Advances in Structural Stability and Dynamics

Advances in Structural Stability and Dynamics

Editors

C. W. Lim
City University of Hong Kong, Hong Kong SAR, P.R. China

Weiyong Wang
Chongqing University, China

Jie Yang
RMIT University, Australia

World Scientific

NEW JERSEY • LONDON • SINGAPORE • BEIJING • SHANGHAI • TAIPEI • CHENNAI

Published by

World Scientific Publishing Co. Pte. Ltd.
5 Toh Tuck Link, Singapore 596224
USA office: 27 Warren Street, Suite 401-402, Hackensack, NJ 07601
UK office: 57 Shelton Street, Covent Garden, London WC2H 9HE

Library of Congress Control Number: 2025013903

British Library Cataloguing-in-Publication Data
A catalogue record for this book is available from the British Library.

ADVANCES IN STRUCTURAL STABILITY AND DYNAMICS

ISBN 978-981-98-1476-3 (hardcover)
ISBN 978-981-98-1477-0 (ebook for institutions)
ISBN 978-981-98-1478-7 (ebook for individuals)

For any available supplementary material, please visit
https://www.worldscientific.com/worldscibooks/10.1142/14356#t=suppl

Desk Editors: Soundararajan Raghuraman/Steven Patt

Typeset by Stallion Press
Email: enquiries@stallionpress.com

Printed in Singapore

https://doi.org/10.1142/9789819814770_fmatter

Preface

This book is a new contribution aimed at providing knowledge on advances in structural stability and dynamics. The contents of this book mainly deal with the stability or dynamics of building and bridge structures. We believe the information presented in this book is useful to design professionals and also researchers or students to undertake structural analysis and design on buildings or bridges. The book is organized into 36 chapters and each chapter is an article selected from the special issue on Structural Engineering Dedicated to the 70th Birthday of Professor Y. B. Yang.

Born in Jinmen, Fujian Province on 22nd August 1954, Professor Y. B. Yang is currently a foreign member of the Austrian Academy of Sciences, an academician of the Chinese Academy of Engineering, Professor Emeritus of National Taiwan University, and Professor and Honorary Dean of School of Civil Engineering of Chongqing University. He is a globally renowned expert in structural mechanics and dynamics research. In appreciation of Professor Yang as one of the founding chief editors of the *International Journal of Structural Stability and Dynamics* and his outstanding contributions to structural mechanics and engineering, a special issue on structural engineering was dedicated to Professor Y. B. Yang on his 70th birthday. This special issue was comprised of invited papers from Professor Y. B. Yang's research collaborators, friends, colleagues, and other well-known experts in the field of structural engineering.

The target audience for this book are engineering professionals, students, and teachers in civil, mechanical, and architectural engineering disciplines, and building officials and regulators. A good knowledge on mechanics and structures is essential when reading this book, while a general background on stability and dynamics is an advantage. It is hoped that the book will enable researchers, practitioners, and students to develop greater insight into structural engineering so that robust structures could be designed through enhancing stability. Although care has been taken to present the research results, some errors might have crept in inadvertently. We would highly appreciate it if these errors and suggestions for improvement are brought to our notice.

C. W. Lim
City University of Hong Kong, Hong Kong SAR, P.R. China

Weiyong Wang
Chongqing University, China

Jie Yang
RMIT University, Australia

https://doi.org/10.1142/9789819814770_fmatter

About the Editors

Professor Jie Yang is a Distinguished Professor in the School of Engineering, RMIT University, Australia. His main research interests include advanced composite and functionally graded structures, mechanical metamaterials, structural stability and dynamics, CNT/graphene-reinforced nanocomposites, smart structures, and control. He has authored 1 book and over 500 publications including 360 journal papers which have so far attracted over 29200 Google Scholar citations with h-index 94. He was the Highly Cited Researcher (Cross Field) in 2019, 2020, 2021, 2022, and 2023 by Clarivate Analytics and was named by the annual Australian Research Magazine as the Global Field Leader in Mechanical Engineering in 2020, Australia's Research Field Leader in Mechanical Engineering in 2019, 2020, 2021, 2022, 2023, in Structural Engineering in 2021, and in Acoustics and Sound in 2023. Prof Yang is the Lead Editor-in-Chief of Engineering Structures and serves on the editorial boards of many other international journals.

Professor Weiyong Wang is the Dean of the Department of Structural Engineering of the College of Civil Engineering at Chongqing University, China. His research has focused on the experimental behavior and analytical modeling of steel structures and composite structures under fire conditions; constitutive modeling of material properties at elevated temperatures, developing guidelines for fire resistance design of steel and composite structures; and evaluating fire performance of high strength steel materials and performance-based fire safety design. He has published 4 books and over 200 peer-reviewed papers in international and Chinese journals. As a major group member, in 2018, he won the Science Research Famous Achievement Award in

Higher Institution (the first prize). In 2020, he won the Science Research Famous Achievement Award of the China Steel Structure Society (the special prize) and the Innovative Talent Award. In 2021, he was supported by the Chongqing Science Fund for Distinguished Young Scholars. In 2022, he won the International Association of Advanced Materials Scientist Award. In 2024, he won the Outstanding Young Scholar Award for High-performance Steel Construction. Prof Wang is an Associate Editor-in-Chief of Progress in Steel Building Structures and serves on the editorial boards of many other international journals.

Ir Professor Lim received a BEng from the University of Technology of Malaysia and MEng and PhD from the National University of Singapore and Nanyang Technological University, respectively. Prior to joining the City University of Hong Kong (CityU), he was a post-doctoral research fellow at the University of Queensland and the University of Hong Kong. Professor Lim is also a Visiting Professor at various universities including the University of Western Sydney (Australia), the University of Johannesburg (South Africa), the University of Canterbury (New Zealand), etc.

Professor Lim receives global recognition for original research in applied and computational mechanics. He was awarded the J.N. Reddy Medal 2020 for "*significant and original contributions to vibration of plates and shells, smart piezoelectric structures, nanomechanics, and symplectic elasticity.*" He published over 410 articles, H-index 69, and over 18,600 citations. He has co-authored one very well-selling title in Engineering Mechanics titled *Symplectic Elasticity*, as recorded in April 2010 by the publisher, World Scientific. As the first and corresponding author in another work, he established one ground-breaking theory and the paper has accumulated over 1,650 citations since first published in 2015. It is also a highly cited paper in *JMPS* as well as the most cited paper since 2004 in the journal. His other work was granted the *IJSS* 2004-2008 most cited article award. He also published a research paper in *Nature Communications.* He was awarded Top Referees in 2009, *Proc. A, The Royal Society.*

Professor Lim holds fellowships in ASME, ASCE, EMI, and HKIE. He was awarded the Tan Chin Tuan Exchange Fellowship by Nanyang Technological University in 2016. Recently, he was elected by secret votes to deliver a Plenary Lecture at the World Congress on Computational Mechanics 2022, the largest biennial meet for computational scientists. WCCM 2020 recorded over 400 mini-symposia and over 5,000 papers accepted. He has delivered over 70 plenary/keynote speeches worldwide. On another scientific occasion, Professor Lim was one of the distinguished speakers at a forum in May 2022 organized by the Chinese

Science Bulletin that was broadcast online on *five different virtual platforms*. It was marvelous to note that the scientific forum accumulatively attracted a remarkable *over 30,000 strong audience* online.

He holds the Chair of Selection Committee (2017-present) for J.S. Rao Medal in Vibration Engineering, GC member of IACM and APACM since 2019, Selection Committee Member of APACM awards, Vice Chair/Chair (2018–2022) for ASCE/EMI Stability Committee, Vice Chair (2021–2023) and Chair (2023-Present) for ASME HK Section, and Technical Committee Member for IFToMM since 2011.

He received multiple research grants totaling millions of dollars, one FE software, and four patents. He serves on over 20 journal editorial boards, notably *JoMMS* (Co-Editor-in-Chief), *JVET* (Managing Editor), *AMM* (International Subject Editor), *JSV* (Subject Editor), *RSOS* (Engineering Subject Editor), etc.

https://doi.org/10.1142/9789819814770_fmatter

Contents

https://doi.org/10.1142/9789819814770_0001

Chapter 1

Optimal Vibration Control Design of Antenna Mast on Super High-Rising Structures Against Multi-Hazards of Earthquake and Wind[#]

Hong-Nan Li[*,†], Can-Hua Liu[*] and Chao Li[*,‡]

[*]*State Lab of Coastal and Offshore Engineering*
Faculty of Infrastructure Engineering
Dalian University of Technology
Dalian 116024, Liaoning, P. R. China

[†]*School of Civil Engineering*
Shenyang Jianzhu University
Shenyang 110168, Liaoning, P. R. China

[‡]*chaoli@dlut.edu.cn*

Antenna mast structures are usually set on top of modern super high-rising structures to meet the requirements of communication and aesthetics, and such buildings are highly sensitive to horizontal loads that can greatly increase the acceleration and displacement responses during their life-cycles owing to the inherent high flexibility and low damping. As a result, the antenna masts with small mass and stiffness may suffer serious whiplash effect under the earthquake or wind excitations. In this paper, a multi-hazard protective system with hybrid isolated and energy-dissipated devices of isolation bearing, viscous damper and mild steel damper is presented for the typical inserted antenna mast structures on super high-rising structures. To determine the optimum parameters of the hybrid system that maximize the structural control efficiency under a single hazard of earthquake or wind load, as well as the coupled conditions of these two hazards, an optimization method based on the genetic algorithm is developed for the presented hybrid control system to resist various hazard scenarios. Objective functions are further proposed to penalize the accelerations and relative displacements at the top of the antenna mast structure. Taking a super-tall TV tower as an example, the OpenSeesPy platform is employed to establish the finite element (FE) model. The numerical results show that the optimization scheme for the hybrid energy-dissipated antenna mast system under a single hazard is not suitable for the other hazard condition, while the optimized results for the multi-hazard condition can give consideration to the effects of both earthquake and wind. Moreover, the sensitive analysis is performed to investigate the effects of each parameter of the hybrid system on the objective functions. It can be concluded that the proposed hybrid system performs well under earthquake, wind and coupled multi-hazards, which is of practical significance for the vibration control of antenna masts on super high-rising structures.

Keywords: Vibration control; antenna mast; super high-rising structures; multi-hazards; genetic algorithm.

[‡]Corresponding author.

[#]This chapter appeared previously on the International Journal of Structural Stability and Dynamics. To cite this chapter, please cite the original article as the following: H.-N. Li, C.-H. Liu & C. Li, *Int. J. Struct. Stab. Dyn.*, **23**, 2340001 (2023), DOI: 10.1142/S0219455423400011.

1. Introduction

Super high-rising structures have witnessed burgeoning development globally in the past decades due to the development of high-strength materials and advanced construction techniques.[1,2] Owing to their high flexibility and low inherent damping, the modern super high-rising structures are highly sensitive to horizontal loads that can greatly increase the acceleration and displacement responses. Therefore, the mushrooming super high-rising structures undoubtedly pose serious challenges for structural safety under natural hazards. A wealth of literature suggests that super high-rising structures are mainly confronted with the two most destructive hazards during their service lives, namely earthquakes and strong winds. Although there is little chance of simultaneous occurrence of earthquake and wind, as generally assumed in almost all the current literature that earthquake and wind should be categorized into the independent hazard type,[3,4] enormous economic losses and catastrophic casualties may be caused once the super high-rising structures are struck by the low-probability but high-consequence scenario of concurrent earthquake and wind. Thus, how to achieve the comfort and safety of super high-rising structures under seismic and wind loads has become a significant concern among designers and researchers in civil engineering.[1,5]

Antenna masts are normally set on the top of super high-rising structures to make the structural appearance more beautiful and meet the functional requirements of communication. However, it is quite easy to lead serious whiplash effect of an antenna mast when the whole building is subjected to earthquake or wind loads, due to the sharp decrease of mass and stiffness of the antenna compared to the main structure.[6,7] Meanwhile, it should be emphasized that the vibration of an antenna mast can in turn affect the dynamic responses of the main structure. From 21 p.m., May 18 to 15 p.m., May 19 of 2021, the 356-meter-tall SEG Plaza building in Shenzhen, China was sharply shaking and all the persons in the building were evacuated. The final survey report from the experts indicated that the resonance of the slender antenna on the super high-rising structure occurred under a specific wind speed and the vibration frequency of antenna was quite close to the high-order frequency of the main structure, thus finally induced large amplitude vibration of the whole building. Moreover, the wind-induced fatigue effect of the connection between the mast structure and the main structure under wind loads is obvious, which may consequently influence the normal use of the structure and property security of the people.[8,9] Although extensive research has been carried out on the mechanism of whiplash effect and wind-induced fatigue of mast structures; however, there is a scarce study about vibration control of antenna mast structures under multi-hazards of earthquake and strong wind.

In recent decades, the technology of seismic isolation and reduction has been extensively applied to the vibration control of super high-rising structures. Previous studies have reported that a variety of energy dissipation systems can effectively control structural vibration,[10–14] such as metallic dampers,[15] friction dampers,[16–18] viscoelastic dampers,[1] viscous dampers,[2] tuned mass dampers (TMD)[10–13,19] and tuned liquid dampers (TLD).[14,20,21] As a useful technology for reducing the structure damage, isolation systems have been used to protect structures for decades.[22–24] Despite their remarkable advances, these vibration control technologies are mainly focused on the case of individual hazard, without sufficiently considering the various effects of earthquake and wind loads. Actually, the wind loads are mainly characterized by low frequency contents concentrated in the range of 0.001–1.0 Hz, while the major energy of earthquake loads usually place in relatively higher frequency zones between 0.1 to 10 Hz.[25] Previous studies by meteorologists revealed that the thermal energy generated before and during the earthquake events can increase the gradient of air temperature, which may finally lead to the occurrence of earthquakes accompanied by strong winds.[26,27] For instance, the northeast and northwest Pacific are known as both earthquake- and cyclone-prone regions, and a recent record has shown that hurricane Otto and the 7.0 magnitude earthquake occurred in Nicaragua on the same day,[28] among other earthquake-wind concurrent events.[27] More importantly, Li *et al.*[29] collected 76 pairs of earthquake and strong wind simultaneously recorded on the same day among 3 389 samples of Dali region with a radius of 300 km from 1971 to 2017. Li *et al.*[30] and Li *et al.*[31] performed a series of investigation on the performance assessment of building structures subjected to multi-hazards of earthquake and wind, and found that the structural acceleration and displacement responses under coupled seismic and wind loads were greater than those under an individual hazard. Furthermore, Vulcano[32] compared the dynamic responses of base-isolated building structures under earthquake and wind loads, the numerical results revealed that the acceleration responses under different hazards were quite different with the increase of isolation ratio of bearings. Xu *et al.*[23] investigated the responses of super high-rising structures equipped with triple friction pendulum isolators under seismic and wind loads, and reported that the optimized design parameters of isolators for one hazard were not appropriate for the other hazard. Thereby, to realize optimal vibration control of antenna mast structure on super high-rising structures, the effect of multi-hazard should be taken into account simultaneously.

This article presents a new hybrid control approach for the traditional inserted antenna mast structures on super high-rising structures to resist multi-hazards of earthquake and wind. The hybrid control system consists of isolation bearings (IBs), viscous dampers (VDs) and mild steel dampers (MSDs), and the detailed parameters of each component are optimized using the genetic algorithm (GA). In Sec. 2, the GA-based optimization method of the proposed hybrid control antenna mast is introduced. Section 3 adopts a super-tall TV tower as an example and a simplified multiple degree-of-freedom flexural-shear (MFS) model of the building is developed using the OpenSees[33] and OpenseesPy[34] platform. The external loads of earthquake and wind hazards are given in Sec. 4. Section 5 reports the optimization results of the hybrid control antenna mast under various hazard conditions with detailed discussions. Finally, a brief summary with key conclusions of the present study is provided in Sec. 6.

2. Optimal Design of the Energy Dissipation Antenna Mast

2.1. *The proposed hybrid control system*

A typical inserted antenna mast structure for the super high-rising structures is shown in Fig. 1(a), the antenna mast structure is connected with the top of the main structure through rigid links, and its bottom is fixed to the core-tube. Large stress will be generated at the connections between the mast structure and the main structure, which are the weak points in the anti-vibration design of the mast structure.[7] Moreover, the relative displacement between the top of main structure and mast structure under horizontal loads is limited due to the existence of fixed constraint, as a result, if the damper devices are directly installed here, they will not exhibit good performances. Jin *et al.*[35] find that the cables can control the vibration of the mast structure, however, the long-term low-frequency wind vibration will lead to fatigue damage and failure of structural members as their characteristics of large flexibility and low natural vibration frequency. On the other hand, the strong nonlinearity of cables will increase the difficulty of vibration control analysis.

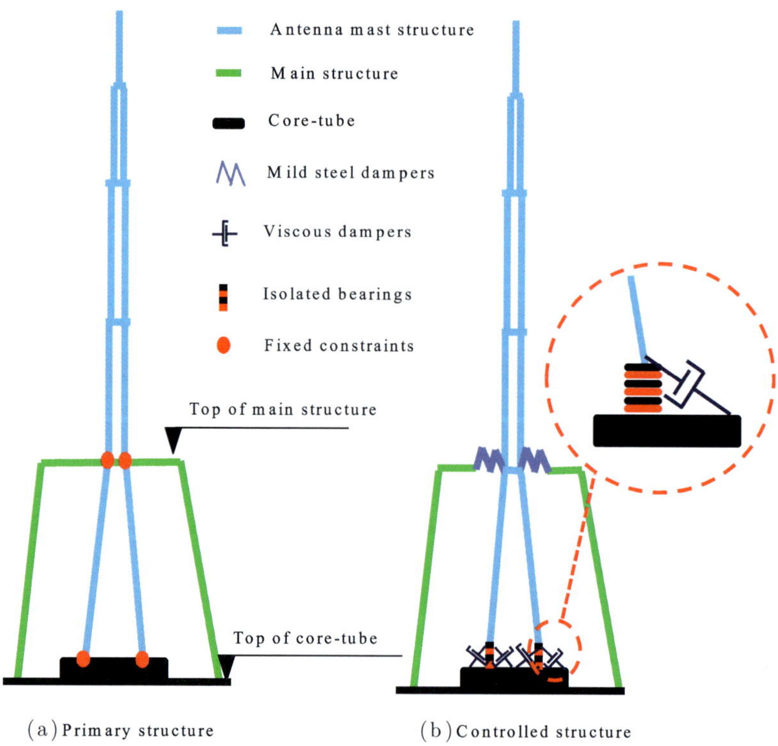

Fig. 1. Primary mast structure and the presented hybrid control system.

In order to control the whiplash effect and suppress the responses of the antenna mast under various dynamic loads, a hybrid control system consisting of IBs, VDs and MSDs is installed between the main structure and the antenna mast. The installation layout of the devices is shown in Fig. 1(b). The IBs and VDs are arranged on top of the core tube to connect with the antenna mast. The MSDs are used to connect the top of the main structure and the antenna. The IBs can prolong the period of the antenna structure and block the secondary excitation of the main structure to the antenna. Fallah *et al.*[36] demonstrated that VDs can reduce the maximum and residual displacements of the IBs. Consequently, the VDs are not only served as energy dissipation devices herein, but also can ensure that the ultimate deformation of bearing does not exceed the limit value. To make the IBs work properly, the connection between the main structure and the antenna is disconnected and equipped with the MSDs, which can further increase the energy dissipation capacity of the antenna and reduce the deformation of the IBs. Since there are multiple parameters to be optimized in the hybrid antenna system consisting of three types of devices, the genetic algorithm is used as the optimization method herein because of its high efficiency.

2.2. *Optimization method*

In this section, a brief description of the proposed optimization process is presented, as shown in Fig. 2. The processes can be roughly divided into the following five steps.

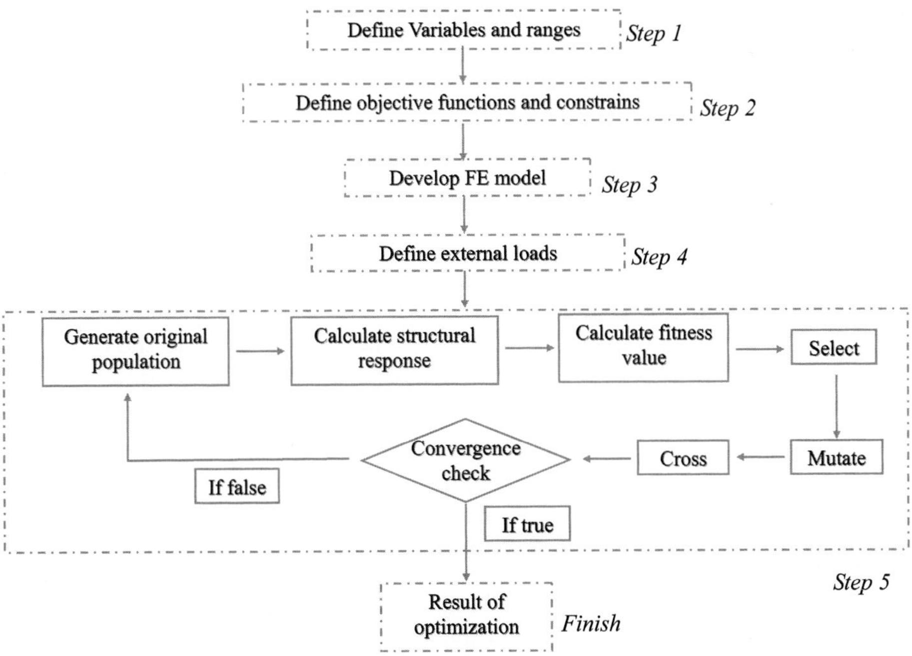

Fig. 2. Flowchart of the optimization framework.

Table 1. The main design variables of the hybrid control antenna system.

No.	Variable	Definition
1	K_b (kN/m)	Pre-yield stiffness of IB
2	Q_d (kN/m)	Characteristic strength of IB
3	α_b	Ratio between post-yield stiffness and initial elastic stiffness of IB
4	C	Viscous damping coefficient of VD
5	α	Velocity exponent of VD
6	K_m (kN/m)	Initial elastic stiffness of MSD
7	u_m (mm)	Yield displacement of MSD
8	α_m	Ratio between post-yield stiffness and initial elastic stiffness of MSD

Step 1. *Define variables and range*

The hybrid control antenna system includes three types of devices; therefore, the key parameters of these devices should be optimized in the design process. The detailed design variables considered in the optimization are stated in Table 1. The value ranges of these parameters can be defined according to the specifications or the products of the manufacturer.

Step 2. *Define objective functions and constrains*

The vibration-reduction rate of the antenna is defined as the variable of the objective functions. All devices should ensure the safety of the structure under ultimate loads, and the control system should ensure that the structural response will not be amplified under all the working conditions, which are the constraints during the whole optimization process.

Step 3. *Develop FE model*

After completing steps 2 and 3, the simplified model of the target super high-rising structure is recommended to conduct the optimization analysis as it is less computationally expensive than refined finite element models.

Step 4. *Define external loads*

The earthquake records are selected from the PEER Strong Motion Database (https://ngawest2.berkeley.edu/) based on the local site conditions and their amplitudes are scaled to match different return periods of earthquakes. Fluctuating wind loads with different return periods are generated considering the correlation between different locations along the height of the structure, and the total wind load duration is 300 s. When considering coupled multi-hazards of earthquake and wind, the seismic ground motion records are loaded at the start of 150 s,[37] in other words, the structure suffers the simultaneous wind and earthquake loads after 150 s until the end of the input seismic motion.

Step 5. *Optimize the design using genetic algorithm*

After determining the design variables, objective functions and constraints; the genetic algorithm is selected as the optimization method to carry out the structural parameter optimization and find the optimal design parameters for the hybrid control antenna structure.

2.3. *Objective functions and design constraints*

Considering that this paper optimizes the vibration control antenna system, the vibration-reduction rates with respect to acceleration and relative displacement responses are chosen as the variables of the objective function. The horizontal displacement at bearings (d_b) under the extreme loads should meet the requirements of the Chinese code,[38] which is the constraint in the whole optimization process.

The objective functions are used to evaluate the performance of the antenna under different loads and various levels of the same hazard, the vibration-reduction rates of peak absolute acceleration and relative displacement of the antenna are considered. The resulting objective functions are:

$$f_c = \sum_{i=1}^{n} (\alpha_d^i \beta_d^i + \alpha_a^i \beta_a^i)(d_b \leq [d_b], \text{extreme cases only}), \tag{1}$$

where the subscript c represents different hazard levels, e.g. the earthquake with an average return period of 50 years, wind loads with an average return period of 10 years, or even simply coupled of the two events. β_d^i and β_a^i are the vibration-reduction rates of peak absolute acceleration and relative displacement of antenna under the ith external loads, respectively. α_d^i and α_a^i are the corresponding weight coefficients, where $\sum_{i=1}^{n}(\alpha_d^i + \alpha_a^i) = 1$. The vibration-reduction rates of peak absolute acceleration and relative displacement vary based on the input design parameters of the hybrid control antenna. $[d_b]$ denotes the ultimate displacements of IBs under extreme case, which means the maximum load intensity considered in each hazard, e.g. the earthquake with a return period of 2 475 years or the multi-hazard case with such earthquake intensity and coupled maximum wind load intensity.

According to the specifications,[38,39] the intensity levels of external loads (I_c) are able to be determined, e.g. the PGA for earthquakes and average velocity for wind loads, respectively. Similarly, the mean annual rate of exceedance probabilities (λ_c) can be obtained by the return periods of various hazard levels. Besides the annual frequency of exceedance, the likelihood of each hazard level to cause damage should also be considered in the objective function f_C. The function F_K is proposed, modified from f_c as

$$F_K = \sum \mu_C f_c', \tag{2}$$

where K denotes different kinds of hazard, e.g. earthquake, wind and coupled multi-hazard of earthquake and wind, f_c' represents objective function of various levels of the same hazard, the importance factors for various levels (μ_c) of the same hazard can be calculated by the following equation:

$$\mu_c = \frac{I_c \lambda_c}{\sum I\lambda}, \tag{3}$$

where $\sum \mu_c = 1$.

3. Description of the Super High-Rising Structure and Numerical Modeling

3.1. *Overview of the example building*

A super-tall TV tower designed to be constructed in Yinchuan, China is adopted as an exemplar in this study. The vertical view and typical story layouts of tower are presented in Fig. 3. According to the Chinese specifications,[38,40] the building is designed as a tube-in-tube structure with a total height of 448.2 m, which has a diameter of 65 m at the bottom and a diameter of 30 m at waist, respectively. The heights of the main structure and the antenna mast are 370 m and 106.9 m, respectively. The antenna mast is inserted into the main structure with 28.7 m and fixed at 341.3 m on the top of the core tube. The main part of the 341.3 m core-tube is an annular tube with the outer diameter of 19.5 m, and the height width ratio is 16.7. Sixteen groups of concrete-filled steel tubular columns are set outside of the core tube, the diameter of columns decreases from 1.3 m at the bottom to 0.9 m at the top. For columns, the material used for the steel is Q355b with yield strength of 355 MPa and ultimate strength of 470 MPa, respectively; and the concrete material is C60–C80.

The installation layout of the hybrid control system to the TV tower is shown in Fig. 4. The isolated lead-rubber bearings and VDs are arranged on the top of the core tube to connect with the antenna mast; the VDs are connected diagonally at the four

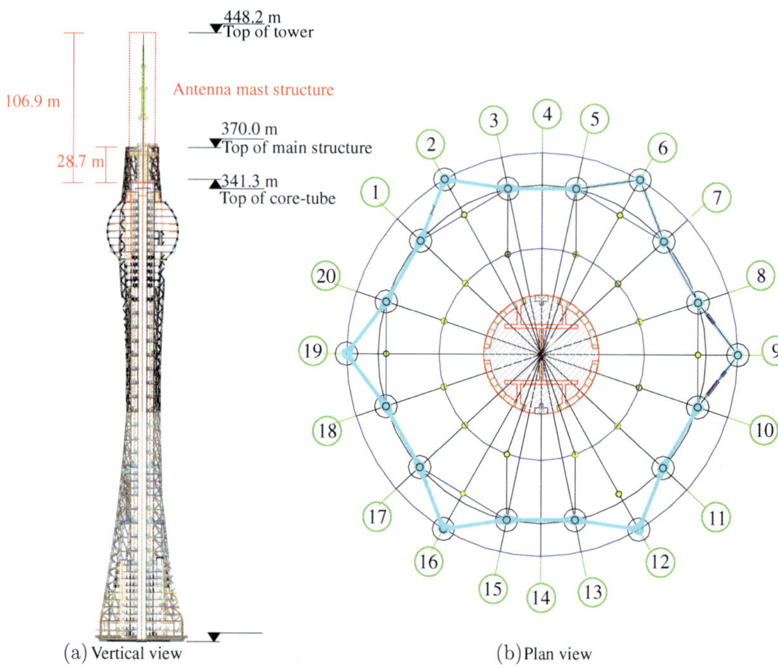

Fig. 3. Vertical view and typical story layout of the tower.

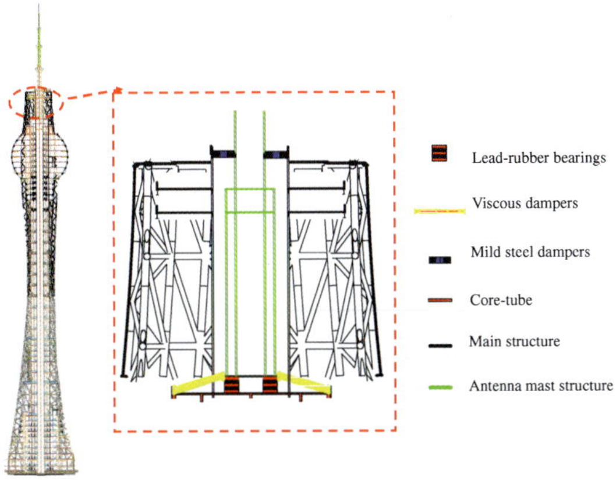

Fig. 4. Tower with hybrid control antenna system.

bottom corners of the antenna. To make the bearings work at full capacity, the antenna is disconnected from the top of the main structure, and MSDs are symmetrically arranged in the X and Y directions to resist loads in the two horizontal directions.

3.2. *Numerical modeling*

The three-dimensional refined finite element (FE) model of the example super high-rising structure is developed using the OpensSees platform,[33] as depicted in Fig. 5(a). All the beams and columns are modeled using the forceBeamColumn elements with elastic sections. The ShellDKGQ elements with ElasticMembranePlate section are adopted to simulate the shear walls and floors. Besides, the pinned beam-column joints are simulated by zeroLength elements, the extreme large stiffness is defined for the translational and torsional directions to represent rigid end, while very small stiffness is assigned to the rotations in X and Y directions to represent the pinned

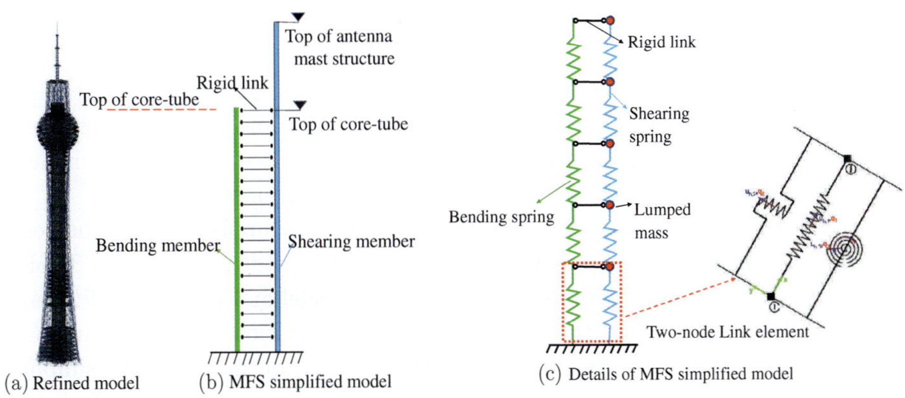

Fig. 5. Illustration of the numerical models.

Table 2. The preceding 12-order natural periods of the tower refined FE model.

Modal	Period (s)	Type
1	8.3062	Translation in Y direction
2	8.1622	Translation in X direction
3	1.9040	Torsion
4	1.8465	Translation in Y direction
5	1.8128	Translation in X direction
6	1.6172	Translation in Y direction
7	1.5825	Translation in X direction
8	0.7963	Translation in Y direction
9	0.7755	Translation in X direction
10	0.5294	Torsion
11	0.4871	Translation in Y direction
12	0.4755	Translation in X direction

connections. There are totally 26 783 nodes, 14 697 forceBeamColumn elements, 16 286 shell elements and 10 450 zeroLength elements in the whole refined FE model. Masses and gravity loads are lumped to the floor nodes. The preceding 12-order natural periods of the tower obtained by the OpenSees platform are summarized in Table 2.

It should be noted that the refined FE model of the super tall building is computationally expensive when performing the dynamic response analysis. In order to balance calculation accuracy and efficiency, a simplified multiple degree-of-freedom flexural-shear (MFS) model is developed in this paper to replace the refined FE model. According to Taghavi and Miranda,[41] such a simplified model can be used to accurately predict the seismic responses of tall buildings. It should also be mentioned that if the design details of a building are available, the parameters of the flexural and shear springs can be determined using the method proposed by Kuang and Huang.[42] As illustrated in Fig. 5(b), the developed MFS model for the exemplar super high-rising structure discretizes each story into a bending spring and a shear spring. The bending spring is utilized to simulate the bending behavior of the shear wall, while the shear spring is used to simulate the shear performance of the frame. These two springs are connected with a rigid link to resist external horizontal loads together. Due to the local degrees of freedom for shear and flexural deformations are coupled, the two-node link elements are ideally situated in the MFS model to simulate the force-displacement relationship of each floor in the structure.

In order to verify the accuracy and rationality of the simplified model, the period and vibration mode are compared with the refined FE model. As listed in Table 3, the first five translational periods of the simplified model and the refined FE model are in good agreement, and the maximum relative deviation is only about 2%. In addition, Fig. 6 shows the first five translational vibration modes of the simplified model and the refined FE model are basically the same, only little difference in higher modes can be observed. The results show that the simplified model is capable

Table 3. The first five translational modal periods of the refined FE model and the MFS model.

Modal	Refined FE model period (s)	MFS model period (s)	Relative error (%)
1	8.1622	8.1585	0.73
2	1.8128	1.8491	2.01
3	1.5825	1.5746	−1.05
4	0.7755	0.7668	−1.12
5	0.4755	0.4851	2.02

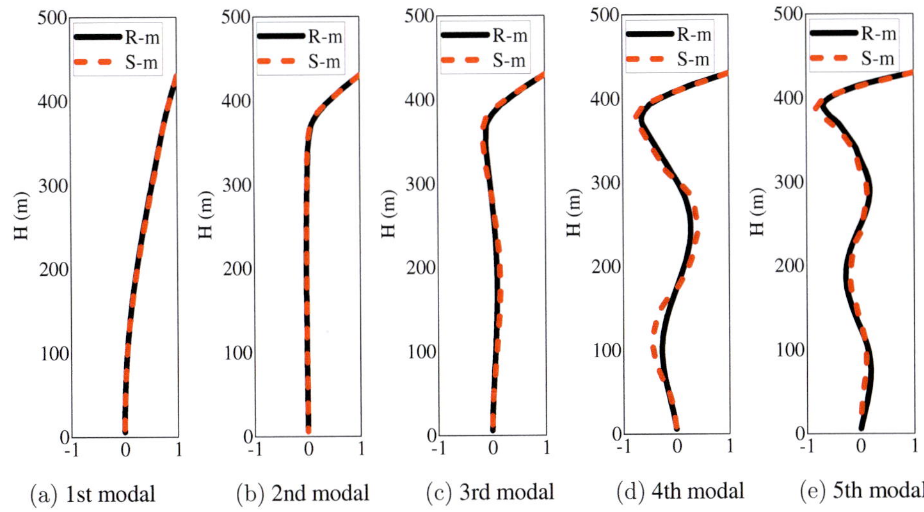

Fig. 6. Modal validation of the simplified model (R-m: Refined FE model; S-m: Simplified model).

of accurately representing the basic dynamic characteristics of the exemplar super high-rising structure.

For the hybrid control antenna system, the VDs are using zeroLength elements with the ViscousDamper material. For the sake of simplicity, the IBs and MSDs are also simulated by zeroLength elements, the Steel 02 material is employed to accurately simulate the hysteretic behaviors of the IBs and MSDs. It should be noted that the uncertainties and fatigue in material properties are not taken into consideration during the dynamic time-history analysis as they are not the key issues focused in the present study.

4. External Loads

4.1. *Earthquake*

To optimize the parameters of the hybrid control antenna system, seven sets of unidirectional ground motions are selected in accordance with site condition of the structure, i.e. site class II in the Chinese design code.[38] All the ground motion records are selected from the PEER Strong Motion Database (https://ngawest2.berkeley.edu/) based on the local site and source properties, including soil type

Table 4. Information of the selected seven ground motion records.

No.	Events name	RSN	Date	Station	R_{rup} (km)	Magnitude
1	Loma Prieta	800	10/18/1989	Salinas-John & Work	32.78	6.93
2	Hector Mine	1 838	10/16/1999	Whitewater Trout Farm	62.91	7.13
3	Loma Prieta	761	10/18/1989	Fremont-Emerson Court	39.85	6.93
4	Kocaeli Turkey	1 170	8/17/1999	Mecidiyekoy	53.43	7.51
5	Chi-Chi Taiwan	1 476	9/20/1999	TCU029	28.04	7.62
6	Tottori Japan	3 961	10/6/2000	TTR004	67.02	6.61
7	Chuetsu-oki	5 196	7/16/2007	NGN004	80.05	6.8

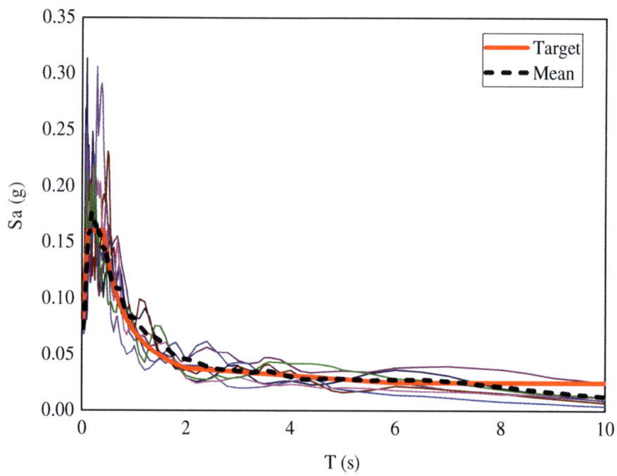

Fig. 7. Comparisons between the response spectra of the selected ground motions and the code response spectrum.

($250\,\mathrm{m/s} \leq \mathrm{VS}_{30} < 500\,\mathrm{m/s}$), epicentral distance (between 10 km and 100 km), and magnitude of the earthquake ($\mathrm{Mw} \geq 6.5$). Table 4 shows the basic information of the selected ground motion records. The comparisons between the response spectrum of each ground motion and the target spectrum of service level earthquake (SLE) defined in the Chinese code[38] are shown in Fig. 7, in which good match results can be obtained. Moreover, the return periods of SLE, design base earthquake (DBE) and maximum considered earthquake (MCE) are respectively 50, 475 and 2 475 years, with PGAs of $0.07\,g$, $0.20\,g$, and $0.40\,g$.

Figure 8 compares the time-history responses at the top of the antenna mast structure and the main structure for the MFS model and the refined FE model under SLE. It is found that the difference between the displacement responses of the two models can be neglected, as a result of the minor errors in higher modes, the acceleration responses between these two models are slightly different. Nevertheless, the error of maximum acceleration is less than 5%, which is acceptable in the following vibration control analysis. Therefore, the developed MFS model is reliable for the accurate computation of the time-history structural responses and optimal design of the proposed hybrid control antenna system.

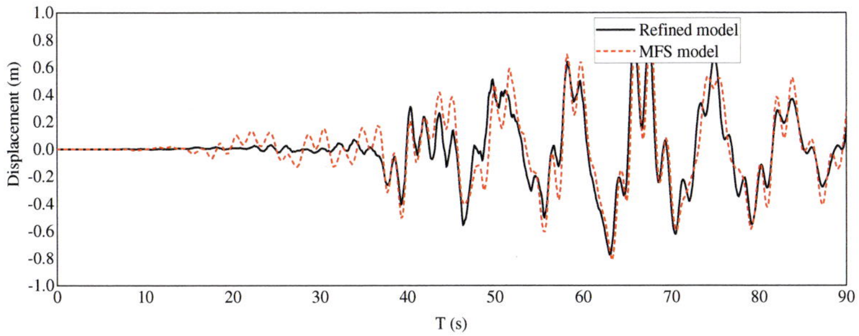

(a) Top displacement time histories for the antenna structure under SLE

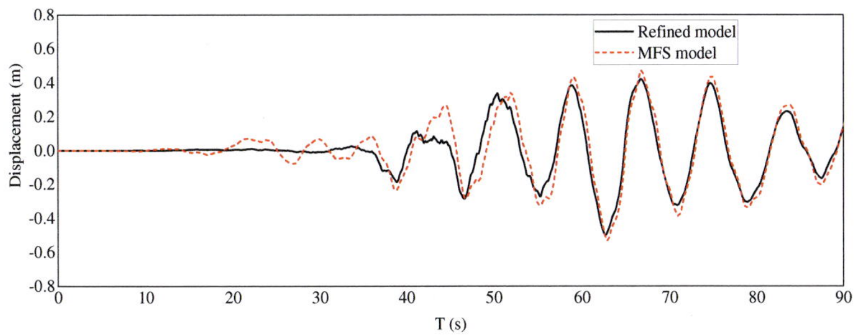

(b) Top displacement time histories for the main structure under SLE

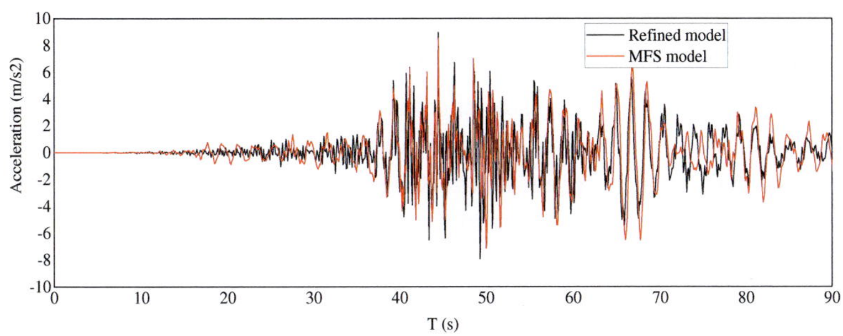

(c) Top acceleration time histories for the antenna structure under SLE

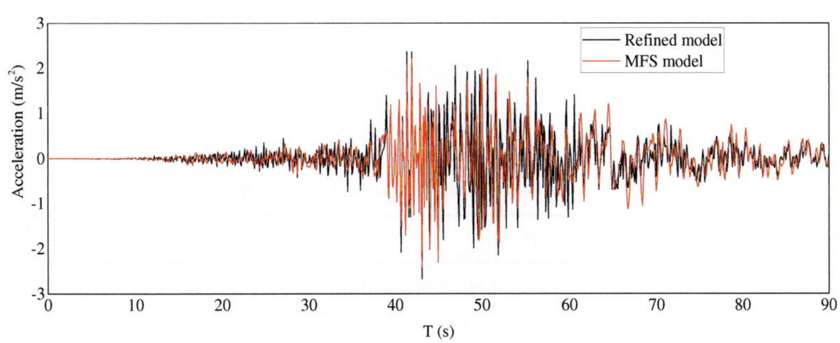

(d) Top acceleration time histories for the main structure under SLE

Fig. 8. Validation of the simplified model based on the seismic responses of structure.

4.2. *Wind load*

The total wind velocity along the height of this super high-rising structure can be decomposed into a constant mean wind velocity $\bar{v}$ and a fluctuating wind velocity $\tilde{v}$. Here, the logarithmic law presented by Tennekes[43] is adopted to describe the variation of mean wind velocity with height z_i, which can be expressed as:

$$\frac{\bar{v_i}(z_i)}{v_{10}} = \frac{\ln(z_i/z_0)}{\ln(10/z_0)}, \tag{4}$$

where v_{10} is the mean wind velocity at an elevation of 10 m above the ground level; and z_0 denotes the roughness length, which is equal to 1.2 m for city exposure.[43] Only one point is selected from each floor of this super high-rising structure to compute the mean wind velocities at different heights based on Eq. (4). The algorithm used to simulate the fluctuating wind velocity is outlined in Appendix A. It is noteworthy that the two-dimensional simplified model incapable of considering the effects of cross-wind. According to the specification,[39] the duration of the generated time histories is set to be 300s with sampling frequency of 10 Hz.[44,45] Finally, the horizontal wind loads are obtained at the 47 points of each floor for the exemplar super high-rising structure (45 floors of the main structure and 2 points for the antenna structure).

5. Optimization of the Vibration Control System

In the optimal design, the ranges of the feasible design variables for IBs, VDs and MSDs are tabulated in Table 5. All the ranges of these parameters are based on practical application. According to the Chinese specification,[46] the recommended lead-rubber bearings (LRBs) have a constant post-yield to pre-yield stiffness ratio of 0.0769, and the characteristic strength and pre-yield stiffness approximately satisfy Eq. (5); consequently, only pre-yield stiffness K_b is required to be determined in the optimization.

$$Q_d = 7.354\times10^{-8} * K_b^2 - 6.1455\times10^{-3} * K_b + 32222. \tag{5}$$

where Q_d(N) and K_b(N/m) are characteristic strength and pre-yield stiffness of LRBs, respectively.

Table 5. The value ranges of the six variables in optimization analysis.

No.	Variable	Instruction	Variation intervals
1	K_b (kN/mm)	pre-yield stiffness of IBs	5–40
2	C (kN* (s/mm))	viscous damping coefficient of VDs	0–20
3	α	velocity exponent of VDs	0–1
4	K_m (kN/mm)	initial elastic stiffness of MSDs	10–500
5	u_m (mm)	yield displacement of MSDs	0.5–2
6	α_m	ratio between post-yield stiffness and initial elastic stiffness of MSDs	0.001–0.02

5.1. *Objective functions*

5.1.1. *Individual hazard*

The vibration-reduction rates regarding the peak absolute acceleration and relative displacement of the antenna structure are taken as the performance indices of the hybrid control system. It should be also noted that the displacement of IBs is limited to three times of the total thickness of the rubber layers under MCE,[38] which is a constraint in the optimization process. In Eq. (1), α_d and α_a are both 0.07143 for each objective function when using seven seismic inputs. According to the specification,[38] the mean annual exceeding probabilities of SLE, DBE and MCE are 0.019608, 0.002105 and 0.000404, respectively. The important factors of the three earthquake levels in the objective function for seismic loads case (F_E) can be obtained by Eq. (3) as follows: $\mu_{\mathrm{SLE}} : \mu_{\mathrm{DBE}} : \mu_{\mathrm{MCE}} = 70 : 22 : 8$.

In accordance with different mean wind velocities at the height of 10 m ($\bar{v}$), the service level wind (SLW, $\bar{v} = 25.3\,\mathrm{m/s}$), design base wind (DBW, $\bar{v} = 32.25\,\mathrm{m/s}$) and maximum considered wind (MCW, $\bar{v} = 34.46\,\mathrm{m/s}$) are considered as different wind load intensity levels, the return periods of SLW, DBW and MCW are 10 years, 50 years and 100 years, respectively. For all the wind load levels, the weighting factor of α_d and α_a are both 0.5 in each objective function. The mean annual exceeding probabilities of SLW, DBW and MCW are 0.1, 0.02 and 0.01, respectively. The important factors of three wind load levels in the objective function for wind loads case (F_W) calculated using Eq. (3) are as follows: $\mu_{\mathrm{SLW}} : \mu_{\mathrm{DBW}} : \mu_{\mathrm{MCW}} = 72 : 18 : 10$.

5.1.2. *Multi-hazards*

The exemplar super high-rising structure is located in Yinchuan City of Ningxia with a seismic fortification intensity of 8-degree. Yinchuan is on the Yinchuan-Hetao seismic belt of China and has experienced earthquakes of magnitude more than 8.0 in history. Besides, strong winds regularly sweep Yinchuan all year round. Although existing literature generally recognizes that earthquake and wind are independent hazard events,[4,47] untold damages may be caused to the society and human beings when the super high-rising structures suffered the low probability but high-consequence condition of concurrent earthquake and wind. Thus, it is of great significance to investigate the performance of such super high-rising structure under the combined effects of earthquake and wind hazards.[31] In order to analyze the performance of the hybrid control system under coupled multi-hazards, 27 sets of "earthquake–wind loads" are formed from the above-mentioned 3 ground motions (No. 2, No. 6 and No. 7 ground motions) and wind loads (3 ground motions $\times$ 3 earthquake levels $\times$3 wind load levels).

The proposed objective functions to evaluate the multi-hazard performance include 9 coupled hazard levels, namely the combination of the three earthquake intensity levels with each of the three wind load levels defined as f_{CSS}, f_{CSD}, f_{CSM}, f_{CDS}, f_{CDD}, f_{CDM}, f_{CMS}, f_{CMD} and f_{CMM}, respectively. For each coupled multi-hazard condition, the weighting factors of α_d and α_a fulfill the restriction of $\alpha_d + \alpha_a = 1$. It is noted that this paper only studies the vibration-reduction effects of the proposed hybrid control system under the coupled earthquake and wind loads. Eujeong *et al.*[47] reported that the multi-hazard of earthquake and wind can also be categorized into independent hazard type that can jointly affect the structures by spatiotemporal coincidence. Therefore, for simplicity, it is considered that the joint occurrence probability of earthquake and wind loads can be obtained by the product of the occurrence probabilities of earthquake and wind loads, for example, $\lambda_{\mathrm{SS}} = \lambda_{\mathrm{SLW}} * \lambda_{\mathrm{SLE}} = 0.0019608$. The intensity indices of earthquake and wind loads are normalized for the computation convenience of the importance factors for various levels of multi-hazard, namely I'_{SLE}, I'_{DBE} and I'_{MCE} are 0.175, 0.5 and 1, and the values of I'_{SLW}, I'_{DBW} and I'_{MCW} are 0.73, 0.94 and 1, respectively. The intensity indices for multi-hazards can also be calculated by the product of the nondimensionalized intensity indices of earthquake and wind loads, namely $I_{\mathrm{SS}} = I'_{\mathrm{SLE}} * I'_{\mathrm{SLW}} = 0.175 * 0.73 = 0.128$, and all the parameter values of μ in the objective functions of the coupled multi-hazards (F_C) can be finally obtained using Eq. (3), as summarized in Table 6.

5.2. *Optimization results and discussions*

Genetic algorithms (GAs) are commonly used to generate high-quality solutions to optimization problems by relying on biologically inspired operators such as mutation, crossover and selection. The GA is adopted to search for the maximum values of F_E, F_W and F_C for the hybrid control antenna system. The parameters of the GA selected in this study are based on "Geatpy" (http://geatpy.com/), which is a genetic and evolutionary algorithm toolbox in Python with high computation capacity. In this paper, the population size, the crossover and mutation probabilities are 150, 0.7 and 0.2, respectively. Figure 9 illustrates the optimization processes on the objective functions of F_E, F_W and F_C, respectively; it can be seen that the objective functions basically converge and achieve the optimal values after 50 generations. Table 7 presents the optimized values of each parameter under earthquakes, wind loads and coupled multi-hazards, respectively.

The vibration-reduction rates of relative displacement and absolute acceleration for three design cases are shown in Table 8, in which Cases 1, 2 and 3 represent the optimized results of the hybrid control antenna under the seismic, wind and multi-hazard loads, respectively. The relative antenna displacement and acceleration time histories for the design schemes of Cases 1, 2 and 3 under MCW, DBE and DD_7 are shown in Fig. 10, where DD_7 denotes coupled multi-hazard of DBW and DBE with the NO. 7 ground motion mentioned in Table 4.

Table 6. The importance factors μ of F_C.

μ_{SS}	μ_{SD}	μ_{SM}	μ_{DS}	μ_{DD}	μ_{DM}	μ_{MS}	μ_{MD}	μ_{MM}
0.504	0.126	0.070	0.158	0.040	0.022	0.058	0.015	0.003

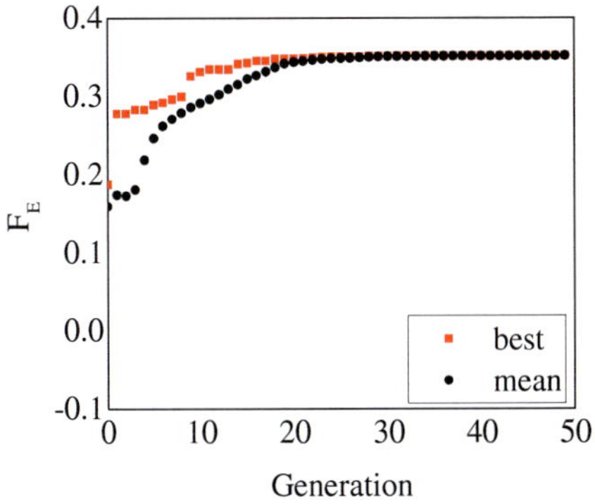

(a) For seismic loads

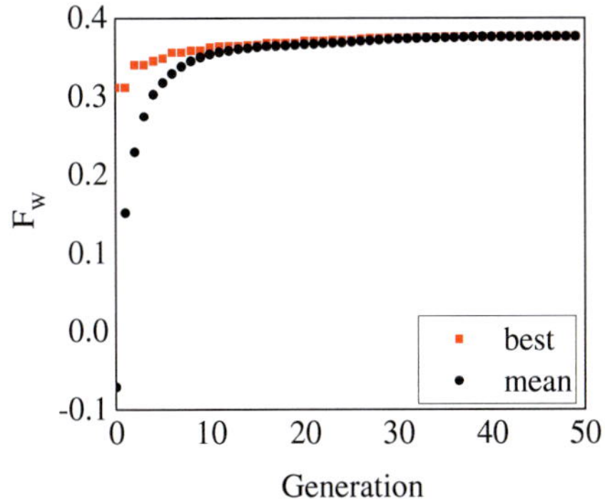

(b) For wind loads

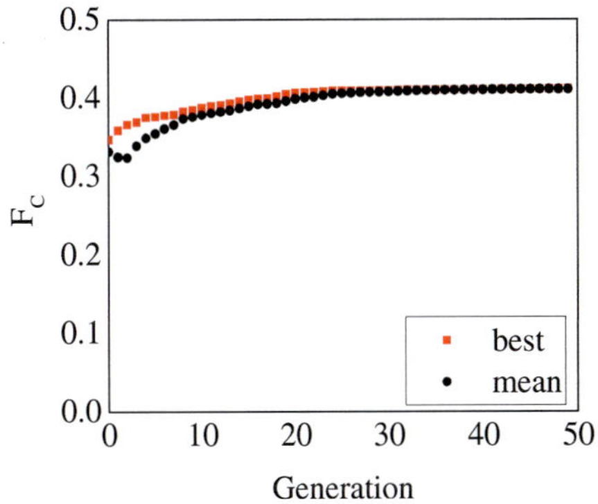

(c) For coupled multi-hazards

Fig. 9. Iterative process during the GA.

Table 7. The optimal parameters for the hybrid control antenna system under various hazards.

Case	Load	K_b (kN/mm)	C kN/(mm/s)$^{\alpha}$	α	K_m (kN/mm)	u_m (mm)	α_m
1	Seismic	21.6476	0.8366	1.00	108.6784	0.5	0.0141
2	Wind	27.0028	2.0807	0.75	138.5852	0.5	0.0039
3	Multi-hazard	26.2883	0.6852	1.00	136.7003	0.5	0.0200

It can be seen from Table 8 that the proposed antenna system performs better for the hazard it is designed for, and when used for a different hazard load, there are differences in the control efficiency regarding the relative displacement and acceleration responses. For different levels of seismic excitations, when using the optimized design parameters of Cases 2 and 3, the mean control effects decrease compared to that of Case 1. For the three levels of wind excitations, when using the optimal design results of Cases 1 and 3, the hybrid antenna system achieves slightly better control effect on displacements but the acceleration responses significantly increase compared to those under the optimal design scheme of Case 2. For the coupled multi-hazard excitations, the design scheme for Case 1 performs the best control effect regarding displacements; while the design of Case 3 shows the best control results on accelerations. However, when the control effects of accelerations and displacements of the nine load levels are comprehensively considered, the design parameters obtained in Case 3 represent the optimal scheme.

It should be noted that the proposed hybrid control system has little control effect on the main structure. For seismic loads, the control effect of top displacement

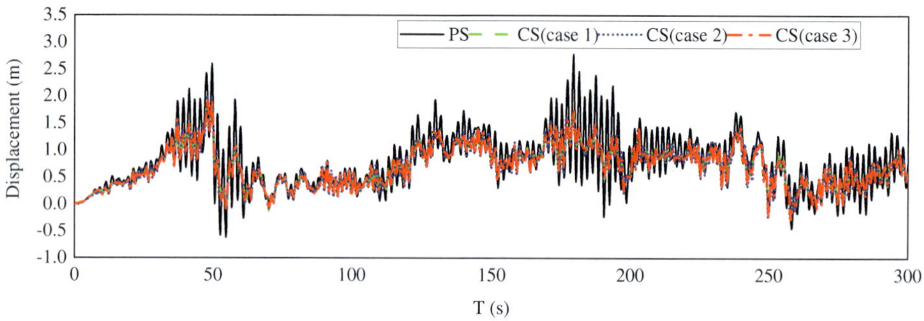

(a) Relative displacement time histories under MCW

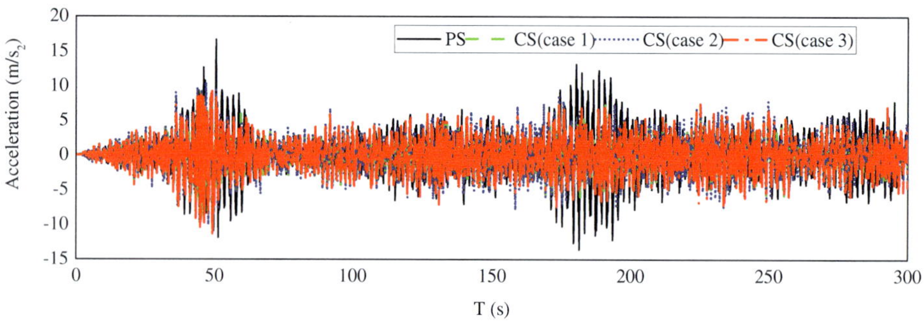

(b) Top acceleration time histories under SLW

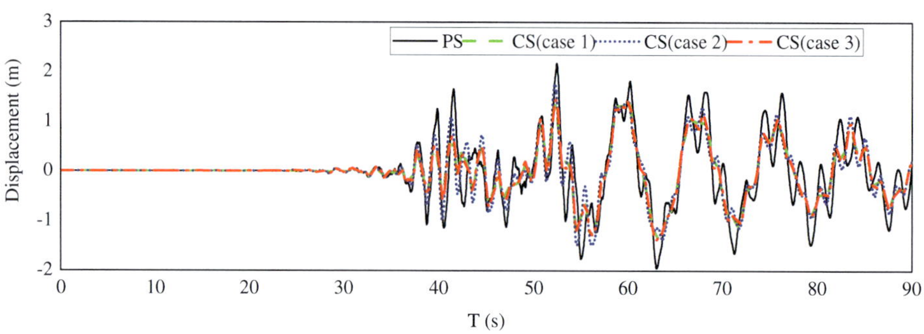

(c) Relative displacement time histories under DBE

Fig. 10. Control effect of the hybrid system using the three design cases under MCW, DBE, DD_7. (PS: primary antenna structure; CS: controlled antenna structure).

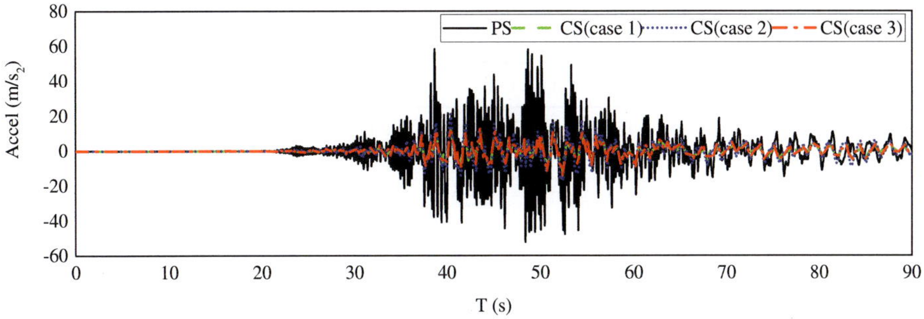

(d) Top acceleration time histories under DBE

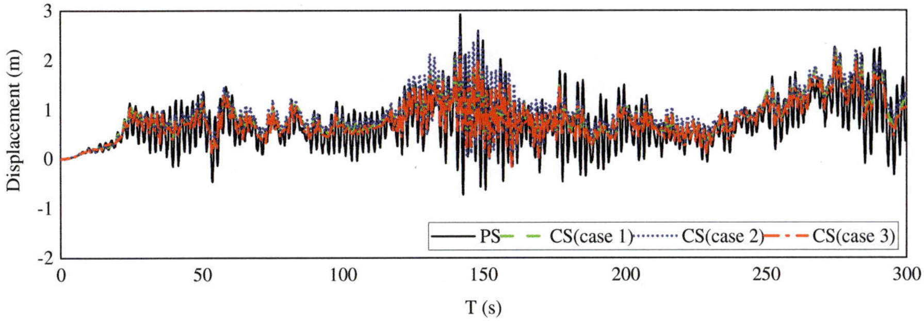

(e) Relative displacement time histories under DD_7

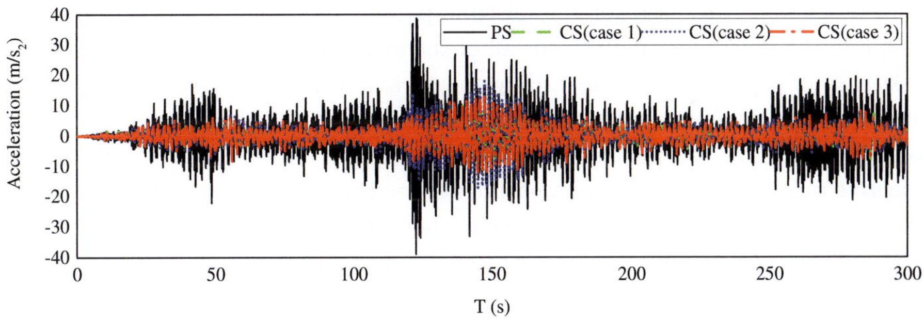

(f) Top acceleration time histories under DD_7

Fig. 10. (*Continued*)

is small and the top vibration reduction ratio of acceleration is up to 6.96%. The influences of the proposed hybrid control system on the main structure's dynamical responses under different levels of wind loads are slight. For the coupled multi-hazard excitations, all the design scheme cases perform a better control effect than the individual hazard excitations, the maximum vibration reduction ratio of acceleration reaches 37.72%, and the change of displacement responses is limited.

Table 8. Control efficiency of the proposed antenna system with different optimal design schemes compared to the primary structure.

		For relative displacement (%)			For acceleration (%)		
Vibration-reduction rates (β)	Hazard level	Design of Case 1	Design of Case 2	Design of Case 3	Design of Case 1	Design of Case 2	Design of Case 3
	SLE	15.64	−5.29	2.88	66.40	60.93	56.16
For seismic (mean)	DBE	35.28	12.50	29.98	75.90	65.75	62.58
	MCE	41.54	22.40	38.02	78.03	78.23	66.76
	SLW	28.34	28.08	27.74	16.26	22.02	9.45
For wind-loads	DBW	20.59	18.19	20.49	11.01	13.33	3.95
	MCW	30.83	28.10	30.15	35.29	37.16	30.24
	SS	30.17	19.78	23.64	46.07	40.07	62.83
	DS	41.65	22.01	35.42	77.61	66.16	73.93
	MS	49.67	32.08	46.35	81.68	71.26	78.62
	SD	26.86	24.66	26.87	65.99	60.23	61.45
For multi-hazard (mean)	DD	32.14	22.56	30.40	51.08	44.05	71.30
	MD	40.32	25.93	38.22	34.77	67.26	72.20
	SM	36.65	26.59	35.26	67.97	66.63	65.22
	DM	40.33	34.51	39.31	77.96	65.50	75.40
	MM	43.45	35.96	41.93	81.63	70.32	79.92

To understand the effect of each design variable on the responses of the antenna structure, sensitivity analysis is also conducted. The design parameters Case 1 (as given in Table 7) are used for the baseline parameter values of the hybrid control system in the sensitivity analysis for all of the excitation cases. Figure 11 shows the sensitivity of F_E, F_W and F_C regarding the parameters of K_b, C, α, K_m, u_m and α_m, from which it can be seen that the influences of the six design parameters on the objective functions of wind and earthquake loads are biased, but the trends of the objective functions with the same parameter are roughly the same. F_E, F_W and F_C decrease as K_b, C and K_m increase, while they increase as α decreases; besides, u_m and α_m have negligible effects on the objective functions. However, it should be noted that the varying trend is not absolutely consistent. When the structure suffered wind loads, the value of objective function decreases when α is larger than 0.6, while increases when α is less than 0.6; the value of objective function decreases at $C > 1.2\,\text{kN/(mm/s)}$ but increases at $C < 1.2\,\text{kN/(mm/s)}$.

In order to quantify the relative importance of design variables in determining the values of objective functions, global sensitivity analysis is further developed to calculate the Sobol sensitivity indices,[48,49] including the frequently used first-order effect (S1) and total effect (ST). S1 represents the first-order effect of the input parameter on the output result alone, while ST represents the sum of the first-order, second-order and other high-order effects of the input parameter on the output results. The difference between S1 and ST (δS) presents the effect of interaction between input parameter and other parameters on the output results. It is worth mentioning that the greater the sensitivity index values are, the more important the

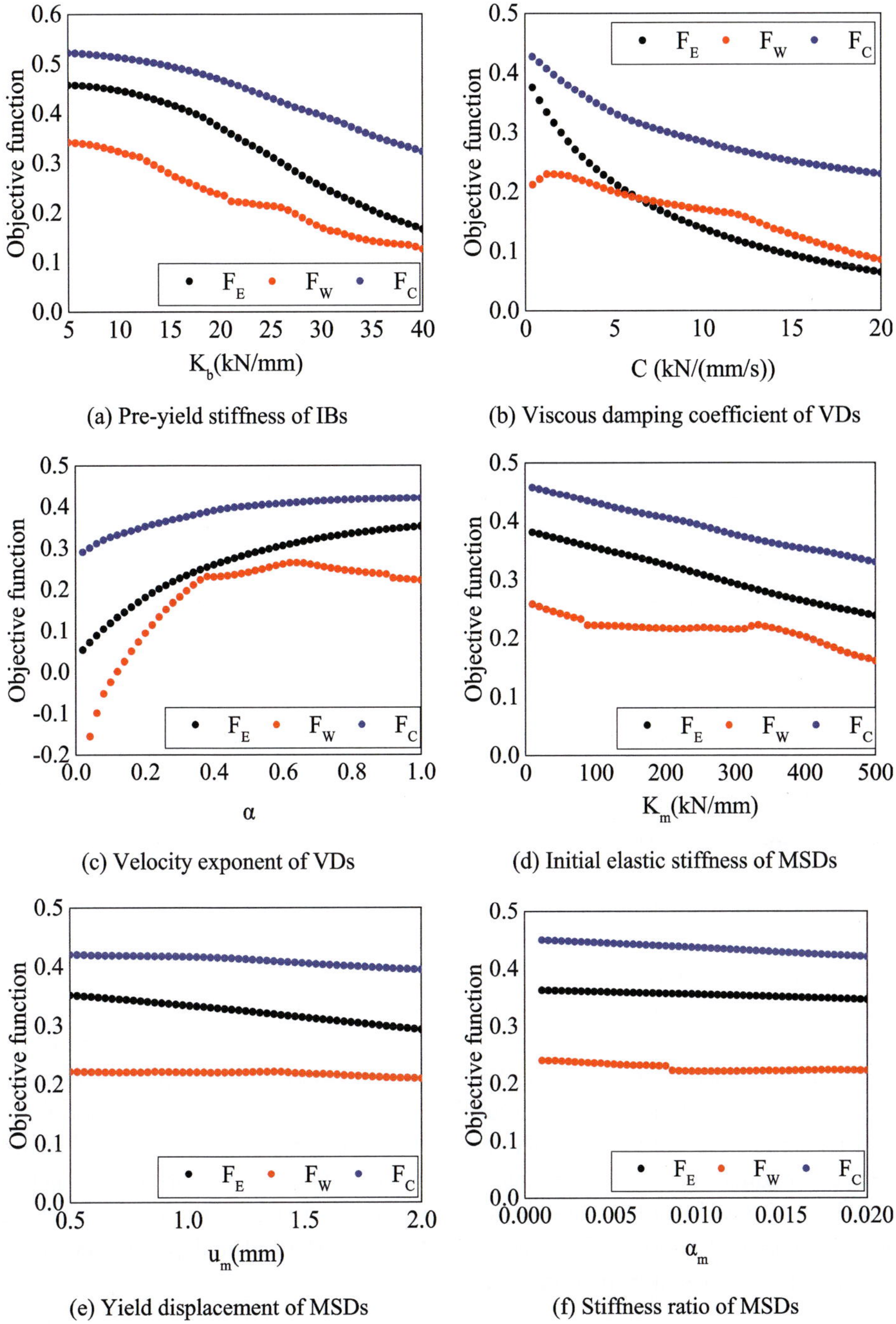

Fig. 11. Sensitivity of F_E, F_W and F_C to each design parameter.

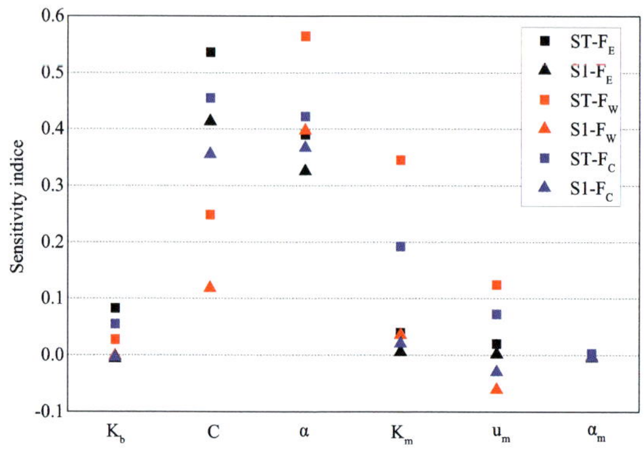

Fig. 12. Global sensitivity indices of F_E, F_W and F_C to each design parameter.

variable is to the objective functions. For the three excitation conditions, in total 3 584 samples of design parameters are generated to calculate the global sensitivity indices, the results are given in Fig. 12. It can be found that the effects of various input parameters on the output results induce considerable differences among F_E, F_W and F_C. For the individual seismic loading condition, the sensitivities of various design variables to the objective function are ranked as $C > \alpha > K_b > K_m > u_m > \alpha_m$, the maximum values of ST, S1 and δS with respect to the parameter of C are 0.536, 0.413 and 0.123, respectively. For the solely wind load case, the sensitivities of design variables to the objective function are ranked as $\alpha > K_m > C > u_m > K_b > \alpha_m$, the ST and S1 values for the parameter of α achieve maximum values of 0.564 and 0.397; while δS is the largest for K_m with a value of 0.309. For the coupled multi-hazard condition, the sensitivity ranking of the design variables is exactly the same with that of the earthquake loading case, indicating that the earthquake load dominates the design when the multi-hazards are taken into account. Besides, the maximum values of ST, S1 and δS for the parameter of C are 0.456, 0.313 and 0.163, respectively. According to these results, it is highlighted the behavior of VDs has the largest impact on the vibration control performance of the antenna system. Moreover, the δS values for all of the three excitation cases are larger than 0.1, which reveals that the interaction between design parameters is an assignable factor affecting the structural vibration control. It should be also noted that for some design parameters, the value of S1 is negative due to the sample size is not large enough, but it can also reflect that the influence of the first-order effect is very limited. Owing to the different dynamic characteristics of the wind and earthquake excitations, some differences existed among the optimal parameters when designed for various hazard conditions.

6. Conclusions

This paper proposes a new hybrid control antenna system for the super high-rising structures, an optimization framework based on GA is developed to obtain the optimal design parameters of the new antenna system. The vibration control performance of the system suffered different intensity levels of seismic excitations, wind loads and multi-hazards are investigated. The main conclusions are drawn as follows:

(1) The developed simplified MFS model of the super-tall TV tower can yield compatible modal periods, vibration modes and dynamic response time histories with the refined FE model, and it is efficient in the optimal design of the presented antenna system.
(2) The hybrid control antenna system achieves the best performance for the external excitations it was designed for, and when used for the different external excitations, the control efficiency will reduce to a certain extent. All the optimal design cases can achieve good performance under different types of hazard loads. (i.e. for the DBW case, the vibration-reduction rates of displacement for three optimal design cases are 20.59%, 18.19%, 20.49%, respectively.)
(3) The influences of all the design parameters on the objective functions of three external excitations are biased, but the trends of the objective functions with the same parameter are roughly the same. (i.e. both F_E, F_W and F_C increase as α increase, but F_W decreases at α larger than 0.6 while increases at α less than 0.6)
(4) It is demonstrated that the viscous damping coefficient C and velocity index α have the greatest influence on the performance of the hybrid control system under all of the three kinds of external excitations, while the pre-yield stiffness of IB and the ratio between post-yield stiffness and initial elastic stiffness of MSD play a negligible role in the hybrid system.
(5) The results of the present study show that seismic load is the dominant external load when the hybrid control antenna system is subjected to various cases of coupled multi-hazards of earthquake and wind considered in this paper. It is suggested that the effect of multi-hazard should be taken into account in the seismic design of the hybrid control antenna system to balance the vibration control effect under various hazard conditions.

Acknowledgment

The authors gratefully acknowledge the financial support from the Dalian Municipal Natural Science Foundation (2019RD01) and National Natural Science Foundation of China (51738007 and 52027811) for carrying out this research.

Appendix A. Simulation of Wind Load Time Histories

The PSD functions of the fluctuating wind velocities at different heights are modeled by the Kaimal spectrum[50]:

$$S(z, f) = \frac{v_*^2}{f} \frac{200x_m}{(1 + 50x_m)^{5/3}}, \tag{A.1}$$

in which

$$x_m = \frac{fz}{\bar{v}_z}, \tag{A.2}$$

$$v_*^2 = kv_{10}^2, \tag{A.3}$$

where f, v_* and x_m are the frequency (Hz), friction velocity and Monin coordinate, respectively; k is the von-Karman's constant, typically set to be 0.4. Due to the spatial correlation effect, the fluctuating wind velocities at different heights are similar although they are different along the building. The cross PSD function of wind velocity at point k and l is represented by

$$S_{v_k v_l}(f) = \sqrt{S_{v_k v_k}(f) S_{v_l v_l}(f)} \mathrm{Coh}(k, l; f), \tag{A.4}$$

where $S_{v_k v_k}$ and $S_{v_l v_l}$ are the PSDs obtained by Eq. (A.4); $\mathrm{Coh}(k, l; f)$ presents the spatial coherency loss function between point k and l, which is given by[51]

$$\mathrm{Coh}(k, l; f) = \exp\left(\frac{Df|k-l|}{\bar{v}}\right) \exp\left(-i\frac{k-l}{v_{\mathrm{app}}}\omega\right)(k > l), \tag{A.5}$$

where $|k - l|$ is the distance between points k and l and $\bar{v}$ is the average of the mean wind velocity at point k and l. The decay constant D and apparent wave velocity v_{app} are assumed to be 0.075 and 10 m/s, respectively.

In the simulation, the TV tower is divided into 48 segments corresponding to each floor. The wind velocity time histories within each segment are assumed to be identical and equal to that simulated at the height of the corresponding floor. The time interval and duration of the generated wind velocities are assigned to be 0.05 s and 300 s, respectively.

Finally, the wind drag force can be obtained by

$$F_i = \frac{1}{2} C_D A_i \rho [\bar{v}_i + v_i(t)]^2, \tag{A.6}$$

where C_D, A_i and ρ are the shape factor, windward area and air density, respectively.

References

1. X. Zhao and L. Qin, Embodied carbon based integrated optimal seismic design for super tall buildings with viscoelastic coupling dampers, *Procedia Eng.* **118** (2015) 223–231.
2. F. Adachi, K. Fujita, M. Tsuji and I. Takewaki, Importance of interstory velocity on optimal along-height allocation of viscous oil dampers in super high-rise buildings, *Eng. Struct.* **56** (2013) 489–500.
3. B. R. Ellingwood, D. V. Rosowsky, Y. Li and J. H. Kim, Fragility assessment of light-frame wood construction subjected to wind and earthquake hazards, *J. Struct. Eng.* **130** (2004) 1921–1930.
4. S. Kameshwar and J. E. Padgett, Multi-hazard risk assessment of highway bridges subjected to earthquake and hurricane hazards, *Eng. Struct.* **78** (2014) 154–166.

5. D. Y. Wang, K. T. Tse, Y. Zhou and Q. X. Li, Structural performance and cost analysis of wind-induced vibration control schemes for a real super-tall building, *Struct. Infrastruct. Eng.* **11** (2015) 990–1011.
6. W. X. Zhong and J. H. Lin, "Whiplash effect" of high buildings caused by earthquake (in Chinese), *J. Vib. Shock* **2** (1985) 5–10.
7. L. J. Zhang, H. Y. Zhang, Y. X. Ji and S. W. Hu, Whiplash effect on hydropower house during an earthquake, *Int. J. Struct. Stab. Dyn.* **14** (2014) 1450007.
8. M. P. Repetto and G. Solari, Wind-induced fatigue collapse of real slender structures, *Eng. Struct.* **32** (2010) 3888–3898.
9. B. Fu, J. B. Zhao, B. Q. Li, J. Yao, A. R. Mouafo Teifouet, L. Y. Sun and Z. Y. Wang, Fatigue reliability analysis of wind turbine tower under random wind load, *Struct. Saf.* **87** (2020) 101982.
10. A. Crewe, Passive energy dissipation systems in structural engineering, *Struct. Saf.* **20** (1998) 197–205.
11. S. Etedali, Z. K. Bijaem, N. Mollayi and V. Babaiyan, Artificial intelligence-based prediction models for optimal design of tuned mass dampers in damped structures subjected to different excitations, *Int. J. Struct. Stab. Dyn.* **21** (2021) 2150120.
12. K. Vanshaj, A. Kumar Shukla and M. Shukla, Seismic response on soil–structure interaction of asymmetric plan buildings with active tuned mass dampers, *Int. J. Struct. Stab. Dyn.* **22** (2022) 2250102. https://doi.org/10.1142/S0219455422501024.
13. F. D. S. Brando, A. K. D. Almeida and L. F. F. Miguel, Optimum design of single and multiple tuned mass dampers for vibration control in buildings under seismic excitation, *Int. J. Struct. Stab. Dyn.* **22** (2022) 2250078. https://doi.org/10.1142/S021945542250078X.
14. T. Konar and A. D. Ghosh, Adaptation of a deep liquid-containing tank into an effective structural vibration control device by a submerged cylindrical pendulum appendage, *Int. J. Struct. Stab. Dyn.* **21** (2021) 2150078. https://doi.org/10.1142/S0219455421500784.
15. B. C. Zhao, B. Lu, X. Zeng and Q. Gu, Experimental and numerical study of hysteretic performance of new brace type damper, *J. Constr. Steel Res.* **183** (2021) 106717.
16. M. N. Eldin, A. J. Dereje and J. Kim, Seismic retrofit of RC buildings using self-centering PC frames with friction-dampers, *Eng. Struct.* **208** (2020) 109925.
17. C. W. Zhang, A. Ali and L. Sun, Investigation on low-cost friction-based isolation systems for masonry building structures: experimental and numerical studies, *Eng. Struct.* **243** (2021) 112645.
18. A. Ghafouri-Nejad, M. Alirezaei, S. M. Mirhosseini and E. Zeighami, Parametric study on seismic response of the knee braced frame with friction damper, *Structures* **32** (2021) 2073–2087.
19. W. Cui, T. Ma and L. Caracoglia, Time-cost "trade-off" analysis for wind-induced inhabitability of tall buildings equipped with tuned mass dampers, *J. Wind Eng. Ind. Aerodyn.* **207** (2020) 104394.
20. Q. H. Wang, H. S. Qiao, D. De Domenico, Z. W. Zhu and Y. Tang, Seismic performance of optimal multi-tuned liquid column damper-inerter (MTLCDI) applied to adjacent high-rise buildings, *Soil Dyn. Earthq. Eng.* **143** (2021) 106653.
21. S. Cammelli, Y. F. Li and S. Mijorski, Mitigation of wind-induced accelerations using tuned liquid column dampers: Experimental and numerical studies. *J. Wind Eng. Ind. Aerodyn.* **155** (2016) 174–181.
22. S. Chowdhury, A. Banerjee and S. Adhikari, Optimal design of inertial amplifier base isolators for dynamic response control of multi-storey buildings, *Int. J. Struct. Stab. Dyn.* (2022), doi: https://doi.org/10.1142/S0219455423500475.
23. Y. Q. Xu, T. C. Becker and T. Guo, Design optimization of triple friction pendulums for high-rise buildings considering both seismic and wind loads, *Soil Dyn. Earthq. Eng.* **142** (2021) 106568.

24. J. Sheikhi, M. Fathi, R. Rahnavard and R. Napolitano, Numerical analysis of natural rubber bearing equipped with steel and shape memory alloys dampers, *Structures* **32** (2021) 1839–1855.
25. H. S. Kim and J. W. Kang, Smart outrigger damper system for response reduction of tall buildings subjected to wind and seismic excitations, *Int. J. Steel Struct.* **17** (2017) 1263–1272.
26. L. V. Kozak, M. I. Dzubenko and V. M. Ivchenko, Temperature and thermosphere dynamics behavior analysis over earthquake epicentres from satellite measurements, *Phys. Chem. Earth* **29** (2004) 507–515.
27. J. Laštovička, J. Baše, F. Hruška, J. Chum, T. Šindelářová, J. Horálek, J. Zedink and V. Krasnov, Simultaneous infrasonic, seismic, magnetic and ionospheric observations in an earthquake epicentre, *J. Atmos. Sol. Terr. Phys.* **72** (2010) 1231–1240.
28. Centre USNH. Hurricane Otto: Category two storm makes Nicaragua landfall. BBCCom 2016. www.bbc.com/news/world-latin-america%0A-38090722.
29. H. N. Li, X. W. Zheng and C. Li, Copula-based approach to construct a joint probabilistic model of earthquakes and strong winds, *Int. J. Struct. Stab. Dyn.* **19** (2019) 1950046.
30. C. Li, Y. Liu and H. N. Li, Fragility assessment and optimum design of a steel–concrete frame structure with hybrid energy-dissipated devices under multi-hazards of earthquake and wind, *Eng. Struct.* **245** (2021) 112878.
31. H. N. Li, Y. Liu, C. Li and X. W. Zheng, Multihazard fragility assessment of steel-concrete composite frame structures with buckling-restrained braces subjected to combined earthquake and wind, *Struct. Des. Tall Spec. Build.* **29** (2020) 1–19.
32. A. Vulcano, Comparative study of the earthquake and wind dynamic responses of base-isolated buildings, *J. Wind Eng. Ind. Aerodyn.* **74–76** (1998) 751–764.
33. S. Mazzoni, F. Mckenna and G. L. Fenves, OpenSees command language manual, in (Pacific Earthquake Engineering Research (PEER) Center, California, USA, 2001).
34. M. J. Zhu, F. Mckenna and M. H. Scott, OpenSeesPy: Python library for the OpenSees finite element framework, *SoftwareX* **7** (2018) 6–11.
35. J. Jin, H. M. Liu, J. Q. Hong and J. S. Zhang, Study on guyed mast vibration control by use of SMA (in Chinese), *Eng. Mech.* **21** (2004) 119–123.
36. N. Fallah and G. Zamiri, Multi-objective optimal design of sliding base isolation using genetic algorithm, *Sci. Iran.* **20** (2013) 87–96.
37. C. Y. Yuan, J. Y. Chen, J. Li and Q. Xu, Fragility analysis of large-scale wind turbines under the combination of seismic and aerodynamic loads, *Renew. Energy* **113** (2017) 1122–1134.
38. GB50011-2016, Code for seismic design of buildings (Architecture and Building Press, Beijing, China, 2016).
39. GB 50009-2012. Load code for the design of building structures (Architecture and Building Press, Beijing, China, 2012).
40. GB 50135-2019. Standard for design of high-rising structures (Architecture and Building Press, Beijing, China, 2019).
41. S. Taghavi and E. Miranda, Approximate floor acceleration demands in multistory buildings. II: applications, *J. Struct. Eng.* **131** (2005) 212–220.
42. J. S. Kuang and K. Huang, Simplified multi-degree-of-freedom model for estimation of seismic. response of regular wall-frame structures, *Struct. Des. Tall Spec. Build.* **20** (2011) 418–432.
43. H. Tennekes, The logarithmic wind profile, *J. Atmos. Sci.* **30** (1973) 234–238.
44. Y. P. Song, J. B. Chen, Y. B. Peng, P. D. Spanos and J. Li, Simulation of nonhomogeneous fluctuating wind speed field in two-spatial dimensions via an evolutionary wavenumber-frequency joint power spectrum, *J. Wind Eng. Ind. Aerodyn.* **179** (2018) 250–259.

45. X. L. Zhao, C. R. Liu, P. G. Wand and G. L. Zhang, Experimental simulation study on dynamic response of offshore wind power pile foundation under complex marine loadings, *Soil Dyn. Earthq. Eng.* **156** (2022) 107232.
46. JG/T 118-2018. Rubber isolation bearing for buildings (Architecture and Building Press, Beijing, China, 2018).
47. E. Choi, J. G. Ha, D. Hahm and M. K. Kim, A review of multihazard risk assessment: Progress, potential, and challenges in the application to nuclear power plants, *Int. J. Disaster Risk Reduct.* **53** (2022) 101933.
48. A. Saltelli, P. Annoni, I. Azzini, F. Campolongo, M. Ratto and S. Tarantola, Variance based sensitivity analysis of model output. Design and estimator for the total sensitivity index, *Comput. Phys. Commun.* **181** (2010) 259–270.
49. A. Saltelli, Making best use of model evaluations to compute sensitivity indices, *Comput. Phys. Commun.* **145** (2002) 280–297.
50. J. C. Kaimal, J. C. Wyngaard, Y. Izumi and O. R. Cote, Spectral characteristics of surface-layer turbulence, *Q. J. R. Meteorol. Soc.* **98** (1972) 563–589.
51. G. Q. Huang, H. L. Liao and M. S. Li, New formulation of Cholesky decomposition and applications in stochastic simulation, *Probabilistic Eng. Mech.* **34** (2013) 40–47.

https://doi.org/10.1142/9789819814770_0002

Chapter 2

Investigation of Spatial-Varying Frequencies Concerning Effects of Moving Mass on a Beam#

Judy P. Yang* and Zhi-Yuan Su

Department of Civil Engineering,
National Yang Ming Chiao Tung University,
Hsinchu 300093, Taiwan
**jpyang@nycu.edu.tw*

J. D. Yau

Department of Architecture, Tamkang University,
New Taipei City 251301, Taiwan

D. S. Yang

Department of Civil Engineering,
Monash University, Clayton, Victoria 3800, Australia

The spatial-varying frequency of a vehicle-bridge interaction (VBI) system subjected to a moving mass is theoretically derived and numerically investigated through a three-dimensional VBI model, in which the effects of moving mass are introduced through the inertial force and centrifugal force in the equation of motion of the bridge. For a large vehicle-to-bridge mass ratio, it has been known that the frequency of a VBI system could change with respect to the location of a moving vehicle. As such, this study derives the analytical solution based on a moving mass-beam system to account for frequency variation and further builds the numerical model with detailed implementation for practical applications. The numerical results show the following findings: (1) The frequency of a VBI system is a function of velocity and location of a moving vehicle. (2) The reduction of spatial-varying frequency ratio for a particular mode decreases with respect to the mode of higher order. (3) The maximum reduction of spatial-varying frequency ratio of the first mode in a moving mass-beam system occurs in the location where the bridge has the maximum deflection as a result of local mode excitation. (4) For the VBI system with high suspension stiffness and large vehicle-to-bridge mass ratio, the

*Corresponding author.
#This chapter appeared previously on the International Journal of Structural Stability and Dynamics. To cite this chapter, please cite the original article as the following: J. P. Yang *et al.*, *Int. J. Struct. Stab. Dyn.*, **23**, 2340002 (2023), DOI: 10.1142/S0219455423400023.

absolute variation of spatial-varying frequency ratio of the first mode can be up to 30–40%.

Keywords: Moving mass; frequency variation; inertial force; centrifugal force; mass ratio; scale effect.

1. Introduction

The fountain head of vehicle-bridge interaction (VBI) can be dated back to the investigation of high-speed trains in the 1990s.[1] With the simplification of connected moving loads by a single vehicle model, the application to analysis of interaction between two sub-systems (i.e. vehicle and bridge) was later developed in 2004.[2,3] Nowadays, the identification of modal parameters such as bridge damping, frequencies, and mode shapes has become one important research direction in the VBI-related topics.[4]

As the VBI system contains two sub-systems, the VBI effect arises from the fact that the moving vehicle excites the bridge while the vibration of bridge turns back and affects the movement of vehicle at the same time. In most previous studies, the VBI effect is not crucial with *a priori* assumptions that the vehicle mass is negligible in comparison with that of the bridge, and the variation of bridge frequencies is therefore ignored in the dynamic analysis. Nevertheless, for vehicles of heavy weights, it can be expected that the instantaneous frequencies of bridge change with time due to the movement of vehicle on the bridge.

The investigation of frequency variation began on the experimental tests in the 2000s. Such a phenomenon was not even mentioned in some studies while the indication of frequency variability could be observed from the results. For instance, an ambient vibration test was conducted on a cable-stayed bridge to investigate the natural frequencies of global modes under normal traffic-induced loads; it was observed that there was 1% change in frequencies within one day.[5] The ambient vibration tests were conducted on three bridges to investigate the effect of vehicle weight on bridge frequencies, and it was shown that the measured natural frequencies differ up to 5.4% for bridges of short spans.[6] On the other hand, a lack of theoretical studies on frequency variation was resolved by the closed-form solutions of a VBI system as proposed by Yang *et al.*,[7] in which a lumped sprung mass is assumed for the vehicle while a simple beam is used for the bridge. The time-varying frequencies of the bridge and the vehicle were solved analytically by the eigenvalue analysis. In this direction, the eigenvalue analysis is further performed on the basis of a two degrees-of-freedom interaction model,[8] where both theoretical and experimental investigations were included.

To account for the inertial force due to moving mass, different approaches were proposed to deal with the moving mass on the bridge. The constitutive equation of an Euler-Bernoulli beam under the excitation of moving mass was analyzed by a control algorithm using different numbers of controlled modes and actuators.[9] The moving finite element formulation was proposed by modelling the moving mass as a moving finite element.[10] Unlike the aforementioned methods using moving mass, a moving internal node element was proposed in the finite element framework to consider the VBI effect for problems of moving loads and moving vehicles.[11]

As motivated by the aforementioned studies, this study conducts the modal analysis to extract the instantaneous frequencies of a VBI system, in which a full trip of a vehicle travelling over the bridge is considered. Since there exists limited research on the theory of spatial-varying frequency in a VBI system subjected to a moving mass, the theoretical derivation of spatial-varying frequency in a VBI system concerning the inertial force and centrifugal force is established for the first time. A three-dimensional analysis is further provided to investigate the frequency variation. The analytical solution derived in this study can be served as a reference in the design of bridges and construction of complex numerical models concerning the effects of moving mass on a beam.

The rest of this study is arranged as follows. The theoretical description of the vehicle-bridge interaction is presented in Sec. 2. The numerical examples along with software implementation in constructing the numerical models are provided in Sec. 3. Section 4 concludes the present study.

2. Theoretical Description of Vehicle-Bridge Interaction

The mathematical model of a VBI system is first introduced as the basic framework of this study. For the VBI system with high suspension stiffness and large vehicle-to-bridge mass ratio, the presence of both inertial force and centrifugal force is further included in the equation of motion of the bridge.

2.1. *Vehicle-bridge interaction system*

The mathematical model of a vehicle-bridge interaction system is depicted in Fig. 1, which consists of a single-axle two-mass vehicle model,[12] and the vehicle moves at a constant speed v on the simply supported bridge of length L. The following parameters are included: sprung mass m_v, unsprung mass m_u, and a suspension unit of a spring constant k_v and damping coefficient c_v.

Under the finite element (FE) framework, by ignoring the effects of moving mass, the equations of motion for a two-mass vehicle and a beam element are derived as

$$\begin{bmatrix} m_v & 0 \\ 0 & m_u \end{bmatrix} \begin{bmatrix} \ddot{q}_v \\ \ddot{q}_u \end{bmatrix} + \begin{bmatrix} c_v & -c_v \\ -c_v & c_v \end{bmatrix} \begin{bmatrix} \dot{q}_v \\ \dot{q}_u \end{bmatrix} + \begin{bmatrix} k_v & -k_v \\ -k_v & k_v \end{bmatrix} \begin{bmatrix} q_v \\ q_u \end{bmatrix} = \begin{bmatrix} 0 \\ f_c - (m_v + m_u)g \end{bmatrix} \tag{1}$$

and

$$\mathbf{m}_b\ddot{\mathbf{q}}_b + \mathbf{c}_b\dot{\mathbf{q}}_b + \mathbf{k}_b\mathbf{q}_b = -\mathbf{N}_c f_c, \tag{2}$$

where $\mathbf{m}_b$, $\mathbf{c}_b$, and $\mathbf{k}_b$ are the mass, damping, and stiffness matrices of the beam element. q_v and q_u are the displacements of the sprung mass and unsprung mass, respectively. $\mathbf{q}_b$ is the displacement vector of the bridge. $\mathbf{N}$ is the vector containing cubic Hermitian interpolation functions, and the subscript c denotes the contact point. The superscript dot denotes the derivative with respect to time throughout the content. The interaction between the vehicle and bridge is through the contact force $f_c = (m_v + m_u)g + m_v\ddot{q}_v + m_u\ddot{q}_u$. Referring to Ref. 12, the coupled VBI system of equations can be established as

$$\mathbf{M}\begin{bmatrix}\ddot{q}_v\\ \ddot{\mathbf{q}}_b\end{bmatrix} + \mathbf{C}\begin{bmatrix}\dot{q}_v\\ \dot{\mathbf{q}}_b\end{bmatrix} + \mathbf{K}\begin{bmatrix}q_v\\ \mathbf{q}_b\end{bmatrix} = \mathbf{F}, \tag{3}$$

where $\mathbf{M}$, $\mathbf{C}$, and $\mathbf{K}$ are the mass, damping, and stiffness matrices of the VBI system while $\mathbf{F}$ is the force vector.

2.2. *Derivation of frequency variation for moving mass-beam system*

To focus on the effects of moving mass, the theoretical derivation of frequency variation in a VBI system is interpreted through a moving mass on the bridge, i.e. a moving mass-beam system. Referring to Fig. 1, for a bridge subjected to a moving mass, the equations of motion for the bridge and moving mass can be, respectively described by the following equations:

$$\bar{m}\ddot{u} + c\dot{u} + EIu'''' = \left[(m_v + m_u)g - m_u\frac{d^2u}{dt^2} - m_v\ddot{u}_v\right]\delta(x - vt), \tag{4}$$

$$m_v\ddot{u}_v + c_v\dot{u}_v + k_v u_v = k_v u(x,t), \tag{5}$$

where an undamped bridge with $c = 0$ is assumed in this study. $(\bullet)' = \frac{\partial(\bullet)}{\partial x}$, $(\dot{\bullet}) = \frac{\partial(\bullet)}{\partial t}$, and $0 \le vt \le L$; $u(x,t)$ is the vertical deflection of the beam; $\bar{m}$ denotes

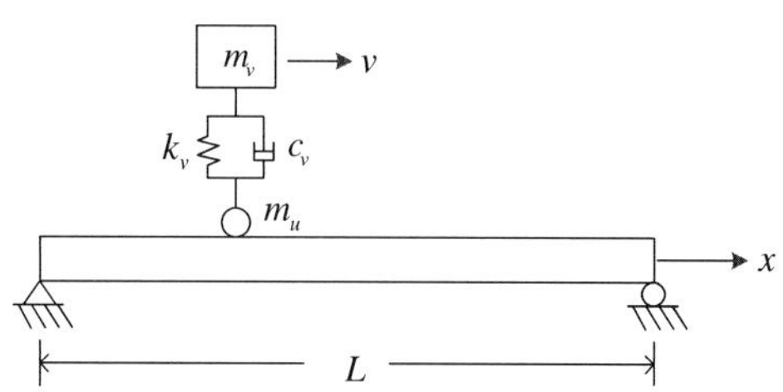

Fig. 1. Mathematical model.

the mass per unit length of the beam; EI is the flexural rigidity of the beam; δ is Dirac delta function; g is the gravitational acceleration; u_v is the displacement of the sprung mass.

For a VBI system with high suspension stiffness and large vehicle-to-bridge mass ratio, the effects of moving mass arise from the second force term in Eq. (4). For simplicity, assuming that the moving mass has mass m (neglecting unsprung mass), the above force term can be modified accordingly as[13,14]

$$m\frac{d^2u}{dt^2} = m\left(\frac{\partial^2 u}{\partial t^2} + 2v\frac{\partial^2 u}{\partial t \partial x} + v^2\frac{\partial^2 u}{\partial x^2}\right), \tag{6}$$

where the three terms on the right-hand side stand for the inertial force, Coriolis force, and centrifugal force. Particularly, the inertial force acts in the direction of the beam deflection; the Coriolis force is related to the rate of inclination of the beam, arising from the dynamic coupling between the velocity of the moving mass and the angular velocity of the bridge; the centrifugal force is related to the curvature of the beam induced by the velocity of the moving mass in the position of action.

For a simple beam, its deflection can be approximated by

$$u(x,t) = \sum_n q_n \sin\left(\frac{n\pi x}{L}\right), \tag{7}$$

where q_n is the modal coordinate, and $\sin(\frac{n\pi x}{L})$ is the mode shape. By substituting Eq. (7) into Eqs. (4) and (6), multiplying $\sin(\frac{l\pi x}{L})$ on both sides of the equation, and performing the integration, the following equation can be reached:

$$\begin{aligned}
&\sum_n \left[\frac{\bar{m}L}{2} + m\sin\left(\frac{l\pi vt}{L}\right)\sin\left(\frac{n\pi vt}{L}\right)\right]\ddot{q}_n \\
&\quad + \sum_n \left[\frac{EIL}{2}\left(\frac{n\pi}{L}\right)^4 - mv^2\left(\frac{n\pi}{L}\right)^2\sin\left(\frac{l\pi vt}{L}\right)\sin\left(\frac{n\pi vt}{L}\right)\right]q_n \\
&\quad + \sum_n 2vm\left[\sin\left(\frac{l\pi vt}{L}\right)\cos\left(\frac{n\pi vt}{L}\right)\left(\frac{n\pi}{L}\right)\right]\dot{q}_n \\
&= mg\sin\left(\frac{l\pi vt}{L}\right).
\end{aligned} \tag{8}$$

The above equation can be arranged in matrix form to obtain the generalized equation of free vibration for a simple beam subjected to a moving mass:

$$\mathbf{M}_m \begin{bmatrix}\ddot{q}_1\\ \vdots \\ \ddot{q}_n\end{bmatrix} + \mathbf{C}_m \begin{bmatrix}\dot{q}_1\\ \vdots \\ \dot{q}_n\end{bmatrix} + \mathbf{K}_m \begin{bmatrix}q_1\\ \vdots \\ q_n\end{bmatrix} = \begin{bmatrix}0\\ \vdots \\ 0\end{bmatrix}, \tag{9}$$

where $\mathbf{M}_m$, $\mathbf{C}_m$, and $\mathbf{K}_m$ are the mass, damping, and stiffness matrices of the VBI system concerning the effects of moving mass, and they are explicitly derived as

$$\mathbf{M}_m = \begin{bmatrix} 1+\rho\sin^2\frac{\pi vt}{L} & \cdots & \rho\sin\left(\frac{\pi vt}{L}\right)\sin\left(\frac{n\pi vt}{L}\right) \\ \vdots & \ddots & \vdots \\ \rho\sin\left(\frac{n\pi vt}{L}\right)\sin\left(\frac{\pi vt}{L}\right) & \cdots & 1+\rho\sin^2\frac{n\pi vt}{L} \end{bmatrix}, \tag{10}$$

$$\mathbf{C}_m = \frac{2\rho\pi v}{L}\begin{bmatrix} \sin\left(\frac{2\pi vt}{L}\right) & \cdots & n\sin\left(\frac{\pi vt}{L}\right)\cos\left(\frac{n\pi vt}{L}\right) \\ \vdots & \ddots & \vdots \\ \sin\left(\frac{n\pi vt}{L}\right)\cos\left(\frac{\pi vt}{L}\right) & \cdots & n\sin\left(\frac{2n\pi vt}{L}\right) \end{bmatrix}, \tag{11}$$

$$\mathbf{K}_m = \begin{bmatrix} \omega_{b1}^2-\varpi_1^2 & \cdots & \rho\left(\frac{\pi v}{L}\right)^2\sin\left(\frac{n\pi vt}{L}\right)\sin\left(\frac{\pi vt}{L}\right) \\ \vdots & \ddots & \vdots \\ \rho\left(\frac{n\pi v}{L}\right)^2\sin\left(\frac{n\pi vt}{L}\right)\sin\left(\frac{\pi vt}{L}\right) & \cdots & \omega_{bn}^2-\varpi_n^2 \end{bmatrix}, \tag{12}$$

where the following parameters are defined: $\varpi_n^2 = \rho(\frac{n\pi v}{L})^2\sin(\frac{l\pi vt}{L})\sin(\frac{n\pi vt}{L})$, the free vibration frequency of the bridge $\omega_{bn} = (\frac{n\pi}{L})^2\sqrt{\frac{EI}{\bar{m}}}$, and the modal mass ratio $\rho = \frac{2m}{\bar{m}L}$.

For demonstration purpose, the first mode is considered in the characteristic equation $|\mathbf{K}_m - \omega^2\mathbf{M}_m| = 0$, which leads to the following:

$$\omega_{b1}^2 - \rho\left(\frac{\pi v}{L}\right)^2\sin^2\left(\frac{\pi vt}{L}\right) - \omega^2\left(1+\rho\sin^2\frac{\pi vt}{L}\right) = 0, \tag{13}$$

where ω is the frequency of moving mass-beam system. By defining the driving frequency implied by the moving mass as $\Omega = \pi v/L$, Eq. (13) can be rendered as

$$\omega_{b1}^2 - \rho\Omega^2\sin^2\left(\frac{\pi v}{L}t\right) - \omega^2\left[1+\rho\sin^2\left(\frac{\pi v}{L}t\right)\right] = 0, \tag{14}$$

which can be further simplified as

$$\frac{\omega}{\omega_{b1}} = \sqrt{\frac{1-\rho S_1^2\sin^2\left(\frac{\pi v}{L}t\right)}{1+\rho\sin^2\left(\frac{\pi v}{L}t\right)}}, \tag{15}$$

where the above expression is defined as the spatial-varying frequency ratio of the first mode for $0 \leq vt \leq L$, with $S_1 = \frac{\Omega}{\omega_{b1}}$ and $S_1 < 1$. As higher modes will be investigated in the numerical examples, $\frac{\omega}{\omega_{bi}}$ and S_i are defined for the ith mode in the following study. Clearly, the ith spatial-varying frequency ratio is defined in terms of the frequency of moving mass-beam system to that of the ith bridge natural frequency to consider the effects of moving mass.

In general, the effect of inertial force is larger than that of the centrifugal force, and the effect of Coriolis force is the least important one. When the mass ratio of vehicle to bridge is larger than 1/10, the effects of moving mass should be taken into consideration.[13] As the effects of moving mass are of major interest herein, only large vehicle-to-bridge mass ratio is investigated in the numerical examples. From theoretical point of view, for a moving mass of high velocity, the above effects of Coriolis force and centrifugal force tend to reduce the dynamic stiffness of the bridge, thereby making the bridge unstable. Nevertheless, it should be noted that these two forces induced by the moving mass can generally be neglected in practice since the bridges are designed to be very stiff in order to meet the requirement of safety.

3. Numerical Examples

To better understand the effects of moving mass arising from the inertial force and centrifugal force in the three-dimensional space, the commercial software ABAQUS/CAE 2020[15] is adopted herein.

3.1. *Implementation and scale effect*

In general, the coupled equation of motion of the VBI system in Eq. (3) is commonly solved by Newmark β method in the finite element programming, while it is solved by Hilber-Hughes-Taylor-α (HHT-α) method as defined in ABAQUS/Standard. More specifically, Eq. (3) is solved by the implicit time integration using moderate dissipation, in which the control of numerical damping associated with the time integration is through HHT-α method, i.e. a modified Newmark β method. The default parameters in HHT-α method are listed as follows: $\alpha = -0.41421$, $\beta = 0.5$, and $\gamma = 0.91421$.

To simulate the moving vehicle on the bridge, the *surface-to-surface contact* setting is adopted to describe the interaction between the master surface (bridge pavement) and slave surface (bottom of unsprung mass). In the contact enforcement, the tangential direction is set as frictionless while the normal direction is set as hard contact to avoid penetration. Note that the bridge span in the present model is exactly the same as the analytical one, unlike the additional span length required in Ref. 16.

The instantaneous frequency of a particular mode for the moving mass-beam system is obtained from the eigenvalue analysis, in which the Lanczos method in ABAQUS is adopted by setting the equivalent density of moving mass. The implementation of the inertial force and centrifugal force in the modal analysis is described herein. In a moving mass-beam system, by assuming small deformation and a linearly elastic beam, the centrifugal force in the third term of Eq. (6) can be calculated by using the relation between the curvature κ and moment M of the beam (i.e. $\kappa = \frac{M}{EI}$) as $mv^2 \frac{M}{EI}$, which varies with the velocity and position of the moving mass. As for the inertial force, upon imposing the gravational force and centrifugal force on the moving mass and then recording the vertical accelerations of moving mass in a one-run trip, the inertial force can be obtained by multiplying the moving mass m with these vertical accelerations, in which the vertical accelerations of moving mass are recorded from 15 positions with equal distance along the beam. Since these two forces are input *equivalently* through the density of moving mass in ABAQUS, the density as a result of gravational force, centrifugal force, and inertial force should be used in combination during calculation in this regard. On the other hand, the centrifugal force imposed on the moving mass in the one-run trip is approximated by using the curvature at midspan of bridge, which varies with velocity. Similarly, in a moving vehicle-beam system, the inertial forces in the sprung mass m_v and unsprung mass m_u are obtained by multiplying the individual mass with the corresponding vertical accelerations recorded in the sprung/unsprung mass.

To investigate the effects of moving mass, the scale effect should be taken into account in the numerical simulation. In the analytical model, the moving mass is assumed to be a mass *point* while the bridge is assumed to be a straight *line* ideally. Therefore, the vertical accelerations of moving mass are the same as the vertical accelerations of the bridge. In contrast, the interaction between the moving mass and beam is through the contact *surface* in the numerical model. For the moving mass-beam system of large vehicle-to-bridge mass ratio, the vehicle-to-bridge dimension ratio of the numerical model in three dimensions is relatively larger than that of the analytical model in two dimensions. Although the moving mass moves on the bridge pavement, the numerical results have shown that the vertical accelerations of moving mass do not equal the vertical accelerations of the upper flange of the box beam as shown from the portion of numerical model in Fig. 2. That is to say, the vertical accelerations of moving mass should be utilized in setting the inertial force in the numerical model.

3.2. *Verification of VBI model*

The parameters of the VBI system used in the analytical model are listed below:[17] For the simply supported bridge, $E = 29$ GPa, $I = 0.2422$ m^4, $\bar{m} = 1{,}536$ kg/m, $L = 30$ m, and $f_{b1} = 3.732$ Hz; for the vehicle moving straightly along the bridge,

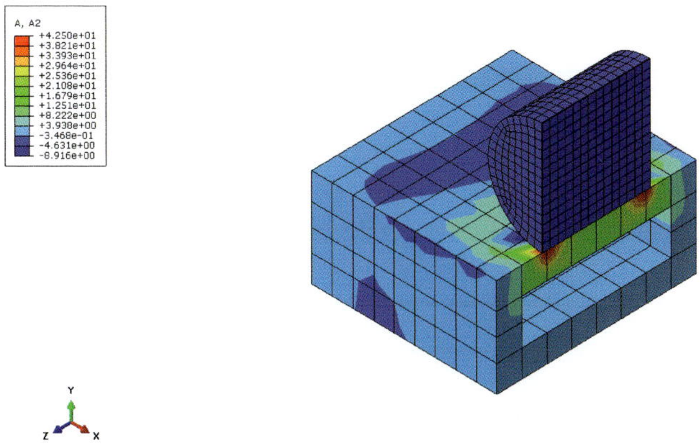

Fig. 2. Illustration of scale effect through the vertical accelerations of moving mass and beam.

$m_v = 1{,}500$ kg, $m_u = 150$ kg, $k_v = 170$ kN/m, $c_v = 0$, and $v = 25$ m/s. To verify the numerical model, the inertial force, Coriolis force, and centrifugal force are neglected in this example. The time interval is set as 0.01 s. The corresponding numerical model is shown in Fig. 3, in which a box cross-section is used for the bridge with width 2.344 m, depth 1.102 m, and constant thickness 0.3 m; the vehicular body has width 1.257 m, height 0.898 m, and length 2.873 m; the wheel has diameter 1.257 m and thickness 1.257 m. The three-dimensional deformable solid brick elements (C3D8, standard) are adopted to mesh the bridge, vehicular body, and wheel with element sizes given by 0.25 m, 0.14 m, and 0.089 m, respectively. In addition to Young's modulus, Poisson's ratio 0.2 is assumed for the bridge. To account for the difference between the analytical model and numerical model, the modified Young's modulus $E = 29.69$ GPa is used herein to match the first bridge frequency in the analytical model as suggested in the reference.[16]

As shown in Figs. 4 and 5, the dynamic responses at midpoint of the bridge and those in the centroid of vehicular body including the displacement, velocity, and acceleration are compared with the closed-form solutions.[17] Obviously, the present numerical model yields results in good agreement with the closed-form solutions while the differences come from different output locations and contact boundary setting between the two-dimensional analytical model and present three-dimensional model. Thus, the correctness of numerical model is verified.

3.3. *Effects of moving mass*

The following parameters are adopted in this example: For the concrete bridge, Young's modulus 24.125 GPa (modified as 24.699 GPa in the numerical model), Poisson's ratio 0.2, density 2,400 kg/m^3, $L = 30$ m, and $f_{b1} = 2.086$ Hz; for the moving mass made of depleted uranium alloy,[18] $m = 15{,}400$ kg, Young's modulus

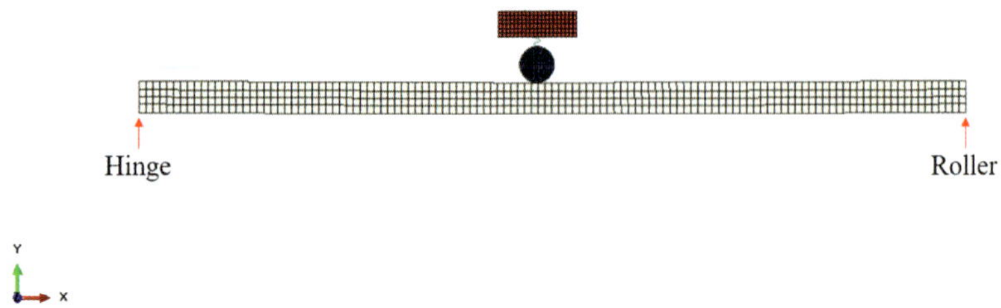

(a)

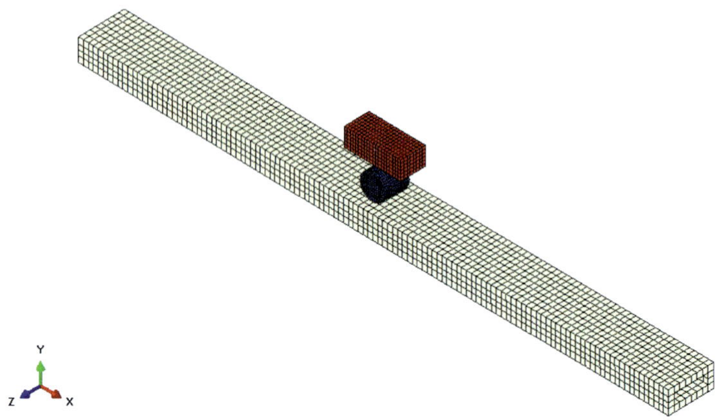

(b)

Fig. 3. A 3D FE model of the VBI system: (a) side view; (b) front view.

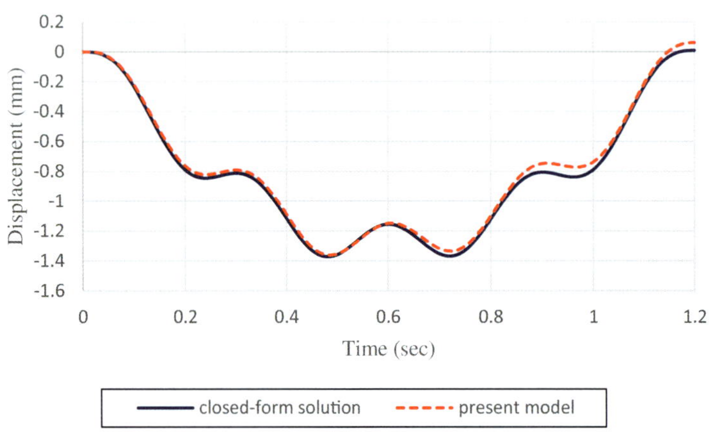

(a)

Fig. 4. Dynamic responses at midpoint of the bridge: (a) displacement, (b) velocity, and (c) acceleration.

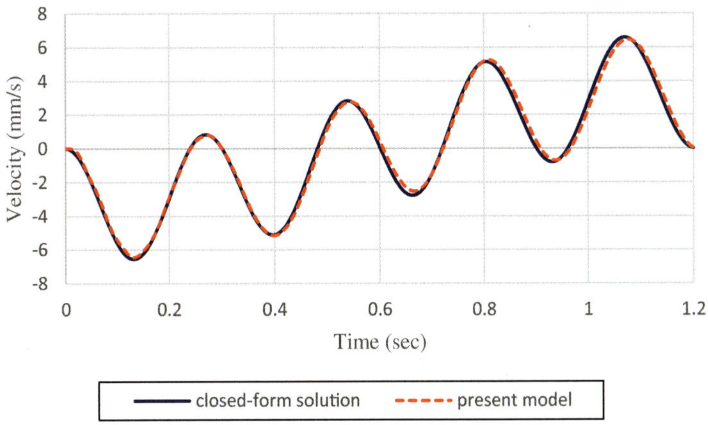

(b)

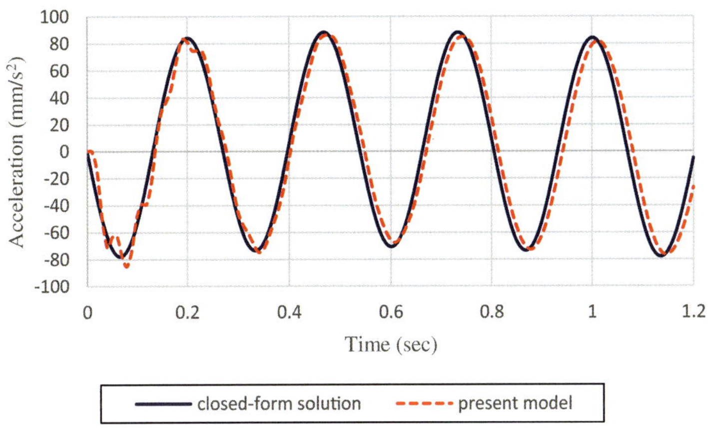

(c)

Fig. 4. (*Continued*)

174 GPa, Poisson's ratio 0.2, and density 9872.451 kg/m^3. The time interval is 0.001 s. As described in Sec. 3.2, the same geometry and boundary conditions of the bridge (including element type and size) are adopted while the geometry and element information of the unsprung mass are used for the moving mass herein. The finite element model of the moving mass-beam system is shown in Fig. 6. Note that the mass ratio of moving mass to bridge is 0.126.

To understand the variation of frequency in the moving mass-beam system, the analytical solution derived in Eq. (15) for the first mode of bridge is interpreted through the three-dimensional plot, as shown in Fig. 7(a); the corresponding result obtained by the numerical simulation is given in Fig. 7(b). By comparison, the trends of spatial-varying frequency ratio in these two sub-figures agree with each other; the verification of numerical simulation is confirmed. It is shown that the first instantaneous frequency of the moving mass-beam system is a function of both velocity and location ($x = vt$) of a moving mass. The decrease in ω/ω_{b1} becomes more obvious with the increase in S_1. In Fig. 8, the spatial-varying frequency ratio ω/ω_{b1} for a moving mass-beam system is examined through various velocities in the range of 0~350 km/hr for a moving mass in different locations of the bridge. The variations of frequency ratios are observed for all velocities of moving mass.

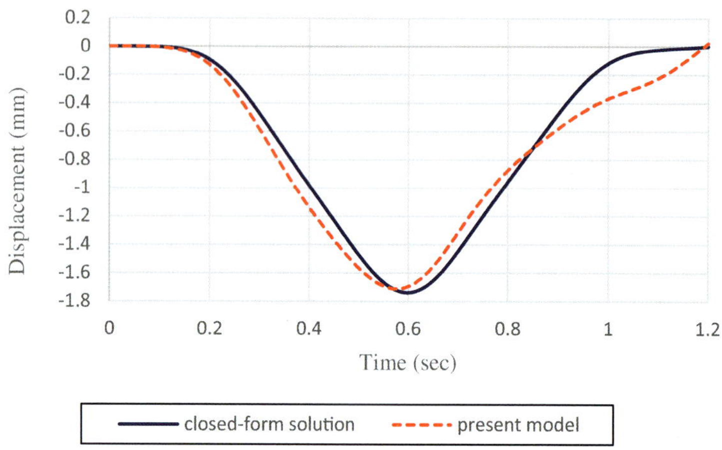

(a)

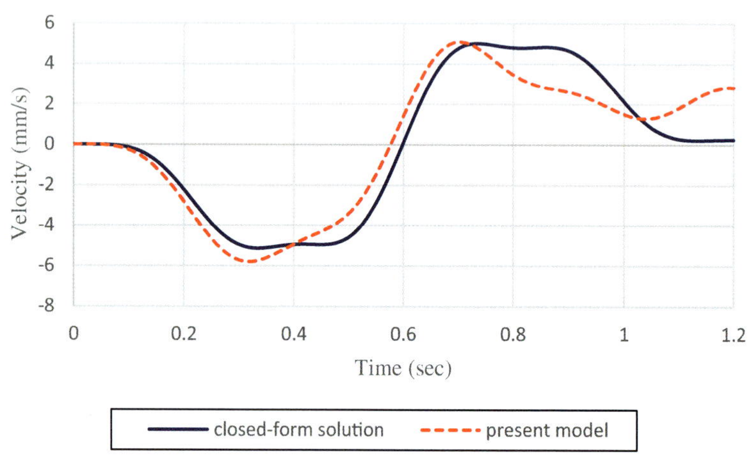

(b)

Fig. 5. Dynamic responses of the vehicular body: (a) displacement, (b) velocity, and (c) acceleration.

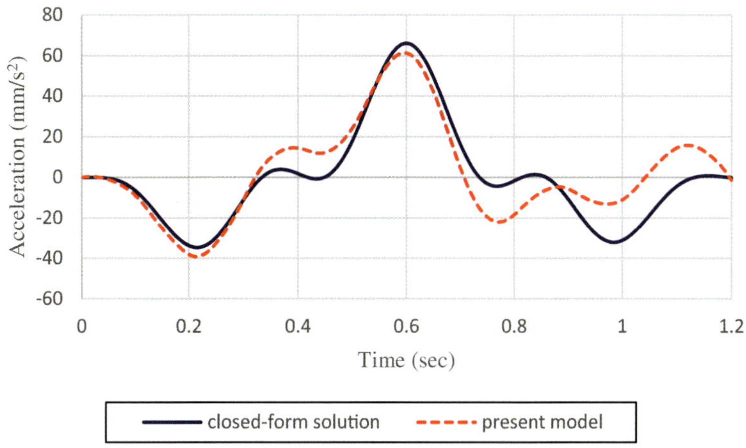

(c)

Fig. 5. (*Continued*)

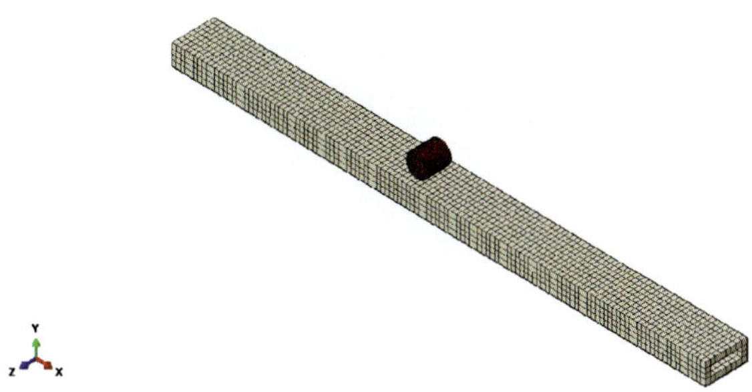

Fig. 6. FEM model of the beam subjected to a moving mass.

For illustration, the maximum reduction of frequency ratio due to moving mass is defined as the minimum frequency ratio divided by the maximum frequency ratio for each velocity; the corresponding dimensionless location of moving mass with the minimum frequency ratio is recorded. As shown in Table 1, the maximum reduction of frequency ratio ω/ω_{b1} varies for different velocities. The maximum reduction of ω/ω_{b1} is 85.5% for velocity of 325 km/hr. Specifically, the reduction of ω/ω_{b1} for 325 km/hr is larger than that of 350 km/hr, which indicates that there is a descending trend in the dynamic stiffness of the system, as also shown from the larger value of maximum deflection of upper flange in the box beam for 325 km/hr depicted in Fig. 9. Thus, it is deduced that the local mode[19] of the upper flange of the beam is excited by the high-speed moving mass.

For the second and third modes of the moving mass-beam system, the spatial-varying frequency ratios are shown in the sub-figures of Fig. 10. In general,

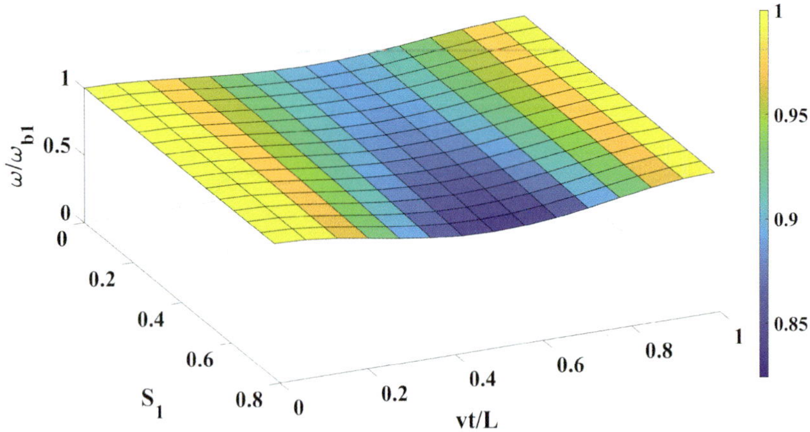

(a)

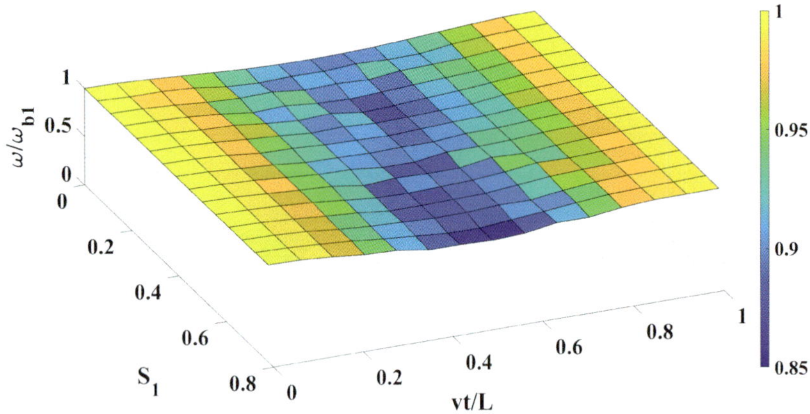

(b)

Fig. 7. Spatial-varying frequency ratios of the first mode for a moving mass-beam system: (a) analytical solution; (b) numerical solution.

similar phenomena to the first mode are observed for these higher modes while some oscillations of frequency ratios are shown. As depicted in Fig. 11, the variations of frequencies for both modes are observed, and the ranges of variations of frequency ratios (ω/ω_{b2} and ω/ω_{b3}) are less than that of the first mode. For visualization, Figs. 8 and 11 are plotted by curve smoothing. From Tables 2 and 3, the maximum reduction of ω/ω_{b2} is 88.8% at a velocity of 350 km/hr, while the maximum reduction of ω/ω_{b3} is 88.4% at a velocity of 325 km/hr. It is concluded that the reduction of spatial-varying frequency ratio decreases with mode of higher order, which is also

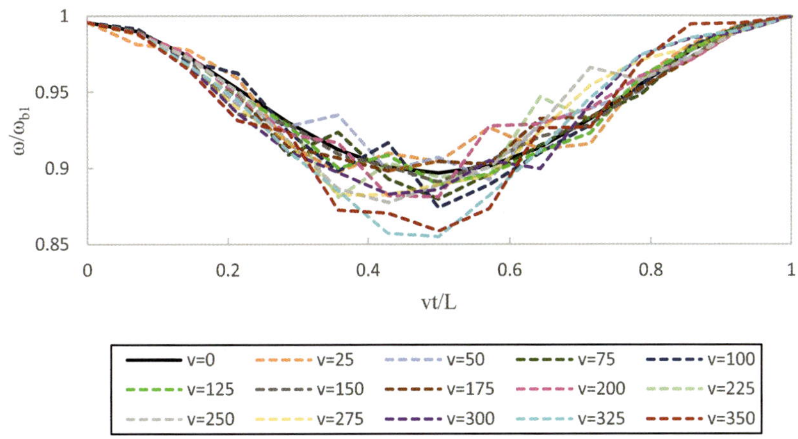

Fig. 8. Frequency ratio ω/ω_{b1} versus various velocities of a moving mass on the bridge. The velocity is in unit of km/hr.

Table 1. Maximum reduction of ω/ω_{b1} versus various velocities of a moving mass.

Velocity (km/hr)	0	25	50	75	100	125	150	175	200	225	250	275	300	325	350
Location x/L	0.50	0.36	0.57	0.50	0.50	0.50	0.50	0.43	0.50	0.36	0.43	0.43	0.43	0.50	0.50
Reduction $\omega/\omega_{b1}(\%)$	89.7	89.8	89.3	88.0	87.5	89.0	89.1	89.9	88.2	88.2	87.8	88.3	88.4	85.5	85.9

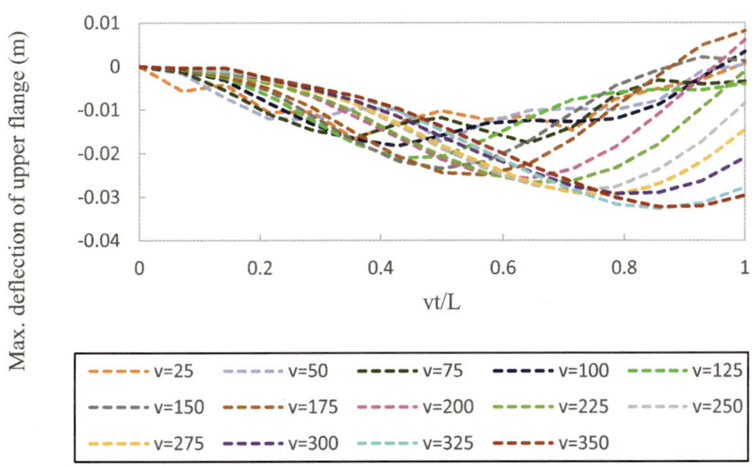

Fig. 9. Maximum deflection of upper flange versus various velocities of a moving mass on the bridge.

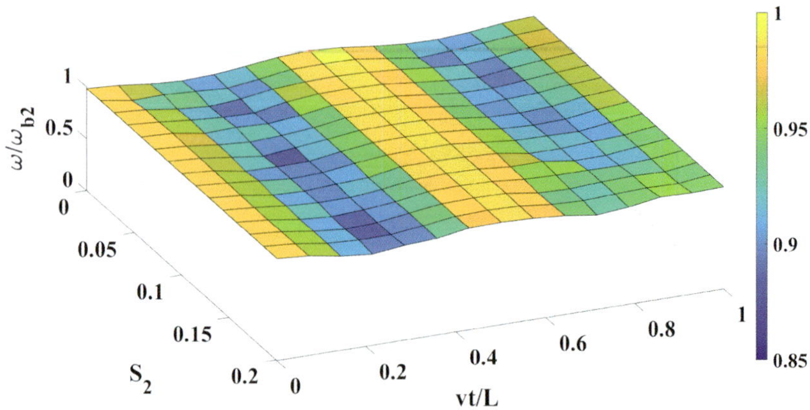

(a)

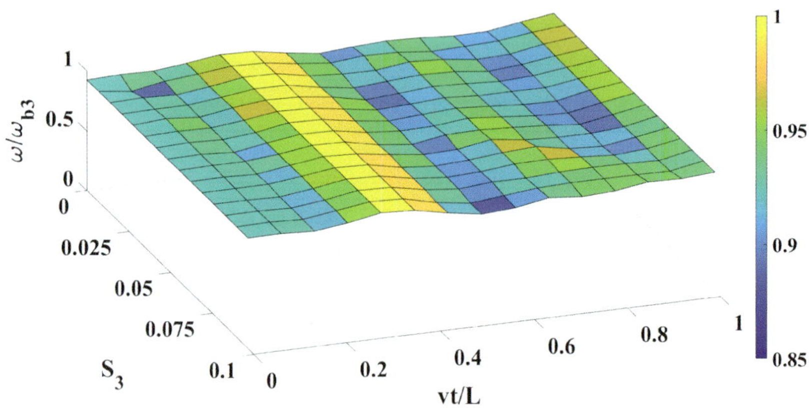

(b)

Fig. 10. Spatial-varying frequency ratios of higher modes for a moving mass-beam system: (a) second mode; (b) third mode.

supported by the decreasing range of spatial-varying frequency-ratio variation for higher modes.

3.4. *Frequency variation in a VBI system*

The tank vehicle has the following parameters: $m_v = 20{,}000$ kg, $m_u = 2{,}000$ kg, Young's modulus 174 GPa, Poisson's ratio 0.2, and $c_v = 0$, sprung mass density 6167.125 kg/m^3, and unsprung mass density 1282.137 kg/m^3. Two scenarios are further considered: (1) The stiffness of tank is chosen as $k_v = 2{,}500$ kN/m so that $\omega_v < \omega_{b1}$, and $f_v = 1.779$ Hz. (2) The stiffness of tank is chosen as $k_v = 5{,}000$ kN/m so that $\omega_v > \omega_{b1}$, and $f_v = 2.516$ Hz. The rest parameters (geometry and

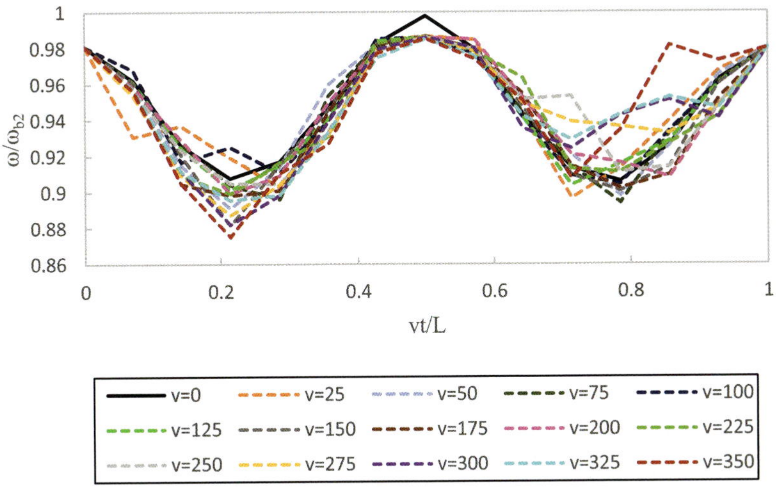

(a)

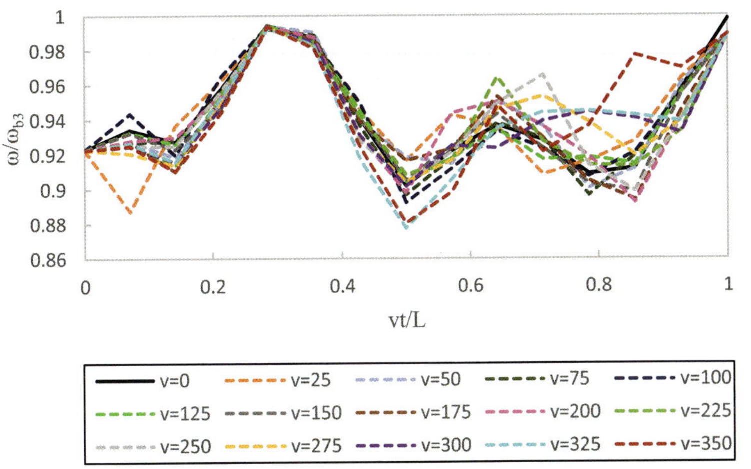

(b)

Fig. 11. Frequency ratios versus various velocities of a moving mass on the bridge: (a) ω/ω_{b2}; (b) ω/ω_{b3}. The velocity is in unit of km/hr.

mesh information) of the VBI system are the same as those described in Sec. 3.2. Note that the mass ratio of vehicle to bridge is 0.179 in this example.

In the first scenario, the corresponding numerical results of spatial-varying frequency of various modes for the VBI system are shown in Fig. 12. It is observed from Fig. 12(a) that the frequency ratio ω/ω_{b1} increases and becomes larger than 1 when the moving vehicle gets closer to the midspan of the bridge, which is also

Table 2. Maximum reduction of ω/ω_{b2} versus various velocities of a moving mass.

Velocity (km/hr)	0	25	50	75	100	125	150	175	200	225	250	275	300	325	350
Location x/L	0.79	0.71	0.21	0.79	0.79	0.21	0.21	0.21	0.21	0.21	0.29	0.21	0.21	0.21	0.21
Reduction $\omega/\omega_{b2}(\%)$	90.8	90.9	90.3	90.7	91.8	91.4	89.4	91.0	91.2	91.2	91.7	90.0	89.5	90.9	88.8

Table 3. Maximum reduction of ω/ω_{b3} versus various velocities of a moving mass.

Velocity (km/hr)	0	25	50	75	100	125	150	175	200	225	250	275	300	325	350
Location x/L	0.50	0.07	0.79	0.79	0.50	0.50	0.86	0.86	0.86	0.50	0.86	0.50	0.50	0.50	0.50
Reduction $\omega/\omega_{b3}(\%)$	90.4	89.4	90.6	90.3	89.8	90.9	90.4	90.0	89.8	91.3	90.5	91.1	90.8	88.4	88.6

clearly presented in Fig. 13(a). As such, the maximum increment of spatial-varying frequency ratio is particularly defined herein to illustrate the variation of frequency, which is defined as the maximum frequency ratio divided by the minimum frequency ratio for each velocity; the corresponding dimensionless location with the maximum frequency ratio is recorded. In Table 4, the maximum increment of ω/ω_{b1} is 147%. To further illustrate the difference between moving mass and moving vehicle, let ω_s denote the instantaneous frequency of sprung mass measured during vehicle passage while ω_v is the static frequency of sprung mass, i.e. $\omega_v = \sqrt{k_v/m_v}$. Unlike ω/ω_{b1}, the frequency ratio ω_s/ω_v decreases and has a minimum around the midspan of

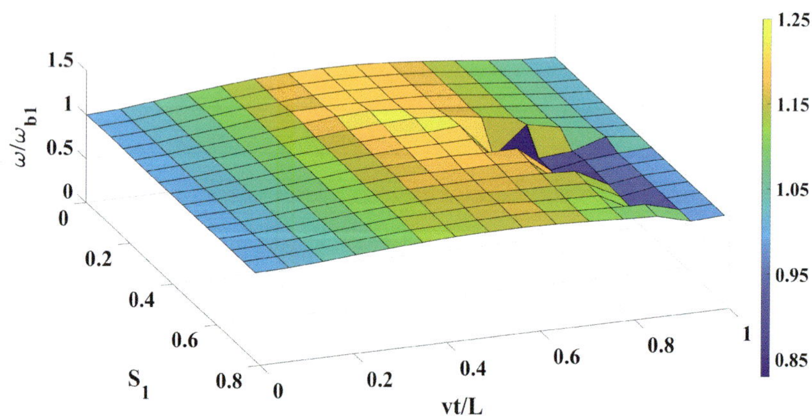

(a)

Fig. 12. Spatial-varying frequency ratios of various modes for a VBI system ($\omega_v < \omega_{b1}$): (a) first mode; (b) second mode; (c) third mode.

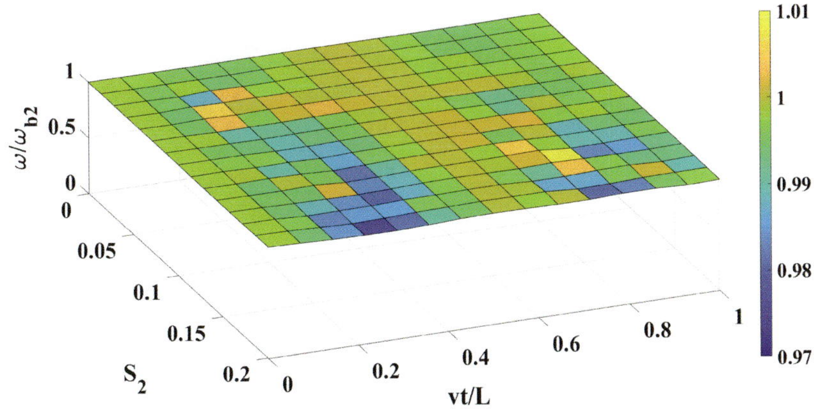

(b)

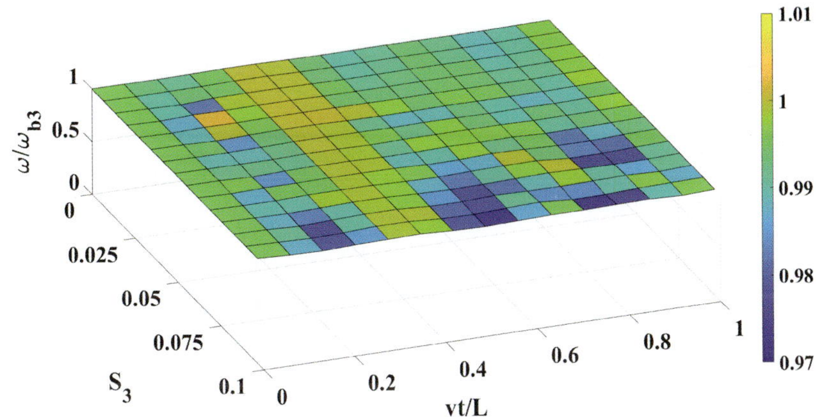

(c)

Fig. 12. (*Continued*)

bridge as shown in Fig. 13(b). For the vehicle with lower frequency ($\omega_v < \omega_{b1}$) as shown in Fig. 13(a), the system becomes softer due to vehicle-bridge interaction, thereby resulting in the sudden drop of ω/ω_{b1} at $\frac{vt}{L} = 0.7$. Recall the impact factor[13] is proportional to $\frac{1}{1-S^2}$ with S the non-dimensional speed parameter proportional to the vehicle velocity v. For the increasing velocity v, the impact factor becomes larger gradually. That is, when the vehicle leaves the bridge in the step-by-step time-history response analysis, as the initial conditions are not zero anymore and the element boundary constraints are changed toward the end support at this moment, the *impact effect* arises due to high velocity and enlargement of system's stiffness, which leads to the sudden increase in ω_s/ω_v starting at $\frac{vt}{L} = 0.7$, as shown in Fig. 13(b).

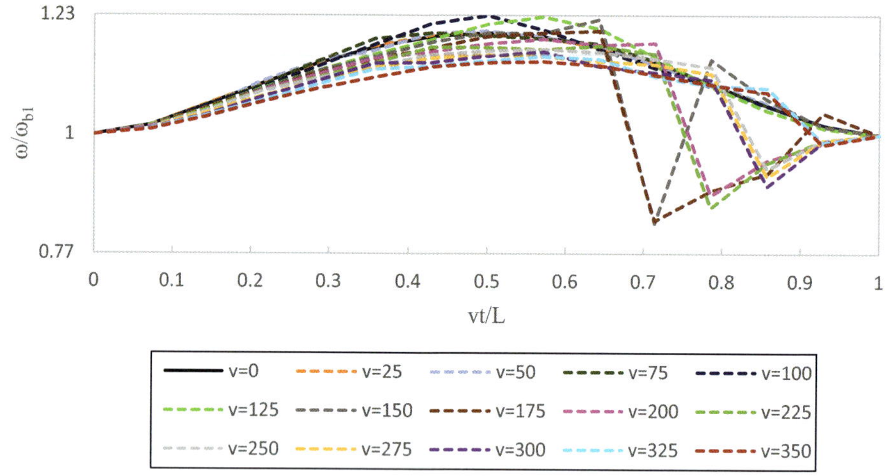

(a)

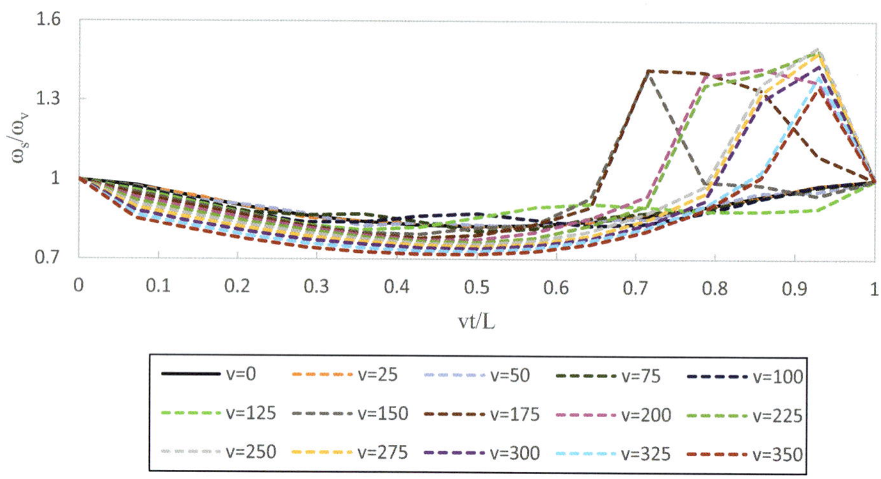

(b)

Fig. 13. Frequency ratios versus various velocities of a moving vehicle on the bridge ($\omega_v < \omega_{b1}$): (a) ω/ω_{b1}; (b) ω_s/ω_v. The velocity is in unit of km/hr.

In the second scenario, the spatial-varying frequency of various modes for the VBI system is shown in Fig. 14. Referring to Figs. 14(a) and 15(a), the frequency ratio ω/ω_{b1} decreases and reaches a minimum around the midspan of bridge, and the maximum reduction of frequency ratio is 72.4% as summarized in Table 5. It is

Table 4. Maximum increment of ω/ω_{b1} versus various velocities of a moving vehicle ($\omega_v < \omega_{b1}$).

Velocity (km/hr)	0	25	50	75	100	125	150	175	200	225	250	275	300	325	350
Location x/L	0.50	0.50	0.50	0.43	0.50	0.57	0.64	0.64	0.57	0.64	0.57	0.50	0.57	0.57	0.57
Increment $\omega/\omega_{b1}(\%)$	120	120	120	120	123	123	147	144	134	136	125	126	129	117	116

found that the trend of first instantaneous frequency of VBI system is close to that of a moving mass-beam system as described in Sec. 3.3; this is mainly due to the fact that the vehicle stiffness is so large that the vibration frequency of sprung mass is larger than the fundamental frequency of bridge, thereby making the moving tank similar to a moving mass. Overall, the reduction of frequency ratio is proportional to the vehicle velocity. For the vehicle with higher frequency ($\omega_v > \omega_{b1}$) as shown in Fig. 15(b), due to impact effect, the sudden peak of ω_s/ω_v is reached at $\frac{vt}{L} = 0.9$ while a delay is observed, due to higher stiffness of vehicle, in comparison with Fig. 13(b).

By comparing Figs. 13(a) and 15(a), the trends of frequency variation of bridge for $\omega_v < \omega_{b1}$ and $\omega_v > \omega_{b1}$ have opposite half-sine forms, which agree well with the phenomena as pointed out in the reference.[7] Unlike the 2D Euler line beam with a solid cross-section, the excitation of higher modes in the current 3D bridge with a box cross-section is particularly observed as shown in Figs. 16 and 17, in

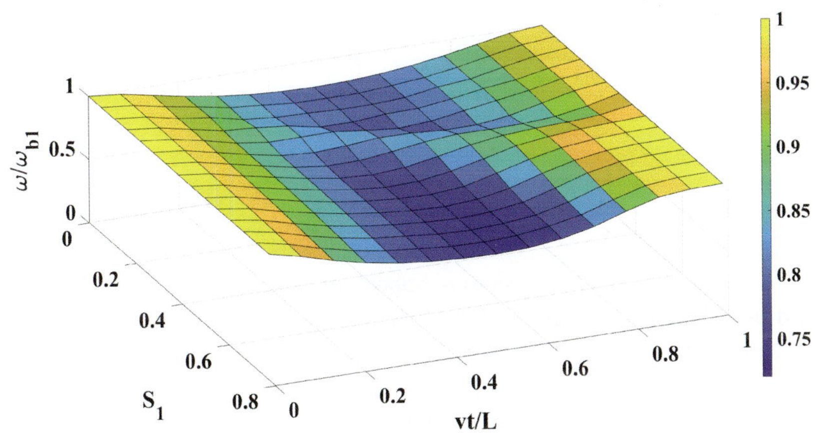

(a)

Fig. 14. Spatial-varying frequency ratios of various modes for a VBI system ($\omega_v > \omega_{b1}$): (a) first mode; (b) second mode; (c) third mode.

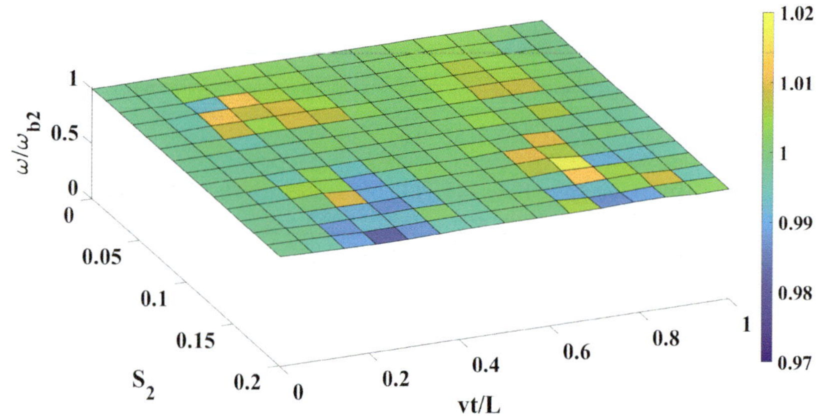

(b)

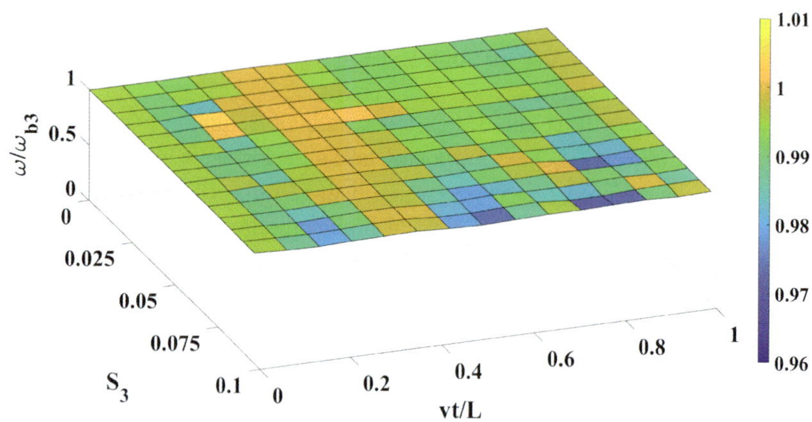

(c)

Fig. 14. (*Continued*)

which the modes of sinusoidal shapes appear only in the first half (in subfigures (a)) and one third (in subfigures (b)) of the bridge while some kinks appear in the rest part of the bridge interrupting the sinusoidal shapes due to the high frequency content arising from the local oscillation of the bridge deck in the 3D model. As shown in Figs. 12(b), 12(c) and 16, for frequency ratios ω/ω_{b2} and ω/ω_{b3} in the first scenario, the corresponding maximum reductions of frequency ratios are 96.9% and 97.1%, respectively, as calculated in Tables 6 and 7. For frequency ratios ω/ω_{b2} and ω/ω_{b3} in the second scenario as shown in Figs. 14(b), 14(c) and 17, the corresponding maximum reductions of frequency ratio are 96.9% and 96.9% as summarized in

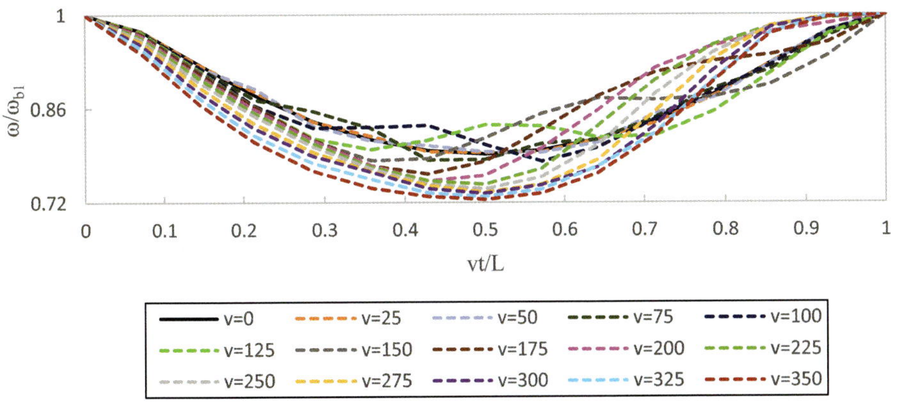

(a)

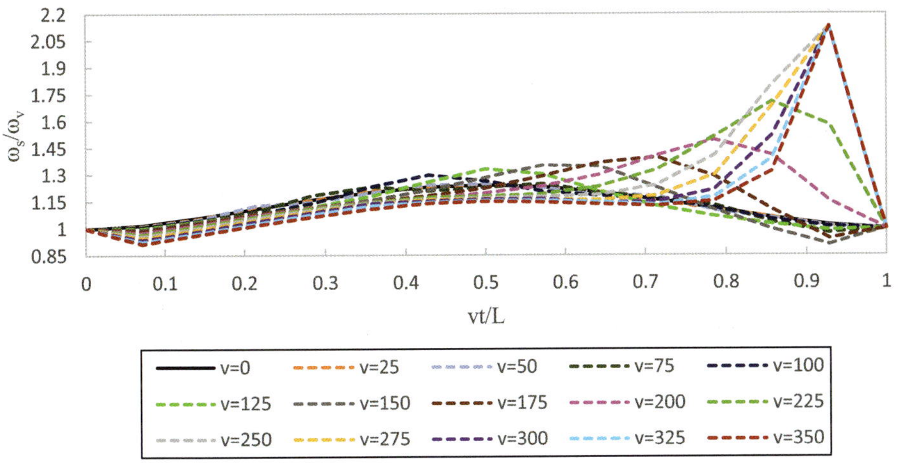

(b)

Fig. 15. Frequency ratios versus various velocities of a moving vehicle on the bridge ($\omega_v > \omega_{b1}$): (a) ω/ω_{b1}; (b) ω_s/ω_v. The velocity is in unit of km/hr.

Tables 8 and 9, respectively. In comparison with the moving mass-beam system, the reductions of second and third spatial-varying frequency ratios in the VBI system (including both scenarios) are smaller than those of a moving mass-beam system, even the total weight of the moving vehicle is heavier than that of the moving mass. As a whole, ω/ω_{b1} makes a major contribution to the variation of instantaneous frequency of the VBI system while higher modes are excited with less influence.

Table 5. Maximum reduction of ω/ω_{b1} versus various velocities of a moving vehicle ($\omega_v > \omega_{b1}$).

Velocity (km/hr)	0	25	50	75	100	125	150	175	200	225	250	275	300	325	350
Location x/L	0.50	0.50	0.57	0.43	0.57	0.36	0.36	0.43	0.43	0.50	0.50	0.50	0.50	0.50	0.50
Reduction $\omega/\omega_{b1}(\%)$	79.2	79.4	79.4	78.3	78.1	79.9	78.2	76.2	75.2	74.7	74.0	73.5	73.3	72.9	72.4

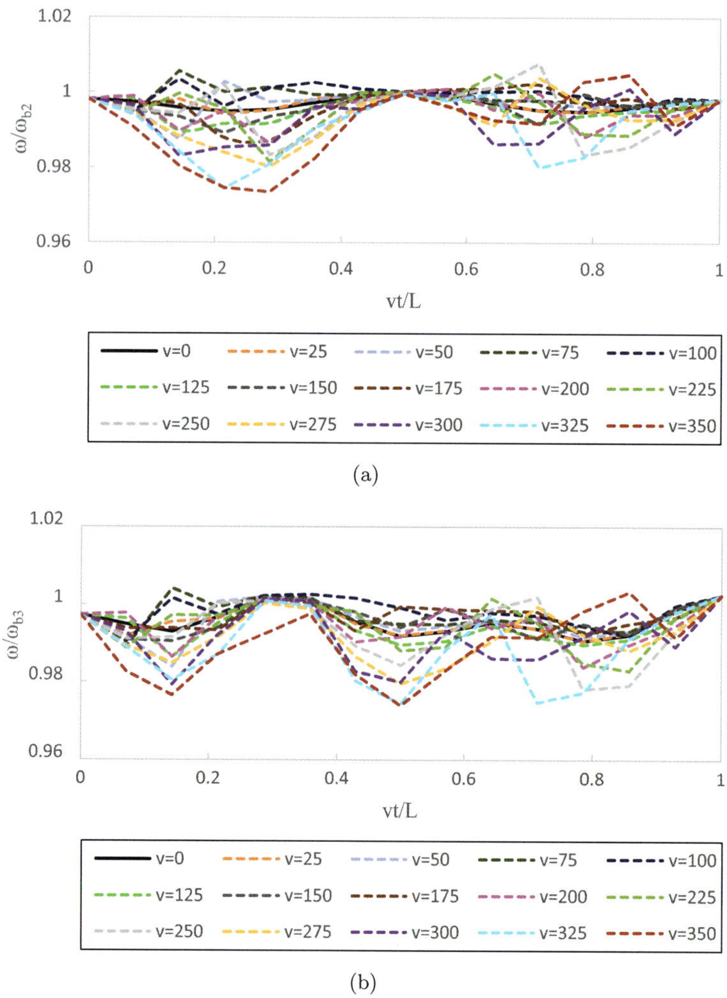

Fig. 16. Frequency ratios versus various velocities of a moving vehicle on the bridge ($\omega_v < \omega_{b1}$): (a) ω/ω_{b2}; (b) ω/ω_{b3}. The velocity is in unit of km/hr.

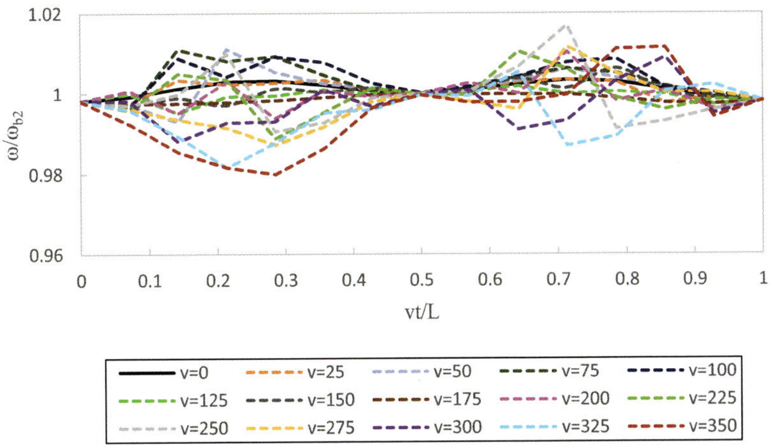

(a)

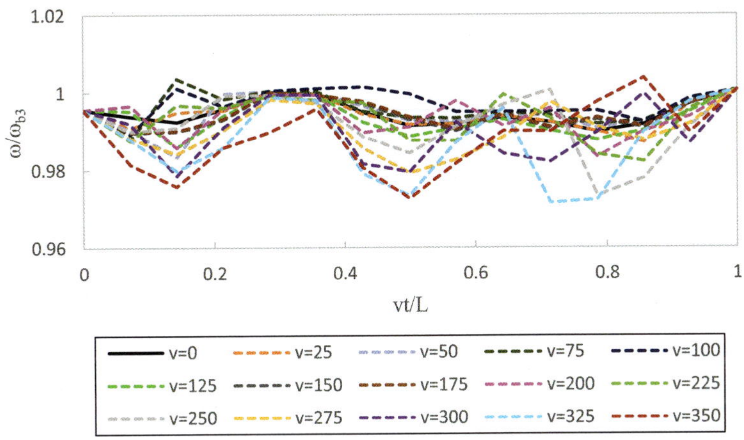

(b)

Fig. 17. Frequency ratios versus various velocities of a moving vehicle on the bridge ($\omega_v > \omega_{b1}$): (a) ω/ω_{b2}; (b) ω/ω_{b3}. The velocity is in unit of km/hr.

Table 6. Maximum reduction of ω/ω_{b2} versus various velocities of a moving vehicle ($\omega_v < \omega_{b1}$).

Velocity (km/hr)	0	25	50	75	100	125	150	175	200	225	250	275	300	325	350
Location x/L	0.79	0.21	0.14	0.07	0.07	0.14	0.21	0.29	0.29	0.29	0.29	0.29	0.14	0.21	0.29
Reduction ω/ω_{b2}(%)	99.5	99.4	98.4	98.9	99.1	99.0	98.8	98.5	98.5	97.7	97.6	97.6	98.2	97.5	96.9

Table 7. Maximum reduction of ω/ω_{b3} versus various velocities of a moving vehicle ($\omega_v < \omega_{b1}$).

Velocity (km/hr)	0	25	50	75	100	125	150	175	200	225	250	275	300	325	350
Location x/L	0.79	0.79	0.14	0.07	0.07	0.14	0.14	0.79	0.79	0.86	0.79	0.50	0.14	0.50	0.50
Reduction $\omega/\omega_{b3}(\%)$	98.8	98.8	98.1	98.5	98.8	98.4	98.8	99.0	98.1	98.1	97.6	97.7	97.7	97.2	97.1

Table 8. Maximum reduction of ω/ω_{b2} versus various velocities of a moving vehicle ($\omega_v > \omega_{b1}$).

Velocity (km/hr)	0	25	50	75	100	125	150	175	200	225	250	275	300	325	350
Location x/L	1.00	0.86	0.14	0.07	0.07	0.14	0.07	0.93	0.29	0.29	0.29	0.29	0.14	0.21	0.29
Reduction $\omega/\omega_{b2}(\%)$	99.5	99.4	98.2	98.5	98.8	99.3	99.1	99.7	98.4	97.9	97.4	97.6	97.9	97.6	96.9

Table 9. Maximum reduction of ω/ω_{b3} versus various velocities of a moving vehicle ($\omega_v > \omega_{b1}$).

Velocity (km/hr)	0	25	50	75	100	125	150	175	200	225	250	275	300	325	350
Location x/L	0.79	0.86	0.14	0.07	0.07	0.14	0.07	0.07	0.79	0.86	0.79	0.50	0.14	0.71	0.50
Reduction $\omega/\omega_{b3}(\%)$	98.9	98.7	98.3	98.4	98.8	98.5	98.9	98.9	98.3	98.2	97.3	97.9	97.8	97.1	96.9

4. Conclusions

In this study, the analytical solution of the spatial-varying frequency in a moving mass-beam system is first derived through the inclusion of both inertial force and centrifugal force in the equation of motion of the bridge. From the numerical results, the following conclusions are drawn:

(1) For large vehicle-to-bridge mass ratios, the frequency of a VBI system is a function of both velocity and location of a moving vehicle. Even for a stationary vehicle, the frequency of the VBI system is a function of vehicle position. The change of spatial-varying frequency ratio ω/ω_{bi} decreases with the increase of mode number (i.e. $i = 1, 2, 3$) as also supported by the decreasing range of variation for the spatial-varying frequency ratio ω/ω_{bi}. It is noted that only the first three modes were investigated for demonstration purpose, including but not limited to, based on the derived analytical solution.

(2) In the moving mass-beam system, due to the excitation of local mode on the upper flange of the box beam, the maximum reduction of ω/ω_{b1} occurs in the location where the bridge has the maximum deflection.

(3) For moving vehicle with $\omega_v < \omega_{b1}$, the first instantaneous frequency of VBI system is larger than the fundamental bridge frequency during vehicle passage for various velocities, and the maximum spatial-varying frequency ratio ω/ω_{b1} is 47% higher than the minimum spatial-varying frequency ratio ω/ω_{b1}. In contrast, the instantaneous frequency of sprung mass is smaller than the natural frequency of sprung mass during vehicle passage for various velocities.

(4) For moving vehicle with $\omega_v > \omega_{b1}$, the first instantaneous frequency of VBI system is less than the fundamental bridge frequency during vehicle passage for various velocities, and the minimum spatial-varying frequency ratio ω/ω_{b1} is 27.6% lower than the maximum spatial-varying frequency ratio ω/ω_{b1}. The trend of first instantaneous frequency of VBI system is close to that of a moving mass-beam system. The reduction of frequency ratio ω/ω_{b1} is generally proportional to the vehicle velocity.

(5) For both $\omega_v < \omega_{b1}$ and $\omega_v > \omega_{b1}$, the impact effect becomes pronounced in the last passage on the bridge. Overall, ω/ω_{b1} has a major contribution to the variation of instantaneous frequency of the VBI system.

Acknowledgments

The authors express their great gratitude to Dr. Xuzhao Lu for software assistance and Dr. Kai-Chun Chang for wording correction. The support of this work under the grant nos. MOST 111-2628-E-A49-009-MY3, NSTC 111-2221-E-032-033, and MOST 110-2923-E-032-001-MY3 by National Science and Technology Council of the Republic of China (Taiwan) is gratefully acknowledged.

References

1. Y. B. Yang, Z. L. Wang, K. Shi, H. Xu and Y. T. Wu, State-of-the-art of vehicle-based methods for detecting various properties of highway bridges and railway tracks, *Int. J. Struct. Stab. Dyn.* **20**(13) (2020) 2041004.
2. Y. B. Yang and J. P. Yang, State-of-the-art review on modal identification and damage detection of bridges by moving test vehicles, *Int. J. Struct. Stab. Dyn.* **18**(2) (2018) 1850025.
3. Z. L. Wang, J. P. Yang, K. Shi, H. Xu, F. Q. Qiu and Y. B. Yang, Recent advances in researches on vehicle scanning method for bridges, *Int. J. Struct. Stab. Dyn.* **22**(15) (2022) 2230005.
4. Y. He, J. P. Yang and Y. F. Li, A three-stage automated modal identification framework for bridge parameters based on frequency uncertainty and density clustering, *Eng. Struct.* **255** (2022) 113891.
5. Q. W. Zhang, L. C. Fan and W. C. Yuan, Traffic-induced variability in dynamic properties of cable-stayed bridge, *Earthq. Eng. Struct. Dyn.* **31** (2002) 2015–2021.
6. C. Y. Kim, D. S. Jung, N. S. Kim, S. D. Kwon and M. Q. Feng, Effect of vehicle weight on natural frequencies of bridges measured from traffic-induced vibration, *Earthq. Eng. Eng. Vib.* **2**(1) (2003) 109–115.

7. Y. B. Yang, M. C. Cheng and K. C. Chang, Frequency variation in vehicle-bridge interaction systems, *Int. J. Struct. Stab. Dyn.* **13**(2) (2013) 1350019.
8. K. C. Chang, C. W. Kim and S. Borjigin, Variability in bridge frequency induced by a parked vehicle, *Smart Struct. Syst.* **13**(5) (2014) 755–773.
9. A. Nikkhoo, F. R. Rofooei and M. R. Shadnam, Dynamic behavior and modal control of beams under moving mass, *J. Sound Vib.* **306** (2007) 712–724.
10. I. Esen, Dynamic response of a beam due to an accelerating moving mass using moving finite element approximation, *Math. Comput. Appl.* **16** (2011) 171–182.
11. D. S. Yang, C. M. Wang and J. D. Yau, Moving internal node element method for dynamic analysis of beam structure under moving vehicle, *J. Sound Vib.* **511** (2021) 116348.
12. J. P. Yang and B. H. Chen, Two-mass vehicle model for extracting bridge frequencies, *Int. J. Struct. Stab. Dyn.* **18**(4) (2018) 1850056.
13. Y. B. Yang, J. D. Yau and Y. S. Wu, *Vehicle-Bridge Interaction Dynamics with Applications to High-Speed Railways* (World Scientific Publishing, 2004).
14. D. S. Yang, C. M. Wang and J. D. Yau, Dynamic stability and response of inclined beams under moving mass and follower force, *Int. J. Struct. Stab. Dyn.* **20**(14) (2020) 2043004.
15. ABAQUS/Standard, *ABAQUS/CAE User's Manual* (Dassault Systemes Simulia Corp, Providence, RI, USA, 2014).
16. X. Lu, C. W. Kim and K. C. Chang, Finite element analysis framework for dynamic vehicle-bridge interaction system based on ABAQUS, *Int. J. Struct. Stab. Dyn.* **20**(3) (2020) 2050034.
17. J. P. Yang, Theoretical formulation of three-mass vehicle model for vehicle-bridge interaction, *Int. J. Struct. Stab. Dyn.* **21**(7) (2021) 2171004.
18. B. Miranbeigi and S. Maleki, Live load distribution factor for tank loading on slab-girder bridges, *KSCE J. Civ. Eng.* **23** (2019) 3420–3430.
19. K. Matsuoka, A. Collina, C. Somaschini and M. Sogabe, Influence of local deck vibrations on the evaluation of the maximum acceleration of a steel-concrete composite bridge for a high-speed railway, *Eng. Struct.* **200** (2019) 109736.

https://doi.org/10.1142/9789819814770_0003

Chapter 3

Drive-By Fleet Monitoring to Detect Bearing Damage in Bridges Using a Moving Reference Influence Function[#]

Eugene J. OBrien*, Daniel P. McCrum† and Shuo Wang‡

School of Civil Engineering
University College Dublin, Belfield, Dublin, Ireland
**eugene.obrien@ucd.ie*
†daniel.mccrum@ucd.ie
‡shuo.wang@ucdconnect.ie

This paper introduces a new bridge damage indicator, the moving reference influence function (MRIF), to detect bridge bearing damage using deflections inferred from vehicle accelerations. Recently, vehicle acceleration has been used to find the apparent profile (AP) of a bridge when a vehicle passes. This AP consists of bridge profile elevations and bridge deflection components. To describe the relationship between these deflection components and load, a MRIF is proposed for the first time in this paper. An error minimization process is used to find the MRIF and the road surface profile on the bridge. The vehicle acceleration signals used in the paper are assumed to be collected from a partially instrumented vehicle fleet. In the fleet, only the first axle acceleration is collected from each vehicle. To simplify the minimization process, both the MRIF and the bridge profile are represented by kernel density functions. The results show that the bridge profile can be accurately obtained and that bridge bearing damage can be identified from the MRIF. Both area and skewness of the MRIF are damage sensitive and can be used together to find the location and severity of bridge bearing damage.

Keywords: Drive-by; vehicle acceleration; fleet monitoring; moving reference influence function; bridge; bearing damage; damage detection; kernel density function.

1. Introduction

As bridges require periodic maintenance interventions, it is essential to understand the condition of these structures during their service lives. Depending on the source of data measurement, structural health monitoring (SHM) methods can be categorized into two groups: direct methods and indirect methods. The direct

‡Corresponding author.
[#]This chapter appeared previously on the International Journal of Structural Stability and Dynamics. To cite this chapter, please cite the original article as the following: E. J. OBrien, D. P. McCrum & S. Wang, *Int. J. Struct. Stab. Dyn.*, **23**, 2340003 (2023), DOI: 10.1142/S0219455423400035.

methods refer to data obtained from sensors positioned directly on the structure itself. Such sensors can directly measure bridge acceleration, strain, deflection or rotation.[1–5] Many bridge modal properties can be obtained from the bridge response to monitor the structure: natural frequency, mode shape and modal curvature.[6] Most frequency-based methods are based on the fact that bridge damage generally causes changes in natural frequencies.[7] However, recent studies have shown that the changes due to damage are relatively small and may be overshadowed by changes due to environmental effects.[8] Mode shapes are found to be less sensitive to environmental effects[9] but are also not very sensitive to local damage in particular.[10] Modal curvature has good potential to detect localized damage but requires large numbers of sensors.[11] Instead of monitoring bridge modal properties, some other methods have been developed, such as non-destructive methods[12] and influence line-based methods.[1] The non-destructive method normally uses radio X-ray, radiography, eddy current or ultrasonics at a localized area of the structure.[12] It relies on detecting the short vibration bursts emitted by the cracks.[13] However, non-destructive methods require the location of the damage and the bridge cannot be in service during the test. The influence line-based method aims to monitor the structure by observing the changes in influence line as bridge damage will cause changes in the influence line.[1] The influence line is defined as the bridge response due to a unit load moving across the bridge. The response can be stress/strain, deflection or rotation. The changes in the influence enable the detection of both damage severity and location.[14,15] The disadvantage of the influence line-based method is that an accurate influence line is difficult to obtain in practice and the changes due to the damage may be conflated with the changes due to the environment e.g. temperature. Generally, for direct methods, a measurement system is required for each target structure and many sensors are required to be installed on each bridge, especially if mode shapes or modal curvature are required.[16] Although the accuracy of direct methods is reasonably good, the installation and maintenance of the sensors, data acquisition and communication electronics are expensive and it would be inordinately expensive to monitor all bridges in a transportation network. Also, there can be challenges with access and with electrical power.

In recent years, indirect methods have been developed as reusable and cost-effective alternatives to direct SHM. Indirect methods, also called "drive-by" methods, use inertial data from vehicles passing over the bridge to infer information about its health condition. As the vehicle passes over the bridge, its vibration is affected by both the vehicle properties (e.g. speed, mass, sprung stiffness, etc.) and the bridge properties (e.g. stiffness, mass, boundary conditions, etc.). Hence, it is possible to obtain information about the condition (e.g. stiffness) of the bridge from passing vehicle measurements.

The concept of drive-by monitoring was first introduced by Yang *et al.*, using vehicle acceleration signals to extract the bridge natural frequency.[17] The change in natural frequency of the bridge was used as an indicator of the structural condition of the bridge. Since then, many different damage indicators have been proposed to monitor the condition of the structure. Yang and Yang divide drive-by monitoring methods into two categories: (1) modal identification methods and (2) damage detection methods.[18] Modal identification refers to the extraction of bridge dynamic properties such as natural frequencies, mode shapes and damping ratios. The challenge for most natural frequency-based methods is that the measured vehicle acceleration contains significant frequency components associated with vehicle frequencies and is perhaps excited by the road profile (pavement roughness).[19] The frequency components, which are close to the bridge frequencies, are difficult to filter out and can significantly affect the accuracy of the inferred bridge natural frequencies. Yang *et al.* developed a method to find bridge mode shapes from a passing vehicle acceleration by analyzing the bridge frequency components.[20] However, the accuracy was shown to be highly sensitive to vehicle speed and does not seem to be feasible in practice. McGetrick *et al.* used vehicle acceleration to infer changes in bridge damping. Their study showed that increasing bridge damping reduces the magnitude of the bridge and vehicle peaks in the power spectral density plot, which allows damage to be identified.[21] However, bridge surface profile tends to prevent damage identification when a rough profile exists. González *et al.* further developed a method to find the damping ratio through a least square minimization process.[22] The results show that different profile roughnesses give similar levels of accuracy. Yang *et al.* proposed a theoretical framework to find the damping in a bridge from an analysis of a passing vehicle's vertical acceleration and the relative displacement between the vehicle and the bridge in the frequency domain.[23] The error in the damping estimation approach was found to be less than 5% when both the vehicle and bridge frequencies are accurately known.

Other drive-by damage detection methods aim to find bridge damage by solving the equations of motion of the vehicle–bridge interaction system. Majumder and Manohar proposed a method to find bridge element stiffness from bridge–vehicle interaction system equations.[24] However, the road profile was not considered in these simulations and is known to be important. Similarly, Bu *et al.* introduced an approach to find stiffness changes in the bridge using a sensitivity analysis of vehicle acceleration signals.[25] Kim and Kawatani used the equation of motion of the vehicle–bridge interaction system in a pseudo-static formulation.[26] They found the normalized change of the bridge stiffness matrix by solving the general inverse problem. Results showed that both location and severity of element stiffness change can be found. In an alternative to directly obtaining the bridge stiffness matrix, other researchers have detected bridge crack size and location by analyzing vehicle vibration response through a continuous wavelet transform.[27,28]

More recently, some researchers have used the inferred bridge profile height to detect damage. When a vehicle is on a bridge, the inferred profile, known as the apparent profile (AP), contains elements of bridge deflection as well as the true (unloaded) profile. OBrien and Keenahan calculated the AP from data generated using a simulated traffic speed deflectometer.[29] They found that the differences between the two APs, damaged minus healthy, can be used to identify local damage. Without using a traffic speed deflectometer to get the AP heights, Keenahan *et al.* further developed an inverse Newmark-beta method to calculate profile height from vehicle axle accelerations.[30] This method requires all vehicle properties such as axle weights, tyre stiffness and spring stiffness. To find all the vehicle properties, a fleet monitoring concept was introduced using cross entropy optimization.[30] With inferred vehicle properties, the bridge AP can be found from a fleet of vehicle accelerations as they pass over the bridge. Yang *et al.* proposed a method to calculate the displacement of the "contact point acceleration response" from a two-axle passing vehicle.[31] This displacement is very similar to the AP proposed in this method. This method requires both axle accelerations and the displacement influence line and is capable of obtaining the bridge profile. Currently, there is no method capable of finding the bridge profile height and the displacement influence line at the same time.

In this paper, a method is proposed to separate the bridge profile from the bridge deflection components. The deflection components are the result of a number of axles acting at different locations on the bridge. The bridge response is represented here with a moving reference influence function (MRIF), defined as the deflection at a moving reference point, x_k, due to a unit axle load at a point, $(x_k - d)$, where d is the distance between the unit load and the measured point. The MRIF is proposed here, for the first time, as a damage sensitive indicator of bridge behavior. Typically, bridge damage can be categorized as global or local. Examples of global damage are foundation settlement, boundary condition changes (bearing damage) and overall uniform reduction in flexural rigidity. Local damage consists of a local loss of flexural rigidity. In this paper, bridge bearing failure is considered as an example of global damage. Many studies show that the damage that occurs in bridge bearings can change the transmission of the force from the main girder to the bearing.[32,33] Kaitlyn simulated bearing damage by gradually increasing the stiffness of the roller support.[34] An in-service highway bridge studied by Matthew *et al.* considered bearings to be in poor condition as they offered restriction to rotational movement due to corrosion and joint leaking.[35] Hence, bearing damage in this paper is represented as an increase in rotational stiffness.

2. Apparent Profile and Moving Reference Deflection

The concept of AP was firstly proposed by OBrien and Keenahan to describe the profile experienced by the instrumented vehicle as it passes over a bridge.[29] The AP can be accurately obtained from vehicle acceleration using a direct integration method (such as inverse Newmark-beta method). However, all vehicle properties are required in a direct integration method (speed, mass, sprung stiffness, etc.). In this study, all relative vehicle properties except axle weights are assumed to be accurately estimated using, for example, the Cross Entropy optimization algorithm derived from on-road vehicle accelerations.[30] Also, it is assumed that vehicle axle weights are known, perhaps from an on-board vehicle weighing system.[36] As all vehicle properties and accelerations are known when the vehicle passes the bridge, direct integration can be used to obtain the AP. Figure 1 shows an example of an AP (due to a two-axle vehicle with total weight 7.6 t, 6.4 m axle spacing and 26.1 m/s velocity on a 20 m span bridge) and the corresponding real bridge profile. More details of the bridge properties are provided in Sec. 5.

As bridge deflection is defined down as negative, the AP is smaller than the real bridge profile in Fig. 1. For scan k, the AP, R_k^a, consists of the bridge surface profile and its deflection components:

$$R_k^a = R_k^0 + \delta_k, \tag{2.1}$$

where R_k^0 denotes the bridge surface profile height and δ_k is the bridge deflection whose location varies in time as the vehicle progresses across the bridge. Figure 2 shows the moving reference deflection due to a two-axle vehicle (the vehicle is represented simply by two individual forces). For a multi-axle vehicle, the superposition of the bridge deflection can be written as

$$\delta_k = \delta_{k,1} + \delta_{k,2} + \cdots + \delta_{k,n}, \tag{2.2}$$

where n is the number of axles that contribute to the deflection.

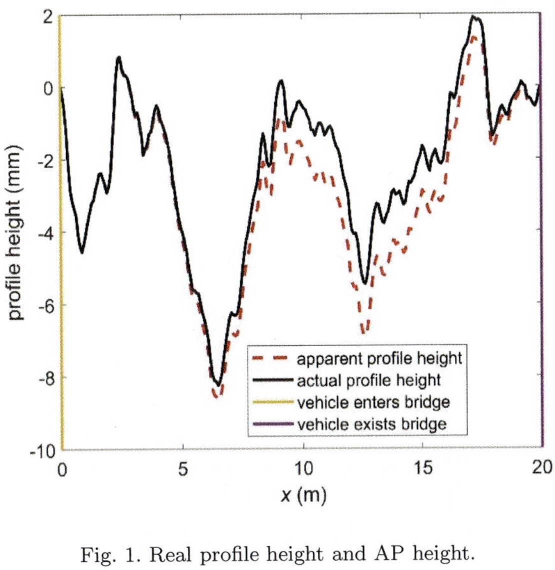

Fig. 1. Real profile height and AP height.

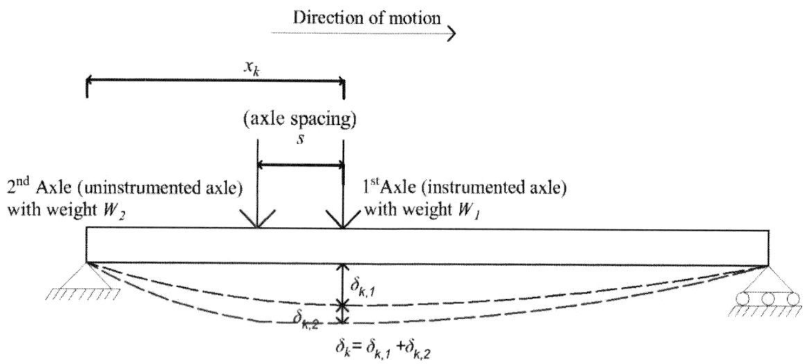

Fig. 2. Moving reference deflection due to a two-axle vehicle for scan k (where xk is the location of the first axle relative to the left support of a simply supported bridge).

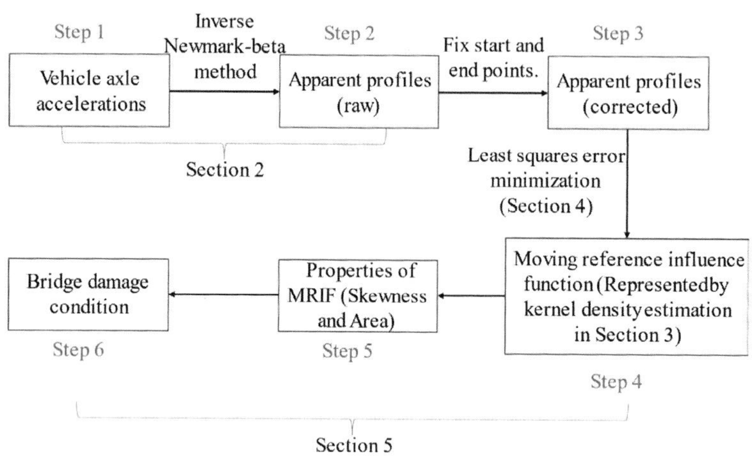

Fig. 3. Flow chart of proposed MRIF procedure to detect bridge damage using measured vehicle accelerations.

Figure 3 shows the concept of using the MRIF, inferred from the AP, to indicate bridge damage. Firstly, vehicle acceleration is used to calculate the AP through the inverse Newmark-beta method (Steps 1 and 2 in Fig. 3 and discussed in Sec. 2 of this paper). Then, APs are corrected to remove drift related to the numerical integration process (Step 3). These corrected APs are used to derive an MRIF through an error minimization process (Steps 3–4), which will be introduced in Sec. 4 of this paper. The properties of this MRIF are used to detect damage in the bridge, as discussed in Sec. 5 of this paper. The principal novelty in this paper is the concept of using MRIF as an indicator of bridge behavior and ultimately of its condition.

3. Moving Reference Influence Function and Its Kernel Density Estimation

In structural mechanics, an influence line is defined as the structural response (deflection, reaction force, bending moment, etc.) at a fixed point due to a unit load moving across the structure. The only variable in this definition is the location of the unit load. Moving reference deflection, on the other hand, has two variables: the location of the unit load and the location of the measurement point. Hence, the MRIF is defined in this paper as the bridge deflection at a (moving) measurement point, x_k, due to a (moving) unit load at a fixed distance, d, to its left. The value of the MRIF is denoted by $J_{xk,d}$, where d is the distance between the unit load and the measured point. By this definition, the relationship between the moving reference deflection components, $\delta_{k,i}$ at scan k due to the nth axle can be written as

$$\delta_{k,n} = W_n J_{x_k,s_n}, \tag{3.1}$$

where W_n is the weight of the nth axle and s_n denotes the distance between the nth and the instrumented axle (first axle in this paper). Adding the contributions of each axle gives

$$\delta_k = W_1 J_{x_k,s_1} + W_2 J_{x_k,s_2} + \cdots + W_n J_{x_k,s_n}. \tag{3.2}$$

3.1. *Kernel density estimation of a function*

Kernel Density Estimation (KDE) is used to smooth an estimated probability distribution of a variable in a population when a limited data sample is available.[37] In this application, KDE is used to reduce the number of parameters needed to define the profile and MRIFs. A simple example is presented here to illustrate the concept. A variable a ($a \epsilon [0, 10]$) and the function $f(a)$ is defined as

$$f(a) = \begin{cases} a, & a \in [0, 6], \\ -1.5a + 15, & a \in [6, 10]. \end{cases} \tag{3.3}$$

Here, the Gaussian function is used as the KDE:

$$g(a) = \sum_{q=1}^{Q} \left(j_q * \exp\left\{ \frac{-(a - x_q)^2}{2\sigma^2} \right\} \right), \tag{3.4}$$

where Q is the total number of Gaussian functions used to estimate $f(a)$, j_q is the kernel value of the qth Gaussian function and x_q is the center of the peak of the qth Gaussian function. The standard deviation of the Gaussian functions is σ. The function $f(a)$ and its KDE, $g(a)$ are plotted in Fig. 4 for different values of Q (3, 5 and 9).

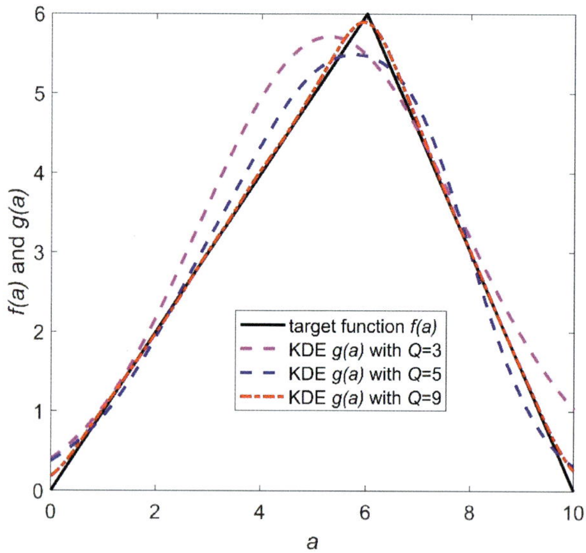

Fig. 4. Function $f(a)$ and its KDE representations, $g(a)$.

It can be seen in Fig. 4 that, as Q increases, the estimate, $g(a)$ fits the target function (black solid line) better, especially at the start, end and peak. However, an excessively large Q will result in a complex KDE and an increase in computation time with little gain in accuracy. Overall, the target function can be accurately estimated by KDE with a moderate value of Q(9 in this example). Another parameter sigma (σ) affects the robustness of the KDE. Figure 5 shows KDEs of $f(a)$ with the same $Q = 9$ but different sigma values (0.6 and 0.4).

From Fig. 5, when σ is 0.4, an unsmooth KDE is inferred as each Gaussian function has a small value when variable a is beyond $[a - 2\sigma, a + 2\sigma]$. When $\sigma = 0.6$ is used, the KDE becomes smooth. Although the KDE with a large σ value is smooth, an error in one Gaussian function (e.g. for No. q Gaussian function, the best j_q value is j_{best} but j_{error} is used in KDE) will affect the overall performance of KDE. Hence, sufficient parameters (σ and Q) selection is important.

3.2. *Kernel density estimation of MRIF*

The example of Sec. 3.1 shows that KDE can estimate a function as the sum of a modest number of Gaussian functions. Compared with the univariate function $f(a)$, the MRIF has two independent variables (x and d): x denotes the location of the axle where measurement takes place and d denotes the distance from that measurement axle to another axle whose weight contributes to the response. Hence, two-dimensional Gaussian functions are used in its estimation. The KDE of the

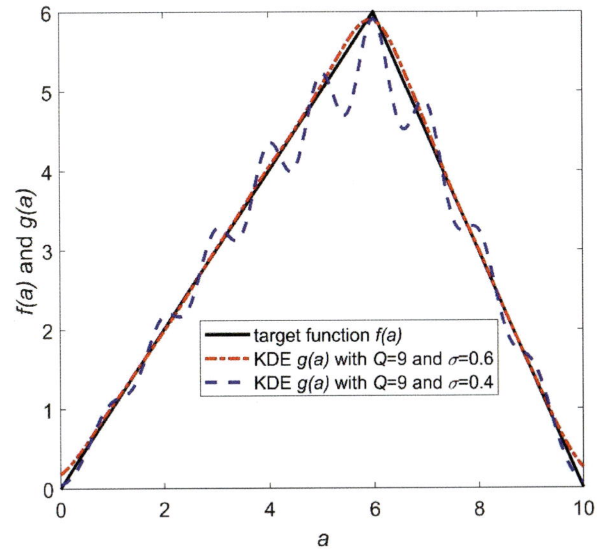

Fig. 5. Function $f(a)$ and its KDE representations, $g(a)$ with different σ values.

MRIF can be written as

$$J_{x,d} = \sum_{q=1}^{Q} \left(j_q * \exp\left\{ \frac{-(x - x_q)^2}{2\sigma_o^2} + \frac{-(d - d_q)^2}{2\sigma_o^2} \right\} \right), \tag{3.5}$$

where j_q is the kernel value of MRIF for the qth Gaussian function. The center of the peak of the qth Gaussian function is denoted by x_q and d_q. In Eq. (3.5), x_q, d_q and σ_o need to be defined in advance according to the bridge span and vehicle axle spacings in the fleet. Hence, in an optimization algorithm to find the MRIF, the kernel values (j_q) are the only unknowns in Eq. (3.5).

4. Obtaining MRIF From Apparent Profiles

OBrien *et al.* developed an algorithm to calculate a bridge influence line from direct strain measurements using an error minimization process.[38] As the moving reference deflection is an implicit bridge deflection measurement (through superposition in Eq. (2.2)) and is contained in the AP, a similar approach will be used in this paper to calculate the MRIF from the AP. Similar to Ref. 38, the error function is defined as the sum of squared differences between the theoretical AP values and the corresponding measurements. The error function, E can be written as

$$E = \sum_{i=1}^{I} \sum_{k=1}^{K} \{R_{i,k}^{a} - R_{i,k}^{a,t}\}^2, \tag{4.1}$$

where K is the total number of scans and I is the total number of vehicles in the fleet. For vehicle i and scan k, the "measurements" ($R^a_{i,k}$) consist of the AP calculated from the measured accelerations. The theoretical AP ($R^{a,t}_{i,k}$), is inferred from Eq. (2.1) and is given by

$$R^{a,t}_{i,k} = R^0_k + W_{i,1} J_{x_k, s_{i,1}} + W_{i,2} J_{x_k, s_{i,2}} + \cdots + W_{i,n} J_{x_k, s_{i,n}}. \tag{4.2}$$

In this study, only two-axle vehicles are investigated. This implies that only two-axle vehicles would be instrumented in the monitoring fleet. In these circumstances, Eq. (4.2) is reduced to three terms and Eq. (4.1) becomes

$$E = \sum_{i=1}^{I} \sum_{k=1}^{K} \{R^a_{i,k} - R^0_k - W_{i,1} J_{x_k, s_{i,1}} - W_{j,2} J_{x_k, s_{i,2}}\}^2. \tag{4.3}$$

Reformulating the MRIF as Q Kernel functions (Eq. (3.5)), this becomes

$$E = \sum_{i=1}^{I} \sum_{k=1}^{K} \left\{ R^a_{i,k} - R^0_k - W_{i,1} \sum_{q=1}^{Q} \left(j_q * \exp\left\{ \frac{-(x_k - x_q)^2}{2\sigma_o^2} + \frac{-(s_{i,1} - d_q)^2}{2\sigma_o^2} \right\} \right) \right.$$
$$\left. - W_{j,2} \sum_{q=1}^{Q} \left(j_q * \exp\left\{ \frac{-(x_k - x_q)^2}{2\sigma_o^2} + \frac{-(s_{i,2} - d_q)^2}{2\sigma_o^2} \right\} \right) \right\}^2. \tag{4.4}$$

In this paper, the first axle is taken to be instrumented so $s_{n,1} = 0$ and $s_{n,2}$ equals the vehicle axle spacing of the ith vehicle, s_i. Hence,

$$E = \sum_{i=1}^{I} \sum_{k=1}^{K} \left\{ R^a_{i,k} - R^0_k - W_{i,1} \sum_{q=1}^{Q} \left(j_q * \exp\left\{ \frac{-(x_k - x_q)^2}{2\sigma_o^2} + \frac{-(d_q)^2}{2\sigma^2} \right\} \right) \right.$$
$$\left. - W_{i,2} \sum_{q=1}^{Q} \left(j_q * \exp\left\{ \frac{-(x_k - x_q)^2}{2\sigma_o^2} + \frac{-(s_i - d_q)^2}{2\sigma_o^2} \right\} \right) \right\}^2. \tag{4.5}$$

The optimization problem is to find the values for j_q and the real profile, R^0_k, which minimize the error function, E. Formulating the profile by KDE gives

$$R^0_k = \sum_{t=1}^{T} \left(r_t * \exp\left\{ \frac{-(x_k - x^r_t)^2}{2\sigma_r^2} \right\} \right), \tag{4.6}$$

where r_t is the kernel value of the tth Gaussian function and x^r_t is the center of the peak of tth Gaussian function. T is the total number of such functions. The standard deviation used in the KDE of the profile is σ_r. Substituting into the error function gives

$$E = \sum_{i=1}^{I}\sum_{k=1}^{K}\left\{R^a_{i,k} - \sum_{t=1}^{T}\left(r_t * \exp\left\{-\frac{(x_k - x^r_t)^2}{2\sigma_r^2}\right\}\right)\right.$$
$$- W_{i,1}\sum_{q=1}^{Q}\left(j_q * \exp\left\{-\frac{(x_k - x_q)^2}{2\sigma_o^2} - \frac{d_q^2}{2\sigma_o^2}\right\}\right)$$
$$\left. - W_{i,2}\sum_{q=1}^{Q}\left(j_q * \exp\left\{-\frac{(x_k - x_q)^2}{2\sigma_o^2} - \frac{(s_i - d_q)^2}{2\sigma_o^2}\right\}\right)\right\}^2. \quad (4.7)$$

In Eq. (4.7), all terms are known except the kernel values of the two KDEs (all j_q and r_t). To minimize this error function, the partial derivatives of the error function are taken with respect to each j_q and r_t, respectively, and set to zero. For example, taking the partial derivative with respect to j_1 gives

$$\frac{\partial E}{\partial j_1} = B_1 - A_{1,1} * j_1 - A_{1,2} * j_2 - \cdots - A_{1,Q} j_Q$$
$$- A_{1,Q+1} r_1 - \cdots - A_{1,Q+T} r_T = 0, \quad (4.8)$$

where the A and B terms are coefficients which are given in Appendix A. Moving all terms containing j_q and r_t to the right-hand side of Eq. (4.8) gives a linear equation in the $(Q+T)$ unknowns. Combining all $(Q+T)$ equations together gives a system of linear equations of the form

$$\begin{Bmatrix} B_1 \\ \vdots \\ B_Q \\ B_{Q+1} \\ \vdots \\ B_{Q+T} \end{Bmatrix} = \begin{bmatrix} A_{1,1} & \cdots & A_{1,Q} & A_{1,Q+1} & \cdots & A_{1,Q+T} \\ \vdots & & \vdots & \vdots & & \vdots \\ A_{Q,1} & \cdots & A_{Q,Q} & A_{Q,Q+1} & \cdots & A_{Q,Q+T} \\ A_{Q+1,1} & \cdots & A_{Q+1,Q} & A_{Q+1,Q+1} & \cdots & A_{Q+1,Q+T} \\ \vdots & & \vdots & \vdots & & \vdots \\ A_{Q+T,1} & \cdots & A_{Q+T,Q} & A_{Q+T,Q+1} & \cdots & A_{Q+T,Q+T} \end{bmatrix} * \begin{Bmatrix} j_1 \\ \vdots \\ j_Q \\ r_1 \\ \vdots \\ r_T \end{Bmatrix} \quad (4.9)$$

which can be simplified as

$$\{\bar{B}\} = [\bar{A}] * \{\bar{j}\}, \quad (4.10)$$

where $\bar{B}$ is a vector of size $\{(Q+T)\times 1\}$, $\bar{A}$ is a matrix of size $[(Q+T)\times(Q+T)]$, and $\bar{j}$ is the vector of unknowns:

$$\{\bar{j}\} = \begin{Bmatrix} j_1 \\ \vdots \\ j_Q \\ r_1 \\ \vdots \\ r_T \end{Bmatrix} \tag{4.11}$$

which can be found by solving the system of equations:

$$\{\bar{j}\} = [\bar{A}]^{-1}\{\bar{B}\}. \tag{4.12}$$

5. Fleet Monitoring Using MRIF

To illustrate the concept of using the MRIF to detect bridge damage, a numerical validation study is presented in this section. A finite element (FE) model of a simply supported bridge and a coupled bridge–vehicle dynamic interaction model were used to generate "simulated measured" vehicle axle accelerations. The bridge model used in this paper represents a 20 m simply supported beam, 4 m wide (representing one lane of a highway bridge). The deck consists of three "Y3" beams and a 200 mm deep (*in situ*) concrete slab. The FE model of the structure contains 20 discretized beam elements. Each element has four degrees of freedom, one rotational and one vertical translational degree of freedom at each end. Two rotational springs are added at each end of the bridge (see Fig. 6) to enable modeling of rotational bearing damage, i.e. frictional resistance to rotation. The stiffness of the bridge end rotational springs is denoted as K_{rot}. The damping is 3% and the viscous damping matrix is generated on the assumption of Rayleigh damping.[39] Table 1 provides the properties of the bridge FE model.

From the Long-Term Pavement Performance database,[41] it is typical for bridges to be normally subject to crossing of more than 1 000 trucks per day. A fleet of 200 two-axle vehicles was simulated crossing the bridge (details of fleet provided later in this section). Currently, many vehicles are instrumented with accelerometers,

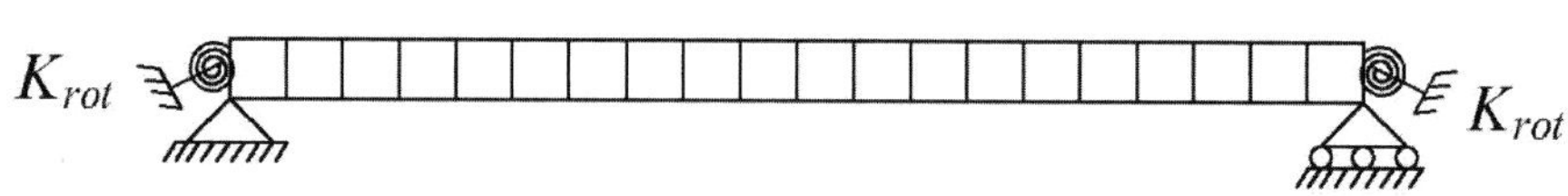

Fig. 6. Bridge FE model with rotational springs at the ends to allow bearing damage to be modeled.[40]

Table 1. Properties of bridge FE model.

Span	20 m
Number of FEs	20
Total degrees of freedom	42
Young's modulus, E	$3 \times 10^{10}\,\mathrm{N/m^2}$
Cross-sectional area	$2.05\,\mathrm{m^2}$
Second moment of area	$0.299\,\mathrm{m^4}$
Damping ratio	3%

and it is anticipated that accelerations may be collected from the general vehicle population in the near future. In the interim, it would be relatively inexpensive to instrument particular types of vehicles (e.g. buses or other public service vehicles). Each two-axle vehicle is represented by two-quarter cars (QCs) with no interaction allowed between axles as shown in Fig. 7.[42] The goal is to demonstrate that monitoring a single axle is enough, even if the response is influenced by loading from more than one axle. The coupled vehicle–bridge system, shown in Fig. 7, has been used in many other studies.[4,5,20–23,28,43–47] Each QC has a total of two degrees of freedom, namely: sprung mass displacement (δ_s) and unsprung mass displacement (δ_u). The class "A" bridge profile is generated according to ISO standard.[46]

Some properties of the 200-vehicle fleet used in the analysis are selected from the Long-Term Pavement Performance Weigh-in-Motion (WIM) database.[41] In this study, only two-axle truck records were taken at random from data for Wisconsin, USA, collected in 2012. Other properties are generated randomly by Monte Carlo simulation using population means and standard deviations from the literature.[46] Table 2 provides a summary of all vehicle fleet properties. Acceleration data is sampled at 1000 Hz from the sprung mass of the first QC. The KDE of profile uses 201 Gaussian functions ($Q = 201$) with 0.1 m as spacing between two center

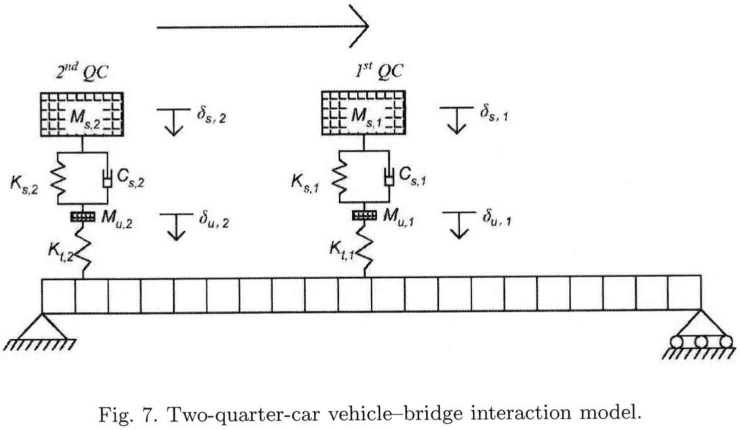

Fig. 7. Two-quarter-car vehicle–bridge interaction model.

Table 2. Vehicle fleet properties.

Property	Symbol/units	Mean value[46]	Standard deviation[46]
Unsprung mass	M_u: kg	700	250
Damping	C_s: Ns/m	2×10^4	5×10^3
Unsprung stiffness	K_s: N/m	4×10^5	1×10^5
Tyre stiffness	K_t: N/m	1.75×10^6	5×10^5
Sprung mass 1	$M_{s,1}$	From database	
Sprung mass 2	$M_{s,2}$		
Vehicle speed	V		
Axle spacing	s_n		

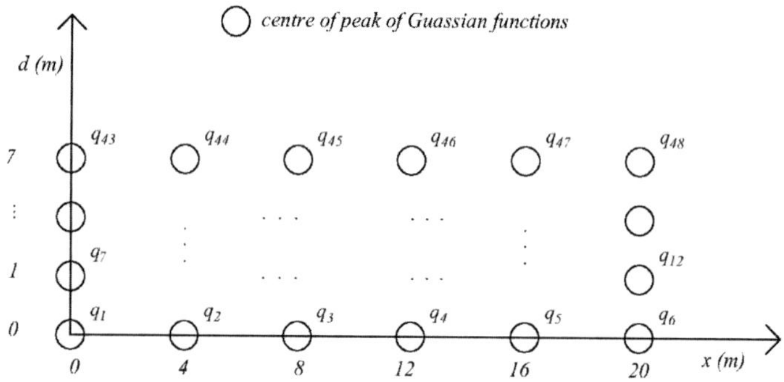

Fig. 8. Distribution of the centers of peak of Gaussian functions in KDE of MRIF.

of peaks of adjacent Gaussian functions ($x^r_{t+1} - x^r_t = 0.1$ m). The KDE of MRIF uses 48 Gaussian functions ($Q = 48$) and the distribution of the centers of peak of Gaussian functions is shown in Fig. 8. Different values of combinations between σ_r and σ_o are applied to calculate MRIFs. The all-inferred MRIF presented in this paper was calculated by using the combination of σ_r and σ_o with the smallest error (the error is calculated through Eq. (4.1)).

Figure 9 is a 3D plot of the MRIF for axle spacing in the range, $d = 4.5$–7 m. As expected, a second axle closer to the first axle measurement point (smaller d) results in greater deflection. Figure 10 is a section through the MRIF at $d = 0$ m, i.e. deflection at a (moving) point due to a unit load at that point. For two-axle trucks or buses, it is unusual for axle spacing to be less than 4.5 m or greater than 7 m. Hence, the inferred MRIF outside of this range is less reliable. Therefore, this paper only uses MRIF in the interval range ($d = 4.5$–7 m) and at the section $d = 0$ m. In Fig. 10, the MRIF inferred from the accelerations of the 200-vehicle fleet is compared to the actual MRIF derived from the static equations governing the beam. While there are some differences due to the location of the kernel points, the general shape of the actual static and inferred MRIFs are similar. Figure 11 shows the actual profile and the profile inferred by the 200-vehicle fleet. It can be

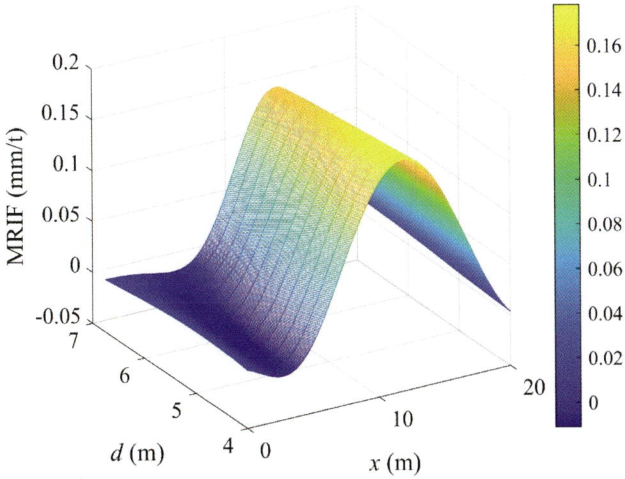

Fig. 9. Inferred MRIF ($d = 4.5$–7 m) from full fleet passing healthy bridge.

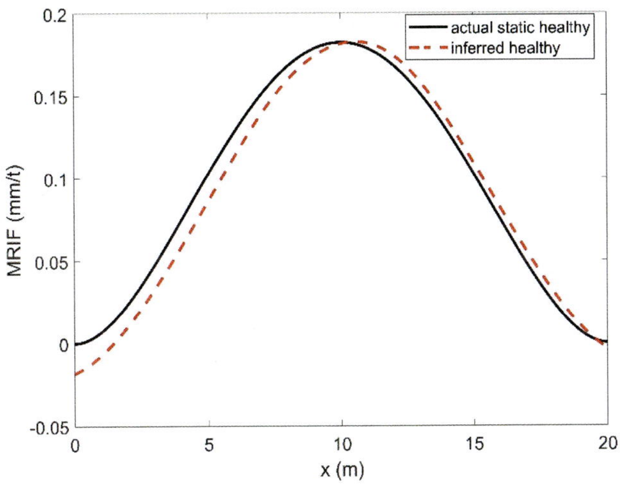

Fig. 10. Inferred section of MRIF ($d = 0$ m) and the section of static MRIF ($d = 0$ m) from full fleet passing healthy bridge.

seen that the inferred profile is a very accurate estimation of the actual profile on the bridge.

To test whether the MRIF is damage sensitive, MRIFs are inferred for different levels of bridge bearing damage (different levels of rotational stiffness), each time using a different fleet of 200 vehicles. Khan *et al.* discuss the modeling of bearing damge by changing the rotational stiffness at the support(s).[40] They define different

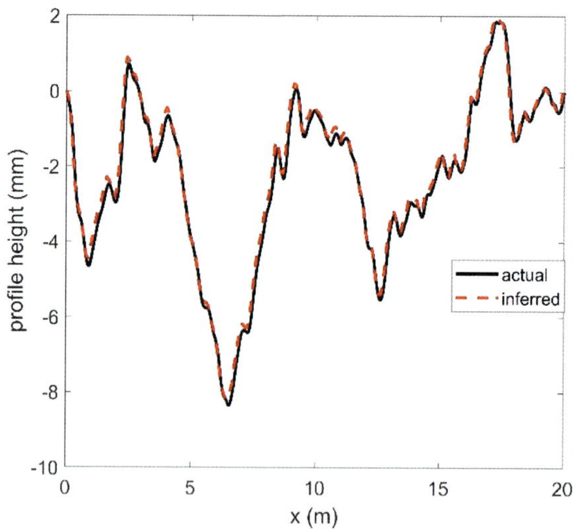

Fig. 11. Inferred profile height (from the full fleet) and real profile height.

damage levels by changing the "degree of fixity" at the support. When the bridge is healthy, clearly K_{rot} is zero and the support is pinned, as intended. As the bearing fails, K_{rot} is assumed to increase. A sensitivity study by Ref. 40 shows a highly nonlinear relationship between damage and spring stiffness. They proposed five levels of damage, as shown in Table 3: healthy (pinned), low, moderate, severe and

Table 3. Different damage conditions in Ref. 40 and this paper.

Ref. 40	K_{rot} (N·m)	0	10^9	10^{10}	10^{11}	10^{12}	10^{13}
	K_{rot} /(4*EI*/*L*)	0	0.343	3.43	34.3	343	3430
	Damage conditon	Healthy (pinned)	Low	Moderate	Severe	Fixed	Fixed
	First natural frequency (Hz)	4.24	5.18	7.81	9.35	9.60	9.64
	Difference in first natural frequency (%)	0	22	84	121	126	127
This paper	K_{rot} (N·m)	0	10^8	10^9	10^{10}	10^{11}	10^{12}
	K_{rot} /(4*EI*/*L*)	0	0.557	5.57	55.7	557	5570
	Damage condition	Healthy (pinned)	Low	Moderate	Severe	Fixed	Fixed
	First natural frequency (Hz)	5.42	5.65	7.15	10.64	12.07	12.25
	Difference in first natural frequency (%)	0	4	32	96	123	126

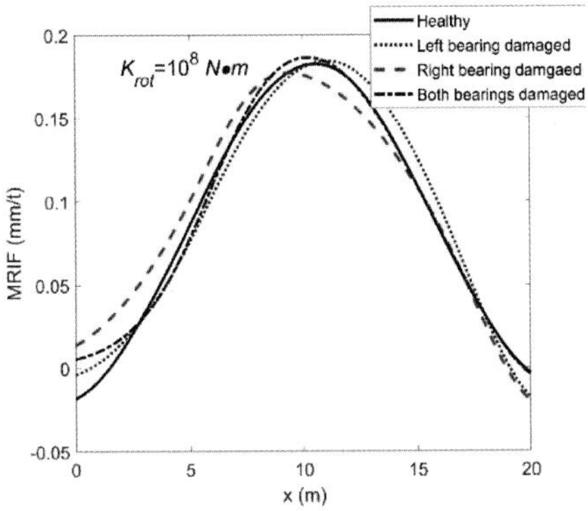

(a)

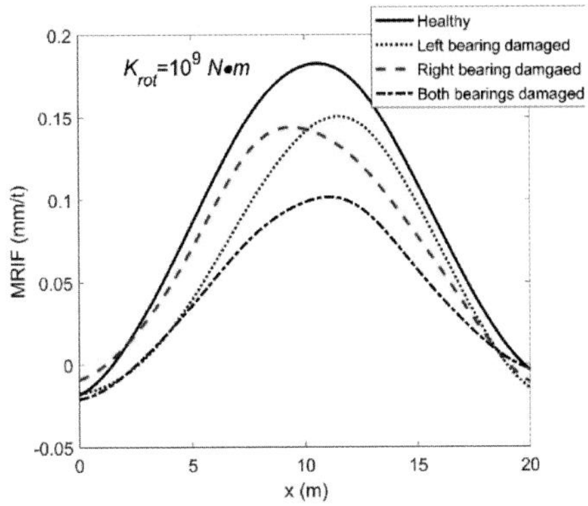

(b)

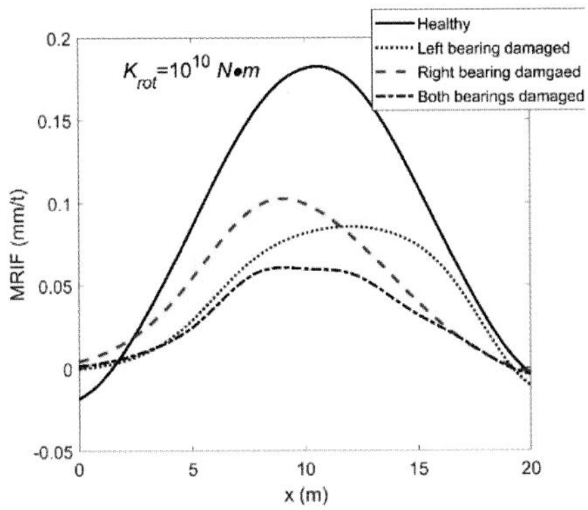

(c)

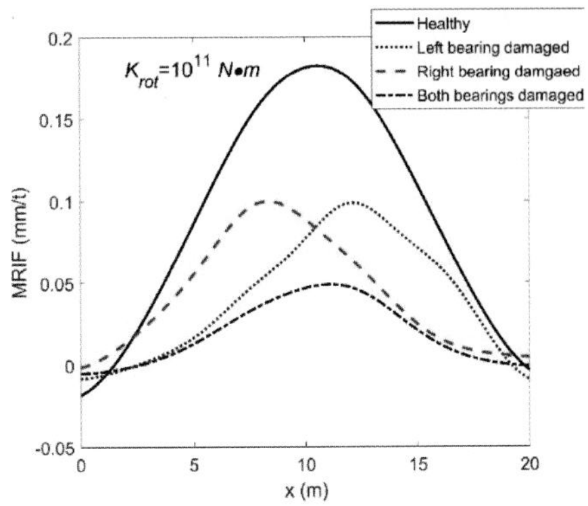

(d)

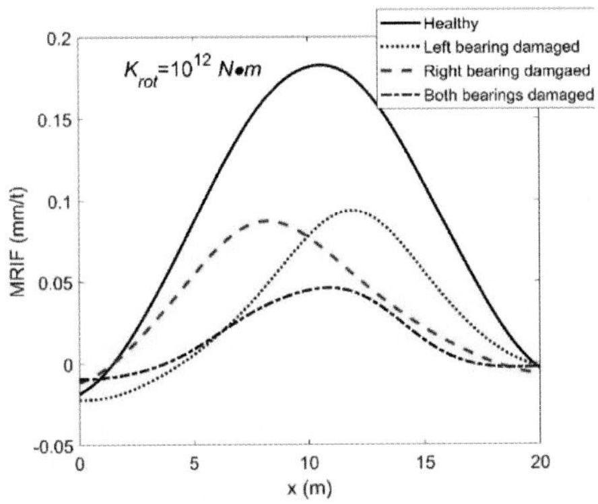

(e)

Fig. 12. Inferred MRIF ($d = 0\,\text{m}$) for different intensities and locations of bearing damage (healthy bridge i.e. $K_{\text{rot}} = 0\,\text{N·m}$ is included in every figure); (a) $K_{\text{rot}} = 10^8\,\text{N·m}$, (b) $K_{\text{rot}} = 10^9\,\text{N·m}$; (c) $K_{\text{rot}} = 10^{10}\,\text{N·m}$; (d) $K_{\text{rot}} = 10^{11}\,\text{N·m}$; and (e) $K_{\text{rot}} = 10^{12}\,\text{N·m}$.

fully fixed. They found that a K_{rot} of 10^{12} N·m corresponded to a fully fixed support. Lower rotational stiffnesses for each damage state were observed in this study (see Fig. 12 for example) as compared to that of Ref. 40, indicating the MRIF approach has a different sensitivity to rotational stiffness from that of the literature. This difference is due to rotational stiffness of a simply supported bridge which is related to it's span and *EI*. Compare $K_{\text{rot}}/(4EI/L)$ between this paper and the literature, similar values are indicated for each damage level.

Figure 12 presents a section through the inferred MRIF ($d = 0$ m) for different damage intensities and locations. Damage in both the left and right bearings is investigated as well as combined damage to both. Similar to the influence line,[48] MRIF should be able to indicate the magnitude of the change in stiffness (damage) in a similar way to the method introduced by Zheng *et al.*[49] In this paper, damage is identified by analyzing the area and the skewness of the MRIF. It is clear from Fig. 12 that bearing damage affects the MRIF at $d = 0$, generally reducing the area under the curve. For this example, most of the changes occur between $K_{\text{rot}} = 10^8$ and 10^{10} — Figs. 12(a)–12(c). For higher values of K_{rot} the reductions in the area under the MRIF are not significant — Figs. 12(c)–12(e). The areas under each MRIF, for $d = 0$, are presented in Table 4. Comparing the left bearing damage (Column 2) and right bearing damage (Column 3), there is no significant difference in area. When both bearings are damaged (Column 4), the area is less than when only one bearing is damaged.

The location of damage also has an effect on the MRIF: damage in the left bearing tends to skew the function toward the right while damage in the right bearing has the opposite effect. Skewness is a statistical concept to describe an asymmetry or distortion from a symmetrical curve in a set of data. It is defined as

$$\text{Sk} = \frac{1}{O}\sum_{i=1}^{O}\left(\frac{X_i - \mu}{\sigma}\right)^3, \tag{5.1}$$

Table 4. Area (unit: mm^2/t) of the section through the MRIF ($d = 0$).

Column 1	Column 2	Column 3	Column 4
K_{rot} (N·m)	Left bearing damaged (% difference to healthy)	Right bearing damaged (% difference to healthy)	Both bearings damaged (% difference to healthy)
0 (healthy)		1.66×10^3 (0%)	
10^8 (Fig. 12(a))	1.63×10^3 (−1.8%)	1.62×10^3 (−2.4%)	1.54×10^3 (−7.2%)
10^9 (Fig. 12(b))	1.25×10^3 (−24.7%)	1.29×10^3 (−22.3%)	0.90×10^3 (−45.8%)
10^{10} (Fig. 12(c))	0.78×10^3 (−53.0%)	0.79×10^3 (−52.4%)	0.48×10^3 (−71.1%)
10^{11} (Fig. 12(d))	0.73×10^3 (−56.0%)	0.64×10^3 (−61.4%)	0.35×10^3 (−78.9%)
10^{12} (Fig. 12(e))	0.69×10^3 (−58.4%)	0.69×10^3 (−58.4%)	0.33×10^3 (−80.1%)

Note: Numbers in the brackets are the percentage differences of area between the damaged and healthy MRIF sections with $d = 0$.

where O is the total amount of data in the set and X_i is a data point. The mean and standard deviation of the data set are μ and σ, respectively. In this paper, the entire three-dimensional MRIF is assessed in relation to skewness. To simplify the process, sections of MRIF with different d values being assessed individually. Each section (value of d) is considered as a distribution curve and the value of the MRIF, $J_{x,d}$, is taken as the frequency. The mean, standard deviation and skewness of the section of MRIF can be obtained through the equations:

$$\mu = \frac{\sum_{k=1}^{K} (x_k \times J_{x_k,d})}{\sum_{k=1}^{K} J_{x_k,d}}, \tag{5.2}$$

$$\sigma = \sqrt{\frac{\sum_{x=1}^{K} \left[(x_k - \mu)^2 \times J_{x_k,d}\right]}{\sum_{k=1}^{K} J_{x_k,d}}}, \tag{5.3}$$

$$\mathrm{Sk} = \frac{\sum_{k=1}^{K} \left[\left(\frac{x-\mu}{\sigma}\right)^3 \times J_{x_k,d}\right]}{\sum_{k=1}^{K} J_{x_k,d}}. \tag{5.4}$$

Table 5 presents the skewness of the MRIF for $d = 0$ and a range of damage states. It can be seen in Table 5 that skewness is close to zero when the bridge is healthy and when both bearings are equally damaged. When only the left bearing is damaged, the skewness is consistently negative and when only the right bearing is damaged, the skewness is consistently positive, though not very much in the case of $K_{\mathrm{rot}} = 10^8$. Clearly skewness of the MRIF can provide a good indication of which bearing is damaged (or more damaged).

It can be seen that bridge bearing damage reduces the area under the MRIF and damage to only one bearing affects its skewness. For completeness, skewness and area can be calculated for the entire three-dimensional MRIF, not just one section ($d = 0$ presented in Tables 4 and 5). To address this, each MRIF is separated into sections with different d. Skewness and area of the MRIF are calculated by summing the weighted skewness and area of each section. The weighting is determined on the basis of the number of vehicles of the fleet used to find it. Hence more common axle spacings will provide more data and sections for these values of d are weighted

Table 5. Skewness of MRIF ($d = 0$).

K_{rot} (N·m)	Left bearing damaged	Right bearing damaged	Both bearings damaged
0 (healthy)		−0.01	
10^8 (Fig. 12(a))	−0.06	0.01	0.00
10^9 (Fig. 12(b))	−0.07	0.04	0.01
10^{10} (Fig. 12(c))	−0.15	0.12	0.03
10^{11} (Fig. 12(d))	−0.24	0.16	−0.01
10^{12} (Fig. 12(e))	−0.14	0.21	−0.04

more heavily. Taking sections at 0.5 m intervals, the weighting at $d = t$ is

$$\rho_t = \frac{n_t}{N}, \tag{5.5}$$

where n_t is the number of vehicles in the fleet with axle spacing between $[t - 0.25, t + 0.25]$ m. The weighted area of the entire MRIF is then calculated as

$$\text{Area} = \frac{1}{2} \times \sum (\rho_t \times A_t) + \frac{1}{2} \times A_0, \tag{5.6}$$

where A_t denotes the area under the section through the MRIF at $d = t$, and A_0 is the area under the section through the MRIF at $d = 0$. All vehicles have one axle with $d = 0$, so this section gets the greatest weighting which is 1/2 (as there are totally $2 \times N$ axles in the fleet and N axles have one axle with $d = 0$). Similarly,

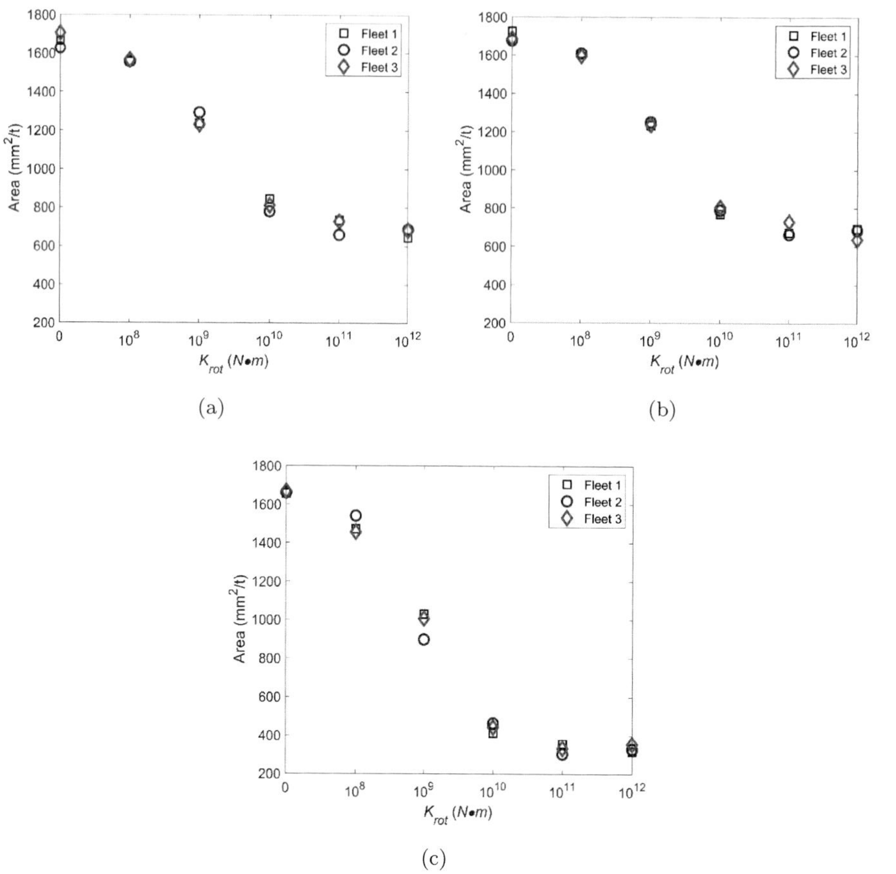

Fig. 13. Weighted area under the entire MRIF for three fleets of healthy and damaged bridges: (a) left bearing damage, (b) right bearing damage, and (c) both bearings damaged.

the skewness of the entire MRIF can be calculated as

$$\text{Skewness} = \frac{1}{2} \times \sum (\rho_t \times S_t) + \frac{1}{2} \times S_0, \tag{5.7}$$

where S_t and S_0 are the skewness for $d = t$ and $d = 0$, respectively. To test the repeatability of using the MRIF as a damage indicator, three fleets of two-axle vehicles (each fleet consisting of 200 vehicles and axle spacings in the range of 4.5–7 m) were run for each damage condition. The results of this analysis are plotted in Fig. 13.

There is good repeatability between the three fleets, each having similar areas for given damage condition. The decreasing trend in area with increasing damage is evident. The trend is generally "S" shaped, i.e. as K_{rot} increases on a log scale,

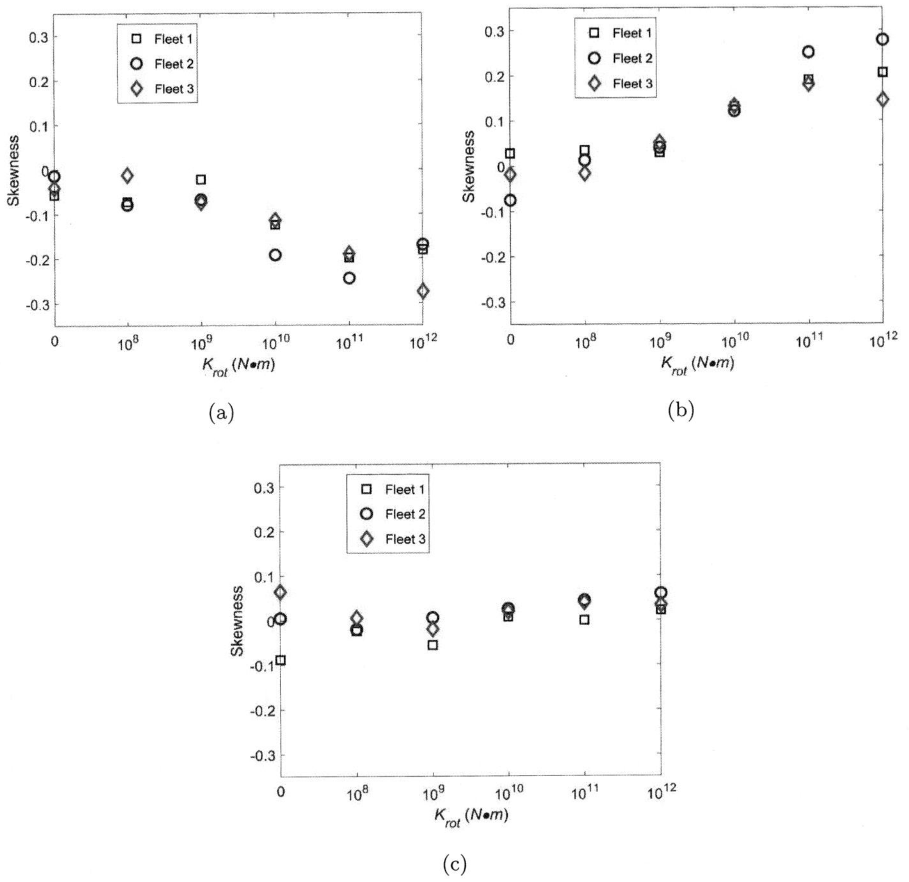

Fig. 14. Skewness of the entire MRIF for three fleets on healthy and damaged bridges: (a) left bearing damage, (b) right bearing damage, and (c) both bearings damaged.

there is little change in area at first, then a reasonably sudden fall and finally little change again.

Figure 14 shows the skewness of the entire MRIF for a range of damage conditions. The results are again repeatable but there is greater variability, particularly for the largest and some of the smallest values of K_{rot}. When K_{rot} is less than 10^9 N·m, it is difficult to identify which bearing is damaged (left, right or both) from the skewness (values close to 0). For greater levels of damage, the skewness values are consistently positive for right bearing damage and negative for left bearing damage (Figs. 14(a) and 14(b)). When both bearings are damaged, the skewness can be positive or negative but is generally of low magnitude.

6. Conclusion

This paper presents a newly developed bridge damage indicator, namely the MRIF. The MRIF and bridge profile height are represented by KDEs. By using accelerations from a partially instrumented two-axle vehicle fleet, profile heights and the MRIF are inferred for a simply supported single span bridge, with good accuracy. Different locations and severity of bearing damage are considered. The results indicate that area under the MRIF reduces as a bridge bearing is damaged. Also, the skewness of the MRIF tends to move away from zero when only one bearing is damaged (positively for right bearing damage and negatively for left bearing damage). The MRIF is capable of detecting the location and severity of bearing damage. Damage that causes a change in the flexural rigidity (local cracks, overall material degradation) of the girder (which is being loaded) will also affect MRIF. Normally, this type of damage will reduce the flexural stiffness and thus will reduce the area under the MRIF curve. Hence, in the context of these two types of damage, the type can be inferred from whether the area under the MRIF increases or decreases. In the unlikely event that both types of damage occur simultaneously, it may not be possible to separate the effects.

Acknowledgments

The authors wish to acknowledge the financial support received from University College Dublin and the Chinese Scholarship Council. The authors also gratefully acknowledge the Federal Highway Administration and the Long-Term Pavement Performance Program for access to WIM data.

Appendix A

The general formulations of the elements in the components of matrix $[\bar{A}]$ are given in Eqs. (A.1)–(A.6) and components of matrix $[\bar{B}]$ are given in Eqs. (31) and (32). We define n_r to represent the row number and n_c to represent the column number in the matrix $[\bar{A}]$ and vector, $\{\bar{B}\}$. When $n_r < Q$ and $n_c < Q$, the coefficients are given by

$$
\begin{aligned}
A_{n_r,n_c} = \sum_{i=1}^{I}\sum_{k=1}^{K}\Bigg(& (W_{i,1}W_{i,1}) * \Bigg[\exp\left\{\frac{-(x_k - x_{n_r})^2 - (d_{n_r})^2}{2\sigma_o^2}\right\} \\
& * \exp\left\{\frac{-(x_k - x_{n_c})^2 - (d_{n_c})^2}{2\sigma_o^2}\right\} + (W_{i,1}W_{i,2}) \\
& * \exp\left\{\frac{-(x_k - x_{n_r})^2 - (d_{n_r})^2}{2\sigma_o^2}\right\} \\
& * \exp\left\{\frac{-(x_k - x_{n_c})^2 - (s_i - d_{n_c})^2}{2\sigma^2}\right\} + (W_{i,1}W_{i,2}) \\
& * \exp\left\{\frac{-(x_k - x_{n_r})^2 - (s_i - d_{n_r})^2}{2\sigma_o^2}\right\} \\
& * \exp\left\{\frac{-(x_k - x_{n_c})^2 - (y_{n_c})^2}{2\sigma_o^2}\right\} + (W_{i,2}W_{i,2}) \\
& * \exp\left\{\frac{-(x_k - x_{n_r})^2 - (s_i - d_{n_r})^2}{2\sigma_o^2}\right\} \\
& * \exp\left\{\frac{-(x_k - x_{n_c})^2 - (s_i - y_{n_c})^2}{2\sigma_o^2}\right\}\Bigg]\Bigg).
\end{aligned} \tag{A.1}
$$

When $n_r < Q$ and $n_c > Q$, the expression is

$$
\begin{aligned}
A_{n_r,n_c} = \sum_{i=1}^{I}\sum_{k=1}^{K}\Bigg(& (W_{i,1}) * \exp\left\{\frac{-(x_k - x_{n_r})^2 - (0 - d_{n_r})^2}{2\sigma_o^2}\right\} + (W_{i,2}) \\
& * \exp\left\{\frac{-(x_k - x_{n_r})^2 - (s_i - d_{n_r})^2}{2\sigma_o^2}\right\}\Bigg) * \exp\left\{\frac{-(x_k - x^r_{n_c-Q})^2}{2\sigma_r^2}\right\}.
\end{aligned} \tag{A.2}
$$

When $n_r > Q$ and $n_c < Q$, it becomes

$$
\begin{aligned}
A_{n_r,n_c} = \sum_{i=1}^{I}\sum_{k=1}^{K}\Bigg(& (W_{i,1}) * \exp\left\{\frac{-(x_k - x_{n_c})^2 - (0 - d_{n_c})^2}{2\sigma_o^2}\right\} + (W_{i,2}) \\
& * \exp\left\{\frac{-(x_k - x_{n_c})^2 - (s_i - d_{n_c})^2}{2\sigma_o^2}\right\}\Bigg) * \exp\left\{\frac{-(x_k - x^r_{n_r-Q})^2}{2\sigma_r^2}\right\}.
\end{aligned} \tag{A.3}
$$

When $n_r > Q$ and $n_c > Q$, the equation becomes

$$A_{n_r,n_c} = \sum_{i=1}^{I}\sum_{k=1}^{K}\left(\exp\left\{\frac{-(x_k - x^r_{n_r-Q})^2}{2\sigma_r^2}\right\} * \exp\left\{\frac{-(x_k - x^r_{n_c-Q})^2}{2\sigma_r^2}\right\}\right). \tag{A.4}$$

When $n_r < Q$, the coefficients of the $\{\bar{B}\}$ vector are

$$B_{n_r} = \sum_{i=1}^{I}\sum_{k=1}^{K} R^a_{i,k} * \left(W_{i,1} * \exp\left\{\frac{-(x_k - x_{n_r})^2}{2\sigma_o^2} + \frac{-(0 - d_{n_r})^2}{2\sigma_o^2}\right\} + W_{i,2} * \exp\left\{\frac{-(x_k - x_{n_r})^2}{2\sigma_o^2} + \frac{-(s_i - d_{n_r})^2}{2\sigma_o^2}\right\}\right). \tag{A.5}$$

When $n_r > Q$, the coefficients become

$$B_{n_r} = \sum_{i=1}^{I}\sum_{k=1}^{K} R^a_{i,k} * \exp\left\{\frac{-(x_k - x^r_{n_r-Q})^2}{2\sigma_r^2}\right\}. \tag{A.6}$$

References

1. C. McGeown, F. Huseynov, D. Hester, P. McGetrick, E. J. OBrien and V. Pakrashi, Using measured rotation on a beam to detect changes in its structural condition, *J. Struct. Integr. Maint.* **6**(3) (2021) 159–166.
2. C. R. Farrar, S. W. Doebling and D. A. Nix, Vibration-based structural damage identification, *Philos. Trans. Roy. Soc. London Ser. A: Math. Phys. Eng. Sci.* **359**(1778) (2001) 131–149.
3. Y. Y. Li, Hypersensitivity of strain-based indicators for structural damage identification: A review, *Mech. Syst. Signal Process.* **24**(3) (2010) 653–664.
4. D. Martinez, A. Malekjafarian and E. J. OBrien, Bridge health monitoring using deflection measurements under random traffic, *Struct. Control Health Monit.* **27**(9) (2020) 2593.
5. S. Wang, E. J. OBrien and D. P. McCrum, A novel acceleration-based moving force identification algorithm to detect global bridge damage, *Appl. Sci.* **11**(16) (2021) 7271.
6. J. R. Casas and J. J. Moughty, Bridge damage detection based on vibration data: Past and new developments, *Front. Built Environ.* **3** (2017) 4.
7. J. R. Casas and A. C. Aparicio, Structural damage identification from dynamic-test data, *J. Struct. Eng.* **120**(8) (1994) 2437–2450.
8. W. Fan and P. Qiao, Vibration-based damage identification methods: A review and comparative study, *Struct. Health Monit.* **10**(1) (2011) 83–111.
9. O. S. Salawu and C. Williams, Bridge assessment using forced-vibration testing, *J. Struct. Eng.* **121**(2) (1995) 161–173.
10. O. Huth, G. Feltrin, J. Maeck, N. Kilic and M. Motavalli, Damage identification using modal data: Experiences on a prestressed concrete bridge, *J. Struct. Eng.* **131**(12) (2005) 1898–1910.
11. Pandey, A., Biswas, M., & Samman, M., Damage detection from changes in curvature mode shapes, *J. Sound Vib.* **145**(2) (1991) 321–332.
12. D. Pines and A. E. Aktan, Status of structural health monitoring of long-span bridges in the United States, *Progr. Struct. Eng. Mater.* **4**(4) (2002) 372–380.

13. M. Vagnoli, R. Remenyte-Prescott and J. Andrews, Railway bridge structural health monitoring and fault detection: State-of-the-art methods and future challenges, *Struct. Health Monit.* **17**(4) (2018) 971–1007.
14. F. Huseynov, C. Kim, E. J. Obrien, J. Brownjohn, D. Hester and K. Chang, Bridge damage detection using rotation measurements — Experimental validation, *Mech. Syst. Signal Process.* **135** (2020) 106380.
15. Z. Chen, Q. L. Cai and S. Zhu, Damage quantification of beam structures using deflection influence lines, *Struct. Control Health Monit.* **25** (2018) e2242.
16. E. P. Carden and P. Fanning, Vibration based condition monitoring: A review, *Struct. Health Monit.* **3**(4) (2004) 355–377.
17. Y. B. Yang, C. W. Lin and J. D. Yau, Extracting bridge frequencies from the dynamic response of a passing vehicle, *J. Sound Vib.* **272**(3–5) (2004) 471–493.
18. Y. B. Yang and J. P. Yang, State-of-the-art review on modal identification and damage detection of bridges by moving test vehicles, *Int. J. Struct. Stab. Dyn.* **18**(2) (2018) 1850025.
19. A. Malekjafarian, P. J. McGetrick and E. J. OBrien, A review of indirect bridge monitoring using passing vehicles, *Shock Vib.* **2015** (2015) 286139.
20. Y. B. Yang, Y. C. Li and K. C. Chang, Constructing the mode shapes of a bridge from a passing vehicle: A theoretical study, *Smart Struct. Syst.* **13**(5) (2014) 797–819.
21. P. J. McGetrick, A. Gonzalez and E. J. OBrien, Theoretical investigation of the use of a moving vehicle to identify bridge dynamic parameters, *Insight-Non-Destr. Test. Cond. Monit.* **51**(8) (2009) 433–438.
22. A. González, E. J. OBrien and P. J. McGetrick, Identification of damping in a bridge using a moving instrumented vehicle, *J. Sound Vib.* **331**(18) (2012) 4115–4131.
23. Y. B. Yang, B. Zhang, Y. Chen, Y. Qian and Y. Wu, Bridge damping identification by vehicle scanning method, *Eng. Struct.* **183** (2019) 637–645.
24. L. Majumder and C. S. Manohar, A time-domain approach for damage detection in beam structures using vibration data with a moving oscillator as an excitation source, *J. Sound Vib.* **268**(4) (2003) 699–716.
25. J. Q. Bu, S. S. Law and X. Q. Zhu, Innovative bridge condition assessment from dynamic response of a passing vehicle, *J. Eng. Mech.* **132**(12) (2006) 1372–1379.
26. C. W. Kim and M. Kawatani, Pseudo-static approach for damage identification of bridges based on coupling vibration with a moving vehicle, *Struct. Infrastruct. Eng.* **4**(5) (2008) 371–379.
27. A. Khorram, F. Bakhtiari-Nejad and M. Rezaeian, Comparison studies between two wavelet based crack detection methods of a beam subjected to a moving load, *Int. J. Eng. Sci.* **51** (2012) 204–215.
28. P. J. McGetrick and C. W. Kim, A parametric study of a drive by bridge inspection system based on the Morlet wavelet, *Key Eng. Mater.* **569** (2013) 262–269.
29. E. J. OBrien and J. Keenahan, Drive-by damage detection in bridges using the apparent profile, *Struct. Control Health Monit.* **22**(5) (2015) 813–825.
30. J. Keenahan, Y. Ren and E. J. OBrien, Determination of road profile using multiple passing vehicle measurements, *Struct. Infrastruct. Eng.* **16**(9) (2020) 1262–1275.
31. Y. B. Yang, B. Q. Wang, Z. L. Wang, K. Shi, H. Xu, B. Zhang and Y. T. Wu, Bridge surface roughness identified from the displacement influence lines of the contact points by two connected vehicles, *Int. J. Struct. Stab. Dyn.* **20**(14) (2020) 2043003.
32. M. Huang and Y. Lei, Bearing damage detection of a reinforced concrete plate based on sensitivity analysis and chaotic moth-flame-invasive weed optimization, *Sensors* **20**(19) (2020) 5488.

33. S.-H. Kim, H.-S. Mha and S.-W. Lee, Effects of bearing damage upon seismic behaviors of a multi-span girder bridge, *Eng. Struct.* **28**(7) (2006) 1071–1080.
34. K. Kliewer and B. Glisic, Normalized curvature ratio for damage detection in beam-like structures, *Front. Built Environ.* **3** (2017) 50.
35. J. W. Matthew and K. D. Janoyan, In-service diagnostics of a highway bridge from a progressive damage case study, *J. Bridge Eng.* **15**(5) (2010) 597–607.
36. G. Radoičić, M. Jovanović and M. Arsić, Experience with an on-board weighing system solution for heavy vehicles, *ETRI J.* **38**(4) (2016) 787–797.
37. B. W. Silverman, *Density Estimation for Statistics and Data Analysis* (Chapman & Hall, London, 1986), p. 26.
38. E. J. OBrien, M. J. Quilligan and R. Karoumi, Calculating an influence line from direct measurements, *Proc. Inst. Civil Eng. — Bridge Eng.* **159**(1) (2006) 31–34.
39. S. Adhikari, *Structural Dynamic Analysis with Generalized Damping Models: Analysis* (Hoboken, New Jersey, 2013).
40. M. A. Khan, D. P. McCrum, E. J. OBrien, C. Bowe, D. Hester, P. J. McGetrick, C. O'Higgins, M. Casero and V. Pakrashi, Re-deployable sensors for modal estimates of bridges and detection of damage-induced changes in boundary conditions, *Struct. Infrastruct. Eng.* **2021** (2021) 1–15.
41. D. Walker and D. Cebon, The metamorphosis of LTPP traffic data, in *6th Int. Conf. Weigh-In-Motion (ICWIM 6)*, Dallas, United States, 4–7 June 2012, pp. 272–281.
42. J. Keenahan, E. J. OBrien, P. J. McGetrick and A. Gonzalez, The use of a dynamic truck–trailer drive-by system to monitor bridge damping, *Struct. Health Monit.* **13**(2) (2014) 143–157.
43. K. Feng, A. González and M. Casero, A kNN algorithm for locating and quantifying stiffness loss in a bridge from the forced vibration due to a truck crossing at low speed, *Mech. Syst. Signal Process.* **154** (2021) 107599.
44. Y. B. Yang and J. D. Yau, Vehicle-bridge interaction element for dynamic analysis, *J. Struct. Eng.* **123**(11) (1997) 1512–1518.
45. E. J. OBrien, P. McGetrick and A. González, A drive-by inspection system via vehicle moving force identification, *Smart Struct. Syst.* **13**(5) (2014) 821–848.
46. D. Martinez, A. Malekjafarian and E. OBrien, Bridge health monitoring using deflection measurements under random traffic, *Struct. Control Health Monit.* **27**(9) (2020) 2593.
47. ISO 8608:1995, Mechanical vibration — Road surface profiles — Reporting of measured data.
48. X. Zheng, D.-H. Yang, T.-H. Yi, H.-N. Li and Z.-W. Chen, Bridge influence line identification based on regularized least-squares QR decomposition method, *J. Bridge Eng.* **24**(8) (2019) 06019004.
49. X. Zheng, T.-H. Yi, D.-H. Yang and H.-N. Li, Stiffness estimation of girder bridges using influence lines identified from vehicle-induced structural responses, *J. Eng. Mech.* **147**(8) (2021) 04021042.

https://doi.org/10.1142/9789819814770_0004

Chapter 4

Study on the EPSD of Wind-Induced Responses of the Sutong Bridge Using Harmonic Wavelets#

Zidong Xu*, Hao Wang† and Kaiyong Zhao‡
Key Laboratory of C&PC Structures of Ministry of Education
Southeast University, Nanjing 211189, P. R. China
**zdxu1993@163.com*
†wanghao1980@seu.edu.cn
‡rl505328@seu.edu.cn

Many long-span bridges are located at typhoon prone regions. With the continuous increase of the bridge span, the typhoon-induced buffeting becomes more and more prominent. In this study, based on the structural health monitoring system installed in the Sutong Bridge, the recorded buffeting responses of the main girder during typhoons Damrey and Haikui were analyzed. The run test method demonstrated that the recorded acceleration responses can be regarded as zero-mean non-stationary random processes. Hence, to capture the energy distribution of the recorded data in the time-frequency domain, the evolutionary power spectral density (EPSD) estimation was conducted using efficient generalized harmonic wavelet (GHW) and filtered harmonic wavelet (FHW), respectively. Compared with the GHW, narrower wavelet bandwidth is required by the FHW to yield a compromise between the time and frequency resolution. For the FHW-based method, the power spectral density amplitudes of the averaging EPSDs are slightly larger for certain major frequency components than those obtained by the Pwelch method. Results show that the non-stationary features of the buffeting of long-span bridges during Typhoon events should be considered. This study can also provide references for non-stationary buffeting analysis of other long-span bridges during extreme wind events.

Keywords: Sutong Bridge; typhoon; harmonic wavelet; structural health monitoring system; non-stationary buffeting response.

†Corresponding author.
#This chapter appeared previously on the International Journal of Structural Stability and Dynamics. To cite this chapter, please cite the original article as the following: Z. Xu, H. Wang & K. Zhao, *Int. J. Struct. Stab. Dyn.*, **23**, 2340004 (2023), DOI: 10.1142/S0219455423400047.

1. Introduction

During the past decades, a large number of long-span cable-stayed bridges have emerged throughout the world.[1] The long-span bridge often becomes more flexible with the increase of its span length, making it a wind-sensitive structure. Thus, the wind-induced vibration is critical to bridge safety and requires in-depth investigation.[2–4]

As one of the recurring aero-elastic dynamic problems, buffeting analysis is indispensable for wind resistance design of long-span cable-stayed bridges.[5–7] In Davenport's doctoral research,[8] a better description of the wind loading was presented for buffeting analysis of long-span bridges and to evaluate the dynamic action on structures. Then, Scanlan[9,10] proposed expressions of the corresponding self-excited aerodynamic forces and developed formulas that allow straightforward calculation of buffeting response. Other researches on bridge buffeting response were also performed in the last half century.[11–15] Nevertheless, most of these researches were mainly based on stationary and Gaussian hypotheses.

A recent field measurement study has uncovered the fact that typhoons actually exhibit obvious non-stationary characteristics.[16] According to the chain of thought that the wind loading on long-span bridge is determined by aerodynamic parameters of wind turbulence,[17] the non-stationary wind characteristics will create non-stationary forces and cause non-stationary features in the buffeting responses of long-span bridges.[18] The development in structural health monitoring system (SHMS) offers the most reliable data for investigation of wind-induced responses during severe typhoon events.[19,20] As a result, the non-stationary features of the buffeting response recorded by SHMS can be obtained based on rational signal processing methods.[21]

The power spectral density (PSD) can be utilized to identify the energy contribution of different frequency contents to the total energy.[22] However, the conventional Fourier analysis fails to capture the evolutionary spectral features.[23] Thus, several approaches were proposed to reflect the transient features of non-stationary processes in both time and frequency domain. For instance, a joint time-frequency analysis proposed by Winger (1932) in the field of quantum mechanics, which was introduced to the field of signal processing by Ville,[24] is known as the Wigner–Ville method (WVM). However, the spectrum generated by WVM lacks physical meaning since there exists negative values in some cases.[25] Alternatively, by applying the Fourier transform to windowed data fragment centered at a time t, the short-term Fourier transform (STFT) can provide local information in both time and frequency domain to establish a spectrum.[26] Nevertheless, due to the fixed width of the time window, all STFT coefficients have the same frequency bandwidth.[27] Therefore, a rational time-frequency resolution is desirable to capture high-frequency components in short-lived trends and low-frequency components in long-lasting trends.[28]

The wavelet-based method was developed to overcome the aforementioned drawbacks of the WVM and the STFT.[29,30] As an efficient digital signal processing algorithm, the wavelet theory is capable of providing multiple resolutions in both time and frequency domains by scaling and shifting the "mother wavelet".[31,32] In this regard, Spanos and Failla (2004) developed a wavelets-based approach to estimate the evolutionary power spectral density (EPSD) of non-stationary processes defined by Priestley.[33] The time-frequency analysis of some typical non-stationary signals in different engineering fields, including wind, earthquake and ocean, were conducted using wavelets-based method.[23,34–36] Among those researches, the harmonic wavelet (HW) introduced by Newland[27,37] was widely used due to its simple structure and property of non-overlapping Fourier transforms.[38–41] It is noted that the various members of the HW family, including generalized harmonic wavelet (GHW) and filtered harmonic wavelet (FHW), were introduced to analyze the EPSD of non-stationary process by Spanos.[25] However, most researches have focused on the EPSD analysis of extreme winds. The non-stationary wind-induced responses remain to be discussed.

In this study, the recorded vibration responses of the main girder of the Sutong Bridge during typhoons Damrey and Haikui were analyzed using wavelets-based methods. Due to the orthogonal property and acceptable computational efficiency of the HW transform, the EPSDs of the buffeting responses are estimated using the GHW and the FHW, respectively. In addition, a comparison of the EPSDs estimated by the two HWs was conducted to show their differences in analysis of the non-stationary buffeting responses.

2. The Recorded Non-Stationary Buffeting Response

2.1. *Accelerometers on Sutong bridge*

As a long-span cable-stayed bridge, the Sutong Bridge (see Fig. 1) connects Suzhou and Nantong in Jiangsu Province. Its main span is 1 088 m, which breaks the world record at its opening in 2008. The bridge, near the entrance of the Yangtze River, is located in the southeastern region of the Asian continent. According to the meteorological survey, the bridge site is dominated by a humid subtropical monsoon climate.[42] Some extreme wind events, such as strong north wind in winter and typhoon climate in summer, may cause strong impact on the bridge.[36] Therefore, the wind-induced responses of this flexible structure require in-depth investigation.

In view of the vital significance of Sutong Bridge in land transportation, an extensive SHMS was installed to provide valuable information for evaluating the structural health conditions on both the constructing and operating periods. There are 15 types of sensors in the sensory subsystem, such as temperature sensor, anemometer, global positioning system, strain gages and accelerometer, etc. These types of sensors were installed on various parts of the bridge to monitor structural

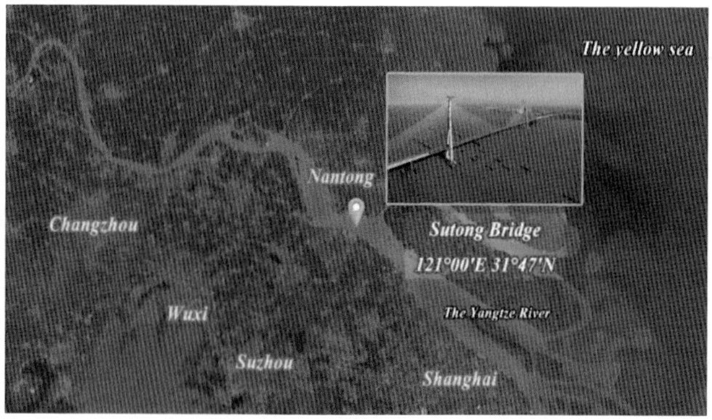

Fig. 1. Location of Sutong Cable-stayed bridge.

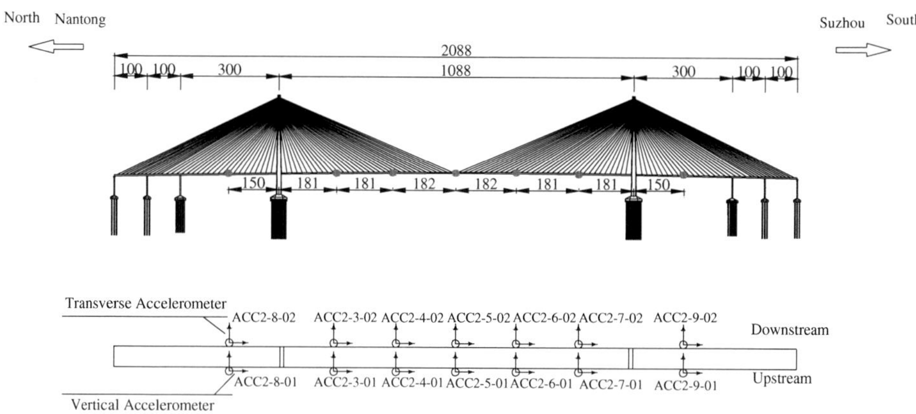

Fig. 2. Layout of the accelerometers on the deck (Unit: m).

behaviors under various excitations. For long-span cable-stayed bridges, the deck vibration may be considered as the main structural wind-induced vibration. Thus, seven pairs of low-frequency bidirectional 8310-type accelerometers, produced by Kistler Instrumental Corporation, were installed on the bridge deck to monitor the wind-induced deck vibrations, as depicted in Fig. 2. Data collected from the anemometer at the middle span of the main girder are used for analysis. The sampling frequency is 1Hz for wind data collection, while the wind direction is defined as 0° facing north, anticlockwise.[19]

As shown in Fig. 2, five pairs of accelerometers were positioned on the main span and one on each side span. Each sensor can accurately measure accelerations ranging from $-2.5\,g$ to $2.5\,g$ (g represents the gravitational acceleration) and work properly with the temperature ranging from −40°C to +85°C. The sampling frequency of the accelerometers is fixed at 20 Hz to obtain more information during the real-time measurements. The transverse distance between the two accelerometers in a pair is 35.4 m, and both lateral and vertical accelerations of the main girder can be obtained by each sensor.[19]

2.2. *Typhoon Damrey and Haikui*

Typhoon Damrey was formed on the northwestern region of the Tropical of Cancer as a tropical storm on July 28, 2012. It is noted that the other Typhoon Saola was developed besides Typhoon Damrey simultaneously. As typhoons Damrey and Saola move towards the eastern coastal China, they shortened the distance between the circulations of their corresponding low-pressure areas and the Fujiwhara effect was enhanced. It landed on Jiangsu at 21:00 local time (UTC+8) on August 2, and passed the Sutong Bridge site from around 10:00 on August 2 to 20:00 on August 3. The moving track of Typhoon Damrey is shown in Fig. 3, while the recorded wind samples are plotted in Fig. 4.

Typhoon Haikui was developed as a tropical depression in the southeast of Iwo Jima late on August 2, 2012. With its ever-increasing intensity, Haikui was upgraded to a typhoon on August 7. It made a landfall on Xiangshan County in Zhejiang, China at 03:20 local time (UTC+8) on August 8. Subsequently, it entered Anhui Province at 20:00 on August 8 and passed through the Sutong Bridge site around 03:00 to 19:00 on August 8. Finally, it weakened into tropical depression again at 12:00 on August 9. Figure 5 shows the moving track of Typhoon Haikui and the location of the Sutong Bridge. The recorded wind samples of Typhoon Haikui are plotted on Fig. 6.

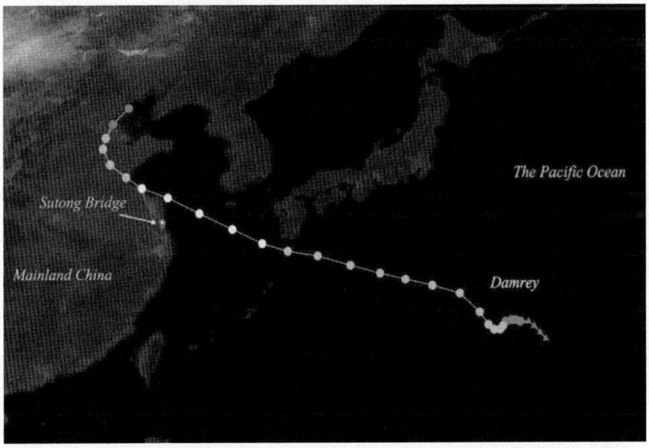

Fig. 3. Moving track of Typhoon Damrey.

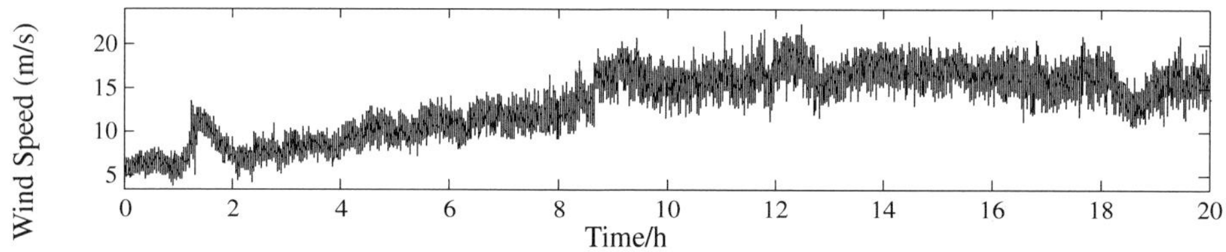

(a) Wind speed

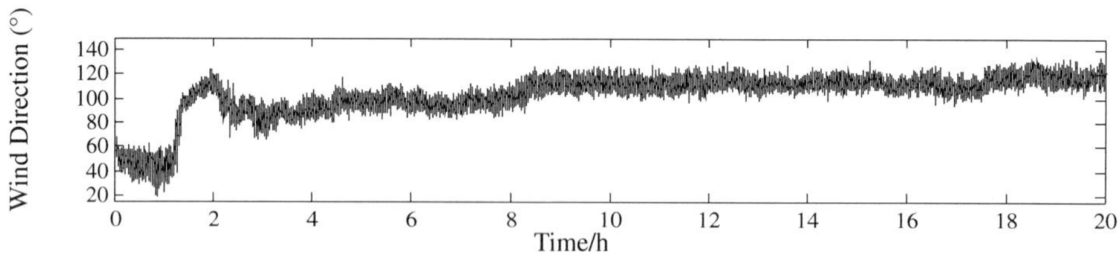

(b) Wind direction

Fig. 4. Wind samples of Typhoon Damrey (2012/08/03T00:00:00-2012/08/03T20:00:00).

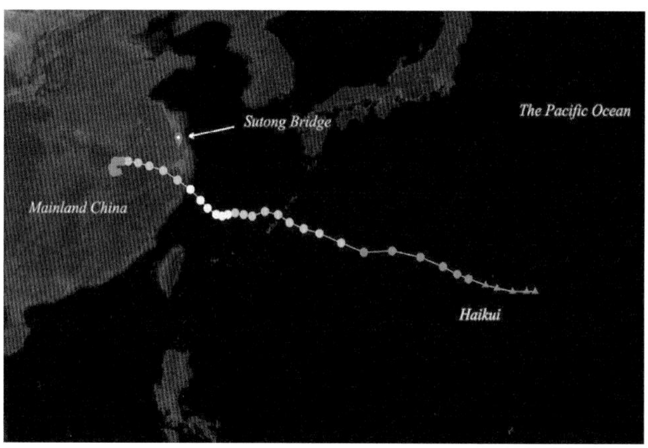

Fig. 5. Moving track of Typhoon Haikui.

2.3. *Wind characteristics of typhoon records*

To characterize the wind field during typhoons, the wind characteristics of the two typhoons are analyzed. Here, a 10-min duration is selected as a basic interval. The mean wind speed and mean wind direction can be obtained using the vector decomposition method.[20] Figure 7 shows the variation trend of the mean wind characteristics of the two typhoons.

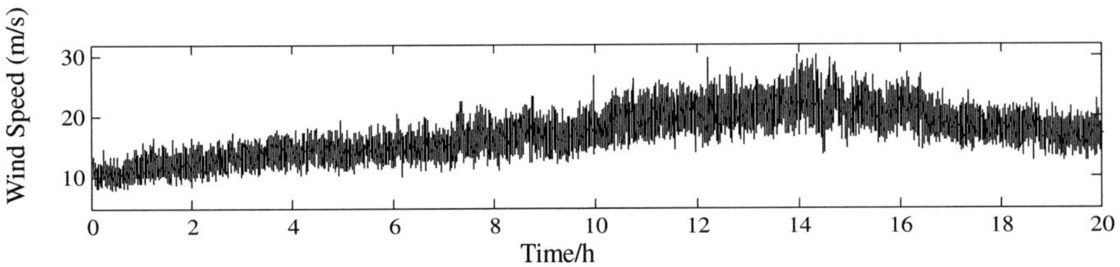

(a) Wind speed

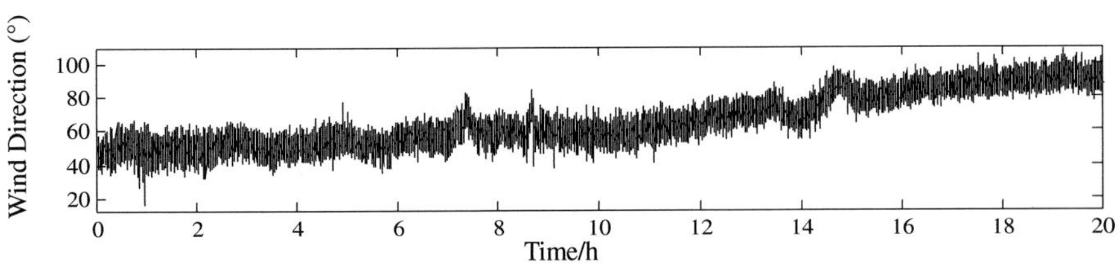

(b) Wind direction

Fig. 6. Wind samples of Typhoon Haikui (2012/08/08T00:00:00-2012/08/08T20:00:00).

In addition, the turbulent wind characteristics, including turbulence intensity (see Fig. 8) and gust factor (see Fig. 9), are also analyzed to provide important wind characteristic references for the wind performance evaluation of the bridge. Note that the gust duration time is taken as 3 s here.[43]

2.4. *Non-stationary buffeting response during typhoons*

During typhoons Damrey and Haikui, the acceleration responses at the key sections were recorded in real time using accelerometers installed on the Sutong Bridge. Due

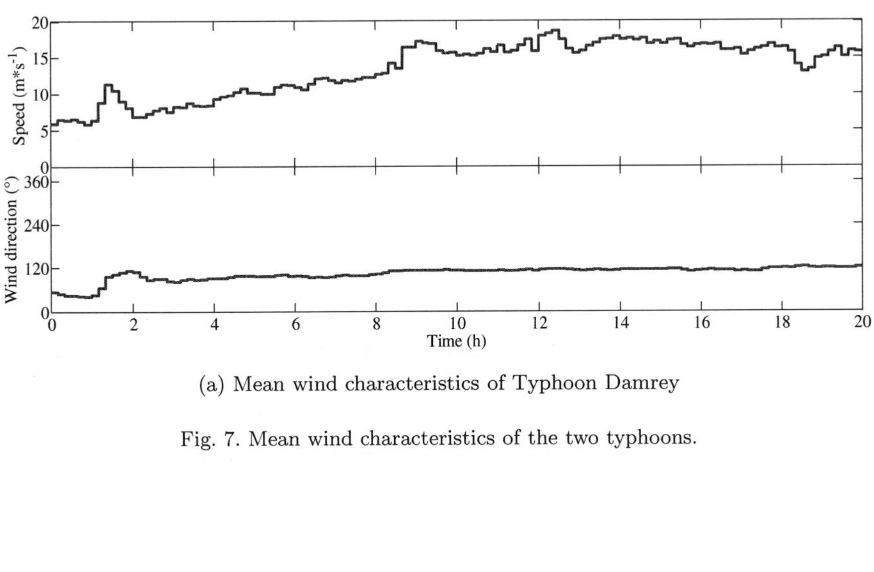

(a) Mean wind characteristics of Typhoon Damrey

Fig. 7. Mean wind characteristics of the two typhoons.

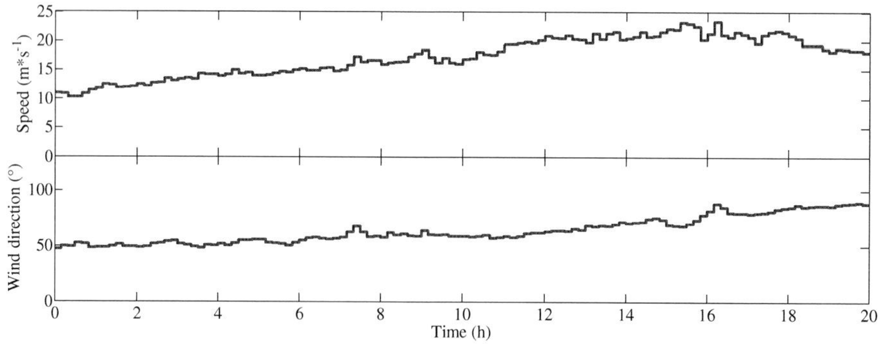

(b) Mean wind characteristics of Typhoon Haikui

Fig. 7. (*Continued*)

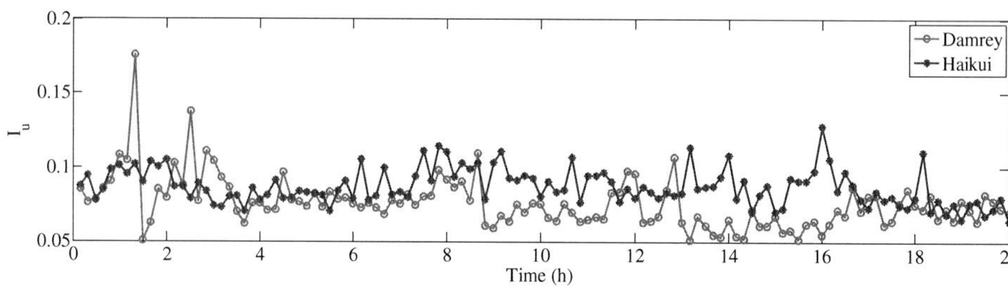

Fig. 8. Turbulence intensity of the two typhoons.

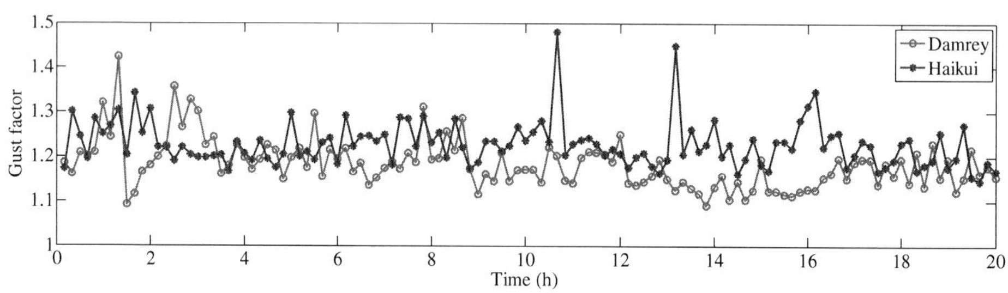

Fig. 9. Gust factor of the two typhoons.

to the influence of the typhoon climate, few cars passed through the bridge than the usual operating hours. In addition, the temperature variation during typhoon has little influence on the modal changes. Hence, the data collected from 07:00 to 09:00 on August 3, 2012 during Typhoon Damrey and from 10:00 to 12:00 on

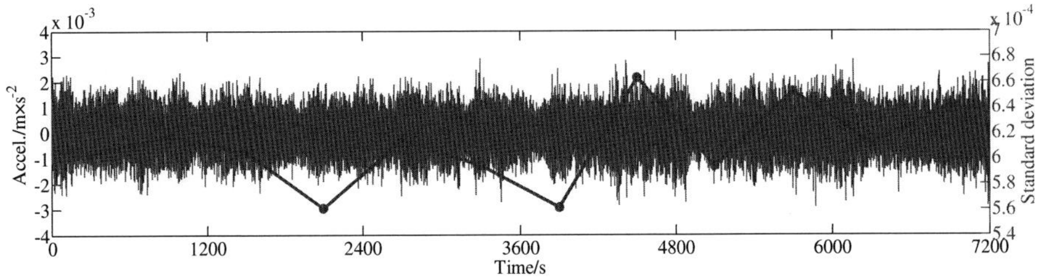

(a) Lateral

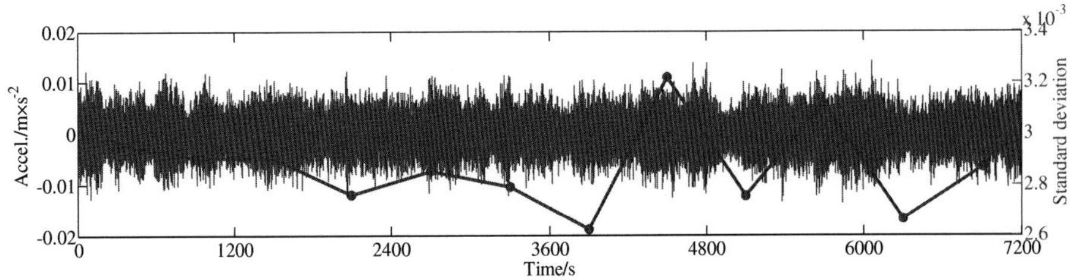

(b) Vertical

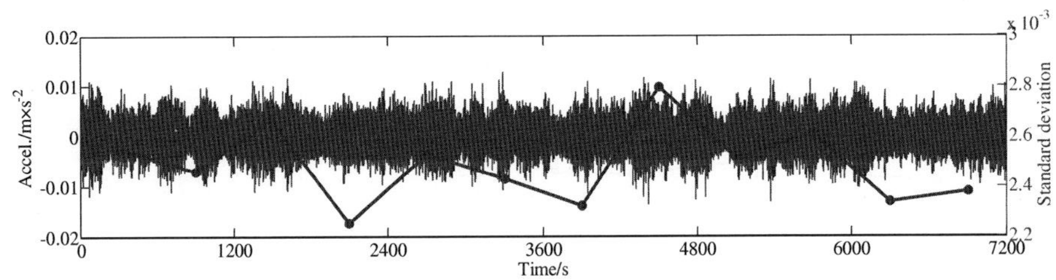

(c) Torsional

Fig. 10. Recorded buffeting response of the main girder during Typhoon Damrey.

August 8, 2012 during Typhoon Haikui can be reasonably considered as wind-induced buffeting responses.

When the two typhoons passed by the Sutong Bridge site, the recorded buffeting responses at mid-span section of the main span have the largest amplitude, as shown in Figs. 10 and 11, and thus selected to be analyzed. Meanwhile, to reflect the non-stationary features, the standard deviation of the buffeting responses is also plotted in Figs. 10 and 11 using the 10-min time interval that is recommended by Chinese code (JTG/T 3360-01-2018). Note that the lateral and vertical responses are the average of the two sets of data recorded by accelerometers ACC2-5-01 and ACC2-5-02, and the torsional responses are the differential values of the two vertical data recorded by the upstream and downstream accelerometers in a pair and divided by

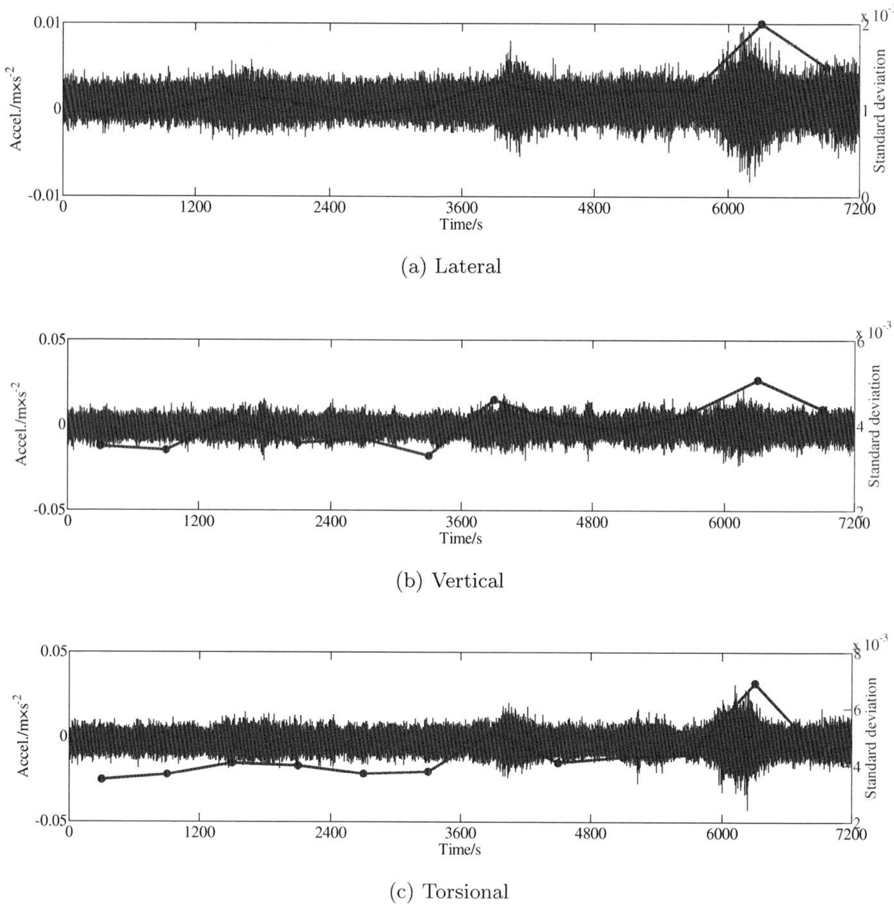

Fig. 11. Recorded buffeting response of the main girder during Typhoon Haikui.

35.4 m (distance between the two sensors). The FIR low-pass filter is utilized to filter the high-frequency components larger than 15 Hz. The recorded lateral, vertical and torsional responses for both typhoons contain 144 000 data points, respectively.

As shown in Figs. 10 and 11, during typhoons Damrey and Haikui, the lateral responses of the main girder at the mid-span section are obviously smaller than the vertical responses, similar to that in the Runyang Suspension Bridge during Typhoon Matsa.[43] It can be observed from the standard deviation that the recorded buffeting responses exhibit time-varying variance. Herein, the run test method is utilized to further evaluate the signal stationarity. The run test method is a widely used non-parametric statistical test. For the recorded buffeting response, the run test method was implemented where each 2-h sample is divided into 144 segments

Table 1. Run test results of the recorded data.

Typhoon	Direction	N_1	N_2	R	$\|z\|$	Results
Damrey	Lateral	71	73	45	4.68	Non-stationary
	Vertical	59	85	47	4.09	Non-stationary
	Torsional	63	81	55	2.87	Non-stationary
Haikui	Lateral	41	103	16	8.98	Non-stationary
	Vertical	58	86	38	5.61	Non-stationary
	Torsional	43	101	24	7.46	Non-stationary

with a significance level of 0.05. The z-score can be defined by Castellan.[44]

$$\begin{cases} \mu = 2\dfrac{N_1N_2}{N_1+N_2}+1 \\ \sigma = \dfrac{2N_1N_2(2N_1N_2-N_1-N_2)}{(N_1+N_2)^2(N_1+N_2-1)}, \\ z = \dfrac{(R-\mu)}{\sigma} \end{cases} \tag{1}$$

where N_1 represents the number of data points with a positive sign; N_2 is the number of data points with a negative sign. R is the number of runs in the time series. With the fixed significance level, the time series will be judged as non-stationary if $|z| \geq 1.96$. More details for the run test method can be found in Ref. 45. The run test results of the recorded responses in Figs. 10 and 11 are summarized in Table 1.

From Table 1, all the calculated z-scores are larger than 1.96, which indicates the non-stationarity of the recorded buffeting responses. Hence, the buffeting responses are selected for in-depth investigation.

3. EPSD Estimation of the Recorded Responses

To investigate the non-stationary features of the recorded responses in time and frequency domain, the EPSDs of the buffeting responses recorded from the mid-span of the girder in each direction for both typhoons were estimated by both GHW and FHW. More details with respect to the EPSD estimation theory using HW are given in the Appendix.

As shown in Figs. 12 and 13, the EPSDs estimated by GHW are similar to those by the FHW. Obviously, the energy of the recorded responses in each direction mainly concentrates on several typical frequencies. For instance, the frequencies of the first peak point at each time instant is 0.186, 0.107 and 0.547 Hz for vertical, lateral and torsional EPSDs of both typhoons, respectively. These frequencies correspond to the first vibration mode in each direction,[46] indicating that the basic modes dominate the buffeting responses of the main girder of the Sutong Bridge during typhoons. This phenomenon indirectly reveals that the turbulence kinetic energy distributions of typhoons mainly concentrate on the fundamental frequency

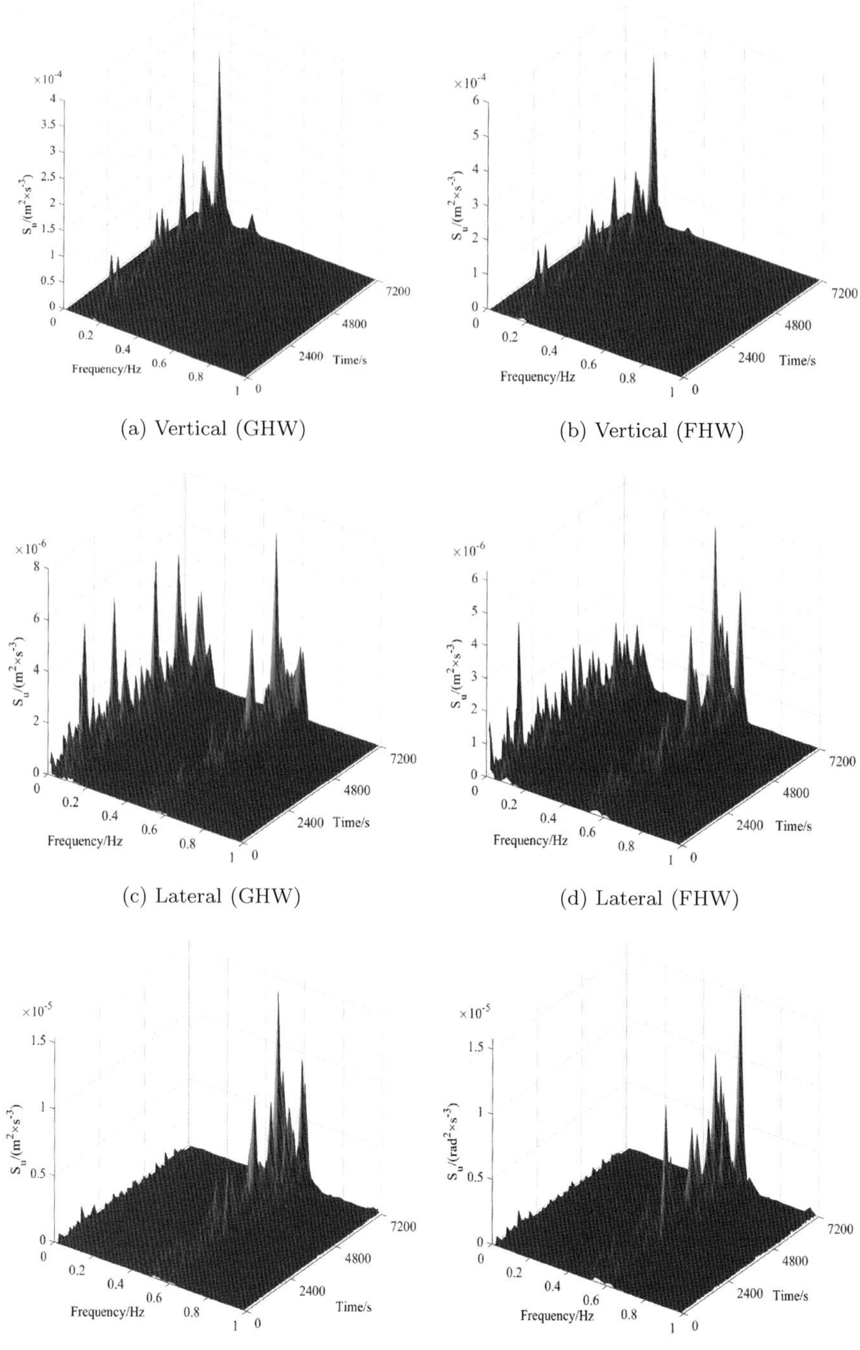

(a) Vertical (GHW)

(b) Vertical (FHW)

(c) Lateral (GHW)

(d) Lateral (FHW)

(e) Torsional (GHW)

(f) Torsional (FHW)

Fig. 12. EPSDs of the recorded buffeting responses using GHW and FHW during Typhoon Damrey.

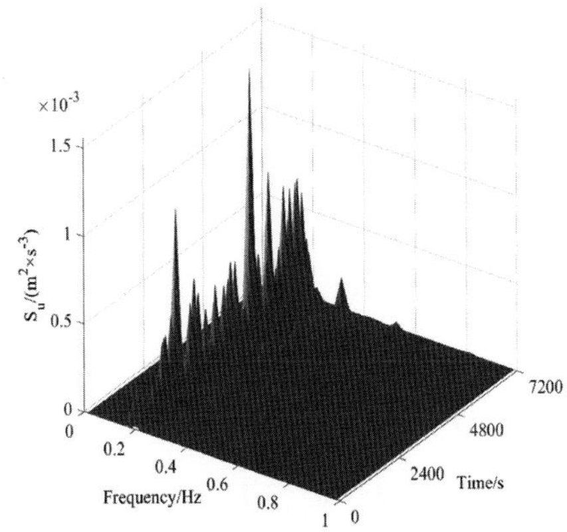

(a) Vertical (GHW)

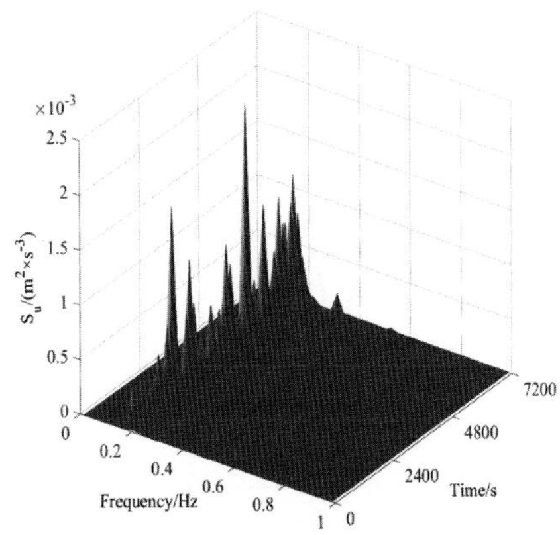

(b) Vertical (FHW)

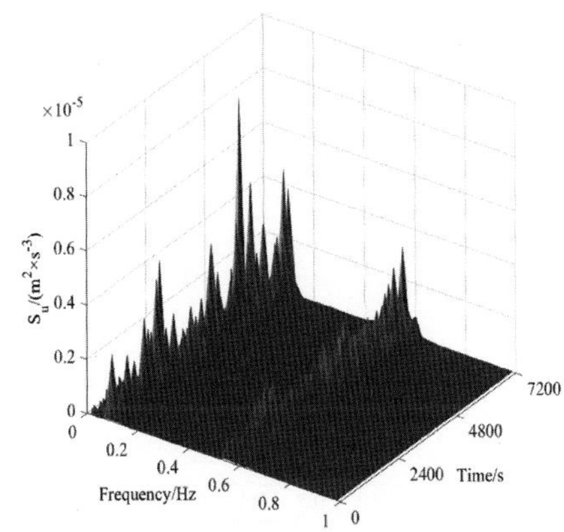

(c) Lateral (GHW)

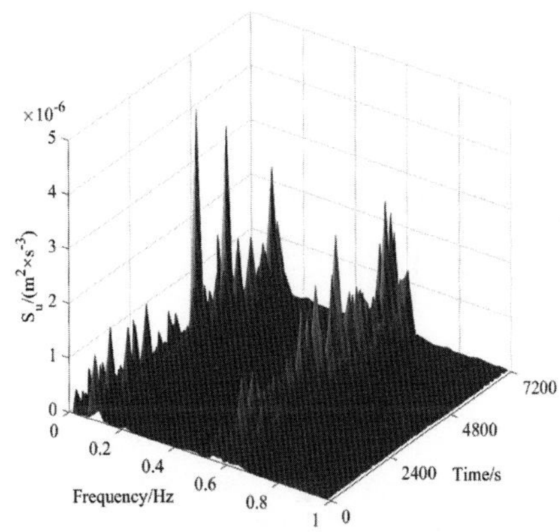

(d) Lateral (FHW)

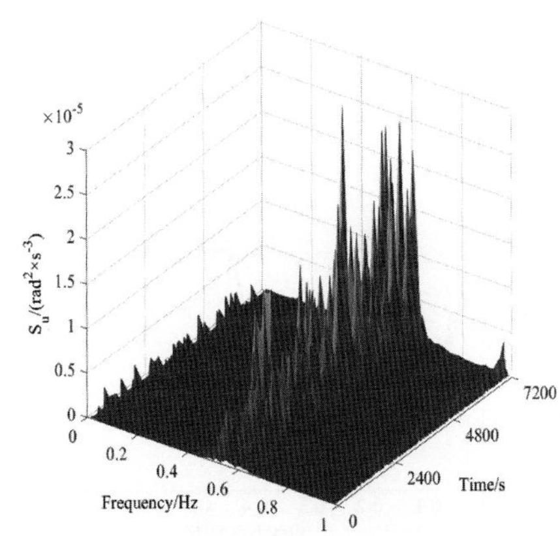

(e) Torsional (GHW)

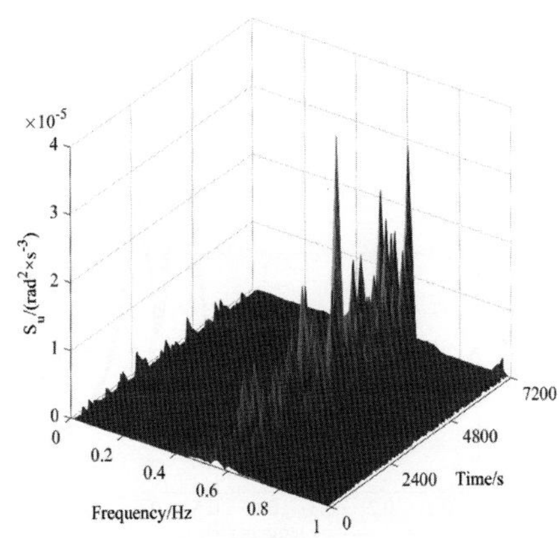

(f) Torsional (FHW)

Fig. 13. EPSDs of the recorded buffeting responses using GHW and FHW during Typhoon Haikui.

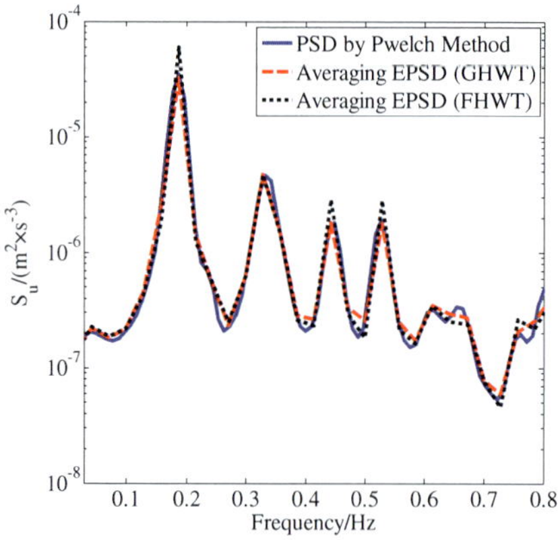

(a) Vertical PSDs during Typhoon Damrey

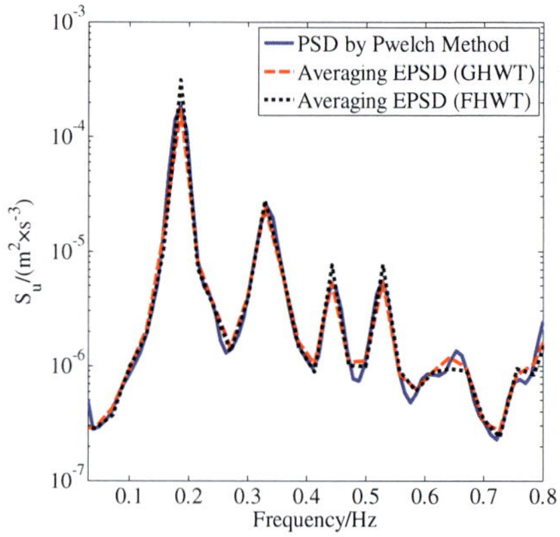

(b) Vertical PSDs during Typhoon Haikui

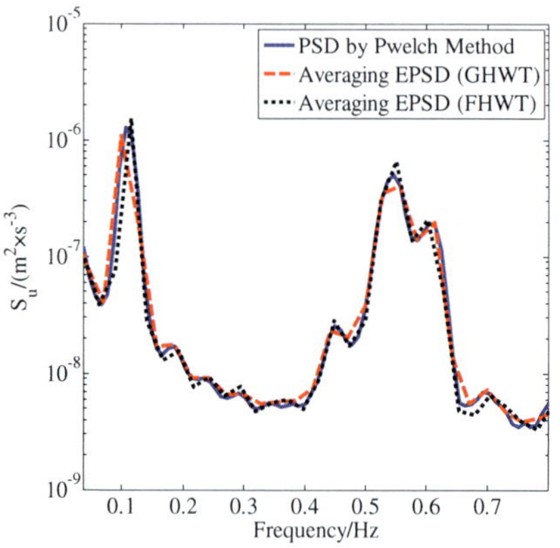

(c) Lateral PSDs during Typhoon Damrey

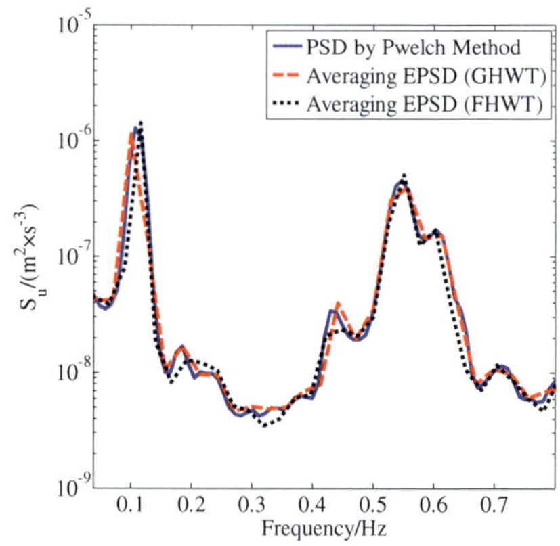

(d) Lateral PSDs during Typhoon Haikui

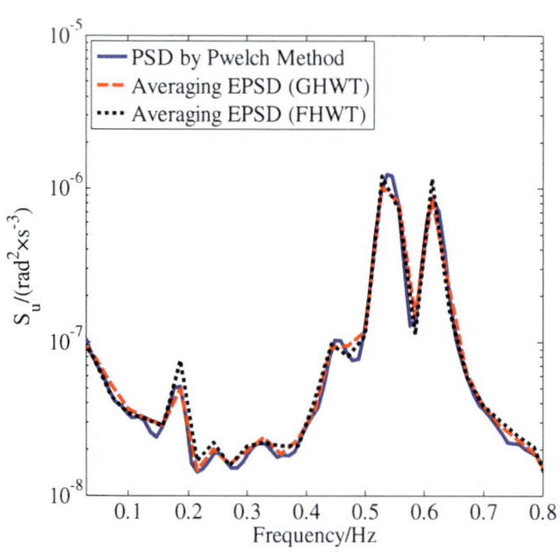

(e) Torsional PSDs during Typhoon Damrey

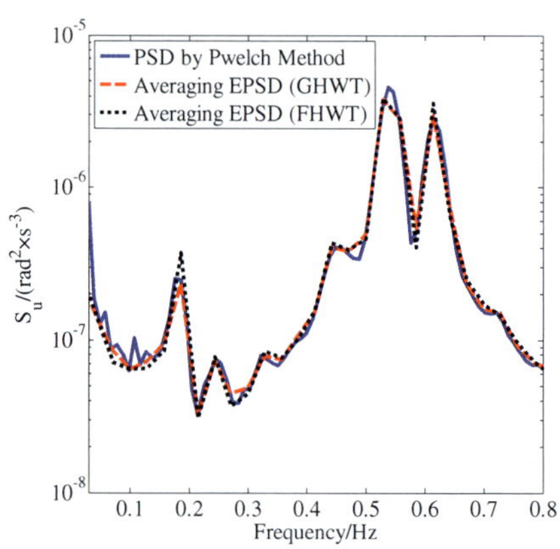

(f) Torsional PSDs during Typhoon Haikui

Fig. 14. Comparison of the averaging EPSD and the PSD by the Pwelch Method.

components. Moreover, the energy contained by each mode at a fixed time instant is different between the two typhoons. In addition, the estimated EPSDs vary with time at fixed frequencies for both typhoons, indicating the non-stationarity of the recorded responses.

To validate the accuracy of estimated EPSDs, the averaging EPSDs and the PSDs estimated by the Pwelch method are compared in Fig. 14. It is noted that the averaging EPSD is obtained by averaging the EPSD along the whole time duration. As shown in Fig. 14, the averaging EPSDs agree well with the PSDs by the Pwelch method, indicating that the estimated EPSDs using HWs are accurate and consistent with the analytical theory by Priestley.[22] However, it can be observed that some peak values of the averaging EPSDs estimated by FHW are slightly larger than that by GHW at certain fixed frequencies in each direction. In addition, the PSDs estimated by the Pwelch method during Typhoon Damrey are similar to those during Typhoon Haikui, indirectly indicating that the frequency components of the stimulus, i.e. typhoons, exhibit analogical features and cause similar response spectra during both typhoons. To reach a meaningful compromise between time and frequency resolution, the frequency band should be divided into more parts for the EPSDs estimated by the FHW than those by the GHW. In general, the time resolution for FHW is better than that of GHW due to the added window function. Therefore, the FHW is recommended to analyze the non-stationary wind-induced response.

4. Conclusions

To investigate the non-stationary features of the recorded buffeting responses of the Sutong Bridge during typhoons, HW-based methods have been utilized to estimate the EPSDs of non-stationary bridge responses. A comparative study between the EPSDs estimated by the GHW and the FHW was conducted. The following conclusions can be drawn:

(1) For the Sutong Bridge during typhoons Damrey and Haikui, the lateral responses of the main girder at mid-span section are much smaller than the vertical ones. Based on the run test method, the recorded buffeting responses in all directions are demonstrated to be non-stationary.
(2) The EPSDs of the recorded buffeting responses estimated by the GHW are similar to those by the FHW for both typhoons. The averaging EPSDs agree well with the PSDs obtained by the Pwelch method. In addition, the fundamental vibrational modes dominate the buffeting response of the main girder of the Sutong Bridge during typhoons.
(3) The frequency band should be divided into more parts for FHW than the GHW to reach a meaningful compromise between time and frequency resolution. In general, the time resolution for FHW is better than that of GHW due to the

added window function, which is recommended to analyze the non-stationary wind-induced response.

(4) The recorded responses of the main girder exhibit non-stationarity during typhoons and the stationary assumption in wind engineering should be switched to a non-stationary analysis scheme.

Appendix A. HW-Based EPSD Estimation Theory

The GHW and FHW are selected to conduct EPSD estimation due to their appealing features. They have simple explicit formula, and possess the property of non-overlapping Fourier transform. Moreover, GHWT and FHWT can be calculated efficiently using FFT technique.[39] The GHWT divides the frequency axis with two indices (m, n) that define the scales at this scheme. In the frequency domain, the GHWT with scale (m, n) at time position k can be defined by Newland.[37]

$$GW_T(\omega) = \begin{cases} \dfrac{1}{(n-m)2\pi} e^{-i\omega k/(n-m)}, & m2\pi \leq \omega \leq n2\pi \\ 0, & \text{elsewhere} \end{cases}, \tag{A.1}$$

where m and n are positive real numbers. Eq. (A.1) is for a GHW centered at time $k/(n-m)$ and frequency $(m+n)\pi$ with bandwidth $(n-m)2\pi$. Additionally, orthogonality can be easily demonstrated for the GHW.

The FHW is an improved version of the GHW, which enhances the time resolution in the wavelet mean square map with a given frequency resolution by filtering the mother wavelet using a Hanning function in the frequency domain.[38] Specifically,

$$FW_T(\omega) = \begin{cases} \dfrac{1}{(n-m)2\pi} \left[1 - \cos\left(\dfrac{\omega - m2\pi}{n-m}\right)\right], & m2\pi \leq \omega \leq n2\pi \\ 0, & \text{elsewhere} \end{cases}. \tag{A.2}$$

For the HW-based EPSD estimation theory, the local spectra estimated by GHW and FHW in the ith frequency and the kth time subdomain, i.e. $m2\pi \leq \omega \leq n2\pi$ and $k/(n-m) \leq t \leq (k+1)/(n-m)$, can be defined by[25]

$$S_{GW}(\omega_i, t_k) = \frac{4E[|g_{(m,n),k}|^2]}{n-m}, \tag{A.3}$$

$$S_{FW}(\omega_i, t_k) = \frac{4E[|g_{F(m,n),k}|^2]}{n-m}, \tag{A.4}$$

where $g_{(m,n),k}$ is the GHW coefficient and $g_{F(m,n),k}$ is the FHW coefficient.

Acknowledgments

This research work was jointly supported by the National Natural Science Foundation of China (Nos. 52208481, 51978155).

References

1. D. Pauselli, D. Salciarini and F. Ubertini, Three-dimensional modeling of soil-structure interaction for a bridge founded on Caissons under seismic conditions, *Appl. Sci.* **12**(21) (2022) 10904.
2. E. Simiu and R. H. Scanlan, *Wind Effects on Structures* (Wiley, 1996).
3. Z. J. Li, M. Q. Feng, L. X. Luo, D. M. Feng and X. L. Xu, Statistical analysis of modal parameters of a suspension bridge based on Bayesian spectral density approach and SHM data, *Mech. Syst. Signal Processing* **98** (2018) 352–367.
4. M. Laura, C. Francesco and F. Antonio, Static and dynamic testing of highway bridges: A best practice example, *J. Civil Struct. Health Monitoring* **10**(1) (2020) 43–56.
5. Z. Q. Chen, Y. Han, X. G. Hua and Y. Z. Luo, Investigation on influence factors of buffeting response of bridges and its aeroelastic model verification for Xiaoguan bridge, *Eng. Struct.* **31**(2) (2009) 417–431.
6. J. Zhu and W. Zhang, Numerical simulation of wind and wave fields for coastal slender bridges, *J. Bridge Eng.* **22**(3) (2016) 04016125.
7. F. Benedettini, M. Dilena and A. Morassi, Vibration analysis and structural identification of a curved multi-span viaduct, *Mech. Syst. Signal Processing* **54** (2015) 84–107.
8. A. G. Davenport, A statistical approach to the treatment of wind loading on tall masts and suspension bridges, Department of Civil Engineering, University of Bristol, UK (1961).
9. R. H. Scanlan, The action of flexible bridges under wind, I: flutter theory, *J. Sound Vib.* **60**(2) (1978) 187–199.
10. R. H. Scanlan, The action of flexible bridges under wind, II: Buffeting theory, *J. Sound Vib.* **60**(2) (1978) 201–211.
11. A. Jain, Coupled flutter and buffeting analysis of long-span bridges, *J. Struct. Eng.* **122**(7) (1996) 716–725.
12. Y. L. Xu, D. K. Sun, J. M. Ko and J. H. Lin, Fully coupled buffeting analysis of tsing ma suspension bridge, *J. Wind Eng. Ind. Aerodyn.* **85**(1) (2000) 97–117.
13. X. Chen and A. Kareem, Advances in modeling of aerodynamic forces on bridge decks, *J. Eng. Mech.* **128**(11) (2002) 1193–1205.
14. F. Tubino and G. Solari, Gust buffeting of long span bridges: Double modal transformation and effective turbulence, *Eng. Struct.* **29**(8) (2007) 1698–1707.
15. W. Zhang, Y. J. Ge and C. S. Cai, Evaluating wind loads on bridge decks using velocity fields, *J. Eng. Mech.* **139**(3) (2012) 339–346.
16. Y. L. Xu and J. Chen, Characterizing nonstationary wind speed using empirical mode decomposition, *J. Struct. Eng.* **130**(6) (2004) 912–920.
17. A. G. Davenport, Buffeting of a suspension bridge by storm winds, *J. Struct. Div.* **88**(3) (1962) 233–270.
18. L. Hu, Y. L. Xu and W. F. Huang, Typhoon-induced non-stationary buffeting response of long-span bridges in complex terrain, *Eng. Struct.* **57**(4) (2013) 406–415.

19. H. Wang, R. M. Hu, J. Xie, T. Tong and A. Q. Li, Comparative study on buffeting performance of sutong bridge based on design and measured spectrum, *J. Bridge Eng.* **18**(7) (2013) 587–600.
20. Q. S. Li, L. H. Zhi, J. Yi, A. To and J. Xie, Monitoring of typhoon effects on a super-tall building in Hong Kong, *Struct. Control Health Monitoring* **21**(6) (2014) 926–949.
21. A. Kareem and T. Kijewski, Time-frequency analysis of wind effects on structures, *J. Wind Eng. Ind. Aerodyn.* **90**(12) (2002) 1435–1452.
22. M. B. Priestley, *Spectral Analysis and Time Series* (Academic Press Inc., New York, 1982).
23. K. Gurley and A. Kareem, Application of wavelet transform in earthquake, wind, and ocean engineering, *Eng. Struct.* **21**(2) (1999) 149–167.
24. J. Ville, Theorie et applications de la notion de signal analytique, *Zootecn. Trop.* **2** (1948) 61–74.
25. P. D. Spanos, J. Tezcan and P. Tratskas, Stochastic processes evolutionary spectrum estimation via harmonic wavelets, *Computer Methods Appl. Mech. Eng.* **194**(12–16) (2005) 1367–1383.
26. L. Cohen, *Time-Frequency Analysis* (Prentice-Hall, Englewood Cliffs, NJ, 1995).
27. D. E. Newland, Harmonic wavelet analysis, *Proc. R. Soc. A Math. Phys. Eng. Sci.* **443**(1917) (1993) 203–225.
28. P. D. Spanos and G. Failla, Evolutionary spectra estimation using wavelets, *J. Eng. Mech.* **130**(8) (2004) 952–960.
29. A. Grossmann and J. Morlet, Decomposition of Hardy functions into square integrable wavelets of constant shape, *SIAM J. Math. Anal.* **15**(4) (1984) 723–736.
30. I. Daubechies, *Ten Lectures on Wavelets* (Society for Industrial and Applied Mathematics, 1992).
31. B. B. Hubbard, *The World According to Wavelets the Story of a Mathematical Technique in the Making* (Wellesley, Massachusetts, USA, 1998).
32. M. R. Taha, A. Noureldin, J. L. Lucero and T. J. Baca, Wavelet transform for structural health monitoring: a compendium of uses and features, *Struct. Health Monitoring* **5**(3) (2006) 267–295.
33. M. B. Priestley, Evolutionary spectra and non-stationary processes, *J. R. Stat. Soc.* **27**(2) (1965) 204–237.
34. T. T. Fujita, *The Downburst, Report of Projects NIMROD and JAWS* (University of Chicago, 1985).
35. G. Huang and X. Chen, Wavelets-based estimation of multivariate evolutionary spectra and its application to nonstationary downburst winds, *Eng. Struct.* **31**(4) (2009) 976–989.
36. H. Wang, T. Wu, T. Y. Tao, A, Q. Li and A. Kareem, Measurements and analysis of non-stationary wind characteristics at Sutong Bridge in Typhoon Damrey, *J. Wind Eng. Ind. Aerodyn.* **151** (2016) 100–106.
37. D. E. Newland, Harmonic and musical wavelets, *Proc. R. Soc. A Math. Phys. Eng. Sci.* **444**(1922) (1994) 605–620.
38. D. E. Newland, Ridge and phase identification in the frequency analysis of transient signals by harmonic wavelets, *J. Vib. Acoustics* **121**(2) (1999) 149–155.
39. P. Tratskas and P. D. Spanos, Linear multi-degree-of-freedom system stochastic response by using the harmonic wavelet transform, *J. Appl. Mech.* **70**(5) (2003) 724–731.

40. C. Cattani, Harmonic wavelets towards the solution of nonlinear PDE, *Computers Math. Appl.* **50**(8) (2005) 1191–1210.
41. P. D. Spanos, F. Kong, J. Li and I. A. Kougioumtzoglou, Harmonic wavelets based excitation-response relationships for linear systems: A critical perspective, *Probab. Eng. Mech.* **44** (2016) 163–173.
42. E. C. C. Choi, Characteristics of typhoons over the South China Sea, *J. Wind Eng. Ind. Aerodyn.* **3**(4) (1978) 353–365.
43. H. Wang, A. Li, T. Guo and J. Xie, Field measurement on wind characteristic and buffeting response of the Runyang Suspension Bridge during typhoon Matsa, *Sci. China Series E: Technol. Sci.* **52**(5) (2009) 1354–1362.
44. N. J. Castellan, *Nonparametric Statistics for the Behavioral Sciences* (McGraw-Hill, 1988).
45. J. S. Bendat and A. G. Piersol, *Random Data: Analysis and Measurement Procedures* (Wiley, NJ, 2010).
46. H. Wang, J. X. Mao, J. H. Huang and A. Q. Li, Modal identification of sutong cable-stayed bridge during typhoon haikui using wavelet transform method, *J. Perform. Constructed Facilities* **30**(5) (2016) 04016001.

https://doi.org/10.1142/9789819814770_0005

Chapter 5

Optimal Design of Funicular Arches Under Equally Spaced Point Loads for Maximum Buckling Load#

C. M. Wang*, W. H. Pan†, Y. P. Zhang*,‡ and Y. Z. Hao*

*School of Civil Engineering, The University of Queensland St Lucia, Queensland 4072, Australia

†Department of Civil Engineering, Zhejiang University Hangzhou, Zhejiang, P. R. China

‡y.zhang16@uq.edu.au

This paper presents the optimal designs of pinned supported funicular arches under equally spaced point loads for maximum in-plane buckling load. Under such loading conditions, the funicular arch shapes comprise straight arch members between the point loads, that is, following the shape of the bending moment diagram of an equivalent simply supported beam under the same loading condition. Two classes of funicular arch optimization problems are considered herein. The first class of funicular arches imposes a constraint on the cross-sectional area to be uniform throughout the entire arch length. The second class of funicular arches allows the cross-sectional area to be different from one straight arch member to another member. To facilitate the buckling analysis, the Hencky bar-chain model (HBM) is adopted. This discrete structural model simplifies the optimization process as the decision variables are the HBM rotational spring stiffnesses that define the cross-sectional areas and the horizontal force that controls the arch shape. Presented herein are new optimal funicular arch shapes under various numbers of equally spaced point loads. By increasing the number of point loads, the optimal solution approaches the solution of a parabolic arch under a uniformly distributed load.

Keywords: Funicular arches; optimal design; elastic buckling; Hencky bar-chain model.

1. Introduction

Arch structures have been used throughout the history of human civilization for spanning long distances such as roofs and bridges. When compared to beam structures, funicular (or momentless) arch structures transmit the applied loads to their base supports by axial compression, which is a very economical means for load

‡Corresponding author.

#This chapter appeared previously on the International Journal of Structural Stability and Dynamics. To cite this chapter, please cite the original article as the following: C. M. Wang *et al.*, *Int. J. Struct. Stab. Dyn.*, **23**, 2340005 (2023), DOI: 10.1142/S0219455423400059.

transmission rather than through bending action. This enables arch structures to carry more loads over long distances when compared to beam girders. As arches are slender in modern construction to minimize materials used and costs, these slender arch structures may fail prematurely by elastic buckling before reaching their maximum compressive strength capacity.

Studies on the optimal design of arches against buckling started in the late 1960s. Wu[1] determined the optimal shape of circular arches for the largest critical buckling pressure by using perturbation methods. He assumed a rectangular cross-section of constant width leaving the depth variation to be calculated. Prompted by Wu's work, Budiansky *et al.*[2] maximized the Rayleigh quotient of the arch for a class of moment of inertia to cross-sectional relationships. They pointed out that Wu's solutions were restricted to arches having a sufficiently large opening angle such that antisymmetrical buckling was critical because shallow archers were associated with symmetrical snapping. Tadjbakhsh and Farshad[3] considered the optimal shape of funicular arches of nonuniform cross-section subjected to a given load. The arch structural weight was minimized for a given buckling load. Farshad[4] continued to work on these funicular arches and investigated a variety of mutual shape optimization problems for arches of nonuniform cross-section and having either given horizontal force or given arc length or both. Amazigo[5] obtained the optimal design of shallow circular arches of nonuniform cross-sections under uniform pressure against snap-buckling. Tadjbakhsh[6] studied the stability and optimum design of arch-type structures with variable cross-sections. Blachut[7] studied the optimal design of funicular arches with piecewise constant cross-section against buckling and vibration. Wroblewski and Zyczkowski[8] treated the multimodal optimization problem of circular arches with nonuniform cross-sections against plane and spatial creep buckling. Zhang and Wang[9] determined the optimal design of circular arches with piecewise varying cross-sections under uniform pressure against buckling. Hernández-Díaz *et al.*[10] obtained the buckling design of fully stressed submerged arches via shape parameterization. Chaves Neto and de Melo[11] optimized shallow arches under uniformly distributed load considering in-plane symmetric buckling. Wang *et al.*[12] determined the optimal design of triangular arches against buckling. Wang *et al.*[13] obtained the optimal design of funicular trapezoidal arches of different member cross-sections and under two vertical point loads for maximum buckling load.

Continuing in this line of investigation, this study focuses on the optimization problem of funicular arches under equally spaced point loads for maximum buckling capacity; a problem that has not been tackled before but yet it is commonly encountered in the form of arch bridges carrying a pavement deck (as shown in Fig. 1). As the number of point loads increases to a large number, the optimal solution approaches that of an optimal parabolic arch under a uniformly distributed load. The Hencky bar-chain model (HBM) will be adopted for the buckling analysis.[14]

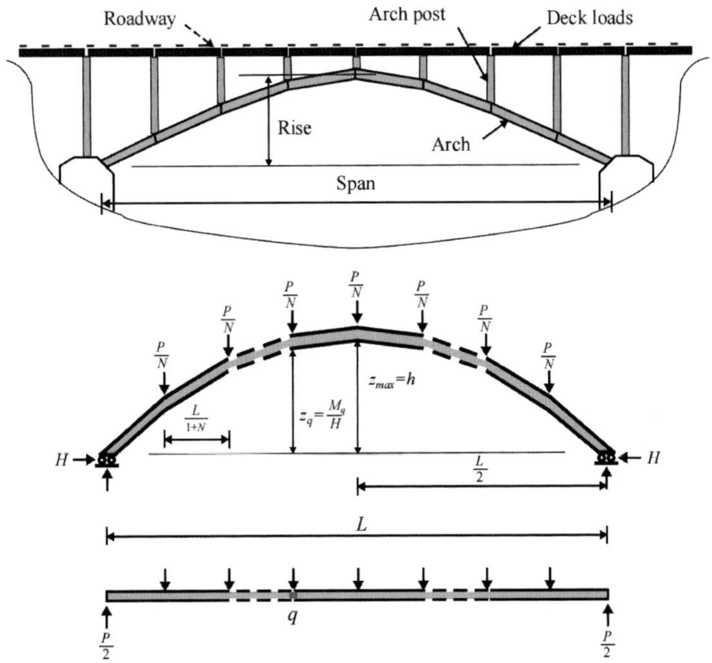

Fig. 1. Funicular arch and equivalent simply supported beam under N equally spaced point loads.

HBM comprises straight rigid bar segments joined to one another by frictionless hinges and elastic rotational springs whose stiffnesses are calibrated appropriately. Based on this HBM, one can readily optimize the varying cross-sectional areas (or varying flexural rigidities) of the arch structure by changing the rotational spring stiffnesses so as to obtain maximum buckling capacity for the arch.

2. Problem Definition

Consider a pinned supported funicular arch of span L. The arch is subjected to N number of equally spaced point loads with magnitude P/N as shown in Fig. 1. In Fig. 1, P is the total vertical load applied on the funicular arch, H is the horizontal reaction at the arch supports and the vertical support reactions are $P/2$ from symmetry.

The geometry of the funicular (or momentless) arches is given by $z(x) = M(x)/H$, where $M(x)$ is the bending moment of a simply supported beam of length L and under the same applied point loads as the arch (see Fig. 1). At any node q under a point load of the equivalent beam, the bending moment is given by

$$M_q = \left[\frac{PL}{N(1+N)}\right]\left[\frac{Nq}{2} - \sum_{k=0}^{q-1} k\right] \quad \text{for } q = 1, 2, \ldots, N. \tag{1}$$

Table 1. Cross-sectional shape constant k for equilateral, square, and circular cross-sections.

Cross-section shape	A	I	k
Equilateral triangle of side length b	$\frac{\sqrt{3}b^2}{4}$	$\frac{\sqrt{3}b^4}{96}$	$\frac{\sqrt{3}}{18}$
Square of side length b	b^2	$\frac{b^4}{12}$	$\frac{1}{12}$
Circular of radius r	πr^2	$\frac{\pi r^4}{4}$	$\frac{1}{4\pi}$

Therefore, the height of an arch node q under a point load is given by

$$z_q = \frac{M_q}{H} = \left[\frac{PL}{N(1+N)H}\right]\left[\frac{Nq}{2} - \sum_{k=0}^{q-1} k\right] \quad \text{for } q = 1, 2, \ldots, N. \tag{2}$$

The arch rise (or apex height) $h_{\max} = z_{(1+N)/2}$ if N is an odd number and $h = z_{N/2}$ if N is an even number.

The cross-sectional shape of the arch is assumed to be geometrically similar (that is of regular polygonal shape or circular shape). Such geometrically similar cross-sections imply that[15]

$$I = kA^2, \tag{3}$$

where I is the second moment of area of the arch cross-section, A is the cross-sectional area, and k is the cross-sectional shape constant. Table 1 shows the k values for a few cross-sectional shapes.

We consider two classes of arch optimization problems in this study. Class 1 problem deals with funicular arches having uniform cross-sectional area throughout the arch, while Class 2 problem considers funicular arches with the nonuniform cross-sectional area. Optimal designs of these two classes of arches are to be determined for maximum buckling load. Detailed descriptions of these two types of the optimization problem are given in Sec. 4. This study focuses on bifurcation in-plane buckling of arches; the arches are assumed to be restrained against out-of-pane buckling. The arch axis compressibility and self-weight are neglected.

3. Hencky Bar-Chain Model for Buckling Analysis

3.1. *Description of HBM for arch considered*

For the elastic in-plane buckling analysis of the funicular arch, the HBM, which is a physical bar-spring model, is adopted.[14] HBM is a discrete structural model that models the arch by rigid segments connected to one another by frictionless hinges with elastic rotational springs. Figure 2 shows the HBM model for a funicular arch under N point loads.

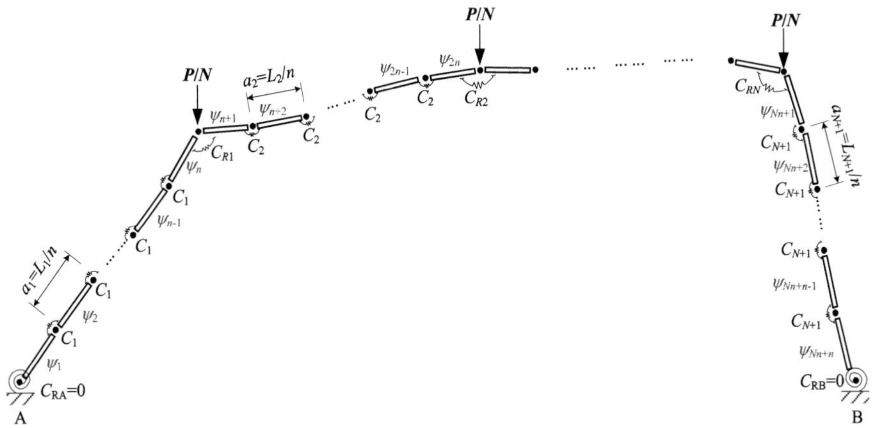

Fig. 2. HBM for funicular arch under N equally spaced point loads.

The HBM of the considered arch has each of its $N+1$ straight members divided into n rigid segments of equal length, i.e. $a_1 = L_1/n$ for the first member, $a_2 = L_2/n$ for the second member, ... and $a_{N+1} = L_{N+1}/n$ for the $(N+1)$th member (see Fig. 2). Note that the number of segments n to be adopted for accurate buckling solutions has to be established from carrying out convergence studies. The rigid segments are connected by frictionless hinges with elastic rotational stiffnesses $C_1 = EI_1/a_1$, $C_2 = EI_2/a_2, \ldots$ and $C_{N+1} = EI_{N+1}/a_{N+1}$ for the first, second, and $(N+1)$th members, respectively. $L_1, L_2, \ldots, L_{N+1}$ are the lengths of the $N+1$ straight arch members and they are given by

$$L_j = \frac{L}{N+1}\frac{\sqrt{H^2 + \frac{P^2}{N^2}\left(\frac{N}{2} - j + 1\right)^2}}{H} \quad \text{for } j = 1, 2, \ldots, N+1. \tag{4}$$

The HBM rotational spring stiffnesses at the pinned supports are $C_{\text{RA}} = C_{\text{RB}} = 0$.[16] The HBM rotational spring stiffnesses at the joints q are given by[14,17]

$$C_{Rq} = \frac{2}{\frac{1}{C_q} + \frac{1}{C_{q+1}}} \quad \text{for } q = 1, 2, \ldots, N. \tag{5}$$

3.2. *Derivation of governing buckling equation for HBM arch using energy approach*

For this funicular arch problem, $(Nn+n)$-order basis deformation vector of segment rotation angles relative to the undeformed arch configuration (positive for clockwise rotations) is selected as $\Psi = (\psi_1, \ldots, \psi_n, \psi_{n+1}, \ldots, \psi_{2n}, \ldots, \psi_{Nn+1}, \ldots, \psi_{Nn+n})^T$.

The elastic strain energy U of HBM is given by[14]

$$U = \frac{1}{2}\begin{bmatrix} C_1((\psi_1-\psi_2)^2+(\psi_2-\psi_3)^2+\cdots+(\psi_{n-1}-\psi_n)^2) \\ +C_2((\psi_{n+1}-\psi_{n+2})^2+(\psi_{n+2}-\psi_{n+3})^2 \\ +\cdots+(\psi_{2n-1}-\psi_{2n})^2)+\cdots+C_{N+1}((\psi_{Nn+1}-\psi_{Nn+2})^2 \\ +(\psi_{Nn+2}-\psi_{Nn+3})^2+\cdots+(\psi_{Nn+n-1}-\psi_{Nn+n})^2) \\ +C_{R1}(\psi_n-\psi_{n+1})^2+C_{R2}(\psi_{2n}-\psi_{2n+1})^2 \\ +\cdots+C_{RN}(\psi_{Nn}-\psi_{Nn+1})^2 \end{bmatrix},$$

$$= \frac{1}{2}\Psi^{\mathrm{T}}[K_{U\Psi}]\Psi, \tag{6}$$

where $[K_{U\Psi}]$ is the elastic stiffness matrix associated with the basis deformation vector and it is given by

$$[K_{U\Psi}] = \begin{bmatrix} C_1[K_{U0}] & & & \\ & C_2[K_{U0}] & & \\ & & \ddots & \\ & & & C_{N+1}[K_{U0}] \end{bmatrix}$$

$$+ \begin{bmatrix} [0_{n-1}] & & & & & & & & & \\ & C_{R1} & -C_{R1} & & & & & & & \\ & -C_{R1} & C_{R1} & & & & & & & \\ & & & [0_{n-2}] & & & & & & \\ & & & & C_{R2} & -C_{R2} & & & & \\ & & & & -C_{R2} & C_{R2} & & & & \\ & & & & & & \ddots & & & \\ & & & & & & & C_{RN} & -C_{RN} & \\ & & & & & & & -C_{RN} & C_{RN} & \\ & & & & & & & & & [0_{n-1}] \end{bmatrix}, \tag{7}$$

where $[K_{U0}]$ is an n-order canonical elastic stiffness matrix for the internal rotational springs in one member:

$$[K_{U0}] = \begin{bmatrix} 1 & -1 & & & \\ -1 & 2 & -1 & & \\ & -1 & \ddots & -1 & \\ & & -1 & 2 & -1 \\ & & & -1 & 1 \end{bmatrix}. \tag{8}$$

The potential energy V due to the equally spaced point loads in the HBM may be expressed as follows[14]:

$$\begin{aligned} V &= -\frac{1}{2}\begin{bmatrix} N_1 a_1(\psi_1^2 + \cdots + \psi_n^2) + N_2 a_2(\psi_{n+1}^2 + \cdots + \psi_{2n}^2) \\ + \cdots + N_{N+1} a_{N+1}(\psi_{Nn+1}^2 + \cdots + \psi_{Nn+n}^2) \end{bmatrix}, \\ &= \frac{1}{2}\Psi^{\mathrm{T}}[K_{V\Psi}]\Psi, \end{aligned} \tag{9}$$

where the geometric stiffness matrix $[K_{V\Psi}]$ is given by

$$[K_{V\Psi}] = -\begin{bmatrix} N_1 a_1[I_n] & & & \\ & N_2 a_2[I_n] & & \\ & & \ddots & \\ & & & N_{N+1} a_{N+1}[I_n] \end{bmatrix}, \tag{10}$$

where $N_1, N_2, \ldots, N_{N+1}$ are the axial forces of the $N+1$ members and they can be formulated as

$$N_j = \sqrt{H^2 + \frac{P^2}{N^2}\left(\frac{N}{2} - j + 1\right)^2} \quad \text{for } j = 1, 2, \ldots, N+1. \tag{11}$$

The total potential energy function is given by

$$\Pi = U + V = \frac{1}{2}\Psi^{\mathrm{T}}([K_{U\Psi}] + [K_{V\Psi}])\Psi. \tag{12}$$

By considering the deformation at the buckled state and the fact that pinned supports A and B do not move from their positions during buckling, the compatibility condition for the basis deformation vector Ψ can be established from kinematic considerations as

$$[K_\mu]\Psi = \begin{Bmatrix} 0 \\ 0 \end{Bmatrix}, \tag{13}$$

where the $2 \times (Nn + n)$-order matrix $[K_\mu]$ is given by

$$[K_\mu] = \begin{bmatrix} a_1 \sin\phi_1 \cdot [1 \cdots 1] & a_2 \sin\phi_2 \cdot [1 \cdots 1] & \cdots & a_{N+1} \sin\phi_{N+1} \cdot [1 \cdots 1] \\ a_1 \cos\phi_1 \cdot [1 \cdots 1] & a_2 \cos\phi_2 \cdot [1 \cdots 1] & \cdots & a_{N+1} \cos\phi_{N+1} \cdot [1 \cdots 1] \end{bmatrix}, \tag{14}$$

where ϕ_1, $\phi_2, \ldots, \phi_{N+1}$ are the inclination angles (positive for clockwise angles) of the $N+1$ members and they are given by

$$\tan\phi_j = \frac{P}{NH}\left(\frac{N}{2} - j + 1\right) \quad \text{for } j = 1, 2, \ldots, N+1. \tag{15}$$

In order to obtain the governing equation for buckling, we minimize the total potential energy subject to the satisfaction of the compatibility conditions (that take care of the boundary conditions). By making use of Lagrange multipliers, we transform the constrained extremum problem into an unconstrained one. Thus, the augmented total potential energy Γ (or Lagrangian) is given by

$$\Gamma(\Psi, \mu_1, \mu_2) = \Pi(\Psi) + \{\mu_1\ \mu_2\} \cdot [K_\mu]\Psi, \tag{16}$$

where μ_1 and μ_2 are the Lagrange multipliers.

The buckling criterion corresponds to the minimum value of the Lagrangian, i.e.

$$\frac{\partial\Gamma(\Psi, \mu_1, \mu_2)}{\partial(\Psi, \mu_1, \mu_2)} = \begin{bmatrix} [K_{U\Psi}] + [K_{V\Psi}] & [K_\mu]^{\mathrm{T}} \\ [K_\mu] & 0 \end{bmatrix} \begin{pmatrix} \Psi \\ \mu_1 \\ \mu_2 \end{pmatrix} = \{0\}. \tag{17}$$

To determine the buckling load of the arch, Eq. (17) has to take on a nontrivial solution of the vector (Ψ, μ_1, μ_2) and this is achieved by setting the determinant of the governing stiffness matrix to zero, i.e.

$$\det \begin{bmatrix} [K_{U\Psi}] + [K_{V\Psi}] & [K_\mu]^{\mathrm{T}} \\ [K_\mu] & 0 \end{bmatrix} = 0. \tag{18}$$

The buckling load P_{cr} is obtained by solving this characteristic equation (18) for the lowest positive root. The eigenvector (Ψ, μ_1, μ_2) in Eq. (17) associated with P_{cr} furnishes the buckling mode of the arch.

4. Optimization Problems

Two classes of optimization problems are tackled herein for funicular arches under equally spaced point loads for maximum buckling load. The first class of optimization problem is to determine the optimal shape of funicular arches with a uniform cross-sectional area. This is achieved by varying C such that EI remains constant while changing a that is calculated by the length of straight arch member between point loads divided by the number n of HBM segments. This problem has only one decision variable and that is the horizontal force H which controls the funicular arch shape. The problem may be mathematically stated as

$$\mathrm{Max}_H\ P_{\mathrm{cr}}, \tag{19}$$

where P_{cr} is determined from Eq. (18). This unconstrained optimization problem only requires the determination of the optimal value of the horizontal force H that maximizes the critical buckling load P_{cr}.

The second class of optimization problem is to determine the optimal shape of funicular arches with nonuniform cross-sectional areas for the straight members for maximum critical buckling load P_{cr}. The arch volume V is kept constant. The rotational spring stiffnesses $C_j = \frac{EI_j}{a_j} = \frac{EkA_j^2}{a_j}$, where $j = 1, 2, \ldots, N$ that defines the cross-sectional areas of the straight arch members are to be appropriately varied. The problem may be mathematically stated as

$$\mathrm{Max}_{H,C_j}\ P_{\mathrm{cr}} \tag{20}$$

subject to the volume constraint

$$\sum A_j L_j = V, \tag{21}$$

where $j = 1, 2, \ldots, N$ and L_j is given by Eq. (4). Owing to the symmetrical design of the funicular arch, the number of rotational spring stiffnesses to optimize reduces to $(N+1)/2$ if N is odd and $N/2+1$ if N is even. The above-constrained optimization problem may be converted to an unconstrained one by expressing a cross-sectional area in terms of the other cross-sectional areas using Eq. (21). By doing so, we not only have an unconstrained optimization problem but also reduce the number of decision variables by one.

5. Results and Discussion

5.1. *Optimal design of funicular arches with a uniform cross-sectional area*

Figures 3–5 show the geometries and variations of the nondimensional buckling load parameter $\beta = P_{\mathrm{cr}}L^2/(EI)$ with respect to the horizontal force parameter $\gamma = P/H$ and the rise-to-span ratio $\varepsilon = h_{\max}/L$ for arches under 3, 7, and 15 point loads, respectively.

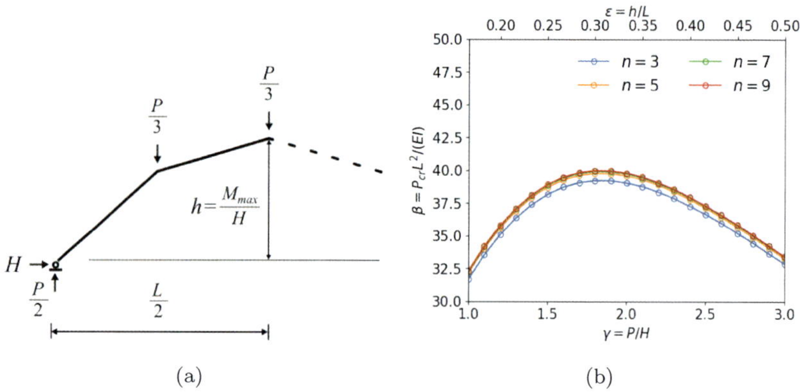

Fig. 3. (a) Geometry and (b) β versus γ, ε for funicular arch under 3 point loads.

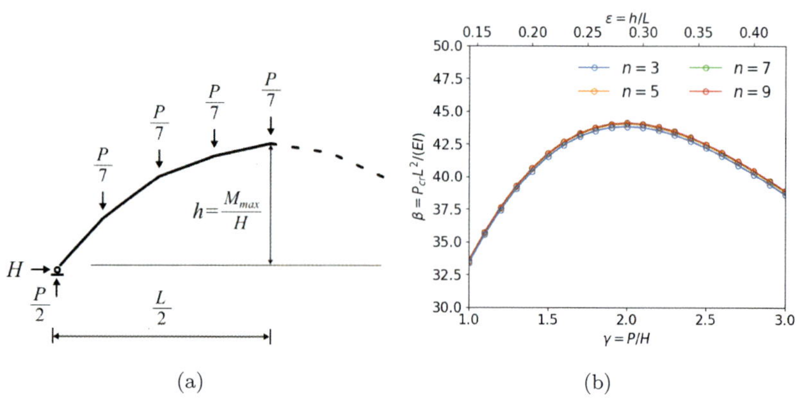

Fig. 4. (a) Geometry and (b) β versus γ, ε for funicular arch under 7 point loads.

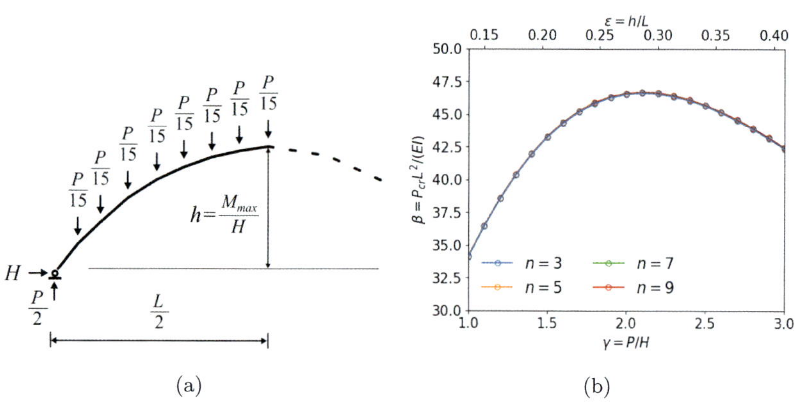

Fig. 5. (a) Geometry and (b) β versus γ, ε for funicular arch under 15 point loads.

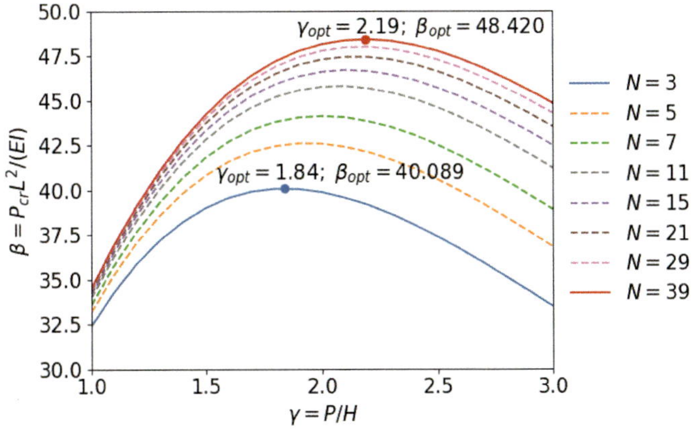

Fig. 6. Variation of β with respect to γ for funicular arches of uniform cross-section and under N point loads.

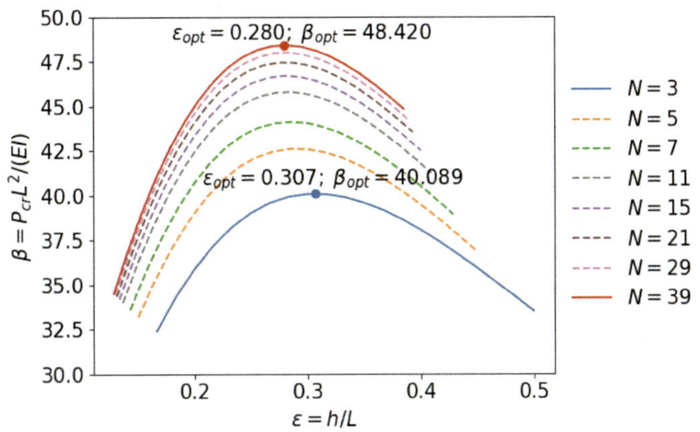

Fig. 7. Variations of β with respect to ϵ for funicular arches of uniform cross-section and under N point loads.

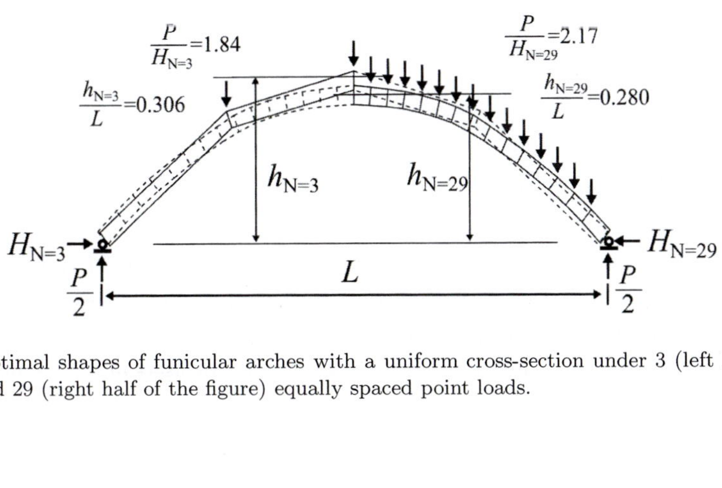

Fig. 8. Optimal shapes of funicular arches with a uniform cross-section under 3 (left half of the figure) and 29 (right half of the figure) equally spaced point loads.

Table 2. Optimal values of horizontal force, arch rise, and buckling loads for funicular arches (of uniform cross-section) under N equally spaced point loads.

Number of point loads	Optimal horizontal force parameter	Optimal rise-to-span ratio	Maximum buckling load parameter
N	$\gamma_{opt} = P/H_{opt}$	$\varepsilon_{opt} = h_{opt}/L$	$\beta_{max} = P_{cr}L^2/(EI)$
3	1.84	0.3067	40.089
5	1.94	0.2910	42.621
7	2.00	0.2857	44.143
11	2.07	0.2823	45.819
15	2.11	0.2813	46.713
21	2.14	0.2806	47.473
29	2.17	0.2803	48.030
39	2.19	0.2802	48.420

It can be seen from Figs. 3–5 that the buckling load parameter increases with increasing number of segments n. For small number of point loads, $n = 9$ is needed for converged buckling loads because of the longer straight arch members between point loads. However, for more point loads ($N = 15$ and above), $n = 5$ will suffice for converged buckling loads since the straight arch members are shorter. The optimal values of γ_{opt} and ε_{opt} for maximum buckling load parameter β_{max} are indicated in the figures and compiled in Table 2 for arches under various numbers of point loads N.

Figures 6 and 7 show the variations of the critical buckling load parameter β with respect to the horizontal force parameter γ and rise parameter ε for various numbers of point loads N. It can be seen from Fig. 6 that the optimal horizontal force parameter somewhat increases and approaches $\gamma = 2.186$ as the number of point loads increases. On the contrary, the optimal rise parameter decreases and approaches $\varepsilon = 0.2803$ as the number of point loads increases (see Fig. 7). Figure 8 compares the optimal shapes of the funicular arches (of uniform cross-section) under 3 and 29 equally spaced point loads.

As the number of point loads increases, the critical buckling load parameter and shape of the funicular arch converge to that of a parabolic arch under a uniformly distributed load. Table 3 shows the comparison of buckling load parameter β between HBM under 39 point loads and the solution of a parabolic arch under uniformly distributed load (with total magnitude of P) for various h/L values. The buckling loads are within 2% of each other.

Interestingly, the optimal design of a constant cross-section arch under uniformly distributed load for the least weight has its rise parameter $\varepsilon = 0.34$ and the horizontal force parameter $\gamma = 2.73$.[18] This shows that the optimal least-weight arch shape is much taller that its counterpart for maximum buckling capacity. So, a trade-off between least weight and maximum buckling capacity designs may be employed for practical arch designs.

Table 3. Comparison of buckling load parameters $\beta = PL^2/(EI)$ of HBM arch under 39 point loads and parabolic arches under uniformly distributed load.

	$h/L = 0.1$	$h/L = 0.3$	$h/L = 0.5$
Funicular HBM arch under 39 point loads	28.35	48.25	37.30
Parabolic arch under uniformly distributed load[21]	29.07	49.47	38.22

5.2. *Optimal design of funicular arches with a nonuniform cross-sectional area*

For this second class of optimization problem where the cross-sectional areas of the arch members between point loads and shape of the arch (defined by the horizontal force) are to be determined for maximum buckling load, we shall adopt the following nondimensional terms to display the results:

$$\tau = \frac{P_{\mathrm{cr}}L}{EV}, \quad \gamma_{\mathrm{opt}} = \frac{P}{H_{\mathrm{opt}}},$$

$$\alpha_i = \frac{A_i L}{V} \quad \text{for } i = 1, 2, \ldots, (N+1)/2.$$

As this is an unconstrained optimization with multiple decision variables γ, α_i, we have used the Nelder and Mead[19] simplex method for seeking the optimal values of the decision variables.

Table 4 presents the optimal funicular arch designs for various numbers N of point loads. The optimal arch design is defined by the optimal horizontal force parameter γ_{opt} and the optimal cross-sectional areas of the arch members $\alpha_{i,\mathrm{opt}}$. The associated rise-to-span ratio $\varepsilon_{\mathrm{opt}}$ and the maximum buckling load parameter $\tau_{\max}$ are also given in the table. It is worth noting that the optimal nonuniform arch design is less tall than its corresponding uniform arch design by comparing $\varepsilon_{\mathrm{opt}}$ in Tables 2 and 4. It is interesting to note that the optimal least-weight arch under uniformly distributed load has a rise-to-span ratio of 0.433,[20] which is about twice taller than the optimal arch design for maximum buckling load.

The left half of Fig. 9 shows the optimal shape and variation of the cross-sectional area of a funicular arch under 21 equally spaced point loads while the right half shows the corresponding optimal shape of arch with a uniform cross-section. It can be seen that the optimal arch with a nonuniform cross-section is slightly shorter than its uniform cross-section counterpart. This is to be expected as the design flexibility of allowing the cross-section to vary leads to a more optimal design characterized by a shorter arch and less material when compared to its arch counterpart with a constant cross-section. It is worth noting that the variation of the cross-section area is similar to that of an optimal circular arch under uniform radial pressure.[9]

Table 4. Optimal funicular arch (of nonuniform cross-section) designs under N point loads.

N	3	5	7	9	11	15	21	29
$\tau_{\max}$	2.4190	2.8462	3.1148	3.2881	3.4186	3.5805	3.7137	3.8084
γ_{opt}	1.3590	1.4438	1.4963	1.5301	1.5538	1.5848	1.6112	1.6307
ε_{opt}	0.2265	0.2165	0.2138	0.2125	0.2119	0.2113	0.2110	0.2109
$\alpha_{1,\text{opt}}$	0.9209	0.8487	0.7737	0.7102	0.6556	0.5724	0.4869	0.4125
$\alpha_{2,\text{opt}}$	0.8667	1.1056	1.124	1.0899	1.0438	0.9496	0.8325	0.7183
$\alpha_{3,\text{opt}}$	—	0.7406	1.0514	1.1420	1.1600	1.1232	1.0298	0.9152
$\alpha_{4,\text{opt}}$	—	—	0.6515	0.9779	1.1072	1.1733	1.1366	1.0445
$\alpha_{5,\text{opt}}$	—	—	—	0.5854	0.9103	1.1326	1.1796	1.1267
$\alpha_{6,\text{opt}}$	—	—	—	—	0.5327	1.0109	1.1720	1.1724
$\alpha_{7,\text{opt}}$	—	—	—	—	—	0.7992	1.1201	1.1882
$\alpha_{8,\text{opt}}$	—	—	—	—	—	0.4567	1.0252	1.1779
$\alpha_{9,\text{opt}}$	—	—	—	—	—	—	0.8832	1.1437
$\alpha_{10,\text{opt}}$	—	—	—	—	—	—	0.6802	1.0864
$\alpha_{11,\text{opt}}$	—	—	—	—	—	—	0.3826	1.0054
$\alpha_{12,\text{opt}}$	—	—	—	—	—	—	—	0.8980
$\alpha_{13,\text{opt}}$	—	—	—	—	—	—	—	0.7585
$\alpha_{14,\text{opt}}$	—	—	—	—	—	—	—	0.5746
$\alpha_{15,\text{opt}}$	—	—	—	—	—	—	—	0.3206

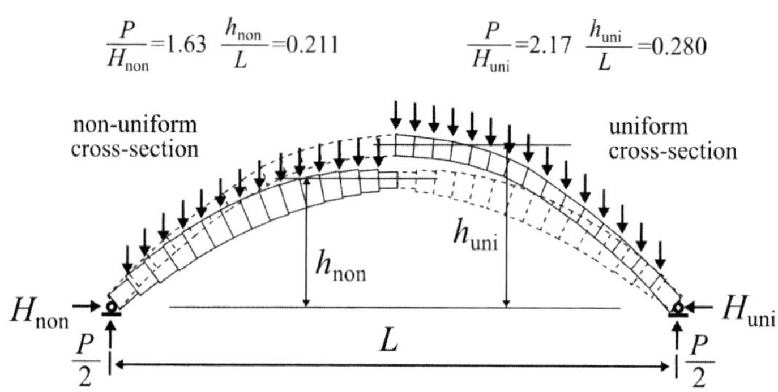

Fig. 9. Optimal shapes for the funicular arch of nonuniform cross-section (left half) and of uniform cross-section (right half) under 29 equally spaced point loads.

6. Concluding Remarks

Presented herein are the optimal designs of pinned supported funicular arches under equally spaced point loads for maximum in-plane buckling load. If the cross-sectional area is constrained to be uniform throughout the arch length, the optimal rise-to-span ratio h/L varies from 0.31 to 0.28 with increasing number of point loads from 3 to 39. However, when the cross-sectional area is allowed to vary from one straight arch member (between point loads) to another member, the optimal rise-to-span ratio h/L varies from 0.227 to 0.221 with increasing number of point loads from 3 to 29. By constraining the cross-section to be uniform, the resulting optimal arch is somewhat taller than its nonuniform cross-section counterpart. By increasing the number of point loads, the optimal solution approaches the solution of a parabolic arch under a uniformly distributed load. When compared to optimal arch designs for the least weight, optimal arch designs against buckling have lower rises.

This study also shows the simplicity of the HBM for the buckling analysis as it avoids deriving and solving the governing differential equation for arch buckling. Its discrete structural feature simplifies the optimization process as the decision variables reduce to only the different HBM rotational spring stiffnesses (that define the cross-sectional areas) and the horizontal force (that controls the arch shape). The HBM and the optimization method can be readily adapted to determine the optimal design of arches under any loading and support conditions including supports on an unequal elevation.

References

1. C. H. Wu, The strongest circular arch — a perturbation approach, *J. Appl. Mech.* **35** (1968) 476–480.
2. B. Budiansky, J. C. Frauenthal and J. W. Hutchinson, On optimal arches, *J. Appl. Mech.* **36** (1969) 880–882.
3. I. Tadjbakhsh and M. Farshad, On conservatively loaded funicular arches funicular arches and their optimal designs, in *Proc. IUTAM Symp. Structural Optimization* (Warsaw, Poland, 1973), pp. 215–228.
4. M. Farshad, On optimal form of arches, *J. Frank. Inst.* **302** (1976) 187–194.
5. J. C. Amazigo, Optimal shape of shallow circular arches against snap-buckling, *J. Appl. Mech.* **45** (1978) 591–594.
6. I. Tadjbakhsh, Stability and optimum design of arch-type structures, *Int. J. Solids Struct.* **17**(6) (1981) 565–574, https://doi.org/10.1016/0020-7683(81)90019-6.
7. J. Blachut, Parametrical optimal design of funicular arches against buckling and vibration, *Int. J. Mech. Sci.* **26**(5) (1984) 305–310.
8. A. Wroblewski and M. Zyczkowski, On multimodal optimization of circular arches against plane and spatial creep buckling, *Struct. Optim.* **1** (1989) 227–234.
9. H. Zhang and C. M. Wang, Hencky bar-chain model for optimal circular arches against buckling, *Mech. Res. Commun.* **88** (2018) 7–11.
10. A. M. Hernández-Díaz, A. Muñoz, J. F. Jiménez-Alonso and A. Sáez, Buckling design of submerged arches via shape parameterization, *Comput. Math. Methods. Med.* **1** (2019) e1057, https://doi.org/10.1002/cmm4.1057.
11. F. G. Chaves Neto and A. M. C. de Melo, Optimization of flat arches under uniformly distributed load considering in-plane buckling, in *Proc. XLI Ibero-Ibero-Latin-American Congress on Computational Methods in Engineering CIL-AMCE 2020*, (ABMEC Foz do Iguaçu, PR, Brazil, 2020).
12. C. M. Wang, W. H. Pan and J. Q. Zhang, Optimal design of triangular arches against buckling, *J. Eng. Mech.* **146**(7) (2020) 04020059.

13. C. M. Wang, W. H. Pan and H. Zhang, Hencky Bar-chain model for buckling analysis and optimal design of trapezoidal arches, in *Modern Trends in Structural and Solid Mechanics 1: Statics and Stability*, Anniversary Volume in honour of Professor Isaac Elishakoff, Chap. 11 (ISTE-Wiley, 2021), pp. 229–247.
14. C. M. Wang, H. Zhang, N. Challamel and W. H. Pan, *Hencky Bar-Chain/Net for Structural Analysis* (World Scientific Publishing Europe Ltd., 2020).
15. V. Thevendran and C. M. Wang, Optimal design of tie-beams, *Int. J. Solids Struct.* **22**(11) (1986) 1343–1356.
16. C. M. Wang, H. Zhang, R. P. Gao, W. H. Duan and N. Challamel, Hencky bar-chain model for buckling and vibration of beams with elastic end restraints, *Int. J. Struct. Stab. Dyn.* **15**(7) (2015) 1540007.
17. W. H. Pan, C. M. Wang and H. Zhang, Hencky bar-chain model for buckling analysis of non-symmetric portal frames. *Eng. Struct.* **182** (2019) 391–402.
18. C. M. Wang, Optimal shape of cables, *J. Eng. Mech.* **110**(11) (1984) 1649–1653.
19. J. A. Nelder and R. Mead, A simplex method for function minimization, *Comput. J.* **7** (1965) 155–162.
20. G. I. N. Rozvany and C. M. Wang, On plane Prager-structures I, *Int. J. Mech. Sci.* **25**(7) (1983) 519–527.
21. W. H. Pan and C. M. Wang, Elastic in-plane buckling of funicular arches, *Int. J. Struct. Stab. Dyn.* **20**(13) (2020) 2041014.

https://doi.org/10.1142/9789819814770_0006

Chapter 6

Internal and External Cancellation Conditions for Free Vibration of Damped Simple Beams Traversed by Successive Moving Loads[#]

Z. L. Wang[*,¶], Z. X. Tan[†,||], L. Chen[*,**], D. S. Yang[‡,††,|||], H. Xu[*,‡‡], K. Shi[*,§§] and Y. B. Yang[*,§,¶¶,|||]

[*]*School of Civil Engineering*
Chongqing University, Chongqing, P. R. China

[†]*State Key Laboratory of Disaster Reduction in Civil Engineering, Tongji University Shanghai, P. R. China*

[‡]*Department of Civil Engineering Monash University, Clayton, VIC 3800, Australia*

[§]*School of Civil Engineering and Architecture Chongqing University of Science and Technology Chongqing, P. R. China*

[¶]*zhlwang@cqu.edu.cn*
[||]*tanzhongxu@tongji.edu.cn*
[**]*LeiChen@stu.cqu.edu.cn*
[††]*ted.yang@monash.edu*
[‡‡]*hxu@cqu.edu.cn*
[§§]*shikang@cqu.edu.cn*
[¶¶]*ybyang@cqu.edu.cn*

In this paper, the internal and external cancellation phenomena for damped beams subjected to multi-moving loads are investigated in detail. To start, the theory for the vibration of a simply supported beam is revisited by including the effect of damping. For the first time, a simple expression is derived for the free vibration of the damped beam under multi-moving loads. Based on the concept of local minimum, two cancellation conditions are identified. One is the *internal cancellation*, which relates to the inherent property of the beam and is conventionally known. The other is the newly formulated *external cancellation* that relates to the number and spacing of moving loads. For comparison, both the resonant condition and the optimal criterion for span length of the bridge are also briefed. By comparing with the classical solution, the present simple expression for the free vibration of the beam is firstly validated. Then the factors affecting the cancellation are investigated against various load cases and damping levels.

[|||]Corresponding authors.
[#]This chapter appeared previously on the International Journal of Structural Stability and Dynamics. To cite this chapter, please cite the original article as the following: Z. L. Wang *et al.*, *Int. J. Struct. Stab. Dyn.*, **23**, 2340007 (2023), DOI: 10.1142/S0219455423400072.

The results show that external cancellation occurs more frequently due to the increase in the number and spacing of the moving loads. The damping of the beam has a leaking effect on cancellation, in that nonzero vibration may occur, but it is also quickly damped out by damping itself.

Keywords: Beam; cancellation; damping; free vibration; moving load; resonance.

1. Introduction

Moving loads-induced vibration on beams has been a hot and challenging topic that has lasted virtually since the mid-nineteenth century.[1,2] Such a problem has become more important than ever due to the widespread construction of high-speed railways, that are often supported mainly by elevated simply supported beams and traversed by trains moving at high speeds. Vibrations induced by the passing moving loads on the beam may affect the fatigue behavior of the beam and entire track system due to the undesired repeated oscillations or frequent resonance in particular. The moving loads-induced problem is becoming more acute due to the use of lightweight materials for bridges and the increase in traffic volume, vehicle weight and speed, especially for those encountered in high-speed railways.[3,4]

Numerous works exist for the vibration of simply supported beams traversed by single or multi-loads. In the books of Timoshenko *et al.*,[1] Fryba[2] and Yang *et al.*,[5] comprehensive studies were conducted on the vibrations of beams subjected to various types of moving loads. Focusing on the phenomena of resonance and cancellation, Yang *et al.*[6] investigated the vibration of simple beams subjected to the passage of high speed trains by an analytical approach. By letting the condition of resonance coincide with that of cancellation, the optimal design criterion for simple beams for suppressing the resonant responses was proposed. The criterion simply states that the bridge span length L should be equal to 0.5 or 1.5 times the car length D for the resonance to be suppressed. In fact, the 1.5-times rule has been mostly adopted by beams for its practicality, which is easy to apply, since it is irrespective of the train speed or bridge frequencies. Subsequently, Ju and Lin[7] studied the high-speed train-induced resonant characteristics of bridges by the finite element method, in which the train–bridge interaction was considered. As for elastically supported beams, the mechanism of resonance and cancellation was studied by Yang *et al.*[8] Museros and Alarcón[9] assessed the contribution of the second bending mode to the resonant behavior of simply supported bridges under the passage of trains of equally spaced loads. Yau and Yang[10] further studied the conditions of resonance and cancellation from the perspective of the vertical acceleration response by the superposition method. Xia *et al.*[11] investigated the effect of the span, lateral and vertical stiffness of the bridge, and vehicle frequencies on the resonance conditions of bridges. Museros *et al.*[12] studied the maximum resonance, cancellation and resonant vertical acceleration of simply supported bridges under moving loads. Kumar *et al.*[13] presented a detailed study of the cancellation phenomenon for simply supported beams subjected to a single moving load, in which the damping effect was included. Cho *et al.*[14] reported that the optimal design criterion for span length of simple beams proposed by Yang *et al.*[6] has been applied to the structural design of the Korea High-Speed Rail (KHSR) bridges. They also extended the criterion to beams with clamped–clamped and clamped–pinned boundary conditions, as well as continuous beams. Recently, Wang *et al.*[15] exploited the soft-end amplification effect on the vibration and frequencies of the bridges with unequal end bearings.

As a matter of fact, numerous investigations exist around the dynamics of beams were induced by moving loads. For instance, Savin[16] derived the analytical expressions for calculating the dynamic amplification factor and the characteristic response spectrum for weakly damped beams subjected to point loads moving at constant speeds. Law and Zhu[17] studied the dynamic behavior of damaged reinforced concrete bridge structures under moving vehicular loads. Extensive researches were also carried out on subjects such as Maglev train/guideway interaction,[18–21] moving load identification,[22,23] moving loads on inclined beams,[24–26] and vehicle scanning method for bridges.[27–32] It is difficult, if not impossible, to have a complete coverage of all the related works on the vibration of beams induced by moving loads.

In the design of railway bridge, *resonance* is the condition to be avoided, as it may cause premature fatigue and failure of the track system, while *cancellation* is the condition to pursue, since it allows unnecessary oscillations to be circumvented. Naturally, a *well-designed bridge* is the one with all the vibrations kept to the minimum in the initial design stage. This was exactly the idea leading to derivation of the optimal design span length for suppressing the resonant responses on beams under the moving loads.[6]

From the aforementioned review, it is clear that abundant research studies have been conducted on the resonance phenomenon of free vibration of simple beams under the moving loads. But only a limited number of research studies has been directed to the cancellation phenomenon of the same problem. For this reason, the paper will be devoted mainly to the phenomenon of cancellation for simple beams under the moving loads.

It should be noted that the cancellation phenomenon often discussed in the literature belongs to the category of *internal cancellation*, which relates to the inherent property of the beam and is totally independent of the moving loads. On the other hand, there is a second category of cancellation called the *external cancellation* that relates to the number and spacing of the moving loads. To the authors' knowledge, the external cancellation has *not* received any attention from previous researchers. In this paper, both the internal and external cancellation will be newly formulated based on the concept of *minimal vibration*, rather than zero vibration as implied by the conventional cancellation. This is primarily to account for the "leaking" effect brought by inclusion of damping of the beam, a factor not well treated in the literature with regard to cancellation.

In this paper, an analytical formulation will be presented for the cancellation phenomena of the *damped* beam subjected to multi-moving loads. For the first time, a *simple* expression is derived in close-form for the free vibration of the damped beam. Based on the concept of minimal vibration, both the internal and external cancellation conditions are identified. To aid in numerical comparison, both the resonance condition and the optimal criterion for bridge span length are briefly introduced as well. In the numerical study, the present simple expression for free vibration of the beam is firstly validated against the classical solution. Then the factors (damping in particular) affecting the internal and external cancellations are assessed, with the resonant response compared for some cases. Finally, the response for the optimized beam with $L/d = 0.5$ and 1.5 are compared with the non-optimized one, by which the "leaking" effect of damping can be observed.

2. Simple Beams Subjected to Successive Moving Loads

This study is focused on the cancellation mechanism of moving loads-induced vibration on simply supported bridges commonly adopted in high-speed railways. The phenomenon of cancellation induced by successively moving loads will be analyzed using the closed-form solutions derived. Particularly, the damping of the beam will be included in the formulation for assessing its effect on the cancellation phenomenon.

2.1. *Basic theory*

For analytical formulation, the mechanical model shown in Fig. 1 is adopted, i.e. a simply supported beam of length L subjected to a sequence of N concentrated loads of magnitude P with constant interval d moving at speed v. The equation of motion for the damped simple beam is

$$\bar{m}\frac{\partial^2 u(x,t)}{\partial t^2} + c\frac{\partial u(x,t)}{\partial t} + EI\frac{\partial^4 u(x,t)}{\partial x^4} = P\sum_{k=1}^{N}\delta[x - v(t-t_k)] \cdot [H(t-t_k) - H(t-t_k-T)], \quad (1)$$

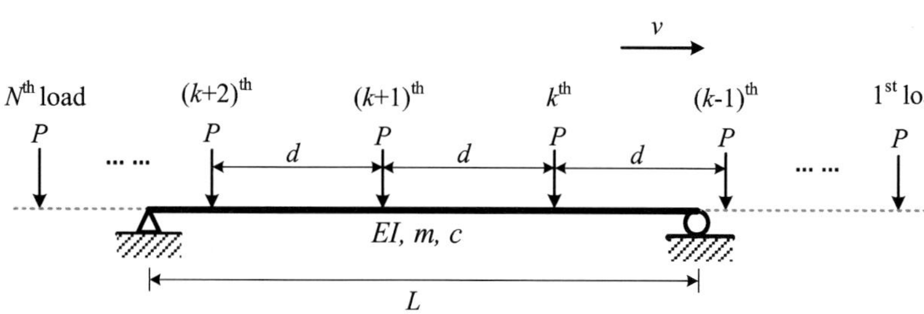

Fig. 1. Simply supported beam subjected to successive uniform moving loads.

where $\bar{m}$ is the mass per unit length of the beam, $u(x,t)$ the vertical displacement, c the damping coefficient, E the elastic modulus, I the moment of inertia of the beam, δ Dirac's delta function, $H(t)$ the unit step function, N the total number of moving loads, $t_k = (k-1)d/v$ the time of the kth load entering the beam, and $T = L/v$ the duration of one load across the beam.

The deflection $u(x,t)$ of the beam can be approximated by the superposition of modal components as

$$u(x,t) = \sum_n q_n(t)\sin\left(\frac{n\pi x}{L}\right), \tag{2}$$

where $\sin(n\pi x/L)$ denotes the nth mode of the simply supported beam and $q_n(t)$ the associated generalized coordinate. Multiplying both sides of Eq. (1) by $\sin(n\pi x/L)$ and integrating from 0 to L, the generalized equation of motion of the damped beam can be obtained as follows:

$$\begin{aligned}\ddot{q}_n(t) + 2\xi_n\omega_n\dot{q}_n(t) + \omega_n^2 q_n(t) = \frac{2P}{\bar{m}L}\sum_{k=1}^{N}\sin\frac{n\pi v(t-t_k)}{L}\\ \cdot[H(t-t_k) - H(t-t_k-T)],\end{aligned} \tag{3}$$

where ξ_n is the damping ratio for the nth mode of the beam and ω_n is the natural frequency, given by

$$\omega_n = \frac{n^2\pi^2}{L^2}\sqrt{\frac{EI}{\bar{m}}}. \tag{4}$$

By imposing zero initial conditions on Eq. (3), the modal coordinate $q_n(t)$ can be solved as follows:

$$\begin{aligned}q_n(t) = \frac{\Delta_{stn}}{1-S_n^2}\sum_{k=1}^{N}\{\sin[\Omega_n(t-t_k)] - S_n^{-\xi_n\omega_n(t-t_k)}\sin[\omega_{dn}(t-t_k)]\}H(t-t_k)\\ -(-1)^n\{\sin[\Omega_n(t-t_k-T)] - S_n^{-\xi_n\omega_n(t-t_k-T)}\\ \times\sin[\omega_{dn}(t-t_k-T)]\}H(t-t_k-T),\end{aligned} \tag{5}$$

in which Δ_{stn} is the static deflection caused by a single load, ω_{dn} is the damped frequency of the beam, Ω_n is the driving frequency, and S_n is the speed parameter, respectively, as

$$\Delta_{stn} = \frac{2P}{\bar{m}L\omega_n^2}, \tag{6a}$$

$$\omega_{dn} = \sqrt{1-\xi_n^2}\omega_n, \tag{6b}$$

$$S_n = \frac{n\pi v}{L\omega_n}, \tag{6c}$$

$$\Omega_n = \frac{n\pi v}{L}. \tag{6d}$$

Equation (5) is the *classical* solution for the vibration of a simple beam. Two typical periods exist in the vibration of the beam in Eq. (5). The first relates to the *forced vibration* when the moving loads are directly acting on the beam, i.e. $t < t_N + T$, and the other the *free vibration* when all the moving loads have passed the beam, i.e. $t \geq t_N + T$. Alternatively, one may also say that the vibration of the beam consists of two parts, one is the *forced vibration* caused by the moving loads, as indicated by the terms containing the driving frequency Ω_n, and the other is the *free vibration* (or self-oscillation of the beam) initially triggered by the moving loads, as indicated by the terms containing the bridge frequency ω_{dn}. When all the moving loads have passed the beam, i.e. for the free vibration period, the forced vibration terminates, while the free vibration continues to prevail until it is eventually damped out. Such a phenomenon will be further analyzed in Sec. 2.2.

It is clear from the classical solution given in Eq. (5) that the successively moving loads contribute to the vibration of the beam by *accumulation* (i.e. via the sign $\sum$), which is related to the resonance condition, since resonance is essentially built up by the summation of wave components. The motivation of this study is to exploit the phenomenon of cancellation triggered by successively moving loads using the closed-form solutions.

2.2. *Simple expression for free vibration of simple beams*

In the following, focus will be placed on the period of free vibration when all the moving loads have passed the beam, i.e. $t \geq t_N + T$, as this is the part relating to the phenomenon of cancellation. The normalized modal coordinate, $q_n(t)/\Delta_{stn}$, for the free vibration $R_n(t)$ can be derived from Eq. (5) as

$$R_n(t) = \frac{q_n(t)}{\Delta_{stn}} = \frac{S_n}{1 - S_n^2} \sum_{k=1}^{N} e^{-\xi_n \omega_n (t - t_k)} \{-\sin[\omega_{dn}(t - t_k)] + (-1)^n e^{\xi_n \omega_n T} \sin[\omega_{dn}(t - t_k - T)]\}. \tag{7}$$

Clearly, the forced vibration part signaled by the driving frequency Ω_n has disappeared in Eq. (7), leaving only the free vibration part containing the self-frequency ω_{dn} of the beam.

In order to eliminate the mathematical form of accumulation in Eq. (7), the following formulas for trigonometric functions are adopted:

$$\vartheta_1 = \sum_{k=1}^{N} e^{\xi_n \omega_n t_k} \sin(\omega_n t_k) = -\frac{\sin(\frac{\omega_n d}{v}) + e^{N \frac{\xi_n \omega_n d}{v}} \sin[(N-1)\frac{\omega_n d}{v}] - e^{(N-1)\frac{\xi_n \omega_n d}{v}} \sin[N \frac{\omega_n d}{v}]}{2[\cos(\frac{\omega_n d}{v}) - \cosh(\frac{\xi_n \omega_n d}{v})]}, \tag{8a}$$

$$\vartheta_2 = \sum_{k=1}^{N} e^{\xi_n \omega_n t_k} \cos(\omega_n t_k)$$
$$= 1 - \frac{\cos(\frac{\omega_n d}{v}) - e^{\frac{\xi_n \omega_n d}{v}} + e^{N\frac{\xi_n \omega_n d}{v}} \cos[(N-1)\frac{\omega_n d}{v}] - e^{(N-1)\frac{\xi_n \omega_n d}{v}} \cos(N\frac{\omega_n d}{v})}{2[\cos(\frac{\omega_n d}{v}) - \cosh(\frac{\xi_n \omega_n d}{v})]}. \tag{8b}$$

The normalized modal coordinate $R_n(t)$ for the free vibration thus reduces to the following *simple* form:

$$R_n(t) = \frac{S_n}{1 - S_n^2} e^{-\xi_n \omega_n t} A_n \sin(\omega_{dn} t + \varphi_n), \quad t \geq t_N + L/v, \tag{9}$$

where $e^{-\xi_n \omega_n t}$ is the attenuation term associated with time t, and A_n the amplitude of the free vibration, given by

$$A_n = \sqrt{\left[(e^{\frac{\xi_n n\pi}{S_n}} - 1)^2 + 4e^{\frac{\xi_n n\pi}{S_n}} \sin^2\left(\frac{n\pi}{2S_n} + \frac{n\pi}{2}\right)\right] \frac{[(e^{N\xi_n \frac{dn\pi}{LS_n}} - 1)^2 + 4e^{N\xi_n \frac{dn\pi}{LS_n}} \sin^2(N\frac{d}{2L}\frac{n\pi}{S_n})]}{[(e^{\xi_n \frac{dn\pi}{LS_n}} - 1)^2 + 4e^{\xi_n \frac{dn\pi}{LS_n}} \sin^2(\frac{d}{2L}\frac{n\pi}{S_n})]}} \tag{10}$$

along with φ_n for the phase angle,

$$\varphi_n = \arctan\left(-\frac{[1 + (-1)^{1+n} e^{T\xi_n \omega_n} \cos(\omega_{dn} T)]\vartheta_1 + (-1)^{1+n} e^{T\xi_n \omega_n} \sin(\omega_{dn} T)\vartheta_2}{(-1)^n e^{T\xi_n \omega_n} \vartheta_1 \sin(\omega_{dn} T) + [1 + (-1)^{1+n} e^{T\xi_n \omega_n} \cos(\omega_{dn} T)]\vartheta_2}\right). \tag{11}$$

It is interesting that the summation sign in Eq. (7) has been eliminated, resulting in a very *simple* expression for calculating the free vibration response of the damped beam traversed by successive moving loads. To the authors' knowledge, the simple expression derived in Eq. (9), as well as the supplementary ones in Eqs. (8) and (10), has not appeared in any previous works, which is the key to deriving the cancellation conditions.

As can be seen from Eq. (10), the amplitude A_n of free vibration of the beam is determined by two terms: the first is the term within the first square bracket, and the other is the remaining fraction term. Aside from the speed parameter S_n, the first term relates only to the damping ratio ξ_n, an inherent attribute of the beam, herein called the *internal term*. On the other hand, the fraction term relates to the number N and spacing d of the external moving loads, which is referred to the *external term*. Another observation of the fraction term is that the numerator differs from the denominator in the involvement of the number of loads N in the numerator. Consequently, one can interpret the denominator as the one representing the contribution by a single load, and the numerator the contribution by the total N

loads. In this sense, the entire fraction term is actually the *normalized contribution* by the total external moving loads.

Based on the above interpretation, the internal term denotes the contribution from the inherent attribute of the beam, and the external term the normalized contribution of the total moving loads. These two terms, when added together, determine the free vibration behavior of the damped beam traversed by successive moving loads. In addition, it is interesting from Eq. (10) that for the damped beam, i.e. with $\xi_n > 0$, the amplitude A_n is always larger than 0, indicating that the amplitude of free vibration for the damped beam will always be positive under any case. Such a phenomenon will be numerically investigated later on.

Correspondingly, the displacement u of the beam over the period of free vibration can be calculated using the normalized modal coordinate $R_n(t)$ in Eq. (9), i.e. by multiplying $R_n(t)$ by Δ_{stn} to obtain $q_n(t)$, and then substituting $q_n(t)$ into Eq. (2), resulting in

$$u(x,t) = \sum_n \frac{\Delta_{stn} S_n}{1 - S_n^2} e^{-\xi_n \omega_n t} A_n \sin(\omega_{dn} t + \varphi_n) \times \sin\left(\frac{n\pi x}{L}\right), \quad t \geq t_N + L/v \tag{12}$$

of which the expression is *simpler* in mathematical form than most existing ones. Besides, the acceleration $\ddot{u}$ of the beam over the period of free vibration can be derived from Eq. (12). By differentiating Eq. (12) twice and neglecting the terms $\xi^2 \omega_n^2$ and $\xi \omega_n \omega_{dn}$ involving light damping ($\xi < 0.05$), the beam's acceleration $\ddot{u}$ over the period of free vibration can be approximated as

$$\ddot{u}(x,t) = -\sum_n \frac{\Delta_{stn} S_n}{1 - S_n^2} \omega_{dn}^2 e^{-\xi_n \omega_n t} A_n \sin(\omega_{dn} t + \varphi_n) \times \sin\left(\frac{n\pi x}{L}\right), \quad t \geq t_N + L/v. \tag{13}$$

2.3. *Cancellation conditions*

In this paper, focus is placed on the cancellation condition for free vibration of the damped beam. For the damped beam with $\xi_n > 0$, it is clear from Eq. (10) that the exponents of all the exponential functions are greater than zero. As such, all the terms of the form $(e^x - 1)^2$ will decrease monotonically as the speed parameter S_n increases. On the other hand, the trigonometric terms of the form $\sin^2 x$ in Eq. (10) are periodic functions with a minimum value of 0. It follows that the cancellation of free vibration, i.e. the *local minimum* of A_n, is dominated by the trigonometric term, i.e. $\sin^2 x$, that consists of two conditions. The first condition to make the internal term in Eq. (10) a local minimum may be met by letting

$$\sin\left(\frac{n\pi}{2S_n} + \frac{n\pi}{2}\right) = 0. \tag{14}$$

Accordingly, the speed parameter for the cancellation condition to occur can be deduced from the internal term as

$$S^{\mathrm{I}}_{c,n} = \frac{n}{2j - n}, \quad j \in Z^{+} \quad \text{and} \quad 2j > n. \tag{15}$$

Herein, the superscript "I" of the speed parameter S denotes the "internal" term, and the subscript c the "cancellation". It is clear that the speed parameter $S^{\mathrm{I}}_{c,n}$ represents the *internal* cancellation of the simple beam, since it only relates to the mode order of the beam, but independent of the external moving loads. It should be noted that the internal cancellation condition in Eq. (15) has been widely recognized in the previous works without the addition of the word "internal".[6,8,10,12,13]

The second cancellation condition to make the external term in Eq. (10) a local minimum can be reached by letting

$$\sin\left(N \frac{d}{2L} \frac{n\pi}{S_n}\right) = 0. \tag{16}$$

Correspondingly, the cancellation speed parameter deduced from the external term is

$$S^{\mathrm{E}}_{c,n} = \frac{nN}{2j} \frac{d}{L}, \quad j \in Z^{+} \quad \text{and} \quad j \neq iN(N \geq 2, i = 1, 2, 3, \ldots), \tag{17}$$

where the superscript "E" of S indicates the "external" term. Clearly, in addition to the mode order of the beam, the speed parameter $S^{\mathrm{E}}_{c,n}$ is related to the nature of external moving loads, as the number N and spacing d of the loads have been included. For this reason, $S^{\mathrm{E}}_{c,n}$ is referred to as the speed parameter for *external* cancellation. It should be reminded that the external cancellation condition is valid only for $N \geq 2$, i.e. when there exist at least two moving loads. Meanwhile, the condition $j \neq iN$ should also be maintained. Otherwise, the denominator in the external term will also be a minimum. It is known that the simultaneous occurrence of minima for both the numerator and denominator does not guarantee the occurrence of minimum for the entire external term. As a side note, the external cancellation condition presented in Eq. (17) is newly formulated herein, which has not been previously reported.

So far, two cancellation conditions have been derived for the free vibration of simple damped beams traversed by successive moving loads. One is the internal cancellation $S^{\mathrm{I}}_{c,n}$ related to the mode number of the beam, and the other is the external cancellation $S^{\mathrm{E}}_{c,n}$ associated with the external moving loads. They can be summarized as follows:

$$S_{c,n} = \begin{cases} \dfrac{n}{2j - n}, & j \in Z^{+} \quad \text{and} \quad 2j > n, \\[2ex] \dfrac{nN}{2j} \dfrac{d}{L}, & j \in Z^{+} \quad \text{and} \quad j \neq iN(N \geq 2, i = 1, 2, 3 \ldots). \end{cases} \tag{18}$$

In a later part of this paper, revelation will be made for the two types of cancellation conditions for the free vibration of thc damped simple beam using some typical examples.

2.4. *Special cases*

2.4.1. *Undamped beam*

For the special case of an undamped beam, the free vibration response can be obtained by substituting $\xi_n = 0$ into Eqs. (9) and (10), i.e.

$$R_n(t) = \frac{S_n}{1 - S_n^2} A_n \sin(\omega_n t + \varphi_n), \quad t \geq t_N + L/v \tag{19}$$

and

$$A_n = \sqrt{\left[4\sin^2\left(\frac{n\pi}{2S_n} + \frac{n\pi}{2}\right)\right] \frac{\sin^2(N \frac{d}{2L} \frac{n\pi}{S_n})}{\sin^2(\frac{d}{2L} \frac{n\pi}{S_n})}}. \tag{20}$$

From Eq. (20), it can be easily deduced that the cancellation conditions for the undamped beam remain *identical* to the damped one. Namely, the cancellation conditions shown in Eq. (18) are applicable to both the undamped and damped beams. The only difference is that *the free vibration for the undamped beam can reach zero under the cancellation conditions*, while *it is nonzero for the damped beam.*

2.4.2. *Beam traversed by a single moving load*

For a single load moving over the simple beam, i.e. $N = 1$, the fraction term in Eq.(10) is equal to 1, and the amplitude of free vibration reduces to

$$A_n = \sqrt{\left[(e^{\frac{\xi_n n\pi}{S_n}} - 1)^2 + 4e^{\frac{\xi_n n\pi}{S_n}} \sin^2\left(\frac{n\pi}{2S_n} + \frac{n\pi}{2}\right)\right]}. \tag{21}$$

Clearly, this indicates that the cancellation effect related to the external moving load on the beam simply vanishes, leaving only the internal cancellation condition as shown in Eq. (15). In turn, this once again confirms that the external cancellation condition is valid only for $N \geq 2$, as was revealed in Eq. (17).

In short, the cancellation conditions presented in Eq. (18) are of general applicability not only to both the damped and undamped beams, but also to both the successively moving loads and single load.

2.5. *Resonance condition*

Although resonance is not the focus of this paper, it would not make sense if no comparison can be made between the resonant response and those of normal speed ("normal" means no cancellation or resonance from here on) or internal or external cancellation speeds. To this end, the resonance condition is also derived from Eq. (10).

As indicated in Sec. 2.4.1, because the form $(e^x - 1)^2$ in Eq. (10) decreases monotonically as the speed parameter S_n increases, this only affects the amplitude of free vibration. It is recognized that damping of the beam will not affect the speed parameter for the occurrence of extreme points corresponding to the cancellation and resonance conditions. Thus, the resonance condition for the damped beam to be derived from Eq. (10) remains *identical* to the one for the undamped beam in Eq. (20).

In this sense, the resonance condition can be obtained from Eq. (20) by maximizing the amplitude A_n or by letting the denominator be equal to zero, i.e.

$$\sin\left(\frac{dn}{2L}\frac{\pi}{S_n}\right) = 0. \tag{22}$$

Accordingly, the speed parameter for resonance condition can be derived from Eq. (10) as follows:

$$S_{r,n} = \frac{n}{2j}\frac{d}{L}, \quad j \in Z^+, \tag{23}$$

where subscript "r" denotes "resonance". For reaching the solution in Eq. (23), the L'Hospital's rule has been employed to yield the following relation:

$$\lim_{\sin\left(\frac{dn}{2L}\frac{\pi}{S_{r,n}}\right)\to 0} \frac{\sin\left(N\frac{dn}{2L}\frac{\pi}{S_{r,n}}\right)}{\sin\left(\frac{dn}{2L}\frac{\pi}{S_{r,n}}\right)} = N. \tag{24}$$

It should be noted that the resonance condition in Eq. (23) has been widely reported in previous works.[6,8,10,12,13] The effect of damping of the beam in this connection will be demonstrated in the numerical study.

3. Criterion for Selection of Span Length or Car Length

In high-speed railway lines, optimal design for bridges with minimal vibration is desired to reduce excessive vibrations, thereby preventing premature fatigue of the track system. The cancellation conditions presented in Eq. (18) are physically meaningful but *cannot* be fully utilized in engineering, as the speed parameter is related to factors such as the train speed v, bridge span L and frequency ω_n, as indicated in Eq. (6c). In contrast, the *criterion for selection of span length* L for bridges aimed at vibration minimization is expected to be practically feasible, since it is *irrespective* of the train speed and bridge frequencies.

As indicated in Eq. (23), the resonant phenomenon with maximum response will occur when the speed parameter equals the first resonant speed $0.5\,d/L$. In practice, the first resonance can be suppressed through the selection of span length

L for the beam, given the car length d, such that either the internal or the external cancellation condition in Eq. (18) is met. Namely, by letting the first resonant speed parameter, $S_{r,1} = 0.5\,d/L$, equal the internal cancellation speed parameter S_c^{I} in Eq. (15), one arrives at the following formula:

$$\frac{L}{d} = j - 0.5, \quad j = 1, 2, 3, \ldots. \tag{25}$$

This is exactly the *criterion for optimal span length* for suppressing the resonant response, which is irrespective of the speed parameter, as was reported by Yang *et al.*[6] While the optimal criterion in Eq. (25) is good for selection of the span length L for the beam, given the car length d, it is equally good for the selection of car length d, given the span length L, although the latter is rarely encountered.

4. Validation of Normalized Modal Coordinate for Free Vibration

As indicated above, the free-to-decay expression presented in Eq. (9) for the first time herein for the normalized modal coordinate of free vibration of the beam is the key to deriving the cancellation conditions. It is a prerequisite that the accuracy of this simple form be assessed prior to further discussion on the cancellation conditions.

The properties close to those used in engineering practice are adopted for the beam: length $L = 30\,\mathrm{m}$, elastic modulus $E = 2.75 \times 10^{10}\,\mathrm{N/m^2}$, moment of inertia $I = 0.50\,\mathrm{m^4}$, per-unit-length mass $m = 2500\,\mathrm{kg/m}$. A total of eight moving loads of equal weight $P = 12\ 000$ kN uniformly spaced at $d = 25\,\mathrm{m}$ is considered.

To verify the accuracy of the simple form in Eq. (9), the midspan vibrational response for the simple beam traversed by eight loads at different speeds calculated from the corresponding Eq. (12) has been plotted in Fig. 2, along with the classical solution obtained by substituting Eq. (5) into Eq. (2). Here part (a) indicates the result for the normal speed of 70 km/h (19.44 m/s), part (b) for the resonance speed of 165 km/h (45.77 m/s), part (c) for the internal cancellation speed of 132 km/h (36.67 m/s), and part (d) for the external cancellation speed of 264 km/h (73.33 m/s). Specifically, solid lines in blue and pink denote, respectively, the responses of the undamped and damped beam calculated by Eq. (12), and dotted lines in gray and cyan the ones by the classical solutions. Although both the forced and free vibrations are given by the classical solution, only the free vibration (red dotted line on right side) will be compared and analyzed, as this is the part relating to the phenomenon of cancellation, which is also the one to be solved by Eq. (12). In the remaining analysis of the paper, only the free-vibration part of the dynamic analysis will be shown for brevity.

As can be seen from Fig. 2, excellent agreement has been achieved between the present and classical solutions for both the undamped and damped beams for all the speeds considered, indicating the *applicability* and high accuracy of the present

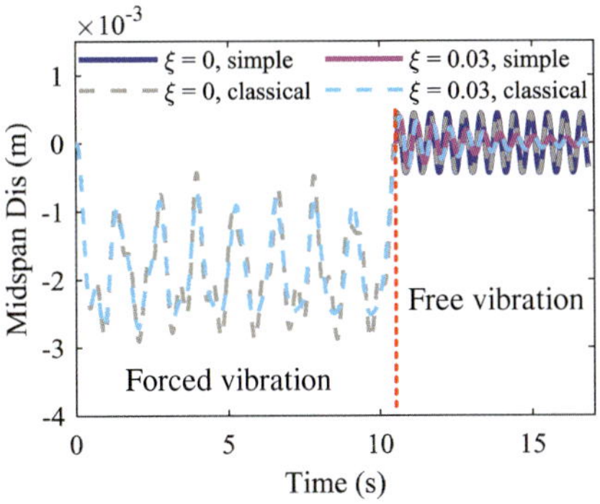

(a) Normal speed: 70 km/h

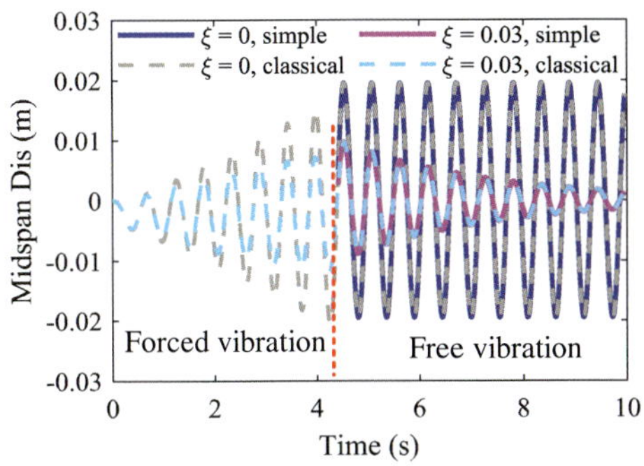

(b) Resonance speed: 167 km/h

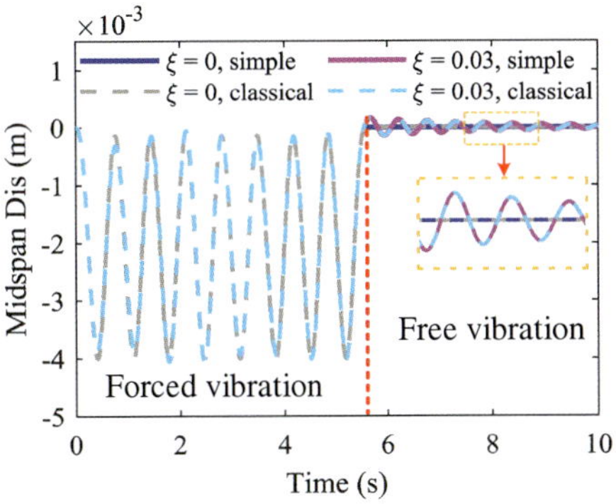

(c) Internal cancellation speed: 132 km/h

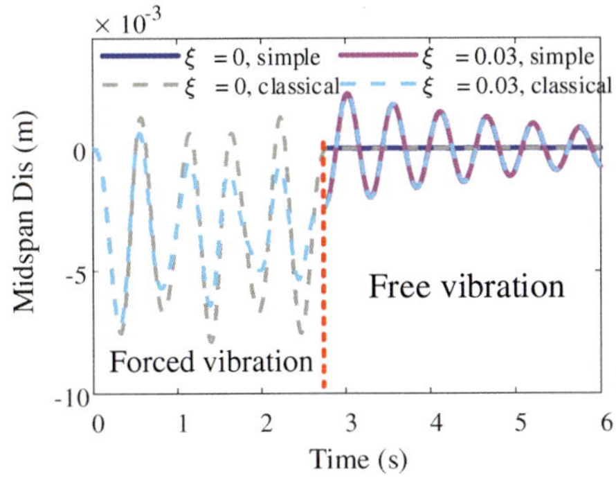

(d) External cancellation speed: 264 km/h

Fig. 2. Vibrational response of beam midspan traversed by eight loads at different speeds.

solution for the free vibration of the beam traversed by successive moving loads. Specifically, for the loads moving over beams at the normal speed (with no cancellation or resonance) of 70 km/h, as shown in Fig. 2(a), the free vibration is stationary without decay for the undamped beam, but is damped out for the damped beam. When the loads are moving at the *resonance* speed of 167 km/h, the free vibration will be significantly amplified by *many times* more than the normal speed, compare Fig. 2(b) with Fig. 2(a) for the amplitudes. Such a resonance phenomenon is likely to exert a detrimental effect on the health status of the railway track. However, when the loads move at the cancellation speeds of 132 and 264 km/h, see Figs. 2(c) and 2(d), the free vibration is totally cancelled out for the undamped beam, but is *nonzero* and *decayed by damping* for the damped beam. The nonzero response is referred to as the "leaking" effect caused by damping on cancellation in this paper. Such a phenomenon originates from the same fact as stated in Secs. 2.2 and 2.4.

Besides, a comparison between the resonance (see Fig. 2(b)) and cancellation (see Figs. 2(c) and 2(d)), especially on the magnitudes of the figures, indicates that the cancellation condition proposed in Eq. (18) does do a great job in minimizing the vibration, although it does not result in zero vibration due to the presence of damping. More investigations on damping effect will be presented in Sec. 5.2.

5. Factors Affecting Cancellation Conditions

As stated in Sec. 2.3, the internal cancellation condition is an inherent property of the beam, while the external cancellation condition is totally related to the external moving loads. In this section, numerical investigation will be conducted on factors affecting the cancellation conditions, with focus placed on the newly formulated external cancellation condition, as the internal one has been widely reported in previous works.[6,10,12,13] Moreover, the effect of damping of the beam on cancellation will also be assessed.

5.1. *Different compositions of moving loads*

As shown in Eq. (17), the external cancellation condition is determined both by the number N and the spacing d of the external moving loads. Both factors of the loads will be assessed for the free vibration of the beam that is simply supported herein.

5.1.1. *Number of moving loads*

To assess the effect of number N of moving loads on the cancellation condition, the normalized amplitudes of the free vibration calculated by Eq. (10) for the beam traversed by the first, third and eighth loads uniformly spaced at $d = 25$ m have

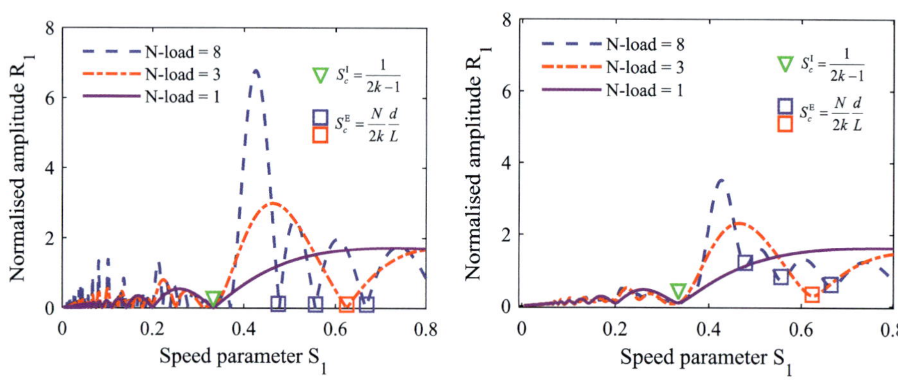

(a) $\xi = 0$ (b) $\xi = 0.03$

Fig. 3. Normalized amplitude of free vibration for beam traversed by different numbers of loads.

Table 1. Cancellation speed parameters for beam traversed by different numbers of loads.

Number of loads	Internal cancellation S_c^{I}	External cancellation S_c^{E}
1	0.333 ($j = 2 \to 0.333$)	—
3	0.333 ($j = 2 \to 0.333$)	0.624 ($j = 2 \to 0.625$)
8	0.333 ($j = 2 \to 0.333$)	0.667 ($j = 5 \to 0.667$); 0.555 ($j = 6 \to 0.556$) 0.477 ($j = 7 \to 0.476$)

Notes: The values outside the parentheses are the cancellation speed parameters extracted from Fig. 3(a), and the parenthesized j-value is the one for Eq. (18) to yield the cancellation speed parameter close to the extracted one.

been plotted in Fig. 3, with part (a) for the undamped beam and part (b) for the damped beam with $\xi = 0.03$. In summary, the cancellation speed parameters extracted from Fig. 3(a) (for the undamped beam) for the range from 0.2 to 0.8 by picking the points of zero amplitude have been listed in Table 1. As can be seen, all the extracted values agree excellently well with those calculated by Eq. (18) with the substitution of the parenthesized j value. This indicates that the cancellation conditions presented in this paper are highly reliable.

Then, let us discuss first the internal cancellation condition. It is clear from Fig. 3(a) and Table 1 that the internal cancellation speed parameter S_c^{I} is the same for all these three different numbers of loads, as marked by the inverse triangular symbols in green. In other words, the internal cancellation speed parameter S_c^{I} given in Eq. (15) and conventionally known as an inherent attribute of beam, which is *independent* of the external moving loads, is revealed by Fig. 3(a).

As for the external speed parameter S_c^{E}, it varies as a function of the number of moving loads. Specifically, for the beam traversed by a single moving load, the external speed parameter is absent, as has been pointed out in Sec. 2.4.2. But when the number of moving loads increases to 3, one external cancellation speed parameter (with zero amplitude) can be observed within the range of 0.4–0.8, as indicated by the red square in Fig. 3(a). When the number of loads increases to 8, two more external cancellation speed parameters (with zero amplitude) can be observed, as marked by the blue squares in Fig. 3(a). Such a phenomenon implies that *the more the external loads travelling over the beam, the more frequent the phenomenon of external cancellation will be induced.*

On the other hand, a comparison between Figs. 3(a) and 3(b) shows that for the beam damped with $\xi = 0.03$, both the internal and external cancellation speed parameters are the same as the undamped beam, indicating that *damping will not affect the speed parameter for the occurrence of cancellation phenomenon.* However, for the damped beam, the amplitude of vibration under cancellation is *nonzero*, as stated in Sec. 2.2, due to the leaking effect. In other words, *for the damped beam, the condition of cancellation should be recognized as the one for the "minimal vibration" to occur*, rather than zero vibration.

It should be added that the effect of damping is *limited* on internal cancellation (see the green triangle in Fig. 3(b)), but significant on external cancellation (see the squares in Fig. 3(b)). The more the number of external moving loads, the more significant such an effect can be observed.

5.1.2. *Spacing between uniform moving loads*

As indicated in Eq. (17), load spacing *d*is another factor that may affect the external cancellation conditions. To assess such an effect, three different load spacings, i.e. $d = 27, 12$, and 5 m, are considered in this section for numerical testing, which should not be regarded as reflective of those used in current practice.

The normalized amplitudes of the free vibration for the beam traversed by eight uniform moving loads of different spacings have been plotted in Fig. 4, with part (a) for the undamped beam and part (b) for the damped beam. Likewise, the

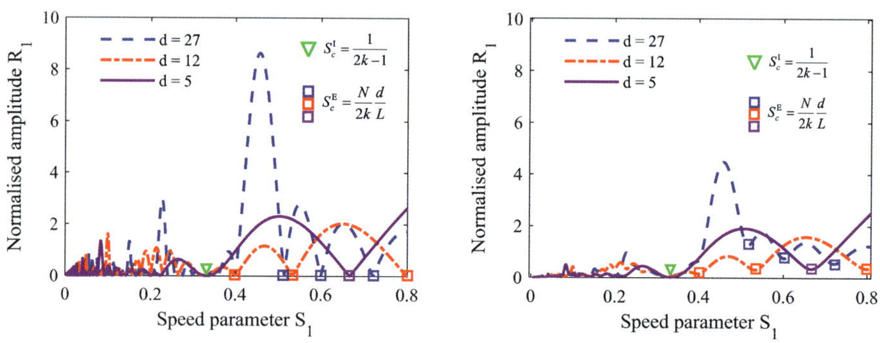

(a) $\xi = 0$ (b) $\xi = 0.03$

Fig. 4. Normalized amplitude of free vibration for beam traversed by eight uniform moving loads of different spacings.

Table 2. Cancellation speed parameters for beam traversed by eight uniform moving loads of different spacings.

Load spacing (m)	Internal cancellation S_c^{I}	External cancellation S_c^{E}
27	0.333 ($j = 2 \rightarrow 0.333$)	0.722 ($j = 5 \rightarrow 0.72$) 0.6 ($j = 6 \rightarrow 0.6$) 0.515 ($j = 7 \rightarrow 0.514$)0.4 ($j = 4 \rightarrow 0.4$)
12	0.333 ($j = 2 \rightarrow 0.333$)	0.8 ($j = 2 \rightarrow 0.8$) 0.531 ($j = 3 \rightarrow 0.533$) 0.402 ($j = 4 \rightarrow 0.4$)
5	0.333 ($j = 2 \rightarrow 0.333$)	0.666 ($j = 1 \rightarrow 0.667$)

Notes: The values outside the parentheses are the cancellation speed parameters extracted from Fig. 3(a), and the parenthesized j-value is the one for Eq. (18) to yield the cancellation speed parameter close to the extracted one.

cancellation speed parameters picked from Fig. 4(a) for the range from 0.3 to 0.8 have been listed in Table 2. It is confirmed that the cancellation speed parameters identified from the figures correspond very well to those calculated from Eq. (18) via substitution of the parenthesized j value. This is an indication of the reliability of the cancellation conditions proposed.

As was expected, the internal cancellation condition S_c^{I} remains unchanged for different load spacings, as indicated by the inversed triangles in Fig. 4. But for the external cancellation conditions, it is clear from Fig. 4 and Table 2 that the *frequency of occurrence* of the cancellation phenomenon decreases as the load spacing decreases. Moreover, the damping effect on the amplitude of free vibration for cancellation conditions also becomes smaller for shorter load spacings, as indicated by the smaller deviation from zero vibration in Fig. 4(b).

5.2. *Various damping ratios of beam*

The above analysis has demonstrated that damping of the beam will not affect the speed parameter for the occurrence of cancellation, but will make the amplitude of vibration under cancellation deviate from zero, as a result of leaking. In this section, the effect of damping of the beam on the cancellation phenomenon will be further assessed from the perspective of both the normalized amplitude and the time history of free vibration. In this regard, the beam is assumed to be traversed by eight uniform moving loads spaced at $d = 25\,\mathrm{m}$, considering three different damping ratios ($\xi = 0, 0.02$, and 0.05).

5.2.1. *Normalized amplitude of free vibration*

To assess the effect of damping of the beam on cancellation, the normalized amplitude of free vibration calculated by Eq. (10) under the moving loads for various damping ratios has been plotted against the speed parameter in Fig. 5. Clearly, the damping effect on the amplitude of free vibration versus speed parameter is sharply demarcated at the internal cancellation point as indicated by the inverse triangular symbol. Specifically, the damping effect is limited on the amplitudes associated with low speed parameters, as located on the left side of the internal cancellation point, while it is prominent on those related to the large speed parameters on the right side.

Moreover, for speed parameters near the external cancellation conditions, see the region within the rectangles, the amplitude decreases as the damping ratio increases. For the special case of zero damping, i.e. $\xi = 0$, the amplitude of free vibration under cancellation virtually goes to zero, a phenomenon same as that stated in Sec. 2.4.1. However, when the speed parameter jumps out of the cancellation regions, the damping effect is precisely the *opposite*. Namely, the larger the damping ratio,

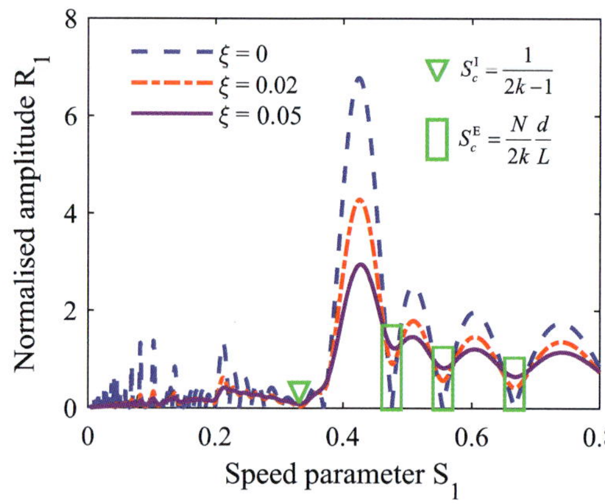

Fig. 5. Normalized amplitude of free vibration of beam with various damping ratios traversed by eight loads.

the smaller the vibration amplitude. Such an effect will reach the maximum for the resonant responses, see the peaks in dashed blue lines in Fig. 5.

5.2.2. *Time-history displacements of free vibration*

Now, let us analyze the damping effect on free vibration of the beam from the perspective of time history. Figure 6 shows the midspan time-history displacements of the beam for various damping ratios ($\xi = 0, 0.02$, 0.05 and 0.1) traversed by eight uniform moving loads at four different speeds, with part (a) for the normal speed (with no cancellation or resonance) of 70 km/h (19.44 m/s), part (b) for the resonance speed of 165 km/h (45.77 m/s), part (c) for the internal cancellation speed of 132 km/h (36.67 m/s), and part (d) for the external cancellation speed of 264 km/h (73.33 m/s). Overall, for all the four speeds, the larger the damping ratio, the faster the attenuation of free vibration, which can be attributed to the term $e^{-\xi_n \omega_n t}$ in Eq. (9).

Some interesting phenomena can be observed around the initial maximum amplitudes, as shown in Fig. 6(a), in that for free vibration of the beam traversed by loads at the normal speed, the initial maximum amplitude will decrease with the increasing damping ratio, though not to a large extent. However, when the loads are moving at the resonant speed, see Fig. 6(b), the free vibration will be amplified *many times* larger than the one for the normal speed (compare the magnitudes in Figs. 6(a) and 6(b)). This kind of vibrations induced by the passing moving loads on the beam is going to severely affect the fatigue behavior of the track system (which is supported by the beam) due to the *undesired repeated oscillations*. For multi-loads moving at the cancellation speeds, as shown in Figs. 6(c) and 6(d), the pure cancellation with zero vibration can be achieved for the undamped beam with

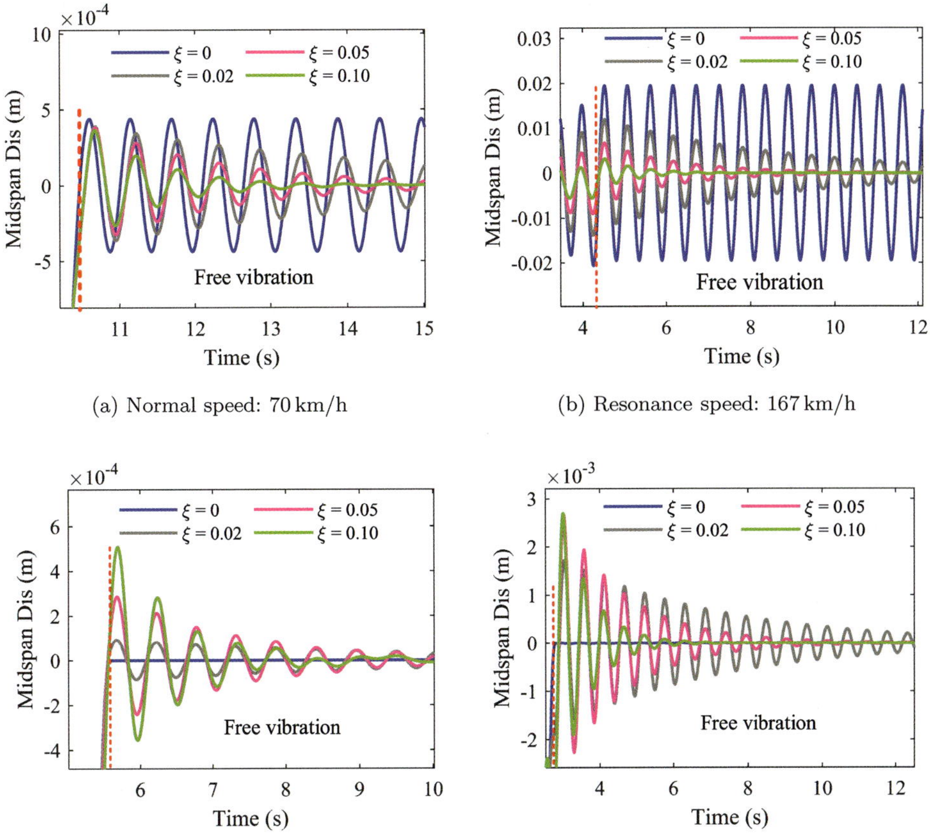

(a) Normal speed: 70 km/h

(b) Resonance speed: 167 km/h

(c) Internal cancellation speed: 132 km/h

(d) External cancellation speed: 264 km/h

Fig. 6. Free vibration at midspan of beam for various dampings traversed by eight loads at different speeds.

$\xi = 0$, as was pointed out in Sec. 2.4.1. But such a phenomenon ceases to exist for the damped beam, and is instead replaced by the pseudo cancellation with nonzero vibration, due to the leaking effect, as indicated in Fig. 6. Despite this fact, the resonance of violent vibration (see Fig. 6(b)) has been greatly suppressed by the minimal vibration under the cancellation condition.

Of more interest is that the initial maximum amplitudes for cancellation of free vibration tend to be larger for larger damping ratios, and the level of increase is quite large, see Figs. 6(c) and 6(d). This is different from the normal speed case in Fig. 6(a), as well as the resonant speed case in Fig. 6(b). For the cases of normal and resonant speeds, the larger the damping ratio, the smaller the vibration amplitude. In fact, for the damped beam, the condition of cancellation should be recognized as the one for the *minimal vibration* to occur, rather than zero vibration, as was previously pointed out in Sec. 4.2.1.

In short, damping of the beam will not affect the speed parameter for the occurrence of cancellation, but will affect the amplitude of free vibration under cancellation. Given that damping always exists in real beams, it is practically impossible to optimize the design of bridge by pure cancellation with zero vibration, but by the concept of minimal vibration.

6. Illustration on Criterion for Optimal Span Length for Bridges

In Sec. 3, the criterion for optimal span length for bridges has been found to be capable of suppressing the resonant response, in a way that is independent of the speed parameter. In this section, the effectiveness of the criterion from the perspectives of both the amplitude and time history of free vibration will be verified. As indicated by the criterion in Eq. (24), for $L/d = 0.5, 1.5, 2.5, \ldots$, the resonant phenomenon can be suppressed. Here, the two values of $L/d = 0.5$ and 1.5 will be taken in the exemplar analysis, since a longer bridge is normally prevented for bridge construction of high-speed railway lines.

6.1. *Amplitude of modal coordinate of free vibration*

The amplitude of modal coordinate of free vibration for the beam with different span lengths traversed by eight loads uniformly spaced at $d = 25\,\text{m}$ have been plotted in Fig. 7, with part (a) for the undamped beam and part (b) for the damped beam. It should be added that for the car length of $d = 25\,\text{m}$, a ratio of L/d of 0.5, 1.2

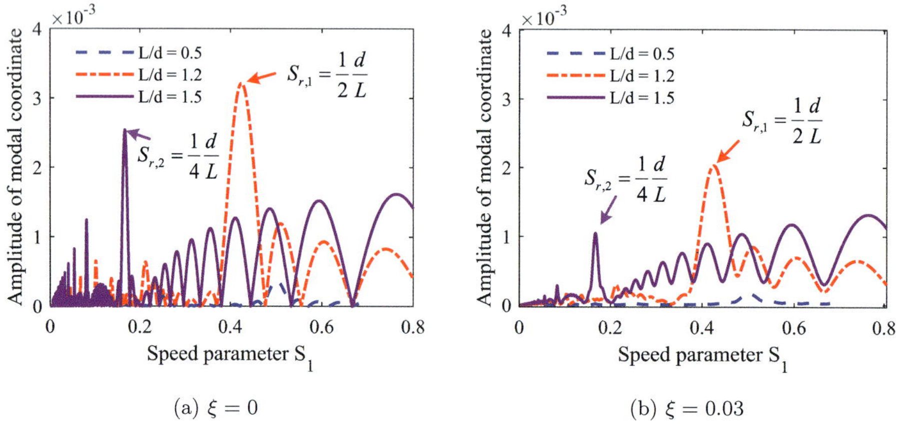

Fig. 7. Amplitude of modal coordinate of free vibration for beam with various span lengths traversed by eight uniform moving loads spaced at 25 m.

(non-optimized), and 1.5 adopted in analysis implies a span length L of 12.5, 30, and 37.5 m, respectively.

For the non-optimized beam of $L/d = 1.2$, the first resonant speed parameter $S_{r,1}$ can be clearly observed by the maximum peak. However, when the span length is optimized to be $L/d = 0.5$ and 1.5, the maximum peak under the resonance condition is largely suppressed, indicating the effectiveness of the criterion for optimization of bridge span length.

Specifically, the resonant response seems to be totally suppressed by the optimized shorter span length, i.e. $L/d = 0.5$. This can be attributed to the fact that the static deflection Δ_{stn} is proportional to the third power of span length L, as implied by ω_n^2 and L in Eq. (6a), the modal coordinate of the free vibration will be significantly reduced by the optimal shorter span length. On the other hand, for the optimized beam of longer span length, i.e. $L/d = 1.5$, the second resonant speed parameter $S_{r,2}$ is observed. Such a phenomenon originates from the fact that the criterion optimization of bridge span length works mainly for suppression of the primary resonance, as indicated by $S_{r,1}$in Eq. (24), but not for the second and higher resonances. However, the amplitudes of these resonant peaks are much smaller than that of the primary one, and even become *negligible* in the presence of damping, as shown in Fig. 7(b). As a side note for Fig. 7(b), the damping of the beam will reduce the resonant response, but will not affect the resonant speed parameters.

6.2. *Time-history responses of free vibration*

Let us verify the effectiveness of the criterion for span length optimization from the perspective of time history. Figure 8 shows the midspan time-history displacements of the beam for different L/d ratios traversed by eight uniform moving loads at the

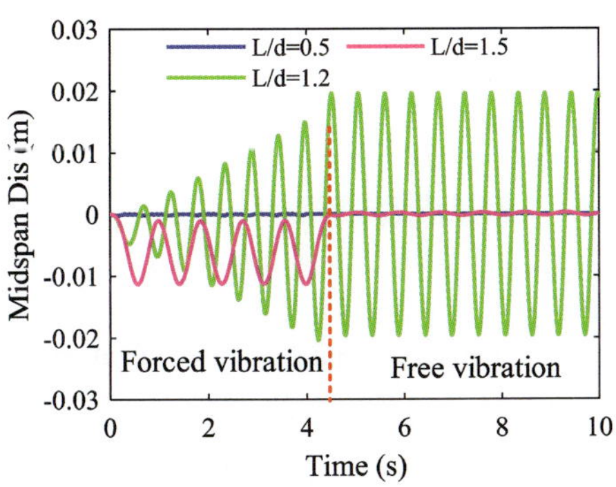

(a) $\xi = 0$

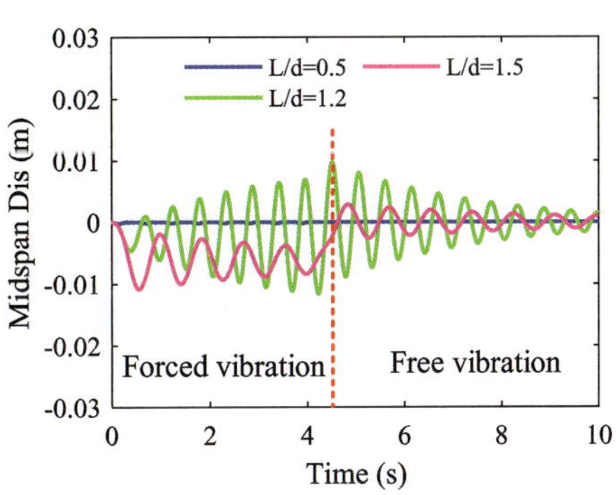

(b) $\xi = 0.03$

Fig. 8. Free displacement vibration at midspan of beam with different span length traversed by eight loads at resonant speed of 165 km/h.

resonant speed of 165 km/h (45.77 m/s), with part (a) for the undamped beam and part (b) for the damped one. As indicated by the green line in Fig. 8(a), resonance occurs for the non-optimized beam of $L/d = 1.2$ for the loads moving at the resonant speed. However, when the span length is optimized to $L/d = 0.5$ and 1.5, the free vibrations of the undamped beam under the resonance condition are almost totally suppressed. This is an indication of the effectiveness of the criterion for optimization of bridge span length.

When the damping of the beam is considered, as shown in Fig. 8(b), the free vibration for the non-optimized beam of $L/d = 1.2$ at the resonance speed gradually decayed, indicating that the damping of the beam is helpful to suppress the resonant response for non-optimized beams. Of interest is that the free vibration of the damped optimized beam (particularly for $L/d = 1.5$) in Fig. 8(b) is slightly enlarged compared with the almost zero vibration of the undamped beam in Fig. 8(a). Such a deviation from zero vibration, though very slight, brought by inclusion of damping of the beam was also revealed in Figs. 3(b) and 4(b), resulting from the leaking effect of damping of the beam. Because of damping, the original perfect (i.e. zero) cancellation has become imperfect (nonzero).

Furthermore, the midspan time-history accelerations of the beam for different L/d ratios traversed by eight uniform moving loads at the resonant speed of 165 km/h computed by Eq. (13) have been plotted in Fig. 9. A comparison of Fig. 9 with Fig. 8 indicates that the acceleration of free vibration induced by different L/d ratios behaves in the same way as that of the displacement. Specifically, the free acceleration vibrations of the undamped beam under the resonance condition are suppressed when the span length is optimized to $L/d = 0.5$ and 1.5, as shown

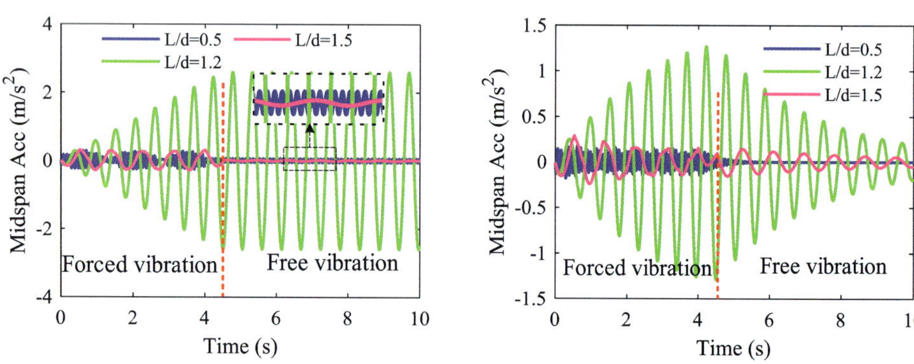

(a) $\xi = 0$ (b) $\xi = 0.03$

Fig. 9. Free acceleration vibration at midspan of beam with different span length traversed by eight loads at resonant speed of 165 km/h.

in Fig. 9(a). When bridge damping is included, deviation from zero vibration also occurs for $L/d = 1.5$, but it is very slight for $L/d = 1.2$, as indicated in Fig. 9(b).

In summary, the criterion given in Eq. (25) for optimization of bridge span length is generally good in suppressing the large-amplitude resonant vibrations. Although damping of the beam will result in nonzero vibration of the beam in the beginning of free vibration, it will quickly be damped out by damping itself. Theoretically, from the viewpoints of both the displacement and acceleration time histories, the optimized beam with shorter span length of $L/d = 0.5$ is more effective than the one with longer span length of $L/d = 1.5$, but the shorter one will increase the overall construction cost greatly. For a balance between the vibration suppression and construction cost, the longer optimal span length of $L/d = 1.5$ is recommended as a better candidate for the design of bridges in high-speed railway lines.

7. Conclusions

This paper presents an analytical and numerical investigation on the cancellation phenomena for damped beams under multi-moving loads. For the first time, a simple form is derived for the free vibration of the damped beam. Based on the concept of minimal vibration, both the internal and external cancellations are identified. The former is conventionally known, but the latter is newly formulated. For comparison, the resonant condition and the optimal criterion for span length of the bridge are also investigated. Based on the theory and numerical analysis presented, the following conclusions are drawn:

(1) The internal cancellation speed parameter S_c^{I} relates to the mode order of the beam, but independent of the external moving loads. On the contrary, the external cancellation speed S_c^{E} relates to the number N and spacing d of the moving loads for a beam of span length L.
(2) The external cancellation will occur more frequently due to the increase in the number and spacing of the moving loads.
(3) The damping of the beam will not affect the speed parameter for occurrence of cancellation phenomenon, but will affect the amplitude of free vibration. Such a phenomenon is referred to as the leaking effect of damping. For damped beams, the cancellation condition should be recognized as the one for minimal vibration to occur, rather than zero vibration.
(4) The criterion for optimization of bridge span length is practically feasible, since it is irrespective of the train speed and bridge frequencies, which has been verified to be capable of suppressing large-amplitude resonant vibrations.
(5) The optimized beam with shorter span length of $L/d = 0.5$ is theoretically more effective than the one with longer span length of $L/d = 1.5$. But the shorter one will increase the overall construction cost greatly. For a balance between vibration suppression and construction cost, the longer optimal span length of

$L/d = 1.5$ is recommended as a better candidate for use in high-speed railway lines.

In short, the cancellation conditions formulated in this paper is of general applicability to both damped and undamped beams, as well as for both multi-moving loads and single load.

Acknowledgments

This study was supported by the following agencies: National Natural Science Foundation of China (Grant Nos. 52008057, 52008315), Chongqing Science and Technology Commission (Grant No. cstc2021yszx-jscxX0001), the fellowship of China Postdoctoral Science Foundation (Grant No. 2021M693738) and Australian Research Council (IH150100006).

References

1. W. Weaver, S. P. Timoshenko and D. H. Young, *Vibration Problems in Engineering*, 5th edn. (Wiley, 1990).
2. L. Frýba, *Vibration of Solids and Structures Under Moving Loads*, 3rd edn. (Thomas Telford, Prague, Czech Republic, 1999).
3. W. M. Zhai, Z. Han, Z. Chen, L. Ling and S. Zhu, Train–track–bridge dynamic interaction: A state-of-the-art review, *Veh. Syst. Dyn.* **57** (2019) 984–1027.
4. R. Zhao, K. Zheng, X. Wei, H. Jia, H. Liao, X. Li, K. Wei, Y. Zhan, Q. Zhang, L. Xiao, L. Zhou, R. Shen, H. Gou, Q. Pu, F. Zhang, Z. Xu and C. Yu, State-of-the-art and annual progress of bridge engineering in 2020, *Adv. Bridge Eng.* **2** (2021) 29.
5. Y. B. Yang, J. D. Yau and Y. S. Wu, *Vehicle–Bridge Interaction Dynamics: With Applications to High-Speed Railways* (World Scientific, Singapore, 2004).
6. Y. B. Yang, J. D. Yau and L. C. Hsu, Vibration of simple beams due to trains moving at high speeds, *Eng. Struct.* **19** (1997) 936–944.
7. S. H. Ju and H. T. Lin, Resonance characteristics of high-speed trains passing simply supported bridges, *J. Sound Vib.* **267** (2003) 1127–1141.
8. Y. B. Yang, C. L. Lin, J. D. Yau and D. W. Chang, Mechanism of resonance and cancellation for train-induced vibrations on bridges with elastic bearings, *J. Sound Vib.* **269** (2004) 345–360.
9. P. Museros and E. Alarcón, Influence of the second bending mode on the response of high-speed bridges at resonance, *J. Struct. Eng.* **131** (2005) 405–415.
10. J. D. Yau and Y. B. Yang, Vertical accelerations of simple beams due to successive loads traveling at resonant speeds, *J. Sound Vib.* **289** (2006) 210–228.
11. H. Xia, N. Zhang and W. W. Guo, Analysis of resonance mechanism and conditions of train–bridge system, *J. Sound Vib.* **297** (2006) 810–822.
12. P. Museros, E. Moliner and M. D. Martínez-Rodrigo, Free vibrations of simply-supported beam bridges under moving loads: Maximum resonance, cancellation and resonant vertical acceleration, *J. Sound Vib.* **332** (2013) 326–345.
13. C. P. S. Kumar, C. Sujatha and K. Shankar, Vibration of simply supported beams under a single moving load: A detailed study of cancellation phenomenon, *Int. J. Mech. Sci.* **99** (2015) 40–47.

14. J. R. Cho, K. Jung, K. Cho, J. W. Kwark, Y. J. Kim and B.-S. Kim, Determination of the optimal span length to minimize resonance effects in bridges on high-speed lines, *Proc. Inst. Mech. Eng. F J. Rail Rapid Transit* **230** (2016) 335–344.
15. Z. L. Wang, Z. X. Tan, H. Yao, K. Shi, H. Xu and Y. B. Yang, Effect of soft-end amplification on elastically supported bridges with bearings of unequal stiffnesses scanned by moving test vehicle, *J. Sound Vib.* **540** (2022) 117308.
16. E. Savin, Dynamic amplification factor and response spectrum for the evaluation of vibrations of beams under successive moving loads, *J. Sound Vib.* **248** (2001) 267–288.
17. S. S. Law and X. Q. Zhu, Dynamic behavior of damaged concrete bridge structures under moving vehicular loads, *Eng. Struct.* **26** (2004) 1279–1293.
18. Y. B. Yang and J. D. Yau, An iterative interacting method for dynamic analysis of the maglev train–guideway/foundation–soil system, *Eng. Struct.* **33** (2011) 1013–1024.
19. Z. L. Wang, Y. L. Xu, G. Q. Li, Y. B. Yang, S. W. Chen and X. L. Zhang, Modelling and validation of coupled high-speed maglev train-and-viaduct systems considering support flexibility, *Veh. Syst. Dyn.* **57** (2019) 161–191.
20. Z. L. Wang, Y. L. Xu, G. Q. Li, S. W. Chen and X. L. Zhang, Dynamic analysis of a coupled system of high-speed maglev train and curved viaduct, *Int. J. Struct. Stab. Dyn.* **18** (2018) 1850143.
21. Y. L. Xu, Z. L. Wang, G. Q. Li, S. Chen and Y. B. Yang, High-speed running maglev trains interacting with elastic transitional viaducts, *Eng. Struct.* **183** (2019) 562–578.
22. X. Q. Zhu and S. S. Law, Moving load identification on multi-span continuous bridges with elastic bearings, *Mech. Syst. Signal Process.* **20** (2006) 1759–1782.
23. Z. W. Chen, W. Yang, J. Li, T. Yi, J. Wu and D. Wang, Bridge influence line identification based on adaptive B-spline basis dictionary and sparse regularization, *Struct. Control Health Monit.* **26** (2019) e2355.
24. D. S. Yang and C. M. Wang, Dynamic response and stability of an inclined Euler beam under a moving vertical concentrated load, *Eng. Struct.* **186** (2019) 243–254.
25. D. S. Yang, C. M. Wang and W. H. Pan, Further insights into moving load problem on inclined beam based on semi-analytical solution, *Structures* **26** (2020) 247–256.
26. D. S. Yang, C. M. Wang and J. D. Yau, Dynamic stability and response of inclined beams under moving mass and follower force, *Int. J. Struct. Stab. Dyn.* **20** (2020) 2043004.
27. Y. B. Yang, C. W. Lin and J. D. Yau, Extracting bridge frequencies from the dynamic response of a passing vehicle, *J. Sound Vib.* **272** (2004) 471–493.
28. Y. Zhang, H. S. Zhao and S. T. Lie, Extracting mode shapes for beams through a passing auxiliary mass, *J. Vib. Acoust.* **141** (2019) 054501.
29. J. Li, X. Zhu, S.-S. Law and B. Samali, A two-step drive-by bridge damage detection using dual Kalman filter, *Int. J. Struct. Stab. Dyn.* **20** (2020) 2042006.
30. J. Li, J. Guo and X. Zhu, Time-varying parameter identification of bridges subject to moving vehicles using ridge extraction based on empirical wavelet transform, *Int. J. Struct. Stab. Dyn.* **21** (2021) 2150046.
31. Y. B. Yang, Z. Li, Z. L. Wang, K. Shi, H. Xu, F. Q. Qiu and J. F. Zhu, A novel frequency-free movable test vehicle for retrieving modal parameters of bridges: Theory and experiment, *Mech. Syst. Signal Process.* **170** (2022) 108854.
32. Z. L. Wang, J. P. Yang, K. Shi, H. Xu, F. Q. Qiu and Y. B. Yang, Recent advances in researches on vehicle scanning method for bridges, *Int. J. Struct. Stab. Dyn.* **22** (2022) 2230005.

https://doi.org/10.1142/9789819814770_0007

Chapter 7

Historical Review on Resonance and Cancellation of Simply Supported Beams Subjected to Moving Train Loads: From Theory to Practice[#]

Y. B. Yang[*,†], J. D. Yau[‡,||,††], S. Urushdaze[§,**,††] and T. Y. Lee[¶]

[*]*School of Civil Engineering, Chongqing University*
Chongqing 400044, China

[†]*School of Civil Engineering and Architecture*
Chongqing University of Science and Technology
Chongqing 401331, China

[‡]*Department of Architecture, Tamkang University*
New Taipei City 25137, Taiwan

[§]*Institute of Theoretical and Applied Mechanics*
Czech Academy of Sciences, 19000 Prague, Czech Republic

[¶]*Department of Civil Engineering*
National Central University, Taoyuan City 32001, Taiwan

[||]*jdyau@mail.tku.edu.tw*
[**]*urushadze@itam.cas.cz*

The advent of railways and especially highspeed railways marks great strides in human transportation history. To guarantee exclusive right-of-way, highspeed railways are often built on equal simply supported beams resting on piers. In this paper, a historical review will be given of the resonance and cancellation phenomena observed for simply supported beams traveled by a set of moving loads, as they are typical of highspeed railways. The phenomenon of resonance was observed by early investigators including Timoshenko, Bolotin, Frýba, Matsuura, etc. However, the phenomenon of cancellation was noted lately in 1997 by Yang *et al.* By letting the conditions of resonance and cancellation coincident, they proposed the *optimal span length* for suppressing the resonance of simple beams, which is equal to 1.5 times the car length. This 1.5 times rule has been verified and adopted in the design of some highspeed railways. In this article, the theoretical solution for the problem will be revisited for unveiling the key parameters such as the resonance speed (in temporal domain) and resonance wavelength (in spatial domain). Then a rather in-depth review will be given of existing works on the resonance and cancellation of

[††]Corresponding authors.
[#]This chapter appeared previously on the International Journal of Structural Stability and Dynamics. To cite this chapter, please cite the original article as the following: Y. B. Yang *et al.*, *Int. J. Struct. Stab. Dyn.*, **23**, 2340008 (2023), DOI: 10.1142/S0219455423400084.

railway bridges from the waves perspective. Some new developments along these lines will also be identified.

Keywords: Bridge; cancellation; CRH; highspeed railway; ICE; moving load; resonance; TGV; wavelength.

1. Historical Aspects

The history of research on moving load problems related to bridge vibrations dates back to the middle of the nineteenth century,[1] when railway bridges were intensively constructed for the lucrative business transport in England. At the time, it was concerned how the train moving at "high" speeds can affect the vibration of bridges.[2] As part of the effort, a dynamic test of a scaffold-supported railroad track (beam) under a moving two-axle vehicle was designed and carried out. From this pioneering experiment, it was shown that the dynamic deflections of the tested beam rose with the increase in moving speeds of the test vehicle.[1,2] This eye-catching experiment confirmed the importance of the moving load effects on the response amplification of the supporting bridge since then. About one century later, Bolotin[3] studied a simply supported beam subjected to a series of masses of equal interval d moving at a constant speed v, and pointed out that the repeated period d/v of the serial loads is a key factor to excitation of the beam to resonance. Subsequently, Frýba[1] indicated that there exists a resonant speed for the regularly spaced trainloads moving over a simply supported beam.

In 1964, the first highspeed railway (HSR) line, the Shinkansen (SKS), connecting Tokyo and Osaka in Japan started its commercial operation. For the SKS bullet train with a car length D (= 25 m) moving at speed v, some research was conducted by Matsuura.[4] He derived a characteristic frequency ($f_b = j \times v/D|_{j=1,2,\ldots}$) for the resonance of the HSR bridges from the response of a simple beam of length L subjected to the equidistant moving loads and implied exciting wavelength. However, the influential key parameter of the critical *maximum* resonant response at some span length ($L = 50\,\text{m}$) was not clearly explained due to lack of knowledge on the resonance wavelength of the bridge excited.

On the other hand, the TGV highspeed trains have been popularly used in Europe. In a field test conducted in France (see Fig. 1), resonance in the form of amplified response was observed on a simply supported bridge of length $L = 38\,\text{m}$ under the passage of the double TGV trainset (power + passenger + power) at 192 km/h, according to Ramondenc (1998).[5] For the car length of $D = 18.7\,\text{m}$ and bridge of frequency $f = 2.8\,\text{Hz}$, Frýba[6] calculated the resonance speed of the train as $v_{\text{res}} = f, D = 52.4$, m/s = 189 km/h, which is close to the train moving speed of 192 km/h. Another field test was conducted on the Tajo Bridge (of length $L = 38\,\text{m}$ and frequency $f = 3.3\,\text{Hz}$) of the Spanish Madrid-Sevilla AV line, according to Goicolea–Ruigómez.[7] Resonance occurred on the bridge when a single AVE trainset (power + passenger + power) with car length $D = 18.7\,\text{m}$ reached the speed of

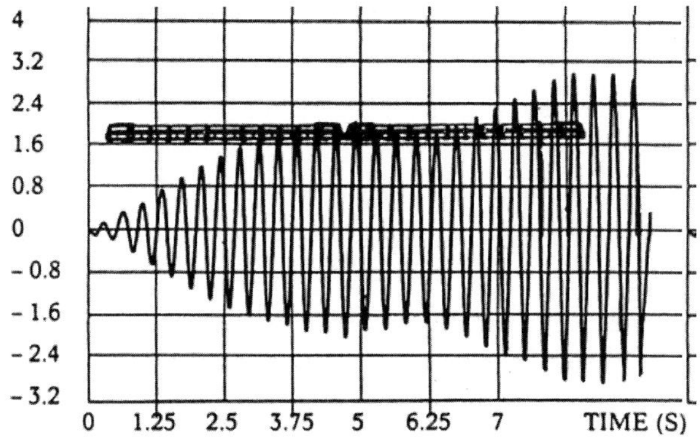

Fig. 1. Resonance observed from the field-test of a 38 m bridge[5,6]

219 km/h, which is close to the resonance speed of $v_{\text{res}} = fD = 222$ km/h. In fact, for both simply supported beams tested above, their span length of $L = 38$ m also roughly meets the critical span length ($L_c = 2D = 37.4$ m) for resonance of the HSR bridges. This is a condition not clearly noted by previous researchers. In the HSR bridge design for speeds up to 350 km/h, the speed $v = fD$ was recognized as the resonance speed by the European Railway Research Institute (ERRI)[8] and used in interpreting the dynamic effects of trainloads moving over simply supported bridges by the 2003 Eurocode.[9] To ensure the health state of the railway tracks for long-time operation, span lengths that are potentially prone to resonance should always be avoided in the design of railway bridges, since excessive vibrations may lead to ballast destabilization, fatigue in mechanical fasteners, and deterioration on the overall quality of supporting track.

Vehicle–bridge interaction (VBI) dynamics was extensively investigated at National Taiwan University since the beginning of 1990s, when Taiwan HSR (THSR) started to design its infrastructure.[10] Yang and co-workers carried out a series of researches on VBI dynamics concerning the impact effects of moving vehicles on highway bridges[11,12] and on the train-induced vibrations of highspeed railway bridges.[13,14] Focusing on the mechanism of vibration of the beam and the repetitive nature of moving trainloads in the spatiotemporal domain, Yang *et al.*[13] derived a closed-form solution for the response of a simply supported beam to a series of equally-spaced moving loads, and used the *speed parameter* to interpret the resonance and cancellation phenomena induced on the beam. The speed parameter defined as $S = \pi v/(\omega L)$, where v = speed, ω = bridge frequency and L = span length, is featured by its nondimensional nature and the merit of wider applications. With this, both the *resonance* and *cancellation* speeds were derived. Several informative conclusions were drawn from Ref. 13 for the train-induced vibration phenomena of railway bridges.

By letting the resonance condition coincide with the cancellation condition (both in terms of the speed parameter), an *optimal design criterion* was derived for the span length of the bridge, for which the resonance will be suppressed all the time.[13] This criterion can be simply put as $(L/D)_{\rm opt} = 1.5$, meaning that the optimal span length L of the simply supported beam is equal to 1.5 times the car (or characteristic) length D, i.e. $L_{\rm opt} = 1.5D$. Focusing on vibration cancellation, Savin[15] pointed out that for a single load moving on a simple beam at constant speed, the residual response of free vibration after the load leaves the beam diminishes under the same cancellation speed mentioned in Ref. 13. Museros and Alarcón[16] examined the dynamic response of simply supported railway bridges under the passage of ICE2 trains in European HSR lines. They found that the resonance is significantly reduced for some 40 m bridges, and that for these bridges, the bridge/coach length ratio L/D (=40/26.4 = 1.52) is quite close to the optimal design length ratio $(L/D)_{\rm opt}$ of 1.5 proposed by Yang *et al.*[13] According to Cho *et al.*,[17] such a design criterion for span length has been applied to the structural design in the Korea High-Speed Rail (KHSR) bridges. They also verified that the optimal design criterion for span length is applicable to vibration reduction of the HSR bridges under the passage of KTX bullet trains in Korea, and concluded that the optimal span length is an economic design means for resonance suppression of the KHSR bridges. It should be emphasized that the suppression of resonance is helpful to ballast stability, as well as running safety for trains traveling over railway bridges.

Over the past two decades, the rapid development of viaduct-based HSR lines in emerging and developed countries has created a large demand for research on the HSR bridge dynamics, especially on the resonance problems of railway bridges. Besides the resonance and cancellation speeds mentioned above, additional interest will be paid in this paper to the *maximum* amplitude of resonance and the *minimum* amplitude of cancellation for a single-span simple beam subjected to moving trainloads, based on the theory reviewed in Sec. 2. To focus on the key parameters involved, the train–bridge interaction system will be neglected from the present theoretical study, in the sense that the train loads will be treated as uniform moving loads, since their moving action on the bridge is mainly of the transient nature and the interaction or inertial effect can be ignored without losing accuracy.

In this short article, the relevant researches that are known to the authors on the *resonance* and *cancellation* of railway bridges have been summarized in Table 1, in the chronological order according to their years of publications. It should be admitted that as continuing researchers on the dynamic problems of the HSR, the authors suffered largely from the lack of access to researches conducted by various national HSR agencies. With reference to Table 1, a brief discussion of some of the cited works is given below.

The book by Frýba,[1] entitled "Vibration of solids and structures under moving loads," is a milestone in the derivation of closed-form solutions for the dynamic

Table 1. List of researches related to resonance and cancellation of railway bridges in chronological order.

Year	Authors	Res.	$v_r = f_b D$	Can.	$(L/D)_{\mathrm{opt}}$	Remarks
1964	Bolotin[3]	•	•			Char. exciting period (d/v.) for resonance
1972	Frýba[1]	•	•			Char. exciting freq. (v/d) for resonance
1976	Matsuura[4]	•	•			Char. freq. $f_b = j * v/D\|_{j=1,2...}$ for resonance
1990	Klasztorny, Langer[38]	•				Numerical studies on resonance and stability
1997	Yang *et al.*[13a]	•[a]	•[a]	•[a]	•[a]	Resonance speed: $v_r = f_b D$ Modal cancellation speed: $v_c = 2f_b L/j\|_{j=3,5,...}$ $L/D = 1.5$ for optimal span length
1998	Ramondenc[5]	•				Resonance measurement from field-test of a 38 m bridge in France
1999	Yau *et al.*[14]	•	•	•		Impact vibrations of railway bridges
2001	Frýba[6]	•	•			Resonance applied to bridge vibration
2001	Savin[15a]	•[a]	•	•		Loading wavelength $= v/f_b$
2001	Yau *et al.*[22a]	•[a]	•	•		Amplification effects of elastic bearings
2001	Yang *et al.*[24a]	•[a]	•	•		Resonance of horizontally curved beam
2001	Yau[39]		•		.	Multiple resonance of continuous beams
2003	Ju, Lin[40a]	•	•			Resonance of high-pier viaducts
2004	Kwark *et al.*[27]	•	•	•	•	Applications of cancellation speeds to bridges with boundary constraints; field-tests
2004	Yang *et al.*[23a]	•[a]	•	•		Supporting elastic bearing to amplify bridge response verified by field tests
2005	Museros, Alarcón[16a]	•[a]	•	•	•	Contribution of the second bending mode
2006	Xia *et al.*[41]	•				Consider different bullet trains: ICE3 (Germany), TGV(France), E500 (Japan), CHT (China)
2006	Yau, Yang[19a]	•[a]	•	•		Sub-resonance of damped short bridges
2007	Vestroni, Vidoli[42a]	•[a]				Loading wavelength
2007	Lacarbonara, Colone[43a]	•[a]				Resonance characteristics of tied-arch bridge

Table 1. (*Continued*)

Year	Authors	Res.	$v_r = f_b D$	Can.	$(L/D)_{\text{opt}}$	Remarks
2012	Goicolea, Antolin[44]		•			Applications and field-test of resonance
2013	Museros *et al.*[28]	•[a]	•	•	•	Cancellation of resonance
2013	Mao, Lu[45a]	•[a]	•	•		Present a resonance indicator in terms of the length ratio of bridge to carriage
2014	Doménech *et al.*[34]		•			VBI effects on tuning resonance
2014	Xia *et al.*[29a]	•[a]	•	•[a]		Conditional cancellation related to the interval between two adjacent moving loads
2014	Arvidsson, Karoumi[46]	•				A survey of 213 papers
2015	Sudheesh-Kumar *et al.*[30a]			•[a]		Cancellation condition of free response (commented by Museros and Moliner in 2017[48])
2015	Yang, Yau[20a]	•[a]	•			Bridge-induced resonance on a moving vehicle
2016	Cho *et al.*[17]		•	•	•	Applications of optimal span length
2017	Yang, Yau[47a]		•	•		Applications of resonance
2018	Yang *et al.*[25a]	•[a]	•	•	•	Torsional resonance and cancellation
2019	Zhai *et al.*[49]	•				Review of train–track–bridge interaction
2019	Yang *et al.*[21a]	•[a]	•	•		Dual resonance of train–bridge system
2020	Martinez-Rodrigo *et al.*[50]		•	•		Resonance and cancellation of a two equal-span continuous beam
2021	Zangeneh *et al.*[51a]	•[a]	•			Maximum resonance of a visco-elastically supported bridge
2021	Auersch[52]	•[a]	•			Resonance prediction in frequency domain
2022	Li *et al.*[31a]	•	•	•[a]		Using the moving load amplitude spectrum to analyze the resonance and two types of cancellation (raised by Xia *et al.* in 2018[36])
2022	Lou *et al.*[35]	•				Shifted resonance of bridges due to VBI effects
2022	Wang *et al.*[53]	•	•			Double flexural-torsional sub-resonance of short bridges (raised by Yang *et al.* in 2018[25])
2023	Xin *et al.*[54]	•[a]	•	•		For a long-span bridge with lower frequencies, cancellation is more significant than resonance

Note: [a]Closed-form solution derived for bridge response. *Res.*, resonance; *Can.*, cancellation.

response of simply supported beams under moving trainloads using simplified train/bridge models. It provided a well-founded survey and theoretical basis for researchers to deal with the mathematical modeling of bridge structures under the action of moving dynamic loads. Aside from the analytical studies using similar approaches, focus was directed by Yang *et al.*[10] to highspeed railways concerning the train–rail–bridge interaction to meet the demands from railway engineers and designers in tackling the vibration problems not encountered by traditional trains of medium or low speeds, such as the riding comfort of train cars[14] or the behavior of trains moving over bridges shaken by earthquakes.[18] Using the finite element method, a range of vehicle models were presented, covering the moving load, moving mass, and sophisticated VBI model, with or without consideration of the rail effects. The studies on the moving load problems[1,10] provided the theoretical basis for vibration analysis of bridge structures under the passage of moving trains.

It is well known that the major parts of highspeed railways existing globally are supported by simply supported beams, elevated from the ground, to keep the exclusive right-of-way for strict maintenance. As can be observed from Table 1, closed-form and analytical solutions of the dynamic response of simply supported bridges opened up a new trend of research on moving load problems for the HSR systems in crossing the millennium. Concerning the resonance of simply supported bridges subjected to moving trains, Yang and co-workers investigated the following problems: (1) a beam subjected to a series of moving loads, for which the train-induced resonance has been derived;[13,19] (2) a moving load traveling over a serious of simple beams, for which the bridge-induced resonance was derived;[20] and (3) a train (i.e. a series of linked railcars) moving over another series of simple beams, for which the wave transmission over the linked railcars under dual resonance was studied.[21] The train-induced resonance has also been extended to the analysis of elastically supported beams[22,23] and horizontally curved beams, for which the phenomena of resonance and cancellation are found to exist as well for the induced centrifugal forces.[24] For the monosymmetic I-beams under the moving loads, it was indicated by Yang *et al.*[25] that the phenomena of resonance and cancellation can also occur in the torsional response. Besides, Yau has included the effect of ground settlement in the analysis of bridge-induced resonance.[26]

From Table 1, it is interesting to note that most researchers have focused on problem of resonance of the HSR bridges. In contrast, only a small portion of the research has been conducted on the cancellation phenomenon, starting from the work by Yang *et al.*[13] See for instance, Kwark *et al.*,[27] Museros *et al.*,[28] Xia *et al.*,[29,36] Kumar *et al.*,[30] and Li *et al.*,[31] Xin *et al.*[54] Specifically, the optimal span length for simple beams proposed by Yang *et al.*[13] was reported by Cho *et al.*[17] to have been applied to some HSR lines in Japan and Korea. Recently, the same optimal span length was also adopted in the design of some newly constructed HSR lines in China. Thanks to the foregoing research effort. Issues related to the dynamics response of the railway–bridge systems in global and the core problem of resonance and cancellation phenomena in particular have been largely adopted in the HSR specifications worldwide.[13,14,32]

In addition to the resonant speeds discussed previously,[4,13,14] an alternative approach will be presented in this study for the resonance wavelength to demonstrate the critical issue of the *maximum* resonance for a beam subjected to moving trainloads. In addition, the optimal design criterion of span length derived by letting the modal cancellation condition coincide with the resonance condition for counteracting or reducing the resonance response will be examined in the illustrated examples. Finally, some considerations on the resonance and cancellation phenomena concerning the selection of optimal span length of HSR bridges will be given.

2. Response of a Simple Beam to a Series of Equidistant Moving Constant Loads

Train-induced resonance on the HSR bridges was designated as a core subject of bridge dynamics since the first commercial operation of the SKS bullet trains in Japan in 1964. Then, numerous research works on the HSR bridge dynamics were conducted in either from the theoretical aspects or field investigations, as partially listed in Table 1 for the works related to the resonance and cancellation phenomena. Among them, practical train-induced resonance phenomena of simply supported bridges by field-experiments were observed and continuously carried out since the late 1990s.[5,7,23] These theoretical and experimental studies have contributed to the design provisions in the 2003 Eurocode[14] and later in the 2014 Code for Design of High Speed Railway[32] of China.

For theoretical formulation of bridge dynamics, an analytical model of a train traveling over a bridge is postulated as a series of equidistant and constant moving loads. Meanwhile, the bridge is modeled as a simply supported beam, for which the Bernoulli–Euler hypothesis applies. The use of a simply supported beam is adequate, since the HSRs are often built on equal simply supported beams resting on piers. A schematic of a simple beam subjected to a sequence of moving point loads P with uniform interval D at a constant speed v is shown in Fig. 2. Depending on the ratio (L/D) of the beam length to the interval of moving loads, there may be two, three, or no wheel loads acting on the beam during the passage of trainloads.[13,19]

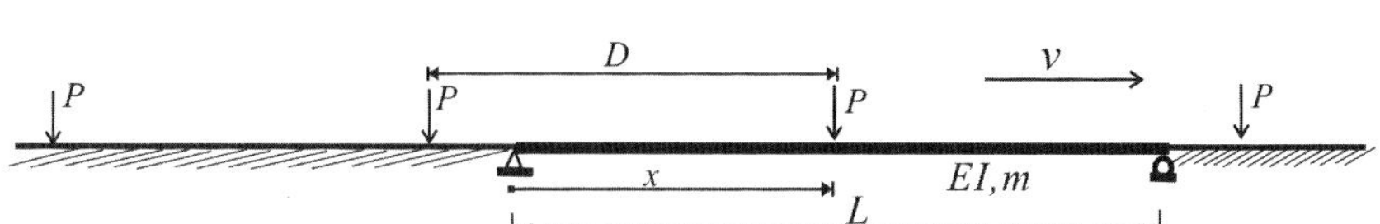

Fig. 2. A simple beam subject to a series of equidistant moving loads.

To obtain closed-form solution for the resonance response, the beam is set to have a span length L greater than the interval D, but not greater than twice the interval, i.e. $D < L < 2D$.[13,19] For this problem, the equation of motion of the beam with zero damping is

$$m\frac{\partial^2 u}{\partial t^2} + EI\frac{\partial^4 u}{\partial x^4} = P\sum_{k=1}^{N} \delta[x - v(t - t_k)] \times [H(t - t_k) - H(t - t_k - L/v)], \quad (1)$$

where m is mass per unit length, $u(x, t)$ is vertical displacement, E is elastic modulus, I is moment of inertia of the beam, δ is Dirac's delta function, $H(t)$ is unit step function, N is Nth moving load on the beam, and $t_k = (k - 1)D/v =$ arriving time of the kth load at the beam. It is assumed that the beam starts to vibrate from rest upon the arrival of the first moving load. Accordingly, the deflection $u(x, t)$ of the simple beam can be expressed in terms of the vibration mode shapes as follows:

$$u(x, t) = \sum_{n=1}^{\infty} q_n(t) \sin\frac{n\pi x}{L}, \quad (2)$$

where $q_n(t)$ denotes the nth modal coordinate of the beam, which can be analytically solved as follows:[19]

$$q_n(t) = \frac{2P}{mL\omega_n^2}\begin{bmatrix} \dfrac{\sin(\Omega_n t) - S_n \sin(\omega_n t)}{1 - S_n^2} H(t - t_{N-1}) \\ -2\dfrac{S_n}{1 - S_n^2} \times \Re_n(\lambda_n, v, t) \times H\left(t - t_{N-1} - \dfrac{L}{v}\right) \end{bmatrix} \quad (3)$$

with

$$S_n = \frac{\pi v}{\omega_n L} = \frac{\lambda_n}{2L}, \quad \lambda_n = \frac{2\pi v}{\omega_n} = \frac{v}{f_n}, \quad \frac{\pi L}{\lambda_n} = \frac{\omega_n L}{2v}, \quad (4a–c)$$

where S_n is nth modal speed parameter, λ_n = loading wavelength,[14,15,55] and $\Re_n(\lambda_n, v, t)$ represents the residual response factor of free vibration accumulated after the series of moving loads have left the beam, i.e.

$$\Re_n(\lambda_n, v, t) = \begin{cases} \cos\dfrac{\pi L}{\lambda_n} \times \left[\sin\left(\omega_n t - \dfrac{\pi L}{\lambda_n}\right) + \sin\left(\omega_n\left(t - \dfrac{t_N}{2}\right) - \dfrac{\pi L}{\lambda_n}\right) \times \Upsilon_n(\lambda_n, D)\right] \\ n = \text{odd}, \\ \sin\dfrac{\pi L}{\lambda_n} \times \left[\cos\left(\omega_n t - \dfrac{\pi L}{\lambda_n}\right) + \cos\left(\omega_n\left(t - \dfrac{t_N}{2}\right) - \dfrac{\pi L}{\lambda_n}\right) \times \Upsilon_n(\lambda_n, D)\right] \\ n = \text{even} \end{cases} \quad (5)$$

with

$$\Upsilon_n(\lambda_n, D) = \frac{\sin((N-2)\pi D/\lambda_n)}{\sin(\pi D/\lambda_n)}. \tag{6}$$

By observing the modal response in Eq. (3), the *underlined* term $S_n/(1-S_n^2)$ can be regarded as an *amplifier* of the residual response $\Re_n(\lambda_n, v, t)$ to reflect the *impact* effect of moving loads on the bridge.[11,12] Moreover, the closed-form expression of the residual response $\Re_n(\lambda_n, v, t)$ reveals several useful messages of vibration for a simple beam under the successive moving loads. The first trigonometric terms $\cos(\pi L/\lambda_n)$ and $\sin(\pi L/\lambda_n)$ in Eq. (5) are related to the modal wavelength ratio of L/λ_n in the spatial domain. The bracketed term of Eq. (5) stands for the residual response of free vibration accumulated in the temporal domain after all the loads have left the beam. Thus, the residual response factor $\Re_n(\lambda_n, v, t)$ is basically a combination of the wave motions and dynamic vibrations in the spatiotemporal domain, from which two design criteria of span length for the HSR bridges can be derived, as will be presented in the following sections.

2.1. *Resonance parameters*

For a railway bridge subjected to moving trainloads, the resonance speed of $v_{r,1} = f_1 D$ is a key issue affecting the bridge design and performance of the HSR lines. As can be seen, the *shorter* the spacing D of the uniform moving loads, the *smaller the resonant speed* of the train will be. By substituting $v_{r,1}$ into Eq. (5), in turn into Eq. (3), and through the modal superposition in Eq. (2), the acceleration response calculated of the beam under vibration is[19]

$$\begin{aligned}\ddot{u}_{r,1}(x,t) = \frac{-2P}{mL}\Bigg\{\sum_{n=1}\Bigg[&\frac{S_{n,r1}}{1-S_{n,r1}^2} \times (S_{n,r1}\sin(\Omega_n t) - \sin(\omega_n t))H(t-t_N) \\ &- \underline{2\frac{S_{n,r1}}{1-S_{n,r1}^2}}\Re_{n,r1}(\lambda_{n,r1}, v, t) \times H\left(t - t_{N-1} - \frac{L}{v_{r,1}}\right)\Bigg] \times \sin\left(\frac{n\pi x}{L}\right)\Bigg\},\end{aligned} \tag{7}$$

where the modal resonance speed parameter $S_{n,r1}$ and loading wavelength $\lambda_{n,r1}$ are, respectively, given as follows:

$$S_{n,r1} = \frac{n\pi v_{r,1}}{\omega_n L} = \frac{D}{2nL}, \quad \lambda_{n,r1} = \frac{D}{n^2} \tag{8a,b}$$

and the corresponding residual response factor $\Re_{n,r1}$ is[19]

$$\Re_{n,r1}(\lambda_{n,r1}, v_{r,1}, t) = \begin{cases} (N-1)\cos\left(\dfrac{\pi L}{D/n^2}\right) \times \sin\left(\omega_n t - \dfrac{\pi L}{D/n^2}\right) & n = \text{odd}, \\ (N-1)\sin\left(\dfrac{\pi L}{D/n^2}\right) \times \cos\left(\omega_n t - \dfrac{\pi L}{D/n^2}\right) & n = \text{even}. \end{cases} \tag{9}$$

Table 2. Critical length ratios $(L/D)_c$.

$(L/D)_c$	$\lvert\cos(\pi L/D)\rvert_{L/D>1} = 1$
$\underline{0.68}$, $1.95 \simeq 2$, $2.97 \simeq 3$, $3.99 \simeq 4$	$L/D = 2, 3, 4$

The preceding equation reveals that once the first mode of the beam is excited in resonance, all the other higher modes ($n > 1$) will be set in resonance as well, as implied by the factor $(N - 1)$. For the *primary* resonance speed of $v_{r,1}(= f_1 D)$ and the fundamental mode of the beam, one can rewrite the *underlined* residual response factor in Eq. (7) as follows:

$$\left|\frac{S_{1,r1}}{1 - S_{1,r1}^2} \times \Re_{1,r1}(\lambda_{1,r1}, v_{r,1}, t)\right|_{S_{1,r1}=D/2L}$$

$$= \left|\frac{(2L/D)}{(2L/D)^2 - 1}\right| \times (N - 1)\left|\cos\left(\frac{\pi L}{D}\right)\right|. \tag{10}$$

By solving the maximum-value problem from Eq. (10), the *critical length ratios* $(L/D)_c$ for the *primary resonance* have been obtained and listed in Table 2. As can be seen, except the critical length ratio of 0.68 for short bridges, where $L/D < 1$,[33] the critical length ratios $(L/D)_c$ can also be estimated from the other condition of $|\Re_{1,r1}|_{\max} = N - 1$, as implied by $|\cos(\pi L/D)| = 1$, for $L/D > 1$ or for longer-span bridges.

For *short bridges* traversed by a train moving at very high (primary resonance) speeds, the *impact* effect may govern the response amplitude of the bridge due to smaller length ratio of $L/D < 1$, which implies larger $S_{r,1}$ but not greater than unity. In this case, larger dynamic amplification factor $S_{r,1}/(1-S_{r,1}^2)$ can be anticipated, as revealed by Eq. (10). In reality, the maximum primary resonance will not occur for the smaller critical length ratio of $L/D < 1$ encountered by short railway bridges. On the other hand, the primary resonance speed ($=f_1 D$) of a short bridge ($L \leqq 30\,\text{m}$)[14,33] is much higher than the operating or design speeds of current bullet trains,[34] only the cases of maximum sub-resonance need to be considered in real HSR lines. However, the resonance does not fully depend on the action time of each moving load on the bridge. Rather, it depends on the "phase" of the residual response induced by each passing load. If the residual response induced by a former (passing) moving load is "in-phase" with that of the following moving load, then resonance will certainly occur, even though they do not simultaneously exist on the bridge.

In contrast, for *long bridges* with length ratio $L/D > 1$ (which implies smaller $S_{r,1}$), the impact effect will be reduced and the in-phase resonant condition of $|\Re_{1,r1}|_{\max} = N - 1$, as implied by $|\cos(\pi L/D)| = 1$, may govern the *maximum* resonance of the bridge. Thus, a simply supported bridge with the length ratio of

$L/D > 1$ will be used in exploration of the resonance phenomenon of HSR bridges in the following study.

Instead of solving the maximum-value problem from Eq. (10) for the critical length ratio $(L/D)_c$, the single term $|\cos(\pi L/D)| = 1$ in Eq. (10) provides a physical means for using the *in-phase* resonance superposition to estimate the critical span length of the simple beam. Such a critical span length may be referred to as the *resonance wavelength* of a simple beam. From the viewpoint of dynamic vibrations and wave motions, when a serious of equally-spaced moving loads pass over a simple beam at the resonance speed ($=f_1 D$), waves of the sinusoidal form will be induced on the beam. If the integral multiple of the loading wavelength (i.e. $nD)_{n=1,2,\ldots}$ happens to be equal to the span length (L) of the primary wave component induced by each moving load, then the in-phase superposition of all these wave components will result in resonance amplification on the beam, which is called the *maximum* resonance. By using the resonance speed ($=f_1 D$) and the critical span length ($L_c = 2D$) in the exemplar analysis, we can explain why a large-amplitude resonance occurred on the 38 m SNCF bridge traversed by the TGV bullet trains in France.[5,6]

2.2. *Extraction of cancelation speeds*

For a simply supported beam with zero damping, the mid-span response is of great interest in the vibration assessment of the structure. By using Eq. (7), together with Eq. (8a,b), the mid-span displacement of the beam can be obtained as

$$u\left(\frac{L}{2}, t\right) = \frac{2P}{mL} \sum_{n=1,3,5,\ldots}^{\infty} \frac{1/\omega_n^2}{1 - S_n^2} \times \begin{bmatrix} (\sin(\Omega_n t) - S_n \sin(\omega_n t))H(t - t_{N-1}) \\ -2S_n \times \Re_n(\lambda_n, v, t)H(t - t_{N-1} - L/v) \end{bmatrix} \sin\frac{n\pi}{2}, \tag{11}$$

where

$$\Re_n(\lambda_n, v, t) = \cos\frac{\pi L}{\lambda_n} \times \left[\sin\left(\omega_n t - \frac{\pi L}{\lambda_n}\right) + \sin\omega_n\left(\omega_n\left(t - \frac{t_N}{2}\right) - \frac{\pi L}{\lambda_n}\right) \times \Upsilon_n(\lambda_n, v, D)\right]. \tag{12}$$

From the first modal cancellation condition of $\cos(\pi L/\lambda_1) = 0$ with $\lambda_1 = v_{c,1}/f_1$ in Eq. (12), the cancellation speed for the fundamental mode of the simple beam can be obtained as

$$v_{c,1} = \left.\frac{2f_1 L}{k}\right|_{k=1,3,5,\ldots}. \tag{13}$$

From the same condition, one can also obtain the following *modal* cancellation conditions for all the odd-modes (n) that results in $\Re_n(\lambda_n, v_{c,1}, t) = 0$:

$$\cos\frac{\pi L}{\lambda_n} = \cos\left(k\frac{n^2\pi}{2}\right)\Bigg|_{\substack{k=1,3,5,\ldots\\ n=1,3,5,\ldots}} = 0. \tag{14}$$

Consequently, when the Nth load left the beam at time $t_E = t_{N-1} + L/v$, one can find from Eq. (11) the following zero responses at the mid-span of the undamped beam:

$$u\left(\frac{L}{2}, t_E\right) = \frac{2P}{mL}\sum_{n=1,3,5,\ldots}^{\infty}\frac{1/\omega_n^2}{1-S_n^2}(\sin(n\pi) - S_n\sin(n^2\pi))\sin\frac{n\pi}{2} = 0, \tag{15a}$$

$$\dot{u}\left(\frac{L}{2}, t_E\right) = \frac{2P}{mL}\sum_{n=1,3,5,\ldots}^{\infty}\frac{1/\omega_n}{1-S_n^2}S_n(\cos(n\pi) - \cos(n^2\pi))\sin\frac{n\pi}{2} = 0, \tag{15b}$$

$$\ddot{u}\left(\frac{L}{2}, t_E\right) = -\frac{2P}{mL}\sum_{n=1,3,5,\ldots}^{\infty}\frac{S_n}{1-S_n^2}(S_n\sin(n\pi) - \sin(n^2\pi))\sin\frac{n\pi}{2} = 0. \tag{15c}$$

As indicated in Eq. (15), even after all the moving loads have left the beam at the cancellation speed, no *residual response* of free vibration occurs at the mid-span of the undamped beam.[48] Obviously, the *modal* cancellation condition of $\cos(\pi L/\lambda_1). = 0$ is a key factor for the selection of optimal span length for response suppression of a simple beam under multi-moving loads.

On the other hand, an observation of the vanishing the dynamic factor $\Upsilon_n(\lambda_n, D) = 0$ from Eqs. (5) and (6) yields another cancellation condition related to the load–interval (D) and number (N) of moving loads.[31,36,56] Such a *special* cancellation may occur only if both the conditions of $\pi D/\lambda_n \notin \mathbb{N}$ (i.e. $\sin(\pi D/\lambda_n) \neq 0$) and $(N_s - 2)D/\lambda_n \in \mathbb{N}$ (i.e. $\sin((N_s - 2)\pi D/\lambda_n) = 0$) are satisfied simultaneously. Here, $\mathbb{N}$ is denoted as the nature numbers and N_s as the *specific group number* of the moving loads that have left the beam.[31] Considering that it is difficult to accommodate trains with different numbers of train cars, the *conditional* cancellation that relates strongly to the wheel–load number N_s will be excluded in searching for the optimal span lengths of a railway bridge against resonance in this study.

2.3. *Optimal design criterion*

The dynamic response expression in Eq. (12) reveals that once the term $\cos(\pi L/\lambda_n)$ related to the modal shapes of the beam[13] vanishes, no residual response (including resonance) will occur at the mid-span of the beam. By letting the primary resonant speed of $v_{r,1} = f_1 D$ be identical to the *modal* cancellation speed of Eq. (11), the *optimal length ratio* $(L/D)_{\text{opt}}$ of the beam under the multi-moving loads can be derived as

$$(L/D)_{\text{opt}} = k/2|_{k=1,3,5,\ldots} \tag{16}$$

which is equivalent to the condition first presented in Ref. 13. This condition is featured by the fact that it is independent of the speed of the moving loads. For the case with length ratio of $L/D > 1$, one can find from Eq. (16) that the optimal length ratio $(L/D)_{\text{opt}}$ is equal to 1.5 (or 2.5) for simply supported beams.[13] It means that even for a train traveling over a simple beam at the resonant speed, the residual (or free) vibrations will be cancelled out after all the moving axle loads of the train have left the beam. This has been referred to as the optimal span length of a simply supported HSR bridge.[13] To demonstrate the effectiveness of the optimal span length ($L_{\text{opt}} = 1.5D$) for real train–bridge systems in resonance suppression, some parametric studies will be carried out in the following sections.

3. Minimum Cancellation and Maximum Resonance

Figure 3 shows a single-span simple beam subjected to a traditional train moving at a constant speed v, of which each car is supported by two bogies. For simplicity, the train is assumed to be composed of 16 identical cars of length D and the train is idealized as a series of moving constant loads with regular intervals, as shown in Fig. 3. Based on the HSR available in the literature,[4–6,10,14,20,29,35,36,55–57] the properties of the train loads are taken as follows: $P = 122.5\,\text{kN}$, $D = 25\,\text{m}$, $d_1 = 15\,\text{m}$ and $d_2 = 2.5\,\text{m}$. For illustration, the empirical structural properties of the mass per unit length (m),[10,13] damping ratio ξ[14] and vertical flexural frequency f_1[14] of the simply supported concrete bridges used for the HSR lines are specified as follows:

$$m = 30 + 0.2L(\text{t/m}), \quad \xi = \begin{cases} 0.015 & L \geq 20\,\text{m}, \\ 0.015 + 0.07(20 - L) & 0 < L < 20\,\text{m}, \end{cases} \tag{17a,b}$$

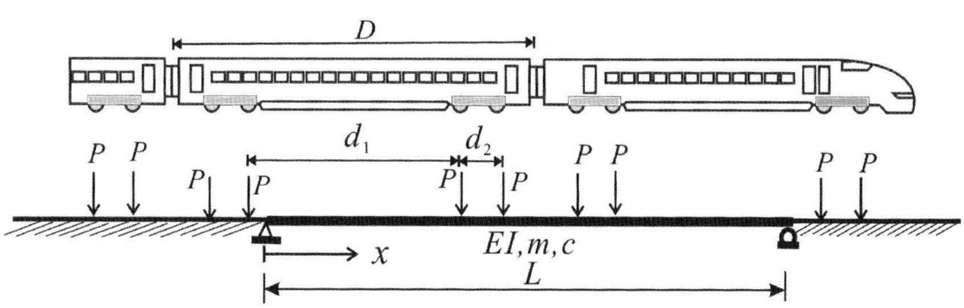

Fig. 3. Schematic of a simple beam to a series of moving trainloads.

$$f_1 = \begin{cases} \dfrac{1}{2}\left(\dfrac{23.58}{L^{0.592}} + \dfrac{94.76}{L^{0.748}}\right) & L \geq 20\,\text{m}, \\ \dfrac{1}{2}\left(\dfrac{80}{L} + \dfrac{94.76}{L^{0.748}}\right) & 0 < L < 20\,\text{m}. \end{cases} \quad (\text{Hz}) \qquad (17\text{c})$$

In Eq. (17c), the empirical frequency f_1 is referred to as the mean value of the upper and lower frequency limits of the concrete railway bridges proposed by Eurocode.[14]

For theoretical demonstration, the VBI effects on a bridge will be neglected and only the primary resonant speed ($v_{r,1} = f_1 D$) are considered in the dynamic analysis of a simple beam subjected to the train loads, as shown in Fig. 3. By ranging the length ratio of the bridge to the car as $1.25 \leq L/D \leq 2.5$, the primary resonance and cancellation speeds have been calculated and listed in Table 3. Particularly, it can be seen that the modal cancellation speeds are identical to the primary resonance ones for the specific (optimal) length ratios of 1.5 and 2.5. For illustration, the maximum acceleration a_{max} computed for the whole beam has been plotted with respect to the axis x/L of the beam and the length ratio L/D in Fig. 4. Such a tri-axial response plot has been referred to as the $a_{\text{max}} - x/L - L/D$ plot.[13]

The tri-axial maximum response plot in Fig. 4 indicates that the resonance amplitudes of acceleration spanned for the length ratios of $1.25 \leq L/D < 1.5$ are much greater than those of $1.5 \leq L/D \leq 2.5$ due to the train-induced impact effects and higher resonance speeds over 400 km/h (see Table 3). Moreover, the *minimum* response amplitudes of cancellation at the optimal length ratios of $(L/D)_{\text{opt}} = 1.5$ and 2.5 are significantly lower than the *maximum* resonance amplitudes at the critical length ratio of $(L/D)_c = 2$, even though the resonance speed of 408 km/h for the optimal span length of $(L/D)_{\text{opt}} = 1.5$ is much higher than that of 333 km/h for the resonance condition at $(L/D)_c = 2$. In addition, some further insights can be obtained from the tri-axial $a_{\text{max}} - x/L - L/D$ response plot given

Table 3. Resonance/cancellation speeds ($v \geqq 200$ km/h) versus length ratio (L/D).

L/D	1.25	1.5	2.0	2.5
L (m)	31.25	37.5	50	62.5
f_1 (Hz)	5.15	4.53	3.70	3.17
$v_{r,1}$ (km/h)	463	408[a]	333[b]	285[a]
$v_{c,1}$ (km/h)	—	408[a]	—	285[a]

[a]Cancellation at primary resonanc
[b]Maximum resonance.

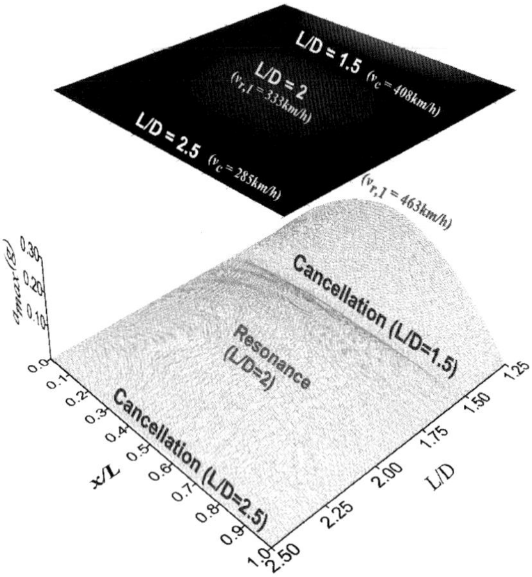

Fig. 4. $a_{\max} - x/L - L/D$ plot of primary resonance for simple beams.

in Fig. 4:

(1) *Minimum* response amplitudes at the primary resonance do take place at the optimal length ratios of $(L/D)_{\text{opt}} = 1.5$ and 2.5, for which the resonance speeds coincide with the theoretical cancellation speeds listed in Table 3;
(2) Under the resonance conditions of $v_{r,1} = f_1 D$ and those implied by $|\cos(\pi L/D)| = 1$, the *maximum* resonance amplitude takes place at the critical length ratio of $(L/D)_c = 2$. Such a critical resonance amplification explains why a large-amplitude resonance vibration took place at the 38 m SNCF bridge traveled by a moving TGV train in France;[5,6]
(3) For a bridge with smaller length ratios $L/D < 1.5$, the train-induced impact effect on the bridge results in significant peak amplification, as indicated by the tri-axial response plot in Fig. 4;
(4) The maximum acceleration amplitudes along the axis x/L of the beam are mostly dominated by the first mode of the simple beams, even though there may exist some minor peaks caused by the higher modes excited.[19]

According to foregoing parametric analysis on the *maximum* resonance and *minimum* cancellation, the aspect length ratio L/D of the bridge to the car dominates how the vibration scenario will be excited on a railway bridge under a moving train. In engineering practice, the definition of the span length L is clear. Mathematically, the car length D is the characteristic distance between two consecutive loads. Since for a train, all the cars are linked by the linking devices, which also add some length.

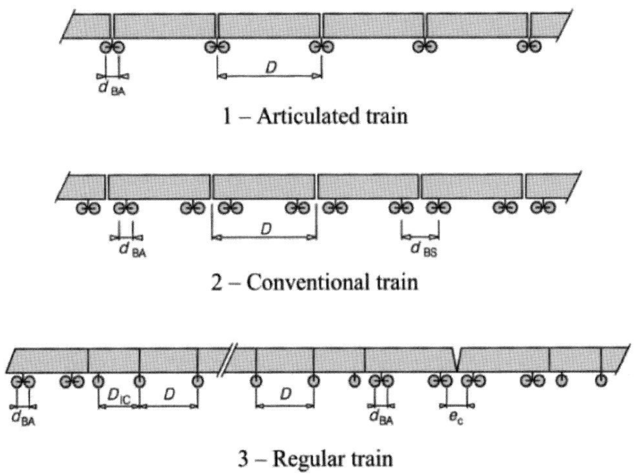

Fig. 5. Different types of highspeed trains according to Eurocode 1.[14]

From the point of mechanics, the distance D should be recognized as the center-to-center distance of each car, or the distance between two front bogies or that between two rear bogies, as defined by Eurocode 1 in Fig. 5. For the various types of typical highspeed trains (conventional, articular and regular coaches)[14] used by different countries in the world, the corresponding optimal and critical lengths calculated of simply supported bridges have been complied for a quick reference in Table 4. It should be noted that the lengths given in the table have been obtained from sources available to the authors, which are not affirmative for the particular car model mentioned herein. Rather, it should be adjusted according to the real car length used by each train model in practice.

According to the length ratio analysis of the bridge to the car shown in Table 4 and real construction of a single-span simply supported bridge, an optimal beam of 40 m length is proposed for the conventional trains, such as CRH380D (China),

Table 4. Optimal (L_{opt}) and critical (L_c) span lengths of simply supported bridges versus bullet trains.[14,32,36]

Train model (D = regular spacing or passenger coach length)	L_{opt} $(L/D)_{opt} = 1.5, 2.5$	L_c $(L/D)_c = 2, 3$
AVE/TGV/Eurostar ($D = 18.7$ m)[a]	28.05 m, 46.75 m	37.4 m, 56.1 m
CRH380D ($D = 26.6$ m)[b]	39.9 m	53.2 m
ICE2 ($D = 26.4$ m)[b]	39.6 m	52.8 m
SKS ($D = 25$ m)[b]	37.5 m	50 m
TALGO AV ($D = 13.14$ m)[c]	19.71, 32.85	26.28, 39.42, 52.56

[a] Articulated train with each two coaches sharing one bogie.
[b] Conventional train with one coach resting on two bogies.
[c] Regular train with one axle per coach.

ICE (Germany) and SKS (Japan) bullet trains, and a beam of 30 m length for the articulated trains such as TGV (France) and Eurostar highspeed trains. In short, for a given HSR system, it is recommended that the optimal length ratio $(L/D)_{\text{opt}} = 1.5$, for which the cancellation effect is enforced, be used for minimizing the vibration of simply supported bridges, but the critical span length ($L_c = 2D$) that may result in the *maximum* resonance be avoided at all times in the design of the HSR bridges.

4. Conclusions

The rapid growth of intercity transportation speeds up the construction and upgrading of highspeed trains, for instance, the new CR450 ultrafast train can run at speeds over 400 km/h in China.[37] Meanwhile, to raise the structural stiffness so as to increase the natural frequencies of HSR bridges and keep them away from the resonance speeds is not always feasible in avoiding or reducing undesired excessive vibrations on train–bridge systems. Given the car length D of a train, the proposed optimal design criterion of $(L/D)_{\text{opt}} = 1.5$ that utilizes the modal cancellation phenomenon to suppress the primary resonance is a simple and feasible means to the design of HSR bridges. In contrast, the critical span length of $L_c = 2D$ (for articulated trains) or $3D$ (for regular trains) should always be avoided for the *maximum* resonance exerted on the bridges, which is harmful to the long-term ballast stability and rail-fastener-track maintenance of a railway system. Concerning future studies on resonance and cancellation problems of railway bridges, focus will be placed on (1) optimal span length of multi-span continuous bridges; (2) suitable optimal length on suppression of sub-resonance of short-span bridges; and (3) proposal of an indicator for railway engineers to select an appropriate bridge span against resonance or sub-resonance in the preliminary design stage.

From the viewpoint of bridge dynamics, the proposed optimal design criterion ($L_{\text{opt}} = 1.5D$) embodies a vibrational connotation of "overcoming resonance by cancellation," which is equivalent to the Chinese proverb of "using the static to overcome the dynamic." (以靜制動) Such an idea can easily improve the performance of railway bridges via vibration suppression.

Acknowledgment

The research reported herein was supported in part by grants from: National Natural Science Foundation of China (Grant No. 52078082), Chongqing Science and Technology Commission (Grant Nos. cstc2021yszx-jscxX0001, cstc2020yszx-jscxX0002, and cstc2019yszx-jcyjX0001), China State Railway Group Co., Ltd. (Grant No. K2019G036), the National Science and Technology Council (Grant No. NSTC 111-2221-E-032-033) in Taiwan and the Taiwan-Czech joint project (Grant No. MOST 110-2923-E-032-001-MY3 and GACR 21-32122J).

References

1. L. Frýba, *Vibration of Solids and Structures under Moving Loads*, 3rd ed. (Thomas Telford Ltd., London, 1999).
2. S. P. Timoshenko, *History of Strength of Materials: With a Brief Account of the History of Theory of Elasticity and Theory of Structures* (McGraw-Hill, New York, 1953).
3. V. V. Bolotin, *The Dynamic Stability of Elastic Systems* (Holden-Day, Inc., San Francisco, 1964).
4. A. Matsuura, A study of dynamic behavior of bridge girder for high speed railway, *Proc. Japan Soc. Civil Eng.* **256** (1976) 35–47 (in Japanese).
5. P. Ramondenc, Vom Einfluß hoher Geschwindigkeiten auf den Entwurf von Eisenbahnbrucken am Beispiel der Stahlbrücken auf der Hochgeschwindigkeitsstrecke (HGS) des TGV Méditerrannée, *Stahlbau* **67**(8) (1998) 652–658 (in German).
6. L. Frýba, A rough assessment of railway bridges for high speed trains, *Eng. Struct.* **23**(5) (2001) 548–556.
7. J. M. Goicolea-Ruigómez, Service limit states for railway bridges in new design codes IAPF and Eurocodes, in *Workshop on Track-Bridge Interaction on High-Speed Railways*, 15–16 October, 2007, Portugal, Porto.
8. ERRI D214, Design of Railway Bridges for Speed up to 350 km/h; Dynamic Loading Effects Including Resonance (Technical report, European Railway Research Institute, 1998).
9. CEN. EN 1991-2, Eurocode 1: Actions on Structures-Part 2: Traffic Loads on Bridges (European Committee for Standardization, Brussels, 2003).
10. Y. B. Yang, J. D. Yau and Y. S. Wu, *Vehicle-Bridge Interaction Dynamics: With Applications to High-Speed Railways* (World Scientific, Singapore, 2004).
11. Y. B. Yang and B. H. Lin, Vehicle-bridge interaction analysis by dynamic condensation method, *J. Struct. Eng.* **121**(11) (1995) 1636–1643.
12. Y. B. Yang, S. S. Liao and B. H. Lin, Impact formulas for vehicles moving over simple and continuous beams, *J. Struct. Eng.* **121**(11) (1995) 1644–1650.
13. Y. B. Yang, J. D. Yau and L. C. Hsu, Vibration of simple beams due to trains moving at high speeds, *Eng. Struct.* **19**(11) (1997) 936–944.
14. J. D. Yau, Y. B. Yang and S. R. Kuo, Impact response of high speed rail bridges and riding comfort of rail cars, *Eng. Struct.* **21**(9) (1999) 836–844.
15. E. Savin, Dynamic amplification factor and response spectrum for the evaluation of vibrations of beams under successive moving loads, *J. Sound Vib.* **248** (2001) 267–288.
16. P. Museros and E. Alarcón, Influence of the second bending mode on the response of high-speed bridges at resonance, *J. Struct. Eng.* **131**(3) (2005) 405–415.
17. J. R. Cho, K. Jung, K. Cho, J. W. Kwark, Y. J. Kim and B. S. Kim, Determination of the optimal span length to minimize resonance effects in bridges on high-speed lines, *Proc. IMechE F: J. Rail Rapid Transit.* **230**(2) (2016) 335–344.
18. Y. B. Yang and Y. S. Wu, Dynamic stability of trains moving over bridges shaken by earthquakes, *J. Sound Vib.* **258**(1) (2002) 65–94.
19. J. D. Yau and Y. B. Yang. Vertical accelerations of simple beams due to successive loads traveling at resonant speeds, *J. Sound Vib.* **289**(1–2) (2006) 210–228.
20. Y. B. Yang and J. D. Yau, Vertical and pitching resonance of train cars moving over a series of simple beams, *J. Sound Vib.* **337** (2015) 135–149.
21. Y. B. Yang, J. D. Yau and S. Urushadze, Wave transmission of linked railcars moving over multi simple beams under dual resonance, *J. Sound Vib.* **452** (2019) 51–57.

22. J. D. Yau, Y. S. Wu and Y. B. Yang, Impact response of bridges with elastic bearings to moving loads, *J. Sound Vib.* **248**(1) (2001) 9–30.
23. Y. B. Yang, C. L. Lin, J. D. Yau and D. W. Chang, Mechanism of resonance and cancellation for train-induced vibrations on bridges with elastic bearings, *J. Sound Vib.* **269**(1–2) (2004) 345–360.
24. Y. B. Yang, C. M. Wu and J. D. Yau, Dynamic response of a horizontally curved beam subjected to vertical and horizontal moving loads, *J. Sound Vib.* **242**(3) (2001) 519–537.
25. Y. B. Yang, M. Li, B. Zhang, Y. T. Wu and J. P. Yang, Resonance and cancellation in torsional vibration of monosymmetric I-sections under moving loads, *Int. J. Struct. Stab. Dyn.* **18**(9) (2018) 1850111.
26. J. D. Yau, Response of a train moving on multi-span railway bridges undergoing ground settlement, *Eng. Struct.* **31** (2009) 2115–2122.
27. J. W. Kwark, E. S. Choi and Y. J. Kim, Dynamic behavior of two-span continuous concrete bridges under moving high-speed train, *Comput. Struct.* **82**(4–5) (2004) 463–474.
28. P. Museros, E. Moliner and M. D. Martinez-Rodrigo, Free vibrations of simply-supported beam bridges under moving loads: Maximum resonance, cancellation and resonant vertical acceleration, *J. Sound Vib.* **332** (2013) 326–345.
29. H. Xia, H. L. Li, W. W. Guo and G. De Roeck, Vibration resonance and cancellation of simply-supported bridges under moving train loads, *J. Eng. Mech.* **140**(5) (2014) 04014015-1–04014015-11.
30. C. P. Sudheesh-Kumar, C. Sujathan and K. Shankar, Vibration of simply supported beams under a single moving load: A detailed study of cancellation phenomenon, *Int. J. Mech. Sci.* **99** (2015) 40–47.
31. J. Li, H. Zhang, D. Zhu and C. Li, A moving load amplitude spectrum for analyzing the resonance and vibration cancellation of simply supported bridges under moving loads, *Euro. J. Mech. A Solids* **92** (2022) 104428.
32. National Railway Administration (NRA), TB 10621-2014, Code for Design of High Speed Railway (Beijing, China, 2015).
33. J. D. Yau, M. D. Martínez-Rodrigo and A. Doménech, An equivalent additional damping approach to assess vehicle-bridge interaction for train-induced vibration of short-span railway bridges, *Eng. Struct.* **188** (2019) 469–479.
34. A. Doménech, P. Museros and M. D. Martínez-Rodrigo, Influence of the vehicle model on the prediction of the maximum bending response of simply-supported bridges under high-speed railway traffic, *Eng. Struct.* **72**(1) (2014) 123–139.
35. P. Lou, J. D. Yau, F. T. K. Au and S. Urushadze, Shifted resonance of railway bridges under trains passing by each other, *Int. J. Struct. Stab. Dyn.* **22** (2022) 2271001.
36. H. Xia, N. Zhang and W. Guo, *Dynamic Interaction of Train-Bridge Systems in High-Speed Railways* (Springer, Germany, 2018).
37. F. Chen, China revs new high-speed train building spree, *Asia Times*, March 8, 2021.
38. M. Klasztorny and J. Langer, Dynamic response of single-span beam bridges to a series of moving loads, *Earthq. Eng. Struct. Dyn.* **19** (1990) 1107–1124.
39. J. D. Yau, Resonance of continuous bridges due to high speed trains, *J. Mar. Sci. Tech.* **9**(1) (2001) 14–20.
40. S. H. Ju and H. T. Lin, Resonance characteristics of high-speed trains passing simply supported bridges, *J. Sound Vib.* **267**(5) (2003) 1127–1141.

41. H. Xia, N. Zhang and W. W. Guo, Analysis of resonance mechanism and conditions of train-bridge system, *J. Sound Vib.* **297** (2006) 810–822.
42. F. Vestroni and S. Vidoli, Closed-form solutions for the structural response to train loads, *J. Sound Vib.* **303** (2007) 691–706.
43. W. Lacarbonara and V. Colone, Dynamic response of arch bridges traversed by high-speed trains, *J. Sound Vib.* **304** (2007) 72–90.
44. J. M. Goicolea-Ruigómez and P. Antolin, Dynamics of high-speed railway bridges: Review of design issues and new research for lateral dynamics, *Int. J. Rail. Tech.* **1**(1) (2012) 27–55.
45. L. Mao and Y. Lu, Critical speed and resonance criteria of railway bridge response to moving trains, *J. Bridge. Eng.* **18** (2013) 131–41.
46. T. Arvidsson and R. Karoumi, Train–bridge interaction — a review and discussion of key model parameters, *Int. J. Rail. Transp.* **2**(3) (2014) 147–186.
47. Y. B. Yang and J. D. Yau, Resonance of high-speed trains moving over a series of simple or continuous beams with non-ballasted tracks, *Eng. Struct.* **143** (2017) 295–305.
48. P. Museros and E. Moliner, Comments on "Vibration of simply supported beams under a single moving load: A detailed study of cancellation phenomenon," *Int. J. Mech. Sci.* **128–129** (2017) 709–713.
49. W. Zhai, Z. Han, Z. Chen, L. Ling and S. Zhu, Train–track–bridge dynamic interaction: A state-of-the-art review, *Veh. Sys. Dyn.* **57**(7) (2019) 984–1027.
50. M. D. Martínez-Rodrigo, A. Andersson, C. Pacoste and R. Karoumian, Resonance and cancellation phenomena in two-span continuous beams and its application to railway bridges, *Eng. Struct.* **222** (2020) 111103.
51. A. Zangeneh, P. Museros, C. Pacoste and R. Karoumi, Free vibration of viscoelastically supported beam bridges under moving loads: Closed-form formula for maximum resonant response, *Eng. Struct.* **244** (2021) 112759.
52. L. Auersch, Resonances of railway bridges analysed in frequency domain by the modal-force-excitation, bridge-transfer and axle-sequence spectra, *Eng. Struct.* **249** (2021) 113282.
53. Y. J. Wang, J. D. Yau, J. Shi and S. Urushadze, Double sub-resonance mechanism in torsional–flexural vibrations of a double-track short bridge under a moving train, *Int. J. Struct. Stab. Dyn.* **22**(16) (2022) 2271005.
54. L. Xin, D. Mu, D. H. Choi, X. Li and F. Wang, General conditions for the resonance and cancellation of railway bridges under moving train loads, *Mech. Syst. Sign. Proc.* **183** (2023) 109589.
55. Q. Li, X. Li and Q. Wu, Resonance analysis of cantilever plates subjected to moving forces by a semi-analytical method, *Int. J. Struct. Stab. Dyn.* **20**(4) (2020) 2050049.
56. Y. F. Duan, S. M. Wang and J. D. Yau, Vector form intrinsic finite element method for analysis of train–bridge interaction problems considering the coach-coupler effect, *Int. J. Struct. Stab. Dyn.* **19**(2) (2019) 1950014.
57. V. Poliakov, N. Zhang, V. Saurin and N. T. Dang, Running safety of a high-speed train within a bridge zone, *Int. J. Struct. Stab. Dyn.* **20**(11) (2020) 2050116.

https://doi.org/10.1142/9789819814770_0008

Chapter 8

Analytical Calculation of Static Deflection of Biperiodic Stepped Euler–Bernoulli Beam[#]

Yuchen Li[*,†], Isaac Elishakoff[*] and Noël Challamel[†,‡]

[*]*Department of Ocean and Mechanical Engineering*
Florida Atlantic University
Boca Raton, FL 33431-0991, USA

[†]*Université de Bretagne Sud*
University of South Brittany
CNRS UMR 6027, IRDL, Centre de Recherche
Rue de Saint Maudé, BP 92116
56321 Lorient Cedex, France
[‡]*noel.challamel@univ-ubs.fr*

In this paper, we investigate the lateral deflection of a simply supported periodic stepped beam under uniform load by using an analytical method. This study considers each element of the biperiodic stepped beam as a Euler–Bernoulli beam. By using the local coordinates alongside with the boundary and continuity conditions, the different coefficients for each element caused by the jump of the bending rigidity are calculated. The continuous deflection problem of the multi-stepped repetitive beam is formulated as a linear first-order difference equation with second member. With these coefficients, the deflection at mid-span of the biperiodic beam is analytically found in exact form. This deflection is satisfactory compared to the results of a finite element model based on beam discretization techniques using Hermitian cubic shape functions. The normalized deflection at mid span converges non-monotonically towards the homogenization beam model based on equivalent homogenized stiffness.

Keywords: Beam theory; biperiodic structures; difference equations; homogenized methods; gradient elasticity; static deflection; scale effects; stepped beams; higherorder homogenization.

1. Introduction

Periodic structures are often present in various fields of engineering, including laminated media or composite materials. Among periodic structures in structural engineering, continuous elastic beams on periodic supports have been extensively

[‡]Corresponding author.
[#]This chapter appeared previously on the International Journal of Structural Stability and Dynamics. To cite this chapter, please cite the original article as the following: Y. Li, I. Elishakoff & N. Challamel, *Int. J. Struct. Stab. Dyn.*, **23**, 2340009 (2023), DOI: 10.1142/S0219455423400096.

studied. Exact solutions may be derived for the bending of continuous elastic beams on periodic supports by solving equivalent linear difference equations.[1] Another class of periodic structures is the so-called biperiodic structure composed of alternate layers of two elastic materials (periodic composite material). For instance, Ref. 2 presents an application of the higher-order asymptotic homogenization method (equivalent continuous model) to study the wave dispersion in periodic composite materials. In this last paper, longitudinal waves in a composite rod and transverse anti-plane shear waves in a fiber-reinforced composite with a square lattice of cylindrical inclusions are studied from equivalent homogenized continuous models and compared to the exact biperiodic composite problem.

Biperiodic stepped beam which belongs to the field of periodic structure is commonly used in civil engineering, mechanical engineering and aeronautical engineering in order to implement different function or control self-weight. Compared to the vibration analysis of biperiodic structures,[3,4] there are less research of static behaviors but there are still research works using different methods in solving stepped beams statics problem. For example, Ref. 5 investigates the application of Galerkin's method[6] to evaluate the deflection of stepped beams. In their study, they deal with the simplest possible problem which is a stepped beam subjected to the uniformly distributed load. It is shown in this paper that the rigorous implementation of the Galerkin method converges to the exact solution, whereas the naïve Galerkin approach does not. Reference 7 verifies modified Rayleigh method which is using a new method for calculating the equivalent moment of inertia of stepped beam as the classic Rayleigh method it is not accurate when the beam is a stepped beam which is also show in Ref. 8. In Ref. 7, they calculate the static deflection of rectangular beam with stepping in width only and rectangular beam with stepping in height only and compare the results of classical Rayleigh method, modified Rayleigh method and finite element method (FEM). They prove that the modified Rayleigh method is better than the classical Rayleigh method for calculating the static deflection of stepped cantilever beam. Reference 9 presents to derivation of the analytical deflection function of a stepped beam using singularity functions made of the Dirac delta function. In their study, the derivation of the deflection function of a stepped beam is achieved in two steps. In the first step, the applied load is formulated and integrated twice to obtain the distribution of bending moment. The moment distribution is then multiplied by the reciprocal of the rigidity distribution to form the right-hand side of a second-order differential equation of beam deflection.

Local coordinate system may be used in the resolution of stepped beam problem as it allows coordinate change along with the jump of the properties. Reference 10 makes a numerical evaluation of high-order modes of a uniform Euler–Bernoulli beam. In this study, using local coordinates significantly simplifies the frequency determinant of vibration equations of the stepped beam, largely eliminates numerical errors in high-order mode evaluation. Reference 11 uses local coordinate systems-based method to analyze high-order modes of vibration of stepped Timoshenko beam. The local coordinate system also simplifies the frequency equation for the vibration of stepped Timoshenko beam, making it possible to obtain high-order modes.

The static behavior of the biperiodic beam asymptotically converges towards the one of the homogenized beam, with a homogenized stiffness which is the harmonic averaging of each weighted stiffness. This has been shown rigorously for rods in Ref. 12 (see also Refs. 13–16). This is shown for beams by Ref. 17 or Ref. 18. Reference 19 deals with the deduction of new homogenized models for the flexural wave in bi-periodic beams. Two higher-order homogenization models are deduced by asymptotic expansion method to describe the free vibration and forced vibration of a periodic beam. This is also known as the so-called Reuss averaging which is the simple arithmetical averaging of the component compliances (see also Ref. 20). The homogenization method is widely used in the research of heterogeneous media as rigidity theorems allow to derive global properties of a deformation from local ones. The homogenization method is also used in composite material research. Reference 21 develops an enhanced asymptotic homogenization method which is from the partial asymptotic expansion and the third-order shear deformation theory. This method can be applied to the model in which the periodic condition through the thickness direction is not assumed.

In this paper, we firstly define a local coordinate system. Then by applying the continuity conditions and the boundary conditions, the deflection functions are found for each element of the repetitive stepped beam from the resolution of a linear difference equation.[21] The deflection in the middle of the beam is compared with those got by a beam finite element model and with a homogenization method in order to check the deflection solution for a large number of bi-periodic elements.

2. Definition of Local Coordinate System and Calculation of Coefficients

The model to be analyzed is shown in Fig. 1. Here we call any blue beam with a stiffness D_1 and a length L_1 or red beam with a stiffness D_2 and a length L_2 an element. We call the combination of a blue beam with a red beam next to it a cell. Totally, there are N cells and $2N$ element.

For each element, the bending moment is calculated as

$$M = \frac{1}{2}qLx - \frac{1}{2}qx^2. \tag{1}$$

The relationship between the local coordinate $x_1^{(i)}$ within blue element and the general coordinate x in longitudinal direction for blue beams is

$$x = x_1^{(i)} + (i-1)(L_1 + L_2). \tag{2}$$

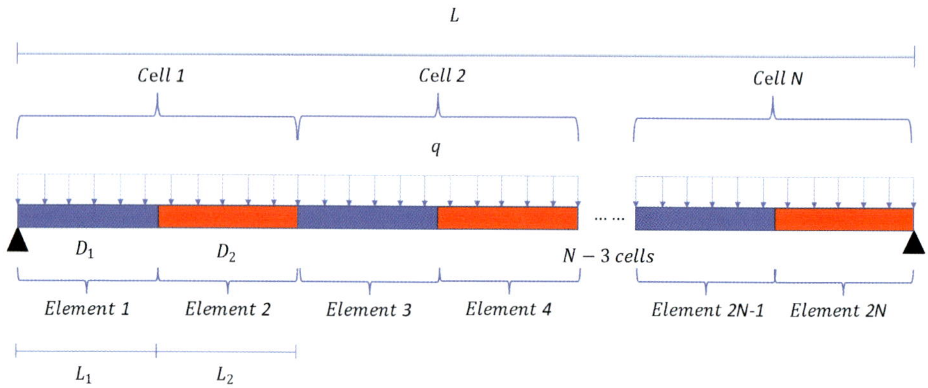

Fig. 1. Biperiodic stepped Euler–Bernoulli beam to be analyzed under a uniform charge.

The relationship between the local coordinate $x_2^{(i)}$ within red element and the general coordinate x in longitudinal direction for red beams is

$$x = x_2^{(i)} + (i-1)(L_1 + L_2) + L_1, \tag{3}$$

where i is the cell serial number.

The expressions of bending moment, slope and deflection for each blue element are given by

$$D_1 w_1^{(i)\prime\prime}(x) = M_1^{(i)} = \frac{1}{2}qLx - \frac{1}{2}qx^2, \tag{4}$$

$$D_1 w_1^{(i)\prime}(x) = \frac{1}{4}qLx^2 - \frac{1}{6}qx^3 + A_1^{(i)}, \tag{5}$$

$$D_1 w_1^{(i)}(x) = \frac{1}{12}qLx^3 - \frac{1}{24}qx^4 + A_1^{(i)}x + A_2^{(i)}, \tag{6}$$

where $A_1^{(i)}$ and $A_2^{(i)}$ are coefficients depending on the cell serial number i.

The expressions of bending moment, slope and deflection for each red element are given by

$$D_2 w_2^{(i)\prime\prime}(x) = M_2^{(i)} = \frac{1}{2}qLx - \frac{1}{2}qx^2, \tag{7}$$

$$D_2 w_2^{(i)\prime}(x) = \frac{1}{4}qLx^2 - \frac{1}{6}qx^3 + B_1^{(i)}, \tag{8}$$

$$D_2 w_2^{(i)}(x) = \frac{1}{12}qLx^3 - \frac{1}{24}qx^4 + B_1^{(i)}x + B_2^{(i)}, \tag{9}$$

where $B_1^{(i)}$ and $B_2^{(i)}$ are coefficients depending on the cell serial number i.

In each cell, with the coordinates given in Eqs. (2) and (3), the continuity condition of slope gives

$$w_1^{(i)\prime}[x = (i-1)(L_1 + L_2) + L_1] = w_2^{(i)\prime}[x = (i-1)(L_1 + L_2) + L_1]. \tag{10}$$

By combining Eq. (10) with Eqs. (5) and (8), one gets

$$A_1^{(i)} = \left(\frac{D_1}{D_2} - 1\right)\left\{\frac{1}{4}qL[(i-1)(L_1 + L_2) + L_1]^2 - \frac{1}{6}q[(i-1)(L_1 + L_2) + L_1]^3\right\} + \frac{D_1}{D_2}B_1^{(i)}. \tag{11}$$

In each cell, with the coordinates given in Eqs. (2) and (3), the continuity condition of deflection gives

$$w_1^{(i)}[x = (i-1)(L_1 + L_2) + L_1] = w_2^{(i)}[x = (i-1)(L_1 + L_2) + L_1]. \tag{12}$$

By combining Eq. (12) with Eqs. (6) and (9):

$$\begin{aligned}&\frac{1}{D_1}\left\{\frac{1}{12}qL[(i-1)(L_1 + L_2) + L_1]^3 - \frac{1}{24}q[(i-1)(L_1 + L_2) + L_1]^4 + A_1^{(i)}[(i-1)(L_1 + L_2) + L_1] + A_2^{(i)}\right\}\\ &= \frac{1}{D_2}\left\{\frac{1}{12}qL[(i-1)(L_1 + L_2) + L_1]^3 - \frac{1}{24}q[(i-1)(L_1 + L_2) + L_1]^4 + B_1^{(i)}[(i-1)(L_1 + L_2) + L_1] + B_2^{(i)}\right\}.\end{aligned} \tag{13}$$

By coupling Eq. (11) with Eq. (13), one obtains

$$A_2^{(i)} = \left(\frac{D_1}{D_2} - 1\right)\left\{\frac{1}{8}q[(i-1)(L_1 + L_2) + L_1]^4 - \frac{1}{6}qL[(i-1)(L_1 + L_2) + L_1]^3\right\} + \frac{D_1}{D_2}B_2^{(i)}. \tag{14}$$

Between cells, with the coordinates given in Eqs. (2) and (3), the continuity condition of slope gives

$$w_1^{(i+1)\prime}[x = i(L_1 + L_2)] = w_2^{(i)\prime}[x = i(L_1 + L_2)]. \tag{15}$$

By combining Eq. (15) with Eqs. (5) and (8):

$$A_1^{(i+1)} = \left(\frac{D_1}{D_2} - 1\right)\left\{\frac{1}{4}qL[i(L_1 + L_2)]^2 - \frac{1}{6}q[i(L_1 + L_2)]^3\right\} + \frac{D_1}{D_2}B_1^{(i)}. \tag{16}$$

Between cells, with the coordinates given in Eqs. (2) and (3), the continuity condition of deflection gives

$$w_1^{(i+1)}[x = i(L_1 + L_2)] = w_2^{(i)}[x = i(L_1 + L_2)]. \tag{17}$$

By combining Eq. (17) with Eqs. (6) and (9):

$$\frac{1}{D_1}\left\{\frac{1}{12}qL[i(L_1+L_2)]^3-\frac{1}{24}q[i(L_1+L_2)]^4+A_1^{(i+1)}[i(L_1+L_2)]+A_2^{(i+1)}\right\}$$
$$=\frac{1}{D_2}\left\{\frac{1}{12}qL[i(L_1+L_2)]^3-\frac{1}{24}q[i(L_1+L_2)]^4+B_1^{(i)}[i(L_1+L_2)]+B_2^{(i)}\right\}. \tag{18}$$

By coupling Eq. (16) with Eq. (18), one obtains

$$A_2^{(i+1)}=\left(\frac{D_1}{D_2}-1\right)\left\{\frac{1}{8}q[i(L_1+L_2)]^4-\frac{1}{6}qL[i(L_1+L_2)]^3\right\}+\frac{D_1}{D_2}B_2^{(i)}. \tag{19}$$

By combining Eq. (11) with Eq. (16), one gets

$$A_1^{(i+1)}-A_1^{(i)}=\left(\frac{D_1}{D_2}-1\right)\left\{\frac{1}{4}qL[i(L_1+L_2)]^2-\frac{1}{6}q[i(L_1+L_2)]^3\right.$$
$$\left.-\frac{1}{4}qL[(i-1)(L_1+L_2)+L_1]^2+\frac{1}{6}q[(i-1)(L_1+L_2)+L_1]^3\right\}. \tag{20}$$

Combining Eq. (14) with Eq. (19), one gets

$$A_2^{(i+1)}-A_2^{(i)}=\left(\frac{D_1}{D_2}-1\right)\left\{\frac{1}{8}q[i(L_1+L_2)]^4-\frac{1}{8}q[(i-1)(L_1+L_2)+L_1]^4\right.$$
$$\left.-\frac{1}{6}qL[i(L_1+L_2)]^3+\frac{1}{6}qL[(i-1)(L_1+L_2)+L_1]^3\right\}. \tag{21}$$

One recognizes in Eqs. (20) and (21) a linear first-order difference equation with constant coefficient and with a second member. We first solve the linear first-order difference equation Eq. (20) (see for instance Appendix A):

$$A_1^{(i)}-A_1^{(1)}=\frac{q}{12}\left(\frac{D_1}{D_2}-1\right)\left\{\left[-2\left(\frac{L}{N}\right)^3+2\left(\frac{L}{N}\right)^2L_1\right]i^3+\left[6\left(\frac{L}{N}\right)^3\right.\right.$$
$$\left.+3\left(\frac{L}{N}\right)^2L-9\left(\frac{L}{N}\right)^2L_1-3\left(\frac{L}{N}\right)LL_1+3\left(\frac{L}{N}\right)L_1^2\right]i^2$$
$$+\left[-6\left(\frac{L}{N}\right)^3-6\left(\frac{L}{N}\right)^2L+13\left(\frac{L}{N}\right)^2L_1+9\left(\frac{L}{N}\right)LL_1\right.$$
$$\left.-9\left(\frac{L}{N}\right)L_1^2-3LL_1^2+2L_1^3\right]i+2\left(\frac{L}{N}\right)^3+3\left(\frac{L}{N}\right)^2L$$
$$\left.-6\left(\frac{L}{N}\right)^2L_1-6\left(\frac{L}{N}\right)LL_1+6\left(\frac{L}{N}\right)L_1^2+3LL_1^2-2L_1^3\right\}. \tag{22}$$

We also solve the linear difference equation Eq. (21) (Appendix A):

$$
\begin{aligned}
A_2^{(i)} - A_2^{(1)} = \frac{q}{24}\left(\frac{D_1}{D_2}-1\right)\Bigg\{&\left[3\left(\frac{L}{N}\right)^4 - 3\left(\frac{L}{N}\right)^3 L_1\right] i^4 + \Bigg[-12\left(\frac{L}{N}\right)^4 \\
&- 4\left(\frac{L}{N}\right)^3 L + 18\left(\frac{L}{N}\right)^3 L_1 + 4\left(\frac{L}{N}\right)^2 LL_1 - 6\left(\frac{L}{N}\right)^2 L_1^2\Bigg] i^3 \\
&+ \Bigg[18\left(\frac{L}{N}\right)^4 + 12\left(\frac{L}{N}\right)^3 L - 39\left(\frac{L}{N}\right)^3 L_1 - 18\left(\frac{L}{N}\right)^2 LL_1 \\
&+ 27\left(\frac{L}{N}\right)^2 L_1^2 + 6\left(\frac{L}{N}\right) LL_1^2 - 6\left(\frac{L}{N}\right) L_1^3\Bigg] i^2 \\
&+ \Bigg[-12\left(\frac{L}{N}\right)^4 - 12\left(\frac{L}{N}\right)^3 L + 36\left(\frac{L}{N}\right)^3 L_1 + 26\left(\frac{L}{N}\right)^2 LL_1 \\
&- 39\left(\frac{L}{N}\right)^2 L_1^2 - 18\left(\frac{L}{N}\right) LL_1^2 + 18\left(\frac{L}{N}\right) L_1^3 + 4LL_1^3 - 3L_1^4\Bigg] i \\
&+ 3\left(\frac{L}{N}\right)^4 + 4\left(\frac{L}{N}\right)^3 L - 12\left(\frac{L}{N}\right)^3 L_1 - 12\left(\frac{L}{N}\right)^2 LL_1 \\
&+ 18\left(\frac{L}{N}\right)^2 L_1^2 + 12\left(\frac{L}{N}\right) LL_1^2 - 12\left(\frac{L}{N}\right) L_1^3 - 4LL_1^3 + 3L_1^4\Bigg\}.
\end{aligned} \tag{23}
$$

One gets $B_1^{(i)}$ from Eqs. (16) and (22):

$$
\begin{aligned}
B_1^{(i)} = \frac{q}{12}\left(1-\frac{D_2}{D_1}\right)\Bigg\{&2\left(\frac{L}{N}\right)^2 L_1 i^3 + \Bigg[-3\left(\frac{L}{N}\right)^2 L_1 - 3\left(\frac{L}{N}\right) L_1 L \\
&+ 3\left(\frac{L}{N}\right) L_1^2\Bigg] i^2 + \Bigg[\left(\frac{L}{N}\right)^2 L_1 + 3\left(\frac{L}{N}\right) LL_1 \\
&- 3\left(\frac{L}{N}\right) L_1^2 - 3LL_1^2 + 2L_1^3\Bigg] i\Bigg\} + \frac{D_2}{D_1} A_1^{(1)}.
\end{aligned} \tag{24}
$$

One gets $B_2^{(i)}$ from Eqs. (19) and (23):

$$
B_2^{(i)} = \frac{q}{24}\left(1-\frac{D_2}{D_1}\right)\Bigg\{-3\left(\frac{L}{N}\right)^3 L_1 i^4 + \Bigg[6\left(\frac{L}{N}\right)^3 L_1 + 4\left(\frac{L}{N}\right)^2 LL_1
$$

$$-6\left(\frac{L}{N}\right)^2 L_1^2\Bigg] i^3+\Bigg[-3\left(\frac{L}{N}\right)^3 L_1-6\left(\frac{L}{N}\right)^2 LL_1+9\left(\frac{L}{N}\right)^2 L_1^2$$
$$+6\left(\frac{L}{N}\right) LL_1^2-6\left(\frac{L}{N}\right) L_1^3\Bigg] i^2+\Bigg[2\left(\frac{L}{N}\right)^2 LL_1-3\left(\frac{L}{N}\right)^2 L_1^2$$
$$-6\left(\frac{L}{N}\right) LL_1^2+6\left(\frac{L}{N}\right) L_1^3+4LL_1^3-3L_1^4\Bigg] i\Bigg\}+\frac{D_2}{D_1}A_2^{(1)}. \tag{25}$$

Now we need to get $A_2^{(1)}$. By coupling the boundary condition $w_1^{(1)}(0)=0$ with Eq. (6), one obtains

$$A_2^{(1)}=0. \tag{26}$$

By combining Eq. (23) with Eq. (26), one gets

$$A_2^{(i)}=\frac{q}{24}\left(\frac{D_1}{D_2}-1\right)\Bigg\{\Bigg[3\left(\frac{L}{N}\right)^4-3\left(\frac{L}{N}\right)^3 L_1\Bigg] i^4+\Bigg[-12\left(\frac{L}{N}\right)^4-4\left(\frac{L}{N}\right)^3 L$$
$$+18\left(\frac{L}{N}\right)^3 L_1+4\left(\frac{L}{N}\right)^2 LL_1-6\left(\frac{L}{N}\right)^2 L_1^2\Bigg] i^3+\Bigg[18\left(\frac{L}{N}\right)^4$$
$$+12\left(\frac{L}{N}\right)^3 L-39\left(\frac{L}{N}\right)^3 L_1-18\left(\frac{L}{N}\right)^2 LL_1+27\left(\frac{L}{N}\right)^2 L_1^2$$
$$+6\left(\frac{L}{N}\right) LL_1^2-6\left(\frac{L}{N}\right) L_1^3\Bigg] i^2+\Bigg[-12\left(\frac{L}{N}\right)^4-12\left(\frac{L}{N}\right)^3 L$$
$$+36\left(\frac{L}{N}\right)^3 L_1+26\left(\frac{L}{N}\right)^2 LL_1-39\left(\frac{L}{N}\right)^2 L_1^2-18\left(\frac{L}{N}\right) LL_1^2$$
$$+18\left(\frac{L}{N}\right) L_1^3+4LL_1^3-3L_1^4\Bigg] i+3\left(\frac{L}{N}\right)^4+4\left(\frac{L}{N}\right)^3 L-12\left(\frac{L}{N}\right)^3 L_1$$
$$-12\left(\frac{L}{N}\right)^2 LL_1+18\left(\frac{L}{N}\right)^2 L_1^2+12\left(\frac{L}{N}\right) LL_1^2$$
$$-12\left(\frac{L}{N}\right) L_1^3-4LL_1^3+3L_1^4\Bigg\}. \tag{27}$$

Equation (27) gives the general solutions of $A_2^{(i)}$. By combining Eq. (25) with Eq. (26), one obtains

$$B_2^{(i)}=\frac{q}{24}\left(1-\frac{D_2}{D_1}\right)\Bigg\{-3\left(\frac{L}{N}\right)^3 L_1 i^4+\Bigg[6\left(\frac{L}{N}\right)^3 L_1+4\left(\frac{L}{N}\right)^2 LL_1$$

$$-6\left(\frac{L}{N}\right)^2 L_1^2\Bigg] i^3+\Bigg[-3\left(\frac{L}{N}\right)^3 L_1-6\left(\frac{L}{N}\right)^2 LL_1+9\left(\frac{L}{N}\right)^2 L_1^2$$
$$+6\left(\frac{L}{N}\right) LL_1^2-6\left(\frac{L}{N}\right) L_1^3\Bigg] i^2+\Bigg[2\left(\frac{L}{N}\right)^2 LL_1-3\left(\frac{L}{N}\right)^2 L_1^2$$
$$-6\left(\frac{L}{N}\right) LL_1^2+6\left(\frac{L}{N}\right) L_1^3+4LL_1^3-3L_1^4\Bigg] i\Bigg\}. \tag{28}$$

Equation (28) gives the general solutions of $B_2^{(i)}$. Now we need to get $A_1^{(1)}$. By combining the boundary condition at the right end $w_2^{(N)}(L)=0$ with Eq. (9):

$$D_2 w_2^{(N)}(L)=\frac{1}{24}qL^4+B_1^{(N)}L+B_2^{(N)}=0. \tag{29}$$

When $i=N$, by combining Eq. (28) with Eq. (29), one gets

$$B_1^{(N)}L=-\frac{1}{24}qL^4-\frac{q}{24}\left(1-\frac{D_2}{D_1}\right)\Bigg\{-3\left(\frac{L}{N}\right)^3 L_1N^4+\Bigg[6\left(\frac{L}{N}\right)^3 L_1$$
$$+4\left(\frac{L}{N}\right)^2 LL_1-6\left(\frac{L}{N}\right)^2 L_1^2\Bigg] N^3+\Bigg[-3\left(\frac{L}{N}\right)^3 L_1-6\left(\frac{L}{N}\right)^2 LL_1$$
$$+9\left(\frac{L}{N}\right)^2 L_1^2+6\left(\frac{L}{N}\right) LL_1^2-6\left(\frac{L}{N}\right) L_1^3\Bigg] N^2+\Bigg[2\left(\frac{L}{N}\right)^2 LL_1$$
$$-3\left(\frac{L}{N}\right)^2 L_1^2-6\left(\frac{L}{N}\right) LL_1^2+6\left(\frac{L}{N}\right) L_1^3+4LL_1^3-3L_1^4\Bigg] N\Bigg\}. \tag{30}$$

When $i=N$, by bringing Eq. (24) back to Eq. (30), it is given that

$$A_1^{(1)}=-\frac{qL^3}{24}\frac{D_1}{D_2}-\frac{q}{24}\left(\frac{D_1}{D_2}-1\right)\Bigg\{-3\left(\frac{L}{N}\right)^3\frac{1}{L}L_1N^4+\Bigg[8\left(\frac{L}{N}\right)^2 L_1$$
$$+6\left(\frac{L}{N}\right)^3\frac{1}{L}L_1-6\left(\frac{L}{N}\right)^2\frac{1}{L}L_1^2\Bigg] N^3+\Bigg[-12\left(\frac{L}{N}\right)^2 L_1-6\left(\frac{L}{N}\right) LL_1$$

$$
\begin{aligned}
&+12\left(\frac{L}{N}\right)L_1^2-3\left(\frac{L}{N}\right)^3\frac{1}{L}L_1+9\left(\frac{L}{N}\right)^2\frac{1}{L}L_1^2-6\left(\frac{L}{N}\right)\frac{1}{L}L_1^3\Bigg]N^2\\
&+\left[4\left(\frac{L}{N}\right)^2L_1+6\left(\frac{L}{N}\right)LL_1-12\left(\frac{L}{N}\right)L_1^2-6LL_1^2+8L_1^3\right.\\
&\left.-3\left(\frac{L}{N}\right)^2\frac{1}{L}L_1^2+6\left(\frac{L}{N}\right)\frac{1}{L}L_1^3-3\frac{1}{L}L_1^4\right]N\Bigg\}.
\end{aligned}
\tag{31}
$$

By combining Eq. (22), with Eq. (31), one obtains

$$
\begin{aligned}
A_1^{(i)}&=\frac{q}{12}\left(\frac{D_1}{D_2}-1\right)\left\{\left[-2\left(\frac{L}{N}\right)^3+2\left(\frac{L}{N}\right)^2L_1\right]i^3+\left[6\left(\frac{L}{N}\right)^3+3\left(\frac{L}{N}\right)^2L\right.\right.\\
&\left.-9\left(\frac{L}{N}\right)^2L_1-3\left(\frac{L}{N}\right)LL_1+3\left(\frac{L}{N}\right)L_1^2\right]i^2+\left[-6\left(\frac{L}{N}\right)^3-6\left(\frac{L}{N}\right)^2L\right.\\
&\left.+13\left(\frac{L}{N}\right)^2L_1+9\left(\frac{L}{N}\right)LL_1-9\left(\frac{L}{N}\right)L_1^2-3LL_1^2+2L_1^3\right]i+2\left(\frac{L}{N}\right)^3\\
&\left.+3\left(\frac{L}{N}\right)^2L-6\left(\frac{L}{N}\right)^2L_1-6\left(\frac{L}{N}\right)LL_1+6\left(\frac{L}{N}\right)L_1^2+3LL_1^2-2L_1^3\right\}\\
&-\frac{qL^3}{24}\frac{D_1}{D_2}+\frac{q}{24}\left(\frac{D_1}{D_2}-1\right)\left\{3\left(\frac{L}{N}\right)^3\frac{1}{L}L_1N^4+\left[-8\left(\frac{L}{N}\right)^2L_1\right.\right.\\
&\left.-6\left(\frac{L}{N}\right)^3\frac{1}{L}L_1+6\left(\frac{L}{N}\right)^2\frac{1}{L}L_1^2\right]N^3+\left[12\left(\frac{L}{N}\right)^2L_1+6\left(\frac{L}{N}\right)LL_1\right.\\
&\left.-12\left(\frac{L}{N}\right)L_1^2+3\left(\frac{L}{N}\right)^3\frac{1}{L}L_1-9\left(\frac{L}{N}\right)^2\frac{1}{L}L_1^2+6\left(\frac{L}{N}\right)\frac{1}{L}L_1^3\right]N^2\\
&+\left[-4\left(\frac{L}{N}\right)^2L_1-6\left(\frac{L}{N}\right)LL_1+12\left(\frac{L}{N}\right)L_1^2+6LL_1^2-8L_1^3\right.\\
&\left.\left.+3\left(\frac{L}{N}\right)^2\frac{1}{L}L_1^2-6\left(\frac{L}{N}\right)\frac{1}{L}L_1^3+3\frac{1}{L}L_1^4\right]N\right\}.
\end{aligned}
\tag{32}
$$

Equation (32) gives the general solutions of $A_1^{(i)}$. By combining Eq. (24), with Eq. (31), one obtains

$$
B_1^{(i)}=\frac{q}{12}\left(1-\frac{D_2}{D_1}\right)\left\{2\left(\frac{L}{N}\right)^2L_1i^3+\left[-3\left(\frac{L}{N}\right)^2L_1-3\left(\frac{L}{N}\right)L_1L\right.\right.
$$

$$+3\left(\frac{L}{N}\right)L_1^2\Bigg]i^2+\left[\left(\frac{L}{N}\right)^2L_1+3\left(\frac{L}{N}\right)LL_1-3\left(\frac{L}{N}\right)L_1^2-3LL_1^2+2L_1^3\right]i\Bigg\}$$

$$-\frac{qL^3}{24}-\frac{q}{24}\left(1-\frac{D_2}{D_1}\right)\Bigg\{-3\left(\frac{L}{N}\right)^3\frac{1}{L}L_1N^4+\Bigg[8\left(\frac{L}{N}\right)^2L_1+6\left(\frac{L}{N}\right)^3\frac{1}{L}L_1$$

$$-6\left(\frac{L}{N}\right)^2\frac{1}{L}L_1^2\Bigg]N^3+\Bigg[-12\left(\frac{L}{N}\right)^2L_1-6\left(\frac{L}{N}\right)LL_1+12\left(\frac{L}{N}\right)L_1^2$$

$$-3\left(\frac{L}{N}\right)^3\frac{1}{L}L_1+9\left(\frac{L}{N}\right)^2\frac{1}{L}L_1^2-6\left(\frac{L}{N}\right)\frac{1}{L}L_1^3\Bigg]N^2+\Bigg[4\left(\frac{L}{N}\right)^2L_1$$

$$+6\left(\frac{L}{N}\right)LL_1-12\left(\frac{L}{N}\right)L_1^2-6LL_1^2+8L_1^3-3\left(\frac{L}{N}\right)^2\frac{1}{L}L_1^2$$

$$+6\left(\frac{L}{N}\right)\frac{1}{L}L_1^3-3\frac{1}{L}L_1^4\Bigg]N\Bigg\}. \tag{33}$$

Equation (33) gives the general solutions of $B_1^{(i)}$. Equation (6) with Eqs. (27) and (32) gives the deflection function of each blue element. Equation (9) with Eqs. (28) and (33) gives the deflection function of each red element.

3. Verification of Deflection Functions by Assuming the Stepped Beam as a Uniform Beams

If $D_1 = D_2$, the beam turns to be a uniform beam. For a uniform beam, the deflection in the mid-span is

$$w\left(\frac{L}{2}\right)=-\frac{5}{384D_1}qL^4. \tag{34}$$

We check the deflection in the mid-span by the using the equations we got in last section 2. If N is an even natural number, the deflection in the mid-span will be between cells $\frac{N}{2}$ and $\frac{N}{2}+1$. For example, if $N=4$, the deflection in the mid-span will be on the border of cell 2 and cell 3 which is shown in Fig. 2.

When N is an even natural number and $L_1 = L_2 = \frac{L}{2N}$, the deflection in the mid-span will be on the border of cell $\frac{N}{2}$ and cell $\frac{N}{2}+1$.

Based on Eqs. (6), (27) and (33), the deflection in the mid-span of the beam is

$$w\left(x=\frac{L}{2}\right)=w_1^{(\frac{N}{2}+1)}\left(x_1^{(\frac{N}{2}+1)}=0\right)$$

$$=\frac{1}{D_1}\left[\frac{1}{128}qL^4+\frac{L}{2}A_1^{(\frac{N}{2}+1)}+A_2^{(\frac{N}{2}+1)}\right], \tag{35}$$

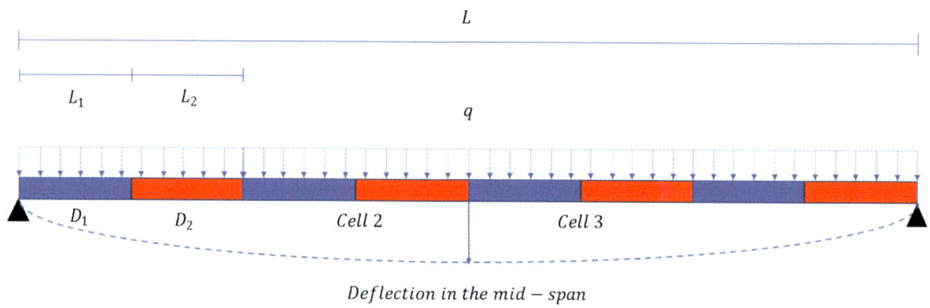

Fig. 2. The deflection in the mid-span when $N = 4$.

where

$$A_1^{(\frac{N}{2}+1)} = \frac{q}{24}\left(\frac{D_1}{D_2}-1\right)\left(-\frac{3}{2}LL_1^2N + 3\frac{1}{L}L_1^4N + L^3 + \frac{3}{2}L^2L_1 - 6L_1^3 + 3LL_1^2\frac{1}{N}\right) - \frac{qL^3}{24}\frac{D_1}{D_2}, \tag{36}$$

$$A_2^{(\frac{N}{2}+1)} = \frac{q}{24}\left(\frac{D_1}{D_2}-1\right)\left(\frac{5}{16}L^3L_1N + \frac{3}{4}L^2L_1^2N + \frac{1}{2}LL_1^3N - \frac{3}{2}L_1^4N - \frac{5}{16}L^4 - \frac{3}{4}L^3L_1 - \frac{3}{4}L^2L_1^2 + 3LL_1^3 + \frac{1}{4}L^3L_1\frac{1}{N} - \frac{3}{2}L^2L_1^2\frac{1}{N}\right). \tag{37}$$

If $D_1 = D_2$, based on Eqs. (35)–(37), the deflection in the middle of beam is

$$w\left(x=\frac{L}{2}\right) = w_1^{(\frac{N}{2}+1)}(x_1^{(\frac{N}{2}+1)}=0) = \frac{1}{D_1}\left(\frac{1}{128}qL^4 - \frac{1}{48}qL^4\right) = -\frac{5}{384D_1}qL^4, \tag{38}$$

which is identical to the deflection in the mid-span of a uniform beam given by Eq. (38).

Let us check the deflection in the mid-span by the using the equations we got in last section. If N is an odd natural number and $L_1 = L_2 = \frac{L}{2N}$, the deflection in the mid-span will be in the cell $\frac{N+1}{2}$. For example, if $N = 3$, the deflection in the mid-span will be in cell 2 which is shown in Fig. 3.

Based on Eqs. (6), (27) and (32), the deflection in the mid-span of the beam is

$$w\left(x=\frac{L}{2}\right) = w_1^{(\frac{N+1}{2})}\left(x_1^{(\frac{N+1}{2})} = \frac{L}{2N}\right) = \frac{1}{D_1}\left[\frac{1}{128}qL^4 + \frac{L}{2}A_1^{(\frac{N+1}{2})} + A_2^{(\frac{N+1}{2})}\right], \tag{39}$$

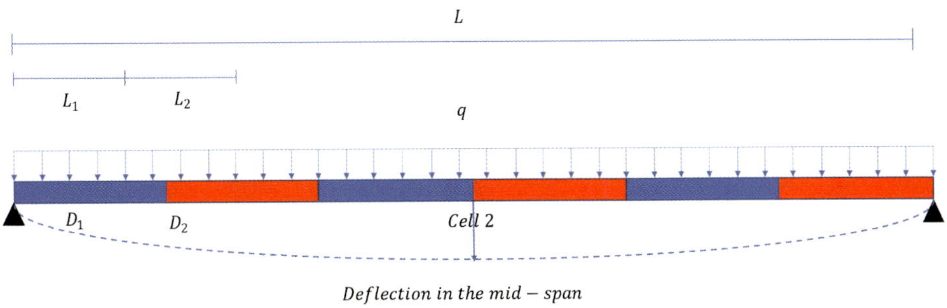

Fig. 3. The deflection in the mid-span when $N = 3$.

where

$$A_1^{(\frac{N+1}{2})} = \frac{q}{24}\left(\frac{D_1}{D_2} - 1\right)\left\{-\frac{3}{2}LL_1^2N + 3\frac{1}{L}L_1^4N + L^3 + 3L^2L_1 - 8L_1^3 - \frac{3}{2}L^3\frac{1}{N} + \frac{15}{2}LL_1^2\frac{1}{N} - 3L^2L_1\frac{1}{N^2} + \frac{1}{2}L^3\frac{1}{N^3}\right\} - \frac{qL^3}{24}\frac{D_1}{D_2}, \quad (40)$$

$$A_2^{(\frac{N+1}{2})} = \frac{q}{24}\left(\frac{D_1}{D_2} - 1\right)\left\{\frac{5}{16}L^3L_1N + \frac{3}{4}L^2L_1^2N + \frac{1}{2}LL_1^3N - \frac{3}{2}L_1^4N - \frac{5}{16}L^4 - \frac{3}{2}L^3L_1 - \frac{3}{2}L^2L_1^2 + 4LL_1^3 + \frac{3}{2}L_1^4 + \frac{3}{4}L^4\frac{1}{N} + \frac{11}{8}L^3L_1\frac{1}{N} - \frac{15}{4}L^2L_1^2\frac{1}{N} - \frac{9}{2}LL_1^3\frac{1}{N} - \frac{3}{8}L^4\frac{1}{N^2} + \frac{3}{2}L^3L_1\frac{1}{N^2} + \frac{9}{2}L^2L_1^2\frac{1}{N^2} - \frac{1}{4}L^4\frac{1}{N^3} - \frac{27}{16}L^3L_1\frac{1}{N^3} + \frac{3}{16}L^4\frac{1}{N^4}\right\}, \quad (41)$$

If $D_1 = D_2$, based on Eqs. (39)–(41), when $N = 3$, the deflection in the middle of beam is

$$w\left(x = \frac{L}{2}\right) = w_1^{(2)}\left(x_1^{(2)} = \frac{L}{2N}\right) = \frac{1}{128D_1}qL^4 - \frac{1}{48D_1}qL^4 = -\frac{5}{384D_1}qL^4, \quad (42)$$

which is identical to the deflection in the middle of a uniform beam given by Eq. (34).

4. Deflection in the Middle of the Beam About N When $D_1 = 2D_2$, $\alpha = 05$

By setting $\frac{L_1}{L_1+L_2} = \alpha$, one gets

$$L_1 = \frac{\alpha L}{N}, \tag{43}$$

$$L_2 = \frac{(1-\alpha)L}{N}. \tag{44}$$

When N is an even number $D_1 = 2D_2$, $\alpha = 0.5$, the deflection in the middle of beam is given by Eqs. (6), (36) and (37):

$$\begin{aligned} w\left(x = \frac{L}{2}\right) &= w_1^{(\frac{N}{2}+1)}\left(x_1^{(\frac{N}{2}+1)} = 0\right) \\ &= \frac{1}{D_1}\left[\frac{1}{128}qL^4 + \frac{L}{2}A_1^{(\frac{N}{2}+1)} + A_2^{(\frac{N}{2}+1)}\right], \end{aligned} \tag{45}$$

where

$$A_1^{(\frac{N}{2}+1)} = \frac{qL^3}{24}\left(\frac{3}{8N} + \frac{3}{16N^3} - 1\right), \tag{46}$$

$$A_2^{(\frac{N}{2}+1)} = \frac{qL^4}{24}\left(-\frac{5}{32} - \frac{3}{16N} - \frac{3}{32N^3}\right). \tag{47}$$

The deflection in the middle of beam can be expressed as

$$w\left(x = \frac{L}{2}\right) = -\frac{5}{256}\frac{qL^4}{D_1}. \tag{48}$$

When N is an odd number, $D_1 = 2D_2$, $\alpha = 0.5$, the deflection in the middle of beam is given by Eqs. (6), (38) and (39):

$$\begin{aligned} w\left(x = \frac{L}{2}\right) &= w_1^{(\frac{N+1}{2})}\left(x_1^{(\frac{N+1}{2})} = \frac{L}{2N}\right) \\ &= \frac{1}{D_1}\left[\frac{1}{128}qL^4 + \frac{L}{2}A_1^{(\frac{N+1}{2})} + A_2^{(\frac{N+1}{2})}\right], \end{aligned} \tag{49}$$

where

$$A_1^{(\frac{N+1}{2})} = \frac{qL^3}{24}\left(-\frac{3}{8N} + \frac{1}{16N^3} - 1\right), \tag{50}$$

$$A_2^{(\frac{N+1}{2})} = \frac{qL^4}{24}\left(-\frac{5}{32} + \frac{3}{16N} - \frac{1}{32N^3}\right). \tag{51}$$

One notices that when N is an odd number, for $\alpha \geq 0.5$, the deflection in the middle of the beam is given by Eqs. (6), (27) and (32). For $\alpha \leq 0.5$, the deflection in the middle of the beam is given by Eqs. (9), (25) and (33). The deflection in the middle of beam can be expressed as

$$w\left(x=\frac{L}{2}\right)=-\frac{5}{256}\frac{qL^4}{D_1}. \tag{52}$$

It is observed, as expected, that the static behavior of the biperiodic beam asymptotically converges towards the one of the homogenized beam which does not depend on n, with a homogenized stiffness which is the harmonic averaging of each weighted stiffness. This has been shown rigorously for biperiodic rods in Refs. 12, 13 or more recently in Ref. 19. This asymptotic formula can be shown for biperiodic beams in Ref. 14. This is also known as the so-called Reuss averaging which is the simple arithmetical averaging of the component compliances. The bending stiffness of the homogenized beam is then equal to

$$\bar{D}=\frac{D_1D_2}{(1-\alpha)D_1+\alpha D_2}, \tag{53}$$

which is also used in Ref. 18 for the lateral vibration problem.

In the specific case where $D_1=2D_2$, $\alpha=0.5$, Eq. (59) gives that

$$\bar{D}=\frac{2}{3}D_1, \tag{54}$$

where $\bar{D}$ is the equivalent stiffness of the stepped beam. In this specific case, the equivalent deflection in the middle span $\bar{w}(x=\frac{L}{2})$ is given by

$$\bar{w}\left(x=\frac{L}{2}\right)=-\frac{5qL^4}{384\bar{D}}=-\frac{5qL^4}{256D_1}, \tag{55}$$

which is the same to the analytical results got in Eqs. (48) and (52).

By setting $w^*_{\text{exact}}(x=\frac{L}{2})=\frac{w(x=\frac{L}{2})}{\bar{w}(x=\frac{L}{2})}$:

$$w^*(N)=1. \tag{56}$$

By verifying with the finite element method (Open source FEM software Code-Aster developed by EDF) by setting $w^*_{\text{FEM}}=\frac{w(x=\frac{L}{2})}{\bar{w}(x=\frac{L}{2})}$, one gets the numerical results presented in Table 1.

The shape functions of each beam element used in Code-Aster are Hermitian cubic functions for the interpolation field:

$$\xi_1(x)=2\left(\frac{x}{L}\right)^3-3\left(\frac{x}{L}\right)^2+1, \tag{57}$$

$$\xi_2(x)=L\left[-\left(\frac{x}{L}\right)^3+2\left(\frac{x}{L}\right)^2-\frac{x}{L}\right], \tag{58}$$

$$\xi_3(x)=-2\left(\frac{x}{L}\right)^3+3\left(\frac{x}{L}\right)^2-\frac{x}{L}, \tag{59}$$

$$\xi_4(x)=L\left[-\left(\frac{x}{L}\right)^3+\left(\frac{x}{L}\right)^2\right]. \tag{60}$$

Table 1. Comparison of analytical results with FEM and homogenization method results when $D_1 = 2D_2$, $\alpha = 0.5$ and $N = 1, 2, 3, 5, 10, 15, 20$.

N	1	2	3	5	10	15	20
w (Exact)	$-0.01953\frac{qL^4}{D_1}$	$-0.01953\frac{qL^4}{D_1}$	$-0.01953\frac{qL^4}{D_1}$	$-0.01953\frac{qL^4}{D_1}$	$-0.01953\frac{qL^4}{D_1}$	$-0.01953\frac{qL^4}{D_1}$	$-0.01953\frac{qL^4}{D_1}$
w_{FEM} (FEM)	$-0.01953\frac{qL^4}{D_1}$	$-0.01953\frac{qL^4}{D_1}$	$-0.01953\frac{qL^4}{D_1}$	$-0.01953\frac{qL^4}{D_1}$	$-0.01953\frac{qL^4}{D_1}$	$-0.01953\frac{qL^4}{D_1}$	$-0.01953\frac{qL^4}{D_1}$
Homogenization method $\bar{w}$				$-0.01953\frac{qL^4}{D_1}$			
w^*_{exact}				1			
w^*_{FEM}				1			

From Table 1, we found that it is a special case that when $\alpha = 0.5$ that all the methods give the same result whatever n is.

5. Deflection in the Middle of the Beam About N When $D_1 = 4D_2, \alpha = \frac{2}{3}$

When N is an even number, $D_1 = 4D_2$, $\alpha = \frac{2}{3}$, the deflection in the middle of beam is given by Eqs. (6), (36) and (37):

$$w\left(x=\frac{L}{2}\right) = w_1^{(\frac{N}{2}+1)}(x_1^{(\frac{N}{2}+1)} = 0)$$

$$= \frac{1}{D_1}\left[\frac{1}{128}qL^4 + \frac{L}{2}A_1^{(\frac{N}{2}+1)} + A_2^{(\frac{N}{2}+1)}\right], \tag{61}$$

where

$$A_1^{(\frac{N}{2}+1)} = \frac{qL^3}{24}\left(\frac{1}{N} + \frac{4}{9N^3} - 1\right), \tag{62}$$

$$A_2^{(\frac{N}{2}+1)} = \frac{qL^4}{24}\left(-\frac{5}{16} - \frac{1}{2N} - \frac{1}{18N^2} - \frac{2}{9N^3}\right). \tag{63}$$

The deflection in the middle of beam can be expressed as

$$w\left(x=\frac{L}{2}\right) = \frac{qL^4}{D_1}\left(-\frac{5}{192} - \frac{1}{432N^2}\right). \tag{64}$$

The normalized deflection for N even with respect to the one of the homogenized beam is plotted in Fig. 4.

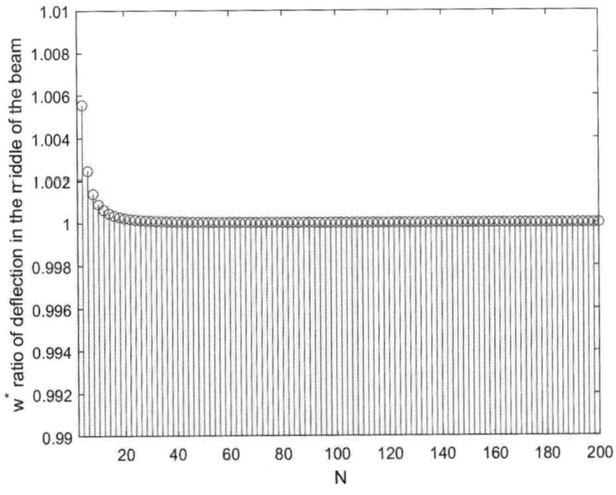

Fig. 4. The ratio of analytical deflection in the mid-span to the deflection got by homogenization method when $D_1 = 4D_2$, $\alpha = \frac{2}{3}$ about N when N is even.

When N is an odd number, $D_1 = 4D_2$, $\alpha = \frac{2}{3}$, the deflection in the middle of beam is given by Eqs. (6), (38) and (39):

$$w\left(x = \frac{L}{2}\right) = w_1^{\left(\frac{N+1}{2}\right)}\left(x_1^{\left(\frac{N+1}{2}\right)} = \frac{L}{2N}\right)$$
$$= \frac{1}{D_1}\left[\frac{1}{128}qL^4 + \frac{L}{2}A_1^{\left(\frac{N+1}{2}\right)} + A_2^{\left(\frac{N+1}{2}\right)}\right], \quad (65)$$

where

$$A_1^{\left(\frac{N+1}{2}\right)} = \frac{qL^3}{24}\left(-\frac{1}{2N} + \frac{1}{6N^3} - 1\right), \quad (66)$$

$$A_2^{\left(\frac{N+1}{2}\right)} = \frac{qL^4}{24}\left(-\frac{5}{16} + \frac{1}{4N} + \frac{5}{72N^2} - \frac{1}{12N^3} + \frac{11}{144N^4}\right). \quad (67)$$

The deflection in the middle of beam can be expressed as

$$w\left(x = \frac{L}{2}\right) = \frac{qL^4}{D_1}\left(-\frac{5}{192} + \frac{5}{1728N^2} + \frac{11}{3456N^4}\right). \quad (68)$$

The normalized deflection for N odd with respect to the one of the homogenized beam is plotted in Fig. 5.

It is observed that unlike the previous case when $\alpha = \frac{1}{2}$, when $\alpha = \frac{2}{3}$, the deflection in the middle depends on N whether it is an odd or and even number which proves that $\alpha = \frac{1}{2}$ is a special case.

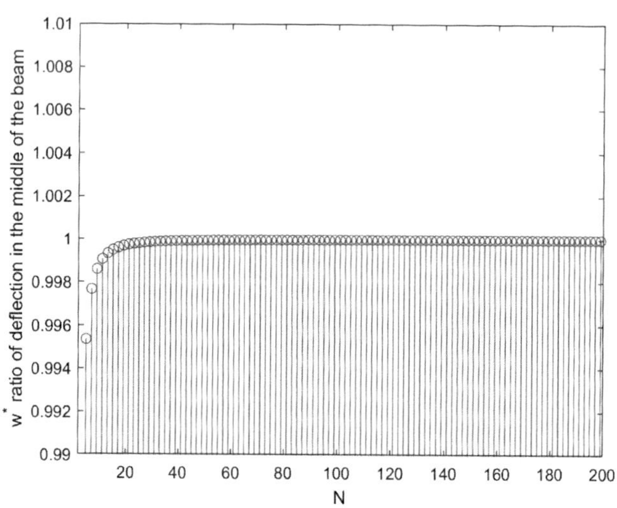

Fig. 5. The ratio of analytical deflection in the mid-span to the deflection got by homogenization method when $D_1 = 4D_2$, $\alpha = \frac{2}{3}$ about N when N is odd.

When $D_1 = 4D_2$, $\alpha = \frac{2}{3}$, the homogenization method gives that

$$\bar{D} = \frac{D_1 D_2}{(1-\alpha)D_1 + \alpha D_2} = \frac{1}{2}D_1, \tag{69}$$

where $\bar{D}$ is the equivalent stiffness of the stepped beam. The equivalent deflection in the middle span $\bar{w}(x = \frac{L}{2})$ is given by

$$\bar{w}\left(x = \frac{L}{2}\right) = -\frac{5qL^4}{384\bar{D}} = -\frac{5qL^4}{192D_1}, \tag{70}$$

which is the same to the analytical result got in Eqs. (48) and (52) when N approaches infinity.

By setting $w^*_{\text{exact}}(x = \frac{L}{2}) = \frac{w(x=\frac{L}{2})}{\bar{w}(x=\frac{L}{2})}$, when N is an even number, one obtains

$$w^*_{\text{exact}}\left(x = \frac{L}{2}\right) = 1 + \frac{4}{45N^2}. \tag{71}$$

When N is an odd number, one obtains

$$w^*_{\text{exact}}\left(x = \frac{L}{2}\right) = 1 - \frac{1}{9N^2} - \frac{11}{90N^4}. \tag{72}$$

By verifying with the finite element method (Open source FEM software Code-Aster developed by EDF) by setting $w^*_{\text{FEM}} = \frac{w(x=\frac{L}{2})}{\bar{w}(x=\frac{L}{2})}$, one gets the numerical results presented in Table 2.

From Table 2, we found that our method and FEM give the same results whatever N is. When $N \geq 20$, the homogenization method also give the same result.

Figure 6 shows the non-monotone convergence of the dimensionless deflection w^*_{exact} towards one of the homogenized biperiodic beams.

This homogenization method is also applied to a higher-order model which is based on the gradient solution:

$$\bar{D}w^{(4)}(x) + D_d w^{(6)}(x) = q, \tag{73}$$

where

$$D_d = \frac{\alpha^2(1-\alpha)^2(D_1 - D_2)^2\bar{D}(\frac{L}{N})^2}{12[(1-\alpha)D_1 + \alpha D_2]^2}. \tag{74}$$

A gradient elasticity beam has been derived in Ref. 18 from a higher-order homogenization method. The higher-order coefficients can be calibrated, using the same procedure that Refs. 2, 12 or 15 followed for deriving the equivalent gradient elasticity rod for a biperiodic continuum rod. For a simply supported higher-order beam

Table 2. Comparison of exact results with FEM results when $D_1 = 4D_2$, $\alpha = \frac{2}{3}$ and $N = 5, 10, 15, 20$.

N	1	2	3	5	10	15	20
w (Exact)	$-0.01997\frac{qL^4}{D_1}$	$-0.02662\frac{qL^4}{D_1}$	$-0.02568\frac{qL^4}{D_1}$	$-0.02592\frac{qL^4}{D_1}$	$-0.02606\frac{qL^4}{D_1}$	$-0.02603\frac{qL^4}{D_1}$	$-0.02604\frac{qL^4}{D_1}$
w_{FEM} (FEM)	$-0.01997\frac{qL^4}{D_1}$	$-0.02662\frac{qL^4}{D_1}$	$-0.02568\frac{qL^4}{D_1}$	$-0.02592\frac{qL^4}{D_1}$	$-0.02606\frac{qL^4}{D_1}$	$-0.02603\frac{qL^4}{D_1}$	$-0.02604\frac{qL^4}{D_1}$
Homogenization method $\bar{w}$				$-0.02604\frac{qL^4}{D_1}$			
w^*_{exact}	0.7667	1.0222	0.9861	0.9953	1.0009	0.9995	1.0000
w^*_{FEM}	0.7667	1.0222	0.9861	0.9953	1.0009	0.9995	1.0000

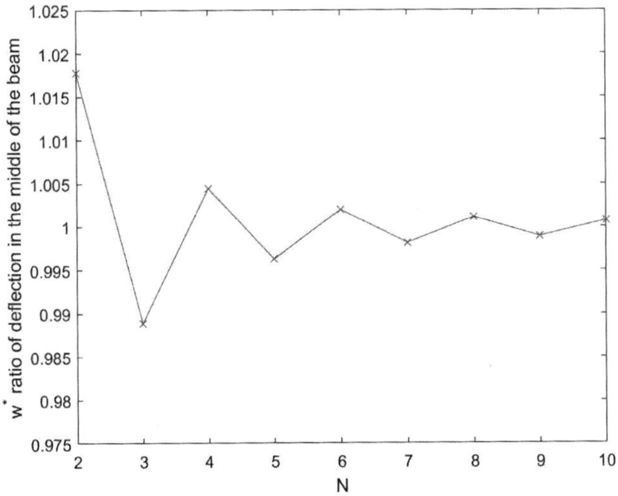

Fig. 6. The ratio of analytical deflection in the mid-span to the deflection got by homogenization method when $D_1 = 4D_2$, $\alpha = \frac{2}{3}$ for $1 \leq N \leq 10$.

under uniform loading, Navier's Fourier series solution is given by

$$w_N(x) = \sum_{m=1}^{\infty} W_{N,m} \sin \frac{m\pi x}{L}, \tag{75}$$

$$q(x) = q = \sum_{m=1}^{\infty} Q_m \sin \frac{m\pi x}{L}, \tag{76}$$

where

$$Q_m = \frac{2q}{L} \int_0^L \sin \frac{m\pi x}{L} dx = -\frac{2q}{L} \left(\frac{L}{m\pi} \right) \cos \frac{m\pi x}{L} \Big|_0^L = \frac{2q}{m\pi}(1 - \cos m\pi), \tag{77}$$

$$Q_m = \frac{4q}{m\pi} \quad \text{for odd } m, \tag{78}$$

$$Q_m = 0 \quad \text{for even } m. \tag{79}$$

By coupling Eq. (73) with Eqs. (75) and (76), one gets

$$W_{N,m} = \frac{Q_m}{\bar{D} \left(\frac{m\pi}{L} \right)^4 \left[1 - \frac{D_d}{\bar{D}} \left(\frac{m\pi}{L} \right)^2 \right]} \quad \text{for odd } m. \tag{80}$$

By coupling Eq. (80) with Eqs. (78) and (79), one obtains

$$w_N(x) = \sum_m \frac{1}{\left[1 - \frac{D_d}{\bar{D}} \left(\frac{m\pi}{L} \right)^2 \right]} \left(\frac{L}{m\pi} \right)^4 \frac{4q}{\bar{D} m\pi} \sin \frac{m\pi x}{L} \quad \text{for odd } m. \tag{81}$$

When $x = \frac{L}{2}$, one gets the deflection in the middle of the beam from Eq. (81):

$$w_N\left(\frac{L}{2}\right) = \sum_m \frac{1}{[1 - \frac{D_d}{\bar{D}}(\frac{m\pi}{L})^2]}\left(\frac{L}{m\pi}\right)^4 \frac{4q}{\bar{D}m\pi}(-1)^{\frac{m-1}{2}} \quad \text{for odd } m. \tag{82}$$

One notices that D_d is a function of N, so $w_N(\frac{L}{2})$ is a function of N.

In the specific case where $D_1 = 4D_2$, $\alpha = \frac{2}{3}$, Eq. (74) gives that

$$D_d = \frac{(\frac{L}{N})^2}{216} D_1. \tag{83}$$

By setting $w^*_{2\text{nd}}(x = \frac{L}{2}) = \frac{w_N(\frac{L}{2})}{\bar{w}(x=\frac{L}{2})}$, one obtains

$$w^*_{2\text{nd}}\left(x = \frac{L}{2}\right) = \sum_m \frac{307.2}{[1 - \frac{1}{108}(\frac{m\pi}{N})^2]}\left(\frac{1}{m\pi}\right)^5 (-1)^{\frac{m-1}{2}}. \tag{84}$$

Equation (84) is a function of N. Typically, $m = 20$ is sufficient for the numerical calculation of the series in the deflection equation. The numerical results are reported in Table 3 for $D_1 = 4D_2$ and $\alpha = 2/3$, for the exact normalized deflection of the biperiodic beam and a comparison with the results issued of the second-order homogenized beam model. Table 4 presents the same comparison with $D_1 = 3D_2$ and $\alpha = 2/3$. It is shown in Fig. 7 and Fig. 8 that the higher-order homogenized beam model accurately approximates the scale effect of the biperiodic beam when the number N of cells is even.

When $D_1 = 3D_2, \alpha = \frac{2}{3}$, one gets.

One notices that the higher-order homogenization method works better when N is an even number. This can be explained as when N is an even number, the total stiffness of the left half is equal to that of the right half. For example, when $N = 4$ shown in Fig. 2, the left half has two blue elements and two red elements and the right half has the same number of blue and red elements. When $N = 3$ shown in Fig. 3, the left half has two blue elements and one red element but the right half has one blue element and two red elements. Therefore, the homogenization method

Table 3. Comparison of exact results with homogenization methods when $D_1 = 4D_2$, $\alpha = \frac{2}{3}$ for $1 \leq N \leq 10$.

N	1	2	3	4	5
w^*_{exact}	0.7667	1.0222	0.9861	1.0056	0.9954
$w^*_{2\text{nd}}$	1.0813	1.0234	1.0100	1.0057	1.0036
	6	7	8	9	10
w^*_{exact}	1.0025	0.9977	1.0014	0.9986	1.0009
$w^*_{2\text{nd}}$	1.0025	1.0018	1.0014	1.0011	1.0009

Table 4. Comparison of exact results with homogenization methods when $D_1 = 3D_2$, $\alpha = \frac{2}{3}$ for $1 \leq N \leq 10$.

N	1	2	3	4	5
w^*_{exact}	0.8133	1.0178	0.9889	1.0044	0.9963
w^*_{2nd}	1.0568	1.0143	1.0063	1.0035	1.0023
	6	7	8	9	10
w^*_{exact}	1.0020	0.9981	1.0011	0.9989	1.0007
w^*_{2nd}	1.0016	1.0011	1.0009	1.0007	1.0006

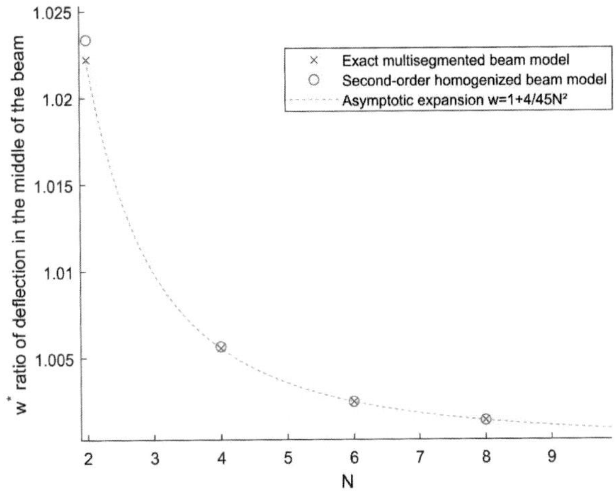

Fig. 7. Comparison of the deflection of the higher-order homogenized beam model with the exact multisegmented beam model when $D_1 = 4D_2$, $\alpha = \frac{2}{3}$ for$N$ is even.

especially the higher-order homogenization method works better when N is an even number.

For a simply supported higher-order beam under uniform loading, Navier Fourier series solution can be used.[22] Equation (73) can be rewritten as

$$\bar{D}w^{(4)}(x) + \varepsilon^2 L^2 \bar{D}w^{(6)}(x) = q \tag{85}$$

with

$$\varepsilon^2 = \frac{\alpha^2(1-\alpha)^2(\frac{D_1}{D_2}-1)^2}{12[(1-\alpha)\frac{D_1}{D_2}+\alpha]^2 N^2}. \tag{86}$$

For a simply supported-simply supported beam, the boundary conditions are given by

$$w(0) = w(L) = 0, \tag{87}$$

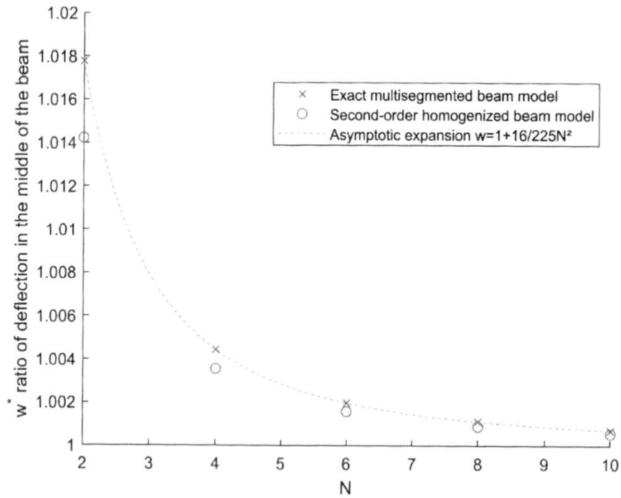

Fig. 8. Comparison of the deflection of the higher-order homogenized beam model with the exact multisegmented beam model when $D_1 = 3D_2$, $\alpha = \frac{2}{3}$ for N is even.

$$w''(0) = w''(L) = 0, \tag{88}$$

$$w^{(4)}(0) = w^{(4)}(L) = 0. \tag{89}$$

By coupling Eq. (86) with Eqs. (87)–(89), one gets

$$w_N = \frac{qx^4}{24\bar{D}} - \frac{qLx^3}{12\bar{D}} - \frac{qL^2\varepsilon^2}{2\bar{D}}x^2 + \frac{qL^3(1+12\varepsilon^2)}{24\bar{D}}x + \frac{qL^4\varepsilon^4}{\bar{D}} - \frac{qL^4\varepsilon^4}{\bar{D}}\cos\left(\frac{x}{L\varepsilon}\right)$$
$$+ \frac{qL^4\varepsilon^4}{\bar{D}}\frac{\cos(\frac{1}{\varepsilon})-1}{\sin(\frac{1}{\varepsilon})}\sin\left(\frac{x}{L\varepsilon}\right) \quad \text{for } \frac{1}{\varepsilon} \neq m\pi, \tag{90}$$

which is equivalent to Eq. (82).

When $x = \frac{L}{2}$, the deflection at mid-span is

$$w_N\left(\frac{L}{2}\right) = \frac{5qL^4}{384\bar{D}} + \frac{qL^4\varepsilon^2}{8\bar{D}} + \frac{qL^4\varepsilon^4}{\bar{D}}\left[1 - \frac{1}{\cos(\frac{1}{2\varepsilon})}\right] \quad \text{for } \frac{1}{\varepsilon} \neq m\pi. \tag{91}$$

Equation (91) expressed with exponential and trigonometric functions is strictly equivalent to the series solution derived previously in Eq. (82).

By setting $w^*_{2\text{nd}}(x = \frac{L}{2}) = \frac{w_N(\frac{L}{2})}{\bar{w}(x=\frac{L}{2})}$, one obtains

$$w^*_{2\text{nd}}\left(x = \frac{L}{2}\right) = 1 + \frac{4}{45N^2} + \frac{8}{1215N^4}\left[1 - \frac{1}{\cos(3\sqrt{3}N)}\right]. \tag{92}$$

One notices that the first two terms of Eq. (82) valid for the second-order homogenized beam model coincide with Eq. (71) of the exact multisegmented beam model.

As the stepped beam is not perfectly symmetrical about the middle, the deflection in the middle is not the maximum deflection alongside the beam while for an equivalent homogenized beam the deflection in the middle is always the maximum deflection alongside the beam.

Supposing that the maximum deflection appears in the elements in the middle where the rotation is zero, one gets

$$D_1 w_1^{(i)\prime}(x) = \frac{1}{4} qLx^2 - \frac{1}{6} qx^3 + A_1^{(i)} = 0, \tag{93}$$

$$D_2 w_2^{(i)\prime}(x) = \frac{1}{4} qLx^2 - \frac{1}{6} qx^3 + B_1^{(i)} = 0. \tag{94}$$

When N is an even number, one gets

$$\begin{aligned} 6Lx^2 - 4x^3 + \left(\frac{D_1}{D_2} - 1\right)\left(-\frac{3}{2} LL_1^2 N + 3\frac{1}{L} L_1^4 N + L^3 + \frac{3}{2} L^2 L_1 \right.\\ \left. - 6L_1^3 + 3LL_1^2 \frac{1}{N}\right) - L^3 \frac{D_1}{D_2} = 0, \end{aligned} \tag{95}$$

$$\begin{aligned} 6Lx^2 - 4x^3 + \left(1 - \frac{D_2}{D_1}\right)\left\{-\frac{3}{2} LL_1^2 N + 3\frac{1}{L} L_1^4 N \right.\\ \left. + \frac{3}{2} L^2 L_1 - 6L_1^3 + 3LL_1^2 \frac{1}{N}\right\} - L^3 = 0. \end{aligned} \tag{96}$$

When $D_1 = 4D_2$, $\alpha = \frac{2}{3}$:

$$6Lx^2 - 4x^3 + L^3 \left(\frac{4}{9N^3} + \frac{1}{N} - 1\right) = 0, \tag{97}$$

$$6Lx^2 - 4x^3 + L^3 \left(\frac{1}{9N^3} + \frac{1}{4N} - 1\right) = 0. \tag{98}$$

By solving Eqs. (87) and (88), one obtains the position of the maximum deflection. Then one gets the maximum deflection by

$$\begin{aligned} w_{\max}^{\left(\frac{N}{2}\right)}(x) = \frac{4q}{D_1}\left[\frac{1}{12} Lx^3 - \frac{1}{24} x^4 + \frac{L^3}{24}\left(-1 + \frac{1}{9N^3} + \frac{1}{4N}\right) x \right.\\ \left. + \frac{L^4}{24}\left(\frac{5}{32} - \frac{1}{8N} - \frac{1}{72N^2} - \frac{1}{18N^3}\right)\right]. \end{aligned} \tag{99}$$

When N is an odd number, one gets

$$6Lx^2 - 4x^3 + \left(\frac{D_1}{D_2} - 1\right)\left(-\frac{3}{2}LL_1^2N + 3\frac{1}{L}L_1^4N + L^3 + 3L^2L_1 - 8L_1^3 - \frac{3}{2}L^3\frac{1}{N} + \frac{15}{2}LL_1^2\frac{1}{N} - 3L^2L_1\frac{1}{N^2} + \frac{1}{2}L^3\frac{1}{N^3}\right) - L^3\frac{D_1}{D_2} = 0. \quad (100)$$

$$6Lx^2 - 4x^3 + \left(1 - \frac{D_2}{D_1}\right)\left(-\frac{3}{2}LL_1^2N + 3\frac{1}{L}L_1^4N - 4L_1^3 + \frac{3}{2}LL_1^2\frac{1}{N}\right) - L^3 = 0. \quad (101)$$

When $D_1 = 4D_2$, $\alpha = \frac{2}{3}$:

$$6Lx^2 - 4x^3 + L^3\left(-1 - \frac{1}{2N} + \frac{1}{6N^3}\right) = 0, \quad (102)$$

$$6Lx^2 - 4x^3 + L^3\left(-1 - \frac{1}{2N} + \frac{1}{18N^3}\right) = 0. \quad (103)$$

By solving Eqs. (92) and (93), one obtains the position of the maximum deflection. Then one gets the maximum deflection by

$$w_{\max}^{(\frac{N+1}{2})}(x) = \frac{q}{D_1}\left[\frac{1}{12}Lx^3 - \frac{1}{24}x^4 + \frac{L^3}{24}\left(-\frac{1}{2N} + \frac{1}{6N^3} - 1\right)x + \frac{L^4}{24}\left(-\frac{5}{16} + \frac{1}{4N} + \frac{5}{72N^2} - \frac{1}{12N^3} + \frac{11}{144N^4}\right)\right]. \quad (104)$$

By setting $w_{\max}^* = \frac{w_{\max}}{\bar{w}(x=\frac{L}{2})}$ and retaking $w_{2\text{nd}}^*(x = \frac{L}{2}) = \frac{w_N(\frac{L}{2})}{\bar{w}(x=\frac{L}{2})}$, one gets the numerical results presented in Table 5.

Similar to the previous table, one notices that the higher-order homogenization method works better when N is an even number.

Table 5. Comparison of maximum deflections of the exact results with those of the homogenization methods when $D_1 = 4D_2$, $\alpha = \frac{2}{3}$ for $1 \le N \le 10$.

N	1	2	3	4	5
$l_{\max}/L$	0.6130	0.4536	0.5537	0.4785	0.5329
$w_{\max}^*$	0.7965	1.0428	0.9930	1.0100	0.9980
$w_{2\text{nd}}^*$	1.0813	1.0234	1.0100	1.0057	1.0036
	6	7	8	9	10
$l_{\max}/L$	0.4859	0.5327	0.4895	0.5185	0.4916
$w_{\max,0}^*$	1.0044	0.9990	1.0024	0.9994	1.0016
$w_{2\text{nd}}^*$	1.0025	1.0018	1.0014	1.0011	1.0009

6. The General Expression of Deflection in the Middle of the Beam About $\frac{D_1}{D_2}$, α and N

When is N an even number, one gets from Eqs. (35)–(37):

$$w\left(x=\frac{L}{2}\right)=w_1^{(\frac{N}{2}+1)}(x_1^{(\frac{N}{2}+1)}=0)$$
$$=\frac{1}{D_1}\left[\frac{1}{128}qL^4+\frac{L}{2}A_1^{(\frac{N}{2}+1)}+A_2^{(\frac{N}{2}+1)}\right] \tag{105}$$

with

$$A_1^{(\frac{N}{2}+1)}=\frac{qL^3}{24}\left(\frac{D_1}{D_2}-1\right)\left[\left(-\frac{3}{2}\alpha^2+\frac{3}{2}\alpha\right)\frac{1}{N}\right.$$
$$\left.+(3\alpha^4-6\alpha^3+3\alpha^2)\frac{1}{N^3}+1\right]-\frac{qL^3}{24}\frac{D_1}{D_2}, \tag{106}$$

$$A_2^{(\frac{N}{2}+1)}=\frac{qL^4}{24}\left(\frac{D_1}{D_2}-1\right)\left\{\frac{5}{16}\alpha-\frac{5}{16}+\left(\frac{3}{4}\alpha^2-\frac{3}{4}\alpha\right)\frac{1}{N}\right.$$
$$\left.+\left(\frac{1}{2}\alpha^3-\frac{3}{4}\alpha^2+\frac{1}{4}\alpha\right)\frac{1}{N^2}+\left(-\frac{3}{2}\alpha^4+3\alpha^3-\frac{3}{2}\alpha^2\right)\frac{1}{N^3}\right\}. \tag{107}$$

By combining Eqs. (105)–(107), one gets

$$w\left(x=\frac{L}{2}\right)=\frac{qL^4}{D_1}\left\{-\frac{5}{384}+\frac{1}{48}\left(\frac{D_1}{D_2}-1\right)\left[\left(\alpha^3-\frac{3}{2}\alpha^2+\frac{1}{2}\alpha\right)\frac{1}{N^2}-\frac{5}{8}+\frac{5}{8}\alpha\right]\right\}. \tag{108}$$

When N is an odd number, the deflection in the middle of beam is given by Eqs. (39)–(41):

$$w\left(x=\frac{L}{2}\right)=w_1^{(\frac{N+1}{2})}\left(x_1^{(\frac{N+1}{2})}=\frac{L}{2N}\right)$$
$$=\frac{1}{D_1}\left[\frac{1}{128}qL^4+\frac{L}{2}A_1^{(\frac{N+1}{2})}+A_2^{(\frac{N+1}{2})}\right], \tag{109}$$

where

$$A_1^{(\frac{N+1}{2})}=\frac{qL^3}{24}\left(\frac{D_1}{D_2}-1\right)\left[\left(-\frac{3}{2}\alpha^2+3\alpha-\frac{3}{2}\right)\frac{1}{N}\right.$$
$$\left.+\left(3\alpha^4-8\alpha^3+\frac{15}{2}\alpha^2-3\alpha+\frac{1}{2}\right)\frac{1}{N^3}+1\right]-\frac{qL^3}{24}\frac{D_1}{D_2}, \tag{110}$$

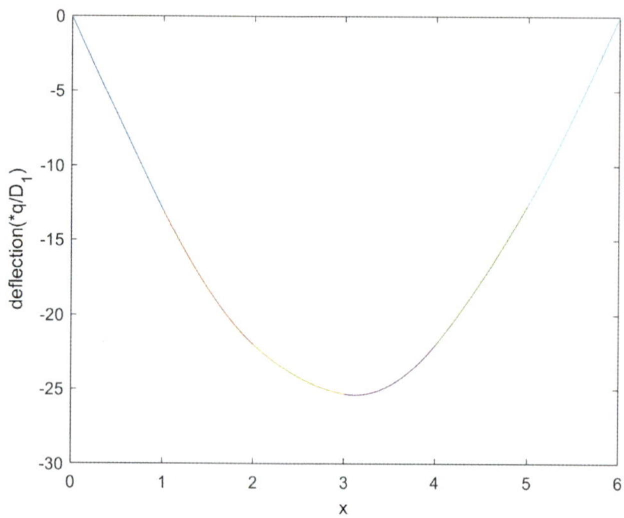

Fig. 9. Deflection curve for a repetitive stepped beam under a uniform load q when $L = 6$, $\frac{D_1}{D_2} = 2$, $\alpha = 0.5$ and $N = 3$.

$$\begin{aligned} A_2^{(\frac{N+1}{2})} &= \frac{qL^4}{24}\left(\frac{D_1}{D_2}-1\right)\left[\frac{5}{16}\alpha - \frac{5}{16} + \left(\frac{3}{4}\alpha^2 - \frac{3}{2}\alpha + \frac{3}{4}\right)\frac{1}{N}\right. \\ &\quad + \left(\frac{1}{2}\alpha^3 - \frac{3}{2}\alpha^2 + \frac{11}{8}\alpha - \frac{3}{8}\right)\frac{1}{N^2} \\ &\quad + \left(-\frac{3}{2}\alpha^4 + 4\alpha^3 - \frac{15}{4}\alpha^2 + \frac{3}{2}\alpha - \frac{1}{4}\right)\frac{1}{N^3} \\ &\quad \left. + \left(\frac{3}{2}\alpha^4 - \frac{9}{2}\alpha^3 + \frac{9}{2}\alpha^2 - \frac{27}{16}\alpha + \frac{3}{16}\right)\frac{1}{N^4}\right]. \end{aligned} \tag{111}$$

The deflection in the middle of beam can be expressed as

$$\begin{aligned} w\left(x = \frac{L}{2}\right) &= \frac{qL^4}{D_1}\left\{-\frac{5}{384} + \frac{1}{48}\left(\frac{D_1}{D_2}-1\right)\left[\left(\alpha^3 - 3\alpha^2 + \frac{11}{4}\alpha - \frac{3}{4}\right)\frac{1}{N^2}\right.\right. \\ &\quad \left.\left. + \left(3\alpha^4 - 9\alpha^3 + 9\alpha^2 - \frac{27}{8}\alpha + \frac{3}{8}\right)\frac{1}{N^4} + \frac{5}{8}\alpha - \frac{5}{8}\right]\right\}. \end{aligned} \tag{112}$$

This analytical method gives not only the solution of the deflection in the mid-span. With the deflection function given by Eqs. (6), (9), (27), (28), (32) and (33), when $\frac{D_1}{D_2} = 2$, $\alpha = 0.5$ and $N = 3$, the deflection curve is plotted in Fig. 9.

7. Conclusion

In this paper, we investigated the static response of a repetitive stepped Euler–Bernoulli beam which may be classified as a continuous biperiodic beam. The exact solution of this problem in term of deflection is obtained via the resolution of a two-step problem. First, we build a local coordinate system. Then, by using this new system of coordinates, along with the continuity conditions (deflection, rotation) and the boundary conditions (simply supported at both ends), the deflection function of the simply supported biperiodic beam is calculated from the resolution of a linear first-order difference equation. Compared to other approximated or asymptotic methods, this analytical method gives an exact closed-form solution of the multi-segmented beam problem. The exact deflection function is numerically compared to FEM-based beam solutions with great accuracy. It is also shown that the exact solution asymptotically converges (for a large number of cells) towards the Homogenization solution based on the so-called Reuss averaging stiffness which is the simple arithmetical averaging of the component compliances. The convergence appears to be non-monotonic for the normalized deflection. This paper is concluded by a parametric study based on the length and stiffness ratio of the biperiodic cell.

Appendix A

Equation (20) is written as

$$A_1^{(i+1)} - A_1^{(i)} = \left(\frac{D_1}{D_2} - 1\right)\left\{\frac{1}{4}qL[i(L_1+L_2)]^2 - \frac{1}{6}q[i(L_1+L_2)]^3 - \frac{1}{4}qL[(i-1)(L_1+L_2)+L_1]^2 + \frac{1}{6}q[(i-1)(L_1+L_2)+L_1]^3\right\} \tag{A.1a}$$

It is noticed that i is the only variable in this equation. Therefore, a series of functions are written:

$$A_1^{(i)} - A_1^{(i-1)} = \left(\frac{D_1}{D_2} - 1\right)\left\{\frac{1}{4}qL[(i-1)(L_1+L_2)]^2 - \frac{1}{6}q[(i-1)(L_1+L_2)]^3 - \frac{1}{4}qL[(i-2)(L_1+L_2)+L_1]^2 + \frac{1}{6}q[(i-2)(L_1+L_2)+L_1]^3\right\}, \tag{A.1b}$$

$$A_1^{(i-1)} - A_1^{(i-2)} = \left(\frac{D_1}{D_2} - 1\right)\left\{\frac{1}{4}qL[(i-2)(L_1+L_2)]^2 - \frac{1}{6}q[(i-2)(L_1+L_2)]^3 - \frac{1}{4}qL[(i-3)(L_1+L_2)+L_1]^2 + \frac{1}{6}q[(i-3)(L_1+L_2)+L_1]^3\right\}, \tag{A.1c}$$

......

$$A_1^{(2)} - A_1^{(1)} = \left(\frac{D_1}{D_2} - 1\right)\left\{\frac{1}{4}qL[(L_1+L_2)]^2 - \frac{1}{6}q[(L_1+L_2)]^3 - \frac{1}{4}qLL_1^2 + \frac{1}{6}qL_1^3\right\}. \tag{A.1d}$$

By summing all equations (A.1a), one gets

$$\begin{aligned} A_1^{(i)} - A_1^{(1)} = {}& \left(\frac{D_1}{D_2} - 1\right)\left\{\frac{1}{4}qL\left\{[(i-1)(L_1+L_2)]^2 + [(i-2)(L_1+L_2)]^2\right.\right. \\ & \left. + \cdots + [2(L_1+L_2)]^2 + (L_1+L_2)^2\right\} - \frac{1}{6}q\{[(i-1)(L_1+L_2)]^3 \\ & + [(i-2)(L_1+L_2)]^3 + \cdots + [2(L_1+L_2)]^3 + (L_1+L_2)^3\} \\ & - \frac{1}{4}qL\{[(i-2)(L_1+L_2)+L_1]^2 + [(i-3)(L_1+L_2)+L_1]^2 \\ & + \cdots + [(L_1+L_2)+L_1]^2 + L_1^2\} + \frac{1}{6}q\{[(i-2)(L_1+L_2)+L_1]^3 \\ & \left. + [(i-3)(L_1+L_2)+L_1]^3 + \cdots + [(L_1+L_2)+L_1]^3 + L_1^3\}\right\}. \end{aligned} \tag{A.2a}$$

By solving the sum of natural numbers (shown in Eq. (A.2b)), sum of square (shown in Eq. (A.2c)) and sum of cube (shown in Eq. (A.2d)) in Eq. (A.2a), one obtains Eq. (22).

$$\sum_{i=1}^{n} i = \frac{(n+1)n}{2}, \tag{A.2b}$$

$$\sum_{i=1}^{n} i^2 = \frac{(2n+1)(n+1)n}{6}, \tag{A.2c}$$

$$\sum_{i=1}^{n} i^3 = \left[\frac{(n+1)n}{2}\right]^2. \tag{A.2d}$$

References

1. F. Bleich, *Buckling Strength of Metal Structures* (McGraw-Hill Inc., 1952).
2. I. V. Andrianov, V. I. Bolshakov, V. V. Danishevs'kyy and D. Weichert, Higher order asymptotic homogenization and wave propagation in periodic composite materials, *Proc. R. Soc. A* **464**(2093) (2008) 1181–1201.
3. L. Banakh and M. Kempner, *Vibrations of Mechanical Systems with Regular Structure* (Springer Science & Business Media, 2010).

4. C. W. Cai, H. C. Chan and J. K. Liu, *Exact Analysis of Bi-periodic Structures* (World Scientific, 2002).
5. I. Elishakoff, A. P. Ankitha and A. Marzani, Rigorous versus naïve implementation of the Galerkin method for stepped beams, *Acta Mech.* **230**(11) (2019) 3861–3873.
6. B. G. Galerkin, Sterzhni i plastiny. Ryady v nekotorykh voprosakh uprugogo ravnovesiya sterzhnei i plastin (Rods and plates series occurring in some problems of elastic equilibrium of rods and plates), Vestnik Inzhenerov i Tekhnikov, Petrograd, Vol. 19, pp. 897–908 (1915) (in Russian) (English Translation: 63-18925, Clearinghouse Fed. Sci. Tech. Info. 1963).
7. L. S. Al-Ansari, Calculating static deflection and natural frequency of stepped cantilever beam using modified Rayleigh method, *Int. J. Mech. Prod. Eng. Res. Dev. (IJMPERD)*, **3**(4) (2013) 107–118.
8. H. J. Abdulsamad, K. J. Wadi, S. K. Al-Raheem and L. S. Alansari, Investigation of static deflection in internal stepped cantilever beam, *J. Mech. Eng. Res. Dev.* **44**(5) (2021) 87–125.
9. P. Cheng, C. Davila and G. Hou, Static, vibration analysis and sensitivity analysis of stepped beams using singularity functions, *J. Struct.* **2014** (2014) 234085.
10. W. Xu, M. Cao, Q. Ren and Z. Su, Numerical evaluation of high-order modes for stepped beam, *J. Vib. Acoust.* **136**(1) (2014) 014503.
11. M. S. Cao, W. Xu, Z. Su, W. Ostachowicz and N. Xia, Local coordinate systems-based method to analyze high-order modes of n-step Timoshenko beam, *J. Vib. Control* **23**(1) (2017) 89–102.
12. J. Fish, W. Chen and G. Nagai, Non-local dispersive model for wave propagation in heterogeneous media: One-dimensional case, *Int. J. Numer. Methods Eng.* **54**(3) (2002) 331–346.
13. N. S. Bakhvalov and G. Panasenko, *Homogenisation: Averaging Processes in Periodic Media: Mathematical Problems in the Mechanics of Composite Materials*, Vol. 36 (Springer Science & Business Media, 2012).
14. T. Zheng and T. Ji, Equivalent representations of beams with periodically variable cross-sections, *Eng. Struct.* **33**(3) (2011) 706–719.
15. I. I. Andrianov, I. V. Andrianov, G. A. Starushenko and E. I. Borodin, Higher order asymptotic homogenization for dynamical problems, *Math. Mech. Solids* **27**(9) (2022) 1672–1686.
16. M. M. Ameen, R. H. J. Peerlings and M. G. D. Geers, Higher-order asymptotic homogenization of periodic materials with low scale separation, in *ECCOMAS Congress 2016, VII European Congress on Computational Methods in Applied Science Engineering*, eds. M. Papadrakakis, V. Papadopoulos, G. Stefanou and V. Plevris, 5–10 June 2016, Crete Island, Greece, pp. 2346–2352.
17. M. M. Ameen, R. H. J. Peerlings and M. G. D. Geers, A quantitative assessment of the scale separation limits of classical and higher-order asymptotic homogenization, *Eur. J. Mech. A Solids* **1** (2018) 89–100.
18. X. Sun, C. Zhou, M. Ichchou, J. P. Lainé and A. M. Zine, Multi-scale homogenization of transversal waves in periodic composite beams, *Int. J. Appl. Mech.* **9**(3) (2017) 1750039.
19. I. V. Andrianov, V. Danishevskyy and J. Awrejcewicz, *Linear and Nonlinear Waves in Microstructured Solids: Homogenization and Asymptotic Approaches* (CRC Press, 2021).

20. C. M. Chen, N. Kikuchi and F. Rostam-Abadi, An enhanced asymptotic homogenization method of the static and dynamics of elastic composite laminates, *Comput. Struct.* **82**(4–5) (2004) 373–382.
21. S. Goldberg, *Introduction to Difference Equations with Illustrative Examples from Economics, Psychology and Sociology* (Dover Publications, New York, 1958).
22. J. N. Reddy, *Mechanics of Laminated Composite Plates and Shells: Theory and Analysis* (CRC Press).

https://doi.org/10.1142/9789819814770_0009

Chapter 9

Lateral Torsional Buckling Resistance and Design of High Strength Steel Beams Subjected to Nonuniform Temperature#

Yanhong Zhang

School of Civil Engineering, Chongqing University
Chongqing 400045, P. R. China

School of Civil Engineering, Tongji University
Shanghai 200092, P. R. China
zyh15723166561@163.com

Linbo Zhang

Department of Civil and Environmental Engineering
University of Waterloo
Waterloo, ON, Canada N2L 3G1
linbo.zhang@uwaterloo.ca

Weiyong Wang*

School of Civil Engineering, Chongqing University
Chongqing 400045, P. R. China

Key Laboratory of New Technology for Construction of Cities in Mountain Area (Ministry of Education)
Chongqing University, Chongqing 400045, P. R. China
wywang@cqu.edu.cn

The mechanical characteristics of high-strength steel (HSS) at high temperatures vary significantly from those of mild steel, so fire design regulations derived from mild steel cannot be transferred to HSS structures. Therefore, this paper investigates the fire resistance of beams made of HSSs, which will facilitate the design and application of HSS structures. Due to the heat dissipation of the concrete floor, steel beams are generally heated from three sides, and the temperature distribution in the steel beam cross-section is nonuniform. For such cases, the fire resistance of HSS beams is assessed using finite element models (FEMs) with nonuniform temperature distribution. A Ramberg–Osgood model was used to predict stress–strain relationships of Q460, Q690, and Q960 HSSs at

*Corresponding author.

#This chapter appeared previously on the International Journal of Structural Stability and Dynamics. To cite this chapter, please cite the original article as the following: Y. Zhang, L. Zhang & W. Wang, *Int. J. Struct. Stab. Dyn.*, **23**, 2340010 (2023), DOI: 10.1142/S0219455423400102.

elevated temperatures, and the model was verified by material tests. In the FEM, the proposed stress–strain relationships of HSSs at elevated temperatures, initial geometric imperfections and residual stress are considered for better accuracy. After verifying the established model against experimental results, the influence of nonuniform temperature distribution, load pattern, steel grade, and slenderness ratio on the resistance of HSS beams is investigated by conducting parametric analyses using the verified FEM. It is found that the critical bending moment of steel beams at elevated temperatures associated with lateral-torsional buckling highly depends on the reduction of elastic modulus of steel. By comparing FEM results with EN 1993-1-2, it is evident that EN 1993-1-2 is unsuitable for assessing the fire resistance of HSS beams. Therefore, the expression specified in EN 1993-1-2 is modified by fitting the finite analysis results to more accurately evaluate the fire resistance of HSS beams considering the nonuniform temperature distribution.

Keywords: High-strength steel; steel beam; nonuniform temperature distribution; lateral-torsional buckling.

1. Introduction

High-strength steels (HSSs) have been gradually used in large engineering projects due to their high strength-to-weight ratio. Steel structures without fire protection are particularly vulnerable to destruction in the event of a fire due to their high thermal conductivity and the rapid degradation of their strength and stiffness when exposed to heat. Without proper fire protection, these structures can become heated up quickly, leading to serious damage or even complete destruction. Therefore, it is essential to ensure adequate fire protection for any steel structures to protect them from the potentially devastating effects of a fire. While there has been a great deal of research related to the fire resistance of mild steel structures, as the HSSs tend to degrade their mechanical properties at elevated temperatures much differently than mild steel does, several significant differences remain to be identified.[1–7] Investigations into the fire safety of HSS columns[8–11] and connections[12,13] illustrate that the current design tailored for mild steel structures is not applicable to HSS structures, thereby hindering the popularization and adoption of HSSs in building structures. Since the progress in the application of HSS structures is being accompanied by a concern about the vulnerability of steel structures under elevated temperatures, researchers have been focusing their attention on assessing HSS structures' fire resistance over the past few years.

Considering that steel beams are a major load-bearing member in steel structures, they can play an influential role in preventing steel structures from collapsing if they are damaged by fire. Material strength, excessive plastic deformation, stability, and fatigue are the possible variables influencing the design of steel beams. If there is no sufficient lateral support, steel beams are prone to fail in lateral-torsional buckling. Bailey *et al.*[14] conducted a numerical analysis of the fire resistance of steel beams with different section dimensions and spans and found that the finite element model (FEM) predicted lower failure temperatures than the current design method.[15] Vila Real *et al.*[16] learned the lateral-torsional buckling behavior of I-shape beams at different temperatures experimentally, and a new proposal was proposed based on the test-supported numerical analysis results, which are safer compared to EN 1993-1-2.[15] In addition, Mesquita *et al.*[17] performed a batch of numerical analysis and experimental tests on steel beams with constant mechanical action and increasing thermal load to capture the critical temperature when lateral-torsional buckling occurred. Saurabh and Avik[18] carried out a parametric study consisting of 2 100 elastic and inelastic beams using a validated numerical model established in the finite element software ABAQUS. According to the parametric study results, they recommended a new proposal to predict the design capacity of monosymmetric steel I-section beams at elevated temperatures, considering different load patterns.

In fire conditions, the current design standard GB51249-2017[19] assumes that the steel beams have a uniform distribution of temperature, which is justifiable due to the high conductivity of steel at elevated temperatures. However, if concrete slabs exist, steel beams can be assumed to be heated from three sides due to the heat dissipation of concrete slabs, and consequently the beam top flange possesses a lower temperature than the web and bottom flange. For such cases, adopting the method stipulated in the standard yields inaccurate results since it neglects the influence of nonuniform temperature distribution. Yin *et al.*[20] analyzed the influence of slenderness ratio, load ratio and type on the later-torsional buckling of steel beams with nonuniform temperature distribution and compared the results of the numerical studies with the predictions obtained from the methods stipulated in BS 5950 Part 8[21] and EN 1993-1-2,[15] both of which were found to be insufficient for analyzing the fire resistance of steel beams with inhomogeneous temperature distribution. George *et al.*[22] investigated the effect of nonuniform heating on the laminated glass-epoxy composite beam experimentally and numerically, and they found that the temperature profile can have a significant impact in determining the critical buckling temperature.

The beams made of HSS usually have smaller sections than those made of mild steel, which allows them to take advantage of the greater strength of HSS. However, the increased slenderness ratio of HSS beams makes them more susceptible to lateral-torsional buckling than mild steel beams. As a result of testing of welded I-section beams made from WGJ (a new type of high performance steel) under concentrated loads by Xu *et al.*,[23] they found that the lateral-rotational buckling is the major mode of failure. Varol and Cashell[24] carried out a numerical simulation on the lateral-torsional buckling performance of unconstrained HSS beams at elevated temperatures. The results show that the buckling curves proposed by Eurocode 3[15] are not applicable for HSS beams. Through experimental and numerical analyses, Luo *et al.*[25] investigated the structural responses of Q690 beams at ambient and elevated temperatures, and propose two practical formulas to assess the load bearing capability of Q690 HSS beams. An experimental study of six Q690 HSS beams after being exposed to elevated temperature by Guo *et al.*[26] revealed that the residual deformation can significantly impact the failure mode and ultimate bearing capacity of Q690 HSS beams. A fire resistance assessment method, based on the equivalent stiffness method, was proposed by Wang *et al.*[27] for assessing Q460 HSS beams with nonuniform temperature distribution for engineering practices. For cold-formed HSS, Li and Ben[28] proposed the stress and strain relationships under different elevated temperatures, which was subsequently incorporated into the numerical analysis of the responses of cold-formed HSS with square and rectangular hollow sections under fire. By comparing the numerical analysis results against AISI S100-16,[29] it was found that the predictions from the current standards are quite conservative. Based on a variety of numerical analysis of Q460, Q690, and Q960 steel beams by Al-Azzani *et al.*,[30] practical design curves were presented to estimate the lateral-torsional buckling of HSS beams in fire.

The aim of this paper is to investigate the fire resistance of HSS beams which are heated from three sides. In order to accurately simulate the behavior of HSSs beams such as Q460, Q690, and Q960 under elevated temperatures, a FEM has been developed that takes into account nonuniform temperature distributions, initial imperfections, and residual stress distributions. This FEM provides a more accurate assessment of the material properties of the aforementioned HSSs, enabling more reliable predictions of their performance under elevated temperatures. Utilizing the verified model, a parametrical examination is conducted to explore the effect of temperature distribution, load type, steel grade and slenderness ratio on the critical moment of HSS beams. Finally, the results obtained from the FEM are compared with the standard EN 1993-1-2,[15] and simple expressions are proposed to evaluate the critical strength of HSS beams in fire design.

2. Finite Element Model

This section introduces the development of a FEM established in ABAQUS, which aims at simulating the steel beam's fire response and is verified against experimental results from Ref. 25. There is an assumption that the beam is simply supported at the end, and that it is not restrained in vertical displacement, lateral displacement, or torsion in any way, and is free to warp and expand. The steel beam is modeled using S4R elements, each node of which has six degrees of freedom.

2.1. *Retention factors for HSSs at elevated temperatures*

The manufacture of HSSs usually involves adding alloying elements or heat treatment methods such as quenching and tempering, which lead the mechanical response of HSS under fire to be different from that of mild steel. Therefore, Wang *et al.*[4–6] carried out experimental studies on Q460QT, Q690QT, and Q960QT steels (the corresponding nominal yield strength is 460 MPa, 690 MPa, 960 MPa, respectively, and the heat treatment method is quenching and tempering, which follows the requirement of GB/T 1591-2018[31]) to attain their elevated temperature mechanical properties. In view of the absence of clearly defined yield plateaus at high temperatures, the yield strength of steel at elevated temperatures is usually reported at strain offsets of some value, such as 0.5, 1.5, or 2.0%. The yield strength at 1.0% strain level was adopted to obtain the fitting equations of yield strength retention factors of Q460 and Q690 steels, while 0.2% offset strain level was adopted for Q960 steel. Table 1 and Fig. 1 exhibit the associated fitting equations proposed to assess the yield strength and elastic modulus retention factors (η_y and η_E, respectively). For comparison purposes, the retention factors specified in EN 1993-1-2 are also presented in Fig. 1.

There are apparent discrepancies in mechanical properties' degradations among different HSSs at elevated temperatures, and Q690 HSS has the lowest yield strength retention factor when $T > 450°\text{C}$ and the lowest elastic modulus retention factor

Table 1. Proposed equations for the mechanical properties retention factors of Q460, Q690 and Q960 steels at elevated temperatures.

Steel	Yield strength	Elastic modulus
Q460[4]	$\eta_y = \frac{f_{y,T}}{f_y} = \begin{cases} 1 & 20°\text{C} \le T \le 450°\text{C} \\ 4.32 \times e^{-T/880} - 1.6 & 450°\text{C} \le T \le 800°\text{C} \end{cases}$	$\eta_E = \frac{E_T}{E} = 1.02 - 0.035e^{T/280}$
Q690[5]	$\eta_y = \frac{f_{y,T}}{f_y} = \frac{1}{1+(T/538)^{10}}$	$\eta_E = \frac{E_T}{E} = \frac{1}{1+(T/534)^{5.5}}$
Q960[6]	$\eta_y = \frac{f_{y,T}}{f_y} = \frac{1}{1.05+1.29\times10^{-4}\times e^{\frac{T}{70}}}$	$\eta_E = \frac{E_T}{E} = \frac{1}{1.02+1.29\times10^{-4}\times e^{\frac{T}{89}}}$

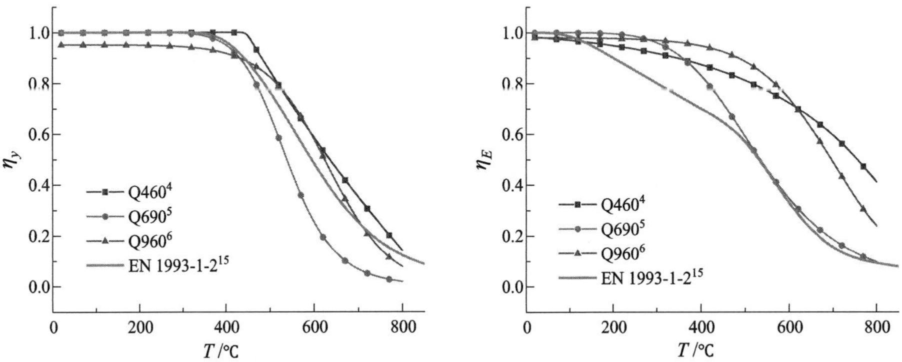

(a) Yield strength (b) Elastic modulus

Fig. 1. Retention factors of mechanical properties of Q460, Q690 and Q960 steels.

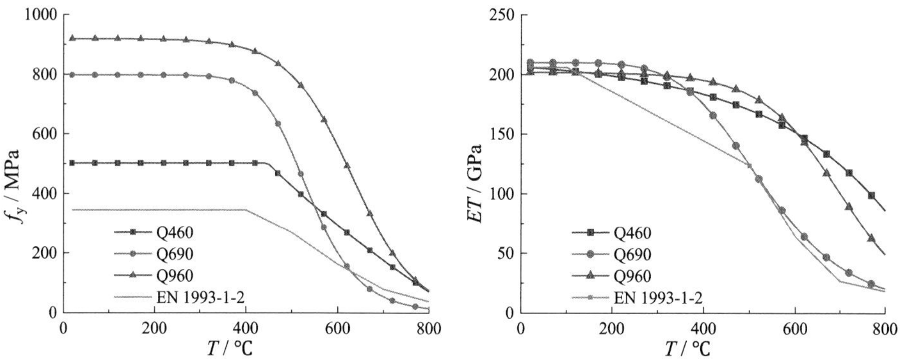

(a) Yield strength

(b) Elastic modulus

Fig. 2. Mechanical properties of Q460, Q690 and Q960 steels.

when $T > 450°C$. When $T > 450°C$, the retention factors of yield strength specified in EN 1993-1-2 are lower than those of Q460 and Q960 HSSs but are higher than those of Q690 HSS. As to the elastic modulus, EN 1993-1-2 provides quite smaller retention factors than those of HSSs, except that the retention factors of elastic modulus in EN 1993-1-2 are close to those of Q690 HSS when $T > 500°C$. Consequently, a simulation of the fire resistance of steel members at elevated temperatures based on the matching mechanical properties of each steel grade is essential in order to generate an accurate assessment of the fire resistance of steel members. Figure 2 illustrates the temperature-dependent evolution of yield strength and modulus of elasticity across steel grades. It is evident that the yield strength of steel increases with steel grade at lower temperatures. However, with increasing temperature, the yield strength of distinct steel grades exhibits notable dissimilarities. Specifically, the yield strength of Q690 steel degrades rapidly, and even falls below that of ordinary steel at $T > 600°C$. Conversely, the modulus of elasticity for different steel grades shows negligible variation at room temperature and follows the same trend as the elastic modulus retention factor.

2.2. *Relationship between stress and strain HSSs at high temperature*

Several models have been developed to describe the constitutive relationship of steel under elevated temperatures. Broadly speaking, these models fall into two categories based on the shape of the model: linear and curve models. Ramberg and Osgood[32] primarily proposed a continuous curve model to describe the relationship between stress and strain of steel at ambient temperature. Subsequently, Hill[33] modified the Ramberg–Osgood model to provide a more accurate representation of the stress–strain relationship in steel that can be applied to simulate their behavior at elevated

Table 2. Fitting coefficients for Q460, Q690, and Q960 steels in the Ramberg–Osgood model.

	Q460			Q690			Q960[6]		
Temperature (°C)	E_T (MPa)	$\sigma_{0.2,T}$ (MPa)	n	E_T (MPa)	$\sigma_{0.2,T}$ (MPa)	n	E_T (MPa)	$\sigma_{0.2,T}$ (MPa)	n
20	212012	528.2	104.7	210000	797.5	67.2	204000	965	538.9
200	174166	492.0	18.4	—	—	—	—	—	—
300	208047	450.6	50.1	191000	706.4	22.1	193060	870.62	46.7
400	197134	453.0	31.4	162500	627.7	12.6	194490	854.83	26.6
450	—	—	—	—	—	—	186510	821.68	18.9
500	152878	427.1	17.3	120500	506.4	16.9	177200	759.7	17.4
550	—	—	—	104000	318.9	10.3	170010	701.5	19.1
600	146853	304.6	27.6	61000	179.9	7.3	159750	631.14	24.6
700	93066	155.1	9.8	22500	47.9	6.4	92530	215.75	13.1
800	59742	76.0	23.7	13950	36.8	6.9	52910	67.11	11.2
900	—	—	—	26750	31.7	22.0	77980	52.18	15.4

temperatures, as shown in Eq. (2.1). The values of parameters in Eq. (2.1) for HSSs are presented in Table 2, in which the values of n for Q460 and Q690 HSSs were obtained by fitting against the existing available test results,[3–5] while the values of coefficient n for Q960 HSS were recommended by Wang *et al.*[6] As shown in Fig. 3, the stress–strain curves of HSSs obtained by fitting Eq. (2.1) are in good agreement with the test data at elevated temperatures.

$$\varepsilon = \frac{\sigma}{E_T} + 0.002\left(\frac{\sigma}{\sigma_{0.2,T}}\right)^n. \tag{2.1}$$

2.3. *Temperature distribution in beam cross-section*

According to EN 1994-1-2,[35] the temperature of each part (top flange, web, and bottom flange, as shown in Fig. 4) of a steel beam without fire protection under fire is calculated by the incremental method as follows:

$$\Delta T_{a,t} = k_{\text{shadow}} \cdot \frac{1}{\rho_a c_a} \cdot \frac{A_i}{V_i} \cdot \dot{h}_{\text{net}} \Delta t, \tag{2.2}$$

where k_{shadow} is the correction factor for the shadow effect, which is determined from the cross-sectional dimensions of the steel beam; ρ_a is the density of steel, which is usually considered to be 7850 kg/m^3; c_a is the specific heat capacity of steel, in J/(kg · °C); A_i and V_i are the exposed surface area and volume of part i of the steel section per unit length; $\frac{A_i}{V_i}$ is the section factor of part $i$related to the sectional shape and size of the member, in m^{-1}; $\dot{h}_{\text{net}}$ is the design value of the net heat flux per unit area, which is determined by

$$\dot{h}_{\text{net}} = \dot{h}_{\text{net,c}} + \dot{h}_{\text{net},r}, \tag{2.3a}$$

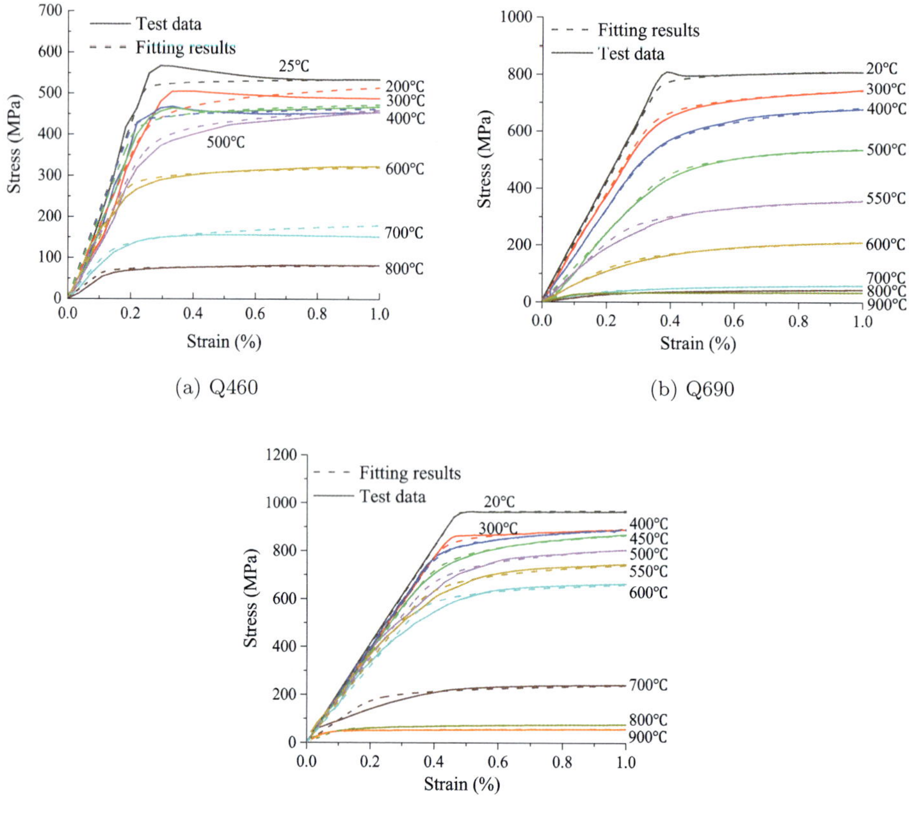

Fig. 3. Fitting results of stress–strain curves of HSSs at elevated temperatures.

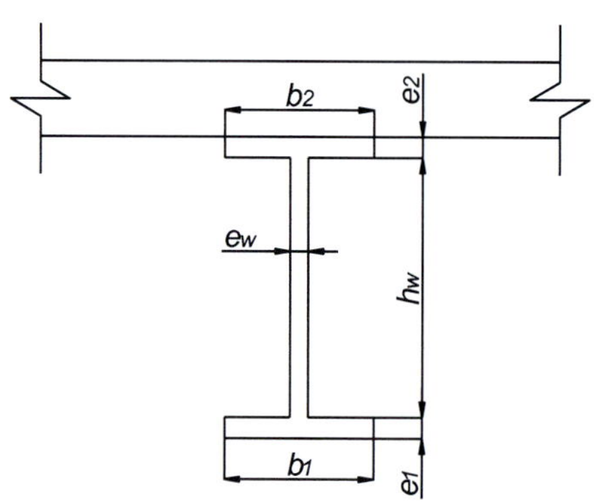

Fig. 4. Elements of a cross-section.

$$\dot{h}_{\text{net},c} = \alpha_c(T_{g,t} - T_{a,t}), \tag{2.3b}$$

$$\dot{h}_{\text{net},r} = 5.67 \times 10^{-8}\varepsilon_m\varepsilon_f[(T_{g,t} + 273)^4 - (T_{a,t} + 273)^4], \tag{2.3c}$$

where α_c is the convective heat transfer coefficient, in $\text{W}/(\text{m}^2 \cdot {}^\circ\text{C})$; ε_m is the emissivity of the fire; ε_f is the emissivity coefficient for steel and concrete related to the surface of the member; $T_{g,t}$ and $T_{a,t}$ are the temperatures of air and steel at time t, respectively.

For the top flange, the section factor $\frac{A_i}{V_i}$ can be calculated as

$$\frac{A_i}{V_i} = \frac{b_1 + 2e_1}{b_1 e_1}. \tag{2.4}$$

For the bottom flange, the section factor $\frac{A_i}{V_i}$ can be calculated as

$$\frac{A_i}{V_i} = \frac{2(b_2 + e_2)}{b_2 e_2}. \tag{2.5}$$

It is recommended that the web temperature can be taken to be the same as the temperature of the bottom flange as long as the beam depth h is less than 500 mm, in accordance with EN1994-1-2. Figure 5 shows the temperature curve of each part of the steel beam with a section of H400 × 200 × 8 × 13 (mm). It can be seen that the temperature of the top flange is significantly lower than that of the bottom flange. The temperature gap between the top and bottom flanges gradually increases with the increase of fire time when $T_{bf} < 700^\circ\text{C}$, while the gap gradually reduces when $T_{bf} < 700^\circ\text{C}$, this is because the increase rate of the bottom flange's temperature slows down due to the increase of steel's specific heat when $T_{bf} < 700^\circ\text{C}$.

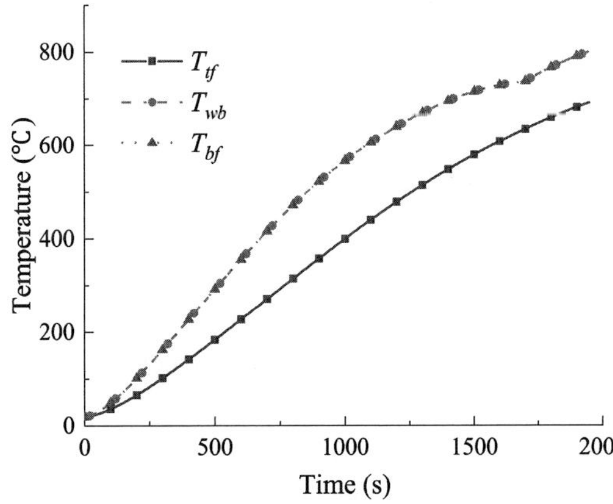

Fig. 5. Variations of T_{tf}, T_{bf} and T_{wb} of the steel beam.

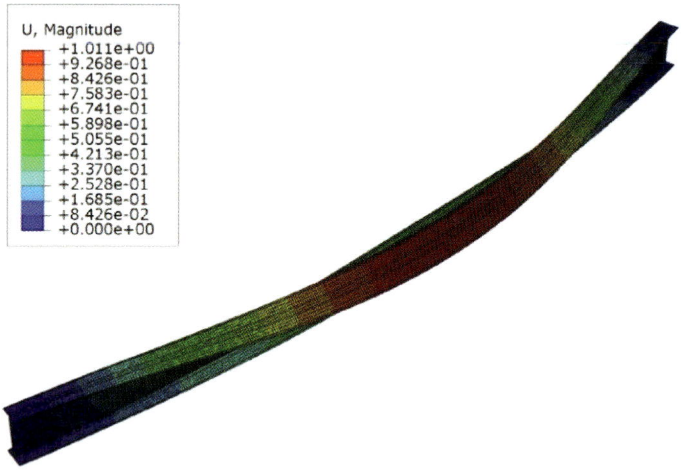

(a) Global geometric imperfection (unit: mm)

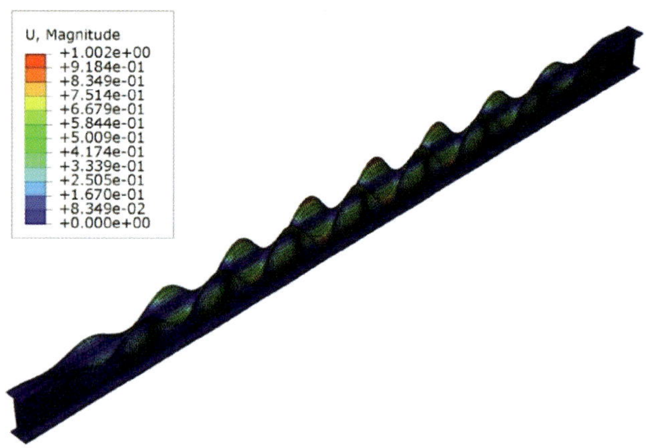

(b) Local geometric imperfections (unit: mm)

Fig. 6. FEM in lateral-torsional buckling.

2.4. *Initial imperfections and residual stresses*

As residual stresses and initial geometric imperfections have unfavorable influences on steel beams' resistance, they should be considered in the fire analysis on steel beams. In the FEM, the initial global and local geometric imperfections are imposed by introducing the buckling modes from the buckling analysis, as shown in Fig. 6. The FEM adopts two modes of the buckling analysis result which depend on the initial geometry of the beam: first, the initial geometric imperfection of the steel beam is considered as the first mode, and second, the modal order of the initial

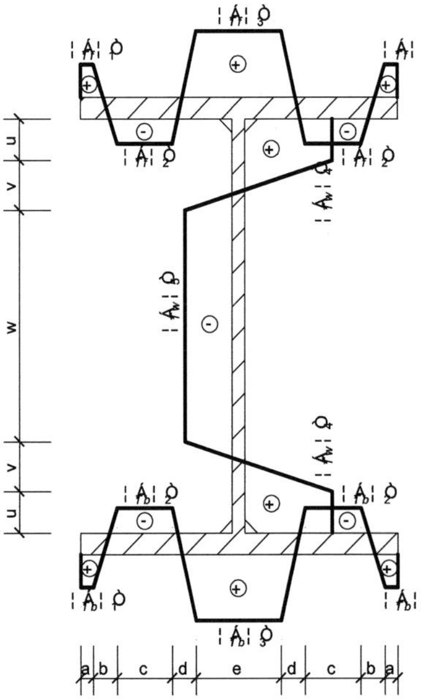

Fig. 7. Residual stress patterns applied to the FEM.[37]

geometric imperfection of the steel beam is considered to be related to its span. According to GB 50205-2001,[36] the original global geometric imperfection is taken as $L/1000$, whereas the initial geometric imperfection value of webs and flanges are $h_w/200$ and $b_f/100$, respectively.

Residual stresses will lead to the elastoplastic state of steel members in advance, which subsequently reduces the bending stiffness. In order to simulate welded high-strength I-sections, the simplified residual stress distribution model described in Ban *et al.*[37] is adopted and implemented as initial stresses of the finite element method, as shown in Fig. 7. The magnitude of the residual tensile stress is related to the yield strength of the steel, and its distribution of such stress is linked to the dimension of the cross-section and welding size. Among them, the distribution range of the maximum residual tensile stress in the middle of the flange (e) is the spacing between the outer weld toes of the welds on both sides of the web. The maximum tensile stress distribution near the flame cutting edge (a) and transition range (b) are taken as 1/10 of the weld toe to the end of the flange. Tensile stress distribution range (u) is taken as the weld toe size. Residual compressive stresses in welded sections are primarily attributed to nonuniform cooling following welding and are not inherently correlated with material strength. The dimensions and distribution of these stresses must satisfy conditions that result in self-balancing residual stresses within the cross-section. While in fire conditions, residual stresses in steel sections can be relieved due to strength deterioration and plastic deformation of steel. Based on the research of residual stress distributions of Q235, Q460, and Q690 steel welded sections, Yang *et al.*[38] proposed the fitting equation of the reduction coefficient of residual stress α_{rT} at elevated temperatures, as shown in Eq. (2.6), which is adopted to obtain the beam's residual stress distribution at elevated temperatures in this study.

$$\alpha_{rT} = \begin{cases} \dfrac{1}{1+(T/550)^{2.05}} & 20^\circ\mathrm{C} \le T \le 400^\circ\mathrm{C} \\ \dfrac{1}{1+(T/460)^{4.66}} & 400^\circ\mathrm{C} < T \le 600^\circ\mathrm{C}\,. \\ \dfrac{1}{1+(T/485)^{5.82}} & 600^\circ\mathrm{C} < T \le 900^\circ\mathrm{C} \end{cases} \tag{2.6}$$

2.5. *Model validation*

Luo *et al.*[25] conducted experiments on simply supported and cantilever Q690 HSS beams at 650°C, the results of which are employed to validate the FEM established in this study. The specimens were made from Q690 HSS, whose yield strength and elastic modulus were measured by tensile tests done on the specimens at ambient temperature, and they were found to have an average yield strength of 768.6 MPa and an average elastic modulus of 206.4 GPa. For simply supported Q690 HSS beams, two sections were selected: H300 × 150 × 10 × 12 (depth × width × flange thickness × web thickness, in mm) and H350 × 150 × 10 × 12, which were named H300 and H350, respectively. In order to prevent local buckling, bearing stiffeners

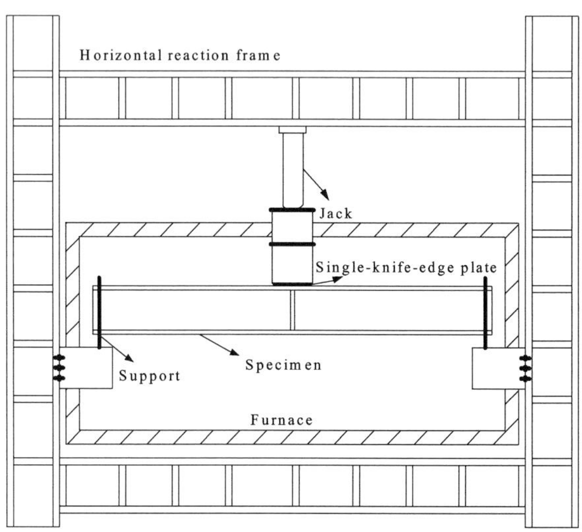

Fig. 8. Test device.[25]

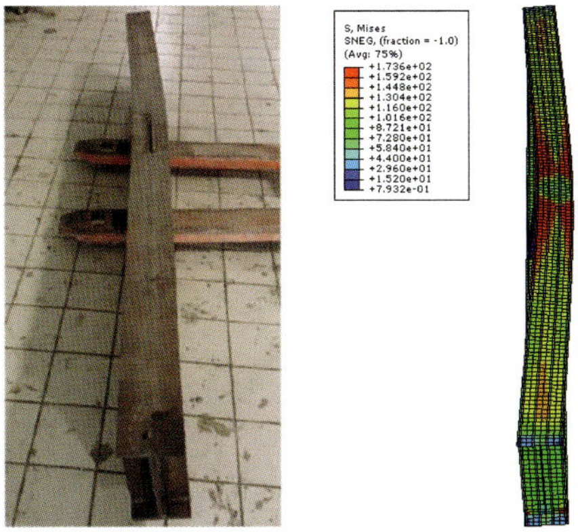

(a) Test (b) FEM (unit: MPa)

Fig. 9. Comparison of the failure mode in the experiment and FEM.

Table 3. Experimental and numerical results of the ultimate load of specimens H300 and H350.

Specimens	Test result[25]	FEM result	Error
H300	91.8kN	91.5 kN	0.3%
H350	102.2kN	104.7 kN	2.4%

are added at the mid-span section and end supports section of the steel beam, with a thickness of 10 mm. The test device for simply supported Q690 steel beams is shown in Fig. 8. There was a concentrated load applied to the beam from a hydraulic jack at the midsection of the beam.

Same as the experiment, lateral torsional buckling mode was obtained in the FEM, as per Fig. 9. A summary of the results of ultimate loads on specimens H300 and H350 based on FEM and tests is presented in Table 3. Based on Table 3, it is evident that the error in estimating the ultimate load by the FEM is no greater than 3%, which is negligible and indicates good agreement between the FEM and experimental results. The curves of load versus lateral deflections at beam mid-span of specimen H350 obtained from the FEM, together with the experimental results, are presented in Fig. 10, whereas the parameter DT2 represents the lateral displacement of the top end flange of the midspan and the parameter DT3 represents the lateral displacement of the bottom end flange of the beam mid-span, respectively. Therefore, the established FEM is verified to predict the lateral-torsional buckling behavior of HSS beams under high temperatures.

3. Parametric Study

Based on the validated model, a study is presented in this section to analyze the effects of temperature, load pattern, steel grade and span on the critical bending

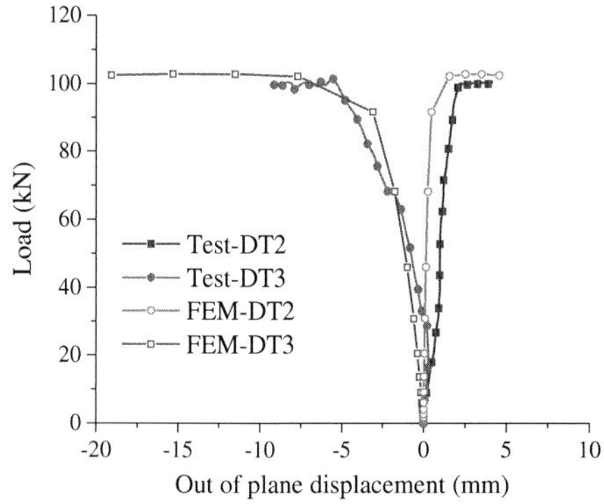

Fig. 10. Comparison of load-displacement curves between the experimental and FEM results.

Table 4. The model information for parameter analysis.

Section (mm)	Temperature level (°C)	Temperature distribution	Steel grade	Span (m)	Load pattern
H400 × 200 × 8 × 13	20, 100, 200, 300, 400, 500, 600 and 700°C	Uniform and nonuniform	Q460, Q690, and Q960	4, 5, 6, 7, 8, 9, and 10 m	Uniform bending moment
			Q960	8 m	A concentrated force and uniformly distributed load

moment of the lateral-torsional buckling (M_{cr}). The model information for parametric analysis is shown in Table 4.

3.1. *Temperature distribution*

Figure 11 illustrates the applied moment and in-plane displacement curves under various levels of elevated temperatures under nonuniform and uniform temperature distributions. The value of M_{cr} is determined by the peak value of the moment-displacement curve obtained from the FEM analysis result. Illustrated in Fig. 12 is the influence of temperature distributions on M_{cr} of HSS beams at elevated temperatures. The temperatures of each part of the steel beams with nonuniform temperature distribution are obtained by Eq. (2.2). The temperatures shown in the parametric study are the temperatures of the bottom flange, T_{bf}, for the reason that the bottom flange possesses the highest temperature in fire. Figure 12 clearly shows that the M_{cr} decreases as temperature increases and considering the effect of the nonuniform temperature distribution yields higher M_{cr}. This is because a beam heated from three sides has a lower top flange temperature and fewer stiffness and strength degradations than the web and bottom flanges. Therefore, the top flange

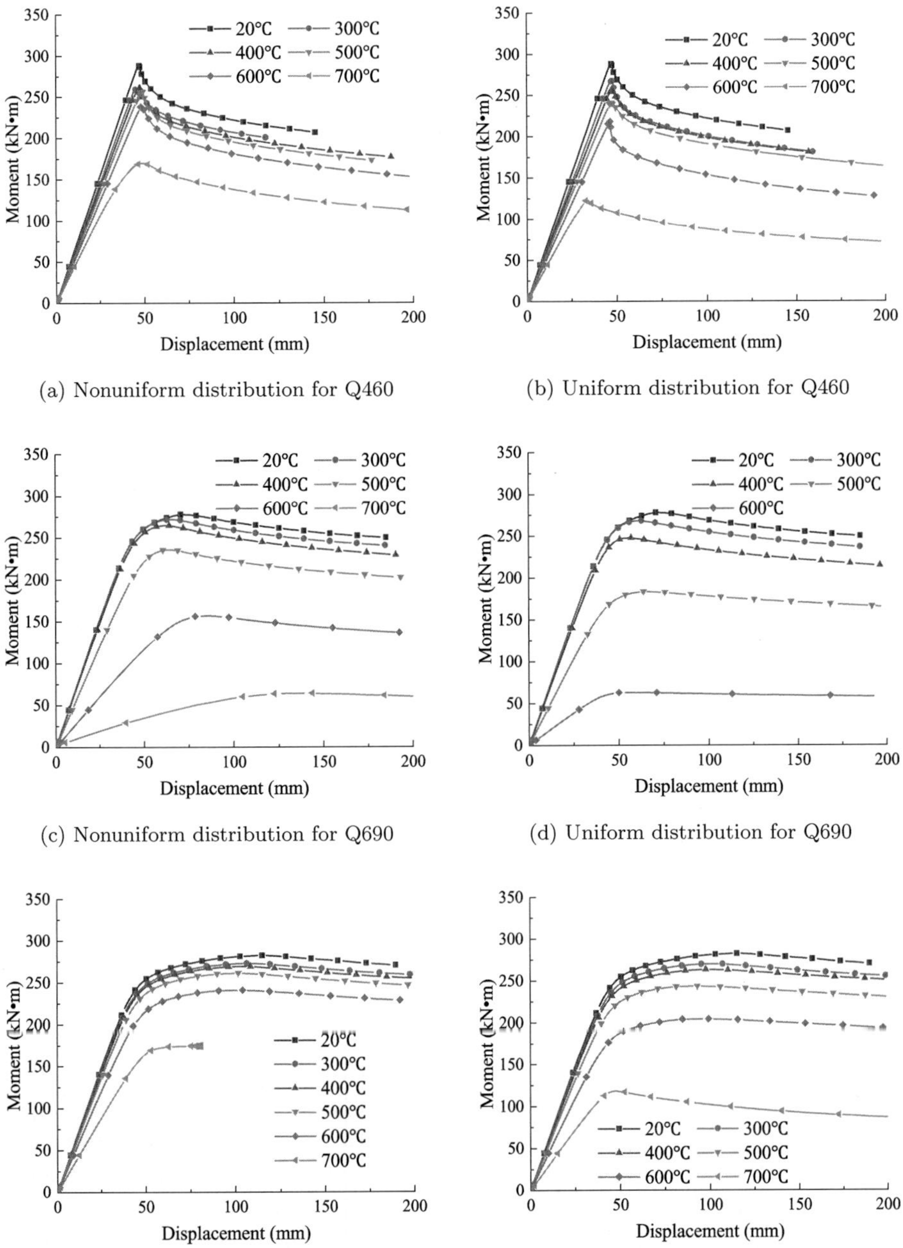

(a) Nonuniform distribution for Q460

(b) Uniform distribution for Q460

(c) Nonuniform distribution for Q690

(d) Uniform distribution for Q690

(e) Nonuniform distribution for Q960

(f) Uniform distribution for Q960

Fig. 11. Moment-displacement curves of HSS beams at elevated temperatures.

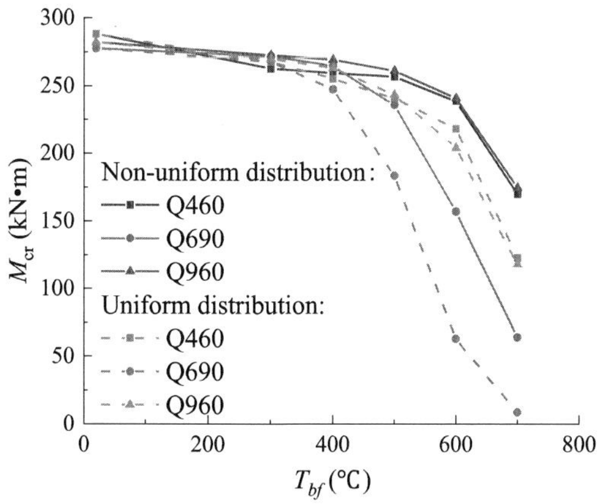

Fig. 12. Critical bending moments of HSS beams with different temperature distributions.

in compression possesses higher strength compared to the steel beam in which the temperature distribution is uniform. Moreover, with an increase in temperature, the discrepancy between M_{cr} calculated from nonuniform and uniform temperature distributions becomes more apparent. That is because the temperature variation between the top and bottom flanges increases with temperature, as discussed in Sec. 2.2, which makes it more apparent that the effect of nonuniform distribution on M_{cr} becomes more obvious as the temperature increases.

The M_{cr} of Q460, Q690, and Q960 HSS beams at ambient temperature are close. This is because the steel beams are in an elastic state when buckling occurs, and thereby M_{cr} depends on the elastic modulus of the steel. Although the yield strength of Q460, Q690 and Q960 are different, their elastic modulus at ambient temperature is almost the same, and thus leading to similar M_{cr} at high temperatures. A noteworthy fact about the Q690 steel beam is that as compared to Q460 and Q960 steel beams, its M_{cr} decreases the most. This is because the elastic modulus of Q690 steel decreases the fastest at elevated temperatures, as shown in Fig. 1, leading to a lower M_{cr} value than that of Q460 and Q960 steel beams at elevated temperatures. When the temperature increases from 20°C to 700°C, the elastic modulus of Q460, Q690, and Q960 steels decreases significantly, with respective drops of 59%, 91%, and 77%. Furthermore, the M_{cr} of Q460, Q690, and Q960 HSS beams with nonuniform temperature distribution decreases by 33%, 77%, and 38%, respectively. On the other hand, the M_{cr} of the same Q460, Q690, and Q960 HSS beams with uniform temperature distribution decreases by 41%, 90%, and 58%, respectively. Therefore, due to the fact that steel elastic modulus is reduced at elevated temperatures, the M_{cr} of HSS beams depends on the decrease in steel elastic modulus, but this is not completely controlled by it. The effect of nonuniform temperature distribution on the M_{cr} of HSS beams is not uniform among different steel grades due to the

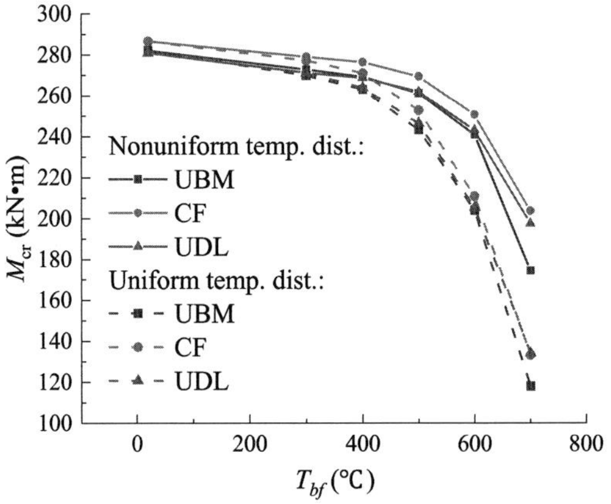

Fig. 13. Critical bending moments of Q960 HSS beams under different load patterns.

obvious discrepancies in material properties reduction factors, as evidenced by the data presented in Fig. 1. This variation has important implications for the design of such beams and must be taken into account in order to ensure the structural integrity of the final product.

3.2. *Load pattern*

There is indeed an effect that the load patterns have on the performance of HSS beams in terms of lateral-torsional buckling. Three load patterns are applied in this paper: uniform bending moment (UBM), a concentrated force (CF) at mid-span, and uniformly distributed load (UDL).

Figure 13 presents the variations of critical bending moments of HSS beams with different load patterns. In Fig. 13, with the increase of T_{bf}, M_{cr} for all load types decreases at almost the same rate when the temperature is raised from 0°C to 500°C and then declines with a larger rate from 500°C to 700°C. Additionally, the M_{cr} of steel beam subjected to UBM degenerates more quickly with temperature rise than the M_{cr} of steel beam afforded CF and UDL. A similar phenomenon can be observed in cases with uniform temperature distribution, as shown in Fig. 13.

3.3. *Steel grade*

Presented in (c) $L = 8\,\mathrm{m}$ (d) $L = 10\,\mathrm{m}$ Fig. 14 is the impact of steel grade on M_{cr}. At ambient temperature, the M_{cr} with lower spans have greater discrepancies among different steel grades, while the discrepancies gradually become insignificant as the span increases, indicating that the influence of steel grade on the M_{cr} of HSS beams decreases as beam span increases. This reason for this is that a shorter

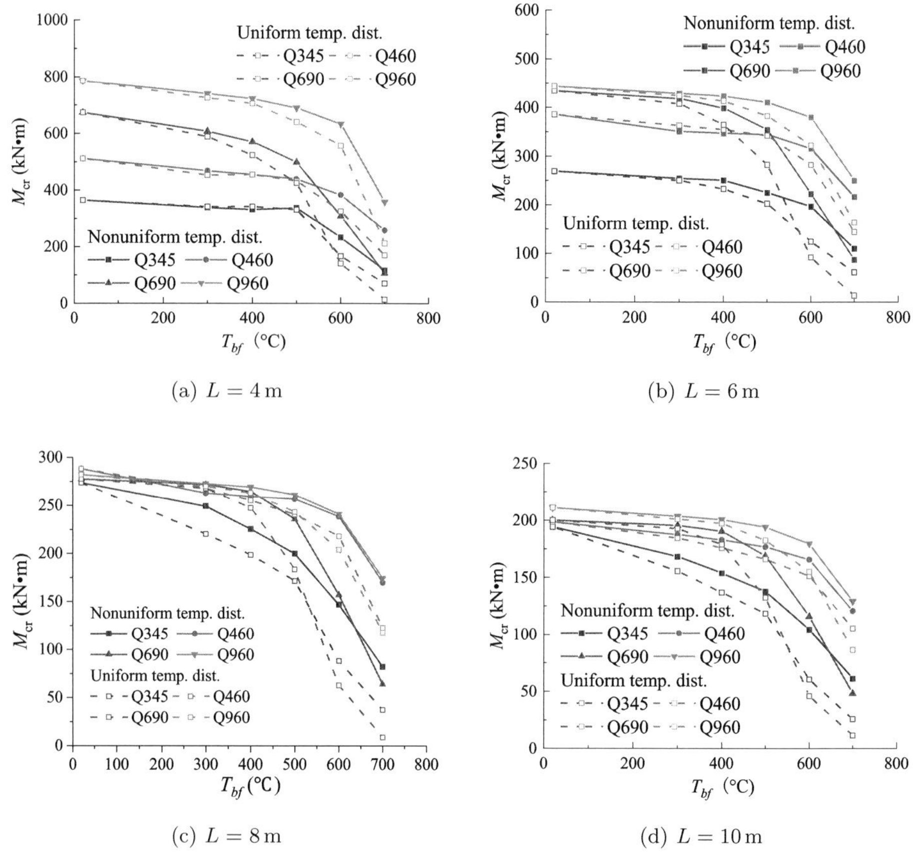

Fig. 14. Influence of the steel grade on the critical bending moment of HSS beams.

beam is more prone to fail in cross-section yielding, and the M_{cr} is primarily controlled by the material strengths. With an increase in the span of beam, the steel beam tends to fail in flexural-torsional buckling and lateral-torsional buckling, and the corresponding M_{cr} of the steel beam is more controlled by its elastic modulus. It is found that beams made from Q690 steel have the fastest decrease in M_{cr} as a result of increased temperature, which is significantly lower than that of beams made of other steels when $T_s \geq 500°C$. This phenomenon can be ascribed to the greater degradations of the elastic modulus of Q690 steel in the range of $T_s \geq 500°C$ compared to Q460 and Q960 steels. Therefore, it is necessary to propose appropriate design equations for evaluating the critical bending moment of beams made of different HSSs. A similar phenomenon can be observed in cases with uniform

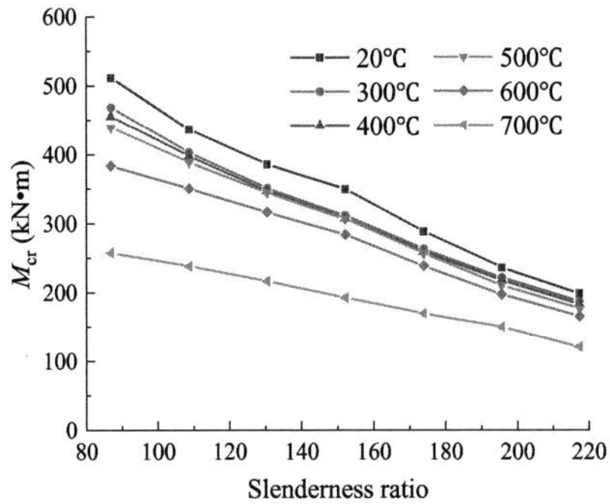

(a) Q460 beams with nonuniform distribution

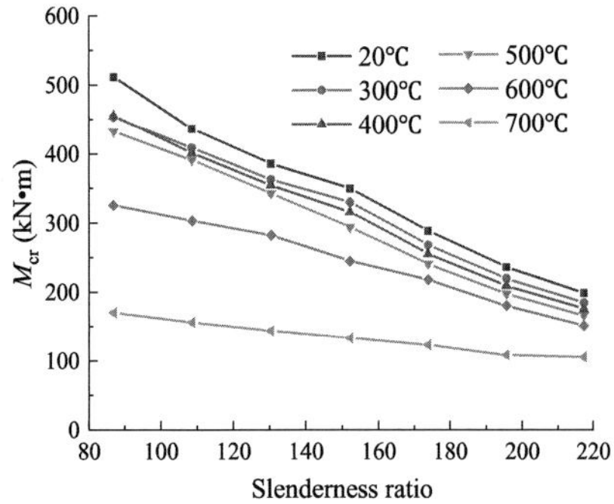

(b) Q460 beams with uniform distribution

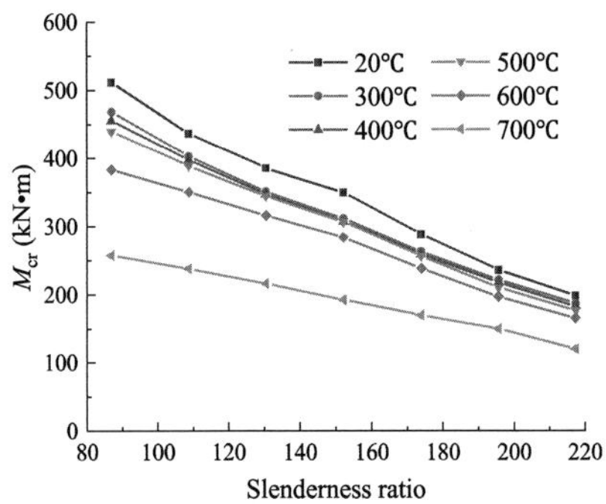

(c) Q690 beams with nonuniform distribution

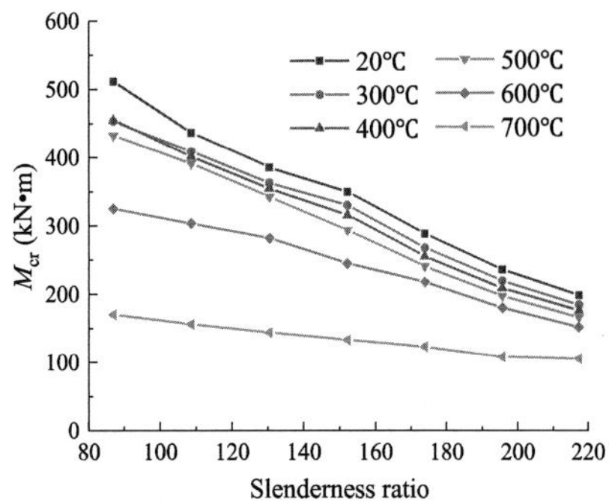

(d) Q690 beams with uniform distribution

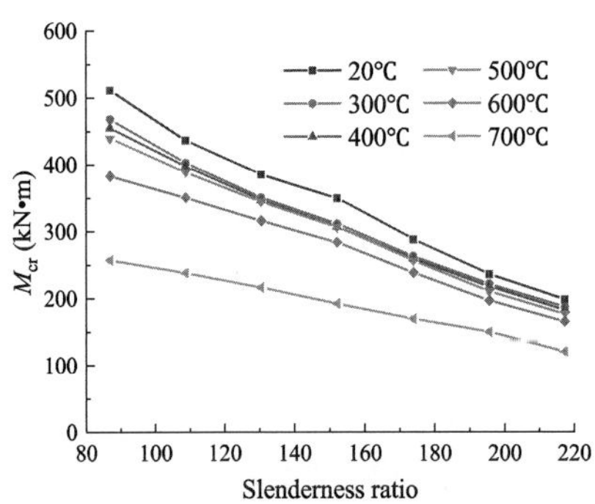

(e) Q960 beams with nonuniform distribution

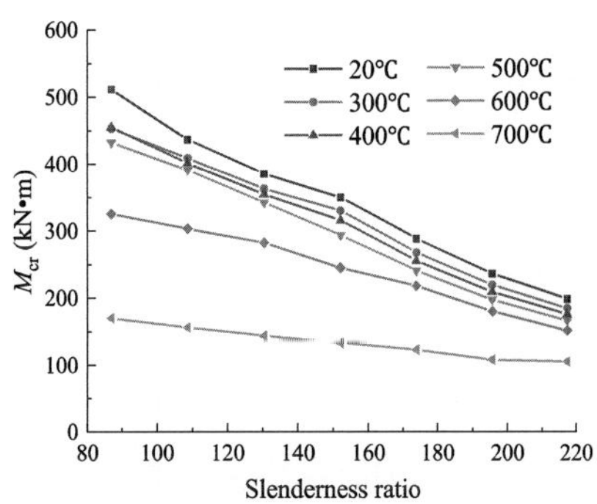

(f) Q960 beams with uniform distribution

Fig. 15. Influence of the slenderness ratio (beam span) on the critical bending moment of HSS beams.

temperature distribution. Furthermore, in order to evaluate the effect of a nonuniform temperature distribution on the performance of HSS structures relative to mild steel structures, the M_{cr} of Q345 steel was computed and compared to that of

HSS steel. The M_{cr} of Q460 and Q960 steels at high temperatures surpasses that of ordinary steels, owing to the fact that the elastic modulus of Q460 and Q960 steels exhibits a lower rate of decay at high temperatures than that of ordinary steels. It is noteworthy that the M_{cr} of Q690 steel beams is lower than that of ordinary steel beams when the temperature achieve 600°C. It can be found from Fig. 2 that when $T > 600$°C, the elastic modulus of Q690 is slightly greater than that of ordinary steel, while its yield strength is lower than that of ordinary steel, thus leading to a lower M_{cr} of Q690.

3.4. *Slenderness ratio*

Presented in Fig. 15 is the influence of slenderness ratio, which is varied by altering the beam span on the M_{cr}. There is no doubt that as the slenderness ratio (beam span) increases, the M_{cr} decreases considerably, as expected. However, the impact of the span on the M_{cr} declines as the temperature rises. There is a specific example that the M_{cr} of Q690 steel beam with nonuniform distribution temperature decreases by 72% at 20°C when the slenderness ratio of the steel beam increases from 86.9 to 217.3, while at 700°C, the reduction of the M_{cr} changes to 44%. On the other hand, the impact of temperature on the M_{cr} of steel beams steadily diminishes as the span increases. For instance, if the temperature of a Q690 steel beam with a span of 4 m increases from 20°C to 700°C, then the M_{cr} decreases by 84%, while if the temperature of the beam with a span of 10 m increases by 76%, then the M_{cr} drops by 74%.

4. Simplified Design Approach

4.1. *Applicability of EN 1993-1-2 to HSS beams*

There is an expression in EN 1993-1-2[15] that describes how to evaluate the temperature resistance of a laterally unsupported steel beam when it is exposed to elevated temperatures, as follows:

$$M_{b,\mathrm{fi,Rd}} = \chi_{\mathrm{LT,fi}} M_{\mathrm{fi},t,\mathrm{Rd}}, \tag{4.1}$$

where $\chi_{\mathrm{LT,fi}}$ is the reduction factor for lateral-torsional buckling in fire design; and $M_{\mathrm{fi},t,\mathrm{Rd}}$ is the design moment resistance, which can be calculated as follows:

$$M_{\mathrm{fi},t,\mathrm{Rd}} = \sum_{i=1}^{n} A_i z_i k_{y,\theta,i} f_{y,i} / \gamma_{M,fi}, \tag{4.2}$$

where A_i is an elemental area of the cross-section with a temperature θ_i. A_i shall be taken as positive if the area is in compression and negative if the area is in tension. $k_{y,\theta,i}$ is the retention factor for the yield strength at temperature θ_i; z_i is the distance from the plastic neutral axis to the centroid of the elemental area A_i;

$f_{y,i}$ is the yield strength at ambient temperature; and $\gamma_{M,fi}$ is the partial factor, taken as unity.

In EN 1993-1-2, the value of $\chi_{LT,fi}$ is determined as follows:

$$\chi_{\mathrm{LT,fi}} = \frac{1}{\phi_{\mathrm{LT},\theta,\mathrm{com}} + \sqrt{[\phi_{\mathrm{LT},\theta,\mathrm{com}}]^2 - [\bar{\lambda}_{\mathrm{LT},\theta,\mathrm{com}}]^2}}, \tag{4.3}$$

with

$$\phi_{\mathrm{LT},\theta,\mathrm{com}} = 0.5[1 + \alpha\bar{\lambda}_{\mathrm{LT},\theta,\mathrm{com}} + (\bar{\lambda}_{\mathrm{LT},\theta,\mathrm{com}})^2], \tag{4.4}$$

in which α is the imperfection factor calculated as

$$\alpha = 0.65\sqrt{235/f_y}. \tag{4.5}$$

The non-dimensional slenderness at elevated temperature $\bar{\lambda}_{\mathrm{LT},\theta,\mathrm{com}}$ is determined as

$$\bar{\lambda}_{\mathrm{LT},\theta,\mathrm{com}} = \bar{\lambda}_{\mathrm{LT}}[k_{y,\theta,\mathrm{com}}/k_{E,\theta,\mathrm{com}}]^{0.5}, \tag{4.6}$$

where $\bar{\lambda}_{\mathrm{LT}}$ is the non-dimensional slenderness at ambient temperature, which can be calculated as

$$\bar{\lambda}_{\mathrm{LT}} = \sqrt{\frac{W_{\mathrm{pl},y} f_y}{M_{\mathrm{cr0}}}}, \tag{4.7}$$

where M_{cr0} is the elastic critical moment for lateral-torsional buckling at ambient temperature, which can be determined from

$$M_{\mathrm{cr0}} = C_1 \frac{\pi}{kL}\sqrt{\mathrm{EI}_y \mathrm{GI}_t}\sqrt{1 + \frac{\pi^2 \mathrm{EI}_w}{(kL)^2 G I_t}}, \tag{4.8}$$

where C_1 is the equivalent uniform moment factor considering for the different load patterns; for the uniform bending moment, $C_1 = 1$. I_y, I_w and I_t are the moments of inertia about the weak axis, warping constant and torsion constant of the section, respectively. k is the effective length factor.

As discussed in Sec. 3.2, the load pattern has a slight effect on the M_{cr} at elevated temperatures. To study the practicality of EN 1993-1-2 towards HSS beams, the critical bending moments of steel beams that made from Q460, Q690 and Q960 HSS with lengths ranging from 4 000 mm to 10 000 mm (corresponding to slenderness ratios ranging from 86.9 to 217.3) under nonuniform bending moment are studied by the verified FEM. The comparison between the results obtained from the FEM and the corresponding curves for EN 1993-1-2 are shown in Fig. 16, indicating that the predictions from EN 1993-1-2 are conservative for Q460 and Q690 HSS beams, especially when $\bar{\lambda}_{\mathrm{LT},\theta,\mathrm{com}} < 1.2$.

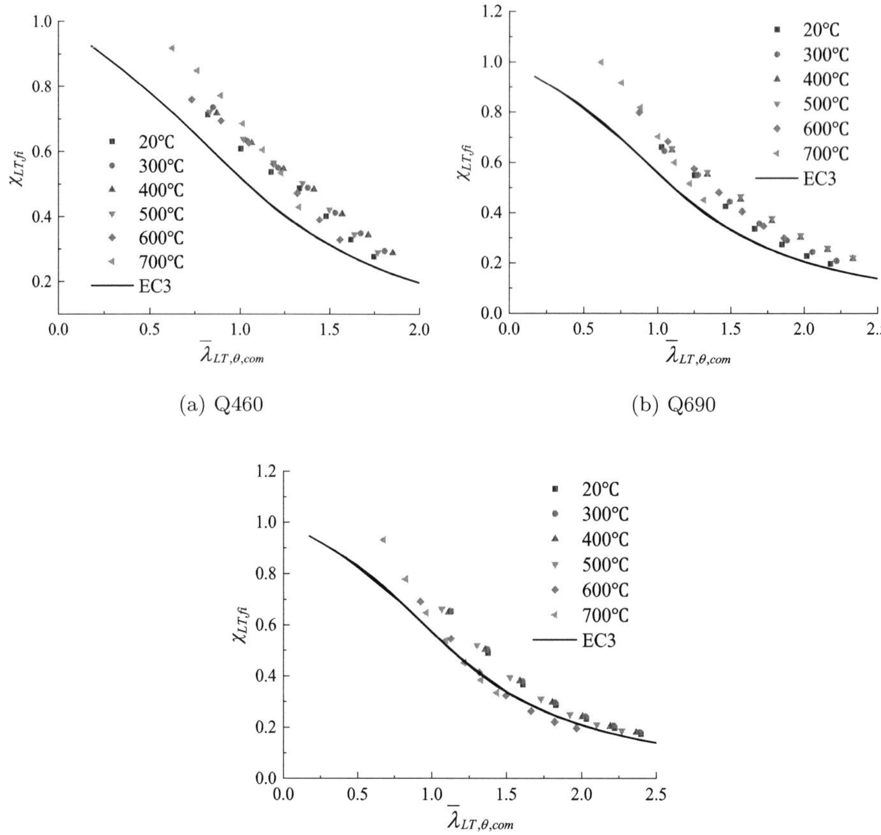

(a) Q460 (b) Q690

(c) Q960

Fig. 16. Comparison results between the FEM and EN 1993-1-2.

4.2. *Proposed design approach*

Based on the parametric study results, the reduction factor for lateral-torsional buckling has been modified, that is, a modified reduction, $\chi_{\text{LT,fi,mod}}$, is proposed to predict the lateral-torsional buckling strength for HSS beams in the fire design:

$$\chi_{\text{LT,fi, mod}} = \frac{h_{\text{FT, mod}}}{\phi_{\text{LT},\theta,\text{com}} + \sqrt{[\phi_{\text{LT},\theta,\text{com}}]^2 - [\bar{\lambda}_{\text{LT},\theta,\text{com}}]^2}}, \tag{4.9a}$$

where $h_{\text{FT,mod}}$ is a coefficient attained by fitting the FEM results.

For Q460 HSS beams:

$$h_{\text{FT, mod}} = \begin{cases} e^{0.255-0.013\bar{\lambda}_{\text{LT},\theta,\text{com}}} & 20°\text{C} \le T_{bf} \le 500°\text{C} \\ e^{0.351-0.100\bar{\lambda}_{\text{LT},\theta,\text{com}}} & 500°\text{C} < T_{bf} \le 700°\text{C} \end{cases}. \tag{4.9b}$$

For Q690 HSS beams:

$$h_{\text{FT, mod}} = \begin{cases} e^{0.454-0.180\bar{\lambda}_{\text{LT},\theta,\text{com}}} & 20°\text{C} \le T_{bf} \le 300°\text{C} \\ e^{0.365-0.039\bar{\lambda}_{\text{LT},\theta,\text{com}}} & 300°\text{C} < T_{bf} \le 500°\text{C} \\ e^{0.456-0.168\bar{\lambda}_{\text{LT},\theta,\text{com}}} & 500°\text{C} < T_{bf} \le 700°\text{C} \end{cases}. \tag{4.9c}$$

For Q960 HSS beams:

$$h_{\text{FT, mod}} = \begin{cases} e^{0.576-0.262\bar{\lambda}_{\text{LT},\theta,\text{com}}} & 20°\text{C} \le T_{bf} \le 500°\text{C} \\ e^{0.600-0.458\bar{\lambda}_{\text{LT},\theta,\text{com}}} & 500°\text{C} < T_{bf} \le 700°\text{C} \end{cases}. \tag{4.9d}$$

The applicable conditions of this approach are as follows: (1) The section of the steel beam is a welded section; (2) The steel material is Q460QT, Q690QT and Q960QT HSS; (3) The slenderness ratio of the steel beam ranges from 86.9 to 217.3,

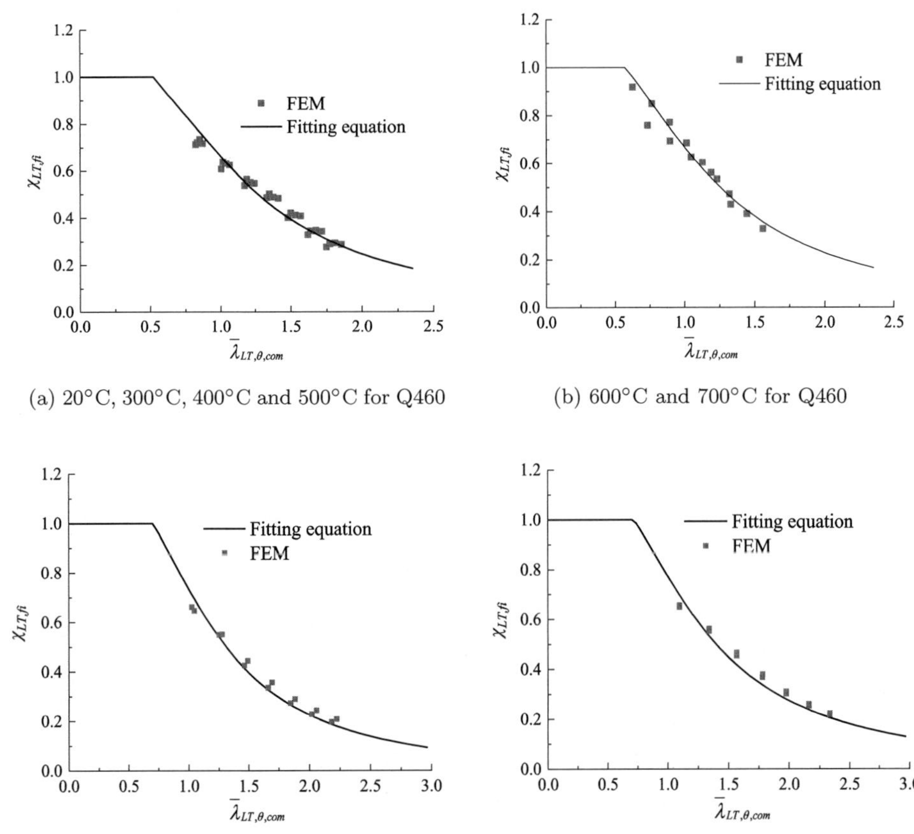

(a) 20°C, 300°C, 400°C and 500°C for Q460

(b) 600°C and 700°C for Q460

(c) 20°C and 300°C for Q690

(d) 400°C and 500°C for Q690

Fig. 17. Comparison results between the FEM and the proposed buckling curve.

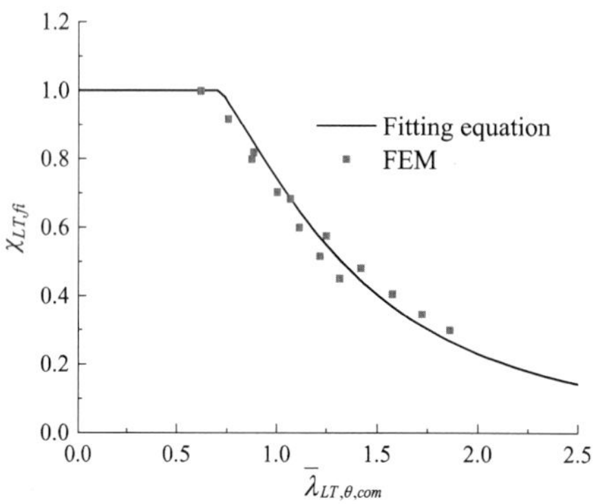

(a) 600°C and 700°C for Q690

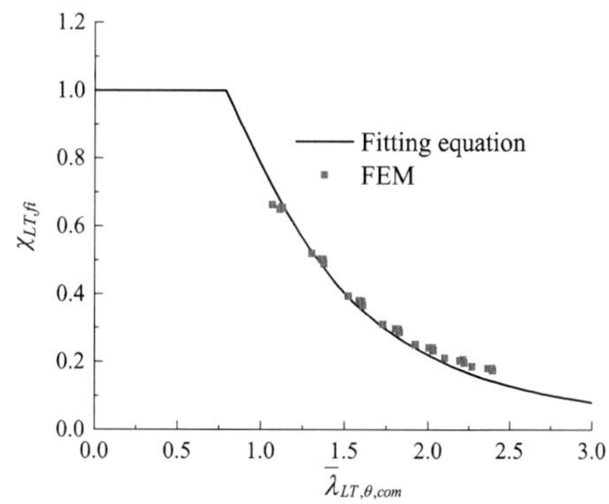

(b) 20°C, 300°C, 400°C and 500°C for Q960

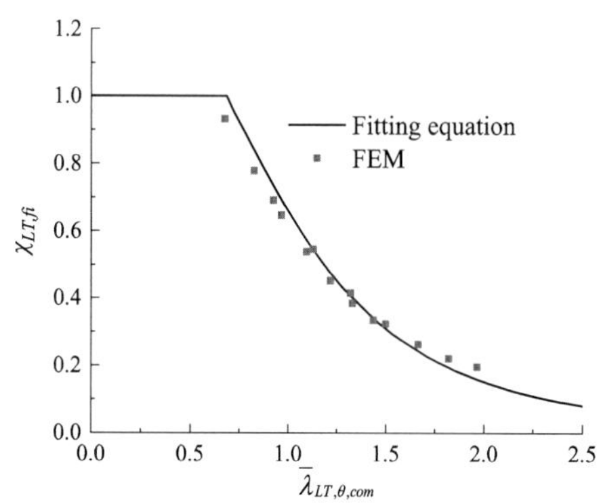

(c) 600°C and 700°C for Q960

Fig. 17. (*Continued*)

and the beam depth h does not exceed 500 mm; (4) Steel beams are insulated with heat-insulating flanges on their tops.

Shown in Fig. 17 are the comparisons between the results obtained from Eq. (4.9) and the FEM. The black line in Fig. 17 in the revised paper is related to Eq. (4.9a), where the $h_{\mathrm{FT,mod}}$ in Eq. (4.9a) is calculated by Eqs. (4.9b)–(4.9d). It can be seen from Fig. 17 that incorporating the proposed modified retention factor enables the expression specified in EN 1993-1-2 to more accurately assess the resistance of Q460, Q690, and Q960 HSS beams exposed to three-side fire.

5. Conclusions

There has been a detailed study carried out on the lateral-torsional buckling behavior of Q460, Q690 and Q960 HSS beams subjected to fire from three sides. By adopting the Ramberg–Osgood model, the stress–strain relationships of Q460 and Q690 HSSs at elevated temperatures are obtained. By incorporating the proposed Ramberg–Osgood model, extensive numerical investigations are performed using the validated FEM to assess the effects of nonuniform temperature distribution, load pattern, steel grade, and slenderness ratio on the M_{cr} of HSS beams. The main conclusions attained from this study are as follows:

(1) With the presence of concrete slabs, the beam top flange has a lower temperature than the other parts in fire scenarios due to the heat dissipation of the concrete. For such cases, it is found that the analysis considering the nonuniform temperature distribution in the beam cross-section yields a higher critical bending moment compared to the analysis assuming the temperature to be uniformly distributed, and the corresponding discrepancies increases as the cross-sectional temperature gradient increases.
(2) The M_{cr} of steel beams associated with lateral-torsional buckling at elevated temperatures depends on the steel elastic modulus degradation with temperature, making Q690 HSS beams have lower resistance than Q460 and Q960 HSS beams at elevated temperatures.
(3) The impact of the span on the M_{cr} of HSS beams declines as the temperature rises. In addition, the M_{cr} of steel beams with lower spans highly depends on the steel strength, while the influence of steel strength on the M_{cr} gradually reduces as the span increases.
(4) For Q460 and Q690 HSS beams with small slenderness ($\bar{\lambda}_{\mathrm{LT},\theta,\mathrm{com}} < 1.2$), the predictions by EN 1993-1-2 are quite conservative. Therefore, a modified retention factor for evaluating the M_{cr} of HSS beams in the fire design is introduced by fitting the FEM results. By incorporating the modified retention factor proposed in this study, it is found that the equation stipulated in EN 1993-1-2 can be employed to more accurately evaluate the resistance of Q460, Q690 and Q960 HSS beams under three-side fire.

Acknowledgments

The authors wish to acknowledge the support of the National Natural Science Foundation of China (Grant No. 52178110), Graduate research and innovation foundation of Chongqing, China (Grant No. CYS21028). Any opinions, findings, conclusions or recommendations expressed in this paper are those of the authors and do not necessarily reflect the views of the sponsors.

References

1. J. Chen, B. Young and B. Uy, Behavior of high strength structural steel at elevated temperatures, *J. Struct. Eng.* **132**(12) (2006) 1948–1954.

2. Y. Z. Li, M. J. Wang, G. Q. Li and B. H. Jiang, Mechanical properties of hot-rolled structural steels at elevated temperatures: A review, *Fire Safety J.* **119** (2021) 103237.
3. G. Q. Li, X. X. Wang, C. Zhang and W. Y. Cai, Creep behavior and model of high-strength steels over 500 MPa at elevated temperatures, *J. Constr. Steel Res.* **168**(5) (2020) 105989.
4. W. Y. Wang, B. Liu and V. Kodur, Effect of temperature on strength and elastic modulus of high-strength steel, *J. Mater. Civil Eng.* **25**(2) (2013) 174–182.
5. W. Y. Wang, K. Wang, V. Kodur and B. Wang, Mechanical properties of high-strength Q690 steel at elevated temperature, *J. Mater. Civil Eng.* **30**(5) (2018) 04018062.
6. W. Y. Wang, Y. H. Zhang, L. Xu and X. Li, Mechanical properties of high-strength Q960 steel at elevated temperature, *Fire Safety J.* **114** (2020) 103010.
7. W. Y. Wang, S. H. Yan and J. P. Liu, Studies on temperature induced creep in high strength Q460 steel, *Mater. Struct.* **50**(1) (2017) 68.
8. W. Y. Wang, L. B. Zhang, Y. Ge and L. Xu, Behaviour of restrained high strength steel columns at elevated temperature, *J. Constr. Steel Res.* **148** (2018) 251–264.
9. G. Q. Li, X. X. Wang, C. Zhang and W. Y. Cai, Creep effect on critical strength of high-strength steel columns exposed to fire, *Eng. Struct.* **220** (2020) 110971.
10. D. Winful, K. A. Cashell, S. Afshan, A. M. Barnes and R. Pargeter, Behaviour of high strength steel columns under fire conditions, *J. Constr. Steel Res.* **150** (2018) 392–404.
11. X. Li, W. Y. Wang and K. Q. Yu, Local buckling of Q690 high-strength steel columns subjected to axial compression at elevated temperature, *Prog. Steel Build. Struct.* **23**(3) (2021) 54–63 (in Chinese).
12. X. H. Qiang, X. Jiang, F. Bijlaard and H. Kolstein, Performance assessment on high strength steel endplate connections in fire, *J. Struct. Fire Eng.* **7**(2) (2016) 168–178.
13. X. H. Qiang, N. D. Wu, Y. F. Luo, X. Jiang and F. Bijlaard, Experimental and theoretical study on high strength steel extended endplate connections after fire, *Int. J. Steel Struct.* **18**(2) (2018) 609–634.
14. C. G. Bailey, I. W. Burgess and R. J. Plank, The lateral-torsional buckling of unrestrained steel beams in fire, *J. Constr. Steel Res.* **36**(2) (1996) 101–119.
15. Eurocode 3: Design of steel structures, Part 1-2: General rules - Structural fire design, (European Committee for Standardization, London, 2005).
16. P. M. M. V. Real, P. A. G. Piloto and J.-M. Franssen, A new proposal of a simple model for the lateral-torsional buckling of unrestrained steel I-beams in case of fire: Experimental and numerical validation, *J. Constr. Steel Res.* **59** (2003) 179–199.
17. L. M. R. Mesquita, P. A. G. Piloto, M. A. P. Vaz and P. M. M. V. Real, Experimental and numerical research on the critical temperature of laterally unrestrained steel I beams, *J. Constr. Steel Res.* **61**(10) (2005) 1435–1446.
18. S. Suman and A. Samanta, A new proposal for design of laterally unsupported monosymmetric steel I-section beams at elevated temperature, *Structures* **38** (2022) 1277–1294.
19. GB 51249-2017, Code for fire safety of steel structures in buildings, China Planning Press, Beijing (2017), (in Chinese).
20. Y. Z. Yin and Y. C. Wang, Numerical simulations of the effects of nonuniform temperature distributions on lateral torsional buckling resistance of steel I-beams, *J. Constr. Steel Res.* **59**(8) (2003) 1009–1033.
21. Structural use of steelwork in building. Part 8, Code of practice for fire resistant design, B. S. Institution, London (1990).

22. N. George and P. Jeyaraj, Nonuniform heat effects on buckling of laminated composite beam: Experimental investigations, *Int. J. Struct. Stab. Dyn.* **18**(12) (2018) 1850153.
23. Y. L. Xu, Y. J. Shi, Y. R. Wu and H. Y. Ban, Experimental and numerical study on lateral-torsional buckling behavior of high performance steel beams, *Int. J. Struct. Stab. Dyn.* **18**(7) (2018) 1850090.
24. H. Varol and K. A. Cashell, Numerical modelling of high strength steel beams at elevated temperature, *Fire Safety J.* **89** (2017) 41–50.
25. J. H. Luo, S. H. Zong, S. J. Zhu, X. N. Guo and M. T. Yu, Mechanical behavior of Q690 high-strength steel beams at room and elevated temperatures, *Struct. Des. Tall. Spec.* **29**(5) (2020) e1712.
26. X. N. Guo, S. H. Zong, M. T. Yu and H. Xu, The bearing capacity of Q690 high-strength steel beams after fire, *Prog. Steel Build. Struct.* **23**(11) (2021) 103–111 (in Chinese).
27. W. Y. Wang, H. Y. Zhou, Y. C. Zhou and L. Xu, A simplified approach for fire resistance design of high strength Q460 steel beams subjected to non-uniform temperature distribution, *Fire Technol.* **54**(2) (2018) 437–460.
28. H. T. Li and B. Young, Cold-formed high strength steel SHS and RHS beams at elevated temperatures, *J. Constr. Steel. Res.* **158** (2019) 475–485.
29. AISI (American Iron and Steel Construction), AISI S100-16: Specification for the design of cold-formed steel structural members, (American Iron and Steel Construction, Washington D.C, 2016).
30. H. Al-Azzani, W. Y. Wang and Y. H. Zhang, Flexural-torsional buckling behavior and design for high-strength steel beams in fire conditions, *Int. J. Struct. Stab. Dyn.* **22**(03N04) (2022) 2240006.
31. GB/T 1591-2019, High strength low alloy structural steels, China Standard Press,Beijing (2020), (in Chinese).
32. W. Ramberg and W. R. Osgood, *Description of Stress-Strain Curves by Three Parameters*, National Advisory Committee for Aeronautics, Technical Report (1943).
33. H. Hill, *Determination of Stress-Strain Relations from Offset Yield Strength Values*, Aluminum Co of America Pittsburgh Pa, Technical Report (1944).
34. W. Y. Wang, R. Yan and L. Xu, Effect of tensile-strain rate on mechanical properties of high-strength Q460 steel at elevated temperatures, *J. Mater. Civ. Eng.* **32**(7) (2020) 04020188.
35. Eurocode 4: Design of composite steel and concrete structures. Part 1-2: General rules — Structural fire design (European Committee for Standardization, London, 2005).
36. GB 50205-2020 34, Standard for acceptance of construction quality of steel structures, China Architecture & Building Press, Beijing (2020), (in Chinese).
37. H. Y. Ban, G. Shi and Y. J. Shi, Experimental and unified model investigations on residual stress within high strength steel welded I-section, *Eng. Mech.* **31**(8) (2014) 83–91 (in Chinese).
38. Z. Y. Yang, W. Y. Wang, J. Zhang and L. Xu, Effect of fire exposure on residual stresses relief in welded high strength Q690 steel sections, *J. Constr. Steel Res.* **177** (2021) 106455.

https://doi.org/10.1142/9789819814770_0010

Chapter 10

Influence of the Relationship Between Excitation Frequency and Natural Frequency on Formulating Structural Random Displacement Response[#]

Qingshan Yang[*,†,‡], Shuai Huang[*], Ling Zhao[*] and Kunpeng Guo[*]

[*]*School of Civil Engineering*
Chongqing University, Chongqing 400044, P. R. China

[†]*Chongqing Key Laboratory of Wind Engineering*
and Wind Energy Utilization
Chongqing 400045, P. R. China
[‡]*qshyang@cqu.edu.cn*

Seismic action and wind excitation are the main sources of excitation to civil engineering structures. The analytical structural responses are similar for both cases, but the simplified formula in design codes on the displacement response under these excitations is quite different. This paper re-visits this difference from under stationary random excitation. The power spectrum density function of the above excitations contains several parameters which define the excitations in frequency domain. The simplified formulas of the displacement variance under different excitations are derived by adjusting these parameters. The responses from these formulas always include the resonance component of the response, whereas the presence of the background component depends on the ratio between the predominant frequency of excitation and the natural frequency of the structure. The influence of this ratio on the displacement covariance and the modal combination rules is then further discussed.

Keywords: Displacement response; frequency domain; civil engineering; random vibration; excitation frequency; natural frequency.

1. Introduction

Civil engineering structures are often subjected to random excitations such as wind, waves, and earthquakes. It is necessary to accurately evaluate the displacement, acceleration, internal force, and other dynamic responses under these actions. The displacement response is a fundamental physical quantity for structural analysis and design. For a linear elastic structure with N degrees-of-freedom, its displacements

[‡]Corresponding author.
[#]This chapter appeared previously on the International Journal of Structural Stability and Dynamics. To cite this chapter, please cite the original article as the following: Q. Yang *et al.*, *Int. J. Struct. Stab. Dyn.*, **23**, 2340011 (2023), DOI: 10.1142/S0219455423400114.

can be expressed by the modal analysis method as

$$\{x(t)\} = \sum_{j=1}^{N} \{\phi_j\} q_j(t), \tag{1}$$

where $\{x(t)\} = \{x_1(t), x_2(t), \ldots, x_N(t)\}$ is the displacement vector, $\{\phi_j\} = \{\phi_{j,1}, \phi_{j,2}, \ldots, \phi_{j,N}\}^T$ and $q_j(t)$ are the jth mode shape and generalized displacement, respectively. The root–mean–square (RMS) value of the displacement response can be obtained from Eq. (1) as

$$\sigma_{x_l} = \sqrt{\sum_{j=1}^{N} \phi_{j,l}^2 \sigma_{q_j}^2 + \sum_{j=1}^{N} \sum_{k=1,k\neq j}^{N} \phi_{j,l}\phi_{k,l}\mathrm{cov}(q_j, q_k)}, \tag{2}$$

where σ_{x_l} denotes the RMS of $x_l(t)$, with $l = 1, 2, \ldots, N$; $\sigma_{q_j}^2$ is the variance of the jth generalized displacement, and $\mathrm{cov}(q_j, q_k)$ is the covariance of the jth and kth generalized displacements. They can be expressed, respectively, as

$$\sigma_{q_j}^2 = \int_0^\infty S_{F_j}(f)|H_j(f)|^2 df, \tag{3}$$

$$\mathrm{cov}(q_j, q_k) = \mathrm{Re}\left[\int_0^\infty S_{F_jF_k}(f) H_j(f) H_k(-f) df\right], \tag{4}$$

where $S_{F_j}(f) = S_{F_jF_j}(f)$ is the single-sided power spectral density (PSD) function of the jth modal generalized force, and $S_{F_jF_k}(f)$ is the single-sided cross-power spectral density function of the jth and kth modal generalized forces. Re[·] indicates the real part of the variable in bracket; $H_j(f) = k_j^{-1}[1 - (f/f_j)^2 + i2\xi_j f/f_j]^{-1}$ is the frequency response function of the jth order mode, and $i = \sqrt{-1}$ is the imaginary unit. f_j and ξ_j are the jth natural frequency and damping ratio respectively. $|H_j(f)|^2$ denotes the transfer function. The RMS of structural displacement can be obtained by substituting the integrals in Eqs. (3) and (4) into Eq. (2).

Since damping of a practical engineering structure is small, approximate formula on the variance of the generalized displacement shown in Eq. (3) from under general random vibration[9,15,16,19] and seismic excitations[17] has been proposed as

$$\sigma_{q_j}^2 = \frac{1}{k_j^2}\frac{\pi f_j}{4\xi_j} S_{F_j}(f_j). \tag{5}$$

However, monographs on wind-induced responses[5,8,20] suggested a different form of the variance as

$$\sigma_{q_j}^2 = \frac{1}{k_j^2}\frac{\pi f_j}{4\xi_j} S_{F_j}(f_j) + \frac{\sigma_{F_j}^2}{k_j^2}. \tag{6}$$

Der Kiureghian[9] studied the displacement response under stationary random excitation, and found that the ratio between the excitation frequency and the structural natural frequency as well as the excitation bandwidth, had a significant influence on the simplified formula. However, there was no explanation on the differences between Eqs. (5) and (6).

This paper re-visits the influence of excitation frequency on the displacement variance. In order to quantify the relative relationship between excitation frequency and structural frequency, a model on the auto-spectral density of excitation is adopted to derive the analytical solution on the displacement variance. Then, the model parameters defining the predominant frequency of excitation and its bandwidth are discussed in detail revealing the source of differences between Eqs. (5) and (6). Then, the background of the simplified formulas of displacement variance under excitations with different bandwidths and predominant frequencies is clarified. The influence of excitation frequency on the covariance of generalized displacement is further discussed. The combination rules for obtaining the standard deviation of displacement response are modified in lieu of findings in this study.

2. Variance of Generalized Displacement

When calculating the variance of generalized displacement according to Eq. (3), the transfer function is expressed as

$$|H_j(f)|^2 = \frac{1}{k_j^2} \frac{1}{[1-(f/f_j)^2]^2 + 4\xi_j^2(f/f_j)^2} \tag{7}$$

which has a significant peak at the natural frequency ($f = f_j$), as shown in Fig. 1. It has the characteristic of narrower bandwidth at a smaller structural damping ratio. Since the damping ratio of building structures is generally small, the curve of the transfer function can be divided into three different subsections as shown. A platform subsection with the curve changes slowly satisfying

$$|H_j(f)|^2 \approx |H_j(0)|^2 = 1/k_j^2, \quad f \leq f_j - \Delta_j \tag{8}$$

within the range of frequency which is less than the natural frequency ($f \leq f_j - \Delta_j$). Δ_j is an increment. The transfer function then increases sharply with frequency

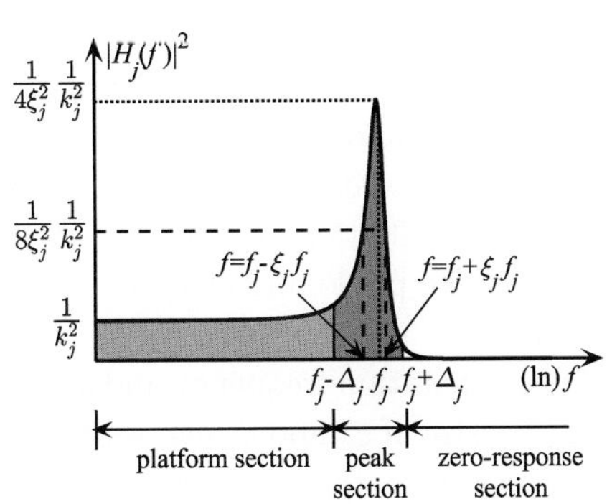

Fig. 1. Segmentation diagram of the transfer.

reaching its maximum value at $f \approx f_j$, with $|H_j(f_j)|^2 = (2\xi_j k_j)^{-2}$. It then decreases rapidly with frequency from f_j, reaching half of the maximum at $f_j \pm \xi_j f_j$. This narrow band frequency range $(f_j - \Delta_j < f < f_j + \Delta_j)$ is called the peak subsection. The increment Δ_j may be taken as multiple of $\xi_j f_j$. It can be seen later that the selection of Δ_j has little effect on the variance.

In the high frequency range,

$$|H_j(f)|^2 \propto f^{-4} \approx 0, \quad f \geq f_j + \Delta_j. \tag{9}$$

The transfer function is noted as small approaching null, indicating the high-frequency responses will be very small. This high-frequency range of transfer function is called the zero-response section.

The value of the transfer function in the peak subsection is noted significantly larger than those in the other two subsections. Although the transfer function in the platform subsection is much smaller than that in the peak subsection, it still contributes to the low-frequency displacement response. Therefore, the variance of generalized displacement is mainly determined from the platform subsection and the peak subsection of the PSD, with contributions depending on the distribution of excitation energy relative to the natural frequency of structure.

The distribution of excitation energy is mainly affected by its predominant frequency and bandwidth. The structural modal damping and natural frequency, in turn, characterize the bandwidth and the predominant frequency of the transfer function. The relationship between the bandwidth of excitation and structural modal damping, predominant frequency of excitation and natural frequency of the structure, as shown in Eq. (10) is adopted for the study.

$$S_{F_j}(f) = \frac{(f/f_0)^{\beta}}{[1-(f/f_0)^2]^2 + 4\xi_0^2(f/f_0)^2} S_0, \tag{10}$$

where f_0 is the predominant excitation frequency which associates with the peak value of excitation spectrum with $f = f_0$. ξ_0 defines the bandwidth characteristics of the load, which is referred to as the bandwidth parameter in this study. β denotes the skewness of the excitation spectrum. S_0 denotes the total excitation energy and is independent of the energy distribution, and it is taken as $S_0 = 1$ in this study. The PSDs with different ξ_0 and β are shown in Fig. 2. It can be seen that a larger ξ_0 corresponds to a wider excitation spectral bandwidth. Parameter β has little influence on the energy distribution. When $\beta = 2$, the excitation spectrum is basically symmetrical about the predominant frequency. The spectrum skews to the left when $\beta < 2$, and skews to the right when $\beta > 2$. Therefore, f_0 and f_j can represent the relative value of excitation frequency and structural natural frequency respectively, while the relative size of ξ_0 and ξ_j can reflect the relationship between the bandwidths of excitation and the transfer function.

The PSD function in Eq. (10) has been widely used in seismic and wind engineering, as the filtered white noise model for seismic ground acceleration,[13,21] and

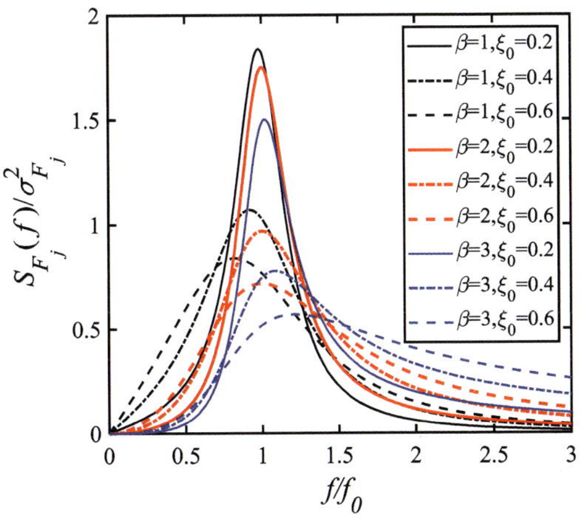

Fig. 2. The effect of the parameters of the spectrum on its shape.

the PSD model of crosswind bending moment and torque at the base of high-rise buildings.[1,14] It should be noted that the bandwidth parameter of most excitations in nature is larger than the damping ratio of the structure (the bandwidth parameter of the transfer function). Ou *et al.*[18] and Jiang *et al.*[12] recommended that the bandwidth parameter of ground acceleration should vary from 0.45 to 1.0. Also, according to the empirical formula in AIJ,[1] the bandwidth parameter of the bending moment at base of high rise building in the crosswind direction varies from 0.27 to 0.86. They indicate that the natural forces acting on a practical engineering structure always exhibit broadband characteristics. Therefore, broadband excitation with $\xi_0 \gg \xi_j$ is considered in this study.

If the excitation spectrum adopts the form of Eq. (10), then the variance of generalized displacement in Eq. (3) can be calculated via the residue theorem of integration or the partial fraction decomposition. Since β has little influence on the distribution of excitation energy, the analytical solution to Eq. (3) is derived by taking $\beta = 2$. The influence of the predominant frequency of broadband excitations on the simplified formula of displacement response is then discussed below.

2.1. *Analytical solution of variance of generalized displacement*

When $\beta = 2$, the PSD function in Eq. (10) can be expressed as

$$S_{F_j}(f) = \frac{(f/f_0)^2}{[1-(f/f_0)^2]^2 + 4\xi_0^2(ff_0)^2} \tag{11}$$

and its variance is given by

$$\sigma_{F_j}^2 = \int_0^\infty S_{F_j}(f)df = \frac{\pi f_0}{4\xi_0}. \tag{12}$$

By substituting Eq. (11) into Eq. (3), we get

$$\sigma_{q_j}^2 = \frac{1}{k_j^2}\int_0^\infty \frac{1}{[1-(f/f_j)^2]^2+4\xi_j^2(f/f_j)^2}\frac{(f/f_0)^2}{[1-(f/f_0)^2]^2+4\xi_g^2(f/f_0)^2}df. \quad (13)$$

Let the ratio of predominant excitation frequency f_0 to the natural frequency f_j be $r(=f_0/f_j)$, and $x=f/f_j$, the variance of the response can be rewritten as

$$\sigma_{q_j}^2 = \frac{f_j r^2}{k_j^2}\int_0^\infty P(x)dx \quad (14)$$

with

$$P(x) = \frac{1}{(1-x^2)^2+(2\xi_j x)^2}\frac{x^2}{(r^2-x^2)^2+(2\xi_0 r x)^2}. \quad (15)$$

According to the residue theorem, the integration in Eq. (14) gives

$$\int_0^\infty P(x)dx = \pi i\sum_{k=1}^{n}\mathrm{Res}_{x=x_k}P(x) = \pi i\sum_{k=1}^{n}[\lim_{x\to x_i}(x-x_i)P(x)], \quad (16)$$

where $\mathrm{Res}_{x=x_k}P(x)$ denotes the residue at the pole x_k of $P(x)$, and $n=4$ is the number of poles of $P(x)$ considered in the upper or lower half plane. All the poles of $P(x)$ are shown, respectively, as

$$\begin{aligned} x_{1,2} &= r\left(\pm\sqrt{1-\xi_0^2}+i\xi_0\right),\\ x_{3,4} &= \pm\sqrt{1-\xi_j^2}+i\xi_j,\\ x_{5,6} &= r\left(\pm\sqrt{1-\xi_0^2}-i\xi_0\right),\\ x_{7,8} &= \pm\sqrt{1-\xi_j^2}-i\xi_j, \end{aligned} \quad (17)$$

where x_1 to x_4 are in the upper half plane, and x_5 to x_8 are in the lower half plane. By substituting Eq. (17) into Eq. (16), we have

$$\sigma_{q_j}^2 = \frac{A}{F}\frac{1}{k_j^2}\frac{\pi f_j}{4\xi_j} + \frac{B}{F}\frac{1}{k_j^2}\frac{\pi f_0}{4\xi_0}, \quad (18)$$

where A, B and F are functions of the frequency ratio r, structural damping ratio ξ_j and excitation bandwidth parameter ξ_0, expressed as

$$A(r,\xi_j,\xi_0) = r^2[(r^2-1)^2+4\xi_0^2r^2-4\xi_j^2], \quad (19)$$

$$B(r,\xi_j,\xi_0) = (r^2-1)^2+4\xi_j^2r^2-4\xi_0^2r^4, \quad (20)$$

$$\begin{aligned} F(r,\xi_j,\xi_0) = {} & [(r^2-1)^2+4\xi_0^2r^2]^2+8\xi_j^2r^2[1-2\xi_0^2\\ & -2(1-\xi_j^2)r^2+(1-2\xi_0^2)r^4], \end{aligned} \quad (21)$$

respectively. So, the bandwidth and predominant frequency of the excitation model are responsible for Eqs. (5) and (6). It is noted that $\frac{1}{k_j^2}\frac{\pi f_j}{4\xi_j} = \int_0^\infty |H_1(f)|^2 df$ and

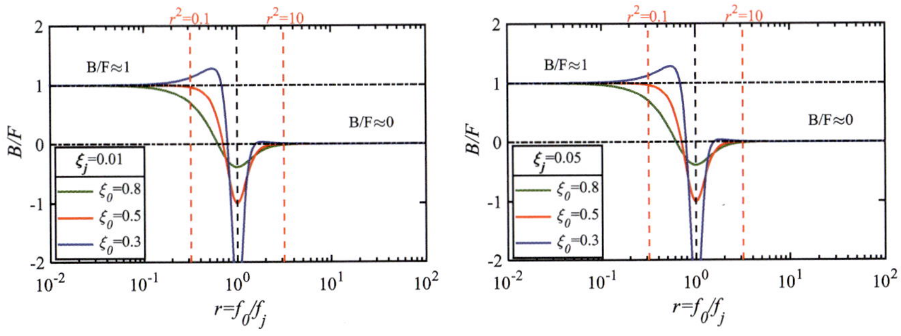

(a) $\xi_j = 0.01$ (b) $\xi_j = 0.05$

Fig. 3. B/F versus frequency ratio and excitation bandwidth.

it represents the area underneath the transfer function curve, and $\frac{1}{k_j^2}\frac{\pi f_0}{4\xi_0} = \frac{\sigma_{F_j}^2}{k_j^2}$ denotes the displacement variance when the excitation is static, i.e. the variance of quasi-static displacement. Therefore, the displacement variance can be expressed as A/F times the area underneath the transfer function curve plus B/F times the variance of the quasi-static displacement.

Since building structures, in general, have small damping, with $\xi_j \ll 1$ and $\xi_j^2 \to 0$, the excitations are therefore relatively broadband with $\xi_g \gg \xi_j$. Then

$$\frac{A}{F} \approx \frac{r^2}{(r^2-1)^2 + 4\xi_0^2 r^2} = S_{F_j}(f_j). \tag{22}$$

This means that for a low damping system, A/F is always the spectral value of excitation at the natural frequency. The coefficient B/F shown in Fig. 3 shows that the structural damping ratio has very small effect on the coefficient B/F.

Then the variance of generalized displacement can be simplified as

$$\sigma_{q_j}^2 \approx \frac{1}{k_j^2}\left[\frac{\pi f_j}{4\xi_j} S_{F_j}(f_j) + \frac{B}{F}\sigma_{F_j}^2\right]. \tag{23}$$

The ratio of predominant excitation frequency to structural natural frequency, f_0/f_j has significant influence on the value of B/F as noted in Fig. 3. The values of B/F within the three ranges of $0 < r^2 \leq 0.1$, $0.1 < r^2 < 10$, and $r^2 \geq 10$ are noted different. But the coefficient will change significantly with the frequency ratio and excitation bandwidth.

2.2. *Influence of predominant excitation frequency on the displacement variance*

The relationship between the PSD of generalized force, the transfer function, and the PSD of displacement is classified into three categories as shown in Fig. 4.

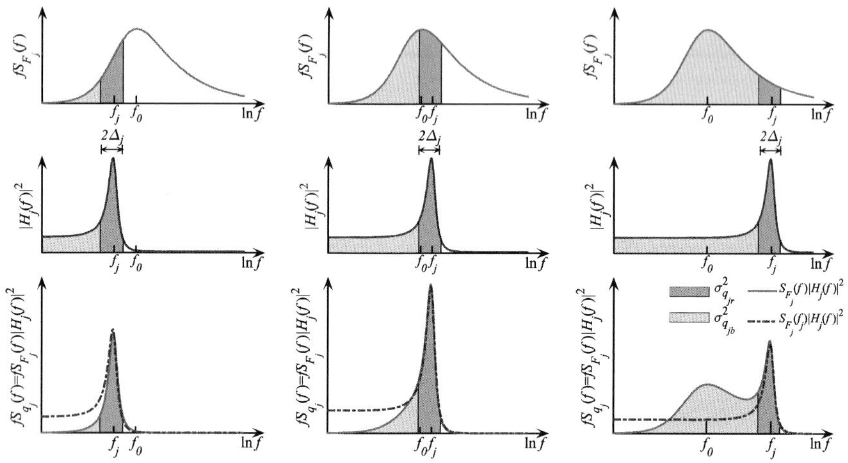

(a) $f_0 \geq f_j + \Delta_j$ (b) $f_j - \Delta_j < f_0 < f_j + \Delta_j$ (c) $f_0 \leq f_j + \Delta_j$

Fig. 4. The relationship between excitation spectrum, transfer function and displacement spectrum.

(1) When the excitation frequency is much higher than the natural frequency

When the predominant excitation frequency is much higher than the natural frequency, the relationship between the PSDs of generalized force and displacement is shown in Fig. 4(a). The predominant frequency of the generalized force is noted within the zero-response section, and the excitation spectrum in the low-frequency range is fairly small relative to that in the peak subsection.

The excitation spectrum changes slowly in the narrow peak subsection, which can be considered as a constant, with

$$S_{F_j}(f) \approx S_{F_j}(f_j), \quad f_j - \Delta_j < f < f_j + \Delta_j. \tag{24}$$

By substituting into Eq. (3), we get

$$\begin{aligned} \sigma_{q_j}^2 &\approx \int_{f_j-\Delta_j}^{f_j+\Delta_j} S_{F_j}(f)|H_j(f)|^2 df \\ &= S_{F_j}(f_j) \times \left[\int_0^{\infty} |H_j(f)|^2 df - \int_0^{f_j-\Delta_j} |H_j(f)|^2 df \right. \\ &\quad \left. - \int_{f_j+\Delta_j}^{\infty} |H_j(f)|^2 df \right]. \end{aligned} \tag{25}$$

The integral of the first term is the area under the transfer function curve, $\int_0^{\infty} |H_j(f)|^2 df = \frac{1}{k_j^2} \frac{\pi f_j}{4\xi_j}$. By substituting Eq. (8) into the second term, $\int_0^{f_j-\Delta_j}$

$|H_j(f)|^2 df = \frac{1}{k_j^2}(f_j - \Delta_j)$ can be obtained. Since $\xi_j \ll 1$, $\pi f_j/4\xi_j \gg f_j - \Delta_j$, the second integral is small and can be ignored. The integral in the high-frequency range of the third term approaches null. Therefore, when the excitation is dominated by high-frequency components, the variance of generalized displacement can be simplified to Eq. (5). The value of Δ_j has little effect on the calculated variance. The displacement response mainly consists of components from excitation in the frequency range $f_j - \Delta_j < f < f_j + \Delta_j$, which is also known as the resonant response. The PSD of generalized force can be approximated as a constant in this range, i.e. the excitation is assumed as a white noise process. The PSD of displacement is shown as the dotted line in Fig. 4(a). It can be seen that the white noise assumption is appropriate within the peak subsection, but it leads to an over-estimate of the displacement in the low-frequency range. However, since the area of the displacement spectrum in low-frequency subsection ($\int_0^{f_j-\Delta_j} |H_j(f)|^2 df$) is much smaller than that in the peak subsection ($\int_0^{\infty} |H_j(f)|^2 df$), this over-estimation is small and considered acceptable. In terms of the analytic solution, it's easy to obtain the approximate formula. For a high frequency excitation with $r^2 > 10$, as shown in Fig. 3(a), $B/F \to 0$. By substituting this relationship into Eq. (23), we have

$$\sigma_{q_j}^2 \approx \frac{1}{k_j^2}\frac{\pi f_j}{4\xi_j} S_{F_j}(f_j). \tag{5}$$

Equation (23) becomes identical to Eq. (5).

(2) When the excitation frequency is much lower than the natural frequency

When the predominant frequency of generalized force is much lower than the natural frequency, the relationship between the PSD of the generalized force and displacement within the low-frequency range is shown in Fig. 4(c). The displacement spectrum has a wide peak in the low-frequency region with a shape similar to that of the excitation spectrum. The variance of displacement is mainly obtained by integrating the power spectrum in two intervals ($0.0 < f < f_j - \Delta_j$) and ($f_j - \Delta_j < f < f_j + \Delta_j$), denoted as $\sigma_{q_{jb}}^2$ and $\sigma_{q_{jr}}^2$, respectively. The generalized excitation spectrum within the narrow band of ($f_j - \Delta_j < f < f_j + \Delta_j$) satisfies the white noise assumption, yielding the resonant response as

$$\sigma_{q_{jr}}^2 \approx \frac{1}{k_j^2}\frac{\pi f_j}{4\xi_j} S_{F_j}(f_j). \tag{26}$$

The PSD of displacement under the white noise assumption is shown by the dotted line in Fig. 4(c). It can be seen that the white noise assumption still has a high accuracy in the peak subsection. However, it significantly underestimates the displacement in the low-frequency range. Therefore, the white noise assumption cannot be used directly for integration in the low-frequency range. Since the transfer function is constant in the low-frequency range ($0.0 < f < f_j - \Delta_j$), then

$$\sigma_{q_{jb}}^2 \approx \int_0^{f_j-\Delta_j} S_{F_j}(f)df/k_j^2 \approx \int_0^{\infty} S_{F_j}(f)df/k_j^2 = \sigma_{F_j}^2/k_j^2. \tag{27}$$

This is the static response caused by the generalized load, and is often called the background response. The variance of generalized displacement is therefore composed of the background and resonant responses, and Eq. (23) is simplified into Eq. (6). From the perspective of analytic solution, with $r^2 < 0.1$, $B/F \to 1$, substituting into Eq. (23) gives

$$\sigma_{q_j}^2 \approx \frac{\sigma_{F_j}^2}{k_j^2} + \frac{1}{k_j^2}\frac{\pi f_j}{4\xi_j}S_{F_j}(f_j). \tag{6}$$

Equation (23) is noted identical to Eq. (6).

(3) When the excitation frequency is close to the natural frequency

When the predominant frequency of generalized force is close to the natural frequency, as shown in Fig. 4(b), B/F gradually decreases with r from a value close to unity to a negative value, reaching a minimum at $r = 1.0$. It then increases to a value close to null. When $B/F = 0$, the two roots of $B/F = 0$ are $r^2 = (1-2\xi_0)^{-1} > 1$ and $r^2 = (1+2\xi_0)^{-1} < 1$. Therefore, $B/F > 0$ within the range of $0.1 < r^2 < (1 + 2\xi_0)^{-1}$ It can be approximate to have $r = 1.0$. Also, $B/F < 0$ within $((1 + 2\xi_0)^{-1} < r^2 < 10.0)$. When $r^2 = 1$, $A = -4(\xi_j^2 - \xi_0^2)$, $B = 4(\xi_j^2 - \xi_0^2)$, and $F = 16(\xi_j^2 - \xi_0^2)^2$. Substituting into Eq. (23) to get

$$\sigma_{q_j}^2 = \frac{1}{k_j^2}\left[\frac{1}{4(\xi_0^2 - \xi_j^2)}\frac{\pi f_j}{4\xi_j} - \frac{1}{4(\xi_0^2 - \xi_j^2)}\frac{\pi f_j}{4\xi_0}\right]. \tag{28}$$

When the generalized force is assumed to be broadband, $\xi_0 \gg \xi_j$, and $\frac{\pi f_j}{4\xi_j} \gg \frac{\pi f_j}{4\xi_0}$. Equation (28) can be approximated as

$$\sigma_{q_j}^2 \approx \frac{1}{k_j^2}\frac{\pi f_j}{4\xi_j}\frac{1}{4\xi_0^2} = \frac{1}{k_j^2}\frac{\pi f_j}{4\xi_j}S_{F_j}(f_j). \tag{29}$$

The displacement variance is also simplified to Eq. (5). It means the background response can be ignored when B/F takes up the maximum negative value, so it can be ignored throughout the range of $(1 + 2\xi_0)^{-1} < r^2 < 10.0$. The displacement power spectrum under the white noise assumption is shown as the dotted line in Fig. 4(b). It is noted that the white noise assumption also gives conservative results in this frequency range.

The above analysis shows that the ratio between the predominant frequency of generalized force and the natural frequency affects significantly the contributions of both the background response and the resonant response. When the ratio approaches unity or beyond, only the resonant response should be considered, and the displacement variance can be calculated according to Eq. (5) with a white

Table 1. Formulas for displacement variances caused by excitations with different bandwidth and predominant frequency.

Theoretical formula	Structures with light damping ($\xi_j \ll 1$)	
	Low frequency excitation ($r^2 < 1/(1+2\xi_0)$)	High frequency excitation ($r^2 \geq 1/(1+2\xi_0)$)
$\sigma^2_{q_j} = \frac{A}{F}\frac{1}{k_j^2}\frac{\pi f_j}{4\xi_j} + \frac{B}{F}\frac{\sigma^2_{F_j}}{k_j^2}$ (18)	$\sigma^2_{q_j} \approx \frac{1}{k_j^2}\frac{\pi f_j}{4\xi_j}S_{F_j}(f_j) + \frac{\sigma^2_{F_j}}{k_j^2}$ (6)	$\sigma^2_{q_j} \approx \frac{1}{k_j^2}\frac{\pi f_j}{4\xi_j}S_{F_j}(f_j)$ (5)

noise assumption on the excitation. When the ratio is small, both the resonant response and the background response should be considered. The displacement variance should be calculated according to Eq. (6). These conclusions are listed in Table 1.

The analytical solution of displacement variance under filtered white noise excitation[9] is shown in the appendix. Since the case with the ratio of frequencies close to or higher than unit was considered, only the simplified formula Eq. (5) has been given. In addition, the coefficients from Der Kiureghian[9] are not exactly correct, and the correct results are also given in the appendix.

3. The Variance of Structural Displacement Variance Under Typical Random Excitations

A building or civil engineering structure is subjected to different random excitations according to their operating environments. Their displacement responses under wind load, seismic action, and wave excitation will be discussed in the following sections.

3.1. *Seismic response*

The filtered white noise model has been used to simulate the PSD of ground acceleration,[13,21] as shown in Eq. (A.1). Since high rise buildings have the first natural frequency varying from 0.1 to 1.0 Hz,[23] it is noted that the predominant frequency (Sec. 2) of earthquakes is, in general, higher than or close to the first natural frequency of high-rise buildings, with a bandwidth parameter (Appendix) much bigger than the damping ratio of structures. Therefore, seismic actions are typically high-frequency, broadband random excitations relative to tall buildings. The corresponding displacement response mainly consists of the resonant response caused by generalized force components in the frequency range of $f_j - \Delta_j < f < f_j + \Delta_j$. The RMS of generalized displacement should be calculated according to Eq. (5).

For some special cases when the predominant seismic frequency is much lower than the natural frequency of structure, the displacement response will include some background responses and it should be calculated according to Eq. (6).

3.2. *Wind-induced response*

Wind excitations on high-rise buildings contain three components in different directions: along wind, crosswind, and torsional. The wind load in the along wind direction is, in general, low-frequency excitation, while the predominant frequency of excitations in the crosswind and torsional directions varies with changes in the mean wind speed.

(1) Along wind displacement response

The along wind fluctuating force can be determined by quasi-steady assumptions, and the power spectrum of wind speed can represent the energy distribution of the generalized force. The along wind excitation is broadband with low-frequency relative to the structure. Therefore, the corresponding structural displacement spectrum has a broadband peak in the low-frequency region similar to the excitation spectrum.[3,24] The variance of the generalized displacement is contributed by both the background and resonant responses, which should be calculated according to Eq. (6).

(2) Crosswind displacement response

The crosswind excitation is affected by vortex shedding, and its predominant frequency is the vortex shedding frequency. According to the Strouhal's Law, the vortex shedding frequency is $f_{\mathrm{vs}} = S_t U/B$, where S_t is the Strouhal number, U is the mean wind speed, and B is lateral dimension of the structure. When the vortex shedding frequency equals the structural natural frequency, the wind speed is called the critical wind speed U_{cr}.

It should be mentioned that the proportion of background and resonant components of the crosswind displacement changes distinctly with the change of wind speed. When the wind speed is much lower than the critical wind speed, the predominant frequency of excitation is much lower than the natural frequency of structure. The excitation energy is concentrated in the platform subsection of the transfer function, resulting in a major contribution from the background response. Both the background and resonant responses are needed to be considered, and the displacement variance is calculated according to Eq. (6).

When the wind speed is close to the critical wind speed, the displacement spectrum has only one distinct peak in the peak subsection of the transfer function, and the background response becomes very small and can be ignored. The displacement variance will be calculated according to Eq. (5). However, complex vortex-induced resonance may occur in this case and the analysis of the displacement responses is not studied in this paper.

When the wind speed exceeds the critical wind speed with a wide margin, the displacement response is mainly dominated by the resonant response.

Since the wind speed is generally lower than the critical wind speed, the calculation of the crosswind response should consider both the background and resonant response. However, when the wind speed is close to or even exceeds the critical wind speed, complex vortex-induced resonance may occur in the structure, and its analysis method is not in the scope of this paper.

3.3. *Response to wave load*

The first natural frequency of an offshore wind turbine is far higher than the predominant frequency of wind excitation and wave excitation,[2,7,10] and therefore, only the background response should be considered in calculating the displacement. However, its natural frequency is lower than the predominant frequency of the seismic acceleration,[2,7] and the background response is very small, so the white noise assumption can be directly used to calculate the displacement.

4. Covariance of Generalized Displacement and the Modal Combination Rules

Discussions in Secs. 2 and 3 have demonstrated that most random excitations in practical engineering are broadband, and its predominant frequency has a distinct influence on the variance of generalized displacement. This section will discuss on the covariance of displacement under broadband excitation. The covariance of generalized displacement of modal orders j and k is shown in Eq. (4).

$$\mathrm{cov}(q_j, q_k) = \int_0^{\infty} \mathrm{Re}[H_j(f)H_k(-f)] \times \mathrm{Re}[S_{F_jF_k}(f)]df. \tag{30}$$

The imaginary component of the cross spectrum of generalized forces, in general, has little influence on the covariance of generalized displacement the real component.[4] Therefore, only the real part of the cross spectrum is considered in the following discussions.

It is also noted that the ratio of predominant excitation frequency to the natural frequency of structure will also affect the covariance similar to that for variance of the generalized displacement as discussed above.

4.1. *When the excitation frequency is close to or exceeds the natural frequency*

Without loss of generality, the case with $f_j \geq f_k$ is discussed. When $f_0 > f_j - \Delta_j$, the covariance of the jth and kth modal displacements contains only the resonant component, i.e. the integral of the cross spectrum within the peak subsection, $\sigma^2_{q_{jkr}}$, can be expressed as

$$\sigma^2_{q_{jkr}} = \int_{f_k-\Delta_k}^{f_j+\Delta_j} \mathrm{Re}[H_j(f)H_k(-f)] \times \mathrm{Re}[S_{F_jF_k}(f)]df. \tag{31}$$

Since the bandwidth of the transfer function is small, the cross-spectral density function of the generalized forces in the peak subsection can be assumed constant, with[4]

$$\sigma^2_{q_{jkr}} \approx \sigma_{q_{jr}} \sigma_{q_{kr}} \gamma_{jkr}, \tag{32}$$

where γ_{jkr} is the correlation coefficient of the resonant response. When the modal frequencies are well separated, the correlation coefficient of the resonant response is very small, and γ_{jkr} tends become nil. When the modal frequencies are close together, the correlation coefficient can be written as

$$\gamma_{jkr} = \alpha_{jkr} \cdot \rho_{jkr} \tag{33}$$

$$\alpha_{jkr} = \frac{\mathrm{Re}[S_{F_j F_k}(f)]|_{f=f_j \text{ or } f=f_k}}{\sqrt{S_{F_j}(f_j) S_{F_k}(f_k)}} \approx \mathrm{coh}_{F_j F_k}(f)|_{f=f_j \text{ or } f=f_k}, \tag{34}$$

$$\rho_{jkr} = \frac{8\sqrt{\xi_j \xi_k}(\beta_{jk}\xi_j + \xi_k)\beta_{jk}^{3/2}}{(1-\beta_{jk}^2)^2 + 4\xi_j\xi_k\beta_{jk}(1+\beta_{jk}^2) + 4(\xi_j^2+\xi_k^2)\beta_{jk}^2}, \tag{35}$$

where $\mathrm{coh}_{F_j F_k}(f)$ is the coherence function of the generalized forces, and $\beta_{jk} = f_j/f_k$ is the frequency ratio between the jth and kth modes. When the structure is subjected to uniform distributed excitations, the generalized forces of different modes are proportional, with $|\mathrm{coh}_{F_j F_k}(f)| \equiv 1.0$, $\gamma_{jk} = \rho_{jkr}$. Therefore, ρ_{jkr} denotes the modal correlation coefficient of the structure under uniform excitations, and α_{jkr} is the reduction coefficient considering the spatial dependence of excitation with $-1.0 \le \alpha_{jkr} \le 1.0$.[4] If the modal damping ratio is the same for all modes with $\xi_j = \xi_k = \xi$, ρ_{jkr} can be further simplified as

$$\rho_{jkr} = \frac{8\xi^2(\beta_{jk}+1)\beta_{jk}^{3/2}}{\left(1-\beta_{jk}^2\right)^2 + 4\xi^2\beta_{jk}(1+\beta_{jk})^2}. \tag{36}$$

The relationship between the correlation coefficient ρ_{jkr} and the natural frequency ratio β_{jk} is shown in Fig. 5. It is noted that ρ_{jkr} decreases with the spacing of modal frequencies. Also a smaller structural damping ratio is associated with a smaller ρ_{jkr}. When the natural frequency ratio satisfies

$$\min\{\beta_{jk}, \beta_{kj}\} < \frac{0.2}{\xi_j + \xi_k + 0.2}, \tag{37}$$

$\rho_{jkr} < 0.1$, and the natural frequencies are considered well separated.[9]

4.2. *When the excitation frequency is much smaller than the natural frequency*

When $f_0 \le f_j - \Delta_j$, the covariance of generalized displacements comes from both the background and resonant responses. The resonant response can be obtained from Eq. (32). The background response is calculated from the integral of the cross

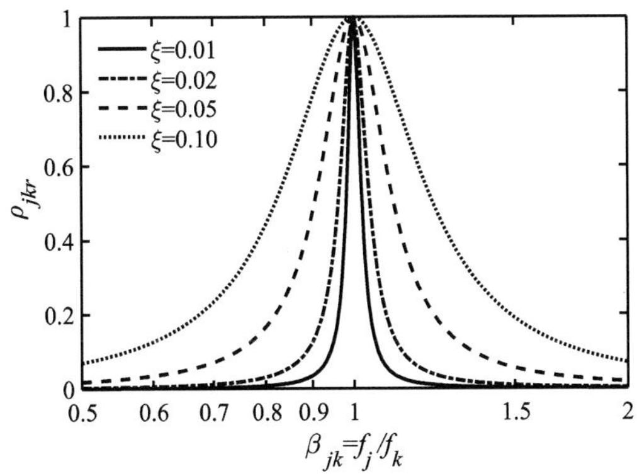

Fig. 5. The relationship between ρ_{jkr} and β_{jk} under uniform excitation.

spectrum of generalized displacements in the platform subsection, $\sigma^2_{q_{jkb}}$, expressed as

$$\sigma^2_{q_{jkb}} = \int_0^{f_k-\Delta_k} \mathrm{Re}[H_j(f)H_k(-f)] \times \mathrm{Re}[S_{F_jF_k}(f)]df. \tag{38}$$

Since the transfer function is approximate uniform in the low-frequency range

$$\begin{aligned}\sigma^2_{q_{jkb}} &\approx \frac{1}{k_jk_k}\mathrm{Re}\left[\int_0^{f_k-\Delta_k} S_{F_jF_k}(f)df\right] \\ &\approx \frac{1}{k_jk_k}\mathrm{Re}\left[\int_0^{\infty} S_{F_jF_k}(f)df\right] = \sigma_{q_{jb}}\sigma_{q_{kb}}\gamma_{jkb},\end{aligned} \tag{39}$$

where $\mathrm{cov}(F_j, F_k) = \mathrm{Re}[\int_0^\infty S_{F_jF_k}(f)df]$ is the covariance of the generalized forces, and γ_{jkb} is the coefficient of background response, which can be expressed as

$$\gamma_{jkb} = \frac{\mathrm{cov}(F_j, F_k)}{\sigma_{F_j}\sigma_{F_k}}. \tag{40}$$

It is shown that the correlation coefficient of background response between different modes is equivalent to the correlation coefficient of generalized forces.

Therefore, when $f_0 \le f_j - \Delta_j$, the covariance of generalized displacement is

$$\mathrm{cov}(q_j, q_k) = \sigma^2_{q_{jkb}} + \sigma^2_{q_{jkr}} = \sigma_{q_{jb}}\sigma_{q_{kb}}\gamma_{jkb} + \sigma_{q_{jr}}\sigma_{q_{kr}}\gamma_{jkr}. \tag{41}$$

The calculation of covariance of generalized displacement is referred to that for the variance of generalized displacement, and it will not be discussed.

4.3. *The modal combination rules*

The modal combination rules for obtaining the standard deviation of displacement are slightly modified to consider both the resonant and background components of the response. By substituting Eqs. (41) and (6) into Eq. (2), we get

$$\sigma_{x_l} = \sqrt{\sigma_{x_{l,b}}^2 + \sigma_{x_{l,r}}^2}, \tag{42}$$

where $\sigma_{x_{l,b}}$ and $\sigma_{x_{l,r}}$ are, respectively, the standard deviations of the background and resonant responses of the structural displacement $x_l(t)$, with

$$\sigma_{x_{l,b}} = \sqrt{\sum_{j=1}^{N} \phi_{j,l}^2 \sigma_{q_{jb}}^2 + \sum_{j=1}^{N} \sum_{k=1,k\neq j}^{N} \phi_{j,l}\phi_{k,l}\sigma_{q_{jb}}\sigma_{q_{kb}}\gamma_{jkb}}, \tag{43}$$

$$\sigma_{x_{l,r}} = \sqrt{\sum_{j=1}^{N} \phi_{j,l}^2 \sigma_{q_{jr}}^2 + \sum_{j=1}^{N} \sum_{k=1,k\neq j}^{N} \phi_{j,l}\phi_{k,l}\sigma_{q_{jr}}\sigma_{q_{kr}}\gamma_{jkr}}. \tag{44}$$

This is the modified Complete Quadratic Combination (CQC) rule.

Since the background response is calculated as a quasi-static displacement generated by the fluctuating forces acting on the structure without dynamic amplification, it can be obtained for all modes of the structure by static analysis of the structure to give

$$x_{l,b}(t) = \sum_{i=1}^{N} \mu_{x_l,i} p_i(t), \tag{45}$$

where $\mu_{x_l,i}$ is the flexibility coefficient for displacement $x_l(t)$, and p_i is the external load corresponding to the degree of freedom $x_i(t)$. The variance of both sides of Eq. (45) is

$$\sigma_{x_{l,b}}^2 = E[x_{l,b}^2(t)] = \sum_{i=1}^{N} \sum_{j=1}^{N} \mu_{x_l,i}\mu_{x_l,j}\mathrm{cov}(p_i, p_j), \tag{46}$$

where $\mathrm{cov}(p_i p_j)$ is the covariance of excitations p_i and p_j.

For the resonant response, if the modal frequencies of structure are well separated satisfying Eq. (37), then $\rho_{jkr} \approx 0 (j \neq k)$, and $\gamma_{jkr} \approx 0$. The cross terms in Eq. (50) can be ignored and the equation can be further simplified as

$$\sigma_{x_{l,r}} \approx \sqrt{\sum_{j=1}^{N} \phi_{j,l}^2 \sigma_{q_j}^2}. \tag{47}$$

This is the modified square–root-of–sum-of–squares (SRSS) rule.

5. Conclusions

The background of the different existing simplified equations for displacement response of a structure in design code when under seismic or wind excitation is studied with an adjustable model on the power spectral density of the excitation.

An analytical solution of displacement under stationary random excitation is derived from which existing simplified equations can be obtained. The ratio between the predominant excitation frequency and natural frequency of structure is responsible for the differences. The following conclusions can be drawn:

(1) The characteristics of frequency and bandwidth of excitations together determine the simplified calculation formula of displacement variance. Table 1 gives the simplified formulas of displacement variance induced by excitations with different frequencies and bandwidths. In practical engineering, the common random excitations are broadband processes. The seismic action is a high-frequency excitation relative to the structural natural frequency, and the corresponding displacement of the structure is mainly a resonant response, so the white noise assumption can be directly used. The along wind force is a low-frequency excitation, and the structural displacement under its action includes both resonant and background responses. The predominant frequency of crosswind force increases with the increase of wind speed. When the wind speed is lower than the critical wind speed, the displacement variance includes resonant response and background response. When the wind speed is close to or exceeds the critical wind speed, only resonant response is included.

(2) The relationship between the excitation predominant frequency and the structural natural frequency also determines the simplified formulas of covariances and the modal combination rules. Under high-frequency loads, the displacement covariances only include resonant response, and the modal correlation coefficients can be determined by the sparsity of modal frequencies, so as to determine whether SRSS or CQC combination is used. Under low-frequency loads, the displacement covariance includes both resonant response and background response. Resonant response is calculated in the same way as high-frequency excitation response, while background response can calculate through the quasi-static analysis, which considered the contribution of all vibration modes.

Appendix

Analytical solution of displacement variance under filtered white noise excitation

The auto-spectral density of filtered white noise is expressed as[13,21]

$$S_{F_j}(f) = \frac{1 + 4\xi_0^2 (f/f_0)^2}{[1 - (f/f_0)^2]^2 + 4\xi_0^2 (f/f_0)^2} S_0 \tag{A.1}$$

where f_0 is the predominant frequency of ground acceleration, ξ_0 is the damping ratio of topsoil, reflecting the bandwidth of ground acceleration, and S_0 is a spectral parameter reflecting the intensity of ground motion. f_g changes from 0.6 to 10 Hz and ξ_g changes from 0.45 to 1 according to the statistical results of Ou *et al.*[18] and

Jiang et al.[12] Then the displacement variance is

$$\sigma_{q_j}^2 = \frac{S_0}{k_j^2}\int_0^\infty \frac{1}{[1-(f/f_j)^2]^2+4\xi_j^2(f/f_j)^2}\frac{1+4\xi_0^2(f/f_0)^2}{[1-(f/f_0)^2]^2+4\xi_0^2(f/f_0)^2}df$$
$$= \frac{f_0 S_0}{k_j^2}\int_0^\infty Q(x)dx \quad \text{(A.2)}$$

$$Q(x) = \frac{r'^4}{(r'^2-x^2)^2+4\xi_j^2 r'^2 x^2}\frac{1+4\xi_0^2 x^2}{[1-x^2]^2+4\xi_0^2 x^2}, \quad \text{(A.3)}$$

where $r' = f_j/f_0 = 1/r$ is the ratio of structural frequency to excitation predominant frequency, and $x = f/f_g$. According to the residue theorem, the integration in Eq. (14) becomes

$$\int_0^\infty Q(x)dx = \pi i \sum_{k=1}^{n} \mathrm{Res}_{x=x_k} Q(x) = \pi i \sum_{k=1}^{n}\left[\lim_{x\to x_i}(x-x_i)Q(x)\right]. \quad \text{(A.4)}$$

The all poles of $Q(x)$ are respectively

$$\begin{aligned} x_{1,2} &= r'(\pm\sqrt{1-\xi_j^2}+i\xi_j)\\ x_{3,4} &= \pm\sqrt{1-\xi_0^2}+i\xi_0\\ x_{5,6} &= r'(\pm\sqrt{1-\xi_j^2}-i\xi_j)\\ x_{7,8} &= \pm\sqrt{1-\xi_0^2}-i\xi_0 \end{aligned} \quad \text{(A.5)}$$

x_1 to x_4 are in the upper half plane, and x_5 to x_8 are in the lower half plane. By substituting Eq. (A.5) into Eqs. (A.4) and (A.2), one can get

$$\sigma_{q_j}^2 = \frac{G_0}{k_j^2}\left[\frac{A}{F}\frac{\pi f_j}{4\xi_j}+\frac{B}{F}\frac{\pi f_0}{4\xi_0}\right] \quad \text{(A.2)}$$

where A, B and F are functions of the frequency ratio r':

$$A = A_0 + 4\xi_j^2 r'^2[2-4\xi_0^2-(3-4\xi_j^2)r'^2-4\xi_0^2 r'^4] \quad \text{(A.3)}$$
$$B = r'^4[1-8\xi_0^2-2(1-2\xi_j^2)r'^2+(1+4\xi_0^2)r'^4] \quad \text{(A.4)}$$
$$F = F_0 + 8\xi_j^2 r'^2[1-2\xi_0^2-2(1-\xi_j^2)r'^2+(1-2\xi_0^2)r'^4] \quad \text{(A.5)}$$
$$A_0 = 1-2(1-4\xi_0^2)r'^2+(1-8\xi_0^2+16\xi_0^4)r'^4+4\xi_0^2 r'^6 \quad \text{(A.6)}$$
$$F_0 = [(r'^2-1)^2+4\xi_0^2 r'^2]^2 \quad \text{(A.7)}$$

The coefficient A_0 given in Der Kiureghian[9] is $A_0 = 1-2(1-4\xi_0^2)r'^2+(1+4\xi_0^2)r'^4$, and other coefficients are the same.

Acknowledgements

This work is funded by the National Natural Science Foundation of China (52221002, 51720105005), Chongqing Science and Technology Bureau (cstc2019yszx-jcyjX0005, cstc2020yszx-jcyjX0007), 111 Project of China (B13002, B18062). This support is gratefully acknowledged.

References

1. Architectural Institute of Japan, *AIJ Recommendations for Loads on Buildings* (AIJ, Tokyo, 2015).
2. L. Arany, S. Bhattacharya, J. H. G. Macdonald and S. J. Hogan, Closed form solution of Eigen frequency of monopile supported offshore wind turbines in deeper waters incorporating stiffness of substructure and SSI, *Soil Dyn. Earthq. Eng.* **83** (2016) 18–32, doi:10.1016/j.soildyn.2015.12.011.
3. B. Chen, H. Zhao, X. Wang and Z. Wang, Wind-induced response and equivalent static wind loads on waiting hall building of regular railway stations, *Int. J. Struct. Stab. Dyn.* **22**(14) (2022) 2250152, doi:10.1142/S0219455422501528.
4. X. Chen and A. Kareem, Coupled dynamic analysis and equivalent static wind loads on buildings with three-dimensional modes, *J. Struct. Eng.* **131**(7) (2005) 1071–1082, doi:10.1061/(ASCE)0733-9445(2005)131:7(1071).
5. D. Claës and S. O. Hansen, *Wind Loads on Structures* (Wiley, New York, 1997).
6. S. H. Crandall and W. D. Mark, Transmission of random vibration, in *Random Vibration in Mechanical Systems*, eds. S. H. Crandall and W. D. Mark, 2nd edn. (Academic Press, Cambridge, 1963), pp. 55–102.
7. R. R. Damiani, H. Song, A. N. Robertson and J. M. Jonkman, Assessing the Importance of Nonlinearities in the Development of a Substructure Model for the Wind Turbine CAE Tool FAST, in *Proceedings of the ASME 2013 32nd International Conference on Ocean, Offshore and Arctic Engineering*, Vol. 8, 9–14 June 2013, Nantes, France, https://doi.org/10.1115/OMAE2013-11434.
8. A. G. Davenport, Gust loading factors, *J. Struct. Div.* **93**(3) (1967) 11–34.
9. A. Der Kiureghian, Structural response to stationary excitation, *J. Eng. Mech. Div.* **106**(6) (1980) 1195–1213, doi:10.1061/JMCEA3.0002659
10. C. Gu, D. Chen, F. Liu, K. Fang, D. Guo and P. Marzocca, Dynamic analysis of flexible wind turbine tower by a transfer matrix method, *Int. J. Struct. Stab. Dyn.* **21**(10) (2021) 2150142, doi:10.1142/S021945542150142X
11. M. Gu and Y. Quan, Across-wind loads of typical tall buildings, *J. Wind Eng. Indus. Aerodyn.* **92**(13) (2004) 1147–1165, doi:10.1016/j.jweia.2004.06.004
12. J. Jiang, Q. Lu and J. Sun, The statistical characterization of strong ground motion specified by response specturm and power spectral density function, *Earthq. Res. Chin.* **9** (1995) 387–402.
13. K. Kanai, Semi-empirical formula for the seismic characteristics of the ground, *Bulletin of the Earthquake Research Institute*, University of Tokyo, **35**(2) (1957) 309–325.
14. A. Kareem, Model for predicting the acrosswind response of buildings, *Eng. Struct.* **6**(2) (1984) 136–141.
15. Y.-K. Lin, *Probabilistic Theory of Structural Dynamics* (McGraw-Hill, New York, 1967).

16. L. D. Lutes and A. S. Shahram, *Random Vibrations: Analysis of Structural and Mechanical Systems* (Butterworth-Heinemann, Oxford, 2004).
17. N. M. Newmark and E. Rosenblueth, *Fundamentals of Earthquake Engineering* (Prentice-Hail Inc, Englewood Cliffs, 1971).
18. J. Ou, D. Niu and X. Du, Random earthquake ground motion model and its parameter determination used in aseismic design, *Earthq. Eng. Eng. Vib.* **11**(3) (1991) 45–54.
19. J. B. Roberts and P. D. Spanos, *Random Vibration and Statistical Linearization* (Courier Corporation, North Chelmsford, 2003).
20. E. Simiu and R. H. Scanlan, *Wind Effects on Structures: Fundamentals and Applications to Design*, Vol. 688 (John Wiley, New York, 1996).
21. H. Tajimi, A statistical method of determining the maximum response of a building structure during an earthquake, in *Proc. 2nd world Conf. Earthquake Engineering, A-2*, 1960, Tokyo, Japan, pp. 781–797.
22. T. von Kármán, Progress in the statistical theory of turbulence, *Proc. Nat. Acad. Sci. United States Am.* **34**(11) (1948) 530–539.
23. P. Xu, C. Xiao and J. Li, Study on relationship between natural vibration periods and heights of structures for high-rise buildings and its reference range, *Int. J. High-Rise Build.* **47** (2014) 1–11, doi:10.21022/IJHRB.2014.3.1.049
24. Q. Yang, H. Pang, M. Liu, Q. Wang, Z. Li and B. Li, Wind-induced impact behavior of base-isolated high-rise buildings with adjacent structures, *Int. J. Struct. Stab. Dyn.* **23**(7) (2022) 2350076, 10.1142/S0219455423500761.

https://doi.org/10.1142/9789819814770_0011

Chapter 11

Static and Dynamic Stability Analyses of Functionally Graded Beam with Inclined Cracks#

Jia-Jia Mao*,‡, Ying-Jie Wang*,§ and Jie Yang†,¶

**Beijing Key Laboratory of Nonlinear Vibrations and Strength of Mechanical Structures Faculty of Materials and Manufacturing Beijing University of Technology, Beijing 100124, China*

†School of Engineering, RMIT University P. O. Box 71,Bundoora, VIC, 3083, Australia

‡jiajia.mao@bjut.edu.cn

§w_yj@emails.bjut.edu.cn

¶j.yang@rmit.edu.au

The focus of this paper is to examine the static and dynamic instabilities of functionally graded beam that contains multiple inclined cracks under the influence of an axial force comprising both static and time-varying harmonic components. The elasticity modulus and mass density of the functionally graded beam are assumed to vary exponentially along its thickness direction. Local stiffness matrix model-based finite element analysis (FEA) is conducted to determine the bending stiffness and tensile stiffness of the section with a crack, and the coupled effect of tensile and bending loadings. Two-node beam elements with three degrees-of-freedom per node are utilized. By combining the Euler–Bernoulli beam theory with Lagrange method, we derive the governing equations that describe the static and dynamic instabilities of a functionally graded beam with multiple inclined cracks. These equations can be solved as eigenvalue problems to obtain the natural frequency and static critical buckling load of the beam. Furthermore, to investigate the dynamic instability of the system, we use the Bolotin method to determine the boundary between the regions of instability and stability based on the same governing equations. By adopting this approach, the study comprehensively investigates the impacts of crack position, inclination angle, and length, as well as elasticity modulus ratio, static and dynamic load factors on both static and dynamic stabilities of a cracked

‡,¶Corresponding authors.

#This chapter appeared previously on the International Journal of Structural Stability and Dynamics. To cite this chapter, please cite the original article as the following: J. J. Mao, Y. J. Wang & J. Yang, *Int. J. Struct. Stab. Dyn.*, **23**, 2340012 (2023), DOI: 10.1142/S0219455423400126.

functionally graded beam to gain valuable insights into the stability and performance of cracked functionally graded structures.

Keywords: Structural stability; critical buckling load; dynamic instability; functionally graded beam; inclined crack.

1. Introduction

Even functionally graded materials (FGMs) are made by mixing two or more materials, and behave gradual variation of material properties due to the composition of each material gradually changes.[1–5] This gradual change in material properties results in a smooth transition between different regions of the material, which allows for better stress distribution and minimizes the occurrence of delamination that are often observed in conventional composites, hence they are a promising candidate for advanced applications in aerospace, automotive, civil engineering, biomedical, and energy sectors. However, these structures usually suffer to complex loads and are involved in crack problems, which degrade the structural stiffness and stability. Therefore, reams of attention are paid from scientific and engineering communities since FGMs were proposed, to discuss the structural stability and dynamic responses of the cracked FGM structures, to insure their safe operation, performance, and durability.

In the last few decades, researchers have developed and refined various models and methods to simulate the decrease in stiffness caused by cracks in materials,[6–13] including the continuous cracked model,[11] local flexibility matrix,[12] lumped flexibility model,[13] to understand the behavior of cracked structures and develop effective strategies to mitigate the negative impact of cracks on material performance and durability. The continuous cracked model represents changes in stiffness and flexibility caused by cracks using a one-dimensional distribution function that is continuous.[14,15] The approximate assumption about the displacement fields and stress fields near cracks in the continuous cracked model may lead to inaccurate and errors for the static and dynamic behaviors related analyses.[11] The successive proposals and developments of lumped flexibility models, as well as the utilization of local flexibility matrices, have facilitated more accurate investigations of the static and dynamic responses of structures containing cracks.[16,17]

The massless rotation spring model,[18] based on the lumped flexibility model, was introduced to simulate the bending stiffness of sections with cracks that are always perpendicular to the structural surfaces, known as mode I cracks. It has been used both for homogeneous,[16,17,19] and FGM structures with mode I cracks[10,12,20–22] to comprehensively investigate their buckling, vibrations, stability and wave propagations. Aydin[23] and Khiem and his groups[24,25] conducted separate discussions on the free vibration characteristics of functionally graded beams, using the Euler–Bernoulli beam theory and the Timoshenko beam theory, respectively, considering arbitrary numbers of surface cracks. With the massless rotating spring model, Yang and his team[12,26–30] investigated the vibration and stability behaviors of the functionally graded beams with edged cracks. Their results show that the gradients can highly enhance the vibration behaviors and stability areas of the functionally graded beam. Though the massless rotation spring model considering the bending stiffness of crack and the continuities of displacements and forces of the sub-structures simultaneously, the effects of tensile flexibility and rotation flexibility of the cracks and their interactions on the stiffness of the cracked structures are ignored.[20]

Dimarogonas *et al.* introduced[31] the local flexibility matrix, which includes bending flexibility, tensile flexibility, and rotational flexibility of the cross-section with cracks and the coupled effect of tensile and bending forces, to obtain the stiffness matrix of the homogeneous structures with cracks. With the local flexibility matrix, Karaagac *et al.*[32–34] investigated the in-plane static and dynamic stabilities, lateral buckling, as well as the free vibration of the cantilever slender beam, and verified their results against experimental data. With the stress intensity factors (SIFs)[35] for the mode I crack on the surface of the FGM plate, the local flexibility matrix was extended to the FGM shaft with cracks by Gayen and Chakraborty.[13] Gayen *et al.*[36,37] further discussed the influences of the varying material distributions of the cracked FGM shafts on their dynamic characteristics. Reynders and Roeck[38] proposed a new technology to locate and quantify the damages of structures through calculating their vibration characteristics based on the local flexibility matrix. The solution to the dynamic behavior of a beam with a single transverse crack has been obtained through the cracked finite element stiffness matrix proposed by Saavedra and Cuitino.[39]

Except for the mode I cracks, inclined cracks are also normal for the FGM structures due to the complex loadings. But there are less literature works investigating the influences of the inclined cracks on the structural bending, structural stability, and dynamic responses of the FGM structures with inclined cracks than that case of mode I cracks. Recently, Duc and his colleagues[40,41] studied how vibrations are affected by a single inclined crack in the cracked FGM plate. In our previous work,[42] we incorporated the local flexibility matrix into functionally graded beams containing multiple inclined cracks to analyze their vibration and wave propagation, and shown that as long as the inclined cracks satisfy some certain conditions, they can be simulated by the equivalent transversal cracks to obtain the actual stiffness of functionally graded beam containing inclined cracks.

Ensuring the safe operation of the cracked FGM structures further, and improving their performance and durability, this paper analyzes the static and dynamic stabilities of the functionally graded beams with multiple inclined cracks through finite element analysis (FEA)-based the local stiffness matrix model, which can simultaneously consider the bending flexibility and tensile flexibility of the section with cracks and the coupled effect of tensile and bending loadings. A beam element with two nodes and three degrees-of-freedom per node is utilized. The properties of the inclined cracked functionally graded beam are varying as exponential laws. By combining the flexibility matrix of the corresponding intact functionally graded beam element with the local additional flexibility matrix resulting from the inclined crack, the stiffness matrix of the functionally graded beam element with an inclined crack can be produced. The governing equations for the stability analysis of the inclined cracked functionally graded beam are formulated with the Euler–Bernoulli beam theory and Lagrange method, which first reduced to the eigenvalue problems to solve the natural frequency and static critical buckling load, and then used to analyze the dynamic stability of the functionally graded beam containing inclined cracks via Bolotin method. The influences of location, angle and length of the multiple inclined cracks on the static critical buckling loads and dynamic stability regions of the inclined cracked functionally graded beam with varying elasticity modulus and boundaries are numerically analyzed.

2. Model and Formulations of Inclined Cracked Functionally Graded Beam

A multiple inclined cracked functionally graded beam with length L is depicted in Fig. 1(a). Taking two inclined cracks as an example, the first crack, inclined at an angle of α_1, is located at $X = L_1$, while the second crack, inclined at an angle of α_2, is located $X = L_2$. a_1 and a_2 represent the lengths of these two inclined cracks, respectively. A time-depending axial load $F(t)$ is acted at $X = L$. The perfect rectangular cross-section with a width of b and a thickness of h of the inclined cracked functionally graded beam is given in Fig. 1(b). The elasticity modulus and mass density of the inclined cracked functionally graded beam are assumed varying as exponential laws, i.e.

$$E(Z) = E_0 e^{\beta_E Z}, \quad \rho(Z) = \rho_0 e^{\beta_\rho Z}, \tag{1}$$

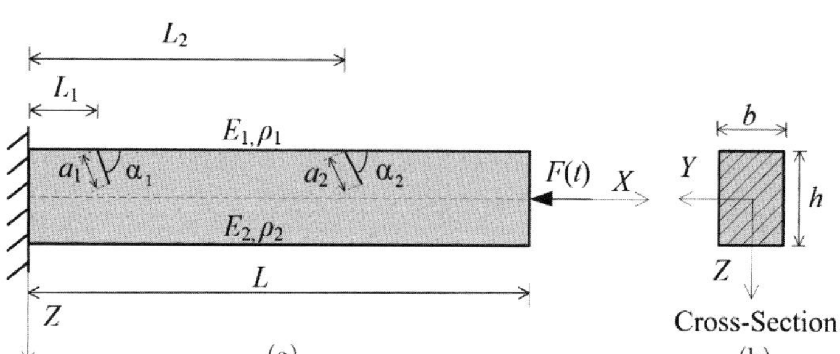

Fig. 1. Functionally graded beam with multiple inclined cracks subjected to time-varying axial load.

where $\beta_E = \frac{\ln r_E}{h}$ and $\beta_\rho = \frac{\ln r_\rho}{h}$ are the gradient indices with

$$r_E = \frac{E_2}{E_1}, \quad r_\rho = \frac{\rho_2}{\rho_1}. \tag{2}$$

Here, the subscripts "0", "1", and "2" are, respectively, the planes of $Z = -h/2$ and $Z = h/2$ of the functionally graded beam with multiple inclined cracks. Poisson's ratio ν of the considered functionally graded beam is supposed as a constant.

According to the Euler–Bernoulli theory, the inclined cracked functionally graded beam has the axial displacement $u(X, Z, t)$ and transversal displacement $w(X, Z, t)$

$$u(X, Z, t) = u_0(X, t) - Z\frac{\partial w_0(X, t)}{\partial X}, \tag{3}$$

$$w(X, Z, t) = w_0(X, t), \tag{4}$$

with the axial and transversal displacements u_0 and w_0 of the arbitrary points on midplane of the inclined cracked functionally graded beam, and time t. The normal stresses for arbitrary points on the midplane of the functionally graded beam with multiple inclined cracks can be expressed by combining them with the linear elastic constitutive equations:

$$\sigma_{XX} = E(Z)\varepsilon_{XX} = E(Z)\left(\frac{\partial u_0}{\partial X} - Z\frac{\partial^2 w_0}{\partial X^2}\right). \tag{5}$$

Hence, the elastic strain energy U_i can be written as

$$U_i = \frac{1}{2}\int_0^L \int_A E(Z)\left[\left(\frac{\partial u_0}{\partial X}\right)^2 - 2Z\frac{\partial u_0}{\partial X}\frac{\partial^2 w_0}{\partial X^2} + \left(Z\frac{\partial^2 w_0}{\partial X^2}\right)^2\right]dAdX. \tag{6}$$

The kinetic energy T can be expressed as

$$T = \frac{1}{2}\int_0^L \int_A \rho(Z)\left[\left(\frac{\partial u_0}{\partial t}\right)^2 - 2\left(\frac{\partial u_0}{\partial t}\right)\left(Z\frac{\partial^2 w_0}{\partial X\partial t}\right) + \left(Z\frac{\partial^2 w_0}{\partial X\partial t}\right)^2 + \left(\frac{\partial w_0}{\partial t}\right)^2\right] \times dAdX. \tag{7}$$

The work U_e done by the external axial load $F(t)$ can be formulated by

$$U_e = -\frac{1}{2}\int_0^L F(t)\left(\frac{\partial w_0}{\partial X}\right)^2 dX. \tag{8}$$

As shown in Fig. 2,[42] the displacement fields u_0 and w_0 of the two-node beam element can be written as

$$u_0(X, t) = N_i^u(X)u_i(t) + N_4^u(X)u_j(t) = \mathbf{N}^u\mathbf{d}^u, \tag{9}$$

$$w_0(X, t) = N_2^w(X)w_i + N_3^w(X)\theta_i + N_5^w(X)v_j + N_6^w(X)\theta_j = \mathbf{N}^w\mathbf{d}^w \tag{10}$$

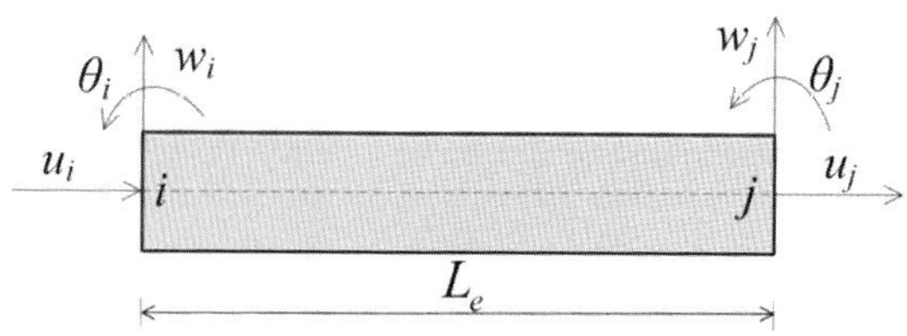

Fig. 2. Two-nodes beam element.

with the Hermitian shape functions for axis and transverse degrees-of-freedom $\mathbf{N}^u$ and $\mathbf{N}^w$ (provided in Appendix A), and the vector $\mathbf{d}(t)$ for the generalized displacements of the two-nodes functionally graded beam element

$$\mathbf{d}(t) = [u_i(t) \quad w_i(t) \quad \theta_i(t) \quad u_j(t) \quad w_j(t) \quad \theta_j(t)]^T, \tag{11}$$

where i and j are, respectively, the left and right nodes of the corresponding element.

By substituting Eqs. (9) and (10) into Eqs. (6)–(8), one yields

$$U_i = \frac{1}{2}\int_0^L \int_A E(Z)\left[\left(\frac{\partial \mathbf{N}^u}{\partial X}\mathbf{d}^u\right)^2 - 2Z\left(\frac{\partial \mathbf{N}^u}{\partial X}\mathbf{d}^u\right)\left(\frac{\partial^2 \mathbf{N}^w}{\partial X^2}\mathbf{d}^w\right) + \left(Z\frac{\partial^2 \mathbf{N}^w}{\partial X^2}\mathbf{d}^w\right)^2\right] dAdX, \tag{12}$$

$$T = \frac{1}{2}\int_0^L \int_A \rho(Z)\left[\left(\mathbf{N}^u\frac{\partial \mathbf{d}^u}{\partial t} - Z\frac{\partial \mathbf{N}^w}{\partial X}\frac{\partial \mathbf{d}^w}{\partial t}\right)^2 + \left(\mathbf{N}^w\frac{\partial \mathbf{d}^w}{\partial t}\right)^2\right] dAdX, \tag{13}$$

$$U_e = -\frac{1}{2}\int_0^L P\left(\frac{\partial \mathbf{N}^w}{\partial X}\mathbf{d}^w\right)^2 dX. \tag{14}$$

According to the Lagrange method, we have

$$\frac{\partial I}{\partial q^e} + \frac{d}{dt}\frac{\partial I}{\partial \dot{q}_k^e} = 0, \quad k = 1, 2, \ldots, 6, \tag{15}$$

where "·" represents derivative of time, and I is the Lagrange function

$$I = T - (U_i + U_e). \tag{16}$$

The matrix form of the governing equations for the stabilities of the functionally graded beam with multiple inclined cracks can be obtained from Eq. (15)

$$\mathbf{M}\ddot{\mathbf{d}}(t) + [\mathbf{K} - F(t)\mathbf{K}_g]\mathbf{d}(t) = \mathbf{0}, \tag{17}$$

where $\mathbf{K_g}$ is the stiffness caused by the structural geometry deformations

$$\mathbf{K}_g = \int_0^{L_e} \left[\left(\frac{\partial \mathbf{N}^w}{\partial X} \right)^T \left(\frac{\partial \mathbf{N}^w}{\partial X} \right) \right] dX. \tag{18}$$

$\mathbf{K}$ and $\mathbf{M}$ are, respectively, the structural stiffness matrix and mass matrix, which can be, respectively, assembled by using the corresponding stiffness matrices of the functionally graded beam element without cracks $\mathbf{K^e}$ and the inclined cracked functionally graded beam element $\mathbf{K^c}$, and the mass matrix $\mathbf{M^e}$ of the functionally graded beam element[42]

$$\mathbf{M^e} = \int_0^{L_e} \mathbf{S}^T \mathbf{D_R} \mathbf{S}^T dX, \tag{19}$$

with

$$\mathbf{S} = \begin{bmatrix} N_1 & 0 & 0 & N_4 & 0 & 0 \\ 0 & N_2 & N_3 & 0 & N_5 & N_6 \\ 0 & \dfrac{\partial N_2}{\partial X} & \dfrac{\partial N_3}{\partial X} & 0 & \dfrac{\partial N_5}{\partial X} & \dfrac{\partial N_6}{\partial X} \end{bmatrix} \tag{20}$$

and

$$\mathbf{D_R} = \int_{-h/2}^{h/2} \begin{bmatrix} \rho(Z) & 0 & -\rho(Z)Z \\ 0 & \rho & 0 \\ -\rho(Z)Z & 0 & -\rho(Z)Z^2 \end{bmatrix} dZ. \tag{21}$$

It is noted that with n elements, the assembled stiffness $\mathbf{K}$, $\mathbf{K_g}$, and mass matrices $\mathbf{M}$ are $3(n+1) \times 3(n+1)$. The left and right boundaries of the functionally graded beam are, respectively, represented by the beginning three rows and columns (i.e., u, w, and θ for the left node of the 1st element) and the last three rows and columns (i.e., u, w, and θ for the right node of the nth element). For the clamped-free (C–F) boundaries, the beginning three rows and columns of $\mathbf{K}$, $\mathbf{K_g}$, and mass matrices $\mathbf{M}$ are set as empty, as the axial displacement, transversal displacement and rotation of the node on the clamped end are constrained. For the clamped–clamped (C–C) boundaries, the beginning three rows and columns and the last three rows and columns are set as empty. For the hinged–hinged (H–H) boundaries, u and w related to the left node of the 1st element and the right node of the nth element are set as empty.

2.1. *Stiffness matrix of intact functionally graded beam element*

As stated in Ref. 42, for the intact functionally graded beam element, its stiffness matrix $\mathbf{K^e}$ is expressed by

$$\mathbf{K^e} = \int_0^{L_e} \mathbf{B}^T \mathbf{D_E} \mathbf{B} dX, \tag{22}$$

where

$$\mathbf{B} = \begin{bmatrix} \frac{\partial N_1}{\partial X} & 0 & 0 & \frac{\partial N_4}{\partial X} & 0 & 0 \\ 0 & \frac{\partial^2 N_2}{\partial X^2} & \frac{\partial^2 N_3}{\partial X^2} & 0 & \frac{\partial^2 N_5}{\partial X^2} & \frac{\partial^2 N_6}{\partial X^2} \end{bmatrix} \tag{23}$$

and

$$\mathbf{D_E} = \int_{-h/2}^{h/2} \begin{bmatrix} E(Z) & -ZE(Z) \\ -ZE(Z) & Z^2E(Z) \end{bmatrix} dZ. \tag{24}$$

2.2. *Stiffness matrix of inclined cracked functionally graded beam element*

As given in Fig. 3, the left node i and the right node j of the functionally graded beam element with length L_e containing an inclined crack are respectively clamped, and loaded by axial force P_1, shearing force P_2 and bending moment P_3. δ_1, δ_2 and δ_3 are the generalized displacements corresponding to P_1, P_2 and P_3, respectively. According to previous studies,[6,42] an inclined crack in a functionally graded beam can be considered equivalent to a crack perpendicular to the top surface. Thus, the additional flexibility matrix of the functionally graded beam element caused by an inclined crack with angle $\alpha(0 \leq \alpha \leq 90°)$ and length a can be calculated by taking the bending stiffness and tensile stiffness of the section with the equivalent crack, as well as the coupled effect of tensile-bending loadings into consideration.[42] The additional strain energy Π_c resulted by the inclined crack can be calculated by the integration of the strain energy release rate G over A_c, the effective area per unit

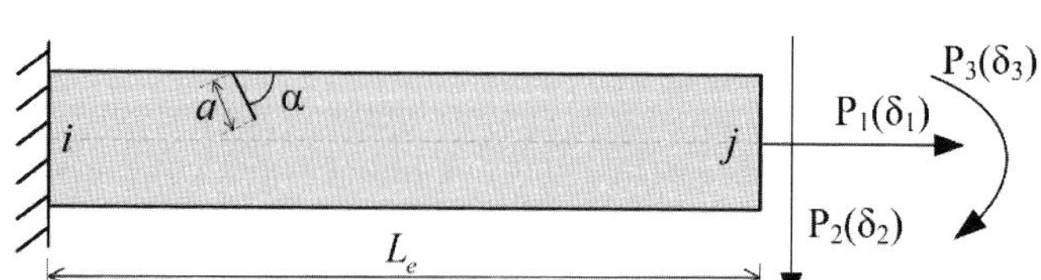

Fig. 3. Functionally graded beam element with an inclined crack.

width of the equivalent crack

$$\Pi_c = \int_{A_c} G dA_c \tag{25}$$

with

$$G = \frac{1}{\tilde{E}(\xi \sin\alpha)}[(K_{IP} + K_{IM})^2 + K_{IIQ}^2], \tag{26}$$

where ξ is a variable which can change from zero to length a of the inclined crack, the SIFs K_{IP}, K_{IIQ} and K_{IM} of the equivalent crack can be separately calculated by

$$K_{\mathrm{IP}} = \frac{P_1}{bh}\sqrt{\pi\xi \sin\alpha} F_1\left(r_E, \frac{\xi \sin\alpha}{h}\right), \tag{27}$$

$$K_{IIQ} = \frac{P_2}{bh}\sqrt{\pi\xi \sin\alpha} F_2\left(r_E, \frac{\xi \sin\alpha}{h}\right), \tag{28}$$

$$K_{IM} = \frac{6P_3}{bh^2}\sqrt{\pi\xi \sin\alpha} F_3\left(r_E, \frac{\xi \sin\alpha}{h}\right) \tag{29}$$

with

$$F_i = \frac{p_1^i + p_2^i \ln(r_E) + p_3^i \ln^2(r_E) + p_4^i \ln^3(r_E) + p_5^i(\frac{\xi \sin\alpha}{h}) + p_6^i(\frac{\xi \sin\alpha}{h})^2}{1 + p_7^i \ln(r_E) + p_8^i \ln^2(r_E) + p_9^i(\frac{\xi \sin\alpha}{h}) + p_{10}^i(\frac{\xi \sin\alpha}{h})^2 + p_{11}^i(\frac{\xi \sin\alpha}{h})^3}, \tag{30}$$

where p_j^1, p_j^2 and p_j^3 $(j = 1, 2, \ldots, 11)$ can be solved through Levenberg–Marquardt method,[43] respectively, corresponding to the axial load, shearing load and bending moment.

The additional displacements caused by the inclined crack can be obtained through the pairs equation

$$\delta_m = \frac{\partial \Pi_c}{\partial P_m}. \tag{31}$$

Therefore, the additional flexibility matrix $\mathbf{c_c}$ is

$$\mathbf{c_c} = \begin{bmatrix} c_{11} & -c_{12} & -c_{13} \\ -c_{21} & c_{22} & c_{23} \\ -c_{31} & c_{32} & c_{33} \end{bmatrix}, \tag{32}$$

where

$$c_{mn} = \frac{\partial \delta_m}{\partial P_n} = \frac{\partial^2 \Pi_c}{\partial P_m \partial P_n}, \quad (m, n = 1, 2, 3). \tag{33}$$

Obviously, $c_{mn} = c_{nm}$. When m equals to n, the corresponding flexibility coefficients c_{mn} are involved to the force P_m, otherwise, they are the coupling flexibility coefficients related to the coupled effect of forces P_m and P_n.

Then, the total flexibility matrix $\mathbf{c_{tot}}$ is expressed as

$$\mathbf{c_{tot}} = \mathbf{c_c} + \mathbf{c_{int}}, \tag{34}$$

where $\mathbf{c_{int}}$ is the flexibility matrix for the undamaged functionally graded beam element

$$\mathbf{c_{int}} = [(\mathbf{LL}^T)^{-1}\,\mathbf{K^e}(\mathbf{LL}^T)^{-1}]^{-1} \tag{35}$$

with

$$\mathbf{L} = \begin{bmatrix} -1 & 0 & 0 & 1 & 0 & 0 \\ 0 & -1 & -L_e & 0 & 1 & 0 \\ 0 & 0 & -1 & 0 & 0 & 1 \end{bmatrix}^T. \tag{36}$$

Hence, the stiffness matrix $\mathbf{K^c}$ for the beam element with one inclined crack can be expressed as

$$\mathbf{K^c} = \mathbf{L}^T \mathbf{c}_{\mathbf{tot}}^{-1} \mathbf{L}. \tag{37}$$

With Eqs. (22) and (37), the stiffness matrix $\mathbf{K}$ in Eq. (17) can be obtained.[40] To investigate the static and dynamic stabilities of functionally graded beams that contain multiple inclined cracks, we express the axial force $F(t)$ as the combination of a static and a time-varying harmonic components

$$F(t) = \alpha P_{\mathrm{cr}} + \beta P_{\mathrm{cr}} \cos(\Omega t), \tag{38}$$

where the static loading factor and dynamic loading factor are denoted by α and β, respectively. P_{cr} represents the static critical buckling load of the inclined cracked functionally graded beam, and the frequency of the axial load is denoted by Ω. Then, the governing equation (17) can be indicated as

$$\mathbf{M\ddot{d}}(t) + [\mathbf{K} - (\alpha P_{\mathrm{cr}} + \beta P_{\mathrm{cr}} \cos(\Omega t))\mathbf{K}_g]\mathbf{d}(t) = \mathbf{0}. \tag{39}$$

3. Solution Procedure

The equation represented by Eq. (39) is a Mathieu–Hill-type system of second-order differential equations with periodic coefficients.[44] Bolotin method can be used to generate periodic solutions at the boundary between regions of the dynamic stability and instability.[45,46] The first-order solution $\mathbf{q}$ with period $2T$, the most practical important one, is given as

$$\mathbf{d}(t) = \mathbf{a_1} \sin\frac{\Omega t}{2} + \mathbf{b_1} \cos\frac{\Omega t}{2}. \tag{40}$$

By substituting Eq. (40) into Eq. (39), one yields

$$\sin\left(\frac{\Omega t}{2}\right)\left[-\left(\frac{\Omega}{2}\right)^2\mathbf{M}+(\mathbf{K}-\alpha P_{\mathrm{cr}}\mathbf{K}_g)\right]\mathbf{a}_1+\frac{1}{2}\beta P_{\mathrm{cr}}\mathbf{K}_g\left[\sin\left(\frac{3\Omega t}{2}\right)\right.$$
$$\left.+\sin\left(\frac{-\Omega t}{2}\right)\right]\mathbf{a}_1+\cos\left(\frac{\Omega t}{2}\right)\left\{-\left(\frac{\Omega}{2}\right)^2\mathbf{M}+(\mathbf{K}-\alpha P_{\mathrm{cr}}\mathbf{K}_g)\right\}\mathbf{b}_1$$
$$+\frac{1}{2}\beta P_{\mathrm{cr}}\mathbf{K}_g\left[\cos\left(\frac{3\Omega t}{2}\right)+\cos\left(\frac{-\Omega t}{2}\right)\right]\mathbf{b}_1=\mathbf{0}. \tag{41}$$

By setting the coefficients of the $\sin(\Omega t/2)$ and $\cos(\Omega t/2)$ terms in Eq. (41) to zero, the equation of motions of the FGM beam with inclined cracks yields as follows:

$$\left|\mathbf{K}-\left(\alpha+\frac{1}{2}\beta P_{\mathrm{cr}}\right)\mathbf{K}_g-\left(\frac{\Omega}{2}\right)^2\mathbf{M}\right|=0, \tag{42}$$

and

$$\left|\mathbf{K}-\left(\alpha-\frac{1}{2}\beta\right)P_{\mathrm{cr}}\mathbf{K}_g-\left(\frac{\Omega}{2}\right)^2\mathbf{M}\right|=0. \tag{43}$$

For given values of α, β and P_{cr}, Eqs. (42) and (43) turn to an eigenvalue problem. Here, with the plus and minus signs, two sets of eigenvalues for Ω can be solved, which represent the dynamic instability and stability regions in the parameter space.

For $\alpha = 0$, $\beta = 0$, and $\omega = \Omega/2$, Eqs. (42) and (43) can be simplified to apply to the case of free vibration of the functionally graded beam with multiple inclined cracks as

$$|\mathbf{K}-\omega^2\mathbf{M}|=0 \tag{44}$$

where the solution is the natural frequency ω of the system.

For $\alpha = 1$, $\beta = 0$, and $\Omega = 0$, Eqs. (42) and (43) can be simplified to apply to the case of static stability of the functionally graded beam containing multiple inclined cracks as

$$|\mathbf{K}-P_{\mathrm{cr}}\mathbf{K}_g|=0, \tag{45}$$

where the solution is the critical buckling load P_{cr} of the system.

4. Numerical Results and Discussions

This paper considers a slender functionally graded beam, which is a plane strain problem, i.e. $\tilde{E}(\xi\sin\alpha) = E(\xi\sin\alpha)$ in Eq. (26). Since the shearing force in a slender functionally graded beam is typically negligible, the elements c_{i2} in the stiffness matrix pertaining to shearing force in $\mathbf{c_c}$ are assumed to be zero. Based on the Levenberg–Marquardt method,[43] p_j^1 and p_j^3 in Eq. (30) are respectively expressed as[47]

$$p_1^1, p_2^1 \dots p_{11}^1 = [0.95868, -0.06038, -0.03295, -0.00466, 0.31604, 0.92944, 0.01004, -0.00746, -0.106047, 0, 0], \tag{46}$$

and

$$p_1^3, p_2^3 \dots p_{11}^3 = [1.17323, -0.35395, 0.02894, -0.00613, 0.66254, 3.072, -0.00141, -0.01173, 1.99173, -0.34962, -3.0982]. \tag{47}$$

To discuss the static and dynamic stabilities of the functionally graded beam with multiple inclined cracks through considering the tensile stiffness and bending stiffness of the cracked sections and the coupled effect of tensile and bending forces parametrically, Aluminum with $E = 70\,\text{Gpa}$, $\nu = 0.33$, and $\rho = 2\,780\,\text{Kg/m}^3$ is selected as the constituent of the top surface of the inclined cracked functionally graded beam. Unless specifically stated, $r_E = r_\rho = r = 5$, $L_1 = 0.1\,L$, $\alpha_1 = \alpha_2 = 90^\circ$, $a_1 = a_2 = 0.3\,h$, and $L_2 = 0.3\,L$. The length of the considered functionally graded beam is $L = 0.3\,\text{m}$ and its thickness is $h = 0.02\,\text{m}$.

4.1. *Model verification*

Before analyzing the stabilities of a functionally graded beam with multiple inclined cracks, we validate the accuracy of our model by comparing various numerical results with published references to ensure the reliability and effectiveness of our mode. First, the critical buckling load ratios $P_{\text{cr1}}/P_{\text{cr0}}$ of the cantilever functionally graded beams with one edge crack at varying locations are presented in Fig. 4. Here, the

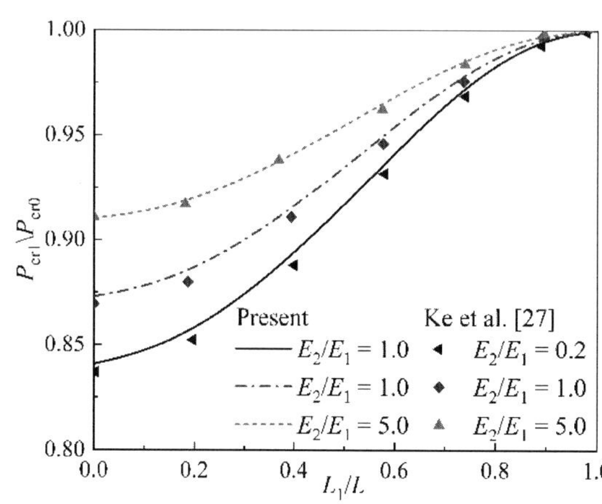

Fig. 4. The critical buckling load ratios of the cantilever functionally graded beams with one edge crack at varying locations.

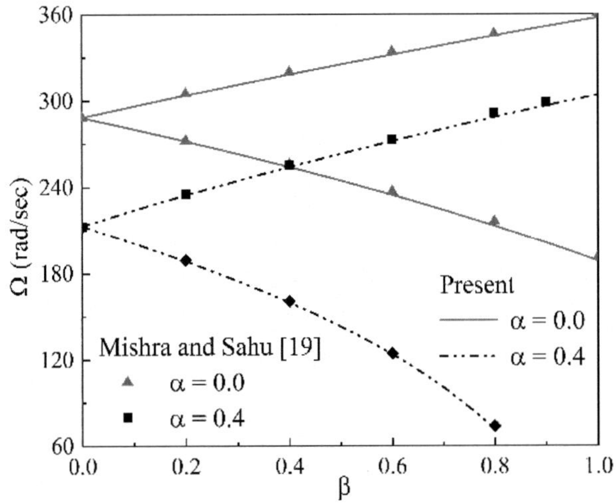

Fig. 5. The dynamic stability of homogenous beam.

length and thickness of the functionally graded beam are 0.6 and 0.1 m, respectively, while the crack depth is 0.02 m. The results for varying elasticity modulus rations are also calculated. P_{cr0} and P_{cr1} are, respectively, the static critical buckling load of the functionally graded beam without crack and that with one edge crack. Results reported by Ke *et al.*[27] are given for comparison. Secondly, the dynamic stabilities of homogenous beam with geometric and material properties of $L = 0.8\,\text{m}$, $b = h = 0.02\,\text{m}$, $E = 210\,\text{GPa}$, $\nu = 0.3$ and $\rho = 7\,800\,\text{kg/m}^3$ investigated by Mishra and Sahu[19] are compared in Fig. 5. As seen, both results for the static and dynamic stabilities have great agreements with the existing solutions.

4.2. *Static stability analyses of inclined cracked functionally graded beam*

In this subsection, the effects of several factors, including inclined angles, elasticity modulus ratios, crack locations, and varying boundary conditions, on the static critical buckling load ratios P_{cr}/P_{cr1} of functionally graded beams with multiple inclined cracks are discussed. Two inclined cracks are present, with the first one positioned at a fixed location of $L_1 = 0.1\,\text{L}$ and the second inclined crack located at varying positions.

Figure 6 shows the influence of the varying locations of the second inclined crack on the critical buckling load ratios P_{cr}/P_{cr1} of the functionally graded beam under different boundary conditions. Three different boundaries are included, C–C, H–H, and clamped-free (C–F). For the most locations of the cracks, they decrease the static stability of the novel beam. Cracks closing to the clamped end and the

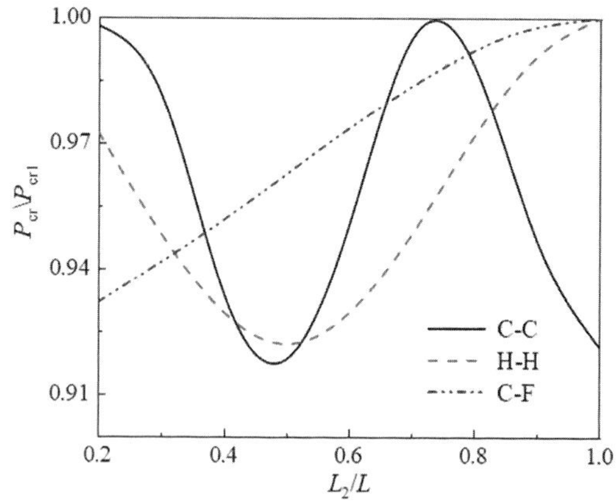

Fig. 6. The influence of the varying locations of the second inclined crack on the critical buckling load ratios of the functionally graded beam under different boundary conditions.

hinged/free ends have totally different influences on the static stability of the functionally graded beam with multiple inclined cracks. When cracks move close to the clamped end, crack can highly reduce the static critical buckling load ratio $P_{\text{cr}}/P_{\text{cr1}}$, but the influences of the cracks diminish when they are near the hinged/free cases. The crack weakens the ability of the clamped end to withstand moments, while the hinged/free ends are unable to withstand moments regardless of whether there is a crack or not. There are certain locations on the functionally graded beam where the inclined cracks have no effect on the critical buckling load ratios. For instance, when the functionally graded beam is under C–C boundary conditions and at the nodes for the first buckling mode, the inclined cracks have no impact on the critical buckling load ratios.

The effect of different elasticity modulus ratios on the static critical buckling load ratios $P_{\text{cr}}/P_{\text{cr1}}$ of the inclined cracked cantilever functionally graded beam with a second inclined crack located at various positions is illustrated in Fig. 7. Remember that E_1 is given in this calculation, increasing E_2/E_1 implies the increasing elasticity modulus for the bottom surface of the functionally graded beam, which can directly enhance the structural stiffness, and increase its stability. As observed in Fig. 4, bigger elasticity modulus ratio is, less the crack can affect the static stability of the functionally graded beam.

Figure 8 illustrates the influences of different inclined angles (a) and varying lengths (b) of the second crack on the static critical buckling load ratios $P_{\text{cr}}/P_{\text{cr1}}$ of the inclined cracked cantilever functionally graded beam with a second inclined crack located at various positions. As expected, both the increasing inclined angle and the rising length of the crack can decrease the static critical buckling load of the functionally graded beam. This is because the ability of the functionally

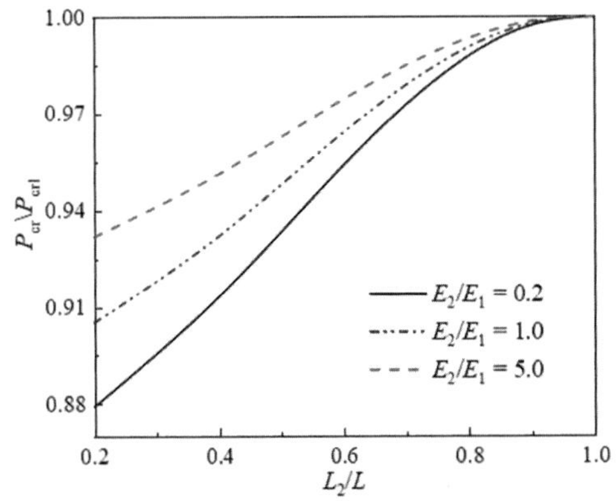

Fig. 7. The effect of different elasticity modulus ratios on the static critical buckling load ratios of the inclined cracked cantilever functionally graded beam with a second inclined crack located at various positions.

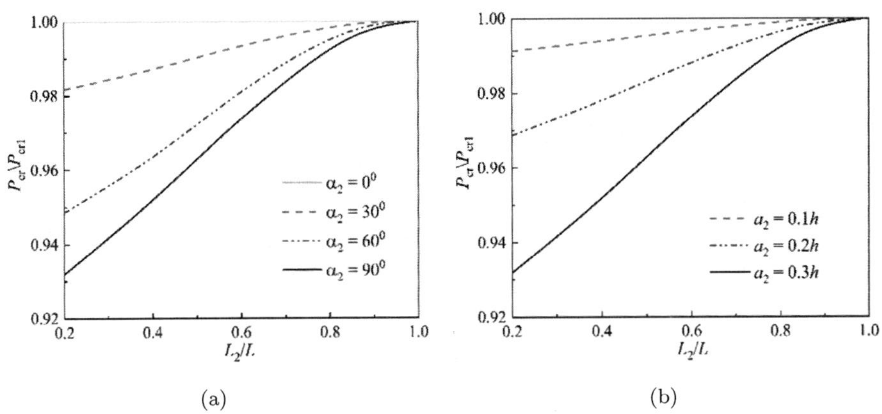

Fig. 8. The influences of different inclined angles (a) and varying lengths (b) of the second crack on the static critical buckling load ratios of the inclined cracked cantilever functionally graded beam with a second inclined crack located at various positions.

graded beam to withstand moments decreases as the second crack inclines more or propagates further.

4.3. *Dynamic stability analyses of functionally graded beam with inclined cracks*

In order to ensure their dynamic stability, a detailed demonstration is provided for the first parametric instability regions of functionally graded beams with multiple inclined cracks. The effects of the boundary conditions, elasticity modulus ratios, inclined angles and crack locations on the boundary frequencies that define the

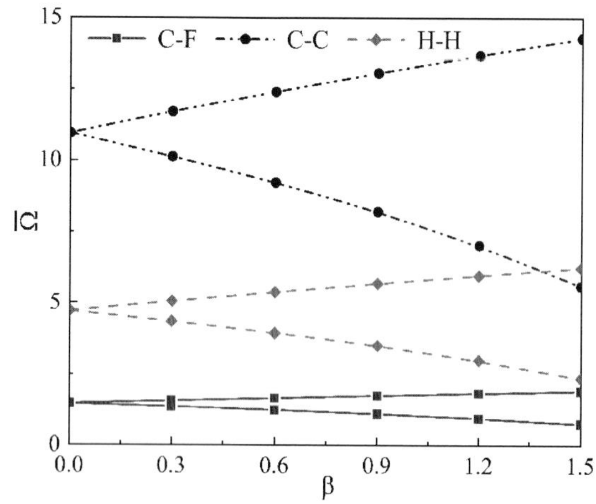

Fig. 9. The influence of boundary condition on the dynamic instability of the inclined cracked cantilever functionally graded beam.

dynamic instability regions are examined. The analysis of dynamic instability uses the dimensionless form of frequency, denoted by $\bar{\Omega} = \Omega/\omega_{CF}$, where ω_{CF} represents the first natural frequency of the homogenous cantilever aluminum beam without any cracks.

For the varying boundary conditions shown in Fig. 9, the more constraints the system has, the higher the stiffness of the inclined cracked functionally graded beam, which leads to a higher dimensionless instability boundary frequency. What's more, the inclined cracked functionally graded beam has the largest instability region with C–C boundaries, while the beam with C–F boundaries has a smaller instability region than that with H–H boundary conditions.

Figure 10 illustrates the influence of the elasticity modulus ratio on the dynamic instability of the inclined cracked cantilever functionally graded beam. Typically, a stiffer structure has a higher instability boundary frequency. However, in Fig. 10, the dimensionless instability boundary frequency does not monotonically increase with increasing elasticity modulus ratio. This is because the elasticity modulus and mass density of the functionally graded beam change synchronously. Specifically, while the increasing elasticity modulus increases the instability boundary frequency, the increasing mass density decreases it.

Figure 11 shows the impact of the static load factor on the dynamic instability of the inclined cracked cantilever functionally graded beam. Here, $\alpha = 0.0$ indicates that no static pressure is acting on the inclined cracked functionally graded beam. As the static load factor α increases, the static pressure rises, resulting in a decrease in both the stiffness and natural frequency of the functionally graded beam. Consequently, the dimensionless instability boundary frequency decreases, and the dynamic instability region expands.

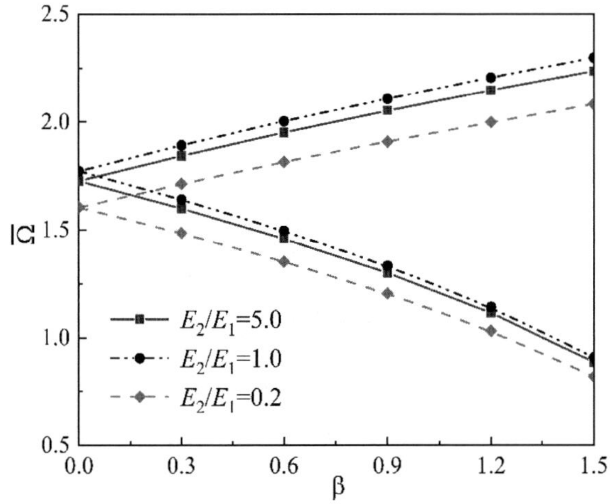

Fig. 10. The influence of the elasticity modulus ratio on the dynamic instability of the inclined cracked cantilever functionally graded beam.

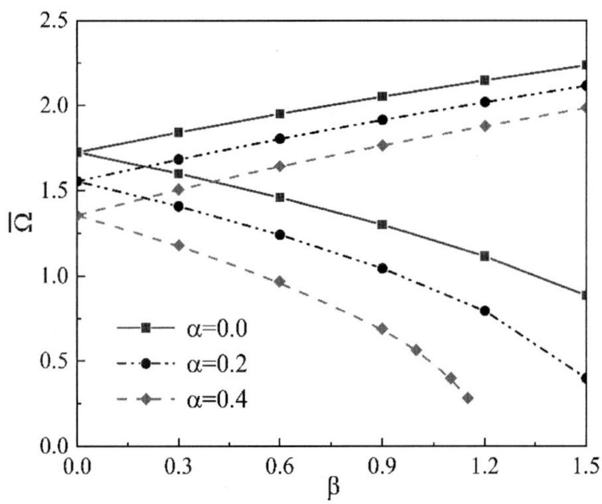

Fig. 11. The influence of static load parameter on the dynamic instability of the inclined cracked cantilever functionally graded beam.

Figure 12 gives the influences of the crack length (a) and the inclined angle (b) of these two inclined cracks on the dynamic instability of the inclined cracked cantilever functionally graded beam. For simplicity, angles of the cracks in Fig. 12(a) are selected as $\alpha_1 = \alpha_2 = 90°$, and lengths of these two inclined cracks in Fig. 12(b) are selected as $a_1 = a_2 = 0.5\,h$. The dimensionless instability boundary frequency

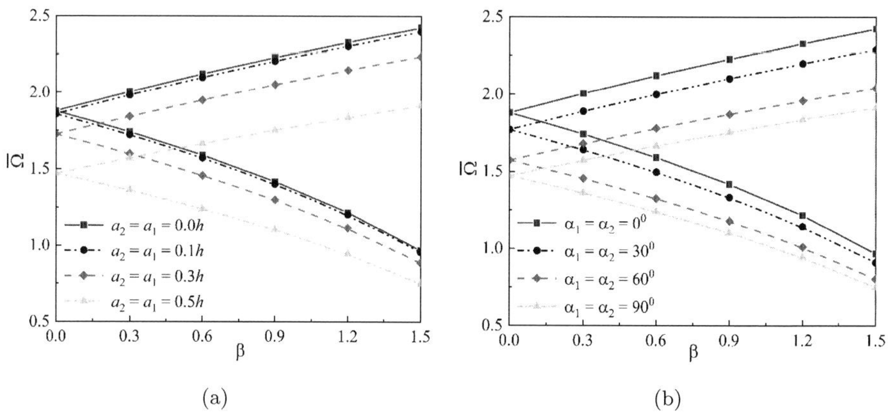

Fig. 12. The influences of crack length (a) and inclined angle (b) of these two inclined cracks on the dynamic instability of the inclined cracked cantilever functionally graded beam.

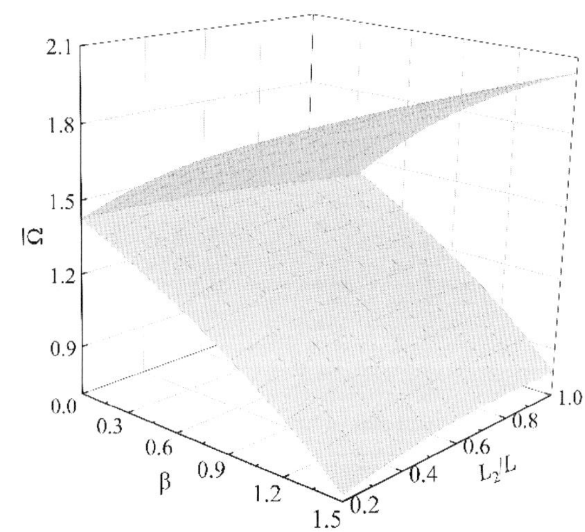

Fig. 13. The variation of dynamic instabilities of the inclined cracked cantilever functionally graded beam with the location of the second inclined crack.

decreases as the crack length increases and the inclined angle rises from 0° to 90°, as both factors reduce the structural stiffness.

Figure 13 depicts the variation of dynamic instabilities of the inclined cracked cantilever functionally graded beam with the location of the second inclined crack. In contrast to the significant impact of crack location on static instability, little effect is observed on the dynamic instability region with the varying crack location. This is because the dynamic instability boundary, as defined in Eq. (42), depends on the difference in stiffness caused by both the structure and geometry. As the crack moves towards the free boundary, both of these factors increase, thereby compensating for each other and resulting in little impact on the dynamic instability region.

5. Conclusion

In this paper, we introduce the novel finite FEA method that utilizes a local additional flexibility matrix to study the static and dynamic stabilities of functionally graded beams containing multiple inclined cracks. It is assumed that the properties of the functionally graded beam containing multiple inclined cracks follow the exponential law in terms of its elasticity modulus and mass densities across its thickness. Two-node beam element with three degree-of-freedom per node is utilized. By considering the bending and tensile flexibilities of the section containing cracks, as well as the coupled effects of bending and tensile forces, the local additional flexibility matrix is utilized to formulate the stiffness matrix for the functionally graded beam element with inclined cracks. The stability analysis of a multiple inclined cracked functionally graded beam can be carried out by applying the Euler–Bernoulli beam theory and Lagrange method to derive governing equations. These equations can be solved as eigenvalue problems to determine both the natural frequency and the static critical buckling load of the beam. To further assess the dynamic stability of the system, Bolotin theory is employed to compute the boundary between the regions of stability and instability based on the same governing equations. Parametric analyses are presented to investigate the effects of crack location, crack inclined angle and crack length on the static critical buckling load and dynamic stability region of the inclined cracked functionally graded beam with different elasticity modulus and boundaries. The results lead to several key conclusions that are relevant to the design, application and analysis of functionally graded beams:

(1) The increasing length and inclined angle of the inclined crack can significantly reduce the stability, both in terms of static and dynamic stabilities, of the functionally graded beam with multiple inclined cracks.
(2) Although increasing the elasticity modulus ratio can increase the static critical buckling load of the inclined cracked functionally graded beam, the synchronous change in mass density can decrease the structural dynamic stability under certain conditions.
(3) The varying position of the inclined crack has a great effect on the static stability of the inclined cracked functionally graded beam, but has little effect on its dynamic instability area.

Acknowledgments

The research was supported by the National Natural Science Foundation of China (Nos. 12172012 and 11802005), and the Australian Research Council under Discovery Project Scheme (Nos. DP210103656 and DP230100548).

Appendix A

The hermit shape functions in Eqs. (9) and (10) are expressed as following forms:

$$N_1^u = -\frac{x}{Le} + 1, \tag{A.1}$$

$$N_2^w = 1 - 3\left(\frac{x}{Le}\right)^2 + 2\left(\frac{x}{Le}\right)^3, \tag{A.2}$$

$$N_3^w = x - 2\frac{x^2}{Le} + \frac{x^3}{Le^2}, \tag{A.3}$$

$$N_4^u = \frac{x}{Le}, \tag{A.4}$$

$$N_5^w = 3\left(\frac{x}{Le}\right)^2 - 2\left(\frac{x^3}{Le^2}\right)^3, \tag{A.5}$$

$$N_6^w = -\frac{x^2}{Le} + \frac{x^3}{Le^2}. \tag{A.6}$$

Appendix B

Nomenclature

Symbol	Meaning	Symbol	Meaning
b, h	Width and thickness of the beam	L, A	Length and area of the beam
α_1, α_2	Angle of the first and second crack	a_1, a_2	Length of the first and second crack
L_1, L_2	Location of the first and second crack	r_E, r_ρ	Ratio of Young's modulus and mass density
$F(t)$	Axial load	$\mathbf{K}^{\mathbf{c}}$	Stiffness matrices of inclined cracked beam element
u, w	Axialand transversal displacements	L_e	Length of beam element
U_i, U_e, T	The elastic strain energy, the work done by the external axial load, kinetic energy	G	Strain energy release rate
$\mathbf{M}$, $\mathbf{K}$, $\mathbf{K}_g$	Overall structural mass, elastic stiffness and geometric stiffness matrix	K_{IP}, K_{IM}, K_{IQ}	SIF corresponding to force P, M, Q

Nomenclature

Symbol	Meaning	Symbol	Meaning
α, β	Static and dynamic loading factor	p_j^1, p_j^2, p_j^3 $(i = 1, 2 \ldots 11)$	Coefficients of stress intensity correction function corresponding to force P, M, Q
$\mathbf{a_1}$, $\mathbf{b_1}$	Coefficient of the first order solution	c_{mn}	Flexibility coefficient
$\mathbf{d}(t)$	Generalized displacement vector	$\mathbf{c_{int}}$	Flexibility matrix of intact beam element
$\mathbf{K^e}, \mathbf{M^e}$	Elastic stiffness and mass matrix of the beam element	F_i $(i = 1, 2, 3)$	Stress intensity correction function corresponding to force P, M, Q
Π_c	Additional strain energy	$\mathbf{c_c}$	Additional flexibility matrix
A_c	Effective area per unit width of the equivalent crack	$\mathbf{c_{tot}}$	Total flexibility matrix
$P_{\mathrm{cr1}}, P_{\mathrm{cr0}}$	Static critical buckling load of the beam with one crack and without crack	$\mathbf{L}$	Transformation matrix
ω_{CF}	The first natural frequency of the intact cantilever aluminum beam	P_{cr}	Static critical buckling load of cracked beam
$\bar{\Omega}$	Dimensionless frequency of cracked FGM beam	Ω	Frequency of the axial load

References

1. S. Suresh, Graded materials for resistance to contact deformation and damage, *Science* **292** (2001) 2447–2451.
2. C. Liu, L. L. Ke, Y. S. Wang, J. Yang and S. Kitipornchai, Buckling and post-buckling of size-dependent piezoelectric Timoshenko nanobeams subject to thermo-electro-mechanical loadings, *Int. J. Struct. Stab. Dyn.* **14** (2014) 1350067.
3. V. Birman, T. Keil and S. Hosder, Functionally graded materials in engineering, in *Structural Interfaces and Attachments in Biology*, eds. S. Thomopoulos, V. Birman and G. M. Genin (Springer Science+Business Media New York, 2013), pp. 19–41.
4. Z. Ni, Y. C. Fan, Z. Y. Hang, J. L. Yang, Y. Wang and C. Feng, Numerical investigation on nonlinear vibration of FG-GNPRC dielectric membrane with internal pores, *Eng. Struct.* **284** (2023) 115928.

5. Z. Ni, F. Zhu, Y. C. Fan, J. L. Yang, Z. Y. Hang and C. Feng, Numerical study on nonlinear vibration of FG-GNPRC circular membrane with dielectric properties, *Mech. Adv. Mater. Struct.* (2023) 1–20.
6. A. Sekhar, Conditioning monitoring of a rotor system having a slant crack in the shaft, *Noise Vib. Worldw.* **30** (1999) 23–31.
7. M. Rezaiee-Pajand and N. Gharaei-Moghaddam, Vibration and static analysis of cracked and non-cracked non-prismatic frames by force formulation, *Eng. Struct.* **185** (2019) 106–121.
8. A. Bose, P. Sathujoda and G. Canale, Natural frequency analysis of a functionally graded rotor-bearing system with a slant crack subjected to thermal gradients, *Int. J. Turbo Jet Engines* (2021).
9. K. J. Bathe, *Finite Element Procedures* (Klaus-Jurgen Bathe, 2006).
10. J. J. Mao, L. J. Guo and W. Zhang, Vibration and frequency analysis of edge-cracked functionally graded graphene reinforced composite beam with piezoelectric actuators, *Eng. Comput.* (2021).
11. Y. Q. Tang, T. J. Li and Q. Y. Yan, Damage analysis of one-dimensional Euler–Bernoulli cracked beam with traveling wave method, *Int. J. Appl. Mech.* **9** (2017) 1750048.
12. D. Gayen and D. Chakraborty, Variation of local flexibility coefficients of functionally graded cracked shaft, *Procedia Eng.* **144** (2016) 1443–1450.
13. M. T. Song, Y. H. Gong, J. Yang, W. D. Zhu and S. Kitipornchai, Nonlinear free vibration of cracked functionally graded graphene platelet-reinforced nanocomposite beams in thermal environments, *J. Sound Vib.* **468** (2020) 115115.
14. W. Huang and N. Sridhar, Fatigue failure risk assessment for a maintained stiffener-frame welded structure with multiple site cracks, *Int. J. Appl. Mech.* **8** (2016) 1650024.
15. G. Long and G. Xu, A combined boundary integral method for analysis of crack problems in multilayered elastic media, *Int. J. Appl. Mech.* **8** (2016) 1650070.
16. A. D. Dimarogonas, Vibration of cracked structures: A state of the art review, *Eng. Fract. Mech.* **55** (1996) 831–857.
17. J. Mackerle, Finite element vibration analysis of beams, plates and shells — A bibliography (1994–1998), *Shock Vib.* **6** (1999) 97–109.
18. P. Rizos, N. Aspragathos and A. Dimarogonas, Identification of crack location and magnitude in a cantilever beam from the vibration modes, *J. Sound Vib.* **138** (1990) 381–388.
19. U. K. Mishra and S. K. Sahu, Parametric instability of beams with transverse cracks subjected to harmonic in-plane loading, *Int. J. Struct. Stab. Dyn.* **15** (2015) 1540006.
20. D. Gayen, R. Tiwari and D. Chakraborty, Static and dynamic analyses of cracked functionally graded structural components: A review, *Compos. B-Eng.* **173** (2019) 106982.
21. B. Panigrahi and G. Pohit, Study of non-linear dynamic behavior of open cracked functionally graded Timoshenko beam under forced excitation using harmonic balance method in conjunction with an iterative technique, *Appl. Math. Model.* **57** (2018) 248–267.
22. L. J. Guo, J. J. Mao, W. Zhang and M. Wu, Stability analyses of cracked functionally graded graphene-platelets reinforced composite beam covered with piezoelectric layers, *Int. J. Struct. Stab. Dyn.* (2023).
23. K. Aydin, Free vibration of functionally graded beams with arbitrary number of surface cracks, *Eur. J. Mech.-A/Solids* **42** (2013) 112–124.

24. N. T. Khiem, N. N. Huyen and N. T. Long, Vibration of cracked Timoshenko beam made of functionally graded material, Shock & Vibration, Aircraft/Aerospace, Energy Harvesting, Acoustics & Optics, in *Proc. 35th IMAC, Conf. and Exposition on Structural Dynamics* (Springer, Cham, 2017), pp. 133–143, https://doi.org/10.1007/978-3-319-54735-0_15.
25. T. V. Lien, N. T. Duc and N. T. Khiem, Free vibration analysis of multiple cracked functionally graded Timoshenko beams, *Lat. Am. J. Solids Struct.* **14** (2017) 1752–1766.
26. M. Song, L. Zhou, W. Karunasena, J. Yang and S. Kitipornchai, Nonlinear dynamic instability of edge-cracked functionally graded graphene-reinforced composite beams, *Nonlinear Dyn.* **109** (2022) 2423–2441.
27. L. L. Ke, J. Yang, S. Kitipornchai and Y. Xiang, Flexural vibration and elastic buckling of a cracked timoshenko beam made of functionally graded materials, *Mech. Adv. Mater. Struct.* **16** (2009) 488–502.
28. M. T. Song, L. Chen, J. Yang, W. D. Zhu and S. Kitipornchai, Thermal buckling and postbuckling of edge-cracked functionally graded multilayer graphene nanocomposite beams on an elastic foundation, *Int. J. Mech. Sci.* **161** (2019) 105040.
29. M. T. Song, Y. H. Gong, J. Yang, W. D. Zhu and S. Kitipornchai, Free vibration and buckling analyses of edge-cracked functionally graded multilayer graphene nanoplatelet-reinforced composite beams resting on an elastic foundation, *J. Sound Vib.* **458** (2019) 89–108.
30. Z. Yang, M. Tam, Y. Zhang, S. Kitipornchai, J. Lv and J. Yang, Nonlinear dynamic response of FG graphene platelets reinforced composite beam with edge cracks in thermal environment, *Int. J. Struct. Stab. Dyn.* **20** (2020) 2043005.
31. A. D. Dimarogonas, S. A. Paipetis and T. G. Chondros, *Analytical Methods in Rotor Dynamics* (Springer Science & Business Media, Germany, 2013).
32. C. Karaagac, H. Ozturk and M. Sabuncu, Crack effects on the in-plane static and dynamic stabilities of a curved beam with an edge crack, *J. Sound Vib.* **330** (2011) 1718–1736.
33. C. Karaagac, H. Ozturk and M. Sabuncu, Effects of an edge crack on the free vibration and lateral buckling of a cantilever laminated composite slender beam, *J. Vib. Control* **19** (2013) 2506–2522.
34. C. Karaagac, H. Öztürk and M. Sabuncu, Free vibration and lateral buckling of a cantilever slender beam with an edge crack: Experimental and numerical studies, *J. Sound Vib.* **326** (2009) 235–250.
35. F. Erdogan and B. Wu, The surface crack problem for a plate with functionally graded properties, *J. Appl. Mech. ASME* **54** (1997) 449–456.
36. D. Gayen, D. Chakraborty and R. Tiwari, Transverse vibration and stability of a cracked functionally graded rotating shaft system, *Adv. Mater. Manuf. Eng.: Proc. ICAMME* **2019** (2020) 625–631.
37. D. Gayen, D. Chakraborty and R. Tiwari, Whirl frequencies and critical speeds of a rotor-bearing system with a cracked functionally graded shaft — finite element analysis, *Eur. J. Mech. A/Solids* **61** (2017) 47–58.
38. E. Reynders and G. De Roeck, A local flexibility method for vibration-based damage localization and quantification, *J. Sound Vib.* **329** (2010) 2367–2383.
39. P. N. Saavedra and L. A. Cuitino, Crack detection and vibration behavior of cracked beams, *Comput. Struct.* **79** (2001) 1451–1459.
40. P. P. Minh, D. T. Manh and N. D. Duc, Free vibration of cracked FGM plates with variable thickness resting on elastic foundations, *Thin-Walled Struct.* **161** (2021) 107425.

41. P. P. Minh and N. D. Duc, The effect of cracks and thermal environment on free vibration of FGM plates, *Thin-Walled Struct.* **159** (2021) 107291.
42. J. J. Mao, Y. J. Wang, W. Zhang, M. Q. Wu, Y. Z. Liu and X. H. Liu, Vibration and wave propagation in functionally graded beams with inclined cracks, *Appl. Math. Model.* **118** (2023) 166–184.
43. J. Fan, Accelerating the modified Levenberg–Marquardt method for nonlinear equations, *Math. Comput.* **83** (2014) 1173–1187.
44. S. C. Mohanty, R. R. Dash and T. Rout, Static and dynamic stability analysis of a functionally graded Timoshenko beam, *Int. J. Struct. Stab. Dyn.* **12** (2012) 1250025.
45. V. Bolotin, *Dynamic Stability of Elastic Systems* (AEROSPACE CORPORATION El Segundo, California, 1962).
46. W. R. Chen, Dynamic stability of linear parametrically excited twisted Timoshenko beams under periodic axial loads, *Acta Mech.* **216** (2010) 207–223.
47. Z. G. Yu and F. L. Chu, Identification of crack in functionally graded material beams using the p-version of finite element method, *J. Sound Vib.* **325** (2009) 69–84.

https://doi.org/10.1142/9789819814770_0012

Chapter 12

Dynamic Analysis of FGM Plates with Variable Delamination Parameters by the Chebyshev–Ritz Method[#]

J. F. Wang[*,§,‖], G. Cao[*], Z. W. Song[†] and S. K. Lai[†,‡,¶,‖]

[*]*Beijing Key Laboratory of Nonlinear Vibrations and Strength of Mechanical Structures Faculty of Materials and Manufacturing Beijing University of Technology Beijing 100124, P. R. China*

[†]*Department of Civil and Environmental Engineering The Hong Kong Polytechnic University Kowloon Hong Kong, P. R. China*

[‡]*Hong Kong Branch of National Rail Transit Electrification and Automation Engineering Technology Research Center The Hong Kong Polytechnic University Kowloon, Hong Kong, P. R. China*

[§]*jf.wang@bjut.edu.cn*
[¶]*sk.lai@polyu.edu.hk*

Delamination is one of the failure modes of functionally graded materials (FGM), resulting in the critical change of vibration characteristics. The Chebyshev polynomials are commonly used as admissible functions to improve the computational efficiency of numerical algorithms and avoid the occurrence of ill-conditioned problems. This paper extends the Chebyshev–Ritz method to the free vibration analysis of delaminated FGM plates, in which the material variation through the plate thickness follows the exponential-law distribution. A plane crack that is considered to be perpendicular to the thickness direction penetrates through the width direction. Based on the region approach, the analysis of FGM plates with a mid-plane delamination is divided into four sub-regions. The kinetic energy and potential energy of each sub-region are derived by the thin plate theory and von Kármán nonlinear strain–displacement relation. The modal functions of the displacement fields of FGM plates can be constructed in accordance with the displacement continuity conditions of the delamination interface and the boundary conditions of such plates. The effects of asymmetric material distribution, delamination length ratio, Young's modulus ratio, and boundary support on the vibration behavior of FGM plates

[‖]Corresponding authors.
[#]This chapter appeared previously on the International Journal of Structural Stability and Dynamics. To cite this chapter, please cite the original article as the following: J. F. Wang *et al.*, *Int. J. Struct. Stab. Dyn.*, **23**, 2340013 (2023), DOI: 10.1142/S0219455423400138.

are investigated. This semi-analytical study provides a reasonable theoretical basis for the behavior prediction and delamination identification of composite structures.

Keywords: Delamination; vibration; Chebyshev polynomials; FGM plate.

1. Introduction

Functionally graded materials (FGM) composed of two or more constituent phases are multifunctional and microscopically inhomogeneous materials with stable and continuous properties.[1–3] It is assumed that the material distribution of advanced composite structures along the structural thickness by following various gradation patterns (e.g. power, sigmoid, and exponential laws) ensures sufficient stiffness, alleviates stress concentration, weakens residual stress, and improves thermal shock resistance.[4] Therefore, various potential applications have been extended from aerospace engineering to mechanical engineering (e.g. fuel cells), biomedical engineering (e.g. orthopedic implants, plasma-treated biomaterials, and biomedical sensors), and even lightweight materials (e.g. lightweight armor).[5]

Defects in composite structures are very common during material processing, component fabrication and in-service operation, including voids, impurities, residual stresses, and curing deformations. All of these may adversely affect system performance and even lead to safety degradation.[6] Although the material variation of FGM structures can reduce and redistribute the processing-induced residual stresses, the interlayer characteristics of multilayered materials still exist due to manufacturing constraints.[7,8] For example, FGM structures that can perform as a thermal barrier coating layer under impact or fatigue loads are not entirely immune to the occurrence of delamination since the crack usually grows along the interfaces.[9] Furthermore, the buckling-driven delamination effect may also occur due to the presence of mechanical and thermal loads.[10] Hence, composite structures are prone to the occurrence of delamination or interlaminar damage, leading to the degradation of mechanical strength and stiffness.

Delamination is a major failure mode of load-bearing structures. Taking FGM-fabricated aircraft wings as an example, the occurrence of delamination is normally perpendicular to the material gradient due to incomplete wetting during the manufacturing process and low-velocity impact.[11] Since delamination is often buried between interconnected layers, visual inspection cannot identify such defects from the structural surface easily, but the vibration behavior (natural frequencies and mode shapes) of defected structures is greatly affected by the change of local flexibility.[12–14] Due to the limitations of experimental environment and equipment sensitivity, theoretical analysis, and numerical simulation are often used to predict the mechanical behaviors of FGM plates.[15] Della and Shu reviewed the dynamic characterization of composite structures with delamination.[11] In general, the analytical model of delaminated plate and shell-type structures can be established by either a region approach or a layerwise approach.[16,17] The latter one is not applicable to multilayered structures due to the reason of heavy computational efforts when the degree of freedom increases with the increase of composite layers. For the region approach, laminates are divided into sub-laminates with a low thickness-to-area ratio, which can be analyzed by the equivalent single layer (ESL) theory separately. In this case, the displacement fields of plates are independent of the number of layers, and thus the computational efficiency of the dynamic analysis of delaminated structures can be improved.[18]

Using the region approach can be traced back to the vibration analysis of delaminated composite beams.[19] Wang and co-workers[20] improved the analysis of an isotropic beam model by considering the interaction of the axial and transverse vibration effects. They assumed that the lateral displacements of the upper and lower parts of the delamination region are free and unconstrained (also called a free model). Kamaloo *et al.*[21] formulated a nonlinear vibration model for the analysis of laminated composite conical shells with a through-width delamination based on the energy method. Shams *et al.*[22] investigated the free vibration behavior of laminated composite beams with a through-width delamination using the dynamic stiffness matrix method. In addition, Tracy and Pardoen[23] also studied the contact effect between various delamination layers on the dynamic responses of simply supported composite beams using a constrained model with independent axial and bending stiffnesses. Recently, Song *et al.*[24] presented two refined models to investigate the natural frequencies of rectangular plates with all-over, central, and edge cracks. Wang *et al.*[25,26] further established an effective analytical model based on the region approach to study the delamination propagation behavior of structural laminates with an arbitrary-shaped delamination. For the analysis of beam structures with a multiple delamination effect, Lee *et al.*[27,28] proposed an analytical solution with the hypothesis of a free model, in which the characteristic equation of the whole delaminated beam was derived by applying a recursive relation to the continuous conditions of each sub-beam. Liu and Shu[29] examined the vibration characteristics of rotating Timoshenko beams with multiple delaminations by considering the effects of free and constrained models. Moreover, Panda *et al.*[30] further investigated the moisture effect on the vibration behavior of woven fiber composite doubly curved panels with a strip delamination effect.

The finite element method (FEM) plays a crucial role in the damage detection of laminated composites.[31] According to the assumption of a free model, various finite delaminated beam element models have been developed.[32,33] In addition, Kiral[34] used the Euler–Bernoulli beam and classical laminated theory to study the free vibration of diverse structural beams with various lay-up sequences. Ganesh *et al.*[35] combined the ESL theory and FEM to investigate the free vibration characteristics of delaminated composite plates. Sharma *et al.*[36] extended the first-order shear deformation theory and FEM to investigate the effects of delamination size and location on the vibration behavior of composite plates with various boundary conditions. Based on the higher-order mid-plane theory and FEM, Hirwani *et al.*[37] further analyzed the bending responses of delaminated composite curved shallow shell panels. Moreover, the vibration behavior of laminated composite plates has been investigated by three-dimensional finite element approaches.[38–40] However, there are only few studies for the vibration behavior of delaminated plate structures, there are ample rooms for further studies.

The Ritz method[41,42] is recognized as an accurate and efficient approach in the static and dynamic behaviors of structures. Selecting appropriate displacement functions to satisfy the geometric and kinematic boundary conditions of plates is crucial for this method.[41] Meirovitch and Kwak[42] compared the computational convergence of the conventional Ritz method and FEM, and demonstrated that the speed of convergence of the former method can be improved vastly by selecting an appropriate class of admissible displacement functions. Although the method is computationally efficient, it may encounter the problem of numerical instability when the admissible functions with higher truncation order. The Chebyshev polynomials as admissible functions in the Ritz method have been proposed to study the vibration analysis of plate and shell structures.[43,44]

In spite of the excellent properties of Chebyshev polynomials,[45] a semi-analytical method is developed to investigate the free vibration of exponential FGM plates with a through-width delamination in this study. Based on the region approach, the FGM plate is divided into four sub-regions, and the kinetic energy and potential energy of each sub-region are derived by the geometrically nonlinear strain-displacement relation. The governing equation of the dynamic system is formulated based on the thin plate theory and the Chebyshev–Ritz method.[41,43–45] The natural frequencies and mode shapes are then obtained by solving the formulated eigenvalue equation. Finally, the influence of various parameters on the dynamic performance of FGM plates is presented.

2. A Four-Region Delamination Model

An exponential FGM plate with a through-width delamination (the dark grey region) is presented in Fig. 1(a). The length and width of the plate are denoted as L and b, respectively. The whole structure is divided into four regions, i.e. Plate i ($i = 1, 2, 3, 4$). In Fig. 1(b), the plate is divided in the x direction as one delamination region and two un-delamination regions. The delamination region consists of two delaminated layers as Plates 2 and 3 with the thicknesses of H_2 and H_3, respectively, which are connected at the ends with the un-delamination regions (i.e. Plates 1 and 4). A delamination crack is assumed at the mid-plane of the plate (i.e. $H_2 = H_3$). h_2 and h_3 are the distances from the mid-plane of Plate 1 to the mid-plane of Plates 2 and 3, respectively. x_1 and x_2 are the horizontal coordinates of the left- and right-end of the delaminated layers, respectively. $L_c/L = (|x_1| + |x_2|)/L$

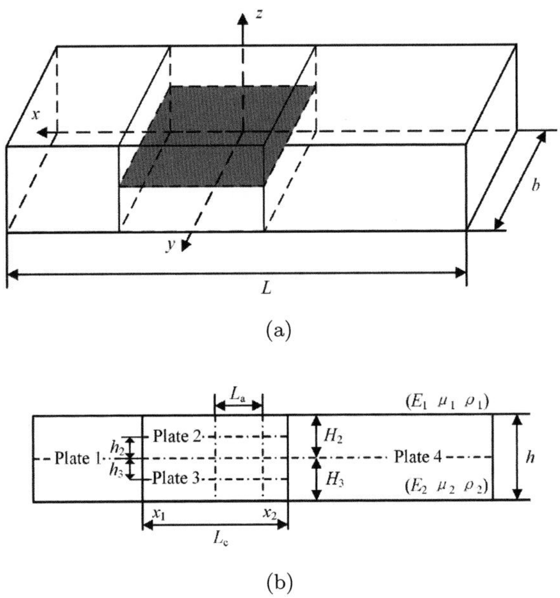

Fig. 1. Schematic of an exponential FGM plate with a through-width delamination.

is the delamination length ratio, and $L_a/L$is the delamination location, in which L_a is the distance from the delamination center. The elastic modulus $E(z)$ and density $\rho(z)$ of the FGM plate follow the exponential law in the thickness direction as follows

$$E(z) = E_0 e^{\beta z}, \quad \rho(z) = \rho_0 e^{\beta z}, \tag{1}$$

where E_0 and ρ_0 are the Young's modulus and mass density of the mid-plane, respectively. β is a constant to characterize the gradual variation of material properties along the plate thickness. The Young's moduli, Poisson's ratios, and densities of the top and bottom layers are expressed as E_i, μ_i and ρ_i ($i = 1$ (top layer); and $i = 2$ (bottom layer)), respectively.

2.1. *Modeling formulation*

According to the classical thin plate theory, the displacement fields on the xy-plane are defined as follows[17]:

$$u = u_0 - z\frac{\partial w}{\partial x}, \tag{2a}$$

$$v = v_0 - z\frac{\partial w}{\partial y}, \tag{2b}$$

$$w = w_0, \tag{2c}$$

where u_0, v_0, and w_0 are the displacements of the mid-plane in the x, y, and z directions, respectively.

Based on the Green strain tensor, we can express the von Kármán strain at a general point by neglecting higher-order terms as follows[46]:

$$\bar{\boldsymbol{\varepsilon}} = \boldsymbol{\varepsilon} + z\boldsymbol{\psi}, \tag{3}$$

where

$$\boldsymbol{\varepsilon} = \begin{Bmatrix} \varepsilon_x \\ \varepsilon_y \\ \gamma_{xy} \end{Bmatrix} = \begin{Bmatrix} \dfrac{\partial u}{\partial x} + \dfrac{1}{2}\left(\dfrac{\partial w_0}{\partial x}\right)^2 \\ \dfrac{\partial v}{\partial y} + \dfrac{1}{2}\left(\dfrac{\partial w_0}{\partial y}\right)^2 \\ \dfrac{\partial u}{\partial y} + \dfrac{\partial v}{\partial x} + \left(\dfrac{\partial w_0}{\partial x}\right)\left(\dfrac{\partial w_0}{\partial y}\right) \end{Bmatrix}, \tag{4}$$

$$\boldsymbol{\psi} = \begin{Bmatrix} \psi_x \\ \psi_y \\ \psi_{xy} \end{Bmatrix} = \begin{Bmatrix} -\dfrac{\partial^2 w_0}{\partial x^2} \\ -\dfrac{\partial^2 w_0}{\partial y^2} \\ -2\dfrac{\partial^2 w_0}{\partial x \partial y} \end{Bmatrix}. \tag{5}$$

The stress–strain behavior of the plate is related to

$$\begin{Bmatrix} \sigma_x \\ \sigma_y \\ \tau_{xy} \end{Bmatrix} = \begin{bmatrix} Q_{11} & Q_{12} & 0 \\ Q_{21} & Q_{22} & 0 \\ 0 & 0 & Q_{66} \end{bmatrix} \begin{Bmatrix} \varepsilon_x \\ \varepsilon_y \\ \gamma_{xy} \end{Bmatrix}, \tag{6}$$

where Q_{kl} $(k,\ l = 1, 2, 6)$ are the transformed reduced stiffness coefficients.

The internal force and moment of the FGM plate can be obtained by integrating the stress components along the thickness direction as follows:

$$\begin{Bmatrix} N_x \\ N_y \\ N_{xy} \\ M_x \\ M_y \\ M_{xy} \end{Bmatrix} = \int_{h/2}^{-h/2} z \begin{Bmatrix} \sigma_x \\ \sigma_y \\ \tau_{xy} \\ z\sigma_x \\ z\sigma_y \\ z\tau_{xy} \end{Bmatrix} dz = \begin{bmatrix} \mathbf{A} & \mathbf{B} \\ \mathbf{B} & \mathbf{D} \end{bmatrix} \begin{Bmatrix} \boldsymbol{\varepsilon} \\ \boldsymbol{\psi} \end{Bmatrix}, \tag{7}$$

where **A**, **B**, and **D** are the tensile stiffness matrix, coupling stiffness matrix, and bending stiffness matrix, respectively. Their inner components are defined as follows:

$$(A_{kl}, B_{kl}, D_{kl}) = \int_{-h/2}^{h/2} Q_{kl}(1, z, z^2)dz, \quad (k, l = 1, 2, 6). \tag{8}$$

By substituting Eqs. (4)–(6) into Eq. (7), the internal force and moment of each region are expressed as follows:

$$\begin{Bmatrix} N_x^{(i)} \\ N_y^{(i)} \\ N_{xy}^{(i)} \end{Bmatrix} = \begin{bmatrix} A_{11}^{(i)} & A_{12}^{(i)} & 0 \\ A_{21}^{(i)} & A_{22}^{(i)} & 0 \\ 0 & 0 & A_{66}^{(i)} \end{bmatrix} \begin{Bmatrix} \dfrac{\partial u_0^{(i)}}{\partial x} + \dfrac{1}{2}\left(\dfrac{\partial w_0^{(i)}}{\partial x}\right)^2 \\ \dfrac{\partial v_0^{(i)}}{\partial y} + \dfrac{1}{2}\left(\dfrac{\partial w_0^{(i)}}{\partial y}\right)^2 \\ \dfrac{\partial u_0^{(i)}}{\partial y} + \dfrac{\partial v_0^{(i)}}{\partial x} + \left(\dfrac{\partial w_0^{(i)}}{\partial x}\right)\left(\dfrac{\partial w_0^{(i)}}{\partial y}\right) \end{Bmatrix}$$
$$+ \begin{bmatrix} B_{11}^{(i)} & B_{12}^{(i)} & 0 \\ B_{21}^{(i)} & B_{22}^{(i)} & 0 \\ 0 & 0 & B_{66}^{(i)} \end{bmatrix} \begin{Bmatrix} -\left(\dfrac{\partial^2 w_0^{(i)}}{\partial x^2}\right) \\ -\left(\dfrac{\partial^2 w_0^{(i)}}{\partial y^2}\right) \\ -2\left(\dfrac{\partial^2 w_0^{(i)}}{\partial x \partial y}\right) \end{Bmatrix} \tag{9}$$

$$\begin{Bmatrix} M_x^{(i)} \\ M_y^{(i)} \\ M_{xy}^{(i)} \end{Bmatrix} = \begin{bmatrix} B_{11}^{(i)} & B_{12}^{(i)} & 0 \\ B_{21}^{(i)} & B_{22}^{(i)} & 0 \\ 0 & 0 & B_{66}^{(i)} \end{bmatrix} \begin{Bmatrix} \dfrac{\partial u_0^{(i)}}{\partial x} + \dfrac{1}{2}\left(\dfrac{\partial w_0^{(i)}}{\partial x}\right)^2 \\ \dfrac{\partial v_0^{(i)}}{\partial y} + \dfrac{1}{2}\left(\dfrac{\partial w_0^{(i)}}{\partial y}\right)^2 \\ \dfrac{\partial u_0^{(i)}}{\partial y} + \dfrac{\partial v_0^{(i)}}{\partial x} \\ + \left(\dfrac{\partial w_0^{(i)}}{\partial x}\right)\left(\dfrac{\partial w_0^{(i)}}{\partial y}\right) \end{Bmatrix}$$
$$+ \begin{bmatrix} D_{11}^{(i)} & D_{12}^{(i)} & 0 \\ D_{21}^{(i)} & D_{22}^{(i)} & 0 \\ 0 & 0 & D_{66}^{(i)} \end{bmatrix} \begin{Bmatrix} -\left(\dfrac{\partial^2 w_0^{(i)}}{\partial x^2}\right) \\ -\left(\dfrac{\partial^2 w_0^{(i)}}{\partial y^2}\right) \\ -2\left(\dfrac{\partial^2 w_0^{(i)}}{\partial x \partial y}\right) \end{Bmatrix} \tag{10}$$

where the superscript i ($i = 1, 2, 3, 4$) represent the i-th plate region, and

$$(A_{kl}^{(i)}, B_{kl}^{(i)}, D_{kl}^{(i)}) = \int_{-h/2}^{h/2} Q_{kl}^{(i)}(1, z_i, z_i^2)dz_i, \quad (k, l = 1, 2, 6), \tag{11}$$

$$Q_{11}^{(i)} = Q_{22}^{(i)} = \frac{E^{(i)}}{1-\nu^2}, \quad Q_{12}^{(i)} = Q_{21}^{(i)} = \frac{\nu E^{(i)}}{1-\nu^2}, \quad Q_{66}^{(i)} = \frac{E^{(i)}}{2(1+\nu)}. \tag{12}$$

The material properties of the plate change continuously along the thickness direction. The elastic moduli of the four regions are expressed as follows:

$$E^{(1)} = E^{(4)} = E_0 e^{\beta z_1}, \quad E^{(2)} = E_0 e^{\beta(z_2+h_2)}, \quad E^{(3)} = E_0 e^{\beta(z_3-h_3)}. \tag{13}$$

The strain energy and kinetic energy of the i-th plate region are obtained as follows:

$$\begin{aligned} U^{(i)} &= \frac{1}{2}\int_{A_i} (\bar{\boldsymbol{\sigma}}^{\mathrm{T}} \cdot \bar{\varepsilon})dA_i \\ &= \frac{1}{2}\int_{A_i} \{\boldsymbol{\varepsilon}^{(i)^{\mathrm{T}}}\mathbf{A}^{(i)}\boldsymbol{\varepsilon}^{(i)} + 2\boldsymbol{\varepsilon}^{(i)^{\mathrm{T}}}\mathbf{B}^{(i)}\boldsymbol{\psi}^{(i)} + \boldsymbol{\psi}^{(i)^{\mathrm{T}}}\mathbf{D}^{(i)}\boldsymbol{\psi}^{(i)}\}dA_i, \end{aligned} \tag{14}$$

$$T^{(i)} = \int_{-h_i/2}^{h_i/2} \rho^{(i)}dz_i \int\int_{A_i} \left[\left(\frac{\partial u^{(i)}}{\partial t}\right)^2 + \left(\frac{\partial v^{(i)}}{\partial t}\right)^2 + \left(\frac{\partial w^{(i)}}{\partial t}\right)^2\right] dA_i, \tag{15}$$

where A_i is the area of each sub-laminate region. The total strain energy and kinetic energy of the whole plate are given by

$$U = \sum_{i=1}^{4} U^{(i)}, \tag{16}$$

$$T = \sum_{i=1}^{4} T^{(i)} \tag{17}$$

and the densities of various plate regions can be expressed as follows:

$$\rho^{(1)} = \rho^{(4)} = \rho_0 e^{\beta z_1}, \quad \rho^{(2)} = \rho_0 e^{\beta(z_2+h_2)}, \quad \rho^{(3)} = \rho_0 e^{\beta(z_3-h_3)}. \tag{18}$$

2.2. *Solution approach*

Displacement components can be represented by various admissible functions, including Chebyshev polynomials, simple polynomials, and trigonometric functions. Using simple polynomials can result in singular mass and stiffness matrices, Singhvi and Kapania[47] demonstrated the superiority of the Chebyshev polynomials for the vibration analysis of beams. Mathematically, the orthogonality properties of the Chebyshev polynomials on the interval [−1, 1] can achieve good convergence and numerical stability.[41] Meanwhile, the Ritz method is recognized as an effective energy-based approach to perform the analysis of structural elements with complex boundary conditions and geometrical configurations.[48,49] In this section, the Chebyshev–Ritz method is extended to investigate the dynamic characteristics of the delaminated FGM plate with

$$P_s(\chi) = \cos[(s-1)\arccos(\chi)], \tag{19}$$

where $P_s(\chi)$ is a one-dimensional Chebyshev polynomial with the truncation order of s. Substituting $\chi = 2x/L$ into Eq. (19) yields the Chebyshev polynomials with respect to x as follows:

$$X_m(x) = \cos\left[(m-1)\arccos\left(\frac{2x}{L}\right)\right]. \tag{20}$$

Similarly, the Chebyshev polynomials as admissible functions in the y direction by assuming $\chi = 2y/b$ can be obtained as follows:

$$Y_n(y) = \cos\left[(n-1)\arccos\left(\frac{2y}{b}\right)\right]. \tag{21}$$

The displacement continuity conditions at the delamination region boundaries are expressed as follows:

$$u^{(1)}|_{x=x_1} = u^{(2)}|_{x=x_1} - h_2\frac{\partial w^{(1)}}{\partial x}\bigg|_{x=x_1} = u^{(3)}|_{x=x_1} + h_3\frac{\partial w^{(1)}}{\partial x}\bigg|_{x=x_1}, \tag{22a}$$

$$u^{(4)}|_{x=x_1} = u^{(2)}|_{x=x_2} - h_2\frac{\partial w^{(4)}}{\partial x}\bigg|_{x=x_2} = u^{(3)}|_{x=x_2} + h_3\frac{\partial w^{(4)}}{\partial x}\bigg|_{x=x_2}, \tag{22b}$$

$$v^{(1)}|_{x=x_1} = v^{(2)}|_{x=x_1} - h_2\frac{\partial w^{(1)}}{\partial y}\bigg|_{x=x_1} = v^{(3)}|_{x=x_1} + h_3\frac{\partial w^{(1)}}{\partial y}\bigg|_{x=x_1}, \tag{23a}$$

$$v^{(4)}|_{x=x_1} = v^{(2)}|_{x=x_2} - h_2\frac{\partial w^{(4)}}{\partial y}\bigg|_{x=x_2} = v^{(3)}|_{x=x_2} + h_3\frac{\partial w^{(4)}}{\partial y}\bigg|_{x=x_2}, \tag{23b}$$

$$w^{(1)}|_{x=x_1} = w^{(2)}|_{x=x_1} = w^{(3)}|_{x=x_1}, \tag{24a}$$

$$w^{(4)}|_{x=x_2} = w^{(2)}|_{x=x_2} = w^{(3)}|_{x=x_2}, \tag{24b}$$

$$\frac{\partial w^{(1)}}{\partial x}\bigg|_{x=x_1} = \frac{\partial w^{(2)}}{\partial x}\bigg|_{x=x_1} = \frac{\partial w^{(3)}}{\partial x}\bigg|_{x=x_1}, \tag{24c}$$

$$\frac{\partial w^{(4)}}{\partial x}\bigg|_{x=x_2} = \frac{\partial w^{(2)}}{\partial x}\bigg|_{x=x_2} = \frac{\partial w^{(3)}}{\partial x}\bigg|_{x=x_2}. \tag{24d}$$

Given that the plate is clamped at $x = -L/2$ and free at other edges (i.e. CFFF case), we have

$$w^{(1)}|_{x=-L/2} = 0, \quad \left.\frac{\partial w^{(1)}}{\partial x}\right|_{x=-L/2} = 0, \tag{25}$$

$$u^{(1)}|_{x=-L/2} = 0, \quad v^{(1)}|_{x=-L/2} = 0. \tag{26}$$

For the plate is clamped at opposite sides (i.e. $x = -L/2$ and $x = L/2$) and free at other two sides (i.e. CFCF case), we obtain

$$w^{(1)}|_{x=-L/2} = w^{(4)}|_{x=L/2} = 0, \quad \left.\frac{\partial w^{(1)}}{\partial x}\right|_{x=-L/2} = \left.\frac{\partial w^{(4)}}{\partial x}\right|_{x=L/2} = 0, \tag{27}$$

$$u^{(1)}|_{x=-L/2} = u^{(4)}|_{x=L/2} = 0, \quad v^{(1)}|_{x=-L/2} = v^{(4)}|_{x=L/2} = 0. \tag{28}$$

The above two cases are widely considered in various engineering applications, e.g. cantilever fans,[50] aircraft wings, and bridge structures. Hence, this work mainly focuses on these two cases. It is worth noting that the present approach is applicable for the vibration analysis of plates with various boundary constraints. The assumed functions of the in-plane displacements with arbitrary boundary conditions are considered as follows:

$$u^{(1)} = Lu^{(1)}(x,y) \sum_{m=0}^{M} \sum_{n=0}^{N} A_{mn}^{(1)} X_m(x) Y_n(y), \tag{29a}$$

$$u^{(2)} = Lu^{(2)}(x,y) \sum_{m=0}^{M} \sum_{n=0}^{N} A_{mn}^{(2)} X_m(x) Y_n(y) + u^{(1)} + H_2^u(x,y), \tag{29b}$$

$$u^{(3)} = Lu^{(3)}(x,y) \sum_{m=0}^{M} \sum_{n=0}^{N} A_{mn}^{(3)} X_m(x) Y_n(y) + u^{(1)} + H_3^u(x,y), \tag{29c}$$

$$u^{(4)} = Lu^{(4)}(x,y) \sum_{m=0}^{M} \sum_{n=0}^{N} A_{mn}^{(1)} X_m(x) Y_n(y), \tag{29d}$$

$$v^{(1)} = Lv^{(1)}(x,y) \sum_{m=0}^{M} \sum_{n=0}^{N} B_{mn}^{(1)} X_m(x) Y_n(y), \tag{30a}$$

$$v^{(2)} = Lv^{(2)}(x,y) \sum_{m=0}^{M} \sum_{n=0}^{N} B_{mn}^{(2)} X_m(x) Y_n(y) + v^{(1)} + H_2^v(x,y), \tag{30b}$$

$$v^{(3)} = Lv^{(3)}(x,y)\sum_{m=0}^{M}\sum_{n=0}^{N} B_{mn}^{(3)} X_m(x)Y_n(y) + v^{(1)} + H_3^v(x,y), \tag{30c}$$

$$v^{(4)} = Lv^{(4)}(x,y)\sum_{m=0}^{M}\sum_{n=0}^{N} B_{mn}^{(1)} X_m(x)Y_n(y), \tag{30d}$$

where M and N are the truncation orders of the Chebyshev polynomials, $Lu^{(i)}(x,y)$ and $Lv^{(i)}(x,y)$ ($i = 1,2,3,4$) are functions related to boundary conditions; $H_2^u(x,y)$, $H_3^u(x,y)$, $H_2^v(x,y)$ and $H_3^v(x,y)$ are obtained from the displacement continuity conditions of the delamination and intact regions as follows:

$$H_3^u(x,y) = h_3\left[(x_1 - x)\left.\frac{\partial w^{(4)}}{\partial x}\right|_{x=x_2} + (x_2 - x)\left.\frac{\partial w^{(2)}}{\partial x}\right|_{x=x_1}\right]/(x_1 - x_2), \tag{31a}$$

$$H_2^u(x,y) = -h_2\left[(x_1 - x)\left.\frac{\partial w^{(4)}}{\partial x}\right|_{x=x_2} + (x_2 - x)\left.\frac{\partial w^{(2)}}{\partial x}\right|_{x=x_1}\right]/(x_1 - x_2), \tag{31b}$$

$$H_3^v(x,y) = h_3\left[(x_1 - x)\left.\frac{\partial w^{(4)}}{\partial y}\right|_{x=x_2} + (x_2 - x)\left.\frac{\partial w^{(2)}}{\partial y}\right|_{x=x_1}\right]/(x_1 - x_2), \tag{32a}$$

$$H_2^v(x,y) = -h_2\left[(x_1 - x)\left.\frac{\partial w^{(4)}}{\partial y}\right|_{x=x_2} + (x_2 - x)\left.\frac{\partial w^{(2)}}{\partial y}\right|_{x=x_1}\right]/(x_1 - x_2). \tag{32b}$$

The out-of-plane displacements are assumed as follows:

$$w^{(1)} = Lw^{(1)}(x,y)\sum_{m=0}^{M}\sum_{n=0}^{N} C_{mn}^{(1)} X_m(x)Y_n(y), \tag{33a}$$

$$w^{(2)} = Lw^{(2)}(x,y)\sum_{m=0}^{M}\sum_{n=0}^{N} C_{mn}^{(2)} X_m(x)Y_n(y) + w^{(1)}, \tag{33b}$$

$$w^{(3)} = Lw^{(3)}(x,y)\sum_{m=0}^{M}\sum_{n=0}^{N} C_{mn}^{(3)} X_m(x)Y_n(y) + w^{(1)}, \tag{33c}$$

$$w^{(4)} = Lw^{(4)}(x,y)\sum_{m=0}^{M}\sum_{n=0}^{N} C_{mn}^{(1)} X_m(x)Y_n(y), \tag{33d}$$

where $Lw^{(i)}(x,y)$ ($i = 1,2,3,4$) are relevant to the supporting boundary conditions.

In the CFFF case, the following functions are assumed as follows:

$$Lu^{(1)}(x,y) = x + \frac{L}{2}, \tag{34a}$$

$$Lu^{(2)}(x,y) = (x - x_1)(x - x_2), \tag{34b}$$

$$Lu^{(3)}(x,y) = (x - x_1)(x - x_2), \tag{34c}$$

$$Lv^{(1)}(x,y) = x + \frac{L}{2}, \tag{35a}$$

$$Lv^{(2)}(x,y) = (x - x_1)(x - x_2), \tag{35b}$$

$$Lv^{(3)}(x,y) = (x - x_1)(x - x_2), \tag{35c}$$

$$Lw^{(1)}(x,y) = \left(x + \frac{L}{2}\right)^2, \tag{36a}$$

$$Lw^{(2)}(x,y) = (x - x_1)^2(x - x_2)^2, \tag{36b}$$

$$Lw^{(3)}(x,y) = (x - x_1)^2(x - x_2)^2. \tag{36c}$$

Similarly, in the CFCF case, we have

$$Lu^{(1)}(x,y) = Lu^{(4)}(x,y) = \left(x + \frac{L}{2}\right)\left(x - \frac{L}{2}\right), \tag{37a}$$

$$Lu^{(2)}(x,y) = (x - x_1)(x - x_2), \tag{37b}$$

$$Lu^{(3)}(x,y) = (x - x_1)(x - x_2), \tag{37c}$$

$$Lv^{(1)}(x,y) = Lv^{(4)}(x,y) = \left(x + \frac{L}{2}\right)\left(x - \frac{L}{2}\right), \tag{38a}$$

$$Lv^{(2)}(x,y) = (x - x_1)(x - x_2), \tag{38b}$$

$$Lv^{(3)}(x,y) = (x - x_1)(x - x_2), \tag{38c}$$

$$Lw^{(1)}(x,y) = Lw^{(4)}(x,y) = \left(x + \frac{L}{2}\right)^2\left(x - \frac{L}{2}\right)^2, \tag{39a}$$

$$Lw^{(2)}(x,y) = (x - x_1)^2(x - x_2)^2, \tag{39b}$$

$$Lw^{(3)}(x,y) = (x - x_1)^2(x - x_2)^2. \tag{39c}$$

The mid-plane displacement components of the plate are assumed as follows:

$$\begin{aligned} &u_0^{(i)}(x,y,t) = u_0^{(i)}(x,y)e^{i\omega t}, \quad v_0^{(i)}(x,y,t) = v_0^{(i)}(x,y)e^{i\omega t}, \\ &w_0^{(i)}(x,y,t) = w_0^{(i)}(x,y)e^{i\omega t}. \end{aligned} \tag{40}$$

By substituting the above modal functions $u_0^{(i)}(x,y,t)$, $v_0^{(i)}(x,y,t)$, $w_0^{(i)}(x,y,t)$ into Eqs. (9), (10), (14), and (15), the potential energy and kinetic energy of the four plate regions with the assumption of small deformation are obtained as follows:

$$
\begin{aligned}
U^{(i)} = \int_{A_i} \Bigg\{ & \left[A_{66}^{(i)} \left(\frac{\partial u_0^{(i)}}{\partial y} + \frac{\partial v_0^{(i)}}{\partial x} \right) - 2B_{66}^{(i)} \left(\frac{\partial^2 w_0^{(i)}}{\partial x \partial y} \right) \right] \left(\frac{\partial u_0^{(i)}}{\partial y} + \frac{\partial v_0^{(i)}}{\partial x} \right) \\
& - 2 \left(\frac{\partial^2 w_0^{(i)}}{\partial x \partial y} \right) \left[B_{66}^{(i)} \left(\frac{\partial u_0^{(i)}}{\partial y} + \frac{\partial v_0^{(i)}}{\partial x} \right) - 2D_{66}^{(i)} \left(\frac{\partial^2 w_0^{(i)}}{\partial x \partial y} \right) \right] \\
& + \left(A_{11}^{(i)} \frac{\partial u_0^{(i)}}{\partial x} + A_{12}^{(i)} \frac{\partial v_0^{(i)}}{\partial y} - B_{11}^{(i)} \left(\frac{\partial^2 w_0^{(i)}}{\partial x} \right)^2 - B_{12}^{(i)} \left(\frac{\partial^2 w_0^{(i)}}{\partial y} \right)^2 \right) \frac{\partial u_0^{(i)}}{\partial x} \\
& - \left(\frac{\partial^2 w_0^{(i)}}{\partial x} \right)^2 \left(B_{11}^{(i)} \frac{\partial u_0^{(i)}}{\partial x} + B_{12}^{(i)} \frac{\partial v_0^{(i)}}{\partial y} - D_{11}^{(i)} \left(\frac{\partial^2 w_0^{(i)}}{\partial x} \right)^2 - D_{12}^{(i)} \left(\frac{\partial^2 w_0^{(i)}}{\partial y} \right)^2 \right) \\
& + \frac{\partial v_0^{(i)}}{\partial y} \left(A_{12}^{(i)} \frac{\partial u_0^{(i)}}{\partial x} + A_{22}^{(i)} \frac{\partial v_0^{(i)}}{\partial y} - B_{21}^{(i)} \left(\frac{\partial^2 w_0^{(i)}}{\partial x} \right)^2 - B_{22}^{(i)} \left(\frac{\partial^2 w_0^{(i)}}{\partial y} \right)^2 \right) \\
& - \left(\frac{\partial^2 w_0^{(i)}}{\partial y} \right)^2 \left(B_{21}^{(i)} \frac{\partial u_0^{(i)}}{\partial x} + B_{22}^{(i)} \frac{\partial v_0^{(i)}}{\partial y} - D_{21}^{(i)} \left(\frac{\partial^2 w_0^{(i)}}{\partial x} \right)^2 \right. \\
& \left. \left. - D_{22}^{(i)} \left(\frac{\partial^2 w_0^{(i)}}{\partial y} \right)^2 \right) \right\} dA_i, \qquad (41)
\end{aligned}
$$

$$
T^{(i)} = \frac{1}{2}\omega^2 \int_{-h_i/2}^{h_i/2} \int_{A_i} \rho^{(i)} ((u_0^{(i)})^2 + (v_0^{(i)})^2 + (w_0^{(i)})^2) \mathrm{d}A_i dz_i. \qquad (42)
$$

By substituting Eqs. (29), (30) and (33) into Eqs. (41) and (42), the total energy of the delaminated FGM plate is given by

$$
\Pi = \sum_{i=1}^{4} (U^{(i)} + T^{(i)}). \qquad (43)
$$

According to the Ritz method, we obtain a set of algebraic equations as follows:

$$
\frac{\partial \Pi}{A_{mn}^{(j)}} = 0, \quad \frac{\partial \Pi}{B_{mn}^{(j)}} = 0, \quad \frac{\partial \Pi}{C_{mn}^{(j)}} = 0, \quad (j = 1, 2, 3), \qquad (44)
$$

where j represents the number of solving domains consisting of two delaminated layers and one un-delamination region. Because of the topological symmetry and displacement continuity, the unknown coefficients of the two un-delamination regions are the same, and thus can be integrated as an un-delamination region. All unknown coefficients in the displacement functions are presented in a vector form as $\mathbf{q} = \{A_{mn}^{(1)}, A_{mn}^{(2)}, A_{mn}^{(3)}, B_{mn}^{(1)}, B_{mn}^{(2)}, B_{mn}^{(3)}, C_{mn}^{(1)}, C_{mn}^{(2)}, C_{mn}^{(3)}\}^{\mathrm{T}}$. Then, Eq. (44) is converted into

$$\frac{\partial \Pi}{q_k} = 0, \quad (k = 1, 2, \ldots, P), \tag{45}$$

where $P = 3 \times 3 \times M \times N$ are the number of coefficients to be determined.

Hence, the problem is transformed into solving a standard eigenvalue equation

$$(\mathbf{K} - \omega^2 \mathbf{M})\mathbf{q} = \mathbf{0}, \tag{46}$$

where the elements of $\mathbf{K}$ and $\mathbf{M}$ are expressed as follows:

$$K_{mn} = \frac{\partial^2 U}{\partial q_m \partial q_n}, \quad M_{mn} = \frac{\partial^2 T}{\partial q_m \partial q_n}. \tag{47}$$

The natural frequencies and mode shapes of the FGM plate can be determined by solving Eq. (46).

3. Results and Discussion

In this section, numerical simulations are carried out to validate the computational efficiency of the present method. The effects of asymmetric material distribution, delamination length ratio, delamination location, Young's modulus ratio, and boundary condition on the vibration characteristics of FGM plates are investigated in detail.

3.1. *Numerical verification*

The geometric dimensions of the exponential FGM plate are assumed as the length-to-width ratio $L/b = 2$, length-to-thickness ratio $L/h = 100$, Poisson's ratio $\mu = 0.33$, and total length $L = 1\,\text{m}$. A numerical example (a beam model in Refs. 10 and 20) is selected to validate the present results. The non-dimensional frequencies Ω_n of the intact and delaminated exponential FGM plates with the CFFF boundary

Table 1. Convergence and comparison of first three-order non-dimensional frequencies of CFFF intact FGM plates.

Young's modulus ratio	Mode number	$M \times N$							Liu and Shu[10]	Error (%)
		2×2	3×3	4×4	5×5	6×6	7×7	8×8		
0.2	1	3.252	3.237	3.224	3.220	3.217	3.215	3.212	3.30	2.4
	2	13.78	13.71	13.67	13.65	13.65	13.65	13.64	/	/
	3	20.25	20.19	20.05	20.03	20.02	20.01	20.00	20.70	3.2
1	1	3.461	3.445	3.431	3.426	3.424	3.420	3.420	3.52	2.7
	2	14.66	14.59	14.56	14.53	14.53	14.52	14.52	/	/
	3	21.55	21.49	21.34	21.32	21.31	21.30	21.28	22.03	3.1

Table 2. Comparison of first-order non-dimensional frequencies of CFFF delaminated FGM plates.

Delamination length ratio	Present	Liu and Shu[10]	Error (%)
0.1	3.326	3.420	2.7
0.2	3.309	3.403	2.8
0.3	3.281	3.360	2.4
0.4	3.227	3.280	1.6
0.5	3.125	3.163	1.2

conditions are presented in Tables 1 and 2, in which $\Omega_n = \omega_n/\sqrt{D_1/\rho_1 h}$ and $D_1 = E_1 h^3/12(1-\nu^2)$. The elastic modulus and density of the top layer are $E_1 = 200$ GPa and $\rho_1 = 2780$ kg/m^3, respectively. It is found that the results decrease gradually and tend to be stable with the deviation of about 1% when the truncation orders M and N are changed from 4 to 8 simultaneously. To improve the computational efficiency, the truncation orders of M and N are assigned as 5 in the subsequent studies.

To investigate the performance of higher-order modes, the admissible functions with higher truncation orders need to be considered. This may lead to expansive computational efforts and complex formula derivations, and even the occurrence of ill-conditioned problems in the Ritz method.[44] In Eqs. (20) and (21), the Chebyshev polynomials are a set of orthogonal functions on the interval [−1,1], which can improve the computational performance of numerical solutions with excellent convergence.[41,44] In Table 1, the present solutions are in good agreement with the available results,[10] and all relative errors are not larger than 3.2%.

A convergence study of the fundamental frequency ratio with different delamination length ratios is depicted in Fig. 2. In this case, Young's modulus ratio (E_2/E_1) is assumed to be 1, i.e. Young's modulus of the top layer is identical to the bottom one. The first-order natural frequencies of the intact and delaminated plates are represented as Ω_{10} and Ω_1, respectively. The first-order natural frequencies of Refs. 10 and 20 are $\Omega_{10} = 22.37$ and $\Omega_{10} = 22.39$, respectively, and the present one is $\Omega_{10} = 21.97$. Furthermore, the natural frequencies of the intact and delaminated plates are obtained in Fig. 2, and the frequency ratio of the intact and delaminated plates is $\omega_1/\omega_{10} = 1$. The present results agree well with the available solutions.

The verification of accuracy is also conducted with the comparison of FEM to validate the vibration characteristics of CFCF plates with a central mid-plane delamination. For simplicity, Young's modulus ratio is set to 1. The four plate regions of the delaminated plate are discretized using C3D8R solid elements with the dimensions of 0.005 m (length) × 0.005 m (width) × 0.0005 m (thickness). They are then assembled to form a layer across the width by creating a binding constraint. In this work, a self-contact interaction is assumed to consider the surface contact in

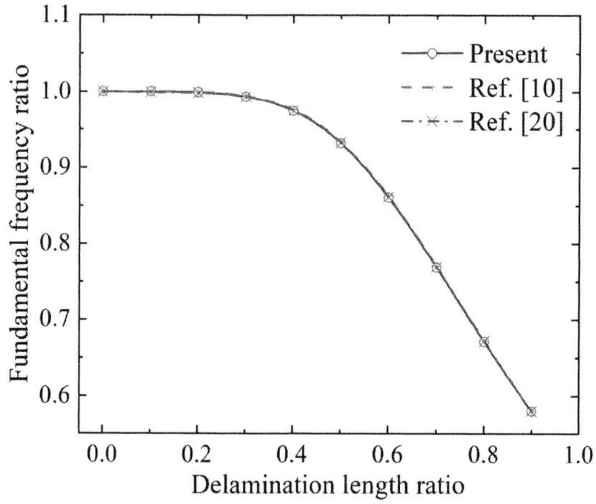

Fig. 2. Fundamental frequency ratio of CFCF delaminated FGM plates.

Table 3. First three-order non-dimensional frequencies of CFCF delaminated FGM plates.

Delamination length ratio	Mode number	Present	FEM	Error (%)
0.5	1	84.54	84.20	0.3
	2	122.51	123.46	0.87
	3	173.77	168.85	2.9
0	1	90.58	90.12	0.52
	2	146.28	146.77	0.33
	3	249.21	248.35	0.35

the normal direction of the upper and lower surfaces to prevent penetration between "Plate 2" and "Plate 3" under vibrational motions. The first three-order natural frequencies with different delaminated length ratios are presented in Table 3. The results are in good agreement with the finite element results, and the maximum relative error is 2.9%.

3.2. *Effect of delamination length ratio*

In this work, delamination is assumed at the mid-plane of the plate, and the mid-point of the delamination area coincides with that of the plate (i.e. $L_a = 0$). The top layer is alumina with $E_1 = 200\,\text{GPa}$, and the bottom layer is aluminum with $E_2 = 70\,\text{GPa}$. The effect of delamination length ratio on the dynamic behavior of FGM plates is investigated. The first-order non-dimensional frequencies of CFFF delaminated exponential FGM plates are presented in Table 2. It is found that the results are consistent with those in Ref. 10 with the maximum relative error of 2.8%.

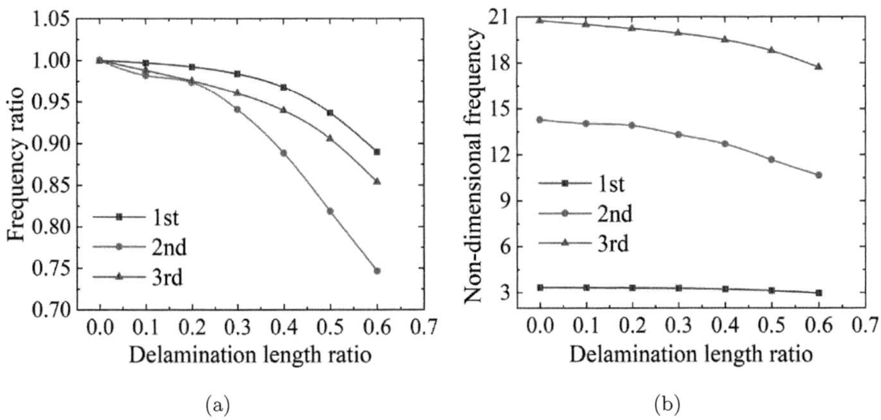

Fig. 3. First three-order modes of CFFF delaminated FGM plates: (a) Frequency ratio and (b) non-dimensional frequency.

The first three-order frequency ratios ω_n/ω_{n0} and frequencies Ω_{n0} of CFFF exponential FGM plates with a central mid-plane delamination are presented in Fig. 3. When the delamination length ratio is less than 0.2, the decrease of the first three-order natural frequencies is not obvious, and the variation of the third-order frequency remains below 5%. All curves tend to decrease when the delamination effect becomes pronounced, and the phenomenon can be explained by the fact that the plate stiffness decreases with the increase of delamination length ratio, resulting in a reduction of natural frequencies. In addition, the second-order natural frequency drops rapidly when the delamination length ratio is close to 0.6, which is consistent with the results of Ref. 51. The presence of delamination also induces a considerable change in the torsional mode of the plate, i.e. second-order vibration mode. Hence, the natural frequencies of the delaminated plate are much lower than that of the intact plate.

Similarly, the results of the CFCF case are presented in Fig. 4. It is observed that the first three-order natural frequencies also decrease with the increase of delamination length ratio, but the third-order frequency drops faster than others. Moreover, the reduction of third-order frequency approaches to 50% when the delamination length ratio is 0.6.

The influence of delamination length ratio for the CFFF and CFCF cases is presented in Figs. 5 and 6, respectively. In these cases, the first five-order modes and their non-dimensional frequencies for the delamination length ratio of 0.2 are similar to those of the intact plate. In the CFFF case, the layers of delamination are in close contact for the first five modes. In the CFCF case, the layers of delamination are still in close contact when the delamination length ratio is below 0.5, but it moves apart (forming two layers as an open state) in the third mode of the

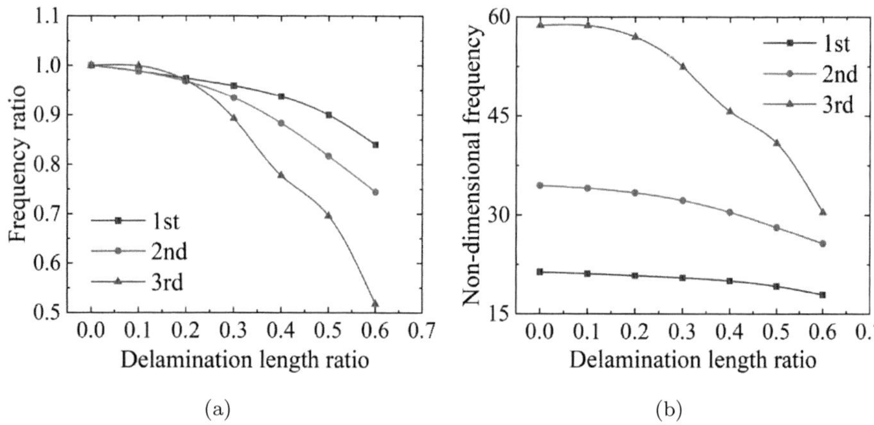

Fig. 4. First three-order modes of CFCF delaminated FGM plates: (a) Frequency ratio and (b) non-dimensional frequency.

CFCF case. Meanwhile, the lateral displacements of the upper and lower delaminated layers in an open state are different because of the asymmetrical material distribution. The transverse deformation of the lower delaminated layer is relatively larger than the upper one due to lower stiffness. A previous study has reported that the delamination interface in an open state is caused by the excessive length of delamination.[52] From the results, it is also found that the vibration frequencies of higher-order modes would decrease evidently with the increase of delamination length ratio.

3.3. *Effect of delamination location*

Owing to the complex stress distribution in various working environments, the occurrence of delamination may appear in FGM structural layers. In this section, the influence of delamination location is investigated on the vibration behavior of FGM plates. The delamination length ratio is set to 0.4 and the delamination location L_a is selected in the range of $[-0.1, 0.3)$.

The effects of delamination location and boundary condition on the first three-order frequency ratios are shown in Fig. 7. In the CFFF case, the second- and third-order frequencies decrease and the first-order frequency increases when the delamination location is close to the fixed edge. In the CFCF case, when the delamination location is close to the mid-plane of the plate, the first two natural frequencies increase while the third-order decreases. When the delamination location is close to both ends of the plate, the first two natural frequencies decrease while the third-order frequency increases contrarily. In the CFCF case, the effect of Young's modulus ratio is also presented in Table 4, and the natural frequencies are totally the

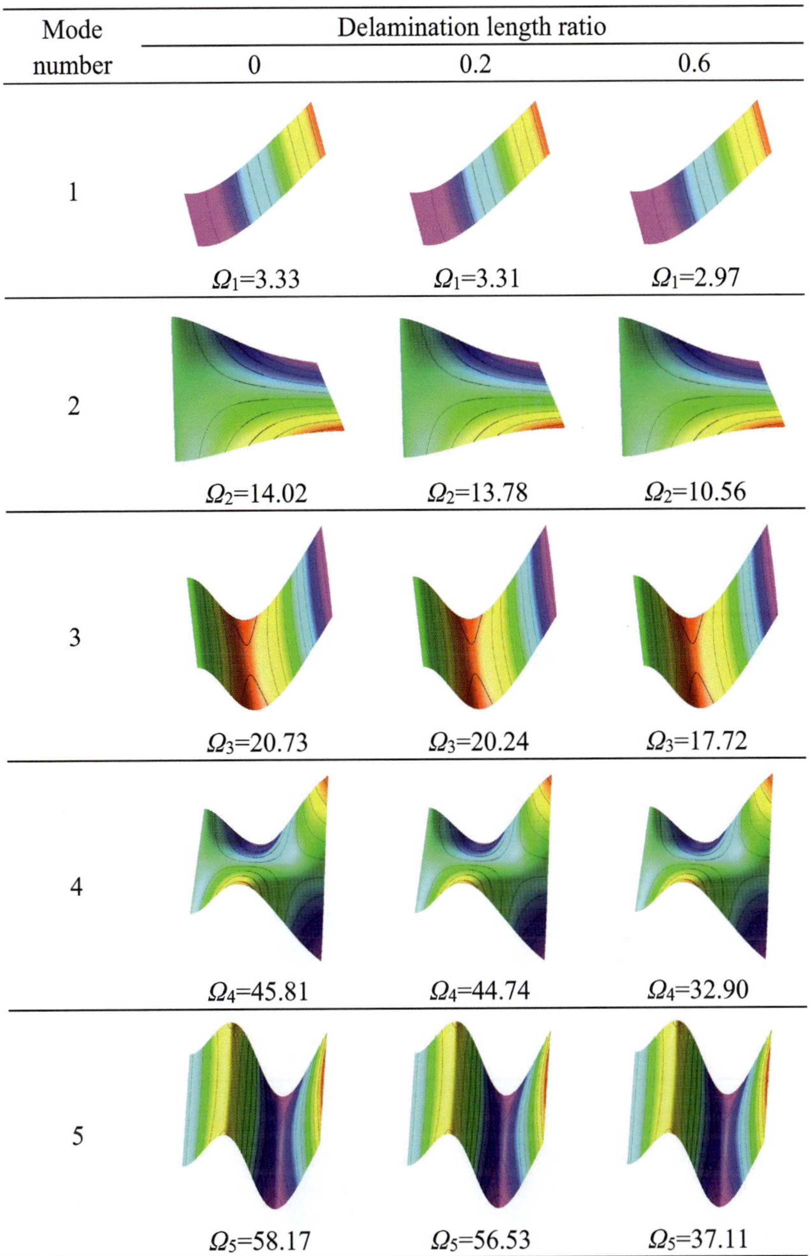

Fig. 5. First five-order vibration modes and their non-dimensional frequencies of CFFF delaminated FGM plates.

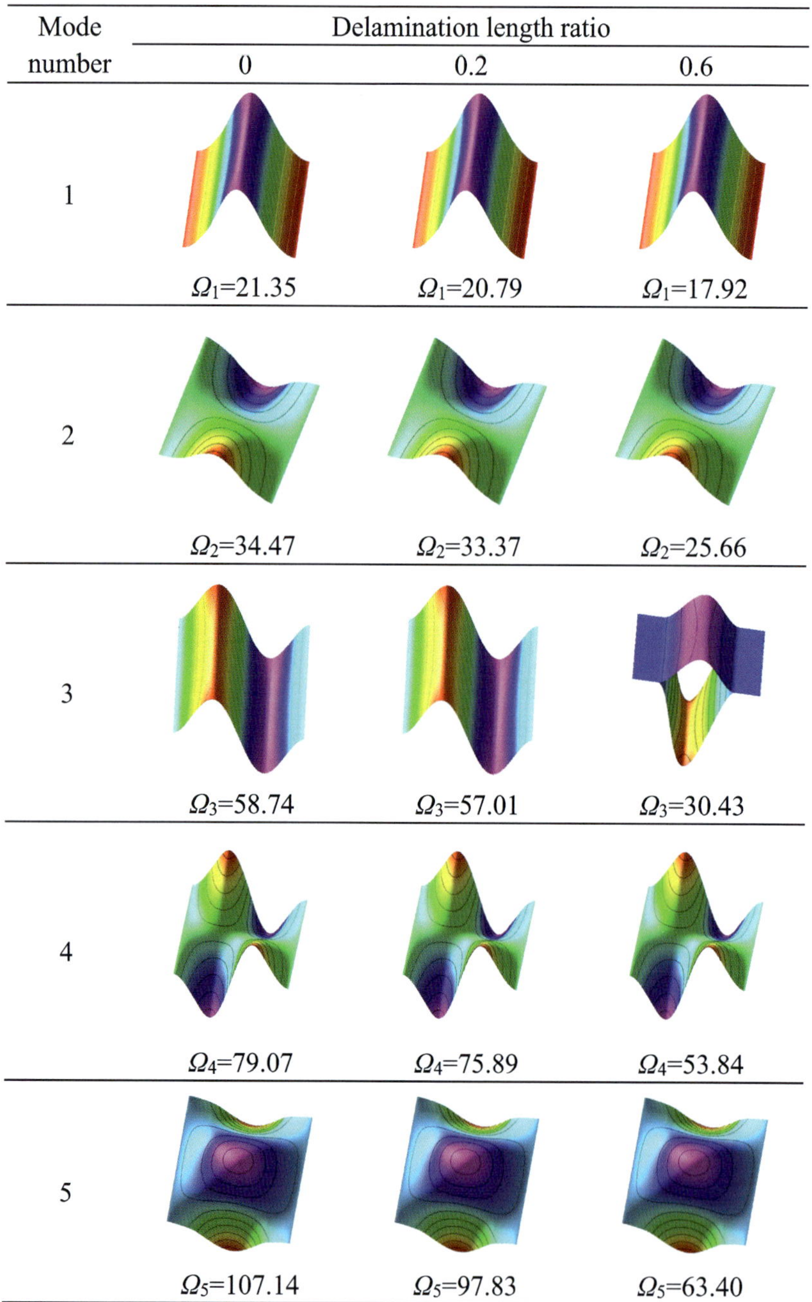

Fig. 6. First five-order vibration modes and their non-dimensional frequencies of CFCF delaminated FGM plates.

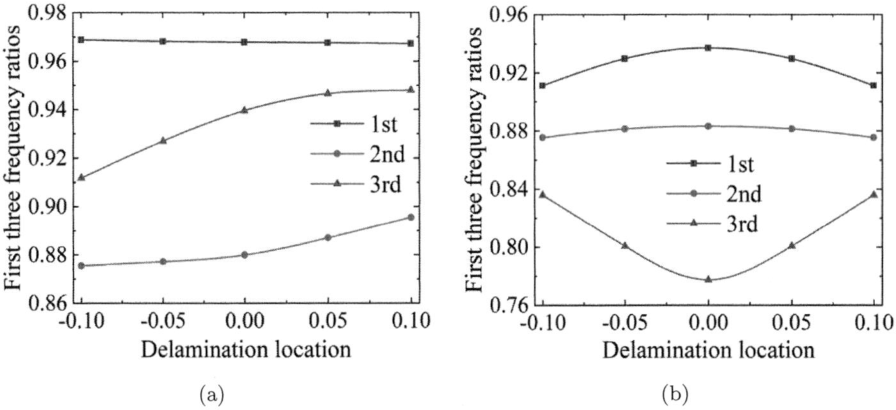

Fig. 7. First three-order frequency ratios of delaminated FGM plates: (a) CFFF case and (b) CFCF case.

Table 4. Influence of Young's modulus and delamination location on the first three-order non-dimensional frequencies of CFCF delaminated FGM plates.

Young's modulus ratio	Delamination location											
	−0.05			0			0.05			0.15		
	Ω_1	Ω_2	Ω_3	Ω_1	Ω_2	Ω_3	Ω_1	Ω_2	Ω_3	Ω_1	Ω_2	Ω_3
0.2	18.30	28.26	45.14	18.39	28.27	44.05	18.30	28.26	45.14	17.61	27.95	48.42
0.4	20.14	30.79	47.39	20.32	30.86	45.97	20.14	30.79	47.39	18.91	30.10	51.18
0.6	20.85	31.75	48.21	21.06	31.84	46.65	20.85	31.75	48.21	19.38	30.77	52.18
0.8	21.12	32.11	48.51	21.35	32.22	46.90	21.12	32.11	48.51	19.55	31.05	52.55
1.0	21.19	32.20	48.59	21.41	32.31	46.97	21.19	32.20	48.59	19.60	31.13	52.65

same when $L_a/L = -0.05$ and 0.05. The results are consistent with those obtained in Fig. 7(b), which is symmetric about the line of $x = 0$.

3.4. *Effect of Young's modulus ratio*

The influence of Young's modulus ratio in the CFFF and CFCF cases is presented in Figs. 8 and 9, respectively. The central mid-plane delamination length ratios are set to 0.2, 0.4, and 0.6. The results show that the natural frequencies increase sharply when the Young's modulus ratio (from top to bottom) is varied from 0 to 0.1, and remain almost constant from 0.1 to 1. Similar trends are observed in both the cases, but the effects of Young's modulus ratio and delamination length ratio are more pronounced in the CFCF case.

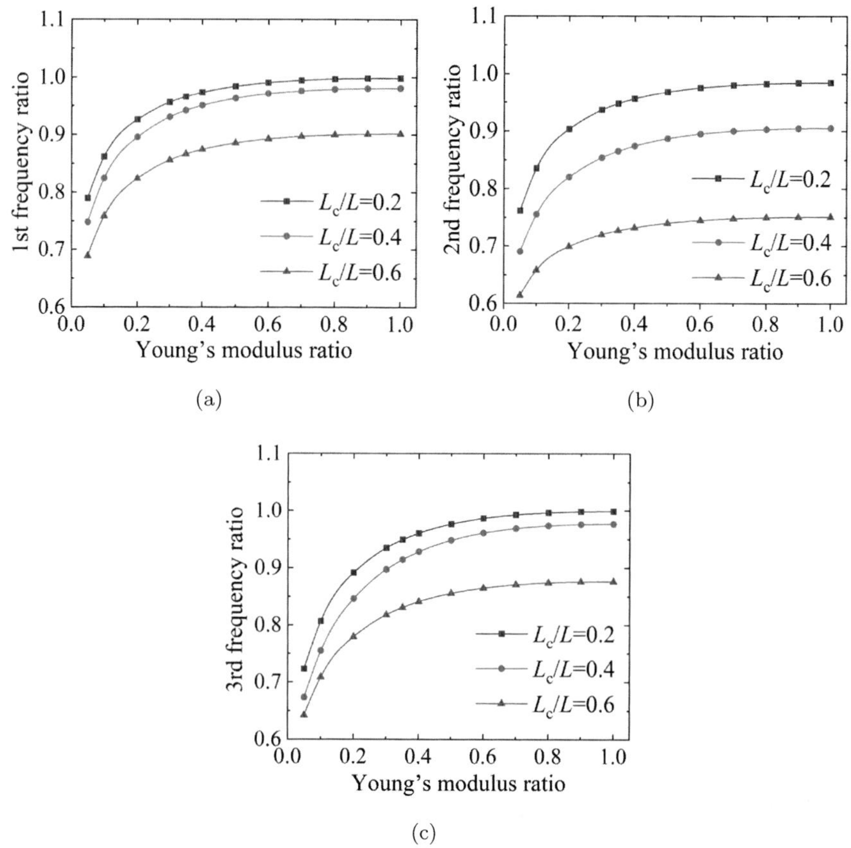

Fig. 8. Effect of Young's modulus ratio on the dynamic performance of CFFF delaminated FGM plates.

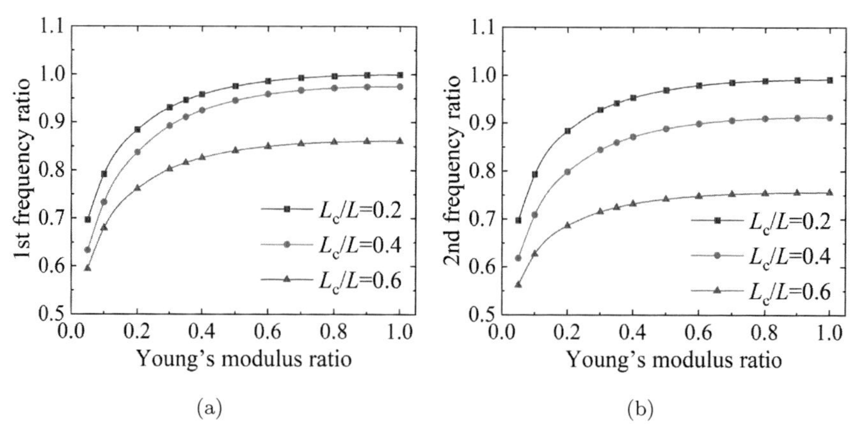

Fig. 9. Effect of Young's modulus ratio on the dynamic performance of CFCF delaminated FGM plates.

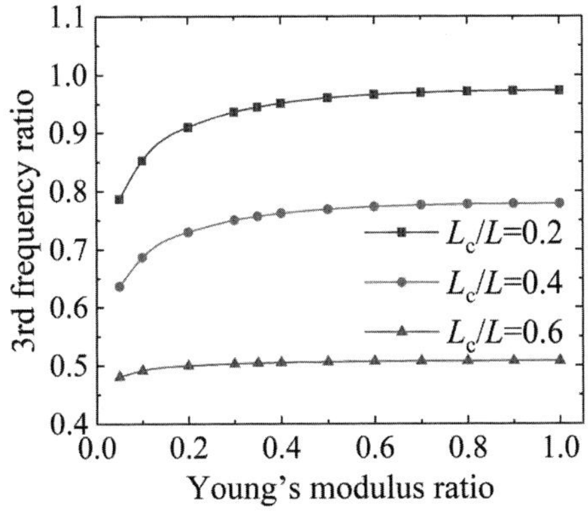

(c)

Fig. 9. (*Continued*)

4. Conclusions

In this work, the Chebyshev polynomials are used as admissible functions by incorporating along with the Ritz method to formulate the governing equation of exponential FGM plates with a through-width delamination. The presence of delamination leads to the reduction of vibration frequencies, which coincides with the variation of mechanical stiffness. When the delamination length exceeds half of the total length, the upper and lower delaminated layers are separated from the transverse direction with asymmetric amplitude responses. However, the layers of delamination remain in close contact under motions. Furthermore, the occurrence of delamination leads to a significant change in the torsional mode of FGM plates. Meanwhile, the natural frequencies of FGM plates increase dramatically with the increase of Young's modulus ratio (top-to-bottom) from 0 to 0.1, and remain constant from 0.1 to 1. The accuracy of the Chebyshev–Ritz method is validated through a series of convergence and comparison studies with respect to the available results. This study offers a feasible approach to predict the delamination effect of composite structures, it is applicable for the analysis of structures with multiple delaminations, complex geometrical dimensions, and various environmental conditions. In addition, further studies will also be extended to the analysis of delamination for the packaging of microelectromechanical devices.

Acknowledgments

The authors gratefully acknowledge the financial support of the Beijing Natural Science Foundation (Grant No. 1222001) and the National Natural Science Foundation of China (Grant No. 12072003). The financial supports from the Innovation and Technology Commission of the Government of the Hong Kong Special Administrative Region to the Hong Kong Branch of National Rail Transit Electrification and Automation Engineering Technology Research Center (K-BBY1) and the Project of Strategic Importance of The Hong Kong Polytechnic University (Project No. 1-ZE0B) are also gratefully acknowledged.

References

1. J. F. Wang, S. H. Cao and W. Zhang, Thermal vibration and buckling analysis of functionally graded carbon nanotube reinforced composite quadrilateral plate, *Eur. J. Mech. A. Solids* **85** (2021) 104105.
2. S. Y. Zhao, Y. Y. Zhang, Y. H. Zhang, J. Yang and S. Kitipornchai, A functionally graded auxetic metamaterial beam with tunable nonlinear free vibration characteristics via graphene origami, *Thin-Walled Struct.* **181** (2022) 109997.
3. S. Y. Zhao, Y. Y. Zhang, Y. H. Zhang, J. Yang and S. Kitipornchai, Vibrational characteristics of functionally graded graphene origami-enabled auxetic metamaterial beams based on machine learning assisted models, *Aerosp. Sci. Technol.* **130** (2022) 107906.
4. P. K. Karsh, T. Mukhopadhyay, S. Chakraborty, S. Naskar and S. Dey, A hybrid stochastic sensitivity analysis for low-frequency vibration and low-velocity impact of functionally graded plates, *Compos. B. Eng.* **176** (2019) 107221.
5. A. Fallah and M. M. Aghdam, Thermo-mechanical buckling and nonlinear free vibration analysis of functionally graded beams on nonlinear elastic foundation, *Compos B. Eng.* **43** (2012) 1523–1530.
6. Y. T. Fu and X. F. Yao, A review on manufacturing defects and their detection of fiber reinforced resin matrix composites, *Compos. C. Open Access* **8** (2022) 100276.
7. N. Zhang, T. Khan, H. M. Guo, S. S. Shi, W. Zhong and W. W. Zhang, Functionally graded materials: An overview of stability, buckling,and free vibration analysis, *Adv. Mater. Sci. Eng.* **2019** (2019) 1354150.
8. F. Sait, N. Saeidi and T. Korkmaz, Finite element analysis of FGM dental crowns using phase-field approach, *J. Mech. Behav. Biomed. Mater.* **138** (2023) 105629.
9. D. H. Li, X. Yang, R. L. Qian and D. Xu, Static and dynamic response analysis of functionally graded material plates with damage, *Mech. Adv. Mater. Struct.* **27** (2020) 94–107.
10. Y. Liu and D. W. Shu, Free vibration of exponential functionally graded beams with a single delamination, *Compos. B. Eng.* **75** (2014) 164–168.
11. C. N. Della and D. W. Shu, Vibration of delaminated composite laminates: A review, *Appl. Mech. Rev.* **60** (2007) 1–20.
12. A. D. Dimarogonas, Vibration of cracked structures: A state of the art review, *Eng. Fract. Mech.* **55** (1996) 831–857.
13. J. Xue, Y. F. Wang and L. H. Chen, Nonlinear vibration of cracked rectangular Mindlin plate with in-plane preload, *J. Sound Vib.* **481** (2020) 115437.
14. Z. F. Zhang, J. W. Pan, W. L. Luo, K. R. Ramakrishnan and H. K. Singh, Vibration-based delamination detection in curved composite plates, *Compos. A. Appl. Sci. Manuf.* **119** (2019) 261–274.
15. J. F. Wang, S. Q. Shi, J. P. Yang and W. Zhang, Multiscale analysis on free vibration of functionally graded graphene reinforced PMMA composite plates, *Appl. Math. Model.* **98** (2021) 38–58.

16. A. K. Noor and W. S. Burton, Assessment of computational models for multilayered composite shells, *Appl. Mech. Rev.* **43** (1990) 67–97.
17. J. N. Reddy and D. H. Bobbins, Theories and computational models for composite laminates, *Appl. Mech. Rev.* **47** (1994) 147–169.
18. S. Kapuria and A. Ahmed, Delamination modeling in doubly curved laminated shells for free vibration analysis using zigzag theory-based facet shell element and hybrid continuity method, *Int. J. Numer. Meth. Eng.* **120** (2019) 1126–1147.
19. R. Ramkumar, S. Kulkarni and R. Pipes, Free vibration frequencies of a delaminated beam, in *34th Annu. Tech. Conf. Proc. Reinforced/Composite Institute*, (Society of Plastics Industry, 1979), pp. 1–5.
20. J. T. S. Wang, Y. Y. Liu and J. A. Gibby, Vibrations of split beams, *J. Sound Vib.* **84** (1982) 491–502.
21. A. Kamaloo, M. Jabbari, M. Y. Tooski and M. Javadi, Nonlinear free vibrations analysis of delaminated composite conical shells, *Int. J. Struct. Stab. Dyn.* **20** (2020) 2050010.
22. S. Shams, A. R. Torabi, M. F. Narab and M. A. A. Atashgah, Free vibration analysis of a laminated beam using dynamic stiffness matrix method considering delamination, *Thin-Walled Struct.* **166** (2021) 107952.
23. J. J. Tracy and G. C. Pardoen, Effect of delamination on the natural frequencies of composite laminates, *J. Compos. Mater.* **23** (1989) 1200–1215.
24. Z. W. Song, S. K. Lai and J. G. Dai, Refined models for free vibration analysis of elastic plates with part-through surface cracks, *Thin-Walled Struct.* **182** (2023) 110312.
25. K. K. Wang, L. B. Zhao, H. M. Hong, J. Y. Zhang and Y. Gong, Parameter studies and evaluation principles of delamination damage in laminated composites, *Chin. J. Aeronaut.* **34** (2021) 62–72.
26. K. K. Wang, L. B. Zhao, H. M. Hong, J. Y. Zhang and N. Hu, An extended analytical model for predicting the compressive failure behaviors of composite laminate with an arbitrary elliptical delamination, *Int. J. Solids Struct.* **185–186** (2020) 439–447.
27. S. Lee, T. Park and G. Z. Voyiadjis, Free vibration analysis of axially compressed laminated composite beam-columns with multiple delaminations, *Compos. B. Eng.* **33** (2002) 605–617.
28. S. Lee, T. Park and G. Z. Voyiadjis, Vibration analysis of multi-delaminated beams, *Compos. B. Eng.* **34** (2003) 647–659.
29. Y. Liu and D. W. Shu, Free vibration analysis of rotating Timoshenko beams with multiple delaminations, *Compos. B. Eng.* **44** (2013) 733–739.
30. H. S. Panda, S. K. Sahu and P. K. Parhi, Effects of moisture on the frequencies of vibration of woven fibre composite doubly curved panels with strip delaminations, *Thin-Walled Struct.* **78** (2014) 79–86.
31. K. Alnefaie, Finite element modeling of composite plates with internal delamination, *Compos. Struct.* **90** (2009) 21–27.
32. A. Zak, M. Krawczuk and W. Ostachowicz, Numerical and experimental investigation of free vibration of multilayer delaminated composite beams and plates, *Comput. Mech.* **26** (2000) 309–315.
33. M. Krawczuk, W. Ostachowicz and A. Zak, Dynamics of cracked composite material structures, *Comput. Mech.* **20** (1997) 79–83.
34. B. G. Kiral, Free vibration analysis of delaminated composite beams, *Sci. Eng. Compos. Mater.* **16** (2009) 209–224.

35. S. Ganesh, K. S. Kumar and P. K. Mahato, Free vibration analysis of delaminated composite plates using finite element method, *Procedia Eng.* **144** (2016) 1067–1075.
36. N. Sharma, P. K. Swain, D. K. Maiti and B. N. Singh, Static and free vibration analyses and dynamic control of smart variable stiffness laminated composite plate with delamination, *Compos. Struct.* **280** (2022) 114793.
37. C. K. Hirwani, S. K. Panda, T. R. Mahapatra, S. K. Mandal, S. S. Mahapatra and A. K. De, Delamination effect on flexural responses of layered curved shallow shell panel-experimental and numerical analysis, *Int. J. Comput. Meth.* **15**(4) (2018) 1850027.
38. P. K. Mishra, A. K. Pradhan, M. K. Pandit and S. K. Panda, The effect of delamination in free vibration responses of adhesively bonded spar wingskin joints, *Mech. Based Des. Struct. Mach.* **51** (2023) 1–18.
39. L. H. Yam, Z. Wei, L. Cheng and W. O. Wong, Numerical analysis of multi-layer composite plates with internal delamination, *Comput. Struct.* **82** (2004) 627–637.
40. Z. Wei, L. H. Yam and L. Cheng, Detection of internal delamination in multi-layer composites using wavelet packets combined with modal parameter analysis, *Compos. Struct.* **64** (2004) 377–387.
41. Y. Kiani, Free vibration of carbon nanotube reinforced composite plate on point supports using Lagrangian multipliers, *Meccanica* **52** (2017) 1353–1367.
42. L. Meirovitch and M. K. Kwak, Convergence of the classical Rayleigh-Ritz method and the finite element method, *AIAA J.* **28** (1990) 1509–1516.
43. D. Chen, J. Yang and S. Kitipornchai, Buckling and bending analyses of a novel functionally graded porous plate using Chebyshev-Ritz method, *Arch. Civ. Mech. Eng.* **19** (2019) 157–170.
44. D. Zhou, F. T. K. Au, Y. K. Cheung and S. H. Lo, Three-dimensional vibration analysis of circular and annular plates via the Chebyshev-Ritz method, *Int. J. Solids Struct.* **40** (2003) 3089–3105.
45. D. Zhou, Y. K. Cheung, S. H. Lo and F. T. K. Au, 3D vibration analysis of solid and hollow circular cylinders via Chebyshev-Ritz method, *Comput. Meth. Appl. Mech. Eng.* **192** (2003) 1575–1589.
46. J. F. Wang, J. P. Yang, L. H. Tam and W. Zhang, Molecular dynamics-based multi-scale nonlinear vibrations of PMMA/CNT composite plates, *Mech. Syst. Signal Process.* **153** (2021) 107530.
47. R. K. Kapania and S. Sarvesh, Comparison of simple and Chebyshev polynomials in Rayleigh—Ritz analysis, *J. Eng. Mech.* **120** (1994) 2126–2135.
48. N. Kharghani and C. Guedes Soares, Influence of different parameters on the deflection of composite laminates containing through-the-width delamination using layerwise HSDT, *Compos. Struct.* **132** (2015) 341–349.
49. N. Kharghani and C. Guedes Soares, Behavior of composite laminates with embedded delaminations, *Compos. Struct.* **150** (2016) 226–239.
50. Y. Kim, S. T. Wereley, C. Chun and Y. Kim, Phase-resolved flow field produced by a vibrating cantilever plate between two endplates, *Am. Inst. Phys.* **16** (2004) 145.
51. S. Keshava Kumar, M. Cinefra, E. Carrera, R. Ganguli and D. Harursampath, Finite element analysis of free vibration of the delaminated composite plate with variable kinematic multilayered plate elements, *Compos. B. Eng.* **66** (2014) 453–465.
52. J. Oh, M. Cho and J. S. Kim, Dynamic analysis of composite plate with multiple delaminations based on higher-order zigzag theory, *Int. J. Solids Struct.* **42** (2005) 6122–6140.

https://doi.org/10.1142/9789819814770_0013

Chapter 13

Linking Mesoscopic and Macroscopic Aspects of Inclined Self-Weight Sandwich Beams with Functionally Graded Porous Cores Under Moving Loads[#]

Da Chen[*,†,‖,**], Shahed Rezaei[‡], Jie Yang[§], Sritawat Kitipornchai[†], Lihai Zhang[*,††] and Philipp L. Rosendahl[¶]

[*]*Department of Infrastructure Engineering*
The University of Melbourne, VIC 3010, Australia

[†]*School of Civil Engineering, The University of Queensland*
St. Lucia, QLD 4072, Australia

[‡]*Mechanics of Functional Materials Division*
Department of Materials Science
Technical University of Darmstadt, Germany

[§]*School of Engineering, RMIT University, P. O. Box 71,*
Bundoora, VIC 3083, Australia

[¶]*Institute of Structural Mechanics and Design*
Department of Civil and Environmental Engineering,
Technical University of Darmstadt, Germany
[‖]*Centre for Infrastructure Engineering and Safety (CIES)*
School of Civil and Environmental Engineering
The University of New South Wales
Sydney, NSW 2052, Australia
[**]*da.chen1@unimelb.edu.au*
[††]*lihzhang@unimelb.edu.au*

The surging interest in porous lightweight structures has been witnessed in recent years to pursue material innovations in broad engineering disciplines for sustainable developments and multifunctional proposes. Functionally graded (FG) porous composites represent a novel way to adjust mechanical characteristics by controlling the porosity distributions. However, the further advance in this field is challenged by the scale gap between mesoscopic and macroscopic aspects of porous structural analysis, i.e. how the local cellular morphologies impact the overall behaviors. The purpose of this paper is to bridge this gap by conducting a theoretical investigation on the performance of inclined self-weight sandwich beams with FG porous cores, where Young's modulus is obtained with representative volume elements (RVEs) in a multiscale modeling study and depends on the cellular morphologies: average cell size and cell wall thickness. The material properties of closed-cell steel foams are adopted in a two-step assessment on target beams, including

[**,††]Corresponding authors.
[#]This chapter appeared previously on the International Journal of Structural Stability and Dynamics. To cite this chapter, please cite the original article as the following: D. Chen *et al.*, *Int. J. Struct. Stab. Dyn.*, **23**, 2340014 (2023), DOI: 10.1142/S021945542340014X.

a static calculation to examine their bending deformations under gravitational loading which are then imported into a forced vibration analysis considering constant and harmonic moving forces. Timoshenko beam theory is used to establish the displacement field, while Ritz and Newmark methods are employed to solve the governing equations in terms of bending, free vibration, and forced vibration. The inclined beams are assumed to rest on a Pasternak foundation, and the corresponding structural responses can be determined based on the specific cell size and cell wall thickness, of which the effects are quantitatively revealed: the stiffness degradation induced from cellular morphologies increases the dynamic deflections, while the corresponding self-weight static deformations are reduced and the fundamental natural frequencies are raised. The influence from geometrical, boundary, and foundation conditions is also discussed to provide a comprehensive overview. This will be valuable for engineers to develop devisable foam-based load-carrying components with enhanced properties.

Keywords: Functionally graded porosity; sandwich composite; inclined beam; moving load; multiscale modeling; metal foam.

1. Introduction

Sustainable developments are now being supported by research communities more strongly than ever, in terms of environmental, human, social, and economic dimensions, etc. Following this trend toward a balanced society, engineers are passionately devoted to the novel lightweight composite components for structural, mechanical, and aerospace applications, with the intention to rationalize the material consumption and to restrict carbon emissions during manufacturing and operation. Metal foams, highlighted by stochastic porosity geometries and emerging large-scale fabrication availability, provide distinctive traits for advancing such lightweight composite forms. The internal cellular morphologies not only reduce the mass density, but also beneficially bring about interesting thermal, biological, acoustic, and electronic attributes, making them even more appealing with multifunctional applicability. A rapidly growing body of publications evidence the potential of metal foams to be adopted in diverse engineering sectors: blast counters and energy absorbers,[1] thermal energy storage systems,[2] bone implant materials and biomedicines,[3] engine brackets,[4] and noise control composites.[5] Moreover, the rapid development in additive manufacturing is able to support the construction of specifically designed porous elements.[6,7]

Lightweight structural components (such as beams, plates, shells, arches, frames) made of metal foams have been widely discussed in recent years for the load carrying purpose, among which the innovative modifications via graded porosities further endow them with devisable characteristics and raise special interests in terms of property enhancement.[8] Grygorowicz and Magnucka-Blandzi.[9] investigated the dynamic stability of sandwich beams with variable foam cores and obtained the related unstable regions and dynamic equilibrium paths. Akbaş *et al.*[10] presented a finite element (FE) model in the examination on forced vibration responses of viscoelastic functionally graded (FG) porous multilayer thick beams under dynamic sine pulse loads, and proved the applicability of this model. Zenkour[11] proposed a refined quasi-3D shear and normal deformation theory for the static analysis of FG porous thick plates and showed the critical role of the porosity factor. Chen *et al.*[12–14] focused on the structural performance improvement caused by non-uniform porosity distributions for FG porous beams and plates, of which the buckling, bending, and vibration traits are disclosed and compared to those from their uniform foam counterparts, emphasizing the superiority of graded porosities. Gao *et al.*[15] considered the material uncertainties in the static buckling assessment of FG porous plates reinforced by graphene platelets, in order to reveal the sensitivities of different parameters.

Most of the existing studies on this topic adopt a rather vague definition of graded porosities by using self-defined coefficients to adjust the variations in elastic moduli, Poisson's ratio, and mass density for the estimation of structural behaviors. The experimental or numerical relations of open- and closed-cell foams are often borrowed to connect these coefficients.[12,16] Carbon nanofillers can also be added into the metal matrix to enhance the component stiffness.[17,18] However, the mesoscopic features of cellular geometries may have a determinant influence on porous structures,[19–21] and how to accurately link them to the overall mechanical performance remains a challenge that requires extensive investigations. In a recent trial,[22] the multiscale modeling is demonstrated to be an effective approach to provide this much-needed knowledge through identifying the quantitative relations between cell size/cell wall thickness and relative density which notably affects the foam properties. In the meantime, although relative density is the key factor to characterize porous structures, the direct impacts from mesoscopic cellular morphologies are important for localized porosity designs but are not fully clarified.

This paper aims to crack this problem via investigating the linkage between mesoscopic and macroscopic aspects of inclined self-weight beams under moving loads, which are highlighted with novel sandwich compositions containing porous skin layers and FG cores. Horizontal/Inclined beams and moving loads form a classic problem reflecting various practical scenarios,[23,24] such as vehicle–bridge or wheel–rail interactions, machine or structural component reactions under traveling force, and skiers passing through snow slabs. The research on this problem is still thriving: Yang *et al.*[25] included the moving mass and moving follower force into the dynamic stability and response examinations on inclined beams, and pursued the analytical solutions for the critical loading conditions; Shokouhifard *et al.*[26] developed an inverse solution method to calculate the moving mass passing through an inclined FG beam. Mohebpour *et al.*[27] presented a numerical assessment on the dynamic deflection of a composite cross-ply laminated beam subjected to a moving mass and deliberated the effects of Coriolis and centrifugal forces. In addition, such beam form may also be conceptually referred to laminated snow slabs in nature: Rosendahl *et al.*[28–30] examined inclined porous snow beams on elastic foundations for the prediction of snow slab avalanche release.

This study adopts the beam theory and multiscale modeling to evaluate the impacts from varying cell sizes and cell wall thicknesses (mesoscopic features) within the FG porous core on the overall bending deformations, free vibration frequencies, and forced vibration deflections (macroscopic responses) of inclined sandwich beams. Both constant and harmonic moving loads are taken into consideration. The structural property changes induced by diverse inclined angles, Pasternak foundation coefficients, slenderness ratio, boundary conditions, and moving force characteristics are weighted in detail for delivering the design guidance of such novel composites. The results from this paper will be beneficial, for example, when designing lightweight high-stiffness components for crane girders subject to moving trolleys/carts which normally carry heavy goods.

2. Formulations of Inclined Sandwich Porous Beams

An example of target inclined sandwich beams is depicted in Fig. 1 and rests on a Pasternak foundation while inclined at an angle of θ. This beam consists of two low-porosity (high-density) skin layers on the top/bottom surfaces and a lightweight FG high-porosity (low-density) core in the middle. The beam elastic properties are dependent on the porosity distributions and can be determined using FE representative volume elements (RVEs) considering the metal foam mesoscopic features (elaborated in Sec. 4). A coordinate system is built on the midplane with x-axis along the length direction, z-axis across the thickness, and the origin point located at the higher/left end of this beam (see Fig. 1). The total beam thickness H_B = core thickness H_C + sum of skin layer thickness $2H_S$. The beam length and width are denoted by L and b, respectively. The core and skin layers are assumed to be adhered together without delamination.

The proposed beam is evaluated under the framework of shear deformable Timoshenko beam theory with a combined static and dynamic analysis, where the static part accounts for the bending deformation subject to uniformly distributed gravitational force, while the dynamic part provides forced vibration solutions under moving loads. Meanwhile, the free vibration natural frequencies may also be acquired. The displacement fields for static and dynamic parts are written as

$$\text{Static:} \begin{cases} u_1^*(x,z) = u_1(x) + z\varphi_1(x), \\ w_1^*(x) = w_1(x), \end{cases} \tag{1}$$

$$\text{Dynamic:} \begin{cases} u_2^*(x,z,t) = u_1(x) + u_2(x,t) + z[\varphi_1(x) + \varphi_2(x,t)], \\ w_2^*(x,t) = w_1(x) + w_2(x,t), \end{cases} \tag{2}$$

where the static displacements are incorporated into their dynamic counterparts, and t reads the time, u_1 (u_2) and w_1 (w_2) are the static (dynamic) axial and transverse displacements, while φ_1 (φ_2) represents the static (dynamic) angle of cross-section rotations. The static deformation occurs under the gravitational load and provides a pre-stressed field for the dynamic analysis which contains time-dependent

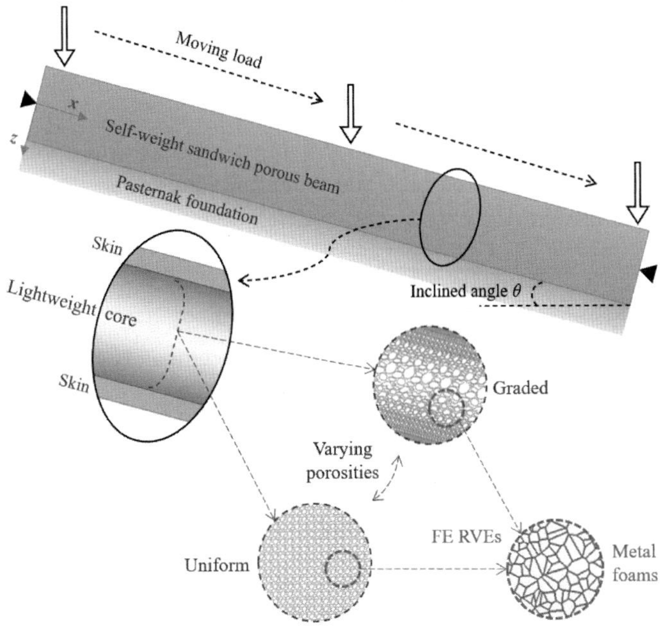

Fig. 1. Schematic diagram of target sandwich porous beam.

components calculated with the moving load. The corresponding linear strains can be expressed as

$$\text{Static:} \begin{cases} \varepsilon_{xx}^{s} = \partial u_1/\partial x + z\partial\varphi_1/\partial x, \\ \gamma_{xz}^{s} = \partial w_1/\partial x + \varphi_1, \end{cases} \tag{3}$$

$$\text{Dynamic:} \begin{cases} \varepsilon_{xx}^{d} = \partial u_1/\partial x + \partial u_2/\partial x + z(\partial\varphi_1/\partial x + \partial\varphi_2/\partial x), \\ \gamma_{xz}^{d} = \partial w_1/\partial x + \partial w_2/\partial x + \varphi_1 + \varphi_2, \end{cases} \tag{4}$$

resulting in the following stresses

$$\text{Normal:} \, \sigma_{xx}^{s} = E^*(z)\varepsilon_{xx}^{s}, \quad \sigma_{xx}^{d} = E^*(z)\varepsilon_{xx}^{d}, \tag{5}$$

$$\text{Shear:} \, \sigma_{xz}^{s} = G^*(z)\gamma_{xz}^{s}, \quad \sigma_{xz}^{d} = G^*(z)\gamma_{xz}^{d}, \tag{6}$$

of which the modulus items $E^*(z)$ and $G^*(z)$ vary on the basis of beam material distributions and are established in Sec. 4. The strain energy of proposed inclined sandwich beams is then written as

$$\text{Static:} \, \Phi_s = \frac{b}{2}\int_0^L \int_{-H_B/2}^{H_B/2} (\sigma_{xx}^{s}\varepsilon_{xx}^{s} + \sigma_{xz}^{s}\gamma_{xz}^{s}) dz dx, \tag{7}$$

$$\text{Dynamic:} \, \Phi_d = \frac{b}{2}\int_0^L \int_{-H_B/2}^{H_B/2} (\sigma_{xx}^{d}\varepsilon_{xx}^{d} + \sigma_{xz}^{d}\gamma_{xz}^{d}) dz dx. \tag{8}$$

The beam kinetic energy under dynamic loads reads

$$T_d = \frac{b}{2}\int_0^L \int_{-H_B/2}^{H_B/2} \rho^*(z)\{[\partial u_2^*(x,z,t)/\partial t]^2 + [\partial w_2^*(x,t)/\partial t]^2\}dzdx, \tag{9}$$

where the mass density item $\rho^*(z)$ is also explained in Sec. 4. The above energy forms are used in Sec. 3 to derive the partial differential equation systems governing the beam bending and vibration behaviors, and the embedded shear correction factor equals to 5/6.

In this study, an elastic Pasternak foundation is adopted to support the target inclined beam and consists of Winkler spring layer and shear layer. The unit foundation reaction force P_f is related to its displacement W_f via[31]

$$P_f = K_l W_f - K_s \partial^2 W_f/\partial x^2, \tag{10}$$

where K_l and K_s represent the stiffness coefficients of Winkler spring layer and shear layer, respectively. The related potential energy stored in the deformed foundation is then given by[16,32–34]

$$\text{Static: } \Phi_s^f = \frac{b}{2}\int_0^L \left\{K_l w_1^{*2}(x) + K_s\left[\frac{\partial w_1^*(x)}{\partial x}\right]^2\right\}dx, \tag{11}$$

$$\text{Dynamic: } \Phi_d^f = \frac{b}{2}\int_0^L \left\{K_l w_2^{*2}(x,t) + K_s\left[\frac{\partial w_2^*(x,t)}{\partial x}\right]^2\right\}dx. \tag{12}$$

3. Bending and Vibration Analyses

Following the formulations in Sec. 2, the bending and vibration analyses of inclined self-weight beams are elaborated in this section, where the bending deflection is examined first. The work done by gravitational loading can be expressed as

$$\text{Static: } \Psi_s = \int_0^L \left[(Q\cos\theta - \varphi_1 Q\sin\theta)w_1 + \frac{1}{2}Qx\sin\theta\left(\frac{\partial w_1}{\partial x}\right)^2\right]dx, \tag{13}$$

where Q represents the evenly distributed force from gravity and $Q = Mg/L$ (M is the total beam mass, $g = 9.8066\,\text{m/s}^2$). In contrast to Ψ_s for the static part, the work Ψ_d done by external loads in the dynamic analysis involves Ψ_{d1} from Q and Ψ_{d2} induced by the moving load P:

$$\text{Dynamic: } \Psi_d = \Psi_{d1} + \Psi_{d2}, \tag{14}$$

$$\Psi_{d1} = \int_0^L \left\{[Q\cos\theta - (\varphi_1+\varphi_2)Q\sin\theta](w_1+w_2) + \frac{1}{2}Qx\sin\theta\left(\frac{\partial w_1}{\partial x} + \frac{\partial w_2}{\partial x}\right)^2\right\}dx, \tag{15}$$

$$\Psi_{d2} = \int_0^L [P\cos\theta - P\sin\theta(\varphi_1 + \varphi_2)] w_2 \delta(x - V_P t) dx$$
$$+ \int_0^L P\sin\theta \left[\frac{1}{2}\left(\frac{\partial w_1}{\partial x} + \frac{\partial w_2}{\partial x}\right)^2 - \frac{1}{2}\left(\frac{\partial w_1}{\partial x}\right)^2 - \frac{H_B}{2}\frac{\partial \varphi_2}{\partial x}\right]$$
$$\times H(x - v_P t) dx, \tag{16}$$

where P is the external load and V_P is its velocity, $\delta(\cdot)$ and $H(\cdot)$ correspond to the Dirac delta function and Heaviside step function, respectively. Equations (15) and (16) also suggest the mentioned load moving direction.

Integrating the derived energy forms (Eqs. (7)–(9), (11)–(16)) results in the energy functionals of the inclined beams

$$\text{Static: } \Lambda_s = \Psi_s - (\Phi_s + \Phi_s^f), \tag{17}$$

$$\text{Dynamic: } \Lambda_d = \Psi_d - (\Phi_d + \Phi_d^f) - T_d. \tag{18}$$

Equations (17) and (18) then can be discretized with the aid of the Ritz method for space domain as per the following trial displacement functions ($1 \le i \le N$, N is the number of employed polynomial items):

$$\frac{u_1(x)}{H_B}, \frac{u_2(x,t)}{H_B} = \sum_{i=1}^{N} \Delta_i^1 \left(\frac{x}{L}\right)^i \left(1 - \frac{x}{L}\right) \quad \text{H–H, C–C, C–H beam,} \tag{19}$$

$$\frac{w_1(x)}{H_B}, \frac{w_2(x,t)}{H_B} = \sum_{i=1}^{N} \Delta_i^2 \left(\frac{x}{L}\right)^i \left(1 - \frac{x}{L}\right) \quad \text{H–H, C–C, C–H beam,} \tag{20}$$

$$\varphi_1(x), \varphi_2(x,t) = \begin{cases} \sum_{i=1}^{N} \Delta_i^3 \left(\frac{x}{L}\right)^{i-1} & \text{H–H beam,} \\ \sum_{i=1}^{N} \Delta_i^3 \left(\frac{x}{L}\right)^i \left(1 - \frac{x}{L}\right) & \text{C–C beam,} \\ \sum_{i=1}^{N} \Delta_i^3 \left(\frac{x}{L}\right)^i & \text{C–H beam,} \end{cases} \tag{21}$$

of which different unknown coefficient sets $\mathbf{\Delta} = \{\{\Delta_i^1\}^T\{\Delta_i^2\}^T\{\Delta_i^3\}^T\}^T$ are employed for static and dynamic calculations ($\mathbf{\Delta}_s$ — static, $\mathbf{\Delta}_d$ — dynamic), and $\mathbf{\Delta}_d$ is time-dependent. These trial functions connect to the beam boundaries with H and C denoting the hinged and clamped ends. Based on the minimum potential energy principle, Eq. (17) is resolved first to gain static displacements (u_1, φ_1, w_1) for the computation of Eq. (18) using Lagrange equations and the average acceleration Newmark method.[13,35] The obtained governing equations (shown as below) are taken to examine the beam bending deflections and vibration histories.

$$\text{Static: } \mathbf{K}_s\boldsymbol{\Delta}_s - \mathbf{P}_s = 0, \tag{22}$$

$$\text{Dynamic: } \mathbf{K}_d\boldsymbol{\Delta}_d + \mathbf{M}\ddot{\boldsymbol{\Delta}}_d - \mathbf{P}_d = 0, \tag{23}$$

where $\mathbf{K}_s$ ($\mathbf{K}_d$) and $\mathbf{P}_s$ ($\mathbf{P}_d$) are the static (dynamic) stiffness matrix and force vector, respectively, $\mathbf{M}$ represents the mass matrix, and the over dot suggests the partial derivative with respect to the time. The deflections from static bending and forced vibration can be determined with $\boldsymbol{\Delta}_s$ and $\boldsymbol{\Delta}_d$ associated to the trial displacement functions.

Additionally, for a comprehensive understanding of the developed sandwich porous beams, a free vibration analysis is carried out to assess their fundamental natural frequencies, where inclined angles and external excitations are ignored. Under the oscillating vibration mode, an eigenvalue equation to identify the natural frequencies ω_f can be written as.[13]

$$\text{Free vibration: } (\mathbf{K}_s - \omega_f^2\mathbf{M})\boldsymbol{\Delta}_f = 0, \tag{24}$$

where $\boldsymbol{\Delta}_f$ is a new coefficient set for trial displacement functions in Eqs. (19)–(21) and differs from those of static bending and forced vibration analyses, i.e. $\boldsymbol{\Delta}_s$ and $\boldsymbol{\Delta}_d$. The lowest eigenvalue of Eq. (24) gives the fundamental natural frequency of target beam.

4. Multiscale Modeling of Beam Properties

4.1. *Multiscale modeling using RVEs*

Multiscale modeling exercises the feasibility to quantitatively build up the relationship between microscopic and macroscopic properties of composite medias, by assuming material heterogeneity and homogeneity in the small and large scales, respectively. In a previous study,[22] Chen *et al.* adopted RVEs for multiscale modeling to mimic the mechanical features of closed-cell metal foams, as illustrated in Fig. 2. These square-shape RVE models are constructed on the basis of the standard uniform distribution and Voronoi tessellation for internal pore positions and cell wall distributions. The key technical aspects are expounded here: (1) the cell wall is of uniform thickness considering the target high-porosity while low-density foams and is simulated with the FE beam element; (2) periodic boundary conditions are applied on the edge node pairs; (3) the RVE compressive stiffness is examined via FE analysis and represents Young's modulus of overall metal foams; (4) results of 655 RVE cases are summarized to disclose the connections between different scales and parameters.

The findings suggest that the relative density has a determinant effect on the obtained foam modulus, while the influences of other factors are rather negligible, including RVE size, internal porosity randomness, compression direction during

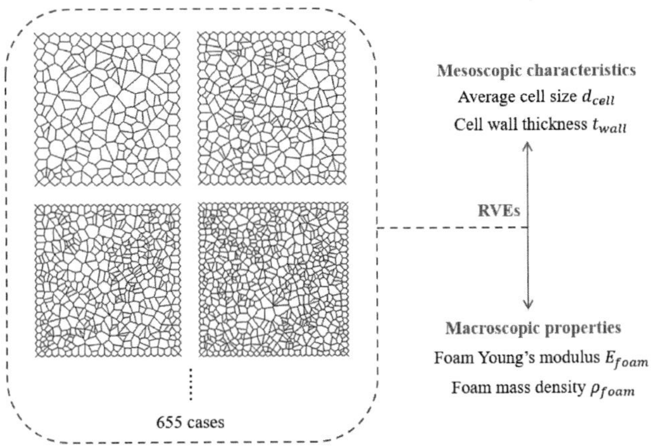

Fig. 2. Multiscale modeling using RVEs.

simulation, and wavy edge setting, etc. The relative density is further revealed to be correlated with averaged cell size and cell wall thickness, leading to the quantitative linkage between macroscopic properties (foam Young's modulus and mass density) and mesoscopic characteristics (RVE cell size and cell wall thickness)[22]:

$$\text{Foam relative density: } \beta_{\text{foam}} = (-\varsigma^2 + \varsigma\sqrt{\varsigma^2 + 4\pi})/2\pi, \tag{25}$$

$$\text{Foam Young's modulus: } E_{\text{foam}} = 0.2032 \cdot E_{\text{matrix}} \cdot \beta_{\text{foam}}, \tag{26}$$

$$\text{Foam mass density: } \rho_{\text{foam}} = \beta_{\text{foam}} \cdot \rho_{\text{matrix}}, \tag{27}$$

of which

$$\varsigma = 3.932 t_{\text{wall}}/d_{\text{cell}}, \tag{28}$$

where E_{foam} and ρ_{foam} are Young's modulus and mass density of foams, while E_{matrix} and ρ_{matrix} denote the corresponding counterparts for cell wall matrix materials, β_{foam} represents the foam relative density, d_{cell} and t_{wall} read the cell size and cell wall thickness, respectively. Note that in this study, E_{foam}, ρ_{foam}, and β_{foam} vary across the beam thickness direction z as per the distributions of d_{cell} and t_{wall}. Meanwhile, E_{matrix} and ρ_{matrix} remain fixed in the multiscale modeling calculation.[22] Due to the nature of Voronoi tessellations and RVEs with beam elements, Eqs. (25)–(28) are only applicable to low-density metal foams ($\beta_{\text{foam}} \leq 13\%$[22]) featured by convex polyhedral cellular configurations.

4.2. *Sandwich porous beam properties*

In this study, the beam lightweight core is highlighted via graded foam densities ($3.95\% \leq \beta_{\text{foam}} \leq 12.45\%$) and can be applied with RVE feature-property relations

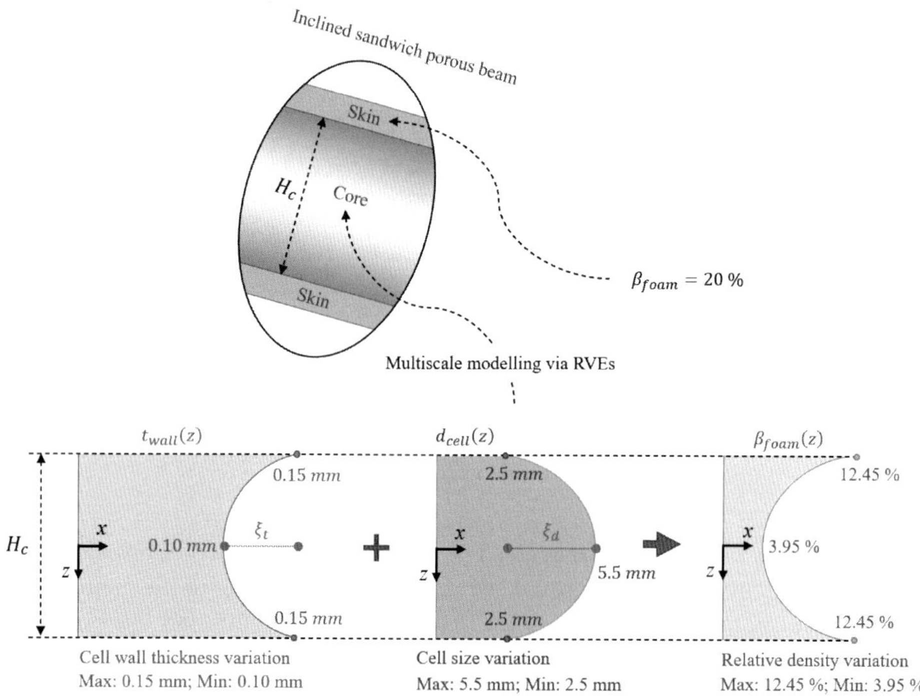

Fig. 3. Variations of cell wall thickness, cell size, and relative density for FG porous cores.

from Sec. 4.1. The corresponding cell wall thickness and average cell size vary along z-axis, as

$$t_{\text{wall}}(z) = t^{s}_{\text{wall}} - (t^{s}_{\text{wall}} - t^{m}_{\text{wall}}) \cdot \xi_t \cdot \cos\left(\pi \frac{z}{H_c}\right), \quad -H_c/2 \leq z \leq H_c/2, \tag{29}$$

$$d_{\text{cell}}(z) = d^{s}_{\text{cell}} - (d^{s}_{\text{cell}} - d^{m}_{\text{cell}}) \cdot \xi_d \cdot \cos\left(\pi \frac{z}{H_c}\right), \quad -H_c/2 \leq z \leq H_c/2, \tag{30}$$

where the combination of $t^{s}_{\text{wall}}(= 0.15\,\text{mm})$ and $d^{s}_{\text{cell}}(= 2.5\,\text{mm})$ results in the maximum relative density as 12.45%, while its minimum value (3.95%) is obtained based on $t^{m}_{\text{wall}}(= 0.10\,\text{mm})$ and $d^{m}_{\text{cell}}(= 5.5\,\text{mm})$. Here we use ξ_t and ξ_d ($0 \leq \xi_t \leq 1$, $0 \leq \xi_d \leq 1$) to control the FG porosity variations. Note that the ranges of relative density, cell wall thickness, and cell size are originated from the RVE parametric discussions in the mentioned work.[22] The joint effect of Eqs. (29) and (30) leads to the symmetrically distributed relative density in porous core about the midplane ($z = 0$), giving rise to the graded beam properties across the thickness direction, as illustrated in Fig. 3.

In the meantime, the top and bottom skins contain relatively high-density uniform foams (relative density $\beta_{\text{foam}} = 20\%$), which may be evaluated via the

closed-cell model with Gaussian random fields (suitable when 15%< β_{foam} < 100%).[36] To summarize, the gradations of relative density, Young's modulus,[22,36] shear modulus (assuming isotropic steel foams),[37] Poisson's ratio,[38,39] and mass density for both skin and core layers are expressed as
Relative density:

$$\beta_{\text{foam}}(z) = \begin{cases} 20\%, & H_c/2 < z \leq H_B/2, \ -H_B/2 \leq z < -H_c/2, \\ \beta_{\text{foam}}(t_{\text{wall}}(z), d_{\text{cell}}(z)), & \text{Eq. (25)}, \ -H_c/2 \leq z \leq H_c/2, \end{cases} \tag{31}$$

Young's modulus:

$$E_{\text{foam}}(z) = \begin{cases} E_{\text{matrix}} \cdot \left(\dfrac{\beta_{\text{foam}} + 0.121}{1.121}\right)^{2.3}, & H_c/2 < z \leq H_B/2, \ -H_B/2 \leq z < -H_c/2, \\ 0.2032 \cdot E_{\text{matrix}} \cdot \beta_{\text{foam}}(\text{z}), & -H_c/2 \leq z \leq H_c/2, \end{cases} \tag{32}$$

Shear modulus:

$$G_{\text{foam}}(z) = \frac{E_{\text{foam}}(z)}{2[1 + v_{\text{foam}}(z)]}, \quad -H_B/2 \leq z \leq H_B/2, \tag{33}$$

Poisson's ratio:

$$v_{\text{foam}}(z) = v_{\text{matrix}} + \frac{3}{2}[1 - \beta_{\text{foam}}(z)]\frac{(1 - 5v_{\text{matrix}})(1 - v^2_{\text{maxtrix}})}{(7 - 5v_{\text{matrix}})}, \quad - H_B/2 \leq z \leq H_B/2, \tag{34}$$

Mass density:

$$\rho_{\text{foam}}(z) = \beta_{\text{foam}}(z) \cdot \rho_{\text{matrix}}, \quad -H_B/2 \leq z \leq H_B/2, \tag{35}$$

The modulus and mass density items in Eqs. (5), (6), and (9) can be further written as $E^*(z) = E_{\text{foam}}(z)/[1 - v^2_{\text{foam}}(z)]$, $G^*(z) = G_{\text{foam}}(z)$, $\rho^*(z) = \rho_{\text{foam}}(z)$. Therefore, the strain and kinetic energies in Eqs. (7)–(9) are reshaped in

$$\Phi_s = \frac{b}{2}\int_0^L [\Gamma_1(\partial u_1/\partial x)^2 + 2\Gamma_2(\partial u_1/\partial x)(\partial \varphi_1/\partial x) + \Gamma_3(\partial \varphi_1/\partial x)^2 + \Gamma_4\varphi_1^2 + 2\Gamma_4\varphi_1(\partial w_1/\partial x) + \Gamma_4(\partial w_1/\partial x)^2]dx, \tag{36}$$

$$\Phi_d = \frac{b}{2}\int_0^L [\Gamma_1(\partial u_1/\partial x + \partial u_2/\partial x)^2 + 2\Gamma_2(\partial u_1/\partial x + \partial u_2/\partial x)(\partial \varphi_1/\partial x + \partial \varphi_2/\partial x)$$
$$+ \Gamma_3(\partial \varphi_1/\partial x + \partial \varphi_2/\partial x)^2 + \Gamma_4(\varphi_1 + \varphi_2)^2 + 2\Gamma_4(\varphi_1 + \varphi_2)$$
$$\times (\partial w_1/\partial x + \partial w_2/\partial x) + \Gamma_4(\partial w_1/\partial x + \partial w_2/\partial x)^2]dx, \tag{37}$$

$$T_d = \frac{b}{2}\int_0^L [\Xi_0(\partial u_2/\partial t)^2 + 2\Xi_1(\partial u_2/\partial t)(\partial \varphi_2/\partial t) + \Xi_2(\partial \varphi_2/\partial t)^2$$
$$+ \Xi_0(\partial w_2/\partial t)^2]dx, \tag{38}$$

where

$$\{\Gamma_1, \Gamma_2, \Gamma_3, \Gamma_4\} = \int_{-H_B/2}^{H_B/2} \left[E^*(z), z \cdot E^*(z), z^2 \cdot E^*(z), \frac{5}{6}G^*(z)\right] dz, \tag{39}$$

$$\{\Xi_0, \Xi_1, \Xi_2\} = \int_{-H_B/2}^{H_B/2} [\rho^*(z), z \cdot \rho^*(z), z^2 \cdot \rho^*(z)]dz. \tag{40}$$

The porosity variations in the FG porous core are identified by ξ_t and ξ_d and connected to the afore-stated updated energy forms. By substituting these forms into the energy functionals of inclined beams (Eqs. (17) and (18)), the governing equation systems (Eqs. (22)–(24)) can be built to measure the corresponding linear performances under gravitational and moving loads, which are validated and discussed in the following section. During the calculation, dimensionless parameters are suggested with the aid of $A_{110} = H_B E_{\text{matrix}} 10^{-6}/(1 - v_{\text{matrix}}^2)$ and $I_{10} = H_B \rho_{\text{matrix}} 10^{-6}$,[12,13] but are not listed here for the purpose of simplification.

5. Results and Discussions

5.1. *Calculation subjects*

In this paper, we adopt closed-cell steel foams ($E_{\text{matrix}} = 210\,\text{GPa}$, $v_{\text{matrix}} = 1/3$, $\rho_{\text{matrix}} = 7850\,\text{kg/m}^3$) as the subject of calculation examples for demonstration and practical applicability.[40] Although the multiscale modeling in Ref. 22 is based on aluminum foams, Eqs. (25)–(28) are derived via elastic analysis and RVEs consisting of classical isotropic beam elements, which may be assigned with various matrix constants such as those from polymers[22] and other metals. Specifically, Eqs. (25), (27), and (28) reflect the RVE geometrical features and are not directly linked to the material parameters. Equation (26) is rather universally applicable as it is obtained from small-deformation examinations. In addition, 155 RVE cases on steel foams are computed with $2.5\,\text{mm} \le d_{\text{cell}} \le 5.5\,\text{mm}$ and $0.04 \le \beta_{\text{foam}} \le 0.13$ (please refer to Sec. 2.3.4 in Ref. 22 for more details), as shown in Fig. 4, where the effectiveness of Eq. (26) is confirmed.

In the following calculations, a base-case sandwich porous beam made of closed-cell steel foams is employed: $H_B = 0.1\,\text{m}$, $b = 0.1\,\text{m}$, $H_C/H_S = 8$. The impacts

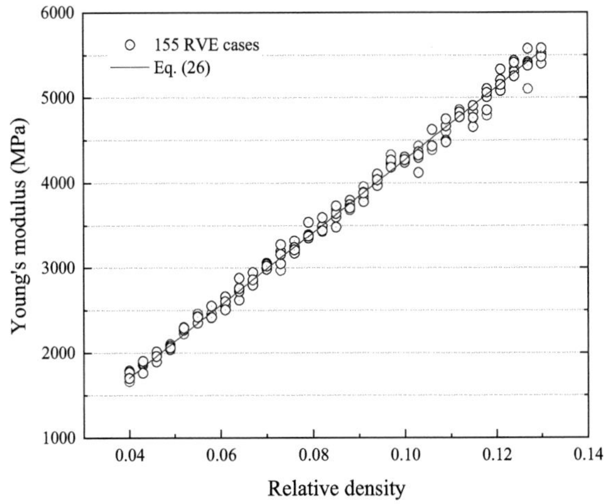

Fig. 4. Young's modulus of closed-cell steel foams: RVEs versus Eq. (26).

from changing parameters on this base-case beam are revealed and discussed, in terms of cell wall thickness (using ξ_t), average cell size (using ξ_d), inclined angle, boundary condition, slenderness ratio (L/H_B), foundation coefficients, etc. The aims of parametric studies are to assess the macroscopic structural performance and to disclose the direct influence from mesoscopic cellular features.

5.2. *Convergence and validation of formulations*

This section provides multiple examples to verify the constructed formulations. Table 1 lists the fundamental natural frequencies of sandwich porous beams, considering different values of N in the displacement trial functions (Eqs. (19)–(21)). It shows that a convergent outcome can be achieved via $N = 10$, which is also confirmed and adopted in the static bending and forced vibration analyses.

We then compared our results to FE simulations in ANSYS Workbench (version 2022 R1) with the aid of 3D solid elements (SOLID186). In the FE model, the entire sandwich beam is divided into 20 homogeneous layers, of which the properties are

Table 1. Convergence analysis on fundamental natural frequencies (Hz).

Boundary conditions	$N = 2$	$N = 4$	$N = 6$	$N = 8$	$N = 10$
H–H	41.761	37.678	37.667	37.667	37.667
C–H	76.814	58.365	58.310	58.310	58.310
C–C	215.534	83.874	83.670	83.669	83.669

Note: $\xi_t = 1$, $\xi_d = 1$, $L/H_B = 20$, without foundation and gravitational load.

Table 2. Validation analysis I on fundamental natural frequencies (Hz).

Boundary conditions	$\xi_t = 1,\ \xi_d = 1$		$\xi_t = 0.5,\ \xi_d = 0.5$		$\xi_t = 0,\ \xi_d = 0$	
			$\bar{K_l} = 0,\ \bar{K_s} = 0$			
	ANSYS	Present	ANSYS	Present	ANSYS	Present
H–H	38.155	37.667	35.796	35.239	32.597	32.021
C–H	58.713	58.310	55.312	54.602	50.533	49.676
C–C	83.113	83.669	78.634	78.437	72.086	71.465
			$\bar{K_l} = 1000,\ \bar{K_s} = 0$			
	ANSYS	Present	ANSYS	Present	ANSYS	Present
H–H	59.777	59.450	55.199	54.841	49.224	48.858
C–H	74.591	74.260	69.456	68.895	62.557	61.879
C–C	94.996	95.472	89.151	88.980	80.969	80.426

Note: $L/H_B = 20$, without gravitational load.

Table 3. Validation analysis II on maximum self-weight bending deflections (mm).

Inclined angles	$\xi_t = 1,\ \xi_d = 1$		$\xi_t = 0.5,\ \xi_d = 0.5$		$\xi_t = 0,\ \xi_d = 0$	
	ANSYS	Present	ANSYS	Present	ANSYS	Present
0°	0.0459	0.0460	0.0513	0.0524	0.0611	0.0632
30°	0.0404	0.0399	0.0452	0.0454	0.0538	0.0547
60°	0.0233	0.0230	0.0261	0.0262	0.0310	0.0316

Note: $L/H_B = 20$, C–C, without foundation.

aligned with the proposed material gradation by using $E^*(z)$, $G^*(z)$, $\rho^*(z)$, $v_{\text{foam}}(z)$ at the mid-layer positions. In this study, we mathematically describe the foundation by using the following dimensionless coefficients:

$$\begin{cases} \bar{K_l} = K_l L^2 / A_{110}, \\ \bar{K_s} = K_s / A_{110}. \end{cases} \tag{41}$$

$\bar{K_l} = 1000$, corresponding to Winkler foundation with $\bar{K_s} = 0$, is approximately realized under the Elastic Support in ANSYS Workbench (stiffness: $A_{110}\bar{K_l}/L^2$, placed on the bottom of beam models). As can be observed, a good match is attained between ANSYS and present results on both foundational natural frequencies in Table 2 and maximum self-weight midspan bending deflections in Table 3 with a variety in boundary conditions, cellular feature settings, foundation coefficients, and inclined angles. Note that the K_s-related verification is not carried out via ANSYS simulations, but is involved in the next validation case where our results are compared to those from open literature.

We also applied our formulations in the free and forced vibration examinations on already-studied horizontal FG beams, given by an existing paper,[41] which are consisting of three layers (nn is the power law index):

$$\begin{cases} VV_1(z) = [(z - h_0)/(h_1 - h_0)]^{nn} & \text{Layer 1 } (h_0 \leq z \leq h_1), \\ VV_2(z) = 1 & \text{Layer 2 } (h_1 \leq z \leq h_2), \\ VV_3(z) = [(z - h_3)/(h_2 - h_3)]^{nn} & \text{Layer 3 } (h_2 \leq z \leq h_3), \end{cases} \tag{42}$$

leading to the layerwise variations in Young's modulus and mass density:

$$\begin{cases} E_i(z) = (E_b - E_t)VV_i(z) + E_t, \\ \rho_i(z) = (\rho_b - \rho_t)VV_i(z) + \rho_t, \end{cases} \tag{43}$$

where $i(= 1, 2, 3)$ indicates the layer number, $E_c = 380\,\text{GPa}$, $\rho_c = 3960\,\text{kg/m}^3$, $E_m = 70\,\text{GPa}$, $\rho_m = 2702\,\text{kg/m}^3$ are employed to determine the beam properties considering either homogeneous hardcore or softcore:

$$\begin{cases} E_b = E_c,\ \rho_b = \rho_c,\ E_t = E_m,\ \rho_t = \rho_m & \text{homogeneous hardcore}, \\ E_b = E_m,\ \rho_b = \rho_m,\ E_t = E_c,\ \rho_t = \rho_c & \text{homogeneous softcore}. \end{cases} \tag{44}$$

Fixed parameters adopted here involve Poisson's ratio = 0.3, beam width = 0.5 m, beam thickness (h) = 1.0 m, nn = 0.5. Three types of layer configurations are presented by 1–0–1 beam ($h_0 = -h/2$, $h_1 = 0$, $h_2 = 0$, $h_3 = h/2$), 1–1–1 beam ($h_0 = -h/2$, $h_1 = -h/6$, $h_2 = h/6$, $h_3 = h/2$), and 2–2–1 beam ($h_0 = -h/2$, $h_1 = -h/10$, $h_2 = 3h/10$, $h_3 = h/2$). Table 4 compares their dimensionless fundamental natural frequencies, expressed as follows, subject to changing slenderness ratio L/h and foundation coefficients which are connected to Eq. (12) via $K_l = T_w D^*/L^4$, $K_s = T_s D^*/L^2$, $D^* = E_m h^3/12$. It is seen that our present results are clearly close to those from Ref. 41.

$$\text{Dimensionless frequency} = \text{Frequency} \cdot (L^2/h)\sqrt{\rho_m/E_m}. \tag{45}$$

In a further trial, let's look at the mid-span deflection histories of these FG beams under a constant moving force (500 kN, 50 m/s). Figure 5 highlights a close match between the published and present results, again verifying the correctness of the constructed formulations. Note that the dimensionless values[41] in this case are written as

$$\text{Dimensionless time} = \text{time}/0.2, \tag{46}$$

$$\text{Dimensionless deflection} = \frac{\text{Deflection}}{\begin{array}{c}\text{Maximum static deflection of aluminum beam}\\ \text{under mid-span point load}\end{array}}. \tag{47}$$

Table 4. Validation analysis III on dimensionless fundamental natural frequencies (Eq. (45)).

Slenderness ratio (L/h)	1–0–1 beam		1–1–1 beam		2–2–1 beam	
	Homogeneous hardcore ($T_w = 0$, $T_s = 0$)					
	Ref. 41	Present	Reference[41]	Present	Reference[41]	Present
5	8.2804	8.2234	8.6518	8.6153	8.7814	8.7481
10	9.3454	9.3246	9.8006	9.7872	9.9725	9.9602
20	9.6884	9.6825	10.1733	10.1695	10.3608	10.3573
	Homogeneous hardcore ($T_w = 100$, $T_s = 100$)					
	Ref. 41	Present	Reference[41]	Present	Reference[41]	Present
5	11.9423	11.8950	12.0896	12.0590	12.1589	12.1310
10	12.9277	12.9079	13.1497	13.1370	13.2562	13.2446
20	13.2713	13.2655	13.5191	13.5154	13.6399	13.6365
	Homogeneous softcore ($T_w = 0$, $T_s = 0$)					
	Ref. 41	Present	Reference[41]	Present	Reference[41]	Present
5	8.4906	8.4039	7.9529	7.8814	7.7094	7.6450
10	10.2146	10.1767	9.6791	9.6468	9.3451	9.3163
20	10.8471	10.8358	10.3277	10.3179	9.9548	9.9461
	Homogeneous softcore ($T_w = 100$, $T_s = 100$)					
	Ref. 41	Present	Reference[41]	Present	Reference[41]	Present
5	12.3998	12.3331	12.1734	12.1206	12.0477	12.0007
10	13.8952	13.8604	13.6362	13.6067	13.4317	13.4054
20	14.5069	14.4958	14.2630	14.2534	14.0246	14.0160

Note: C–C, $nn = 0.5$, without gravitational load.

5.3. *Parametric study on the self-weight static bending*

This section discusses the static bending deflections of sandwich porous beams, caused by the gravitational load which is calculated from the overall beam mass and changes according to their porosity variations. As illustrated in Fig. 3, "$\xi_t = 0$" stands for a uniform cell wall thickness ($t_{\text{wall}} = 0.15\,\text{mm}$) for porous core, and "$\xi_t = 1$" indicates an FG core where $t_{\text{wall}} = 0.15\,\text{mm}$ on the top and bottom surfaces while $t_{\text{wall}} = 0.1\,\text{mm}$ on the mid-height plane. The design of core cell size follows a similar pattern: "$\xi_d = 0$" and "$\xi_d = 1$" suggest constant ($d_{\text{cell}} = 2.5\,\text{mm}$) and graded ($d_{\text{cell}} = 2.5 \sim 5.5\,\text{mm}$) distributions, respectively. Increments in ξ_t and ξ_d represent the general transformation from uniform porosities to FG ones that reach lower cell wall thickness and larger cell size at $z = 0$. In the meantime, the relative density of skin layers remains the same (20%).

Figure 6 evaluates the effects of ξ_t, ξ_d, and inclined angles on the maximum self-weight static deflections. It is noticed in Fig. 6(a) that an increasing ξ_t reduces

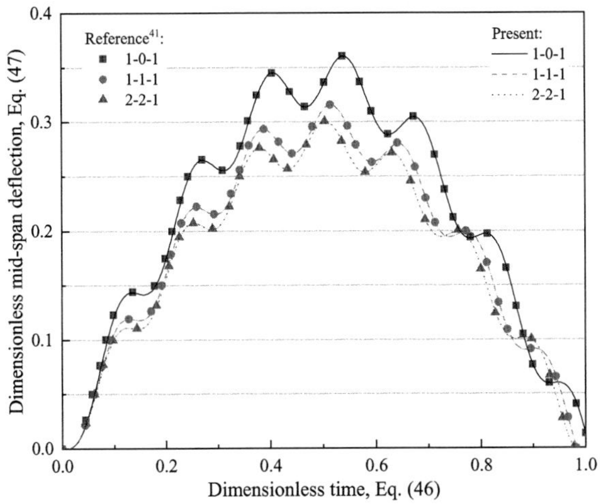

Fig. 5. Validation analysis IV on forced vibrations of FG beams (homogeneous hardcore, H–H, $L/h = 10$, $nn = 0.5$, without foundation).

the cell wall thickness around the core center and consequently the corresponding stiffness of sandwich porous beams, of which the deflection, however, tends to be smaller due to a decreased gravitational load. And the rising cell size ($\xi_d > 0$) on the FG core also results in lower bending deflections. The influence of ξ_d is rather more evident in the current setting, as evidenced by a deflection drop around 18–21% when ξ_d is raised from 0 to 1, compared to that of ξ_t (8–11%). Moreover, an FG porous beam graded at both cell wall thickness and size with $\xi_t = \xi_d = 1$ has a deflection of 0.0399 mm, 27.1% lower than that from its fully uniform counterpart

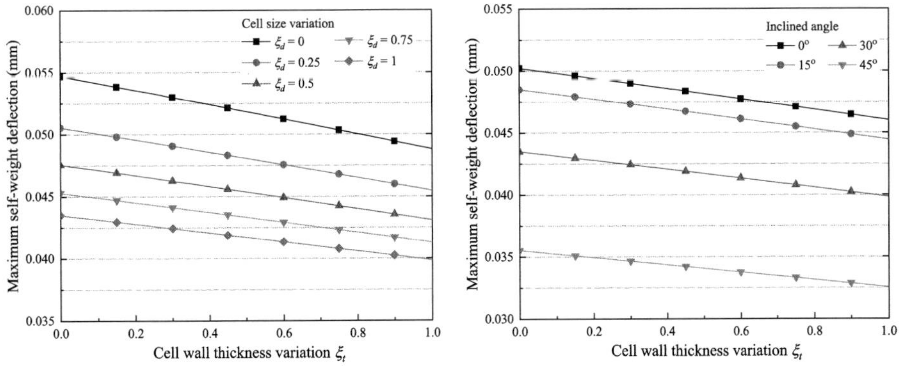

(a) Varying ξ_t and ξ_d (inclined angle 30°) (b) Varying ξ_t and inclined angle ($\xi_d = 1$)

Fig. 6. Maximum self-weight static deflections (mm, C–C, $L/H_B = 20$, without foundation).

Table 5. Maximum self-weight static deflections (mm).

Cell wall thickness variation ξ_t	$\bar{K}_l = 0,\ \bar{K}_s = 0$	$\bar{K}_l = 1000,\ \bar{K}_s = 0$	$\bar{K}_l = 1000,\ \bar{K}_s = 1000$
0	0.0435	0.0335	0.0091
0.25	0.0426	0.0328	0.0089
0.5	0.0417	0.0320	0.0087
0.75	0.0408	0.0313	0.0084
1	0.0399	0.0305	0.0082

Note: C–C, $\xi_d = 1$, inclined angle 30°, $L/H_B = 20$.

(0.0547 mm, $\xi_t = \xi_d = 0$). On the other hand, the maximum self-weight deflection reduces under an increase in the inclined angle, as shown in Fig. 6(b), related to the varying force components along the beam thickness and axial directions.

Table 5 further assesses the role of elastic Pasternak foundation on the beam static bending performance. As expected, the foundation significantly restrains the structural deformation. The inclusion of Winkler spring layer ($\bar{K}_l = 1000$, $\bar{K}_s = 0$) brings down the maximum deflection by roughly 23%, while the combination of Winkler spring and shear layers ($\bar{K}_l = \bar{K}_s = 1000$) is able to contribute to a total reduction of about 79%. These percentage numbers vary slightly according to ξ_t.

5.4. *Parametric study on the free vibration*

The beam free vibration is characterized by its natural frequencies, which are critical features for the structural design in engineering fields. Figure 7 depicts the fundamental natural frequency versus cell size variation curves, given different cell

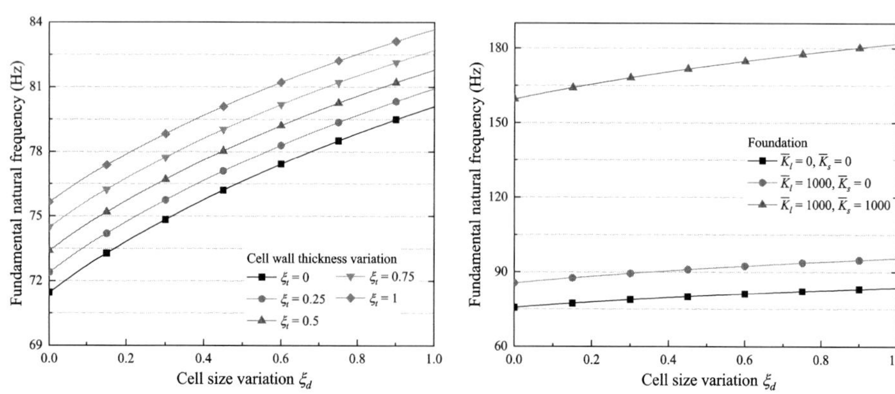

(a) Varying ξ_t and ξ_d (without foundation) (b) Varying ξ_d and foundation settings ($\xi_t = 1$)

Fig. 7. Fundamental natural frequencies (Hz, C–C, $L/H_B = 20$).

Table 6. Fundamental natural frequencies (Hz) with varying ξ_d and boundary conditions.

Cell size variation ξ_d	H–H	C–H	C–C
0	53.68	68.76	88.26
0.25	56.09	71.64	91.75
0.5	58.02	73.94	94.51
0.75	59.62	75.84	96.78
1	60.98	77.45	98.70

Note: $\xi_t = 1$, $L/H_B = 20$, $\bar{K}_l = 100$, $\bar{K}_s = 100$.

Table 7. Fundamental natural frequencies (Hz) with varying ξ_d and slenderness ratio.

Cell size variation ξ_d	$L/H_B = 20$	$L/H_B = 30$	$L/H_B = 40$
0	88.26	45.60	29.74
0.25	91.75	47.54	31.07
0.5	94.51	49.08	32.12
0.75	96.78	50.35	33.00
1	98.70	51.44	33.74

Note: C–C, $\xi_t = 1$, $\bar{K}_l = 100$, $\bar{K}_s = 100$.

wall thickness distributions (Fig. 7(a)) and foundation coefficient values (Fig. 7(b)). It is interesting to observe that although the porous beam obviously grows softer when ξ_d and ξ_t get higher, because of larger cell sizes and thinner cell walls on the core mid-height plane, a rising frequency is obtained instead based on the integrated changes in stiffness and mass matrices (see Eq. (24)). Furthermore, similar to the conclusions in Sec. 5.3, the effect of ξ_d is more evident than that from ξ_t, and the foundation stiffens the beam dramatically, especially when a shear layer is added.

Tables 6 and 7 are focused on the fundamental natural frequencies under different boundary conditions and slenderness ratios, in terms of graded cell sizes on the FG porous core, where a fixed elastic foundation is adopted with $\bar{K}_l = \bar{K}_s = 100$. It is found out that clamped ends raise the frequency which can be slashed by the beam length elongation. The cell size variation appears to be more impactful for less stiff beams with hinged supports and/or a higher slenderness ratio.

5.5. *Parametric study on the forced vibration*

5.5.1. *Constant moving force*

We then look at the forced vibration performance of self-weight inclined sandwich porous beams under a constant force $P = 5000\,\mathrm{N}$ moving at a fixed velocity of $2/3\,\mathrm{m/s}$ from the higher ($x = 0$) to lower ($x = L$) end (see Fig. 1). The adopted time step length is $0.003\,\mathrm{s}$ for Newmark method. Figure 8 discloses the mid-span

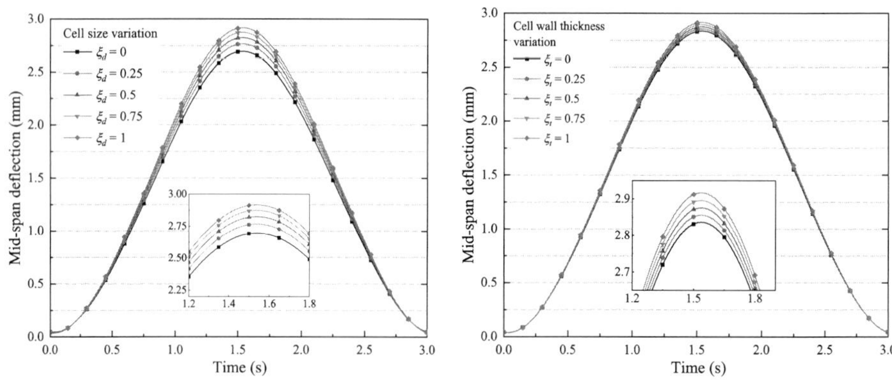

(a) Varying $\xi_d(\xi_t = 1)$ (b) Varying $\xi_t(\xi_d = 1)$

Fig. 8. Mid-span deflection histories under constant moving force (mm, C–C, inclined angle 30°, $L/H_B = 20$, without foundation).

deflection histories of an example beam, where Figs. 8(a) and 8(b) elaborate on the differences brought by variations in cell size and cell wall thickness, respectively. The initial deflection at $t = 0$ is identified via the self-weight static bending in Sec. 5.3. We may see that the deflection increases with a beam stiffness reduction induced from graded porosity distributions, and the growths of peak values are maximized at 8.2% and 2.8% when ξ_d and ξ_t are raised (from 0 to 1), indicating that the influence of cell wall thickness is rather minor in this example. Moreover, the joint effect of inclined angles and cellular geometrical features is also worth investigating. Figure 9 suggests that varying ξ_d and ξ_t affect the forced vibration behavior of target porous beam with changing inclined angle in a similar way, and its deflection decreases once inclined at a larger angle. The FG core with $\xi_t = \xi_d = 1$ gives rise to an increment of about 12.4% in the maximum mid-span vibration amplitude, compared with the case for $\xi_t = \xi_d = 0$ (uniform core). On the other hand, the foundation again has demonstrated its marked impact in adjusting the structural performance, as presented in Table 8. The outcome divergence turns out to be milder if a stiffer foundation is included: 12.4%, 9.5%, and 2.4% ($\xi_t = \xi_d = 0$ versus $\xi_t = \xi_d = 1$) correspond to "$\bar{K}_l = \bar{K}_s = 0$", "$\bar{K}_l = 1000$ & $\bar{K}_s = 0$", and "$\bar{K}_l = \bar{K}_s = 1000$", respectively.

5.5.2. *Harmonic moving force*

Finally, we apply a harmonic moving force $P = 2500[1+\sin(\omega_P t-\pi/2)]$ N (frequency ω_P, velocity V_P m/s) which shares the same peak value (5000 N) and direction with the constant force in Sec. 5.5.1. Figure 10 describes the mid-span forced vibration deflection histories with respect to the cellular variations, and Table 9 tabulates

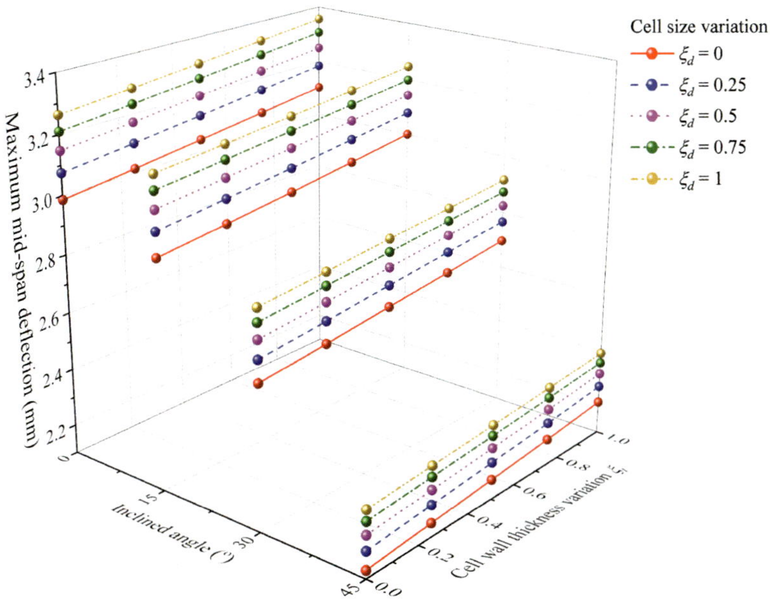

Fig. 9. Maximum mid-span deflections under constant moving force (mm, C–C, $L/H_B = 20$, without foundation).

Table 8. Maximum mid-span deflections under constant moving force (mm).

Cell wall thickness variation ξ_t	Cell size variation ξ_d				
	$\xi_d = 0$	$\xi_d = 0.25$	$\xi_d = 0.5$	$\xi_d = 0.75$	$\xi_d = 1$
	$\bar{K}_l = 0,\ \bar{K}_s = 0$				
0	2.5958	2.6725	2.7359	2.7899	2.8370
0.5	2.6445	2.7182	2.7792	2.8311	2.8762
1	2.6963	2.7666	2.8248	2.8743	2.9171
	$\bar{K}_l = 1000,\ \bar{K}_s = 0$				
0	2.0659	2.1134	2.1523	2.1853	2.2136
0.5	2.0961	2.1415	2.1787	2.2100	2.2371
1	2.1280	2.1709	2.2062	2.2359	2.2615
	$\bar{K}_l = 1000,\ \bar{K}_s = 1000$				
0	0.5808	0.5842	0.5870	0.5893	0.5913
0.5	0.5830	0.5862	0.5889	0.5911	0.5931
1	0.5852	0.5883	0.5908	0.5929	0.5948

Note: C–C, inclined angle 30°, $L/H_B = 20$.

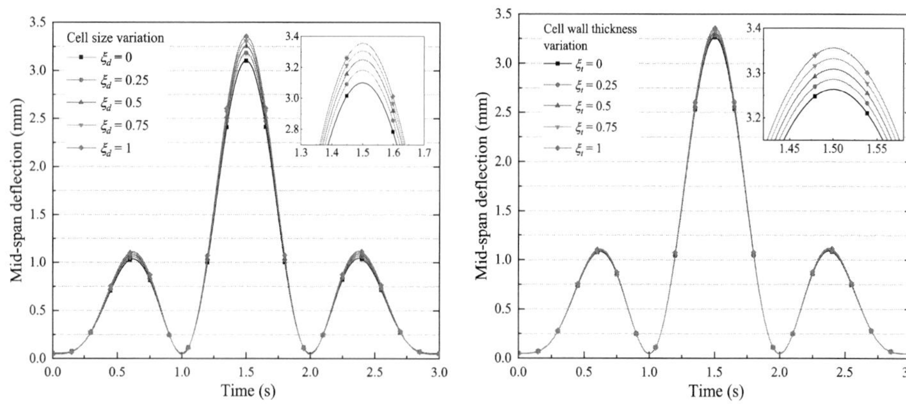

(a) Varying $\xi_d(\xi_t = 1)$

(b) Varying $\xi_t(\xi_d = 1)$

Fig. 10. Mid-span deflection histories under harmonic moving force (mm, $\omega_P = 2\pi$, $V_P = 2/3$, C–C, inclined angle 0°, $L/H_B = 20$, without foundation).

Table 9. Maximum mid-span deflections under harmonic moving force (mm).

Boundary conditions	Inclined angle (°)					
	0	30	0	30	0	30
	$\bar{K}_l = 0,\ \bar{K}_s = 0$		$\bar{K}_l = 1000,\ \bar{K}_s = 0$		$\bar{K}_l = 1000,\ \bar{K}_s = 1000$	
C–C	3.356	2.913	2.603	2.258	0.685	0.594
C–H	5.814	4.769	3.657	3.009	0.748	0.625
H–H	13.043	10.903	5.347	4.462	0.822	0.689

Note: $\omega_P = 2\pi$, $V_P = 2/3$, $L/H_B = 20$, $\xi_t = \xi_d = 1$.

the maximum values in terms of different boundary conditions, inclined angles, and foundation settings when $\xi_t = \xi_d = 1$. The conclusions are close to what obtained from the previous discussions, i.e. higher ξ_t and ξ_d lead to larger vibration amplitudes, which, in turn, may be restrained with increased inclined angle and foundation coefficients, as well as the fixed end supports, related to the changes in stiffness matrix and force vector.

It is also important to distinguish how the beam responses shift according to the frequency ω_P and velocity V_P of the applied harmonic moving force. Figure 11(a) reveals the force amplitude versus time curves for a constant load ($\omega_P = 0$) and four harmonic ones ($\omega_P = 0.5\pi$, π, 2π, 3π) featured by the equivalent peak value 5000 N and moving velocity 2/3 m/s. The corresponding mid-span deflection histories are described in Fig. 11(b), indicating similar maximum deformations (around 3.36 mm) from $\omega_P = 0$ and 2π, while those for $\omega_P = 0.5\pi$, π, 3π are 3.03, 2.75, 3.25 mm, respectively. The deflection under the constant force at any time is larger or equals to

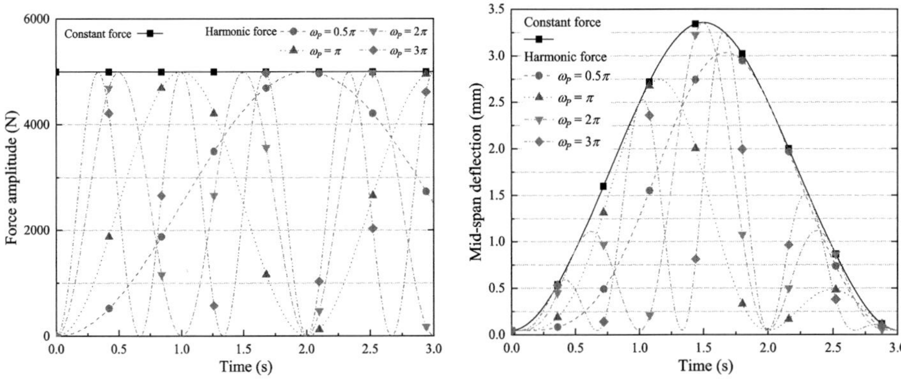

(a) Force amplitude histories

(b) Mid-span deflection histories

Fig. 11. Force amplitude and mid-span deflection histories under constant and harmonic moving force (mm, $V_P = 2/3$, C–C, $\xi_t = \xi_d = 1$, inclined angle 0°, $L/H_B = 20$, without foundation).

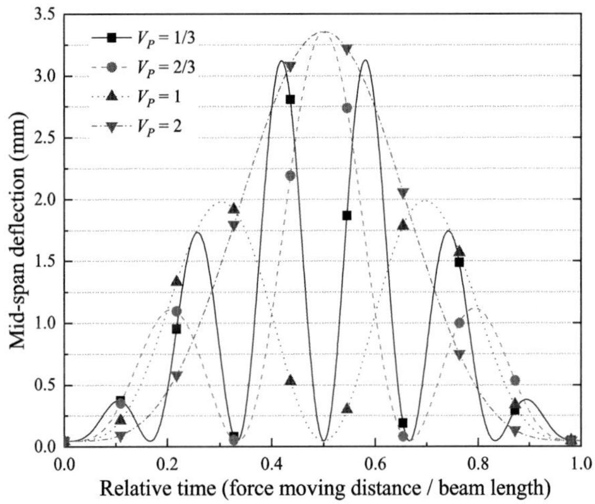

Fig. 12. Mid-span deflection histories under harmonic moving force with varying V_P (mm, $\omega_P = 2\pi$, C–C, $\xi_t = \xi_d = 1$, inclined angle 0°, $L/H_B = 20$, without foundation).

the counterparts induced by the adopted harmonic loads (limited to the mentioned specific frequencies). It is noted that ω_P used here is significantly lower than the fundamental natural frequency of the examined beam (see Sec. 5.4). Meanwhile, the effect of V_P (m/s) is assessed in Fig. 12, where a constant ω_P (2π) is employed. It is fascinating to find out that $V_P = 2/3$ and 2 m/s result in a close peak beam deformation (around 3.36 mm), while $V_P = 1/3$ m/s gives a slightly milder response (peak 3.13 mm) and the lowest maximum deflection (1.99 mm) comes from the case with $V_P = 1$ m/s. This is due to the fact that the beam/column vibration behavior heavily depends on the external excitation frequency and amplitude, as well as the load moving velocity, which has also been explored in existing works.[13,42,43]

6. Conclusions

To build up an explicit linkage between the mesoscopic and macroscopic aspects of porous structures, this paper combines the multiscale modeling and structural assessment in the examination on the bending and vibration performances of inclined sandwich beams containing FG metal foam cores, where the variations in non-uniform distributions of cell wall thickness and average cell size show certain levels of influence on the self-weight static bending responses, free vibration fundamental natural frequencies, and forced vibration deflection histories. The key findings are

(i) The reduction of cell wall thickness and the increment in cell size on the mid-thickness plane can lead to graded relative densities within the FG porous core, and decrease the maximum self-weight bending deformation with a smaller gravitational load.
(ii) Although the structural stiffness is degraded under thinner cell wall thickness and larger cell size, the beam foundational natural frequencies are found to be higher, due to the joint effects from stiffness and mass matrices.
(iii) The forced vibration deflections are closely associated to cell wall thickness and cell size distributions, as well as the moving force velocities and harmonic load frequencies.
(iv) The foundation evidently reduces the bending and vibration deformations and raises the beam natural frequencies. The impact from the shear layer is rather outstanding.
(v) The other examined parameters and settings also have considerable effects on the structural performances of the proposed beam form, including inclined angle, boundary condition, and slenderness ratio.

This paper successfully evaluates the macroscopic reactions of sandwich beams with respect to the mesoscopic cellular morphologies of their FG porous cores. The presented strategy will be useful to develop advanced porous structural components in a practical manner with the straightforward metal foam design guidance.

To further extend this study, 3D models will be established in our future research and link cellular characteristics with the variation of Poisson's ratio. They can also be used for both low- and high-density foams, while the current 2D models are only applicable to the low-density ones. Moreover, coupled mesoscopic and macroscopic analyses will be explored to include damages and large deformations by evaluating the behaviors of local cell walls during the overall beam bending and vibration assessments.

Acknowledgments

Chen acknowledges the financial support from the Australian Research Council (ARC DECRA DE220100876).

References

1. L. Jing and L. Zhao, Blast resistance and energy absorption of sandwich panels with layered gradient metallic foam cores, *J. Sandw. Struct. Mater.* **21**(2) (2017) 464–482.
2. B. V. S. Dinesh and A. Bhattacharya, Effect of foam geometry on heat absorption characteristics of PCM-metal foam composite thermal energy storage systems, *Int. J. Heat Mass Transf.* **134** (2019) 866–883.
3. R. Oriňaková *et al.*, Novel trends and recent progress on preparation methods of biodegradable metallic foams for biomedicine: A review, *J. Mater. Sci.* **56**(25) (2021) 13925–13963.
4. A. Bisht, V. K. Patel and B. Gangil, Future of metal foam materials in automotive industry, in *Automotive Tribology*, eds. J. K. Katiyar *et al.* (Springer, 2019), pp. 51–63.
5. L. Yuvaraj *et al.*, Investigation of light weight closed cell metallic foam for noise reduction in automobile application, *Mater. Today: Proc.* **54** (2022) 359–365.
6. S. H. Siddique *et al.*, Lessons from nature: 3D printed bio-inspired porous structures for impact energy absorption — a review, *Addit. Manufact.* **58** (2022) 103051.
7. G. K. Meenashisundaram *et al.*, Binder jetting additive manufacturing of high porosity 316L stainless steel metal foams, *Materials* **13**(17) (2020) 3744.
8. H. Wu, J. Yang and S. Kitipornchai, Mechanical analysis of functionally graded porous structures: A review, *Int. J. Struct. Stab. Dyn.* **20**(13) (2020) 2041015.
9. M. Grygorowicz and E. Magnucka-Blandzi, Mathematical modeling for dynamic stability of sandwich beam with variable mechanical properties of core, *Appl. Math. Mech.* **37**(10) (2016) 1361–1374.
10. Ş. D. Akbaş *et al.*, Dynamic analysis of viscoelastic functionally graded porous thick beams under pulse load, *Eng. Comput.* **38**(1) (2022) 365–377.
11. A. M. Zenkour, Quasi-3D refined theory for functionally graded porous plates: Displacements and stresses, *Phys. Mesomech.* **23**(1) (2020) 39–53.
12. D. Chen, J. Yang and S. Kitipornchai, Elastic buckling and static bending of shear deformable functionally graded porous beam, *Comp. Struct.* **133** (2015) 54–61.
13. D. Chen, J. Yang and S. Kitipornchai, Free and forced vibrations of shear deformable functionally graded porous beams, *Int. J. Mechan. Sci.* **108** (2016) 14–22.
14. D. Chen, J. Yang and S. Kitipornchai, Buckling and bending analyses of a novel functionally graded porous plate using Chebyshev–Ritz method, *Arch. Civ. Mech. Eng.* **19**(1) (2019) 157–170.
15. K. Gao *et al.*, Probabilistic stability analysis of functionally graded graphene reinforced porous beams, *Aerosp. Sci. Technol.* **98** (2020) 105738.
16. K. Gao *et al.*, Nonlinear free vibration of functionally graded graphene platelets reinforced porous nanocomposite plates resting on elastic foundation, *Comp. Struct.* **204** (2018) 831–846.
17. J. Yang, D. Chen and S. Kitipornchai, Buckling and free vibration analyses of functionally graded graphene reinforced porous nanocomposite plates based on Chebyshev–Ritz method, *Comp. Struct.* **193** (2018) 281–294.
18. S. Kitipornchai, D. Chen and J. Yang, Free vibration and elastic buckling of functionally graded porous beams reinforced by graphene platelets, *Mater. Des.* **116** (2017) 656–665.

19. D. Chen, S. Kitipornchai and J. Yang, Dynamic response and energy absorption of functionally graded porous structures, *Mater. Des.* **140** (2018) 473–487.
20. D. Chen *et al.*, AI enhanced finite element multiscale modeling and structural uncertainty analysis of a functionally graded porous beam, preprint (2022), arXiv: 2211.01970.
21. K. S. Verma *et al.*, Influences of cell size, cell wall thickness and cell circularity on the compressive responses of closed-cell aluminum foam and its FEA analysis, *Int. J. Met.* **16**(5) (2021) 1–16.
22. D. Chen *et al.*, Multiscale modeling of functionally graded porous beams: Buckling and vibration analyses, *Eng. Struct.* **266** (2022) 114568.
23. H. Ouyang, Moving-load dynamic problems: A tutorial (with a brief overview), *Mech. Syst. Signal Process.* **25**(6) (2011) 2039–2060.
24. A. Mamandi, M. H. Kargarnovin and D. Younesian, Nonlinear dynamics of an inclined beam subjected to a moving load, *Nonlinear Dyn.* **60**(3) (2010) 277–293.
25. D. Yang, C. Wang and J. Yau, Dynamic stability and response of inclined beams under moving mass and follower force, *Int. J. Struct. Stab. Dyn.* **2**(14) (2020) 2043004.
26. V. Shokouhifard *et al.*, Inverse dynamic analysis of an inclined FGM beam due to moving load for estimating the mass of moving load based on a CGM, *Iran. J. Sci. Technol. Trans. Mech. Eng.* **44**(3) (2020) 543–556.
27. S. R. Mohebpour, M. Vaghefi and M. Ezzati, Numerical analysis of an inclined cross-ply laminated composite beam subjected to moving mass with consideration the Coriolis and centrifugal forces, *Europ. J. Mech. — A/Solids* **59** (2016) 67–75.
28. P. L. Rosendahl and P. Weißgraeber, Modeling snow slab avalanches caused by weak-layer failure — Part 1: Slabs on compliant and collapsible weak layers, *Cryosphere* **14**(1) (2020) 115–130.
29. P. L. Rosendahl and P. Weißgraeber, Modeling snow slab avalanches caused by weak-layer failure — Part 2: Coupled mixed-mode criterion for skier-triggered anticracks, *Cryosphere* **14**(1) (2020) 131–145.
30. P. Weißgraeber and P. L. Rosendahl, A closed-form model for layered snow slabs, *Cryosphere Discuss.* **2022** (2022) 1–37.
31. Z. Celep, K. Güler and F. Demir, Response of a completely free beam on a tensionless Pasternak foundation subjected to dynamic load, *Struct. Eng. Mech.* **37**(1) (2011) 61.
32. Ş. D. Akbaş, Static analysis of a fiber reinforced composite beam resting on Winkler–Pasternak Foundation, *Int. J. Eng. Appl. Sci.* **12**(3) (2020) 88–98.
33. M. T. A. Robinson and S. Adali, Buckling of non-uniform and axially functionally graded non-local Timoshenko nanobeams on Winkler–Pasternak foundation, *Comp. Struct.* **206** (2018) 95–103.
34. S. Kumar, Vibration analysis of Non-uniform axially functionally graded beam resting on Pasternak foundation, *Mater. Today: Proc.* **62** (2022) 619–623.
35. D. Chen *et al.*, Impact response of inclined self-weighted functionally graded porous beams reinforced by graphene platelets, *Thin-Walled Struct.* **179** (2022) 109501.
36. A. P. Roberts and E. J. Garboczi, Elastic moduli of model random three-dimensional closed-cell cellular solids, *Acta Mater.* **49**(2) (2001) 189–197.
37. M. F. Ashby *et al.*, *Metal Foams: A Design Guide* (Butterworth-Heinemann, Boston, 2000).
38. J. Kováčik, L. Marsavina and E. Linul, Poisson's ratio of closed-cell aluminium foams, *Materials* **11**(10) (2018) 1904.

39. E. Gregorová *et al.*, Processing, microstructure and elastic properties of mullite-based ceramic foams prepared by direct foaming with wheat flour, *J. Europ. Ceram. Soc.* **36**(1) (2016) 109–120.
40. B. Smith *et al.*, Steel foam for structures: A review of applications, manufacturing and material properties, *J. Construct. Steel Res.* **71** (2012) 1–10.
41. W. Songsuwan, M. Pimsarn and N. Wattanasakulpong, Dynamic responses of functionally graded sandwich beams resting on elastic foundation under harmonic moving loads, *Int. J. Struct. Stab. Dyn.* **18**(9) (2018) 1850112.
42. N. T. Khiem and P. T. Hang, Frequency response of a beam-like structure to moving harmonic forces, *Vietnam J. Mech.* **38**(4) (2016) 223–238.
43. D. Chen *et al.*, Vibration absorber for spring-mass system using a hanging heavy column with rotationally restrained end, *J. Eng. Mech.* **146**(4) (2020) 06020001.

https://doi.org/10.1142/9789819814770_0014

Chapter 14

Nonlinear Stability of Three-Layer Circular Shallow Arches with Elastic Interlayer Bonding#

Christoph Adam*, Ivan Paulmichl† and Thomas Furtmüller‡

Universität Innsbruck
Unit of Applied Mechanics, Technikerstr. 13,
6020 Innsbruck, Austria
**christoph.adam@uibk.ac.at*
†ivan.paulmichl@uibk.ac.at
‡thomas.furtmueller@uibk.ac.at

In this paper, stability-prone circular shallow arches composed of three symmetrically arranged flexibly bonded layers with fixed and hinged supports at both ends are examined. Based on the differential equations of equilibrium and a series expansion of the governing kinematic variables, analytical expressions for the limit points and bifurcation points are derived. Solutions for the nonlinear equilibrium path are also provided. Comparison with the results of much more complex numerical analyses with 2D finite continuum elements show high accuracy of these analytical expressions. The application examples indicate the importance of considering the flexibility of the interlayers in the stability analysis. With the assumption of a rigid bond between the layers, the stability limit is overestimated by up to 100% in the examples considered.

Keywords: Instability; interlayer slip; limit-point buckling; bifurcation buckling.

1. Introduction

When the compressive force in a structural member such as a shallow arch exceeds a certain threshold, the member becomes unstable, it significantly changes its geometry and may lose its ability to support the applied loads. This undesirable structural behavior has been recognized for a long time, and accordingly, numerous analytical, semi-analytical, and numerical theories have been developed in the last decades for predicting the stability limits and the nonlinear pre- and post-equilibrium paths of stability-prone structures. Yang and co-workers are among those who have made and

*Corresponding author.
#This chapter appeared previously on the International Journal of Structural Stability and Dynamics. To cite this chapter, please cite the original article as the following: C. Adam, I. Paulmichl & T. Furtmüller, *Int. J. Struct. Stab. Dyn.*, **23**, 2340016 (2023), DOI: 10.1142/S0219455423400163.

continued to make significant contributions to various problems in structural stability.[1–10] In Ref. 1, Yang and Kuo studied the effect of curvature on the stability of curved beams. During the same period, they derived analytical expressions for out-of-plane buckling of angled frames subjected to various types of moment loads.[3,4] Lateral buckling of cantilevered circular arches under various end moments has been examined in Ref. 9. Yang[6] showed that the out-of-plane buckling of such frame structures and the buckling of curved beams[5] can be resolved by the same suitable general straight-beam elements. An extension of the straight-beam approach for analyzing lateral buckling of curved beams including warping effects was recently presented by Yang *et al.*[11] Yang and Kuo[7] improved the tracing of post-buckling paths of structures with multiple loops. In another paper,[2] Yang and Shieh proposed the generalized displacement control method to numerically solve structural instability problems with multiple limit points and snap-back points. Recently, building on an earlier review paper by Yang *et al.*[10] Yang *et al.*[8] provided a comprehensive insight into the state-of-the-art research on nonlinear, post-buckling, and elasto-plastic analyses of framed structures and curved beams.

Shallow arches may exhibit unstable behavior under transverse loading. In the case of in-plane instability in the form of limit-point buckling (snap-through), the structure jumps to a remote equilibrium position. In the bifurcation buckling mode, by contrast, at least two equilibrium paths intersect at the so-called bifurcation point. In a series of papers, Bradford and co-workers[12–14] investigated shallow arches of different shapes (such as circular and parabolic axis) with different boundary conditions subjected to distributed and concentrated loads. Zhou *et al.*[15] studied the effect of arbitrary asymmetric geometric imperfections on their nonlinear instability for shallow arches. The instability behavior of elastically supported shallow arches may become even more complex, as reported in more recent studies (e.g. Refs. 16–18]). In most papers, homogeneous arches are considered, while snap-through and in-plane bifurcation of inhomogeneous and layered curved structures have been analyzed less frequently (e.g. Refs. 19 and 20). As an example, two papers by Kiss[21,22] are mentioned, in which shallow arches made of functionally graded material (FGM) are examined. As long as the layers of multilayer slender beam-like structures are rigidly connected, the kinematic hypothesis of Euler–Bernoulli theory can be used for the formulation of governing equations. The resulting equations differ from those of the homogeneous structure only by effective structural parameters (e.g. effective bending stiffness).

When layers are flexibly bonded (i.e. partial layer interaction), then under load a relative displacement occurs at the interfaces, referred in the literature to as interlayer slip. In such a case, the kinematic hypotheses of the Euler–Bernoulli theory can only be assumed to be valid separately for each layer. The governing equations are then much more complex than those of a homogeneous slender structure. While the static response and dynamic response of linear and nonlinear straight and precurved multilayered beams with interlayer slip have been extensively addressed in the literature,[23–27] research on the stability of shallow arches with flexible bond was nonexistent until recently. In a newly published paper,[28] the stability of two-layer half-sine shallow arches with interlayer slip was considered. This involved an investigation of the effect of crown height, boundary conditions, and loading on the limit loads and the pre- and post-buckling equilibrium branches. In another recent paper,[29] analytical expressions for the limit points, bifurcation points, and load–displacement paths were derived for symmetrically layered three-layer half-sine shallow arches with flexible bond composite subjected to half-sine loads.

In the present contribution, for the first time, the instability behavior of circular shallow arches composed of three symmetrically arranged, elastically bonded layers and immovably and soft-hinged supported at both ends is explored. Building on the differential equations of equilibrium and boundary conditions in terms of the governing kinematic variables, analytical expressions are derived to determine the limit points, bifurcation points, and pre- and post-buckling paths of equilibrium of these structural members. This is achieved by an infinite series expansion of the deflection and the application of Galerkin's method. The comparison of the results obtained in several application examples with the outcomes of Finite Element (FE) analyses with plane stress continuum elements show the great accuracy of the presented theory.

2. Governing Equations

The considered shallow arch shown in Fig. 1 with a circular member axis is composed of three layers that are elastically bonded in the circumferential direction. The layered structure is symmetrical, i.e. the two face layers have the same geometry and are made of the same material. Therefore, the member axis coincides with the neutral axis of the middle layer. In this section, the layers are denoted from top to bottom with the subscripts $i = 1, 2, 3$. The arch is fixed at both ends and linked to the surrounding by a hinge in the member axis. The x-axis with the origin in the left support follows the member axis and is therefore curvilinear, y is the out-of-plane coordinate and the radial in-plane coordinate is denoted by z. Furthermore, a local (η_i, ζ_i) coordinate system parallel to y and z coordinates is located in the neutral axis of each of the three layers, see Fig. 1. In the following, the in-plane instability of this component under the radial load per unit length

$$p(x) = p_0 f(x) \tag{2.1}$$

is investigated, where p_0 is the load amplitude and $f(x)$ is the distribution function of the load normalized to the maximum 1.

Instead of the radius of curvature R, which is much larger than the member length l for shallow arches, its reciprocal $k = R^{-1}$ is used in the following to describe the initial member curvature, which is in the case of circular members independent of

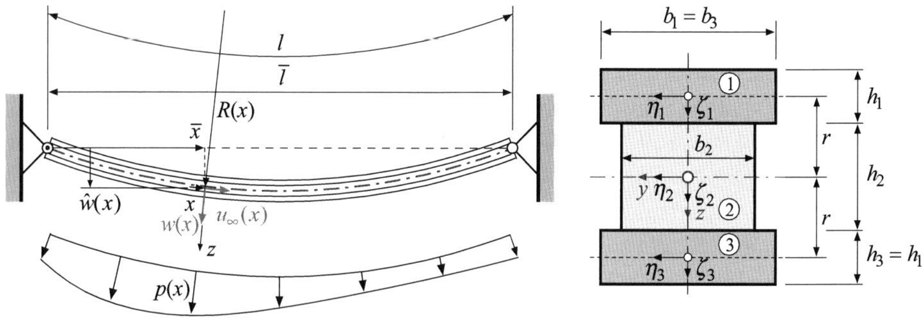

Fig. 1. Immovably soft-hinged supported three-layer circular shallow arch with partial layer interaction subjected to the radial load $p(x)$, modified from Ref. 29.

x, i.e. k = const. The radial and tangential displacement components of the member axis $w(x)$ and $u_\infty(x)$ are much smaller than the radius of curvature, i.e. $|w| \ll R$ and $|u_\infty| \ll R$. Due to the elastic bonding of the layers in the circumferential direction, the layers are displaced relative to each other in the circumferential direction. This relative displacement, referred to as interlayer slip, between the top layer and the middle layer is denoted by $\Delta u_1(x)$ and between the middle layer and the bottom layer by $\Delta u_2(x)$. In the radial direction, a rigid bond is assumed such that the radial displacement of each fiber can be assumed to be equal to $w(x)$.

The interlayer slips yield a discontinuous displacement field in tangential direction. Therefore, the kinematic hypothesis of Euler–Bernoulli beams is applied for each layer separately. The cross-sectional rotation $\chi(x)$ of each layer is however the same, and since the shallow arch is immovably supported at both ends, it can be approximated by[29,30] $\chi(x) \approx -w_{,x}$, i.e. the contribution from the axial deformation is negligible. Here and in the following $(.)_{,x}$ denotes the differentiation of (.) with respect to x. The discontinuous tangential displacement field of the symmetrically layered shallow arches can thus be described layerwise as follows[29]:

$$u_i(x, \zeta_i) \approx u_i^{(0)} - \zeta_i w_{,x}, \quad i = 1, 2, 3. \tag{2.2}$$

Here, $u_i^{(0)}$ is the tangential displacement of the neutral axis of the i-th layer (see Fig. 2 and Ref. 29)

$$u_1^{(0)} = u_\infty + w_{,x} r - \Delta u_1, \quad u_2^{(0)} = u_\infty, \quad u_3^{(0)} = u_\infty - w_{,x} r + \Delta u_2 \tag{2.3}$$

with r as the distance between the member axis and the neutral axis of the top/bottom layer.

Stability problems of shallow arches are inherent to moderately large displacements, therefore the longitudinal strains must be formulated nonlinearly. That is, the axial strains at the centroid of the ith layer are composed of the contributions

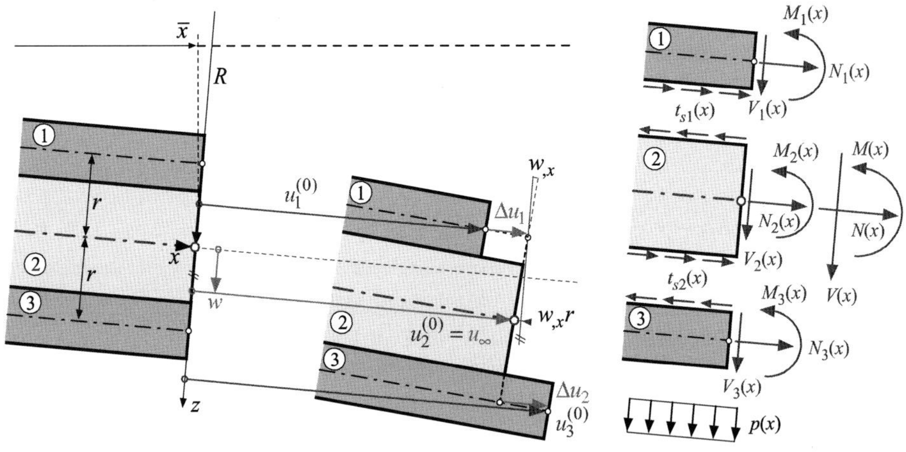

Fig. 2. (a) Cross-section at x in its initial and deformed state (modified from Ref. 29). (b) Internal forces.

due to the longitudinal displacement $u_{i,x}^{(0)}$ and the curvature of the member axis wk as well as the nonlinear part $\frac{1}{2}w_{,x}^2$ [28]

$$e_i(x) \approx u_{i,x}^{(0)} + wk + \frac{1}{2}w_{,x}^2, \quad i = 1, 2, 3. \tag{2.4}$$

For the longitudinal strain at the distance ζ_i from the ith centroid, the bending strain $\epsilon_{bi} \approx -\zeta_i w_{,xx}$ must be added

$$\epsilon_i(x, \zeta_i) = e_i - \zeta_i w_{,xx}, \quad i = 1, 2, 3. \tag{2.5}$$

Assuming linear-elastic material behavior, the longitudinal normal stresses $\sigma_i(x, \zeta_i)$ in the ith layer are then obtained by multiplying $\epsilon_i(x, \zeta_i)$ with the Young's modulus of the ith layer E_i, i.e. $\sigma_i(x, \zeta_i) = E_i\epsilon_i(x, \zeta_i)$. Layerwise integration of the normal stresses over the cross-sectional area A_i leads to the normal force in the ith layer[29]

$$N_i(x) = E_i A_i e_{i,x}, \quad i = 1, 2, 3, \tag{2.6}$$

with the extensional stiffness of the ith layer $E_i A_i$ and $E_3 A_3 = E_1 A_1$. The layerwise bending moments are obtained by integrating the normal stresses multiplied by ζ_i over A_i,[31]

$$M_i(x) = -E_i J_i w_{,xx}, \quad i = 1, 2, 3, \tag{2.7}$$

where $E_3 J_3 = E_1 J_1$, and J_i is the area moment of inertia about the η_i axis of the ith layer. Statically equivalent reduction of the layerwise axial forces N_i and bending moments M_i into the beam axis at x results in the overall normal force N and the overall bending moment $M(x)$,[29]

$$N = \sum_{i=1}^{3} N_i = EA_e \left(u_{\infty,x} + wk + \frac{1}{2} w_{,x}^2 \right) - E_1 A_1 (\Delta u_{1,x} - \Delta u_{2,x}), \quad (2.8)$$

$$M(x) = \sum_{i=1}^{3} M_i - (N_1 - N_3) r = -EJ_\infty w_{,xx} + E_1 A_1 r (\Delta u_{1,x} + \Delta u_{2,x}). \quad (2.9)$$

$EA_e = 2E_1A_1 + E_2A_2$ is the overall extensional stiffness, $EJ_\infty = EJ_0 + 2r^2 E_1 A_1$ the overall bending stiffness of the member with rigidly bonded layers and $EJ_0 = 2E_1J_1 + E_2J_2$ the bending stiffness of the member without any bond between the layers.

The shear tractions $t_{s1}(x)$ and $t_{s2}(x)$ in the upper interlayer and lower interlayer are related to the corresponding interlayer slip via the slip modulus K_s, which is the same in both interlayers, as follows[23]:

$$t_{s1}(x) = K_s \Delta u_1, \quad t_{s2}(x) = K_s \Delta u_2. \quad (2.10)$$

Since the radial stiffness of the interlayer is assumed to be infinite, the peel stress has no effect on the response and is therefore not considered.

2.1. *Differential equations of equilibrium and boundary conditions*

The differential equations of equilibrium and boundary conditions are most efficiently found by the principle of minimum total potential energy,[31] which states that the variation of the total potential energy is zero, i.e. $\delta\Pi = 0$. Application of this principle to the potential energy of the shallow arch and the load $p(x)$[29]

$$\Pi = \frac{1}{2} \sum_{i=1}^{3} E_i \int_{V_i} \epsilon_i^2 \, dV + \frac{1}{2} K_s \int_l ({\Delta u_1}^2 + {\Delta u_2}^2) dx - \int_l pw \, dx \quad (2.11)$$

and with the steps described in Ref. 29, the following differential equations of equilibrium expressed in terms of the internal forces are obtained[29]

$$N_{,x} = 0, \quad N_{1,x} + t_{s1} = 0, \quad -N_{3,x} + t_{s2} = 0, \quad -M_{,xx} - N w_{,xx} + Nk - p = 0. \quad (2.12)$$

This principle also provides the boundary conditions. For the present case of a hinged supported member in which the interlayer slip is not constrained at the ends (referred to as soft-hinged support) and fixed in all directions, the boundary conditions read[29]

$$u_\infty(x_r) = 0, \quad N_1(x_r) = 0, \quad N_3(x_r) = 0, \quad w(x_r) = 0, \quad M(x_r) = 0 \quad (2.13)$$

with $x_r = 0, l$. According to the first of Eqs. (2.12), the overall normal force N is constant over x. Therefore, and considering the boundary conditions $u_\infty(x_r) = 0$,

Eq. (2.8) can be advantageously replaced by the following integral relation:[29]

$$N = \frac{EA_e}{l}\int_0^l \left(wk + \frac{1}{2}w_{,x}^2\right)dx - \frac{E_1A_1}{l}\int_0^l (\Delta u_{1,x} - \Delta u_{2,x})dx. \tag{2.14}$$

Substituting the internal forces defined above into Eq. (2.12), recasting the resulting equations as described in Ref. 29 and taking into account that in the circular shallow arch $k_{,x} = 0$, leads to the following set of differential equations of equilibrium, expressed in terms of the four governing kinematic variables $w, u_\infty, \Delta u_1, \Delta u_2$,

$$\Delta u_{1,xx} + \Delta u_{2,xx} - \frac{K_s}{E_1A_1}(\Delta u_1 + \Delta u_2) - 2w_{,xxx}r = 0, \tag{2.15}$$

$$\Delta u_{1,xx} - \Delta u_{2,xx} - \kappa^2(\Delta u_1 - \Delta u_2) = 0, \tag{2.16}$$

$$u_{\infty,xx} + w_{,x}k + w_{,x}w_{,xx} - \frac{K_s}{E_2A_2}(\Delta u_1 - \Delta u_2) = 0, \tag{2.17}$$

$$EJ_\infty w_{,xxxx} - E_1A_1r(\Delta u_{1,xxx} + \Delta u_{2,xxx}) - N(w_{,xx} - k) = p \tag{2.18}$$

with

$$\kappa = \left(\frac{K_sEA_e}{E_1A_1E_2A_2}\right)^{1/2}. \tag{2.19}$$

The same procedure applied to the boundary conditions equation (2.13) leads to[29]

$$w(x_r) = 0, \quad w_{,xx}(x_r) = 0, \tag{2.20}$$

$$\Delta u_{1,x}(x_r) + \Delta u_{2,x}(x_r) = 0, \tag{2.21}$$

$$u_\infty(x_r) = 0, \tag{2.22}$$

$$2\left(u_{\infty,x}(x_r) + w(x_r)k(x_r) + \frac{1}{2}w_{,x}^2(x_r)\right) - (\Delta u_{1,x}(x_r) - \Delta u_{2,x}(x_r)) = 0. \tag{2.23}$$

In the following analyses, it is assumed that the circular arch is curved upwards (i.e. against the positive z-direction). The curvature k, which is constant for the circular arch, is related to the crown height $\hat{w}_0$ as follows:[32]

$$k = -\frac{8\hat{w}_0}{4\hat{w}_0^2 + l^2}. \tag{2.24}$$

3. Series Expansion of the Response and the Equilibrium Equations

To solve the boundary value problem described by Eqs. (2.15)–(2.18) and Eqs. (2.20)–(2.23), the deflection is expanded into an infinite series of the weighted shape functions $\phi_i(x)$

$$w(x) = \sum_{i=1}^{\infty} q_i \phi_i(x), \tag{3.1}$$

which represent the buckling modes of the corresponding straight beam with interlayer slip. In the present case, where the member is soft-hinged supported at both ends, these are the sine half-wave and a multiple thereof[23]

$$\phi_i(x) = \sin(\lambda_i x), \quad \lambda_i = \frac{i\pi}{l}, \quad i = 1, \ldots, \infty. \tag{3.2}$$

Thus, this series expansion satisfies the boundary conditions defined in Eq. (2.20) in w and $w_{,xx}$. Note that from now on (from Eq. (3.1)) the subscript i denotes the summations index in the series expansion of the response variables.

For shallow arches with other boundary conditions, the buckling modes of the corresponding straight beam are also advantageously used as shape functions, which, however, generally represent more complex trigonometric functions. As an alternative, orthogonal shape functions determined according to the Gram–Schmidt[33] process can be used, or the shape functions given in Quatu,[34] as for example it was done in Ref. 28 for two-layer shallow arches with interlayer slip. The disadvantage is, however, that determining the buckling loads becomes more complex.

The three other kinematic variables Δu_1, Δu_2 and u_∞ must also be expressed as a function of the unknown generalized coordinates q_i.[25] To this end, in the differential equilibrium equations (2.15)–(2.17) as well as in the boundary condition (2.23) the derivatives of the deflection w are replaced by the series representation equation (3.1) and treated as known quantities. Then, these equilibrium equations together with the boundary conditions, Eqs (2.21)–(2.23), are solved analytically for Δu_1, Δu_2 and u_∞. In the present problem, this was done with the software package Mathematica.[35] The results are the following nonlinear series:

$$\Delta u_1(x) = \sum_{i=1}^{\infty} q_i \left(\frac{r\lambda_i^3 \cos(\lambda_i x)}{\lambda_i^2 + \frac{K_s}{E_1 A_1}} - \frac{\beta}{2}\left(\frac{1}{2}\lambda_i^2 q_i - \chi_i\right)\right), \tag{3.3}$$

$$\Delta u_2(x) = \sum_{i=1}^{\infty} q_i \left(\frac{r\lambda_i^3 \cos(\lambda_i x)}{\lambda_i^2 + \frac{K_s}{E_1 A_1}} + \frac{\beta}{2}\left(\frac{1}{2}\lambda_i^2 q_i - \chi_i\right)\right), \tag{3.4}$$

$$\begin{aligned} u_\infty(x) = &-\frac{1}{4}\sum_{i=1}^{\infty}\sum_{j=1}^{\infty}\left(\frac{1}{\lambda_{j-i} + \delta_{ji}}\sin(\lambda_{j-i}x) + \frac{1}{\lambda_{i+j}}\sin(\lambda_{i+j}x)\right)\lambda_i\lambda_j q_i q_j \\ &+ \sum_{i=1}^{\infty}\frac{q_i}{\lambda_i}\left(\frac{\theta}{4}\lambda_i^3 q_i - \frac{8\hat{w}_0}{4\hat{w}_0^2 + l^2}\left(\cos(\lambda_i x) - \frac{1}{2}(1-(-1)^i)\gamma \right.\right. \\ &\left.\left. - \frac{1}{2}(1+(-1)^i)\right)\right), \end{aligned} \tag{3.5}$$

where δ_{ji} is the Kronecker delta and the coefficients

$$\chi_i = \frac{2(1-(-1)^i)}{\lambda_i l} \frac{8\hat{w}_0}{4\hat{w}_0^2 + l^2}, \quad i = 1, \ldots, \infty, \tag{3.6}$$

capture the effect of the initial curvature of the shallow arch on the response. The functions depending on x

$$\beta(x) = \frac{EA_e l \sinh\left(\frac{1}{2}\kappa(l-2x)\right)}{4E_1A_1 \sinh\left(\frac{\kappa l}{2}\right) + E_2A_2\kappa l \cosh\left(\frac{\kappa l}{2}\right)}, \tag{3.7}$$

$$\theta(x) = \frac{2E_1A_1\left((l-2x)\sinh\left(\frac{\kappa l}{2}\right) - l\sinh\left(\frac{1}{2}\kappa(l-2x)\right)\right)}{4E_1A_1 \sinh\left(\frac{\kappa l}{2}\right) + E_2A_2\kappa l \cosh\left(\frac{\kappa l}{2}\right)}, \tag{3.8}$$

$$\gamma(x) = \frac{4E_1A_1 \sinh\left(\frac{1}{2}\kappa(l-2x)\right) + E_2A_2\kappa(l-2x)\cosh\left(\frac{\kappa l}{2}\right)}{4E_1A_1 \sinh\left(\frac{\kappa l}{2}\right) + E_2A_2\kappa l \cosh\left(\frac{\kappa l}{2}\right)} \tag{3.9}$$

have been introduced as abbreviations.

To express the total normal force N also as a function of the unknown generalized coordinates q_i, the series expansions of the kinematic variables w, Δu_1 and Δu_2, Eqs (3.1), (3.3) and (3.4), are differentiated with respect to x and inserted into the integral form for N, Eq. (2.14). Integration over the length l leads to

$$N = \frac{\psi}{4}\sum_{i=1}^{\infty} q_i(\lambda_i^2 q_i - 2\chi_i),$$
$$\psi = \frac{EA_e E_2A_2\kappa l \cosh\left(\frac{\kappa l}{2}\right)}{4E_1A_1 \sinh\left(\frac{\kappa l}{2}\right) + E_2A_2\kappa l \cosh\left(\frac{\kappa l}{2}\right)}. \tag{3.10}$$

The nonlinear coupled equations for the generalized coordinates q_i are obtained from the remaining fourth differential equilibrium equation (2.18) by applying the Galerkin method. According to the Galerkin method,[31] Eq. (2.18) is multiplied by the ith shape function ϕ_i and this expression is integrated over the length l,

$$\int_0^l (EJ_\infty w_{,xxxx} - E_1A_1 r(\Delta u_{1,xxx} + \Delta u_{2,xxx})$$
$$- N(w_{,xx} - k) - p)\phi_i(x)dx = 0, \quad i = 1, \ldots, \infty. \tag{3.11}$$

Substituting the above series representations of the kinematic variables (or their derivatives with respect to x) into Eq. (3.11) and performing the integration for each of the shape functions finally leads to the following equations,

$$h_i = (\lambda_i^2 q_i - \chi_i)N + c_i q_i - \frac{2}{l}p_0 f_i = 0, \quad 1, \ldots, \infty, \tag{3.12}$$

with the series representation of the overall axial force N according to Eq. (3.10), and the generalized stiffness coefficient c_i of the structure associated with the ith shape function and the ith load distribution coefficient f_i

$$c_i = \lambda_i^4(\lambda_i^2 + \alpha^2)\left(\frac{\alpha^2}{EJ_\infty} + \frac{\lambda_i^2}{EJ_0}\right)^{-1}, \quad \alpha^2 = \frac{K_s EJ_\infty}{E_1 A_1 EJ_0} \tag{3.13}$$

$$f_i = \int_0^l f(x)\phi_i(x)dx, \quad i = 1, \ldots, \infty. \tag{3.14}$$

Depending on the result of a convergence study, the first few equations are considered and, accordingly, the above series are approximated by a finite number of members in the analysis.

4. Stability and Instability Analysis

4.1. *Primary equilibrium path*

The starting point for determining the stable and unstable equilibrium paths is Eq. (3.12), which is solved for the ith generalized coordinate

$$q_i = \frac{2p_0 f_i + l\chi_i N}{l(c_i + \lambda_i^2 N)}, \quad 1, \ldots, \infty. \tag{4.1}$$

Substituting q_i into the series representation of the overall normal force N, Eq. (3.10), leads to a quadratic equation for load amplitude p_0 as a function of N

$$\frac{4}{l^2}C_1 p_0^2 - \frac{4}{l}(C_4 - C_2 N)p_0 + N(C_3 N - 2C_5 - 1) = 0, \tag{4.2}$$

with the coefficients depending on N

$$C_1 = \frac{\psi}{4}\sum_{i=1}^{\infty}\frac{\lambda_i^2 f_i^2}{(c_i + \lambda_i^2 N)^2}, \quad C_2 = \frac{\psi}{4}\sum_{i=1}^{\infty}\frac{\lambda_i^2 f_i \chi_i}{(c_i + \lambda_i^2 N)^2}, \quad C_3 = \frac{\psi}{4}\sum_{i=1}^{\infty}\frac{\lambda_i^2 \chi_i^2}{(c_i + \lambda_i^2 N)^2},$$

$$C_4 = \frac{\psi}{4}\sum_{i=1}^{\infty}\frac{f_i \chi_i}{c_i + \lambda_i^2 N}, \quad C_5 = \frac{\psi}{4}\sum_{i=1}^{\infty}\frac{\chi_i^2}{c_i + \lambda_i^2 N}, \tag{4.3}$$

and the two solutions

$$p_0(N) = \frac{l}{2C_1}(C_4 - C_2 N \pm \sqrt{C_4^2 + C_1 N(1 + 2C_5 - C_3 N) + C_2 N(C_2 N - 2C_4)}). \tag{4.4}$$

Evaluating these relationships leads to the primary equilibrium path for the overall normal force in the form p_0 over N.

To find the primary equilibrium path of the other response variables, p_0 from Eq. (4.4) is substituted into the generalized coordinates q_i, Eq. (4.1), which thus become a function of N only, i.e. $q_i(N)$. For example, if the relation for the radial displacement $w(x)$ is sought in the form $p_0(w)$, the $q_i(N)$ are substituted into the series expansion of w, Eq. (3.1), making w also a function of N: $w(N)$. Simultaneous evaluation of $w(N)$ and $p_0(N)$ (Eq. (4.4)) then leads to the relation $p_0(w)$.

The procedure for finding the representations $p_0(\Delta u_1)$, $p_0(\Delta u_2)$, etc., is analogous by substituting $q_i(N)$ into the corresponding series expansions, here Eqs. (3.3) and (3.4).

4.2. *Buckling analysis*

At the stability limit, the tangent stiffness matrix of the structure is singular. The components of the tangent stiffness matrix $\mathbf{K}$ are obtained by the successive partial derivation of the equilibrium equations h_i, Eq. (3.12), with respect to the generalized coordinates q_j as follows:[15]

$$K_{ij} = \frac{\partial h_i}{\partial q_j} = (c_i + \lambda_i^2 N)\delta_{ij} + \frac{\psi}{2}(\lambda_i^2 q_i - \chi_i)(\lambda_j^2 q_j - \chi_j), \tag{4.5}$$

with N according to Eq. (3.10). The determinant of $\mathbf{K}$ is zero at the stability limit, leading to the following buckling condition:

$$\det \mathbf{K} = \prod_{i=1}^{\infty}(c_i + \lambda_i^2 N^{(c)}) + \frac{\psi}{2}\sum_{j=1}^{\infty}\left(\frac{(\chi_j - \lambda_j^2 q_j^{(c)})^2}{c_j + \lambda_j^2 N^{(c)}}\prod_{i=1}^{\infty}(c_i + \lambda_i^2 N^{(c)})\right) = 0. \tag{4.6}$$

Evaluating this buckling equation together with the equilibrium equations h_i, $i = 1, 2, \ldots$, Eq. (3.12) yields the critical loads (both for limit point buckling and bifurcation buckling) and the corresponding critical generalized coordinates $q_i^{(c)}$.

4.2.1. *Limit loads and limit points*

For $N^{(c)} \neq -c_i/\lambda_i^2$, in Eq. (4.6) the expression $\prod_{i=1}^{\infty}(c_i + \lambda_i^2 N^{(c)})$ cancels, and the buckling equations simplifies considerably

$$\det \mathbf{K} = 1 + \frac{\psi}{2}\sum_{i=1}^{\infty}\frac{(\chi_i - \lambda_i^2 q_i^{(lc)})^2}{c_i + \lambda_i^2 N^{(lc)}} = 0. \tag{4.7}$$

This equation captures the critical loads for limit point buckling, the so-called limit loads.[15] The superscript (lc) denotes variables in the limit state of limit point buckling. The case $N^{(c)} = -c_i/\lambda_i^2$ is related to bifurcation buckling and can be treated separately, as shown in Sec. 4.2.2.

Substituting the generalized coordinate according to Eq. (4.1) into Eq. (4.7) yields a nonlinear equation with the limit loads $p_0^{(lc)}$ and the corresponding critical overall normal forces $N^{(lc)}$ as unknowns

$$\frac{4}{l^2}D_1(p_0^{(lc)})^2 - \frac{4}{l}(2C_2 - D_2 N^{(lc)})p_0^{(lc)} + N^{(lc)}(D_3 N^{(lc)} - 4C_3) + 2C_5 + 1 = 0, \tag{4.8}$$

with the coefficients depending on $N^{(lc)}$

$$D_1 = \frac{\psi}{2}\sum_{i=1}^{\infty}\frac{\lambda_i^4 f_i^2}{(c_i+\lambda_i^2 N^{(lc)})^3}, \quad D_2 = \frac{\psi}{2}\sum_{i=1}^{\infty}\frac{\lambda_i^4 f_i \chi_i}{(c_i+\lambda_i^2 N^{(lc)})^3},$$
$$D_3 = \frac{\psi}{2}\sum_{i=1}^{\infty}\frac{\lambda_i^4 \chi_i^2}{(c_i+\lambda_i^2 N^{(lc)})^3}. \tag{4.9}$$

Since the limit states are points on the equilibrium path, the simultaneous numerical solution of Eqs. (4.2) and (4.8) yields the limit loads $p_0^{(lc)}$ and the corresponding critical overall normal forces $N^{(lc)}$. Then, the pairs of values $p_0^{(lc)}$ and $N^{(lc)}$ are substituted into Eq. (4.1) to obtain the critical generalized coordinates $q_i^{(lc)}$, which in turn, when substituted into the series equations (3.1), (3.3)–(3.5), yield the limit points $w^{(lc)}, \Delta u_1^{(lc)}, \Delta u_2^{(lc)}, u_\infty^{(lc)}$ of the shallow arch with interlayer slip.

4.2.2. *Bifurcation loads and bifurcation points*

The shallow arch may also become unstable due to bifurcation buckling. With bifurcation buckling, the shallow arch changes from a pre-buckling equilibrium path to an orthogonal path where the overall normal force N remains constant.[21] Therefore, in the symmetrical shallow arch bifurcation buckling is possible only if the radial load $p(x) = p_0 f(x)$ is also symmetrically distributed, i.e. $f(x) = f(l-x)$.

As has been shown in earlier publications,[21,36,37] during bifurcation buckling the symmetrical arch deforms antimetrically in the form of an antimetric buckling mode of the corresponding straight member under compression. With hinged–hinged boundary conditions, as it is here the case, the second buckling mode (i.e. the lowest antimetric mode) of a beam column (and thus of the considered shallow arch) corresponds to the sine wave,[29] which is the second shape function $\phi_2(x)$. However, it is possible that in addition to bifurcation buckling in the second buckling mode, bifurcation buckling may occur in the fourth buckling mode ($\phi_4(x)$) at a larger normal force, or even in the sixth buckling mode ($\phi_6(x)$), and so on.[15]

The corresponding overall bifurcation normal forces $N_n^{(bc)}$ at the bifurcation points correspond to the normal force leading to buckling (in the nth buckling mode) of the corresponding straight member with interlayer slip (with the same set-up and boundary conditions as the actual shallow arch).[29] Thus, the overall bifurcation normal forces $N_n^{(bc)}$ do not depend on the shape of the initial deflection of the shallow arch[29] (i.e. they are the same for a circular arch and a sinusoidal arch with the same structure, for example). Since for the series expansion of w, Eq. (3.1), the buckling modes of the corresponding hinged–hinged symmetrically layered straight member were chosen as shape functions, for the hinged–hinged symmetrically layered shallow arch with interlayer slip, the overall bifurcation normal forces $N_n^{(bc)}$ are[29]

$$N_n^{(bc)} = -\frac{c_n}{\lambda_n^2}, \quad n = 2, 4, 6, \ldots \tag{4.10}$$

where c_n is the generalized stiffness of the structure associated with the nth buckling mode, see Eq. (3.13). With the overall bifurcation normal force $N_n^{(bc)}$ known, the corresponding symmetric bifurcation buckling load $p_n(x)^{(bc)} = p_{0n}^{(bc)} f(x) = p_{0n}^{(bc)} f(l-x)$ can be determined analytically, as will be shown in what follows. Note that the bifurcation buckling load $p_{0n}^{(bc)}$ and the corresponding bifurcation point depend on the shape of the initial deformation and its amplitude $\hat{w}_0$, unlike $N_n^{(bc)}$.

Since the response of the shallow arch is symmetric up to the bifurcation point, only the generalized coordinates $q_{in}^{(bc)}$ associated with the symmetric shape functions ($i = 1, 3, 5, \ldots$) are nonzero and given by Eq. (4.1), the others are zero, $q_{in}^{(bc)} = 0$ ($i = 2, 4, 6, \ldots$). The nonzero generalized coordinates $q_{in}^{(bc)}$ ($i = 1, 3, 5, \ldots$) are substituted into the series expression for the overall normal force, Eq. (3.10), which is subsequently replaced by the bifurcation buckling normal force $N_n^{(bc)}$, Eq. (4.10).[15] This gives a quadratic equation for the bifurcation load $p_{0n}^{(bc)}$

$$\frac{\lambda_n^4}{l^2} F_1^{(n)} (p_{0n}^{(bc)})^2 - \frac{\lambda_n^2}{l} F_2^{(n)} p_{0n}^{(bc)} + c_n \left(\frac{1}{2} F_3^{(n)} + \frac{1}{\lambda_n^2} \right) = 0, \quad n = 2, 4, 6, \ldots, \tag{4.11}$$

with the coefficients

$$\begin{aligned}
F_1^{(n)} &= \psi \sum_{i=1}^{\infty} \frac{\lambda_{2i-1}^2 f_{2i-1}^2}{(\lambda_n^2 c_{2i-1} - \lambda_{2i-1}^2 c_n)^2}, \\
F_2^{(n)} &= \psi \left(c_n \sum_{i=1}^{\infty} \frac{\lambda_{2i-1}^2 f_{2i-1} \chi_{2i-1}}{(\lambda_n^2 c_{2i-1} - \lambda_{2i-1}^2 c_n)^2} + \sum_{i=1}^{\infty} \frac{f_{2i-1} \chi_{2i-1}}{\lambda_n^2 c_{2i-1} - \lambda_{2i-1}^2 c_n} \right), \\
F_3^{(n)} &= \psi \left(\frac{c_n}{2} \sum_{i=1}^{\infty} \frac{\lambda_{2i-1}^2 \chi_{2i-1}^2}{(\lambda_n^2 c_{2i-1} - \lambda_{2i-1}^2 c_n)^2} + \sum_{i=1}^{\infty} \frac{\chi_{2i-1}^2}{\lambda_n^2 c_{2i-1} - \lambda_{2i-1}^2 c_n} \right).
\end{aligned} \tag{4.12}$$

Solving the quadratic equation (4.11) leads to the two bifurcation loads for the nth bifurcation buckling mode

$$p_{0n}^{(bc)(1)} = \frac{l}{2F_1^{(n)} \lambda_n^3} \left(F_2^{(n)} \lambda_n + \sqrt{(F_2^{(n)})^2 \lambda_n^2 - 2F_1^{(n)} c_n (2 + F_3^{(n)} \lambda_n^2)} \right), \tag{4.13}$$

$$p_{0n}^{(bc)(2)} = \frac{l}{2F_1^{(n)} \lambda_n^3} \left(F_2^{(n)} \lambda_n - \sqrt{(F_2^{(n)})^2 \lambda_n^2 - 2F_1^{(n)} c_n (2 + F_3^{(n)} \lambda_n^2)} \right). \tag{4.14}$$

Substituting Eqs. (4.13) and (4.10) into Eq. (4.1) leads to the generalized coordinates $q_i^{(bc)(1)}$ and $q_i^{(bc)(2)}$ at the bifurcation points. Substituting these generalized coordinates into the series representation of the kinematic variables (Eqs. (3.1), (3.3)–(3.5)), the response $w_n^{(bc)}, \Delta u_{1n}^{(bc)}, \Delta u_2^{(bc)}, u_{\infty n}^{(bc)}$ of the shallow arch at the bifurcation points is obtained.

The smallest so-called critical crown height at which bifurcation buckling can occur is obtained by zeroing the root in Eq. (4.13) (or Eq. (4.14)).

4.3. *Post-bifurcation equilibrium paths*

In the following, the response in the nth post-bifurcation equilibrium path is analyzed. For reasons of simplicity, however, the index n is not specified for the resulting quantities, unless absolutely necessary. The response is divided into the previously determined response at the bifurcation point and the response to be computed relative to that point[21,37]

$$\begin{aligned} w &= w^{(bc)} + \Delta w, \quad u_\infty = u_\infty^{(bc)} + \Delta u_\infty, \\ \Delta u_1 &= \Delta u_1^{(bc)} + \Delta(\Delta u_1), \quad \Delta u_2 = \Delta u_2^{(bc)} + \Delta(\Delta u_2). \end{aligned} \tag{4.15}$$

The relative response is induced by the load $\Delta p(x)$[29]

$$\Delta p(x) = p(x) - p^{(bc)}(x) = \Delta p_0 f(x), \quad \Delta p_0 = p_0 - p_0^{(bc)}. \tag{4.16}$$

The radial displacement Δw is composed of a symmetric part $\Delta w^{(s)}$ and the antimetric part $\Delta w^{(a)}$ with the shape of the bifurcation buckling mode[37]

$$\Delta w = \Delta w^{(s)} + \Delta w^{(a)}, \quad \Delta w^{(s)} = \sum_{i=1}^{\infty} \Delta q_{2i-1}\phi_{2i-1}, \quad \Delta w^{(a)} = \Delta q_n \phi_n, \tag{4.17}$$

with Δq_i denoting the generalized coordinates governing the response relative to the bifurcation point. Between the two bifurcation points the overall normal force is constant, i.e.[21,30]

$$N = N^{(bc)}, \quad \Delta N = 0. \tag{4.18}$$

Consequently, the upper and lower interlayer slips relative to their value at the bifurcation point are equal[29]

$$\Delta(\Delta u_1) = \Delta(\Delta u_2). \tag{4.19}$$

Before expressing $\Delta(\Delta u_1)(= \Delta(\Delta u_2))$ as a function of the generalized coordinates $\Delta q_i, i = 1, 3, 5, \ldots$ and Δq_2, the differential equation of equilibrium equation (2.15) must be appropriately formulated for the post-bifurcation path. To this end, in Eq. (2.15) the separation of the kinematic variable equation (4.15) is inserted and then the differential equation of equilibrium for the bifurcation point is subtracted. In the resulting equation

$$\Delta(\Delta u_1)_{,xx} - \frac{K_s}{E_1 A_1}\Delta(\Delta u_1) - \Delta w_{,xxx} r = 0. \tag{4.20}$$

Equation (4.17) is substituted and solved together with the boundary conditions $\Delta(\Delta u_1)_{,x}(0) = \Delta(\Delta u_1)_{,x}(l) = 0$ (see Eq. (2.21))

$$\Delta(\Delta u_1)(x) = \Delta(\Delta u_2)(x) = \sum_{i=1}^{\infty} \frac{r\lambda_{2i-1}^3 \cos(\lambda_{2i-1}x)}{\lambda_{2i-1}^2 + \frac{K_s}{E_1 A_1}}\Delta q_{2i-1} + \frac{r\lambda_n^3 \cos(\lambda_n x)}{\lambda_n^2 + \frac{K_s}{E_1 A_1}}\Delta q_n. \tag{4.21}$$

The basis for the series representation of Δu_∞ in terms of Δq_i is the relation for the normal force equation (2.8), from which the normal force $N^{(bc)}$ at the bifurcation point is subtracted. Since $\Delta N = 0$, after dividing by EA_e, the following differential equation is obtained:

$$\Delta u_{\infty,x} + \Delta w k + \Delta w_{,x} w_{,x}^{(bc)} + \frac{1}{2}\Delta w_{,x}{}^2 = 0. \tag{4.22}$$

Substituting the series representations for Δw and $w^{(bc)}$ and solving this equation together with the boundary conditions $\Delta u_\infty(0) = 0$ (Eq. (2.22)) yields

$$\begin{aligned}
\Delta u_\infty(x) = {} & \frac{8\hat{w}_0}{4\hat{w}_0^2 + l^2}\left(\sum_{i=1}^{\infty} \frac{1-\cos(\lambda_{2i-1}x)}{\lambda_{2i-1}}\Delta q_{2i-1} + \frac{1-\cos(\lambda_n x)}{\lambda_n}\Delta q_n\right) \\
& - \frac{1}{2}\sum_{i=1}^{\infty}\sum_{j=1}^{\infty}\left(\frac{\sin(\lambda_{2(j-i)}x)}{\lambda_{2(j-i)} + \delta_{ji}} + \frac{\sin(\lambda_{2(i+j-1)}x)}{\lambda_{2(i+j-1)}}\right)\lambda_{2i-1}\lambda_{2j-1} \\
& \times \left(q_{2i-1}^{(bc)} + \frac{1}{2}\Delta q_{2i-1}\right)\Delta q_{2j-1} - \frac{\lambda_n \Delta q_n}{2}\sum_{i=1}^{\infty}\left(\frac{\sin(\lambda_{2i-n-1}x)}{\lambda_{2i-n-1}}\right. \\
& \left. + \frac{\sin(\lambda_{2i+n-1}x)}{\lambda_{2i+n-1}}\right)\lambda_{2i-1}(q_{2i-1}^{(bc)} + \Delta q_{2i-1}) - \frac{x}{2}\sum_{i=1}^{\infty}\lambda_{2i-1}^2 \\
& \times \left(q_{2i-1}^{(bc)} + \frac{1}{2}\Delta q_{2i-1}\right)\Delta q_{2i-1} - \frac{1}{4}\left(\frac{\sin(\lambda_{2n}x)}{\lambda_{2n}} + x\right)\lambda_n^2 \Delta q_n^2.
\end{aligned} \tag{4.23}$$

Finally, the series expansions equations (4.17) and (4.21) for Δw and $\Delta(\Delta u_1)$ and Eq. (4.10) for $N^{(bc)}$ are substituted into differential equation of equilibrium

$$EJ_\infty \Delta w_{,xxxx} - 2E_1 A_1 r \Delta(\Delta u_1)_{,xxx} - N^{(bc)}\Delta w_{,xx} = \Delta p_0 f(x), \tag{4.24}$$

which follows from Eq. (2.18) by subtracting the response at the bifurcation point. To this equation, the Galerkin method is applied, see Eq. (3.11). The resulting expression is solved for the ith generalized coordinate for the symmetric part of the response

$$\Delta q_i = \frac{2\lambda_n^2}{l\lambda_i^2}\left(\frac{EJ_\infty\lambda_i^2 + EJ_0\alpha^2}{\lambda_n^2\lambda_i^2 EJ_0 EJ_\infty(\lambda_i^2+\alpha^2) - c_n(EJ_\infty\lambda_i^2 + EJ_0\alpha^2)}\right) f_i \Delta p_0,$$
$$i = 1,3,5\ldots, \quad n = 2,4,6\ldots \tag{4.25}$$

To find Δq_n (which governs the antimetric response part), the separation of the kinematic variables (Eq. (4.15)) is inserted into the integral representation of the overall normal force (Eq. (2.14)), and then the overall normal force at the bifurcation point is subtracted. From $\Delta N = 0$ follows for the remaining terms

$$\Delta N = \frac{EA_e}{l}\int_0^l \left(\Delta wk + \frac{1}{2}\Delta w_{,x}^2 + w_{,x}^{(bc)}\Delta w_{,x}\right) dx = 0. \tag{4.26}$$

Substituting the series expansion of Δw, Eq. (4.17), into this relation leads after some algebra to a quadratic equation for Δq_n with the two solutions

$$\Delta q_n = \pm\frac{1}{\lambda_n}\sqrt{-\sum_{i=1}^{\infty}\left(\frac{\Delta q_{2i-1}}{\lambda_{2i-1}}\left(\lambda_{2i-1}^3(\Delta q_{2i-1} + 2q_{2i-1}^{(bc)}) - \frac{8}{l}\frac{8\hat{w}_0}{4\hat{w}_0^2 + l^2}\right)\right)}, \tag{4.27}$$

which reflect the fact that the shallow arch can bifurcate either upwards or downwards.[29]

5. Application and Validation

5.1. *Example* 1

In a first example, the stability limits and the equilibrium path of a shallow arch with the following dimensions and material parameters are investigated: Length $l = 1.0$ m, layer thicknesses $h_1 = 0.005$ m, $h_2 = 0.01$ m, width $b = 0.1$ m, crown height $\hat{w}_0 = 0.025$ m, Young's modulus of the two face layers $E_1 = 7.0 \cdot 10^{10}$ N/m^2, Young's modulus of the middle layer $E_2 = 1.0 \cdot 10^{10}$ N/m^2, slip modulus $K_s = 1.0 \cdot 10^9$ N/m^2. The load is uniformly distributed over the length l, i.e. $f(x) = 1$ and $p(x) = p_0$, as shown in Fig. 3. The evaluation of Eqs. (4.2) and (4.8) leads to the two limit loads for this structure: $\bar{p}_0^{(lc)(1)} = p_0^{(lc)(1)} l^3/EJ_\infty = 1.978$ and $\bar{p}_0^{(lc)(2)} = p_0^{(lc)(2)} l^3/EJ_\infty = 1.024$ in normalized format. There are no real solutions for the bifurcation loads. Therefore, the considered structure represents a limit point problem. The primary stable equilibrium path and unstable equilibrium path are determined as described in Sec. 4.1. The corresponding series expansions of the response variables were approximated by the first five symmetric shape functions.

To assess the accuracy of the proposed beam theory and its results, additional comparative FE analyses assuming a plane stress state are performed with the software suite Abaqus.[38] Poisson's ratio is required as an additional material parameter, which is uniformly assumed to be $\nu = 0.3$. The three layers are discretized

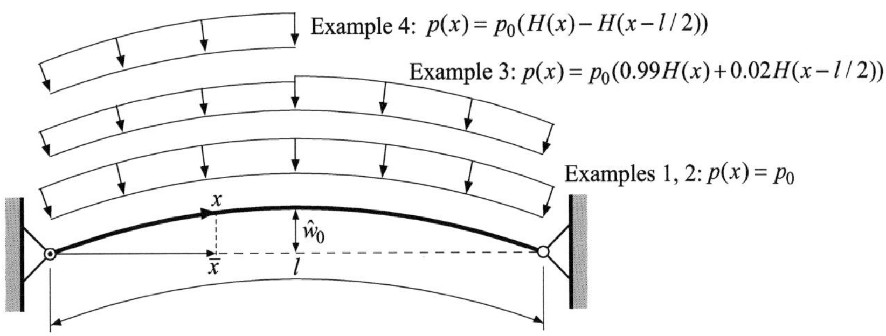

Fig. 3. Example problems.

with plane stress continuum elements of type CPS8R (8-noded quadrilateral 2D elements with reduced integration). In the FE model, a thickness of $2.0 \cdot 10^{-6}$ m is assumed for each of the two interlayers and discretized with cohesive elements of type COH2D4. The longitudinal stiffness corresponds to the slip modulus in the beam model. For the transverse stiffness, which is infinite in the beam model, the value $1.0 \cdot 10^{13}\,\mathrm{N/m^2}$ is used. The hinged support is realized by kinematic coupling of the outer surface of the middle layer to an additional node. In total, the FE model has 108.146 degrees of freedom. In comparison, only the first three symmetric shape functions are required to predict the nonlinear response with the presented beam model. The equilibrium path of the FE model was found by applying the Riks arc-length algorithm implemented in Abaqus.

In Fig. 4(a), the normalized load amplitude $\bar{p}_0 = p_0 l^3/EJ_\infty$ is plotted with a blue solid line versus the normalized radial deflection $w(l/2)/l$ at the center of the arch with flexible bonds. As can be seen, with increasing load $\bar{p}_0$ the equilibrium path becomes increasingly nonlinear. The first limit point ($\bar{p}_0^{(lc)(1)}$) is located at point A, at which the arch snaps to the remote equilibrium position at point B. Between the two limit points A and C ($\bar{p}_0^{(lc)(2)}$) there is the unstable equilibrium path, which can only be traced in a displacement-controlled analysis. The red-dotted line with cross-shaped markers shows the result of the plane stress FE analysis. As can be seen, the midspan deflection from the proposed beam theory and the FE analysis are almost identical. The figure to the right, Fig. 4(b), shows the normalized radial deflection $w(x)/l$ of the beam with interlayer slip over the member length at the points indicated by A, B, C, D in Fig. 4(a) (beam theory). Figure 4(a) also shows with the thin solid black line the load–displacement diagram for the case where the three layers are rigidly bonded, i.e. no interlayer slip occurs and the kinematics is captured by the Euler–Bernoulli theory. The comparison of the solutions demonstrates the influence of the flexible layer bond on the response. While the shallow arch with interlayer slip becomes unstable at $\bar{p}_0^{(lc)(1)} = 1.978$

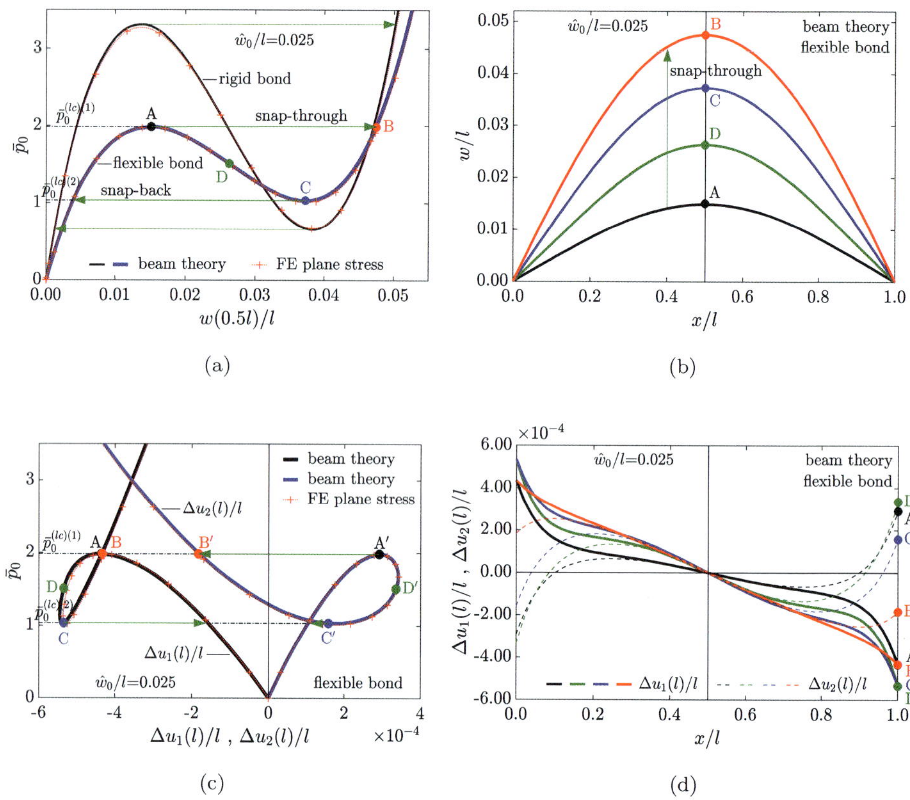

Fig. 4. Shallow arch with crown height $\hat{w}_0/l = 0.025$ subjected to a uniformly distributed load. Left column: load over deformation curves. Right column: distribution of deformation variables over x/l at discrete load levels. (a) and (b) Radial displacement; (c) and (d) interlayer slips. Nondimensional representation.

(limit point A), this is the case for the rigidly bonded at $\bar{p}_0^{(lc)(1)} = 3.306$, i.e. at a load that is 67% larger.

In Fig. 4(c), $\bar{p}_0$ is plotted against the normalized interlayer slips $\Delta u_1(l)/l$ and $\Delta u_2(l)/l$ at the right support ($x = l$). The black solid line refers to the upper interlayer slip $\Delta u_1(l)/l$, and the blue solid line to the lower interlayer slip $\Delta u_2(l)/l$. The interlayer slips that correspond to the deflections denoted by A, B, C, D in Fig. 4(a) are indicated as A, B, C, D (upper interlayer slip) and A′, B′, C′, D′ (lower interlayer slip), respectively. The proposed beam theory reproduces excellently these two response quantities in comparison to the FE solution. Only at $\Delta u_1(l)/l$ in the domain of the "loop" a small deviation is observed. The two interlayer slips as a function of the normalized member axis x/l in Fig. 4(d) illustrate that these two

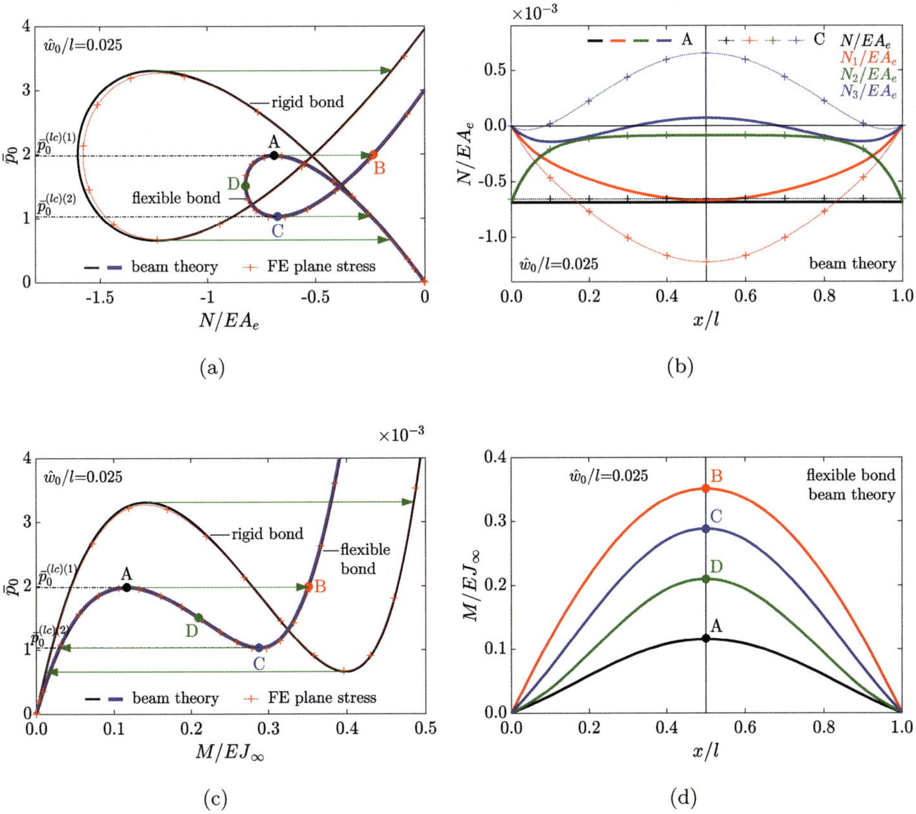

Fig. 5. Shallow arch with crown height $\hat{w}_0/l = 0.025$ subjected to a uniformly distributed load. Left column: load over internal force curves. Right column: distribution of internal forces over x/l at discrete load levels. (a) and (b) Normal forces; (c) and (d) overall bending moments. Nondimensional representation.

response quantities become increasingly different towards the supports due to the geometric nonlinearity.

In Fig. 5, the corresponding internal forces are depicted. Figure 5(a) shows with a solid blue line the normalized load amplitude $\bar{p}_0$ against the normalized overall normal force N/EA_e as a result of the proposed semi-analytical solutions of the discussed beam theory. Up to the first limit point A, the compressive overall normal force increases and then abruptly decreases as the structure snaps through point B. The maximum overall normal force in compression is less than half that of the shallow arch with rigid bond (solid black line). The result of the much more elaborate FE plane stress analysis (red-dotted line with cross-shaped markers) is virtually identical to that of the beam theory presented. Figure 5(b) shows the distribution of the normalized overall normal force and the layer-wise normal forces at the limit points A and C. It can be seen that according to the boundary conditions at the two supports, the overall normal force is completely transferred to the middle layer. While in the equilibrium positions A and C only compression occurs in the normal force of the top layer along x/l, the normal force of the bottom layer exhibits both compression and tension. The normal forces in the two face layers differ from each other at the two limit points A and C, while the total normal force and the normal force in the middle layer are the same in A and C.

The normalized load amplitude $\bar{p}_0$ versus the normalized total moment M/EJ_∞ in Fig. 5(c) shows a similar pattern as the load-deflection diagram in Fig. 4(a). Once more it can be seen how important it is to consider the flexibility of the interlayers (blue solid line) to realistically predict the response, i.e. here the overall bending moment. That is, assuming a rigid bond between the layers, the overall bending moment at the first limit point A is significantly overestimated (black solid line). From Fig. 5(d), it can be seen that the distribution of the overall bending moment over x/l at the two limit points as well as in the unstable branch at point D deviates substantially from a quadratic parabola, while at point B, which is reached after snap-through, the moment resembles a quadratic parabola, because in this position the normal force due to the geometric nonlinearity is comparatively small.

5.2. *Example* 2

In the second example, the normalized crown height is increased to $\hat{w}_0/l = 0.035$, corresponding to 3.5% of the member length. All other parameters and the load (uniform load) remain unchanged. Due to the increase of the crown height, a stability problem with bifurcation buckling (with $n = 2$) is present. Equations (4.13) and (4.14) yield two bifurcation loads $\bar{p}_0^{(bc)(1)} = 3.961$ and $\bar{p}_0^{(bc)(2)} = 0.443$, Eqs. (4.2) and (4.8) yield the limit loads $\bar{p}_0^{(lc)(1)} = 4.096$ and $\bar{p}_0^{(lc)(2)} = 0.443$. As observed, the second bifurcation load and the second limit load are virtually identical. In this example, the series expansions of the response variables were approximated by the first nine shape functions.

Figure 6(a) shows blue lines with the normalized load $\bar{p}_0$ over the normalized deflection w/l at $x = l/2$ of the shallow arch with interlayer slip found by the presented solution based on the beam theory. As observed, the response shows the characteristic behavior typical of a bifurcation buckling instability problem. The solid blue lines from $\bar{p}_0 = 0$ to the first bifurcation point A ($\bar{p}_0^{(bc)(1)}$), and from the second bifurcation point C ($\bar{p}_0^{(bc)(2)}$) upwards represent the stable response paths. The blue solid line between the bifurcation points A ($\bar{p}_0^{(bc)(1)}$) and C ($\bar{p}_0^{(bc)(1)}$) corresponds to the hypothetical equilibrium path for limit point buckling. The post-bifurcation equilibrium path between A and C illustrated with the dashed blue graph is at midspan a straight line between points A and C, but due to bifurcation buckling the distribution of the deflection over x/l is nonsymmetric, as can be seen

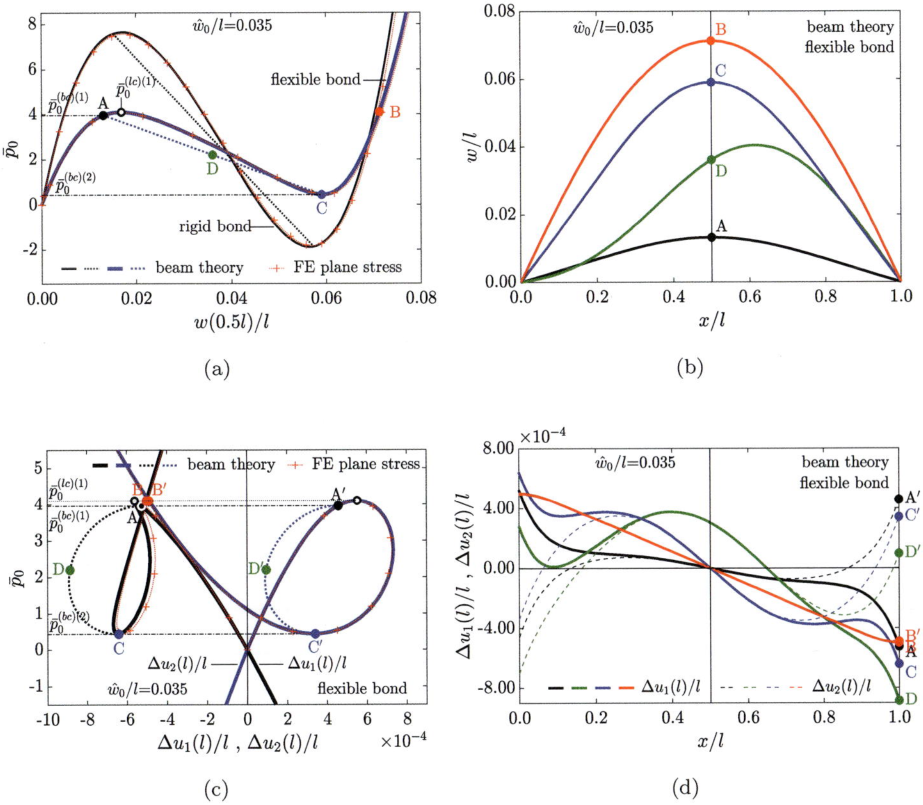

Fig. 6. Shallow arch with crown height $\hat{w}_0/l = 0.035$ subjected to a uniformly distributed load. Left column: load over deformation curves. Right column: distribution of deformation variables over x/l at discrete load levels. (a) and (b) Radial displacement; (c) and (d) interlayer slips. Nondimensional representation.

in Fig. 6(b) at the position D. At both bifurcation points A and C and beyond the post-bifurcation equilibrium path at position B, the response is symmetric, see Fig. 6(b). With the FE plane stress analysis (red-dotted line with cross-shaped markers), only the limit point equilibrium path (snap-through) was reproduced. In such a numerical analysis, the equilibrium path after bifurcation can be found only by imposing a small perturbation on the uniformly distributed load,[15] as shown in the following example.

The comparison of the stability limits at point A of the present and the previous example shows that by increasing the crown height from 2.5% to 3.5% of the member length, the normalized critical load doubles from 1.978 to 3.961. The solution for the structure with rigidly bond layers, also shown in Fig. 6(a), demonstrates that neglecting the flexibility of the interlayers overestimates the critical load by about two times.

The load-interlayer slip diagrams $\bar{p}_0, \Delta u_1(l)/l$ (black lines) and $\bar{p}_0, \Delta u_2(l)/l$ (blue lines) presented in Fig. 6(c) demonstrate the difference of the post-bifurcation equilibrium branch (dashed lines) and the limit point equilibrium branch (solid lines) between stability limits A and C. The nonsymmetric response in the post-bifurcation branch of the response is also reflected in the interlayer slips, shown in Fig. 6(d) for positions D and D′, respectively.

The corresponding internal forces are shown in Fig. 7. In the plot of the normalized load amplitude $\bar{p}_0$ versus the normalized overall normal force N/EA_e in Fig. 7(a), the constant normal force in the post-bifurcation buckling branch A–C is apparent. Figure 7(b) with the overall and layerwise normal forces shows that

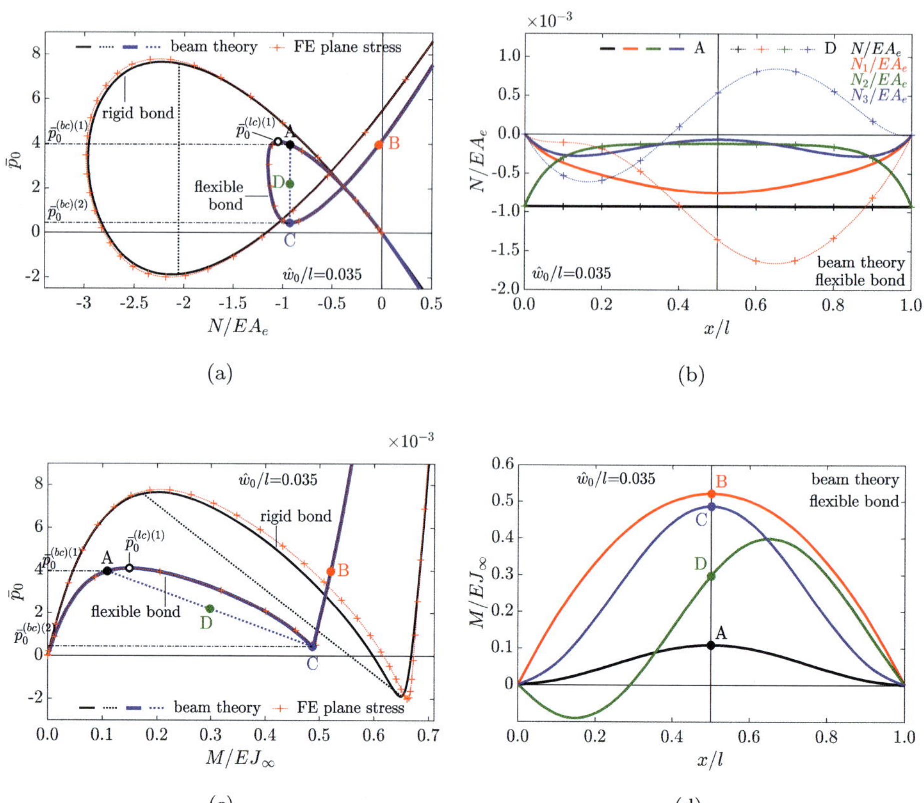

Fig. 7. Shallow arch with crown height $\hat{w}_0/l = 0.035$ subjected to a uniformly distributed load. Left column: load over internal force curves. Right column: distribution of internal forces over x/l at discrete load levels. (a) and (b) Normal forces; (c) and (d) overall bending moments. Nondimensional representation.

in the post-bifurcation buckling path (here in position D), the normal force in the two face layers is nonuniform over x/l, while at bifurcation point A the distribution is symmetric. In Fig. 7(c), the normalized load amplitude $\bar{p}_0$ over the normalized overall moment M/EJ_∞ is presented, next to it in Fig. 7(d) the distribution of M/EJ_∞ over x/l in the positions A, B, C and D.

5.3. *Example* 3

The same structure with crown height (i.e. $\hat{w}_0/l = 0.035$) is considered. However, the load is disturbed and is therefore slightly asymmetrically distributed according to $f(x) = 0.99H(x)+0.02H(x-l/2)$, where $H(x)$ is the Heaviside step function, see

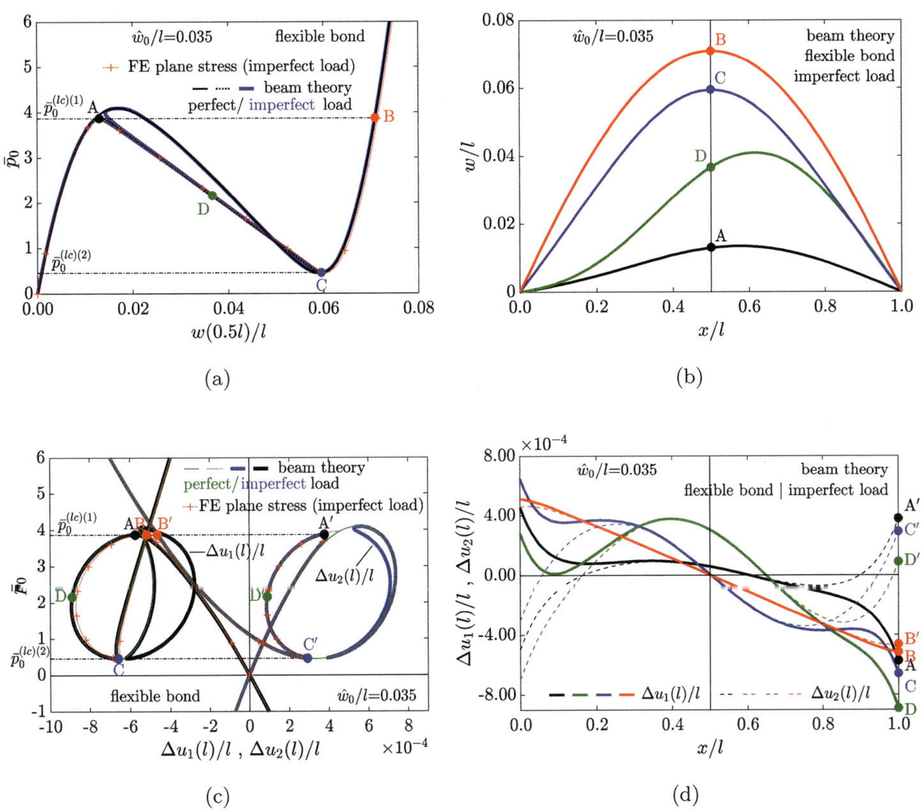

Fig. 8. Shallow arch with crown height $\hat{w}_0/l = 0.035$ subjected to a perturbed (imperfect) distributed load. Left column: load over deformation curves. Right column: distribution of deformation variables over x/l at discrete load levels. (a) and (b) Radial displacement; (c) and (d) interlayer slips. Nondimensional representation.

Fig. 3. In this example problem, nine shape functions were necessary for convergence of the considered response quantities.

Due to the asymmetry of the load (in the subsequent figures denoted as imperfect load), this example is a limit point buckling problem. The resulting normalized load amplitude $\bar{p}_0$-radial midspan deflection $w(l/2)/l$ diagram illustrated in Fig. 8 by blue solid lines shows that there are two unconnected primary equilibrium paths, which approximate the primary and post-bifurcation equilibrium branch of the shallow arch subjected to the (perfectly) uniformly distributed load, depicted by black lines. The limit load $\bar{p}_0^{(lc)(1)} = 3.869$ at limit point A with zero tangent is slightly smaller than the bifurcation load $\bar{p}_0^{(bc)(1)} = 3.961$ of the member subjected to the perfect load. As observed, the FE analysis only delivered the primary post-buckling

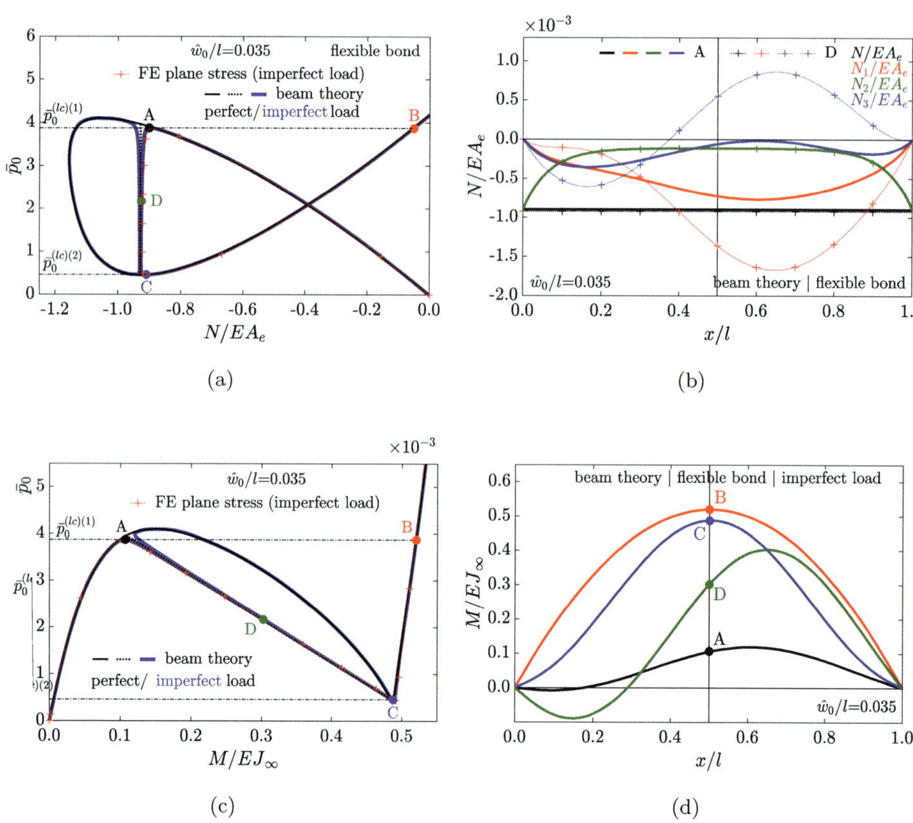

Fig. 9. Shallow arch with crown height $\hat{w}_0/l = 0.035$ subjected to a perturbed (imperfect) distributed load. Left column: load over internal force curves. Right column: distribution of internal forces over x/l at discrete load levels. (a) and (b) Normal forces; (c) and (d) overall bending moments. Nondimensional representation.

equilibrium path between points A and C (red-dotted line with cross-shaped markers), i.e. the unconnected equilibrium path was not found by the Riks arc-length algorithm. Figure 8(b) shows the asymmetric deflection at point D (between the two branch points A and C), as for the arch subjected to perfect loading. Due to the disturbed distributed loading, the response at the limit point A is slightly asymmetric, in contrast to the perfectly loaded structure (see Fig. 6(b)).

The plots $\bar{p}_0$ over the interlayer slips $\Delta u_1(l)/l$ and $\Delta u_2(l)/l$, respectively, depicted in Fig. 8(c) show how complex the perturbed loading response becomes when a bifurcation buckling problem is approximated. Figure 8(d) shows the distribution of these response quantities for the four positions A, B, C, D and A′, B′, C′, D′, respectively.

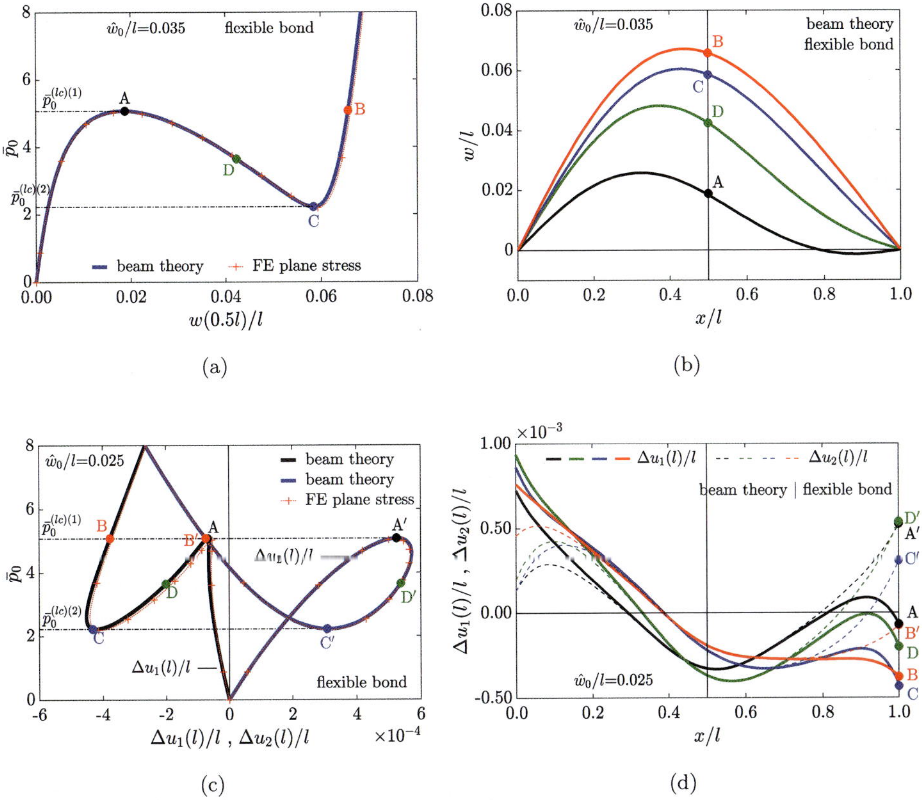

Fig. 10. Shallow arch with crown height $\hat{w}_0/l = 0.035$ subjected to a load distributed over left half. Left column: load over deformation curves. Right column: distribution of deformation variables over x/l at discrete load levels. (a) and (b) Radial displacement; (c) and (d) interlayer slips. Nondimensional representation.

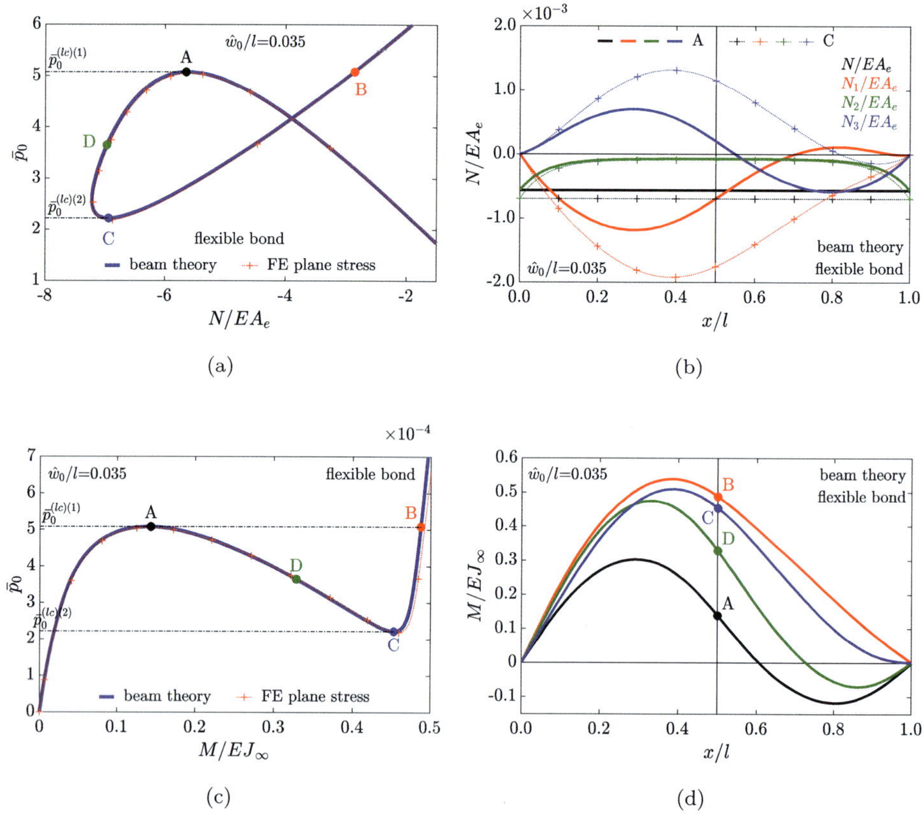

Fig. 11. Shallow arch with crown height $\hat{w}_0/l = 0.035$ subjected to a load distributed over left half. Left column: load over internal force curves. Right column: distribution of internal forces over x/l at discrete load levels. (a) and (b) Normal forces; (c) and (d) overall bending moments. Nondimensional representation.

The internal forces depicted in Fig. 9 also show that the bifurcation buckling problem of the perfectly loaded shallow arch is very well approximated by the member subjected to the imperfect load. In particular, Fig. 9(a) clearly shows the two unconnected equilibrium paths.

5.4. *Example* 4

In the last example, the same shallow arch as in the two previous examples is considered (i.e. $\hat{w}_0/l = 0.035$), but the load is uniformly distributed only over the left half of the beam: $f(x) = H(x) - H(x - l/2)$, see Fig. 3. In the series approximation of the response, nine shape functions are considered. The normalized limit loads are $\bar{p}_0^{(lc)(1)} = 5.071$ and $\bar{p}_0^{(lc)(2)} = 2.215$.

The response of the structure is depicted in Figs. 10 and 11. As can be seen in Figs. 10(a), 10(c), 11(a) and 11(c), the asymmetric load causes the structure to snap to the remote equilibrium point B after reaching the limit point A. The limit load $\bar{p}_0^{(lc)(1)}$ at limit point A is 53% larger than the bifurcation load $\bar{p}_0^{(bc)(1)}$ at bifurcation point A of the structure under the uniform load over the entire member. However, when the critical resultant of the applied load instead of the critical load amplitude $\bar{p}_0$ is considered, it becomes apparent that in the asymmetrically loaded structure the stability limit is reached at a critical resultant load 23.3% lower. The equilibrium paths of the interlayer slip are less complex than for the imperfect uniform load, as can be seen from Fig. 10(c). Figures 10(b), 10(d), 11(b) and 11(d) illustrate the nonsymmetric distribution of the response over x/l due to the asymmetric load. Also in this example problem, the results of the proposed beam theory match excellently the response of the FE plane stress analysis.

6. Summary and Conclusions

Novel semi-analytical solutions for the stability limits and the pre- and post-buckling equilibrium paths of symmetrically composed circular three-layer shallow arches with elastically bonded layers were presented. Several examples demonstrated the importance of considering the interlayer slip when predicting the response of the structures under consideration. When the layers are rigidly bonded, as is generally assumed, the stability limits are significantly overestimated. To validate the presented semi-analytical solutions, computationally much more expensive Finite Element analyses with 2D continuum elements were conducted. The comparison of the results found in these analyses with the ones of the presented beam theory shows the high accuracy of the presented solution equations. With the equations presented here, the stability behavior of the considered shallow arches with interlayer slip can be estimated rapidly and accurately.

References

1. Y.-B. Yang and S. Kuo, Effect of curvature on stability of curved beams, *J. Struct. Eng.* **113**(6) (1987) 1185–1202.
2. Y.-B. Yang and M.-S. Shieh, Solution method for nonlinear problems with multiple critical points, *AIAA J.* **28**(12) (1990) 2110–2116.
3. Y.-B. Yang and S.-R. Kuo, Out-of-plane buckling of angled frames, *Int. J. Mech. Sci.* **33**(1) (1991) 55–67.
4. Y.-B. Yang and S.-R. Kuo, Buckling of frames under various torsional loadings, *J. Eng. Mech.* **117**(8) (1991) 1681–1697.
5. Y.-B. Yang, S.-R. Kuo and J.-D. Yau, Use of straight-beam approach to study buckling of curved beams, *J. Struct. Eng.* **117**(7) (1991) 1963–1978.
6. Y.-B. Yang, Recent researches on buckling of framed structures and curved beams, *J. Construct. Steel Res.* **26**(2) (1993) 193–210, Special South East Asian Issue.

7. S.-R. Kuo and Y.-B. Yang, Tracing postbuckling paths of structures containing multi-loops, *Int. J. Numer. Methods Eng.* **38**(23) (1995) 4053–4075.
8. Y.-B. Yang, J.-D. Yau and L.-J. Leu, Recent developments in geometrically nonlinear and postbuckling analysis of framed structures, *Appl. Mech. Rev.* **56** (2003) 431–449.
9. Y.-B. Yang and Y. Z. Liu, Lateral buckling of cantilevered circular arches under various end moments, *Int. J. Struct. Stabil. Dyn.* **20**(07) (2020) 2071005.
10. Y. B. Yang, A. Chen and S. He, Research on nonlinear, postbuckling and elasto-plastic analyses of framed structures and curved beams, *Meccanica* **56**(6) (2021) 1587–1612.
11. Y. B. Yang, P. X. Jiang and Y. Z. Liu, Straight-beam approach for analyzing lateral buckling of thin-walled curved i-beams, *Int. J. Struct. Stab. Dyn.* **22**(11) (2022) 2271002.
12. M. A. Bradford, B. Uy and Y.-L. Pi, In-plane elastic stability of arches under a central concentrated load, *J. Eng. Mech.* **128**(7) (2002) 710–719.
13. M. A. Bradford, Y.-L. Pi, G. Yang and X.-C. Fan, Effects of approximations on non-linear in-plane elastic buckling and postbuckling analyses of shallow parabolic arches, *Eng. Struct.* **101** (2015) 58–67.
14. Y.-L. Pi and M. Bradford, Nonlinear analysis and buckling of shallow arches with unequal rotational end restraints, *Eng. Struct.* **46** (2013) 615–630.
15. Y. Zhou, W. Chang and I. Stanciulescu, Non-linear stability and remote unconnected equilibria of shallow arches with asymmetric geometric imperfections, *Int. J. Nonlinear Mech.* **77** (2015) 1–11.
16. J. Cail and J. Feng, Effect of support stiffness on stability of shallow arches, *Int. J. Struct. Stab. Dyn.* **10**(05) (2010) 1099–1110.
17. Y.-L. Pi, M. Bradford and F. Tin-Loi, Nonlinear analysis and buckling of elastically supported circular shallow arches, *Int. J. Solids Struct.* **44**(7) (2007) 2401–2425.
18. F. Kaviani and H. R. Mirdamadi, Snap-through and bifurcation of nano-arches on elastic foundation by the strain gradient and nonlocal theories, *Int. J. Struct. Stab. Dyn.* **13**(05) (2013) 1350022.
19. H. Babaei and M. Reza Eslami, Nonlinear snap-through instability of FGM shallow micro-arches with integrated surface piezoelectric layers based on modified couple stress theory, *Int. J. Struct. Stab. Dyn.* **19**(08) (2019) 1950088.
20. M. Faraji Oskouie, M. Zargar and R. Ansari, Dynamic snap-through instability of hygro-thermally excited functionally graded porous arches, *Int. J. Struct. Stab. Dyn.* **23**(03) (2023) 2350030.
21. L. Kiss, Sensitivity of FGM shallow arches to loading imperfection when loaded by a concentrated radial force around the crown, *Int. J. Nonlinear Mech.* **116** (2019) 62–72.
22. L. P. Kiss, Nonlinear stability analysis of FGM shallow arches under an arbitrary concentrated radial force, *Int. J. Mech. Mater. Des.* **16**(1) (2020) 91–108.
23. U. A. Girhammar and D. H. Pan, Exact static analysis of partially composite beams and beam-columns, *Int. J. Mech. Sci.* **49**(2) (2007) 239–255.
24. S. Schnabl, I. Planinc and G. Turk, Buckling loads of two-layer composite columns with interlayer slip and stochastic material properties, *J. Eng. Mech.* **139**(8) (2013) 961–966.
25. C. Adam, D. Ladurner and T. Furtmüller, Moderately large deflection of slightly curved layered beams with interlayer slip, *Arch. Appl. Mech.* **92**(5) (2022) 1431–1450.

26. C. Adam, D. Ladurner and T. Furtmüller, Free and forced small flexural vibrations of slightly curved slender composite beams with interlayer slip, *Thin Walled Struct.* **180** (2022) 109857.
27. C. Adam, D. Ladurner and T. Furtmüller, Moderately large vibrations of flexibly bonded layered beams with initial imperfections, *Compos. Struct.* **299** (2022) 116013.
28. C. Adam, D. Ladurner and T. Furtmüller, In-plane instability of shallow layered arches with interlayer slip, *Acta Mech.* **233**(9) (2022) 3813–3828.
29. C. Adam, D. Ladurner and T. Furtmüller, In-plane buckling of flexibly bonded three-layer pinned-fixed half-sine shallow arches, *Int. J. Nonlinear Mech.* **151** (2023) 104369.
30. Y.-L. Pi, M. Bradford and B. Uy, In-plane stability of arches, *Int. J. Solids Struct.* **39**(1) (2002) 105–125.
31. F. Ziegler, *Mechanics of Solids and Fluids*, 2nd edn. (Springer, New York, 1995).
32. J. W. Harris and H. Stöcker, *Handbook of Mathematics and Computational Science*, 1st edn. (Springer-Verlag, New York, 1998).
33. R. Bhat, Natural frequencies of rectangular plates using characteristic orthogonal polynomials in Rayleigh–Ritz method, *J. Sound Vib.* **102**(4) (1985) 493–499.
34. M. Qatu, In-plane vibration of slightly curved laminated composite beams, *J. Sound Vib.* **159**(2) (1992) 327–338.
35. Mathematica v. 11.2.0, Wolfram Research, Inc. (2017).
36. M. Bateni and M. Eslami, Non-linear in-plane stability analysis of fgm circular shallow arches under central concentrated force, *Int. J. Nonlinear Mech.* **60** (2014) 58–69.
37. H. Lu, A. Liu, Y.-L. Pi, M. A. Bradford, J. Fu and Y. Huang, Localized loading and nonlinear instability and post-instability of fixed arches, *Thin Walled Struct.* **131** (2018) 165–178.
38. Simulia (Dassault Systèmes), Abaqus Standard 3DEXPERIENCE R2016x (2016).

https://doi.org/10.1142/9789819814770_0015

Chapter 15

Vibration Suppression for an Elastically Supported Nonlinear Beam Coupled to an Inertial Nonlinear Energy Sink[#]

Zhi-Yong Chang[*,§], Jie Chen[*,†,¶,**] and Qiu-Sheng Li[†,‡,||,**]

[*]*Beijing Key Laboratory of Nonlinear Vibrations and Strength of Mechanical Structures Beijing 100124, P. R. China*

[†]*Architecture and Civil Engineering Research Center City University of Hong Kong Shenzhen Research Institute Shenzhen 518057, P. R. China*

[‡]*Department of Architecture and Civil Engineering City University of Hong Kong Kowloon, Hong Kong*

[§]*zychang@emails.bjut.edu.cn*

[¶]*jchen@bjut.edu.cn*

[||]*bcqsli@cityu.edu.hk*

This paper investigates the vibration suppression of an elastically supported nonlinear cantilever beam attached to an inertial nonlinear energy sink (NES). The nonlinear terms introduced by the NES are transferred as the external excitations acting on the beam. The governing equations of the nonlinear beam with an inertial NES are derived according to the Lagrange equations and the assumed mode method. The linear and nonlinear frequencies of the beam are numerically obtained by the Rayleigh–Ritz method and the direct iteration method, respectively. The frequencies are verified by the results of the finite element analysis and literature. The responses of the beam under shock excitations and harmonic excitations are numerically solved by the fourth-order Runge–Kutta method. The suppression effect of the inertial NES on the transverse vibration of the beam is evaluated through the values of amplitude reduction and energy dissipation. In addition, a parametric analysis of the inertial NES is conducted to improve the vibration reduction effect of the NES on the beam.

Keywords: Inertial nonlinear energy sink; elastic support; vibration control; geometric nonlinearity, nonlinear frequency.

[**]Corresponding authors.

[#]This chapter appeared previously on the International Journal of Structural Stability and Dynamics. To cite this chapter, please cite the original article as the following: Z.-Y. Chang, J. Chen & Q.-S. Li, *Int. J. Struct. Stab. Dyn.*, **23**, 2340017 (2023), DOI: 10.1142/S0219455423400175.

1. Introduction

Vibrations in engineering practices are generally unintentional, unavoidable, and harmful to structural safety. Traditionally, research in the field of vibration suppression has primarily focused on introducing additional linear vibration absorbers, which have proven to be effective in reducing structural vibration due to their structural simplicity and low cost.[1–4] However, linear vibration absorbers only reduce the vibration near the natural frequency of structures, and their working frequency band is narrow. In 2001, nonlinear vibration absorbers, termed nonlinear energy sinks (NESs), were first proposed by Vakakis[5] and have since gained much attention due to the advantages of their small attached mass, wide working frequency band, and targeted energy transfer characteristics.[6–10]

In recent years, many studies have concentrated on complex dynamic phenomena due to the introduction of nonlinearity. Gendelman[11] proved that the existence of purely nonlinear stiffness enables the primary structure to irreversibly transfer energy to the nonlinear array in one direction. Zang and Chen[12] studied the chaotic, bifurcation and other complex dynamic behaviors of a single-degree-of-freedom oscillator caused by the introduction of NES. Nili Ahmadabadi and Khadem[13] demonstrated that the high-amplitude periodic response of a forced two-degree-of-freedom (2DOF) oscillatory system can be eliminated by using NES. Tsakirtzis *et al.*[14] analyzed the complex dynamics and target energy transfer characteristics in a linear oscillator coupled to multi-degrees-of-freedom essentially nonlinear attachments. Georgiades and Vakakis[15] investigated the dynamic response of a linear beam with a local NES under shock excitation. The numerical results showed that the NES can absorb most of the impact energy in a unidirectional and irreversible manner. Zhang *et al.*[16] studied the suppression effect of NES on excessive vibration of a composite laminated plate under wind loadings. Increasing interest in NESs has motivated various new types of NESs for different applications, such as vibro-impact-type NESs,[17,18] tuned bistable NESs,[19] pendulum-type NESs,[20] rotating NESs,[21] and hysteretic NESs.[22] However, the mass of the additional shock absorber has become a focal point for optimal design. In 2002, Smith[23] proposed a special mechanical element named the inerter that can provide inertial parameters several times its own mass. Based on this advantage, inertial devices have been widely applied for vibration reduction in engineering structures such as buildings,[24] aircraft,[25] and vehicles.[26] Shi *et al.*[27] investigated the dynamic characteristics of a stay cable with an inerter damper installed close to one end of the cable. Zhang *et al.*[28] proposed a novel archetype of an inertial enhanced NES for vibration reduction of vibrating systems. The results showed that there exists an optimal interval for the parameter values of the inerter. Zhang *et al.*[29,30] proved that the inertial NES can suppress the multimodal transverse vibration of a beam.

In engineering applications, most of the primary structures should be modeled as beam or plate structures with elastic boundary conditions rather than classical boundaries, such as simply supported or clamped. Furthermore, the effect of the geometric nonlinearity of the primary structures cannot be ignored due to the large vibration amplitude. Mao *et al.*[31] first studied the nonlinear response of flexible structures under general nonlinear support conditions. The differential quadrature element method was adopted to analyze the boundary nonlinearity in detail. Zang *et al.*[32] investigated the steady state response of an asymmetrically elastically supported viscoelastic beam coupled to a lever-type nonlinear energy sink. In their study, the recurring closed detached response was observed for the first time in a continuous system. Proposing a new vertical elastic support boundary, Ding *et al.*[33] studied the steady state responses and the force transmission of the nonlinear vibration of beams supported by vertical springs under periodic excitation. Kani *et al.*[34] first analyzed the energy transfer between the nonlinear primary system and the NES by considering more than one mode of the beam. Parseh *et al.*[35] studied the dynamics of a nonlinear beam coupled to an NES under harmonic excitation. The approximate solution of the steady-state vibration of the system was obtained by combining the complex averaging method with the arc-length continuation algorithm. Zhang *et al.*[36] investigated the vibration suppression effect of an NES on laminated composite beams in thermal and humid environments.

The problem of energy transfer in vibration systems is another topic of vibration control. Based on the advantages of targeted energy transfer in nonlinear vibration systems, many recent studies have focused on the vibration reduction and energy acquisition of NES systems. In dynamical and acoustical systems, strong nonlinearity can be viewed as the scattering of energy across frequencies.[37] Chen *et al.*[38] investigated the effects of an internal NES on the dynamic response of a truss core sandwich plate. The results showed that a reasonable design of the NES parameters can lead to a relatively high energy dissipation rate. Based on a modified snap-through mechanism, Lu *et al.*[39] found that a bistable energy harvester with only horizontal springs has a better harvesting performance than a harvester with both horizontal and vertical springs under random excitation. Fang *et al.*[40] proposed a novel apparatus consisting of an NES and giant magnetostrictive material to realize vibration control and energy harvesting.

As reviewed above, previous studies have focused on the applications of the NES to beams with classical boundary conditions and ignored structural geometric nonlinearity. The investigation of the inertial NES on a nonlinear beam model with the elastic support boundary condition can promote the engineering application of NESs on the vibration reduction of structures.

So, in this paper, the dynamic model of an elastically supported nonlinear cantilever beam attached to an inertial nonlinear energy sink is established by the Lagrange equations of motion. The linear frequencies of the beam are obtained by the Rayleigh–Ritz method and verified by the finite element method. The nonlinear frequencies are derived by the iterative method and compared with the results of the literature. The transient and steady-state responses of the beam are numerically solved by the fourth-order Runge–Kutta method. The effects of the NES on the vibration reduction of beams subjected to shock excitation and harmonic excitation are evaluated. In addition, to achieve more effective vibration reduction, a detailed parametric analysis of the NES is carried out.

2. Equations of Motion

The nonlinear cantilever beam with an intermediate spring is attached with an inertial nonlinear energy sink, as shown in Fig. 1. The external force Γ is imposed on x_a. x and t are the axial coordinate of the beam and time, respectively. The spring with linear stiffness k_s is attached at an arbitrary position x_s of the beam. $w(x,t)$ and $v(t)$ denote the displacements of the transverse vibration of the beam and NES, respectively. b_m is the inertance of the NES mounted at the right end of the beam. k_{nes} and c_{nes} are the coefficients of the cubic nonlinear stiffness and the linear damping of the NES, respectively.

The geometric nonlinearity of the beam can be obtained by the von Karman strain-displacement equation as

$$\varepsilon_x = \frac{\partial u}{\partial x} + \frac{1}{2}\left(\frac{\partial w}{\partial x}\right)^2, \tag{1}$$

where ε_x, u and w represent the longitudinal strain and axial and vertical displacements, respectively.

According to the Euler-Bernoulli beam theory, the axial displacement can be expressed as

$$u = -z\frac{\partial w}{\partial x}. \tag{2}$$

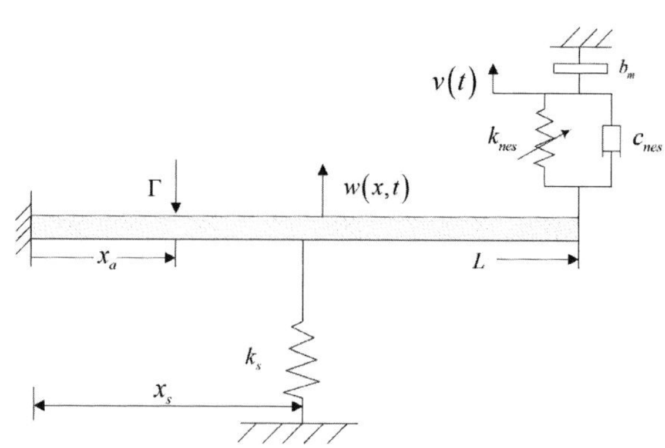

Fig. 1. Dynamic model of a nonlinear beam with an inertial NES.

The kinetic energy of the coupled beam and NES system is obtained as

$$T = \frac{1}{2}\rho A \int_0^L \left(\frac{\partial w}{\partial t}\right)^2 dx + \frac{1}{2}b_m \dot{v}^2, \tag{3}$$

where ρ and A represent the material density and the cross-sectional area of the beam, respectively.

The strain energy of the coupled system is given as

$$U = \frac{1}{8}\int_0^L EA\left(\frac{\partial w}{\partial x}\right)^4 dx + \frac{1}{2}\int_0^L EI\left(\frac{\partial^2 w}{\partial x^2}\right)^2 dx + \frac{1}{2}k_s w_s^2 + \frac{1}{4}k_{\text{nes}}[w(x,t) - v(t)]^4 \delta(x - x_d), \tag{4}$$

where E and I denote the elasticity modulus of the beam and the area moment of inertia, respectively.

The Rayleigh's dissipation function of the coupled system can be written as[35]

$$D = \frac{1}{2}\int_0^L c\left(\frac{\partial w}{\partial t}\right)^2 dx + \frac{1}{2}c_{\text{nes}}[\dot{w}(x,t) - \dot{v}(t)]^2 \delta(x - x_d), \tag{5}$$

where c and x_d are the viscous damping coefficient of the beam and the displacement of the NES, respectively. δ is the Dirac delta function applied to transfer the nonlinear term and damping term of the NES into the external excitations acting on the beam.[41,42]

The equations of motion are obtained by using the Lagrange equation with Rayleigh's dissipation function and the assumed mode method. The Lagrange equation can be expressed as

$$\frac{d}{dt}\left(\frac{\partial T}{\partial \dot{q}}\right) - \frac{\partial T}{\partial q} + \frac{\partial U}{\partial q} + \frac{\partial D}{\partial \dot{q}} = Q, \tag{6}$$

where Q represents the generalized force.

In the assumed mode method, the displacement of the transverse vibration of the beam can be assumed in the form of components or matrices

$$w(x,t) = \sum_{n=1}^{N} \varphi_n(x) q_n(t) = \boldsymbol{\varphi}^T(x)\mathbf{q}(t). \tag{7}$$

Substituting Eqs. (1)–(5) and (7) into Eq. (6), the governing equations of the clamped-elastically supported beam are obtained as

$$\begin{aligned}&\rho A \int_0^L (\varphi\varphi^T)\ddot{q}dx + EI\int_0^L \frac{\partial^2 \varphi}{\partial x^2}\frac{\partial^2 \varphi^T}{\partial x^2} q dx + k_s \varphi_s \varphi_s^T q + c\int_0^L \varphi\varphi^T \dot{q} dx \\ &+ \frac{EA}{2}\int_0^L \left(q^T \frac{\partial \varphi}{\partial x}\frac{\partial \varphi^T}{\partial x} q\right)\left(\frac{\partial \varphi}{\partial x}\frac{\partial \varphi^T}{\partial x}\right) q dx \\ &+ [k_{\text{nes}}(q^T\varphi - v)^3 + c_{\text{nes}}(\dot{q}^T\varphi - \dot{v})]\varphi_i(d) = \Gamma\varphi_i(x_a).\end{aligned} \tag{8}$$

The governing equation of the NES installed on the beam is expressed as

$$b_m \ddot{v} + k_{\text{nes}} \left(v - \sum_{j=1}^{n} q_j(t)\varphi_j(d) \right)^3 + c_{\text{nes}} \left(\dot{v} - \sum_{j=1}^{n} \dot{q}_j(t)\varphi_j(d) \right) = 0. \tag{9}$$

3. Comparative Verification

In this section, the linear and nonlinear frequencies of the nonlinear beam are solved to verify the correctness of the present solution methodology. Without the external force, damping and the force of the NES, the matrix form of the nonlinear governing Eqs. (8) and (9) can be rewritten as

$$(\mathbf{K}_L + \mathbf{K}_S + \mathbf{K}_{NL})\mathbf{q} - \omega^2 \mathbf{M}\mathbf{q} = 0, \tag{10}$$

where $\mathbf{M}$ is the mass matrix and $\mathbf{K}_L$ and $\mathbf{K}_{NL}$ are the linear and nonlinear stiffness matrices, respectively. $\mathbf{K}_S$ is the stiffness matrix provided by the vertical linear spring. The expression of the above matrices can be found in Appendix A.

The geometric and material parameters of the aluminum alloy beam are presented in Table 1.[29] The first four frequencies of the beam for different spring locations are compared with the results obtained by the finite element analysis software ABAQUS, as shown in Fig. 2. Good agreement can be found for the beam with different spring locations. Besides, the spring position has a more obvious influence on the first two frequencies of the beam. Figure 3 depicts the first mode shapes of the beam with different spring locations by ABAQUS software. A special mode shape of the beam can be found for the spring mounted at the right end of the beam. For this condition, the deformation of the right end of the beam is not the largest, similar to the elastic support boundary condition. In the following study, the spring is installed at the $L/2$ of the beam if there is no special study on the spring location parameter x_s.

The nonlinear frequency of the nonlinear beam can be solved from Eq. (10) by a direct iterative algorithm[43] via the following steps:

Table 1. Geometric and material parameters of an aluminum alloy beam.

Item	Notation	Value
Young's modulus of the beam	E	68.9 GPa
Density of the beam	ρ	$2800\,\text{kg}\cdot\text{m}^{-3}$
Length of the beam	L	0.5 m
Width of the beam	b	0.02 m
Height of the beam	h	0.01 m
Damping	c	$5\,\text{N}\cdot\text{s/m}$
Cross-sectional area	A	$2\times10^{-4}\,\text{m}^2$
Cross-sectional moment of inertia	I	$1.67\times10^{-9}\,\text{m}^4$
Vertical spring stiffness at the right end	k_s	$46025.2\,\text{N}\cdot\text{m}^{-1}$

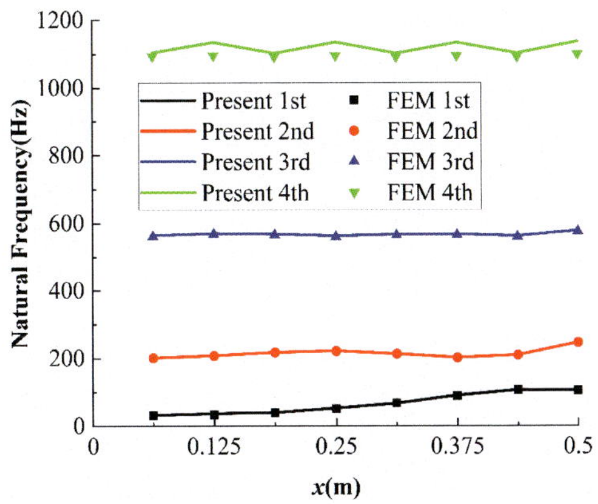

Fig. 2. First four frequencies of the nonlinear beam for different spring locations.

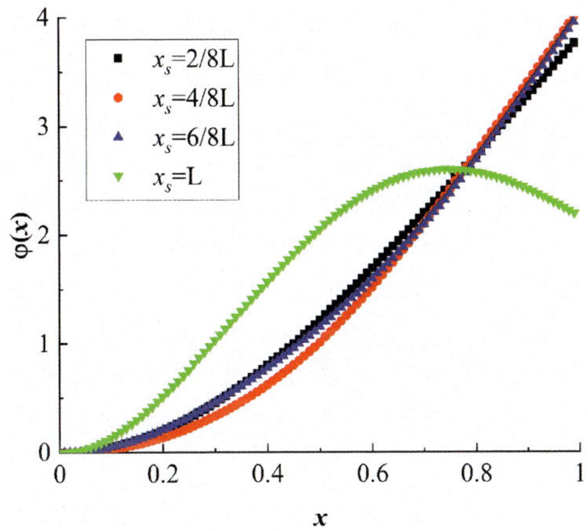

Fig. 3. The first mode shape for different spring locations.

Step 1. Calculate the linear eigenvalue and related eigenvector by neglecting the nonlinear matrices $\mathbf{K}_{NL}$ from the following linear eigenvalue equation:

$$(\mathbf{K}_L + \mathbf{K}_S)\mathbf{q} - \omega^2\mathbf{M}\mathbf{q} = 0. \tag{11}$$

The obtained linear eigenvalue λ_i $(i = 1)$ and eigenvector q_i $(i = 1)$ are regarded as an initial approximation solution for the next step.

Step 2. Apply the given vibration amplitude $W_{\max}$ and vibration mode q_i to solve the amplified eigenvector

$$\bar{q}_i = \frac{W_{\max}}{w_{\max}(q_i)} q_i, \tag{12}$$

where $w_{\max}(q_i)$ represents the amplitude of q_i.

Table 2. Comparison of the nonlinear frequency ratio ω_{nl}/ω_l for clamped support isotropic beams.

$W_{\max}/\sqrt{I/A}$	Present	Ke *et al.*[42]	Marur *et al.*[43]
1	1.0412	1.0303	1.0300
2	1.1437	1.1152	1.1147
3	1.2808	1.2419	1.2420
4	1.4418	1.3983	1.3987

Step 3. Use the amplified eigenvector $\bar{q}_i$ obtained in Step 2 to compute the nonlinear matrices $\mathbf{K}_{NL}$. Then, the new nonlinear eigenvalue λ_{i+1} and eigenvector q_{i+1} can be obtained from the updated Eq. (10).

Step 4. Repeat Step 2 to Step 3 until the relative error between eigenvalues obtained from two consecutive iterations is within 0.1%.

Table 2 shows the nonlinear frequency ratio ω_{nl}/ω_l of clamped isotropic beams obtained in this work and the results of Ke *et al.*[44] and Marur *et al.*[45] The geometric parameters used in this example are $L/h = 100$ and $h = 0.3$. Here, $\Theta = \sqrt{I/h}$ represents the radius of gyration of the beam. The present results are in good agreement with the results of reference.[44,45]

Figure 4 shows the effect of the different spring locations x_s on the nonlinear frequency ratio ω_{nl}/ω_l of the beam. As the vibration amplitude $w_{\max}/\Theta$ increases, the nonlinear frequency ratio of different spring locations increases. This means that the nonlinear beam has a hard spring characteristic. At a given vibration amplitude $w_{\max}$, the nonlinear frequency ratio increases as x_s increases, except for $x_s = 6L/8$. The reason is that the spring location is near the node of the second-order mode of the beam.

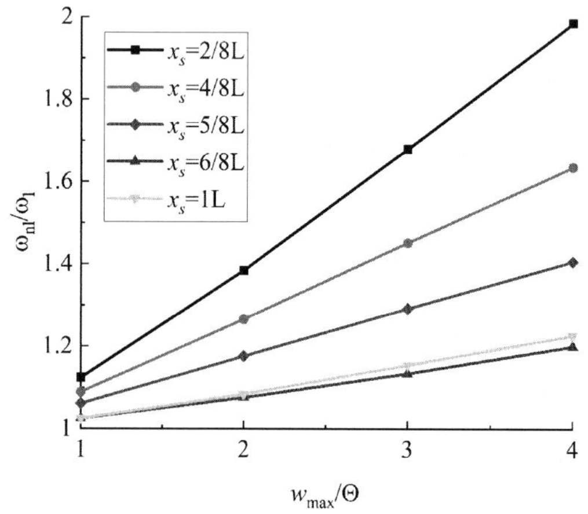

Fig. 4. Effect of the spring location on the nonlinear frequency ratio ω_{nl}/ω_l.

4. Beam Response Under Shock Excitation

This section focuses on the influence of NES's parameters on the beam response under shock excitation. The response of the beam can be numerically solved by the Runge-Kutta method based on Eqs. (8) and (9). The parameters of the inertial NES are given as follows: the mass of the inerter $m_{\text{nes}} = 0.0028\,\text{kg}$, nonlinear stiffness $k_{\text{nes}} = 1 \times 10^7\,\text{N} \times \text{m}^{-3}$, and linear damping $\mu = 1\,\text{N} \times \text{s} \times \text{m}^{-1}$. In this paper, the ballscrew inerter mentioned in reference[46] is adopted. The inertance of the inerter is expressed as

$$b_m = \gamma m_{\text{nes}}, \tag{13}$$

where m_{nes} represents the mass of the inerter of the NES, and γ denotes the inertance coefficient. The range of inertance coefficient values is 60–240.

The nonlinear beam is imposed by the following half-sine pulse[38]:

$$\Gamma = \begin{cases} f\sin(2\pi t/T), & 0 \le t \le T/2, \\ 0, & t > T/2, \end{cases} \tag{14}$$

where $T = 0.4/\pi$.

The vibration energy in the primary structures is absorbed and dissipated by the NES damper to reduce the vibration amplitude, and the efficiency of the NES can be calculated by the following expression:

$$\eta = \frac{c_{\text{nes}} \int_0^t [\varphi_1(x_d)\dot{q}_1 + \varphi_2(x_d)\dot{q}_2 - \dot{v}]^2 dt}{\int_0^T \Gamma[\varphi_1(x_a)q_1 + \varphi_2(x_a)q_2]dt}. \tag{15}$$

Figure 5 presents the influence of the location of NES on the NES efficiency under different external shock excitation locations. As the NES location gradually

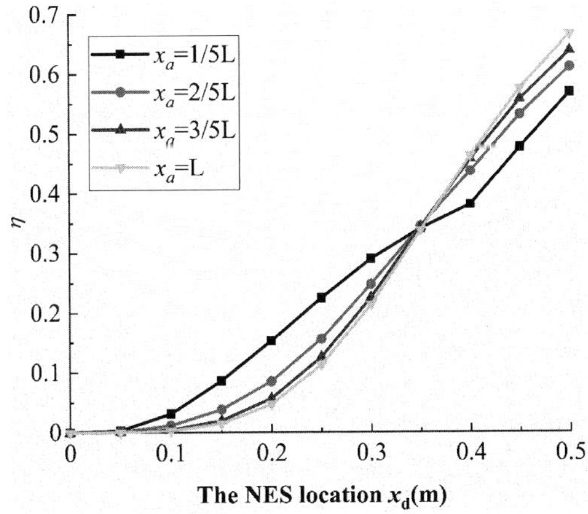

Fig. 5. Variation of the energy dissipation rate with the NES location for different external shock excitation locations.

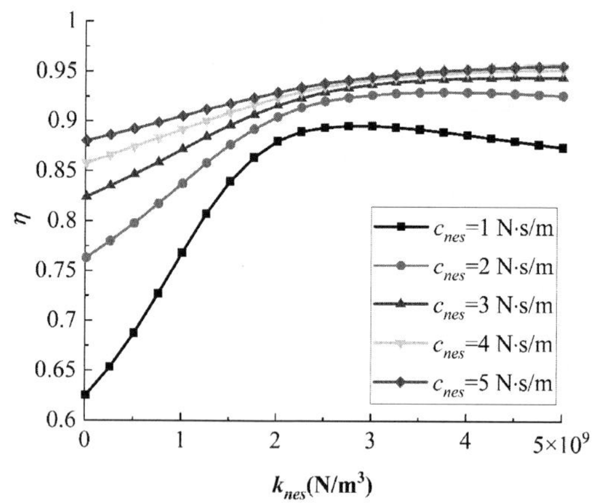

Fig. 6. Variation of the energy dissipation rate with the nonlinear stiffness for different linear damping.

approaches the free end of the beam, the energy dissipation rate of the NES gradually increases. Besides, the performance of the NES has the best vibration reduction performance when it is installed at the right end of the beam. Therefore, in the following parametric study, the NES is installed at the right end of the beam. When the NES is mounted at 0.35 m, the efficiency of the NES does not change with the location of the external shock excitation because this point is the node of the second-order mode of the beam vibration.

Figure 6 illustrates the effect of the nonlinear stiffness of the NES on the energy dissipation rate for different damping. Herein, the inertial mass is set to 0.28 kg. When the damping is chosen to be $c_{\text{nes}} = 1\,\text{N}\cdot\text{s/m}$, the energy dissipation efficiency of the NES increases first and then decreases with increasing nonlinear stiffness of the NES. In contrast, the energy dissipation efficiency gradually increases as the nonlinear stiffness rises for other values of damping. Furthermore, at a given nonlinear stiffness, k_{nes} the energy dissipation efficiency increases as damping c_{nes} increases.

As the damping c_{nes} continuously increases, the energy dissipation efficiency will no longer increase and decrease slowly, as observed in Fig. 7. At a given damping c_{nes} that is lower than $10\,\text{N}\cdot\text{s/m}$, the energy dissipation efficiency increases as the nonlinear stiffness k_{nes} increases. For other damping c_{nes}, the effect of the nonlinear stiffness k_{nes} on the variation of the energy dissipation efficiency is not obvious. Hence, the effect of nonlinear stiffness on the energy dissipation efficiency of the NES is only more pronounced when the damping is taken to a relatively small value.

Figure 8 demonstrates the influence of the inertial mass b_m on the energy dissipation rate. The nonlinear stiffness and damping are set to be $1 \times 10^7\,\text{N/m}^3$ and

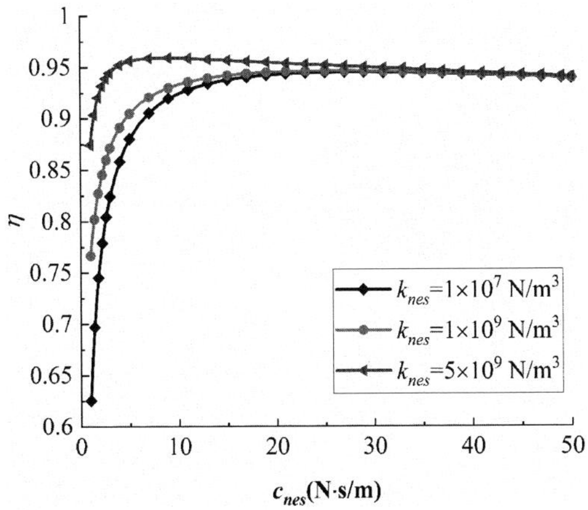

Fig. 7. Variation of the energy dissipation rate with the linear damping for different nonlinear stiffnesses.

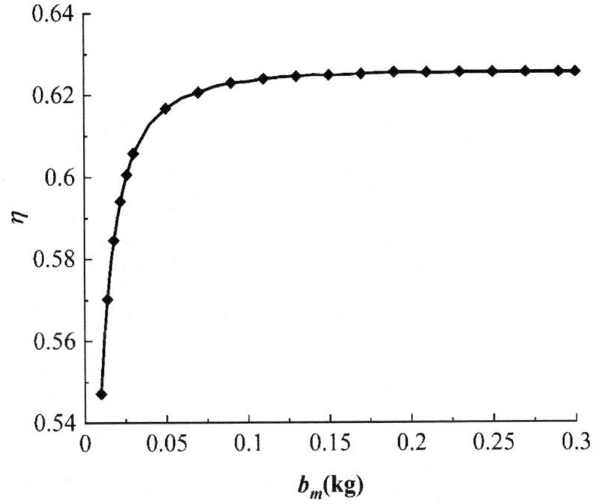

Fig. 8. Variation of the energy dissipation rate with inertial mass.

1 N·s/m, respectively. As the inertial mass parameter b_m increases, the energy dissipation efficiency of the NES increases initially and then remains almost unchanged when the inertial mass is more than 0.1.

5. Beam Response Under Harmonic Excitation

In this section, the nonlinear beam is exposed to a harmonic excitation that can be described as $\Gamma = A\cos(\omega t)$. To evaluate the suppression effect of the NES on the beam vibration, the derived governing equations are investigated in both time and

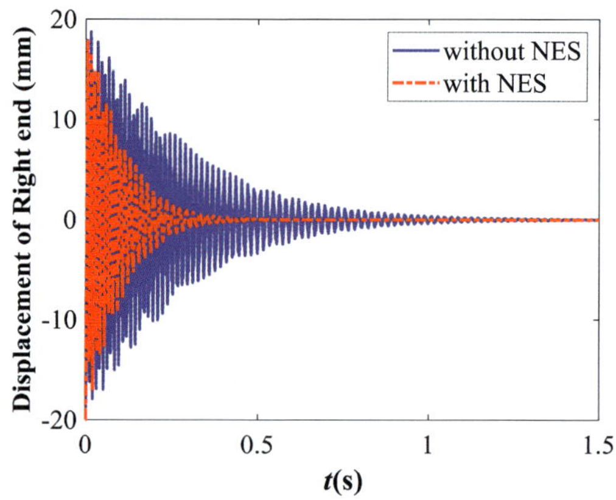

Fig. 9. Vibration suppression effect of the free vibration response at the right end of the beam.

frequency domains using the fourth-order Runge-Kutta method. In the following study, the external amplitude is set to $A = 20\,\mathrm{N/m}$. The initial values are given as

$$q_1 = 0.01, \quad \dot{q}_1 = 0, \quad q_i(i = 1, 2, 3, 4, \ldots, K) = 0, \quad \dot{q}_i = 0, \quad v = \dot{v} = 0. \quad (16)$$

Based on the parameters in Table 1, the free vibration suppression effect of the right point of the beam with and without the NES is depicted in Fig. 9. This indicates that the free vibration response of the beam decreases more quickly to zero after the introduction of the NES.

Figure 10 shows the steady-state response for different external excitation frequencies by the sweep method.[29] It shows that by introducing an NES, the values

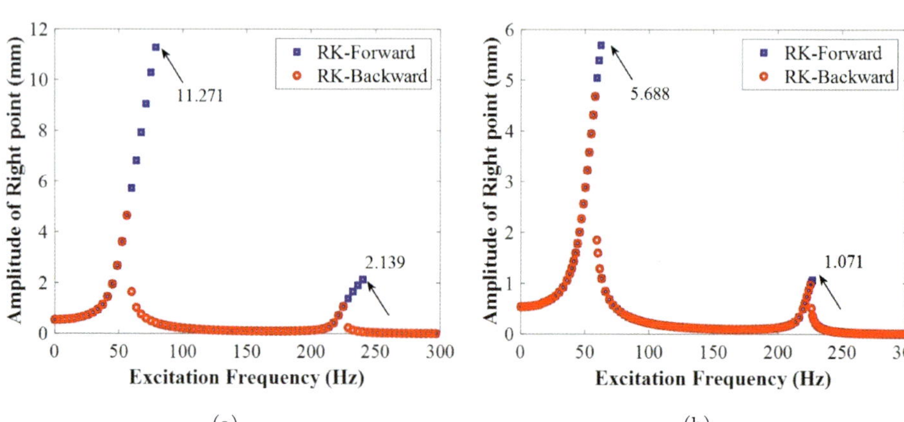

Fig. 10. The steady state response of the right end of the nonlinear beam: (a) without an NES and (b) with an NES.

of the first two primary resonance peaks at the right end point of the beam are reduced by 49.53% and 49.93%, respectively.

To further improve the performance of the NES on the beam vibration suppression, three parameters of the NES, including the inertial mass b_m, nonlinear stiffness k_{nes} and damping c_{nes}, are analyzed in detail. The influence of different damping and nonlinear stiffness on the free vibration response at the right point of the beam is given in Figs. 11 and 12, respectively. With a slight increase in damping, the ability of the NES to suppress the free vibration is significantly improved, while the influence of the nonlinear stiffness coefficient is relatively small.

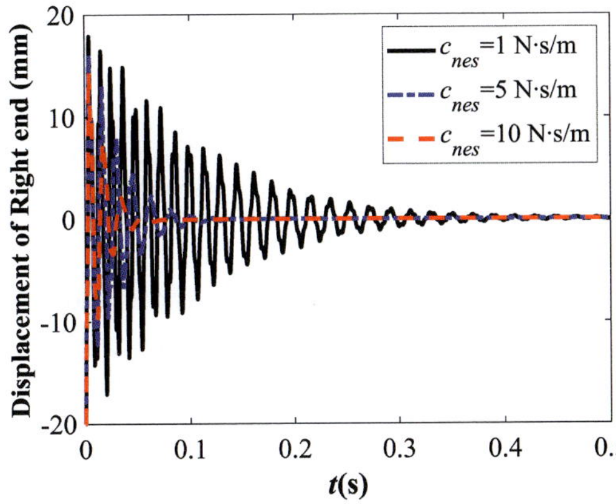

Fig. 11. Free vibration of the right end of the beam with different linear damping.

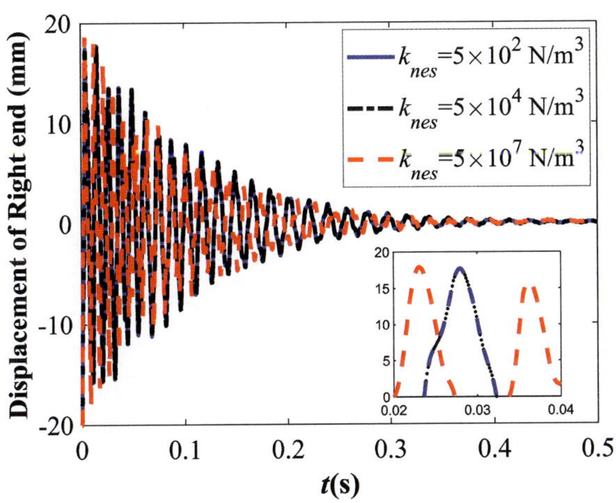

Fig. 12. Free vibration of the right end of the beam with different nonlinear stiffnesses.

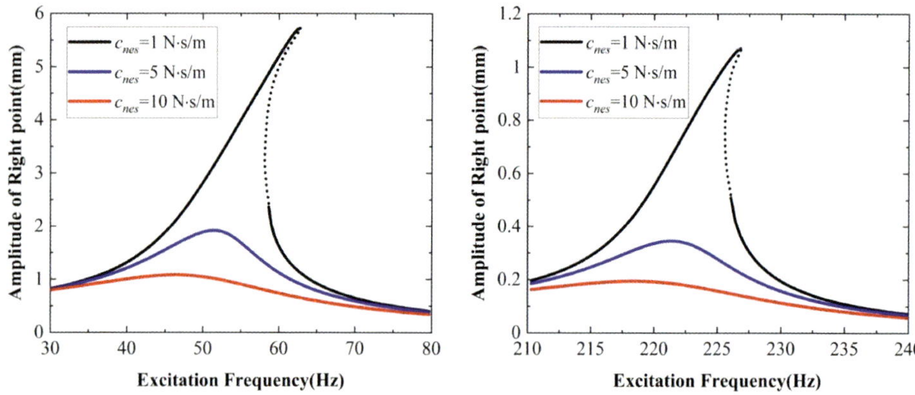

(a) (b)

Fig. 13. The steady state response of the right end of the beam for different linear damping: (a) the first primary resonance and (b) the second primary resonance.

Figure 13 plots the influence of damping on the steady-state response of the right point of the beam. As the damping increases, the peak values of the first- and second-order amplitudes of the right point of the beam decrease significantly. In addition, the amplitude peak of the frequency response curve gradually decreases and shifts to the left. This demonstrates that the hardening characteristics decrease and eventually disappear.

The influences of the nonlinear stiffness and inertia parameters on the steady-state response of the right point of the beam are presented in Figs. 14 and 15. As

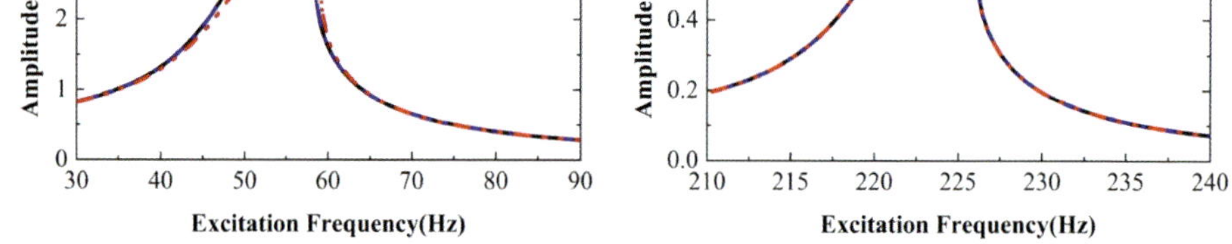

(a) (b)

Fig. 14. The steady state response of the right end of the beam for different nonlinear stiffnesses: (a) the first primary resonance and (b) the second primary resonance.

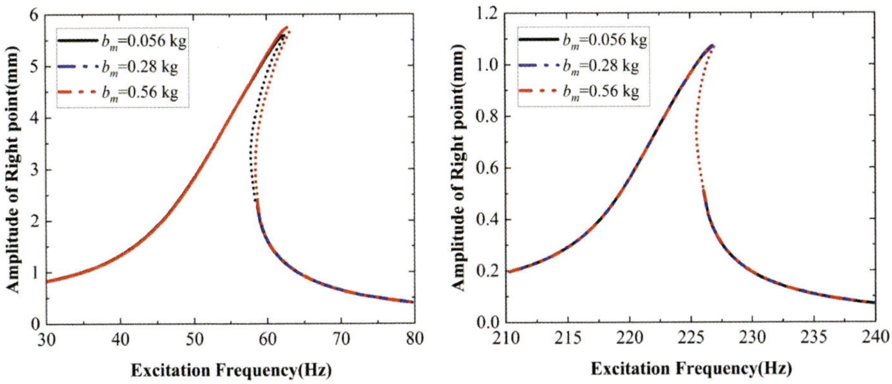

(a) (b)

Fig. 15. The steady state response of the right end of the nonlinear beam for different inertial mass: (a) the first primary resonance and (b) the second primary resonance.

shown in Fig. 14(a), as the nonlinear stiffness increases, the peak of the first-order primary resonance shifts to the lower right. This indicates that the increase in the nonlinear stiffness can enhance the vibration suppression performance of the NES and strengthen the beam's nonlinear hardening characteristics. However, the effect of the nonlinear stiffness of the NES on the second-order primary resonance of the beam is not evident, as observed from Fig. 14(b). Figure 15 shows that the change of the inertial parameters has no noticeable effect on either of the first two orders of the primary resonance response.

6. Conclusion

In this paper, the inertial NES is mounted on an elastically supported nonlinear beam to suppress the multimodal vibration of the beam. Based on the Lagrange equation and the assumed mode method, the motion equations of the beam-NES system are derived. The linear frequency and nonlinear frequencies of the beam are numerically solved by the Rayleigh-Ritz method and iterative method, respectively. In addition, applying the fourth-order Runge-Kutta method, the transient and steady-state responses of the beam are solved numerically in terms of both the amplitude variation rate and energy transfer rate, and a detailed parametric analysis of the NES is carried out. The main conclusions of this study are drawn as follows:

(1) NES can be used to effectively suppress the transverse vibration of nonlinear elastically supported beams under harmonic excitation and shock excitation.

(2) Appropriate parametric design of the NES can effectively improve its vibration suppression performance, and the values of damping of the NES are more sensitive than other parameters.
(3) Elastic supporting structures are more common in engineering practice. Inertial NES and boundary installation solve the problem of excessive additional mass and inappropriate installation.

Acknowledgments

The work described in this paper was fully supported by the National Natural Science Foundation of China (Grant No. 12102015), the General Program of Science and Technology Development Project of Beijing Municipal Education Commission (Project No. KM202110005030), grants from the Fundamental Research Program of Shenzhen Municipality (Project No: JCYJ20170818094653701) and the Tianjin Natural Science Foundation (Grant No. 19JCZDJC32300).

Appendix A

The expression of the mass matrix $\mathbf{M}$ is given as

$$\mathbf{M} = \rho A \int_0^L (\varphi \varphi^T) dx,$$

where

$$\begin{aligned}
\varphi_i &= [\varphi_1, \varphi_2]^T; \\
\varphi_1 &= \cos(1.8751x/L) - \mathrm{ch}(1.8751x/L) \\
&\quad - (0.7341)[\sin(1.8751x/L) - \mathrm{sh}(1.8751x/L)]; \\
\varphi_2 &= \cos(4.6941x/L) - \mathrm{ch}(4.6941x/L) \\
&\quad - (1.0185)[\sin(4.6941x/L) - \mathrm{sh}(4.6941x/L)].
\end{aligned}$$

The expression of the linear stiffness matrix $\mathbf{K}_L$ is given as

$$\mathbf{K}_L = EI \int_0^L \frac{\partial^2 \varphi}{\partial x^2} \frac{\partial^2 \varphi^T}{\partial x^2} dx.$$

The expression of the elastic stiffness matrix $\mathbf{K}_S$ is given as

$$\mathbf{K}_S = k_s \varphi_s \varphi_s^T.$$

The expression of the nonlinear stiffness matrix $\mathbf{K}_{NL}$ is given as

$$\mathbf{K}_{NL} = \frac{EA}{2} \int_0^L \left(q^T \frac{\partial \varphi}{\partial x} \frac{\partial \varphi^T}{\partial x} q \right) \left(\frac{\partial \varphi}{\partial x} \frac{\partial \varphi^T}{\partial x} \right) dx.$$

References

1. L. Zuo and S. A. Nayfeh, The two-degree-of-freedom tuned-mass damper for suppression of single-mode vibration under random and harmonic excitation, *J. Vib. Acoust.* **128** (2006) 56–65.
2. O. F. Tigli, Optimum vibration absorber (tuned mass damper) design for linear damped systems subjected to random loads, *J. Sound Vib.* **331** (2012) 3035–3049.
3. C. L. Davis and G. A. Lesieutre, An actively tuned solid-state vibration absorber using capacitive shunting of piezoelectric stiffness, *J. Sound Vib.* **232** (2000) 601–617.
4. C. H. Loh and P. Y. Lin, Kalman filter approach for the control of seismic-induced building vibration using active mass damper systems, *Struct. Des. Tall Build.* **6** (1997) 209–224.
5. A. F. Vakakis, Inducing passive nonlinear energy sinks in vibrating systems, *J. Vib. Acoust.* **123** (2001) 324–332.
6. J. R. Xue, Y. W. Zhang, H. Ding and L. Q. Chen, Vibration reduction evaluation of a linear system with a nonlinear energy sink under a harmonic and random excitation, *Appl. Math. Mech.* **41** (2019) 1–14.
7. A. Blanchard, L. A. Bergman and A. F. Vakakis, Vortex-induced vibration of a linearly sprung cylinder with an internal rotational nonlinear energy sink in turbulent flow, *Nonlinear Dyn.* **99** (2019) 593–609.
8. Y. Wei, S. Wei, Q. Zhang, X. Dong, Z. Peng and W. Zhang, Targeted energy transfer of a parallel nonlinear energy sink, *Appl. Math. Mech.* **40** (2019) 621–630.
9. D. M. McFarland, G. Kerschen, J. J. Kowtko, Y. S. Lee, L. A. Bergman and A. F. Vakakis, Experimental investigation of targeted energy transfers in strongly and nonlinearly coupled oscillators, *J. Acoust. Soc. Am.* **118** (2005) 791–799.
10. X. Li, K. Liu, L. Xiong and L. Tang, Development and validation of a piecewise linear nonlinear energy sink for vibration suppression and energy harvesting, *J. Sound Vib.* **503** (2021) 116104.
11. O. V. Gendelman, Targeted energy transfer in systems with external and self-excitation, *Proc. Inst. Mech. Eng. C: J. Mech. Eng. Sci.* **225** (2011) 2007–2043.
12. J. Zang and L. Q. Chen, Complex dynamics of a harmonically excited structure coupled with a nonlinear energy sink, *Acta Mech. Sin.* **33** (2017) 801–822.
13. Z. N. Ahmadabadi and S. E. Khadem, Annihilation of high-amplitude periodic responses of a forced two degrees-of-freedom oscillatory system using nonlinear energy sink, *J. Vib. Control* **19** (2012) 2401–2412.
14. S. Tsakirtzis, P. N. Panagopoulos, G. Kerschen, O. Gendelman, A. F. Vakakis and L. A. Bergman, Complex dynamics and targeted energy transfer in linear oscillators coupled to multi-degree-of-freedom essentially nonlinear attachments, *Nonlinear Dyn.* **48** (2006) 285–318.
15. F. Georgiades and A. F. Vakakis, Dynamics of a linear beam with an attached local nonlinear energy sink, *Commun. Nonlinear Sci. Num. Simul.* **12** (2007) 643–651.
16. Y. W. Zhang, H. Zhang, S. Hou, K. F. Xu and L. Q. Chen, Vibration suppression of composite laminated plate with nonlinear energy sink, *Acta Astronaut.* **123** (2016) 109–115.
17. O. V. Gendelman and A. Alloni, Dynamics of forced system with vibro-impact energy sink, *J. Sound Vib.* **358** (2015) 301–314.
18. M. Farid, O. V. Gendelman and V. I. Babitsky, Dynamics of a hybrid vibro-impact nonlinear energy sink, *ZAMM J. Appl. Math. Mech.* **101** (2019) e201800341.

19. G. Habib and F. Romeo, The tuned bistable nonlinear energy sink, *Nonlinear Dyn.* **89** (2017) 179–196.
20. M. Farid and O. V. Gendelman, Tuned pendulum as nonlinear energy sink for broad energy range, *J. Vib. Control* **23** (2016) 373–388.
21. M. A. Al Shudeifat, N. E. Wierschem, L. A. Bergman and A. F. Vakakis, Numerical and experimental investigations of a rotating nonlinear energy sink, *Meccanica* **52** (2016) 763–779.
22. G. C. Tsiatas and A. E. Charalampakis, A new hysteretic nonlinear energy sink (HNES), *Commun. Nonlin. Sci. Numer. Simul.* **60** (2018) 1–11.
23. M. C. Smith, Synthesis of mechanical networks: The inerter, *IEEE Trans. Autom. Control* **47** (2002) 1648–1662.
24. A. Giaralis and F. Petrini, Wind-induced vibration mitigation in tall buildings using the tuned mass-damper-inerter, *J. Struct. Eng.* **143** (2017) 04017127.
25. Y. Li, J. Z. Jiang and S. Neild, Inerter-based configurations for main-landing-gear shimmy suppression, *J. Aircraft* **54** (2017) 684–693.
26. Y. Shen, L. Chen, X. Yang, D. Shi and J. Yang, Improved design of dynamic vibration absorber by using the inerter and its application in vehicle suspension, *J. Sound Vib.* **361** (2016) 148–158.
27. X. Shi and S. Zhu, Dynamic characteristics of stay cables with inerter dampers, *J. Sound Vib.* **423** (2018) 287–305.
28. Y. W. Zhang, Y. N. Lu, W. Zhang, Y. Y. Teng, H. X. Yang, T. Z. Yang and L. Q. Chen, Nonlinear energy sink with inerter, *Mech. Syst. Sig. Process.* **125** (2019) 52–64.
29. Z. Zhang, H. Ding, Y. W. Zhang and L. Q. Chen, Vibration suppression of an elastic beam with boundary inerter-enhanced nonlinear energy sinks, *Acta Mech. Sin.* **37** (2021) 387–401.
30. Z. Zhang, Z. Q. Lu, H. Ding and L. Q. Chen, An inertial nonlinear energy sink, *J. Sound Vib.* **450** (2019) 199–213.
31. X. Y. Mao, H. Ding and L. Q. Chen, Vibration of flexible structures under nonlinear boundary conditions, *J. Appl. Mech.* **84** (2017) 11.
32. J. Zang, R. Q. Cao and Y. W. Zhang, Steady-state response of a viscoelastic beam with asymmetric elastic supports coupled to a lever-type nonlinear energy sink, *Nonlinear Dyn.* **105** (2021) 1327–1341.
33. H. Ding, M. H. Zhu and L. Q. Chen, Nonlinear vibration isolation of a viscoelastic beam, *Nonlinear Dyn.* **92** (2018) 325–349.
34. M. Kani, S. E. Khadem, M. H. Pashaei and M. Dardel, Vibration control of a nonlinear beam with a nonlinear energy sink, *Nonlinear Dyn.* **83** (2015) 1–22.
35. M. Parseh, M. Dardel, M. H. Ghasemi and M. H. Pashaei, Steady state dynamics of a non-linear beam coupled to a non-linear energy sink, *Int. J. Non-Linear Mech.* **79** (2016) 48–65.
36. Y. W. Zhang, S. Hou, Z. Zhang, J. Zang, Z. Y. Ni, Y. Y. Teng and L. Q. Chen, Nonlinear vibration absorption of laminated composite beams in complex environment, *Nonlinear Dyn.* **99** (2020) 2605–2622.
37. A. F. Vakakis, Passive nonlinear targeted energy transfer, *Philos. Trans. A Math. Phys. Eng. Sci.* **376** (2018) 20170132.
38. J. Chen, W. Zhang, M. Yao, J. Liu and M. Sun, Vibration reduction in truss core sandwich plate with internal nonlinear energy sink, *Compos. Struct.* **193** (2018) 180–188.

39. Z. Lu, K. Li, H. Ding and L. Chen, Nonlinear energy harvesting based on a modified snap-through mechanism, *Appl. Math. Mech.* **40** (2018) 167–180.
40. Z. W. Fang, Y. W. Zhang, X. Li, H. Ding and L. Q. Chen, Integration of a nonlinear energy sink and a giant magnetostrictive energy harvester, *J. Sound Vib.* **391** (2017) 35–49.
41. H. Ding, Z. Q. Lu and L. Q. Chen, Nonlinear isolation of transverse vibration of pre-pressure beams, *J. Sound Vib.* **442** (2019) 738–751.
42. W. Zhang, Z. Y. Chang and J. Chen, Vibration reduction for an asymmetric elastically supported beam coupled to an inertial nonlinear energy sink, *J. Vib. Eng. Technol.* (2022) 1–13, https://doi.org/10.1007/s42417-022-00666-x.
43. Y. Wang and J. Chen, Nonlinear free vibration of rotating functionally graded graphene platelets reinforced blades with variable cross-sections, *Eng. Anal. Bound. Elem.* **144** (2022) 262–278.
44. L. L. Ke, J. Yang and S. Kitipornchai, Nonlinear free vibration of functionally graded carbon nanotube-reinforced composite beams, *Compos. Struct.* **92** (2010) 676–683.
45. S. R. Marur and G. Prathap, Non-linear beam vibration problems and simplifications in finite element models, *Comput. Mech.* **35** (2004) 352–360.
46. M. Chen, C. Papageorgiou, F. Scheibe, F. C. Wang and M. Smith, The missing mechanical circuit element, *IEEE Circ. Syst. Mag.* **9** (2009) 10–26.

https://doi.org/10.1142/9789819814770_0016

Chapter 16

High-Frequency Resonances of Train-Track Coupled System Due to Multiple Wheels Interference#

Yaojun Ge* and Simian Lei†

State Key Laboratory of Disaster Reduction in Civil Engineering
Tongji University, 1239 Siping Road
Shanghai 200092, P. R. China
**yaojunge@tongji.edu.cn*
†1810723@tongji.edu.cn

Qi Li

Department of Bridge Engineering
Tongji University, 1239 Siping Road
Shanghai 200092, P. R. China
liqi_bridge@tongji.edu.cn

A significant resonance pattern at high frequencies around 400 Hz to 1000 Hz has been well observed from measured rail accelerations. Such resonances are key issues for rail noise and corrugation. Although this pattern has been noted and discussed in various papers in terms of wave reflections among multiple wheels, the aim of this study is to develop novel dynamic model and reveal generation mechanism of this resonance phenomenon with the train-track coupled system. An infinite Timoshenko beam with continuous supports is adopted for modeling the track system, and the point and transfer receptances of the rail for a moving excitation are explicitly deduced by the residue theorem together with a Fourier transform-based method. A frequency-domain method is then established to calculate the power spectral densities (PSD) of the train-track coupled system responses to stochastic irregularities with the moving vehicle model with multiple wheels. It is found that the high-frequency resonances occur when the reflected waves generated from multiple wheels on the rail are exactly in phase with the initial track irregularity. The load speed has negligible influence on the wheel–rail interaction in the lower frequency range and a small influence in the upper frequency range only for very high train speeds.

Keywords: High-frequency resonance; infinite Timoshenko beam; point receptance; train–track couple system; multi-wheel moving excitation; track irregularity.

*Corresponding author.
#This chapter appeared previously on the International Journal of Structural Stability and Dynamics. To cite this chapter, please cite the original article as the following: Y. Ge, S. Lei & Q. Li, *Int. J. Struct. Stab. Dyn.*, **23**, 2340018 (2023), DOI: 10.1142/S0219455423400187.

1. Introduction

High-frequency resonance around 400 Hz to 1000 Hz has been well observed from field measurement of rail accelerations, which are key issues for rail noise and corrugation.[1–3] Various studies found this resonance phenomenon can be attributed to the wheel-rail interactions between irregular profiles on wheels and rail contact surfaces and the vibration induced by multiple wheel moving on the rail with interference.[4–6] The key to the question is how to develop high-precision dynamic model[7,8] and reveal intrinsic generation mechanism of this resonance in train-track coupled system.[9]

Wheel–rail interactions generated in the railway system are primarily caused by irregular profiles on the wheel and rail contact surfaces, among which the general track geometry irregularities are the main source. It is found the responses of the train-track coupled system induced by track irregularities may present a series of peaks in the spectra at different frequencies associated with various mechanisms.[10,11] These peaks may lead to higher ambient noise at specific frequencies up to thousands of Hertz,[12,13] negatively influencing the comfort of passengers and residential areas with trains passing by. Moreover, the dominant wheel-rail contact forces generated in high frequencies prove crucial in the process of railhead corrugation,[14] leading to increased interactions in turn. In view of these detrimental effects, a great many attempts have been made to investigate the generation mechanism of these peaks in the vibrations of train-track system and wheel-rail interaction forces.

Due to the effects of multiple wheels on a rail, the vibration induced by multiple wheel-rail interactions is observed to display more peaks than that of a single wheel-rail interaction.[15] In an earlier study by Igeland,[14] a series of peaks around 600–800 Hz were found in the spectra of wheel-rail contact forces in the model of a bogie running on a discretely supported rail. Wu and Thompson[10] first brought forward the concepts of active wheel and passive wheel for evaluating the track response excited by multiple wheels. An interesting peak around 600 Hz was then observed in the active wheel-rail interaction force as well as in the point receptance of a rail with additional passive wheels attached.[10,16] Wu and Thompson[17] then developed forward this research and studied these peaks in detail by using a track model with multiple wheels attached. Unlike the results from a single-wheel model, four main peaks were shown to appear in the spectra of contact forces around 550–1200 Hz. In order to explain these additional peaks, the influence of wave reflections between multiple wheels was qualitatively studied. Such cases also happened to Ding *et al.*[18] when investigating the effects of wave reflections between multiple wheels on curve squeal.

The primary step in studying these resonances is to evaluate the point receptance of the rail, which is crucial in dealing with wheel–rail interactions either in the time domain or frequency domain.[19] An infinite Timoshenko beam on discrete supports is often adopted to consider the effect of shear deformation and rotation inertia of the rail at frequencies above 500 Hz.[20] Sheng *et al.*[21] developed a wavenumber-based method in order to calculate the response of the track to a harmonic load of high frequency and moving at high speed. Note that the point receptance of such a discretely supported model is not completely harmonic, raising troubles in analyzing train–track dynamics in the frequency domain.[21,22] However, the discreteness effect can be disregarded if the interested frequency does not fall in the vicinity of the pinned–pinned resonance.[23] With a continuously supported model, the track becomes a linear time invariant system and gives a stationary response with a certain frequency in the moving frame when subjected to a moving harmonic load.[24] It should be mentioned that a moving excitation model[10,16,17,25] is often applied to calculate wheel-rail interactions, in which wheels are regarded as stationary and irregularities are strapped between the wheels and the rail as a relative excitation.[17] Unlike the moving wheel model, the effect of load motion on the rail receptance is not considered in such a moving roughness model.[26] This effect on the point receptance of an Euler–Bernoulli beam was later discussed by Thompson,[27] showing that the wavelength ahead of the load is shorter than that behind the load.

Recently, peaks around the frequencies of 400 Hz to 1000 Hz in track vibrations have been once more observed during several field measurements made on an elevated railway with ballastless track. Although this phenomenon has been revisited in several papers, it still needs quantitative phase analyzes and detailed demonstrations of wave interference process. The aim of this study is to theoretically investigate the generation mechanism of this phenomenon both from mathematical and physical points of view. A typical model of an infinite Timoshenko beam with continuous supports is adopted to model the track system, and its point and transfer receptances are explicitly derived by means of the Fourier transform-based method and the residue theorem. Then, the power spectral densities (PSDs) of the system response are analytically derived and expressed as three parts: the frequency response function (FRF) of the vehicle subsystem, the dynamic stiffness matrix of equivalent wheel-rail contact forces and the PSD for the track irregularities. Based on the concepts of active and passive wheels,[10] the multiple peaks in the PSDs of the wheel-rail interaction forces are investigated from a mathematical point of view by extreme value analysis of the response function and from a physical point of view based on the wave propagation theory. In the following section, the effect of load motion on the response of train-track coupled system is discussed, based on the model of four wheels running on a Timoshenko rail, considering the influence of different types of pad stiffness and train speed.

2. Field Measurement

A field measurement was made at an urban rail transit viaduct in Shanghai, China, where the track is supported by U-shaped girders, as shown in Fig. 1(a). The nominal fastener stiffness is 40 MN/min the vertical direction, with a fastener spacing of 0.6 m. The rail vibration was measured using a piezoelectric accelerometer mounted on the outboard rail foot shown schematically in Fig. 1(b), with a sampling frequency of 10 kHz. Figure 2 shows the time history and a PSD estimate of the vertical acceleration of the rail during the passage of a three-vehicle train running at 60 km/h. The PSD is obtained by using Welch's method with overlapping time segments, resulting in a frequency resolution of 4.883 Hz.

As can be seen from Fig. 2(b), in addition to random variations caused by the excitation due to the track irregularities, there are several peaks in the PSD which can be divided into two categories. In the low-frequency region, a strong

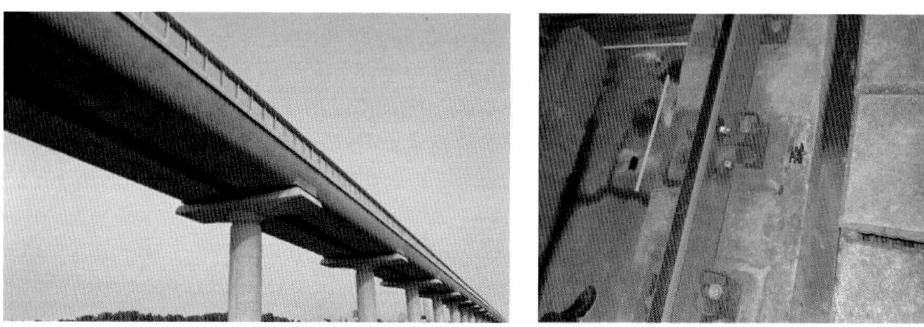

(a) (b)

Fig. 1. Photo of field test: (a) an urban rail transit viaduct in Shanghai and (b) piezoelectric accelerometer mounted on the outboard rail foot.

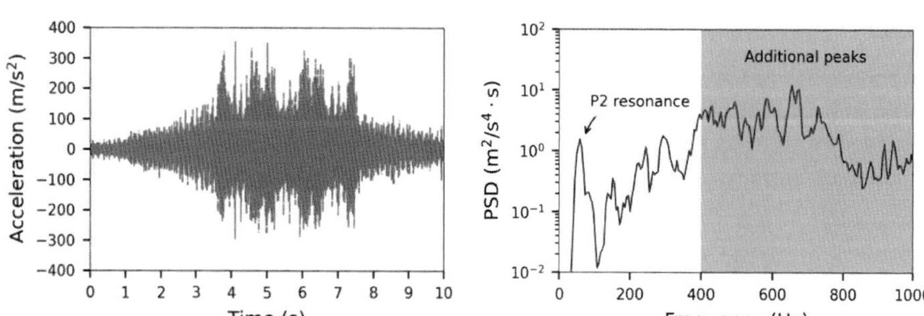

(a) (b)

Fig. 2. Measured vertical accelerations of the rail from field test: (a) time history and (b) PSD.

peak occurs at around 50 Hz that corresponds to the so-called P2 resonance of the wheel mass on the track stiffness.[27,28] A peak at around 170 Hz corresponds to the resonance of the rail mass bouncing on the pad stiffness.[27] In the frequency range from 400 Hz to 1000 Hz, there are additional peaks which may be due to the presence of multiple wheels on the rail.[10,16] The generation mechanism of these additional peaks is investigated in the following sections.

3. Point Receptance of Infinite Timoshenko Beam

The track is usually supported by periodic sleepers, yet the point receptance of it is not purely harmonic, causing difficulties in dealing with train-track dynamics in the frequency domain. In fact, the discreteness effect can be neglected if dominating frequencies don't fall in the vicinity of the pinned–pinned resonance,[20,29] especially if the rail pads are relatively soft.[15,30,31] In addition, it was shown by Wu and Thompson[10] that the effects of multiple wheel–rail interactions may also suppress the pinned-pinned mode. As the shear deformation and rotational inertia of the rail should be considered for frequencies over 500 Hz,[20] a continuously supported Timoshenko beam model subjected to a moving harmonic load[24] is used in this study to obtain the point and transfer receptances of the track in the moving frame of reference, initially without wheels mounted on it.

3.1. *Track model*

The vibration of a Timoshenko beam is described by the vertical displacement w_r (directed positive downwards) and the rotation angle ϕ_r of the cross-section (clockwise), as shown in Fig. 3. For a harmonic load $Qe^{i\omega_0 t}$ moving at a speed V, with amplitude Q and angular frequency ω_0, the kinetic equations for an infinite Timoshenko beam with continuous supports[27] can be rewritten in the moving coordinate system attached to the load, i.e. $\xi = x - Vt$,

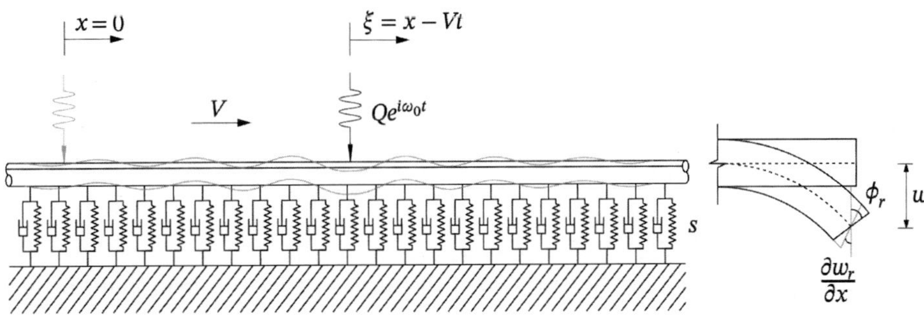

Fig. 3. An infinite Timoshenko beam on an elastic foundation showing the vibration generated by a moving harmonic load.

$$\begin{cases} GA\kappa\dfrac{\partial}{\partial\xi}\left(\phi_r - \dfrac{\partial w_r}{\partial\xi}\right) + sw_r + \rho A\left(\dfrac{\partial^2 w_r}{\partial t^2} + V^2\dfrac{\partial^2 w_r}{\partial\xi^2} - 2V\dfrac{\partial^2 w_r}{\partial\xi\partial t}\right) = Q\delta(\xi)\mathrm{e}^{i\omega_0 t}, \\ GA\kappa\left(\phi_r - \dfrac{\partial w_r}{\partial\xi}\right) - EI\dfrac{\partial^2\phi_r}{\partial\xi^2} + \rho I\left(\dfrac{\partial^2\phi_r}{\partial t^2} + V^2\dfrac{\partial^2\phi_r}{\partial\xi^2} - 2V\dfrac{\partial^2\phi_r}{\partial\xi\partial t}\right) = 0, \end{cases} \tag{1}$$

where ρA is the rail mass per unit length, ρI is the rail rotational inertia, EI is the rail bending stiffness, GA is the rail shear stiffness, κ is the rail shear parameter, and s is the support stiffness per unit length. Damping is introduced in the supports by making the stiffness complex with the form $s(1 + i\eta)$ (this is suppressed in the equations for simplicity).

3.2. *Solution based on Fourier-transform method*

By taking a Fourier transform, Eq. (1) can be transformed from the spatial domain to the wavenumber domain

$$\begin{cases} -GA\kappa ik\Phi + [GA\kappa k^2 + s - \rho A(kV + \omega_0)^2]W = Q, \\ GA\kappa ikW + [GA\kappa + EIk^2 - \rho I(kV + \omega_0)^2]\Phi = 0. \end{cases} \tag{2}$$

Solving Eq. (2) yields the complex amplitudes W and Φ, and then, the response $w_r(\xi, t)$ at position ξ and time t can be obtained by performing the inverse Fourier transform with respect to k and ω

$$w_r(\xi, t) = \frac{Q\mathrm{e}^{i\omega_0 t}}{2\pi}\int_{-\infty}^{\infty}\frac{M(k)}{N(k)}\mathrm{e}^{-ik\xi}dk, \tag{3}$$

where

$$M(k) = GA\kappa + EIk^2 - \rho I(kV + \omega_0)^2, \tag{4}$$

$$N(k) = [GA\kappa k^2 + s - \rho A(kV + \omega_0)^2][GA\kappa + EIk^2 - \rho I(kV + \omega_0)^2] - G^2A^2\kappa^2k^2. \tag{5}$$

It is worth pointing out that the dynamic response is stationary and harmonic if the observation point moves with the load. Therefore, the train-track coupled system retains the linear superposition characteristic and the frequency-domain method becomes possible.

3.3. *Track receptance*

The transfer receptance of the track at a distance ξ from the load can be then obtained as

$$H(\xi, \omega_0) = \frac{1}{2\pi}\int_{-\infty}^{\infty}\frac{M(k)}{N(k)}\mathrm{e}^{-ik\xi}dk. \tag{6}$$

The integral in Eq. (6) can be evaluated by using contour integration. For $\xi \geq 0$, the integral is equal to $-2\pi i$ times the sum of the residues of the poles in the lower half plane, whereas for $\xi < 0$, the integral is equal to $2\pi i$ times the sum of the residues of the poles in the upper half plane. Hence, the receptance is given by

$$H(\xi, \omega_0) = \begin{cases} -i \sum_{n=3}^{4} \text{Res}\left[\frac{M(z)}{N(z)}\text{e}^{-i\xi z}, k_n\right] \\ \quad = -i\left[\frac{M(k_3)\text{e}^{-i\xi k_3}}{N'(k_3)} + \frac{M(k_4)\text{e}^{-i\xi k_4}}{N'(k_4)}\right] (\xi \geq 0), \\ i \sum_{n=1}^{2} \text{Res}\left[\frac{M(z)}{N(z)}\text{e}^{-i\xi z}, k_n\right] \\ \quad = i\left[\frac{M(k_1)\text{e}^{-i\xi k_1}}{N'(k_1)} + \frac{M(k_2)\text{e}^{-i\xi k_2}}{N'(k_2)}\right] (\xi < 0), \end{cases} \tag{7}$$

where k_n are the zeros of $N(k)$ and $\text{Im}(k_{1,2}) > 0, \text{Im}(k_{3,4}) < 0$. To illustrate the contour integration, use is made of typical parameters given in Table 1 for the rail, which are also applied in the following analysis. It is noted that the support stiffness of $s = 66.67\,\text{MN/m}^2$ corresponds to a vertical fastener stiffness of 40 MN/m at a fastener spacing of 0.6 m. Damping is introduced in the rail pads and the rail, as listed in Table 1. Figure 4(a) shows the distribution of the four poles k_n on the complex plane, for load speeds ranging from 0 m/s to 100 m/s and a frequency of 300 Hz. The decay rates for the track waves corresponding to k_3 and k_4 at a load speed of 40 m/s are shown in Fig. 4(b).

As is shown in Fig. 4, k_2 and k_4 are the poles with the largest real part (positive or negative). The real part of the poles represents the wavenumber, and the imaginary part corresponds to decay with distance. It can be seen from Fig. 4(b) that the decay rates for waves k_3 and k_4 both remain at a high level at frequencies below 200 Hz. However, at frequencies above 250 Hz, the rate of attenuation of the wave corresponding to the pole k_4 is far lower than that of the pole k_3, indicating that the transfer receptance for the track at sufficient distance from the load is dominated by the k_4 component at high frequencies.

Table 1. Parameters for railway track based on Timoshenko beam model.[27,32]

Parameter	Definition	Unit	Value
GA	Rail shear stiffness	MN	617
EI	Rail bending stiffness	MNm^2	6.42
η_r	Damping loss factor of rail	/	0.02
ρA	Rail mass per unit length	kg/m	60
ρI	Rail rotational inertia	Kgm	0.240
κ	Rail shear parameter	/	0.4
s	Support stiffness per unit length of foundation	MN/m^2	66.67
η_s	Damping loss factor of foundation	/	0.15

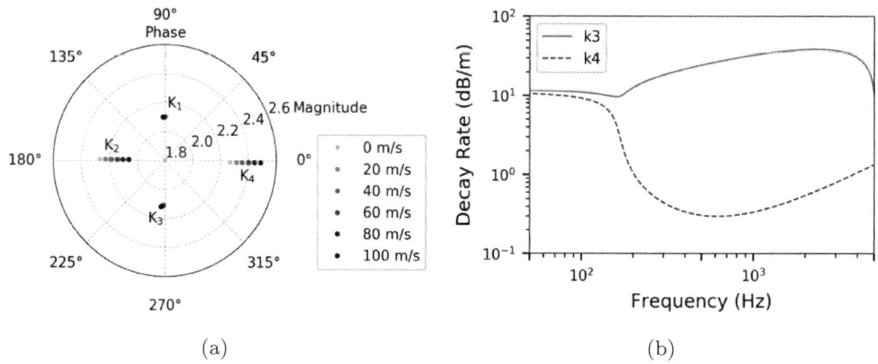

Fig. 4. Poles of $N(k)$ on the complex plane and decay rates for the track waves using the parameters in Table 1: (a) distribution of four poles at a frequency of 300 Hz for various load speeds and (b) decay rates at a load speed of 40 m/s.

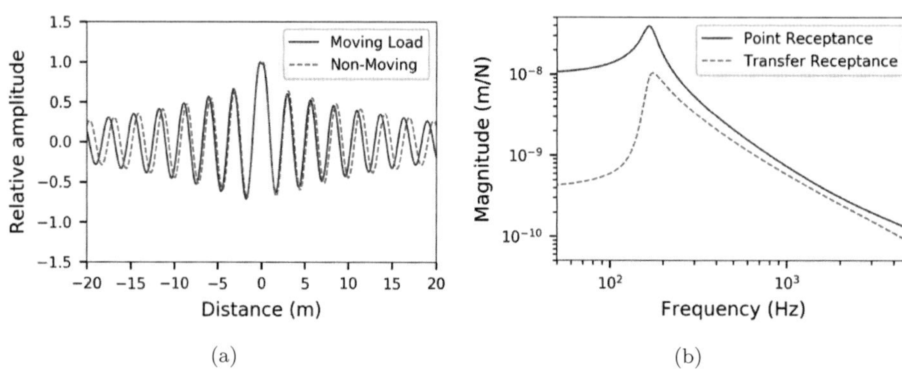

Fig. 5. The response of the rail: (a) instantaneous wave propagation along the rail at 300 Hz for a non-moving load and a load moving to the right with a speed of 40 m/s and (b) the point receptance and the transfer receptance magnitude at $\xi = 2.5\,\text{m}$ for a load moving to the right at 40 m/s.

The effect of the load speed on the poles is also shown in Fig. 4(a). The pole k_4 is related to the wave propagating in the same direction as the load speed and the pole k_2 is related to the wave in the opposite direction. As the load speed increases, the poles k_2 and k_4 migrate slightly to the right, meaning that the wavelength for k_2 wave increases and the wavelength for k_4 wave decreases. This is the basis of the Doppler effect.[33] Figure 5(a) illustrates the effect of the load motion on the wave propagation along the rail, for a speed of 40 m/s and a frequency of 300 Hz. It is shown that the wave ahead of the load is compressed compared to the static case, whereas the wave behind the load is lengthened.

The point receptance of the rail and the transfer receptance at $\xi = 2.5\,\mathrm{m}$ for this case are also shown in Fig. 5(b). It can be seen that magnitude of the transfer receptance is far lower than that of the point receptance at frequencies below 200 Hz, due to the high decay rates in this frequency region. A peak is found at around 200 Hz in both point and transfer receptances, corresponding to the resonance of the rail mass vibrating on the pad stiffness.[27]

4. Dynamics of Train-Track Coupled System

In this section, the receptances obtained in Sec. 3 are employed to derive the spectral density matrices for the response of the track and the vehicle (see Fig. 6) excited by random roughness or irregularities.

4.1. *Train-track coupled model*

The vehicle is assumed to be symmetric about the track centreline. It is represented by a half vehicle model consisting of a half car body, two half bogie frames, four wheels and several flexible spring-damper elements connecting them. The car body, bogie frames, and wheels are regarded as rigid bodies, and only vibration in the vertical direction is considered.[7] It is noted that, for vertical motion, the rigid model for the wheel is reasonable for frequencies below 1 kHz. At higher frequencies flexible modes in the radial direction occurring above 2 kHz should be considered.[27] Considering the natural frequencies of bridges are much lower than that of the wheel bouncing on the track, and the frequencies of the peaks of interest in this study are rather high, therefore the bridge structure is disregarded in the model, and this will have little effect on the stochastic response of the vehicle apart from its mean

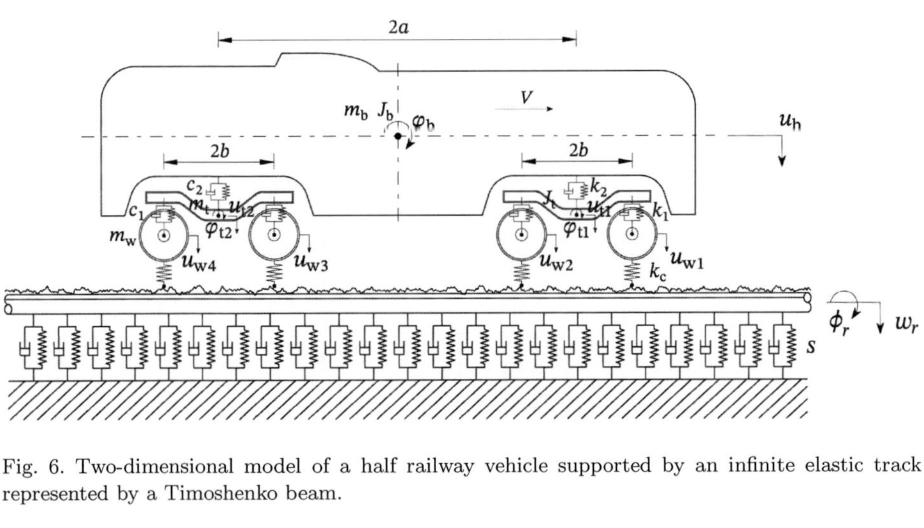

Fig. 6. Two-dimensional model of a half railway vehicle supported by an infinite elastic track represented by a Timoshenko beam.

Table 2. Parameters for a metro used in urban rail transit viaduct in Shanghai.

Parameter	Definition	Unit	Value
m_b	Mass of the half car body	kg	2.045×10^4
J_b	Pitch moment of inertia for the half car body	$\text{kg} \cdot \text{m}^2$	8.994×10^5
m_t	Mass of the half bogie frame	kg	2.223×10^3
J_t	Pitch moment of inertia for the half bogie frame	$\text{kg} \cdot \text{m}^2$	1.629×10^3
m_w	Mass of the wheel	kg	8.94×10^2
a	Half distance between the secondary suspensions	M	7.85
b	Half distance between the two wheels of each bogie	M	1.25
k_1	Vertical stiffness of the primary suspension per wheel	N/m	1.22×10^6
c_1	Vertical damping of the primary suspension per wheel	$\text{N} \cdot \text{s/m}$	1.5×10^4
k_2	Vertical stiffness of the secondary suspension per bogie	N/m	3.4×10^5
c_2	Vertical damping of the secondary suspension per bogie	$\text{N} \cdot \text{s/m}$	2.0×10^4
k_c	Equivalent vertical stiffness of wheel-track contact per wheel	N/m	1.5×10^9

response.[34] Table 2 lists the definitions of the main parameters used to describe a half vehicle along with their values that relate to a metro train in Shanghai.

Based on the aforementioned assumptions, the vehicle model has a total of 10 degrees of freedom (DOFs), including seven translational and three rotational DOFs. The equation of motion for the vehicle subsystem can be expressed as

$$\mathbf{m}\ddot{\mathbf{u}} + \mathbf{c}\dot{\mathbf{u}} + \mathbf{k}\mathbf{u} = -\mathbf{f}, \tag{8}$$

where **m, c,** and **k** are the mass, damping, and stiffness matrices for the vehicle, respectively. **u** is the displacement vector of the vehicle. **f** is the vector of the wheel-rail interaction forces. The negative sign in Eq. (8) represents the fact that the forces act upward on the wheels.

The wheel-rail contact model is an essential element for analyzing the dynamic performance of the entire system. A linearized contact spring is adopted to simplify the system as a linear stationary stochastic vibration problem (see Fig. 7), which

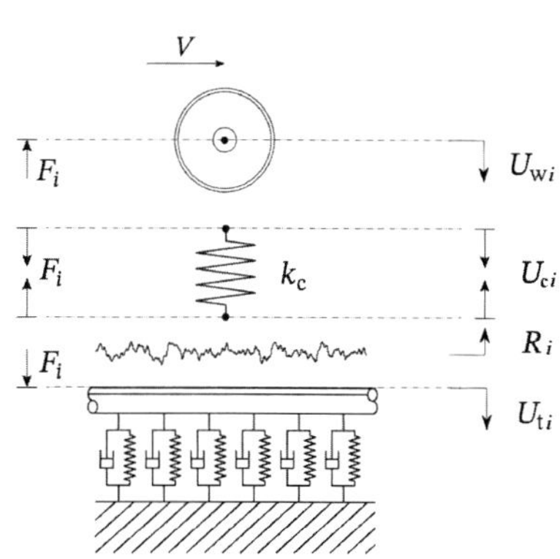

Fig. 7. Schematic diagram for mechanical analysis of the decomposed wheel-rail system.

is reasonable for typical levels of excitation. Thus, the interaction forces can be determined by

$$f_i = k_c u_{ci}, \tag{9}$$

where u_{ci} is the relative compression across the ith contact spring ($i = 1, 2, 3, 4$) and k_c is the linearized contact stiffness.

Only vertical track irregularities are taken into consideration in this study. These are expressed by the US FRA vertical alignment irregularity PSD function $S_r(\Omega)$[35]

$$S_r(\Omega) = \frac{kA_r\Omega_c^2}{\Omega^2(\Omega^2 + \Omega_c^2)}, \tag{10}$$

where Ω represents the spatial angular frequency of track irregularities; k represents the safety factor, which ranges from 0.25 to 1.0; A_r represents the roughness constant; Ω_c represents the cutoff spatial angular frequency. The track irregularity of class 6 is used in this study for illustration purposes with $k = 1, A_r = 0.0339\,\text{cm}^2 \cdot \text{rad/m}$, and $\Omega_c = 0.8245\,\text{rad/m}$. Note that as track irregularities are broadband, the behavior of the peaks of interest won't change substantially except their magnitude if a different type of irregularities is considered.

4.2. *Frequency response of coupled system*

To facilitate mechanical analysis, the original coupled model can be decomposed into several subsystems, the vehicle, the contact springs, the irregularity, and the rail, as shown schematically in Fig. 7. The displacement response of the vehicle can be determined by the frequency-domain method

$$\mathbf{U} = -\mathbf{H}_v\mathbf{F}, \tag{11}$$

where $\mathbf{H}_v = (-\mathbf{m}\omega^2 + i\omega\mathbf{c} + \mathbf{k})^{-1}$ is defined as the frequency response function (FRF) matrix of the vehicle subsystem. It is noted that interactions between the wheels and the track occur at 4 DOFs. Therefore, the vehicle system can be reduced to 4 DOF by using the condensation technique. The response amplitudes of the wheels and the interaction forces are obtained by

$$\mathbf{U}_w = \mathbf{I}_T\mathbf{U} = \{U_{w1} \quad U_{w2} \quad U_{w3} \quad U_{w4}\}^T, \tag{12}$$

$$\mathbf{F}_w = \mathbf{I}_T\mathbf{F} = \{F_1 \quad F_2 \quad F_3 \quad F_4\}^T, \tag{13}$$

where $\mathbf{I}_T$ is the transfer matrix for reducing the 10-DOF system to a 4-DOF one.

The response amplitudes of the wheels can then be derived from the interaction forces

$$\mathbf{U}_w = -\mathbf{H}_w\mathbf{F}_w, \tag{14}$$

where the reduced FRF matrix of the wheels $\mathbf{H}_\mathrm{w}$ can be determined as $\mathbf{H}_\mathrm{w} = \mathbf{I}_\mathrm{T}\mathbf{H}_\mathrm{v}\mathbf{I}_\mathrm{T}^\mathrm{T}$ by using the condensation technique.

The relative compressions across the contact springs can be derived by

$$\mathbf{U}_c = \mathbf{H}_c \mathbf{F}_w, \tag{15}$$

where $\mathbf{H}_c$ is defined as the FRF matrix of the contact subsystem. As the four linearized contact springs are unconnected, $\mathbf{H}_c$ is determined as a diagonal matrix, i.e. $\mathbf{H}_c = \mathrm{diag}\{1/k_c \; 1/k_c \; 1/k_c \; 1/k_c\}$.

As is demonstrated above, when the observation point moves with the vehicle, the response of the contact points on the track will be stationary and harmonic. Therefore, the frequency-domain method can be applied to obtain the response of the contact points on the track through

$$\mathbf{U}_t = \mathbf{H}_t \mathbf{F}_w, \tag{16}$$

where $\mathbf{H}_t$ is the FRF matrix of the track subsystem and, by means of the superposition principle, can be determined from the point and transfer receptances, i.e. Eq. (7),

$$\mathbf{H}_t = \begin{bmatrix} H(0) & H(2b) & H(2a) & H(2a+2b) \\ H(-2b) & H(0) & H(2a-2b) & H(2a) \\ H(-2a) & H(-2a+2b) & H(0) & H(2b) \\ H(-2a-2b) & H(-2a) & H(-2b) & H(0) \end{bmatrix}. \tag{17}$$

The track irregularities are assumed to be a spatial stochastic process. If the running of the vehicle is taken into consideration, the irregularities under each contact point become temporal variables. To ensure the wheel-rail contact is maintained, the various displacements are related by

$$\mathbf{U}_w - \mathbf{U}_t + \mathbf{R}_w = \mathbf{U}_c, \tag{18}$$

where $\mathbf{R}_w$ is the amplitude vector of irregularities at certain frequency (positive for an asperity). Substituting Eqs. (14)–(16) into Eq. (18), the interaction force $\mathbf{F}_w$ is obtained as

$$\mathbf{F}_w = (\mathbf{H}_w + \mathbf{H}_t + \mathbf{H}_c)^{-1} \mathbf{R}_w = \mathbf{K}_d \mathbf{R}_w, \tag{19}$$

where $\mathbf{K}_d = (\mathbf{H}_w + \mathbf{H}_t + \mathbf{H}_c)^{-1}$ is a dynamic stiffness matrix, the inverse of the combined receptance matrix, which converts spatial geometric irregularities into equivalent wheel-rail interaction forces. In the meantime, the displacement response of the contact points on the track can be easily obtained through Eq. (16) as $\mathbf{U}_t = \mathbf{H}_t \mathbf{K}_d \mathbf{R}_w$.

4.3. *Response PSD*

The spectral density matrix for the wheel-rail contact force $\mathbf{S}_{\mathbf{f_w}}$ and for the response of the contact points on the track $\mathbf{S}_{\mathbf{U_t}}$ can be then obtained as

$$\mathbf{S}_{\mathbf{f_w}} = \lim_{T\to\infty} \frac{\mathbf{F}_w^* \mathbf{F}_w^T}{2T} = \mathbf{K}_d^* \mathbf{S}_r \mathbf{K}_d^T, \tag{20}$$

$$\mathbf{S}_{\mathbf{U_t}} = \lim_{T\to\infty} \frac{\mathbf{U}_t^* \mathbf{U}_t^T}{2T} = \mathbf{H}_t^* \mathbf{K}_d^* \mathbf{S}_r \mathbf{K}_d^T \mathbf{H}_t^T, \tag{21}$$

where the superscript $*$ denotes the conjugate and superscript T denotes the transpose operation. $\mathbf{S}_r$ is defined as the spectral density matrix for the irregularities of contact points given by

$$\mathbf{S}_r = \lim_{T\to\infty} \frac{\mathbf{R}_w^* \mathbf{R}_w^T}{2T} = \mathcal{B}(\omega, V)\frac{1}{V} S_r\left(\frac{\omega}{V}\right), \tag{22}$$

where $S_r(\Omega)$ is the spatial-PSD function of track irregularities in Eq. (10); $\mathcal{B}(\omega, u)$ is a phase-lag matrix, describing relationships of time delay among the excitations. The use of the phase-lag matrix allows different types of excitations to be represented. For example, at low frequencies, excitations at each wheel can be regarded as strictly correlated, and $\mathcal{B}(\omega, V)$ can be expressed as

$$\mathcal{B}(\omega, V) = \begin{bmatrix} 1 & e^{i\frac{\omega}{V}2b} & e^{i\frac{\omega}{V}2a} & e^{i\frac{\omega}{V}2(a+b)} \\ e^{-i\frac{\omega}{V}2b} & 1 & e^{i\frac{\omega}{V}2(a-b)} & e^{i\frac{\omega}{V}2a} \\ e^{-i\frac{\omega}{V}2a} & e^{-i\frac{\omega}{V}2(a-b)} & 1 & e^{i\frac{\omega}{V}2b} \\ e^{-i\frac{\omega}{V}2(a+b)} & e^{-i\frac{\omega}{V}2a} & e^{-i\frac{\omega}{V}2b} & 1 \end{bmatrix} \tag{23}$$

of which the derivation can be found in Ref. 36. In contrast, at high frequencies, an uncorrelated excitation model is more reasonable, for which $\mathcal{B}(\omega, u)$ can be obtained as $\mathcal{B}(\omega, V) = \mathrm{diag}\{1\ 1\ 1\ 1\}$. For the active–passive wheel model which has been developed in Ref. 10. the irregularity excitation is only applied on one wheel at a time. If the first wheel is treated as the active wheel, $\mathcal{B}(\omega, u)$ can be expressed as $\mathcal{B}(\omega, V) = \mathrm{diag}\{1\ 0\ 0\ 0\}$.

5. Mechanism Analysis

5.1. *Multiple peaks in response PSD*

The frequency-domain method presented in Sec. 4 is first applied to analyze the vibration of the train–track coupled system introduced in Fig. 6 with the uncorrelated irregularity excitations at four wheels. The PSDs of vertical contact forces from the first and second wheels are shown in Fig. 8(a), for a passage speed of 40 m/s. Moreover, Fig. 8(b) shows the PSDs of contact forces for a track coupled with four wheels with the irregularity excitation only on the third wheel. This model represents a pair of bogies at the adjacent ends of two vehicles, with $b = 1.25$ m

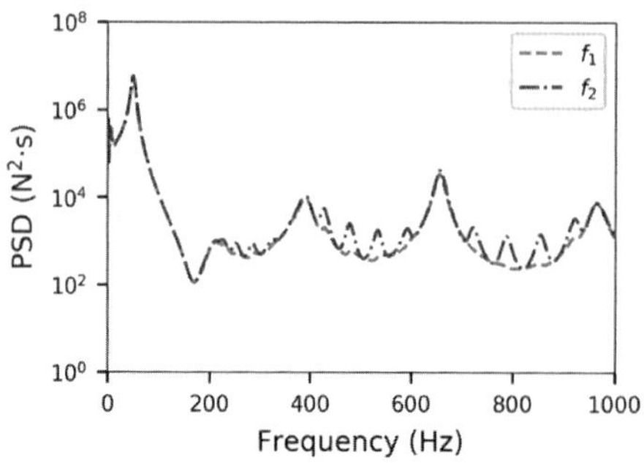

(a)

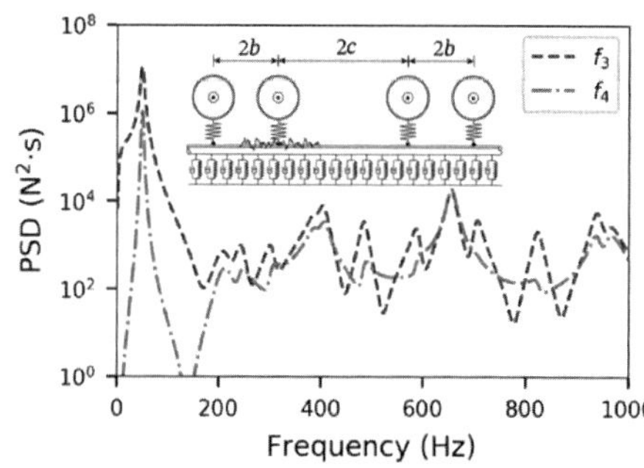

(b)

Fig. 8. PSD functions of wheel-rail contact forces for the system of: (a) a half vehicle on the rail subjected to uncorrelated irregularity excitations at all four wheels and (b) four wheels on the rail with track irregularity only on the third wheels.

and $c = 3.55\,\mathrm{m}$. In the latter case, the bogies and car body are omitted from the model; this only affects the results at frequencies below 10 Hz. As can be seen from Fig. 8, similar response peaks can be observed at frequencies above 200 Hz in the contact forces obtained from both models. The middle wheels (blue lines) present more peaks than the side wheels (red lines). These results broadly resemble the experimental results in Fig. 2, especially when the four adjacent wheels are included in the model. Note that the discrepancies between the field test results and simulation results will be discussed in Sec. 5.4.

5.2. *Influence of two wheels on wheel–rail interactions*

In order to investigate the multiple peaks observed both in the field measurement and the simulations in Fig. 8, a simplified model including active and passive wheel/rail interactions is applied in the following sections. The simplified model consists of an infinite continuously supported Timoshenko beam for the rail together with two longitudinally arranged wheels linked by a hinged massless bar, as shown in Fig. 9. Track irregularities are assumed to exist under the right-hand wheel (the active wheel), using the US FRA class 6 spectrum in Eq. (10), and the other wheel is excited only by the rail vibration; this is treated as the passive wheel. Responses of this simplified model can be easily calculated based on previous derivations in Sec. 4, in which certain functions need to be adjusted and their expressions are given in Appendix A for the sake of brevity.

The response PSDs of the active and passive contact forces are presented in Fig. 10(a), for a passage speed of 40 m/s. As a comparison, the response PSD of wheel-rail contact forces for a single wheel running on the rail at 40 m/s is shown in Fig. 10(b).

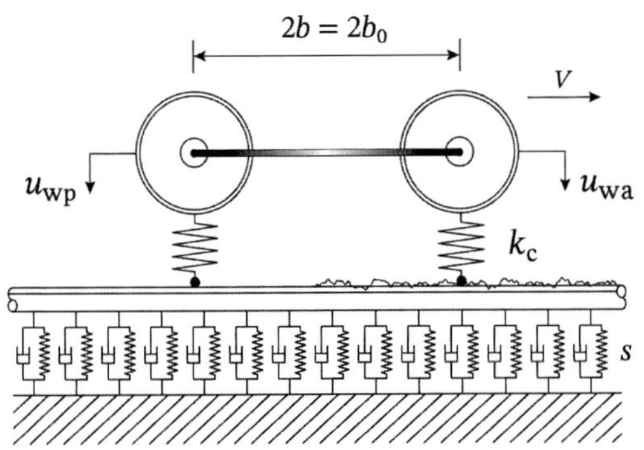

Fig. 9. An infinite Timoshenko beam model of the rail with two wheels linked by a hinged bar with irregularities only existing under the active wheel.

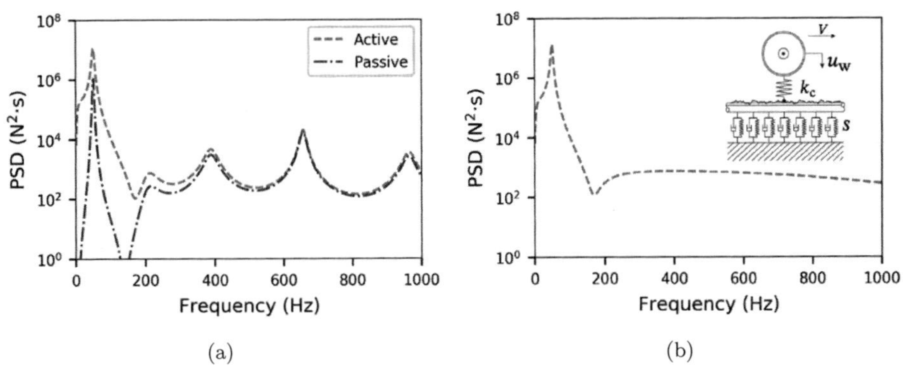

Fig. 10. PSD functions of wheel-rail contact forces for the system of (a) the active and passive wheels on the rail and (b) a single wheel on the rail.

Similar to Fig. 8, it can be observed from Fig. 10(a) that besides the original resonance peak around 50 Hz there is a set of peaks at 400 Hz and above. In contrast, these additional peaks do not occur in Fig. 10(b), with only the peak corresponding to the P2 resonance of the wheel on the rail appearing. Therefore, the effects of multiple wheels on the rail have changed the pattern of wheel/rail interactions, leading to multiple peaks in the response.

5.3. *Contact force expressions*

These multiple peaks can be explained from a mathematical perspective. Assume the active wheel is subjected to a particular component of track irregularities with a constant wavelength. Utilizing the frequency-domain method established above, the interaction forces $\mathbf{F}_w = [F_a\ F_p]^T$ can be derived from Eq. (19), in which F_a

represents contact forces of the active wheels and F_{p} for the passive wheels. Note that the irregularity under the passive wheels is zero, so the irregularity vector can be expressed as $\mathbf{R}_w = [R\ 0]^T$. Besides, the FRF matrices of the subsystem are all second-order matrices. The active and passive contact forces can then be determined by

$$F_a = \frac{\left[H(0,\,\omega) + \dfrac{1}{k_c} - \dfrac{1}{m_w\omega^2}\right] R}{\left[H(0,\,\omega) + \dfrac{1}{k_c} - \dfrac{1}{m_w\omega^2}\right]^2 - H(2b,\,\omega)H(-2b,\,\omega)}, \tag{24}$$

$$F_p = \frac{-H(-2b,\,\omega)R}{\left[H(0,\,\omega) + \dfrac{1}{k_c} - \dfrac{1}{m_w\omega^2}\right]^2 - H(2b,\,\omega)H(-2b,\,\omega)}, \tag{25}$$

where the element $H(\xi, \omega)$ represents the receptance of the rail, given by Eq. (7). It can be seen that the contact forces are complicated complex functions, especially as the rail receptance $H(\xi, \omega)$ is a complex integral, but the contact force will reach a local maximum if the modulus of the denominator in Eq. (24) or Eq. (25) has a local minimum. Figure 11 gives the modulus of the denominator in contact forces at different frequencies.

As is expected, the curve has local minima at around 50, 380, 635 and 950 Hz, exactly the places where the multiple peaks occur in Fig. 10(a). This kind of explanation makes sense from a mathematical viewpoint, but it fails to provide an intuitive physical interpretation and still cannot answer the question why these functions have local extrema at certain frequencies.

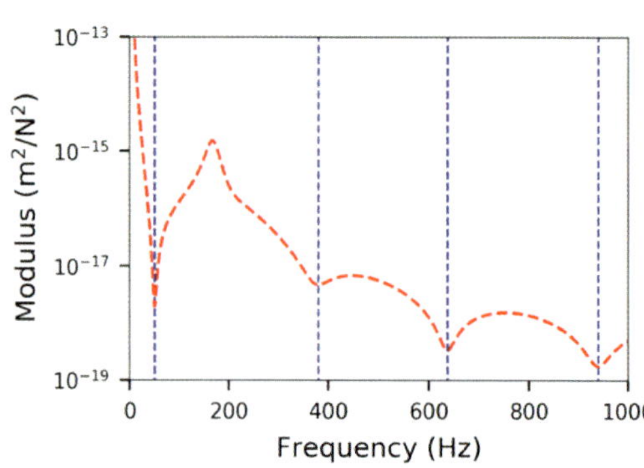

Fig. 11. The modulus of the denominator in contact force expressions at different frequencies.

5.4. *Explanation based on wave propagation theory*

In fact, the vibration of the rail consists of waves, which propagate along the rail and can be reflected by the wheels; the responses are formed of the superposition of these waves. To show the wave propagation effect, a series of models are considered in which the wheelbase $2b$ of a bogie is varied. Figure 12 gives the PSDs of the active and passive contact forces and the phase differences between them. Results are shown for cases with $b = 0.5b_0, b_0, 1.5b_0$ and $2b_0$, in which b_0 is the original half wheelbase of 1.25 m.

These results show that, when the responses reach the local maxima, the two contact forces are either in phase or 180° out of phase, signifying intensification of the waves. For the active wheel, for example, the wave excited by the active wheel radiates outwards along the rail, and is partially reflected by the passive wheel. If the reflected wave is in phase with the irregularities, the excitation for the active wheel will be enhanced, and so will be the response.

In this case, the active wheel is initially excited by irregularities and generates the wheel-rail contact force F_{a_1}, defined as the first-order active contact force with

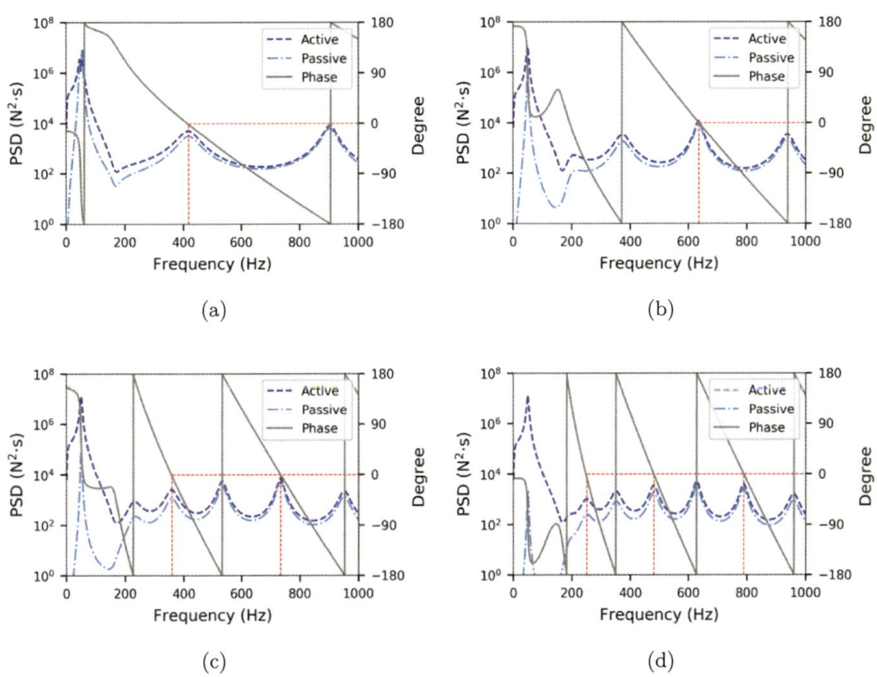

Fig. 12. PSDs and phase differences of the active and passive contact forces for wheelbase of (a) $b = 0.5b_0$, (b) $b = b_0$, (c) $b = 1.5b_0$ and (d) $b = 2b_0$.

a subscript 1 and derived as

$$F_{a_1} = K_d R_0, \tag{26}$$

where R_0 is the irregularity (positive downward) and K_d is the dynamic stiffness function for a single wheel-rail contact

$$K_d = \frac{1}{-\frac{1}{m_w \omega^2} + H(0,\omega) + \frac{1}{k_c}}, \tag{27}$$

where $H(0,\omega)$ is the point receptance of the rail at the contact point ($\xi = 0$). It is worth pointing that, for frequencies between 100 Hz and 1000 Hz, $|H(0,\omega)|$ is much larger than the other two terms and therefore

$$K_d \approx \frac{1}{H(0,\omega)}. \tag{28}$$

The incident wave $w_i(\xi, t)$ excited by F_{a_1} propagates (in the negative ξ direction) to the passive contact point, and the vibration of this point $w_i(-2b, t)$ can be obtained from Eq. (26) by multiplying by a transfer receptance $H(-2b, \omega)$

$$wi(-2b, t) = H(-2b,\, \omega) F_{a_1}. \tag{29}$$

As discussed in Sec. 3, the transfer receptance at frequencies above 300 Hz is dominated by the k_2 (or k_4) component at sufficient distance from the excitation point. Therefore, by ignoring the rapidly decaying component (k_1 term), from Eq. (7) $H(-2b,\, \omega)$ can be simplified as

$$H(-2b,\, \omega) \approx i \frac{M(k_2)}{N'(k_2)} e^{ik_2 2b}. \tag{30}$$

The vibration $w_i(-2b, t)$ can be seen as an equivalent track irregularity that causes a secondary interaction force at the passive wheel. This can be written as

$$R_1 = w_i(-2b, t) = \eta_i R_0. \tag{31}$$

Here, η_i describes the changes of the magnitude and phase of R_1 from the initial irregularity R_0

$$\eta_i = \frac{R_1}{R_0} = H(-2b,\, \omega) K_d \approx \frac{H(-2b,\, \omega)}{H(0,\, \omega)}. \tag{32}$$

The equivalent irregularity R_1 will excite the passive wheel and generate a first-order passive contact force F_{p_1}. This is given by

$$F_{p_1} = K_d R_1. \tag{33}$$

Comparing Eqs. (26), (32) and (33) it can be seen that $F_{p_1} = \eta_i F_{a_1}$. The force F_{p_1} will then excite a reflected wave $w_r(\xi, t)$ in the rail which will propagate back to the active wheel. It will excite the active wheel and can again be seen as an equivalent irregularity R_2. R_2 can be obtained as

$$R_2 = H(2b,\, \omega) F_{p_1} = \eta_r R_0, \tag{34}$$

where $H(2b,\, \omega)$ is a transfer receptance at $\xi = 2b$ and can be simplified as

$$H(2b,\, \omega) \approx -i\frac{M(k_4)}{N'(k_4)}\mathrm{e}^{-ik_4 2b} \tag{35}$$

and η_r describes the changes of the magnitude and phase of R_2 from the initial irregularity R_0

$$\eta_r = \frac{R_2}{R_0} = H(2b,\, \omega)H(-2b,\, \omega)K_d^2. \tag{36}$$

In this process, the magnitude and phase of the reflection R_2 generated from the initial irregularity R_0^{26} undergo a series of changes

$$R_0 \overset{\times K_d}{\longrightarrow} F_{a_1} \overset{\times H(-2b,\omega)}{\longrightarrow} R_1 \overset{\times K_d}{\longrightarrow} F_{p_1} \overset{\times H(2b,\omega)}{\longrightarrow} R_2. \tag{37}$$

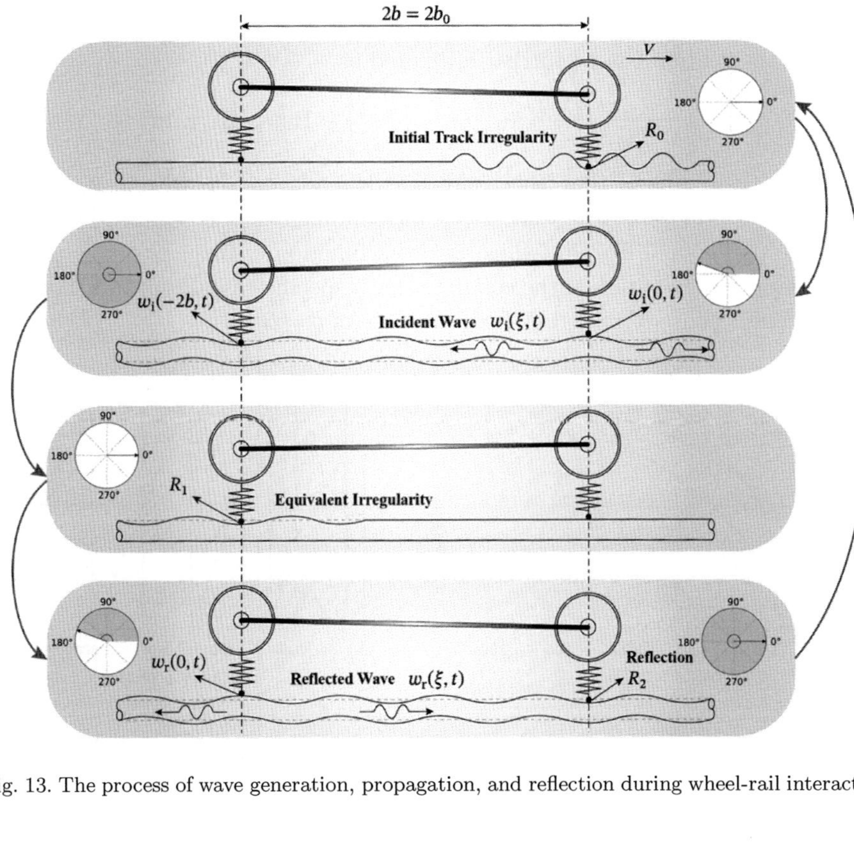

Fig. 13. The process of wave generation, propagation, and reflection during wheel-rail interactions.

Figure 13 depicts the process of wave generation, propagation, and reflection during wheel-rail interactions. After being excited by initial irregularities R_0, the incident wave $w_i(\xi, t)$ propagates outward from the active contact point, and serves as an equivalent track irregularity R_1 at the passive wheel. In the same way, the equivalent irregularity R_1 excites a reflected wave $w_r(\xi, t)$, and it will propagate back to the active point as a reflection R_2. If the reflection R_2 and the initial irregularity R_0 are in phase,

$$\mathrm{Arg}[\eta_r] = 2n\pi \tag{38}$$

for integer values of n, the excitation for the active wheel will be enhanced due to the constructive interference.

It is noted that k_2 is approximately equal to $-k_4$ because the speed of bending waves is much greater than the train speed. In this case, $H(2b, \omega)$ is approximately equal to $H(-2b, \omega)$ and $\eta_\mathrm{r} \approx \eta_\mathrm{i}^2$. Therefore, Eq. (38) can be equivalently expressed as

$$\mathrm{Arg}(\eta_i) = n\pi. \tag{39}$$

Equation (39) also represents the condition on the phase difference between F_{a_1} and F_{p_1} in the case of intensification. By substituting Eqs. (32) and (7), the condition of wave intensification can be further evaluated as

$$\mathrm{Arg}\left[\frac{M(k_4)N'(k_3)}{M(k_3)N'(k_4) + M(k_4)N'(k_3)}\right] - \mathrm{Re}(k_4)2b = n\pi. \tag{40}$$

This qualitatively shows that the condition depends on the wheelbase and the wavenumbers of the rail. Note that as the wavenumbers are determined by roots of the quartic polynomial Eq. (5), which is also a function of frequency, it is possible to give a numerical solution of the condition for a specific case. Figure 14 shows the magnitude and phase of η_i and η_r for wheelbase of $b = b_0$.

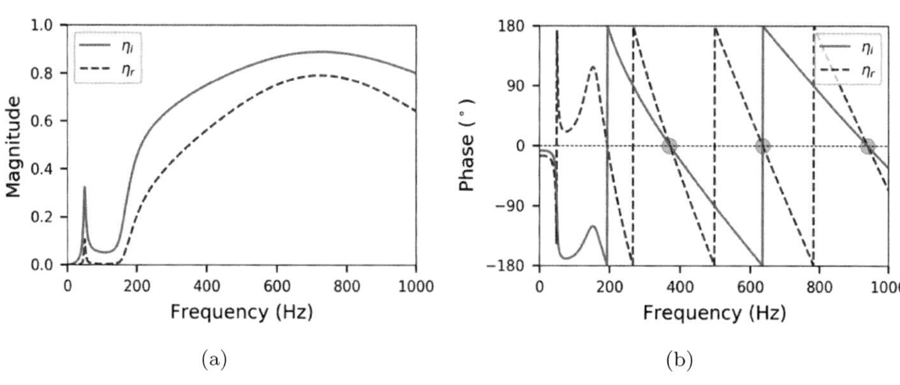

Fig. 14. The coefficients η_i and η_r for wheelbase of $b = b_0$: (a) magnitude and (b) phase.

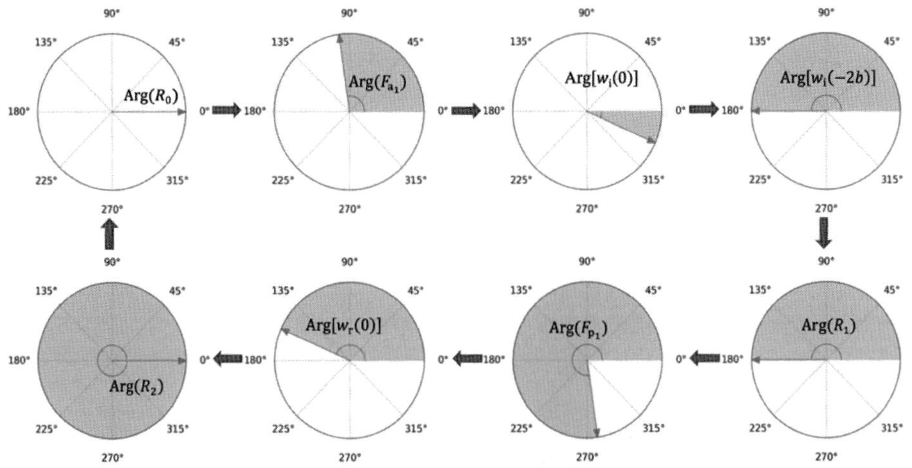

Fig. 15. The phase changes in the case of intensification between the initial irregularity and the reflection at 635 Hz.

It can be seen that the magnitudes of the two coefficients are both less than 1, due to the effect of wave attenuation and the influence of the wheel and contact receptances in K_d. The phase of η_i reaches 0 or π at around 380, 635 and 950 Hz, where the phase of η_r is 2π. This is consistent with the results in Fig. 12(b) in which the peak in the contact forces occurs when the phase difference between them is either 0 or π.

Figure 15 shows the phase changes of the intermediate variables from the initial irregularity to the reflection in the case of intensification at 635 Hz.

The phase of the reflection returns to 2π, and it will continue to propagate and get reflected between the two wheels. In fact, as long as the reflection R_2 has the same phase as the initial irregularity, the excitation will be continuously intensified. Thus, the total equivalent irregularity becomes

$$R_{\text{total}} = \sum_{n=0}^{\infty} |\eta_r|^n R = \frac{R}{1 - |\eta_r|}. \tag{41}$$

Figure 14(a) shows that $|\eta_r|$ is larger than 0.6 at frequencies above 400 Hz, so R_{total} will be more than $2.5R$ according to Eq. (41). This shows that wave interference will have a significant effect on the intensification of excitations. In particular, $|\eta_r|$ reaches the peak (around 0.8) at 700 Hz, and R_{total} exceeds $5R$ in this case. This explains that there is maximum amplification at the peak around 635 Hz, compared to other additional peaks in Fig. 12(b).

As the number of wheels increases, there will be more reflected waves between multiple wheels, leading to a more complicated condition of wave intensification and an increase in consequent peaks in the response. This can be justified by comparing the two-wheel results of Fig. 10 to the four-wheel results of Fig. 8, and also the field test results of a three-vehicle train in Fig. 2(b). Note that the field measurement was conducted at a fixed point on the rail, and there may be additional peak energies in excitations at certain wavelengths, e.g. due to wheel flats or combined irregularities of wheel and rail, etc. These facts may lead to unexpected changes in the peak patterns of the field test results, e.g. peaks from 200 Hz to 400 Hz in Fig. 2(b).

The wave interference phenomenon is insignificant at low frequencies because waves are strongly attenuated when traveling in the rail in this frequency region.[17] According to Fig. 4(b), the decay rates at frequencies below 200 Hz are above 10 dB/m. Therefore, the wheel–rail interactions at low frequencies are not affected by the wave reflections.

6. Influence of Load Speed on Wheel–Rail Interactions

Usually, a moving irregularity model is applied to solve train-track coupled dynamics in the frequency domain, in which the train is assumed to be stationary and irregularities are pulled between wheels and the rail as a relative excitation. As discussed in Sec. 3, the wavelength of the wave propagating in the same direction as the moving excitation is shorter than that propagating in the opposite direction due to effects of the load motion on the poles, namely the Doppler effect. The difference between the wavelengths of the waves heading in each direction gradually increases as the load speed increases.

In this section, it is investigated whether the load speed has an effect on the peaks in the PSD of wheel-rail contact forces. Different values of pad stiffness are also considered. The PSDs of the wheel-rail contact forces are calculated by using the moving load approach and the moving irregularity approach and compared for

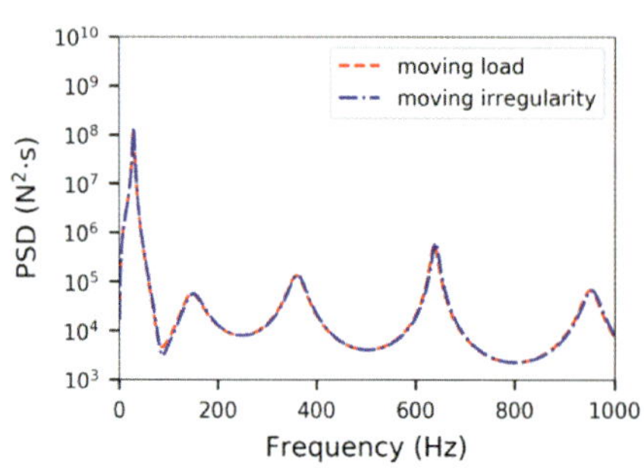

(a)

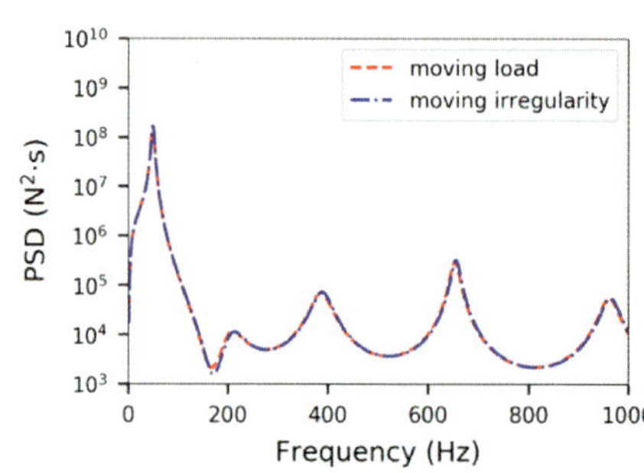

(b)

Fig. 16. PSDs of the active contact forces at $V = 100\,\mathrm{m/s}$ by using the moving load model (red line) and the moving irregularity model (blue line) with pad stiffness of (a) $10\,\mathrm{MN/m}$ and (b) $40\,\mathrm{MN/m}$.

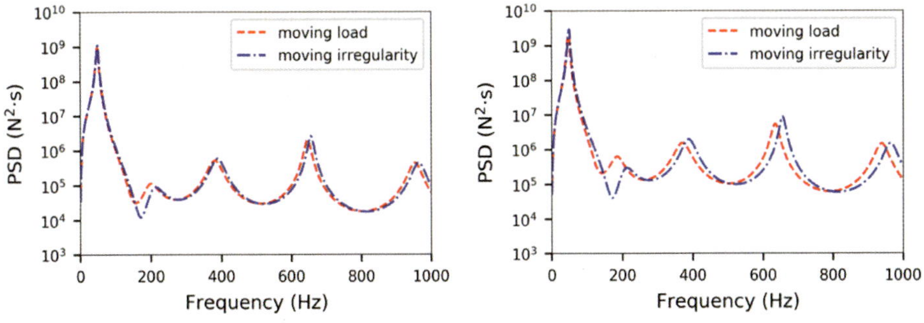

(a) (b)

Fig. 17. PSDs of the active contact forces with pad stiffness of 40 MN/m at a fastener spacing of 0.6 m by using the moving load model (red line) and the moving irregularity model (blue line) at the train speed of (a) 200 m/s and (b) 300 m/s.

different fastener stiffness and train speeds. The investigation is still based on the simplified model shown in Fig. 9 with $b = 1.25$ m.

Figure 16 shows the force PSD for two values of pad stiffness (10 MN/m and 40 MN/m) for a fastener spacing of 0.6 m, and a train speed of 100 m/s; Fig. 17 shows results for two higher values of train speed (200 m/s and 300 m/s) with the pad stiffness of 40 MN/m for a fastener spacing of 0.6 m. Note that these high speeds are considered in order to illustrate Doppler effect on wheel–rail interactions despite their impracticability for the current trains in operation.

It can be seen from Fig. 16 that differences between the contact forces obtained using the moving load model and the moving irregularity model are small, showing that Doppler effect has little influence on wheel-rail interactions for these cases. In contrast, as is shown in Fig. 17, the difference is slightly increased with the increase of train speed, and Doppler effect can be observed for a train speed of 300 m/s. These peaks for the moving load model are seen to move to the lower frequency range, compared to that of the moving irregularity model. It is noted that the shift in the peaks in the upper frequency region is greater than in the low-frequency region. However, since practical train speeds are usually below 100 m/s, these results show that the moving irregularity model is sufficient in most practical cases.

Comparing the cases for different pad stiffness in Fig. 16, the position of the peaks is unaffected but the height of the peaks is increased for the softer rail pads. This is related to the lower decay rate that such a track exhibits.

7. Conclusions

This study has investigated the phenomenon and the mechanism of multiple peaks in the high-frequency rail vibration spectra. A frequency-domain method has been adopted to calculate the PSD responses of the train-track coupled system excited by multi-wheel moving excitations and stochastic irregularities. The track is modeled as an infinite Timoshenko beam with continuous supports. Based on the residue theorem, the point and transfer receptances of the rail are explicitly determined using a Fourier transform-based method.

The results from the train-track dynamic interaction model show the multiple peaks in the responses of the system, around 400–1000 Hz, occur with the presence of multiple wheelsets on the rail. Based on the wave propagation theory, it is found that the multiple peaks occur when the bending rail waves reflected from multiple wheels on the rail are exactly in phase with the initial irregularity. This suggests the high-frequency rail response to multiple wheels cannot be simply treated as the combination of individual responses to a series of single wheels by using superposition principle, as conventionally done.[10] Instead, the wave interference between multiple wheels shall be considered. The consequent peak responses would be responsible for typical wear patterns on the railhead.[17] Although the model has been based on a moving excitation, the influence of the Doppler effect on the wheel-rail contact forces is found to be small for a practical range of parameter values, where the train speed is 100 m/s or below, the effect of load motion is insignificant, and a moving irregularity model can be used to solve train-track coupled dynamics.

Acknowledgments

The study was supported by the National Natural Science Foundation of China (Grant Nos. 52278520, 51978527 and 52178432), and the State Key Laboratory of Disaster Reduction in Civil Engineering (Grant No. SLDRCE19-A-04) and the China Scholarship Council (No. 202106260170).

Appendix A

$$\mathbf{H}_{\mathrm{w}} = -\mathrm{diag}\left\{\frac{1}{m_{\mathrm{w}}\omega^2} \quad \frac{1}{m_{\mathrm{w}}\omega^2}\right\}, \tag{A.1}$$

$$\mathbf{H}_{\mathrm{c}} = \mathrm{diag}\left\{\frac{1}{k_{\mathrm{c}}} \quad \frac{1}{k_{\mathrm{c}}}\right\}, \tag{A.2}$$

$$\mathbf{H}_{\mathrm{t}} = \begin{bmatrix} H(0) & H(2b) \\ H(-2b) & H(0) \end{bmatrix}, \tag{A.3}$$

$$\mathbf{B}(\omega, V) = \mathrm{diag}\{1 \quad 0\}. \tag{A.4}$$

References

1. Y. B. Yang and J. D. Yau, Vertical and pitching resonance of train cars moving over a series of simple beams, *J. Sound Vib.* **337** (2015) 135–149.

2. Q. Zeng, Y. B. Yang and E. G. Dimitrakopoulos, Dynamic response of high speed vehicles and sustaining curved bridges under conditions of resonance, *Eng. Struct.* **114** (2016) 61–74.
3. Y. B. Yang and J. D. Yau, Resonance of high-speed trains moving over a series of simple or continuous beams with non-ballasted tracks, *Eng. Struct.* **143** (2017) 295–305.
4. Y.-S. Wu, Y.-B. Yang and J.-D. Yau, Three-dimensional analysis of train-rail-bridge interaction problems, *Veh. Syst. Dyn.* **36** (2001) 1–35.
5. Y.-S. Wu and Y.-B. Yang, Steady-state response and riding comfort of trains moving over a series of simply supported bridges, *Eng. Struct.* **25** (2003) 251–265.
6. H. H. Hung, G. H. Chen and Y. B. Yang, Effect of railway roughness on soil vibrations due to moving trains by 2.5d finite/infinite element approach, *Eng. Struct.* **57** (2013) 254–266.
7. Y.-B. Yang, J. Yau, Z. Yao and Y. Wu, *Vehicle-Bridge Interaction Dynamics: With Applications to High-Speed Railways* (World Scientific, 2004).
8. Y. B. Yang and C. W. Lin, Vehicle-bridge interaction dynamics and potential applications, *J. Sound Vib.* **284** (2005) 205–226.
9. Y. B. Yang, C. L. Lin, J. D. Yau and D. W. Chang, Mechanism of resonance and cancellation for train-induced vibrations on bridges with elastic bearings, *J. Sound Vib.* **269** (2004) 345–360.
10. T. X. Wu and D. J. Thompson, Vibration analysis of railway track with multiple wheels on the rail, *J. Sound Vib.* **239** (2001) 69–97.
11. Y. Yang, B. Zhang, T. Wang, H. Xu and Y. Wu, Two-axle test vehicle for bridges: Theory and applications, *Int. J. Mech. Sci.* **152** (2019) 51–62.
12. M. Sun, X. S. Fang, D. X. Mao, X. Wang and Y. Li, Broadband acoustic ventilation barriers, *Phys. Rev. Appl.* **13** (2020) 8.
13. X. Wang, D. Mao, W. Yu and Z. Jiang, Acoustic performance of balconies having inhomogeneous ceiling surfaces on a roadside building facade, *Build. Environ.* **93** (2015) 1–8.
14. A. Igeland, Railhead corrugation growth explained by dynamic interaction between track and bogie wheelsets, *Proc. Inst. Mech. Eng. F J. Rail Rapid Transit* **210** (1996) 11–20.
15. G. Fang, Y. Wang, Z. Peng and T. Wu, Theoretical investigation into the formation mechanism and mitigation measures of short pitch rail corrugation in resilient tracks of metros, *Proc. Inst. Mech. Eng. F: J. Rail Rapid Transit* **232** (2018) 2260–2271.
16. T. Wu and D. Thompson, The effects on railway rolling noise of wave reflections in the rail and support stiffening due to the presence of multiple wheels, *Appl. Acoust.* **62** (2001) 1249–1266.
17. T. Wu and D. Thompson, Behaviour of the normal contact force under multiple wheel/rail interaction, *Veh. Syst. Dyn.* **37** (2002) 157–174.
18. B. Ding, G. Squicciarini and D. Thompson, Effect of rail dynamics on curve squeal under constant friction conditions, *J. Sound Vib.* **442** (2019) 183–199.
19. X. Sheng, T. Zhong and Y. Li, Vibration and sound radiation of slab high-speed railway tracks subject to a moving harmonic load, *J. Sound Vib.* **395** (2017) 160–186.
20. K. L. Knothe and S. L. Grassie, Modeling of railway track and vehicle track interaction at high-frequencies, *Veh. Syst. Dyn.* **22** (1993) 209–262.
21. X. Sheng, C. J. C. Jones and D. J. Thompson, Responses of infinite periodic structures to moving or stationary harmonic loads, *J. Sound Vib.* **282** (2005) 125–149.

22. X. Sheng and M. H. Li, Propagation constants of railway tracks as a periodic structure, *J. Sound Vib.* **299** (2007) 1114–1123.
23. J. C. O. Nielsen and A. Igeland, Vertical dynamic interaction between train and track — influence of wheel and track imperfections, *J. Sound Vib.* **187** (1995) 825–839.
24. X. Sheng, Generalization of the fourier transform-based method for calculating the response of a periodic railway track subject to a moving harmonic load, *J. Mod. Transp.* **23** (2015) 12–29.
25. T. X. Wu and Y. R. Wang, Modeling of wheel-track interaction with rail vibration damper and its application for suppressing short pitch rail corrugation, in *Noise and Vibration Mitigation for Rail Transportation Systems* (Springer, 2015), pp. 361–368.
26. T. Wu and D. Thompson, An investigation into rail corrugation due to micro-slip under multiple wheel/rail interactions, *Wear* **258** (2005) 1115–1125.
27. D. Thompson, *Railway Noise and Vibration: Mechanisms, Modelling and Means of Control* (Elsevier, Oxford, 2008).
28. Q. Li and D. J. Wu, Analysis of the dominant vibration frequencies of rail bridges for structure-borne noise using a power flow method, *J. Sound Vib.* **332** (2013) 4153–4163.
29. S. L. Grassie, R. W. Gregory, D. Harrison and K. L. Johnson, The dynamic response of railway track to high frequency vertical excitation, *J. Mech. Eng. Sci.* **24** (1982) 77–90.
30. J. Egana, J. Vinolas and M. Seco, Investigation of the influence of rail pad stiffness on rail corrugation on a transit system, *Wear* **261** (2006) 216–224.
31. H. Ilias, The influence of railpad stiffness on wheelset/track interaction and corrugation growth, *J. Sound Vib.* **227** (1999) 935–948.
32. Q. Li and D. J. Thompson, Prediction of rail and bridge noise arising from concrete railway viaducts by using a multilayer rail fastener model and a wavenumber domain method, *Proc. Inst. Mech. Eng. F J. Rail Rapid Transit* **232** (2018) 1326–1346.
33. T. Mazilu, Prediction of the interaction between a simple moving vehicle and an infinite periodically supported rail — green's functions approach, *Veh. Syst. Dyn.* **48** (2010) 1021–1042.
34. S. Lei, Y. Ge, Q. Li and D. J. Thompson, Frequency-domain method for non-stationary stochastic vibrations of train-bridge coupled system with time-varying characteristics, *Mech. Syst. Signal Proc.* **183** (2023) 109637.
35. V. Garg and R. V. Dukkipati, *Dynamics of Railway Vehicle Systems* (Elsevier, Toronto, 2012).
36. S. M. Lei, Y. J. Ge and Q. Li, Effect and its mechanism of spatial coherence of track irregularity on dynamic responses of railway vehicles, *Mech. Syst. Signal Proc.* **145** (2020) 21.

https://doi.org/10.1142/9789819814770_0017

Chapter 17

Thermal Vibration Analysis of Sandwich Cylindrical Shells with Porous FGM Surface Layers#

Chen Liang*, Zhenyu Chen†, Guifeng Wang* and C. W. Lim*,‡

**Department of Architecture and Civil Engineering*
City University of Hong Kong Tat Chee Avenue
Kowloon, Hong Kong SAR, P. R. China

†Department of Mechanical Engineering
National University of Singapore
9 Engineering Drive 1, Singapore 117575, Singapore
‡bccwlim@cityu.edu.hk

This study aims at investigating the thermal vibration characteristics of sandwich cylindrical shells consisting of two surface layers crafted from functionally graded materials (FGMs) and a central metal core layer. The sandwich cylindrical shells with FGMs surface layers, with and without porosity, are modelled by using the Kirchhoff–Love shell theory. A porosity function composed of three distinct parts is introduced, including the core-to-thickness ratio, porosity volume fraction, and porosity distribution function. Through the function, the significant effects of porosity that varies with the mixing degree of constituent materials can be analyzed. The material properties are assumed to be temperature-dependent and they show continuous graded variation along the thickness direction. A theoretical approach for analyzing thermal strain energy in the cylindrical shells subjected to thermal environments is established by incorporating Green's nonlinear strains. The governing equations are derived by applying Hamilton's principle. Subsequently, analytical solutions for the system's natural frequencies are determined. Further, to validate the analytical results, a comparative analysis is conducted, drawing upon numerical simulations and other data available in the open literature. Additionally, the thermal vibration characteristics of the composite shell structures are examined in a comprehensive study with respect to various parameters such as porosity type, porosity volume fraction, core-to-thickness ratio, power-law exponent, and temperature changes.

Keywords: Cylindrical shell; functionally graded; Kirchhoff–Love shell; porosity; sandwich; thermal stresses; vibration.

1. Introduction

Functionally graded materials, or FGMs in short, emerged as one of the most prominent composite materials with its advent in 1984 as proposed by some research

‡Corresponding author.

#This chapter appeared previously on the International Journal of Structural Stability and Dynamics. To cite this chapter, please cite the original article as the following: C. Liang *et al.*, *Int. J. Struct. Stab. Dyn.*, **23**, 2340019 (2023), DOI: 10.1142/S0219455423400199.

scientists in Japan.[1] Originally, FGMs were invented as an outstanding composite material characterized by excellent thermal resistance properties and were intended for use in applications in aerospace engineering and fusion reactor designs.[1–5] One of the distinct properties of FGMs is the continuously varying material properties, achieved by gradually changing the volume fraction of constituent materials. The exceptional properties and excellent characteristics of desirable materials can be leveraged by FGMs with a unique design. For instance, FGMs consisting of metal and ceramic components have been extensively studied. The ceramic counterparts of FGMs possess outstanding heat resistance while the metal counterparts provide toughness capability and strong ductility, respectively leading to the abilities for thermal loading bearing and fracture prevention.[6,7]

Because of the continuity of materials properties, FGM sandwich structures can effectively relieve the significant interfacial shear stress concentration on the surface–core layers,[8] and thus attracting much attention of academic and industrial communities. For these reasons, some researchers paid specific attention to this type of unique composite structure.[9–12] By using the four-unknown shear and normal deformation theory, Thai *et al.*[13] conducted static, buckling, and dynamic analyses of FGM sandwich plates (FGMSP). Fazzolari[14] utilized the quasi-3D Ritz models and the Hierarchical Trigonometric Ritz Formulation to analyze the stability characteristics of FGMSP. Nguyen and Nguyen[15] applied the higher-order shear deformation theory (HSDT) to study the vibration and buckling response of FGM sandwich beams (FGMSB). Neves *et al.*[16] presented static analysis of FGMSP by employing a hyperbolic theory, in which the effects of warping and Murakami's zig–zag function were considered. Based on the first-order shear deformation theory (FSDT), vibration and stability of FGMSB were investigated by Kahya and Turan.[17] Liu *et al.*[18] analyzed the vibration characteristics of FGMSP by applying the refined higher-order sandwich panel theory. Li *et al.*[19] analyzed nonlinear vibration of FGMSB, considering thermal conditions and auxetic porous copper core effects. The meshless global collocation method based on a plate spline radial basis function is presented by Xiang *et al.*[20] to analyze the vibration response of FGMSP. In the works of Katili and Katili,[21,22] the improved Felippa Bergan Triangular element and Bergan Wang Quadrilateral plate element were developed to analyze FGMSP by using a unified and integrated approach. Garg *et al.*[23] used the higher-order zigzag theory to study the bending characteristics of FGMSP. Fu *et al.*[24] used the differential quadrature method and HSDT to study the vibration of FGMSP. Liu *et al.*[25] used the differential quadrature method to analyze the bending behaviors of FGMSP, in which the auxetic core is tunable by adjusting the Poisson's ratio. Mohammadzadeh and Noh[26] presented nonlinear dynamic analyses of FGMSP by considering elastic foundation and blast loads. Based on the sinusoidal shear deformation plate theory, the buckling and vibration of FGMSP were investigated by Zenkour.[27] To the best of the authors' knowledge and from a comprehensive review,[28] there exist very few research works on the dynamics and stability of sandwich cylindrical shells made of FGMs (FGMSCS). Consequently, it is significant to conduct a timely study to establish the dynamic properties of FGMSCS.

Several preparation technologies, such as the nonpressure sintering technique, self-propagating high temperature synthesis, and multi-step continuous infiltration method can be used to manufacture FGMs.[29] Unlike well-tailored foam composite structures,[30–37] porosity in FGMs is generated due to inherent limitations of the current fabrication techniques. Moreover, considerable discrepancies in solidification temperature between ceramic and metal elements can lead to the formation of porosities and micro-voids during the manufacturing process.[38] In a study conducted by Wattanasakulpong *et al.*,[39] a similar porosity phenomenon was observed during an experiment with FGMs. Several scholars investigated the impacts of porosity on the properties of FGM sandwich structures.[40–44] In view of the significant effects of porosity that may vary with the mixing degree of constituent materials and because there exist few theoretical models for FGM sandwich structures, a new porosity function is established via a theoretical model in this paper. The porosity effects on FGM sandwich structures can be accurately prescribed by composing several structural and geometrical functions.

For decades, composite structures have received considerable attention for their thermal behaviors, particularly for FGM structures under thermal conditions.[45–54] For instance, Do and Lee[55] analyzed thermal buckling of FGMSP based on an advanced mesh-free method combined with radial point interpolation. Considering three types of thermal loads, Fazzolari and Carrera[56] studied thermal stability of FGMSP. In the work of Shao *et al.*,[57] the thermal transient vibration behaviors of FGMSP were investigated by considering several types of temperature field distribution conditions. Jalali *et al.*[58] conducted a thermal stability analysis of circular FGMSP via FSDT. On the basis of the four-variable refined plate theory, Li *et al.*[59] presented a thermomechanical bending analysis of FGMSP. Considering thermal conditions, Shi *et al.*[60,61] presented free and transient vibration of two types of combined composite structures, the FGM cylindrical–conical shells with annular plates and the FGM conical–cylindrical coupled shells. Vaghefi[62] applied a three-dimensional model to study thermo-elastoplasticity of FGMSP. The thermal buckling characteristics of FGM plates were studied by Do *et al.*,[63] in which the effects of crack length were analyzed by using a theoretical analysis and finite element method. Considering the coupling fields of acoustic and thermal conditions, Wang *et al.*[64] developed a formulation to predict the fatigue and nonlinear random response of laminated composite panels with elastic boundary conditions. Wang and Shen[65] studied the nonlinear dynamic behaviors of FGMSP, where the effects of thermal environments and elastic foundations were considered. In the work of

Sun *et al.*,[66] torsional bifurcation buckling behaviors of FGM cylindrical shells considering thermal effects were investigated. Further considering the coupling effects of compressive and thermal loadings, Sun *et al.* presented analytical bifurcation buckling solutions for FGM cylindrical shells.[67] Later, Sun *et al.*[68] accurately analyzed buckling behaviors of grid-stiffened FGM cylindrical shells and considered the similar coupling effects. For cylindrical shell composite structures, an analytical method for thermal strain energy is presented to investigate the thermal dynamic response by referring to numerical and experimental results and some theoretical models of the plate theory.[69–72]

In this work, a thermal strain energy analytical method for FGMSCS composed of two FGM surface layers and metal core is proposed. The Kirchhoff–Love shell theory is adopted and thermal vibration response is reported. The response of FGM surface layers for sandwich cylindrical shells with or without porosity is analyzed. A porosity function that considers different mixing degrees of constituent materials is introduced by combining three distinct components, i.e. porosity volume fraction, core-to-thickness ratio, and porosity distribution function. Further, a theoretical framework for thermal vibration analysis on FGMSCS is presented and analytical natural frequency solutions are reported. Furthermore, the present analytical model and theoretical solution methods are verified by comparing with numerical simulations and published benchmarks. In addition, the impacts of different system parameters on the vibration characteristics of FGMSCS subjected to thermal conditions are elucidated.

2. Theoretical Methodology and Modeling

A sandwich cylindrical shell, described in a coordinate system (x, θ, z) on the shell middle surface as shown in Fig. 1, is considered. The thickness, length and middle-surface radius of the shell are h, L, and r, respectively. The sandwich cylindrical shell consists of Layers 1, 2, and 3, labeled from the inner to outer shell surface. The respective thicknesses are h_1, h_2, and h_3, while the z-direction ordinates of the different surface layers are z_0, z_1, z_2, and z_3. In the present study, a metal core layer and FGM inner and outer surface layers are considered. FGMSCS that incorporates porosity is denoted by Type-1 shell, while Type-2 shell designates FGMSCS without porosity, and it is used as a reference for comparison.

2.1. *Material properties of FGMSCS*

The constituent material properties P_i of FGMSCS are functions of temperature that can be described as[73]

$$P_i(T) = P_0(P_{-1}T^{-1} + 1 + P_1T + P_2T^2 + P_3T^3), \tag{1}$$

where P_{-1}, P_0, P_1, P_2, and P_3 are temperature coefficients, and T is temperature in Kelvin.

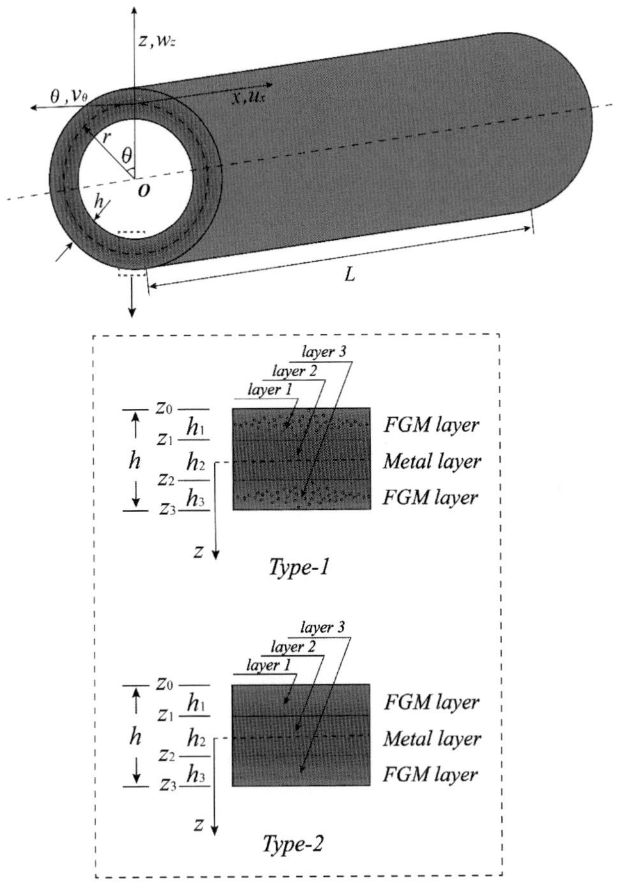

Fig. 1. Sandwich cylindrical shells where FGM surface layers are with or without porosity.

The material properties $P^{(N)}(z, T)$ are temperature-dependent for each layer and they can be determined by individual volume fractions $V_i^{(N)}(T)$ and material properties $P_i(T)$ as

$$P^{(N)}(z,T) = \sum_{i=1}^{M} P_i(T)V_i^{(N)}(z) - \frac{f^{(N)}(z)}{2}\sum_{i=1}^{M} P_i(T), \tag{2}$$

where the number of types of component materials is M; and the layer number is N ($N = 1, 2, 3$). Considering all the constituent materials, the summation of volume fractions should be unit,[74,75] namely,

$$\sum_{i=1}^{M} V_i^{(N)}(z) = 1. \tag{3}$$

A porosity function $f^{(N)}(z)$ is introduced and it is composed of the porosity distribution function $\varphi^{(N)}(z)$ and porosity volume fraction β . The porosity function can be defined as

$$f^{(N)}(z) = \beta\varphi^{(N)}(z). \tag{4}$$

For FGMSCS composed of two types of materials, metal and ceramic, the volume fraction of ceramic $V_c^{(N)}(z)$ and the porosity distribution function $\varphi^{(N)}(z)$ for Type-1 and Type-2 shells can be expressed as

Type-1

$$\begin{cases} V_c^{(1)}(z) = \left(\dfrac{z - z_1}{z_0 - z_1}\right)^p, & \varphi^{(1)}(z) = 1 - \dfrac{|4z + h(1+\zeta)|}{2h - h(1+\zeta)}, \quad z \in [z_0, z_1], \\ V_c^{(2)}(z) = 0, & \varphi^{(2)}(z) = 0, \qquad z \in [z_1, z_2], \\ V_c^{(3)}(z) = \left(\dfrac{z - z_2}{z_3 - z_2}\right)^p, & \varphi^{(3)}(z) = 1 - \dfrac{|4z - h(1+\zeta)|}{2h - h(1+\zeta)}, \quad z \in [z_2, z_3]. \end{cases} \tag{5}$$

Type-2

$$\begin{cases} V_c^{(1)}(z) = \left(\dfrac{z - z_1}{z_0 - z_1}\right)^p, & \varphi^{(1)}(z) = 0, \ z \in [z_0, z_1], \\ V_c^{(2)}(z) = 0, & \varphi^{(2)}(z) = 0, \ z \in [z_1, z_2], \\ V_c^{(3)}(z) = \left(\dfrac{z - z_2}{z_3 - z_2}\right)^p, & \varphi^{(3)}(z) = 0, \ z \in [z_2, z_3], \end{cases} \tag{6}$$

where $p(0 \le p < \infty)$ is the power-law exponent and $\zeta = h_2/h$ is the core-to-thickness ratio. On the other hand, Young's modulus $E^{(N)}(z)$, Poisson's ratio $\mu^{(N)}(z)$, mass density $\rho^{(N)}$(z), and the thermal expansion coefficient $\alpha^{(N)}(z)$ for each layer of the cylindrical shells are given by

$$\begin{aligned} E^{(N)}(z,T) = {} & E_m(T) + [E_c(T) - E_m(T)]V_c^{(N)}(z) \\ & - \frac{f^{(N)}(z)}{2}[E_c(T) + E_m(T)], \end{aligned} \tag{7}$$

$$\begin{aligned} \mu^{(N)}(z,T) = {} & \mu_m(T) + [\mu_c(T) - \mu_m(T)]V_c^{(N)}(z) \\ & - \frac{f^{(N)}(z)}{2}[\mu_c(T) + \mu_m(T)], \end{aligned} \tag{8}$$

$$\begin{aligned} \rho^{(N)}(z,T) = {} & \rho_m(T) + [\rho_c(T) - \rho_m(T)]V_c^{(N)}(z) \\ & - \frac{f^{(N)}(z)}{2}[\rho_c(T) + \rho_m(T)], \end{aligned} \tag{9}$$

$$\alpha^{(N)}(z,T) = \alpha_m(T) + [\alpha_c(T) - \alpha_m(T)]V_c^{(N)}(z) - \frac{f^{(N)}(z)}{2}[\alpha_c(T) + \alpha_m(T)], \tag{10}$$

where E_m and E_c denote Young's modulus; μ_m and μ_c denote Poisson's ratio; ρ_m and ρ_c denote mass density; α_m and α_c denote the thermal expansion coefficient, respectively, for metal and ceramic.

2.2. *Kirchhoff–Love shell model*

Based on the Kirchhoff–Love shell theory, the global displacement field functions $u_x(x,\theta,z,t)$, $v_\theta(x,\theta,z,t)$, and $w_z(x,\theta,z,t)$ of an arbitrary point in the cylindrical shell are expressed as

$$u_x(x,\theta,z,t) = u(x,\theta,t) - z\frac{\partial w(x,\theta,t)}{\partial x}, \tag{11}$$

$$v_\theta(x,\theta,z,t) = v(x,\theta,t) - z\frac{\partial w(x,\theta,t)}{r\partial \theta}, \tag{12}$$

$$w_z(x,\theta,z,t) = w(x,\theta,t), \tag{13}$$

where t is time; while $u(x,\theta,t)$, $v(x,\theta,t)$, $w(x,\theta,t)$ are the displacement components of a point on the midplane. The stain–displacement relations are given by

$$\begin{Bmatrix} \varepsilon_x \\ \varepsilon_\theta \\ \gamma_{x\theta} \end{Bmatrix} = \begin{Bmatrix} \varepsilon_x^0 \\ \varepsilon_\theta^0 \\ \gamma_{x\theta}^0 \end{Bmatrix} + z\begin{Bmatrix} \varepsilon_x^1 \\ \varepsilon_\theta^1 \\ \gamma_{x\theta}^1 \end{Bmatrix}, \tag{14}$$

where the reference surface strains ε_x^0, ε_θ^0, $\gamma_{x\theta}^0$ and the surface curvatures ε_x^1, ε_θ^1, $\gamma_{x\theta}^1$ are defined as

$$\begin{Bmatrix} \varepsilon_x^0 \\ \varepsilon_\theta^0 \\ \gamma_{x\theta}^0 \end{Bmatrix} = \begin{Bmatrix} \dfrac{\partial u}{\partial x} \\ \dfrac{1}{r}\left(\dfrac{\partial v}{\partial \theta} + w\right) \\ \dfrac{\partial v}{\partial x} + \dfrac{\partial u}{r\partial \theta} \end{Bmatrix}, \quad \begin{Bmatrix} \varepsilon_x^1 \\ \varepsilon_\theta^1 \\ \gamma_{x\theta}^1 \end{Bmatrix} = \begin{Bmatrix} -\dfrac{\partial^2 w}{\partial x^2} \\ -\dfrac{1}{r^2}\left(\dfrac{\partial^2 w}{\partial \theta^2} - \dfrac{\partial v}{\partial \theta}\right) \\ -\left(\dfrac{2}{r}\dfrac{\partial^2 w}{\partial x \partial \theta} - \dfrac{1}{r}\dfrac{\partial v}{\partial x}\right) \end{Bmatrix}. \tag{15}$$

The constitutive relation for a cylindrical shell is stated as

$$\begin{Bmatrix} \sigma_x^{(N)} \\ \sigma_\theta^{(N)} \\ \tau_{x\theta}^{(N)} \end{Bmatrix} = \begin{bmatrix} Q_{11}^{(N)} & Q_{12}^{(N)} & 0 \\ Q_{21}^{(N)} & Q_{22}^{(N)} & 0 \\ 0 & 0 & Q_{33}^{(N)} \end{bmatrix} \begin{Bmatrix} \varepsilon_x \\ \varepsilon_\theta \\ \gamma_{x\theta} \end{Bmatrix} \tag{16}$$

in which the stress components are $\sigma_x^{(N)}$, $\sigma_\theta^{(N)}$, $\tau_{x\theta}^{(N)}$, respectively; $Q_{ij}^{(N)}$ $(i,\ j = 1,2,3)$ are the reduced stiffness and they are functions of temperature T and z as follows

$$Q_{11}^{(N)} = Q_{22}^{(N)} = \frac{E^{(N)}(z,T)}{1-[\mu^{(N)}(z,T)]^2}, \tag{17}$$

$$Q_{12}^{(N)} = Q_{21}^{(N)} = \frac{\mu^{(N)}(z,T)E^{(N)}(z,T)}{1-[\mu^{(N)}(z,T)]^2}, \tag{18}$$

$$Q_{33}^{(N)} = G^{(N)}(z,T) = \frac{E^{(N)}(z,T)}{2[1+\mu^{(N)}(z,T)]}. \tag{19}$$

The resultants forces and moments are given by

$$\begin{Bmatrix} N_x \\ N_\theta \\ N_{x\theta} \end{Bmatrix} = \sum_{N=1}^{3} \int_{z_{N-1}}^{z_N} \begin{Bmatrix} \sigma_x^{(N)} \\ \sigma_\theta^{(N)} \\ \tau_{x\theta}^{(N)} \end{Bmatrix} dz, \quad \begin{Bmatrix} M_x \\ M_\theta \\ M_{x\theta} \end{Bmatrix} = \sum_{N=1}^{3} \int_{z_{N-1}}^{z_N} \begin{Bmatrix} \sigma_x^{(N)} \\ \sigma_\theta^{(N)} \\ \tau_{x\theta}^{(N)} \end{Bmatrix} z\mathrm{d}z. \tag{20}$$

Substituting Eqs. (14)–(19) into Eq. (20) yields

$$N_x = A_{11}\frac{\partial u}{\partial x} - A_{12}\frac{\partial^2 w}{\partial x^2} + A_{13}\frac{1}{r}\left(\frac{\partial v}{\partial \theta} + w\right) - A_{14}\frac{1}{r^2}\left(\frac{\partial^2 w}{\partial \theta^2} - \frac{\partial v}{\partial \theta}\right), \tag{21}$$

$$N_\theta = A_{13}\frac{\partial u}{\partial x} - A_{14}\frac{\partial^2 w}{\partial x^2} + A_{11}\frac{1}{r}\left(\frac{\partial v}{\partial \theta} + w\right) - A_{12}\frac{1}{r^2}\left(\frac{\partial^2 w}{\partial \theta^2} - \frac{\partial v}{\partial \theta}\right), \tag{22}$$

$$N_{x\theta} = B_{11}\left(\frac{\partial v}{\partial x} + \frac{\partial u}{r\partial \theta}\right) - B_{12}\left(\frac{2}{r}\frac{\partial^2 w}{\partial x\partial \theta} - \frac{1}{r}\frac{\partial v}{\partial x}\right), \tag{23}$$

$$M_x = C_{11}\frac{\partial u}{\partial x} - C_{12}\frac{\partial^2 w}{\partial x^2} + C_{13}\frac{1}{r}\left(\frac{\partial v}{\partial \theta} + w\right) - C_{14}\frac{1}{r^2}\left(\frac{\partial^2 w}{\partial \theta^2} - \frac{\partial v}{\partial \theta}\right), \tag{24}$$

$$M_\theta = C_{13}\frac{\partial u}{\partial x} - C_{14}\frac{\partial^2 w}{\partial x^2} + C_{11}\frac{1}{r}\left(\frac{\partial v}{\partial \theta} + w\right) - C_{12}\frac{1}{r^2}\left(\frac{\partial^2 w}{\partial \theta^2} - \frac{\partial v}{\partial \theta}\right), \tag{25}$$

$$M_{x\theta} = D_{11}\left(\frac{\partial v}{\partial x} + \frac{\partial u}{r\partial \theta}\right) - D_{12}\left(\frac{2}{r}\frac{\partial^2 w}{\partial x\partial \theta} - \frac{1}{r}\frac{\partial v}{\partial x}\right), \tag{26}$$

where coefficients A_{ij}, B_{ij}, C_{ij} and D_{ij} $(i,\ j\ =\ 1,2,3,4)$ are presented in Appendix A.

The strain energy due to bending moment and normal force can be expressed as

$$\begin{aligned} S &= \frac{1}{2}\int_0^L \int_0^{2\pi} \sum_{N=1}^{3} \int_{z_{N-1}}^{z_N} \left(\sigma_x^{(N)}\varepsilon_x + \sigma_\theta^{(N)}\varepsilon_\theta + \tau_{x\theta}^{(N)}\gamma_{x\theta}\right) r\mathrm{d}z\mathrm{d}\theta\mathrm{d}x \\ &= \frac{1}{2}\int_0^L \int_0^{2\pi} (N_x\varepsilon_x^0 + M_x\varepsilon_x^1 + N_\theta\varepsilon_\theta^0 + M_\theta\varepsilon_\theta^1 + N_{x\theta}\gamma_{x\theta}^0 + M_{x\theta}\gamma_{x\theta}^1) r\mathrm{d}\theta\mathrm{d}x, \end{aligned} \tag{27}$$

while the kinetic energy can be determined as

$$K = \frac{1}{2}\int_0^L \int_0^{2\pi} \sum_{N=1}^{3} \int_{z_{N-1}}^{z_N} \left[\left(\frac{\partial u_x}{\partial t}\right)^2 + \left(\frac{\partial v_\theta}{\partial t}\right)^2 + \left(\frac{\partial w_z}{\partial t}\right)^2\right] \rho^{(N)}(z,T)\, r\mathrm{d}z\mathrm{d}\theta\mathrm{d}x$$

$$= \frac{1}{2}\int_0^L \int_0^{2\pi} \left\{ I_1 \left[\left(\frac{\partial u}{\partial t}\right)^2 + \left(\frac{\partial v}{\partial t}\right)^2 + \left(\frac{\partial w}{\partial t}\right)^2\right] + I_2\left[-2\frac{\partial u}{\partial t}\frac{\partial^2 w}{\partial x \partial t} - 2\frac{\partial v}{\partial t}\frac{\partial^2 w}{r\partial\theta\,\partial t}\right]\right.$$

$$\left. + I_3\left[\left(\frac{\partial^2 w}{\partial x \partial t}\right)^2 + \left(\frac{\partial^2 w}{r\partial\theta\,\partial t}\right)^2\right]\right\} r\mathrm{d}\theta\mathrm{d}x, \tag{28}$$

where the coefficients I_1, I_2 and I_3 are defined as

$$\begin{Bmatrix} I_1 \\ I_2 \\ I_3 \end{Bmatrix} = \sum_{N=1}^{3} \int_{z_{N-1}}^{z_N} \begin{Bmatrix} 1 \\ z \\ z^2 \end{Bmatrix} \rho^{(N)}(z,T)\mathrm{d}z. \tag{29}$$

2.3. *Thermal analysis for FGMSCS*

In this study, we consider FGMSCS that is subjected to initial-final temperature variation conditions. It is hypothesized that the FGMSCS is placed in a thermal environment for a sufficiently long period of time, indicating that temperature has risen uniformly from an initial T_0 to a final T_z. The temperature difference ΔT_1 is

$$\Delta T_1 = T_z - T_0. \tag{30}$$

The stresses owing to temperature variation can be expressed as

$$\begin{Bmatrix} \sigma_x^{T(N)} \\ \sigma_\theta^{T(N)} \\ \tau_{x\theta}^{T(N)} \end{Bmatrix} = -\begin{Bmatrix} Q_{11}^{(N)} & Q_{12}^{(N)} & 0 \\ Q_{21}^{(N)} & Q_{22}^{(N)} & 0 \\ 0 & 0 & Q_{33}^{(N)} \end{Bmatrix} \begin{Bmatrix} 1 \\ 1 \\ 0 \end{Bmatrix} \alpha^{(N)}(z,T)\Delta T, \tag{31}$$

where the thermal stress components are $\sigma_x^{T(N)}$, $\sigma_\theta^{T(N)}$, $\tau_{x\theta}^{T(N)}$. The temperature rise from the reference temperature T_f can be expressed as

$$\Delta T = T_z - T_f \tag{32}$$

in which the reference temperature for a cylindrical shell in the stress-free state is $T_f = 298\,\mathrm{K}$.

With reference to the experiment and numerical analysis[69,70] and theoretical analyses[71,72] for plates in thermal environments, a new analytical model that applies Green's nonlinear strains is introduced here. It is used to derive the thermal strain energy and the thermal effects on sandwich cylindrical shells with more accurate solutions. The common thermal stresses of thermal conditions are adopted.[76] For the cylindrical shell structures, Green's nonlinear normal strains are given by

$$\varepsilon_{xnl} = \frac{1}{2}\left[\left(\frac{\partial u}{\partial x} - z\frac{\partial^2 w}{\partial x^2}\right)^2 + \left(\frac{\partial v}{\partial x} - z\frac{\partial^2 w}{r\partial x\partial\theta} + z\frac{\partial v}{r\partial x}\right)^2 + \left(\frac{\partial w}{\partial x}\right)^2\right], \tag{33}$$

$$\varepsilon_{\theta nl} = \frac{1}{2}\left[\left(\frac{\partial u}{r\partial\theta} - z\frac{\partial^2 w}{r\partial x\partial\theta}\right)^2 + \left[\frac{1}{r}\left(\frac{\partial v}{\partial\theta} + w\right) - \frac{z}{r^2}\left(\frac{\partial^2 w}{\partial\theta^2} - \frac{\partial v}{\partial\theta}\right)\right]^2 + \left(\frac{1}{r}\frac{\partial w}{\partial\theta}\right)^2\right]. \tag{34}$$

The thermal strain energy is given by

$$\begin{aligned} S^T &= \frac{1}{2}\int_0^L\int_0^{2\pi}\sum_{N=1}^{3}\int_{z_{N-1}}^{z_N}(2\sigma_x^{T(N)}\varepsilon_{xnl} + 2\sigma_\theta^{T(N)}\varepsilon_{\theta nl} + \tau_{x\theta}^{T(N)}\gamma_{x\theta nl})r\mathrm{d}z\mathrm{d}\theta\mathrm{d}x \\ &= \frac{1}{2}\int_0^L\int_0^{2\pi}\left\{A_x^T\left[\left(\frac{\partial u}{\partial x}\right)^2 + \left(\frac{\partial v}{\partial x}\right)^2 + \left(\frac{\partial w}{\partial x}\right)^2\right]\right. \\ &\quad + B_x^T\left[-2\frac{\partial u}{\partial x}\frac{\partial^2 w}{\partial x^2} - 2\frac{\partial v}{\partial x}\left(\frac{\partial^2 w}{r\partial x\partial\theta} - \frac{\partial v}{r\partial x}\right)\right] \\ &\quad + C_x^T\left[\left(\frac{\partial^2 w}{\partial x^2}\right)^2 + \left(\frac{\partial^2 w}{r\partial x\partial\theta} - \frac{\partial v}{r\partial x}\right)^2\right] \\ &\quad + A_\theta^T\left[\frac{1}{r^2}\left(\frac{\partial u}{\partial\theta}\right)^2 + \frac{1}{r^2}\left(\frac{\partial v}{\partial\theta} + w\right)^2 + \frac{1}{r^2}\left(\frac{\partial w}{\partial\theta}\right)^2\right] \\ &\quad + B_\theta^T\left[-2\frac{\partial u}{r\partial\theta}\frac{\partial^2 w}{r\partial x\partial\theta} - 2\left(\frac{\partial v}{r\partial\theta} + \frac{w}{r}\right)\left(\frac{\partial^2 w}{r^2\partial\theta^2} - \frac{\partial v}{r^2\partial\theta}\right)\right] \\ &\quad \left. + C_\theta^T\left[\left(\frac{\partial^2 w}{r\partial x\partial\theta}\right)^2 + \left(\frac{\partial^2 w}{r^2\partial\theta^2} - \frac{\partial v}{r^2\partial\theta}\right)^2\right]\right\}r\mathrm{d}\theta\mathrm{d}x \end{aligned} \tag{35}$$

in which coefficients A_x^T, B_x^T, C_x^T, A_θ^T, B_θ^T, and C_θ^T under thermal conditions are presented in Appendix B.

2.4. *Solution procedure*

To determine natural frequencies of FGMSCS in thermal conditions via the Kirchhoff–Love shell theory, the Hamilton principle is applied, as

$$\int_0^t [\delta K - (\delta S + \delta S^T)]\mathrm{d}t = 0. \tag{36}$$

Substituting Eqs. (27), (28), and (35) into Eq. (36), the governing equations can be obtained as

$$\frac{\partial N_x}{\partial x} + \frac{\partial N_{x\theta}}{r\partial\theta} + A_x^T \frac{\partial^2 u}{\partial x^2} + A_\theta^T \frac{\partial^2 u}{r^2 \partial\theta^2} - B_x^T \frac{\partial^3 w}{\partial x^3} - B_\theta^T \frac{\partial^3 w}{r^2 \partial x \partial\theta^2}$$
$$= I_1 \frac{\partial^2 u}{\partial t^2} - I_2 \frac{\partial^3 w}{\partial x \partial t^2}, \tag{37}$$

$$\frac{\partial N_{x\theta}}{\partial x} + \frac{1}{r}\frac{\partial M_{x\theta}}{\partial x} + \frac{1}{r}\frac{\partial N_\theta}{\partial\theta} + \frac{1}{r^2}\frac{\partial M_\theta}{\partial\theta} + A_x^T \frac{\partial^2 v}{\partial x^2} + A_\theta^T \left(\frac{1}{r^2}\frac{\partial^2 v}{\partial\theta^2} + \frac{1}{r^2}\frac{\partial w}{\partial\theta}\right)$$
$$+ B_x^T \left(\frac{2}{r}\frac{\partial^2 v}{\partial x^2} - \frac{1}{r}\frac{\partial^3 w}{\partial x^2 \partial\theta}\right) + B_\theta^T \left(\frac{1}{r^3}\frac{\partial w}{\partial\theta} + \frac{2}{r^3}\frac{\partial v^2}{\partial\theta^2} - \frac{1}{r^3}\frac{\partial^3 w}{\partial\theta^3}\right)$$
$$+ C_x^T \left(\frac{1}{r^2}\frac{\partial^2 v}{\partial x^2} - \frac{1}{r^2}\frac{\partial^3 w}{\partial x^2 \partial\theta}\right) + C_\theta^T \left(\frac{1}{r^4}\frac{\partial^2 v}{\partial\theta^2} - \frac{1}{r^4}\frac{\partial^3 w}{\partial\theta^3}\right)$$
$$= I_1 \frac{\partial^2 v}{\partial t^2} - I_2 \frac{\partial^3 w}{r\partial\theta \partial t^2}, \tag{38}$$

$$\frac{\partial^2 M_x}{\partial x^2} + \frac{2}{r}\frac{\partial^2 M_{x\theta}}{\partial x \partial\theta} + \frac{1}{r^2}\frac{\partial^2 M_\theta}{\partial\theta^2} - \frac{N_\theta}{r} + A_x^T \frac{\partial^2 w}{\partial x^2} + A_\theta^T \left(\frac{1}{r^2}\frac{\partial^2 w}{\partial\theta^2} - \frac{1}{r^2}\frac{\partial v}{\partial\theta} - \frac{w}{r^2}\right)$$
$$+ B_x^T \left(\frac{1}{r}\frac{\partial^3 v}{\partial x^2 \partial\theta} + \frac{\partial^3 u}{\partial x^3}\right) + B_\theta^T \left(\frac{2}{r^3}\frac{\partial^2 w}{\partial\theta^2} - \frac{1}{r^3}\frac{\partial v}{\partial\theta} + \frac{1}{r^3}\frac{\partial^3 v}{\partial\theta^3} + \frac{1}{r^2}\frac{\partial^3 u}{\partial x \partial\theta^2}\right)$$
$$+ C_x^T \left(\frac{1}{r^2}\frac{\partial^3 v}{\partial x^2 \partial\theta} - \frac{1}{r^2}\frac{\partial^4 w}{\partial x^2 \partial\theta^2} - \frac{\partial^4 w}{\partial x^4}\right)$$
$$+ C_\theta^T \left(\frac{1}{r^4}\frac{\partial^3 v}{\partial\theta^3} - \frac{1}{r^4}\frac{\partial^4 w}{\partial\theta^4} - \frac{1}{r^2}\frac{\partial^4 w}{\partial x^2 \partial\theta^2}\right)$$
$$= I_1 \frac{\partial^2 w}{\partial t^2} + I_2 \left(\frac{1}{r}\frac{\partial^3 v}{\partial\theta \partial t^2} + \frac{\partial^3 u}{\partial x \partial t^2}\right) - I_3 \left(\frac{\partial^4 w}{\partial x^2 \partial t^2} + \frac{1}{r^2}\frac{\partial^4 w}{\partial\theta^2 \partial t^2}\right). \tag{39}$$

Further substituting Eqs. (21)–(26) into Eqs. (37)–(39) yields

$$A_{11}\frac{\partial^2 u}{\partial x^2} - A_{12}\frac{\partial^3 w}{\partial x^3} + A_{13}\frac{1}{r}\left(\frac{\partial w}{\partial x} + \frac{\partial^2 v}{\partial x \partial\theta}\right) + A_{14}\frac{1}{r^2}\left(\frac{\partial^2 v}{\partial x \partial\theta} - \frac{\partial^3 w}{\partial x \partial\theta^2}\right)$$
$$+ B_{11}\left(\frac{1}{r^2}\frac{\partial^2 u}{\partial\theta^2} + \frac{\partial^2 v}{r\partial x \partial\theta}\right) + B_{12}\frac{1}{r^2}\left(\frac{\partial^2 v}{\partial x \partial\theta} - 2\frac{\partial^3 w}{\partial x \partial\theta^2}\right) A_x^T \frac{\partial^2 u}{\partial x^2}$$
$$+ A_\theta^T \frac{1}{r^2}\frac{\partial^2 u}{\partial\theta^2} - B_x^T \frac{\partial^3 w}{\partial x^3} - B_\theta^T \frac{1}{r^2}\frac{\partial^3 w}{\partial x \partial\theta^2} = I_1 \frac{\partial^2 u}{\partial t^2} - I_2 \frac{\partial^3 w}{\partial x \partial t^2}, \tag{40}$$

$$
\begin{aligned}
&A_{11}\left(\frac{1}{r^2}\frac{\partial^2 v}{\partial\theta^2}+\frac{1}{r^2}\frac{\partial w}{\partial\theta}\right)+A_{12}\left(\frac{1}{r^3}\frac{\partial^2 v}{\partial\theta^2}-\frac{1}{r^3}\frac{\partial^3 w}{\partial\theta^3}\right)+A_{13}\frac{\partial^2 u}{r\partial x\partial\theta}-A_{14}\frac{\partial^3 w}{r\partial x^2\partial\theta}\\
&\quad+B_{11}\left(\frac{1}{r}\frac{\partial^2 u}{\partial x\partial\theta}+\frac{\partial^2 v}{\partial x^2}\right)+B_{12}\left(\frac{1}{r}\frac{\partial^2 v}{\partial x^2}-\frac{2}{r}\frac{\partial^3 w}{\partial x^2\partial\theta}\right)\\
&\quad+C_{11}\left(\frac{1}{r^3}\frac{\partial w}{\partial\theta}+\frac{1}{r^3}\frac{\partial^2 v}{\partial\theta^2}\right)+C_{12}\frac{1}{r^4}\left(\frac{\partial^2 v}{\partial\theta^2}-\frac{\partial^3 w}{\partial\theta^3}\right)+C_{13}\frac{1}{r^2}\frac{\partial^2 u}{\partial x\partial\theta}\\
&\quad-C_{14}\frac{1}{r^2}\frac{\partial^3 w}{\partial x^2\partial\theta}+D_{11}\left(\frac{1}{r^2}\frac{\partial^2 u}{\partial x\partial\theta}+\frac{1}{r}\frac{\partial^2 v}{\partial x^2}\right)\\
&\quad+D_{12}\left(\frac{1}{r^2}\frac{\partial^2 v}{\partial x^2}-\frac{2}{r^2}\frac{\partial^3 w}{\partial x^2\partial\theta}\right)+A_x^T\frac{\partial^2 v}{\partial x^2}+A_\theta^T\left(\frac{1}{r^2}\frac{\partial w}{\partial\theta}+\frac{1}{r^2}\frac{\partial^2 v}{\partial\theta_2}\right)\\
&\quad+B_x^T\left(\frac{2}{r}\frac{\partial^2 v}{\partial x^2}-\frac{1}{r}\frac{\partial^3 w}{\partial x^2\partial\theta}\right)+B_\theta^T\left(\frac{1}{r^3}\frac{\partial w}{\partial\theta}+\frac{2}{r^3}\frac{\partial^2 v}{\partial\theta^2}-\frac{1}{r^3}\frac{\partial^3 w}{\partial\theta^3}\right)\\
&\quad+C_x^T\left(\frac{1}{r^2}\frac{\partial^2 v}{\partial x^2}-\frac{1}{r^2}\frac{\partial^3 w}{\partial x^2\partial\theta}\right)+C_\theta^T\frac{1}{r^4}\left(\frac{\partial^2 v}{\partial\theta^2}-\frac{\partial^3 w}{\partial\theta^3}\right)\\
&=I_1\frac{\partial^2 v}{\partial t^2}-I_2\frac{\partial^3 w}{r\partial\theta\partial t^2},
\end{aligned}
\tag{41}
$$

$$
\begin{aligned}
&A_{11}\left(-\frac{w}{r^2}-\frac{1}{r^2}\frac{\partial v}{\partial\theta}\right)+A_{12}\left(\frac{1}{r^3}\frac{\partial^2 w}{\partial\theta^2}-\frac{1}{r^3}\frac{\partial v}{\partial\theta}\right)-A_{13}\frac{1}{r}\frac{\partial u}{\partial x}+A_{14}\frac{1}{r}\frac{\partial^2 w}{\partial x^2}\\
&\quad+C_{11}\left(\frac{1}{r^3}\frac{\partial^2 w}{\partial\theta^2}+\frac{1}{r^3}\frac{\partial^3 v}{\partial\theta^3}+\frac{\partial^3 u}{\partial x^3}\right)+C_{12}\left(\frac{1}{r^4}\frac{\partial^3 v}{\partial\theta^3}-\frac{1}{r^4}\frac{\partial^4 w}{\partial\theta^4}-\frac{\partial^4 w}{\partial x^4}\right)\\
&\quad+C_{13}\left(\frac{1}{r^2}\frac{\partial^3 u}{\partial x\partial\theta^2}+\frac{1}{r}\frac{\partial^2 w}{\partial x^2}+\frac{1}{r}\frac{\partial^3 v}{\partial x^2\partial\theta}\right)+C_{14}\left(\frac{1}{r^2}\frac{\partial^3 v}{\partial x^2\partial\theta}-\frac{2}{r^2}\frac{\partial^4 w}{\partial x^2\partial\theta^2}\right)\\
&\quad+D_{11}\left(\frac{2}{r^2}\frac{\partial^3 u}{\partial x\partial\theta^2}+\frac{2}{r}\frac{\partial^3 v}{\partial x^2\partial\theta}\right)+D_{12}\left(\frac{2}{r^2}\frac{\partial^3 v}{\partial x^2\partial\theta}-\frac{4}{r^2}\frac{\partial^4 w}{\partial x^2\partial\theta^2}\right)\\
&\quad+A_x^T\frac{\partial^2 w}{\partial x^2}+A_\theta^T\left(\frac{1}{r^2}\frac{\partial^2 w}{\partial\theta^2}-\frac{w}{r^2}-\frac{1}{r^2}\frac{\partial v}{\partial\theta}\right)+B_x^T\left(\frac{\partial^3 u}{\partial x^3}+\frac{1}{r}\frac{\partial^3 v}{\partial x^2\partial\theta}\right)\\
&\quad+B_\theta^T\left(\frac{2}{r^3}\frac{\partial^2 w}{\partial\theta^2}-\frac{1}{r^3}\frac{\partial v}{\partial\theta}+\frac{1}{r^3}\frac{\partial^3 v}{\partial\theta^3}+\frac{1}{r^2}\frac{\partial^3 u}{\partial x\partial\theta^2}\right)\\
&\quad+C_x^T\left(\frac{1}{r^2}\frac{\partial^3 v}{\partial x^2\partial\theta}-\frac{1}{r^2}\frac{\partial^4 w}{\partial x^2\partial\theta^2}-\frac{\partial^4 w}{\partial x^4}\right)\\
&\quad+C_\theta^T\left(\frac{1}{r^4}\frac{\partial^3 v}{\partial\theta^3}-\frac{1}{r^4}\frac{\partial^4 w}{\partial\theta^4}-\frac{1}{r^2}\frac{\partial^4 w}{\partial x^2\partial\theta^2}\right)\\
&=I_1\frac{\partial^2 w}{\partial t^2}-I_2\left(-\frac{1}{r}\frac{\partial^3 v}{\partial\theta\partial t^2}-\frac{\partial^3 u}{\partial x\partial t^2}\right)-I_3\left(\frac{1}{r^2}\frac{\partial^4 w}{\partial\theta^2\partial t^2}+\frac{\partial^4 w}{\partial x^2\partial t^2}\right).
\end{aligned}
\tag{42}
$$

For the current work, the simply supported boundary condition for cylindrical shell are considered, as

$$v = w = N_x = M_x = 0, \quad \text{at } x = 0 \text{ and } x = L \tag{43}$$

which are satisfied by the following displacement functions:

$$u = U\cos\left(\frac{m\pi x}{L}\right)\cos(n\theta)\cos(\omega t), \tag{44}$$

$$v = V\sin\left(\frac{m\pi x}{L}\right)\sin(n\theta)\cos(\omega t), \tag{45}$$

$$w = W\sin\left(\frac{m\pi x}{L}\right)\cos(n\theta)\cos(\omega t), \tag{46}$$

where U, V, and W are constants that denote the amplitude of thermal vibration; n and m are the circumferential and axial wave numbers; and ω (rad/s) is the natural angular frequency.

The substitution of Eqs. (44)–(46) into Eqs. (40) and (42) leads to a generalized eigenvalue equation for the thermal vibration problem, and it can be expressed in a matrix form as

$$([K] - \omega^2[M])\begin{Bmatrix} U \\ V \\ W \end{Bmatrix} = \{0\}, \tag{47}$$

where the expressions of K_{ij} and M_{ij} $(i, j = 1, 2, 3)$ are presented in Appendix C. The natural frequency can be obtained by solving the nontrivial solutions of Eq. (47), as

$$\mathrm{Det}[K_{3\times3} - \omega^2 M_{3\times3}] = 0. \tag{48}$$

Analytically, Eq. (48) can be solved to yield three eigenvalue solutions for the natural frequencies and the smallest eigenvalue is of primary significance in this work.

3. Results and Discussion

To establish credibility of the theoretical derivation and analytical model proposed in this study, a comparison is presented with reference to the results of the published data and benchmarks for homogenous simply supported circular cylindrical shells.[77–79] In this respect and for comparison purposes, a homogenous cylindrical shell can be obtained from the present model by setting: $\Delta T = 0$, $p = 0$, $\zeta = 0$, $\beta = 0$, $r = 0.1\,\text{m}$, $h = 0.247 \times 10^{-3}\,\text{m}$, $L = 0.2\,\text{m}$, $E = 71.02 \times 10^9\,\text{Pa}$, $\rho = 2796\,\text{kg/m}^3$, and $\mu = 0.31$. The comparation presented in Table 1 indicates a high level of agreement between the new model established in this study and the published benchmarks.[77–79]

Further, another detailed study is presented to authenticate the new theoretical model for FGMSCS with and without porosity proposed here. As stated above, there

Table 1. Comparison of natural frequencies (Hz) of a homogeneous circular cylindrical shell under simply supported boundary condition.

Mode		Natural frequency (Hz)			
m	n	Present	Ref. 77	Ref. 78	Ref. 79
1	7	484.663	484.6	484.6	484.55
1	8	489.667	489.6	489.6	489.55
1	9	546.296	546.2	546.2	546.2
1	6	553.431	553.3	553.3	553.33
1	10	636.886	636.8	636.8	636.81
1	5	722.202	722.1	722.1	722.13
1	11	750.722	750.7	750.7	750.66
1	12	882.265	882.2	882.2	882.23
2	10	968.315	968.1	968.1	968.09
2	11	983.556	983.4	983.4	983.34

Table 2. Temperature-dependent material coefficients.[6]

Materials		P_0	P_{-1}	P_1	P_2	P_3
Stainless steel	E(Pa)	201.04×10^9	0	3.079×10^{-4}	-6.534×10^{-7}	0
	μ	0.3262	0	-2.002×10^{-4}	3.797×10^{-7}	0
	ρ(kg/m^3)	8166	0	0	0	0
	α(1/K)	12.330×10^{-6}	0	8.086×10^{-4}	0	0
Silicon nitride	E(Pa)	348.43×10^9	0	-3.070×10^{-4}	2.160×10^{-7}	-8.946×10^{-11}
	μ	0.2400	0	0	0	0
	ρ(kg/m^3)	2 370	0	0	0	0
	α(1/K)	5.8723×10^{-6}	0	9.095×10^{-4}	0	0

are very few studies available that deal with the surface layers' porosity effects of FGMSCS. As explained earlier, the porosity effects vary extensively with the mixing degree of component materials. The new analytical model and solutions for FGMSCS here are verified by comparing with FEM numerical simulations using COMSOL Multiphysics. The geometric parameters adopted are $r = 1$ m, $h = 0.05r$, $L = 5r$, and $\zeta = 0.4$. The thermal coefficients and material properties are very sensitive to temperature changes, and they are tabulated in Table 2 for metal (stainless steel) and ceramic (silicon nitride).[6] Table 3 shows very good consistency and agreement between analytical solutions and FEM simulations, where the power-law exponent is $p = 1$ for both types of shells. The porosity volume fraction is $\beta = 0.1$ for Type-1 shell, and it is $\beta = 0$ for Type-2 shell. Hence, the present analytical model and solution methodology are validated.

Subsequently, a systematic parameter study is performed to show the thermal vibration response of FGMSCS with and without porosity under thermal environments. The material coefficients are shown in Table 2. The geometric parameters $r = 1$ m, $h = 0.05r$, and $L = 5r$ are set for two types of FGMSCS.

Table 3. Comparison of natural frequencies (Hz) of simply supported FGM cylindrical sandwich shells.

Mode		Type-1 shell		Type-2 shell	
m	n	Present	COMSOL	Present	COMSOL
1	1	190.511	190.58	191.155	191.23
1	2	90.176	89.75	90.750	90.32
1	3	128.199	127.39	129.808	128.98
1	4	230.771	229.10	233.851	232.12
1	5	368.163	364.41	373.103	369.17
1	6	536.729	529.07	543.928	535.89
2	1	469.869	470.93	471.353	472.42
2	2	252.839	253.43	253.938	254.53
2	3	195.134	194.64	196.829	196.32
2	4	260.778	258.86	264.011	262.02
2	5	389.305	385.09	394.446	390.05
2	6	555.573	547.32	562.990	554.33
3	1	680.174	684.77	682.505	687.12
3	2	428.128	434.03	429.920	435.84
3	3	310.309	315.84	312.375	317.92
3	4	324.584	327.19	328.070	330.61
3	5	431.034	429.48	436.490	434.75
3	6	589.795	583.02	597.568	590.38

The natural frequency response of sandwich cylindrical shells for two types of FGM surface layers with and without porosity subjected to the different temperature loadings ΔT are presented in Tables 4 and 5, where $\zeta = 0.4$, $p = 1$, and $\beta = 0.1$. An increase in temperature change leads to decreasing the natural frequency of both types of shells. As observed, the fundamental natural frequency of Type-2 shell is larger than that of Type-1 shell when the temperature starts to rise initially. However, as the temperature continues to rise, the situation is reversed. In addition, it is interesting to note that the fundamental mode shape for both types of shell changes from $m = 1$, $n = 2$ to $m = 1$, $n = 3$ with a higher temperature change. This finding indicates that the fundamental mode shape of FGMSCS does not always exist in a specific wavenumber but rather it varies with variations of temperature change.

The impact of power-law exponent p on natural frequency for two types of FGMSCS under thermal effects and with $\zeta = 0.4$, $\beta = 0.1$, and $\Delta T = 50\,\mathrm{K}$ are shown in Table 6. From the table and for both types of FGMSCS, a decrease of natural frequency occurs with increasing power-law exponent. In the meantime, it is observed that the fundamental natural frequency of Type-2 shell is smaller than the corresponding value of Type-1 shell. On the contrary, the natural frequencies of other modes of Type-2 shell are higher than those of Type-1 shell, except for the fundamental natural frequency. In addition, the fundamental natural frequency always corresponds to mode $m = 1$, $n = 2$ for varying power-law exponent.

Table 4. Natural frequencies (Hz) of Type-1 shell subjected to different temperature loadings ΔT.

Mode		ΔT					
m	n	0 K	20 K	40 K	60 K	80 K	100 K
1	1	190.511	189.452	188.344	187.187	185.981	184.725
1	2	90.176	83.223	75.456	66.600	56.152	42.978
1	3	128.199	116.544	103.295	87.745	68.363	39.941
1	4	230.771	219.038	206.373	192.602	177.483	160.648
1	5	368.163	356.392	343.960	330.804	316.845	301.983
1	6	536.729	524.773	512.282	499.226	485.568	471.269
2	1	469.869	468.436	466.918	465.314	463.624	461.848
2	2	252.839	249.326	245.661	241.842	237.863	233.719
2	3	195.134	186.474	177.186	167.173	156.304	144.398
2	4	260.778	249.590	237.622	224.760	210.853	195.690
2	5	389.305	377.610	365.285	352.274	338.509	323.906
2	6	555.573	543.610	531.122	518.081	504.452	490.200

Table 5. Natural frequencies (Hz) of Type-2 shell subjected to different temperature loadings ΔT.

Mode		ΔT					
m	n	0 K	20 K	40 K	60 K	80 K	100 K
1	1	191.155	190.062	188.920	187.728	186.486	185.193
1	2	90.750	83.533	75.447	66.183	55.159	40.995
1	3	129.808	117.787	104.089	87.949	67.661	36.961
1	4	233.851	221.767	208.711	194.497	178.863	161.412
1	5	373.103	360.987	348.185	334.628	320.232	304.889
1	6	543.928	531.628	518.774	505.333	491.266	476.530
2	1	471.353	469.884	468.329	466.688	464.961	463.147
2	2	253.938	250.298	246.500	242.542	238.417	234.119
2	3	196.829	187.871	178.251	167.863	156.564	144.150
2	4	264.011	252.480	240.135	226.853	212.471	196.761
2	5	394.446	382.407	369.714	356.307	342.111	327.037
2	6	562.990	550.683	537.833	524.407	510.371	495.684

The influence of porosity volume fraction β on the natural frequency for both types of FGMSCS with $\zeta = 0.4$, $p = 1$, and $\Delta T = 50\,\mathrm{K}$ and subjected to the thermal loadings is presented in Table 7. For $\beta = 0$, the natural frequency of Type-2 shell can be referred. For increasing β, the corresponding natural frequency refers to Type-1 shell. It can be observed that the fundamental natural frequency of FGMSCS under thermal effects reaches its maximum for $\beta = 0.2$, thus indicating that a change of fundamental natural frequency does not always indicate monotonous increasing in β. However, the nonfundamental natural frequencies decrease monotonically with increasing β. Thus, it implies that the natural frequency of Type-2 shell is always

Table 6. Natural frequencies (Hz) for two types of FGMSCS with respect to different power-law exponents p under thermal effects.

Mode		$P = 0.5$		$P = 1$		$P = 5$	
m	n	Type-1	Type-2	Type-1	Type-2	Type-1	Type-2
1	1	202.159	202.345	187.772	188.330	163.347	164.452
1	2	78.320	78.003	71.188	70.990	58.418	58.368
1	3	106.253	106.615	95.873	96.396	75.148	75.956
1	4	217.509	219.271	199.639	201.763	164.661	167.472
1	5	365.626	369.040	337.477	341.506	282.868	288.061
1	6	546.677	552.057	505.827	512.129	426.896	434.943
2	1	501.704	502.164	466.127	467.519	405.720	408.489
2	2	262.828	263.127	243.771	244.541	211.137	212.601
2	3	187.271	187.837	172.277	173.161	144.848	146.234
2	4	251.383	253.270	231.311	233.620	192.680	195.760
2	5	388.609	392.189	358.869	363.104	301.366	306.833
2	6	566.952	572.516	524.673	531.194	443.058	451.385

Table 7. Natural frequencies (Hz) of FGMSCS with respect to different porosity volume fraction β under thermal effects.

Mode		β				
m	N	0	0.1	0.2	0.3	0.4
1	1	188.330	187.772	187.160	186.488	185.747
1	2	70.990	71.188	71.264	71.218	71.046
1	3	96.396	95.872	95.159	94.245	93.116
1	4	201.763	199.639	197.339	194.842	192.121
1	5	341.506	337.477	333.269	328.846	324.169
1	6	512.129	505.827	499.332	492.594	485.558
2	1	467.519	466.127	464.606	462.939	461.104
2	2	244.541	243.771	242.914	241.961	240.902
2	3	173.161	172.277	171.255	170.083	168.749
2	4	233.620	231.311	228.833	226.161	223.271
2	5	363.104	358.869	354.454	349.821	344.930
2	6	531.194	524.673	517.957	510.995	503.729

higher than the corresponding value of Type-1 shell except the fundamental natural frequency. Similarly, the fundamental natural frequency always corresponds to the mode of $m = 1$, $n = 2$ for increasing β.

Table 8 examines the influence of core-to-thickness ratio ζ on the natural frequency for both types of FGMSCS with $p = 1$, $\beta = 0.1$, $m = 1$, and $\Delta T = 50\,\text{K}$ under thermal loadings. The natural frequency for both types of shells exhibits a monotonically decreasing trend for increasing ζ. This is because an increase in ζ results in reduced contents of component materials of silicon nitride in both types of FGMSCS, thus indicating a lower strength of composite structure. In addition,

Table 8. Natural frequencies (Hz) of FGMSCS with respect to different core-to-thickness ratio ζ under thermal effects.

ζ	Type	$n = 1$	$n = 2$	$n = 3$	$n = 4$	$n = 5$	$n = 6$
0.1	1	209.923	82.235	111.587	226.394	379.431	566.550
	2	209.926	81.119	109.741	225.120	378.697	566.385
0.2	1	202.001	78.357	106.203	217.155	364.895	545.480
	2	202.285	77.612	105.301	217.287	366.171	548.077
0.3	1	194.641	74.687	100.988	208.274	350.967	525.321
	2	195.104	74.246	100.877	209.540	353.835	530.078
0.4	1	187.772	71.188	95.873	199.639	337.477	505.827
	2	188.330	70.990	96.396	201.763	341.506	512.129
0.5	1	181.333	67.828	90.795	191.151	324.272	486.779
	2	181.917	67.814	91.787	193.840	329.012	493.985
0.6	1	175.274	64.579	85.693	182.717	311.212	467.979
	2	175.825	64.691	86.972	185.657	316.184	475.407
0.7	1	169.551	61.413	80.499	174.246	298.165	449.244
	2	170.021	61.593	81.865	177.095	302.848	456.155
0.8	1	164.127	58.308	75.142	165.647	285.003	430.393
	2	164.474	58.495	76.360	168.019	288.819	435.970
0.9	1	158.970	55.238	69.533	156.824	271.592	411.248
	2	159.159	55.368	70.323	158.275	273.884	414.569

the variation of ζ does not affect the fundamental mode shape for both types of FGMSCS.

4. Concluding Remarks

In this work, the thermal vibration characteristics of FGM sandwich cylindrical shells (FGMSCS) with and without porosity, and with simply supported boundary condition is investigated. The mixing degree of component materials leads to changes in porosity, and it is accurately described by an analytical porosity function. A theoretical model with analytical solution methodology considering Green's nonlinear strains is established to derive thermal strain energy of FGMSCS exposed to thermal effects. The governing equations are derived via Hamilton's principle and analytical solutions are presented. The effects of porosity type, porosity volume fraction, core-to-thickness ratio, power-law exponent, and temperature change are solved and discussed in detail. The key findings from the present investigation are summarized as follows:

(1) A higher temperature loading results in a smaller natural frequency for the two types of shells considered. In addition, the fundamental mode shape for both types of shells changes with a higher temperature change, thus indicating that

the fundamental mode shape of FGMSCS does not always exist in a specific wavenumber but rather it varies with variations of temperature change.

(2) An increase in the power-law exponent leads to decreasing natural frequency for both types of shells. The fundamental natural frequency of Type-1 shell is slightly higher than the corresponding value of Type-2 shell, while the opposite occurs for nonfundamental natural frequencies.

(3) For increasing porosity volume fraction, the fundamental natural frequency increases to a maximum and then decreases. However, the nonfundamental natural frequency of FGMSCS decreases monotonically.

(4) Increasing core-to-thickness ratio results in reduced silicon nitride as one of the component materials, and the reduction in natural frequency of both types of shells.

(5) The fundamental mode shape for both types of FGMSCS is not affected by the variations of power-law exponent, porosity volume fraction, and core-to-thickness ratio.

Acknowledgments

The work described in this paper was supported by the City University of Hong Kong (Project No. ARG 9667253).

Appendix A

$$A_{11} = \sum_{N=1}^{3} \int_{z_{N-1}}^{z_N} \frac{E^{(N)}(z)}{1 - [\mu^{(N)}(z)]^2} \mathrm{d}z, \quad A_{12} = \sum_{N=1}^{3} \int_{z_{N-1}}^{z_N} \frac{E^{(N)}(z)}{1 - [\mu^{(N)}(z)]^2} z \, \mathrm{d}z,$$

$$A_{13} = \sum_{N=1}^{3} \int_{z_{N-1}}^{z_N} \frac{E^{(N)}(z)\mu^{(N)}(z)}{1 - [\mu^{(N)}(z)]^2} \mathrm{d}z, \quad A_{14} = \sum_{N=1}^{3} \int_{z_{N-1}}^{z_N} \frac{E^{(N)}(z)\mu^{(N)}(z)}{1 - [\mu^{(N)}(z)]^2} z \, \mathrm{d}z$$

$$B_{11} = \sum_{N=1}^{3} \int_{z_{N-1}}^{z_N} \frac{E^{(N)}(z)}{2[1 + \mu^{(N)}(z)]} \mathrm{d}z, \quad B_{12} = \sum_{N=1}^{3} \int_{z_{N-1}}^{z_N} \frac{E^{(N)}(z)}{2[1 + \mu^{(N)}(z)]} z \, \mathrm{d}z,$$

$$C_{11} = A_{12}, \quad C_{12} = \sum_{N=1}^{3} \int_{z_{N-1}}^{z_N} \frac{E^{(N)}(z)}{1 - [\mu^{(N)}(z)]^2} z^2 \, \mathrm{d}z, \quad C_{13} = A_{14},$$

$$C_{14} = \sum_{N=1}^{3} \int_{z_{N-1}}^{z_N} \frac{E^{(N)}(z)\mu^{(N)}(z)}{1 - [\mu^{(N)}(z)]^2} z^2 \, \mathrm{d}z, \quad D_{11} = B_{12},$$

$$D_{12} = \sum_{N=1}^{3} \int_{z_{N-1}}^{z_N} \frac{E^{(N)}(z)}{2[1 + \mu^{(N)}(z)]} z^2 \, \mathrm{d}z.$$

Appendix B

$$A_x^T = \sum_{N=1}^{3} \int_{z_{N-1}}^{z_N} -\frac{E^{(N)}(z)}{1-\mu^{(N)}(z)}\,\alpha^{(N)}(z,T)\Delta T \mathrm{d}z,$$

$$B_x^T = \sum_{N=1}^{3} \int_{z_{N-1}}^{z_N} -\frac{E^{(N)}(z)}{1-\mu^{(N)}(z)}\,\alpha^{(N)}(z,T)\Delta T\, z\mathrm{d}z,$$

$$C_x^T = \sum_{N=1}^{3} \int_{z_{N-1}}^{z_N} -\frac{E^{(N)}(z)}{1-\mu^{(N)}(z)}\,\alpha^{(N)}(z,T)\Delta T\, z^2\mathrm{d}z,$$

$$A_\theta^T = A_x^T, \quad B_\theta^T = B_x^T, \quad C_\theta^T = C_x^T.$$

Appendix C

$$K_{11} = -\frac{(A_{11}+A_x^T)m^2\pi^2}{L^2} - \frac{A_\theta^T n^2}{r^2} - \frac{B_{11}n^2}{r^2},$$

$$K_{12} = \frac{mn\pi[A_{14}+B_{12}+(A_{13}+B_{11})r]}{Lr^2},$$

$$K_{13} = \frac{m\pi[(A_{14}+2B_{12}+B_\theta^T)L^2n^2 + A_{13}L^2r + (A_{12}+B_x^T)m^2\pi^2r^2]}{L^3r^2},$$

$$K_{21} = \frac{mn\pi[C_{13}+D_{11}+(A_{13}+B_{11})r]}{Lr^2},$$

$$K_{22} = -\frac{1}{L^2r^4}\begin{pmatrix} C_{12}L^2n^2 + C_\theta^T L^2n^2 + A_{12}L^2n^2r + 2B_\theta^T L^2n^2r + C_{11}L^2n^2r \\ + A_{11}L^2n^2r^2 + A_\theta^T L^2n^2r^2 + C_x^T m^2\pi^2r^2 + D_{12}m^2\pi^2r^2 \\ + B_{12}m^2\pi^2r^3 + 2B_x^T m^2\pi^2r^3 + D_{11}m^2\pi^2r^3 \\ + A_x^T m^2\pi^2r^4 + B_{11}m^2\pi^2r^4 \end{pmatrix},$$

$$K_{23} = -\frac{1}{L^2r^4}\begin{pmatrix} C_{12}L^2n^3 + C_\theta^T L^2n^3 + B_\theta^T L^2nr + C_{11}L^2nr + A_{12}L^2n^3r \\ + B_\theta^T L^2n^3r + A_{11}L^2nr^2 + A_\theta^T L^2nr^2 + C_{14}m^2n\pi^2r^2 \\ + C_x^T m^2n\pi^2r^2 + 2D_{12}m^2n\pi^2r^2 + A_{14}m^2n\pi^2r^3 \\ + 2B_{12}m^2n\pi^2r^3 + B_x^T m^2n\pi^2r^3 \end{pmatrix},$$

$$K_{31} = \frac{1}{L^4r^4}\begin{pmatrix} B_\theta^T L^3mn^2\pi r^2 + C_{13}L^3mn^2\pi r^2 + 2D_{11}L^3mn^2\pi r^2 \\ + A_{13}L^3m\pi r^3 + B_x^T Lm^3\pi^3r^4 + C_{11}Lm^3\pi^3r^4 \end{pmatrix}$$

$$K_{32} = -\frac{1}{L^4r^4}\begin{pmatrix} C_{12}L^4n^3 + C_\theta^T L^4n^3 + A_{12}L^4nr + B_\theta^T L^4nr + B_\theta^T L^4n^3r \\ + C_{11}L^4n^3r + A_{11}L^4nr^2 + A_\theta^T L^4nr^2 + C_{14}L^2m^2n\pi^2r^2 \\ + C_x^T L^2m^2n\pi^2r^2 + 2D_{12}L^2m^2n\pi^2r^2 + B_x^T L^2m^2n\pi^2r^3 \\ + C_{13}L^2m^2n\pi^2r^3 + 2D_{11}L^2m^2n\pi^2r^3 \end{pmatrix},$$

$$K_{33} = -\frac{1}{L^4r^4}\begin{pmatrix} C_{12}L^4n^4 + C_\theta^T L^4n^4 + A_{12}L^4n^2r + 2B_\theta^T L^4n^2r + C_{11}L^4n^2r \\ + A_{11}L^4r^2 + A_\theta^T L^4r^2 + A_\theta^T L^4n^2r^2 + 2C_{14}L^2m^2n^2\pi^2r^2 \\ + C_x^T L^2m^2n^2\pi^2r^2 + C_\theta^T L^2m^2n^2\pi^2r^2 + 4D_{12}L^2m^2n^2\pi^2r^2 \\ + A_{14}L^2m^2\pi^2r^3 + C_{13}L^2m^2\pi^2r^3 + A_x^T L^2m^2\pi^2r^4 \\ + C_{12}m^4\pi^4r^4 + C_x^T m^4\pi^4r^4 \end{pmatrix}.$$

$$M_{11} = I_1, \quad M_{12} = 0, \quad M_{13} = -I_2\frac{m\pi}{L}, \quad M_{21} = 0, \quad M_{22} = I_1, \quad M_{23} = I_2\frac{n}{r},$$

$$M_{31} = -I_2\frac{m\pi}{L}, \quad M_{32} = I_2\frac{n}{r}, \quad M_{33} = I_1 + I_3\left(\frac{m\pi}{L}\right)^2 + I_3\left(\frac{n}{r}\right)^2.$$

References

1. M. Koizumi, FGM activities in Japan, *Compos. Part B Eng.* **28** (1997) 1–4.
2. M. Yamanouchi, M. Koizumi, T. Hirai and I. Shiota, in *Proc. the first Int. Sym. Functionally Gradient Materials*, 8-9 October 1990, Sendai, Japan.
3. T. Hirai and L. Chen, Recent and prospective development of functionally graded materials in Japan, *Mater. Sci. Forum* **308** (1999) 509–514.
4. S. H. Chan, Performance and emissions characteristics of a partially insulated gasoline engine, *Int. J. Therm. Sci.* **40** (2001) 255–261.
5. S. Uemura, The activities of FGM on new application, *Mater. Sci. For.* **423** (2003) 1–10.
6. H.-S. Shen, *Functionally Graded Materials: Nonlinear Analysis of Plates and Shells* (CRC Press, London, 2016).
7. Y. Q. Wang, C. Liang and J. W. Zu, Wave propagation in functionally graded cylindrical nanoshells based on nonlocal Flügge shell theory, *Eur. Phys. J. Plus* **134** (2019) 233.
8. Q. Li, V. P. Iu and K. P. Kou, Three-dimensional vibration analysis of functionally graded material sandwich plates, *J. Sound Vib.* **311** (2008) 498–515.
9. S. Natarajan and G. Manickam, Bending and vibration of functionally graded material sandwich plates using an accurate theory, *Finite Elem. Anal. Des.* **57** (2012) 32–42.
10. L. Iurlaro, M. Gherlone and M. Di Sciuva, Bending and free vibration analysis of functionally graded sandwich plates using the refined zigzag theory, *J. Sandw. Struct. Mater.* **16** (2014) 669–699.
11. S. Pandey and S. Pradyumna, Analysis of functionally graded sandwich plates using a higher-order layerwise theory, *Compos. Part B Eng.* **153** (2018) 325–336.
12. W. Li, H. Ma and W. Gao, A higher-order shear deformable mixed beam element model for accurate analysis of functionally graded sandwich beams, *Compos. Struct.* **221** (2019) 110830.

13. C. H. Thai, A. M. Zenkour, M. Abdel Wahab and H. Nguyen-Xuan, A simple four-unknown shear and normal deformations theory for functionally graded isotropic and sandwich plates based on isogeometric analysis, *Compos. Struct.* **139** (2016) 77–95.
14. F. A. Fazzolari, Stability analysis of FGM sandwich plates by using variable-kinematics Ritz models, *Mech. Adv. Mater. Struct.* **23** (2016) 1104–1113.
15. T. K. Nguyen and B. D. Nguyen, A new higher-order shear deformation theory for static, buckling and free vibration analysis of functionally graded sandwich beams, *J. Sandw. Struct. Mater.* **17** (2015) 613–631.
16. A. M. A. Neves *et al.*, Static analysis of functionally graded sandwich plates according to a hyperbolic theory considering Zig-Zag and warping effects, *Adv. Eng. Softw.* **52** (2012) 30–43.
17. V. Kahya and M. Turan, Vibration and stability analysis of functionally graded sandwich beams by a multi-layer finite element, *Compos. Part B Eng.* **146** (2018) 198–212.
18. M. Liu, Y. Cheng and J. Liu, High-order free vibration analysis of sandwich plates with both functionally graded face sheets and functionally graded flexible core, *Compos. Part B Eng.* **72** (2015) 97–107.
19. C. Li, H.-S. Shen and J. Yang, Nonlinear vibration behavior of FG sandwich beams with auxetic porous copper core in thermal environments, *Int. J. Struct. Stab. Dyn.* **23** (2023) 2350144.
20. S. Xiang, G. Kang, M. Yang and Y. Zhao, Natural frequencies of sandwich plate with functionally graded face and homogeneous core, *Compos. Struct.* **96** (2013) 226–231.
21. A. M. Katili and I. Katili, Improving Felippa Bergan triangular element by using UI approach for analysis of isotropic and FGM sandwich plates, *Compos. Struct.* **312** (2023) 116823.
22. A. M. Katili and I. Katili, Improving the Bergan Wang quadrilateral plate element by using the UI approach for analysis of isotropic and FGM sandwich plates, *Thin-Walled Struct.* **183** (2023) 110328.
23. A. Garg, H. D. Chalak and A. Chakrabarti, Bending analysis of functionally graded sandwich plates using HOZT including transverse displacement effects, *Mech. Based Des. Struct. Mach.* **50** (2022) 3563–3577.
24. T. Fu, Z. Chen, H. Yu, Z. Wang and X. Liu, Free vibration of functionally graded sandwich plates based on nth-order shear deformation theory via differential quadrature method, *J. Sandw. Struct. Mater.* **22** (2020) 1660–1680.
25. B. L. Liu, S. Li and Y. S. Li, Bending of FGM sandwich plates with tunable auxetic core using DQM, *Eur. J. Mech. A/Solids* **97** (2023) 104838.
26. B. Mohammadzadeh and H. C. Noh, Analytical method to investigate nonlinear dynamic responses of sandwich plates with FGM faces resting on elastic foundation considering blast loads, *Compos. Struct.* **174** (2017) 142–157.
27. A. M. Zenkour, A comprehensive analysis of functionally graded sandwich plates: Part 2-Buckling and free vibration, *Int. J. Solids Struct.* **42** (2005) 5243–5258.
28. A. Garg, M.-O. Belarbi, H. D. Chalak and A. Chakrabarti, A review of the analysis of sandwich FGM structures, *Compos. Struct.* **258** (2021) 113427.
29. Y. Q. Wang, Y. H. Wan and Y. F. Zhang, Vibrations of longitudinally traveling functionally graded material plates with porosities, *Eur. J. Mech. A/Solids* **66** (2017) 55–68.

30. J. W. Yan, W. Zhang, S. K. Lai, J. F. Wang and J. J. Mao, Large amplitude vibration and bistable jump of functionally graded graphene-platelet reinforced porous composite plates, *Waves Random Complex Media* (2022) 1–29.
31. J. Yang, D. Chen and S. Kitipornchai, Buckling and free vibration analyses of functionally graded graphene reinforced porous nanocomposite plates based on Chebyshev-Ritz method, *Compos. Struct.* **193** (2018) 281–294.
32. D. Chen, J. Yang and S. Kitipornchai, Nonlinear vibration and postbuckling of functionally graded graphene reinforced porous nanocomposite beams, *Compos. Sci. Technol.* **142** (2017) 235–245.
33. C. Liang and Y. Q. Wang, Wave propagation in three-dimensional graphene aerogel cylindrical shells resting on Winkler–Pasternak elastic foundation, *J. Mech. Mater. Struct.* **15** (2020) 435–455.
34. Y. Q. Wang, C. Liang and J. W. Zu, Examining wave propagation characteristics in metal foam beams: Euler–Bernoulli and Timoshenko models, *J. Brazilian Soc. Mech. Sci. Eng.* **40** (2018) 565.
35. Y. Q. Wang and C. Liang, Wave propagation characteristics in nanoporous metal foam nanobeams, *Results Phys.* **12** (2019) 287–297.
36. Y. Q. Wang, C. Ye and J. W. Zu, Nonlinear vibration of metal foam cylindrical shells reinforced with graphene platelets, *Aerosp. Sci. Technol.* **85** (2019) 359–370.
37. M. W. Teng and Y. Q. Wang, Nonlinear forced vibration of simply supported functionally graded porous nanocomposite thin plates reinforced with graphene platelets, *Thin-Walled Struct.* **164** (2021) 107799.
38. J. Zhu, Z. Lai, Z. Yin, J. Jeon and S. Lee, Fabrication of ZrO_2-NiCr functionally graded material by powder metallurgy, *Mater. Chem. Phys.* **68** (2001) 130–135.
39. N. Wattanasakulpong, B. Gangadhara Prusty, D. W. Kelly and M. Hoffman, Free vibration analysis of layered functionally graded beams with experimental validation, *Mater. Des.* **36** (2012) 182–190.
40. C. Liang and Y. Q. Wang, A quasi-3D trigonometric shear deformation theory for wave propagation analysis of FGM sandwich plates with porosities resting on viscoelastic foundation, *Compos. Struct.* **247** (2020) 112478.
41. Y. Liu, S. Su, H. Huang and Y. Liang, Thermal-mechanical coupling buckling analysis of porous functionally graded sandwich beams based on physical neutral plane, *Compos. Part B Eng.* **168** (2019) 236–242.
42. B. Adhikari, P. Dash and B. N. Singh, Buckling analysis of porous FGM sandwich plates under various types nonuniform edge compression based on higher order shear deformation theory, *Compos. Struct.* **251** (2020) 112597.
43. B. Bouderba and H. M. Berrabah, Bending response of porous advanced composite plates under thermomechanical loads, *Mech. Based Des. Struct. Mach.* **50** (2022) 3262–3282.
44. Y. Ni *et al.*, An accurate model for free vibration of porous magneto-electro-thermo-elastic functionally graded cylindrical shells subjected to multi-field coupled loadings, *J. Intell. Mater. Syst. Struct.* **32** (2021) 2006–2023.
45. K. M. Liew, S. Kitipornchai, X. Z. Zhang and C. W. Lim, Analysis of the thermal stress behaviour of functionally graded hollow circular cylinders, *Int. J. Solids Struct.* **40** (2003) 2355–2380.
46. J. Sun, X. Xu and C. W. Lim, Stress waves and dynamic buckling of functionally graded cylindrical shells under combined axial impact and thermal load, *Acta Mech.* **226** (2015) 1323–1339.

47. X. K. Xia and H. S. Shen, Vibration of post-buckled sandwich plates with FGM face sheets in a thermal environment, *J. Sound Vib.* **314** (2008) 254–274.
48. Z. Zhao *et al.*, Thermo-electro-mechanical size-dependent buckling response for functionally graded graphene platelet reinforced piezoelectric cylindrical nanoshells, *Int. J. Struct. Stab. Dyn.* **20** (2020) 1–28.
49. X. Z. Zhang, S. Kitipornchai, K. M. Liew, C. W. Lim and L. X. Peng, Thermal stresses around a circular hole in a functionally graded plate, *J. Therm. Stress.* **26** (2003) 379–390.
50. A. M. Zenkour and N. A. Alghamdi, Thermomechanical bending response of functionally graded nonsymmetric sandwich plates, *J. Sandw. Struct. Mater.* **12** (2010) 7–46.
51. P. Liu *et al.*, Transient thermal shock fracture analysis of functionally graded piezoelectric materials by the extended finite element method, *Int. J. Solids Struct.* **51** (2014) 2167–2182.
52. C. Li, C. Zhu, C. W. Lim and S. Li, Nonlinear in-plane thermal buckling of rotationally restrained functionally graded carbon nanotube reinforced composite shallow arches under uniform radial loading, *Appl. Math. Mech. (English Ed.)* **43** (2022) 1821–1840.
53. D. Li, Z. Deng, G. Chen, H. Xiao and L. Zhu, Thermomechanical bending analysis of sandwich plates with both functionally graded face sheets and functionally graded core, *Compos. Struct.* **169** (2017) 29–41.
54. J. Sun, X. Xu and C. W. Lim, Accurate symplectic space solutions for thermal buckling of functionally graded cylindrical shells, *Compos. Part B Eng.* **55** (2013) 208–214.
55. V. N. Van Do and C. H. Lee, Thermal buckling analyses of FGM sandwich plates using the improved radial point interpolation mesh-free method, *Compos. Struct.* **177** (2017) 171–186.
56. F. A. Fazzolari and E. Carrera, Thermal stability of FGM sandwich plates under various through-the-thickness temperature distributions, *J. Therm. Stress.* **37** (2014) 1449–1481.
57. D. Shao, W. Liang, W. Wu and Y. Tao, A unified transient vibration analysis of FGM Sandwich plates in thermal environment based on a further refined zigzag plate theory, *Int. J. Struct. Stab. Dyn.* **22** (2022) 1–51.
58. S. K. Jalali, M. H. Naei and A. Poorsolhjouy, Thermal stability analysis of circular functionally graded sandwich plates of variable thickness using pseudo-spectral method, *Mater. Des.* **31** (2010) 4755–4763.
59. D. Li, Z. Deng and H. Xiao, Thermomechanical bending analysis of functionally graded sandwich plates using four-variable refined plate theory, *Compos. Part B Eng.* **106** (2016) 107–119.
60. X. Shi, P. Zuo, R. Zhong, C. Guo and Q. Wang, Thermal vibration analysis of functionally graded conical-cylindrical coupled shell based on spectro-geometric method, *Thin-Walled Struct.* **175** (2022) 109138.
61. X. Shi, P. Zuo, R. Zhong, C. C. Guo and Q. Wang, Vibration analysis of combined functionally graded cylindrical-conical shells coupled with annular plates in thermal environment, *Compos. Struct.* **294** (2022) 115738.
62. R. Vaghefi, Thermo-elastoplastic analysis of functionally graded sandwich plates using a three-dimensional meshless model, *Compos. Struct.* **242** (2020) 112144.
63. T. Van Do, D. Hong Doan, N. Chi Tho and N. Dinh Duc, Thermal buckling analysis of cracked functionally graded plates, *Int. J. Struct. Stab. Dyn.* **22** (2022) 1–24.

64. Y. Wang *et al.*, Nonlinear random responses and fatigue prediction of elastically restrained laminated composite panels in thermo-acoustic environments, *Compos. Struct.* **229** (2019) 111391.
65. Z. X. Wang and H. S. Shen, Nonlinear dynamic response of sandwich plates with FGM face sheets resting on elastic foundations in thermal environments, *Ocean Eng.* **57** (2013) 99–110.
66. J. Sun, X. Xu and C. W. Lim, Torsional buckling of functionally graded cylindrical shells with temperature-dependent properties, *Int. J. Struct. Stab. Dyn.* **14** (2014) 1–23.
67. J. Sun, X. Xu and C. W. Lim, Buckling of functionally graded cylindrical shells under combined thermal and compressive loads, *J. Therm. Stress.* **37** (2014) 340–362.
68. J. Sun, C. W. Lim, X. Xu and H. Mao, Accurate buckling solutions of grid-stiffened functionally graded cylindrical shells under compressive and thermal loads, *Compos. Part B Eng.* **89** (2016) 96–107.
69. M. K. Rath and S. K. Sahu, Vibration of woven fiber laminated composite plates in hygrothermal environment, *JVC/Journal Vib. Control* **18** (2012) 1957–1970.
70. M. Sit and C. Ray, A third order nonlinear model to study the dynamic behaviour of composite laminated structures under thermal effect with experimental verification, *Compos. Struct.* **212** (2019) 106–117.
71. Y. W. Kim, Temperature dependent vibration analysis of functionally graded rectangular plates, *J. Sound Vib.* **284** (2005) 531–549.
72. N. Wattanasakulpong, G. B. Prusty and D. W. Kelly, Free and forced vibration analysis using improved third-order shear deformation theory for functionally graded plates under high temperature loading, *J. Sandw. Struct. Mater.* **15** (2013) 583–606.
73. Y. S. Touloukian, *Thermophysical Properties of High Temperature Solid Materials.* (Macmillian, New York, 1966).
74. C. T. Loy, K. Y. Lam and J. N. Reddy, Vibration of functionally graded cylindrical shells, *Int. J. Mech. Sci.* **41** (1999) 309–324.
75. S. C. Pradhan, C. T. Loy, K. Y. Lam and J. N. Reddy, Vibration characteristics of functionally graded cylindrical shells under various boundary conditions, *Appl. Acoust.* **61** (2000) 111–129.
76. C. Liang, Z. Yaw and C. W. Lim, Thermal strain energy induced wave propagation for imperfect FGM sandwich cylindrical shells, *Compos. Struct.* **303** (2023) 116295.
77. F. Pellicano, Vibrations of circular cylindrical shells: Theory and experiments, *J. Sound Vib.* **303** (2007) 154–170.
78. Z. Qin, F. Chu and J. Zu, Free vibrations of cylindrical shells with arbitrary boundary conditions: A comparison study, *Int. J. Mech. Sci.* **133** (2017) 91–99.
79. Y. Q. Wang, C. Ye and J. W. Zu, Vibration analysis of circular cylindrical shells made of metal foams under various boundary conditions, *Int. J. Mech. Mater. Des.* **15** (2019) 333–344.

https://doi.org/10.1142/9789819814770_0018

Chapter 18

Patch Loading Design Recommendations for Titanium–Clad Bimetallic Steel Plate Girder#

Xuhong Zhou, Zeqiao Luo, Yu Shi and Xuanyi Xue*

School of Civil Engineering, Chongqing University
Chongqing 400045, P. R. China

Key Laboratory of New Technology for Construction of Cities in Mountain Area (Chongqing University)
Ministry of Education, Chongqing 400045, P. R. China
xuexuanyi@126.com

For steel structures service in corrosive environments, corrosion reduces the mechanical properties and effective thickness of the structural steel, which has a significant effect on the durability of steel structures. Titanium–clad bimetallic steel (TCBS), a bimetallic material composed of titanium alloy and carbon steel with excellent corrosion resistance has been suggested to solve the corrosion issue. The TCBS consisting of the TA1 titanium cladding and Q235 carbon steel substrate was introduced in the plate girder as the web to improve durability in this study. The stress–strain properties and failure performance of TCBS and Q235 carbon steel were discussed. Notably, the separation between the cladding and the substrate in the TCBS was observed during the necking stage in the tensile coupon test. Welding method between TCBS web and Q235 carbon steel flange was suggested for processing the plate girder, which was validated to be appropriate according to the experimental results. No crack in welding and TCBS web of plate girder was observed during the patch loading test. A numerical parametric study on the ultimate resistance of TCBS plate girder subjected to patch loading was conducted using the finite element software ABAQUS, in which three different key parameters were included. A comparison between the numerical and predicted results indicated that the design equations suggested in EN 1993–1–5 were too conservative to be directly used for predicting the ultimate resistance of the TCBS plate girder under patch loading, which was caused by the strain–hardening effect and yield plateau of the structural steel. The current design equations were modified to effectively predict the resistance properties of TCBS plate girders.

Keywords: Titanium–clad bimetallic steel; plate girder; design recommendation; numerical analysis; patch loading.

*Corresponding author.
#This chapter appeared previously on the International Journal of Structural Stability and Dynamics. To cite this chapter, please cite the original article as the following: X. Zhou *et al.*, *Int. J. Struct. Stab. Dyn.*, **23**, 2340020 (2023), DOI: 10.1142/S0219455423400205.

1. Introduction

In the service process, structural and reinforcing steels are in contact with erosion factors in the service environment. Steel reacts with corrosion factors and enters an oxidation state, which is called corrosion.[1–3] Corrosion weakens the steel member section and reduces the resistance performance, which also affects the failure mode. To improve the service time of steel structures, it is important to improve their corrosion resistance. Remarkably, the price of structural steel with good corrosion resistance is typically significantly higher than that of carbon steel, such as the stainless steel. If corrosion–resistant steel is widely used in a project, it will significantly increase the direct costs of building structures. The progress in bimetallic material technology provides a feasible engineering solution to the aforementioned corrosion issue.[4,5] Bimetallic materials are composed of two metals with different properties[6–8] that exhibit outstanding corrosion resistance and mechanical properties. Presently, the titanium–clad bimetallic steel (TCBS) consisting of the titanium cladding and carbon steel substrate performs an excellent corrosion resistance, which has garnered significant attention.[9] Based on an investigation conducted by Shi *et al.*,[9] the mechanical properties of TCBS are suitable for steel structures. The thickness of titanium cladding is clearly smaller than that of carbon steel substrate. The above bimetallic construction form improves the corrosion resistance and effectively reduces the amount of alloy elements, thus controlling the price of TCBS. Therefore, it is meaningful to introduce the TCBS into steel structures, which enhances the durability significantly.

The plate girder analyzed in this study is a welded I-section steel girder composed of flange, web, and transverse stiffener. The web slenderness of the steel plate girders is generally large, which can effectively improve the spacing between the top and bottom flanges. Therefore, the ability of the section to resist the bending moment is effectively improved. Consequently, the plate girder exhibits a strong span capacity, which is often used in bridge and long–span building structures. It is worth noting that a web with a large height–thickness ratio may exhibit the local buckling when the load is relatively small. Nevertheless, the local buckling of the web is not the limit state of the plate girder. After the local buckling of the web, the plate girder develops a stable post-buckling capacity, which helps plate girder continue to resist the load. Many scholars have studied the resistance capacity of plate girders considering the post-buckling capacity.[10–18] Thus, the resistant performance of plate girders developing post-buckling capacity has become a research hotspot. In practical engineering, plate girders frequently encounter patch loading, e.g. wheel loads acting on crane beams, purlin loads acting on main beams, and launch loads during the launch of plate and box girders. Roberts and Rockey[19] proposed a collapse mechanism solution for predicting the ultimate resistance of a plate girder subjected to patch loading, which was characterized by four plastic hinges occurring in the loaded flange and three yield lines occurring in the upper half of the web.

Johansson and Lagerqvist[20,21] presented a new approach to determine the ultimate resistance of unstiffened plate girders under patch loading, which was related to the critical load of plate girder. Remarkably, this method was also adopted by the European standard, *Eurocode 3 — Design of steel structures — Part 1–5: Plate structural elements* (EN 1993–1–5),[22] has been used until now. Chacón *et al.*[23] numerically investigated the effect of flange strength on the bearing performance of plate girders, revealing that a certain correction factor should be added to obtain good accuracy when calculating the ultimate resistance of hybrid plate girders using the EN 1993–1–5 procedure.[22] In addition, Santos *et al.*[24,25] conducted experimental and numerical studies on the bearing performance of stainless steel plate girders under unilateral and symmetrical local loads and proposed a revised design equation. Furthermore, Graciano *et al.*[26] numerically studied the patch–loading resistance behavior of aluminum beams, which considered the influence of larger deformation, initial imperfection, and the hardening effect of aluminum alloys. However, there is no research on TCBS plate girder.

In this study, TCBS is introduced into the plate girder to enhance the corrosion resistance. When the web is made of TCBS, the deformation in the web brings higher requirements for the metallurgical bonding performance of TCBS. Therefore, the resistance performance of TCBS plate girders under patch loading was clarified through experimental and numerical studies. In Sec. 2, the design equations in EN 1993–1–5[22] for the patch loading resistance of plate girders under patch loading were introduced. In Sec. 3, the details of the ultimate patch–loading experiment of TCBS plate girders were introduced. The mechanical properties of TA1 titanium, Q235 carbon steel, and TCBS were discussed. A welding method between TCBS and Q235 carbon steel was suggested and validated to be suitable for TCBS plate girders. In Sec. 4, a comprehensive parameter study on TCBS plate girders under patch loading was presented, where three different key parameters were considered. In Sec. 5, the design equations in EN 1993–1–5[22] were modified to effectively predict the ultimate resistance of the TCBS plate girders under patch loading.

Notations

a	Length of web
A	Percentage elongation after fracture
b_f	Width of flange
e	Spacing between two transverse stiffeners at the end of plate girder
E	Young's modulus
f_u	Ultimate stress of steel
f_y	Yield stress of steel
f_{yf}	Yield stress of flange
f_{yw}	Yield stress of web
F_{cr}	Critical buckling resistance

F_{EN}	Ultimate resistance obtained from EN 1993–1–5
F_{exp}	Ultimate resistance obtained from experimental results
F_{FEM}	Ultimate resistance obtained from numerical analysis results
F_{pre}	Predicted ultimate resistance computed using the proposed formula
F_y	Yield resistance
h_w	Depth of web
K_F	Critical buckling factor
l_y	Effective loaded length
S	Standard deviation
S_s	Load length
t_b	Thickness of base metal
t_c	Thickness of cladding metal
t_f	Thickness of flange
t_s	Thickness of transverse stiffener
t_w	Thickness of web
α_{F_0}	Imperfection coefficient
γ_{M_1}	Partial safety factor
ε_{pl}	Plastic strain
$\varepsilon_{\mathrm{nom}}$	Nominal strain
λ_F	Slenderness parameter
λ_{F_0}	Plateau length of resistance reduction function
σ_{true}	True stress
σ_{nom}	Nominal stress
χ_F	Resistance reduction factor

2. Patch Loading Resistance Model for Carbon Steel Plate Girder

Many design codes have suggested equations for the resistance of stainless steel and carbon steel plate girders subjected to patch loading. However, no design equation has been proposed for TCBS plate girders subjected to patch loading. In EN 1993–1–5,[22] the ultimate resistance of unstiffened carbon steel plate girders should be calculated using

$$F_{EN} = \chi_F F_y / \gamma_{M1}, \tag{1}$$

where χ_F is the resistance reduction coefficient, γ_{M_1} is partial safety factor, and F_y is the yield resistance, which is given as:

$$F_y = f_{yw} t_w l_y. \tag{2}$$

The effective loaded length l_y can be obtained from

$$l_y = S_s + 2t_f\left(1 + \sqrt{m_1 + m_2}\right). \tag{3}$$

The dimensionless parameters m_1 and m_2 can be calculated using Eqs. (4) and (5):

$$m_1 = \frac{f_{yf} b_f}{f_{yw} t_w}, \tag{4}$$

$$m_2 = \begin{cases} 0.02\left(\frac{h_w}{t_f}\right)^2 & \lambda_F > 0.5 \\ 0 & \lambda_F \leq 0.5 \end{cases}. \tag{5}$$

Additionally, the resistance reduction coefficient χ_F is given by

$$\chi_F = \frac{0.5}{\lambda_F}, \tag{6}$$

where λ_F is the general slenderness parameter calculated as

$$\lambda_F = \sqrt{\frac{F_y}{F_{cr}}}, \tag{7}$$

where

$$F_{cr} = 0.9 K_F E \frac{t_w^3}{h_w}, \tag{8}$$

$$K_F = 6 + 2\left(\frac{h_w}{a}\right)^2. \tag{9}$$

It should be emphasized that Müller[27] harmonized the shape of the resistance reduction function of all structural compression members in EN 1993–1–5,[22] which can be expressed as

$$\chi_F = \frac{1.0}{\varphi_F + \sqrt{\varphi_F^2 - \lambda_F}} \leq 1.0, \tag{10}$$

where

$$\varphi_F = \frac{1}{2}[1 + \alpha_{F0}(\lambda_F - \lambda_{F0}) + \lambda_F], \tag{11}$$

where α_{F0} is the imperfection coefficient and λ_{F0} is the plateau length. According to the research results of Mirambell *et al.*,[28] when partial safety factor $\gamma_{M1} = 1.0$, the plateau length $\lambda_{F0} = 0.5$ and the imperfection coefficient $\alpha_{F0} = 0.75$. Based on Eq. (11) and the above factors, Eq. (10) became Eq. (6), as suggested in EN 1993–1–5.[22]

3. Experimental Study

3.1. *Design of TCBS plate girder*

The TA1 titanium and Q235 carbon steel were used to produce TCBS via a hot–rolling process. Owing to its outstanding corrosion resistance, TA1 titanium was considered to be the cladding. Q235 carbon steel was considered to be the substrate, which was supposed to provide resistance properties. It is worth noting that the thicknesses of TA1 titanium and Q235 carbon steel remained almost constant in the TCBS plate, whose dimensions were 1500 mm × 7300 mm (width × length). The thicknesses of the cladding materials TA1 titanium (t_c) and Q235 carbon steel (t_b) were 1.4 mm and 6.0 mm, respectively. The chemical compositions of TA1 titanium (cladding material) and Q235 carbon steel (base material) were presented in Ref. 9

The plate girder was composed of a web, transverse stiffener, and flange (Fig. 1).[11,13] The height–thickness ratio of the web is relatively large to increase the moment resistance capacity of the plate girder.[10] There was no longitudinal stiffener in the plate girders used in this study. Notably, the web of the plate girder was divided into several panels by using transverse stiffeners and flanges. To investigate the resistance of plate girder subjected to patch loading, a standard element in the plate girder was selected to perform the experiment, as shown in Fig. 2. The aforementioned design method for the specimen was recommended in Refs. 11, 29 and 30. Remarkably, the current researches on TCBS were focused on material properties and microstructure. Therefore, to explore the application of

Fig. 1. Plate girder in bridge.[11]

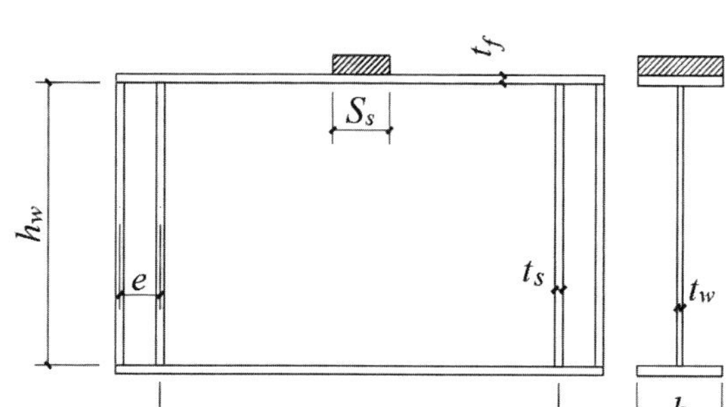

Fig. 2. Plate girder subjected to patch loading (notation).

TCBS in steel members and structures, two plate girder specimens were designed in this paper. The specimen numbers and dimensions were presented in Table 1. Based on the girder components with natural corrosion, the web exhibited more severe corrosion than the flange.[31] Hence, in this experiment, TCBS was used in the web and Q235 carbon steel was used in the flanges and transverse stiffeners.

The tensile coupon test was performed in accordance with the requirements in GB/T 228.1–2010.[32] The dimensions of TCBS and Q235 tensile coupon specimens are displayed in Fig. 3. The mechanical properties of TCBS and Q235 were obtained, as shown in Table 2 and Fig. 4. There is no clear yield plateau in the stress–strain curve of TCBS, which is similar to that of HSS and stainless steel.[10,11,33–36] There is a yield plateau in the stress–strain curve of Q235 mild steel. It should be emphasized that for structural steel without a clear yield plateau, the stress corresponding to a residual strain of 0.2% was taken as the yield stress.[32,37,38] Remarkably, the elastic segment of TCBS is evidently smaller than that of Q235 carbon steel. When the stress is relatively low, plastic strain occurs in TCBS. When the yield stress is attained, the stress gradually increases with the increase of strain, which is the strain-hardening effect. When the ultimate stress is attained, the stress decreases

Table 1. Dimensions of the plate girder specimens.

No.	h_w(mm)	t_w(mm)	a (mm)	b_f(mm)	t_f(mm)	t_s(mm)	e (mm)	S_s(mm)
P1	1000	7.4	1780	150	20	20	100	150
P2								300

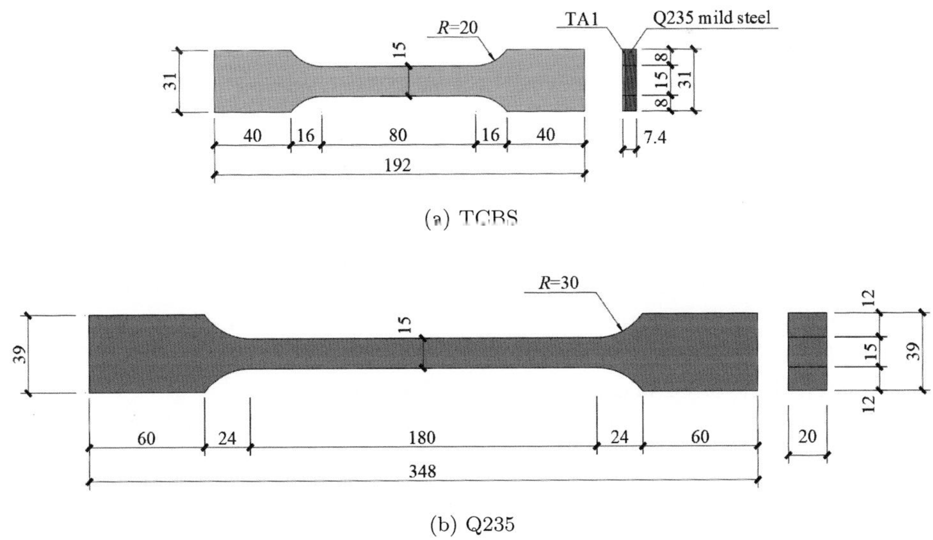

Fig. 3. Dimensions of tensile coupon specimens.

Table 2. Mechanical properties of TCBS.

Type	Location	Thickness (mm)	f_y (MPa)	f_u (MPa)	E (GPa)	A(%)
TCBS	Web	7.4	191.7	379.2	182	37.95
Q235 mild steel	Flange and stiffener	20	314.8	447.7	207	35.20

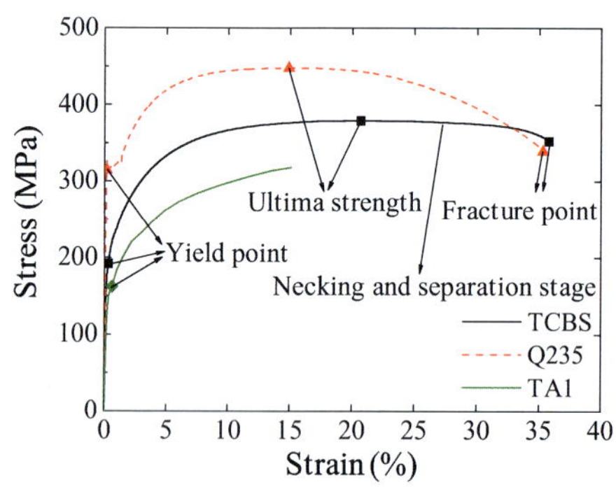

Fig. 4. Nominal stress–strain curves of TCBS, Q235 carbon steel, and TA1.[39]

(a) Fracture of Q235 specimen

(b) Fracture of TCBS specimen

Fig. 5. Fractures of TCBS and Q235 carbon steel.

gradually with an increase in strain. After the peak point, the descending rate of stress of Q235 mild steel is larger than that of TCBS, indicating that TCBS exhibits better ductile properties than Q235 mild steel. Additionally, there is also no yield plateau in the stress–strain curve of TA1 titanium[39] (Fig. 4). The strength of TCBS is lower than that of Q235 carbon steel and higher than that of TA1 titanium. It is proved that the stress–strain properties of TCSB are influenced by the cladding ratio $t_c/(t_c + t_b)$.[40] Hence, it is reasonable that the strength properties of TCBS are between those of Q235 carbon steel and TA1 titanium.

After the static tensile test, the fracture surfaces of TCBS and Q235 carbon steel are shown in Fig. 5. A clear necking phenomenon was observed in TCBS and Q235 carbon steel. In the necking stage, the separation between the cladding and substrate was observed in TCBS.

Transverse stiffeners, which were supposed to simulate the fixed effect from the adjacent elements in the plate girder, were used to create the rigid boundary of the web. Fillet weld was selected for welding the web and transverse stiffeners to flanges (Fig. 6). The E43 welding rod was used in this experiment. The tensile strength of E43 was greater than 430 MPa, meeting the requirements of GB/T 5117-2012.[41] The melting point of TA1 titanium is significantly higher than that of Q235 carbon steel, resulting in more difficult welding between TCBS web and Q235 flanges. Consequently, considering that the thickness of the Q235 carbon steel substrate was evidently greater than that of the TA1 titanium cladding, TA1 titanium was polished before welding at the weld position (Fig. 6).

3.2. *Experimental program and results*

Based on current investigations, for the plate girder under patch loading, the deformation was mainly concentrated in the area near the loading position of the flange and web. Some scholars have suggested reversing the loaded specimen to perform the second loading.[42–45] As the deformation of the first load was concentrated near the loading area, it can be considered that the effects of the existing

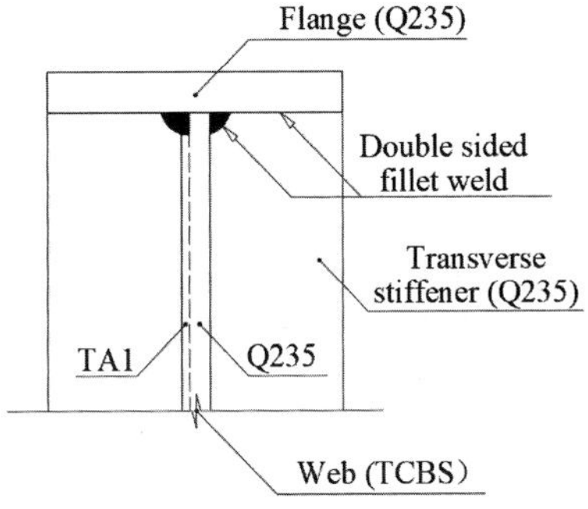

Fig. 6. Welding form.

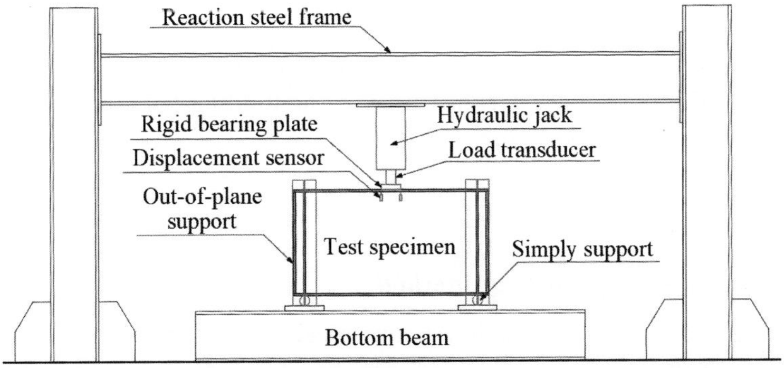

Fig. 7. Test setup.

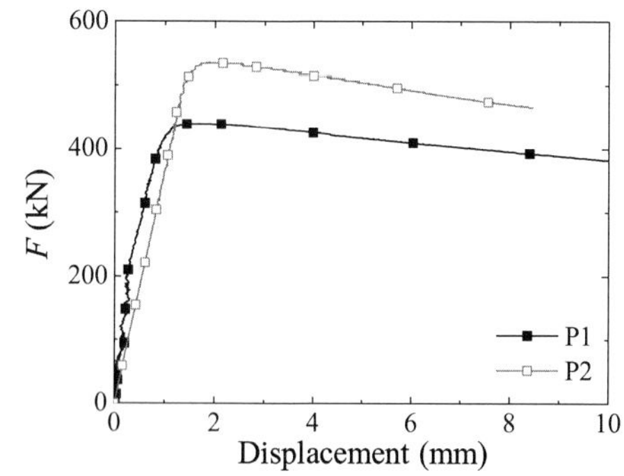

Fig. 8. Load-deflection curves.

deformation on the test results of the second load were relatively small. In this study, the aforementioned experimental method was selected.

A monotonic loading test was conducted to clarify the mechanical response of the TCBS plate girder subjected to transverse patch loading. The test setup is shown in Fig. 7. In this experiment, a simple support boundary condition was selected. Out-of-plane clamping was applied to the support to prevent the overall buckling. The abovementioned boundary conditions are recommended in Refs. 46 and 47. The loaded length S_s is considered as a variable. Patch loading was applied to the mid-span area. S_s was predetermined to be 150 mm and 300 mm, respectively. During the ultimate loading test, the displacement was controlled at 0.1 mm/min before the ultimate resistance and 0.5 mm/min after the peak point.

After the patch loading experiment, the load-deflection curves of the specimens were obtained, as shown in Fig. 8. It is revealed that the ultimate resistance of TCBS plate girder increased significantly with the increment in loading length. The failure modes of the TCBS plate girder were presented in Fig. 9, from which it can be seen that the deformation of the web panel under the patch loading was mainly concentrated in the area near the load action. The flange near the edge of the loading block exhibited significant plastic deformation. It is worth noting that no separation was observed in the bonding interface between the cladding and the substrate of TCBS. No crack was observed in the welding between the TCBS web and the carbon steel flange.

4. Numerical Study

4.1. *Finite element modeling method*

In this paper, to develop the nonlinear numerical model of the TCBS plate girder, the finite element software ABAQUS[48] was selected. S4R shell element was used in the plate girder model (Fig. 10). To simulate the load transmitted to the upper

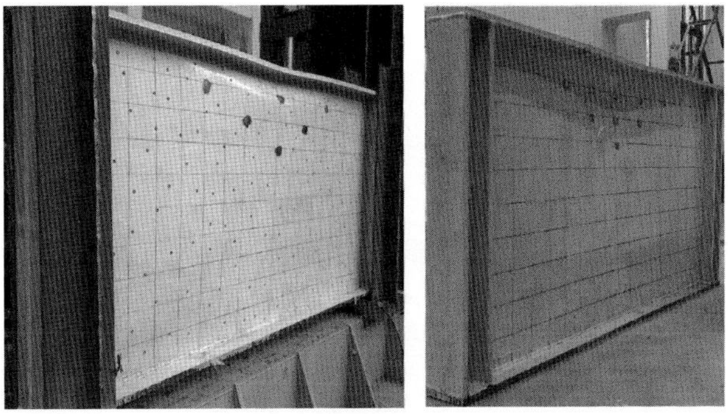

(a) P1 girder

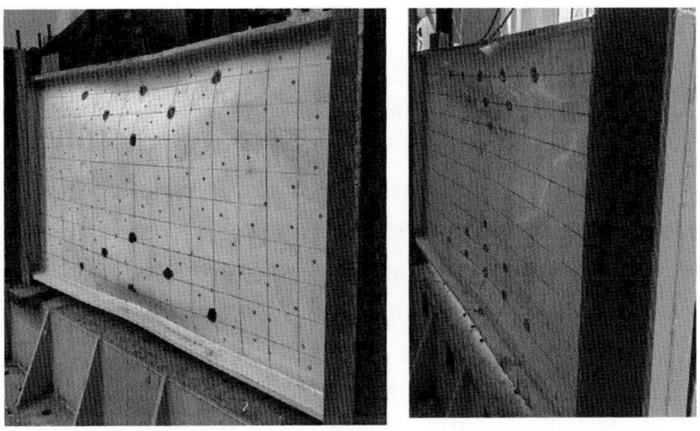

(b) P2 girder

Fig. 9. Failure modes of P1 and P2 girder. (The left image is a front view and the right image is the other view).

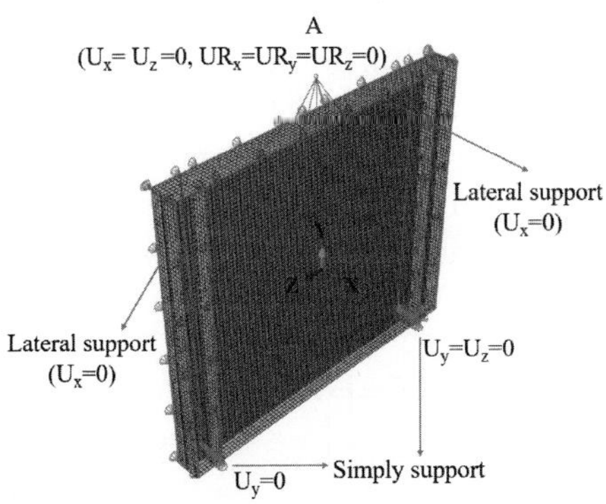

Fig. 10. Boundary conditions applied in the numerical model.

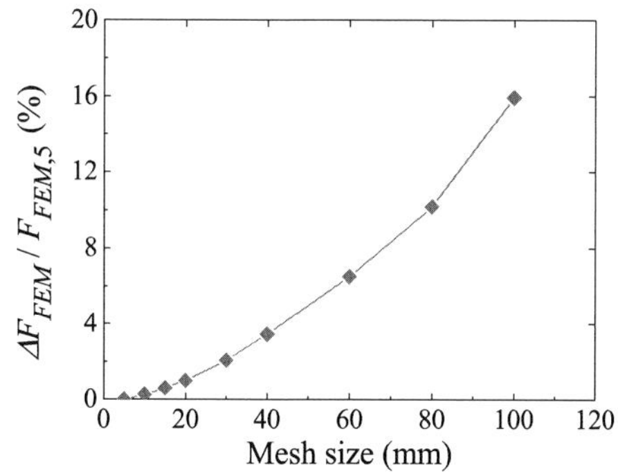

Fig. 11. Convergence study results.

flange of the plate girder through the rigid loading block, all nodes in the area of equal width between the upper flange and the rigid loading block were coupled to the reference Point A. Furthermore, the reference point A was only allowed to be displaced in the vertical direction, i.e. the degrees of freedom U_x, U_z, UR_x, UR_y, and UR_z were restricted. Subsequently, displacement was applied at Point A to introduce patch loading. To simulate the simply supported boundary condition, the translational degrees of freedom in the Y–direction of the left supported area were restrained (i.e. $U_y = 0$), and the translational degrees of freedom in the Y and Z directions of the right supported area were restrained (i.e. $U_y = U_z = 0$). To avoid lateral torsional buckling, the translational degrees of freedom in the X–direction of the edges of the whole upper flange and transverse stiffener were restricted (i.e. $U_x = 0$).

The true stress–plastic strain properties of Q235 carbon steel and TCBS were obtained using Eqs. (12) and (13), respectively. The nominal σ_{nom}–$\varepsilon_{\mathrm{nom}}$ curves are shown in Fig. 4. A convergence study on the P2 plate girder was conducted to determine the appropriate mesh element size, where 5, 10, 15, 20, 30, 40, 60, 80, and 100 mm were included. The ultimate resistance of the numerical model with a 5 mm mesh element size was considered as the benchmark. The dimensionless ratio $\Delta F_{\mathrm{FEM}}/F_{\mathrm{FEM},5}$ was selected to quantify the numerical calculation accuracy, where $\Delta F_{\mathrm{FEM}} = F_{\mathrm{FEM},i} - F_{\mathrm{FEM},5}$ (Fig. 11). After convergence analysis, a 10 mm mesh element size was used in the parameter study to balance the calculation accuracy and calculation cost.

$$\sigma_{true} = \sigma_{nom}(1 + \varepsilon_{nom}), \tag{12}$$

$$\varepsilon_{pl} = \ln(1 + \varepsilon_{nom}) - \frac{\sigma_{true}}{E}. \tag{13}$$

Based on the research on plate girders in Refs. 10, 11, 13, 35 and 49, the initial imperfection caused by processing and transportation affects the resistance performance significantly. In this study, the elastic buckling mode is selected as the initial

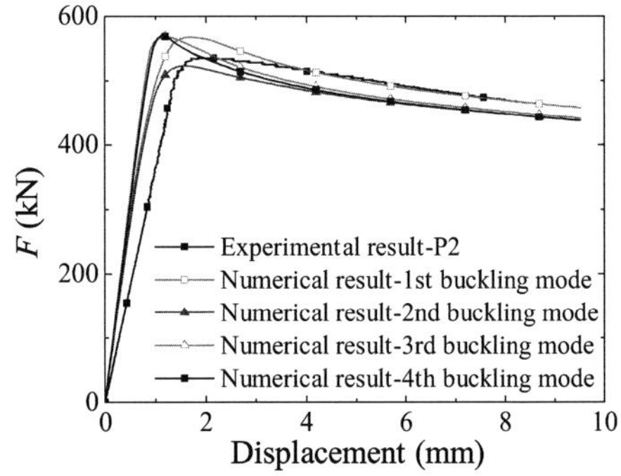

Fig. 12. Comparison of load–displacement curves obtained from numerical results with different imperfection shapes and P2 test results.

Table 3. Comparison of ultimate strength between F_{FEM} and F_{exp}.

Imperfection shape	1	2	3	4
$F_{\text{FEM}}/F_{\text{exp}}$	1.06	0.98	1.06	1.06

Table 4. Comparison of ultimate strength between F_{FEM} and F_{exp}.

Imperfection amplitudes	$h_w/50$	$h_w/100$	$h_w/200$	$h_w/300$
$F_{\text{FEM}}/F_{\text{exp}}$	0.84	0.92	0.98	1.00

imperfection shape. The amplitude is used to determine the magnitude of the initial imperfection. Four different initial imperfection shapes and amplitudes were selected to clarify their effects. For the P2 plate girder numerical model with different initial imperfection shapes, $h_w/200$ was considered as the amplitude. After numerical analysis, the comparisons of the ultimate resistance and load–displacement curves between the experimental results and the numerical results with different imperfection shapes are shown in Fig. 12 and Table 3. The result indicated that the second buckling mode was appropriate for the numerical simulation of the TCBS plate girder under patch loading. When the second buckling mode was selected, different amplitudes were considered, e.g. $h_w/50, h_w/100, h_w/200$, and $h_w/300$. The numerical study results shown in Table 4 indicate that both $h_w/200$ and $h_w/300$ are appropriate for the numerical simulation of TCBS plate girders under patch loading. Considering the requirements of EN 1993–1–3[22] and previous studies,[10,11,13,35,49] $h_w/200$ was considered to be the amplitude of the initial imperfection.

Table 5. Comparison of ultimate resistance between numerical and experimental results.

No.	Girders	h_w (mm)	t_w (mm)	a (mm)	b_f (mm)	t_f (mm)	S_s (mm)	f_{yw} (MPa)	f_{yf} (MPa)	F_{exp} (kN)	F_{FEM} (kN)	$F_{\mathrm{FEM}}/F_{\mathrm{exp}}$
1	P1	1000	7.4	1780	150	20	150	191.7	314.8	439.6	456.5	1.04
2	P2	1000	7.4	1780	150	20	300	191.7	314.8	535.1	523.5	0.98
3	PL-550-1	600	8	600	190	10	80	650	617	940.6	908.1	0.97
4	PL-550-2	500	8	600	180	10	80	650	617	943.3	901.3	0.96
5	PL-550-3	400	8	600	350	10	80	650	617	962.1	905.3	0.94
6	PL-550-4	300	8	600	130	10	80	650	617	949.0	898.2	0.95
7	pli 4301:1	240	4	1000	120	12	40	297	285	176.0	186.2	1.06
8	pli 4301:2	240	4	1000	120	12	80	297	285	196.0	212.7	1.09
9	pli 4301:3	316	4	1400	120	12	40	297	285	168.0	180.4	1.07
10	pli 4301:4	440	4	1623	120	12	40	297	285	169.0	181.2	1.07
11	A1	500	4	500	120	8	50	305	–	165.0	171.3	1.04
12	A2	500	4	500	120	8	150	323.5	321	215.0	225.9	1.05
13	A11	500	4	500	120	8	100	305	–	199.0	199.4	1.00
14	A12	500	4	500	120	8	25	323.5	321	154.6	159.2	1.03
15	A13	500	4	500	120	8	150	324	325	230.0	225.9	0.98
16	CI5/1	500	3.2	500	150	8	0	258	258	90	90.5	1.01
17	CI5/2	500	3.2	500	150	8	0	258	258	92	90.5	0.98
18	CII5/1	500	4.3	500	150	8	0	278	278	147	136.5	0.93
19	CII5/2	500	4.3	500	150	8	0	278	278	137	136.5	1.00

Note: Nos. 3 – 6 were obtained from Ref.50, Nos. 7–10 were obtained Ref.51, Nos. 11–15 were obtained from Ref. 44, and Nos. 16–19 were obtained from Ref. 52.

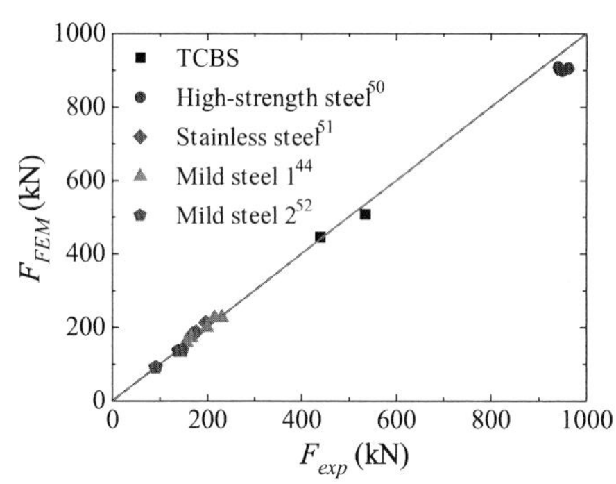

Fig. 13. Experimental vs. numerical analysis results.

4.2. *Numerical model validation*

To validate the accuracy of the aforementioned numerical modeling method, two TCBS plate girders in this study, four high-strength steel plate girders,[50] four stainless steel plate girders,[51] and nine carbon steel plate girders[44,52] were selected. In this study, the ultimate resistance of TCBS plate girders under patch loading was mainly focused. A comparison of the ultimate resistance between the numerical and

Table 6. Determination of key parameters.

Parameters	t_w	h_w/t_w	a/h_w	S_s/h_w
Value	7.4	100	0.75	0
		150	1.0	0.1
		200	1.25	0.2
		250	1.5	0.3
		300	1.75	0.4
			2.0	0.5

experimental results is presented in Table 5 and Fig. 13. The abovementioned results indicate that the numerical modeling method used in this study can be used to predict the ultimate resistance of plate girders with different structural steel, including the TCBS plate girder. It is worth noting that a similar numerical modeling method, which was used in Refs. 10, 11, 13 and 35, was validated comprehensively.

4.3. *Parameter study*

4.3.1. *General statement*

The patch loading resistance of the slender TCBS plate girder was investigated experimentally, as discussed in Sec. 3. However, the experimental results are very limited. To expand the database, the numerical modeling method validated in Sec. 4.2 was selected. Based on the design equations discussed in Sec. 2 and previous investigations on plate girders,[24,49,53] $S_s/h_w, a/h_w$, and h_w/t_w were considered in the parameter study. The following geometry was applied in the current numerical model: $t_w = 7.4$ mm, $h_w/t_w = 100 - 300$, $a/h_w = 0.75 - 2.0, S_s/h_w = 0 - 0.5, b_f =$ 150 mm, $t_f = t_s = 20$ mm. Table 6 summarizes various variable combinations, thereby a total of 180 numerical models were performed. It is noteworthy that t_f/t_w influenced the resistance performance of the plate girder under patch loading to a certain extent. However, for a plate girder with a relatively large h_w/t_w, the effects of t_f/t_w on the ultimate resistance are relatively small.[49] Therefore, t_f/t_w was not considered in the following parameter study. To effectively quantify the effects of key parameters on the resistance performance of the TCBS plate girder under patch loading, resistance reduction factor χ_F in Eq. (1) was selected. However, the resistance reduction factor obtained from the numerical results should be calculated using Eq. (14), where F_y is given by Eqs. (2)–(5).

$$\chi_F = F_{FEM}/F_y. \tag{14}$$

4.3.2. *Effects of S_s/h_w*

Based on the numerical results, the effects of relative load length S_s/h_w on the ultimate resistance $F_{\rm FEM}$ of the numerical models are illustrated in Fig. 14. With

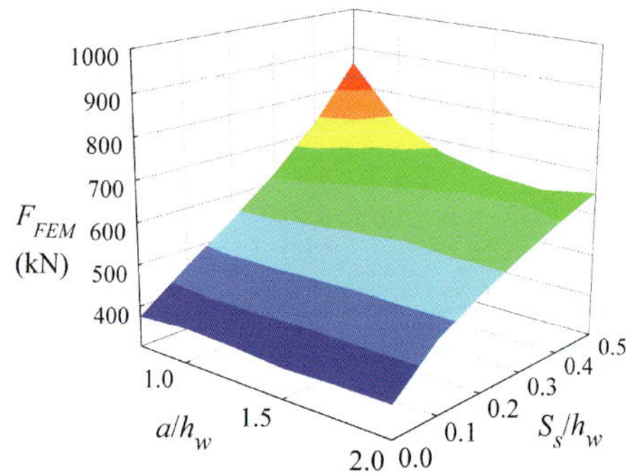

Fig. 14. Variation of ultimate resistance $F_{\rm FEM}$ with different $S_s/h_w (h_w/t_w = 200)$.

the increment in S_s/h_w, $F_{\rm FEM}$ increased gradually. This finding was consistent with stainless steel and mild steel plate girders.[24,45] With a decrease in web aspect ratio a/h_w, the abovementioned effects of S_s/h_w on $F_{\rm FEM}$ became more significant. To clarify the effects S_s/h_w on resistant performance of TCBS plate girder comprehensively, the effects of S_s/h_w on resistance reduction factor χ_F in Eq. (14) for the TCBS plate girder with different a/h_w and web slenderness h_w/t_w values are shown in Fig. 15. For the numerical model with $a/h_w = 0.75$, the effects of S_s/h_w on χ_F were different from those of the numerical model in which a/h_w was not lower than 1.0. With an increase in $S_s/h_w, \chi_F$ of the numerical model with $a/h_w = 0.75$ first decreased and then increased gradually, especially when $h_w/t_w = 100$. For the numerical model with $a/h_w = 0.75$ and $h_w/t_w = 100$, when the S_s/h_w ranged from 0.3 to 0.5, χ_F increased from 0.744 to 1.124. When a/h_w was not lower than 1, with an increase in $S_s/h_w, \chi_F$ of the numerical model decreased gradually. It is easy to see that when a/h_w was the same, variation trends of χ_F curves with different h_w/t_w were almost the same. When a/h_w was not lower than 1, a linear negative correlation between S_s/h_w and χ_F was observed.

4.3.3. *Effects of a/h_w*

After the numerical analysis, the variation trends of the ultimate resistance $F_{\rm FEM}$ of the TCBS plate girder with different web aspect ratios a/h_w and web slenderness h_w/t_w are shown in Fig. 16. The results indicated that with an increase in a/h_w, the $F_{\rm FEM}$ of TCBS plate girders decreased gradually, which showed a good agreement with stainless steel, mild steel, and longitudinally stiffened TCBS plate girders.[49,54–56] This was because the deformation of the web was concentrated near the loading area under the patch load. The transverse stiffeners provided the more significant reinforcement for the web when a/h_w was small. However, an increase in h_w/t_w reduced the adverse effects of a/h_w on $F_{\rm FEM}$. Moreover, the effects of a/h_won the variation trends of resistance reduction factor χ_F are shown in Fig. 17.

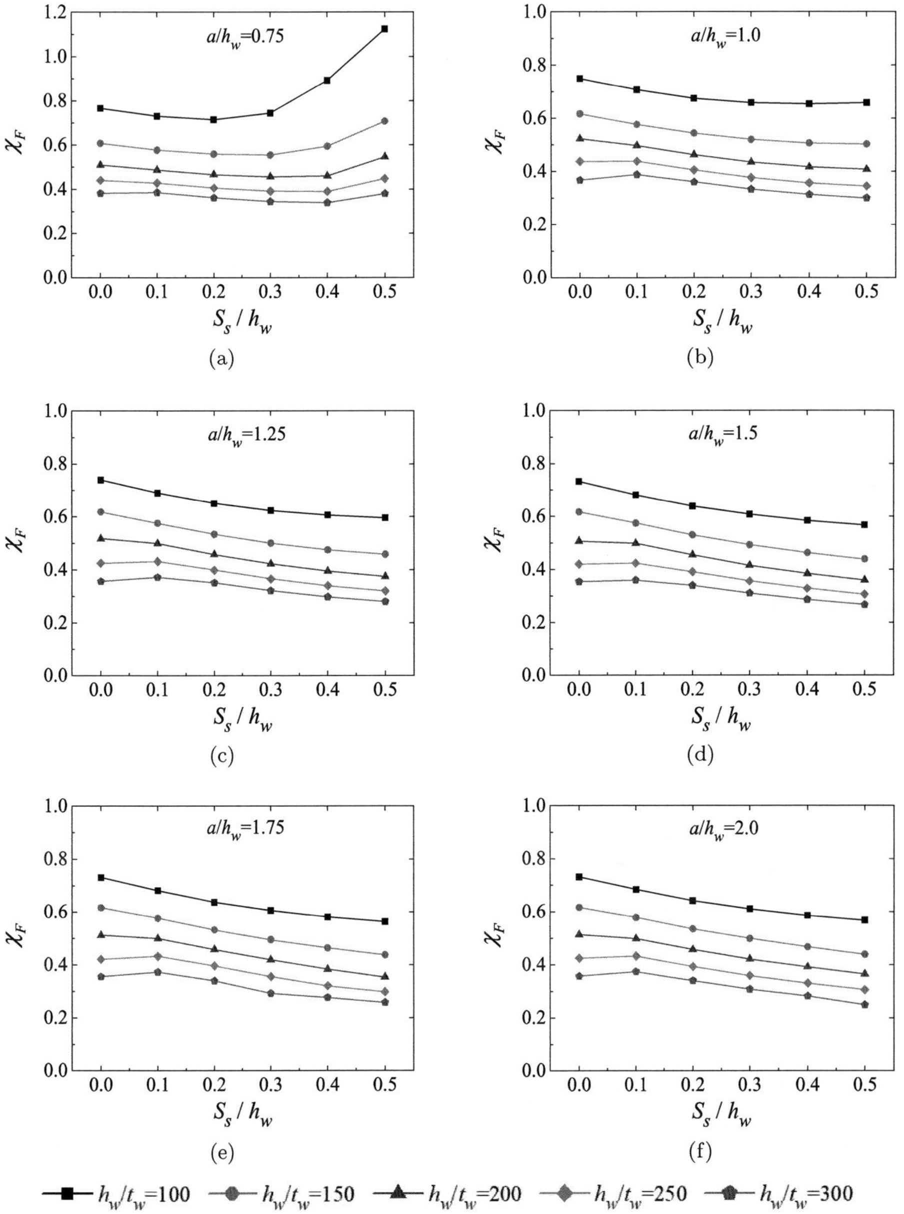

Fig. 15. Effects of S_s/h_w on χ_F of TCBS plate girders.

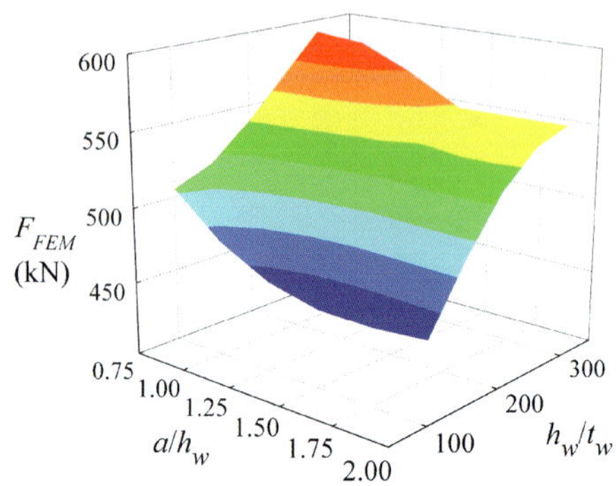

Fig. 16. Variation of ultimate resistance F_{FEM} with different $a/h_w(S_s/h_w = 0.2)$.

When the relative load length S_s/h_w was relatively small (i.e. $S_s/h_w = 0, 0.1$, and 0.2), the influence of a/h_w on χ_F could be ignored. However, when S_s/h_w and a/h_w remained unchanged, χ_F increased significantly as the web slenderness h_w/t_w decreased. Furthermore, it is worth noting that when S_s/h_w was large and a/h_w ranged from 0.75 to 1.00, χ_F of TCBS plate girders diminished. Nevertheless, when a/h_w ranged from 1.00 to 2.00, the effects of a/h_w on χ_F were negligible.

4.3.4. *Effects of* h_w/t_w

The numerical results for ultimate resistance F_{FEM} of the TCBS plate girder with different web slenderness h_w/t_w and relative loading lengths S_s/h_w are shown in Fig. 18. When $S_s/h_w = 0$, the effects of h_w/t_w on F_{FEM} of the TCBS plate girder could be ignored. When S_s/h_w was large, with increments in h_w/t_w, the F_{FEM} of TCBS plate girder increased, and a linear relationship was observed. However, the research results of Graciano *et al.*[49] showed that the ultimate resistance of stainless steel plate girders decreased significantly with the increase of h_w/t_w. A notable difference between these two studies was the way to obtain different h_w/t_w. In this section, the t_w was a constant. Therefore, a larger h_w/t_w was achieved by increasing h_w. However, in Ref. 49, the larger h_w/t_w was obtained through decreasing in t_w. Most notably, when h_w/t_w remained constant, F_{FEM} increased significantly with increasing S_s/h_w. Furthermore, the curves of the resistance reduction factor χ_F vary with h_w/t_w, as shown in Fig. 19. It is worth noting that with an increase in h_w/t_w, χ_F of the TCBS plate girder evidently decreased. However, when S_s/h_w was relatively small (i.e. $S_s/h_w = 0$, 0.1, and 0.2), the influence of the web aspect ratio a/h_w on χ_F could be ignored. When S_s/h_w was relatively large ($S_s/h_w = 0.3, 0.4$, and 0.5), χ_F of the TCBS plate girder with $a/h_w = 0.75$ was clearly higher than that of other TCBS plate girders with different a/h_w. Furthermore, when

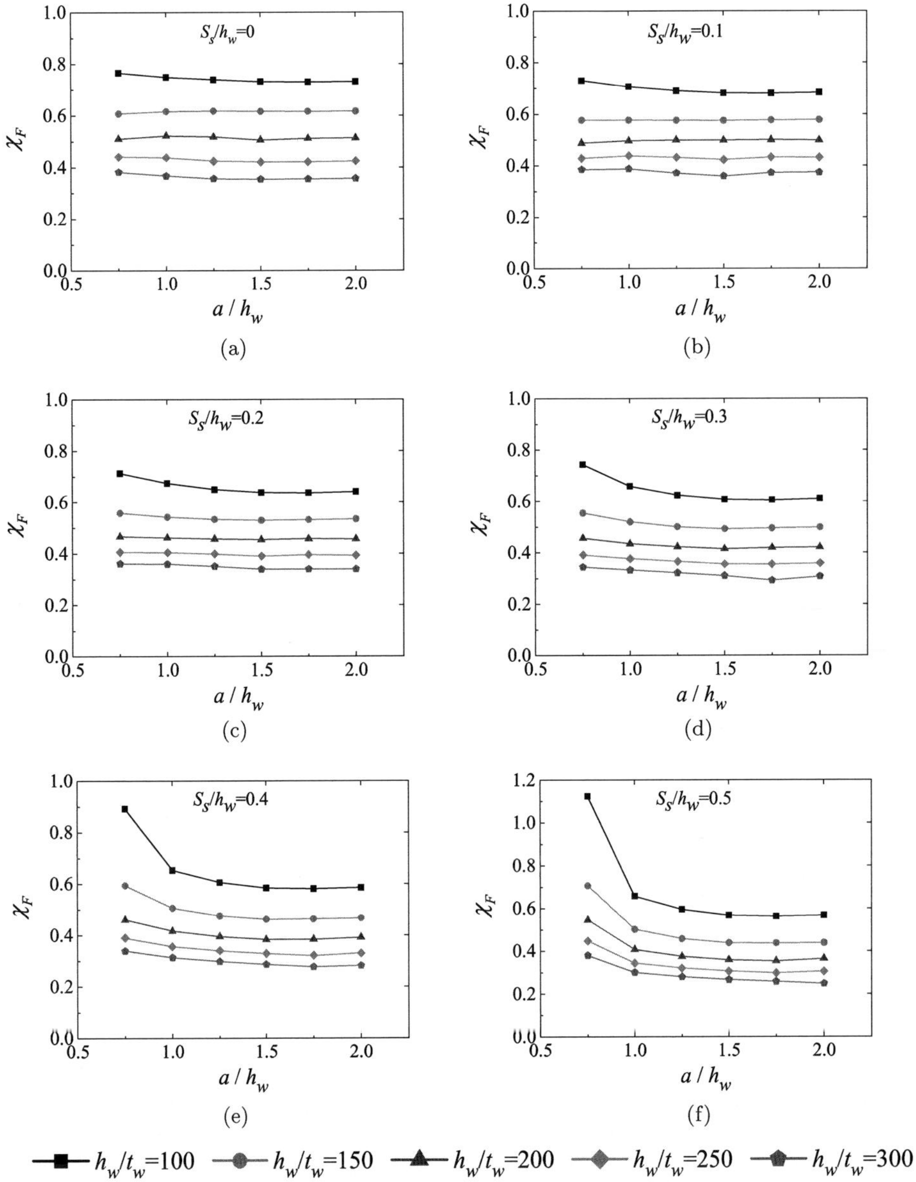

Fig. 17. Effects of a/h_w on χ_F of TCBS plate girders.

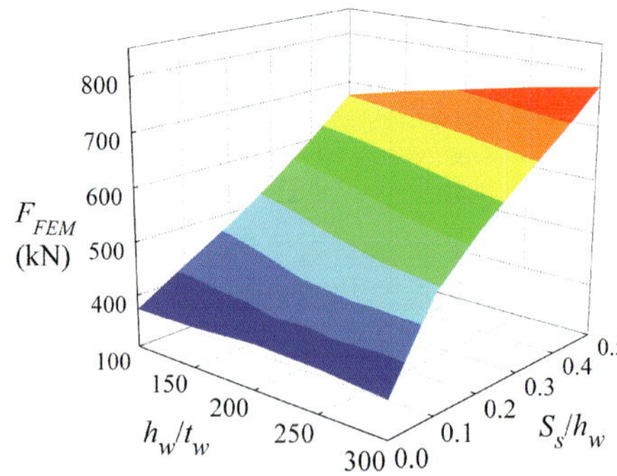

Fig. 18. Variation of ultimate resistance F_{FEM} with different h_w/t_w ($a/h_w = 1.0$).

$a/h_w = 0.75$ and h_w/t_w ranged from 100 to 150, there was an obvious reduction in χ_F.

5. Design Suggestions

Based on the numerical analysis results obtained from the abovementioned parameter study, the ultimate resistance obtained from the numerical analysis results and the calculated results from EN 1993–1–5[22] were compared to validate the applicability (Fig. 20). In this study, the partial safety factor γ_{M1} was predetermined as 1.0. The numerical analysis result F_{FEM} was always greater than the calculated result F_{EN} from European code. The mean value and standard deviation of $F_{\mathrm{FEM}}/F_{\mathrm{EN}}$ were 1.66 and 0.21, respectively. The abovementioned results revealed that the design equations in EN 1993–1–5[22] were too conservative to be used directly for predicting the ultimate resistance of the TCBS plate girder under patch loading. It is worth noting that research on the ultimate patch loading resistance of stainless steel plate girders reached a similar conclusion.[49] Based on the investigation conducted in Ref.35, the yield plateau and strain–hardening effect in the stress–strain curve of structural steel had adverse and beneficial effects on the resistance performance of the plate girder, respectively. TCBS and stainless steel exhibited a noteworthy strain–hardening effect, and no yield plateau was observed. The design equations in EN 1993–1–5[22] were proposed based on the research on carbon steel plate girders. Therefore, the design equations in EN 1993–1–5[22] underestimated the ultimate resistance of the TCBS and stainless-steel plate girders.

It is worth noting that the patch loading resistance model in EN 1993–1–5[22] was simplified from the calculation methodology proposed by Lagerqvist and Johansson.[21,57] In Ref.21, Lagerqvist systematically studied the elastic buckling coefficient of a plate girder under patch loading and considered the influence of relative load length and the fixed effect of flanges. The calculation formula for the elastic critical

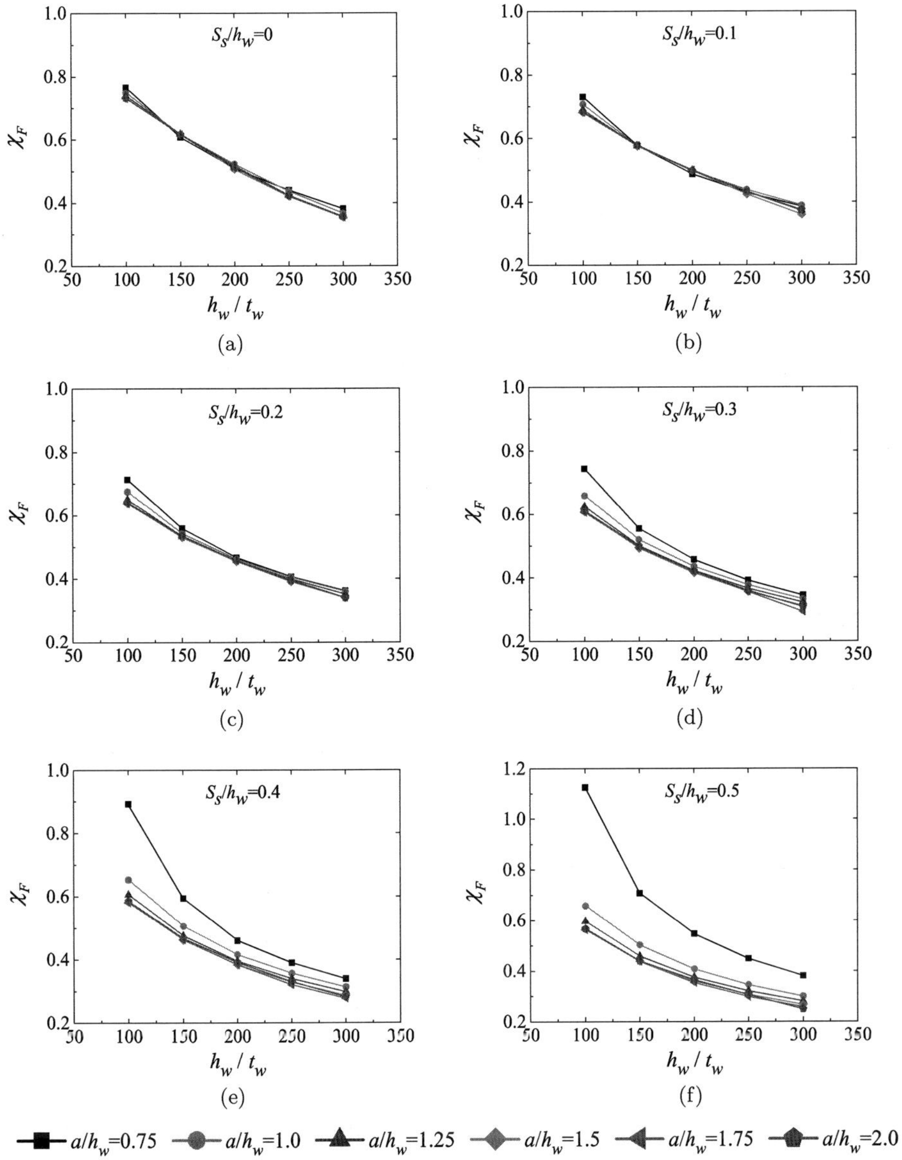

Fig. 19. Effects of h_w/t_w on χ_F of TCBS plate girders.

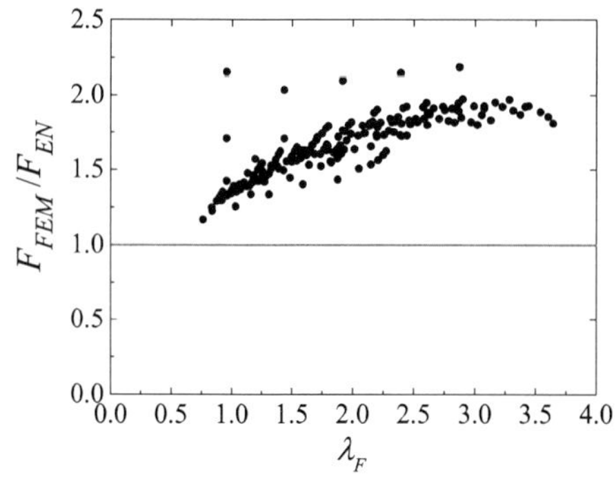

Fig. 20. Comparison of ultimate resistance between F_{FEM} and F_{EN}.

buckling factor is as follows:

$$K_F = \left(1 + \frac{S_s}{2h_w}\right)\left(5.3 + 1.9\left(\frac{h_w}{a}\right)^2 + 0.4\sqrt[4]{\beta}\right) \text{with} \beta = \frac{b_f t_f^3}{h_w t_w^3}. \tag{15}$$

In Refs.21 and 57, Lagerqvist suggested that the resistance reduction factor χ_F should be calculated using Eq. (16). However, when calculating the slenderness parameter λ_F using Eqs. (7)–(9), Eq. (9) was replaced by Eq. (17) to calculate the critical buckling factor K_F. Although Eq. (17) considered the fixed action of the flanges, it did not consider the influence of the relative load length. It is believed that the effects of relative load length on ultimate resistance of carbon steel plate girder were relatively small.

$$\chi_F = 0.06 + \frac{0.47}{\lambda_F}, \tag{16}$$

$$K_F = 5.82 + 2.1\left(\frac{h_w}{a}\right)^2 + 0.46\sqrt[4]{\beta}. \tag{17}$$

In this study, the patch–loading resistance model of EN 1993–1–5[22] was selected. To make EN 1993–1–5[22] applicable to different structural steel plate girders, many scholars suggested modifying the calculation formula of the strength reduction factor.[25,26,49,56] Few scholars suggested modifying the calculation formula of the effective loaded length.[29,58] Notably, for the effective loaded length, the difference between the revised calculation formula and the formula in EN 1993-1-5[22] was relatively small. Therefore, Eq. (6) for calculating the resistance reduction factor χ_F was modified to predict the resistance properties of TCBS plate girders. In Fig. 21, Eq. (16), as proposed by Lagerqvist *et al.*,[21,57] and Eq. (18), as suggested by Ren,[53] were compared to validate their applicability. In Fig. 21, FEM–1 and FEM–2 referred to the numerical analysis results. Remarkably, the ordinates χ_F of FEM–1 and FEM–2 were the same and were fully discussed in Sec. 4. However, the λ_F of FEM–1 and FEM–2 were different. As discussed in Sec. 2, K_F was

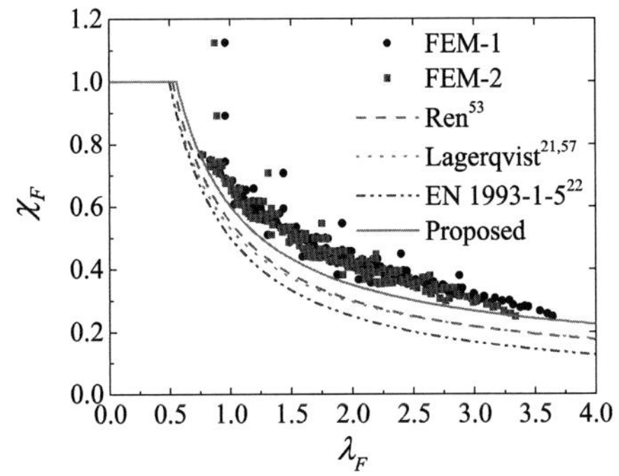

Fig. 21. Resistance reduction factor χ_F obtained using different calculation methods.

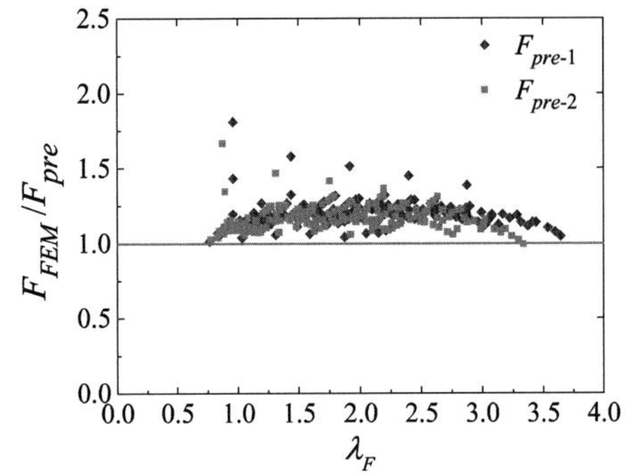

Fig. 22. Comparison of ultimate resistance between F_{FEM} and F_{pre}.

an important parameter for determining λ_F. Thereby, when calculating the λ_F of FEM–1 and FEM–2, Eqs. (9) and (15) were used for calculating the critical buckling factor K_F, respectively. The comparison results indicate that the error between FEM–1 and FEM–2 was relatively small. Additionally, the applicability of Eqs. (16) and (18) is better than that of the design equations (i.e. Eq. (6)) suggested in EN 1993–1–5.[22] However, the accuracy of Eqs.(16) and (18) remained unsatisfactory. Based on the numerical results, Eq. (19) was suggested to calculate the resistance reduction factor for predicting the resistance properties of the TCBS plate girder subjected to patch load. A comparison of χ_F between the numerical results and results from Eq. (19) is shown in Fig. 21.

$$\chi_F = 0.05 + \frac{0.5}{\lambda_F}, \tag{18}$$

$$\chi_F = 0.1 + \frac{0.5}{\lambda_F}. \tag{19}$$

Figure 22 shows a comparison between the ultimate resistance calculated by numerical analysis and the predicted values $F_{\text{pre}-i}$. $F_{\text{pre}-1}$ refers to the predicted ultimate resistance obtained according to Eqs. (1)–(9), however, Eq. (6) was replaced by Eq. (19) to calculate the resistance reduction factor. But, different from $F_{\text{pre}-1}$, Eq. (15) was used to calculate the critical buckling factor K_F in the determination of $F_{\text{pre}-2}$. The mean value and standard deviation of $F_{\text{FEM}}/F_{\text{pre}-1}$ were 1.20 and 0.091, respectively. Moreover, the mean value and standard deviation of $F_{\text{FEM}}/F_{\text{pre}-2}$ were 1.17 and 0.083, respectively. Therefore, using Eq. (15) slightly increases the accuracy of the predictive equation of the ultimate resistance. Considering the convenience and accuracy of calculations, the simplified Eq. (9), as suggested in EN 1993–1–5,[22] can be used to appropriately determine the critical buckling factor of the TCBS plate girder under patch loading. Therefore, it is believed that the proposed Eq. (19) can be used to predict the resistance reduction factor of the TCBS plate girder under patch loading effectively.

6. Conclusions

A TCBS consisting of the TA1 titanium cladding and Q235 carbon steel substrate was introduced in the plate girder as the web to improve durability. In this study, the ultimate resistance of a TCBS plate girder under patch loading was investigated experimentally and numerically. The main conclusions of this study are as follows.

(1) An ultimate resistance experiment was performed on the TCBS plate girder under patch loading. The stress–strain properties and failure performance of TCBS and Q235 carbon steel were discussed. The separation between the cladding and substrate in the TCBS was observed. The welding between TCBS and Q235 carbon steel was suggested and validated to be appropriate based on the test results.
(2) Through numerical analysis, a parameter study was conducted on the ultimate resistance of the TCBS plate girder subjected to patch loading, where S_s/h_w, a/h_w, and h_w/t_w were included. With the increment in S_s/h_w, F_{FEM} increased significantly. With a decrease in a/h_w, the above effects of S_s/h_w on F_{FEM} became more significant. With an increase in a/h_w, F_{FEM} of the TCBS plate girder decreased gradually. An increase in h_w/t_w reduced the adverse effects of a/h_w on F_{FEM}. When $S_s/h_w = 0$ was applied, the effects of h_w/t_w on F_{FEM} of the TCBS plate girder were negligible. When $S_s/h_w = 0.5$, with an increase in h_w/t_w, F_{FEM} of the TCBS plate girder increased, and a linear relationship was observed.
(3) Based on the numerical results, the effects of S_s/h_w, a/h_w, and h_w/t_w on χ_F of TCBS plate girder were investigated. A comparison between the numerical and predicted results indicated that the design equations in EN 1993–1–5[22] were too conservative to be directly used for predicting the ultimate resistance of

the TCBS plate girder under patch loading, which was caused by the strain–hardening effect and yield plateau of structural steel. The design equation for χ_F was modified to predict the resistance properties of the TCBS plate girders.

Acknowledgments

This study was financially supported by the Chongqing Science and Technology Commission grant number cstc2021yszx–jcyj0003 and the Special Funding for Research Projects of Postdoctoral Researchers in Chongqing grant number XmT20200011.

References

1. J. Hua, F. Wang, L. Huang, N. Wang and X. Xue, Experimental study on mechanical properties of corroded stainless-clad bimetallic steel bars, *Constr. Build. Mater.* **287** (2021) 123019.
2. J. Hua, F. Wang, N. Wang, L. Huang, L. Hai, Y. Li, X. Zhu and X. Xue, Experimental and numerical investigations on corroded stainless-clad bimetallic steel bar with artificial damage, *J. Build. Eng.* **44** (2021) 102779.
3. J. Hua, H. Fan, X. Xue, F. Wang, Z. Chen, L. Huang and N. Wang, Tensile and low-cycle fatigue performance of bimetallic steel bars with corrosion, *J. Build. Eng.* **43** (2021) 103188.
4. J. Hua, F. Wang, Z. Yang, X. Xue, L. Huang and Z. Chen, Low-cycle fatigue properties of bimetallic steel bars after exposure to elevated temperature, *J. Constr. Steel Res.* **187** (2021) 106959.
5. J. Hua, H. Fan, W. Yan, N. Wang, X. Xue and L. Huang, Seismic resistance of the corroded bimetallic steel bar under different strain amplitudes, *Constr. Build. Mater.* **319** (2022) 126088.
6. J. Hua, F. Wang, Y. Xiang, Z. Yang, X. Xue, L. Huang and N. Wang, Mechanical properties of stainless-clad bimetallic steel bars exposed to elevated temperatures, *Fire Saf. J.* **127** (2022) 103521.
7. J. Hua, Z. Yang, X. Xue, L. Huang and N. Wang, Bond properties of bimetallic steel bar in seawater sea-sand concrete at different ages, *Constr. Build. Mater.* **323** (2022) 126539.
8. J. Hua, Z. Yang, F. Wang, X. Xue, N. Wang and L. Huang, Relation between the metallographic structure and mechanical properties of a bimetallic steel bar after fire, *J. Mater. Civ. Eng.* **34** (2022) 04022193, doi:10.1061/(ASCE)MT.1943-5533.0004351.
9. Y. Shi, Z. Luo, X. Zhou, X. Xue and J. Li, Post-fire mechanical properties of titanium–clad bimetallic steel in different cooling approaches, *J. Constr. Steel Res.* **191** (2022) 107169.
10. Y. Xiao, X. Y. Xue, F. F. Sun and G. Q. Li, Postbuckling shear capacity of high-strength steel plate girders, *J. Constr. Steel Res.* **150** (2018) 475–490.
11. X. Y. Xue, X. Zhou, Y. Shi and Y. Xiang, Ultimate shear resistance of S600E high-strength stainless steel plate girders, *J. Constr. Steel Res.* **179** (2021) 106535.
12. Y. D. Kim and D. W. White, Transverse stiffener requirements to develop shear-buckling and postbuckling resistance of steel I-Girders, *J. Struct. Eng.* **140** (2014) 04013098.

13. Y. Xiao, X. Y. Xue, F. F. Sun and G. Q. Li, Intermediate transverse stiffener requirements of high-strength steel plate girders considering postbuckling capacity, *Eng. Struct.* **196** (2019) 109289.
14. A. Hayatdavoodi and N. E. Shanmugam, Web buckling and ultimate strength of composite plate girders subjected to shear and bending, *Int. J. Struct. Stab. Dyn.* **15** (2015) 1450047.
15. M. A. Alikhanifard, A. Rahai and P. Tehrani, New Formulations for prediction of buckling loads in steel plate girders through linear and nonlinear stability analysis, *Int. J. Struct. Stab. Dyn.* **23** (2023) 2350005.
16. M. Kaloud and M. Zornerova, The post-buckled behavior in thin-walled construction and its partial "erosion" under repeated loading, *Int. J. Struct. Stab. Dyn.* **11** (2011) 805–827.
17. Z. Y. Wen, X. Wei, L. Xiao and L. J. Dai, Shear performance of corrugated steel webs with local uniform defects, *Int. J. Struct. Stab. Dyn.* **22** (2022) 2250153.
18. A. Rossi, M. Hosseinpour, C. H. Martins, and Y. Sharifi, A new formula for predicting lateral distortional buckling strength of I-Beams subjected to different loading conditions, *Int. J. Struct. Stab. Dyn.* **22** (2022) 2250129.
19. T. M. Roberts and K. C. Rockey, A mechanism solution for predicting the collapse loads of slender plate girders when subjected to in-plane patch loading, *Proc. Inst. Civ. Eng.* **67** (1979) 155–175.
20. B. Johansson and O. Lagerqvist, Resistance of plate edges to concentrated forces, *J. Constr. Steel Res.* **32** (1995) 69–105.
21. O. Lagerqvist, *Patch Loading Resistance of Steel Girders Subjected to Concentrated Forces* (Luleå University of Technology, Division of Steel Structures, Luleå, 1995).
22. EN 1993-1-5, *Eurocode 3 – Design of Steel Structures – Part 1–5: Plated Structural Elements* (European Committee for Standardisation, Brussels, 2006).
23. R. Chacón, E. Mirambell and E. Real, Influence of flange strength on transversally stiffened girders subjected to patch loading, *J. Constr. Steel Res.* **97** (2014) 39–47.
24. G. B. dos Santos, L. Gardner and M. Kucukler, Experimental and numerical study of stainless steel I-sections under concentrated internal one-flange and internal two-flange loading, *Eng. Struct.* **175** (2018) 355–370.
25. G. B. dos Santos and L. Gardner, Design recommendations for stainless steel I-sections under concentrated transverse loading, *Eng. Struct.* **204** (2020) 109810.
26. C. Graciano, N. Loaiza and E. Casanova, Numerical study and design of extruded aluminum beams subjected to concentrated loads, *Thin-Walled Struct.* **155** (2020) 106917.
27. C. Müller, Zum Nachweisebener Tragwerkeaus Stahlgegenseitliches Auswei-chen, Doctoral Thesis, RWTH Aachen, Lehrstuhlfür Stahlbau, Shaker Verlag No.47.
28. E. Mirambell, R. Chacón, U. Kuhlmann and B. Braun, Statistical evaluation of the new resistance model for steel plate girders subjected to patch loading, *Steel Constr.* **5** (2012) 10–15.
29. R. Chacón, M. Bock and E. Real, Longitudinally stiffened hybrid steel plate girders subjected to patch loading, *J. Constr. Steel Res.* **67** (2011) 1310–1324.
30. A. Arabzadeh and M. Varmazyari, Strength of I-girders with Delta stiffeners subjected to eccentric patch loading, *J. Constr. Steel Res.* **65** (2009) 1385–1391.
31. T. Nakai, H. Matsushita, N. Yamamoto and H. Arai, Effect of pitting corrosion on local strength of hold frames of bulk carriers (1st report), *Mar. Struct.* **17** (2004) 403–432.

32. GB/T 228.1-2010, Metallic materials—Tensile testing—Part 1: Method of test at room temperature, Standard Press of China, Beijing, China, 2010.
33. F. F. Sun, X. Y. Xue, Y. Xiao, Y. M. Le and G. Q. Li, Effect of welding and complex loads on the high-strength steel T-stub connection, *J. Constr. Steel Res.* **150** (2018) 76–86.
34. X. Zhou, X. Xue, Y. Shi and J. Xu, Post-fire mechanical properties of Q620 high-strength steel with different cooling methods, *J. Constr. Steel Res.* **180** (2021) 106608.
35. J. Hua, X. Xue, Q. Huang, Y. Shi and W. Deng, Post-fire performance of high-strength steel plate girders developing post-buckling capacity, *J. Build. Eng.* **52** (2022) 104442.
36. J. Hua, F. Wang, X. Xue, Z. Ding, Y. Sun and L. Xiao, Ultra-low cycle fatigue performance of Q690 high-strength steel after exposure to elevated temperatures, *J. Build. Eng.* **57** (2022) 104832.
37. H. Zhou, W. Wang, K. Wang and L. Xu, Mechanical properties deterioration of high strength steels after high temperature exposure, *Constr. Build. Mater.* **199** (2019) 664–675.
38. W. Wang, T. Liu and J. Liu, Experimental study on post-fire mechanical properties of high strength Q460 steel, *J. Constr. Steel Res.* **114** (2015) 100–109.
39. L. Hai, H. Ban and Y. Shi, Preliminary numerical investigation on uniaxial monotonic properties of titanium-clad bimetallic steel, in *Proc. 17th Colloq. Inst. Struct. Stab. Fatigue, China Steel Constr. Soc.* (Shaanxi, Xi'an, 2021) pp. 223–227.
40. X. Liu, R. Bai, B. Uy and H. Ban, Material properties and stress-strain curves for titanium-clad bimetallic steels, *J. Constr. Steel Res.* **162** (2019) 105756.
41. GB/T 5117-2012, Covered electrodes for manual metal arc welding of non-alloy and fine grain steels, Standard Press of China, Beijing, China, 2012.
42. N. Marković and N. Hajdin, A contribution to the analysis of the behaviour of plate girders subjected to patch loading, *J. Constr. Steel Res.* **21** (1992) 163–173.
43. L. M. Gil-Martín, B. Šćepanović, E. Hernández-Montes, M. A. Aschheim and D. Lučić, Eccentrically patch-loaded steel I-girders: The influence of patch load length on the ultimate strength, *J. Constr. Steel Res.* **66** (2010) 716–722.
44. N. Markovic and S. Kovacevic, Influence of patch load length on plate girders. Part I: Experimental research, *J. Constr. Steel Res.* **157** (2019) 207–228.
45. S. Kovacevic and N. Markovic, Experimental study on the influence of patch load length on steel plate girders, *Thin-Walled Struct.* **151** (2020) 106733.
46. R. Chacón, E. Mirambell and E. Real, Transversally stiffened plate girders subjected to patch loading. Part 1. Preliminary study, *J. Constr. Steel Res.* **80** (2013) 483–491.
47. R. Chacón, E. Mirambell and E. Real, Influence of designer-assumed initial conditions on the numerical modelling of steel plate girders subjected to patch loading, *Thin-Walled Struct.* **47** (2009) 391–402.
48. Abaqus Simulia v.6.14, Dassault systems, 2016.
49. C. Graciano, N. Loaiza and E. Casanova, Resistance of slender austenitic stainless steel I- girders subjected to patch loading, *Structures.* **20** (2019) 924–934.
50. K. L. Xu, Local buckling behaviour and design method of high strength steel welded I-section beams, PhD Thesis, Beijing: Tsinghua University (2017) (in Chinese).
51. E. Unosson, Patch loading of stainless steel girders–Experiments and finite element analyses, Lulea University of Technology (2003).
52. S. Aleksić, M. Rogač and D. Lučić, Analysis of locally loaded steel plate girders: Model for patch load resistance, *J. Constr. Steel Res.* **89** (2013) 153–164.

53. T. Ren, Elastic buckling and ultimate strength of web plates in I-girders under patch loading and shear, PhD Thesis, Zhejiang: Zhejiang University (2005) (in Chinese).
54. S. Kovacevic, N. Markovic, D. Sumarac and R. Salatic, Unfavorable geometric imperfections in steel plate girders subjected to localized loads, *Thin-Walled Struct.* **161** (2021) 107412.
55. Z. Luo, Y. Shi, X. Xue and T. Gao, Experimental and numerical investigation on patch loading capacity of longitudinally stiffened hybrid titanium-clad bimetallic steel plate girder, *Eng. Fail. Anal.* **146** (2023) 107117.
56. Z. Luo, Y. Shi, X. Xue, L. Xu and H. Zhang, Design recommendations on longitudinally stiffened titanium-clad bimetallic steel plate girder, *J. Constr. Steel Res.* **201** (2023) 107748.
57. O. Lagerqvist and B. Johansson, Resistance of I-girders to concentrated loads, *J. Constr. Steel Res.* **39** (1996) 87–119.
58. R. Chacón, E. Mirambell and E. Real, Hybrid steel plate girders subjected to patch loading, Part 2: Design proposal, *J. Constr. Steel Res.* **66** (2010) 709–715.

https://doi.org/10.1142/9789819814770_0019

Chapter 19

A Comparative Study of Dynamic Responses of Coastal Long-Span Bridge Under Typhoon with Different Crossing Paths[#]

Jian Guo[*,‡], Chengjie Zhong[†], Kaijiang Ma[†] and Yufeng Shen[†]

[*]*National Key Laboratory of Bridge Intelligent and Green Construction Southwest Jiaotong University Chengdu, Sichuan 610031, P. R. China*

[†]*College of Civil Engineering Zhejiang University of Technology Hangzhou, Zhejiang 310023, P. R. China*

[‡]*guoj@swjtu.edu.cn*

Coastal long-span bridges are intensively constructed along the western Pacific coast. While the coast is frequently disturbed by typhoons. Given that a coastal long-span bridge will experience a number of typhoons during its service life, it is important to have an understanding of the wind fields formed by typhoons with different crossing paths and corresponding structural dynamic responses. Wind and acceleration data of the Xihoumen Bridge under the influence of onshore, offshore, and landfalling typhoons were acquired by the structural health monitoring system (SHMS). Wind field features of typhoons with three types of paths were compared. The landfalling Typhoon In-fa was used to investigate the differences of wind-induced structural dynamic responses under the influence of each typhoon structural region. The covariance-driven stochastic subspace identification (SSI-COV) was used to identify the modal parameters of the bridge. The results revealed that wind field features at bridge site varied with typhoon crossing paths. The SSI-COV accurately identified the structural frequencies of first 15 vertical modes using the acceleration data under the influence of spiral rainband. The first eyewall has the most detrimental wind conditions to the bridge because the stiffening girder vibrated violently in this region.

Keywords: Typhoon influence; cyclone structure; coastal long-span bridge; structural health monitoring system (SHMS); wind field features; structural dynamic responses; stochastic subspace identification (SSI).

[‡]Corresponding author.

[#]This chapter appeared previously on the International Journal of Structural Stability and Dynamics. To cite this chapter, please cite the original article as the following: J. Guo *et al.*, *Int. J. Struct. Stab. Dyn.*, **23**, 2340021 (2023), DOI: 10.1142/S0219455423400217.

1. Introduction

Typhoon is a kind of tropical cyclone generated on the West Pacific Ocean and the South China Sea each summer and autumn.[1,2] As one of the most destructive natural disasters, typhoons are unpredictable and have a wide range of influence.[3,4] The western Pacific coast is one of the coastal long-span bridge intensively constructed areas with a high typhoon risk.[5] Due to the large span of the coastal bridge, low structural stiffness and damping ratio make bridges sensitive to wind.[6] Affected by the complex wind conditions of typhoons, many coastal long-span bridges exhibited excessive vibrations.[7,8] Potential structural damage and failure threaten the health and safety of bridges.[9,10] To improve the wind resistance of coastal bridges, investigations on wind field features formed by typhoons and corresponding structural dynamic responses are deemed essential. Noticing that different typhoons have different effects on bridge structures, it is of great significance to reveal the wind field and dynamic response characteristics under typhoons with different crossing paths.

In response to the extreme consequences caused by tropical cyclones, extensive research has been conducted. The cyclone structure, track, and model are vital for disaster prevention and mitigation.[11–13] For civil engineering, the near-surface wind field characteristics under cyclone influence are the basis of wind resistance practice. Based on the data measured by a wind tower, Xia *et al.*[14] studied the wind characteristics of Typhoon Maria in the near-ground boundary layer. The wind profile parameter was found to be larger than the recommended value in the Chinese code. The turbulence characteristic parameters and their relationships were also revealed. Using the evolutionary power spectral density, Huang *et al.*[15] compared the wind characteristics of three similar typhoons based on stationary and nonstationary models. They discovered the wind characteristics calculated by the nonstationary model were more centralized and stable than those obtained by the stationary model and recommended the nonstationary model for wind characteristic analysis of typhoons. Cao *et al.*[16,17] investigated the wind characteristics of Typhoon Maemi at a meteorological station, including wind speed, wind direction, turbulence intensity, gust factor, etc. Compared with normal wind, gust factor of the typhoon has no significant differences. However, wind field characteristic parameters are sensitive to local surface roughness, which emphasizes the importance of field measurement. When the typhoon passes through the observation position, the unique wind condition of typhoon center can be observed. In the typhoon center, time history of wind speed presents an M-shaped curve, and wind direction changes drastically.[18] This phenomenon is the result of the joint influence of the calm wind condition in typhoon center and the typhoon movement.

Wind field characteristics vary widely with topographic features and altitude. The data measured by meteorological stations and wind towers far away from engineering sites are unsuitable for engineering practice. The development and application of structural health monitoring system (SHMS) make up for the shortage of wind data at the engineering site. With SHMS, environmental information (e.g. humidity, temperature, wind speed, wind direction, etc.) and structural condition (e.g. displacement, strain, acceleration, damage, etc.) can be measured and transmitted in real time.[19–21] SHMS makes the analysis of wind field characteristics and structural responses more convenient, accurate, and efficient. Based on the long-term monitoring of a suspension bridge, Aksel *et al.*[22] investigated the wind characteristics, dynamic responses, and their relationship. They discovered that the dynamic responses of the Hardanger Bridge were governed by wind-induced vibration when the mean wind speed exceeded 8 m/s. The mean wind speed and turbulence intensity are critical to describing the dynamic responses. Using the wind field data in a storm, they predicted the dynamic response of the bridge deck, and the result achieved an agreement with the measured data in case of lateral and torsional responses.[23] Guo *et al.*[24] compared the power spectral density of a stiffening girder under typhoons and normal wind. They found that most of the vibration energy concentrates in the low-frequency band (0–1.5625 Hz) under typhoons, while it has a wider frequency band (0–4.6875 Hz) under normal wind. Although SHMS has been widely used in bridge engineering, this method of obtaining bridge status information still has some shortcomings. For instance, on-site installation of the measurement equipment is costly and inconvenient,[25] which makes SHMS only applicable to long-span bridges. There are some more economical and convenient field measurement methods. A vehicle can be used as the exciter and receiver of bridge vibrations, and the dynamic properties of the bridge can be calculated by the signals obtained by the vehicle.[26,27] Zhang *et al.*[28] used a noncontact sensing technology including microwave radar, optical camera, and video equipment to measure the displacements of a bridge.

The wind field features of typhoons vary from hurricanes, monsoons, and thunderstorms.[29,30] Even the wind fields formed by typhoons differ due to the unique track, intensity, and unpredictable evolution process of each typhoon.[15,19] However, differences in wind field features and bridge dynamic response characteristics caused by typhoons with different crossing paths were seldom reported. Therefore, there is an urgent need to find these differences. It will help bridge managers cope with typhoons with different paths. A typhoon can be divided into four regions radially (typhoon eye, eyewall, spiral rainband, and peripheral gale area). The airflow state in each typhoon structural region is different. Accordingly, the wind field features vary with the typhoon structural regions influencing the observation position. Analysis of wind features and corresponding dynamic responses region by region is needed. However, the specificity of the wind field in each typhoon structural region is not clear. There are few comparative studies on wind-induced responses of bridges in different typhoon structural regions.

In the present study, the measured data were observed by the SHMS installed on the Xihoumen Bridge. Three typhoons (Typhoon Meranti, Talim, and In-fa) were used to find the wind field differences between onshore, offshore, and landfalling typhoons. Since the bridge site has gone through various stages of the landfalling typhoon, Typhoon In-fa was chosen to investigate the bridge response differences between typhoon structural regions. Using the acceleration of the stiffening girder, the wind-induced responses are analyzed to find the most detrimental typhoon structural region to the bridge. In addition, the SSI-COV method is applied for modal parameter identification of the bridge under typhoon influence. The research is supposed to provide references for the bridge wind resistance design and protect bridge structures from wind-induced damage and failure.

The engineering background and details of the three typhoons are described in Sec. 2. Section 3 compares the wind field features including mean wind speed, wind direction, turbulence intensity, gust factor, and power spectral density of typhoons with three typical crossing paths. Section 4 identifies the bridge natural frequency and analyzes the structural dynamic responses under the influence of each typhoon structural region. Section 5 summarizes the results of the analysis.

2. Engineering Background and Typhoon Overview

2.1. *Engineering background*

Xihoumen Bridge is a two-span continuous float steel suspension bridge with a tower height of 236.486 m (Fig. 1). The main span reaches 1 650 m, which makes it the longest steel-box suspension in the world when it was completed. The bridge is located in a high wind speed area of the western Pacific coast. The basic wind speed at the bridge site reaches 41.12 m/s. Since the bridge opened to traffic in 2009, an average of 4.1 typhoons have affected Zhoushan area annually, of which seven typhoons have passed through or landed in Zhoushan area, namely Fung-wong in 2014, Ampil, Jongdari and Runbia in 2018, Mitag in 2019, In-fa in 2021, and Muifa in 2022. Therefore, wind resistance is crucial for the safety and health of the bridge. To optimize the aerodynamic shape, the bridge girder was designed as a twin-box form (Fig. 2).

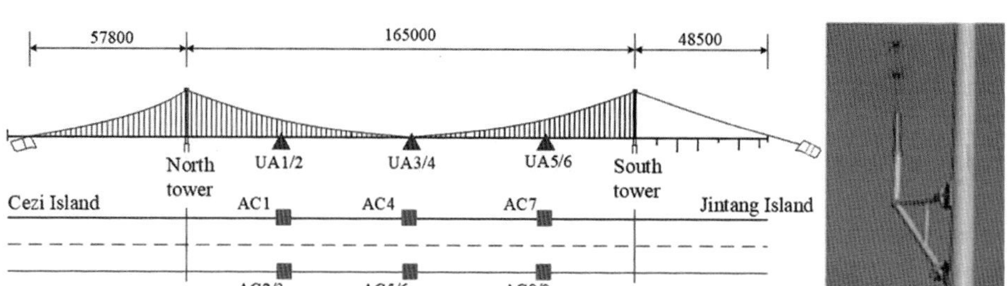

Fig. 1. Layout of sensors installed on bridge and 3D ultrasonic anemometers.

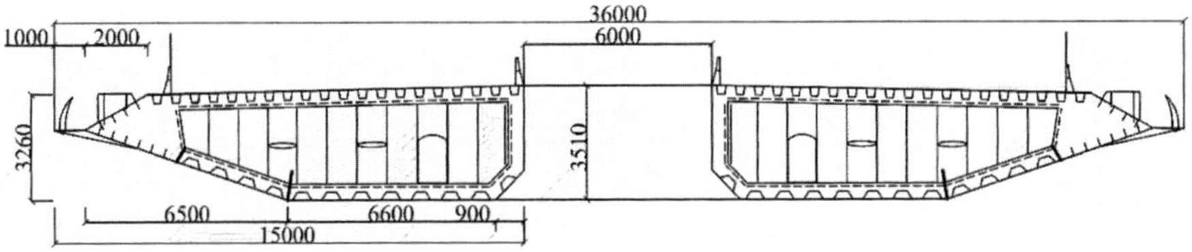

Fig. 2. Cross section of stiffening girder.

Table 1. Details of anemometers and accelerometers.

Sensor	Anemometers			Sensor	Accelerometers		
	Location	Direction	Frequency (Hz)		Location	Direction	Frequency (Hz)
UA1	1/4 span, E	3D	32	AC1	1/4 span, E	Vertical	50
UA2	1/4 span, W			AC2	1/4 span, W	Lateral	
UA3	1/2 span, E			AC3	1/4 span, W	Vertical	
UA4	1/2 span, W			AC4	1/2 span, E	Vertical	
UA5	3/4 span, E			AC5	1/2 span, W	Lateral	
UA6	3/4 span, W			AC6	1/2 span, W	Vertical	
				AC7	3/4 span, E	Vertical	
				AC8	3/4 span, W	Lateral	
				AC9	3/4 span, W	Vertical	

Note: E and W refer to the east and west sides of the stiffening girder. 3D refers to the bridge's longitudinal, lateral, and vertical directions.

With the purpose of measuring and transmitting wind speed and stiffening girder acceleration data in real-time, six 3D ultrasonic anemometers (UA1–UA6) and nine accelerometers (AC1–AC9) were deployed as the components of SHMS. The sensor layout is shown in Fig. 1, and the details of these sensors are listed in Table 1.

2.2. *Overview of three typhoons with typical crossing paths*

Typhoon In-fa is the sixth named tropical cyclone generated in the Pacific in 2021. At 2:00 on July 18, Typhoon In-fa rose from a tropical depression to a tropical storm, and continuously approached the coast of China. It landed in Putuo District, Zhoushan City at about 12:30 on July 25 with a central air pressure of 965 hPa and a maximum wind speed of 38 m/s. Subsequently, it moved northwestward, and its intensity began to decrease. Eventually, Typhoon In-fa dissipated at 20:00 on July 30. Figure 3 illustrates the path of Typhoon In-fa.

Different typhoons form wind fields with different features at the bridge site due to unique paths, intensities, landing points, and evolution processes. And the responses of bridge structures change with the wind field features. To analyze the impact of typhoons with different crossing paths on the bridge site, Typhoon Meranti and Talim are used for comparison. The paths of Typhoon Meranti and Talim

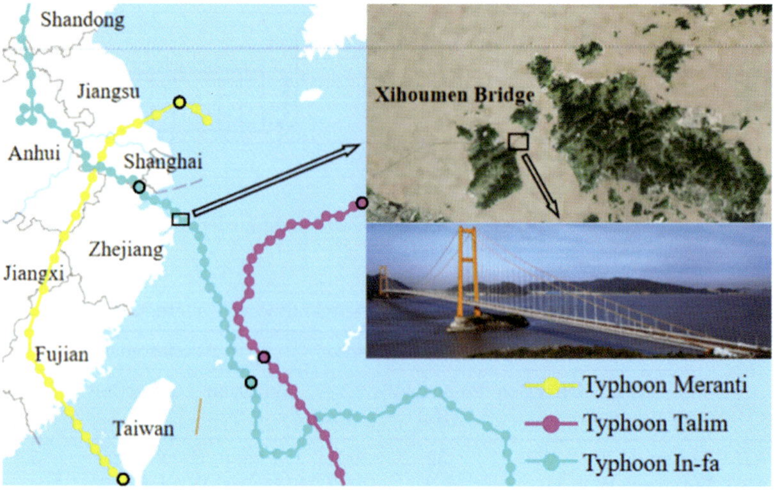

Fig. 3. Paths of Typhoon Meranti, Talim and In-fa (*Note*: Dots with black circles represent the beginning and end of the observation).

Table 2. Details of Typhoon Meranti, Talim and In-fa.

Name	Lifecycle	Duration of observation (h)	Landing point	Minimum distance (km)
Meranti	9.09–9.17, 2016	72 (9.14 00:00–9.17 00:00)	Fujian	277
Talim	9.09–9.18, 2017	72 (9.14 00:00–9.17 00:00)	—	320
In-fa	7.18–7.30, 2021	59 (7.24 00:00–7.26 11:00)	Zhejiang	2

are also illustrated in Fig. 3. Typhoon Meranti landed in Zhangzhou, Fujian, then moved northward and returned to the sea. The intensity of Typhoon Meranti kept weakening after its landfall. The central pressure rose to 1 002 hPa when it was only 277 km from the bridge site. In the process of moving to the mainland, the moving direction of Typhoon Talim changed sharply, making it fail to land in China. Figure 3 illustrates typhoons with three typical crossing paths: onshore typhoon (Meranti), offshore typhoon (Talim), and close landfalling typhoon (In-fa). The lifecycles, durations of observation, landing points, and minimum distances from typhoon centers to the bridge site of these typhoons are listed in Table 2.

3. Wind Field Feature Analysis

3.1. *Wind speed and direction*

The mid-span of the bridge is sensitive to the wind effect. Therefore, the raw data collected by the anemometers at 1/2 span are applied to the analysis. The measured wind velocities are referred to as the along-bridge (u_x), cross-bridge (u_y), and vertical (u_z) components. The mean wind speed U and wind direction θ are determined by the following equations:

$$U = \sqrt{\overline{u_x^2} + \overline{u_y^2}}, \tag{1}$$

$$\theta = \begin{cases} \arccos \dfrac{\overline{u_x}}{U}, & \overline{u_y} \geq 0, \\ 360 - \arccos \dfrac{\overline{u_x}}{U}, & \overline{u_y} < 0, \end{cases} \tag{2}$$

where $\overline{u_x}$ and $\overline{u_y}$ are the mean value of u_x and u_y with a time interval of 10 min.

The 10 min–mean wind speed and wind direction during three typhoons are illustrated in Fig. 4. The time history of mean wind speed and direction are diverse for three typhoons. For onshore typhoon, the wind speed at bridge site kept rising slowly with the approach of Typhoon Meranti and reached the maximum value of 18.83 m/s when the typhoon center was only 277 km from the bridge site. Then the wind speed decreased with the leaving of the typhoon. For offshore Typhoon

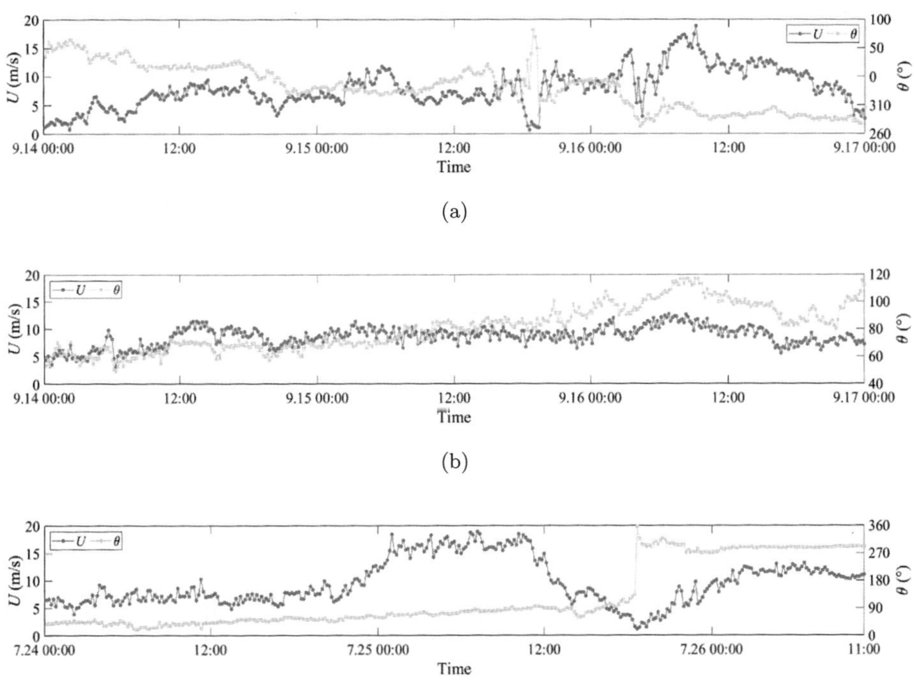

Fig. 4. Time history of mean wind speed and direction: (a) Meranti, (b) Talim, and (c) In-fa.

Talim, the wind speed was more stable than the other typhoons with the range from 3.32 to 11.99 m/s. Typhoon In-fa landed in Zhoushan City where the bridge is located. The wind speed at bridge site was quite variable. When the typhoon was far from the bridge (00:00–17:10 on July 24, more than 358 km), the wind speed was less than 10.5 m/s and stable. Before the typhoon landed, the wind speed rose and reached the maximum of 18.90 m/s at 7:10 on July 25. In the typhoon center (18:40 on July 25, 3 km), the bridge site was almost in a calm wind condition. Then the wind speed rose again with the leaving of the typhoon center.

In general, the wind direction at the bridge site changed slowly and continuously with the movement of typhoons. While for Typhoon In-fa, the wind direction turned sharply from 132.6° to 356.6° when the typhoon center passed the bridge site. The airflow in the typhoon structure moves circumferentially. The change in the relative position between the bridge site and the typhoon center made the bridge site affected by the circumferential airflow on the other side of the typhoon center suddenly, which is the reason for the rapid change in wind direction at the bridge site.

Figure 5 shows the probability density distribution, mean values, and standard deviations of instantaneous wind speed (σ_U) and direction (σ_θ) in three typhoons. The offshore Typhoon Talim has the smallest standard deviations of wind speed and direction, which denotes the bridge was continuously affected by the strong wind from the same direction. Affected by the ground roughness, the onshore Typhoon Meranti has higher standard deviations than the offshore typhoon. Experienced the whole lifecycle of typhoon, wind conditions at the bridge site are the most complex under the influence of the landfalling Typhoon In-fa.

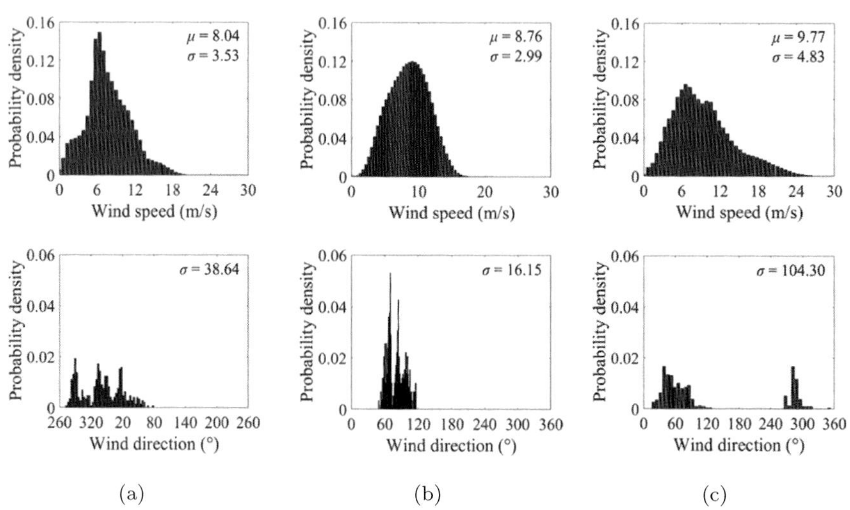

Fig. 5. Probability density distributions of instantaneous wind speed and direction: (a) Meranti, (b) Talim and (c) In-fa.

3.2. *Turbulence intensity*

Turbulence intensity reflects the turbulence component in wind and can be defined as the ratio of the standard deviation of fluctuating wind to mean wind speed:

$$I_i = \frac{\sigma_i}{U}, \quad i = u, v, w, \tag{3}$$

where I_i are the turbulence intensities in along-wind, cross-wind, and vertical directions; σ_i are the standard deviations of the fluctuation component in three directions. The along-wind (u), cross-wind (v), and vertical (w) fluctuating wind velocity can be calculated as follows:

$$u = u_x \cos\theta + u_y \sin\theta - U, \tag{4}$$

$$v = -u_x \sin\theta + u_y \cos\theta, \tag{5}$$

$$w = u_z - \overline{u_z}, \tag{6}$$

where $\overline{u_z}$ is the mean value of u_z with a time interval of 10 min. The turbulence intensities of three typhoons are illustrated in Fig. 6.

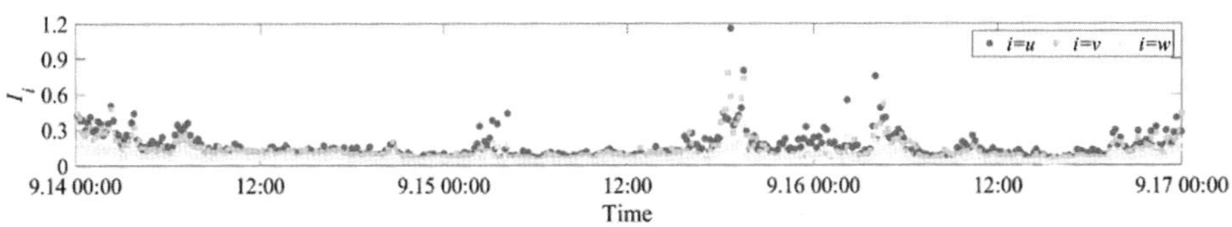

(a)

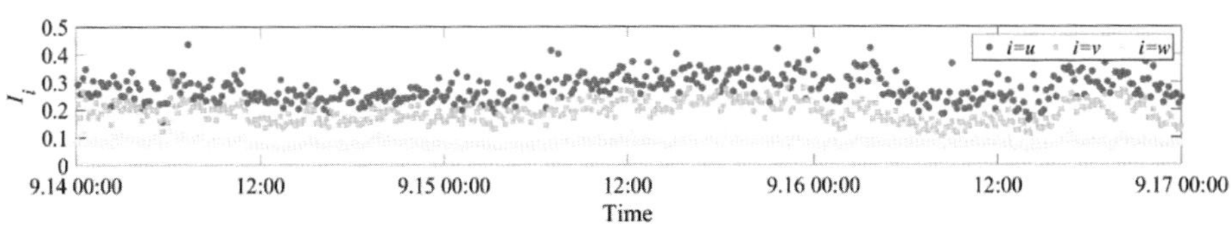

(b)

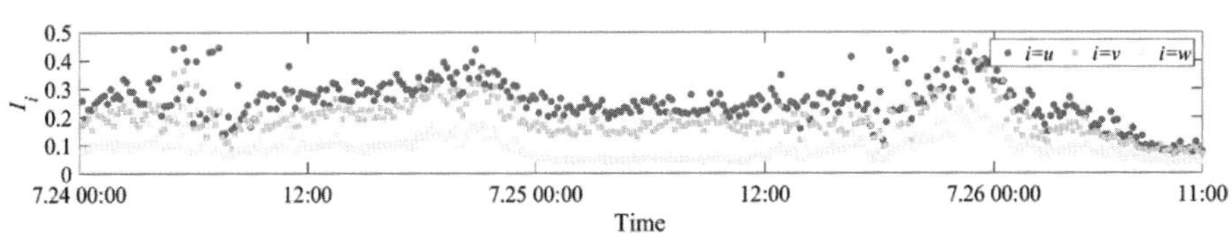

(c)

Fig. 6. Turbulence intensities in three directions of three typhoons: (a) Meranti, (b) Talim and (c) In-fa.

Table 3. Mean turbulence intensities and their ratios of three typhoons.

	Meranti	Talim	In-fa before landing	In-fa after landing
I_u	0.1596	0.2766	0.2641	0.0879
I_v	0.1301	0.1881	0.1981	0.0809
I_w	0.0722	0.0794	0.0934	0.0437
Ratio	1:0.815:0.452	1:0.680:0.287	1:0.750:0.353	1:0.920:0.497

Comparing Figs. 6(a) and 6(b), the turbulence intensities in three directions of Typhoon Meranti are lower than those of Typhoon Talim. After Typhoon In-fa landed on the mainland at about 8:00 on July 26, the turbulence intensity decreased and kept lower than the value when it was on the sea surface, that is, the turbulence component of offshore typhoons is less than onshore typhoons. In some design specifications, the longitudinal turbulence intensity is calculated according to the height of the structure and the condition of exposure.[31,32] The turbulence intensities in the other two directions could be calculated by the proportional relation among them. The three-direction turbulence intensity ratio given by China's wind-resistant design specification is $I_u : I_v : I_w$ = 1:0.88:0.50. However, the recommended value applies to normal wind and varies from the ratio of typhoons. Table 3 illustrated the mean values of I_u, I_v, I_w, and their ratios in three typhoons. To compare the turbulence intensity of onshore typhoons and offshore typhoons, Typhoon In-fa is divided into two periods (before landing and after landing). For onshore typhoons, the ratios of three-direction turbulence intensity are closer to the recommended value of specification. While for offshore typhoons, the values of $I_v : I_u$ and $I_w : I_u$ are much smaller. Thus, smaller ratios are required to calculate the lateral and vertical turbulence intensities of offshore typhoons.

3.3. *Gust factor*

The gust factor is defined as the ratio of the maximum gust wind speed to the mean wind speed and is used to describe the turbulence of wind. The calculation formulas are as follows:

$$G_u = \frac{\max(U_{t_g})}{\overline{U}} = 1 + \frac{\max(u_{t_g})}{\overline{U}}, \tag{7}$$

$$G_i = \frac{\max(i_{t_g})}{\overline{U}}, \quad i = v, w, \tag{8}$$

where G_u, G_v, and G_w are the gust factor in longitudinal, lateral, and vertical directions; U_{tg} is the longitudinal gust wind speed in the gust duration ($t_g = 3\,\mathrm{s}$); u_{tg}, v_{tg}, and w_{tg} are the three-direction mean fluctuating wind speeds in t_g. The three-direction gust factors of three typhoons are illustrated in Fig. 7.

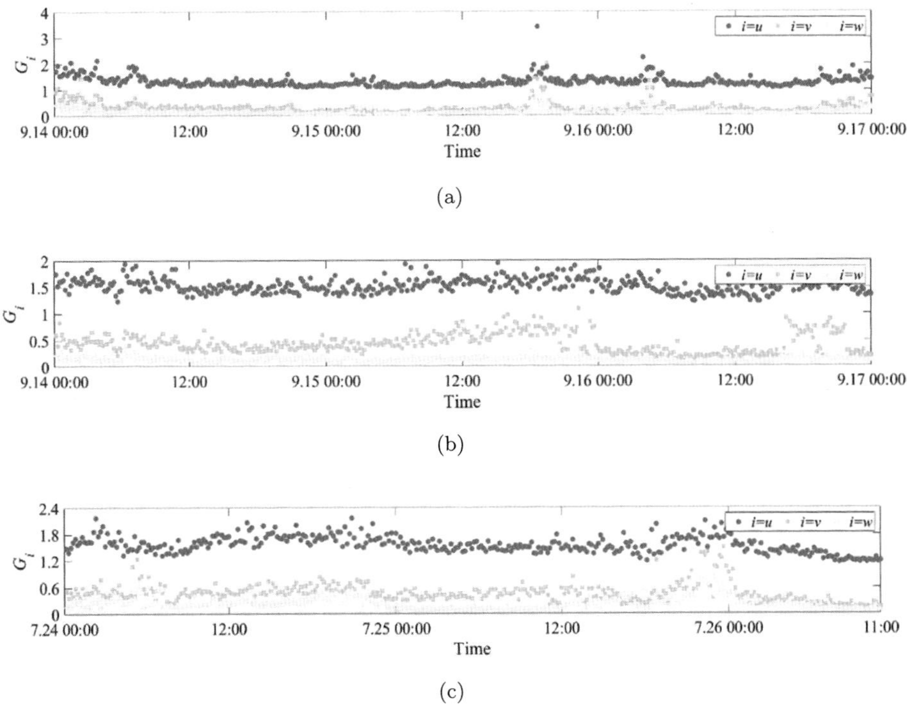

Fig. 7. Gust factors in three directions of three typhoons: (a) Meranti, (b) Talim and (c) In-fa.

As shown in Fig. 7, the tendencies of the gust factor are similar to the turbulence intensity. The relationship between gust factor and turbulence intensity has been studied for many years. Different models and model coefficients have been proposed.[16,33,34] The most commonly used model was proposed by Mackey, Choi and Lam[34]:

$$G_u(t_g, T) = 1 + k_1 I_u^{k_2} \ln\left(\frac{T}{t_g}\right), \tag{9}$$

where T is the time interval of mean wind speed (10 min); k_1 and k_2 are the model coefficients. The measured data of three typhoons are fitted based on the relational model, as shown in Fig. 8. Then the root mean square error (RMSE) is used to evaluate the fitting accuracy and find the optimum coefficients. RMSE is calculated as follows:

$$\mathrm{RMSE} = \sqrt{\frac{1}{n}\sum_{i=1}^{n}(G_{ui} - G'_{ui})^2}, \tag{10}$$

where G_{ui} is the measured gust factor; G'_{ui} is the gust factor calculated by the relational model with the measured turbulence intensity; n is the number of samples.

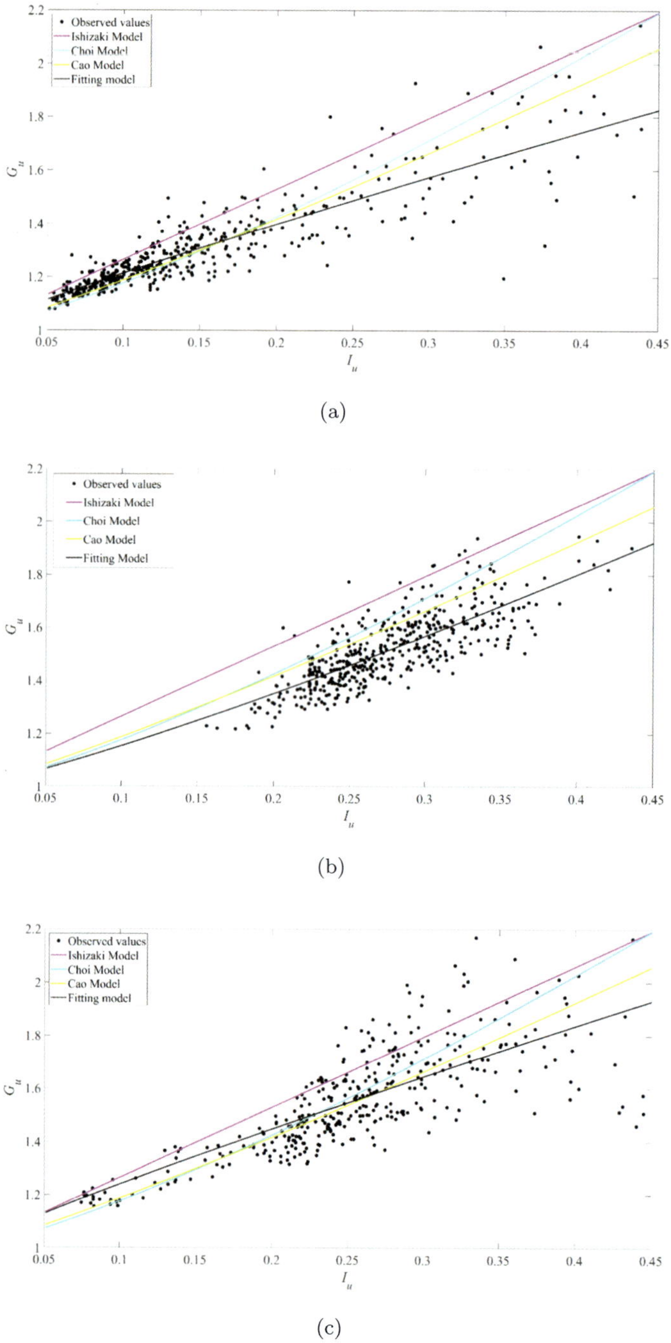

Fig. 8. Relationship between gust factor and turbulence intensity of three typhoons: (a) Meranti, (b) Talim, and (c) In-fa.

Table 4. Model coefficients and RMSE of three typhoons.

Model	Ishizaki	Choi	Cao	Meranti	Talim	In-fa
k_1	0.5	0.62	0.5	0.32	0.45	0.36
k_2	1.0	1.27	1.15	0.90	1.19	0.90
$\mathrm{RMSE}_{\mathrm{Meranti}}$	0.1761	0.1792	0.1347	0.0961	—	—
$\mathrm{RMSE}_{\mathrm{Talim}}$	0.2306	0.1560	0.1219	—	0.0848	—
$\mathrm{RMSE}_{\mathrm{In\text{-}fa}}$	0.1802	0.1452	0.1292	—	—	0.1233

The lower RMSE value means better fitting results. The model coefficients proposed by Ishizaki, Choi, and Cao and RMSE values for three typhoons are listed in Table 4. Based on the above model, three groups of coefficients are obtained by fitting the measured data of three typhoons with the lowest RMSE as the target. The optimum coefficients of each typhoon are also illustrated in Table 4.

As summarized in Table 4, there are significant differences in model coefficients between the offshore typhoon and the onshore typhoon. The coefficients of offshore Typhoon Talim are close to those of Cao's suggestion where k_2 exceeds 1.0, and the fitting curve warps upward. However, for onshore Typhoon Meranti and landfalling Typhoon In-fa, k_2 is less than 1.0, and the curve bends downward. The coefficients k_1 of three typhoons are smaller than 0.5, particularly for onshore typhoon, which is much smaller. The k_1 of landfalling typhoon is between k_1 values of the onshore typhoon and the offshore typhoon.

3.4. *Turbulence power spectral density*

Turbulence power spectral density (PSD) is the distribution density of pulsation kinetic energy in frequency and is used to describe the contribution of the kinetic energy of vortices with different scales to wind fluctuation. In the empirical turbulence PSD formulations, the Kaimal spectrum and the von Karman spectrum are frequently used. The Kaimal spectrum and von Karman spectrum for longitudinal wind are defined as Eqs. (11) and (12), respectively.

$$\frac{nS_u(n,z)}{u_*^2} = \frac{200f_z}{(1+50f_z)^{5/3}}, \tag{11}$$

$$\frac{nS_u(n,z)}{\sigma_u^2} = \frac{4f_z}{(1+70.8f_z^2)^{5/6}}, \tag{12}$$

$$u_*^2 = \frac{\sigma_u^2}{6}, \tag{13}$$

$$f_z = \frac{nz}{U(z)}, \tag{14}$$

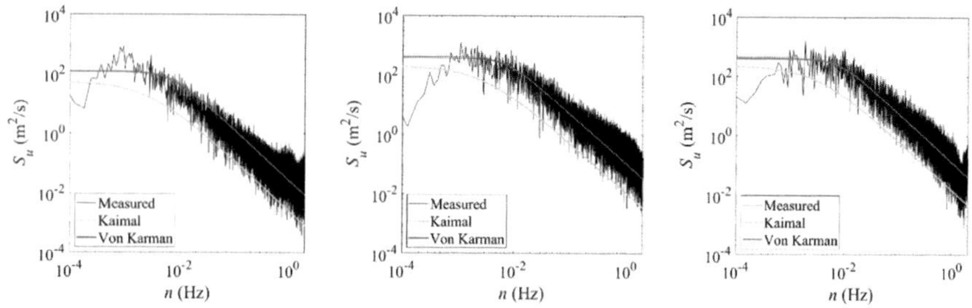

(a) (b) (c)

Fig. 9. Comparison of the measured longitudinal spectra of three typhoons: (a) Meranti, (b) Talim, and (c) In-fa.

where n is the frequency; $S_u(n, z)$ is the power spectral density; $u*$ is the friction wind speed and can be calculated by Eq. (13); f_z is the Monin coordinate and is expressed as Eq. (14); $U(z)$ is the mean wind speed at the height of z.

The spectra of three typhoons are shown in Fig. 9. Compared with the Kaimal spectrum, the measured spectra of three typhoons are in better agreement with the von Karman spectrum. The spectrum energy of turbulent wind with low frequency is lower than the corresponding value of the von Karman spectrum. In general, the spectra of three different types of typhoons are similar.

4. Dynamic Responses of Stiffening Girder During Landfalling Typhoon

As a synoptic-scale system, typhoon has a complex structure. In the vertical direction, a typhoon can be divided into three layers. From low altitude to high altitude, they are convergent layer, middle layer, and divergent layer. In the northern hemisphere, air moves counterclockwise and flows into typhoon structure at convergent layer, spirals up at the middle layer, and flows out clockwise at divergent layer. Most civil engineering structures are in convergent layer. Therefore, the wind features at the convergent layer are the research priority. In the horizontal direction, a typhoon can be divided into four regions radially: typhoon eye, eyewall, spiral rainband, and peripheral gale area as shown in Fig. 10. Each region has unique wind features that excite different structure responses.

The processes of Typhoon In-fa moving on the sea, landing, and moving on the mainland were recorded by the SHMS. The bridge experienced the influence of all the typhoon structural regions. Accordingly, this typhoon is used to find the differences in wind-induced dynamic responses under the influence of different typhoon structural regions.

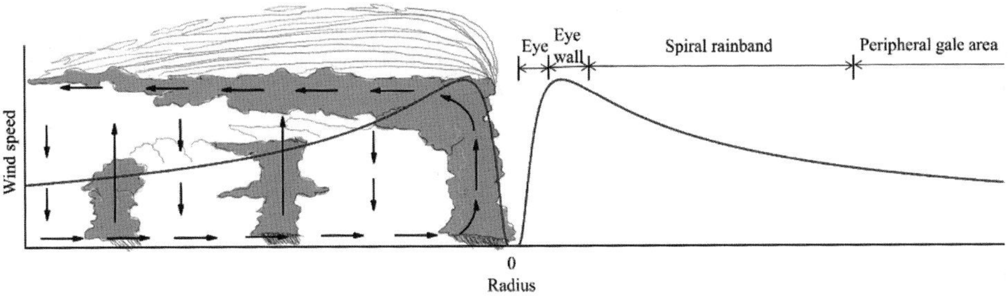

Fig. 10. Air flow condition and structure regions of typhoon.

4.1. *Acceleration responses under Typhoon In-fa*

With nine accelerometers deployed on the stiffening girder, the vertical, lateral, and torsional accelerations at 1/4 span, 1/2 span, and 3/4 span can be obtained. Taking the acceleration at 1/2 span as an example, the vertical acceleration is the average of the signals from AC4 and AC6. The lateral acceleration is equal to the signals from AC5. The torsional acceleration can be calculated by dividing the difference of vertical signals by the distance between AC4 and AC6 (33 m). As presented in Fig. 11(a), typhoon structural regions are divided according to the wind speed characteristics in each region.[35] The vertical, lateral, and torsional accelerations are presented in Figs. 11(b)–11(d), respectively.

The variation tendencies of the vertical, lateral, and torsional accelerations are similar and consistent with the time history of wind speed. The wind speed peaked at the first eyewall. The accelerations in the first eyewall were much larger than the accelerations under the influence of the other typhoon regions. Weakened in landing, the wind speed of the second eyewall is lower than the first eyewall, and the accelerations are also smaller in this region. Due to the calm wind condition in the typhoon eye, the acceleration approaches 0 in the typhoon center.

4.2. *Relationship between wind features and dynamic responses*

The root mean square (RMS) values of vibrational accelerations at 1/2 span were calculated with a time interval of 10 min. The relationships between RMS of acceleration and mean wind speed, direction, and turbulence intensity in each typhoon structural region are presented in Figs. 12–14, respectively.

RMS values under the influence of most typhoon structural regions are less than $0.02\,\text{m/s}^2$. However, nearly all the RMS values in the first eyewall exceed $0.02\,\text{m/s}^2$, indicating that the stiffening girder vibrated intensively in this typhoon structural region. The RMS of acceleration increases with the mean wind speed. Their relationship can be fitted by a quadratic function, as shown in Eq. (15).

$$\text{RMS}_i = a_i U^2, \quad i = z, y, \phi, \tag{15}$$

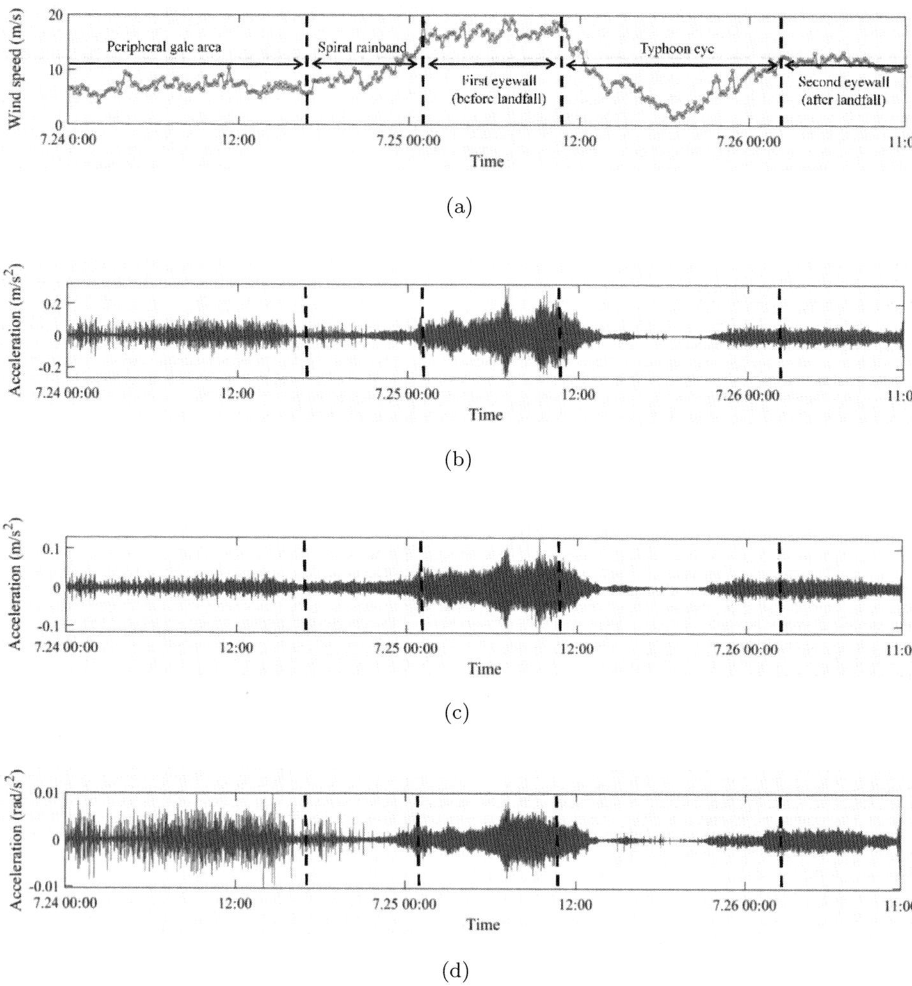

Fig. 11. Accelerations in each typhoon structural region: (a) Typhoon structural regions,(FFT) (b) vertical, (c) lateral and (d) torsional.

where RMS_z, RMS_y, RMS_ϕ are the RMS values of vertical, lateral, and torsional accelerations; $a_z = 1.320 \times 10^{-4}$, $a_y = 5.597 \times 10^{-5}$, and $a_\phi = 3.832 \times 10^{-6}$. With this model and wind data of typhoons, the bridge dynamic responses can be preliminarily predicted.

As shown in Fig. 13, the RMS values under the influence of each typhoon structural region are concentrated in a narrow wind direction range, except the typhoon eye. The wind direction had a mutation in typhoon eye. It is consistent with the time history of wind condition shown in Fig. 4(c). According to Fig. 14, the turbulence intensity in the first eyewall is distributed in the range from 0.196 to 0.282,

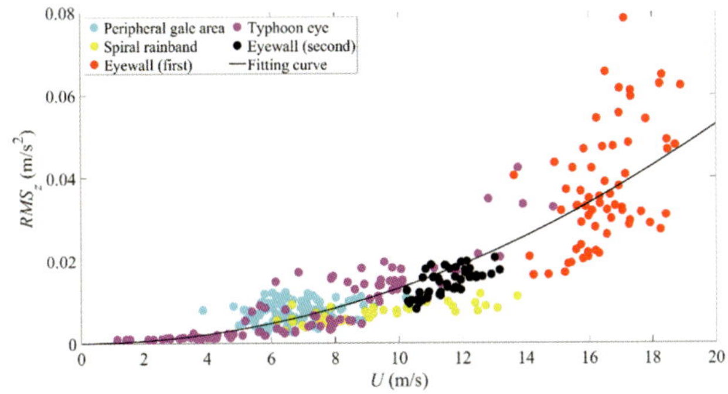

(a)

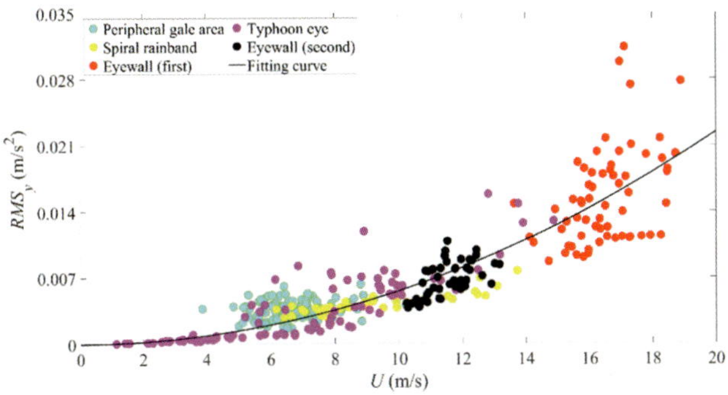

(b)

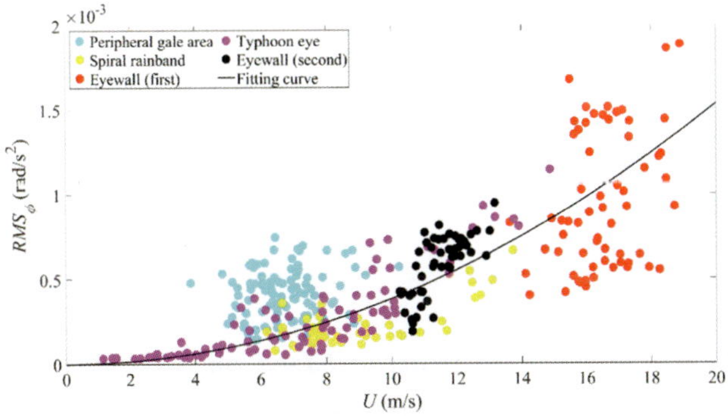

(c)

Fig. 12. Relationship between RMS of acceleration and mean wind speed: (a) Vertical, (b) lateral, (c) torsional.

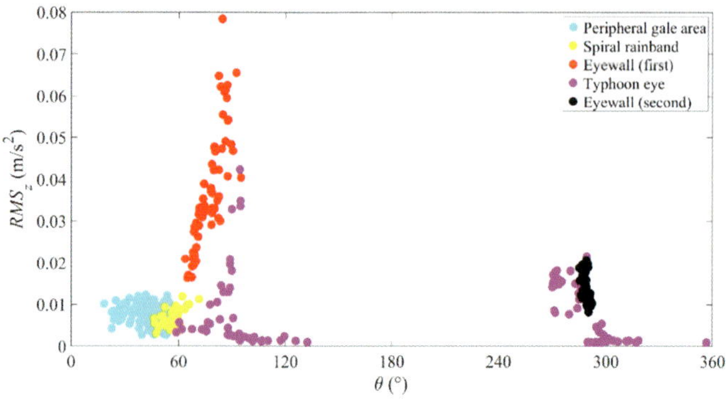

Fig. 13. Relationship between RMS of acceleration and wind direction.

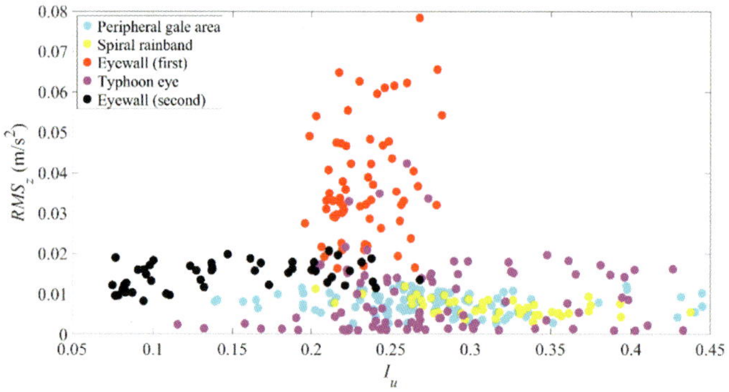

Fig. 14. Relationship between RMS of acceleration and turbulence intensity.

which is more concentrated than those in the other regions. The stiffening girder vibrated severely only within this range. Low and high turbulence intensities failed to excite serious vibrations. Although turbulence intensity in other typhoon structural regions also can reach this range, low-speed wind in these regions cannot excite the intensive vibrations of the bridge.

4.3. *Modal parameters identification based on SSI-COV*

Stochastic subspace identification (SSI) applies to the random responses under the turbulent wind,[36] which can be used to analyze the bridge responses under the influence of typhoons. Covariance-driven stochastic subspace identification (SSI-COV) is applied to identify the modal parameters of the bridge. The first step of the SSI-COV, similar to the data-driven stochastic subspace identification (SSI-DATA), is to construct the Hankel matrix according to the structural system with

n degrees of freedom, as shown in Eq. (16).[37]

$$\begin{aligned} \mathbf{x_{k+1}} &= \mathbf{A}\mathbf{x}_k + \boldsymbol{\omega}_\mathbf{k}, \\ \mathbf{y_k} &= \mathbf{C}\mathbf{x_k} + \boldsymbol{\nu}_\mathbf{k}, \end{aligned} \tag{16}$$

where $\mathbf{x_k} \in \mathbf{R}^{n\times 1}$ is the state vector; $\mathbf{y_k} \in \mathbf{R}^{l\times 1}$ is the response vector; $\mathbf{A} \in \mathbf{R}^{n\times n}$ is the system matrix; $\mathbf{C} \in \mathbf{R}^{l\times n}$ is the output matrix; $\boldsymbol{\omega}_\mathbf{k}$ and $\boldsymbol{\nu}_\mathbf{k}$ are the input and output noise sequences, respectively. The noise sequences are assumed to be zero mean samples of a white noise process.

$$E\left[\begin{pmatrix} \boldsymbol{\omega}_\mathbf{p} \\ \boldsymbol{\nu}_\mathbf{p} \end{pmatrix} \begin{pmatrix} \boldsymbol{\omega}_\mathbf{q}^T & \boldsymbol{\nu}_\mathbf{q}^T \end{pmatrix}\right] = \begin{pmatrix} \mathbf{Q} & \mathbf{S} \\ \mathbf{S}^T & \mathbf{R} \end{pmatrix} \delta_\mathbf{pq}, \tag{17}$$

where $\delta_\mathbf{pq}$ is the Kronecker delta.

The Hankel matrix can be divided into a past block and a future block.[38]

$$\mathbf{H} = \frac{1}{\sqrt{n}} \left[\begin{array}{cccc} y_0 & y_1 & \cdots & y_{n-1} \\ y_1 & y_2 & \cdots & y_n \\ \vdots & \vdots & \ddots & \vdots \\ y_{i-1} & y_i & \cdots & y_{i+n-2} \\ \hline y_i & y_{i+1} & \cdots & y_{i+n-1} \\ y_{i+1} & y_{i+2} & \cdots & y_{i+n} \\ \vdots & \vdots & \ddots & \vdots \\ y_{2i-1} & y_{2i} & \cdots & y_{2i+n-2} \end{array}\right] = \left[\frac{\mathbf{Y}_{0|i=1}}{\mathbf{Y}_{i|2i-1}}\right] = \left[\frac{\mathbf{Y}_p}{\mathbf{Y}_f}\right], \tag{18}$$

where n is the number of matrix columns; y_i is the acceleration data at the ith measurement point; $\mathbf{Y}_p$ is the past response block; and $\mathbf{Y}_f$ is the future response block. The essential difference between the two SSI methods is the calculation method of the correlation between the $\mathbf{Y}_p$ and the $\mathbf{Y}_f$. The SSI-DATA calculates the projection matrix by projecting $\mathbf{Y}_f$ onto $\mathbf{Y}_p$.

$$\mathbf{P}_i = \mathbf{Y}_f/\mathbf{Y}_p = \mathbf{Y}_f\mathbf{Y}_p^T(\mathbf{Y}_p\mathbf{Y}_p^T)^{\dagger}\mathbf{Y}_p, \tag{19}$$

where $()^{\dagger}$ is the pseudo-inverse of a matrix. The SSI-COV expresses the correlation by calculating the covariance matrix.

$$\begin{aligned} \boldsymbol{\Lambda}_\mathbf{i} &= \mathbf{Y_f}\mathbf{Y}_\mathbf{p}^T = E[\mathbf{y}_{k+i} \quad \mathbf{y}_k^T], \\ \mathbf{G} &= E[\mathbf{x}_{k+1} \quad \mathbf{y}_k^T]. \end{aligned} \tag{20}$$

According to the assumption, $\mathbf{\Lambda_i}$ can be expressed as $\mathbf{\Lambda_i} = \mathbf{CA^{i-1}G}$. The Toeplitz matrix can be expressed as

$$\mathbf{T} = \begin{pmatrix} \mathbf{\Lambda}_i & \mathbf{\Lambda}_{i-1} & \cdots & \mathbf{\Lambda}_1 \\ \mathbf{\Lambda}_{i+1} & \mathbf{\Lambda}_i & \cdots & \mathbf{\Lambda}_2 \\ \vdots & \vdots & \ddots & \vdots \\ \mathbf{\Lambda}_{2i-1} & \mathbf{\Lambda}_{2i-2} & \cdots & \mathbf{\Lambda}_i \end{pmatrix} = \begin{pmatrix} \mathbf{CA^{i-1}G} & \mathbf{CA^{i-2}G} & \cdots & \mathbf{CG} \\ \mathbf{CA^{i}G} & \mathbf{CA^{i-1}G} & \cdots & \mathbf{CAG} \\ \vdots & \vdots & \ddots & \vdots \\ \mathbf{CA^{2i-2}G} & \mathbf{CA^{2i-3}G} & \cdots & \mathbf{CA^{i-1}G} \end{pmatrix}$$

$$= \begin{pmatrix} \mathbf{C} \\ \mathbf{CA} \\ \vdots \\ \mathbf{CA^{i-1}} \end{pmatrix} (\mathbf{A^{i-1}G} \quad \cdots \quad \mathbf{AG} \quad \mathbf{G}) = \mathbf{O_i \Gamma_i}, \tag{21}$$

where $\mathbf{O_i}$ is the extended observability matrix; and $\Gamma_\mathbf{i}$ is the reversed extended stochastic controllability matrix. They can be obtained by SVD decomposition of the Toeplitz matrix.

$$\mathbf{T} = \mathbf{U} \sum \mathbf{V}^T = (\mathbf{U_1} \quad \mathbf{U_2}) \begin{pmatrix} \sum_1 & 0 \\ 0 & \sum_2 = 0 \end{pmatrix} \begin{pmatrix} \mathbf{V}_1^T \\ \mathbf{V}_2^T \end{pmatrix}, \tag{22}$$

$$\begin{aligned} \mathbf{O_i} &= \mathbf{U_i} \sum_1^{1/2}, \\ \mathbf{\Gamma_i} &= \sum_1^{1/2} \mathbf{V}_1^T. \end{aligned} \tag{23}$$

The system matrix and the output matrix can be expressed as

$$\begin{aligned} \mathbf{A} &= \mathbf{O}(1:l,:), \\ \mathbf{C} &= \mathbf{O}(1:l,:)^\dagger \mathbf{O}(l+1:2l,:). \end{aligned} \tag{24}$$

Then the modal parameters can be obtained by eigenvalue decomposition of the system matrix $\mathbf{A}$.

$$\mathbf{A} = \phi \widehat{\mathbf{\Lambda}} \phi^{-1}, \tag{25}$$

where $\widehat{\mathbf{\Lambda}} = \text{diag}\{\widehat{\lambda}_i\}$ is the eigenvalue matrix of $\mathbf{A}$; ϕ is the eigenvector of $\mathbf{A}$. The eigenvalues of the continuous-time system λ_i can be calculated by Eq. (26).

$$\lambda_i = \frac{\ln(\widehat{\lambda}_i)}{\Delta t}, \tag{26}$$

where Δt is the time interval of sampling. The relationship between the circular frequency, damping ratio, and the eigenvalue can be expressed as Eq. (27).

$$\lambda_i, \overline{\lambda_i} = -\xi_i \omega_i \pm j\omega\sqrt{1-\xi_i^2}, \tag{27}$$

where $j = \sqrt{-1}$. The natural frequency f_i and damping ratio ξ_i at ith mode can be expressed as Eq. (28).

$$\begin{aligned} f_i &= \frac{|\lambda_i|}{2\pi}, \\ \xi_i &= \frac{\lambda_i + \overline{\lambda_i}}{2\sqrt{\lambda_i \overline{\lambda_i}}}. \end{aligned} \tag{28}$$

Owing to the wind speed with a wide range in this region that can fully excite the structural vibration at different frequencies, the accelerations of the bridge under the influence of spiral rainband were used for modal parameter identification. The natural frequencies of the Xihoumen Bridge calculated by SSI-COV are illustrated and compared with the consequences obtained by the other methods[39–42] in Table 5.

The frequencies identified by these methods have a high consistency with each other numerically. The slight difference between the values may be caused by the influence of temperature, humidity, vehicles, and other factors on the bridge status.[43,44] However, the correspondence between frequencies and mode orders is different. Taking the results of PP and finite element simulation as the reference standard,

Table 5. Natural frequencies of Xihoumen Bridge.

Mode order	Methods				
	Peak picking (PP)	Finite element simulation	Bayesian fast Fourier transform (FFT)	SSI-DATA	SSI-COV
1	0.098	0.101	0.0948	0.0953	0.0948
2	0.134	0.133	0.1328	0.1328	0.1317
3	0.183	0.184	0.1825	0.1825	0.1819
4	0.232	0.229	0.2767	0.2301	0.2279
5	0.275	0.274	—	0.2757	0.2750
6	0.342	0.324	—	0.3237	0.3272
7	0.378	0.375	—	0.4351	0.3578
8	0.433	0.428	—	0.4918	0.4351
9	0.494	0.482	—	—	0.4910
10	—	—	—	—	0.5511
11	—	—	—	—	0.6852
12	—	—	—	—	0.7633
13	—	—	—	—	0.8220
14	—	—	—	—	0.9922
15	—	—	—	—	1.0643

Note: Only vertical mode frequencies of the Xihoumen Bridge are presented. "—" indicates that the value is not listed in the researches.

the fourth and seventh mode frequencies are not identified by some methods. The fourth and seventh mode frequencies are close to the fifth and sixth mode frequencies, respectively. Closely spaced modes like these are difficult to distinguish. The excitation condition has an impact on the accuracy of structural modal parameter identification. The wind with a wide speed range from 6 to 14 m/s in the spiral rainband fully excited the bridge vibrations, which is conducive to the identification of modal parameters.

4.4. *Power spectral density of vibrational acceleration*

During the impact of Typhoon In-fa, the bridge was closed to traffic at about 20:00 on July 24. Therefore, there were no effects from traffic loading, and the vibrations were only excited by the wind. The wind-induced vibrations of suspension bridges are mainly at the frequency of 0–1 Hz.[23,29] The spectra of the vertical acceleration measured at 1/2 span of the bridge shown in Fig. 15 verified this range. Accordingly, the vibrations at this frequency range are analyzed. To compare the dynamic responses in each typhoon structural region, the power spectra of vertical accelerations are presented in Fig. 16.

The spectra of vertical accelerations are similar to each other in the typhoon eye and two eyewalls of Typhoon In-fa. Most vibrational energy is concentrated in the fundamental frequency, and less energy is owned by higher-order vibrations. The power spectra of the bridge under the influence of the peripheral gale area and spiral rainband are more irregular. Located at a wave node, acceleration data 1/2 span cannot reflect the power spectral density of the third, fifth, seventh, and ninth modes. The accelerometers AC1-3 and AC7-9 were not strictly installed at 1/4 span and 3/4 span, so the influence of wave nodes is avoided. The vibrational energy in the first eyewall is the largest. Therefore, the first eye wall is the most notable typhoon structural region in typhoon-resistance practice.

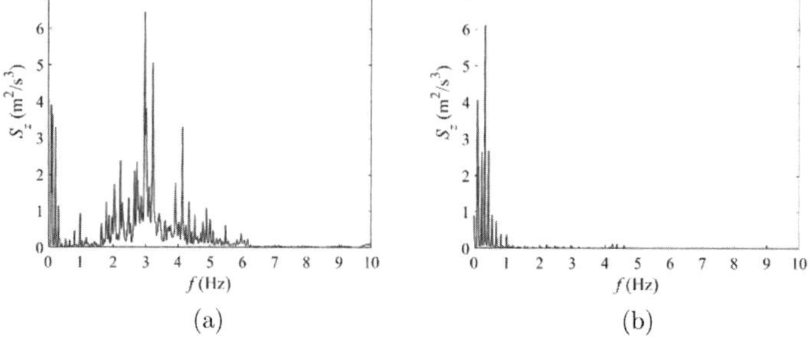

Fig. 15. Power spectra of vertical acceleration at 1/2 span: (a) Open to traffic and (b) closed to traffic.

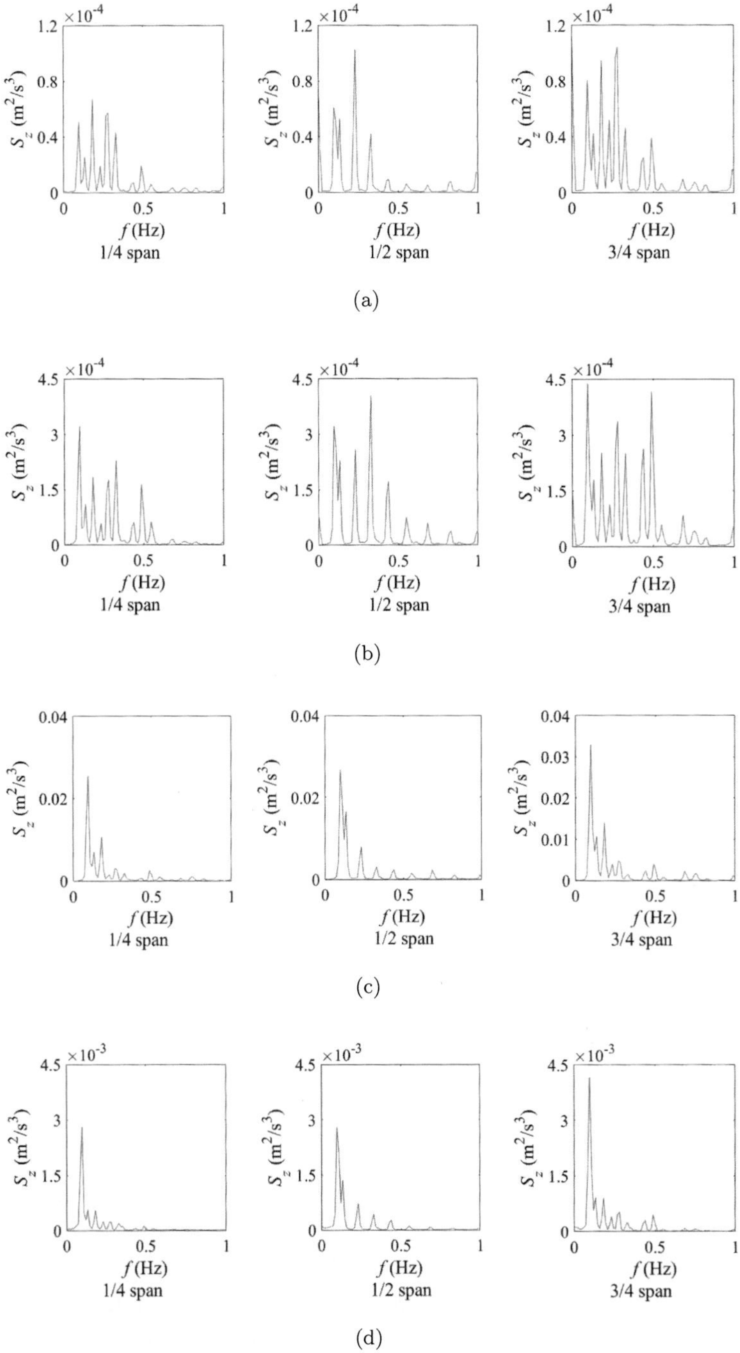

Fig. 16. Power spectra in each structural region: (a) Peripheral gale area, (b) spiral rainband, (c) first eyewall, (d) typhoon eye, and (e) second eyewall.

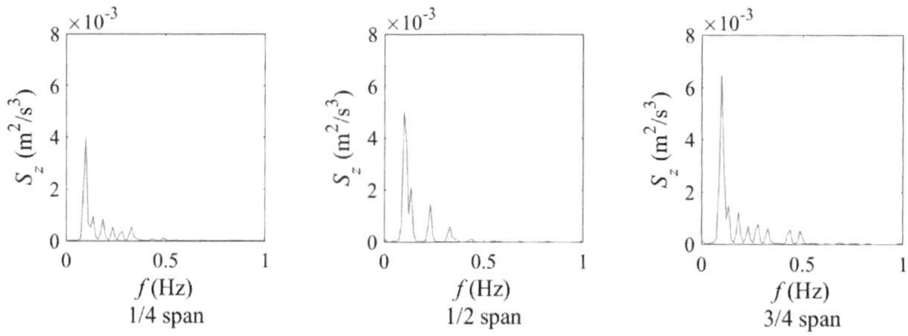

(e)

Fig. 16. (*Continued*)

5. Conclusions

The wind field features of onshore, offshore, and landfalling typhoons are compared using the data recorded by the SHMS of Xihoumen Bridge. Based on the obtained acceleration responses, the structural frequency of the bridge is identified by the SSI-COV. To determine the most detrimental typhoon structural region to the bridge, the RMS of acceleration and vibration energy under the influence of each typhoon structure region are compared. Conclusions are summarized as follows:

(1) Among the typhoons with three typical crossing paths, the landfalling typhoon owns the highest standard deviations of wind speed and direction ($\sigma_U = 4.83$, $\sigma_\theta = 104.30$), due to the influence of different wind conditions caused by typhoon structural regions. The standard deviations of the onshore typhoon ($\sigma_U = 3.53$, $\sigma_\theta = 38.64$) are higher than those of the offshore typhoon ($\sigma_U = 2.99$, $\sigma_\theta = 16.15$) as a result of being influenced by surface roughness.
(2) The turbulence intensity of the offshore typhoon is higher than that of the landfalling typhoon and onshore typhoon. Onshore typhoon ratios of I_v to I_u and I_w to I_u are higher than those for offshore typhoons and are close to the values suggested by China's wind-resistant design specification. The relational model proposed by Mackey, Choi, and Lam can be used for the conversion between turbulence intensity and gust factor, while the modal parameters vary with the crossing paths of typhoons. Three typhoons with different paths have similar turbulence power spectra that fit with the von Karman spectrum well.
(3) A quadratic function applies to express the relationship between the RMS of acceleration and wind speed. The stiffening girder vibrated severely at the turbulence intensity range from 0.196 to 0.282 under the influence of the first eyewall. Low and high turbulence intensities cannot excite serious vibrations.
(4) The vibration responses fully excited by the complex wind condition in spiral rainband are applicable for identifying modal parameters by SSI-COV. Using

the acceleration data under the influence of spiral rainband, the bridge's frequencies (0–1.1 Hz) of 15 vertical modes are identified accurately, including low frequencies and closely spaced modes.

(5) Under the influence of typhoon eye and eyewall, the vibrational energy is mainly concentrated in the fundamental frequency, and less energy is held by higher-order vibrations. The stiffening girder's vibration acceleration and energy reach their peak value under the impact of the first eyewall. Thus, bridge managers need to pay close attention to the first eyewall.

Acknowledgments

This research is supported by the National Natural Science Foundation of China (Grant No. U22A20231, 51578506).

References

1. Y. C. He, P. W. Chan and Q. S. Li, Wind characteristics over different terrains, *J. Wind. Eng. Ind. Aerodyn.* **120** (2013) 61–69.
2. S. Schmidt, C. Kemfert and P. Höppe, Tropical cyclone losses in the USA and the impact of climate change: A trend analysis based on data from a new approach to adjusting storm losses, *Environ. Impact. Asses. Rev.* **29**(6) (2009) 359–369.
3. S. H. Li and H. P. Hong, Typhoon wind hazard estimation for China using an empirical track model, *Nat. Hazard.* **82**(2) (2016) 1009–1029.
4. S. H. Li and H. P. Hong, Use of historical best track data to estimate typhoon wind hazard at selected sites in China, *Nat. Hazard.* **76**(2) (2015) 1395–1414.
5. X. Xu, D. Sun and T. Guo, A systemic analysis of typhoon risk across China, *Nat. Hazard.* **77**(1) (2015) 461–477.
6. L. Hu, Y. Xu, Q. Zhu, A. Guo and A. Kareem, Tropical storm-induced buffeting response of long-span bridges: Enhanced nonstationary buffeting force model, *J. Struct. Eng.* **143**(6) (2017) 04017027.
7. Y. Xu, L. Hu and A. Kareem, Conditional simulation of nonstationary fluctuating wind speeds for long-span bridges, *J. Eng. Mech.* **140**(1) (2014) 61–73.
8. Z. W. Guo, S. Lin and Q. Ni, Advances in active control of wind-induced vibration of long-span suspension bridges, *Int. J. Struct. Stab. Dyn.* **21**(12) (2022) 2230002.
9. Q. Yang, R. Gao, F. Bai, T. Li and Y. Tamura, Damage to buildings and structures due to recent devastating wind hazards in East Asia, *Nat. Hazard.* **92** (2018) 1321–1353.
10. Y. Ding, P. Wang, X. Liu, X. Zhang, L. Hong and Z. Cao, Risk assessment of highway structures in natural disaster for the property insurance, *Nat. Hazard.* **140** (2020) 2663–2685.
11. G. Holland, A revised hurricane pressure-wind model, *Mon. Weather. Rev.* **136**(9) (2007) 3432–3445.
12. J. P. Terry and G. Gienko, Developing a new sinuosity index for cyclone tracks in the tropical South Pacific, *Nat. Hazards.* **59**(2) (2011) 1161–1174.
13. F. Jakobsen and H. Madsen, Comparison and further development of parametric tropical cyclone models for storm surge modelling, *J. Wind. Eng. Ind. Aerodyn.* **92**(5) (2004) 375–391.

14. D. D. Xia, L. M. Dai, H. F. Wang and H. T. Hu, A field measured based wind characteristics analysis of a typhoon in near-ground boundary layer, *Atmosphere* **12**(7) (2021) 873.
15. P. Huang, W. Xie and M. Gu, A comparative study of the wind characteristics of three typhoons based on stationary and nonstationary models, *Nat. Hazard.* **101** (2020) 785–815.
16. S. Y. Cao, Y. Tamura, N. Kikuchi, M. Saito, I. Nakayama and Y. Matsuzaki, Wind characteristics of a strong typhoon, *J. Wind. Eng. Ind. Aerodyn.* **97**(1) (2009) 11–21.
17. Y. Matsuzakia, S. Y. Cao, Y. Tamura, N. Kikuchi, M. Saito and I. Nakayama, A case study of gust factor of a strong typhoon, *J. Wind. Eng. Ind. Aerodyn.* **138** (2015) 52–60.
18. L. L. Song, W. C. Chen, B. L. Wang, S. Q. Zhi and A. J. Liu, Characteristics of wind profiles in the landfalling typhoon boundary layer, *J. Wind. Eng. Ind. Aerodyn.* **149** (2016) 77–88.
19. Z. L. Ye, N. Li and F. J. Zhang, Wind characteristics and responses of Xihoumen Bridge during typhoons based on field monitoring, *J. Struct. Health. Monit.* **9** (2019) 1–20.
20. Y. Q. Ni and Y. X. Xia, Strain-based condition assessment of a suspension bridge instrumented with structural health monitoring system, *Int. J. Struct. Stab.* **16**(4) (2016) 1640027.
21. G. H. Zhai, Y. Narazaki, S. Wang, S. A. V. Shajihan and B. F. Spencer, Synthetic data augmentation for pixel-wise steel fatigue crack identification using fully convolutional networks, *Smart Struct. Syst.* **29**(1) (2022) 237–250.
22. A. Fenerci, O. Øiseth and A. Rønnquist, Long-term monitoring of wind field characteristics and dynamic response of a long-span suspension bridge in complex terrain, *Eng. Struct.* **147** (2017) 269–284.
23. A. Fenerci and O. Øiseth, Strong wind characteristics and dynamic response of a long-span suspension bridge during a storm, *J. Wind Eng. Ind. Aerodyn.* **172** (2018) 116–138.
24. J. Guo, C. J. Hu, M. J. Zhu and Y. Q. Ni, Monitoring-based evaluation of dynamic characteristics of a long span suspension bridge under typhoons, *J. Civ. Struct. Health. Monit.* **11** (2021) 397–410.
25. Y. B. Yang, C. W. Lin and J. D. Yau, Extracting bridge frequencies from the dynamic response of a passing vehicle, *J. Sound. Vib.* **272**(3–5) (2004) 471–493.
26. C. W. Lin and Y. B. Yang, Use of a passing vehicle to scan the fundamental bridge frequencies: An experimental verification, *Eng. Struct.* **27**(13) (2005) 1865–1878.
27. Y. B. Yang and K. C. Chang, Extracting the bridge frequencies indirectly from a passing vehicle: Parametric study, *Eng. Struct.* **31**(10) (2009) 2448–2549.
28. J. Zhang, L. M. Zhou, Y. D. Tian, S. S. Yu, W. J. Zhao and Y. Y. Cheng, Vortex-induced vibration measurement of a long-span suspension bridge through noncontact sensing strategies, *Comput. Aided Civ. Inf.* **37**(12) (2022) 1617–1633.
29. L. X. Li, A. Kareem, Y. Q. Xiao, L. L. Song and C. Y. Zhou, A comparative study of field measurements of the turbulence characteristics of typhoon and hurricane winds, *J. Wind. Eng. Ind. Aerodyn.* **140** (2015) 49–66.
30. J. Y. He, Q. S. Li, M. Asce, P. W. Chan, L. Li, C. Lu, L. Zhang and H. L. Yang, Characteristics and vertical profiles of mean wind and turbulence for typhoon, monsoon, and thunderstorm winds, *J. Struct. Eng.* **147**(11) (2021) 04021188.

31. V. A. Reston, ASCE 7-10, Minimum Design Loads for Buildings and other Structures (American Society of Civil Engineering, Reston, VA, 2010).
32. Professional Standard of the People's Republic of China, *Wind-Resistant Design Specification for Highway Bridges* (China Communications Co., Ltd., Beijing, 2018).
33. H. Ishizaki, Wind profiles, turbulence intensities and gust factors for design in typhoon-prone regions, *J. Wind. Eng. Ind. Aerodyn.* **13** (1983) 55–66.
34. E. Choi, Gradient height and velocity profile during typhoons, *J. Wind. Eng. Ind. Aerod.* **13** (1983) 31–41.
35. K. A. Emanuel, An air-sea interaction theory for tropical cyclones. Part I: Steady-state maintenance, *J. Wind. Eng. Ind. Aerod.* **43** (1986) 585–604.
36. T. Janesupasaeree, Determination of flutter derivatives of bridge decks by covariance-driven stochastic subspace identification, *Int. J. Struct. Stab.* **11**(1) (2011) 73–99.
37. Y. F. Shao, C. Q. Miao, B. W. Li and Q. D. Wu, Simultaneous de-noising and enhancement method for long-span bridge health monitoring data based on empirical mode decomposition and fractal conservation law, *Meas. Sci. Technol.* **30**(6) (2019) 065103.
38. B. Peeters and G. D. Roeck, Reference based stochastic subspace identification in civil engineering, *Inverse. Probl. Sci. Eng.* **8**(1) (2000) 47–74.
39. Z. W. Huang, Y. Z. Li, X. G. Hua and Z. Q. Chen, Automatic identification of bridge vortex-induced vibration using random decrement method, *Appl. Sci.* **9**(10) (2019) 2049.
40. X. L. Chu, H. N. Sinh, W. Cui, L. Zhao and Y. J. Ge, Life-cycle assessment for flutter probability of a long-span suspension bridge based on field monitoring data, *Comput.* **2106** (2021) 10694.
41. H. Li, S. J. Laima, J. P. Ou, X. F. Zhao, W. S. Zhou, Y. Yan, N. Li and Z. Q. Liu, Investigation of vortex-induced vibration of a suspension bridge with two separated steel box girders based on field measurements, *Eng. Struct.* **33**(6) (2011) 1894–1907.
42. Y. X. Yang, T. T. Ma and Y. J. Ge, Evaluation on bridge dynamic properties and VIV performance based on wind tunnel test and field measurement, *Wind Struct.* **20**(6) (2015) 719–737.
43. Y. B. Yang, M. C. Cheng and K. C. Chang, Frequency variation in vehicle-bridge interaction systems, *Int. J. Struct. Stab.* **13**(2) (2011) 1350019.
44. Y. Xia, H. Hao, G. Zanardo and A. Deeks, Long term vibration monitoring of an RC slab: Temperature and humidity effect, *Eng Struct.* **28**(3) (2006) 441–452.

https://doi.org/10.1142/9789819814770_0020

Chapter 20

Random Dynamic Analysis and Running Safety Assessment of Long-Span Cable-Stayed Bridge Under High-Speed Trains Subjected to Earthquake Excitations and Track Irregularities[#]

Jian-Feng Mao[*,†,‡,§], Xiao-Tong Chen[*,¶], Zhi-Wu Yu[*,†,‡,||,‡‡], Li-Zhong Jiang[*,†,‡,**] and Aliyu Yusuf Moyi[*,††]

[*]*School of Civil Engineering, Central South University Changsha 410075, P. R. China*

[†]*National Engineering Research Center of High Speed Railway Construction Technology Changsha 410075, P. R. China*

[‡]*China Railway Group Limited Beijing 100039, P. R. China*

[§]*jfmao1@csu.edu.cn*

[¶]*chenxiaotong@csu.edu.cn*

[||]*zhwyu@csu.edu.cn*

[**]*lzhjiang@csu.edu.cn*

[††]*yali9998@yahoo.com*

Running trains on a long-span bridge under earthquake is an inevitable problem for the high-speed railway in seismic regions. This paper aims to investigate the seismically random dynamic behavior of train–cable stayed bridges interaction system using the probability density evolution method (PDEM). The system simultaneously involves the multi-effect of random multi-point earthquake excitations with traveling waves and random track irregularity. The multi-point earthquake excitations, composed of lateral and vertical positions, are generated by stochastic harmful functions (SHFs) and uniformly modulated into nonstationary random processes. The motion equation of such a system is established by coupling the 38-degree vehicles and bridges through the refined random wheel/rail contact relationship and accounting for the phase-lags of seismic travelling waves between pier excitations. PDEM is proven to be applicable to such time-dependent systems and is then used to transform the random excitations into a series of deterministic representative excitations with initial probability. By solving for the corresponding deterministic probability responses, various nonstationary random responses, including the time-dependent probability density evolution functions of random responses, mean

[‡‡]Corresponding author.

[#]This chapter appeared previously on the International Journal of Structural Stability and Dynamics. To cite this chapter, please cite the original article as the following: J.-F. Mao *et al.*, *Int. J. Struct. Stab. Dyn.*, **23**, 2340022 (2023), DOI: 10.1142/S0219455423400229.

values curves and standard deviations, can be obtained easily. A case study based on the cable-stayed bridge is then presented, and the results show the effectiveness and accuracy of the proposed method by comparison with Monte Carlo Simulation. Additionally, the influence of train speed and seismic intensity on the safety and reliability of trains is discussed.

Keywords: Multi-points earthquake excitation; probability density evolution method; train–bridge systems; stochastic harmful functions; high speed railway.

1. Introduction

Due to the high stability of track irregularity on the bridge, the bridge is widely used in the high-speed railway in China, which aims to get a smoother track to ensure the safety and stability of moving trains. Due to the high proportion of bridges and complex geological conditions, the probability of earthquakes occurring when vehicles move on the bridges is highly increased, e.g. during the Niigata Earthquake on 23rd October 2004, a Shinkansen high-speed train derailed when crossing an elevated bridge at 200 km/h. This establishes the importance of investigating the safety of train–bridge systems under the action of earthquakes. The high-accuracy calculations that aim to estimate the seismic bridge dynamic behaviors are greatly important, especially for a long-span cable-stayed bridge with large seismic displacement responses.

As a matter of fact, the complex time-dependent random characteristics of train–bridge coupled systems under nonstationary earthquake excitations are difficult to be reproduced. When solving such nonstationary random vibration problems of time-dependent systems, the conventional response spectrum method[1,2] is inconvenient, so the time-history methods[3–18] are usually used, in which sufficient representative samples of track irregularity and seismic accelerations are constructed as deterministic loads to calculate the structural responses, e.g. Yang and Wu[7] analyzed the dynamic stability of train–bridge coupled systems by using as input excitations the earthquake records. Xia *et al.*[9] investigated the dynamic responses of bridge–vehicle systems by using the nonuniform seismic accelerations input from different foundations. Fryba and Yau[8,10] used the modified Kobe earthquake records to study the influences of train speeds and wave passage effects on the responses of suspended bridges. Du *et al.*[11] presented a finite element method-based framework for dynamic analysis of coupled bridge–train systems under nonuniform seismic ground motion. Montenegro *et al.*[13] presented the running safety assessment of trains moving over bridges subjected to moderate earthquakes. Jin *et al.*[14] presented a numerical investigation on the derailment of vehicles on simply-supported bridges, where a total of 18 ground motion records were used to account for the uncertainty of ground motions. Nishimura *et al.*[16] investigated vehicle safety in terms of dynamic stability and the possibility of derailment directly caused by the track excitations of great earthquakes. Zhang *et al.*[17] proposed a numerical solution for the dynamic responses of a train–bridge interaction system subjected to multi-support seismic loads. Ju[18] developed a simple nonlinear moving wheel element to deal with the contact and separation modes of the wheel/rail interaction when the high-speed trains moved on multi-span simply supported bridges under earthquakes. Gao *et al.*[19] conducted a numerical analysis of the dynamic response of long-span bridges under different incident angles of seismic waves and different train speeds. Guo *et al.*[20] performed the safety assessment of high-speed trains on the simply supported bridges under low-level earthquakes utilizing spectral intensity index. Feng *et al.*[21] presented a track irregularity spectrum after the action of earthquakes. Huang and Cheng[22] developed a numerical model for predicting the seismic responses of vehicle-guideway bridge systems for low-medium-speed maglev trains. Then, the point estimation method was used to discuss the running safety performance of trains under different ground motion (GM) intensities, different train speeds, and different pier heights. Liu *et al.*[23] applied the point estimation method to analyze the vibration of train–bridge coupled system under earthquake and determined the ground motion intensity threshold for the bridge based on the maximum probability of running safety.

The presented investigations mostly based on the deterministic analysis are the approximation of random dynamic responses of train–bridge system subjected to earthquake. The precision and reliability rely heavily on the quality and number of the selected samples. Therefore, developing a random dynamic theory of train–bridge coupled systems under earthquake excitations based on traditional deterministic models is significant to precisely obtain the dynamic reliability and statistical results. Actually, many random theories have been employed to investigate the randomness of the train–bridge system. Typically, in the early time, Monte Carlo simulation (MCS) is employed as the traditional approach for stochastic structural simulation. But now, it is mainly adopted as a validation method with other algorithms due to its huge computational cost. Later, in Refs. 24–26, the pseudo excitation method (PEM) is employed as an efficient algorithm for a train–bridge coupled system to obtain its random characteristics of earthquakes and track irregularities represented by spectra. Even though PEM is considered to be more appropriate for lateral vibration estimation due to its zero-mean characteristic, as it ignored the spatial variability of excitations along the whole railway lines,[27] not to mention the dynamic amplification coupling effect of train–bridge interaction mainly in the vertical vibration under the combinations of vehicle loads, track irregularities and earthquakes. Fortunately, the deficiency will be remedied by the probability density evolution method (PDEM),[28] a new stochastic approach developed by Li and Chen[29] with its evolutionary criterion based on the physical randomness of systems. It may provide a new perspective for tackling stochastic systems. Inspired by this, Yu and Mao[30–33] successfully and systematically employed PDEM to investigate the stochastic characteristic of train–bridge coupled systems and established the corresponding probabilistic models, which can simultaneously involve the multi-effect of random track irregularities and stochastic structural uncertainties. These models have been verified to be reliable and have high efficiency for train–bridge interaction. Meanwhile, Xu and Zhai[34,35] established similar probabilistic models for the temporal–spatial stochastic analysis of vehicle–track coupled systems based on PDEM under track irregularities or earthquakes. The PDEM-application research studies for railway train-substructure systems are attracting increasing attention. Jin and Liu[36] used the PDEM to derive the probability density function (PDF) for the extreme values of wheel unloading under the action of an earthquake and track unevenness.

In this paper, based on the probability density evolution method (PDEM), the seismically random dynamic behavior of train–cable stayed bridges system is explored under the multi-effect of random multi-point earthquake excitations with traveling waves and random track irregularity. First, the 38-degree-vehicles models and the cable-stayed bridges model are established with the refined random wheel/rail contact relationship. Then, the multi-point earthquake excitations in the lateral and vertical positions are generated by stochastic harmful functions (SHFs) and uniformly modulated into nonstationary random processes. Thus, the random excitations are transformed into a series of deterministic representative excitations with initial probability. Third, by solving the corresponding deterministic probability responses, various nonstationary random responses, including the time-dependent probability density evolution functions of random responses, mean values curves and standard deviations are obtained. Finally, a case study based on the cable-stayed bridge is presented and the proposed method's effectiveness and accuracy are verified with Monte Carlo Simulation. Also, the influence of train speed and seismic intensity on the safety and reliability of trains is discussed.

2. Modelling of Train–Bridge Coupled System

The dynamic analysis model of the train–bridge coupling system subjected to multi-point earthquake excitations can be regarded as a spatial dynamic system composed of the vehicle and track–bridge subsystem, which interact with each other through refined wheel/rail contact interaction. In this model, the randomness of excitations is mainly composed of the nonuniform earthquake excitations imposed on the bridge piers and the random track irregularities between the wheel and rail.

In order to simplify the complex train–bridge system, the following assumptions are made for the modeling: (1) The train runs across the bridge at a constant speed. (2) The components of each vehicle are regarded as rigid components, and the vehicle suspension systems are simulated using linear springs and dampers. (3) The wheel/rail interactions do not slide due to the high wheel/rail friction coefficient.

2.1. *Train model*

Similar to the vehicle models presented in Refs. 31 and 37–39, the train model here is composed of two motor cars in the front and rear and several trailer cars in the

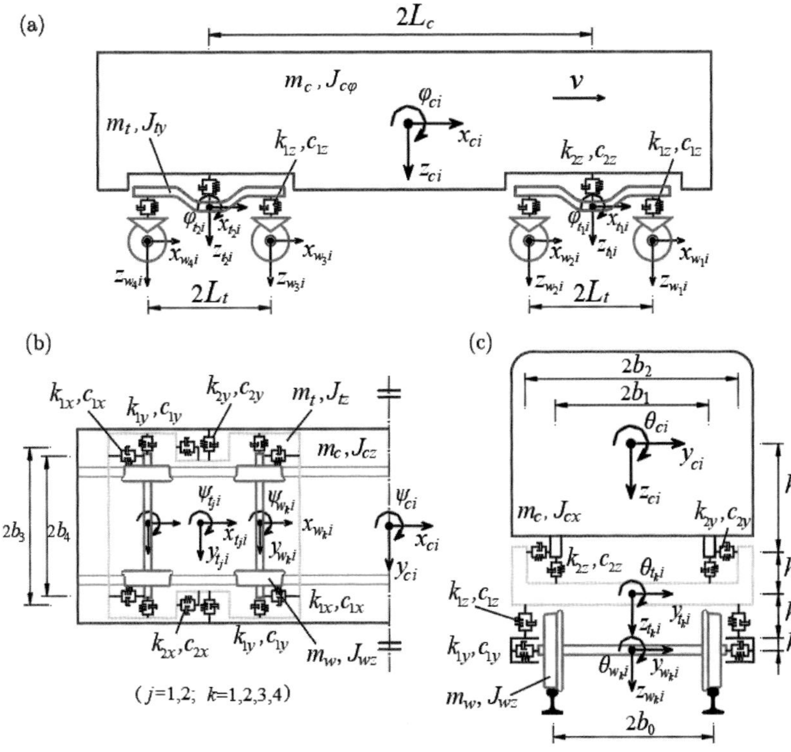

Fig. 1. The vehicle model: (a) lateral view; (b) top viewand (c) front view.

middle, all moving at a constant speed. As shown in Fig. 1, the train model consists of locomotives and/or passenger cars, each of which is a complex multi-degrees of freedoms (DOFs) dynamic system composed of one car body, two bogies, and four-wheel sets, with several suspension spring-dampers.

The vehicle model is established as a four-axle vehicle with 38 DoFs,[31] as shown in Fig. 1, and each car body and bogie has six DOFs, namely, the longitudinal, lateral, vertical, roll, yaw and pitch displacement, respectively. And each wheel-set has five DOFs, namely the longitudinal, lateral, vertical, roll and yaw displacement for each vehicle. The linear springs and viscous dashpots are applied for the connections between the car body, bogies and wheel-sets. All the train parameters can be found in Fig. 1 or Ref. 31, and the symbols in the figure are defined in Table A.1 in Appendix A, or as needed. The train model here, does not consider the randomness of vehicle parameters.

The equation of motion of the train can be expressed as

$$\mathbf{M}_v\{\ddot{\mathbf{X}}_v\} + \mathbf{C}_v\{\dot{\mathbf{X}}_v\} + \mathbf{K}_v\{\mathbf{X}_v\} = \mathbf{F}_v(\boldsymbol{\xi}^r, t), \tag{1}$$

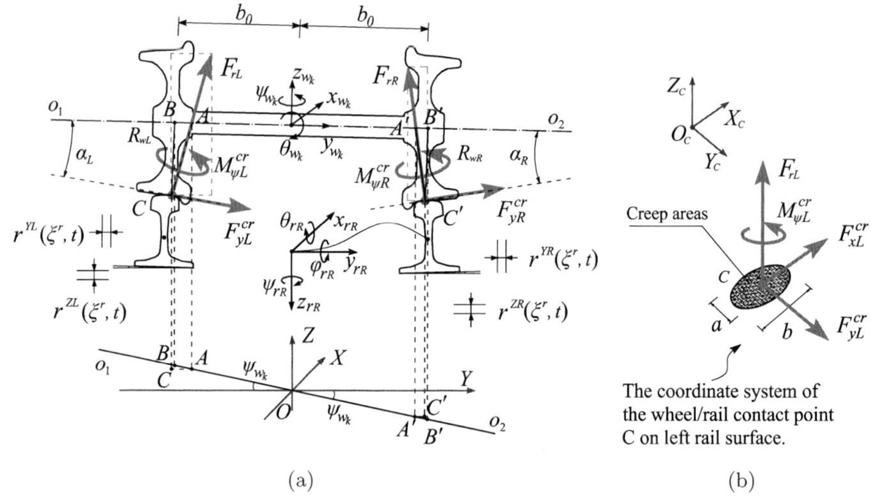

Fig. 2. The 3D wheel/rail interaction model.[31]

where $\mathbf{M}_v$, $\mathbf{C}_v$ and $\mathbf{K}_v$ are the mass, damping and stiffness matrices of the train model, $\mathbf{X}_v$, $\dot{\mathbf{X}}_v$ and $\ddot{\mathbf{X}}_v$ are the corresponding displacement vector, velocity vector and acceleration vector. $\mathbf{F}_v(\boldsymbol{\xi}^r, t)$ is the random vector of wheel/rail forces subjected to random track irregularities and wheel/rail interactions. The symbol $\boldsymbol{\xi}^r$ in the force vector $\mathbf{F}_v$ represents the uniform point design of random frequencies and random phases used to generate representation of irregularity samples from track spectrum. Details of these matrices and vectors can be found in Ref. 31.

As the details of the wheel/rail contact model are shown in Fig. 2, the wheel/rail interactions consist of the contact forces in the normal plane, formulated using Hertz theory,[40] and in the tangential plane, formulated using Kalker linear rolling contact theory,[41] the reference to the detailed information of the wheel/rail interaction is Refs. 4 and 31.

Therefore, the random vector $\mathbf{F}_v(\boldsymbol{\xi}^r, t)$ with order $38N_v \times 1$ involving the excitations of random track irregularities can be written as

$$\mathbf{F}_v(\boldsymbol{\xi}^r, t) = \{\mathbf{F}_{v1}(\boldsymbol{\xi}^r, t)\mathbf{F}_{v2}(\boldsymbol{\xi}^r, t)\cdots\mathbf{F}_{vN_v}(\boldsymbol{\xi}^r, t)\}^{\mathrm{T}}, \tag{2}$$

where $\mathbf{F}_{vi}(\boldsymbol{\xi}^r, t)$ denotes the random load vector for each vehicle. N_v is the total number of vehicles in the train model. The detailed expression can be written as

$$\mathbf{F}_{vi}(\boldsymbol{\xi}^r, t) = \left\{ \begin{matrix} \begin{bmatrix} \mathbf{0}_{6\times1} & \mathbf{0}_{6\times1} & \mathbf{0}_{6\times1} \end{bmatrix}^{\mathrm{T}} \\ \begin{bmatrix} \mathbf{F}_{w_1 i}(\boldsymbol{\xi}^r, t) & \mathbf{F}_{w_2 i}(\boldsymbol{\xi}^r, t) & \mathbf{F}_{w_3 i}(\boldsymbol{\xi}^r, t) & \mathbf{F}_{w_4 i}(\boldsymbol{\xi}^r, t) \end{bmatrix}^{\mathrm{T}} \end{matrix} \right\}, \tag{3}$$

where the submatrix $\mathbf{F}_{w_k i}(\boldsymbol{\xi}^r, t)$ $(k = 1, 2, 3, 4; i = 1, 2, \ldots, N_v)$ denotes the random force vector of the kth wheel-set in ith vehicle. Respectively, these vectors can be expressed as

$$\mathbf{F}_{w_k i}(\boldsymbol{\xi}^r, t) = \left\{ \begin{array}{c} \sum\limits_{p=L,R} F^{cr}_{xp,i}(\boldsymbol{\xi}^r, t) \\ \sum\limits_{p=L,R} \left((-1)^p F^z_{rp,i}(\boldsymbol{\xi}^r, t)\, \lambda^p_k(t) + \varsigma^p_{wk,i} \dot{r}^{Yp}(\boldsymbol{\xi}^r, t) \right) \\ + \sum\limits_{p=L,R} F^{\mathrm{cr}}_{yp,i}(\boldsymbol{\xi}^r, t) \\ F^z_{rL,i}(\boldsymbol{\xi}^r, t) + F^z_{rR,i}(\boldsymbol{\xi}^r, t) - G_{\mathrm{axle}} \\ 0 \\ \sum\limits_{p=L,R} M^{cr}_{\psi p,i}(\boldsymbol{\xi}^r, t) \end{array} \right\}^{\mathrm{T}}, \tag{4}$$

$$F^z_{rp,i}(\boldsymbol{\xi}^r, t) = k_{\mathrm{rail}} \left(z_{wki} + (-1)^I b_0 \theta_{wki} - z^p_{rki} - r^{Zp}(\boldsymbol{\xi}^r, t) \right); \quad p = L|_{I=1}, R|_{I=2}, \tag{5}$$

where the superscripts 'L' and 'R' denote the left and right rail; $r^{Zp}(\boldsymbol{\xi}^r, t)$ and $r^{Yp}(\boldsymbol{\xi}^r, t)$ $(p = L, R)$, respectively, represent the random vertical and lateral track irregularity samples at left/right rail; $\dot{r}(\cdot)$ is the first derivative of each track irregularity; $\lambda^L_k(t)$ and $\lambda^R_k(t)$ denote the wheel/rail contact angle slope of left/right rail at kth wheel-set; $\varsigma^p_{wk,i}(p = L, R)$ denotes the lateral creep coefficient between the kth wheel-set with the left/right rail;[4] F^{cr}_{xp}, F^{cr}_{yp} and $M^{\mathrm{cr}}_{zp}(p = L, R)$ denote the tangential contact forces transmitted between the wheel-sets and the rails; $F^z_{rp,i}(\boldsymbol{\xi}^r, t)$ $(p = L, R)$ denotes the vertical wheel/rail forces at left/right rail; z_{wki} and θ_{wki} denote the vertical and roll displacement at kth wheel-set of ith vehicle; $z^p_{rki}(p = L, R)$ denotes the vertical displacement of rail under the kth wheel-set of ith vehicle; b_0 is the half width of wheel-set, k_{rail} is the equivalent contact stiffness coefficient between the wheel and rail which is linearized using the Hertz nonlinear contact theory.[42] G_{axle} is the wheel axle load due to the static vehicle system. The rail parameters are defined in Table A.2 in Appendix A, or as needed.

For more details on the matrices of vehicle dynamic motion, please refer to Ref. 31.

2.2. *Bridge model*

As shown in Fig. 2, the cable-stayed bridge model is composed of several subsystems, of which the stiffening girder and other substructures are modeled as linear elastic 3D Bernoulli–Euler beams, not considering the nonlinearity. The random motion equation of the bridge system under random earthquake excitations can be written

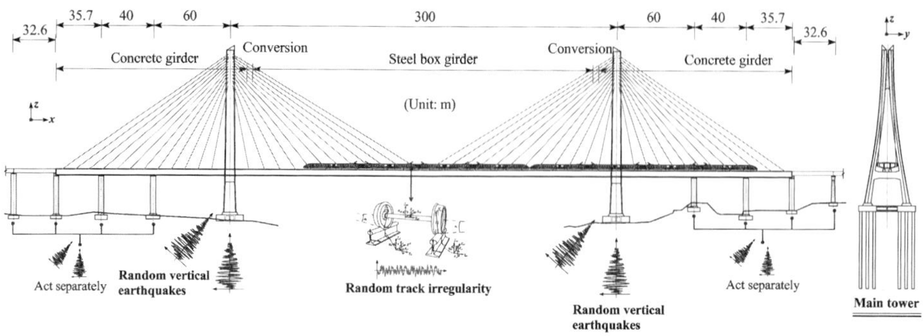

Fig. 3. Dynamic model of train-bridge coupled system with random track irregularities and random multi-point earthquake excitations with traveling waves.

in the following partitioned form

$$\begin{bmatrix} \mathbf{M}_{bb} & \mathbf{M}_{bs} \\ \mathbf{M}_{bs}^{\mathrm{T}} & \mathbf{M}_{ss} \end{bmatrix} \begin{Bmatrix} \ddot{\mathbf{X}}_b \\ \ddot{\mathbf{X}}_s \end{Bmatrix} + \begin{bmatrix} \mathbf{C}_{bb} & \mathbf{C}_{bs} \\ \mathbf{C}_{bs}^{\mathrm{T}} & \mathbf{C}_{ss} \end{bmatrix} \begin{Bmatrix} \dot{\mathbf{X}}_b \\ \dot{\mathbf{X}}_s \end{Bmatrix} + \begin{bmatrix} \mathbf{K}_{bb} & \mathbf{K}_{bs} \\ \mathbf{K}_{bs}^{\mathrm{T}} & \mathbf{K}_{ss} \end{bmatrix} \begin{Bmatrix} \mathbf{X}_b \\ \mathbf{X}_s \end{Bmatrix} = \begin{Bmatrix} \mathbf{F}_b(\boldsymbol{\xi}^r, t) \\ \mathbf{F}_s(\boldsymbol{\xi}^s, t) \end{Bmatrix}, \tag{6}$$

where $\mathbf{X}_s$ represents the support displacements of the cable-stayed bridge, i.e. those at its ends and the bottoms of the piers and main towers of the cable-stayed bridge; $\mathbf{X}_b$ represents all of the other bridge displacements; the superscript 'T' denotes the transpose matrix; $\mathbf{M}_{bb}$, $\mathbf{C}_{bb}$ and $\mathbf{K}_{bb}$ denote the mass, damping and stiffness matrices of the cable-stayed bridge system corresponding to $\mathbf{X}_b$; $\mathbf{M}_{ss}$, $\mathbf{C}_{ss}$ and $\mathbf{K}_{ss}$ denote the mass, damping and stiffness matrices of the cable-stayed bridge system corresponding to $\mathbf{X}_s$; $\mathbf{M}_{bs}$, $\mathbf{C}_{bs}$ and $\mathbf{K}_{bs}$ denote the mass, damping and stiffness interaction matrices of the cable-stayed bridge system between $\mathbf{X}_b$ and $\mathbf{X}_s$; $\mathbf{F}_b(\boldsymbol{\xi}^r, t)$ represents the force vector exerted on the cable-stayed bridge by the running train and random track irregularities excitations; $\mathbf{F}_s(\boldsymbol{\xi}^s, t)$ represents the random load vector subjected to random multi-point earthquake excitations with travelling waves that act on the cable-stayed bridge. The symbol $\boldsymbol{\xi}^s$ in the force vector $\mathbf{F}_s$ presents the uniform point design of random frequencies and random phases used to generate representation of earthquake excitation samples from power spectrum density (PSD).

The random load vector, $\mathbf{F}_b(\boldsymbol{\xi}^r, t)$, which is subjected to the influence of random track irregularities and the wheel axle load with an order of $N_b \times 1$, can be expressed in details as

$$\mathbf{F}_b(\boldsymbol{\xi}^r, t) = -\sum_{i=1}^{N_v} \sum_{k=1}^{4} \left(F_{rL,i}^z(\boldsymbol{\xi}^r, t)\, \boldsymbol{\Phi}_{ik}^{ZL} + F_{rR,i}^z(\boldsymbol{\xi}^r, t)\, \boldsymbol{\Phi}_{ik}^{ZR} \right) \tag{7}$$

$$
\begin{aligned}
&-\sum_{i=1}^{N_v}\sum_{k=1}^{4}\left(F_{xL,i}^{cr}\left(\boldsymbol{\xi}^r,t\right)\boldsymbol{\Phi}_{ik}^{XL}+F_{xR,i}^{z}\left(\boldsymbol{\xi}^r,t\right)\boldsymbol{\Phi}_{ik}^{XR}\right)\\
&+\sum_{i=1}^{N_v}\sum_{k=1}^{4}\left(0.5G_{axle}b_0^{-1}\left(\lambda_k^L(t)r^{YL}\left(\boldsymbol{\xi}^r,t\right)\boldsymbol{\Phi}_{ik}^{YL}\right.\right.\\
&\left.\left.+\lambda_k^R(t)r^{YR}\left(\boldsymbol{\xi}^r,t\right)\boldsymbol{\Phi}_{ik}^{YR}\right)\right)\\
&+\sum_{i=1}^{N_v}\sum_{k=1}^{4}\left(\varsigma_{wk}^{22L_{\dot{r}}^{YL}}\left(\boldsymbol{\xi}^r,t\right)\boldsymbol{\Phi}_{ik}^{YL}+\varsigma_{wk}^{22R_{\dot{r}}^{YR}}\left(\boldsymbol{\xi}^r,t\right)\boldsymbol{\Phi}_{ik}^{YR}\right.\\
&\left.-\sum_{p=L,R}F_{yp,i}^{cr}\left(\boldsymbol{\xi}^r,t\right)\boldsymbol{\Phi}_{ik}^{Yp}\right),
\end{aligned}
$$

where N_b denotes the DOFs of the displacement vector $\mathbf{X}_b$; $\boldsymbol{\Phi}_{ik}^{Xp}$, $\boldsymbol{\Phi}_{ik}^{Yp}$ and $\boldsymbol{\Phi}_{ik}^{Zp}(p = L, R)$ represents the decomposition-index vectors of wheel–rail force at the kth wheel-set of ith vehicle in the X, Y and Z position, which decompose the wheel/rail forces in Eq. (7) in the element into nodal forces.

2.3. *Train–bridge coupled system subjected to random excitations*

Considering the wheel/rail interactions in Fig. 3, the coupled equations of motions for the train–cable-stayed bridge system can be obtained by combining Eqs. (1) and (6) together. So, the coupled equation can be established and rewritten as follows:

$$
\begin{bmatrix}\mathbf{M}_v & 0 & 0\\ 0 & \mathbf{M}_{bb} & \mathbf{M}_{bs}\\ 0 & \mathbf{M}_{bs}^{\mathrm{T}} & \mathbf{M}_{ss}\end{bmatrix}\begin{Bmatrix}\ddot{\mathbf{X}}_v\\ \ddot{\mathbf{X}}_b\\ \ddot{\mathbf{X}}_s\end{Bmatrix}+\begin{bmatrix}\mathbf{C'}_v & \mathbf{C}_{vb} & 0\\ \mathbf{C}_{vb}^{\mathrm{T}} & \mathbf{C'}_{bb} & \mathbf{C}_{bs}\\ 0 & \mathbf{C}_{bs}^{\mathrm{T}} & \mathbf{C}_{ss}\end{bmatrix}\begin{Bmatrix}\dot{\mathbf{X}}_v\\ \dot{\mathbf{X}}_b\\ \dot{\mathbf{X}}_s\end{Bmatrix}
+\begin{bmatrix}\mathbf{K'}_v & \mathbf{K}_{vb} & 0\\ \mathbf{K}_{vb}^{\mathrm{T}} & \mathbf{K'}_{bb} & \mathbf{K}_{bs}\\ 0 & \mathbf{K}_{bs}^{\mathrm{T}} & \mathbf{K}_{ss}\end{bmatrix}\begin{Bmatrix}\mathbf{X}_v\\ \mathbf{X}_b\\ \mathbf{X}_s\end{Bmatrix}=\begin{Bmatrix}\mathbf{F'}_v\left(\boldsymbol{\xi}^r,t\right)\\ \mathbf{F'}_b\left(\boldsymbol{\xi}^r,t\right)\\ \mathbf{F}_s\left(\boldsymbol{\xi}^s,\mathbf{t}\right)\end{Bmatrix}, \quad (8)
$$

where $\mathbf{C}_{vb}$ and $\mathbf{K}_{vb}$ represent the damping and stiffness matrices of the wheel/rail interaction, respectively; $\mathbf{C'}_v$ and $\mathbf{K'}_v$ represent the damping and stiffness matrices of the train system when considering the wheel/rail interaction, respectively; $\mathbf{C'}_{bb}$ and $\mathbf{K'}_{bb}$ represent the damping and stiffness matrices of the cable-stayed bridge system subjected to the wheel/rail interaction, respectively; and $\mathbf{F'}_v\left(\boldsymbol{\xi}_q^r,t\right)$ and $\mathbf{F'}_b\left(\boldsymbol{\xi}_q^r,t\right)$, respectively denotes the random load vector acting on the vehicles and bridge systems involving the excitations of random track irregularities and the wheel axle load.

For more convenient expression, the random dynamic equation of Eq. (8) can be derived in the form as follows:

$$\begin{bmatrix} \mathbf{M}_{VB} & \mathbf{M}_{VBs} \\ \mathbf{M}_{VBs}^{\mathrm{T}} & \mathbf{M}_{ss} \end{bmatrix} \begin{Bmatrix} \ddot{\mathbf{X}}_{VB} \\ \ddot{\mathbf{X}}_{s} \end{Bmatrix} + \begin{bmatrix} \mathbf{C}_{VB} & \mathbf{C}_{VBs} \\ \mathbf{C}_{VBs}^{\mathrm{T}} & \mathbf{C}_{ss} \end{bmatrix} \begin{Bmatrix} \dot{\mathbf{X}}_{VB} \\ \dot{\mathbf{X}}_{s} \end{Bmatrix} + \begin{bmatrix} \mathbf{K}_{VB} & \mathbf{K}_{VBs} \\ \mathbf{K}_{VBs}^{\mathrm{T}} & \mathbf{K}_{ss} \end{bmatrix} \begin{Bmatrix} \mathbf{X}_{VB} \\ \mathbf{X}_{s} \end{Bmatrix} = \begin{Bmatrix} \mathbf{0} \\ \mathbf{F}_{s}(\boldsymbol{\xi}^{s},\mathbf{t}) \end{Bmatrix} + \begin{Bmatrix} \mathbf{F}_{VB}(\xi^{r},t) \\ \mathbf{0} \end{Bmatrix}, \tag{9}$$

where the displacement vector is $\{\mathbf{X}_{VB}\} = \{\mathbf{X}_v \quad \mathbf{X}_b\}^{\mathrm{T}}$, the random load vector of the train–cable stayed bridge interaction system involving the excitations of random track irregularities and the wheel axle load is $\mathbf{F}_{VB}(\boldsymbol{\xi}^r,t) = \{\mathbf{F}'_v(\boldsymbol{\xi}^r,t) \quad \mathbf{F}'_b(\boldsymbol{\xi}^r,t)\}^{\mathrm{T}}$, and the respective mass, damping and stiffness matrices are written as follows:

$$\mathbf{M}_{VB} = \begin{bmatrix} \mathbf{M}_v & \mathbf{0} \\ \mathbf{0} & \mathbf{M}_b \end{bmatrix}, \quad \mathbf{C}_{VB} = \begin{bmatrix} \mathbf{C}'_v & \mathbf{C}_{vb} \\ \mathbf{C}_{vb}^{\mathrm{T}} & \mathbf{C}'_b \end{bmatrix}, \quad \mathbf{K}_{VB} = \begin{bmatrix} \mathbf{K}'_v & \mathbf{K}_{vb} \\ \mathbf{K}_{vb}^{\mathrm{T}} & \mathbf{K}'_b \end{bmatrix},$$

$$\mathbf{M}_{VBs} = [\mathbf{0},\mathbf{M}_{bs}]^{\mathrm{T}}, \quad \mathbf{C}_{VBs} = [\mathbf{0},\mathbf{C}_{bs}]^{\mathrm{T}}, \quad \mathbf{K}_{VBs} = [\mathbf{0},\mathbf{K}_{bs}]^{\mathrm{T}}.$$

2.3.1. *Random excitations subject to earthquake*

In order to easily understand the dynamic characteristic of Eq. (9), using the superposition principle of linear systems, the Eq. (9) can be calculated separately by adding these different excitations $\mathbf{F}_s(\boldsymbol{\xi}^s,t)$ and $\mathbf{F}_{VB}(\boldsymbol{\xi}^r,t)$. Therefore, the random dynamic equation of motion of such cable-stayed bridge under high-speed trains subjected to multi-point earthquakes can be extracted from Eq. (9) as

$$\begin{bmatrix} \mathbf{M}_{VB} & \mathbf{M}_{VBs} \\ \mathbf{M}_{VBs}^{\mathrm{T}} & \mathbf{M}_{ss} \end{bmatrix} \begin{Bmatrix} \ddot{\mathbf{X}}_{VB}^{e} \\ \ddot{\mathbf{X}}_{s}^{e} \end{Bmatrix} + \begin{bmatrix} \mathbf{C}_{VB} & \mathbf{C}_{VBs} \\ \mathbf{C}_{VBs}^{\mathrm{T}} & \mathbf{C}_{ss} \end{bmatrix} \begin{Bmatrix} \dot{\mathbf{X}}_{VB}^{e} \\ \dot{\mathbf{X}}_{s}^{e} \end{Bmatrix} + \begin{bmatrix} \mathbf{K}_{VB} & \mathbf{K}_{VBs} \\ \mathbf{K}_{VBs}^{\mathrm{T}} & \mathbf{K}_{ss} \end{bmatrix} \begin{Bmatrix} \mathbf{X}_{VB}^{e} \\ \mathbf{X}_{s}^{e} \end{Bmatrix} = \begin{Bmatrix} \mathbf{0} \\ \mathbf{F}_{s}(\boldsymbol{\xi}^{s},t) \end{Bmatrix}, \tag{10}$$

where the superscript 'e' shows that the random response is caused by earthquake excitations.

Normally, the absolute displacement vector $\mathbf{X}_{VB}^{e}$ can be decomposed into two parts,[43] which are expressed as follows:

$$\mathbf{X}_{VB}^{e} = \mathbf{X}_{VB}^{eI} + \mathbf{X}_{VB}^{ed}, \tag{11}$$

where $\mathbf{X}_{VB}^{eI}$ is the quasi-static displacement vector caused by earthquakes, and $\mathbf{X}_{VB}^{ed}$ is the dynamic displacement vector caused by inertial forces subjected to earthquakes.[24,44] So putting the quasi-static displacement vector $\mathbf{X}_{VB}^{eI}$ into Eq. (10) satisfies

$$\mathbf{X}_{VB}^{eI} = -\mathbf{K}_{VB}^{-1}\mathbf{K}_{VBs}\mathbf{X}_s^e \equiv \mathbf{R}_{VBs}\mathbf{X}_s^e. \tag{12}$$

Substituting Eqs. (11) and (12) into Eq. (10), and it can be rewritten as

$$\begin{aligned}&\mathbf{M}_{VB}\ddot{\mathbf{X}}_{VB}^{ed} + \mathbf{C}_{VB}\dot{\mathbf{X}}_{VB}^{ed} + \mathbf{K}_{VB}\mathbf{X}_{VB}^{ed}\\ &\quad = -(\mathbf{M}_{VB}\mathbf{R}_{VBs} + \mathbf{M}_{VBs})\ddot{\mathbf{X}}_s^e - (\mathbf{C}_{VB}\mathbf{R}_{VBs} + \mathbf{C}_{VBs})\dot{\mathbf{X}}_s^e.\end{aligned} \tag{13}$$

Due to the high-speed train running across the cable-stayed bridge, all of the matrices in Eq. (13) are time-dependent.

Making reference to Ref. 24, the dynamic Eq. (13) cannot be reduced to the conventional form where $\mathbf{X}_s^e$ represents uniform ground displacements,[45] since Eq. (13) implies that the damping forces are proportional to absolute velocity vector $\left\{\dot{\mathbf{X}}_{VB}^{eT} \quad \dot{\mathbf{X}}_s^{eT}\right\}^{\mathrm{T}}$. So to avoid the inconsistency on this, the damping forces should be assumed to be proportional to the relative velocity vector $\left\{\dot{\mathbf{X}}_{VB}^{eT} \quad 0\right\}^{\mathrm{T}}$ in Eq. (13). This has been proven in Section 30.2.4 in Ref. 46.

Additionally, the earthquake displacement vector $\mathbf{X}_s^e$ here corresponds to the random load vector $\mathbf{F}_s(\boldsymbol{\xi}^s, t)$, which is subjected to random multi-point earthquake excitations with traveling waves in Eq. (9). So when considering the influence of random parameters $\boldsymbol{\xi}^s$ in the earthquake excitations, the following equation for the train–bridge system under random earthquakes can be obtained

$$\begin{aligned}&\mathbf{M}_{VB}\ddot{\mathbf{X}}_{VB}^{ed} + \mathbf{C}_{VB}\dot{\mathbf{X}}_{VB}^{ed} + \mathbf{K}_{VB}\mathbf{X}_{VB}^{ed}\\ &\quad = -(\mathbf{M}_{VB}\mathbf{R}_{VBs} + \mathbf{M}_{VBs})\ddot{\mathbf{X}}_s^e(\boldsymbol{\xi}^s, t) = \mathbf{P}_{\ddot{\mathbf{X}}^e}\ddot{\mathbf{X}}_s^e(\boldsymbol{\xi}^s, t).\end{aligned} \tag{14}$$

One thing to note is that the damping matrices on the right hand of dynamic Eq. (14) have been neglected. This is considered to be reasonable as Clough and Penzien[45] have pointed out that the damping terms on the right hand side of this kind of equation, like Eq. (14), can be neglected because they make little contribution to the effective load of a relatively lowly damped system, of which the train–bridge system here is considered to be. By taking the earthquake excitations $\ddot{\mathbf{X}}_s^e(\boldsymbol{\xi}^s, t)$ to be nonuniform, Eq. (14) can be used to study the random dynamic characteristic of train–cable stayed bridge system under multi-point wave passage effort.

2.3.2. *Random excitation subjected to track irregularities*

If only the random excitation of track irregularities is considered in Eq. (9), there will be no earthquake displacement, in this case, $\mathbf{X}_s^e = \mathbf{0}$. Therefore, the random dynamic equation of motion for the cable-stayed bridge under high-speed train due to random track irregularities can be written as

$$\mathbf{M}_{VB}\ddot{\mathbf{X}}_{VB}^{r} + \mathbf{C}_{VB}\dot{\mathbf{X}}_{VB}^{r} + \mathbf{K}_{VB}\mathbf{X}_{VB}^{r} = \mathbf{F}_{VB}\left(\boldsymbol{\xi}^{s}, t\right), \tag{15}$$

where the superscript 'r' represents that the random responses are caused by the excitations of random track irregularities.

When combining the random excitation of earthquake and track irregularities, the random dynamic equation of motion of the train–bridge system can be rewritten as

$$\mathbf{M}_{VB}\ddot{\mathbf{X}}_{VB} + \mathbf{C}_{VB}\dot{\mathbf{X}}_{VB} + \mathbf{K}_{VB}\mathbf{X}_{VB} = \mathbf{P}_{\ddot{\mathbf{X}}^{e}}\ddot{\mathbf{X}}_{s}^{e}\left(\boldsymbol{\xi}^{s}, t\right) + \mathbf{F}_{VB}\left(\boldsymbol{\xi}^{r}, t\right). \tag{16}$$

In the next section, the point design method is introduced to generate the representation excitation samples of the earthquakes $\ddot{\mathbf{X}}_{s}^{e}\left(\boldsymbol{\xi}^{s}, t\right)$ and the track irregularities in the force vector of $\mathbf{F}_{VB}\left(\boldsymbol{\xi}^{r}, t\right)$, and the random dynamic Eq. (16) is solved by using PDEM.

3. Uncertainty Parameter Design and the Representative Excitation Samples Generation

After establishing the dynamic equation of motion of the train–bridge coupled system, the next step is to generate the random track irregularity and nonstationary earthquake excitations. As there exist a large number of multi-dimensional uncertainty parameters, the point design of which is always considered to be the key technology to improve the computational efficiency and accuracy when taking the solutions in the stochastic processes.

In this study, a stochastic harmonic function (SHF)[47] with a nonstationary stochastic process,[48] of which the power spectrum density (PSD) of the generated target samples which is precisely proven to be equivalent to the original one,[28] was employed to generate the representation of random nonstationary earthquake samples by power spectrum density that involves the random spatial frequencies and phases, as well as the stationary track irregularity samples.

Here, as mentioned in Sec. 2.3, there exist two kinds of uncertainty parameters $\boldsymbol{\xi}^{s}$ and $\boldsymbol{\xi}^{r}$, which both represent the uniform point set vectors of random frequencies and random phases used to generate the representation track irregularity samples and the representation earthquake samples from the power spectrum.

3.1. *Uniform point design of multi-dimensional parameters*

The difficulty of uniform point design of high dimensional uncertainty parameters has troubled researchers for decades, and the higher parameter dimensions will distinctly increase the point set deviation and decrease the calculation accuracy. Herein, considering the dimension of the parameters within the allowable range of accuracy, the number theoretical method (NTM)[49,50] in multi-dimensional hypercube, of which the point set design methods involving good point (GP), good lattice point (GLP), the Hua–Wang method,[51] is introduced to get the discrete point design of uncertainty parameters set.

Assume that there is a random vector $\mathbf{P}_S$, in an S-dimensional hypercube $\mathbf{C}^S = [0, \ \ 1]^S$, generated using the good point method (GP),[52] which is derived as

$$\mathbf{P}_S = \left\{\boldsymbol{\chi}_q = [\chi_{q,1}, \chi_{q,2}, \ldots, \chi_{q,S}] \in \mathbf{C}^S, q = 1, 2, \ldots, n_{pt}\right\} \tag{17}$$

where the point set$\boldsymbol{\chi}_q$ in $\mathbf{C}^S$ is uniformly designed. The S-dimensional hypercube point set in Eq. (18) is generated by using the square root method of sequence.[49]

$$\chi_{q,i} = q\wp^{\frac{i}{S+1}} - \int \left(q\wp^{\frac{i}{S+1}}\right), \tag{18}$$

where $\chi_{q,i} \in (0, 1)$ $(i = 1, 2, \ldots, S; q = 1, 2, \ldots, n_{pt})$ is the uniformly discrete point set in the S-dimensional hypercube, $\wp$ is one of the prime numbers, $\text{int}(\cdot)$ denotes the integer part of $q\wp^{\frac{i}{S+1}}$. Thus, the point vector $\boldsymbol{\chi}_q$ can be used to form the point design of uncertainty parameters in the following section.

Without loss of generality, we can take the symbol $\boldsymbol{\Theta}$ to represent the random state vector space that includes the uncertainty parameters. Thus, there exists

$$\boldsymbol{\Theta} = \left\{\boldsymbol{\xi}_q = [\xi_{q,1}, \xi_{q,2}, \ldots, \xi_{q,N}] \in \{\boldsymbol{\xi}^s, \boldsymbol{\xi}^r\}, q = 1, 2, \ldots, n_{pt}\right\}, \tag{19}$$

where $\xi_{q,i}(i = 1, 2, \cdots, N)$ denotes one of these uncertainty parameters of random frequencies and random phases used to generate representation excitation samples. n_{pt} is the total number of representation samples, and N is the total dimension of random parameters.

In detail, Eq. (19) can be expressed as follows:

$$\{\boldsymbol{\xi}^s, \boldsymbol{\xi}^r\} = \left\{\begin{matrix} \left\{\Omega^{s,I}_{q,1}, \Omega^{s,I}_{q,1}, \cdots, \Omega^{s,I}_{q,N_1}, \phi^{s,I}_{q,1}, \phi^{s,I}_{q,1}, \cdots, \phi^{s,I}_{q,N_1}\right\} \in \boldsymbol{\xi}^s_q; I = h, z \\ \left\{\Omega^{r,J}_{q,1}, \Omega^{r,J}_{q,1}, \cdots, \Omega^{r,J}_{q,N_2}, \phi^{r,J}_{q,1}, \phi^{r,J}_{q,1}, \cdots, \phi^{r,J}_{q,N_2}\right\} \in \boldsymbol{\xi}^r_q; J = z, \theta, y, \varphi \end{matrix}\right\}, \tag{20}$$

where $q = 1, 2, \cdots, n_{pt}$, $N = 2 \times N_1 + 4 \times N_2$, there are random frequencies $\Omega^s_{q,i} \in \boldsymbol{\xi}^s$ and random phases $\phi^s_{q,i} \in \boldsymbol{\xi}^s$ $(i = 1, 2, \ldots, N_1)$ with uniform probability distributions for the earthquake functions, as well as random spatial frequencies $\Omega^r_{q,i} \in \boldsymbol{\xi}^r$ and random phases $\phi^r_{q,i} \in \boldsymbol{\xi}^r$ $(i = 1, 2, \ldots, N_2)$ with uniform probability distributions for track irregularities, all of which can be assumed to have uniform probability distributions. In Eq. (20), the superscripts 'h' and 'z' denote lateral and vertical earthquake excitations, and the superscripts 'z', 'θ', 'y' and 'φ' separately denote the vertical profile, cross-level, lateral alignment and lateral track-distance track irregularities.

Normally, it is impossible to get a good solution on the point design of the high dimensional uncertainty parameters in Eq. (20) unless through the approximate processing. Meanwhile, since the power spectra of earthquake and track irregularity are independent of each other, one can assume that each kind of these representation excitation sample generation uses the same point design in Eqs. (17) and (18). Referring to Ref. 47, the stochastic harmonic function (SHF) for the stochastic evolutionary process, which contains random frequencies, random phases, and random amplitudes that are determined by frequencies, could reproduce the exact target auto-correlation function and the power spectrum density functions, and its minimum dimension used for the frequency truncation to ensure enough precision for calculation is 10. In this study, the number for the dimension of uncertainty parameters was at least ≥ 50, which would be accurate enough both in the representational track irregularity samples generation[30] and the earthquake samples generation.

Therefore, the discrete representative points of uncertainty parameters $\boldsymbol{\xi}_q \in \boldsymbol{\Theta}$, which are expressed in Eqs. (19) and (20), can be rewritten as

$$\xi_{q,k} = \hbar_k(\boldsymbol{\Omega}, \boldsymbol{\phi}, \chi_{q,i}) \in \{\boldsymbol{\xi}^s, \boldsymbol{\xi}^r\}, \tag{21}$$

where $k = 1, 2, \ldots, N$; $i = 1, 2, \ldots, S$. $\hbar_k(\boldsymbol{\Omega}, \boldsymbol{\phi}, \chi_{q,i})$ denote the conversion function of the random variables that are related to random frequencies $\boldsymbol{\Omega}$ and random phases $\boldsymbol{\phi}$. Further, there is

$$\hbar_k(\boldsymbol{\Omega}, \boldsymbol{\phi}, \chi_{q,i}) = F^{-1}_{\boldsymbol{\Omega},\boldsymbol{\phi}}(\chi_{q,i}), \tag{22}$$

where $F^{-1}_{\boldsymbol{\Omega},\boldsymbol{\phi}}(\cdot)$ denotes the inverse function of the distribution function of the random variable Ω or $\boldsymbol{\phi}$, this method is often referred to as the inverse transform method.[53]

Therefore, referring to Ref. 28, take $\mathbf{U}$ as the unit vector of interested stochastic dynamic response, the joint probability density function for the random vectors $\boldsymbol{\xi}_q$ is $p_\Theta(\mathbf{U}, \boldsymbol{\xi}_q, t)$, and the initial probability of each designed point set is P_q. Then, there exists

$$\sum_{q=1}^{n_{pt}} P_q(\mathbf{U}, t) = \sum_{q=1}^{n_{pt}} \int_{\boldsymbol{\xi}_q \in \boldsymbol{\Theta}} p_{\mathbf{U}\Theta}(u, \boldsymbol{\xi}_q, t) \mathrm{d}\boldsymbol{\xi}$$
$$= \int_{\bigcup_{q=1}^{n_{pt}} \boldsymbol{\xi}_q = \boldsymbol{\Theta}} p_{\mathbf{U}\Theta}(u, \boldsymbol{\xi}_q, t) \mathrm{d}\boldsymbol{\xi} = \int_{\boldsymbol{\Theta}} p_{\mathbf{U}\Theta}(u, \boldsymbol{\xi}_q, t) d\boldsymbol{\xi} = \mathbf{1}. \tag{23}$$

The initial conditions of each design are correspondingly partially discretized as

$$p_{\mathbf{U}\Theta}(\boldsymbol{u}, \xi_q, t)|_{t=t_0} = \delta(\boldsymbol{u} - \boldsymbol{u}_0) p_{\mathbf{U}\Theta}(\xi_q, t)|_{t=t_0} = \delta(\boldsymbol{u} - \boldsymbol{u}_0) P_q, \tag{24}$$

where $\delta(\cdot)$ is the Dirac delta function.

3.2. *Random representational samples of earthquakes and track irregularities*

Based on PDEM, the stochastic harmonic function (SHF) for nonstationary stochastic processes[47,48] is employed to generate the representational nonstationary random earthquake samples with initial probability information, of which the PSD is reproduced to match the exact target PSD.

Referring to Ref. 54, considering a two-dimensional nonstationary representative stochastic process of earthquake samples in the lateral and horizontal positions, the following can be written

$$\ddot{x}_s^I(\boldsymbol{\xi}^s, t) = A_I(\boldsymbol{\Omega}, t) g^I(\boldsymbol{\xi}^s, t) \quad I = h, z, \tag{25}$$

where $\ddot{x}_s^h(\boldsymbol{\xi}^s, t)$ and $\ddot{x}_s^z(\boldsymbol{\xi}^s, t)$ denote the lateral and vertical nonstationary earthquake excitations. $A_I(\omega, t)$, $I = h, z$ denote the modulating functions that involve frequency ω and time t. And $g^I(\boldsymbol{\xi}^s, t)$, $I = h, z$ denote the stationary earthquake excitations that are generated by SHF, written as

$$g^I(\boldsymbol{\xi}^s, t) = \sum_{i=1}^{N} \sqrt{2S_I(\Omega_{q,i}^{s,I})\Delta\Omega_{q,i}^{s,I}} \cos(\Omega_{q,i}^{s,I} t + \phi_{q,i}^{s,I}) \quad I = h, z, \tag{26}$$

where $q = 1, 2, \cdots, n_{pt}$, $S_h(\Omega_\Theta^{s,h})$ and $S_z(\Omega_\Theta^{s,z})$ are set as PSDs of the lateral and vertical earthquake. Referring to Ref. 48, $\sqrt{2S_I(\Omega_{q,i}^{s,I})\Delta\Omega_{q,i}^{s,I}}$, $I = h, z$ are the random amplitude functions, $\Omega_{\Theta,i}^{s,I} \in \left[\Omega_{q,i-1}^{s,I(p)}, \Omega_{q,i}^{s,I(p)}\right]$ is mutually exclusive, it satisfied $\bigcup_{i=1}^{N} \left[\Omega_{q,i-1}^{s,I(p)}, \Omega_{q,i}^{s,I(p)}\right] = \left[\Omega_L^{s,I}, \Omega_u^{s,I}\right]$ with $\Omega_L^{s,I}$ and $\Omega_u^{s,I}$ as the upper and lower frequencies, and $\mu\left\{\left[\Omega_{q,i-1}^{s,I(p)}, \Omega_{q,i}^{s,I(p)}\right] \cap \left[\Omega_{q,\varphi-1}^{s,I(p)}, \Omega_{q,\varphi}^{s,I(p)}\right]\right\} = 0$ for $i \neq \varphi$; $\Delta\Omega_{q,i}^{s,I} = \Omega_{q,i}^{s,I(p)} - \Omega_{q,i-1}^{s,I(p)}$, and $\phi_{q,i}^{s,I}$, $I = h, z$ independent uniformly distributed random variable over $[0, 2\pi]$. For details, combined with Eqs. (17) to (20), the uniformly designed random frequencies and phases can be written as

$$\begin{cases} \Omega_{q,i}^{s,I} = \Omega_{q,i-1}^{s,I(p)} + [\Omega_{q,i}^{s,I(p)} - \Omega_{q,i-1}^{s,I(p)}]\chi_{q,i} \\ \phi_{q,i}^{I} = 2\pi\chi_{q,i+N_1} \end{cases} \quad I = h, z \tag{27}$$

in which, $i = 1, 2, \cdots, N_1$ with $S = 2N_1$.

Therefore, the representative nonstationary stochastic functions of earthquake excitations with $g^h(\boldsymbol{\xi}^s, t)$ and $g^z(\boldsymbol{\xi}^s, t)$ in Eq. (25) having the mean value equal close to zero are:

$$E\left[g^I(\boldsymbol{\xi}^s, t)\right] \cong 0 \quad I = h, z \tag{28}$$

Assume that phase-lags of earthquake excitations exist between the bridge piers, so that the earthquake acceleration vector $\ddot{\mathbf{X}}_s^e(\boldsymbol{\xi}^s, t)$ in Eq. (14) can be written as

$$\begin{aligned} \ddot{\mathbf{X}}_s^e(\boldsymbol{\xi}^s, t) &= \left[\left\{\ddot{\mathbf{x}}_s^h(\boldsymbol{\xi}^s, t)\right\}^{\mathrm{T}} \quad \{\ddot{\mathbf{x}}_s^z(\boldsymbol{\xi}^s, t)\}^{\mathrm{T}}\right]^{\mathrm{T}} \\ &= \operatorname{diag}\left[\mathbf{A}_h(\Omega, t) \quad \mathbf{A}_z(\Omega, t)\right] \left[\left\{\ddot{\mathbf{g}}_s^h(\boldsymbol{\xi}^s, t)\right\}^{\mathrm{T}} \quad \{\ddot{\mathbf{g}}_s^z(\boldsymbol{\xi}^s, t)\}^{\mathrm{T}}\right]^{\mathrm{T}}, \end{aligned} \tag{29}$$

where

$$\mathbf{A}_h(\Omega, t) = \mathrm{diag}[\, A_h(\Omega, t - t_1)\; A_h(\Omega, t - t_2) \cdots A_h(\Omega, t - t_M)\,],$$

$$\ddot{\mathbf{g}}^I_s(\boldsymbol{\xi}^s, t) = [\ddot{g}^I_s(\boldsymbol{\xi}^s, t - t_1) \quad \ddot{g}^I_s(\boldsymbol{\xi}^s, t - t_2) \quad \cdots \quad \ddot{g}^I_s(\boldsymbol{\xi}^s, t - t_M)], \quad I = h, z,$$

where M is the number of bridge piers in the cable-stayed bridge.

Based on the point design from Eqs. (17)–(18) and the similar method used in Ref. 31, the random frequencies $\Omega^J_{q,i} \in \boldsymbol{\xi}^r$ and random phases $\phi^J_{q,i} \in \boldsymbol{\xi}^r$, $J = z, \theta, y, \varphi$, that are both used to generate the representative track irregularity samples can be designed as follows:

$$\begin{cases} \Omega^{r,J}_{q,i} = \left(\left(\Omega^{r,J(p)}_{q,i-1}\right)^{-1} + \left[\left(\Omega^{r,J(p)}_{q,i-1}\right)^{-1} - \left(\Omega^{r,J(p)}_{q,i}\right)^{-1} \right] \chi_{q,i} \right)^{-1} & J = z, \theta, y, \varphi, \\ \phi^{r,J}_{q,i} = 2\pi \chi_{q,i+N_1}, \end{cases} \tag{30}$$

where $\Omega^{r,J(p)}_{q,i-1}$, $J = z, \theta, y, \varphi$, are the dividing frequencies for the track spectrum that satisfies $\Omega^{r,J}_L < \Omega^{r,J(p)}_{q,1} < \Omega^{r,J(p)}_{q,2} < \cdots < \Omega^{r,J(p)}_{q,N_1-1} < \Omega^{r,J}_u$, $J = z, \theta, y, \varphi$, and $\Omega^{r,J}_u$ and $\Omega^{r,J}_L$ are the upper and lower cut-off spatial frequencies, respectively. Thus, the random track irregularity is written as

$$\gamma_J(\boldsymbol{\xi}^r, t) = \sum_{i=1}^{N_1} \sqrt{2S_{r,J}(\Omega^{r,J}_{q,i})\Delta\Omega^{r,J}_{q,i}} \cos(\Omega^{r,J}_{q,i} V t + \phi^{r,J}_{q,i}) \quad J = z, \theta, y, \varphi, \tag{31}$$

where $\gamma_z(\boldsymbol{\xi}^r_q,t)$, $\gamma_\theta(\boldsymbol{\xi}^r_q,t)$, $\gamma_y(\boldsymbol{\xi}^r_q,t)$ and $\gamma_\varphi(\boldsymbol{\xi}^r_q,t)$ denote the random track irregularities in the vertical, cross-level, lateral and track-distance positions, respectively.

Therefore, the vertical track irregularity of the left and right rail is $r^{Zp}(\boldsymbol{\xi}^r, t)\, r^{Zp}(\boldsymbol{\xi}^r,t) = \gamma_z(\boldsymbol{\xi}^r,t) \pm \gamma_\theta(\boldsymbol{\xi}^r,t)b_0$, $p = L, R$, and the lateral track irregularity of the left and right rail is $r^{Y(L,R)}(\boldsymbol{\xi}^r_q,t) = \gamma_y(\boldsymbol{\xi}^r_q,t) \pm \gamma_v(\boldsymbol{\xi}^r_q,t)/2$. And b_0 is half the width of the lateral track.

4. Solution and Dynamic Reliability Assessment

As mentioned before, the probability density evolution method (PDEM) proposed by Li and Chen as a new stochastic approach[28] can capture the probability evolution information with its evolutionary criterion based on the physical randomness of systems.

Without a loss of generality, similar to Refs. 28 and 33, take $\mathbf{U} = (u_1, u_2, \cdots, u_{nU})^T$ to be the connection with the random dynamic response vector $\mathbf{X}_{VB}$. Therefore, the state vector of the random train–bridge coupled system in Eq. (16) can be written as

$$\mathbf{U} = \mathbf{G}(\mathbf{X}_{VB}, \dot{\mathbf{X}}_{VB}, \ddot{\mathbf{X}}_{VB}, \boldsymbol{\xi}_q \in \boldsymbol{\Theta}, t), \tag{32}$$

$$u_l = G_l(\mathbf{X}_{VB}, \dot{\mathbf{X}}_{VB}, \ddot{\mathbf{X}}_{VB}, \boldsymbol{\xi}_q \in \boldsymbol{\Theta}, t), \quad (l = 1, 2, \ldots, n_U), \tag{33}$$

where $\mathbf{G}(\cdot)$ denotes the transfer operator. u_l, $l = 1, 2, \cdots, n_U$ is the sub-element of the state vector $\mathbf{U}$, and n_U is the total number of sub-elements.

Likewise, the velocity of state vector $\mathbf{U}$ can be assumed as

$$\dot{\mathbf{U}} = g(\mathbf{X}_{VB}, \dot{\mathbf{X}}_{VB}, \ddot{\mathbf{X}}_{VB}, \boldsymbol{\xi}_q \in \boldsymbol{\Theta}, t), \tag{34}$$

$$\dot{u}_l = g_l(\mathbf{X}_{VB}, \dot{\mathbf{X}}_{VB}, \ddot{\mathbf{X}}_{VB}, \boldsymbol{\xi}_q \in \boldsymbol{\Theta}, t) \quad (l = 1, 2, \ldots, n_U), \tag{35}$$

where $\mathbf{g}(\cdot)$ denotes the transfer operator. Obviously, $\mathbf{g}(\cdot) = \partial \mathbf{G}(\cdot)/\partial t$.

Therefore, based on Eq. (31), set $p_{\mathrm{U}\Theta}(\boldsymbol{u}, \boldsymbol{\xi}_q, t)$ as the joint probability density function of $\mathbf{U}$ which yields

$$p_{\mathrm{U}\Theta}(\boldsymbol{u}, \boldsymbol{\xi}_q, t) = \delta(u - G_l(\mathbf{X}_{VB}, \dot{\mathbf{X}}_{VB}, \ddot{\mathbf{X}}_{VB}, \boldsymbol{\xi}_q \in \boldsymbol{\Theta}, t)) p_{\Theta}(\boldsymbol{\xi}_q). \tag{36}$$

Based on the *Reynold* transport theorem and the related derivation, noting the arbitrariness of $D_t \times D_\Theta$ and regarding the train–cable-stayed bridge coupled system as a probabilistic conservative system, the generalized probability density evolution equation[28] is obtained as

$$\frac{\partial p_{U\boldsymbol{\Theta}}(u, \boldsymbol{\xi}_q, t)}{\partial t} + \dot{\boldsymbol{u}}(\boldsymbol{\xi}_q, t) \frac{\partial p_{U\boldsymbol{\Theta}}(u, \boldsymbol{\xi}_q, t)}{\partial u} = 0, \tag{37}$$

where the initial condition reads

$$p_{\mathrm{U}\Theta}(u, \boldsymbol{\xi}_q, t)|_{t=0} = \delta\left(u - G_l(\mathbf{X}_{VB0}, \dot{\mathbf{X}}_{VB0}, \ddot{\mathbf{X}}_{VB0}, \boldsymbol{\xi}_q \in \boldsymbol{\Theta}, t)\right) p_{\Theta}(\boldsymbol{\xi}_q). \tag{38}$$

Then, the partial differential equations composed of Eqs. (36) and (37) can be solved by the bilateral difference method using the total variation diminishing (TVD)[29] and the solution of joint probabilistic function $p_{U\boldsymbol{\Theta}}(\boldsymbol{u}, \xi_q, t)$ can be obtained. The joint probability density function of the random system response is given as

$$p_{\mathrm{U}}(\boldsymbol{u}, t) = \int_{\mathrm{U}\boldsymbol{\Theta}} p_{U\boldsymbol{\Theta}}(\boldsymbol{u}, \boldsymbol{\xi}_q, t) d\boldsymbol{\xi}_q. \tag{39}$$

Based on the time-variant joint probability density functions of random dynamic responses of the cable-stayed bridge under the high speed train subjected to excitations involving random earthquake and random track irregularities, it is convenient to get their reliability assessment which will greatly improve the comprehensibility of analytical results.

According to the probability density function $p_{\mathrm{U}}(\boldsymbol{u}, t)$ in Eq. (38), the mean values $E(\boldsymbol{u}, t)$ of the random dynamic response is written as

$$E(\boldsymbol{u}, t) = \int_{-\infty}^{\infty} u p_{\mathrm{U}}(\boldsymbol{u}, t) du. \tag{40}$$

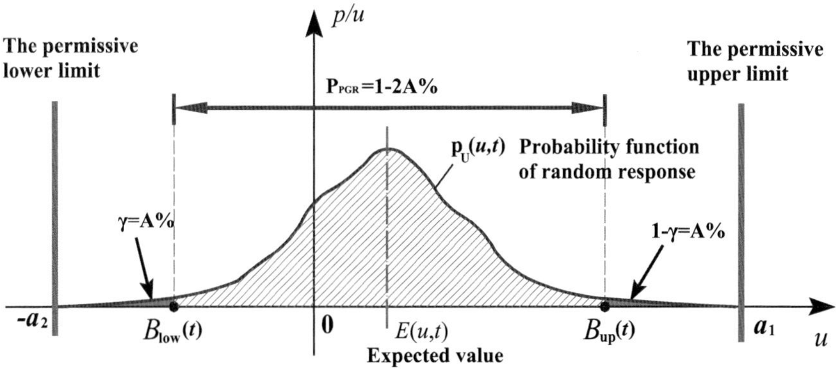

Fig. 4. The estimation of uncertainty bounds using quantiles of the probability distribution.

When the upper boundary of random dynamic responses $\boldsymbol{u}_{\text{Limit}}$ is considered, similar to Ref. 55, the train–bridge system dynamic reliability can be written as

$$R_{\text{U}\Theta}(t) = \text{P}\left\{|u_l(t)| < u_{\text{Limit}}, l = 1, 2, \cdots, n_U, t = [0, T]\right\}, \tag{41}$$

where P$\{\cdot\}$ denotes the time-varying dynamic probability function of the random responses $u_l(t)$. T is the total time period of random response.

Obviously, it is easy to obtain the time-varying reliability function $R_{\text{U}\Theta}(t)$ under the upper boundary conditions, which is written as

$$R_{\text{U}\Theta}(t) = \int_{-u_{\text{Limit}}}^{u_{\text{Limit}}} p_{\text{U}}(\boldsymbol{u}, t)du = \int_{\text{U}\Theta}\int_{-u_{\text{Limit}}}^{u_{\text{Limit}}} p_{U\Theta}(\boldsymbol{u}, \xi_q, t)\text{d}u\text{d}\xi_q. \tag{42}$$

Therefore, the time-varying dynamic reliability function $R_{\text{U}\Theta}(t)$ is calculated, of which the minimum value during $t \in [0, T]$ can be considered as the system dynamic reliability. The calculated $R_{\text{U}\Theta}(t)$ is the instantaneous reliability of the structure.

The probabilistic guarantee bounds for the response can be more easily understood through Fig. 4, where $\boldsymbol{A}\%$ is the quantile, $\mathbf{p}_{PGR}$ is the probability guarantee rate, $B_{\text{up}}(t)$ and $B_{\text{low}}(t)$ are the upper and lower bounds of the response probability guarantee rate, respectively, and the shaded part of the figure represents the probability guarantee rate of the response.

5. Case Studies

5.1. *Basic model information*

Consider the CRH3 high-speed train traversing across the cable-stay bridge mentioned in Sec. 2.2, in which the bridge model consists of the main span with a length of 35.5 + 40 + 60 + 300 + 60 + 40 + 35.5 = 571 m and separately two spans of 32 m prestressed simply supported concrete bridge, as shown in Fig. 3. As the calculation of train–bridge coupled interaction is solved based on the program developed on MATLAB, a finite element model based on MIDAS is established to verify the veracity of the cable-stayed bridge model, as shown in Fig. B.1 in Appendix B. The material parameters of its deck are shown in Table A.3 in Appendix A. And the geometry characteristics of major sections of the concrete or steel girders are shown in Table A.4 in Appendix A.

The cable-stayed bridge was assumed to have a Rayleigh damping matrix. The girder, concrete tower and piers were discretized using 3D Bernoulli–Euler beam elements, with the deck sub-divided into the identical elements of each length of 1.5 m. The first 10 modes of the cable-stayed bridge were included for verification with the MATLAB model. The first vertical and lateral natural frequencies of the deck were 0.911 rad/s (0.147 Hz) and 3.36 rad/s (0.535 Hz), respectively. The CRH3 high-speed train consists of eight locomotive cars (denoted as 'T') and eight trailer cars (denoted as 'M'), whose marshalling is $2 \times (1T + 2M + 2T + 2M + 1T)$. Their main parameters of the high-speed train refer to Ref. 31. For better simulation of the train motion when it arrives at the bridge, the train was assumed to start moving from a location 150 m before the left-hand end of the cable-stayed bridge.

Assume that the earthquake wave propagates in the subsoil from one end of the bridge to the other while the train is traversing the bridge. The lateral and vertical earthquake components have the forms given in Eq. (48), and the parameters of the modulation functions are given in Eq. (49). The power spectral density (PSD) functions take the Clough–Penzien model.[45]

$$S_h(\omega) = \frac{\omega_{gh}^4 + 4\zeta_{gh}^2\omega_{gh}^2\omega^2}{(\omega_{gh}^2 - \omega^2)^2 + 4\zeta_{gh}^2\omega_{gh}^2\omega^2} \cdot \frac{\omega^4}{(\omega_{fh}^2 - \omega^2)^2 + 4\zeta_{fh}^2\omega_{fh}^2\omega^2} S_{0h}, \tag{43}$$

$$S_v(\omega) = \frac{\omega_{gv}^4 + 4\zeta_{gv}^2\omega_{gv}^2\omega^2}{(\omega_{gv}^2 - \omega^2)^2 + 4\zeta_{gv}^2\omega_{gv}^2\omega^2} \cdot \frac{\omega^4}{(\omega_{fv}^2 - \omega^2)^2 + 4\zeta_{fv}^2\omega_{fv}^2\omega^2} S_{0v}, \tag{44}$$

where ζ_{gh} and ζ_{gv} denote the damping ratios of the soil; ω_{gh} and ω_{gv} denote the predominant frequencies of the soil; ω_{fh} and ω_{fv}, ζ_{fh} and ζ_{fv} denote the filter parameters hindering the low-frequency components of earthquake acceleration, valued by $\omega_{fI} = 0.1\omega_{gI}$ and $\zeta_{fI} = \zeta_{gI}$, $I = h, v$, respectively. S_{0h} and S_{0v} denote the spectral intensity of the earthquake acceleration process, which is related to the peak ground acceleration (PGA) and can be respectively defined as follows.:[56]

$$S_{0h} = \frac{\bar{a}_{h,\max}^2}{\gamma_h^2 \left[\pi\omega_{gh}(2\zeta_{gh} + 1/2\zeta_{gh})\right]}, \quad S_{0v} = \frac{\bar{a}_{v,\max}^2}{\gamma_v^2 \left[\pi\omega_{gv}(2\zeta_{gv} + 1/2\zeta_{gv})\right]}, \tag{45}$$

where $\bar{a}_{h,\max}$ and $\bar{a}_{v,\max}$ denote the mean values of the random earthquake peak ground acceleration (PGA); γ_h and γ_v denotes the peak factors.

A nonuniform modulation function[48,57] is taken here to generate nonstationary earthquake excitations:

$$A(\omega, t) = \frac{\exp(-at) - \exp(-c\omega t - bt)}{\exp(-at^*) - \exp(-c\omega t^* - bt^*)}, \quad \omega > 0, t > 0, \tag{46}$$

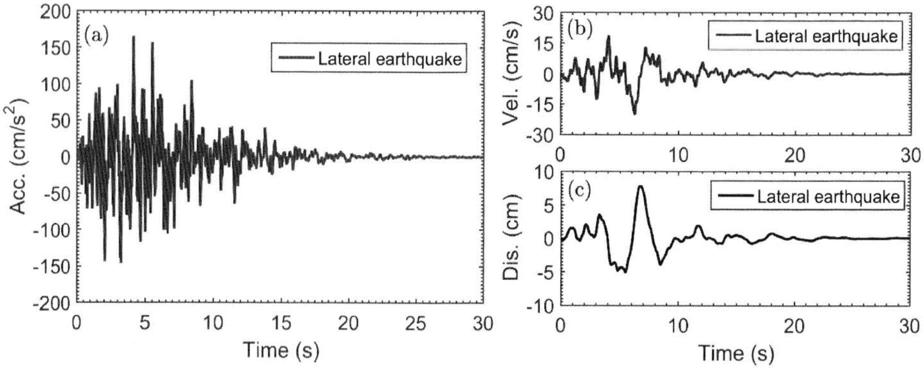

Fig. 5. Typical representative lateral earthquake samples at PGA = 196.2cm/s^2. (a)acceleration; (b) velocity; and (c) displacement.

where

$$t^* = \frac{\ln(c\omega + b) - \ln a}{c\omega + (b - a)}. \tag{47}$$

In this case, they take the values: $a = 0.25\text{s}^{-1}$, $b = a + 0.001 = 0.251\text{s}^{-1}$, $c = 0.005$, $\overline{a}_{v,\max} = 0.4\overline{a}_{h,\max}$, $\overline{a}_{h,\max} = 0 \sim 196.2\text{cm/s}^2$, $\omega_{gv} = \omega_{gh}$, $\omega_{gh} = 17.95\,\text{rad/s}, \zeta_{gh} = \zeta_{gv} = 0.72, \gamma_h = \gamma_v = 2.8$, the frequency region is $\omega \in [1, 100]\,\text{rad/s}$, the earthquake duration time is $T = 30\,\text{s}$.

Based on the point design method in Sec. 3 and the solution with PDEM in Sec. 4, the total number of representation samples $n_{\text{pt}} = 400$ is used to generate the representative samples. An example of the typical representative lateral earthquake samples, which include the acceleration, velocity and displacement based on the PSD of Clough–Penzien model in Eq. (42), is shown in Fig. 5. Besides, the track spectrum functions of generated track irregularities in Eq. (30) use the China high-speed railway track spectrum,[4] where their spatial frequency lies between $0.01 \times 2\pi \sim 1.0 \times 2\pi\,(\text{rad/m})$. The train used here is the China Railway High-Speed (CRH) train, which is composed of eight vehicles with $M + 2T + 2M + 2T + M$ as its marshalling pattern at a speed of $250\,\text{km/h}$. The vehicle parameters are listed in Table A.1 in the Appendix A.

5.2. *Verification of computational efficiency and accuracy*

The random bridge deflection and acceleration at mid-span are considered the most important factors in the bridge safety assessment. Similar to the previous investigations,[31,33] the calculated results both on the bridge deflection at mid-span of the cable-stayed bridge and the vehicle acceleration at the center of gravity ($v = 250\,\text{km/h}, \text{PGA} = 50\,\text{cm/s}^2$) obtained by MCM in 9 999 runs are compared

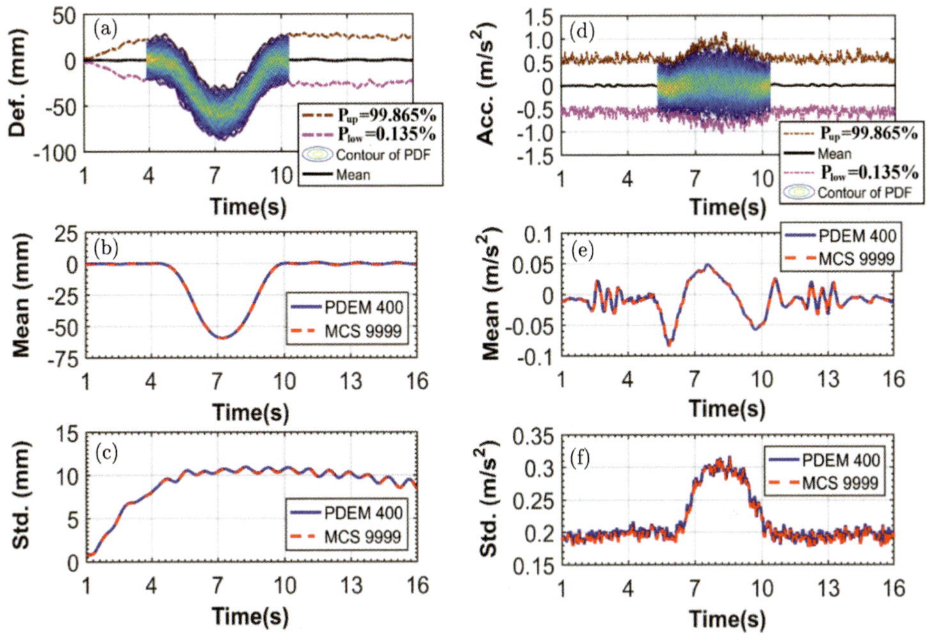

Fig. 6. Verification of PDEM and MCS both on the bridge deflection at mid span of the cable stayed bridge and the vehicle acceleration at the center of gravity ($v = 250\,\text{km/hPGA} = 50\,\text{cm/s}^2$).

with the results obtained by PDEM with 400 runs for the purpose of verifying the application reliability of PDEM, which are shown in detail in Fig. 6.

The probability density function (PDF) contours of the random bridge deflection at mid-span are marked from 1.0 to 16.0 s. It is worth mentioning that the PDF flows smoothly with time, and the contours' cover area represents the randomness of bridge responses with upper and lower ranges. As shown in Figs. 6(a) and 6(d), the integral value of the probability density function at each time step is always equal to 1, which means that the train–bridge coupled responses remain conservative, with no probability loss.

Figures 6(b) and 6(c) show the mean value curves and the standard deviation curves obtained from the PDF of the random bridge deflection, as well as the vehicle acceleration curves shown in Figs. 6(e) and 6(f). Obviously, both of the mean value curves in Figs. 6(b) and 6(e) or the standard deviations in Figs. 6(c) and 6(f) obtained by PDEM in 400 runs are nearly coincident with the results obtained by MCM in 9 999 runs.

The Monte Carlo method is a valid stochastic analysis method in which the frequency of sample values represent the probability. However, in achieving the same accuracy, the probabilistic density evolution method has higher computational efficiency than the Monte Carlo method, probably by 1–2 orders of magnitude.

These results are considered as a convincing example of verification for the proposed method.

5.3. *Response of vehicle and bridge under seismic acceleration*

According to the dynamic analysis work of the bridge and the actual engineering data collected by the research workers,[58–61] the lateral response of the train–bridge system is more significant compared to the vertical response when the train–bridge system encounters seismic excitation, which is due to the greater lateral seismic acceleration and the weaker lateral restraint of the structural system compared to the vertical restraint.

Figures 7 and 8 show the lateral and vertical seismic acceleration time curves for five different peak earthquake accelerations for a typical sample, respectively. Where the peak lateral and vertical seismic accelerations for each working condition satisfy $\overline{a}_{v,\max} = 0.4\overline{a}_{h,\max}$, the subsequent structural dynamic analysis is based on the probability density evolution of the samples under the action of these five operating conditions in this paper.

Under the excitation of random multi-point earthquakes and random track irregularities, the probability PDF contours of the lateral vehicle acceleration show

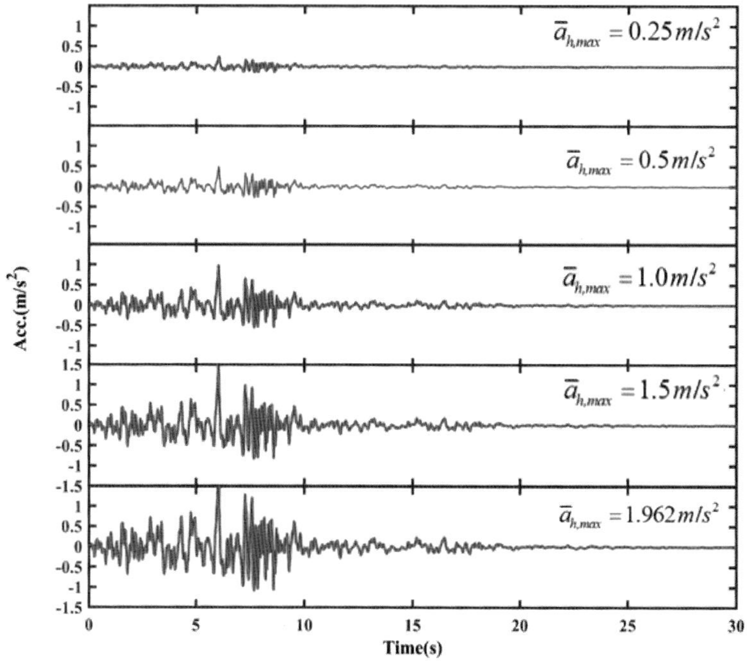

Fig. 7. Time history of the lateral seismic acceleration.

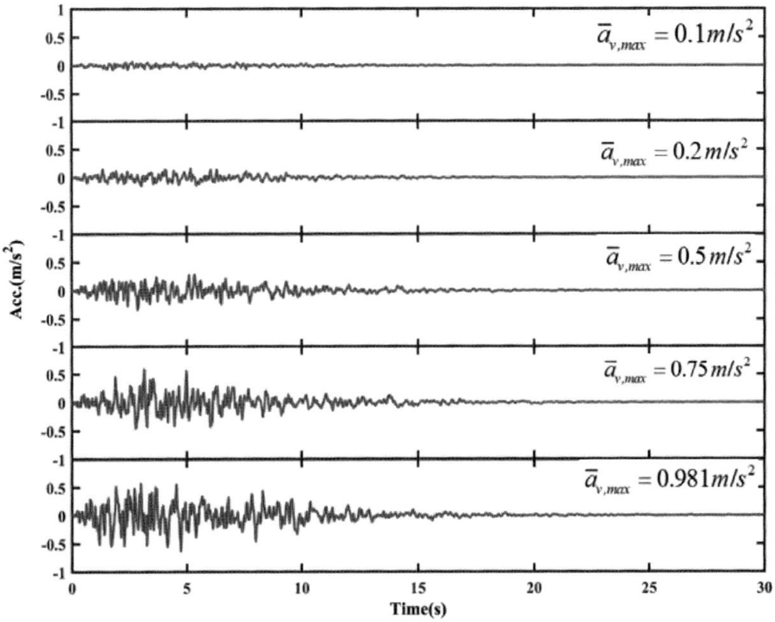

Fig. 8. Time history of the vertical seismic acceleration.

different shapes in Figs. 9(a)–9(f). In reality, the range of the PDF contour indicates the upper and lower limits of random lateral vehicle acceleration. The 99.865% upper and 0.235% lower limit of probability guarantee rate are used for comparison with the PDF contours; the comparison shows that most of the PDF contours fall within the 99.865% upper and 0.235% lower limit of probability guarantee rate. The closer the response is to the mean curve, the larger its probability density value will be. A sample value curve is to be compared with the mean curves, as seen in a sample value curve floating above and below the mean. It is evident that the mean curve of lateral vehicle acceleration in Fig. 9 is a curve almost equal to zero. The mean curve has a slight disturbance when the train passes over the bridge at roughly 3.9–10.2 s, this reveals that a coupling effect occurs between the train and the bridge.

In addition, as shown in Figs. 9(a)–9(f), there is a significant variation in the PDF contours of lateral vehicle acceleration. Where $\bar{a}_{h,\max}$ and $\bar{a}_{v,\max}$ denote the mean values of the random earthquake peak ground acceleration (PGA). The lateral mean values of the random earthquake PGA in Figs. 10(a)–10(f) are 0, 0.25, 0.5, 1, 1.5 and 1.962m/s^2, respectively, and the amplitude of the lateral vehicle acceleration is 0.5, 0.88, 2.34, 3.91, 4.95 and 6.48m/s^2. This indicates that the mean value of random earthquake PGA has a great effect on lateral vehicle acceleration and that the latter increases as the former increases.

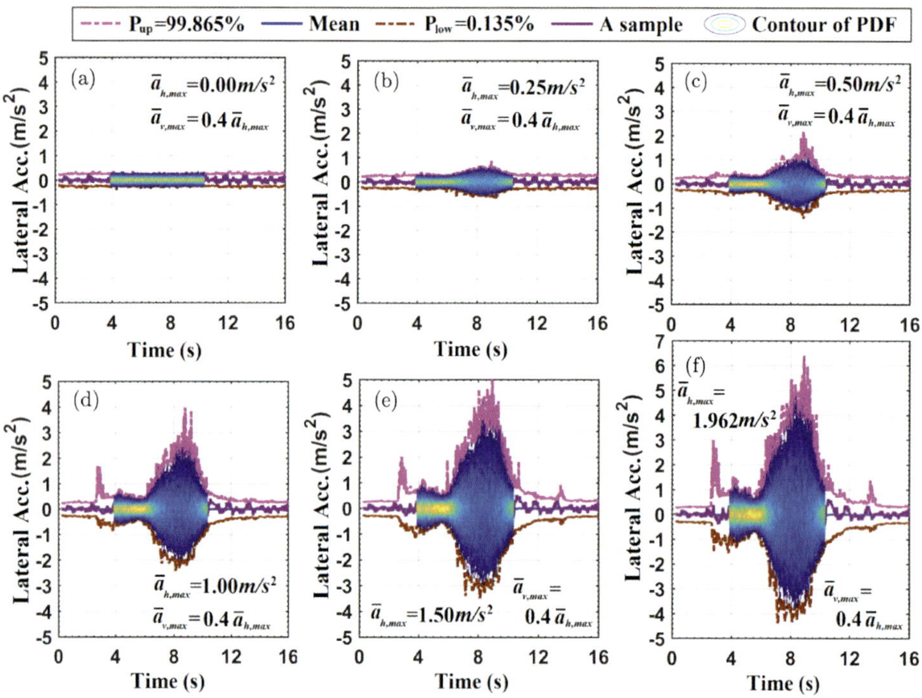

Fig. 9. Comparative analysis of the random lateral vehicle acceleration at the center of the fourth car body under different mean values of the random earthquake PGA ($v = 250\,\text{km/h}$).

Several comparisons similar to the lateral vehicle acceleration response are shown in Figs. 10–12. Figure 10 markedly demonstrates the considerable diversity of bridge lateral acceleration amplitudes under diverse mean values of the earthquake PGA. In Fig. 10(a), the values of the lateral bridge acceleration appear almost zero, whereas there is a significant fluctuation only for a short period of time when the train crosses the bridge according to the magnification due to the influence of the train–bridge coupled interaction. Moreover, the amplitude of the lateral bridge acceleration from the upper limit to the lower limit in Fig. 10(b) at 3.9–10.2 s is 1.10–1.04m/s^2. Nevertheless, the amplitudes of the upper and lower limits are 2.0 and -1.96m/s^2 in Fig. 10(c), respectively. Against the response of the lateral vehicle acceleration analyzed above, a comparable conclusion can be drawn that the lateral bridge acceleration at the mid-span of the cable-stayed bridge also increases with the increase of the earthquake PGA.

5.4. *Wheel–rail contact position under seismic acceleration*

As one of the most important responses that directly act on the running safety of wheel/rail interaction, the wheel/rail contact point position under the action

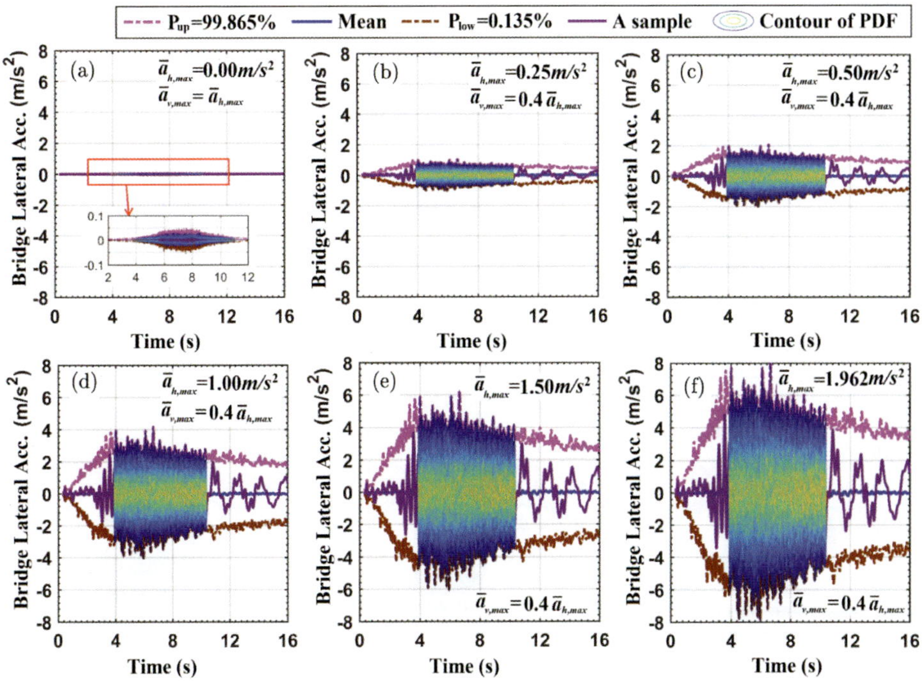

Fig. 10. Comparative analysis of the random lateral bridge acceleration at the mid-span of cable-stayed bridge under different mean values of the random earthquake PGA ($v = 250\,\text{km/h}$).

of earthquake attracts more attention. The comparative analysis of the left and right random wheel/rail contact point position under different mean values of the random earthquake PGA is shown in Figs. 11 and 12, respectively. Based on the wheel–rail contact model mentioned in Sec. 2.1 and the wheel–rail contact diagram shown in Fig. 11(c), it is explicit that the wheel–rail contact point position indicates the distance from the wheel–rail contact point to the inside edge of the rail. In Fig. 11(a), the wheel–rail contact point position ranges from 27.5 mm to 42.3 mm when the earthquake PGA $\bar{a}_{h,\max}$ is 0, and the difference reaches 14.8 mm in between the upper and lower limit values. From the Fig. 11(b), one can clearly know that the position of the wheel–rail contact point is distributed in a wide range from 14.8 to 45.2 mm when the earthquake PGA $\bar{a}_{h,\max}$ is 0.5m/s^2. A large difference between the two limit values reaches 30.4 mm, for that reason, the earthquake PGA intensifies the lateral vehicle–bridge coupling vibration, resulting in an increase in the relative lateral displacement of the wheel-set and rail, so the interval of the contact point position becomes larger.

It is worth mentioning that the contact point position does not continue to increase as the earthquake PGA increases due to the limitation of the wheel rim.

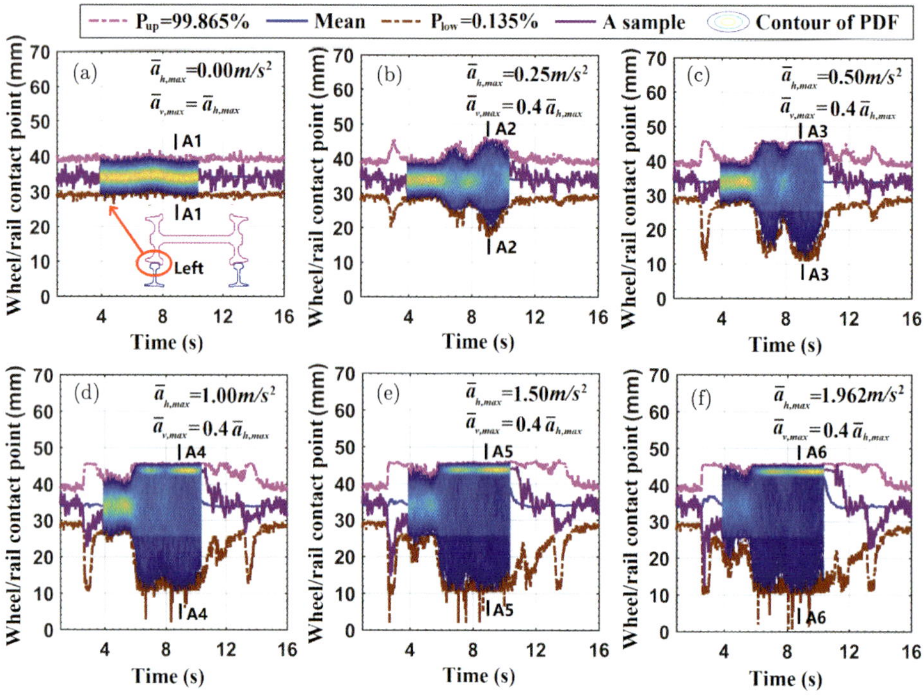

Fig. 11. Comparative analysis of the left random wheel/rail contact point position under different mean values of the random earthquake PGA ($v = 250\,\text{km/h}$).

As illustrated in Figs. 11(c)–11(f), the range of the wheel–rail contact point position always remains between 10 and 46 mm, although the earthquake PGA is constantly rising. A similar conclusion is found for the right wheel pair in Fig. 12. In addition, due to the effect of random track irregularity, the position of the left and right wheel–rail contact points is not the same at the same moment considering the super elevation of outer rail and track turning radius (Fig. 13).

In order to better demonstrate the advantages of probability density evolution method, the position simulation of wheel-rail random contact point is taken as follow. According to Figs. 11 and 12, the probability density functions of the left and right wheel–rail contact position at different earthquake PGAs were plotted at the train running time of 9.1 s as shown in Fig. 13. Taking the wheel–rail contact point location on the left rail surface shown in Fig. 13(a) as an example, it is clear that these probability density function curves named A1–A6 roughly show a trend of normal distribution.

The larger the PDF of a curve at a contact location, the greater the probability of a contact point at that location. As the earthquake PGA increases, the mean value of the distribution functions decreases. This indicates that the enhanced vehicle–bridge

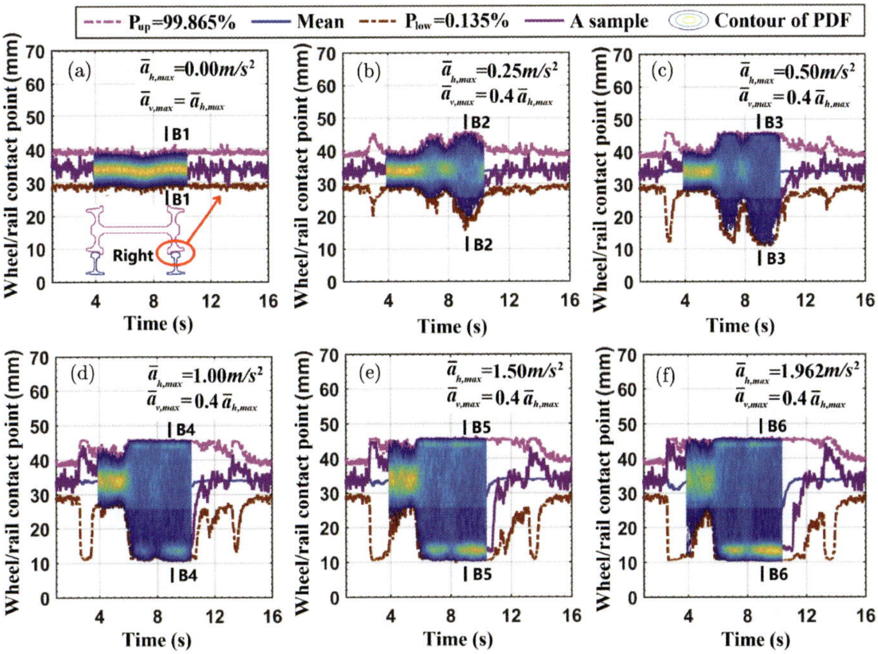

Fig. 12. Comparative analysis of the right random wheel/rail contact point position under different mean values of the random earthquake PGA ($v = 250\,\text{km/h}$).

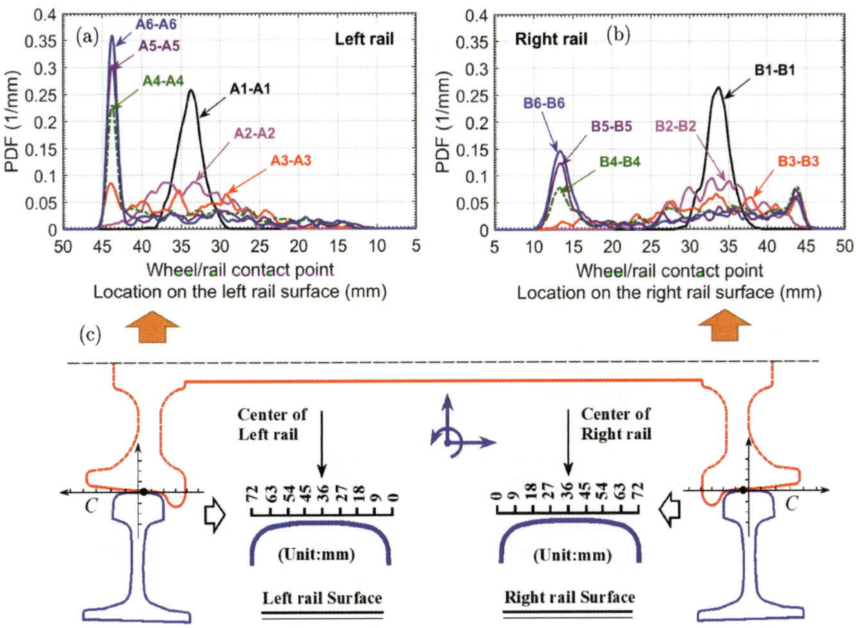

Fig. 13. PDF of the wheel railwheel rail contact point C in Fig. 14 at different earthquake PGAs.

coupling vibration results in a larger range of contact points, and the position of the wheel–rail contact point is mainly concentrated on the left side. The same conclusion was reached for the right wheel–rail. The total probability that is equivalent to the area enclosed by each probability density function curve and the horizontal axis is 1, this is in accordance with the basic principle of the probability density evolution method. When no old random source disappears and no new random source is created by the stochastic process, the total probability is conserved.

5.5. *Train speed analysis and safety and reliability assessment*

Based on the probabilistic reliability assessment method described in Chap. 4, the maximum values of the bridge and vehicle response were studied. All cases were calculated for a train speed ranging from 150 km/h to 350 km/h with quake and no quake, where "Quake" indicates that the lateral PGA is 1.5 m/s^2 (Figs. 14 and 15).

The maximum values of displacement and acceleration of the bridge at mid-span, vehicle acceleration, derailment factors, unloading rate, and lateral wheel force for different train speeds are shown in Tables 1 and 2. As safety factors, the derailment factor and the unloading rate are chosen as controlling indexes in many countries, and are defined as

$$\text{Derailment factor} = \frac{Q}{P}, \tag{48}$$

$$\text{Unloading rate} = \frac{P_s - P}{P_s}, \tag{49}$$

where Q and P are the lateral and vertical wheel–rail interaction forces for an individual wheel, and P_s is the static weight of that individual wheel. The safety

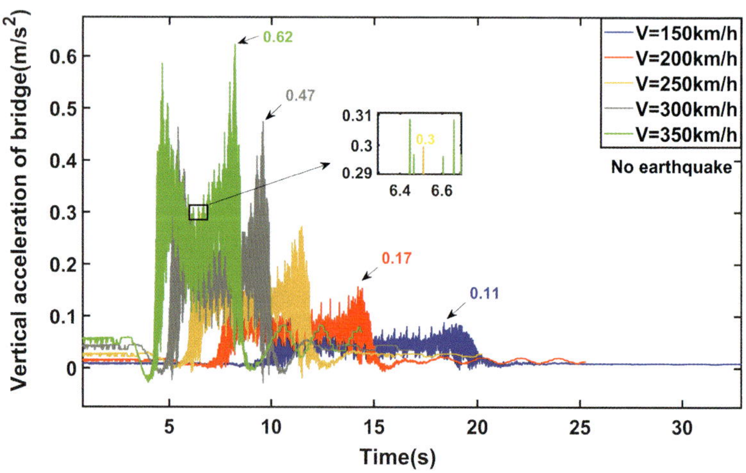

Fig. 14. Vertical acceleration in the span of a bridge in the absence of seismic action.

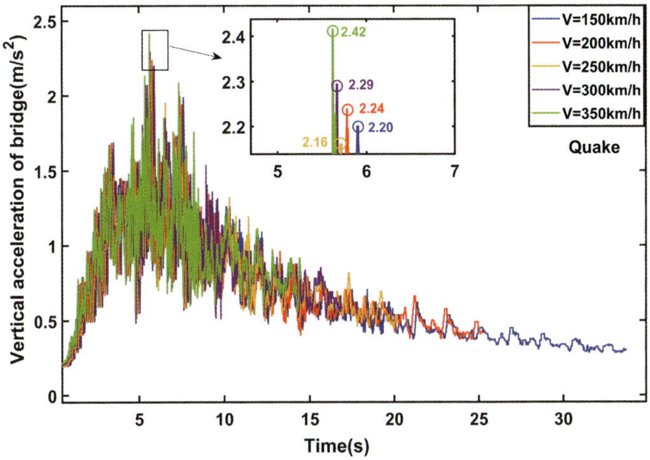

Fig. 15. Vertical acceleration in the span of a bridge under earthquake action.

Table 1. Maximum dynamic response of the bridge with different vehicle speeds.

	Displacement of bridge (mm)				Acceleration of bridge (m/s^2)			
	Lateral		Vertical		Lateral		Vertical	
Train speed (km/h)	No quake	Quake	No quake	Quake	No quake	Quake	No quake	Quake
150	1.77	256.41	69.26	130.68	0.02	8.69	0.11	2.20
200	1.88	256.09	68.65	158.40	0.03	8.69	0.17	2.24
250	1.88	249.39	68.55	159.10	0.05	8.67	0.30	2.16
300	1.87	242.99	69.32	150.88	0.07	8.53	0.47	2.29
350	1.87	238.17	73.38	150.55	0.08	8.65	0.62	2.42

Table 2. Maximum dynamic response of vehicle at different vehicle speeds.

							Acceleration of vehicle (m/s^2)			
	Derailment factors		Unloading rate		Lateral wheel Force (kN)		Lateral		Vertical	
Train speed (km/h)	No quake	Quake	No quake	Quake	No quake	Quake	No quake	Quake	No quake	Quake
150	0.31	0.36	0.10	0.24	20.78	29.51	0.18	1.23	0.44	0.87
200	0.34	0.41	0.14	0.31	22.79	36.36	0.20	1.82	0.58	1.35
250	0.33	0.61	0.19	0.45	23.02	47.88	0.22	2.64	0.68	1.60
300	0.32	0.72	0.25	0.51	21.20	50.00	0.27	2.81	0.80	1.48
350	0.32	1.05	0.29	0.56	21.03	61.15	0.29	3.27	0.91	1.60

evaluation criteria used for derailment factor, unloading rate and lateral wheel force are

$$Q/P \leq 0.8, \tag{50}$$

$$\frac{P_s - P}{P_s} \leq 0.8, \tag{51}$$

$$\begin{cases} Q \leq 29 + 0.3P_{\mathrm{st}} \text{ (danger limit)}, \\ Q \leq 19 + 0.3P_{\mathrm{st}} \text{ (tolerance limit)}, \end{cases} \tag{52}$$

where P_{st} is the static wheel load, the value of P_{st} is taken as 146 kN for the high-speed trains.

In addition, the maximum vertical acceleration histories of the bridge at mid-span with and without earthquake are shown in Figs. 14 and 17, respectively.

The lateral and vertical displacements of the bridge at mid-span are insensitive to changes in vehicle speed. However, the lateral and vertical acceleration of the bridge at mid-span without seismic effects increases with increasing train speed and not significantly with seismic effects. The reason is that the contribution of seismic excitation to the bridge response is greater than the contribution of vehicle speed.

Similarly, same conclusions can be drawn from the car body responses listed in Table 2. In the absence of seismic effects, the derailment factors and lateral wheel force do not vary significantly with car speed, and there is a significant increase in the car body lateral and vertical acceleration, and the individual car body responses vary more significantly with car speed under seismic effects. This is because excessive vehicle speed under seismic action will aggravate the axle vibration.

The maximum values of the derailment factors, unloading rate and lateral wheel force for the body under different lateral PGA are shown in Figs. 16–18, respectively.

It is clear that all the three indicators as mentioned above increase with increasing seismic acceleration. When the lateral earthquake acceleration exceeds $0.75\,\mathrm{m/s^2}$, the derailment factor has exceeded the permissible limit of 0.8. The unloading rate is always within the safe range. According to the limit value calculated by Eq. (51), it can be seen that when the lateral PGA exceeds $1.5\,\mathrm{m/s^2}$, the wheel–rail transverse force exceeds the permissible limit and the safety limit. It is very unsafe for trains to travel under such seismic intensities.

Assuming that only two failure modes, the lateral and the vertical acceleration of the bridge exists in the system, the instantaneous reliability of the structure can be calculated by the PDEM according to the reliability analysis method introduced in Chap. 4, and the time course curve of the system reliability is given in Fig. 19.

It can be seen in Fig. 19 that the structural reliability is 1 at the initial moment, which indicates that the structure is very reliable before the train goes on the bridge,

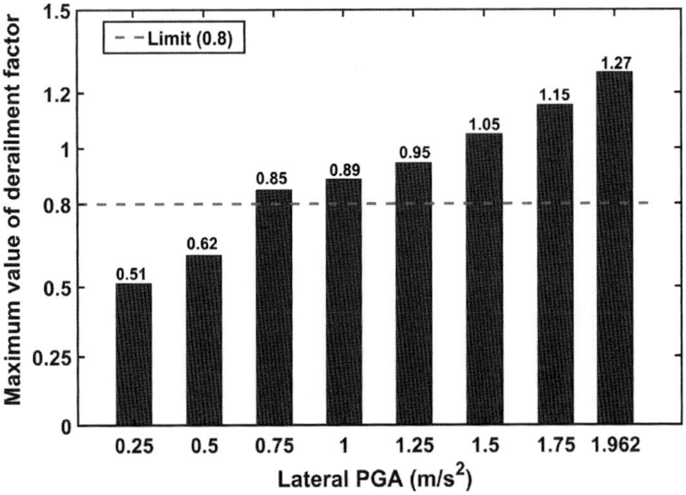

Fig. 16. Maximum derailment factor under different seismic intensities.

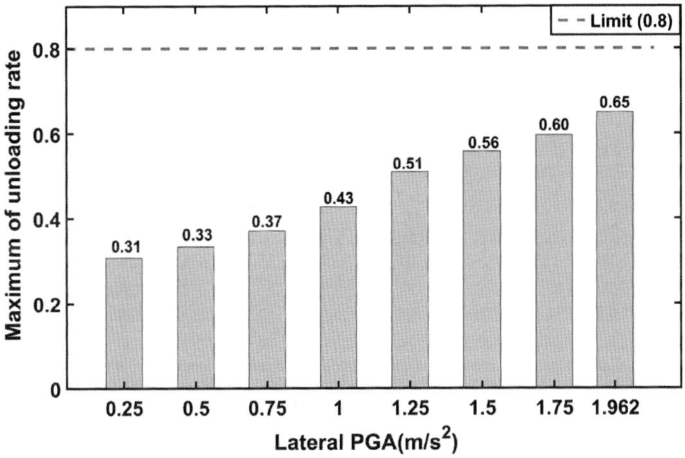

Fig. 17. Maximum unloading rate under different seismic intensities.

but as the train runs to the middle of the bridge span, the random vibration of the structure will increase, and the structural reliability will be significantly reduced. It is also noted that the reliability of the structure decreases significantly with the increase of the peak earthquake acceleration. The reliability of the structure is only 0.73 when the PGA increases to $0.75\,\mathrm{m/s^2}$, which proves that the acceleration response of the bridge has a certain probability of exceeding the limit value at this time.

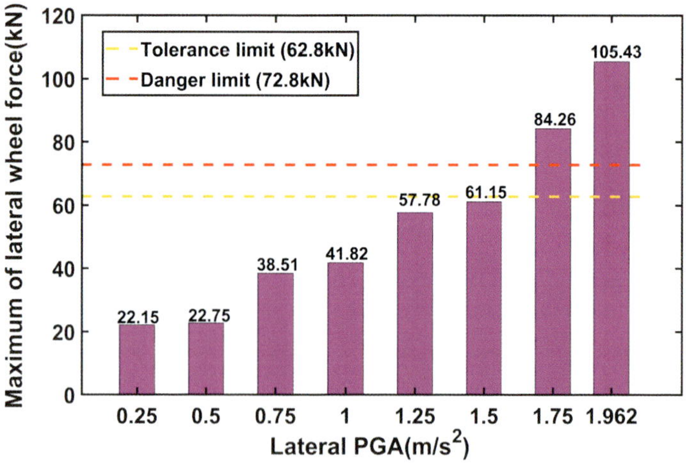

Fig. 18. Maximum lateral wheel force under different seismic intensities.

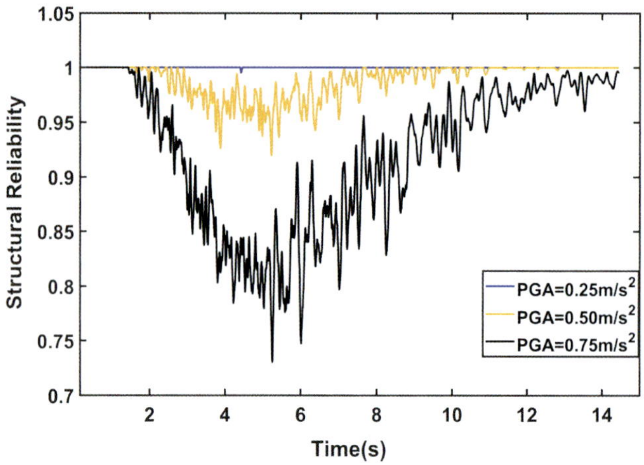

Fig. 19. Instantaneous reliability of the structure under different earthquake accelerations.

6. Conclusions

A refined model of the vehicle–bridge coupling vibration for a long-span cable-stayed bridge is established based on the PDEM involving stochastic vibration theory, in which the multi-effect of random multi-point earthquake excitations with traveling waves and random track irregularity are simultaneously considered. A case study is propounded for the detailed numerical analysis. Based on these cases, the conclusions are as follows:

(1) The PDEM is applicable to the nonstationary stochastic dynamic analysis of the vehicle–bridge system for high-speed trains, and can improve the operational efficiency by 1–2 orders of magnitude over the Monte Carlo method with the same accuracy.
(2) The lateral acceleration of the vehicle and bridge increases significantly under earthquake action, even more than 10 times the normal operation under intense earthquake action, which indicates that the earthquake action has a considerable effect on the vehicle–bridge coupling vibration.
(3) When the earthquake PGA is less than $0.5\,\mathrm{m/s^2}$, the relative lateral displacement of the wheel and rail will increase with the increase of the earthquake PGA, which is also a reflection of the earthquake affecting the lateral vehicle–bridge coupling vibration.
(4) The derailment factor, unloading rate and lateral wheel force increase markedly with the PGA. When the earthquake acceleration reaches $0.5\,\mathrm{m/s^2}$, the lateral wheel force exceeds the limit when it is greater than 1.5, then the train should stop. The reliability of the structure is affected by the earthquake, and will decrease with the increase of the earthquake intensity. The minimum instantaneous reliability is only 0.73 when the earthquake PGA $=$ 0.75, and the structure is unsafe.

Acknowledgments

Joint support was received from National Natural Science Foundation of China (Grant No. 52078485); Joint Fund of National Natural Science Foundation of China (Grant No. U1934217); the Science and Technology Research and Development Program Project of China railway group limited (Project No. 2020-Special-02, 2021-Special-08 and 2021-Major-16).

Appendix A

Table A.1. Major parameters of the vehicle used in this study.

Vehicle parameters	Units	Tractor	Trailer
M_c (mass of the body)	kg	48 000	44 000
J_{cx} (roll mass moment of the body)	kg m^2	115 000	100 000
J_{cy} (pitch mass moment of the body)	kg m^2	2 700 000	2 700 000
J_{cz} (yaw mass moment of the body)	kg m^2	2 700 000	2 700 000
M_t (mass of the bogie)	kg	3 200	2 400
J_{tx} (roll mass moment of the bogie)	kg m^2	3 200	2 400
J_{ty} (pitch mass moment of the bogie)	kg m^2	6 800	6 800
J_{tz} (yaw mass moment of the bogie)	kg m^2	7 200	7 200
m_w (mass of the wheel-set)	kg	2 400	2 400
J_{wx} (roll mass moment of the wheel-set)	kg m^2	1 200	1 200
J_{wz} (yaw mass moment of the wheel-set)	kg m^2	1 200	1 200

(*Continued*)

Table A.1. (*Continued*)

Vehicle parameters	Units	Tractor	Trailer
k_{1x} (longitudinal stiffness of the first suspension system, per side)	kN/m	9 000	15 000
k_{1y} (lateral stiffness of the first suspension system, per side)	kN/m	1 040	700
k_{1z} (vertical stiffness of the first suspension system, per side)	kN/m	3 000	5 000
k_{2x} (longitudinal stiffness of the second suspension system, per side)	kN/m	240	280
k_{2y} (lateral stiffness of the second suspension system, per side)	kN/m	400	300
k_{2z} (vertical stiffness of the second suspension system, per side)	kN/m	480	560
c_{1x} (longitudinal damping of the first suspension system, per side)	kN s/m	50	0
c_{1y} (lateral damping of the first suspension system, per side)	kN s/m	50	50
c_{1z} (vertical damping of the first suspension system, per side)	kN s/m	30	30
c_{2x} (longitudinal damping of the second suspension system, per side)	kN s/m	60	120
c_{2y} (lateral damping of the second suspension system, per side)	kN s/m	60	60
c_{2z} (vertical damping of the second suspension system, per side)	kN s/m	30	25
L (full length of the vehicle)	m	24.775	24.775
L_c (half-distance of the two bogies)	m	17.375/2	17.375/2
L_t (half-distance of two wheel-sets)	m	1.25	1.25
b_1 (half-span of the second vertical suspension system)	m	1.00	1.00
b_2 (half-span of the first vertical suspension system)	m	0.95	0.95
b_3 (half-span of the second horizontal suspension system)	m	1.00	1.00
b_4 (half-span of the first horizontal suspension system)	m	0.95	0.95
b_0 (half-span of the wheel-set)	m	1.496/2	1.496/2
e(lateral distance from the wheel-set to the bridge center)	m	2.50	2.50
h (vertical distance from the rail to the bridge center)	m	1.80	1.80
h_1 (height of the body above the second suspension system)	m	0.80	0.80
h_2 (height of the second suspension system above the bogie)	m	0.30	0.20
h_3 (height of the bogie above the wheel-set)	m	−0.05	0.10
h_4 (height of the wheel-set above the bridge centroid)	m	2.30	2.30
R_{w0} (initial rolling radius of the wheel)	m	0.92/2	0.92/2

Table A.2. Major parameters of the rail used in this study.

Rail parameters	Units	Value
E_r (Elastic modulus)	N/m^2	2.059×10^{11}
I_r (Sectional moment of inertia)	m^4	3.217×10^{-5}
$\overline{m}_r$ (Mass per unit length)	kg/m	60.64

Table A.3. Material characteristics of each component.

Components	Materials	Modulus of elasticity(N/m^2)	Density (kg/m^3)	Poisson rate
Concrete girder	C60 concrete	3.60×10^{10}	2 600	0.2
Steel girder	Q345qD steel	2.10×10^{11}	7 850	0.3
Concrete tower	C50 concrete	3.45×10^{10}	2 600	0.2
Piers	C35 concrete	3.15×10^{10}	2 600	0.2
Cable-stayed cable	steel tendon	2.05×10^{11}	8 321	0.3

Table A.4. Geometry characteristics of major sections.

Girder sections	A (m^2)	I_y (m^4)	I_z or I_x (m^4)	I_ρ (m^4)
Steel box girder: A–A	1.393	7.370	63.830	1.813
Steel box girder: B–B	0.922	4.606	48.935	1.615
Concrete girder section: C–C	62.09	89.92	1269.12	311.98
Concrete girder section: D–D	64.14	90.66	1331.37	315.35
Concrete tower section (Middle)	21.18	38.06	140.5	103.88

Appendix B

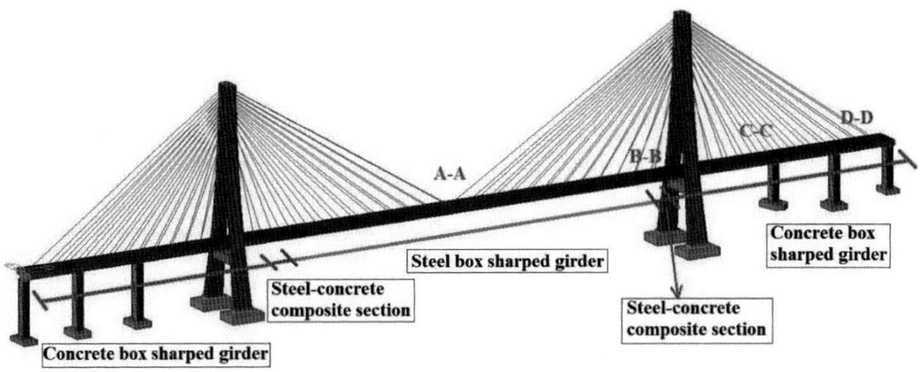

Fig. B.1. Midas model of the cable-stayed bridge with different girder sections.

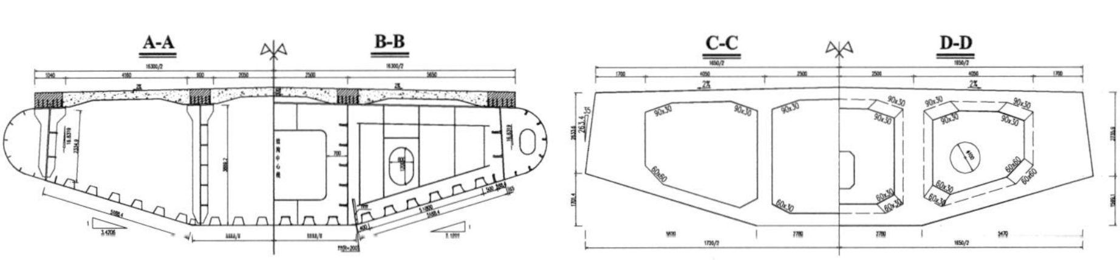

Fig. B.2. Typical cross sections of the bridge girder in Fig. B.

References

1. E. Herediazavoni and E. H. Vanmarcke, Seismic random-vibration analysis of multisupport-structural systems, *J. Eng. Mech. ASCE* **120**(5) (1994) 1107–1128.
2. A. D. Kiureghian and A. Neuenhofer, Response spectrum method for multisupport seismic excitations, *Earthq. Eng. Struct. Dyn.* **21**(8) (1992) 713–740.
3. Q. Y. Zeng and X. R. Guo, *Train–Bridge Time-Variant System Analysis Theory and Application* (China Railway Press, Beijing, 1999).
4. W. M. Zhai, *Vehicle-Track Coupling Dynamics*, 4th edn. (Science Press, Beijing, 2015).

5. H. Xia, N. Zhang and W. W. Guo, *Coupling Vibrations of Train–bridge System* (Science Press, Beijing, 2014).
6. H. Xia, N. Zhang and W. Guo, *Dynamic Interaction of Train–bridge Systems in High-Speed Railways* (Springer, Berlin, Heidelberg, 2018).
7. Y. B. Yang and Y. S. Wu, Dynamic stability of trains moving over bridges shaken by earthquakes, *J. Sound Vib.* **258**(1) (2002) 65–94.
8. J. D. Yau and L. Fryba, Response of suspended beams due to moving loads and vertical seismic ground excitations, *Eng. Struct.* **29**(12) (2007) 3255–3262.
9. H. Xia, Y. Han, N. Zhang and W. Guo, Dynamic analysis of train–bridge system subjected to non-uniform seismic excitations, *Earthq. Eng. Struct. Dyn.* **35**(12) (2006) 1563–1579.
10. L. Fryba and J. Yau, Suspended bridges subjected to moving loads and support motions due to earthquake, *J. Sound Vib.* **319**(1–2) (2009) 218–227.
11. X. T. Du, Y. L. Xu and H. Xia, Dynamic interaction of bridge-train system under non-uniform seismic ground motion, *Earthq. Eng. Struct. Dyn.* **41**(1) (2012) 139–157.
12. M. F. Liu, T. P. Chang and D. Y. Zeng, The interactive vibration behavior in a suspension bridge system under moving vehicle loads and vertical seismic excitations, *Appl. Math. Model.* **35**(1) (2011) 398–411.
13. P. A. Montenegro, R. Calcada, N. V. Pouca and M. Tanabe, Running safety assessment of trains moving over bridges subjected to moderate earthquakes, *Earthq. Eng. Struct. Dyn.* **45**(3) (2016) 483–504.
14. Z. Jin, S. Pei, X. Li, H. Liu and S. Qiang, Effect of vertical ground motion on earthquake-induced derailment of railway vehicles over simply-supported bridges, *J. Sound Vib.* **383** (2016) 277–294.
15. Y. Li, S. Zhu, C. S. Cai, C. Yang and S. Qiang, Dynamic response of railway vehicles running on long-span cable-stayed bridge under uniform seismic excitations, *Int. J. Struct. Stab. Dyn.* **16** (2016) 1550055, doi: 10.1142/S0219455415500054
16. K. Nishimura, Y. Terumichi, T. Morimura and K. Sogabe, Development of vehicle dynamics simulation for safety analyses of rail vehicles on excited tracks, *J. Comput. Nonlin. Dyn.* **4**(1) (2009) 011001, doi: 10.1115/1.3007901.
17. N. Zhang, H. Xia and G. De Roeck, Dynamic analysis of a train–bridge system under multi-support seismic excitations, *J. Mech. Sci. Technol.* **24**(11) (2010) 2181–2188.
18. S. H. Ju, Nonlinear analysis of high-speed trains moving on bridges during earthquakes, *Nonlinear Dyn.* **69**(1–2) (2012) 173–183.
19. X. Gao, P. Duan and H. Qian, Dynamic response analysis of long-span continuous bridge considering the effect of train speeds and earthquakes, *Int. J. Struct. Stab. Dyn.* **20**(6) (2020) 2040013, doi: 10.1142/S0219455420400131.
20. W. Guo, Y. Wang, H. Liu, Y. Long, L. Jiang and Z. Yu, Running safety assessment of trains on bridges under earthquakes based on spectral intensity theory, *Int. J. Struct. Stab. Dyn.* **21**(14) (2021) 2140008, doi: 10.1142/S0219455421400083.
21. Y. Feng, Y. Hou, L. Jiang, W. Zhou, J. Yu and W. Guo, Stochastic transverse earthquake-induced damage track irregularity spectrum considering the uncertainty of track-bridge system, *Int. J. Struct. Stab. Dyn.* **21**(14) (2021) 2140004, doi: 10.1142/S0219455421400046.
22. F. Huang, B. Cheng and N. Teng, Dynamic performance of vehicle-guideway bridge systems for low-medium-speed maglev trains under earthquakes, *Int. J. Struct. Stab. Dyn.* **21**(12) (2021) 2150166, doi:10.1142/S0219455421501662.

23. X. Liu, L. Jiang, P. Xiang, Y. Feng, Z. Lai and W. Zhou, Running safety of high-speed railway train on bridge during earthquake considering uncertainty parameters of bridge, *Int. J. Struct. Stab. Dyn.* **22**(10) (2022) 2241008, doi:10.1142/S0219455422410085.
24. Z. Zhang, Y. Zhang, J. Lin, Y. Zhao, W. P. Howson and F. W. Williams, Random vibration of a train traversing a bridge subjected to traveling seismic waves, *Eng. Struct.* **33**(12) (2011) 3546–3558.
25. Z. C. Zhang, J. H. Lin, Y. H. Zhang, Y. Zhao, W. P. Howson and F. W. Williams, Non-stationary random vibration analysis for train bridge systems subjected to horizontal earthquakes, *Eng. Struct.* **32**(11) (2010) 3571–3582.
26. Z. Zeng, X. He, Y. Zhao, Z. Yu, L. Chen, W. Xu and P. Lou, Random vibration analysis of train-slab track-bridge coupling system under earthquakes, *Struct. Eng. Mech.* **54**(5) (2015) 1017–1044.
27. L. Xu and W. Zhai, Stochastic analysis model for vehicle-track coupled systems subject to earthquakes and track random irregularities, *J. Sound Vib.* **407** (2017) 209–225.
28. J. Li and J. B. Chen, *Stochastic Dynamic of Structures* (Wiley, Singapore, 2009).
29. J. Li and J. B. Chen, The principle of preservation of probability and the generalized density evolution equation, *Struct. Saf.* **30**(1) (2008) 65–77.
30. Z. Yu, J. Mao, F. Guo and W. Guo, Non-stationary random vibration analysis of a 3d train–bridge system using the probability density evolution method, *J. Sound Vib.* **366** (2016) 173–189.
31. Z. W. Yu and J. F. Mao, Probability analysis of train-track-bridge interactions using a random wheel/rail contact model, *Eng. Struct.* **144** (2017) 120–138.
32. J. Mao, Z. Yu, Y. Xiao, C. Jin and Y. Bai, Random dynamic analysis of a train–bridge coupled system involving random system parameters based on probability density evolution method, *Probabilist. Eng. Mech.* **46**(Supp C) (2016) 48–61.
33. Z. W. Yu and J. F. Mao, A stochastic dynamic model of train-track-bridge coupled system based on probability density evolution method, *Appl. Math. Model.* **59** (2018) 205–232.
34. L. Xu and W. Zhai, Stochastic analysis model for vehicle-track coupled systems subject to earthquakes and track random irregularities, *J. Sound Vib.* **407** (2017) 209–225.
35. L. Xu and W. Zhai, A new model for temporal — spatial stochastic analysis of vehicle — track coupled systems, *Vehicle Syst. Dyn.* **55**(3) (2017) 427-448.
36. Z. Jin, W. Liu and S. Pei, Probabilistic evaluation of railway vehicle's safety on bridges under random earthquake and track irregularity excitations, *Eng. Struct.* **266** (2022) 114527.
37. W. W. Guo, H. Xia, G. De Roeck and K. Liu, Integral model for train-track-bridge interaction on the sesia viaduct: dynamic simulation and critical assessment, *Comput. Struct.* **112–113** (2012) 205–216.
38. Z. P. Zeng, Y. G. Zhao, W. T. Xu, Z. W. Yu, L. K. Chen and P. Lou, Random vibration analysis of train–bridge under track irregularities and traveling seismic waves using train-slab track-bridge interaction model, *J. Sound Vib.* **342** (2015) 22–43.
39. W. M. Zhai, H. Xia, C. B. Cai, M. M. Gao, X. Li, X. G. Guo, N. Zhang and K. Y. Wang, High-speed train-track-bridge dynamic interactions-part I: Theoretical model and numerical simulation, *Int. J. Rail Transp.* **1** (2013) 1–24.
40. K. L. Johnson, *Contact Mechanics* (Cambridge University Press, UK, 1985).
41. J. J. Kalker, *Three-Dmensional Elastic Bodies in Rolling Contact* (Kluwer Academic Publishers, The Netherlands, 1990).

42. G. Chen and W. M. Zhai, A new wheel/rail spatially dynamic coupling model and its verification, *Veh. Syst. Dyn.* **41**(4) (2004) 301-322.
43. E. L. Wilson, *Static and Dynamic Analysis of Structures: A Physical Approach With Emphasis on Earthquake Engineering* (Computer and Structures Inc, Berkley, California, 2004).
44. H. Xia, Y. L. Xu and T. Chan, Dynamic interaction of long suspension bridges with running trains, *J. Sound Vib.* **237**(2) (2000) 263–280.
45. R. W. Clough and J. Penzien, *Dynamics of Structures* (McGraw-Hill, New York, 1993).
46. J. H. Lin and Y. H. Zhang, Seismic random vibration of long-span structures, in *Vibration and Shock Handbook*, ed. C. De Silva (CRC Press, Boca Raton, FL, 2005).
47. J. B. Chen, W. L. Sun, J. Li and J. Xu, Stochastic harmonic function representation of stochastic processes, *J. Appl. Mech.-Trans. ASME* **80**(0110011) (2013) 1–11.
48. J. Chen, F. Kong and Y. Peng, A stochastic harmonic function representation for non-stationary stochastic processes, *Mech. Syst. Signal Pr.* **96** (2017) 31–44.
49. K. T. Fang, *Number-Theoretic Methods in Statistics* (Chapman and Hall, England, 1994).
50. J. Li and J. Chen, The number theoretical method in response analysis of nonlinear stochastic structures, *Comput. Mech.* **39**(6) (2007) 693–708.
51. L. K. Hua and Y. Wang, *Applications of Number Theory to Numerical Analysis* (Springer, Heridelberg Berlin, New York, 1981).
52. K. T. Fang, Uniform design: Application of number-theoretic methods in experimental design, *Acta Math. Appl. Sin.* **3**(4) (1980) 363–372.
53. L. Zhangjun, *Structural Dynamics* (China Building Industry Press, 2021).
54. G. Deodatis, Non-stationary stochastic vector processes: Seismic ground motion applications, *Probabilis. Eng. Mech.* **11**(3) (1996) 149–167.
55. J. Li and J. B. Chen, Dynamic response and reliability analysis of structures with uncertain parameters, *Int. J. Numer. Meth. Eng.* **62**(2) (2005) 289–315.
56. H. Seya, M. E. Talbott and H. H. M. Hwang, Probabilistic seismic analysis of a steel frame structure, *Probabilist. Eng. Mech.* **8**(2) (1993) 127–136.
57. G. Deodatis and M. Shinozuka, Auto-regressive model for nonstationary stochastic processes, *J. Eng. Mech.* **114**(11) (1988) 1995–2012.
58. Y. Yang and Y. Wu, Dynamic stability of trains moving over bridges shaken by earthquakes, *J. Sound Vib.* **258**(1) (2002) 65–94.
59. Z. Zeng, Y. Zhao, W. Xu, Z. Yu, L. Chen and P. Lou, Random vibration analysis of train–bridge under track irregularities and traveling seismic waves using train-slab track-bridge interaction model, *J. Sound Vib.* **342** (2015) 22–43.
60. Y. Li, S. Zhu, C. S. Cai, C. Yang and S. Qiang, Dynamic response of railway vehicles running on long-span cable-stayed bridge under uniform seismic excitations, *Int. J. Struct. Stab. Dyn.* **16**(5) (2016) 1550005, doi:10.1142/S0219455415500054.
61. W. Gong, Z. Zhu, Y. Liu, R. Liu, Y. Tang and L. Jiang, Running safety assessment of a train traversing a three-tower cable-stayed bridge under spatially varying ground motion, *Railw. Eng. Sci.* **28**(2) (2020) 184–198.

https://doi.org/10.1142/9789819814770_0021

Chapter 21

Turbulence Effects on Vortex-Induced Dynamic Response of a Twin-Box Bridge and Ride Comfort of the Vehicle[#]

Guo-Qing Zhang[*], You-Lin Xu[*,‡], Bin Wang[*] and Qing Zhu[†]
[*]*Department of Bridge Engineering*
Southwest Jiaotong University
Chengdu, P. R. China
[†]*Department of Bridge Engineering*
Tongji University, Shanghai, P. R. China
[‡]*ceylxu@polyu.edu.hk*

The rapid growth of suspension bridges' span makes vortex-induced vibration (VIV) appears more and more frequently, and once it occurs the closure of the bridge results in considerable economic losses. Investigating the dynamic behavior of the bridge experiencing VIV and the vehicles running on it is thus imperative for providing a reliable guidance for the managers to make operation decisions. Nevertheless, most of the existing studies focus on VIV of bridges subjected to smooth winds, but a certain level of turbulence always exists in reality. The effects of turbulence on vortex-induced dynamic response of the bridge and ride comfort of the vehicles are not clear. This study thus develops a coupled vortex-vehicle-bridge system applicable to the multi-mode lock-in regions of a twin-box deck subjected to both vortex-induced forces and buffeting forces in a turbulent flow. The system is then applied to a real long suspension bridge with three types of vehicles subjected to either smooth or turbulent winds. The results from the case study show that the increasing turbulence mitigates vortex-induced responses of both the bridge and the vehicles and reasonably improves the vehicles' ride comfort. However, the buffeting forces induced by turbulent wind component should not be ignored when turbulence intensity becomes high.

Keywords: Turbulence effect; vortex-induced force; buffeting force; vortex-vehicle-bridge system; dynamic response; ride comfort.

1. Introduction

Benefitting from the use of new materials with less mass but greater strength, the crossing ability of long-span bridges keeps increasing in the last decades but the

[‡]Corresponding author.
[#]This chapter appeared previously on the International Journal of Structural Stability and Dynamics. To cite this chapter, please cite the original article as the following: G.-Q. Zhang *et al.*, *Int. J. Struct. Stab. Dyn.*, **23**, 2340023 (2023), DOI: 10.1142/S0219455423400230.

stiffness and damping of the bridges decrease inevitably, which results in vortex-induced vibration (VIV) more frequently under normal wind conditions. In recent years, VIV events have been reported in the Xihoumen Bridge,[1] the Yi Sun-Shin Bridge,[2] the Humen Bridge[3] and others, arising public's great concern. In fact, based on wind tunnel tests, several semi-empirical vortex-induced forces (VIF) models[4–9] have been proposed and some of them have been applied in predicting vortex-induced responses (VIR) of long-span bridges. The mechanisms of VIV and the measures to reduce VIV have also been investigated. For instance, the mechanism of VIV of a rectangular cylinder was discussed in detail in terms of VIF, aeroelastic response, vibration energy, aerodynamic damping ratio, and flow visualization by Yan *et al.*[10] Li *et al.*[11] studied the surface pressure distribution and aerodynamic nonlinearities of a typical open section through wind tunnel tests and numerical simulations, and they found that the presence of high-order harmonic components makes the degree of aerodynamic nonlinearity continue to increase, but the influence of the third-order and above components on VIV can be ignored. Chen and Li[12] presented a fluid-structure coupling model to investigate the VIV of a circular cylinder subjected to a uniform cross-flow. Xu *et al.*[13] assessed the dynamic stability of a VIV system by combining analytical and experimental approaches. The results indicate that the experimentally observed hysteresis phenomenon corresponds to the occurrence of saddle-node bifurcation of the VIV system. Wang *et al.*[14] proposed suspended flexible sheets (SFSs) with high strength to suppress VIV and the maximum reduction ratio for VIV amplitude approximately reached 2/3. Hasheminejad *et al.*[15] introduced the concept of energy-regenerative damping for semi-active VIV suppression of an elliptical cylinder. However, most of the existing studies pay attention to VIV of the bridges subjected to smooth winds. As a certain level of turbulence always exists in the reality, revealing the effects of turbulence on vortex-induced dynamic responses of the bridge and the ride comfort of the vehicles is thus imperative for providing a reliable guidance for the managers to make operation decisions.

By conducting wind tunnel tests on a rectangular cylinder under smooth winds and turbulent winds of 8% turbulence intensity, Shang[16] found that turbulence could significantly restrict VIV and reduce the span-wise correlation of VIF. Trush *et al.*[17] also confirmed by wind tunnel experiments that increasing turbulence intensity of incoming flow could cause a reduction in the VIR of stay cables. The wind tunnel tests conducted by Fu *et al.*[18] showed that vortex-induced aeroelastic effects on wind-induced responses and unsteady aerodynamic forces are significant, and an increase in turbulence intensity may result in a weak correlation of the unsteady aerodynamic forces and thus suppress vortex-induced aeroelastic effects. Liu *et al.*[19] analyzed the cable vibration data and wind field characteristics of the Jingyue Yangtze River Bridge, collected by the structural health monitoring system, and found that the cable vibration increased with higher wind speed when turbulence intensity is large but it decreased with higher wind speed of lower turbulence intensity. Utsunomiya *et al.*[20] investigated the VIR of a single box deck in the slowly fluctuating wind field generated by the active-controlled wind tunnel system and demonstrated the decrease of VIR of the bridge deck with increasing wind fluctuation. Wu *et al.*[21] investigated the VIV of a cable-stayed bridge without backstays through the segmental model test and the scaled full-aeroelastic model wind tunnel test. The VIV of the bridge, which can be clearly identified in the smooth flow, disappeared in turbulent flow and demonstrated that turbulence could impair the correlation of vortices and the variation of aerodynamic forces.

When it comes to twin-box decks, Zhu *et al.*[22] introduced a turbulence-induced positive damping term into the VIF model proposed by Zhu *et al.*[23] to make it available for turbulent winds and validated it using a newly-developed wind tunnel test technique[8] that could measure VIF directly rather than VIR. The comparative results showed that the proposed VIF model could effectively predict the VIR of a twin-box deck under different turbulent winds and different structural damping ratios. Based on the proposed model, Zhu *et al.*[24] also analyzed VIV of a real twin-box bridge under non-uniformly distributed turbulent winds mode by mode. The results showed that turbulence could reduce VIV, and the reduction effect was larger on lower-order modes of vibration than higher-order modes of vibration.

Nevertheless, although the changes that turbulence winds bring to VIF were disposed by the VIF model proposed by Zhu *et al.*,[22] the buffeting forces induced by turbulent winds and acting on the bridge as well as the vehicles were not included. Moreover, the previous studies,[25,26] which attempted to evaluate the ride comfort of the vehicle running on the vortex-excited bridge, adopted either the semi-empirical VIF model developed by Ehsan and Scanlan[27] or the simple harmonic VIF model proposed by Ruscheweyh and Sedlacek.[28] Since these two VIF models were developed based on the wind tunnel tests under smooth winds, turbulence effects on VIR of the bridge and ride comfort of the vehicles have not been appropriately considered until now.

In this connection, this study aims to investigate the turbulence effects on dynamic behavior of a twin-box bridge at lock-in and the vehicles running on it as well as their ride comfort by establishing a coupled vortex-vehicle-bridge (VVB) system that takes all of the wind loads, including static wind forces, buffeting forces, and VIF, into consideration. Sections 2 and 3 introduce the equations of motion of a twin-box bridge and the vehicles running on it, while Sec. 4 couples the two subsystems via considering road roughness and the contact forces. The solving procedure of the system is also given in Sec. 4. Section 5 introduces the criteria adopted to evaluate the vehicles' ride comfort. Section 6 discusses the results of a case study and the conclusions are remarked in Sec. 7.

2. Equations of Motion for VIV of a Twin-Box Deck in Turbulent Winds

Twin-box decks are adopted widely for long-span suspension bridges in recent years as it is believed that the twin-box deck can enhance the stability of the bridges against flutter.[29] However, some of these bridges suffer from VIV even under normal wind speeds. The field observations also show that the majority of the VIV events occur in the form of the deck's vertical vibration.[30] Therefore, this study focuses on the vortex-induced vertical vibration of the twin-box deck only.

In turbulent winds, wind forces acting on a bridge deck can be classified into three parts: static wind forces, buffeting forces and self-excited forces, which are induced by the mean part of fluctuating winds, turbulence component, and aeroelastic interaction between wind and deck motion respectively. Since most of the existing studies focus on predicting the peak amplitude of VIR, at which VIF dominates the bridge's vibration, VIF is usually regarded as the only force exciting the bridge at lock-in region. Based on the results of wind tunnel tests, several semi-empirical models for VIF have been developed,[5,6,27] but most of them concern smooth winds only until Zhu *et al.*[22] proposed a VIF model available in turbulent winds for twin-box decks. According to the works of Ehsan and Scanlan[27] and Zhu *et al.*,[22] the pure-vortex-induced force is negligible to the steady-state VIR, so that VIF can be classified as self-excited forces since the remaining terms in the VIF models are all induced by the bridge's motion. Although buffeting forces act on the bridge towers and cables as well, these forces generally result in the horizontal vibration of the bridge. Since this study concerns the vortex-induced vertical vibration of the twin-box deck, only the buffeting forces acting on the deck in the vertical direction are considered.

With all the above discussions, the equation of motion of the vortex-excited bridge deck in turbulent winds can be written in terms of the finite element method (FEM) as

$$M_b\ddot{Z}_b + C_b\dot{Z}_b + K_bZ_b = \tilde{F}_{VI} + F_{bu}^{BL} + F_{st}^{BL} - F_{wz}, \tag{1}$$

where M_b, C_b, and K_b are the mass matrix, damping matrix, and stiffness matrix of the deck in the vertical direction; Z_b is the deck's displacement vector; and $\tilde{F}_{VI}$, F_{bu}^{BL}, F_{st}^{BL}, F_{wz} represent the VIF, buffeting forces, static wind forces and the contact forces from the vehicles, respectively. The minus sign before F_{wz} indicates its inverse acting direction compared with the others.

During VIV, a long suspension bridge usually vibrates in a single mode of vibration with a specific natural frequency.[27,31] Therefore, the deck subsystem can be treated as a generalized single degree of freedom (SDOF) system and Eq. (1) can be simplified by using the modal analysis method. That is, the displacement of the deck at time t can be expressed as

$$Z_b(t) = \phi_n q(t), \tag{2}$$

where ϕ_n is the n-order model shape of the deck; and $q(t)$ is the generalized coordinate for the n-order mode of vibration of the bridge.

By applying Eq. (2) to Eq. (1) with the Rayleigh damping assumption, the generalized equation of motion for the n-order mode of vibration of the deck can be obtained as

$$M_n(\ddot{q}(t) + 2\omega_n\xi_n\dot{q}(t) + \omega_n^2 q(t)) = \int_0^L (\tilde{F}_{VI} + F_{bu}^{BL} + F_{st}^{BL} - F_{wz})\phi_n dx, \quad (3)$$

where L is the length of the deck; $M_n = \phi_n^T M_b \phi_n$ represents the n-order generalized mass of the bridge; ω_n and ξ_n are the n-order circular frequency and damping ratio of the bridge, respectively.

Based on the work of Zhu *et al.*,[24] the VIF acting on the deck in turbulent winds throughout the entire lock-in regions of multi-mode VIVs can be expressed as

$$\tilde{F}_{VI} = \frac{\rho D^2}{m}[Y_1(1 - \varepsilon\dot{\eta}^2 - \varepsilon_t w_r^2)\dot{\eta} + Y_2\eta], \quad (4)$$

while the VIF model in smooth winds can be expressed as

$$\tilde{F}_{VI} = \frac{\rho D^2}{m}[Y_1(1 - \varepsilon\dot{\eta}^2)\dot{\eta} + Y_2\eta], \quad (5)$$

where Y_1, Y_2, and ε are the parameters to-be-determined by wind tunnel tests under smooth incoming winds; ε_t is the to-be-determined parameter related to the turbulence-induced nonlinear damping term; $w_r = \frac{w}{U}$ is the dimensionless vertical turbulence component as w and U are the vertical fluctuating wind speed and the mean wind speed, separately; D is the deck depth; m is the deck's mass per length; ρ is the air density; $\eta = y/D$ is the dimensionless vertical displacement; and y is the deck's vertical displacement.

This VIF model counts the aerodynamic damping caused by w rather than the other two turbulent components as w will change the relative speed between the wind and the deck that vibrates in the vertical direction. The accuracy of this VIF model has already been validated by the wind tunnel tests conducted by Zhu *et al.*[22] The span-wise correlation of VIF along the bridge deck at lock-in was also investigated by Zhu *et al.*,[32] and their results will be used in this study.

Buffeting forces induced by turbulent winds were not considered in the previous studies. According to the quasi-steady theory, vertical buffeting forces acting on the deck can be expressed as[33]

$$F_{bu}^{BL} = \frac{1}{2}\rho U^2 B\left[2C_{BL}\frac{u}{U}\gamma_1 + \left(C'_{BL} + \frac{H}{B}C_{BD}\right)\frac{w}{U}\gamma_2\right], \quad (6)$$

where C'_{BL} is the slope of C_{BL}; C_{BL} is the lift coefficient; u is the wind velocity fluctuation in the mean wind direction; γ_1 and γ_2 are the aerodynamic admittance functions taken as unit in the following analysis; B is the width of the deck; and H is the height of the deck.

For line-like structure, static wind forces usually include three components: drag, lift and moments.[6] The lift component concerned in this study can be expressed as

$$F_{st}^{BL} = \frac{1}{2}\rho U^2 BC_{BL}. \tag{7}$$

The contact forces from the vehicles running on the deck are explained in the following subsections.

3. Equations of Motion for the Vehicles in Turbulent Winds

In this study, the three typical vehicles, including sedan car, light truck, and heavy truck, running on a long span bridge,[34,35] are modeled as several rigid bodies connected by a series of springs and dampers, as shown in Figs. 1 and 2. Heavier trucks of five or six-axles, which often result in more significant bridge responses, are not considered in this study because their detail parameters are not available to the writers. Nevertheless, the procedure and computer program developed in this

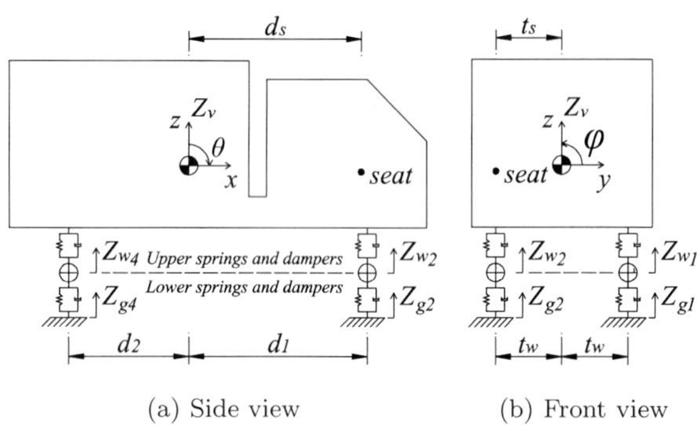

Fig. 1. Model of the two-axle vehicle (sedan car and light truck).

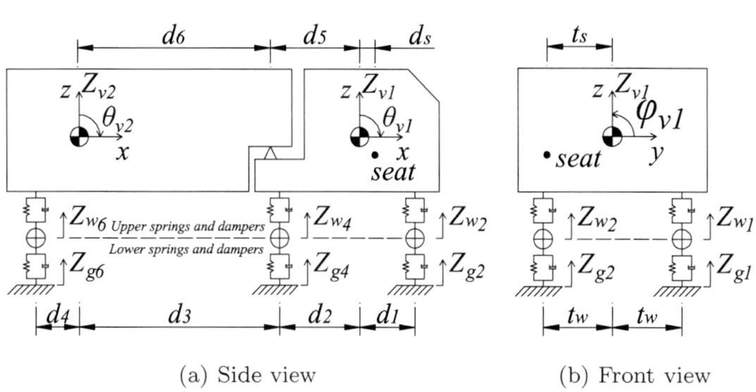

Fig. 2. Model of the three-axle vehicle (heavy truck).

study can be easily used to analyze their dynamic responses and comfort levels once the information of heavier trucks is available. Note that the pitching motion of the trailer of the heavy truck depends on other degrees of freedom, which can be defined as

$$\theta_{b2} = (Z_{b2} - Z_{b1} - d_5\theta_{b1})/d_6. \tag{8}$$

Most previous studies on wind-vehicle-bridge (WVB) systems often assume that the wheels of the vehicle keep touch with the deck without separation. Thus, the equations of motion of the vehicle are usually established at its static equilibrium position under self-weight so that the gravity does not need to be included in the equations of motion, except for calculating the contact forces of the vehicle acting on the deck. To consider the case where vehicle jumps from the deck, the gravity is regarded as an external force in this study no matter whether one or more wheels jump from the deck or not. The equations of motion for the vehicle are established at a special position where the tires touch with the deck with zero forces and the springs and dampers of the vehicle keep their initial state as the gravity does not act.

Similar to what acting on the deck, wind forces acting on the vehicle can also be classified into three parts: static wind forces, buffeting forces and self-excited forces. Considering the vehicle width is usually small and the maximum wind speed, beyond which the vehicle should be stopped, is not very high, the aeroelastic stiffness and aeroelastic damping of the vehicle are generally weak, which leads to the omission of self-excited forces.[36] Furthermore, according to the ride comfort evaluation stipulated in the standard,[37] the ride comfort of the vehicle is determined by the vehicle's vibration in the vertical, lateral, pitching, and rolling direction. It can be reasonably assumed that the vehicle's vibration in the lateral direction will be obviously smaller than that in the vertical direction when VIV of the bridge is concerned. This is because the former is excited mainly by the lateral turbulence but the latter is excited by the deck's vibration, turbulence as well as the road roughness. Therefore, this study considers only the lift force, rolling moment, and pitching moment acting on the vehicle. The equations of motion of the two-axle vehicle can be derived as

$$\begin{aligned} m_v\ddot{Z}_v = {} & F_{st}^{VL} + F_{bu}^{VL} - m_v g + K_s(-4Z_v + Z_{w1} + Z_{w2} + Z_{w3} + Z_{w4} \\ & + 2\theta d_1 - 2\theta d_2) + C_s(-4\dot{Z}_v + \dot{Z}_{w1} + \dot{Z}_{w2} + \dot{Z}_{w3} \\ & + \dot{Z}_{w4} + 2\dot{\theta} d_1 - 2\dot{\theta} d_2), \end{aligned} \tag{9a}$$

$$\begin{aligned} I_y\ddot{\theta} = {} & F_{st}^{VP} + F_{bt}^{VP} + m_v g h\theta + K_s(2Z_v(d_1 - d_2) - d_1(Z_{w1} + Z_{w2}) \\ & + d_2(Z_{w3} + Z_{w4}) - 2\theta(a^2 + b^2)) + C_s(2\dot{Z}_v(d_1 - d_2) \\ & - d_1(\dot{Z}_{w1} + \dot{Z}_{w2}) + d_2(\dot{Z}_{w3} + \dot{Z}_{w4}) - 2\dot{\theta}(a^2 + b^2)), \end{aligned} \tag{9b}$$

$$I_x\ddot{\varphi} = F_{st}^{VR} + F_{st}^{VR} + m_v gh\varphi + K_s t_w(Z_{w1} - Z_{w2} + Z_{w3} - Z_{w4} - 4\varphi t_w)$$
$$+ C_s t_w(\dot{Z}_{w1} - \dot{Z}_{w2} + \dot{Z}_{w3} - \dot{Z}_{w4} - 4\dot{\varphi} t_w), \tag{9c}$$

$$m_{w1}\ddot{Z}_{w1} = F_{zw1} - m_{w1}g + K_s(Z_v - Z_{w1} - \theta d_1 + \varphi t_w) + C_s(\dot{Z}_v - \dot{Z}_{w1}$$
$$- \dot{\theta} d_1 + \dot{\varphi} t_w) + K_t(-Z_{w1} + Z_{g1}) + C_t(-\dot{Z}_{w1} + \dot{Z}_{g1}), \tag{9d}$$

$$m_{w2}\ddot{Z}_{w2} = F_{zw2} - m_{w2}g + K_s(Z_v - Z_{w2} - \theta d_1 - \varphi t_w) + C_s(\dot{Z}_v - \dot{Z}_{w2}$$
$$- \dot{\theta} d_1 - \dot{\varphi} t_w) + K_t(-Z_{w2} + Z_{g2}) + C_t(-\dot{Z}_{w2} + \dot{Z}_{g2}), \tag{9e}$$

$$m_{w3}\ddot{Z}_{w3} = F_{zw3} - m_{w3}g + K_s(Z_v - Z_{w3} + \theta d_2 + \varphi t_w) + C_s(\dot{Z}_v - \dot{Z}_{w3}$$
$$- \dot{\theta} d_2 + \dot{\varphi} t_w) + K_t(-Z_{w3} + Z_{g3}) + C_t(-\dot{Z}_{w3} + \dot{Z}_{g3}), \tag{9f}$$

$$m_{w4}\ddot{Z}_{w4} = F_{zw4} - m_{w4}g + K_s(Z_v - Z_{w4} + \theta d_2 - \varphi t_w) + C_s(\dot{Z}_v - \dot{Z}_{w4}$$
$$+ \dot{\theta} d_2 - \dot{\varphi} t_w) + K_t(-Z_{w4} + Z_{g4}) + C_t(-\dot{Z}_{w4} + \dot{Z}_{g4}), \tag{9g}$$

where m_v and m_{wi} $(i = 1, 2, 3, 4)$ are the mass of the vehicle body and the ith wheel; I_y and I_x are the mass moment of inertia of the vehicle body about the y-axis and x-axis, separately; F_i^j $(i = st, bu$ and $j = VL, VP, VR)$ represents the static or buffeting component of the left force, pitching moment, and rolling moment; F_{zwi} $(i = 1, 2, 3, 4)$ represents the external force acting on the ith wheel without considering the gravity and it will be zero in this study; g is the gravity acceleration; K_s and C_s are the vertical stiffness and damping of the upper springs and dampers used to model the suspension system of the vehicle; K_t and C_t are the vertical stiffness and damping of the lower springs and dampers used to model the tires.

Organizing the equations in state space for programing leads to

$$\dot{X} = M^{-1}NX + M^{-1}PU \tag{10}$$

in which X and U are the state variable vector and the input variable vector, respectively, which can be expressed as

$$X = [Z_v \quad \varphi \quad \theta \quad Z_{w1} \quad Z_{w2} \quad Z_{w3} \quad Z_{w4} \quad \dot{Z}_v \quad \dot{\varphi} \quad \dot{\theta} \quad \dot{Z}_{w1} \quad \dot{Z}_{w2} \quad \dot{Z}_{w3} \quad \dot{Z}_{w4}]^T, \tag{11a}$$

$$U = \begin{bmatrix} U_{11} \\ U_{21} \\ U_{31} \end{bmatrix}, \tag{11b}$$

$$U_{11} = [F_{st}^{VL} + F_{bu}^{VL} - m_v g \quad F_{st}^{VR} + F_{bu}^{VR} \quad F_{st}^{VP} + F_{bu}^{VP}]^T, \tag{11c}$$

$$U_{21} = [F_{zw1} - m_{w1}g \quad F_{zw2} - m_{w2}g \quad F_{zw3} - m_{w3}g \quad F_{zw4} - m_{w4}g]^T, \tag{11d}$$

$$U_{31} = [Z_{g1} \quad Z_{g2} \quad Z_{g3} \quad Z_{g4} \quad \dot{Z}_{g1} \quad \dot{Z}_{g2} \quad \dot{Z}_{g3} \quad \dot{Z}_{g4}]^T. \tag{11e}$$

What must be emphasized is that a period of time, during which the gravity is activated and the vehicle vibrates until it reaches the equilibrium position, is needed before the vehicle enters the bridge. More details about Eq. (10) and the equations of motion of the three-axle vehicle can be found in Appendix.

For a vehicle moving on the deck along a straight line at constant velocity V_T, the concerned static wind forces and buffeting forces can be deduced following the formula proposed by Li *et al.*[36]

$$F_{st}^{VL} = \frac{1}{2}\rho A C_{VL}(\beta)\bar{V}_R^2, \tag{12a}$$

$$F_{st}^{VR} = \frac{1}{2}\rho A h C_{VR}(\beta)\bar{V}_R^2, \tag{12b}$$

$$F_{st}^{VP} = \frac{1}{2}\rho A h C_{VP}(\beta)\bar{V}_R^2, \tag{12c}$$

$$F_{bu}^{VL} = \frac{1}{2}\rho A\bar{V}_R^2\left(C_{VL}(\beta)\frac{2Uu}{\bar{V}_R^2} + (C'_{VL}(\beta) + C_{VD}(\beta))\frac{w}{U}\right), \tag{12d}$$

$$F_{bu}^{VR} = \frac{1}{2}\rho A h\bar{V}_R^2\left(C_{VR}(\beta)\frac{2Uu}{\bar{V}_R^2} + C'_{VR}(\beta)\frac{w}{U}\right), \tag{12e}$$

$$F_{bu}^{VP} = \frac{1}{2}\rho A h\bar{V}_R^2\left(C_{VP}(\beta)\frac{2Uu}{\bar{V}_R^2} + C'_{VP}(\beta)\frac{w}{U}\right), \tag{12f}$$

where $\bar{V}_R^2 = U^2 + V_T^2$; A is the area of the vehicle's side surface; h is the height of the center of gravity of the vehicle body; $\beta = \arctan(V_T/\sqrt{(U+u)^2 + w^2})$ is the relative yaw angle of the vehicle; C_{VL}, C_{VD}, C_{VR}, C_{VP} are the lift, drag, rolling, and pitching coefficients, that can be obtained from the work of Wang *et al.*[38]

$$C_{Vi}(\beta) = C_{i0} + C_{i1}\beta + C_{i2}\beta^2 + C_{i3}\beta^3, \tag{13}$$

where C_{ij} ($i = L, D, R, P$ and $j = 0, 1, 2, 3$) are the fitting constants.

C'_{Vi} ($i = L, R, P$) is the differential between the angle of the wind's incidence $\alpha = \arctan(w/(U+u))$, which represents the influence of the vertical turbulence on the forces acting on the vehicle. Because of its tiny values compared with the horizontal flow, the differentials are set as 0 in this study.

4. Coupled VVB System and Its Solving Procedure

Based on the equations introduced in the last two subsections, the coupled VVB system can be realized by adding the contact forces between the vehicle tires and the deck into the equations. Before calculating the contact forces, the simulation of road roughness that cannot be ignored between the two subsystems needs to be introduced.

4.1. *Road roughness*

Road roughness is one of the most important factors that affects the vibration of the vehicle. By assuming it as a homogeneous and ergodic random field satisfying a Gaussian distribution with a zero mean,[39,40] the spectral representation method proposed by Shinozuka and Deodatis[41] can be employed to simulate it. After the lateral width of the deck is divided into N_l segments, the deck surface roughness of the jth segment at the longitudinal location X can be obtained by

$$r_j(X) = \sqrt{2(\Delta\Omega)} \sum_{m=1}^{j} \sum_{k=1}^{N_\Omega} \sqrt{G(\Omega_{mk})} R(\Omega_{mk}) \cos(\Omega_{mk} + \theta_{mk}), \tag{14}$$

where $\Delta\Omega$ is the interval of spatial frequency of the deck surface; N_Ω is the total number of the selected frequencies; Ω_{mk} is the kth spatial frequency for the mth segment; θ_{mk} is the phase corresponding to Ω_{mk}, which is considered as a random phase angle distributed from 0 to 2π.

$$R(\Omega_{mk}) = \begin{cases} 0 & 1 \le j < m \le N_l, \\ \gamma_e(\Omega_{mk})^{|j-m|} & m = 1,\ m \le j \le N_l, \\ \gamma_e(\Omega_{mk})^{|j-m|}\sqrt{1-\gamma_e(\Omega_{mk})^2} & 2 \le m \le j \le N_l, \end{cases} \tag{15}$$

where $\gamma_e(\Omega_{mk})$ is the lateral coherence coefficient between the left and right wheel paths. Bogsjo[42] compared different coherence models and recommended an exponentially decreasing coherence model

$$\gamma_e(\Omega_{mk}) = \exp(-\rho_r d_w \Omega_{mk}), \tag{16}$$

where ρ_r is a parameter in the range of $3.1 \sim 5.5$[42] and taken as 4 in this study; and d_w is the lateral distance between the left and right wheel paths;

$G(\Omega_{mk})$ is the power spectral density (PSD) function that can be expressed as[43]

$$G(\Omega_{mk}) = G(\Omega_0)\left(\frac{\Omega}{\Omega_0}\right)^{-2} \quad (\Omega_1 < \Omega < \Omega_2), \tag{17}$$

where $\Omega_0 = 0.1\,\mathrm{m}^{-1}$ is the reference spatial frequency of the deck surface; $G(\Omega_0)$ is the roughness coefficient whose value should be chosen according to the deck surface condition, such as A, B, and C in ISO 8608[43]; Ω_1 and Ω_2 are the lower and upper cut-off frequencies, respectively.

4.2. *Contact forces between the vehicle tires and the deck*

The geometric contact boundaries of the tires of a moving vehicle are formed from the surface profile consisting of the displacements of the deck and the road roughness. This surface profile Z_p at the location (X, Y) can be expressed as

$$Z_p(X, Y) = Z_b(X, Y) + Z_g(X, Y), \tag{18}$$

where Z_g is the road roughness height; and Z_b is the vertical displacement of the deck.

The time derivative of the surface profile is

$$\dot{Z}_p(X, Y) = \dot{Z}_b(X, Y) + \frac{\partial Z_b(X, Y)}{\partial X}\dot{X} + \frac{\partial Z_g(X, Y)}{\partial X}\dot{X}, \tag{19}$$

where $\dot{X}$ represents the vehicle speed.

The normal contact force between the tire and the deck is generated in consideration of the deformation and damping effect of the tire in the vertical direction. Assuming that the tire is linearly elastic,[44] the contact force at the ith tire can be calculated by

$$\tilde{F}_{wzi} = K_t(Z_{pi} - Z_{wi}) + C_t(\dot{Z}_{pi} - \dot{Z}_{wi}), \tag{20}$$

where Z_{pi} and Z_{wi} are the surface profile at the contact position of the ith tire and the displacement of the tire, respectively.

It is worthy noting that the contact force calculated by Eq. (20) will become negative when the tire jumps from the deck and result in a deviation from the true vibration of the system. To simulate the jumping accurately, it should be set to zero. Therefore, the contact force should be expressed as

$$F_{wzi} = \begin{cases} \tilde{F}_{wzi} & \tilde{F}_{wzi} > 0 \\ 0 & \tilde{F}_{wzi} \leq 0. \end{cases} \tag{21}$$

4.3. *Numerical solving procedure*

The solving procedure for the VVB system at lock-in region can be summarized in Fig. 3. In the first phase, the wind spectrum proposed by Busch and Panofsky,[45] which can be expressed as Eq. (22), is adopted to simulate the turbulent wind field through the spectral representation method, and the aerodynamic coefficients of the deck and the vehicles are collected to calculate the wind forces acting on the deck and the vehicle according to the equations introduced in Secs. 2 and 3. After that, the Newmark-Beta method and the Four-order Runge-Kutta approach are selected to solve the deck and the vehicles' equations, respectively. It is emphasized that both of the two subsystems need enough time to reach their steady state before they are coupled by inserting the contact forces into the equations because of the

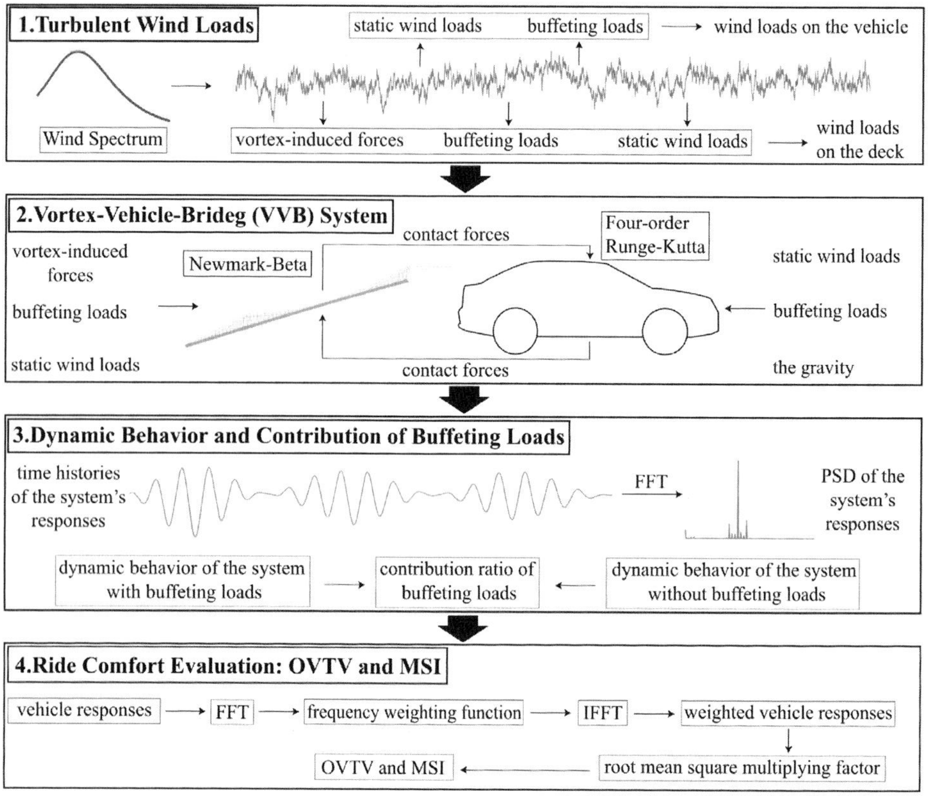

Fig. 3. Solving procedure for the VVB system at lock-in.

effects of the gravity on the vehicles and the initial excitation on the deck. Considering that the contact forces depend on the dynamic responses of the two subsystems, it should be determined via an iterative approach at each time step. In the third phase, the dynamic behavior of the two subsystems can be obtained and the contribution of the buffeting loads, which are ignored in most of the existing studies, can be uncovered by comparing the dynamic behavior of the system with and without the action of the buffeting loads. Finally, the ride comfort of the vehicles is evaluated according to ISO 2631-1[37] for investigating the influences of several factors, including driving speed, road roughness as well as vehicle type, on it in turbulent wind field.

$$\frac{nS_w(n)}{\sigma_w^2} = \frac{2.15f}{1 + 11.16f^{5/3}}, \quad f = \frac{nz}{U}, \tag{22}$$

where $S_w(n)$ is the vertical wind spectrum; σ_w is the standard deviation of the vertical turbulence component; U is the mean wind speed at height z.

5. Ride Comfort Evaluation

To the best knowledge of the authors, there is no international standard specialized for the ride comfort of the vehicles running on the bridge experiencing VIV. During VIV, not only the drivers' physiological status but also their psychology are affected and need to be considered when evaluating their comfort, which is quite different from the drivers of the vehicles running on the ground. While further detailed studies on this topic wait for consideration, this study adopts the ride comfort evaluation method suggested by the International Standard Organization (ISO) in ISO 2631-1[37] for the time being.

According to the standard, drivers' discomfort is caused by the acceleration transmitted to them. Since a driver is supported by the seat, backrest, and floor surfaces, the accelerations at these positions should be taken into account, which can be obtained based on the surfaces' distances from the centroid of the vehicle after the dynamic responses of the vehicle are calculated via the VVB system. To reflect the varying tolerance and sensitivity of human bodies to the outside vibration with different frequencies, the standard suggests two ride comfort indexes, overall vibration total value (OVTV) and motion sickness incidence (MSI), with different factors. While the former quantifies the overall vibration transmitted to the drivers, the latter indicates the percentage of the drivers who may vomit in such condition.[37] Before the root-mean-squares (RMS) of the accelerations are combined for all locations and directions to calculate the indexes, they need to be weighted according to their frequencies. The weighting factors are drawn in Fig. 4, in which W_k, W_d, W_e and W_c are related to vertical vibration at the seat as well as vertical and lateral vibrations at the floor, lateral vibration at the seat and floor, rotational

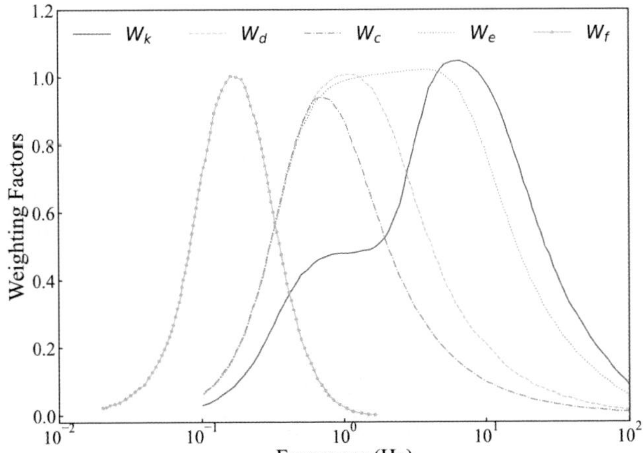

Fig. 4. Weighting curve for different participating axes.[29]

Table 1. Multiplying factors for each vibration measure in evaluating OVTV.[37]

Multiplying factor	Value	Location	Direction
M_k	1.00	Seat	Vertical
M_d	1.00	Seat	Lateral
M_e	0.40	Seat	Pitching
M_e	0.20	Seat	Rolling
M_c	0.40	Backrest	Vertical
M_d	0.50	Backrest	Lateral
M_k	0.40	Floor	Vertical
M_k	0.25	Floor	Lateral

Table 2. Comfort criteria based on OVTV value.[37]

OVTV value (m/s^2)	Comfort level
<0.315	Not uncomfortable
0.315–0.630	A little uncomfortable
0.500–1.000	Fairly uncomfortable
0.800–1.600	Uncomfortable
1.250–2.500	Very uncomfortable
>2.000	Extremely uncomfortable

vibration at the seat, and vertical vibration at the backrest, respectively, for calculating OVTV, while W_f is employed for all positions and directions for calculating MSI. Except the weighting factors, several multiplying factors are also adopted in evaluating OVTV to count the drivers' sensitivity and tolerance to the vibrations from different locations and directions, which are listed in Table 1, but there is no multiplying factor in evaluating MSI. Comfort criteria in terms of the OVTV value are listed in Table 2 while the MSI value directly indicates the percentage of the drivers who may vomit.

From the weighting factors and multiplying factors shown above, the differences between OVTV and MSI can be summarized as: (1) OVTV is more sensitive to vibration with high frequency but MSI is affected by lower frequency; (2) OVTV attaches more importance to the seat's vibration while MSI ignores the different effects of the vibration in different positions and directions. Both of them will be adopted in this study to evaluate the drivers' ride comfort comprehensively.

6. Case Study and Discussions

The Xihoumen Bridge, located on the east coast of China, is a suspension bridge with a main span of 1650 m and a side span of 578 m (see Fig. 5). The bridge adopts flat streamlined and separate double box section with its total width 36 m, height 3.5 m and the gap between the two boxes is 6 m (see Fig. 6). According to the records from Li *et al.*,[1] 37 VIV events have been observed since 2009 and all

Fig. 5. Layout of the Xihoumen Bridge.

Fig. 6. Cross section of the Xihoumen Bridge.

Table 3. Frequencies of the first six vertical modes of the bridge.[24]

Mode order	Frequency (Hz)
3	0.097
6	0.133
8	0.182
14	0.228
19	0.260
22	0.323

of them occurred in the vertical direction. More specifically, seven of the events occurred with the third vertical bending mode of the bridge deck under the mean wind speeds closing to 5.9 m/s. Twenty-three of them occurred with the six vertical bending mode under the mean wind speeds of approximate 10.5 m/s. The turbulence intensity during the VIV events varied from 2.5% to 7.8%. The deck's maximum displacement and acceleration reached 23.89 cm and 61.82 cm/s^2, respectively. The first six vertical modes of the bridge are listed in Table 3, in which the mode order is the order accounting all the modes of the bridge.

Using the newly developed VIF measurement equipment, Zhu *et al.*[22] carried out extensive wind tunnel tests and worked out a VIF model for the twin-box deck based on the directly measured VIF. Table 4 lists the identified parameters by Zhu *et al.*,[24] among which Y_1, Y_2 and ε were fitted from the time histories of the VIF in smooth flow but ε_t was fitted from the VIF in different turbulent winds so that the parameters are available for the wind fields of various turbulence intensities. The damping ratios of the bridge in the first six vertical modes are around 0.5% according to the field measurement, and 0.5% is adopted for all modes in this study. Other aerodynamic coefficients of the deck can be found from the work of Ge.[46]

In this section, the dynamic behavior of the twin-box deck experiencing VIV and the vehicles in turbulent winds will be investigated mode by mode to reveal

Table 4. VIF parameters of the Xihoumen Bridge.[24]

Reduced wind speed	Y_1	Y_2	ε	ε_t
6.026	1.63	3.998	85.7	115.1
6.39	9.95	4.801	161.0	115.1
6.733	35.10	4.645	285.2	115.1
7.077	50.02	4.726	351.7	115.1
7.428	48.50	4.983	422.6	115.1
7.778	41.10	5.445	593.4	115.1
8.135	30.65	5.883	949.8	115.1
8.493	18.24	6.520	1671.0	115.1
8.823	1.74	7.104	8464.3	115.1
9.1868	12.50	8.726	4817.0	115.1

the turbulence effects on them. Then, the ride comfort of the vehicles under various scenarios will be evaluated for exploring its influencing factors.

6.1. *Turbulence effects on the twin-box deck at lock-in regions*

By decoupling and neglecting the vehicle subsystem, the solving procedure introduced in Sec. 4.3 can be employed to analyze the vortex-induced dynamic response of the twin-box deck without the influence of vehicles. The coming mean wind speed increases from 2.8 m/s to 13 m/s to cover the first six lock-in regions of the deck and the turbulence intensity ranging from 0% to 20% are considered.

Figures 7 and 8 show the maximum displacement and acceleration responses of the deck against different mean wind speeds. It can be seen that with the increase of turbulence intensity, both the maximum displacement and acceleration responses of the deck within the lock-in regions decrease continuously until reaching a level similar to that outside the lock-in regions. The width of lock-in region of the deck for each of the six modes seems to be affected slightly by turbulence intensity. Different from the maximum displacement response whose amplitude seems to be stable within the lock-in regions of different modes, higher order VIV induced by higher wind speed results in higher acceleration response.

To show the effects more clearly, several mean wind speeds, marked by the chain lines in Figs. 7 and 8, are selected and the corresponding maximum displacement and acceleration responses of the deck against turbulence intensity are displayed in Figs. 9 and 10 separately. The left subpicture of Fig. 9 illustrates the variation of the deck's maximum displacement response against turbulence intensity when the wind cannot induce VIV of the bridge deck (i.e. outside the lock-in regions). It can be seen that the deck's maximum displacement response increases with increasing turbulent intensity. This is because outside the lock-in regions, only the buffeting forces exist and they increase with increasing turbulence intensity. The right subpicture of Fig. 9 shows the variation of the deck's maximum displacement response against turbulence intensity when VIV occurs. What can be found from this subpicture

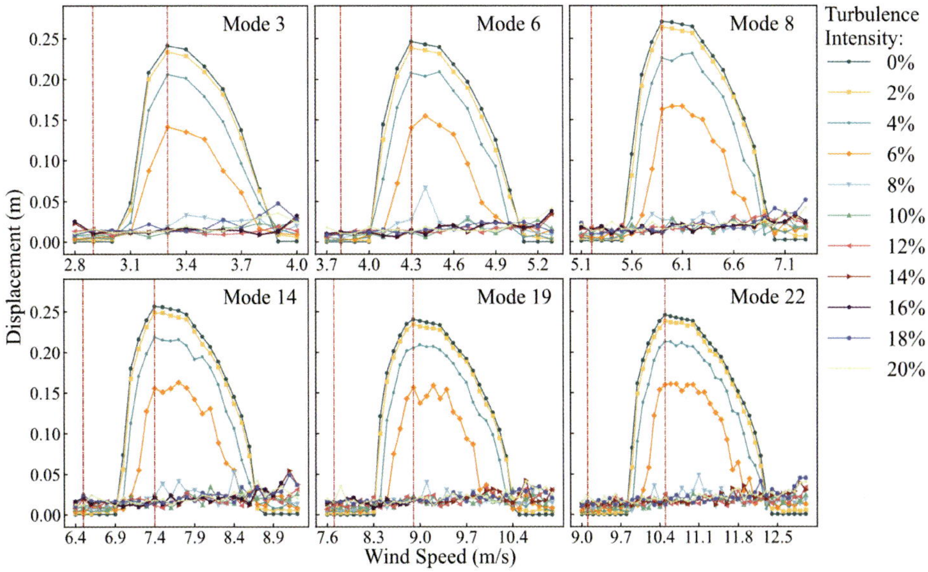

Fig. 7. Max vertical displacement of the twin-box deck in turbulent wind field.

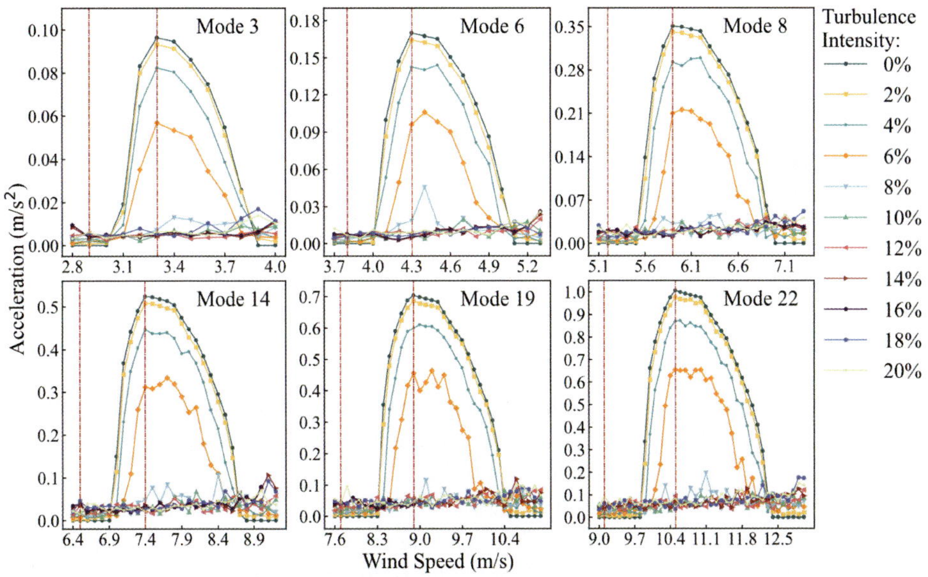

Fig. 8. Max vertical acceleration of the twin-box deck in turbulent wind field.

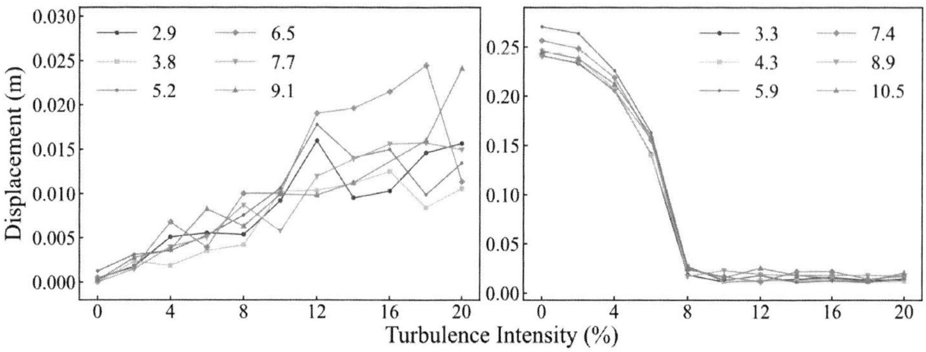

Fig. 9. Variation of the maximum vertical displacement of the twin-box deck against turbulence intensity and wind speed (outside and inside the lock-in regions).

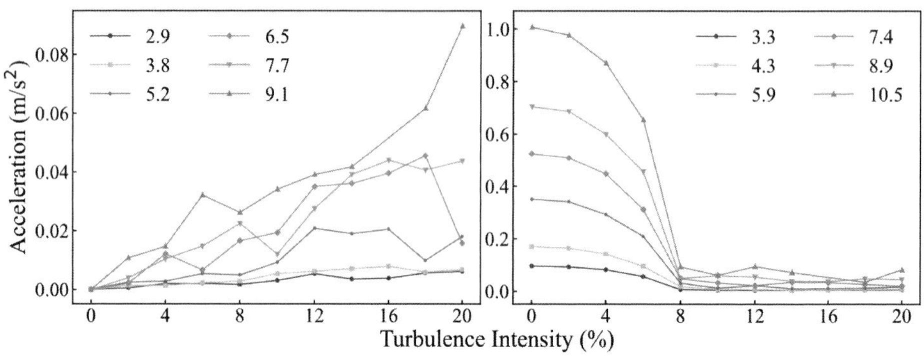

Fig. 10. Variation of the maximum vertical acceleration of the twin-box deck against turbulence intensity and wind speed (outside and inside the lock-in regions).

is that the rising turbulence intensity restrains the deck's maximum displacement response and that the reduction effect keeps increasing before the turbulence intensity reaches 8%. After that turbulence intensity, the deck's maximum displacement response maintains in a relatively low level even though the turbulence intensity keeps rising. The above trend is almost similar for all the concerned wind speeds despite the deck vibrating in different modes.

The left subpicture of Fig. 10 shows that not only higher turbulence intensity but also higher wind speed results in higher acceleration response when VIV does not appear (outside the lock-in regions). The right subpicture of Fig. 10 illustrates that within the lock-in regions, the maximum acceleration response of the deck vibrating at different modes decreases to a similar low level when the turbulence intensity reaches 8%, but the reduction effect that turbulence brings for high order VIV is more considerable than that for low order VIV. It is not hard to find that

8% turbulence intensity is a critical turbulence intensity for the Xihoumen Bridge. Before this value, VIV can be identified although the deck's maximum displacement and acceleration response are reduced with increasing turbulence intensity. After this value, VIV seems to disappear as the deck's vibration is similar to that outside the lock-in regions.

To explore the turbulence effect further, the response time histories at the designated node of the maximum model shape are extracted. Figures 11 and 12 give

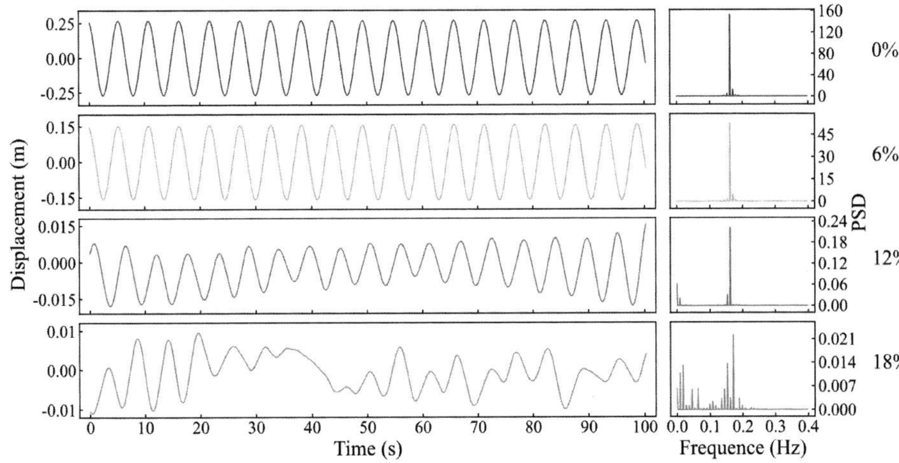

Fig. 11. Vertical displacement responses of the twin-box deck at several turbulence intensities (mean wind speed of 5.9 m/s).

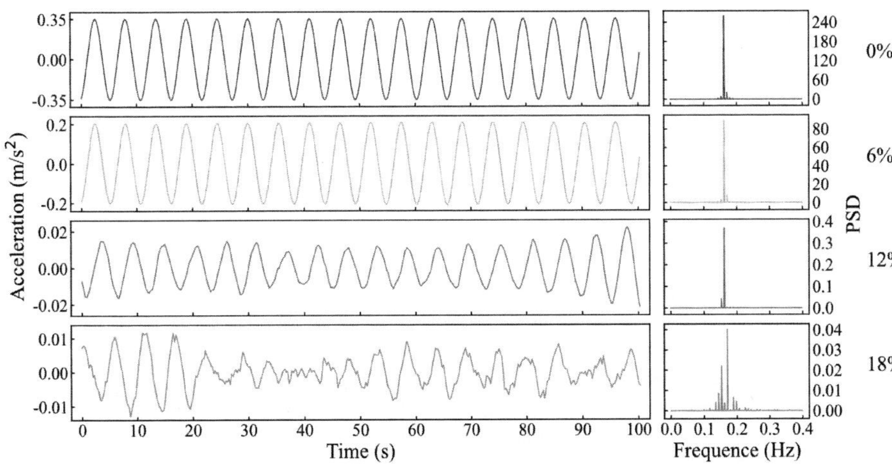

Fig. 12. Vertical acceleration responses of the twin-box deck at several turbulence intensities (mean wind speed of 5.9 m/s).

the time histories of the node's displacement and acceleration as well as their PSD when the mean wind speed is 5.9 m/s and the turbulence intensity varies from 0% to 18%. The figures show that not only the amplitude of the deck's vibration keeps reducing with the ascending turbulence intensity but also its pattern changes considerably. When the deck vibrates in wind field of relatively low turbulence intensity, its pattern closes to a harmonic vibration. However, when the turbulence intensity rises to a relatively high level, the deck's vibration shows a degree of randomness especially for the deck's acceleration response. The PSD curves further prove this point that the vibration energy concentrates at the VIV frequency, 0.182 Hz, when the turbulence intensity is less than 8% but its distribution becomes wider as the turbulence intensity further increases.

6.2. *Dynamic responses of the vehicle running on the deck*

In this subsection, the dynamic response of the sedan car running on the deck subjected to VIV in turbulent wind field is analyzed through the VVB system to uncover the turbulence effects. The road condition is supposed as 'smooth' to exclude the influence of road roughness and the driving speed is set as 80 km/h.

Figures 13 and 14 give the maximum vertical displacement and acceleration responses of the vehicle, respectively. It can be seen that with the increase of

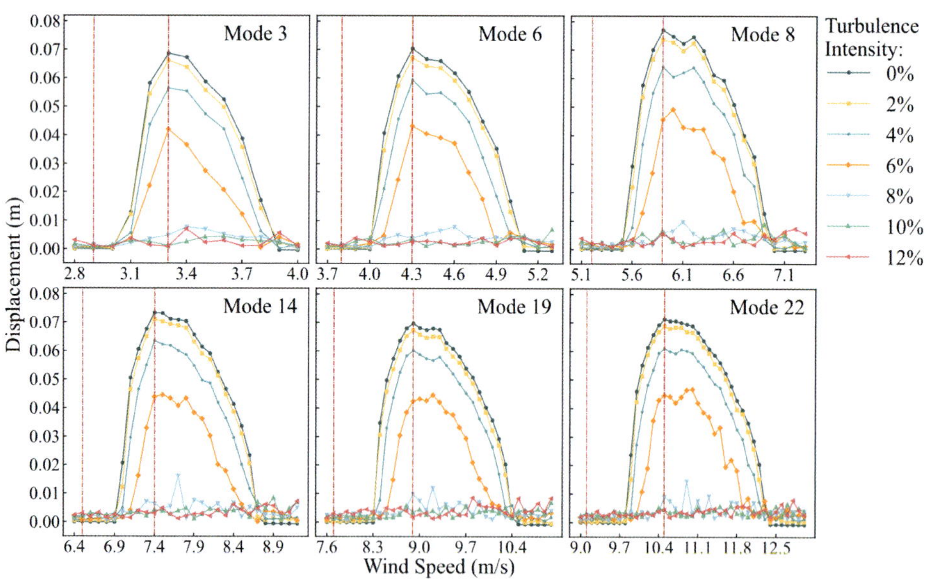

Fig. 13. Maximum vertical displacement of the vehicle in turbulent wind field.

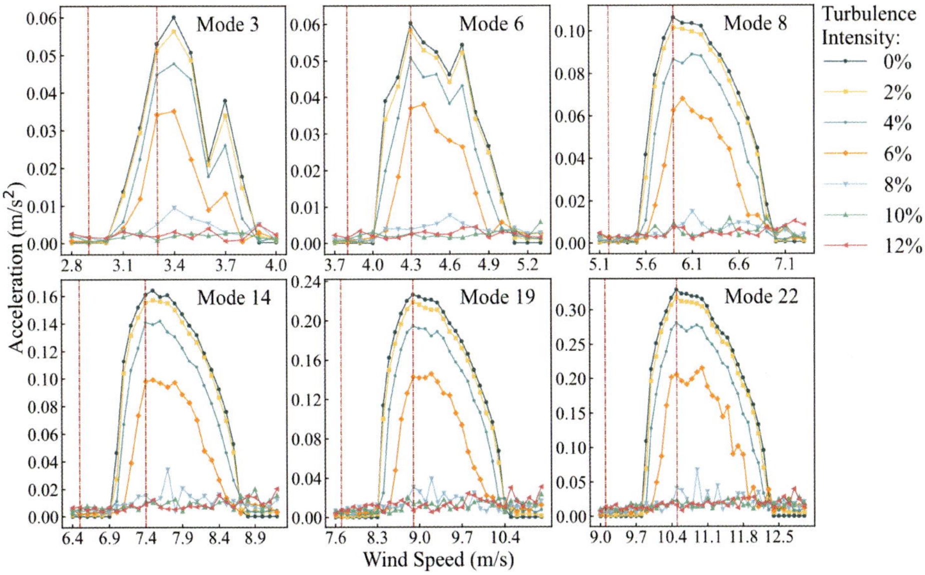

Fig. 14. Maximum vertical acceleration of the vehicle in turbulent wind field.

turbulence intensity, both the maximum displacement and acceleration responses of the vehicle within the lock-in regions of the bridge deck decrease continuously until reaching a level similar to that outside the lock-in regions of the bridge deck. However, the width of the lock-in region for different modes is affected only slightly. It is noted that the static displacement of the vehicle, induced by the gravity as well as the static wind forces, is excluded in the figures.

To explore more details, the same mean wind speeds as selected in the last subsection and marked by the chain lines in Figs. 13 and 14 are selected and the corresponding displacement and acceleration responses against turbulence intensity are displayed in Figs. 15 and 16. From the figures, it can be seen that outside the lock-in regions, the displacement response of the vehicle increases with increasing turbulence intensity but it is less affected by wind speed. The maximum displacements response of the vehicle within the lock-in regions among different modes of vibration of the bridge and excited by different wind speeds decreases with increasing turbulence intensity but it is also less affected by wind speed. It is noted that the wind speed mentioned here is the actual wind speed rather than the reduced wind speed that is actually constant when the maximum responses refer to. For acceleration response of the vehicle, it increases with increasing turbulence intensity and wind speed outside the lock-in regions but it decreases with increasing turbulence intensity and increases with increasing wind speed inside the lock-in regions.

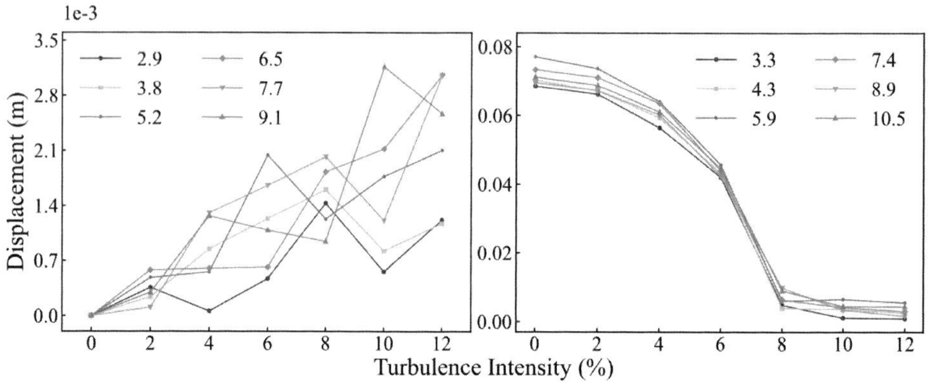

Fig. 15. Variation of the maximum vertical displacement of the vehicle against turbulence intensity and wind speed (outside and inside the lock-in regions).

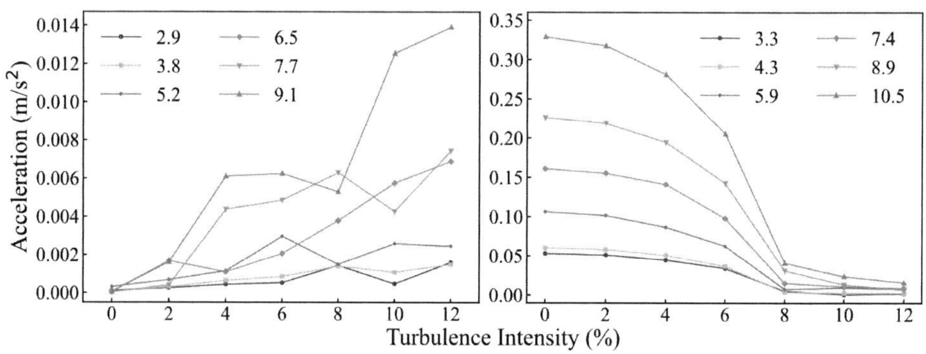

Fig. 16. Variation of the maximum vertical acceleration of the vehicle against turbulence intensity and wind speed (outside and inside the lock-in regions).

The time histories of the sedan car's displacement and acceleration responses when the wind speed is 5.9 m/s are shown in Figs. 17 and 18, in which the location of the vehicle is counted by the vehicle's distance from the north end of the deck. The corresponding PSD curves are drawn in the right subpictures of Figs. 17 and 18. From these figures, it can be found that the vehicle's vibration seems to be dominated by the deck's vertical vibration. When the vehicle moves to the 1/2 side span as well as the 1/4 and 3/4 main span, where the deck's mode shape is relatively large as the 8th mode of vibration of the deck is excited, it vibrates with large displacement and acceleration, but both of them reduce significantly when the vehicle moves to the ends of the deck or the tower, 578 m from the deck's north end. A special point, 1 403 m from the deck's north end, is noticed since the vehicle vibrates slightly when it closes to this point. The reason behind this is that the

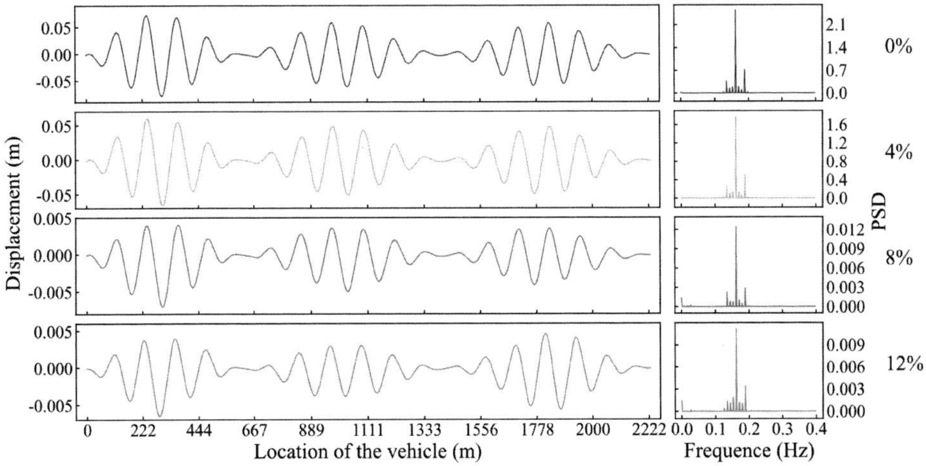

Fig. 17. Vertical displacement responses of the vehicle at several turbulence intensities (mean wind speed of 5.9 m/s).

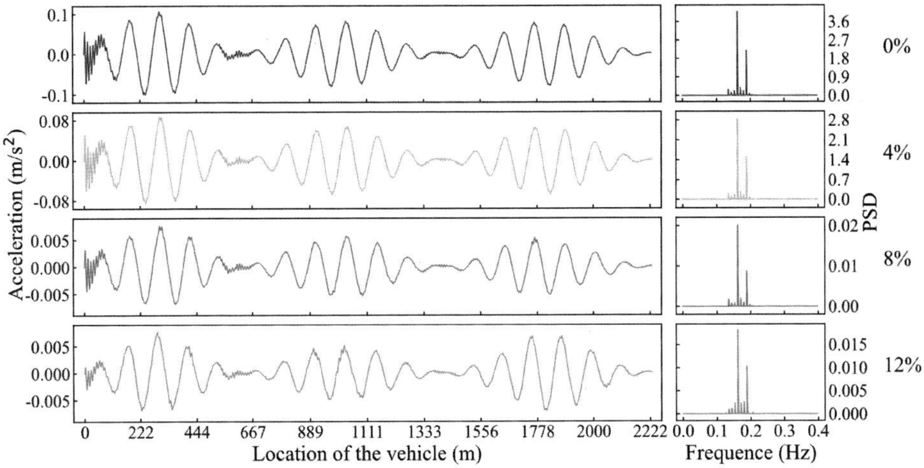

Fig. 18. Vertical acceleration responses of the vehicle at several turbulence intensities (mean wind speed of 5.9 m/s).

point is the stagnation point of the deck when the 8th mode of vibration is excited. Furthermore, different from the deck, whose vibration time histories are drawn in Figs. 11 and 12, the pattern of the vehicle's vibration seems not to be affected by the increasing turbulence despite its reducing amplitude. The similar PSD patterns shown by the right subpictures of Figs. 17 and 18 also prove this point and demonstrate that the vehicle's vibration is dominated by the component having the similar frequency with the deck even though turbulence exists.

6.3. *Effects of buffeting forces*

As discussed in Secs. 2 and 3, not only a nonlinear damping term in the VIF model but also the buffeting forces due to the turbulent wind component are considered in this study. For evaluating the effects of buffeting forces on the VVB system's dynamic responses, which are often ignored in the existing studies, this subsection compares the vibration of the deck and the sedan car with and without the action of buffeting forces, while the road roughness is supposed as 'smooth' and the driving speed is set as 80 km/h. What should be noted is that the VIF model's nonlinear damping term induced by turbulence is always included whether or not the buffeting forces are included.

To facilitate the quantification of the effects of buffeting forces, the following coefficient is defined:

$$Con = \frac{R_{wb} - R_{nb}}{R_{wb}}, \tag{23}$$

where Con is the buffeting force effect ratio; R_{wb} and R_{nb} represent the amplitude of the vibration of the deck or the vehicle with and without the buffeting forces, respectively.

By the definition, the meanings of different Con values can be summarized as follows: 1 represents that the vibration is totally determined by the buffeting forces and VIF does not exist; 0 indicates the vibration is totally determined by the VIF and there is no the effect of buffeting forces; the values lower than 0 mean that the buffeting forces restrain the VIR; the values higher than 1 mean the amplitude of the VIR is smaller than 0, which is theoretically impossible; and the values between 0 and 1 represent the extent of buffeting force effects on the vibration of either deck or vehicle.

What can be found from Figs. 19 and 20, which show the "Con" values of the deck's displacement and acceleration responses, is that the buffeting forces totally determine the deck's vibration when turbulence intensity exceeds 8% or wind speed is beyond the lock-in regions. For wind conditions with turbulence intensities lower than 8%, the VIV of the bridge deck induced by certain wind speeds, such as 5.7 m/s to 6.6 m/s of 6% turbulence intensity for Mode 8, dominates the deck's vibration and yields a "plateau region" around the zero "Con" value although the buffeting forces still bring slight fluctuations as shown by the deviations of Con from zero in the plateau region. The fluctuation of the "Con" value around zero value is more obvious for the cases with higher turbulence intensity. In the "descending region" or the "ascending region" of the "Con" value, such as 6.9 ~ 7.2 m/s and 8.4 ~ 8.8 m/s with 6% turbulence intensity for Mode 14, both the VIF and the buffeting forces play role in exciting the deck even though the turbulence intensity drops to 2%. This indicates that within the descending region or the ascending region, the buffeting forces should not be ignored. As the "Con" value in the "plateau region" is also

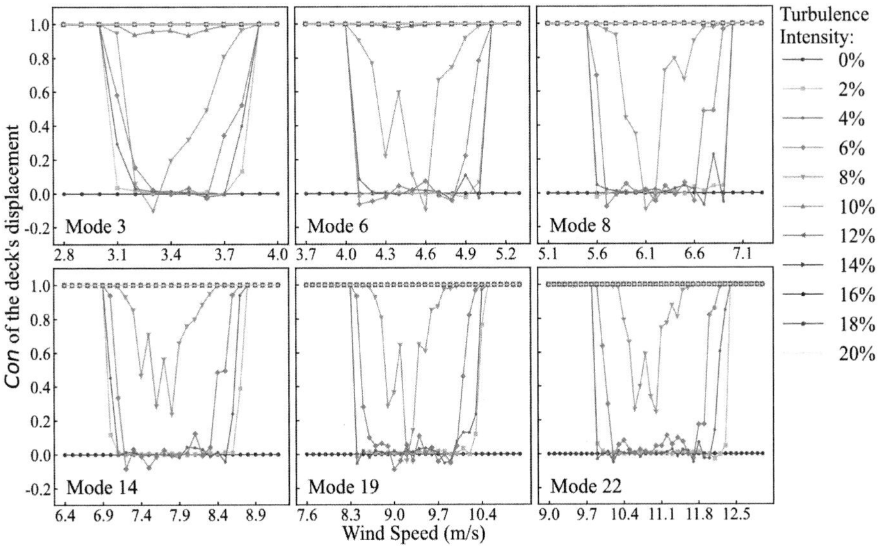

Fig. 19. Effects of buffeting forces on deck's displacement responses.

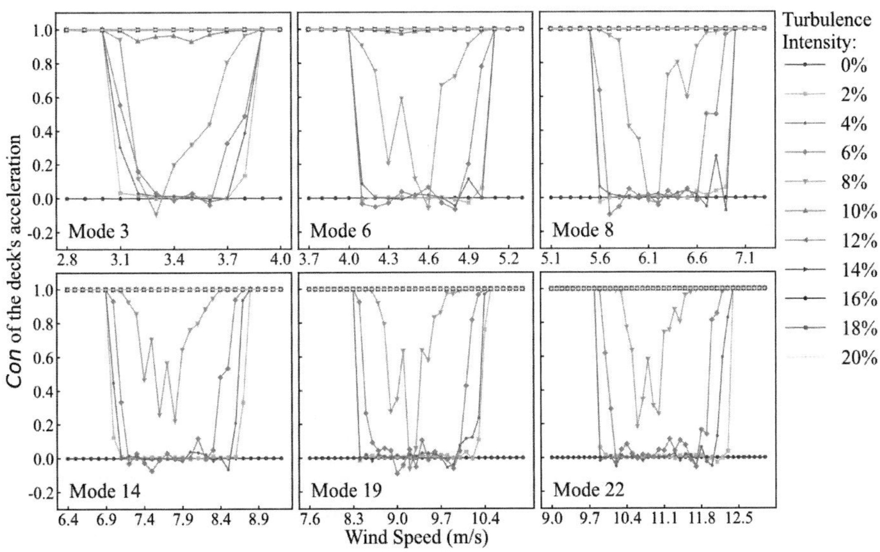

Fig. 20. Effects of buffeting forces on deck's acceleration responses.

fluctuating, it will be more accurate to include the buffeting forces in prediction for all wind conditions.

Figures 21 and 22 display the "*Con*" value of the vehicle's displacement and acceleration response, respectively. The similar trends with the deck can be

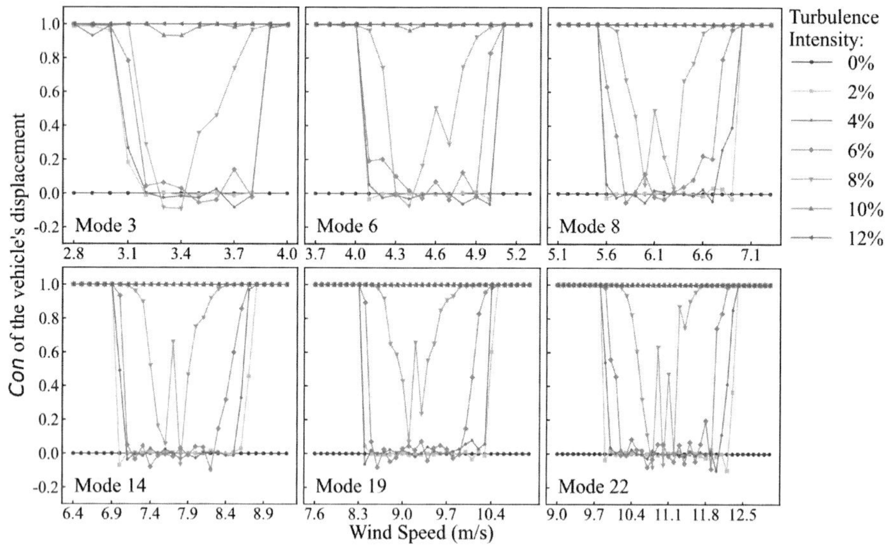

Fig. 21. Effects of buffeting forces on vehicle's displacement responses.

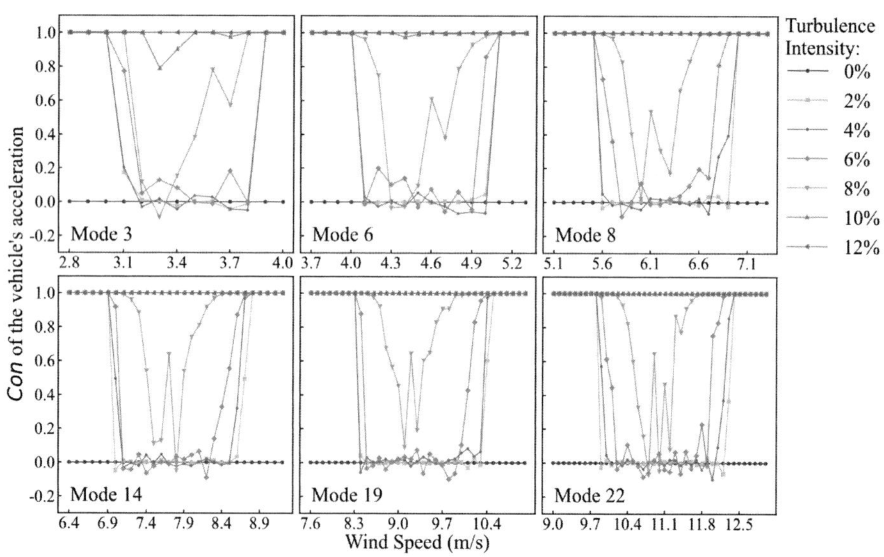

Fig. 22. Effects of buffeting forces on vehicle's acceleration responses.

observed, which indicates that the buffeting forces play the same role in determining the dynamic response of the vehicle. In summary, although the buffeting forces contribute slightly to the VVB system's vibration when the wind speed falls in the "plateau region" and the turbulence intensity is lower than 8%, they should not be ignored at least in other wind conditions, such as the wind speed in the "ascending region" and the "descending region" with weak turbulence as well as the wind with the critical and above turbulence intensity.

6.4. *Ride comfort of the vehicles*

Ride comfort is the key factor that the managers have to consider when making the operation decisions during VIV of the bridge. In order to explore the turbulence effects on ride comfort of the vehicles, this subsection first considers the ride comfort of a sedan car running on the deck experiencing VIV in turbulent wind field according to ISO 2631-1.[37] After that, the influences of other factors, including driving speed, road condition and vehicle type, are analyzed for a turbulence intensity of 4%.

6.4.1. *Turbulence effects on the sedan car's ride comfort*

Let the sedan car run on the deck with smooth surface to evaluate its ride comfort according to ISO 2631-1[37] without considering the effects of road roughness. Figure 23 gives the vehicle's OVTV within the multi-mode lock-in regions when the driving speed is 80 km/h. From the figure, it can be found that while the higher order mode of the deck is excited by higher wind speed, which reasonably results in ascending OVTV, the increasing turbulence intensity brings lower OVTV, which

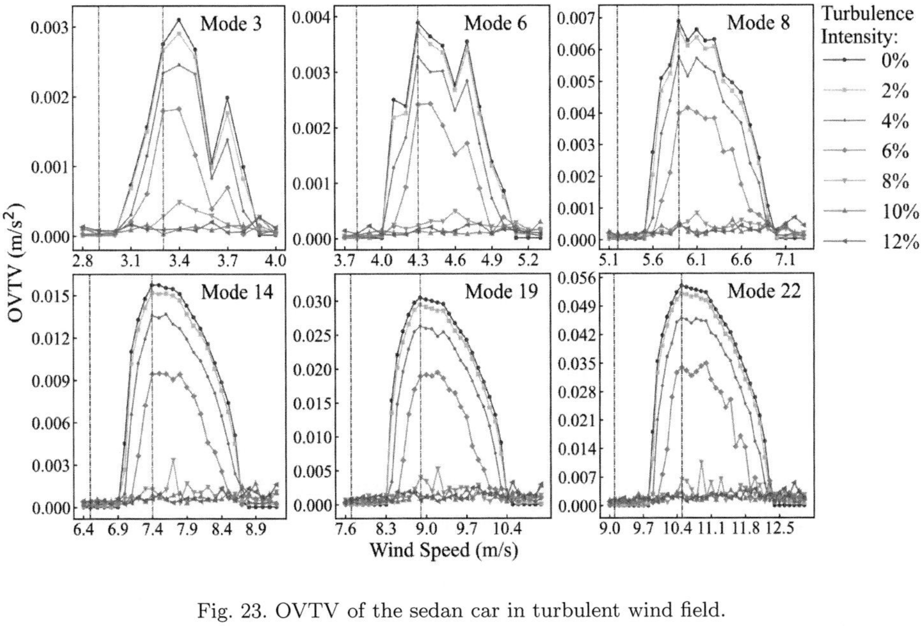

Fig. 23. OVTV of the sedan car in turbulent wind field.

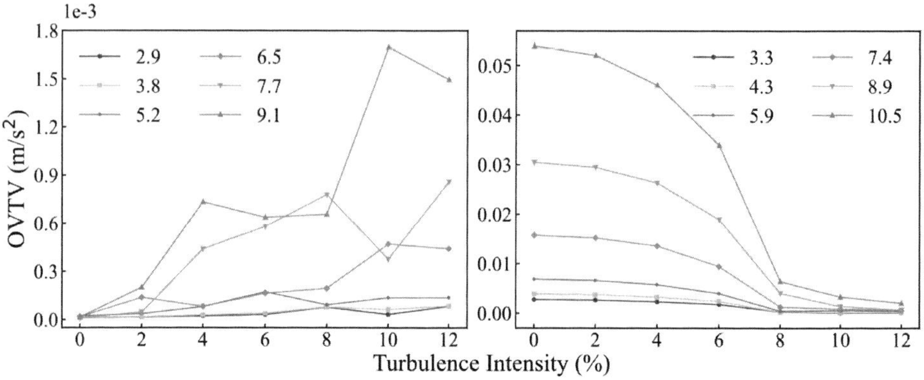

Fig. 24. Variation of the sedan car's OVTV against turbulence intensity and wind speed (outside and inside the lock-in regions).

confirms that the turbulence existing in the reality can improve the ride comfort significantly. To display such a phenomenon more clearly, some wind speeds, marked by the chain lines in Fig. 23, are selected and the corresponding OVTV against turbulence intensity is drawn in Fig. 24. The left subpicture of Fig. 24 shows that both higher wind speed and stronger turbulence can deteriorate the vehicle's ride comfort but the OVTV values are so small that the driver will not be uncomfortable according to the criteria listed in Table 2. The right subpicture of Fig. 24 confirms that although higher order VIV brings higher OVTV, the increasing turbulence makes more significant influence on high order VIV. After the turbulence intensity reaches 8%, the OVTV within all the six lock-in regions drops to the approximate level lower than 0.01. This trend is similar to the trend shown in Fig. 16 for the deck's acceleration, as OVTV is determined by the vehicle's acceleration. The same trends can also be observed in Figs. 25 and 26 for the MSI values of the vehicle.

From the phenomenon discussed above, the turbulence effects on the sedan car's ride comfort can be summarized that during VIV, rising turbulence intensity can considerably improve the vehicle's ride comfort until OVTV and MSI value drops to the level similar to that when only the buffeting occurs. Furthermore, what must be emphasized is that although turbulence effects on the vehicle's ride comfort can be revealed by OVTV and MSI values of the vehicle in different turbulent wind fields, their absolute values are so small that the driver's feeling is believed as 'no uncomfortable' according Table 2 and only 4.2% of the drivers will be sick even when the maximum amplitudes of the deck's displacement and acceleration reach 0.27 m and 1.0 m/s^2, respectively. However, in reality such large amplitudes of the deck's displacement and acceleration responses will already obstruct the drivers' sight, cause discomfort to the drivers, and give rise to the drivers' concern about the safety of the bridge. Therefore, both of the indexes used above according to the

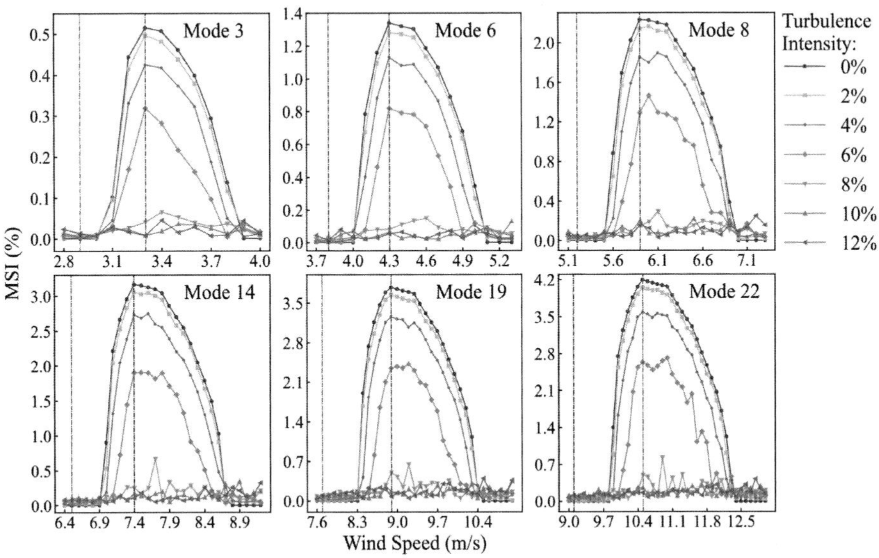

Fig. 25. MSI of the sedan car in turbulent wind field.

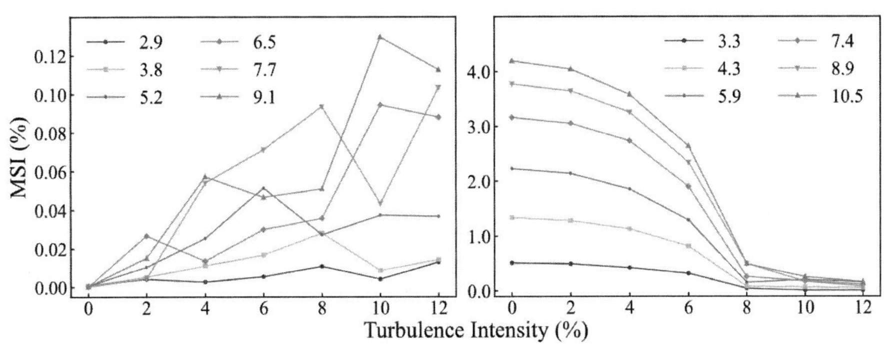

Fig. 26. Variation of the sedan car's MSI against turbulence intensity and wind speed (outside and inside the lock-in regions).

standard may not be suitable for evaluating the ride comfort of the vehicles running on the bridge deck experiencing VIV. A new evaluation standard, taking not only the drivers' physiological status but also their psychology into consideration, is needed and worthy to be investigated in the near future.

6.4.2. *Ride comfort with various driving speeds*

Assuming that the deck is smooth and the turbulence intensity is 4%, Fig. 27 gives the sedan car's OVTV against driving speed and wind speed for outside and inside

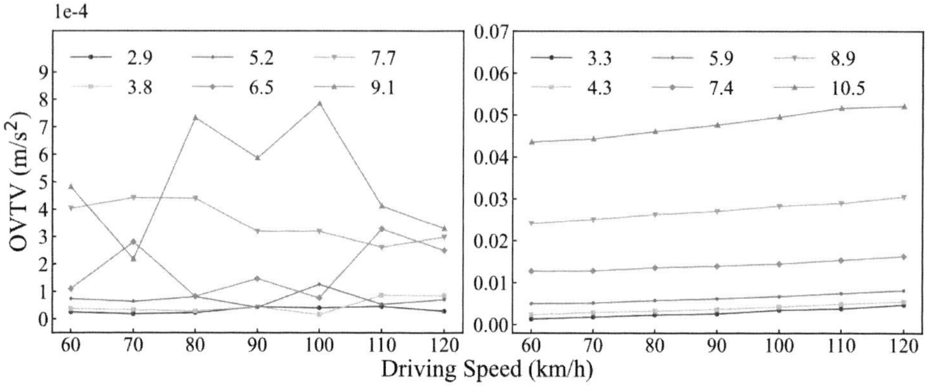

Fig. 27. Variation of the sedan car's OVTV against driving speed and wind speed (outside and inside the lock-in regions).

the lock-in regions. The left subpicture of Fig. 27 demonstrates that the OVTV value is quite small and seems not to be impacted by driving speed when VIV does not occur. However, the right one of Fig. 27 illustrates that the OVTV value increases moderately with the rising driving speed when VIV occurs. Furthermore, the right subpicture also shows that the amplification effect from the ascending driving speed is more obvious when higher order VIV appears, which can be explained by a higher wave number of the deck when it vibrates in higher order mode of vibration.

Figure 28 shows the sedan car's MSI value against driving speed and wind speed outside and inside the lock-in regions. It can be seen that the MSI value is quite small outside the lock-in region, compared with that inside the lock-in region. Increasing

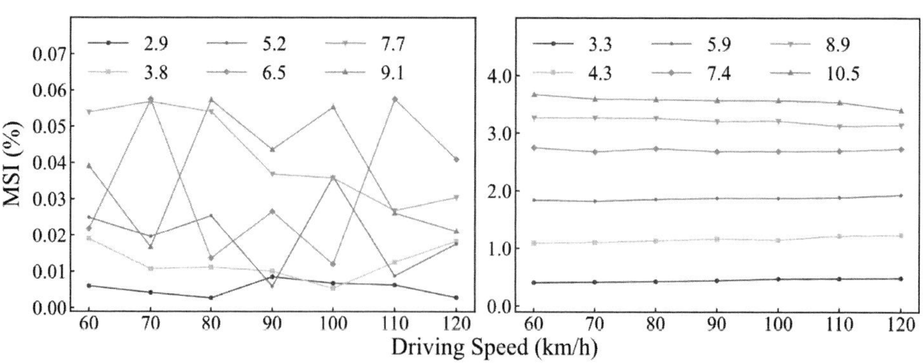

Fig. 28. Variation of the sedan car's MSI against driving speed and wind speed (outside and inside the lock-in regions).

Table 5. The selected four cases and their weighting factors.

Case no.	Wind speed (m/s)	Driving speed (km/h)	Deck's frequency	Vehicle's frequency	RMS	Weighting factors W_k	W_d	W_c	W_e	W_f
1	3.3	60	0.097	0.105	0.010	0.032	0.066	0.070	0.068	0.738
2	3.3	120	0.097	0.120	0.013	0.042	0.088	0.090	0.088	0.862
Difference = 2(Case 2 − Case 1)/(Case 2 + Case 1)					26.1%	27.0%	28.6	25.0%	25.6%	15.5%
3	10.5	60	0.323	0.359	0.112	0.310	0.622	0.630	0.623	0.490
4	10.5	120	0.323	0.389	0.119	0.339	0.684	0.690	0.684	0.415
Difference = 2(Case 4 − Case 3)/(Case 4 + Case 3)					6.1%	8.9	9.5%	9.1%	9.3%	−16.6%

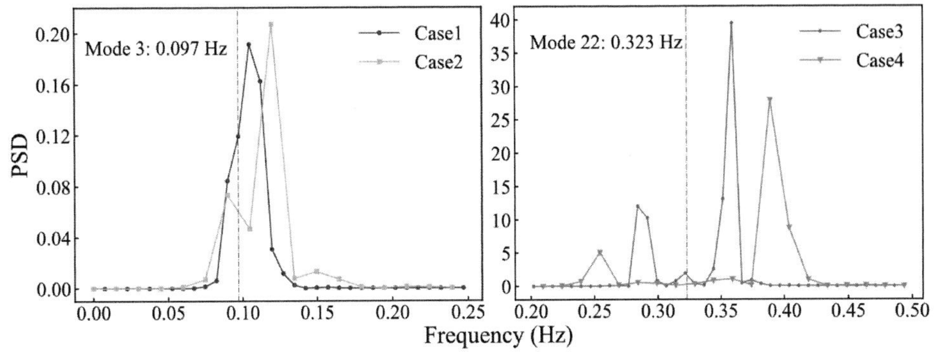

Fig. 29. PSD of the sedan car's acceleration for the selected four cases.

driving speed does not make more drivers vomit no matter whether VIV occurs or not. To explore the reason behind this, the four cases listed in Table 5 are selected, the PSD of the vehicle's acceleration responses for the four cases are worked out and displayed in Fig. 29, and the peak frequencies of the vehicle and the corresponding weighting factors are listed in Table 5. It can be found from Table 5 and Fig. 29 that increasing driving speed results in the increase of both the vibration frequency of the vehicle and the root mean squares (RMS) value of the vehicle's acceleration response even though the deck vibrates still in the same mode of vibration. It should be noted that the vehicle's frequency mentioned here is the main frequency of the vehicle's response, which is obtained via the fast Fourier transform (FFT), rather than its inherent natural frequencies. The change in the vehicle's frequency moderately causes the change of the weighting factors according to Fig. 4. For Case 1 and Case 2 listed in Table 5, the increase of the weighting factors for calculating OVTV, including W_k, W_d, W_c, and W_e, approximates 26% but that for MSI, W_f, is only 15.5%, indicating that OVTV's increasing rate is higher than MSI's increasing rate even though the RMS of the vehicle's acceleration rises by the same degree.

For Case 3 and Case 4 listed in Table 5, while the weighting factors for OVTV keep going up with the increasing driving speed, the weighting factors for MSI decrease considerably. This can explain why there is an obvious decrease in the MSI curve at wind speed of 10.5 m/s while the OVTV curve keeps increasing with increasing driving speed. In summary, faster driving speed brings not only higher RMS of vehicle response but also higher frequency of the vehicle. Because of the sensitivity of the two indexes in terms of the weighting factors to vehicle vibration frequency, the driving speed's effects on the two indexes are different.

6.4.3. *Effects of road roughness on ride comfort*

To reveal the effects of road roughness on the vehicle's ride comfort, a sedan car running on the deck with 80 km/h in the wind field of 4% turbulence intensity is considered in this subsection. It can be seen from Fig. 30 that a dramatic increase of OVTV is observed when the road condition deteriorates. The overlapped lines in each subpicture of Fig. 30 illustrate that road roughness determines the OVTV value no matter whether VIV occurs or not. The OVTV value up to 5.0, which is much higher than that in the smooth road condition, also confirms the dominant role of road condition in determining the vehicle's OVTV.

Figure 31 displays the variation of the vehicle's MSI value with road condition for both outside and inside the lock-in regions. The left subpicture of Fig. 31 shows that the vehicle's MSI is also determined by the road condition when VIV does not occur, but the right subpicture of Fig. 31 demonstrates that the VIV's role affecting the driver's sickness cannot be ignored. The reason behind this phenomenon can be explained by the fact that MSI value is more sensitive to lower frequency but OVTV value is more sensitive to higher frequency. It can be found from Table 5

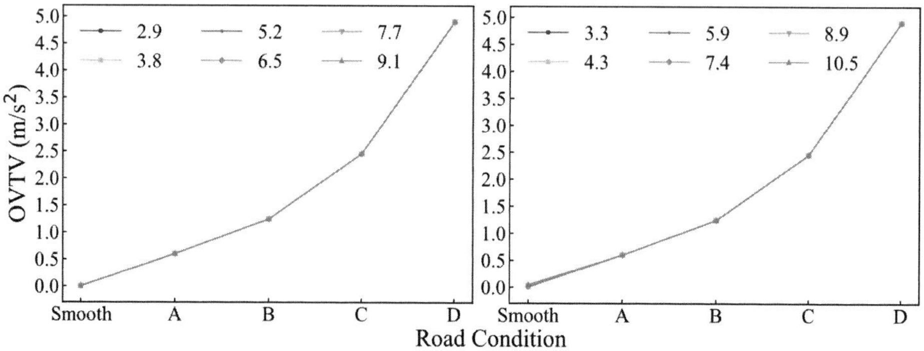

Fig. 30. Variation of the sedan car's OVTV against road roughness and wind speed (outside and inside the lock-in regions).

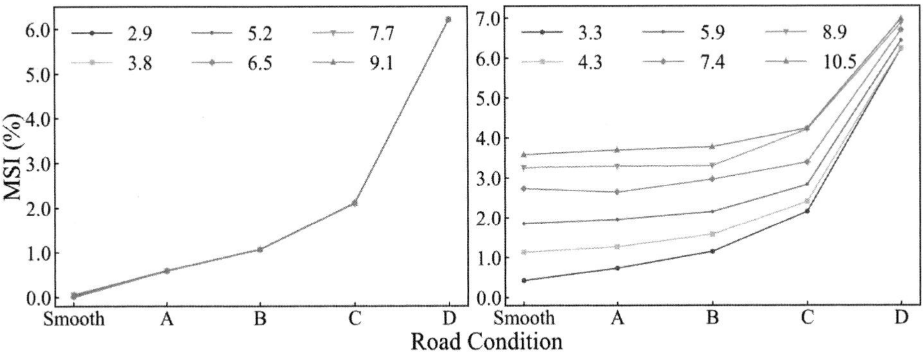

Fig. 31. Variation of the sedan car's MSI against road roughness and wind speed (outside and inside the lock-in regions).

that the frequency of the vehicle vibration induced by VIV ranges from 0.1 Hz to 0.4 Hz for the Xihoumen Bridge. According to Fig. 4, all the weighting factors for OVTV within this range are small and this range is far away from the regions of the maximum weighting factors of OVTV, which are within 1–10 Hz, so that OVTV is not sensitive to VIV. This condition results in the overlapped lines in the right subpicture of Fig. 30 and the difference which the VIV brings to OVTV can be neglected, compared with that from the road condition. However, this range covers the frequency at which the weighting factor for MSI reaches 1 at 0.2 Hz and the weighting factor for MSI drops to 0.2 when the vehicle's frequency exceeds 0.5 Hz. These two factors make MSI more sensitive to VIV than road roughness and increase the distances between the neighboring curves in the right subpicture of Fig. 31, in which the vertical distances between the curves at the same road condition represent the change caused by VIV, while the increase in the given curve represents the change caused by the road condition. Therefore, compared with OVTV and considering the real situation of VIV, MSI is more suitable for evaluating the drivers' ride comfort when VIV occurs.

6.4.4. *Ride comfort of different vehicle types*

It is easy to understand that the ride comfort of different types of vehicles will be different owing to different dynamic properties of the vehicles. In order to uncover their ride comfort of different types of vehicles in turbulent wind field, the three typical vehicles, including sedan car, light truck and heavy truck, are taken into consideration in this subsection. The OVTV and MSI values of the vehicles, when the road condition is supposed as 'smooth' and turbulence intensity is set as 4%,

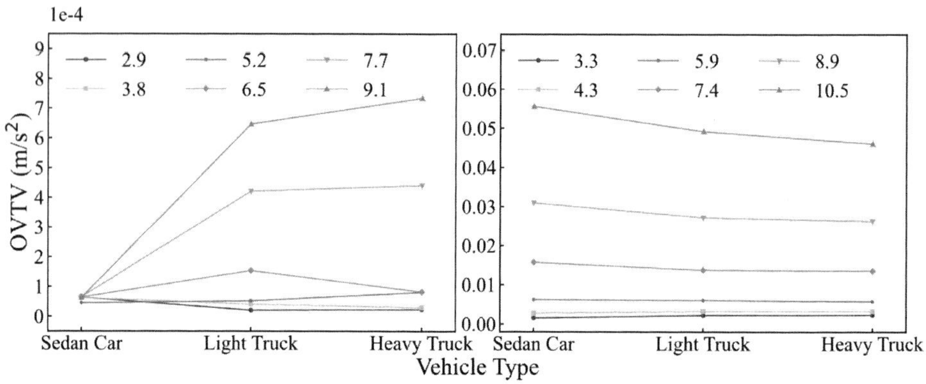

Fig. 32. Variation of OVTV against vehicle type and wind speed (outside and inside the lock-in regions).

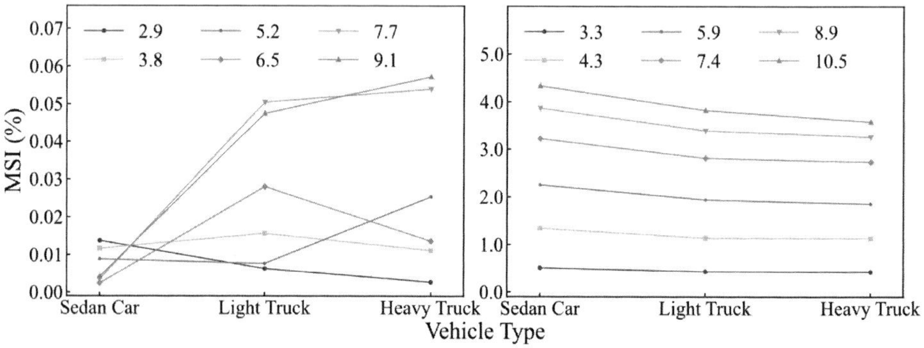

Fig. 33. Variation of MSI against vehicle type and wind speed (outside and inside the lock-in regions).

are calculated and the results are plotted in Figs. 32 and 33 for both outside and inside the lock-in regions.

In the left subpictures of the two figures without VIV, the OVTV and MSI values of the two trucks tend to increase with increasing wind speed while that of the sedan car keeps relatively stable. This is because higher turbulent wind forces act on the trucks of larger side area. However, all the OVTV and MSI values in the left subpictures of the two figures are so small that the difference can be ignored when there is no VIV. From the right subpictures of the two figures, it can be found that heavier vehicle has slightly better ride comfort, compared with the lighter vehicle, when higher order VIV occurs by higher wind speed. The reasons behind this phenomenon can be explained by the fact that the fluctuation of the contact forces due to VIV dominates the vehicle's behavior rather than the buffeting forces and the light car has less gravity force, making it more susceptible to the fluctuating contact forces.

7. Concluding Remarks

The coupled VVB system applicable to the multi-mode lock-in regions of the deck in turbulent wind field has been established and applied to a real long suspension bridge, and the vortex-induced dynamic behavior of the bridge and the ride comfort of the vehicles have been explored in this study. Concluding remarks from this study can be made as follows:

(1) Different from the displacement response of the bridge deck whose amplitude seems to be similar within the lock-in regions of different modes of vibration, higher order VIV induced by higher wind speed results in higher acceleration response of the bridge deck.
(2) With the increase of turbulence intensity, both the maximum displacement and acceleration responses of the bridge deck within the lock-in regions decrease continuously until reaching a level similar to that outside the lock-in regions.
(3) The increasing turbulence intensity not only restrains the vibration amplitude but also changes the vibration pattern of the bridge deck, in which the displacement time history of the bridge deck is far from a quasi-harmonic VIV curve when turbulent intensity is higher.
(4) There is a critical turbulence intensity of 8% for the Xihoumen Bridge. Before this critical value, the VIV of the bridge deck could occur and higher turbulent intensity yields less VIR. When the turbulence intensity exceeds this value, the VIV of the bridge deck is very weak and the buffeting dominates the bridge vibration.
(5) With the increase of turbulence intensity, both the maximum displacement and acceleration responses of the vehicle within the lock-in regions of the bridge deck decrease continuously until reaching a level similar to that outside the lock-in regions of the bridge deck. The increasing turbulence intensity thus improves the ride comfort of the vehicles running on the deck during VIV.
(6) The buffeting would dominate the coupled system's vibration when the critical turbulence intensity exceeds or the wind speed is beyond the lock-in regions. The VIF would dominate the system's vibration when the maximum VIV occurs and the turbulence intensity is less than its critical value. Within the ascending range and the descending range of the lock-in region, both VIF and buffeting forces play important role in exciting the system's vibration.
(7) According to the two indexes stipulated in ISO 2631-1[29] for evaluating the ride comfort of vehicles running on the ground, the drivers of the vehicles running on the bridge deck at lock-in do not feel any uncomfortable even when the

amplitude of the deck's vibration ups to 0.27 m. This is in contrast with the actual situation when the bridge will be closed to traffic. Thus, a new evaluation standard, taking not only the drivers' physiological status sbut also their psychology into consideration, is needed and worthy to be investigated in the near future.

(8) Higher driving speed would result in higher frequency of vehicle vibration and higher acceleration response of the vehicle when it runs on the deck at lock-in. Since the different sensitivities of the two indexes, OVTV and MSI, to vehicle vibration frequency, the MSI value of the vehicle would be different from the OVTV value. The results from this study show that MSI may be more suitable for evaluating the vehicle's ride comfort when it running on the real deck at lock-in.

(9) Heavier vehicle would have a better ride comfort than lighter vehicle when the vehicle runs on the bridge during VIV. The road condition (roughness) would impair the vehicle ride comfort significantly no matter whether VIV of the bridge occurs or not.

Acknowledgments

The works described in this paper are financially supported by the Changjiang Scholars Program of the Ministry of Education of China (SWJTU-WB0100111012209) to which the authors are most grateful. Any opinions and conclusions presented in this paper are entirely those of the authors.

Appendix

The equations of motion of the three-axle vehicle are listed below:

$$\begin{aligned} m_{v1}\ddot{Z}_{v1} = {} & F_{st1}^{VL} + F_{bu1}^{VL} - m_{v1}g - m_{v2}gd_1/(d_4+d_6) + K_{s1}(-2Z_{v1} + Z_{w1} \\ & + Z_{w2} + 2\theta_{v1}d_1) + K_{s2}(-2Z_{v1} + Z_{w3} + Z_{w4} - 2\theta_{v1}d_2) \\ & + C_{s1}(-2\dot{Z}_{v1} + \dot{Z}_{w1} + \dot{Z}_{w2} + 2\dot{\theta}_{v1}d_1) \\ & + C_{s2}(-2\dot{Z}_{v1} + \dot{Z}_{w3} + \dot{Z}_{w4} - 2\dot{\theta}_{v1}d_2), \end{aligned} \tag{A.1}$$

$$\begin{aligned} I_{yv1}\ddot{\theta}_{v1} = {} & F_{st1}^{VP} + F_{bu1}^{VP} + m_{v1}gh\theta_{v1} - m_{v2}gd_5d_4/(d_4+d_6) \\ & + K_{s1}d_1(2Z_{v1} - Z_{w1} - Z_{w2} - 2\theta_{v1}d_1) + K_{s2}d_2(-2Z_{v1} + Z_{w3} \\ & + Z_{w4} - 2\theta_{v1}d_2) + C_{s1}d_1(2\dot{Z}_{v1} - \dot{Z}_{w1} - \dot{Z}_{w2} - 2\dot{\theta}_{v1}d_1) \\ & + C_{s2}d_2(-2\dot{Z}_{v1} + \dot{Z}_{w3} + \dot{Z}_{w4} - 2\dot{\theta}_{v1}d_2), \end{aligned} \tag{A.2}$$

$$I_{xv1}\ddot{\varphi}_{v1} = F_{st1}^{VR} + F_{bu1}^{VR} + m_{v1}gh\varphi_{v1} + K_{s1}t_w(Z_{w1} - Z_{w2} - 2\varphi_{v1}t_w) + K_{s2}t_w(Z_{w3} - Z_{w4} - 2\varphi_{v1}t_w) + C_{s1}t_w(\dot{Z}_{w1} - \dot{Z}_{w2} - 2\dot{\varphi}_{v1}t_w) + C_{s2}t_w(\dot{Z}_{w3} - \dot{Z}_{w4} - 2\dot{\varphi}_{v1}t_w), \quad (A.3)$$

$$m_{v2}\ddot{Z}_{v2} = F_{st2}^{VL} + F_{bu2}^{VL} - m_{v2}g + K_{s3}(-2Z_{v2}(d_6 + d_4)/d_6 + Z_{w5} + Z_{w6}) + C_{s3}(-2\dot{Z}_{v2}(d_6 + d_4)/d_6 + \dot{Z}_{w5} + \dot{Z}_{w6}), \quad (A.4)$$

$$I_{xv2}\ddot{\varphi}_{v2} = F_{st2}^{VR} + F_{bu2}^{VR} + m_{v2}gh\varphi_{v2} + K_{s3}t_w(Z_{w5} - Z_{w6} - 2\varphi_{v2}t_w) + C_{s3}t_w(\dot{Z}_{w5} - \dot{Z}_{w6} - 2\dot{\varphi}_{v2}t_w), \quad (A.5)$$

$$m_{w1}\ddot{Z}_{w1} = F_{zw1} - m_{w1}g + K_{s1}(Z_{v1} - Z_{w1} - \theta_{v1}d_1 + \varphi_{v1}t_w) + C_{s1}(\dot{Z}_{v1} - \dot{Z}_{w1} - \dot{\theta}_{v1}d_1 + \dot{\varphi}_{v1}t_w) + K_{t1}(-Z_{w1} + Z_{g1}) + C_{t1}(-\dot{Z}_{w1} + \dot{Z}_{g1}), \quad (A.6)$$

$$m_{w2}\ddot{Z}_{w2} = F_{zw2} - m_{w2}g + K_{s1}(Z_{v1} - Z_{w2} - \theta_{v1}d_1 - \varphi_{v1}t_w) + C_{s1}(\dot{Z}_{v1} - \dot{Z}_{w2} - \dot{\theta}_{v1}d_1 - \dot{\varphi}_{v1}t_w) + K_{t1}(-Z_{w2} + Z_{g2}) + C_{t1}(-\dot{Z}_{w2} + \dot{Z}_{g2}), \quad (A.7)$$

$$m_{w3}\ddot{Z}_{w3} = F_{zw3} - m_{w3}g + K_{s2}(Z_{v1} - Z_{w3} + \theta_{v1}d_2 + \varphi_{v1}t_w) + C_{s2}(\dot{Z}_{v1} - \dot{Z}_{w3} + \dot{\theta}_{v1}d_2 + \dot{\varphi}_{v1}t_w) + K_{t2}(-Z_{w3} + Z_{g3}) + C_{t2}(-\dot{Z}_{w3} + \dot{Z}_{g3}), \quad (A.8)$$

$$m_{w4}\ddot{Z}_{w4} = F_{zw4} - m_{w4}g + K_{s2}(Z_{v1} - Z_{w4} + \theta_{v1}d_2 - \varphi_{v1}t_w) + C_{s2}(\dot{Z}_{v1} - \dot{Z}_{w4} + \dot{\theta}_{v1}d_2 - \dot{\varphi}_{v1}t_w) + K_{t2}(-Z_{w4} + Z_{g4}) + C_{t2}(-\dot{Z}_{w4} + \dot{Z}_{g4}), \quad (A.9)$$

$$m_{w5}\ddot{Z}_{w5} = F_{zw5} - m_{w5}g + K_{s3}(Z_{v2}(d_6 + d_4)/d_6 - Z_{w5} + \varphi_{v2}t_w) + C_{s3}(\dot{Z}_{v2}(d_6 + d_4)/d_6 - \dot{Z}_{w5} + \dot{\varphi}_{v2}t_w) + K_{t3}(-Z_{w5} + Z_{g5}) + C_{t3}(-\dot{Z}_{w5} + \dot{Z}_{g5}), \quad (A.10)$$

$$m_{w6}\ddot{Z}_{w6} = F_{zw6} - m_{w6}g + K_{s3}(Z_{v2}(d_6 + d_4)/d_6 - Z_{w6} - \varphi_{v2}t_w) + C_{s3}(\dot{Z}_{v2}(d_6 + d_4)/d_6 - \dot{Z}_{w6} - \dot{\varphi}_{v2}t_w) + K_{t3}(-Z_{w6} + Z_{g6}) + C_{t3}(-\dot{Z}_{w6} + \dot{Z}_{g6}), \quad (A.11)$$

where m_{v1}, m_{v2} and m_{wi} $(i = 1, 2, 3, 4)$ are the mass of the tractor, trailer and ith wheel, respectively; I_{yv1}, I_{xv1}, and I_{xv2} are the mass moment of inertia of the tractor about the y-axis and x-axis, and that of the trailer about the x-axis, respectively; F_{ik}^{j} $(i = st, bu,\ j = VL, VP, VR$ and $k = 1, 2)$ represents the static or buffeting force, pitching moment, and rolling moment acting on the tractor and trailer, separately; F_{zwi} $(i = 1, 2, 3, 4)$ represents the external force acting on the ith wheel without considering the gravity; g is the gravity acceleration; K_{si} and C_{si} $(i = 1, 2, 3)$ are the vertical stiffness and damping of the upper springs and dampers used to model the ith suspension system of the vehicle; K_{ti} and C_{ti} $(i = 1, 2, 3)$ are the vertical stiffness and damping of the lower springs and dampers used to model the tires on the ith axis.

For the two-axle vehicle, the variables in Eq. (10) can be listed as:

$$M = \mathrm{diag}[1 \quad 1 \quad 1 \quad 1 \quad 1 \quad 1 \quad 1 \quad m_v \quad I_x \quad I_y \quad m_{w1} \quad m_{w2} \quad m_{w3} \quad m_{w4}], \tag{A.12}$$

$$N = \begin{bmatrix} [0]_{7\times 7} & N_{12} \\ N_{21} & N_{22} \end{bmatrix}, \tag{A.13}$$

$$N_{12} = \mathrm{diag}[1 \quad 1 \quad 1 \quad 1 \quad 1 \quad 1 \quad 1], \tag{A.14}$$

$$N_{21} = \begin{bmatrix} -4K_s & 0 & 2K_s(d_1 - d_2) & K_s & K_s & K_s & K_s \\ 0 & m_v gh - 4K_s t_w^2 & 0 & t_w K_s & -t_w K_s & t_w K_s & -t_w K_s \\ 2K_s(d_1 - d_2) & 0 & m_v gh - 2K_s(d_1^2 + d_2^2) & -d_1 K_s & -d_1 K_s & d_2 K_s & d_2 K_s \\ K_s & t_w K_s & -d_1 K_s & -(K_s + K_t) & 0 & 0 & 0 \\ K_s & -t_w K_s & -d_1 K_s & 0 & -(K_s + K_t) & 0 & 0 \\ K_s & t_w K_s & d_2 K_s & 0 & 0 & -(K_s + K_t) & 0 \\ K_s & -t_w K_s & d_2 K_s & 0 & 0 & 0 & -(K_s + K_t) \end{bmatrix}, \tag{A.15}$$

$$N_{22} = \begin{bmatrix} -4C_s & 0 & 2C_s(d_1-d_2) & C_s & C_s & C_s & C_s \\ 0 & -4C_s t_w^2 & 0 & t_w C_s & -t_w C_s & t_w C_s & -t_w C_s \\ 2C_s(d_1-d_2) & 0 & -2C_s(d_1^2+d_2^2) & -d_1 C_s & -d_1 C_s & d_2 C_s & d_2 C_s \\ C_s & t_w C_s & -d_1 C_s & -(C_s+C_t) & 0 & 0 & 0 \\ C_s & -t_w C_s & -d_1 C_s & 0 & -(C_s+C_t) & 0 & 0 \\ C_s & t_w C_s & d_2 C_s & 0 & 0 & -(C_s+C_t) & 0 \\ C_s & -t_w C_s & d_2 C_s & 0 & 0 & 0 & -(C_s+C_t) \end{bmatrix}, \tag{A.16}$$

$$P = \begin{bmatrix} [0]_{7\times 7} & [0]_{7\times 4} & [0]_{7\times 4} \\ P_{21} & P_{22} & P_{23} \end{bmatrix}, \tag{A.17}$$

$$P_{21} = \mathrm{diag}[1 \quad 1 \quad 1 \quad 1 \quad 1 \quad 1 \quad 1], \tag{A.18}$$

$$P_{22} = \begin{bmatrix} 0 & 0 & 0 & 0 \\ 0 & 0 & 0 & 0 \\ 0 & 0 & 0 & 0 \\ K_t & 0 & 0 & 0 \\ 0 & K_t & 0 & 0 \\ 0 & 0 & K_t & 0 \\ 0 & 0 & 0 & K_t \end{bmatrix}, \tag{A.19}$$

$$P_{23} = \begin{bmatrix} 0 & 0 & 0 & 0 \\ 0 & 0 & 0 & 0 \\ 0 & 0 & 0 & 0 \\ C_t & 0 & 0 & 0 \\ 0 & C_t & 0 & 0 \\ 0 & 0 & C_t & 0 \\ 0 & 0 & 0 & C_t \end{bmatrix}. \tag{A.20}$$

For the three-axle vehicle, the variables in Eq. (10) can be listed as

$$X = \begin{bmatrix} X_{11} \\ X_{21} \end{bmatrix}, \tag{A.21}$$

$$X_{11} = [Z_{v1} \quad \varphi_{v1} \quad \theta_{v1} \quad Z_{v2} \quad \varphi_{v2} \quad Z_{w1} \quad Z_{w2} \quad Z_{w3} \quad Z_{w4} \quad Z_{w5} \quad Z_{w6}]^T, \tag{A.22}$$

$$X_{21} = [\dot{Z}_{v1} \quad \dot{\varphi}_{v1} \quad \dot{\theta}_{v1} \quad \dot{Z}_{v2} \quad \dot{\varphi}_{v2} \quad \dot{Z}_{w1} \quad \dot{Z}_{w2} \quad \dot{Z}_{w3} \quad \dot{Z}_{w4} \quad \dot{Z}_{w5} \quad \dot{Z}_{w6}]^T, \tag{A.23}$$

$$U = \begin{bmatrix} U_{11} \\ U_{21} \\ U_{31} \end{bmatrix}, \tag{A.24}$$

$$\begin{aligned} U_{11} = [&F_{st1}^{VL} + F_{bu1}^{VL} - m_{v1}g - m_{v2}gd_4/(d_4 + d_6) \\ &F_{st1}^{VR} + F_{bu1}^{VR} \quad F_{st1}^{VP} + F_{bu1}^{VP} - m_{v2}gd_5d_4/(d_4 + d_6) \\ &F_{st2}^{VL} + F_{bu2}^{VL} - m_{v2}g \quad F_{st2}^{VR} + F_{bu2}^{VR}]^T, \end{aligned} \tag{A.25}$$

$$\begin{aligned} U_{21} = [&F_{zw1} - m_{w1}g \quad F_{zw2} - m_{w2}g \quad F_{zw3} - m_{w3}g \\ &F_{zw4} - m_{w4}g \quad F_{zw5} - m_{w5}g \quad F_{zw6} - m_{w6}g]^T, \end{aligned} \tag{A.26}$$

$$U_{31} = [Z_{g1} \quad Z_{g2} \quad Z_{g3} \quad Z_{g4} \quad Z_{g5} \quad Z_{g6} \quad \dot{Z}_{g1} \quad \dot{Z}_{g2} \quad \dot{Z}_{g3} \quad \dot{Z}_{g4} \quad \dot{Z}_{g5} \quad \dot{Z}_{g6}]^T, \tag{A.27}$$

$$M = \mathrm{diag}[M_{11} \quad M_{12}], \tag{A.28}$$

$$M_{11} = \mathrm{diag}[1 \quad 1 \quad 1 \quad 1 \quad 1 \quad 1 \quad 1 \quad 1 \quad 1 \quad 1 \quad 1], \tag{A.29}$$

$$M_{22} = \mathrm{diag}[m_{v1} \quad I_{xv1} \quad I_{yv1} \quad m_{v2} \quad I_{xv2} \quad m_{w1} \quad m_{w2} \quad m_{w3} \quad m_{w4} \quad m_{w5} \quad m_{w6}], \tag{A.30}$$

$$N = \begin{bmatrix} [0]_{11\times5} & [0]_{11\times6} & N_{13} & N_{14} \\ N_{21} & N_{22} & N_{23} & N_{24} \\ N_{31} & N_{32} & N_{33} & N_{34} \end{bmatrix}, \tag{A.31}$$

$$N_{13} = \begin{bmatrix} \mathrm{diag}[1 \quad 1 \quad 1 \quad 1 \quad 1] \\ [0]_{6\times5} \end{bmatrix}, \tag{A.32}$$

$$N_{14} = \begin{bmatrix} [0]_{5\times6} \\ \mathrm{diag}[1 \quad 1 \quad 1 \quad 1 \quad 1 \quad 1] \end{bmatrix}, \tag{A.33}$$

$$N_{21} = \begin{bmatrix} -2(K_{s1}+K_{s2}) & 0 & 2(K_{s1}d_1 - K_{s2}d_2) & 0 & 0 \\ 0 & m_{v1}gh - 2t_w^2(K_{s1}+K_{s2}) & 0 & 0 & 0 \\ 2(K_{s1}d_1 - K_{s2}d_2) & 0 & m_{v1}gh - 2(K_{s1}d_1^2 + K_{s2}d_2^2) & 0 & 0 \\ 0 & 0 & 0 & -2K_{s3}(d_6+d_4)/d_6 & 0 \\ 0 & 0 & 0 & 0 & m_{v2}gh - 2K_{s3}t_w^2 \end{bmatrix}, \tag{A.34}$$

$$N_{22} = \begin{bmatrix} K_{s1} & K_{s1} & K_{s2} & K_{s2} & 0 & 0 \\ t_wK_{s1} & -t_wK_{s1} & t_wK_{s2} & -t_wK_{s2} & 0 & 0 \\ -d_1K_{s1} & -d_1K_{s1} & d_2K_{s2} & d_2K_{s2} & 0 & 0 \\ 0 & 0 & 0 & 0 & K_{s3} & K_{s3} \\ 0 & 0 & 0 & 0 & t_wK_{s3} & -t_wK_{s3} \end{bmatrix}, \tag{A.35}$$

$$N_{23} = \begin{bmatrix} -2(C_{s1}+C_{s2}) & 0 & 2(C_{s1}d_1 - C_{s2}d_2) & 0 & 0 \\ 0 & m_{v1}gh - 2t_w^2(C_{s1}+C_{s2}) & 0 & 0 & 0 \\ 2(C_{s1}d_1 - C_{s2}d_2) & 0 & m_{v1}gh - 2(C_{s1}d_1^2 + C_{s2}d_2^2) & 0 & 0 \\ 0 & 0 & 0 & -2C_{s3}(d_6+d_4)/d_6 & 0 \\ 0 & 0 & 0 & 0 & m_{v2}gh - 2C_{s3}t_w^2 \end{bmatrix}, \tag{A.36}$$

$$N_{24} = \begin{bmatrix} C_{s1} & C_{s1} & C_{s2} & C_{s2} & 0 & 0 \\ t_wC_{s1} & -t_wC_{s1} & t_wC_{s2} & -t_wC_{s2} & 0 & 0 \\ -d_1C_{s1} & -d_1C_{s1} & d_2C_{s2} & d_2C_{s2} & 0 & 0 \\ 0 & 0 & 0 & 0 & C_{s3} & C_{s3} \\ 0 & 0 & 0 & 0 & t_wC_{s3} & -t_wC_{s3} \end{bmatrix}, \tag{A.37}$$

$$N_{31} = \begin{bmatrix} K_{s1} & t_w K_{s1} & -d_1 K_{s1} & 0 & 0 \\ K_{s1} & -t_w K_{s1} & -d_1 K_{s1} & 0 & 0 \\ K_{s2} & t_w K_{s2} & d_2 K_{s2} & 0 & 0 \\ K_{s2} & -t_w K_{s2} & d_2 K_{s2} & 0 & 0 \\ 0 & 0 & 0 & K_{s3}(d_6 + d_4)/d_6 & t_w K_{s3} \\ 0 & 0 & 0 & K_{s3}(d_6 + d_4)/d_6 & -t_w K_{s3} \end{bmatrix}, \tag{A.38}$$

$$N_{32} = \mathrm{diag}[-K_{s1} - K_{t1} \quad -K_{s1} - K_{t1} \quad -K_{s2} - K_{t2} \quad -K_{s2} - K_{t2} \\ -K_{s3} - K_{t3} \quad -K_{s3} - K_{t3}], \tag{A.39}$$

$$N_{33} = \begin{bmatrix} C_{s1} & t_w C_{s1} & -d_1 C_{s1} & 0 & 0 \\ C_{s1} & -t_w C_{s1} & -d_1 C_{s1} & 0 & 0 \\ C_{s2} & t_w C_{s2} & d_2 C_{s2} & 0 & 0 \\ C_{s2} & -t_w C_{s2} & d_2 C_{s2} & 0 & 0 \\ 0 & 0 & 0 & C_{s3}(d_6 + d_4)/d_6 & t_w C_{s3} \\ 0 & 0 & 0 & C_{s3}(d_6 + d_4)/d_6 & -t_w C_{s3} \end{bmatrix}, \tag{A.40}$$

$$N_{34} = \mathrm{diag}[-C_{s1} - C_{t1} \quad -C_{s1} - C_{t1} \quad -C_{s2} - C_{t2} \quad -C_{s2} - C_{t2} \\ -C_{s3} - C_{t3} \quad -C_{s3} - C_{t3}], \tag{A.41}$$

$$P = \begin{bmatrix} [0]_{11\times 11} & [0]_{11\times 6} & [0]_{11\times 6} \\ P_{21} & P_{22} & P_{23} \end{bmatrix}, \tag{A.42}$$

$$P_{21} = \mathrm{diag}[1 \quad 1 \quad 1 \quad 1 \quad 1 \quad 1 \quad 1 \quad 1 \quad 1 \quad 1 \quad 1], \tag{A.43}$$

$$P_{22} = \begin{bmatrix} 0 & 0 & 0 & 0 & 0 & 0 \\ 0 & 0 & 0 & 0 & 0 & 0 \\ 0 & 0 & 0 & 0 & 0 & 0 \\ 0 & 0 & 0 & 0 & 0 & 0 \\ 0 & 0 & 0 & 0 & 0 & 0 \\ K_{t1} & 0 & 0 & 0 & 0 & 0 \\ 0 & K_{t1} & 0 & 0 & 0 & 0 \\ 0 & 0 & K_{t1} & 0 & 0 & 0 \\ 0 & 0 & 0 & K_{t1} & 0 & 0 \\ 0 & 0 & 0 & 0 & K_{t1} & 0 \\ 0 & 0 & 0 & 0 & 0 & K_{t1} \end{bmatrix}, \tag{A.44}$$

$$P_{23} = \begin{bmatrix} 0 & 0 & 0 & 0 & 0 & 0 \\ 0 & 0 & 0 & 0 & 0 & 0 \\ 0 & 0 & 0 & 0 & 0 & 0 \\ 0 & 0 & 0 & 0 & 0 & 0 \\ 0 & 0 & 0 & 0 & 0 & 0 \\ C_{t1} & 0 & 0 & 0 & 0 & 0 \\ 0 & C_{t1} & 0 & 0 & 0 & 0 \\ 0 & 0 & C_{t1} & 0 & 0 & 0 \\ 0 & 0 & 0 & C_{t1} & 0 & 0 \\ 0 & 0 & 0 & 0 & C_{t1} & 0 \\ 0 & 0 & 0 & 0 & 0 & C_{t1} \end{bmatrix}. \quad \text{(A.45)}$$

References

1. H. Li, S. Laima, Q. Zhang, N. Li and Z. Liu, Field monitoring and validation of vortex-induced vibrations of a long-span suspension bridge, *J. Wind Eng. Ind. Aerodyn.* **124** (2014) 54–67.
2. Y. C. Hwang, S. Kim and H. K. Kim, Cause investigation of high-mode vortex-induced vibration in a long-span suspension bridge, *Struct. Infrastruct. E* **16** (2020) 84–93.
3. Y. Ge, L. Zhao and J. Cao, Case study of vortex-induced vibration and mitigation mechanism for a long-span suspension bridge, *J. Wind Eng. Ind. Aerodyn.* **220** (2022) 104866.
4. G. Diana, F. Resta, M. Belloli and D. Rocchi, On the vortex shedding forcing on suspension bridge deck, *J. Wind Eng. Ind. Aerodyn.* **94** (2006) 341–363.
5. A. Larsen, A generalized-model for assessment of vortex-induced vibrations of flexible structures, *J. Wind Eng. Ind. Aerodyn.* **57** (1995) 281–294.
6. E. Simiu and R. H. Scanlan, *Wind Effects on Structures: Fundamentals and Applications to Design* (John Wiley, USA, 1996).
7. T. Wu and A. Kareem, Vortex induced vibration of bridge decks: Volterra series-based model, *J. Eng. Mech.* **139** (2013) 1831–1843.
8. L. D. Zhu, X. L. Meng and Z. S. Guo, Nonlinear mathematical model of vortex-induced vertical force on a flat closed-box bridge deck, *J. Wind Eng. Ind. Aerodyn.* **122** (2013) 69–82.
9. P. Hu, G. Hu, Y. Han, F. Zhang and Y. Tang, Vortex-induced force model and reliability analysis of a steel box girder with projecting slab in a cable-stayed bridge, *Int. J. Struct. Stab. Dyn.* **22**(13) (2022) 2250139.
10. B. Yan, Y. Yuan, D. Li, K. Li, Q. Yang, D. Lin and G. Piccardo, 2D and 2.5D numerical simulations for vortex-induced vibration (VIV) of a 1.5:1 rectangular cylinder, *Int. J. Struct. Stab. Dyn.* **22**(7) (2022) 2250033.
11. J. W. Li, Z. X. Ma, X. R. Qiao, J. M. Hao, G. H. Zhao and W. Feng, Aerodynamic nonlinear energy evolution characteristics of vertical vortex-induced vibration of open bridge girder, *Int. J. Struct. Stab. Dyn.* **22**(14) (2022) 2250165.

12. J. Chen and Q. S. Li, Nonlinear dynamics of a fluid-structure coupling model for vortex-induced vibration, *Int. J. Struct. Stab. Dyn.* **19**(7) (2019) 1950071.
13. K. Xu, Y. Ge, L. Zhao and X. Du, Experimental and numerical study on the dynamic stability of vortex-induced vibration of bridge decks, *Int. J. Struct. Stab. Dyn.* **18**(3) (2018) 1850033.
14. B. Wang, M. Zhang, F. Xu and B. Su, An emergency aerodynamic measure to suppress vortex-induced vibration of bridge decks: Suspended flexible sheets, *Int. J. Struct. Stab. Dyn.* (2023) 2350141, https://doi.org/10.1142/S0219455423501419.
15. S. M. Hasheminejad, A. H. Rabiee and M. Jarrahi, Semi-active vortex induced vibration control of an elastic elliptical cylinder with energy regeneration capability, *Int. J. Struct. Stab. Dyn.* **17**(9) (2017) 1750107.
16. D. Shang, *Experimental Study on Model Scale Effect of Vortex-Induced Vibrations for 1:6 Rectangular Cross-Sections* (Hunan University, 2015) (in Chinese).
17. A. Trush, S. Pospisil, S. Kuznetsov and H. Kozmar, Wind-tunnel experiments on vortex-induced vibration of rough bridge cables, *J. Bridge Eng.* **22**(10) (2017) 06017001.
18. G. Fu, Y. Quan, M. Gu and C. Feng, Experimental investigation of unsteady aerodynamic forces and vortex-induced aeroelastic effects of a slender structure in uniform and turbulent flows, *Int. J. Struct. Stab. Dyn.* **23**(4) (2023) 2350040.
19. Z. Liu, Z. Zhu, W. Chen and Z. Chen, Observation of vortex-induced vibration and wind characteristics of cables across the Yangtze river bridge, *J. Railway Sci. Eng.* **17**(7) (2020) 1760–1768 (in Chinese).
20. H. Utsunomiya, F. Nagao, M. Noda and E. Tanaka, Vortex-induced oscillation of a bridge in slowly fluctuating wind, *J. Wind Eng. Ind. Aerodyn.* **89** (2001) 1689–1699.
21. Y. Wu, X. Wu, J. Li, H. Xin, Q. Sun and J. Wang, Investigation of vortex-induced vibration of a cable-stayed bridge without backstays based on wind tunnel tests, *Eng. Struct.* **250** (2022) 113436.
22. Q. Zhu, Y. L. Xu, L. D. Zhu and B. Y. Chen, A semi-empirical model for vortex-induced vertical forces on a twin-box deck under turbulent wind flow, *J. Fluids Struct.* **71** (2017) 183–198.
23. L. D. Zhu, L. Du, X. Meng and Z. Guo, Nonlinear mathematical models of vortex-induced vertical force and torque on a centrally-slotted box deck, in *Proc. 14th Int. Conf. Wind Eng.* (Porto Alegre, Brazil, 2015).
24. Q. Zhu, Y. L. Xu, L. D. Zhu and H. Li, Vortex-induced vibration analysis of long-span bridges with twin-box decks under non-uniformly distributed turbulent winds, *J. Wind Eng. Ind. Aerodyn.* **172** (2018) 31–41.
25. J. Zhu, Z. Xiong, H. Xiang, X. Huang and Y. Li, Ride comfort evaluation of stochastic traffic flow crossing long-span suspension bridge experiencing vortex-induced vibration, *J. Wind Eng. Ind. Aerodyn.* **219** (2021) 104794.
26. H. Yu, B. Wang, G. Zhang, Y. Li and X. Chen, Ride comfort assessment of road vehicle running on long-span bridge subjected to vortex-induced vibration, *Wind Struct.* **31** (2020) 393–402.
27. F. Ehsan and R. H. Scanlan, Vortex-induced vibrations of flexible bridges, *J. Eng. Mech.* **116** (1990) 1392–1411.
28. H. Ruscheweyh and G. Sedlacek, Crosswind vibrations of steel stacks — Critical comparison between some recently proposed codes, *J. Wind Eng. Ind. Aerodyn.* **30** (1988) 173–183.

29. Y. L. Xu, *Wind Effects on Cable-Supported Bridges* (John Wiley & Sons Inc, Singapore, 2013).
30. S. Xu, R. Ma, D. Wang, A. Chen and H. Tian, Prediction analysis of vortex-induced vibration of long-span suspension bridge based on monitoring data, *J. Wind Eng. Ind. Aerodyn.* **191** (2019) 312–324.
31. M. Mashnad and N. P. Jones, A model for vortex-induced vibration analysis of long-span bridges, *J. Wind Eng. Ind. Aerodyn.* **134** (2014) 96–108.
32. Q. Zhu, B. Y. Chen, L. D. Zhu and Y. L. Xu, Investigation on characteristics and spanwise correlation of vortex-induced forces on a twin-box deck using newly-developed wind-tunnel test technique, *J. Wind Eng. Ind. Aerodyn.* **164** (2017) 69–81.
33. R. H. Scanlan and N. P. Jones, Aeroelastic analysis of cable-stayed bridges, *J. Struct. Eng.* **116** (1990) 279–297.
34. S. R. Chen and J. Wu, Dynamic performance simulation of long-span bridge under combined loads of stochastic traffic and wind, *J. Bridge Eng.* **15** (2010) 219–230.
35. S. R. Chen and J. Wu, Modeling stochastic live load for long-span bridge based on microscopic traffic flow simulation, *Comput. Struct.* **89** (2011) 813–824.
36. Y. Li, S. Qiang, H. Liao and Y. L. Xu, Dynamics of wind–rail vehicle–bridge systems, *J. Wind Eng. Ind. Aerodyn.* **93** (2005) 483–507.
37. ISO 2631-1, Mechanical vibration and shock: Whole-body vibration, *International Organization for Standardization* (1997).
38. B. Wang, Y. L. Xu and Y. Li, Dynamic analysis of wind-vehicle-bridge systems using mutually-affected aerodynamic parameters, *Wind Struct.* **20**(2) (2015) 191–211.
39. K. M. A. Kamash and J. D. Robson, The application of isotropy in road surface modelling, *J. Sound Vib.* **57**(1) (1978) 89–100.
40. L. Sun and X. J. Deng, Predicting vertical dynamic loads caused by vehicle-pavement interaction, *J. Transp. Eng.* **124** (1998) 470–478.
41. M. Shinozuka and G. Deodatis, Simulation of stochastic processes by spectral representation, *Appl. Mech. Rev.* **44** (1991) 191–204.
42. K. Bogsjo, Coherence of road roughness in left and right wheel-path, *Veh. Syst. Dyn.* **46** (2008) 599–609.
43. ISO 8608, Mechanical vibration–road surface profiles–reporting of measured data, *International Organization for Standardization* (1995).
44. N. Zhang and H. Xia, Dynamic analysis of coupled vehicle–bridge system based on inter-system iteration method, *Comput. Struct.* **114–115** (2013) 26–34.
45. N. E. Busch and H. A. Panofsky, Recent spectra of atmospheric turbulence, *Q. J. Roy. Meteor. Soc.* **94** (1968) 132–148.
46. Y. Ge, *Wind Resistance of Long Span Suspension Bridge* (CC Press, China, 2011).

https://doi.org/10.1142/9789819814770_0022

Chapter 22

Study and Application of Steel and Composite Walls with Corrugated Steel Plates in China#

Yan-Lin Guo*, Chen-Bao Wen and Hao-Jun Sun

Department of Civil Engineering
Tsinghua University, Beijing 100084, P. R. China
gyl@tsinghua.edu.cn

Steel and composite walls with corrugated steel plates have found their wide application in China due to their excellent structural performance. Extensive research has been conducted over the past few decades to investigate their behavior and load-bearing capacity. This paper summarized numerous experimental and numerical studies on the corrugated steel plate shear walls (CSPSWs), double corrugated steel plate shear walls (DCSPSWs), and concrete-infilled double steel corrugated-plate walls (CDSCWs). First, the load-bearing mechanism and design methods of CSPSWs and DCSPSWs under monotonic and horizontal cyclic loads were elucidated. Subsequently, this paper elaborates the load-bearing capacity and corresponding design methods including the sectional strength capacity, the overall stability, and the wall panel stability of CDSCWs under axial compression, compressive-bending combination, and compressive-bending-shear combination. The proposed design methods of CSPSWs, DCSPSWs, and CDSCWs have successfully implemented in practical high-rise building structures and some of them have been adopted by the Chinese engineering construction standards T/CECS 290 and T/CECS 624.

Keywords: Corrugated steel plate; shear wall; stability; design method; review.

1. Introduction

In seismically active regions, the design for lateral force resistance in structures is of utmost importance. Both braced frames and shear walls are commonly employed structural systems in high-rise buildings.[1–5] The shear wall system is favored by engineers due to its superior integrity, exceptional lateral stiffness, substantial bearing capacity, and commendable seismic performance. While reinforced concrete shear walls have conventionally served as cyclic load-resisting systems, their application is limited by concrete's susceptibility to brittle failure and lack of ductility.

*Corresponding author.
#This chapter appeared previously on the International Journal of Structural Stability and Dynamics. To cite this chapter, please cite the original article as the following: Y.-L. Guo, C.-B. Wen & H.-J. Sun, *Int. J. Struct. Stab. Dyn.*, **23**, 2340024 (2023), DOI: 10.1142/S0219455423400242.

With the promotion of steel structures, steel plates have been adopted in shear wall structural systems.

In the 1970s, thin flat steel plate shear walls (FSPSWs) began to be studied and applied in high-rise buildings.[6–11] However, the low out-of-plane stiffness of the thin flat plates leads to early buckling and then induces the post-buckling diagonal tension field. This can cause severe pinching hysteresis loops under cyclic loading and potentially harm the frame column. Some researchers proposed alternative shear walls, such as stiffened steel plate shear walls and buckling-restrained steel plate shear walls.[12] The stiffeners could prevent the shear wall from premature buckling. The ductility and energy dissipation are enhanced, but the fabrication cost is also increased.

For the above reasons, corrugate steel plate shear walls (CSPSWs) have been proposed, primarily due to their remarkable out-of-plane geometric stability.[13–16] These walls offer excellent lateral load-resisting performance without the need for stiffeners or thicker plates. The processing and welding can be efficiently carried out in factory settings. Then, they can be directly connected using high-strength bolts on site, enabling industrial production, and modular construction. CSPSWs exhibit significant ductility and enhanced energy dissipation. Various types of CSPSWs have been proposed, including horizontal- and vertical-CSPSWs, stiffened CSPSWs, and double CSPSWs (DCSPSWs). Numerous studies have been conducted to assess their lateral and cyclic performance, and also corresponding design methods have been proposed.

Besides, steel and concrete composite walls can take advantage of the two materials due to the combined effect.[17,18] They can be employed as bearing walls, retaining walls, and shear walls to provide lateral stiffness and strength support for high-rise buildings. An efficient composite wall known as concrete-infilled double steel corrugated-plate walls (CDSCWs) has been introduced, which consists of two bolt-connected corrugated steel plates infilled with concrete.[19–21] Thorough investigations have been conducted on the sectional load-bearing capacity and global stability capacity of CDSCWs. They present high load-bearing capacity efficiency and excellent seismic performance. In China, the design methods of CSPSWs, DCSPSWs, and CDSCWs were established and adopted by the engineering construction standards titled "Technical specification for structures with corrugated steel plate components" (T/CECS 624-2019).[22]

This paper presents state-of-the-art knowledge on steel and composite walls with corrugated steel plates, including analytical, experimental, and numerical studies, as well as strength design recommendations. First, this paper summarizes the load-bearing mechanism and design methods of CSPSWs, DCSPSWs, and CDSCWs under monotonic loads and horizontal cyclic loads. The effects of various parameters, such as dimensions, openings, and corrugation shape, are introduced in this paper. In particular, the load-bearing capacity of CDSCWs is focused on the sectional strength capacity, overall stability, and wall panel stability. The key design formulas are provided for promoting the application of steel and composite walls with corrugated steel plates.

2. Corrugated Steel Plate Shear Walls (CSPSWs)

2.1. *Application and type of CSPSWs*

CSPSWs are usually embedded within a steel boundary frame to resist the lateral loads resulting from horizontal seismic or wind loads. They resist the lateral loads by shear yielding rather than by the post-buckling diagonal tension field as FSPSWs. CSPSWs can achieve high shear buckling strength and lateral stiffness without the need to increase the thickness of the plates, resulting in a more cost-effective design. CSPSWs have a lower demand on the stiffness and strength of the frame. This is because CSPSWs experience minimal development of the diagonal tension field after buckling, thereby allowing their initial buckling state to be considered as their ultimate load-capacity state. Besides, the transportation and installation of CSPSWs are less likely to induce significant initial deformations, facilitating better control of construction errors and initial imperfections.

Various corrugation shapes are employed in practice for CSPSWs, as shown in Fig. 1. The thickness of the corrugated plates may reach a maximum value of 8 mm, which is suitable for structural designs of high-rise buildings up to 150 m in height. The sinusoidal corrugation has very smooth curves, resulting in minimal residual stress induced during the rolling–forming manufacture. However, the curved shape poses challenges for connecting the CSPSWs with surrounding frame beams and columns. The trapezoidal corrugation with straight lines offers more convenient bolted or welded connections, making it a preferred choice for structural design engineers. Sun *et al.*[23] conducted a comparison of the material utilization efficiency

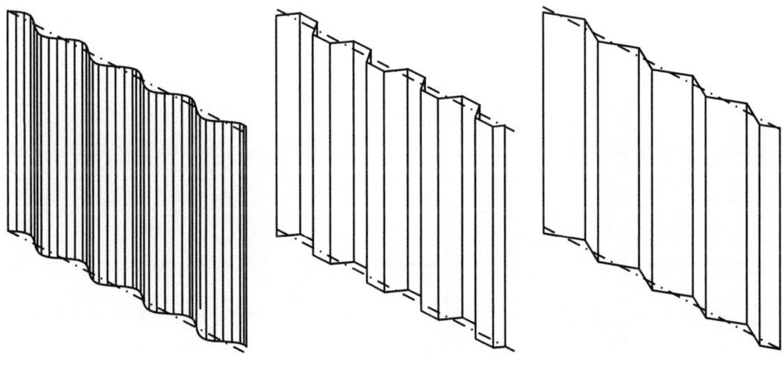

(a) Sinusoidal (b) Trapezoidal (c) Triangular

Fig. 1. The corrugation shapes of CSPSWs.

between the trapezoidal and sinusoidal corrugations. They found that trapezoidal corrugation exhibits higher buckling loads due to the presence of more material located farther away from the neutral axis of the section. As for triangular corrugation, it suffers from poor material utilization efficiency and induces large residual stress, resulting in its limited usage in practice.

CSPSWs can be divided into horizontal-CSPSWs (H-CSPSWs) and vertical-CSPSWs (V-CSPSWs), as shown in Fig. 2. The axial compression stiffness in the direction perpendicular to the corrugation direction is negligible due to the accordion effect. Hence, H-CSPSWs can release vertical loads efficiently, approximating a pure shear state.

Recently, CSPSWs have been used in public and residential buildings in China (see Fig. 3). One notable example is Building Two, Block 81, Hangzheng Storage (2019), situated in Hangzhou, Zhejiang Province. This building comprises 18 above-ground floors and two underground floors, with a height of 76.28 m. The entire structure is a steel frame-shear wall structure system with H-CSPSWs as the

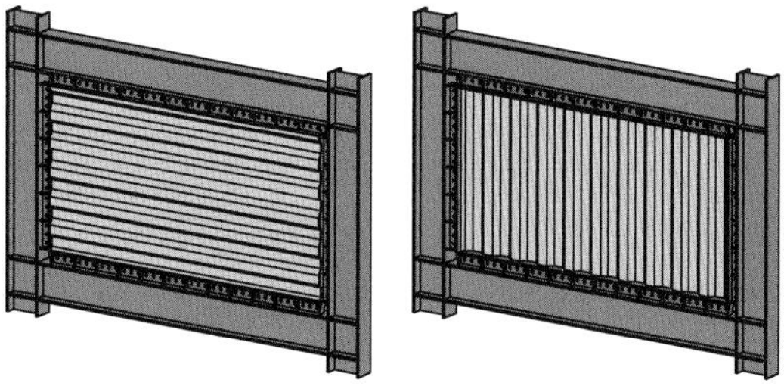

(a) Horizontally CSPSWs (b) Vertically CSPSWs

Fig. 2. CSPSWs with different corrugation directions.

(a) (b)

Fig. 3. Photograph of the CSPSWs in practice engineering structures.

primary lateral load-resisting elements. Another noteworthy structure is Building 16, West Block of Jiangbei District, Xuhui, a residential building located in Ningbo, Zhejiang. It has 27 above-ground floors and one underground floor, and the building height is 79.110 m. The adoption of CSPSWs greatly reduces material consumption while demonstrating excellent load-bearing capacity.

2.2. *Elastic shear buckling of corrugated steel plates*

2.2.1. *Elastic bending rigidities*

The corrugated steel plate is usually regarded as an orthotropic plate. Their flexural rigidity constants are the basic parameters. Back in 1959, Timoshenko.[24] presented the expressions of rigidity constants of the corrugated plate, denoted as follows:

$$D_x = EI_x/\lambda, \tag{2.1}$$

$$D_y = \frac{\lambda}{q} \cdot \frac{Et^3}{12(1-\upsilon^2)}, \tag{2.2}$$

$$H = \frac{q}{\lambda} \cdot \frac{Et^3}{12(1+\upsilon)}. \tag{2.3}$$

where D_x and D_y are the flexural rigidity constants with the bending moment direction perpendicular and parallel to the corrugation line, respectively, and H is the torsional rigidity constant. I_x is the moment of inertia of the cross-section of one repeating corrugation section about its neutral axis. The dimension parameters are presented in Fig. 4. The flexural rigidity constant D_x is usually much larger than D_y. D_x of the trapezoidally corrugated plate can be obtained accurately, expressed by Eq.(2.4).

$$D_x = \frac{Eta^2}{2\lambda}\left(d_1 + \frac{d_2}{3\cos\gamma}\right). \tag{2.4}$$

As for sinusoidal corrugation, Timoshenko[24] suggested the formula for the expansion length q and the flexural rigidity constant D_x. However, it is found that the calculation formulas can lead to discrepancy, especially when the corrugation

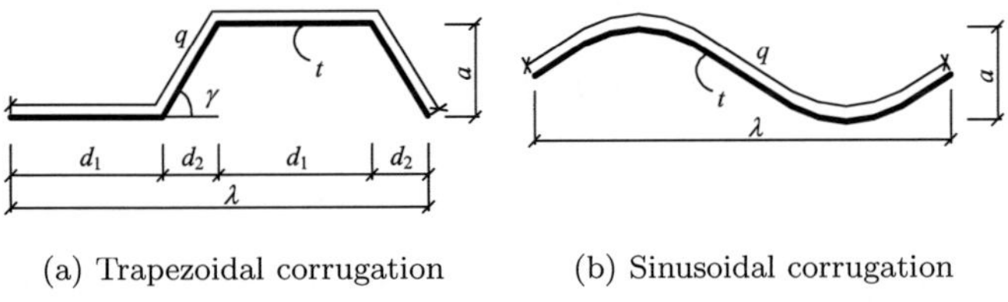

(a) Trapezoidal corrugation (b) Sinusoidal corrugation

Fig. 4. Two typical corrugation shapes.

ratio a/λ is large. Briassoulis,[25] Liew,[26] and Zhang[27] *et al.* modified the formulas with different forms. Dou *et al.*[28] also recommended a more accurate prediction for q and D_x, based on numerical integral, as Eqs. (2.5) and (2.6) show. The comparison indicated that Eq. (2.6) could provide accurate calculation results in various corrugation dimensions.

$$q = \lambda \cdot \sqrt{1 + 16.3\left(\frac{a}{2\lambda}\right)^{1.92}}, \tag{2.5}$$

$$D_x = \frac{Eta^2}{4} \cdot \sqrt{\left[0.250 + 2.35\left(\frac{a}{2\lambda}\right)^2\right] \Big/ \left[1 + 0.342\left(\frac{a}{2\lambda}\right)^2\right]}. \tag{2.6}$$

2.2.2. *Elastic buckling load*

The elastic shear buckling behavior of corrugated steel plates was studied back in the 1960s by Easley and McFarland.[29,30] The elastic global shear buckling stress was derived using the minimum potential energy theorem and the Rayleigh–Ritz method, as Eq. (2.7) shows. Parameter b is the width of H-CSPSWs, but as for V-CSPSWs, it should be replaced with the height h of the wall. The global shear buckling coefficient k_g equals 36 for simply supported metal shear diaphragms and 68.4 for clamped supported diaphragms.

$$\tau_{\mathrm{cr}} = k_g \sqrt[4]{D_x^3 D_y} \Big/ (tb^2). \tag{2.7}$$

Easley[31] also tested 46 corrugated web beams and carried out finite elements numerical analyses. They suggested that k_g should be taken as 31.6 and 59.2 for simply supported and clamped edges, respectively. Since then, the shear buckling stress formulas for the corrugated webs of beams have been studied experimentally and numerically by many researchers.[32–37] However, the boundary condition and dimension differences between CSPSWs and the corrugated webs were not considered. Tong and Guo[15] derived the same formula as Eq. (2.7) and accordingly, the shear buckling coefficient was fitted as Eq.(2.8). The coefficient k_g was related to the converted aspect ratio β and the constant of the equivalent orthotropic plate θ.

$$k_g = (7 + 20\theta)\beta^2 + 8\beta + 61.2 + 29.5\theta, \tag{2.8}$$

$$\beta = b/h \cdot \sqrt[4]{D_y/D_x}, \tag{2.9}$$

$$\theta = H/\sqrt{D_x D_y}. \tag{2.10}$$

The elastic shear buckling loads of the corrugated plates were well-studied based on the orthotropic plate theory. It is worth noting that the parameter of corrugation repeating number significantly influences elastic shear buckling loads, which is neglected in the theoretical analyses but investigated by Zhao,[38] Dou,[28] and Wang[39] based on the FE analyses. Qiu and Zhao[38] investigated the elastic shear buckling behavior of corrugated plates considering constraints from neighboring subpanels and boundary frame members. Dou *et al.*[28] studied the elastic shear buckling load of sinusoidally corrugated plates. Besides, Wang *et al.*[39] proposed formulas for large-scale corrugated webs that are used in bridge girders. Wen *et al* .[40] proposed a highly accurate formula, expressed by Eq. (2.11), where m is the corrugation number. Then, elastic shear buckling loads can be utilized to obtain the normalized slenderness ratio by Eq. (2.12), which is a crucial parameter in strength design of CSPSWs and naturally influences their load-bearing capacity.

$$k_g = 50 + 207\beta/m + 102/(\beta m) + 1610\theta\beta^3, \tag{2.11}$$

$$\lambda_n = \sqrt{f_{\mathrm{vy}}/\tau_{\mathrm{cr}}}, \tag{2.12}$$

2.3. *Shear behavior and design of horizontal-CSPSWs (H-CSPSWs)*

2.3.1. *Experimental study*

The experimental studies on H-CSPSWs started from the light-gauge steel plate with a relatively small thickness of 0.3–2.0 mm. Mo *et al.*[41] tested five reinforced concrete framed shear wall specimens made of corrugated steel plates. Berman *et al.* and Berman and Bruneau[42,43] compared the hysteretic behavior of FSPSWs, H-CSPSWs, and braced frames. It was found that H-CSPSWs could achieve significant ductility and energy dissipation while minimizing the demands placed on the surrounding frame. Emami *et al.*[14] conducted experimental studies and revealed that although the ultimate strength of the FSPSW specimen was larger than the H-CSPSW specimen, the energy dissipation capacity, ductility ratio, and initial stiffness of the H-CSPSW specimen were larger. Shimizu *et al.*[44] performed cyclic loading tests of simple shear panels with pure shear loading. Dou *et al.*[45] proposed the flat-CSPSWS and tested their cyclic behaviors. Many experimental studies on the cyclic performance of H-CSPSWs were conducted and corresponding design recommendations were proposed.[46–51]

The plates with a small thickness are easily profiled into the corrugation shape. However, the potential benefits of corrugations are not fully realized due to premature buckling. Therefore, many researchers adopted the plates with a large thickness (i.e. $t = 2.0-30.0$ mm) in CSPSWs.[52–56] Shon *et al.*[52] performed cyclic loading tests to evaluate the shear hysteresis and energy dissipation of the CSPSWs with the plate thickness being 3.2 mm. The test specimens failed by the global buckling and had a greater accumulated energy dissipation capacity. Qiu *et al.*[53] revealed that the lateral load-resisting mechanism in CSPSWs was shear yielding or inelastic shear buckling.

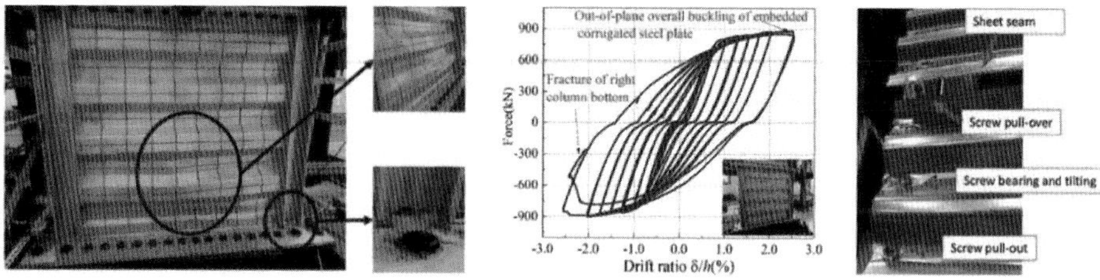

(a) Test conducted by Jin *et al.*[56] ($t = 5.0$ mm)

(b) Test results of Jin *et al.*[56]

(c) Test results of Yu *et al.*[47]

Fig. 5. The test results of CSPSWs.

Many cyclic tests on H-CSPSWs have been conducted in the last decade, as listed in in Table 1. Some of the results are shown in Fig. 5. Typically, H-CSPSWs with a small thickness are fastened to the boundary elements using bolts. Consequently, the end-failure modes are unavoidable and also limit the load-bearing capacity of H-CSPSWs.[44] Besides, they are prone to form tension field like flat plates.[14] However, corrugated plates with a large thickness are commonly welded to the boundary element. In this case, the shear load-carrying mode is the primary mode, while the tension field is not easily formed. Furthermore, yielding or inelastic shear buckling contributes to a more stable hysteretic curve with slight pinching.

2.3.2. *Parametric analysis*

The corrugation shape determines the flexural rigidities of H-CSPSWs. Edalati *et al.*[57] found that trapezoidally corrugated plates had higher load-bearing capacity than the sinusoidally corrugated plates. While Zhao *et al.*[58] found that the corrugation shape had little influence on the hysteretic behavior of H-CSPSWs when the trapezoidal and sinusoidal corrugations had the similar expansion length and corrugation depth.

The corrugation angle and depth also attract the interest of researchers.[16,44,56,57,59–64] They found that the corrugation angle had a negligible effect on the ultimate strength but a significant and positive influence on the residual strength. Edalati[57] Kalali,[63] Dou,[64] and Jin *et al.*[56] also focused on the shape of sinusoidal corrugation. A balance should be achieved between the load-bearing capacity and the steel consumption in the design.

The walls with rectangular openings that serve as windows or doors are commonly used to fulfill building functions. Hence, some researchers performed experimental and numerical studies on H-CSPSWs with openings. Bahrebar *et al.*[60] and Farzampour *et al.*[65] studied the structural performance of H-CSPSWs with openings. The results indicated that H-CSPSWs are more suitable for accommodating openings under cyclic loadings compared to FSPSWs.

Table 1. CSPSWs with various corrugation, loading, and study types.

Corrugation direction	Type of the study	Author (Year)	Loading type	Thickness (mm)	Remarks
Only H-CSPSWs	Experimental	Mo *et al.*[41] (2000)	Cyclic	0.4–0.8	The ductility can be improved if the thickness of the corrugated plate is appropriate to the frame.
		Berman *et al.*[42,43] (2005)	Cyclic	0.75	Compared the hysteretic behavior of FSPSWs, CSPSWs, and braced frames.
		Shimizu *et al.*[44] (2013)	Monotonic Cyclic	0.4–0.8	Proposed strength formulas and cyclic hysteresis models for end-failure action of the corrugated steel sheets.
		Yu *et al.*[47] (2016)	Cyclic	1.73	By having circular holes, the shear walls could fail in sheathing ruptures instead of screw failures.
		Zhang *et al.*[50] (2018)	Cyclic	0.4–0.8	Created openings in the field of the corrugated sheets in order to improve the ductility.
	Experimental Numerical	Vigh *et al.*[46] (2013)	Cyclic	——	Proposed design recommendations based on FEMA P-695.
		Zhang *et al.*[48] (2017)	Monotonic Cyclic	0.6 $\sim$ 1.8	Found that gravity load at the service load level led to an increase in shear strength and initial stiffness.
		Cao *et al.*[70] (2018)	Cyclic	2.47–2.93	Used the PFI theory and strip model for the design of a CSPSW.
		Jin *et al.*[71] (2021)	Cyclic	6.0–20.0	Established theoretical equations for predicting plastic global buckling behavior of CSPSWs.
		Yu *et al.*[55] (2022)	Cyclic	3.0	Performed quasi-static cyclic tests on specially shaped CSPSWs and FCSPSWs.
	Numerical	Edalati *et al.*[57] (2014)	Monotonic	1.5–20.0	Trapezoidal plates had better structural performance than sinusoidal plates.
		Kalali *et al.*[16] (2015)	Cyclic	1.25–6.0	H-CSPSWs could provide higher stiffness and energy absorption than FCSPSWs.
		Bahrebar *et al.*[60] (2016)	Cyclic	1.25–4.0	Investigated the cyclic behavior of centrally-perforated H-CSPSWs.
		Kalali *et al.*[63] (2016)	Cyclic	1.25–6.0	Compared the effect of the corrugation geometrical variations of sinusoidal corrugated plates.
		Tong and Guo[66] (2018)	Monotonic	4.0–8.0	Proposed a design method for stiffened H-CSPSWs.

(Continued)

Table 1. *(Continued).*

Corrugation direction	Type of the study	Author (Year)	Loading type	Thickness (mm)	Remarks
		Shariati *et al.*[72] (2019)	Cyclic	2.25	Studied corrugated low-yield point steel plate shear walls having circular openings.
		Feng *et al.*[67] (2021)	Monotonic	2.0–8.0	Introduced a theoretical method of determining the number of steel strips for stiffened CSPSWs.
		Cao *et al.*[62] (2022)	Monotonic	3.0–30.0	Derived the geometric parameters of three buckling modes and the demand of the frame stiffness.
		Zhao *et al.*[73] (2022)	Cyclic	4.0–16.0	Proposed a performance-based seismic design (PBSD) method for H-CSPSWs.
Both H-CSPSWs and V-CSPSWs	Experimental	Emami *et al.*[14] (2013)	Cyclic	1.25	Compared the structural performance between CSPSWs and FSPSWs
		Shon *et al.*[52] (2017)	Cyclic	3.2	Performed cyclic loading tests to evaluate the structural performance of CSPSWs.
		Qiu *et al.*[53] (2018)	Cyclic	2.0	Revealed the load-resisting mechanism of CSPSWs was shear yielding or inelastic shear buckling.
	Experimental Numerical	Wang *et al.*[54] (2019)	Cyclic	3.0	Compared the lateral force bearing capacity, lateral stiffness, ductility, and energy dissipation.
		Wang *et al.*[74] (2021)	Cyclic	2.0–6.0	Assessed the matching relationship between out-of-plane and in-plane stiffness of the frames.
		Jin *et al.*[56] (2022)	Cyclic Monotonic	2.0–5.0	Proposed an assembled multi-grid CSPSW and conducted experimental and numerical studies on it.
	Numerical	Emami *et al.*[75] (2014)	Cyclic	1.25	Compared the test results and FE analysis results.
		Farzampour *et al.*[65] (2015)	Monotonic	1.5–4.0	Proposed an ultimate strength prediction procedure for CSPSWs with rectangular openings.
		Zhao *et al.*[58] (2017)	Monotonic Cyclic	5.0	Found that CSPSWs were less sensitive to the gravity loads and boundary frame stiffness.
		Dou *et al.*[64] (2018)	Monotonic	4.0–10.0	Proposed equations to predict the maximum shear resistance and post-buckling behavior.

Table 1. (*Continued*)

Corrugation direction	Type of the study	Author (Year)	Loading type	Thickness (mm)	Remarks
		Fadhil *et al.*[59] (2018)	Cyclic	1.25	Investigated the effect of corrugation angle and direction on the performance of CSPSWs.
		Deng *et al.*[76] (2020)	Cyclic	0.4–2.0	Develop a high-efficiency analysis model (HEAM) and equivalent cross braces for CSPSWs.
		Dou *et al.*[61] (2021)	Monotonic	4.0–10.0	Established design formulas for the ultimate strength and residual strength of CSPSWs.
		Ghodratian *et al.*[77] (2021)	Cyclic	0.5–2.0	Explored the cyclic performance of CSPSWs with beam-only-connected infill plates.
		Qiu *et al.*[38] (2022)	Monotonic	2.0–30.0	Proposed simplified load-drift curves for CSPSWs with different dominant shear buckling modes.
		Zheng *et al.*[78] (2022)	Cyclic	10.0	Compared the seismic performance of CSPSWs with different placement angles.
Only V-CSPSWs	Experimental	Ding *et al.*[79] (2017)	Cyclic	1.6	Investigated the effects of the opening on the structural performance of CSPSWs.
	Experimental Numerical	Hosseinzadeh *et al.*[49] (2017)	Cyclic	1.25	Investigated the effect of variation in the angle of the trapezoidal plate.
		Gilvaee *et al.*[51] (2021)	Cyclic	1.25	Investigated the influence of an opening in CSPSWs experimentally and numerically.
	Numerical	Farzampour *et al.*[80] (2018)	Monotonic	1.5	Proposed analytical equations to estimate the strength of CSPSWs with a reduced beam section.
		Behzad *et al.*[81] (2021)	Monotonic	1.25	Studied the CSPSWs with stiffeners in the form of box and plate sections around the opening.
		Vaziri *et al.*[82] (2021)	Monotonic	3.5–15.0	Estimated the force-displacement curve of based on the tension field angle and PFI method.
		Zheng *et al.*[83] (2022)	Cyclic	2.0–10.0	Studied the asymmetric diagonal stiffened beam-only-connected CSPSWs.
		Wen *et al.*[40,84] (2023)	Monotonic	4.0–8.0	Proposed prefabricated CSPSWs and studied the behavior under compression and shear loads.

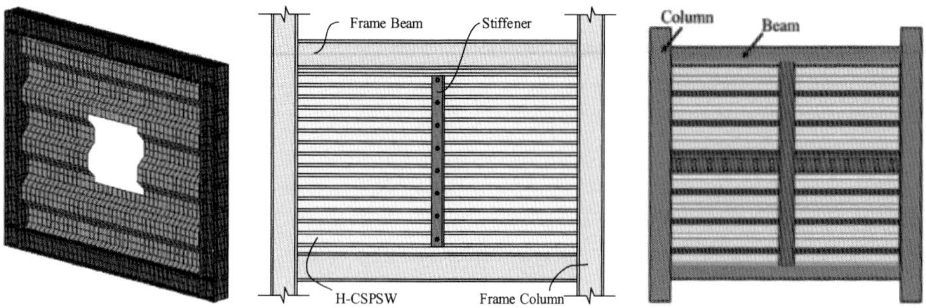

(a) H-CSPSWs with an opening (b) Stiffened H-CSPSWs (c) Multi-grid H-CSPSWs

Fig. 6. Different types of H-CSPSWs.

In some long-span frame structures, the width of H-CSPSWs can become substantial, potentially leading to a decrease in load-bearing capacity. To address this issue, Tong and Guo[66] proposed the stiffened H-CSPSWs as depicted in Fig. 6. The numerical results indicated that the stiffening system could restrain the out-of-plane displacements of the corrugated steel plate. Feng *et al.*[67] studied the H-CSPSW with multiple vertical stiffeners. Jin *et al.*[56] studied the behavior of an assembled multi-grid H-CSPSWs.

2.3.3. *Design method*

In 1996, Elgaaly[31] proposed the inelastic shear buckling stress for corrugated steel webs. Then, the design manual for bridges with corrugated steel webs[68] and Chinese specification T/CECS 290-2022[69] both proposed different shear strength formulas. The definition of the stability coefficient is expressed by Eq. (2.13). For different corrugation shapes, and the formula form of the stability coefficient is similar.

$$\varphi = \tau_u / f_{\mathrm{vy}}. \tag{2.13}$$

However, the dimensions of H-CSPSWs are quite different from those of corrugated steel webs. Tong and Guo[66] fitted the shear buckling curves for unstiffened and stiffened H-CSPSWs based on the numerical results, which were also adopted by T/CECS 624-2019.[22] The rigidity ratio η reflects the restraining effect of the stiffening system. The ultimate strength factor could be expressed by Eqs.(2.16) and (2.18) for the case $\eta < \eta_{0,p}$ and the case $\eta \geq \eta_{0,p}$, respectively.

$$\eta = 2EI_s / D_y b, \tag{2.14}$$

$$\eta_{0,p} = (750 - 3850\theta) \cdot \beta + 760\theta - 175, \tag{2.15}$$

$$\varphi = \begin{cases} 1 - 0.148\lambda_n^2 & \lambda_n \le 0.8, \\ \left[\Phi - \sqrt{\Phi^2 - 4\lambda_n^2}\right] \Big/ 2\lambda_n^2 & 0.8 < \lambda_n \le 1.05, \\ 0.637/\lambda_n & \lambda_n > 1.05, \end{cases} \tag{2.16}$$

$$\Phi = 0.5 + 0.68\lambda_n + \lambda_n^2, \tag{2.17}$$

$$\varphi = \begin{cases} 1 - 0.137\lambda_n^2 & \lambda_n \le 0.8. \\ 0.73/\lambda_n & \lambda_n > 0.8. \end{cases} \tag{2.18}$$

Dou *et al.*[64] investigated the shear resistance and post-buckling behavior of H-CSPSWs with sinusoidal corrugation. The design formula for predicting the maximum shear strength was expressed by Eq. (2.19). As for trapezoidally H-CSPSWs, Dou *et al.*[61] found that when $\lambda_n \le 0.35$, the ultimate strength factor $\varphi > 0.95$, indicating full shear yielding could be ensured.

$$\varphi = \begin{cases} 1 - 0.20\lambda_n^2 & \lambda_n \le 1.0, \\ 1.60/\left(1 + \lambda_n^2\right) & \lambda_n > 1.0. \end{cases} \tag{2.19}$$

Qiu *et al.*[38] also conducted numerical studies and obtained the design formula expressed by Eq. (2.20). Fig. 7 depicts different formulas for predicting the ultimate shear strength. When the normalized slenderness ratio λ_n is below 0.50, the H-CSPSWs are approximately subjected to full shear yielding failure.

$$1/\tau_u^{2.5} = 1/f_{\mathrm{vy}}^{2.5} + 1/\tau_{\mathrm{cr}}^{2.5}. \tag{2.20}$$

The design methods for openings in H-CSPSWs have been investigated by some researchers. Farzampour *et al.*[65] proposed a simplified approach, suggesting that the ultimate shear strength of H-CSPSWs with an opening can be determined by

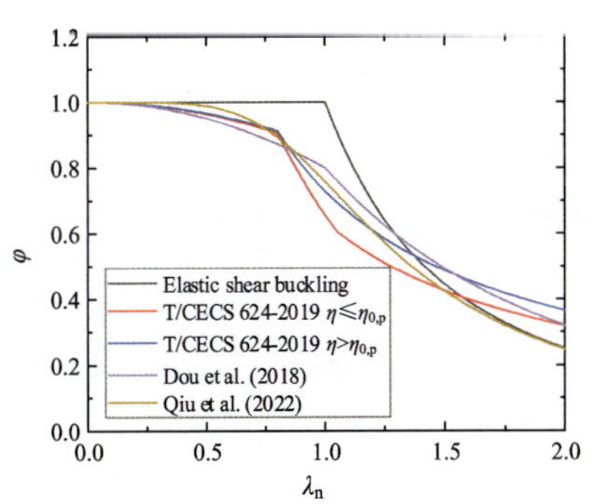

Fig. 7. Comparison of different formulas for predicting the shear strength.

multiplying a reduction factor $(1 - b_{op}/b)$, regardless of opening location. However, the validity of this simplification may require further investigation.

The interaction effect between the wall and the frame is another crucial aspect to consider. In the case of CSPSWs with a larger thickness, the interaction behavior between the plate and the frame differs due to the shear yielding failure mode, as opposed to the tension field failure mode. Therefore, there is a need for future studies to enhance the design methods considering the wall–frame interaction effect for H-CSPSWs.

2.4. *Shear behavior and design of vertical CSPSWs (V-CSPSWs)*

2.4.1. *Experimental study*

The shear behavior of V-CSPSWs also attracts the attention of many researchers. Emami *et al.*[14] conducted a comparative study on the cyclic behavior of V-CSPSWs and H-CSPSWs, finding that the seismic performance of the corrugated specimens was not influenced by the orientation of the corrugations. Ding *et al.*[79] and Gilvaee *et al.*[51] conducted cyclic tests on V-CSPSWs with rectangular openings. Hosseinzadeh *et al.*[49] experimentally investigated the effect of the corrugation angle on the shear behavior.

Recently, some researchers have also presented experiments on V-CSPSWs with a thickness of 2.0–6.0 mm. Shon *et al.*[52] compared the energy dissipation curves of H-CSPSW and V-CPSW specimens and observed similar results. Qiu *et al.*[53] performed an experiment with CSPSWs of width and height both measuring 1 100 mm on each floor. The test results for the H-CSPSW specimen and V-CSPSW specimen were nearly identical, as depicted in Fig. 8. Wang *et al.*[54,74] also conducted experimental comparisons of the cyclic behavior between H-CSPSWs and V-CSPSWs.

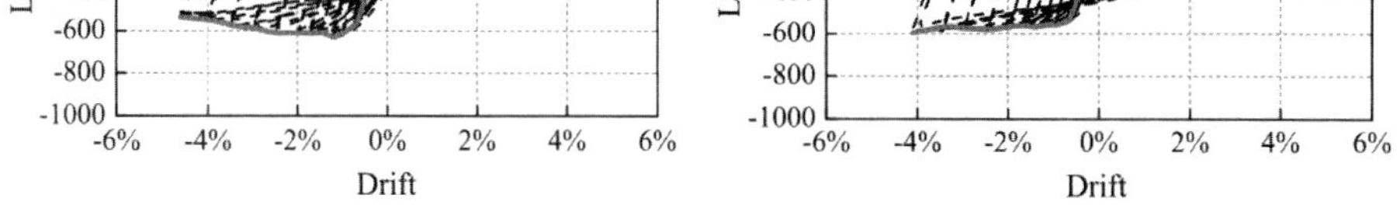

(a) V-CSPSW specimen (b) H-CSPSW specimen

Fig. 8. The hysteretic curves in the experiment conducted by Qiu *et al.*[53] ($t = 2.0$ mm).

2.4.2. *Parametric analysis*

In addition to the experimental studies, several numerical studies have been conducted to investigate the impact of the corrugation direction.[38,56,58,59,61,64,77,78] Zhao *et al.*[58] observed that V-CSPSWs exhibited comparable elastic lateral stiffness to H-CSPSWs, with slightly higher lateral strength, unloading stiffness. Dou *et al.*[61,64] discovered that V-CSPSWs demonstrated superior shear resistance compared to H-CSPSWs when the width of the wall (b) exceeded the height of the wall (h). Qiu *et al.*[38] performed nonlinear pushover analyses on a V-CSPSW model and observed minimal effect of corrugation direction on lateral behavior, as the elastic buckling loads were nearly identical, as shown in Fig. 9. Fadhil *et al.*[59] compared H-CSPSWs and V-CSPSWs with different corrugation angles. Ghodratian *et al.*[77] introduced CSPSWs with beam-only-connected infill plates, finding that detaching columns from the infill plate had a minor impact on responses for V-CSPSWs.

In summary, the shear behavior of V-CSPSWs is primarily influenced by the height, rather than the width. Additionally, the constraints applied to the corrugated edges of the plate are of greater importance, highlighting the need for effective restraint of the top and bottom edges in V-CSPSWs and the left and right edges in H-CSPSWs.

Hosseinzadeh *et al.*[49] investigated the cyclic behavior of V-CSPSWs with different angles of the trapezoidal plate. Fadhi *et al.*[59] found that the corrugation angle should be 30° in V-CSPSWs to achieve the maximum initial stiffness and highest ductility. Dou *et al.*[61,64] examined the impact of corrugation length, angle, and depth. The effect of the geometrical parameters on V-CSPSWs is similar to that on H-CSPSWs.

Similar to H-CSPSWs, Farzampour *et al.*[65] and Ding *et al.*[79] investigated the shear-carrying capacity of V-CSPSWs with a rectangular opening. The vicinity of the openings is typically a weak area in the wall, prone to premature failure. To

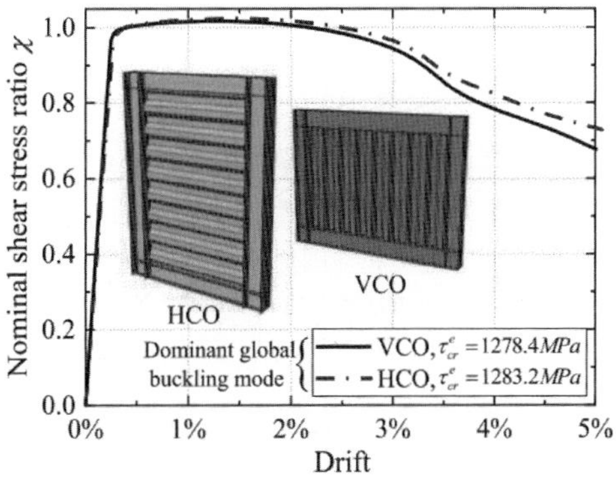

Fig. 9. Effects of corrugation direction on the lateral behavior (Qiu *et al.*[38]).

address this issue, Behzad *et al.*[81] proposed the use of arranged stiffeners around the openings.

Besides, Jin *et al.*[56] conducted numerical investigations on the shear resistance of V-CSPSWs with both longitudinal and transverse connecting stiffeners. The results demonstrated that multi-grid V-CSPSWs exhibited excellent out-of-plane stability. Wen *et al.*[40] conducted numerical studies on the global shear stability capacity of V-CSPSWs with vertical boundary elements and proposed the corresponding design method.

2.4.3. *Design method*

Qiu *et al.*[38] introduced identical design formulas for both V-CSPSWs and H-CSPSWs, as expressed in Eq. (2.20). Wen *et al.*[40] proposed the ultimate strength factor φ could be calculated by Eq. (2.21).

$$\varphi = 1/(1 + 0.8\lambda_n^{3.2})^{1/1.6}. \tag{2.21}$$

The design method for V-CSPSWs has not been provided in T/CECS 624-2019.[22] The numerical results of Qiu *et al.*[38] and Wen *et al.*[40] indicated that the formulas were accurate enough, which may be utilized in the engineering practice.

2.5. *Compressive-shear behavior of V-CSPSWs*

Researchers have examined the compressive-shear behavior of FSPSWs to address the challenges posed by axial compression. Zhang and Guo.[85] investigated the behavior of FSPSWs with pre-compression. Xu *et al.*[86] studied the elastic–plastic buckling of vertically stiffened FSPSWs under compression and shear loads.

Due to the accordion effect, the axial compression can be efficiently released in H-CSPSWs. However, for V-CSPSWs, the compression load from the gravity load or overturning moment of the whole structure is unavoidable and must be

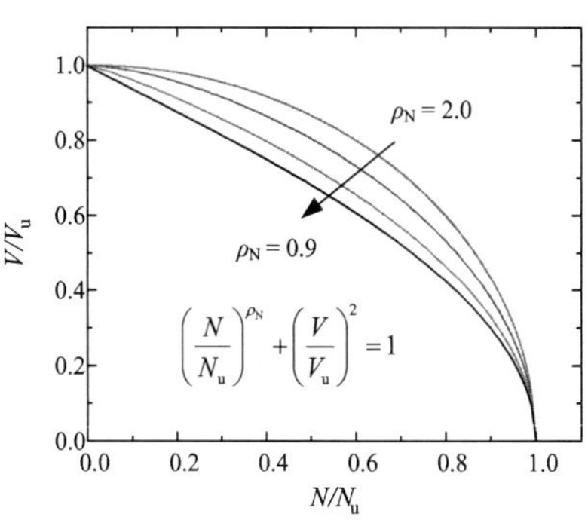

Fig. 10. Correlation design formula proposed by Wen *et al.*[84]

concerned. Zhao *et al.*[58] evaluated the effect of the gravity load. They found that the V-CSPSWs were less sensitive to the gravity load effects than FSPSWs. Wen *et al.*[84] investigated the compression-shear behaviors of V-CSPSWs. The correlation curve was proposed for design. The parameter ρ_N reflected the effect of the stability of the wall on the compression-shear behavior.

Nonetheless, Wen *et al.*[84] assumed the axial compression to be distributed uniformly and proposed the design method, as shown in Fig. 10 and Eqs. (2.22)–(2.24). It needs more investigation into the behavior and design when the non-uniform compression and bending loads are applied on the V-CSPSWs.

$$\left(\frac{N}{N_u}\right)^{\rho_\mathrm{N}} + \left(\frac{V}{V_u}\right)^2 = 1, \tag{2.22}$$

$$\rho_\mathrm{N} = \frac{2 + 5.4\lambda_\mathrm{sc}^{7.5}}{1 + 6\lambda_\mathrm{sc}^{7.5}}, \tag{2.23}$$

$$\lambda_\mathrm{sc} = \frac{h}{2\pi}\sqrt{\frac{f_y \omega t}{D_x}}. \tag{2.24}$$

3. Double Corrugated Steel Plate Shear Walls (DCSPSWs)

3.1. *Application and type of DCSPSWs*

As previously mentioned, CSPSWs have been extensively studied and widely employed as a lateral load-resisting system integrated within frames. However, there is an increasing demand for higher shear resistance, and CSPSWs may not fully meet the requirements for enhanced load-bearing capacity. Hence, double corrugated steel plate shear walls (DCSPSWs) have been proposed as a novel type of shear walls. DCSPSW consists of two identical trapezoidal corrugated steel plates that are directly connected using high-strength bolts. Typically, there are three different types of DCSPSWs: Parallel DCSPSWs with horizontal corrugation, parallel DCSPSWs with vertical corrugation, and orthogonal DCSPSWs, as illustrated in Fig. 11.

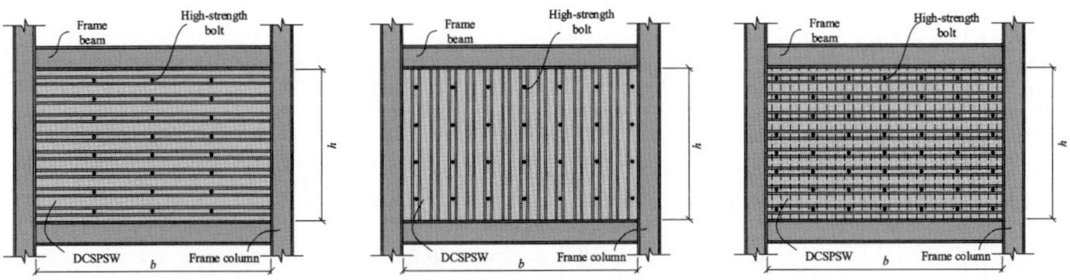

(a) Parallel DCSPSWs with horizontal corrugation (b) Parallel DCSPSWs with vertical corrugation (c) Orthogonal DCSPSWs

Fig. 11. Diagram of three types of DCSPSWs.

In parallel DCSPSWs, the corrugated plates are assembled in a parallel configuration, while in orthogonal DCSPSWs, they are assembled in an orthogonal configuration. The high-strength bolts are arranged at the valleys of the corrugations, connecting two plates to form a whole part. In parallel DCSPSWs with horizontal corrugation, the vertical load can also be efficiently released. Parallel DCSPSWs with vertical corrugation and orthogonal DCSPSWs can partially share the vertical compression load. Orthogonal DCSPSWs can be regarded as flat plates with modified rigidity constants. The presence of cavities formed by the two corrugated plates significantly enhances the stiffness of DCSPSWs. As an alternative to CSPSWs, DCSPSWs demonstrate great potential for application in multi-story or high-rise building structures.

3.2. *Behavior and design of parallel DCSPSWs with horizontal corrugation*

3.2.1. *Elastic shear buckling*

Tong *et al.*[87–90] first presented parallel DCSPSWs with horizontal corrugation and investigated their buckling behavior. The buckling behavior of DCSPSWs is affected by the spacing of the bolts. The vertical spacing of the bolts was taken as the corrugation length. When the horizontal spacing d_b is very large, the strength of DCSPSWs can be regarded as a simple superposition of two single plates. For one single plate (see Fig. 12), the rigidity constants were derived, expressed by Eqs. (3.1)–(3.3).

$$D_{x1} = \frac{E}{\lambda} \cdot \left[\frac{(d_1 + d_3)\, t^3}{12} + \left(\frac{2}{3} d_0 + d_1 - \frac{(d_0 + d_1)^2}{q} \right) a^2 t \right], \tag{3.1}$$

$$D_{y1} = \frac{\lambda}{q} \cdot \frac{Et^3}{12(1 - \upsilon^2)}, \tag{3.2}$$

$$H_1 = \frac{q}{\lambda} \cdot \frac{Et^3}{12(1 + \upsilon)}. \tag{3.3}$$

When d_b is small enough, the two plates can form cavities to work together. The rigidity constants can be expressed by Eqs. (3.4)–(3.6).

$$D_x = \frac{E}{6\lambda} \cdot [(d_1 + d_3)t^3 + 4(2d_0 + 3d_1)ta^2], \tag{3.4}$$

$$D_y = \frac{\lambda}{d_3 + \dfrac{d_0^2 + 2d_0 d_1}{2d_0 + 3d_1}} \cdot \frac{Et^3}{6\,(1 - \upsilon^2)}, \tag{3.5}$$

$$H = \frac{2Et}{\lambda q\,(1 + \upsilon)} \cdot a^2\,(d_1 + d_2)^2\,. \tag{3.6}$$

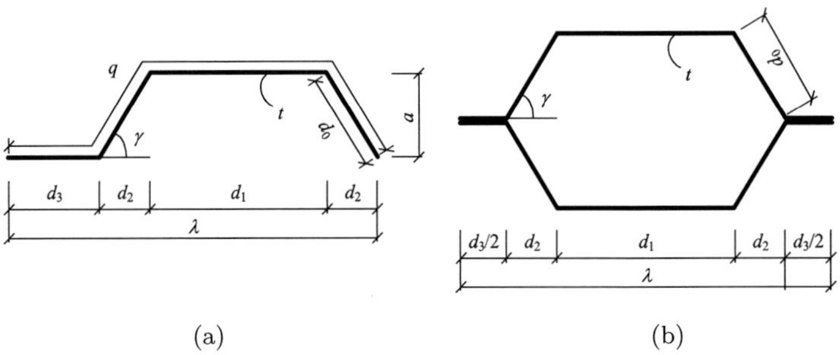

Fig. 12. Dimensions of the corrugation of DCSPSWs.

Tong *et al.*[87] presented the formulas to predict the buckling coefficient, expressed by Eq. (3.7), where ω is the reduction factor considering discrete bolt connections. The ratio of the bolt spacing and the width η was introduced to reflect the constraint strength under different bolt spacing. Then, the normalized slenderness ratio can also be obtained after calculating the elastic buckling stress by Eq. (2.12).

$$k = \omega k_0. \tag{3.7}$$

$$\omega = 1 - \frac{\eta^{1.1} \cdot (1 - \omega_1)}{\eta^{1.1} + 0.08 \cdot (1 - \eta)^{1.1}}, \tag{3.8}$$

$$\eta = d_b / b. \tag{3.9}$$

3.2.2. *Shear behavior and design method*

The monotonic and cyclic tests have been conducted by Tong *et al.*[88,90] to investigate the behavior of DCSPSWs. The tests revealed that DCSPSWs with high aspect ratios were prone to global buckling failure. The cyclic tests demonstrated that DCSPSWs exhibited stable and well-rounded hysteretic curves. The main failure modes were the low-cycle fatigue failure of the boundary columns and the bucking of the corrugated plate. It was also found that the effect of connecting bolts was insignificant for DCSPSWs with small aspect ratios. In addition, increasing the corrugation depth and angle could increase the shear resistance.

Based on the numerical studies, the ultimate strength factor could be predicted by Eq. (3.10), which has also been employed by T/CECS 624-2019.[22]

$$\varphi = \begin{cases} 1 - 0.233\lambda_n^2 & \lambda_n \le 0.67, \\ 0.6/\lambda_n & \lambda_n > 0.67. \end{cases} \tag{3.10}$$

The post-ultimate ductile behavior was also analyzed by Tong *et al.*[88–90] and the corresponding predicting formulas for the shear-resistant curve were fitted. These predictions could be adopted for performing the push-over elastoplastic analysis of the DCSPSWs. However, the vertical spacing of the bolts in the study of Tong *et al.*[88–90] was the corrugation length. More studies may be needed for the DCSPSWs with relatively large vertical spacing. The stress analyses and design for the bolts may also be investigated experimentally and numerically.

3.3. *Behavior and design of parallel DCSPSWs with vertical corrugation*

Deng *et al.*[91] and Ghodratian-Kashan *et al.*[92,93] conducted numerical and experimental studies on parallel DCSPSWs with vertical corrugation. The cyclic tests on DCSPSWs of Deng *et al.*[91] showed that the local and global shear buckling was the main failure mode of DCSPSWs. Owing to the combined effect, the ultimate shear resistance of DCSPSWs was higher than that of a single-corrugated plate with the same steel volume.

The experiments conducted by Ghodratian-Kashan *et al.*[93] showed that DCSPSWs with vertical corrugation exhibited a spindle-shaped hysteresis curve, indicating favorable cyclic performance even in the case of beam-only connections. Ghodratian-Kashan *et al.*[92,93] also developed equations to predict the ultimate strength of DCSPSWs by considering the tension field and interactive shear buckling stress. These equations provide valuable insights into the shear behavior of DCSPSWs and can assist in their design and analysis.

However, the studies conducted by Deng *et al.*[91] and Ghodratian-Kashan *et al.*[92,93] focused on DCSPSWs with relatively small thickness, which may not provide sufficient lateral resistance and limit their practical application. Additionally, there is still a lack of research on parallel DCSPSWs with vertical corrugation, and the existing design methods for such systems require further improvement. Future investigations could consider the behavior of DCSPSWs with larger thicknesses, ensuring their effective utilization in structural applications.

3.4. *Behavior and design of orthogonal DCSPSWs*

3.4.1. *Elastic shear buckling*

The behavior of orthogonal DCSPSWs was studied systematically by Tong *et al.*[94] First, the rigidity constants of orthogonal DCSPSWs were proposed and are expressed as follows:

$$D_x = D_y = D_{x1} + \frac{Et^3\lambda}{12\left(1-\nu^2\right)} \frac{\left(\frac{2}{3}d_0 + d_1\right)q + 3(d_0+d_1)^2}{\left(\frac{2}{3}d_0 + d_1\right)q^2 - (d_0+d_1)^2 q}, \tag{3.11}$$

$$H = 1.82\sqrt{\frac{t}{a}} \cdot \frac{Eta^2}{1+\upsilon} \frac{(d_1+d_2)^2}{\lambda(q+\lambda)}. \tag{3.12}$$

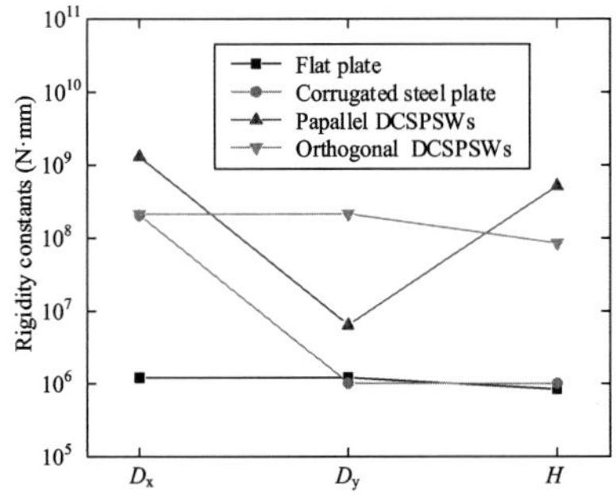

Fig. 13. The comparison of the rigidity constants.

As for orthogonal DCSPSWs, the two flexural rigidity constants in two orthogonal directions, D_x and D_y, are equal. The rigidity constants of the flat plate, corrugated steel plate, parallel DCSPSW, and orthogonal DCSPSWs are compared in Fig. 13. It is evident that the flexural rigidity constant D_y of orthogonal DCSPSWs is much larger than that of others. The shear buckling modes and the buckling loads of orthogonal DCSPSWs are related to the arrangement of bolts. Tong *et al.*[94] proposed the formulas to calculate the shear buckling stress of orthogonal DCSPSWs based on the numerical results.

3.4.2. *Shear behavior and design method*

In the parametric analyses conducted by Tong *et al.*[94] it was observed that orthogonal DCSPSWs exhibited good shear-carrying capacity and ductility. The shear-carrying capacity of the orthogonal DCSPSWs was superior to that of the parallel DCSPSWs when the dimensions and bolt spacing were kept constant. Most examples could achieve the shear-yielding mechanism, indicating the excellent shear-bearing performance of DCSPSWs. A relatively conservative design formula, expressed by Eq. (3.13), was proposed to predict the shear-carrying capacity of orthogonal DCSPSWs.

$$\varphi = \begin{cases} 1 & \lambda_n \le 0.3, \\ -0.0588\lambda_n + 1.018 & 0.3 < \lambda_n < 2.0, \\ 0.9 & \lambda_n \ge 2.0. \end{cases} \tag{3.13}$$

Zhu *et al.*[95] numerically investigated the shear-resistant capacity of orthogonal DCSPSWs with two boundary elements. It is suggested that the ultimate strength factor can be calculated by the following equation.

$$\varphi = \begin{cases} 1/\left(1 + 1.2\lambda_n^{3.0}\right)^{1/1.5} & \lambda_n \le 0.8, \\ 0.65/\sqrt{\lambda_n} & \lambda_n > 0.8. \end{cases} \tag{3.14}$$

Besides, Tong *et al.*[94] investigated the effect of the thickness, aspect ratio, and bolt spacing on the cyclic performance of orthogonal DCSPSWs. Increasing the number of bolts could significantly improve the cumulative energy dissipation and ductility.

4. Concrete-Infilled Double Steel Corrugated-Plate Wall (CDSCWs)

4.1. *Application and type of CDSCWs*

Traditional composite shear walls are composed of two outer flat steel plates and infilled concrete, using connectors to achieve the composite action, such as shear studs, bolts, fastener screws, hooks, and tie rods. The outer steel plates can provide restraint to the infilled concrete and improve the ductility and deformation capacity of the concrete. The steel plates are anchored to the infilled concrete through connectors. Thus, the infilled concrete can restrain steel plates by one side, preventing or delaying the out-of-plane buckling of the steel plates. Besides, steel plates are used as the framework of concrete pouring construction, which can improve construction efficiency to save time and costs.

Extensive experimental and numerical studies on composite shear walls have been conducted.[17,18] Traditional outer steel plates are flat steel plates or skinned profiled steel sheets which are often less than 2.0 mm thick. Therefore, the contributions to the load-bearing capacity of the profiled steel sheets are neglected, and the load-bearing capacity is limited. Besides, the dense shear connectors are inevitable in flat steel plates to achieve excellent seismic performance, which produces a complex fabrication process.

Due to the limitations of traditional composite shear walls, Guo *et al.*[19] proposed a new type of CDSCWs. As depicted in Fig. 14, CDSCW consists of two corrugated steel plates with infilled concrete and two vertical boundary elements. The corrugated plates are directly connected using bolts and the boundary elements are concrete-filled steel tubes (CFSTs). Besides, Tong *et al.*[96,97] also proposed multi-cell multi-celled corrugated-plate CFST walls to combine the corrugated plate and concrete together. Unlike CSPSWs or DCSPSWs, CDSCWs are arranged vertically in a continuous manner, without any interruptions by beams along the wall height. They are not only subjected to axial compression and in-plane bending moments, but also to horizontal shear forces.

It is well known that CDSCWs have better load-bearing capacity and seismic performance than the traditional composite shear walls, owing to the high out-of-plane flexural stiffness of corrugated steel plates. The compression loads can be transmitted effectively and continuously. As Fig. 15 shows, CDSCWs have been employed in several engineering buildings by Zhejiang Zhongnan Construction Group Steel Structure Co. Ltd. These walls offer numerous advantages, including high load-bearing capacity, excellent seismic performance, flexible layout

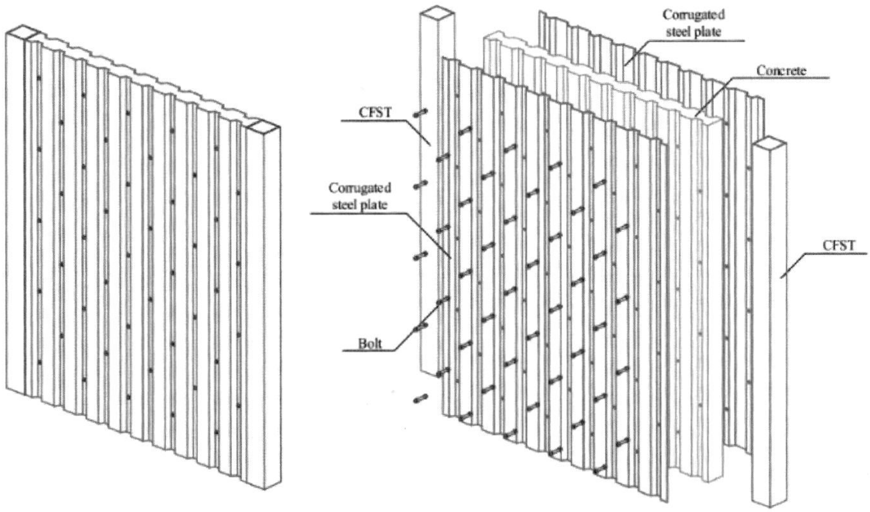

(a) Assembled diagram (b) Exploded diagram

Fig. 14. Composition diagrams of a CDSCW.

(a) (b)

Fig. 15. Photograph of the CDSCWs in practical engineering structures.

options, efficient construction processes, and superior engineering quality. Therefore, CDSCWs have become increasingly favored by engineers, leading to their widespread adoption as both axial compressive and lateral resisting structural systems in new buildings. Numerous experimental and numerical studies have been conducted, Guo *et al.*[19–21,98–104] have proposed a comprehensive design method that has been incorporated into T/CECS 624-2019.[22] Recent cyclic tests on CDSCWs highlighted the significance of boundary elements in enhancing the seismic performance of CDSCWs.[98,105] These findings suggest that further improvements can be made in the earthquake-resistant design of CDSCWs.

CDSCWs can be placed flexibly at the different positions of the buildings. According to architectural and structural design demands, there are different types of cross-sections, such as I-section, T-section, and L-section. The high-strength bolts could also be arranged in aligned and staggered ways to obtain different load-bearing performances.

4.2. *Sectional load-bearing capacity and design*

4.2.1. *Under uniform compression*

Zhu and coworkers[106] investigated the load-bearing mechanism and strength design formulas of CDSCWs under uniform compression. The sectional load-bearing capacity was analyzed using the concept of full-section plasticity. When the vertical spacing of the bolts is large, or the thickness of the steel corrugated plates is small, the local buckling of the steel plates may occur prior to yielding. Therefore, the influence of local buckling on section strength should be considered.

Through a series of FE eigenvalue buckling analyses, it was determined that local buckling failure often occurred between two rows of bolts. A parametric study indicated that the vertical bolt spacing d_1 and the plate thickness t_s greatly affected the buckling load. The equations of the elastic local buckling load were fitted based on the numerical results obtained, expressed by Eqs. (4.1)–(4.4), where E_s was the elastic modulus of steel, I_s was the out-of-plane moment of inertia of the corrugated plate, and s was the unfolded length of the corrugated plate.

$$\sigma_s = k_s \frac{\pi^2 E_s I_s}{d_1^2 s t_s}, \tag{4.1}$$

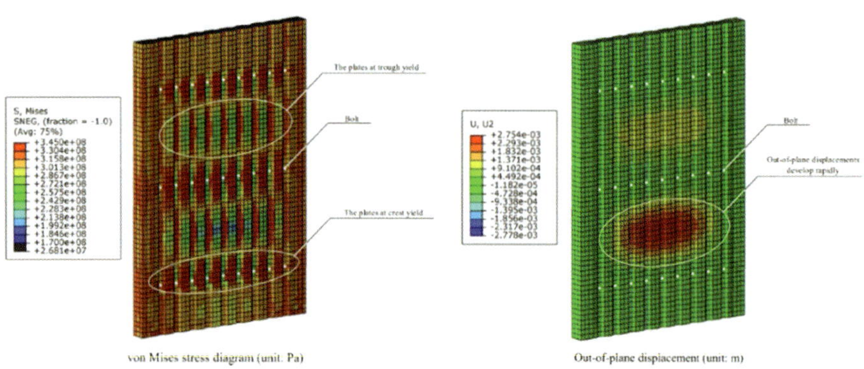

(a) (b)

Fig. 16. Stress and deformation diagrams of the typical example.

$$k_s = 0.480 + 0.285 \ln \beta_1 + 0.314\beta_2, \tag{4.2}$$

$$\beta_1 = d_1/d_2, \tag{4.3}$$

$$\beta_2 = t_s/d_2. \tag{4.4}$$

The ultimate load-bearing performance of the corrugated steel plate in CDSCWs under uniform compression was analyzed by Zhu and coworkers,[106] Fig 16 depicted the stress and deformation at the peak load of the typical example. Due to the lateral restraining effect, the plates can only bulge away from the infilled concrete.

Based on abundant FE results, the ultimate load-bearing capacity of the corrugated steel plates under uniform compression could be calculated as follows:

$$N_{\mathrm{us}} = \varphi_s f_y A_s, \tag{4.5}$$

$$\varphi_s = \begin{cases} 0.97 & 0 \leq \lambda s \leq 0.5, \\ -0.042\lambda s^3 - 0.056\lambda s^2 + 0.083\lambda s + 0.95 & 0.5 < \lambda s \leq 1.75. \end{cases} \tag{4.6}$$

$$\lambda_s = \sqrt{f_y/\sigma_s}. \tag{4.7}$$

It was further observed that following the occurrence of local buckling in the corrugated plates, the infilled concrete would experience damage. Therefore, the ultimate compressive strain of entire CDSCWs could be defined as the ultimate compressive strain of the steel plates. The sectional load-bearing capacity of CDSCWs could be expressed as follows:

$$N_u = N_{\mathrm{us}} + N_{\mathrm{uc}} + N_{\mathrm{cfst}} = \varphi_s f_y A_s + \sigma_c(\varepsilon_c) A_c + N_{\mathrm{cfst}}, \tag{4.8}$$

$$\varepsilon_c = \varepsilon_u = \varepsilon_{\mathrm{us}} = \varphi_s f_y / E_s. \tag{4.9}$$

Zhu and coworkers[20,106] also performed experimental studies on the sectional load-bearing capacity of CDSCWs. The test results agreed well with the numerical simulation results and also verified the safety and validity of the proposed formulas of the sectional load-bearing capacity of CDSCWs.

4.2.2. *Under combined compression and in-plane bending loads*

As Fig. 17 depicts, the sectional load-bearing capacity under combined compression and in-plane bending loads is also developed based on the full-section plastic failure mode. The sectional load-bearing capacity of the corrugated steel plate is affected by local buckling and transition zone of the tension and compression, which is reflected by a reduction factor φ_{sp}. The contribution of concrete in the tension area is neglectable. The reduction of the transition zone on the load-bearing capacity of the concrete is reflected by the factor β_c. The parameter χ reflects the position of the neutral axis.

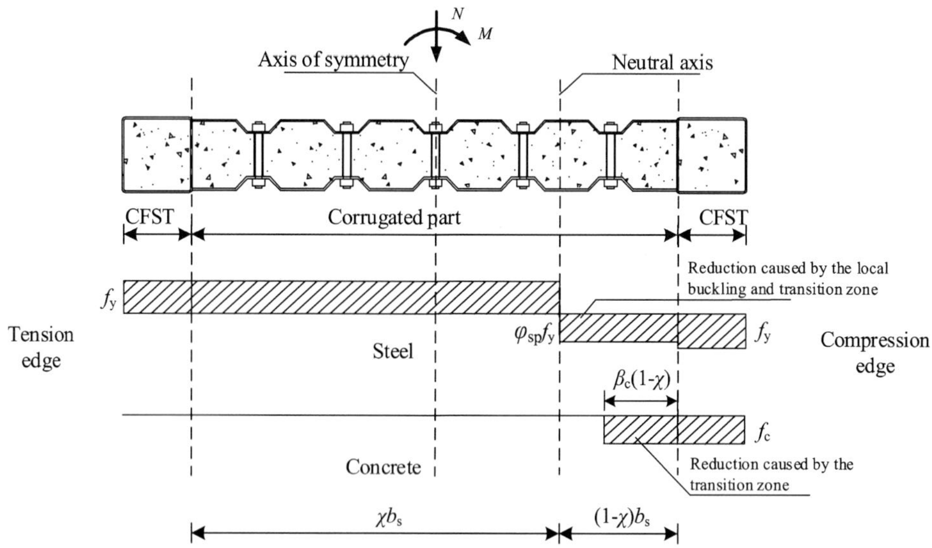

Fig. 17. Stress distribution diagram under combined compression and in-plane bending loads.

The ultimate load-bearing capacity of CDSCWs could be obtained by superposing the contribution of the corrugated plate, concrete, and the boundary elements, expressed as follows:

$$N_u = N_{\text{SCP}} + N_{\text{Con}} + N_{\text{CFST}}, \tag{4.10}$$

$$M_u = M_{\text{SCP}} + M_{\text{Con}} + M_{\text{CFST}}. \tag{4.11}$$

Zhu and coworkers[20,106] conducted numerical studies based on the validated FE model. The reduction factor φ_{sp} was fitted expressed by Eq. (4.12), and β_c was suggested to be 0.4.

$$\varphi_{\text{sp}} = \begin{cases} \varphi_p & \chi > 0.5, \\ 2(\varphi_p - \varphi_s)\chi + \varphi_s & 0 < \chi \leq 0.5 \\ \varphi_s & \chi \leq 0, \end{cases} \tag{4.12}$$

$$\varphi_p = 1.775 \left(\beta_1/\beta_2^2\right)^{-0.112}. \tag{4.13}$$

Then, the ultimate load-bearing capacity of the steel plate and concrete could be calculated as follows:

$$N_{\text{SCP}} = \varphi_{\text{sp}}(1-\chi)N_{\text{sw}} - \chi N_{\text{sw}}, \tag{4.14}$$

$$M_{\text{SCP}} = \frac{1}{2}\chi(1-\chi)(1+\varphi_{\text{sp}})N_{\text{sw}}b_s, \tag{4.15}$$

$$N_{\mathrm{Con}} = \begin{cases} \beta_c(1-\chi)N_{\mathrm{cw}} & \chi > 0, \\ \dfrac{b_s}{b_c}\left[\beta_c - \dfrac{\sigma_c(\varepsilon_c)}{f_c}\right] N_{\mathrm{cw}}\chi + \beta_c N_{\mathrm{cw}} & -\dfrac{b_c}{b_s} \le \chi \le 0, \\ \dfrac{\sigma_c(\varepsilon_c)}{f_c} N_{\mathrm{cw}} & \chi < -\dfrac{b_c}{b_s}, \end{cases} \tag{4.16}$$

$$M_{\mathrm{Con}} = \begin{cases} \dfrac{1}{2}\beta_c(1-\chi)(1-\beta_c+\beta_c\chi)N_{\mathrm{cw}}b_s & \chi > 0, \\ \dfrac{1}{2}\beta_c(1-\beta_c)\left(\dfrac{b_s}{b_c}\chi + 1\right) N_{\mathrm{cw}}b_s & -\dfrac{b_c}{b_s} \le \chi \le 0, \\ 0 & \chi < -\dfrac{b_c}{b_s}. \end{cases} \tag{4.17}$$

An experimental study was conducted to investigate the sectional load-bearing capacity of CDSCWs under combined compression and in-plane bending load. The local buckling of the corrugated steel plate was observed in the test, as illustrated in Fig. 18. Point E in Fig 19 was the test value of the ultimate load-bearing capacity of the specimen. The restraining effect of bolts and concrete on the corrugated steel plate was considered. It was found that the design formulas were conservative and accurate enough.

4.2.3. *Reduction caused by the shear load*

The compression and the bending loads were the governing factors in the studies of Zhu and coworkers.[20,106] Therefore, the influence of the shear load was considered by introducing the reduction factors ρ_N and ρ_M, as shown in Fig. 20. The shear load was assumed to be subjected by the steel plates parallel to the CDSCW plane.

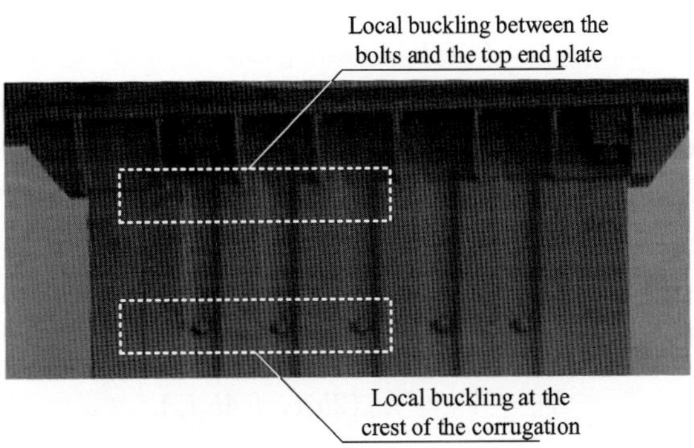

Fig. 18. Local buckling of the corrugated steel plate in the test.

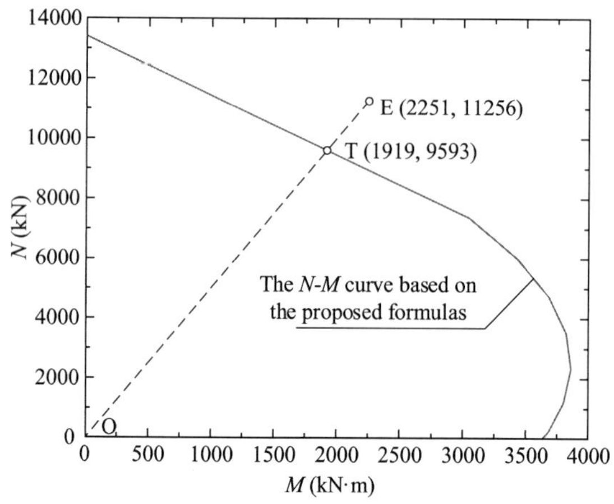

Fig. 19. Comparison between the test results and the proposed formulas.

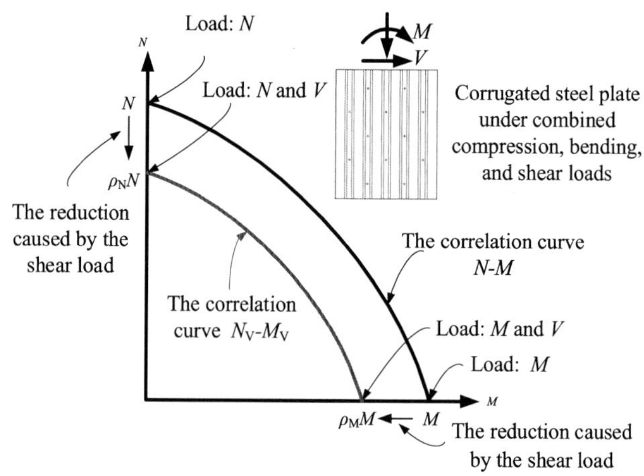

Fig. 20. The correlation curve of the load-bearing capacity of the corrugated steel plate.

Zhu and coworkers[20,106] developed an FE model to investigate the load-bearing capacity of the corrugated steel plate under combined loads. The FE results indicated that the corrugated steel plates were subjected to full shear yielding failure in the common size range of engineering applications. Hence, the ultimate shear capacity of the plate could be estimated by the shear yield capacity.

$$V_u = V_y = f_{vy}(2b_s t_s + 4b_c t_c). \tag{4.18}$$

Then, Zhu *et al.*[20,106] fitted the correlation curves, and the reduction factors ρ_{sw} and ρ_{sc} could be expressed by Eqs. (4.19) and (4.20), respectively. The sectional load-bearing capacity could be calculated by the proposed formulas as follows:

$$\rho_{\text{sw}} = \rho_{\text{NV}} = \rho_{\text{MV}} = \left[1 - (V/V_u)^2\right]^{2/3}, \tag{4.19}$$

$$\rho_{\text{sc}} = \begin{cases} 1 & V/V_u \leq 0.5, \\ 1 - (2V/V_u - 1)^2 & V/V_u > 0.5. \end{cases} \tag{4.20}$$

4.3. *Global stability capacity of CDSCWs with I-section*

4.3.1. *Under uniform compression*

For CDSCWs having slender ratios in their out-of-plane direction, overall instability failure may occur and becomes a major concern in structural design. Therefore, the global stability capacity of CDSCWs with *I*-section should be examined. Guo *et al.*[19] studied the overall instability performance of CDSCWs experimentally and numerically. The corresponding strength design recommendations were established for the convenient engineering application in structural design and construction.

The elastic buckling load was analyzed first, as shown in Fig. 21. The steel-to-concrete ratio ρ_w, defined as the ratio of the steel sectional area and concrete sectional area, was introduced to fit the design formulas. Then, the elastic buckling load and the normalized slenderness ratio of CDSCWs could be obtained as Eqs. (4.21) and (4.24).

$$N_{\text{cr}} = \pi^2[0.86\rho_w^{-0.054}(E_{\text{sw}}I_{\text{sw}} + E_{\text{cw}}I_{\text{cw}}) + 2(E_{\text{sc}}I_{\text{sc}} + E_{\text{cc}}I_{\text{cc}})]/h_s^2, \tag{4.21}$$

$$\rho_w = A_{\text{sw}}/A_{\text{cw}}, \tag{4.22}$$

$$N_y = A_{\text{sw}}f_{\text{yw}} + A_{\text{cw}}f_{\text{cw}} + 2A_{\text{cc}}f_{\text{cc}} + 2A_{\text{sc}}f_{\text{yc}}, \tag{4.23}$$

$$\lambda_n = \sqrt{N_y/N_{\text{cr}}}. \tag{4.24}$$

The elastic–plastic analyses were carried out to investigate the global stability of CDSCWs with *I*-section. According to the FE numerical results, the φ_N–λ_n curve for estimating the overall compressive stability was attained (see Fig. 22) and is expressed as follows:

$$N_u = \varphi_{\text{N}}N_y, \tag{4.25}$$

$$\varphi_{\text{N}} = \begin{cases} 1 - 0.835\lambda_n^2 + 0.37\lambda_n^3 & 0 \leq \lambda_n \leq 0.8, \\ 1\Big/\left[\phi + \sqrt{\phi^2 - \lambda_n^2}\right] & \lambda_n > 0.8, \end{cases} \tag{4.26}$$

$$\phi = 0.50\left[1 + 0.90(\lambda_n - 0.46) + \lambda_n^2\right]. \tag{4.27}$$

Besides, Guo *et al.*[19] also conducted the overall instability experimental tests of CDSCWs under axial compression. A 4.5 m-high *I*-section CDSCW specimen was tested to estimate its overall flexural instability performance. The experimental results yielded an overall stability coefficient of 0.968 for the specimen, indicating a high load-bearing capacity of CDSCWs. The simulation also validated the FE nonlinear analyses.

4.3.2. *Under combined compression and in-plane bending loads*

Guo *et al.*[19] also investigated the global stability capacity of CDSCW with I-section under combined compression and in-plane bending loads. The formulas for predicting the global stability capacity were established based on the numerical results. Then the correlation curve $N/N_u - M/M_u$ could be obtained, where M_u was the

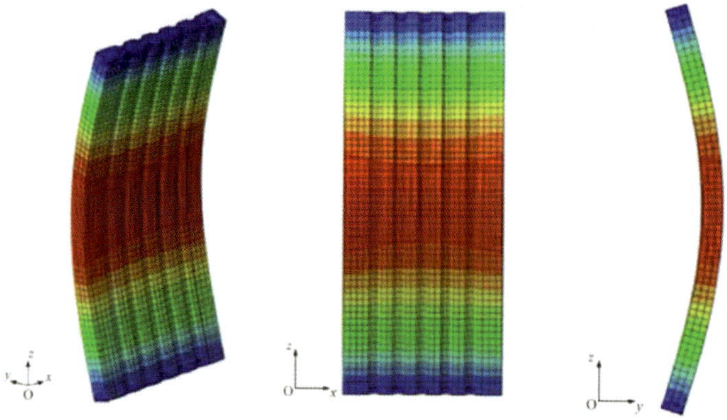

Fig. 21. The first-order elastic buckling modes of CDSCWs.

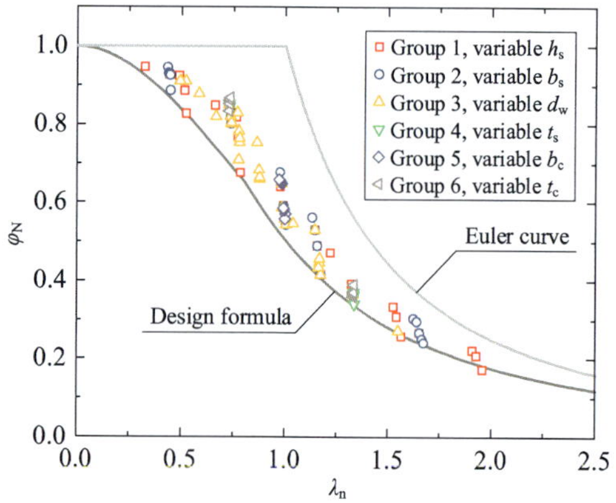

Fig. 22. Lower envelope of numerical examples of CDSCWs under compression.

stability capacity under in-plane bending only.

$$M_u = \varphi_{\mathrm{M}} M_y, \tag{4.28}$$

$$\varphi_{\mathrm{M}} = -\frac{0.18}{10^3}\frac{h_s b_s}{d_w^2} + 1.04 \leq 1, \tag{4.29}$$

$$(N/N_u)^2 + M/M_u \leq 1. \tag{4.30}$$

4.4. *Global stability capacity and design of wall panel*

4.4.1. *Under uniform compression*

Considering the CDSCWs consisting of several wall panels in T-section or L-section, as shown in Fig. 23, the global stability capacity of wall panels of CDSCWs is another concern for structural designers. In CDSCWs with T-section or L-section, the wall panel is susceptible to global buckling under compression and bending. The CFST at the intersection of wall panels provides lateral restraint, simplifying the flange wall panel as three-side simply supported and one-side panels. Similarly, the web panels of CDSCWs with C-section or Z-section can be simplified as four-side simply supported panels.

In the engineering construction standard T/CECS 624-2019,[22] the design formulas for predicting the global stability capacity of wall panels have been proposed. First, the elastic buckling load of the panel should be calculated. Then, by introducing the normalized slenderness ratio, the stability coefficients of wall panels could be obtained. As for three-side simply supported and one-side free wall panels, the

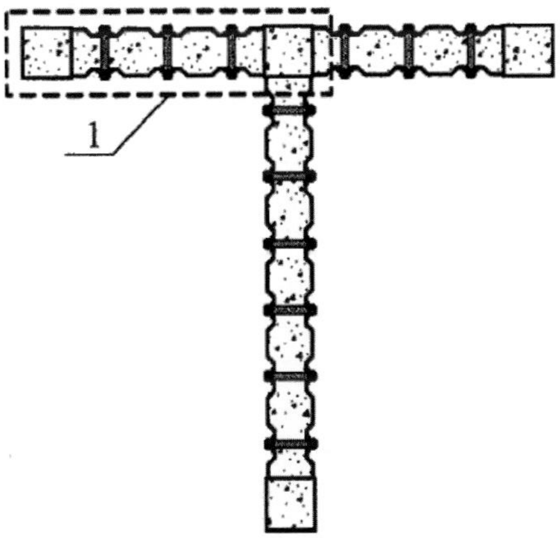

(a) Three-side simply supported and one-side free

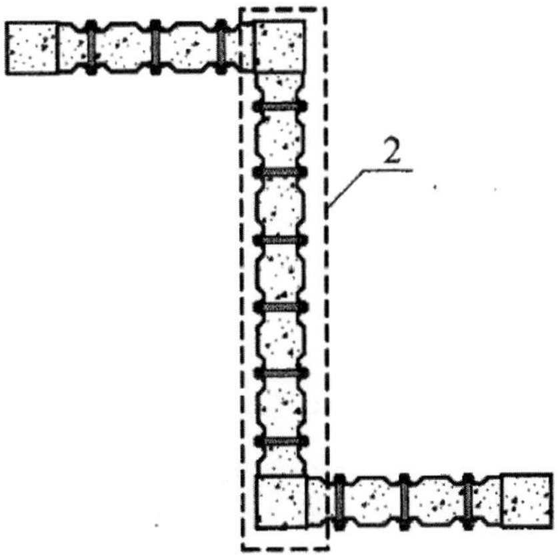

(b) Four-side simply supported

Fig. 23. The boundary conditions of the wall panels.

stability coefficient is expressed by Eqs. (4.31) and (4.32). Similarly, the stability coefficient of four-side simply supported and one-side free wall panels is expressed by Eqs. (4.33) and (4.34).

$$\varphi_{\mathrm{pf}} = \begin{cases} 0.9 & 0 \leq \lambda_n \leq 0.4, \\ \dfrac{1}{\phi + \sqrt{\phi^2 - \lambda_n^2}} & \lambda_n > 0.4, \end{cases} \tag{4.31}$$

$$\phi = 0.50[1 + 1.15(\lambda_n - 0.32) + \lambda_n^2], \tag{4.32}$$

$$\varphi_{\mathrm{pf}} = \begin{cases} 1 & 0 \leq \lambda_n \leq 0.35, \\ 1\Big/[\phi + \sqrt{\phi^2 - \lambda_n^2}] & \lambda_n > 0.35, \end{cases} \tag{4.33}$$

$$\phi = 0.50[1 + 2(\lambda_n - 0.35) + \lambda_n^2]. \tag{4.34}$$

4.4.2. *Under combined compression and in-plane bending loads*

The FE numerical analyses revealed that the wall panels were not prone to flexural instability failure under the pure in-plane bending moments. Therefore, the global stability capacity under the pure bending load M_u was calculated by the yield load-bearing capacity M_y. Then, the correlation curve N/N_u–M/M_u under the combined compression and in-plane bending loads was obtained.

As for the three-side simply supported and one-side free wall panels, the correlation curve N/N_u–M/M_u could be expressed by Eq. (4.35) when the free side was under compression. When the simply supported side was under compression, the correlation curve N/N_u–M/M_u was calculated by Eq. (4.36). In addition, the global stability capacity of the four-side simply supported wall panels should be checked by Eq. (4.37).

$$1.3(N/N_u)^2 - 0.3N/N_u + M/M_u \leq 1, \tag{4.35}$$

$$N/N_u + M/M_u \leq 1, \tag{4.36}$$

$$1.7(N/N_u)^2 - 0.7N/N_u + M/M_u \leq 1. \tag{4.37}$$

5. Summary and Conclusions

This paper provides a summary of previous research on steel and composite walls utilizing corrugated steel plates, specifically focusing on CSPSWs, DCSPSWs, and CDSCWs. This paper highlights the significant structural advantages offered by corrugated steel plates, including high load-bearing capacity, effective energy dissipation, and favorable ductility. The following conclusions can be drawn:

(1) Unlike FSPSWs, CSPSWs primarily experience shear-yielding or inelastic buckling failure, resulting in stable hysteresis curves and effective energy dissipation.

The load-bearing capacity of CSPSWs can be evaluated using the ultimate strength factor, which is inversely correlated with the normalized slenderness ratio.

(2) The V-CSPSWs can bear vertical compression along with the frame columns. However, the axial compression causes a reduction in the shear-carrying capacity of V-CSPSWs. There is limited research on the compression-shear behavior of V-CSPSWs, suggesting the need for future experimental studies.

(3) Since the bolted connections enable the two steel plates to work collaboratively, DCSPSWs could exhibit a higher load-bearing capacity and better seismic performance than that of two individual corrugated steel plates. However, more studies on DCSPSWs should be conducted to supplement the data of DCSPSWs, especially focusing on the compression-shear behavior of orthogonal DCSPSWs.

(4) CDSCWs demonstrate exceptional load-bearing and seismic performance due to the synergistic effects of corrugated plates, concrete, and bolt connections. The design strength recommendations for all failure modes of CDSCWs are currently available and considered conservative. However, there is still room for further investigation into the seismic performance of CDSCWs, particularly in understanding the enhancing effect of vertical boundary elements.

Acknowledgments

This study was supported by research grants from the National Natural Science Foundation of China (No. 51878376). The authors wish to express their gratitude to the sponsor.

References

1. J. Bai, W. Li, H. Chen, M. Feng and Y. H. Wang, A stiffness ratio-based seismic design for reinforced concrete frames with buckling-restrained braces, *Int. J. Struct. Stab. Dyn.* **22** (2022) 1–22.
2. J. He *et al.*, Hysteretic behavior and multi-wave buckling of assembled buckling-restrained braces wrapped with fiber clothes, *Int. J. Struct. Stab. Dyn.* **22** (2022) 1–31.
3. Y. Cao, Y. Zhou, G. Zhong, X. Deng and J. Takagi, An investigation of the hysteretic behavior of a buckling-restrained brace subsystem frame with out-of-plane eccentricity, *Int. J. Struct. Stab. Dyn.* **23** (2022) 1–28.
4. W. Wang, J. Ma, H. Al-Azzani and Y. Xing, Experimental study on seismic performance of steel frame sheathing with concrete and plasterboard composite wall, *Int. J. Struct. Stab. Dyn.* **22** (2022) 1–29.
5. L. He, X. Sun, H. Bu and Z. Tang, Seismic performance of MRSF structures damped with steel slit shear panels, *Int. J. Struct. Stab. Dyn.* **22** (2022) 1–17.
6. R. G. Troy and R. M. Richard, Steel plate shear walls resist lateral load, cut costs, *Civ. Eng.* **49** (1979) 53–55.

7. E. W. Tromposch and G. L. Kulak, Cyclic and static behaviour of thin panel steel plate shear walls, Structural Engineering Report (1987).
8. M. Elgaaly, V. Caccese and C. Du, Postbuckling behavior of steel–plate shear walls under cyclic loads, *J. Struct. Eng.* **119** (1993) 588–605.
9. J. W. Berman, Seismic behavior of code designed steel plate shear walls, *Eng. Struct.* **33** (2011) 230–244.
10. A. Razzazzadeh and F. R. Mashiri, Analysis of steel plate shear walls, *Steel Compos. Struct.* (2012) 549–554, doi:10.3850/978-981-08-6218-3 ss-we025.
11. R. Purba and M. Bruneau, Experimental investigation of steel plate shear walls with in-span plastification along horizontal boundary elements, *Eng. Struct.* **97** (2015) 68–79.
12. Y. L. Guo and J. S. Zhu, J. S. Research progress of shear walls: Types and design methods, *Gongcheng Lixue/Eng. Mech.* **37** (2020) 19–33.
13. M. Gholizadeh and Y. Yadollahi, Comparing steel plate shear wall behavior with simple and corrugated plates, *Appl. Mech. Mater.* **147** (2012) 80–85.
14. F. Emami, M. Mofid and A. Vafai, Experimental study on cyclic behavior of trapezoidally corrugated steel shear walls, *Eng. Struct.* **48** (2013) 750–762.
15. J. Z. Tong and Y. L. Guo, Elastic buckling behavior of steel trapezoidal corrugated shear walls with vertical stiffeners, *Thin-Walled Struct.* **95** (2015) 31–39.
16. H. Kalali, M. Hajsadeghi, T. Zirakian and F. J. Alaee, Hysteretic performance of SPSWs with trapezoidally horizontal corrugated web-plates, *Steel Compos. Struct.* **19** (2015) 277–292.
17. S. J. Hilo *et al.*, A state-of-the-art review on double-skinned composite wall systems. *Thin-Walled Struct.* **97** (2015) 74–100.
18. J. Mo, B. Uy, D. Li, H. T. Thai and H. Tran, A review of the behaviour and design of steel–concrete composite shear walls, *Structures* **31** (2021) 1230–1253.
19. Y. L. Guo, J. S. Zhu, M. Z. Wang, X. Yang and P. Zhou, Overall instability performance of concrete-infilled double steel corrugated-plate wall, *Thin-Walled Struct.* **130** (2018) 372–394.
20. J. S. Zhu, Y. L. Guo, M. Z. Wang, X. Yang and B. L. Zhu, Strength design of concrete-infilled double steel corrugated-plate walls under uniform compressions, *Thin-Walled Struct.* **141** (2019) 153–174.
21. M. Z. Wang, Y. L. Guo, J. S. Zhu, X. Yang and J. Z. Tong, Sectional strength design of concrete-infilled double steel corrugated-plate walls with T-section, *J. Constr. Steel Res.* **160** (2019) 23–44.
22. China Association for Engineering Construction Standardization Committee, T/CECS 624-2019: Technical Specification for Structures with Corrugated Steel Plate Components, China Architecture & Building Press (Beijing, China, 2019).
23. H.-J. Sun, Y.-L. Guo, C.-B. Wen and J.-Q. Zuo, Local and global buckling prevention design of corrugated steel plate shear walls, *J. Build. Eng.* **68** (2023) 106055.
24. S. Timoshenko and S. Woinowsky-Krieger, *Theory of Plates and Shells* (Springer, Berlin, Heidelberg, 1959).
25. D. Briassoulis, Equivalent orthotropic properties, *Comput. Struct.* **23** (1986) 129–138.
26. K. Liew, L. Peng and S. Kitipornchai, Buckling analysis of corrugated plates using a mesh-free Galerkin method based on the first-order shear deformation theory, *Comput. Mech.* **38**, 61–75 (2006).

27. Z. Qing-Lin, *Study on the Stability Behavior of H-Shaped Members with Sinusoidal Webs* (Tsinghua University, China, 2008).
28. C. Dou, Z. Q. Jiang, Y. L. Pi and Y. L. Guo, Elastic shear buckling of sinusoidally corrugated steel plate shear wall, *Eng. Struct.* **121** (2016) 136–146.
29. J. T. Easley, Buckling formulas for corrugated metal shear diaphragms, *J. Struct. Div.* **101** (1975) 1403–1417.
30. J. T. Easley and D. E. McFarland, Buckling of light-gage corrugated metal shear diaphragms, *J. Struct. Div.* **95** (1969) 1497–1516.
31. M. Elgaaly, R. W. Hamilton and A. Seshadri, Shear strength of beams with corrugated webs, *J. Struct. Eng.* **122** (1996) 390–398.
32. R. G. Driver, H. H. Abbas and R. Sause, Shear behavior of corrugated web bridge girders, *J. Struct. Eng.* **132** (2006) 195–203.
33. J. Moon, J. Yi, B. H. Choi and H. E. Lee, Shear strength and design of trapezoidally corrugated steel webs, *J. Constr. Steel Res.* **65** (2009) 1198–1205.
34. R. Sause and T. N. Braxtan, Shear strength of trapezoidal corrugated steel webs *J. Constr. Steel Res.* **67** (2011) 223–236.
35. M. F. Hassanein and O. F. Kharoob, Behavior of bridge girders with corrugated webs: (I) Real boundary condition at the juncture of the web and flanges, *Eng. Struct.* **57** (2013) 554–564.
36. M. F. Hassanein and O. F. Kharoob, Behavior of bridge girders with corrugated webs: (II) Shear strength and design, *Eng. Struct.* **57** (2013) 544–553.
37. M. Leblouba, M. T. Junaid, S. Barakat, S. Altoubat and M. Maalej, Shear buckling and stress distribution in trapezoidal web corrugated steel beams, *Thin-Walled Struct.* **113** (2017) 13–26.
38. J. Qiu, Q. Zhao, C. Yu and Z. Wang, Lateral behavior of trapezoidally corrugated wall plates in steel plate shear walls, Part 2: Shear strength and post-peak behavior, *Thin-Walled Struct.* **174** (2022) 109103.
39. S. Wang, Y. Zhang, T. Luo and Y. Liu, Elastic critical shear buckling stress of large-scale corrugated steel web used in bridge girders, *Eng. Struct.* **244** (2021) 112757.
40. C. B. Wen, Y. L. Guo, J. Q. Zuo and H. J. Sun, Global shear stability capacity of trapezoidally corrugated steel plate shear walls with boundary elements, *J. Build. Eng.* **72** (2023) 106553.
41. Y. L. Mo and S. F. Perng, Behavior of framed shearwalls made of corrugated steel under lateral load reversals, *Adv. Struct. Eng.* **3** (2000) 255–262.
42. J. W. Berman, O. C. Celik and M. Bruneau, Comparing hysteretic behavior of light-gauge steel plate shear walls and braced frames, *Eng. Struct.* **27** (2005) 475–485.
43. J. W. Berman and M. Bruneau, Experimental investigation of light-gauge steel plate shear walls, *J. Struct. Eng.* **131** (2005) 259–267.
44. N. Shimizu, R. Kanno, K. Ikarashi, K. Sato and K. Hanya, Cyclic behavior of corrugated steel shear diaphragms with end failure, *J. Struct. Eng.* **139** (2013) 796–806.
45. C. Dou, Y. Xiao, C. Xie and N. Yang, Hysteretic experimental study and lateral performance analysis of flat-corrugated steel plate shear walls, *Eng. Mech.* **39** (2022) 1–13.
46. L. G. Vigh, G. G. Deierlein, E. Miranda, A. B. Liel and S. Tipping, Seismic performance assessment of steel corrugated shear wall system using non-linear analysis, *J. Constr. Steel Res.* **85** (2013) 48–59.

47. C. Yu and G. Yu, Experimental investigation of cold-formed steel framed shear wall using corrugated steel sheathing with circular holes, *J. Struct. Eng.* **142** (2016) 1–9.
48. W. Zhang, M. Mahdavian, Y. Li and C. Yu, Experiments and simulations of cold-formed steel wall assemblies using corrugated steel sheathing subjected to shear and gravity loads, *J. Struct. Eng.* **143** (2017) 1–13.
49. L. Hosseinzadeh, F. Emami and M. Mofid, Experimental investigation on the behavior of corrugated steel shear wall subjected to the different angle of trapezoidal plate, *Struct. Des. Tall Spec. Build.* **26** (2017) e1390.
50. W. Zhang, M. Mahdavian and C. Yu, Different slit configuration in corrugated sheathing of cold-formed steel shear wall, *J. Constr. Steel Res.* **150** (2018) 430–441.
51. M. Y. Gilvaee and M. Mofid, Experimental/numerical evaluation of steel trapezoidal corrugated infill panels with an opening, *Appl. Sci.* **11** (2021) 3275.
52. S. Shon, M. Yoo and S. Lee, An experimental study on the shear hysteresis and energy dissipation of the steel frame with a trapezoidal-corrugated steel plate, *Materials (Basel)* **10** (2017) 261.
53. J. Qiu, Q. Zhao, C. Yu and Z. Li, Experimental studies on cyclic behavior of corrugated steel plate shear walls, *J. Struct. Eng.* **144** (2018) 1–12.
54. W. Wang *et al.*, Experimental study of the hysteretic behaviour of corrugated steel plate shear walls and steel plate reinforced concrete composite shear walls, *J. Constr. Steel Res.* **160** (2019) 136–152.
55. Y. Yu, C. Hu, F. Zhao and L. Jiang, Research on the specially-shaped corrugated steel plate shear walls with horizontal corrugation, *J. Constr. Steel Res.* **188** (2022) 107012.
56. S. Jin, Q. Wang, J. Zhou and J. Bai, Experimental and numerical investigation of assembled multi-grid corrugated steel plate shear walls, *Eng. Struct.* **251** (2022) 113544.
57. S. A. Edalati, Y. Yadollahi, I. Pakar, A. Emadi and M. Bayat, Numerical study on the performance of corrugated steel shear walls, *Wind Struct.* **19** (2014) 405–420.
58. Q. Zhao, J. Sun, Y. Li and Z. Li, Cyclic analyses of corrugated steel plate shear walls, *Struct. Des. Tall Spec. Build.* **26** (2017) 1–17.
59. H. Fadhil, A. Ibrahim and M. Mahmood, Effect of corrugation angle and direction on the performance of corrugated steel plate shear walls, *Civ. Eng. J.* **4** (2018) 2667.
60. M. Bahrebar, M. Z. Kabir, T. Zirakian, M. Hajsadeghi and J. B. P. Lim, Structural performance assessment of trapezoidally-corrugated and centrally-perforated steel plate shear walls, *J. Constr. Steel Res.* **122** (2016) 584–594.
61. C. Dou, X. Cheng, Y.-Y. Zhao and N. Yang, Shear resistance and design of infill panels in corrugated-plate shear walls, *J. Struct. Eng.* **147** (2021) 04021179.s
62. Q. Cao, J. Huang, D. Wang, L. Zhao and Z. Wu, Research on seismic performance of corrugated steel plate shear walls with inelastic buckling subjected to lateral loads, *Int. J. Steel Struct.* **22** (2022) 816–832.
63. H. Kalali, T. G. Ghazijahani, M. Hajsadeghi, T. Zirakian and F. J. Alaee, Numerical study on steel shear walls with sinusoidal corrugated plates, *Lat. Am. J. Solids Struct.* **13** (2016) 2502–2514.
64. C. Dou, Y. L. Pi and W. Gao, Shear resistance and post-buckling behavior of corrugated panels in steel plate shear walls, *Thin-Walled Struct.* **131** (2018) 816–826.
65. A. Farzampour and J. A. Laman, Behavior prediction of corrugated steel plate shear walls with openings, *J. Constr. Steel Res.* **114** (2015) 258–268.

66. J.-Z. Tong and Y.-L. Guo, Shear resistance of stiffened steel corrugated shear walls, *Thin-Walled Struct.* **127** (2018) 76–89.
67. L. Feng, T. Sun and J. Ou, Method of determining the minimum number of stiffeners for stiffened corrugated steel walls, *Structures* **34** (2021) 3487–3500.
68. Design manual for PC bridges with corrugated steel webs, Research committee for hybrid structures with corrugated steel web (Tokyo, Japan, 1998).
69. T/CECS 290-2022, *Technical Specification for Steel Corrugated-Plate Structures* (China Planning Press, 2022).
70. Q. Cao and J. Huang, Experimental study and numerical simulation of corrugated steel plate shear walls subjected to cyclic loads, *Thin-Walled Struct.* **127** (2018) 306–317.
71. H. J. Jin *et al.*, Simplified inelastic global buckling theory of non-buckling corrugated steel shear walls for seismic energy dissipation, *Thin-Walled Struct.* **159** (2021) 107209.
72. M. Shariati *et al.*, Numerical study on the structural performance of corrugated low yield point steel plate shear walls with circular openings, *Steel Compos. Struct.* **33** (2019) 569–581.
73. Q. Zhao, J. Qiu, Y. Zhao and C. Yu, Performance-based seismic design of corrugated steel plate shear walls, *KSCE J. Civ. Eng.* **26** (2022) 3486–3503.
74. W. Wang, Q. Luo, Z. Sun, B. Wang and S. Xu, Relation analysis between out-of-plane and in-plane failure of corrugated steel plate shear wall, *Structures* **29** (2021) 1522–1536.
75. F. Emami and M. Mofid, On the hysteretic behavior of trapezoidally corrugated steel shear walls, *Struct. Des. Tall Spec. Build.* **23** (2014) 94–104.
76. E. F. Deng, L. Zong, H. P. Wang, F. W. Shi and Y. Ding, High efficiency analysis model for corrugated steel plate shear walls in modular steel construction, *Thin-Walled Struct.* **156** (2020) 106963.
77. S. M. Ghodratian-Kashan and S. Maleki, Cyclic performance of corrugated steel plate shear walls with beam-only-connected infill plates, *Adv. Civ. Eng.* **2021** (2021) 5542613.
78. L. Zheng *et al.*, Seismic performance of steel corrugated plate structural walls with different corrugation inclinations, *J. Constr. Steel Res.* **192** (2022) 107248.
79. Y. Ding *et al.*, Cyclic tests on corrugated steel plate shear walls with openings in modularized-constructions, *J. Constr. Steel Res.* **138** (2017) 675–691.
80. A. Farzampour, I. Mansouri, C. H. Lee, H. B. Sim and J. W. Hu, Analysis and design recommendations for corrugated steel plate shear walls with a reduced beam section, *Thin-Walled Struct.* **132** (2018) 658–666.
81. S. Behzad, T. Farshid, K. Mohammad and E. Torki, Stiffened trapezoidally - corrugated plates in open - section shear walls, *Int. J. Steel Struct.* **21** (2021) 1644–1661.
82. E. Vaziri, M. Gholami and M. Gorji Azandariani, The wall–frame interaction effect in corrugated steel plate shear walls systems, *Int. J. Steel Struct.* **21** (2021) 1680–1697.
83. H. Zheng, Z. Zhang, L. Jiang, Y. Hu and W. Wang, Cyclic performance and shear resistance design of asymmetric diagonal stiffened beam-only-connected corrugated steel plate shear walls, *Arch. Civ. Mech. Eng.* **22** (2022) 183.
84. C. Wen, Y. Guo, J. Zuo and X. Zhao, Strength design of prefabricated corrugated steel plate shear walls under combined compression and shear loads, *J. Build. Eng.* **65** (2023) 105790.

85. X. Zhang and Y. Guo, Behavior of steel plate shear walls with pre-compression from adjacent frame columns, *Thin-Walled Struct.* **77** (2014) 17–25.
86. Z. Xu, G. Tong and L. Zhang, Design of vertically stiffened steel plate walls under combined uniaxial compression and shear loads, *Structures* **26** (2020) 348–361.
87. J. Z. Tong, Y. L. Guo and J. Q. Zuo, Elastic buckling and load-resistant behaviors of double-corrugated-plate shear walls under pure in-plane shear loads, *Thin-Walled Struct.* **130** (2018) 593–612.s
88. J. Z. Tong, Y. L. Guo, J. Q. Zuo and J. K. Gao, Experimental and numerical study on shear resistant behavior of double-corrugated-plate shear walls, *Thin-Walled Struct.* **147** (2020) 106485.
89. J. Z. Tong, Y. L. Guo and W. H. Pan, Ultimate shear resistance and post-ultimate behavior of double-corrugated-plate shear walls, *J. Constr. Steel Res.* **165** (2020) 105895.
90. J. Z. Tong, R. M. Wu, Z. Y. Xu and Y. L. Guo, Subassemblage tests on seismic behavior of double-corrugated-plate shear walls, *Eng. Struct.* **276** (2023) 115341.
91. R. Deng, J. Yang, Y. H. De Wang, Q. Q. Li and J. K. Tan, Cyclic shear performance of built-up double-corrugated steel plate shear walls: Experiment and simulation, *Thin-Walled Struct.* **181** (2022) 110077.
92. S. M. Ghodratian-Kashan and S. Maleki, Numerical investigation of double corrugated steel plate shear walls, *J. Civ. Eng. Constr.* **10** (2021) 44–58.
93. S. M. Ghodratian-Kashan and S. Maleki, Experimental investigation of double corrugated steel plate shear walls, *J. Constr. Steel Res.* **190** (2022) 107138.
94. J. Z. Tong, Theoretical and experimental study on design methods for steel corrugated plate shear walls, dissertation, Tsinghua University (2018).
95. B. L. Zhu, J. Z. Tong, J. X. Zhang and Y. L. Guo, Shear-resistant capacity and design of orthogonal double-corrugated-plate shear walls, *Thin-Walled Struct.* **187** (2023) 110759.
96. J. W. Zhang *et al.*, Experimental evaluation on seismic performance of multi-celled corrugated-plate CFST walls, *J. Constr. Steel Res.* **201** (2023) 107743.
97. J. Z. Tong *et al.*, Flexural tests and behavior of multi-celled corrugated-plate CFST members, *J. Build. Eng.* **49** (2022) 104051.
98. J. S. Zhu, Y. L. Guo, M. Z. Wang, X. Yang and Y. L. Pi, Seismic performance of concrete-infilled double steel corrugated-plate walls: Experimental research, *Eng. Struct.* **215** (2020) 110601.
99. M. Z. Wang, Y. L. Guo, J. S. Zhu and X. Yang, Flexural buckling of axially loaded concrete-infilled double steel corrugated-plate walls with T-section, *J. Constr. Steel Res.* **166** (2020) 105940.
100. Y.-Q. Zhou *et al.*, Numerical and experimental studies on sectional load capacity of concrete-infilled double steel corrugated-plate walls under combined compression and in-plane bending, *Thin-Walled Struct.* **159** (2021) 107250.
101. M. Z. Wang, Y. L. Guo and P. Zhou, Load-carrying mechanism of truss-confined buckling-restrained braces: Numerical and theoretical analyses, *Eng. Struct.* **212** (2020) 110500.
102. M. Z. Wang, Y. L. Guo, X. Yang and J. S. Zhu, Interaction equations of composite walls with T-section under axial compression and biaxial bending, *Eng. Struct.* **229** (2021) 111667.

103. M. Z. Wang, Y. L. Guo, X. Yang and J. S. Zhu, Compressive buckling behaviour of steel corrugated-plates in contact with a rigid medium, *Compos. Struct.* **262** (2021) 113356.
104. M. Z. Wang, Y. L. Guo, J. S. Zhu and X. Yang, Flexural-torsional buckling and design recommendations of axially loaded concrete-infilled double steel corrugated-plate walls with T-section, *Eng. Struct.* **208** (2020) 110345.
105. C. B. Wen, Y. L. Guo, H. J. Sun, Z. H. Tian and X. He, Experimental and numerical study on seismic performance of concrete-infilled double steel corrugated-plate walls, *J. Build. Eng.* **68** (2023) 106171.
106. J. S. Zhu, Theoretical and experimental study on design methods for I-section concrete-infilled double steel corrugated-plate walls, thesis, Tsinghua University (2020).

https://doi.org/10.1142/9789819814770_0023

Chapter 23

Shaking Table Test of a Split-Foundation RC Frame Structure with One Lower Story Compared with a Conventional RC Frame Structure[#]

Baolong Jiang[*], Yingmin Li[*,†], Yangyang Tang[*], Shuyan Ji[‡,§] and Liping Liu[*,†]

[*]*School of Civil Engineering*
Chongqing University, Chongqing, P. R. China

[†]*MOE Key Laboratory of New Technology*
for Construction of Cities in Mountain Area
Chongqing University, Chongqing, P. R. China

[‡]*School of Management Science and Real Estate*
Chongqing University, Chongqing, P. R. China
[§]*jishuyan@cqu.edu.cn*

Split-foundation (SF) building structures, as a kind of hill-side buildings, are widely used in mountainous regions due to the scarcity of plain areas. In order to experimentally investigate the seismic responses of the SF structure, shaking table test was carried out on two 1/8-scale models: a split-foundation RC frame structure supported by foundations at two different levels and a conventional RC frame structure supported by foundations at one single level. The failure patterns, dynamic characteristics, acceleration responses, displacement responses, and torsion responses of the two models were assessed and compared under seismic ground motions. The test results indicate that the two models were severely damaged after the last intensity level excitation and showed an intermediate failure mechanism. The damage to the SF model was uneven and special with relatively severe damage of the upper grounded columns. The difference in natural frequencies between the two models resulted from the effect of the lower story on lateral stiffness, and the effect was relatively dramatic in the transverse direction. The difference in structural stiffness led to larger acceleration amplification factors of the SF model and smaller acceleration amplification factors in the longitudinal direction. Although the restriction at upper ground level created a certain effect on displacement responses of the SF model especially in the first story, the curve shapes of displacement responses of the two models were similar from the first floor to fourth floor. In addition, the torsion responses of the SF model, especially in the first story (upper ground story), were obviously much larger than those of the conventional model. The torsional effect cannot be ignored in the SF structure which is a torsionally sensitive structure.

Keywords: Mountain area; hill slope; shaking table test; split-foundation building; RC frame; seismic performance.

[§]Corresponding author.
[#]This chapter appeared previously on the International Journal of Structural Stability and Dynamics. To cite this chapter, please cite the original article as the following: B. Jiang *et al.*, *Int. J. Struct. Stab. Dyn.*, **23**, 2340025 (2023), DOI: 10.1142/S0219455423400254.

1. Introduction

In mountain areas, many buildings are constructed on hill slopes with irregular structural configurations. To suit the slope's geometry, foundations of these building structures rest on different ground levels, forming hill-side buildings.[1] As a special kind of frame structures supported by foundations at two different levels, the split-foundation (SF) frame structures are different from conventional frame structures on flat ground, and are widely used in mountainous regions of many countries such as China, India and Nepal (Fig. 1).[1,2] However, the SF buildings in mountainous regions are often located in seismically vulnerable zones. The SF building structures are highly irregular in plane and elevation, leading to severe damages to this type of buildings during earthquakes according to previous studies and earthquake disasters.[3–7] Therefore, it is necessary to specially study the seismic response of SF structures.[8]

A number of researchers have contributed in the area of the seismic performance of SF structures. Li *et al.*[8] reviewed the special problem existing in seismic behavior and design, presenting an overview of the progress in research regarding seismic behavior of structures on slopes. According to the research, the special specification for structural design was also recommended. Singh *et al.*[5] and Narayanan *et al.*[9] noticed the different performance of hill buildings during the 2011 Sikkim earthquake, and carried out an analytical study to investigate the peculiar seismic behavior of hill buildings. Singh *et al.*[10] presented seismic response of hill buildings subjected to bi-directional excitation, and the primary failure mode of shear failure of short column was observed. Surana *et al.*[11] reviewed the seismic characterization

(a) (b)

Fig. 1. Typical structural configurations of split-foundation (SF) buildings: (a) a SF building in India,[1] (b) a SF building in China.[2]

and vulnerability of building stock in hilly regions. Li *et al.*[2] presented the fragility curves of two reinforced concrete frames on steep slopes under both near-fault and far-field ground motions. Xu *et al.*[12,13] studied the parameter's influence of different numbers of stories and ground columns on the seismic performance of the step-back structure along the slope, and the seismic performance and damage pattern of the split-foundation frame buildings in across-slope direction are studied. Three methods of controlling vertical structural stiffness to reduce the vertical structural irregularities were contrasted by Tang *et al.*[14] Jiang *et al.*[15] investigated the seismic performance of interior beam-column joints under axial beam tension of step back RC frame structures on hill slopes. Surana *et al.*[1] presented the floor acceleration demand in hill-side building by numerical study, and suggested simple floor spectral amplification functions for seismic design of acceleration-sensitive nonstructural components. Xu *et al.*[16] simulated the overall failure mechanism and failure path of split-foundation RC frames along the slope in OpenSees. The comparative analysis of seismic performance between three-dimensional step-back frame structure and common frame structure was performed by Tang *et al.*[17] Surana *et al.*[18] assessed the effect of different code design levels, ground-motion intensity measure and structural configuration type on damage probabilities of different damage states and mean damage ratio for the maximum considered earthquake hazard level. Stability of slope with differently configured buildings has been studied by Paul *et al.*[19] Nonlinear 2D finite element model taking into account the coupled behavior of building, foundation and slope was built and analyzed by Raj *et al.*[20] The interaction has a significant effect not only on the slope stability but also on the building response. Tang *et al.*[21] presented the seismic response of structures presumed to be built on the sites under uniform and nonuniform strong seismic excitation. Li *et al.*[22] performed a quasi-static test and finite element analysis to investigate the seismic performance of RC split-foundation plane frame structures with steel braces. Tang *et al.*[23] conducted a nonlinear time history analysis to study the redistribution of internal force and the development of the plastic hinge of the split-foundation frame structures with two lower stories.

Previous research mainly focused on finite element analysis of SF building structures with multi-lower stories. Only a few studies have built SF structural models with one lower story, considering a larger range of stiffness ratio. And few experimental studies, especially shaking table tests, on the seismic performance of SF building structures were conducted. In view of this, to further investigate the seismic performance SF building structures in a straightforward manner, shaking table tests of a split-foundation RC frame structure with one lower story (Fig. 2) and a conventional RC frame structure were conducted in this study. The seismic performance of the two structures was analyzed and compared under different input ground accelerations.

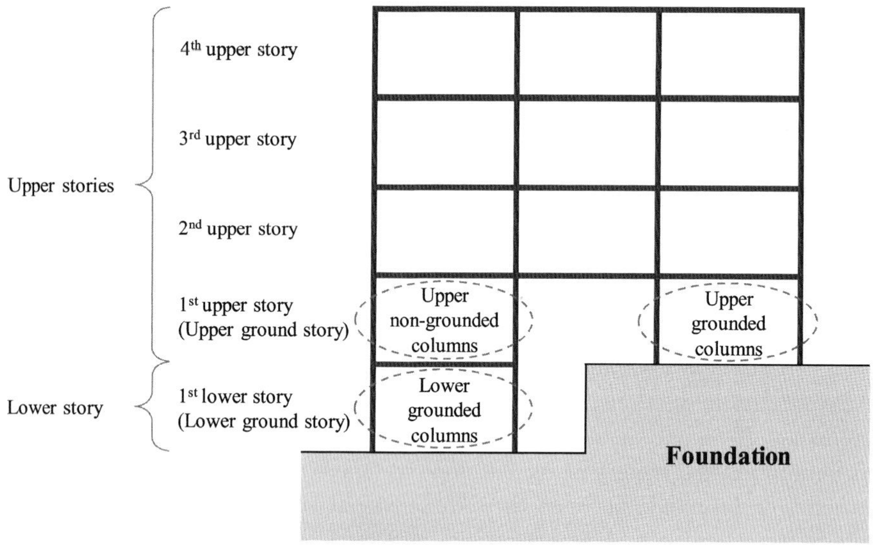

Fig. 2. A schematic diagram of the test SF frame structure with one lower story (model 1).

2. Design of Experimental Structures

2.1. *Model structure design*

The structural models used in the shaking table test were two 1/8-scale RC frame structures designed according to the seismic design code[24] with a seismic fortification intensity of 8 (0.20 g) and a site class of II. The two models were tested at the

Table 1. Main similitude scale factors.

Type	Parameters	Similitude relationship	Scale factors
Geometric characteristics	Length	S_l	1/8
	Linear displacement	S_l	1/8
	Angular displacement	S_θ	1
	Area	$S_A = S_l^2$	1/64
Material characteristics	Strain	S_ε	1
	Elastic modulus	S_E	0.33
	Stress	$S_\sigma = S_E$	0.33
	Mass	$S_m = S_\sigma \cdot (S_l)^2/S_a$	2.734×10^{-3}
Loading	Concentrate load	$S_F = S_\sigma \cdot (S_l)^2$	5.156×10^{-3}
	Area load	$S_p = S_\sigma$	0.33
	Moment	$S_M = S_\sigma \cdot (S_l)^3$	6.45×10^{-4}
Dynamic characteristics	Time	$S_T = (S_l)^{0.5} \cdot (S_a)^{-0.5}$	0.257
	Frequency	$S_f = (S_l)^{-0.5} \cdot (S_a)^{0.5}$	3.884
	Velocity	$S_v = (S_l \cdot S_a)^{0.5}$	0.486
	Horizontal acceleration	S_a	1.89

same time on the shaking table for comparison: model 1 (M1) is a 5-story split-foundation (SF) RC frame structure supported by a foundation at two different levels, and model 2 (M2) is a 4-story conventional RC frame structure supported by foundations at one single level.

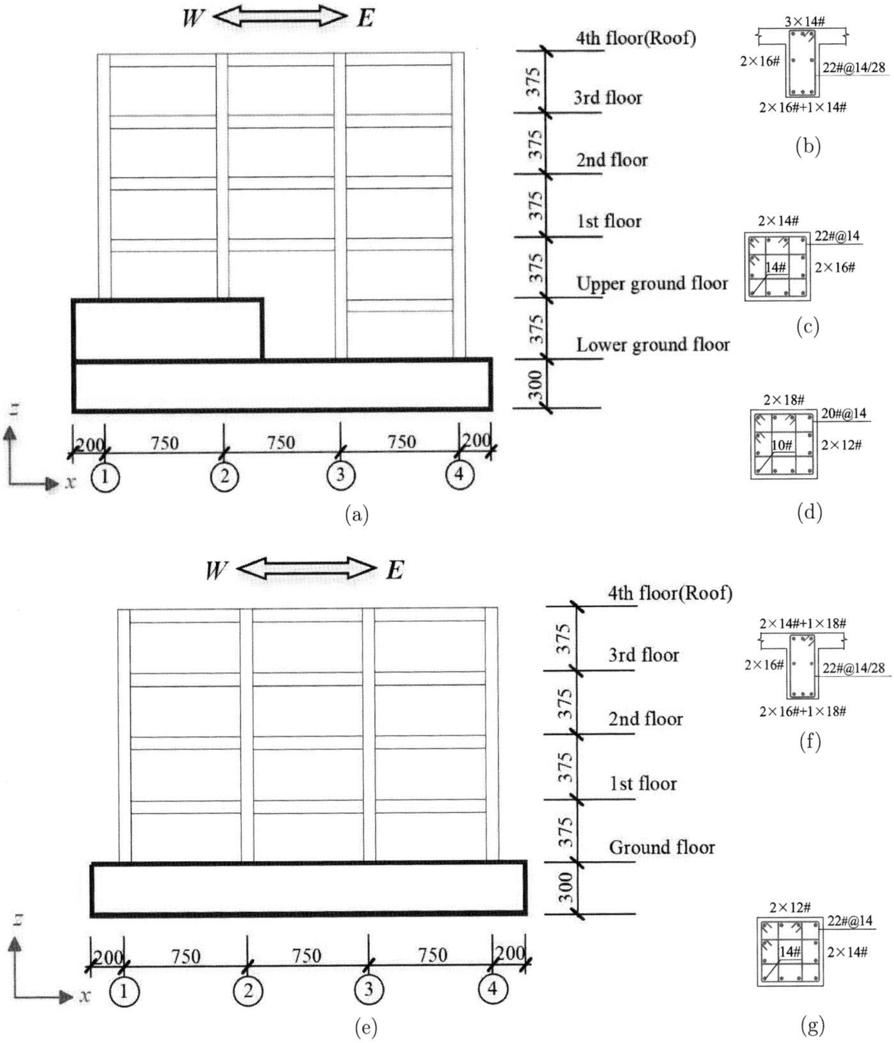

Fig. 3. Layout and reinforcements of the main parts of each model (unit: mm): (a) elevation of model 1, (b) beam reinforcements of model 1, (c) lower grounded column reinforcements of model 1, (d) upper grounded column reinforcements of model 1, (e) elevation of model 2, (f) beam reinforcements of model 2, (g) column reinforcements of model 2, (h) plan layout and measurement points of the lower story of model 1, (i) plan layout and measurement points of model 2 (upper stories of model 1).

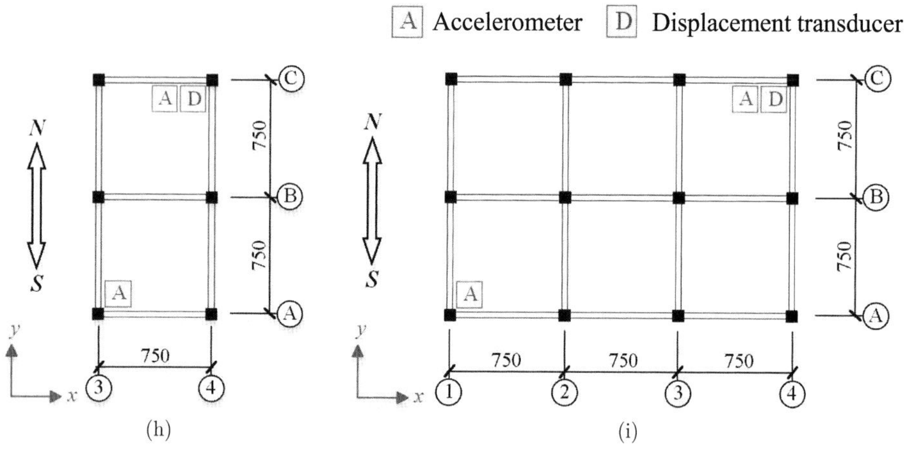

Fig. 3. (*Continued*)

The main similitude scale factors of the two models are shown in Table 1. According to the dimension and capacity of the shaking table and the model material, three controlling basic similitude factors (the length scale factor S_l, the stress scale factor S_σ, and the acceleration scale factor S_a) were chosen. The other scale factors were then derived from the three controlling factors.

The model structures were fabricated according to the similitude relationship, and the layout of the building is shown in Fig. 3. Model 2 was a 4-story RC frame with three bays in the x direction (longitudinal direction, EW direction, or along-slope direction) and two bays in the y direction (transverse direction, NS direction, or across-slope direction). As a SF structure, model 1 consisted of one lower story and four upper stories. The lower story of model 1 was designed with one bay in the x direction and two bays in the y direction, while the configuration of the upper stories was the same as model 1. The plan dimension of each frame model was 2 325 mm × 1 575 mm, and the model structures remained a constant story height of 37.5 mm. The cross-sectional dimensions of each column and each beam were 75 mm × 75 mm and 37.5 mm × 75 mm, respectively. The floor slab thickness was 17.5 mm. The reinforcement details of the main parts of the structures are shown in Fig. 3, and Fig. 4 shows the completed test model structures on the shaking table.

Steel and lead mass blocks were used as additional mass for the test model structures. The total mass of model 1 was 8 118 kg including the frame of 1 015 kg, added mass of 2 163 kg, and the model base of 4 940 kg. The total mass of model 2 is 10 334 kg including the frame of 1 115 kg, added mass of 2 359 kg, and the model base of 6 860 kg.

(a) (b)

Fig. 4. Test model structures on the shaking table: (a) model 1 and (b) model 2.

Table 2. Material properties of the galvanized iron wire.

Grade	Diameter (mm)	Elastic modulus (MPa)	Yield strength (MPa)	Ultimate strength (MPa)
22#	0.64	2.02×10^5	345	422
20#	0.80	2.04×10^5	367	433
18#	1.20	1.26×10^5	302	413
16#	1.60	1.42×10^5	328	428
14#	2.20	2.08×10^5	335	394
12#	2.80	2.61×10^5	240	327
10#	3.42	2.01×10^5	340	447

2.2. *Structure materials*

Microconcrete was used to simulate the concrete of the prototype structures with a compressive strength of 9.09 MPa and elastic modulus of 8 410 MPa measured at the time of testing. Galvanized iron wires were used to simulate the steel reinforcements. Thus the RC structural members of model structures were designed and constructed according to the similitude factors. The measured mechanical properties of the galvanized iron wires are provided in Table 2.

3. Testing Program

3.1. *Shaking table facility*

The test was performed on a three-dimensional and six-DOF earthquake simulation shaking table in the multi-functional shaking table laboratory at Chongqing

University, China. The shaking table was manufactured by MTS Systems Corporation with dimensions of 6.1 m × 6.1 m and a capacity of 6×10^4 kg. The maximum horizontal and vertical accelerations of the shaking table are 1.5 and 1.0 g, respectively.

3.2. *Seismic motion selection*

According to the design response spectrum of Chinese code[22] and the dynamic characteristics of the structures, five seismic ground motions were selected, including four natural records (Wenchuan, Iwate-Miyagi Nairiku, Northridge and Taft) and one artificial ground motion (ACC). The acceleration response spectra of the selected ground motions are compared with the design response spectrum in Fig. 5.

3.3. *Instrumentation*

The instrumentation consisted of accelerometers and displacement transducers, and the arrangement of the key measurement points is shown in Fig. 3. A total of 50 accelerometers were installed on each floor and the base slab to monitor the acceleration responses in horizontal or vertical directions. Eight displacement transducers were arranged on the key floors and the base slab to measure the horizontal displacements. To prevent damage to the displacement transducers caused by the possible collapse of the models during testing, displacement transducers were detached during the last three intensity levels of ground motions.

3.4. *Loading protocol*

Eight seismic intensity levels with the input PGA ranging from 0.11 to 1.48 g were considered, and thus a total of 48 loading cases, including white noise, were

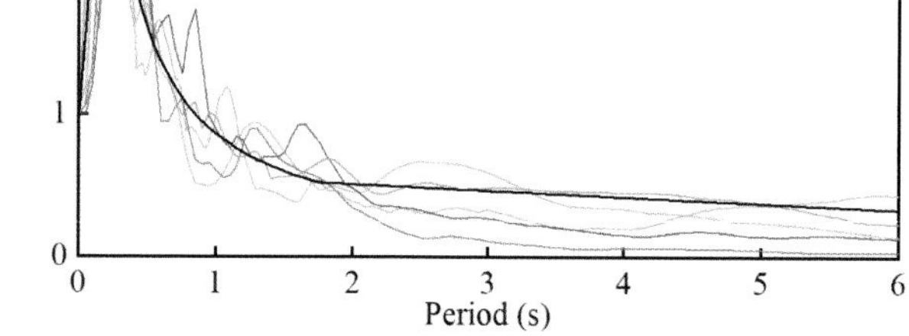

Fig. 5. Comparison of acceleration response spectra.

Table 3. The loading sequence.

Intensity level	PGA (g)	Case	Ground motion	Direction
A	0.11 g	1	White noise	
		2–5	Wenchuan	x, y, xy, yx
		6–9	ACC	x, y, xy, yx
		10–13	Iwate-Miyagi Nairiku	x, y, xy, yx
		14–17	Northridge	x, y, xy, yx
		18–21	Taft	x, y, xy, yx
		22–23	Wenchuan	xyz, yxz
		24–25	Iwate-Miyagi Nairiku	xyz, yxz
		26–27	Northridge	xyz, yxz
		28–29	Taft	xyz, yxz
		30	White noise	
B	0.34 g	31–32	ACC	x, xy
		33	Iwate-Miyagi Nairiku	xyz
		34	Northridge	xyz
		35	White noise	
C	0.48 g	36	ACC	xy
		37	Iwate-Miyagi Nairiku	xyz
		38	White noise	
D	0.62 g	39	Iwate-Miyagi Nairiku	xyz
		40	White noise	
E	0.80 g	41	Iwate-Miyagi Nairiku	xyz
		42	White noise	
F	1.07 g	43	Iwate-Miyagi Nairiku	xyz
		44	White noise	
G	1.23 g	45	Iwate-Miyagi Nairiku	xyz
		46	White noise	
H	1.48 g	47	Iwate-Miyagi Nairiku	xyz
		48	White noise	

conducted throughout the test. The model structures were subjected to unidirectional shaking in each horizontal direction of the models followed by bidirectional shaking and three-directional shaking. Table 3 lists the loading sequences of the shaking table test. In Table 3, the terms "x", "y", "xy", "yx", "xyz" and "yzx" represent different loading cases with different loading directions. In these cases, each letter of the term represents one of the loading directions, and the first letter represents the main loading direction.

4. Test Results and Analysis

4.1. *Test phenomena*

The structure models exhibited no visible cracks when the peak ground acceleration (PGA) was 0.11 g. Then, with the increasing load, the cracks gradually emerged and developed.

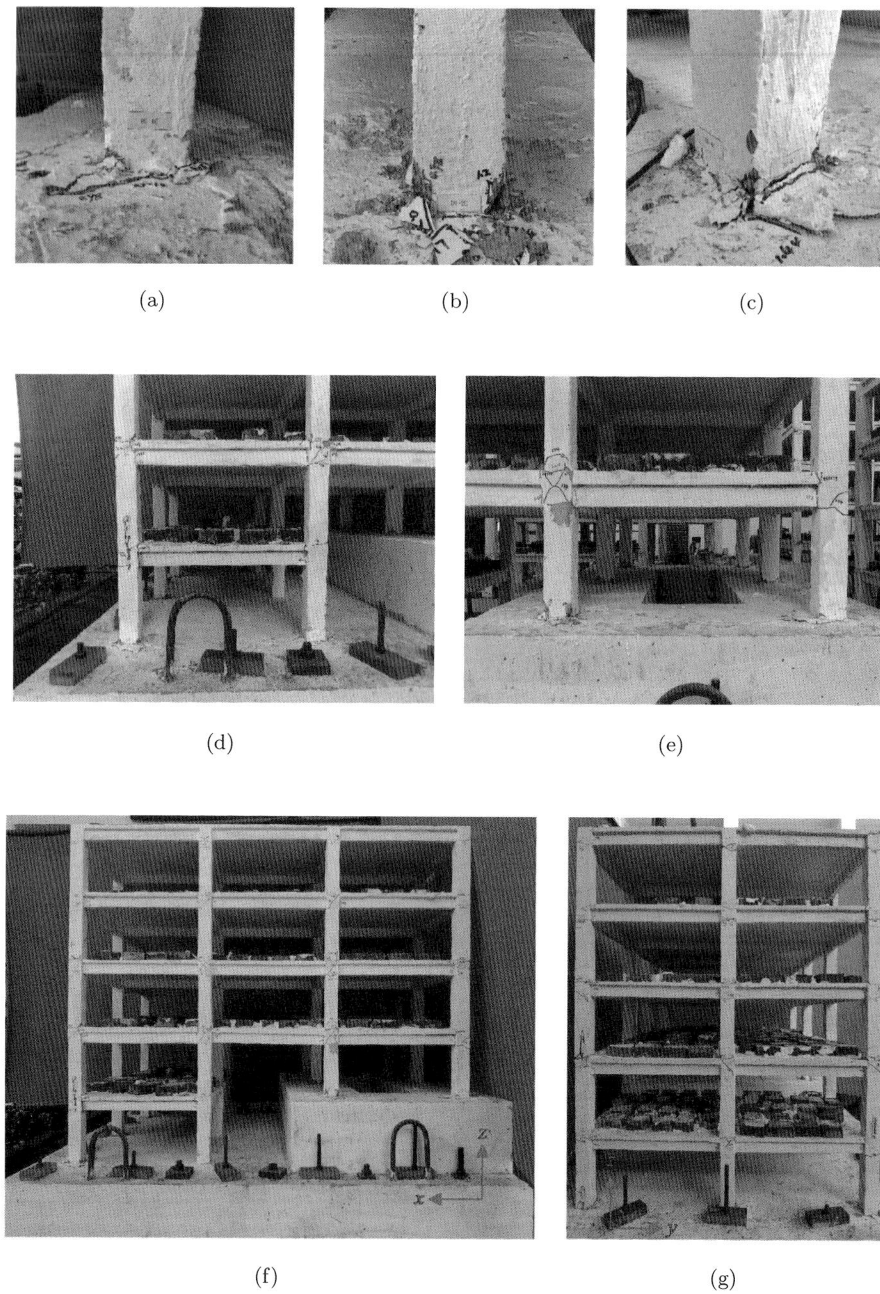

Fig. 6. Damage to model 1: (a) bottom of the upper grounded corner column, (b) bottom of the upper grounded exterior column, (c) bottom of the lower grounded corner column, (d) the lower ground side, (e) the upper ground side, (f) north elevation and (g) east elevation.

For model 1, after excitations with a PGA of 0.34 g, cracks occurred at column ends and beam ends from the first to third upper stories as shown in Figs. 6(a) and 6(b). Then the cracks in the second and third stories tended to develop more rapidly on the lower ground side of model 1. After excitation with a PGA of 0.62 g, the ends of upper grounded columns began to crack, and the crack width increased after excitation with a PGA of 0.80 g. Meanwhile, the bottoms of the lower grounded columns developed horizontal cracks as shown in Fig. 6(c), and the ends of non-grounded columns in the first story cracked. Then more cracks were observed, developing at the ends of beams and columns. After excitation with a PGA of 1.23 g, cracks extended through some beams and columns in the lower ground story as shown in Fig. 6(d), and upper grounded columns were seriously damaged with concrete spalling as shown in Fig. 6(e). After excitation with a PGA of 1.48 g, there was significant concrete spalling at the ends of column on the third and fourth stories. As shown in Figs. 6(f) and 6(g), the final damage to model 1 was not uniformly distributed with more severe damage to the upper ground side. However, in general, the failure mechanism of model 1 was an intermediate mechanism[25] (or partial-beam mechanism[26]) instead of story mechanism or beam mechanism.

For model 2, after excitations with a PGA of 0.34 g, minor cracks started to appear at some column ends in the third and fourth stories. Then the ends of beams in the first and second stories began to crack. After excitation with a PGA of 0.62 g, the cracks developed at the ends of members in each story, and some horizontal cracks were observed at the bottom of columns on the first story, as shown in Figs. 7(a) and 7(b). The beam ends connected with corner columns in the first and second stories cracked, and some beam-column joints in the first story suffered diagonal cracks, as shown in Figs. 7(c) and 7(d). An obvious horizontal crack developed on most column ends in the second and third stories. After excitation with a PGA of 1.48 g, the concrete of some columns was crushed at the bottom of the first story, and there were obvious concrete fragments falling off the bottom of the second story and the top of the third story. Finally, the damage distribution of model 2 was more symmetrical and uniform than that of model 1, and model 2 also showed an intermediate failure mechanism with some column yielding, as depicted in Figs. 7(e) and 7(f).

4.2. *Dynamic characteristics*

The dynamic properties of the test structures are identified under white noise excitations conducted before and after each intensity level test. The frequencies and damping ratios of the models after each intensity level test can be obtained by analyzing the transfer function derived from the acceleration signals at the model base and roof under white noise. The fundamental natural frequencies in each horizontal direction were picked at the first peaks of the frequency response curves.

Fig. 7. Damage to model 2: (a) bottom of the ground floor corner column, (b) bottom of the ground floor exterior column, (c) corner beam-column joint on the first floor, (d) exterior beam-column joint on the first floor, (e) north elevation and (f) west elevation.

Table 4. First natural frequency in each direction (unit: Hz).

Case (white noise)	Model 1		Model 2	
	x	y	x	y
1	9.27	7.32	9.81	9.09
30	8.69	7.19	9.60	8.87
35	7.28	6.35	7.91	7.65
38	6.45	5.66	6.70	6.39
40	5.84	5.16	6.23	5.88
42	4.76	4.08	4.96	4.72
44	4.11	3.29	4.36	4.03
46	3.45	2.84	3.62	3.32
48	2.78	2.35	3.00	2.77

The corresponding damping ratios were calculated using the half-power bandwidth method.[27]

Table 4 shows the first natural frequency in both horizontal directions. The initial natural frequencies of model 1 in the x and y direction were 9.27 and 7.32 Hz, and those of model 2 were 9.81 Hz and 9.09 Hz, respectively. Due to the additional lower story, the differences between the initial frequencies of the two models were 5.5% and 19.55% in the x and y direction, respectively. The greater difference of 19.55% in the y direction is attributed to the different restriction conditions between the two directions of model 1. In the y direction of model 1, the frame was restricted in plane of only two axes (axis 1 and 2) at the upper ground level, while it was restricted in plane of all three axes (axis A to C) in the x direction. As a result, the natural frequencies of model 1 were much closer to those of model 2 in the x direction than in the y direction, indicating that the frame with additional lower story trends to lose less lateral stiffness in along-slope direction (x direction) than in across-slope direction (y direction).

The natural frequencies of models gradually decreased with increasing input of PGA, as shown in Fig. 8(a), indicating the deterioration of the structural stiffness resulting from the accumulation of structural damage. After the first intensity level test, the structures began to be damaged, leading to a constantly significant reduction in frequency. After excitation with a PGA of 0.8 g, both the models were severely damaged, and the rate of frequency reduction was slower. Generally, the frequencies were reduced more rapidly in the x direction than in the y direction. Specially, the first natural frequency in the x direction of model 1 decreased more rapidly than those of model 2, and by contrast in the y direction, those of model 1 decreased more slowly. This indicated more serious damage to model 1 in the x direction leading to a relatively rapid increase in the stiffness degradation. Compared to model 2, the additional lower story of model 1 tends to cause earlier damage propagation in x direction (along-slope direction) and slower damage propagation in y direction (across-slope direction).

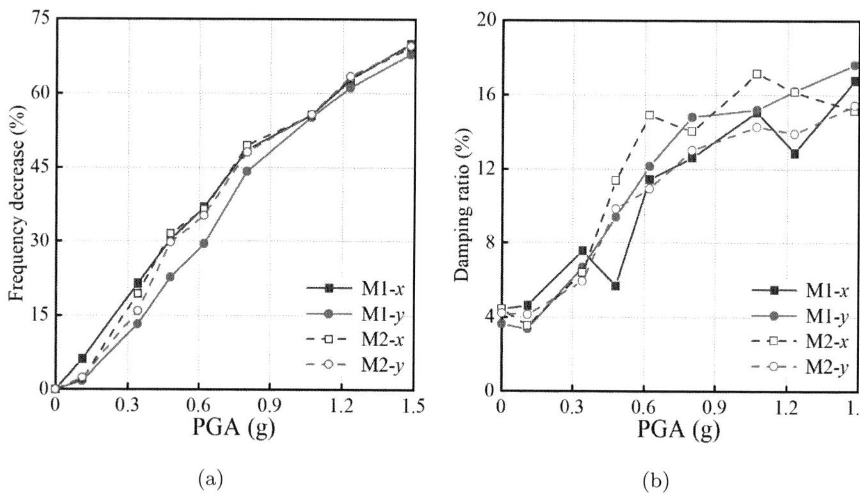

Fig. 8. Variations in natural frequencies and damping ratios of the two models after ground motions of each intensity level: (a) decrease in natural frequencies and (b) damping ratios.

The damping ratio, one of the dynamic characteristics, was used to present the magnitude of structural damping, reflecting the energy dissipation during the test. The first damping ratios in both directions of the two models are shown in Fig. 8(b), and roughly increased with increasing input PGA. The initial damping ratios in both directions of the two models were all around 4% within the reasonable range for reinforced concrete structures.[28] Then, due to the accumulation of structural damage, the damping ratios generally increased after the first intensity level excitations, and the growth rate of the damping ratios was slowed down after excitation with the PGA of 0.8 g.

4.3. *Acceleration responses*

The acceleration amplification factor, determined as the ratio of the peak acceleration achieved at each floor to the peak acceleration of the input motion, was used to analyze the structural acceleration response during the test. It is an important parameter of the structural seismic response which is related to the spectral characteristics of the seismic ground motions and the structural dynamic properties.[29–31]

The acceleration amplification factors of the two models are depicted in Fig. 9. The general variation trend of the acceleration amplification factors of the two models was similar. The acceleration amplification factors in the x direction and y direction increased roughly with the increase of the building height. Accordingly, the largest acceleration amplification during different intensity levels of excitation occurred on the roof. Moreover, the acceleration amplification factors of the two models were generally reduced with the increase in the input PGA due to the

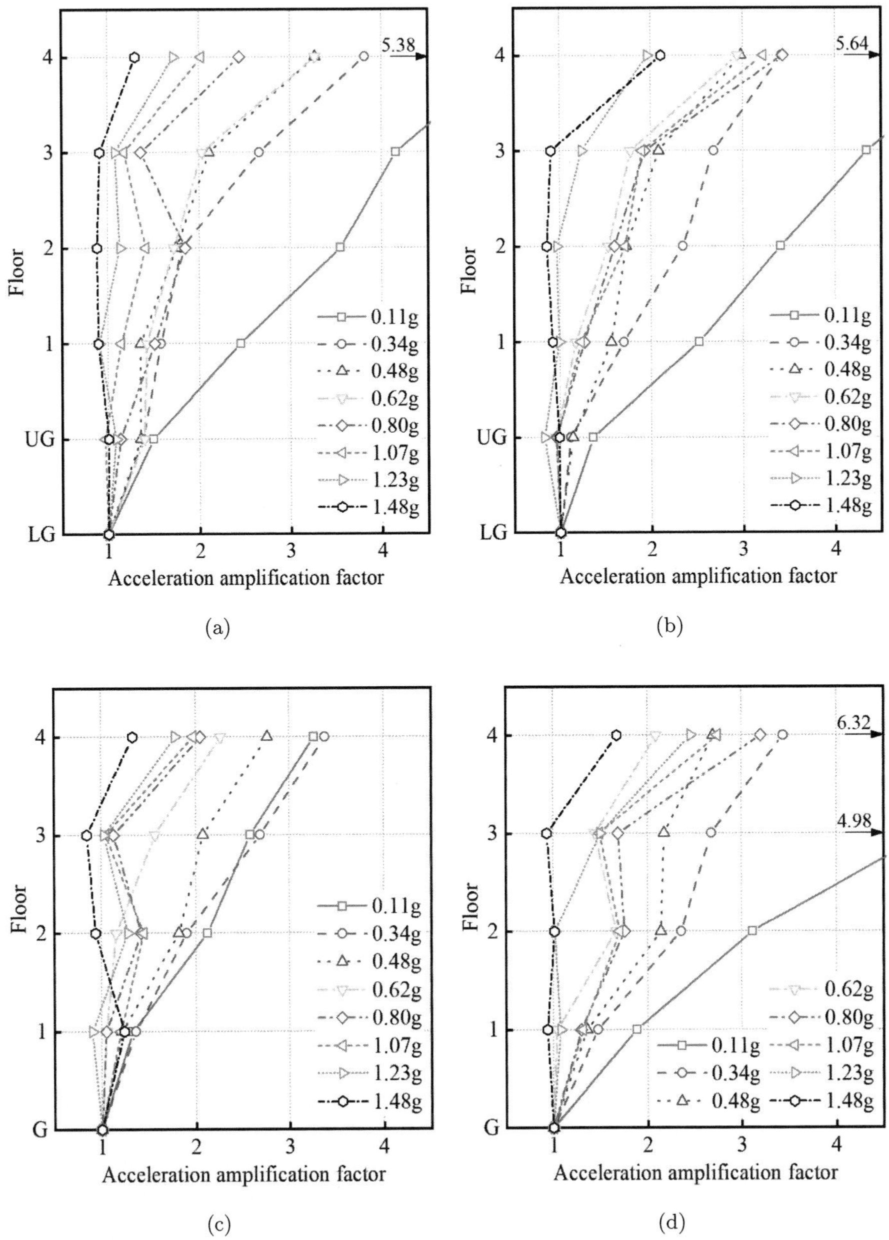

Fig. 9. Acceleration amplification factors of the two models: (a) model 1 in the x direction, (b) model 1 in the y direction, (c) model 2 in the x direction and (d) model 2 in the y direction.

accumulation of structural damage, the deterioration of the structural stiffness and the increase in the damping ratio. Specifically, during the input PGA of 1.48 g, the acceleration amplification factors of many floors were even less than 1, indicating that the seismic responses of the two severely damaged structures were reduced evidently.

In general, the acceleration amplification factors of each model in the x direction were smaller than those in the y direction, mainly due to the difference in stiffness. The comparison between the two models showed that, during the input PGA ranging from 0.11 to 1.07 g, the acceleration amplification factors of model 1 overall were slightly larger than those of model 2. This was because the lower story of model 1 made the accelerations of upper stories further amplified with the increase of structural height at one side.

4.4. *Displacement responses*

Figure 10 shows the peak floor displacements of the two models in the x and y directions. The two structures deformed in a similar overall trend under different intensity level ground motions, and the shear mode deformation dominated the peak floor displacements of the two models. The peak displacement of each floor in both directions increased with increasing PGA and along with the structural height, and the shapes of displacement curves of model 1 were similar to those of model 2, especially from the first floor. Also, the peak floor displacements of the two models were smaller in the x direction than in the y direction, especially under the last four intensity level excitations. The general displacement responses of model 1 were slightly larger than those of model 2, especially in the y direction. These were mainly due to the difference in structural stiffness between the models and between the directions. Combining the previous frequency results (Table 4), it can be observed that the deformation is smaller for a more rigid model or in a more rigid direction, indicating that the stiffer the structure, the smaller its deformation.

Interstory drift ratio (IDR), defined as the relative translational displacement between two consecutive floors divided by the story height, is an important engineering response quantity and indicator of structural performance. Figure 11 shows the maximum interstory drift ratios (IDRs) of the two models in the x direction and y direction. Similarly, the maximum IDR increased with increasing PGA, and the shapes of IDR curves of model 1 were similar to those of model 2 from the first to fourth floors. Along with the building height, the IDRs of the two models generally increased from the bottom and decreased in top stories. Moreover, it is worthwhile to note that the restriction at the upper ground level created a certain effect on displacement responses of model 1 especially in the first story, as depicted in Figs. 10(a) and 10(b) and Figs. 11(a) and 11(b).

With the first input PGA of 0.11 g, the maximum IDRs of the two models were around 0.2%, and the structures remained in a virtually elastic stage (less than

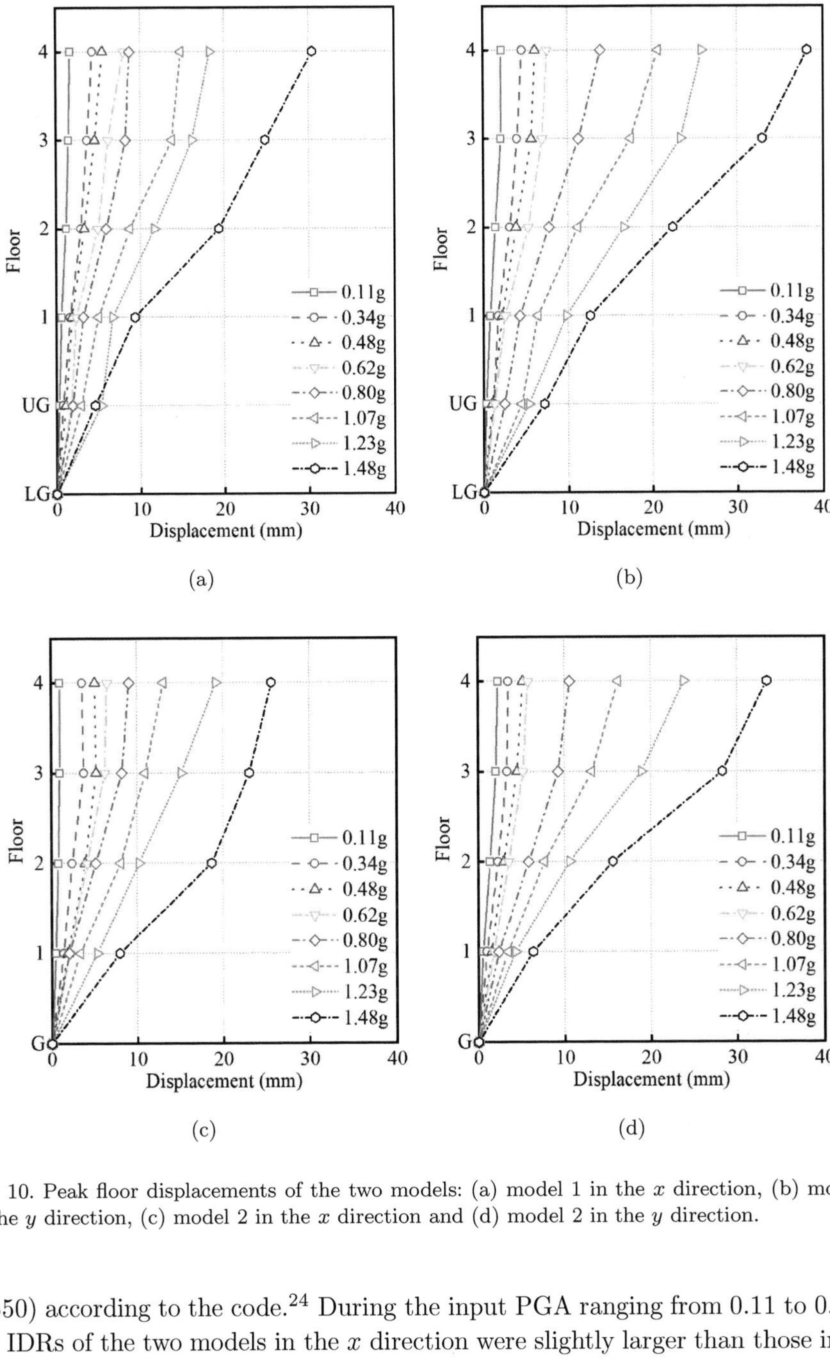

Fig. 10. Peak floor displacements of the two models: (a) model 1 in the x direction, (b) model 1 in the y direction, (c) model 2 in the x direction and (d) model 2 in the y direction.

1/550) according to the code.[24] During the input PGA ranging from 0.11 to 0.62 g, the IDRs of the two models in the x direction were slightly larger than those in the y direction, and conversely became the smaller ones during the last four intensity level excitations. This indicates the two models suffered more serious damage in the

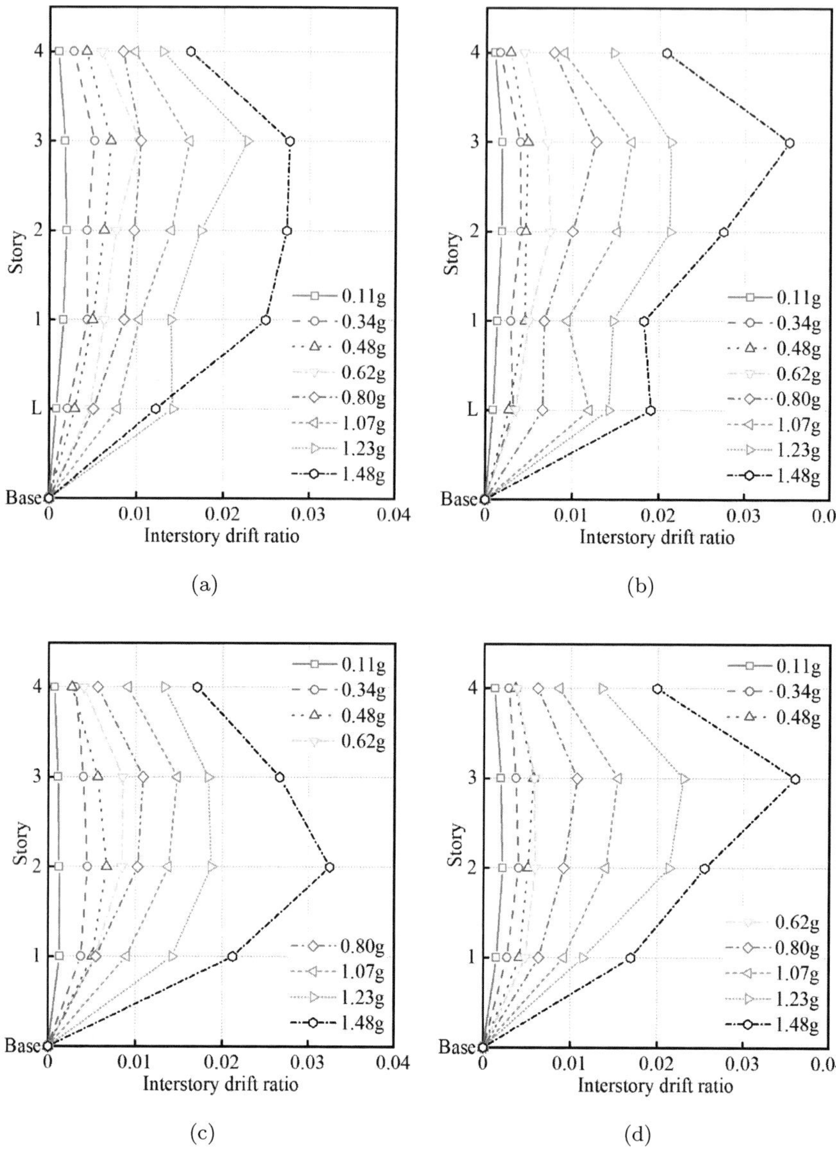

Fig. 11. Maximum interstory drift ratios of the two models: (a) model 1 in the x direction, (b) model 1 in the y direction, (c) model 2 in the x direction, (d) model 2 in the y direction.

x direction at earlier stages. Under the last two level excitations, a great amount of cumulative damage to the two models led to relatively large IDRs exceeding 2% which is just the collapse limit (1/50) of a RC frame according to the Chinese code.[24] During each intensity level test, the largest IDR among all the stories of each model was concentrated in the second and third stories, the IDR of which obviously increased more rapidly with increasing PGA. In addition, for model 1, the IDR of the lower story also increased relatively rapidly under strong ground motions. It shows that the damage to the two models was mainly concentrated in the second and third stories, and in this case, the damage to model 1 began to develop from the upper stories into the lower story at the last test stages. The findings were demonstrated earlier in the test phenomena.

4.5. *Torsion responses*

To measure and analyze the horizontal torsion responses, this study utilized the torsion angle, which is defined as the difference between the translational displacements along two edges in one direction at each floor divided by the normal distance between the two edges. In addition, the interstory torsion angle was defined as the relative torsion angle between two consecutive floors. Figure 12 shows the torsion angle and interstory torsion angle of the two models. It can be seen that both torsion angle and interstory torsion angle of model 1 were obviously much larger than those of model 2. This indicates that the torsional effect cannot be ignored in the SF structure which is a torsionally sensitive structure.

The torsion angles of the two models roughly increased with increasing PGA and along with the structural height. For model 1, the torsion angle increased rapidly after the PGA of 0.62 g, and finally reached 0.0084 rad at the roof. The upper ground floor contributed most to the torsion angles at the upper floors so that the torsion angles at the upper floors increased slowly before the PGA of 1.07 g. However, for model 2, the torsion angles were much smaller and steadily increased in an approximately linear distribution along with the structural height. Before the PGA of 1.07 g, the torsion angles of model 2 were around 0.001 rad within a limited range, but increased rapidly during the last three level test.

The interstory torsion angles of the two models also increased with increasing PGA in general. Before the last two level test, there was not a big difference in the interstory torsion angle between the two models from the second to fourth stories. During the last two level tests, the interstory torsion angles increased dramatically, which of model 1 were much larger on the whole. Specially, the key characteristic of interstory torsion angles of model 1 was the sudden change in the lower and first stories, and the maximum interstory torsion angle among the stories occurred at the first story (upper ground story) in most cases. It shows that torsion responses of model 1 were remarkable especially in the lower and first stories due to the vertical geometric change. The irregular geometric change resulted in a significant misalignment between the center of mass and center of rigidity, causing an inherent eccentricity in the SF structures.[32] When the interstory torsion angles of model 1 were associated with the interstory drift ratios of the lower story (Fig. 11), it can be inferred that the torsional effect on the interstory drift ratio of model 1 mainly

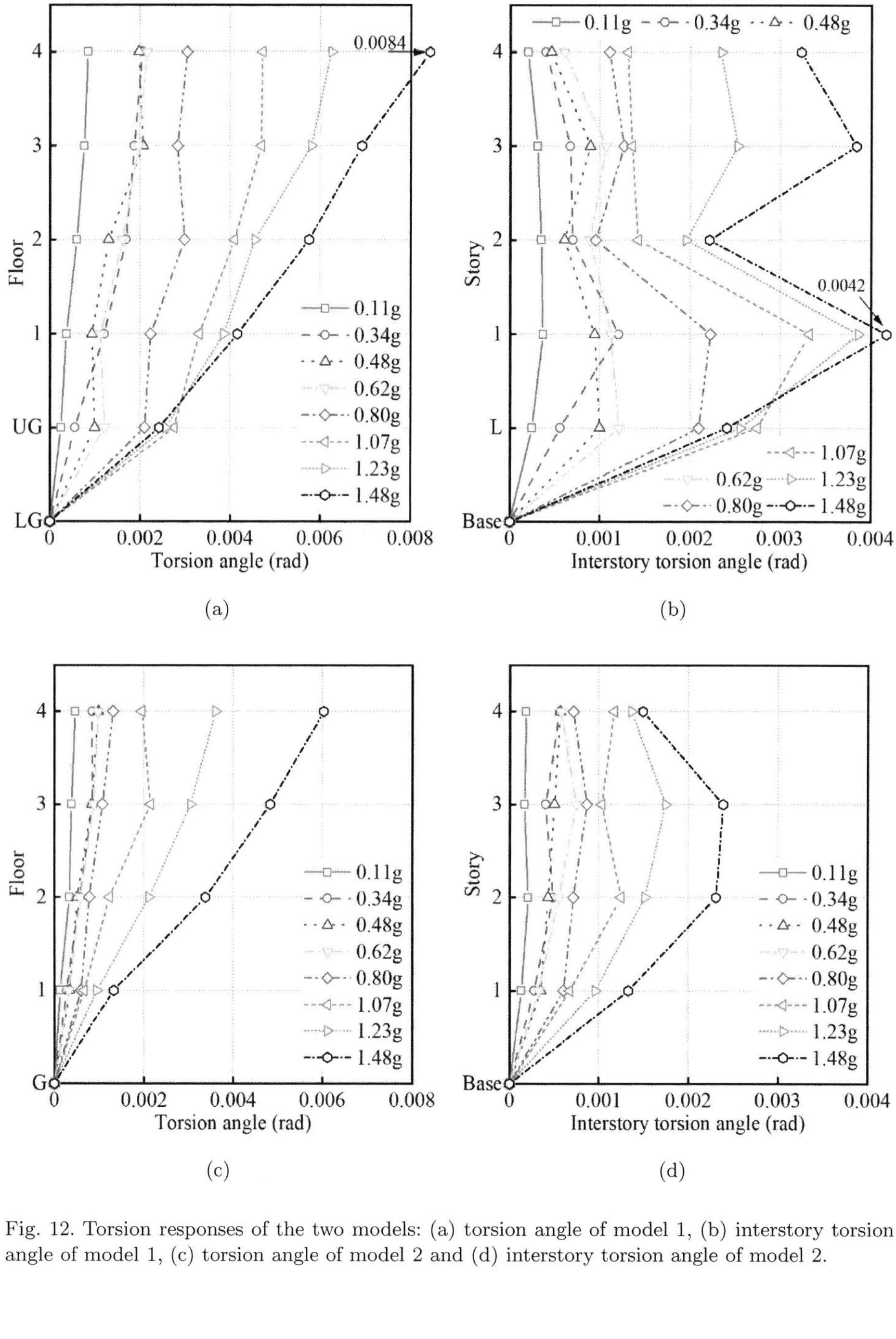

Fig. 12. Torsion responses of the two models: (a) torsion angle of model 1, (b) interstory torsion angle of model 1, (c) torsion angle of model 2 and (d) interstory torsion angle of model 2.

reflected in the y direction, leading to a relatively obvious amplification effect on the interstory drift ratio of the lower story in the y direction.

5. Conclusions

Shaking table test was carried out on two 1/8-scale models: a split-foundation RC frame structure supported by the foundation at two different levels and a conventional RC frame structure supported by foundations at one single level. The structural dynamic characteristics and seismic performance of the SF frame structure were investigated and were compared with those of the conventional frame structure. The following conclusions can be drawn from this test:

(1) The two model structures started to crack after excitations with a PGA of 0.34 g. For the SF model, cracks occurred in the lower story when the PGA reached 0.48 g and the concrete spalled after ground motions with a PGA of 1.07 g. The concrete of the conventional model started spalling during the PGA of 1.23 g. After the last intensity level excitation with a PGA of 1.48 g, the two model structures were already severely damaged. Finally, both models generally showed an intermediate failure mechanism. However, the final damage to model 1 was more severe on the upper ground side, and the final damage distribution of model 2 was more uniform and symmetrical.

(2) The initial natural frequencies of the SF model in the x and y direction were 9.27 Hz and 7.32 Hz, and those of the conventional model were 9.81 Hz and 9.09 Hz, respectively. The difference in natural frequencies between the two models resulted from the effect of the lower story on lateral stiffness, and the effect was relatively dramatic in the y direction. The natural frequencies of models gradually decreased with increasing PGA, and more serious damage to the SF model in the x direction led to a relatively rapid increase in stiffness degradation. The initial damping ratios in both directions of the two models were all around 4%, and then the damping ratios roughly increased during the whole test.

(3) The acceleration amplification factors of the two models roughly increased along with the structural height and were generally reduced with increasing PGA. Due to severe damage to the models, the acceleration amplification effect was no longer obvious at the last stages. The difference in structural stiffness led to larger acceleration amplification factors of the SF model and smaller acceleration amplification factors in the x direction.

(4) Displacement responses of the two models generally increased with increasing PGA. The peak floor displacements of the two models increased along with the structural height, the trend of which showed a shear mode deformation. The curve shapes of displacement responses of the two models were similar from the first to fourth floors. The restriction at the upper ground level created

a certain effect on displacement responses of the SF model especially in the first story. During each intensity level test, the largest interstory drift ratio among all the stories of each model was concentrated in the second story or third story, and specially the interstory drift ratio of the lower story also increased relatively rapidly at the last test stages. This was well agreed with damage development and distribution in the test phenomena.

(5) Torsion responses of the two models increased with increasing PGA in general, and the torsion responses of the SF model were obviously much larger than those of the conventional model. The upper ground floor of the SF model contributed most to the torsion angles at the upper floors, and the maximum interstory torsion angle occurred at the first story (upper ground story) in most cases. The torsional effect cannot be ignored in the SF structure which is a torsionally sensitive structure.

The seismic responses of the split-foundation RC frame structure and the conventional structure were analyzed through the shaking table test in this study. Overall, the seismic responses of the split-foundation structure are obviously different from the conventional regular structure on flat ground resulting from the irregular and asymmetric configuration of the split-foundation structure. However, the number of models in this study is limited and many different configurations of split-foundation structures were not included in this paper. Therefore, the influence of different configurations and improvements on the seismic performance of split-foundation structures will be considered in further study.

Acknowledgments

The research in this paper was supported by the National Natural Science Foundation of China (Grant Nos. 51638002, 51878101 and 52278484).

References

1. M. Surana, Y. Singh and D. H. Lang, Effect of irregular structural configuration on floor acceleration demand in hill-side buildings, *Earthquake Engng. Struct. Dyn.* **47**(10) (2018) 2032–2054.
2. Y. M. Li, J. Xu and G. J. Wang, Seismic fragility analysis of RC frames on steep slopes under near-fault and far-field ground motions, in *Proc. 16th European Conf. Earthquake Engineering*, Thessaloniki, Greece (June 2018), p. 11401.
3. Y. M. Li and L. P. Liu, *Damage and Thinking of Building in Wenchuan Earthquake* (Chongqing University Press, Chongqing, 2008) (in Chinese).
4. L. P. Wang, Y. M. Li, N. N. Zheng, L. P Liu, J. Han and J. W. Liu, Seismic damage investigation on typical slope building in Wenchuan earthquake, *J. Xi'an Univ. of Arch. Tech.* (Natural Science Edition) **41**(6) (2009) 822–826 (in Chinese).
5. Y. Singh, P. Gade, D. H. Lang and E. Erduran, Seismic behavior of buildings located on slopes-an analytical study and some observations from Sikkim earthquake of 18th

September 2011, in *Proc. 15th World Conference on Earthquake Engineering*, Lisbon, Portugal (September 2012).
6. X. C. Lin, H. Y. Zhang, H. F. Chen, H. Chen and J. Q. Lin, Field investigation on severely damaged aseismic buildings in 2014 Ludian earthquake, *Earthq. Eng. Eng. Vib.* **14**(1) (2015) 169–176.
7. Z. Qu and Y. Q. Yang, Seismic damages to owner-built dwellings in the 2015 earthquake sequence in Nepal, *Earthq. Eng. Eng. Dyn.* **35**(4) (2015) 51–59 (in Chinese).
8. Y. M. Li, Y. Y. Tang, B. L. Jiang, M. H. Dai and Q. Xie, State-of-the-art of study on seismic behavior of structures in mountainous terrain and establishment of the specification for structural design, in *Proc. 16th World Conf. Earthquake Engineering*, Santiago, Chile (January 2017).
9. A. R. V. Narayanan, R. Goswami and C. V. R. Murty, Performance of RC buildings along hill slopes of Himalayas during 2011 Sikkim Earthquake, in *Proc. 15th World Conf. Earthquake Engineering*, Lisbon, Portugal (September 2012).
10. Y. Singh, V. R. Yeluguri and D. H. Lang, Seismic response of hill buildings subjected to bi-directional excitation, in *Proc. 10th U.S. National Conf. Earthquake Engineering*, Anchorage, Alaska (July 2014).
11. M. Surana, Y. Singh and D. H. Lang, Seismic characterization and vulnerability of building stock in hilly regions, *Nat. Hazards Rev. ASCE* **19**(1) (2018) 04017024.
12. G. Xu and A. Q. Li, Seismic performance and improvements of stepback steel frames, *J. Earthq. Eng.* **25**(2) (2021) 163–187.
13. G. Xu, R. J. Zhang and A. Q. Li, Seismic performance and improvements of split-foundation buildings in across-slope direction, *Adv. Struct. Eng.* **23**(4) (2020) 749–763.
14. Y. Y. Tang, Y. M. Li, B. L. Jiang, S. Y. Ji, X. Y. Zhou and G. J. Wang, Study of controlling vertical structural stiffness for the frame structure supported by foundations with different elevations, in *Proc. 16th World Conference of Earthquake Engineering*, Santiago, Chile (January 2017).
15. B. L. Jiang, S. Y. Ji, Y. M. Li, Z. K. Chen and Y. Y. Tang, Influence of axial beam tension on seismic performance of RC beam-column joints, in *Proc. 16th World Conf. Earthquake Engineering*, Santiago, Chile (January 2017).
16. J. Xu, S. Y. Ji, Y. M. Li, X. W. Wang and G. J. Wang, Seismic failure modes identification of split-foundation RC frame structures, in *Proc. 17th World Conf. on Earthquake Engineering*, Sendai, Japan (September 2020).
17. Y. Y. Tang, Y. M. Li, S. Y. Ji and L. P. Liu, Comparative analysis of seismic performance between step-back frame structure and common frame structure, in *Proc. 17th World Conf. on Earthquake Engineering*, Sendai, Japan (September 2020).
18. M. Surana, A. Meslem, Y. Singh and D. H. Lang, Analytical evaluation of damage probability matrices for hill-side RC buildings using different seismic intensity measures, *Eng. Struct.* **207** (2020) 110254.
19. D. K. Paul and S. Kumra, Stability analysis of slope with building loads, *Soil Dyn. Earthq, Eng.* **16**(6) (1997) 395–405.
20. D. Raj and Y. Singh, Pseudostatic analysis of a coupled building-foundation-slope system for seismic and gravity actions, in *Proc. 4th Geo-China Int. Conf.* Jinan, China (July 2016), pp. 99–107.
21. Y. Y. Tang, Y. M. Li, Z. M. Huang, L. P. Liu and Y. S. Ding, Analysis of local site effect on seismic response of frame structure supported by foundations with different elevations, in *Proc. 1st South Asia Conf. Earthquake Engineering*, Karachi, Pakistan (February 2019).

22. R. Li, L. Liu, Y. Yin, C. Zhang and Y. Li, Seismic performance of RC split-foundation frame structures with steel braces in the mountainous area, *Int. J. Struct. Stabil. Dyn.* **22**(03n04) (2022) 2240011.
23. Y. Tang, Y. Li, B. Jiang, L. Wang and S. Ji, Study on redistribution of internal force and plastic hinge development of mountainous building structures with foundations at two different elevations, *Buildings* **13**(4) (2023) 909.
24. GB 50011-2010, *Code for Seismic Design of Buildings* (China Architecture & Building Press, Beijing, 2010).
25. J. P. Moehle, J. D. Hooper and C. D. Lubke, Seismic design of reinforced concrete special moment frames: a guide for practicing engineers, in *NEHRP Seismic Design Technical Brief No. 1, NIST GCR 8-917-1* (National Institute of Standards and Technology, Gaithersburg, MD, USA, August 2008).
26. F. Jalayer, P. Franchin and P. E. Pinto, A scalar damage measure for seismic reliability analysis of RC frames, *Earthquake Engng. Struct. Dyn.* **36**(13) (2007) 2059–2079.
27. K. Yuan, D. Gan, J. Guo and W. Xu, Hybrid geotechnical and structural seismic isolation: Shake table tests, *Earthquake Engng. Struct. Dyn.* **50**(12) (2021) 3184–3200.
28. N. Satake, K. Suda, T. Arakawa, A. Sasaki and Y. Tamura, Damping evaluation using full-scale data of buildings in Japan, *J. Struct. Eng. ASCE* **129**(4) (2003) 470–477.
29. Y. M. Li, B. L. Jiang, M. L. Zhang, B. B. Chen and X. Y. Li, Shaking table test on model structure of Chongqing Gaoke Sun Constellation Mansion, *J. Build. Struct.* **40**(3) (2019) 142–151 (in Chinese).
30. T. C. Hutchinson, X. Wang, G. Hegemier, P. Kamath and B. Meacham, Earthquake and postearthquake fire testing of a midrise cold-formed steel-framed building. I: Building response and physical damage, *J. Struct. Eng. ASCE* **147**(9) (2021) 04021125.
31. J. Y. Xue, R. Guo, D. Xu, L. J. Qi and Z. He, Shaking table test on 1/2-scale model of column-and-tie timber structure filled with wooden walls, *J. Struct. Eng. ASCE* **146**(12) (2020) 04020282.
32. Y. Y. Tang, Y. M. Li, B. L. Jiang and S. Y. Ji, Study on control parameters of torsional effect of the structure supported by foundations with different elevations, *Earthq. Eng. Eng. Dyn.* **42**(6) (2022) 122–128 (in Chinese).

https://doi.org/10.1142/9789819814770_0024

Chapter 24

Bridge Evaluation Based on Identified Influence Lines and Influence Surfaces: Multiple-Scenario Application[#]

Xu Zheng[*,†], Ting-Hua Yi[*,‡,¶], Dong-Hui Yang[*], Hong-Nan Li[*] and Yu Zhou[§]

[*]*School of Civil Engineering*
Dalian University of Technology, Dalian, P. R. China

[†]*Key Laboratory of Damage Diagnosis*
for Engineering Structures of Hunan Province
Hunan University, Changsha, P. R. China

[‡]*School of Civil and Transportation Engineering*
Beijing University of Civil Engineering and Architecture
Beijing, P. R. China

[§]*School of Civil Engineering*
Anhui Jianzhu University, Hefei 230601, P. R. China
[¶]*yth@dlut.edu.cn; yth@bucea.edu.cn*

Bridge influence lines (BILs) and bridge influence surfaces (BISs) are inherent static parameters of bridges which can be extracted from moving vehicle-induced bridge responses. Compared with dynamic parameters, these parameters are directly related to the stiffness and internal forces in each cross-section of a bridge therefore can be considered as an effective bridge metamodel. To accelerate the engineering practice of BIL- and BIS-based bridge evaluation, this paper first briefly reviews the current BIL and BIS field test and identification methods. Then, the bridge evaluation guidelines of China and the United States are introduced as the basis of the evaluation methods. Engineering application scenarios for bridge evaluation, including permit load verification, performance degradation checking, and load carrying capacity evaluation, are summarized, and a multiple-scenario bridge evaluation method is established. At the end of this paper, an evaluation example of a four-span continuous bridge is presented to illustrate the application procedure and verify the effectiveness of the proposed method. The outcomes of this paper provide a promising application method of field test BILs and BISs, which may help bridge engineers more effectively use these parameters in engineering practice.

Keywords: Bridge evaluation; influence line; influence surface; load testing; moving load; continuous bridge.

[¶]Corresponding author.
[#]This chapter appeared previously on the International Journal of Structural Stability and Dynamics. To cite this chapter, please cite the original article as the following: X. Zheng *et al.*, *Int. J. Struct. Stab. Dyn.*, **23**, 2340026 (2023), DOI: 10.1142/S0219455423400266.

1. Introduction

In the near future, many additional bridges will require major rehabilitation projects, as most of the bridges in the United States were designed for a lifespan of 50 years, and 42% of bridges in the United States are at least 50 years old.[1] In China, there are also nearly 1 million highway bridges in service in 2022.[2] The evaluation and maintenance of these bridges will become a major challenge for bridge management. To answer performance questions about an existing bridge, structural health monitoring (SHM),[3,4] bridge nondestructive inspection (NDI),[5] and bridge load testing[6] are three main approaches. However, the SHM system is too expensive for implementation in such a large number of bridges,[7] and the NDI method can only detect the damage of bridge components rather than capture global bridge performance.[8] For these reasons, load testing is still the most reliable way to evaluate the actual behavior of a large number of in-service bridges.

Compared with dynamic tests that require a large number of sensors[9] and obtain local damage insensitive indicators,[10] diagnostic load tests are widely used worldwide as a low-cost and efficient method for bridge evaluation.[11,12] Deferring repair or replacement through diagnostic load testing can deliver cost savings and have significant paybacks for both bridge owners and users. When implementing diagnostic load testing, single-calibration trucks should pass through the tested bridge at a crawl speed. The vehicle-induced bridge response can be used for verifying the bridge model, which is based on model updating technology,[13] or simply calculating the adjustment factor.[14] In addition to this application scenario, vehicle-induced bridge responses during diagnostic load testing also contain another important feature of bridges: bridge influence lines (BILs) and bridge influence surfaces (BISs). Bridge influence lines are defined as a function of the bridge response and the longitudinal location of the load when a unit-concentrated load crosses a bridge.[15] A bridge influence surface is the extension of the BIL, which is defined as a function of the bridge response and the spatial location of the load when a unit-concentrated load travels throughout the entire bridge deck.[16] As inherent properties of the bridge, these two features are unrelated to the magnitude of the live load, which provides a promising future for bridge evaluation.

Combining bridge diagnostic load test data and vehicle parameters, BILs and BISs can be calculated through various identification algorithms. To accurately identify BILs and BISs, dynamic fluctuations induced by moving vehicles should be removed first.[17,18] After extracting the quasi-static component, Obrien[19] proposed a matrix method to extract BILs in the spatial domain. Based on this work, a frequency domain method and regularization[20,21] methods were proposed for improving identification efficiency and accuracy. Zheng[16] summarized this research work and accurately identified BISs by considering the spatial distribution of vehicle loads. Khuc *et al.*[22] and Zhou[23] proposed a BIS identification method during bridge operation combined with computer vision technology. Current bridge evaluation methods based on BIL and BIS are mainly aimed at the damage detection of bridge components. Ono *et al.*[24] applied the gray relation coefficient and deflection influence line method and successfully detected the stiffness loss of a deck slab. Alamdari *et al.*[25,26] applied a BIL-based method to a cable-stayed bridge, and bridge damage was successfully identified. Cai[27] researched the sensitivity of different BILs to damage and verified the method through a simply supported beam. Zhou *et al.*[28] derived the formula of an inclination BIL for damage identification, and the proposed index was successfully applied to identify the damage of a truss bridge. Khuc *et al.*[22] proposed a bridge support damage detection method based on the variation in identified BISs. Zheng *et al.*[29] derived the relationship between BILs and girder bridge stiffness and successfully identified the rotational stiffness of the support and flexural stiffness of the beam based on the proposed formula.

For the influence line and influence surface-based bridge evaluation method, Sun *et al.*[30] extracted the actual influence line based on diagnostic load testing data. Based on the calculated adjustment factor, the rating factor of an unfavorable bridge girder was significantly improved. Based on the features of an extracted influence line, Wang *et al.*[31] proposed a capacity evaluation index based on the maximum amplitude and total area of influence lines for bridge evaluation. These pioneering efforts successfully applied influence lines to bridge performance evaluation with a base-state finite element model. However, in engineering practice, these models are sometimes difficult to obtain. In addition, an accurate evaluation of load-carrying capacity is always necessary to combine bridge inspection data such as crack and sectional defect information. Therefore, how to efficiently combine BILs and BISs in bridge evaluation needs to be systematically summarized.

Based on the current developments and shortcomings of BIL- and BIS-based bridge evaluation methods, this paper briefly reviews the field testing and identification methods for BILs and BISs. Then, the engineering application scenario for bridge evaluation, including permit load verification, performance degradation checking, and load-carrying capacity evaluation, is summarized. Based on the possible application scenario, the concept of a multiple-scenario bridge evaluation method is established. The application procedure and effectiveness of the proposed method are verified through a four-span continuous bridge evaluation example at the end of this paper.

2. BIL and BIS Testing and Identification Method

The accurate identification of BILs and BISs is the basis for reliable bridge evaluation. The basic testing and identification process of BILs and BISs is briefly introduced in this section.

2.1. *Field test procedure*

High-quality field test data are fundamental to the accurate identification of BILs and BISs. A field test of BILs and BISs includes the testing of truck parameters and bridge responses during the truck passage. The truck parameters include the axle weight (wheel weight), axle distance (wheel distance), and position during the test. The vehicle position can be tested by a global navigation satellite system (GNSS). To match the sampling frequency of bridge responses, real-time kinematic (RTK) technology should be used as a supplement to GNSS during positioning. To capture more response points during the truck passage, the velocity of the truck should be as small as possible, and the truck routine should fully cover the tested bridge. Various forms of bridge responses can be tested for BIL and BIS testing, mainly including inclinations, strains, and deflections of the bridge. For BIL testing, the calibration truck is only required to run through a representative lane of the tested bridge. For BIS testing, every vehicle lane of the bridge should be covered. The basic procedure of the BIL and BIS field tests is summarized in Fig. 1.

2.2. *Identification method*

The raw data of the bridge response and test vehicle should be preprocessed before being applied to BIL and BIS identification. Data preprocessing can be mainly divided into three steps. First, the bridge response during the period when the truck is not on the bridge should be eliminated, and the truck-induced bridge responses should be extracted. Then, the truck positioning data should be converted to bridge coordinates (with one bridge support as the origin and the longitudinal direction of the bridge as the positive direction), and then the time history of bridge response should be converted to vehicle location-related bridge responses. Finally, to extract the linear-elastic portion of the bridge response, the residual of the bridge response should be removed by linear revision.

After data preprocessing, the response vector **R** can be constructed, and the dimension of the vector should correspond to the following equation:

$$p = \frac{l + D}{d} + 1, \tag{1}$$

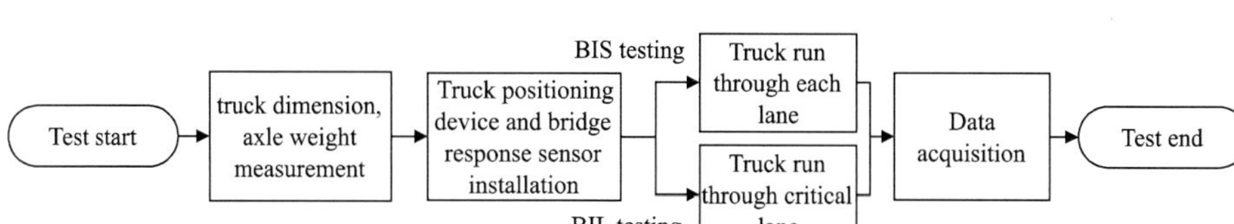

Fig. 1. Flowchart of the BIL and BIS field tests.

where p is dimension of response vector $\mathbf{R}$. l is the total length of the tested bridge. d is the spatial distance between two continuous bridge response points. D is distance between the front and rear axles of the calibration trucks.

After constructing the response vector $\mathbf{R}$, the information matrix of calibration truck $\mathbf{L}$ can be constructed based on filling the $p \times q$ zero matrix $\mathbf{L}_0$, and the column number q can be determined by

$$q = p - \frac{D}{d}. \tag{2}$$

Then, the matrix $\mathbf{L}_0$ should be filled according to the following rules:

$$L_{k+C_i,k} = A_i, \tag{3}$$

$$C_i = \frac{D_i}{d}, \tag{4}$$

where k is number between $1 \sim q$; A_i is the ith axle weight of calibration trucks (take the front axle as the first axle); C_i is the column number difference between A_i and A_1; D_i is the distance between the ith axle and the first axle.

After establishing vector $\mathbf{R}$ and matrix $\mathbf{L}$, the problem of influence line identification can be converted to solve the linear equations (Eq. (5)) and extract the influence coefficient vector $\mathbf{\Phi}$.

$$\mathbf{R} = \mathbf{L}\mathbf{\Phi}. \tag{5}$$

To solve Eq. (5) more accurately, inverse problem-solving methods such as the least square method, curve fitting method, and regularization method can be used.[15] If the spatial effect of the bridge is considered, the influence line identification problem becomes the influence surface identification problem. In addition to refined testing of the target influence coefficient,[16] an easier way to identify a BIS is to fit the influence line in each lane to construct the influence surface.

3. Multiple-Scenario Bridge Evaluation Method

3.1. *Governing codes and guidelines*

China and the United States have the largest bridge infrastructures in the world. To manage these bridges rationally, the bridge evaluation codes of these two countries are of the greater reference value. Due to the early start of bridge construction in the United States, bridge evaluation codes in the United States have also been developed over a longer period. The first guideline for bridge evaluation is the AASHTO-1988 Manual for bridge rating through load testing.[32] In this code, three rating methods, including allowable stress rating (ASR), load factor rating (LFR), and load-resistance factor rating (LRFR), were introduced. After this version of the code, The American Association of State Highway and Transportation Officials (AASHTO) published the Manual for Bridge Evaluations,[33] which is still the governing code for bridge evaluation in the United States. This code recommends diagnostic load testing as the preferred method of the load test. After diagnostic load testing, the rating factor of the bridge can be revised based on the adjustment factor K, which can be calculated by $K = 1 + K_a K_b$. K_b should be assigned a value between 0 and 1.0 to indicate the level of test benefit that is expected at the rating load level,[33] and K_a accounts for the benefit derived from the load test, which can be determined by

$$K_a = \frac{\varepsilon_C}{\varepsilon_T} - 1, \tag{6}$$

where ε_T is the maximum member strain measured during the load test. ε_C is the corresponding calculated strain due to the test vehicle at its position on the bridge, which produces ε_T.

The development of the Chinese bridge evaluation code is quite different from that of the US code. Because the US bridge design can choose any version of specifications, the evaluation method is a general method that is suitable for multiple design theories. In China, since the selection of the latest version of design codes for bridge design is mandatory, bridge evaluation codes should also be based on the latest design codes. The timeline of the design and evaluation code of the bridge is summarized in Table 1.

Based on the Chinese bridge evaluation code, the load-carrying capacity evaluation formula of the reinforced concrete bridge is shown as follows:

$$\gamma_0 S \leq R(f_d, \xi_c a_{dc}, \xi_s a_{ds}) Z(1 - \xi_e), \tag{7}$$

where S and $R(\cdot)$ account for the theoretical load effect and resistance of the structure. ξ_c and ξ_s account for the section reduction of reinforced concrete and steel, respectively, and ξ_e accounts for the potential for further bridge degradation. The values of ξ_c, ξ_s and ξ_e can be determined by field inspection. γ_0 represents the

Table 1. Timeline of bridge design and evaluation code in China.

Bridge design code	Bridge evaluation code
1985: Code for design of highway reinforced concrete and prestressed concrete bridges and culverts[34]	1988: Highway old bridge load bearing capacity identification methods (for trial implementation)[35]
2004: Code for design of highway reinforced concrete and prestressed concrete bridges and culverts[36]	2011: Specification for inspection and evaluation of load-bearing capacity of highway bridges[37]
2018: Specifications for design of highway reinforced concrete and prestressed concrete bridges and culverts[38]	New Specification will soon be available

importance of the tested structure. Z accounts for the benefit derived from the load test, which can be determined from a table in the code.[37]

3.2. *Basic scenarios for bridge evaluation*

Based on the aforementioned summary, to accurately evaluate bridge load-carrying capacity, the following preparations are required: (1) Field inspection results, which include the defect situation of bridge section and bridge environment. (2) Bridge design information, which is used to calculate resistance of critical section. (3) Field load testing data and a finite element model, which account for the benefit derived from the load test. However, in engineering practice, due to the different requirements of bridge owners and users, this information is not always available. Therefore, how to satisfy the requirements of bridge owners and users and give reasonable bridge evaluation results is important for the application of BIL- and BIS-based bridge evaluation methods. The basic scenarios for bridge evaluation are summarized as follows:

Permit load verification: This project intended to evaluate the structure's ability to carry multiple superheavy transports that were required to cross the structure. For the passage capacity evaluation of heavy trucks and vehicles, bridge model and field inspection data are not available, but the BIS or BIL of the bridges should be obtained.

Performance degradation checking: For the bridges for which proof load tests have been conducted, the BIL and BIS tests can be used to reproduce the previous load test conditions based on the virtual loading of BILs and BISs. The performance degradation can be checked through a comparison of virtual loading and proof load test results.

Load carrying capacity evaluation: For bridges with detailed field inspection results, load testing data, design information, and finite element models, BIL and BIS tests can provide similar information as diagnostic load test, which accounts for the benefit derived from load testing.

3.3. *Bridge evaluation method based on BILs and BISs*

For permit load verification, the heavy vehicle passage capacity can be guaranteed if the following relations can be satisfied:

$$f_d \geq \max(P(x) * \mathrm{DIL}(x)), \tag{8}$$

$$R_d \geq \max(P(x) * \varphi(x)), \tag{9}$$

where f_d and R_d account for the displacement and resistance limit of the bridge; $P(x)$ is the Dirac-delta function consisting of the concentrated load of the heavy vehicle; and $*$ is the convolution of the two functions. $\mathrm{DIL}(x)$ is the displacement

Table 2. Multiple-scenario bridge evaluation method based on BIL and BIS.

Scenario	Available information	Evaluation method	Objective
Scenario 1	Design information	Vehicle-induced bridge response calculation based on BIL and BIS	Permit load verification
Scenario 2	Load testing results	Comparison between proof load testing and BIL or BIS calculation results	Performance degradation checking
Scenario 3	Design information Field inspection results	Influence line and FEM comparison to account benefit of bridge performance	Load-carrying capacity evaluation

influence line function and $\varphi(x)$ is the internal forces influence line function, which can be calculated by strain influence lines and section information.

To check performance degradation, bridge performance can be considered degraded if the following equation is satisfied:

$$S_s < \sum_{i=1}^{n} P_i \varphi_i, \tag{10}$$

where P_i is the ith concentrated load applied to the bridge during load testing; φ_i is the influence coefficient in the BIL or BIS corresponding to the ith concentrated load; and S_s is the real bridge response during the proof load test.

For load-carrying capacity evaluation, the load-carrying capacity of the bridge satisfies the requirement if the following equation is satisfied.

$$\gamma_0 S \leq R(\cdot) Z_v, \tag{11}$$

where $R(\cdot)$ is the bridge resistance based on the design code and field inspection results. For the reinforced concrete bridge, $R(\cdot)$ can be calculated as Eq. (7) based on Chinese code.[37] Z_v is the adjustment factor that accounts for the benefit derived from the field test BIL and BIS, which can be calculated based on different guidelines.[33,37]

Based on the previous summary, the multiple-scenario bridge evaluation method based on the influence line can be summarized, as shown in Table 2.

4. Four-Span Continuous Bridge Evaluation Examples

To show the implementation procedure of the proposed bridge evaluation method, a field test example is proposed in this section. The basic situation and field test conditions are introduced first, and the bridge is evaluated based on the identified BILs and BISs in different scenarios.

4.1. *Basic situation of the bridge*

The bridge to be tested is one of a pair of four-span variable section continuous girder bridges, as shown in Fig. 2. The total length of the bridge is 260 m, and the main girder is a reinforced concrete box girder with a maximum height of 4.8 m in the mid-span section and a minimum height of 2 m in the support section, the specific geometry of which is shown in Fig. 3.

4.2. *Field test conditions*

To comprehensively evaluate the performance of the tested bridge, various forms of load testing are conducted, as shown in Fig. 4. The load testing procedure mainly

Fig. 2. South three spans of the tested bridge (imaged by Yu and Zhou).

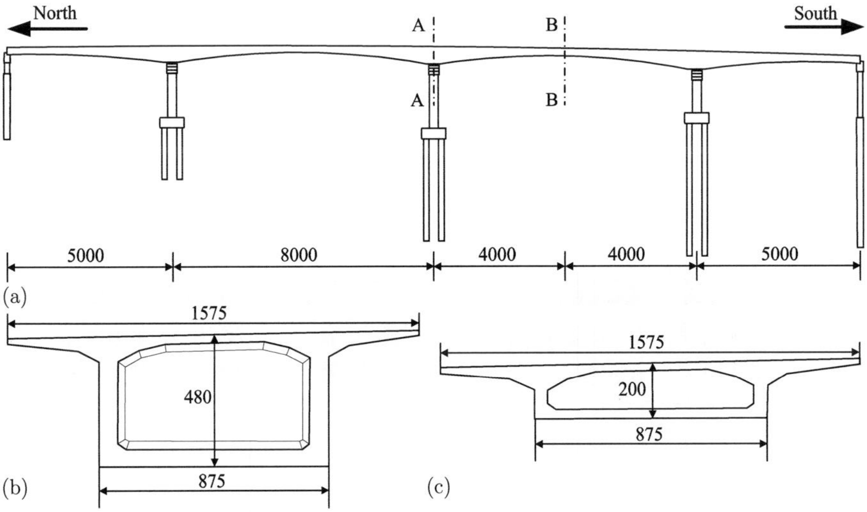

Fig. 3. Front and elevation views of the tested bridge: (a) side view; (b) detail of Section A; and (c) detail of Section B. (Unit: cm).

(a) (b) (c)

Fig. 4. Field test conditions: (a) load case 1: single truck moving; (b) load case 2: three trucks moving; and (c) load case 3: static proof load testing. (Imaged by Xu Zheng).

contains three load cases. In load case 1, a single calibration truck was used to test the BIS by passing through each lane of the bridge at crawl speed and repeating the loading twice for each lane, as shown in Fig. 4(a). Vehicle-mounted GNSS-RTK equipment was used for the real-time positioning of the loading vehicle, and a reference station was placed in a fixed station to provide real-time kinematics, as shown in Fig. 4(b). The positioning frequency is 20 Hz. The axle distance and axle weight of the truck were measured before the test. The axle weight measurement results were 4.41 t (front), 10.64 t, 9.89 t, and 10.14 t (rear). The gross weight of the truck is 35.08 t. The axle distance measurements of the test vehicle were 1.9 m, 4.2 m, and 1.4 m, respectively. The wheel distance in each axle is 2.0 m. In load case 2, three calibration trucks ran side-by-side to test the BIL, with the positioning system mounted only on the middle green truck. In load case 3, the load trucks are placed near Section B to conduct proof load testing. This condition can be divided into two subconditions of six truck loading and nine truck loading, and the specific arrangement of trucks can be seen in Fig. 5.

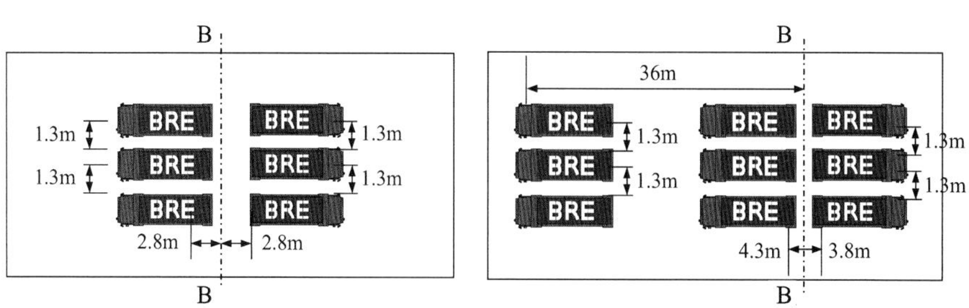

(a) (b)

Fig. 5. Field test conditions of static load testing: (a) six trucks loading; and (b) nine trucks loading.

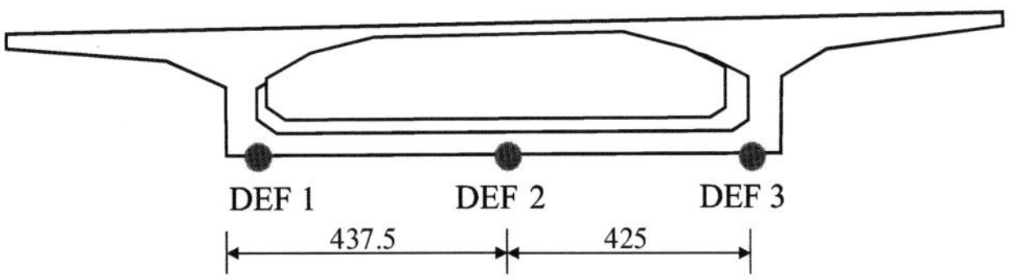

Fig. 6. Deflection measurement points in Section B (unit: cm).

The bridge responses were collected simultaneously during the test. A non-contact distributed image deflectometer with a sampling frequency of 200 Hz was applied in this test to remotely measure the deflection of the bridge, and the measurement points of the tested bridge are shown in Fig. 6.

4.3. *BIL and BIS identification*

Combining the bridge response data and truck positioning data in load case 2, the BILs at each measurement point can be identified, as shown in Fig. 7. It can be seen in Fig. 7 that the bridge support positions generally correspond well to the influence line position, and the sensor position corresponds well to the peak of the influence line, indicating that the identification results are reliable.

Combining the bridge responses and truck positioning data in load case 1, BISs of each measurement point can be identified, as shown in Fig. 8. The influence lines corresponding to each wheel of the vehicle are identified[14] and BISs are identified through linear interpolation.

4.4. *Multiple-scenario bridge evaluation*

Due to the restriction, only the deflection BILs and BISs are considered in this paper for bridge evaluation. It is recommended to incorporate the strain into the

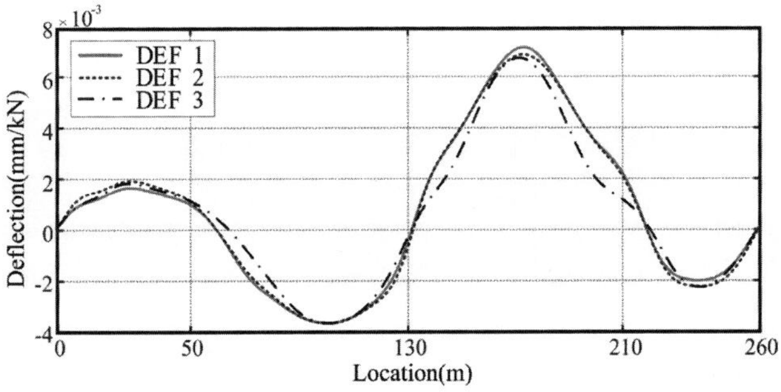

Fig. 7. BIL identification results.

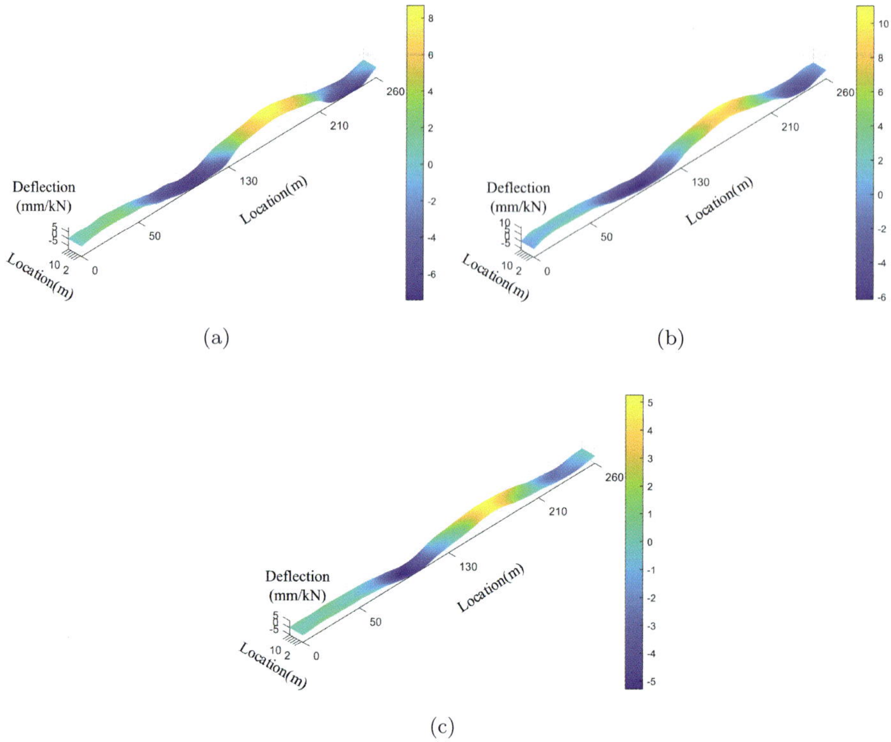

Fig. 8. BIS identification results: (a) DEF 1; (b) DEF 2; and (c) DEF 3.

process for further evaluation of the internal forces if available. A brief application for the proposed method is introduced as follows.

4.4.1. *Permit load verification*

When the heavy vehicle shown in Fig. 9 applies to this bridge, the calculated bridge responses based on BIL are shown in Table 3, and the vehicle-induced displacements

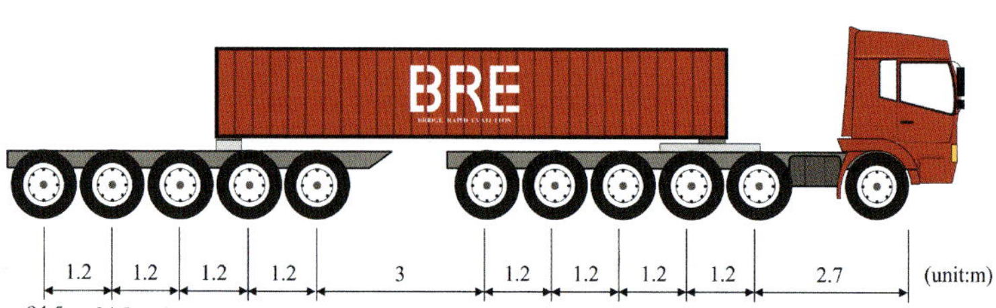

Fig. 9. Permit truck loading configuration.

Table 3. Permit truck-induced load effect.

Sensor	Deflection 1	Deflection 2	Deflection 3	Deflection limit	Result
Deflection	6.80 mm	6.62 mm	6.41 mm	133 mm	OK

Table 4. Comparison between static loading and BIL & BIS calculation results.

	Six trucks loading — Fig. 5(a)			Nine trucks loading — Fig. 5(b)		
	DEF 1	DEF 2	DEF 3	DEF 1	DEF 2	DEF 3
Static loading (mm)	15.63	16.69	15.42	21.41	20.00	19.78
BIS calculation (mm)	15.14	15.96	16.91	18.07	19.40	18.71
Error	3.1%	4.4%	9.7%	15.6%	3.0%	5.4%
BIL calculation (mm)	14.65	14.76	13.45	17.25	17.61	15.01
Error	6.3%	11.6%	12.8%	19.4%	11.9%	23.7%

Notes: Average error: 6.4% (BIS calculation), 14.3% (BIL calculation); standard deviation error: 4.9% (BIS calculation), 6.2% (BIL calculation).

are less than the design limit of the bridge (1/600 of bridge length[38]). To check the internal forces induced by this vehicle, strain influence lines can be used if available.

4.4.2. *Performance degradation checking*

To illustrate the performance degradation checking method, the load pattern of static load testing, as shown in Fig. 5, is applied to the BILs and BISs, and a comparison is made between the calculated bridge responses and static load testing results. The comparison results are shown in Table 4.

Since all three loading conditions of this bridge were implemented on the same day, the bridge did not actually degrade. The comparison results show that the actual response matches well with the virtual loading response, and the BIS calculation is more accurate than the BIL calculation. In the above calculations, the loading trucks are all assumed to be symmetrically along the centerline of the bridge. However, it is difficult to control the transverse location of trucks during the field testing. This is the main source of error between calculation and static loading.

4.4.3. *Load-carrying capacity evaluation*

Considering that the bridge is newly built and has no defects, take the comprehensive modification factor ξ_c, ξ_s, which accounts for the defect of the tested bridge, as 1. Based on the comparison between the field-testing BIS and the finite element model of the bridge, the verification factor η, which accounts for the benefit derived from the load, can be calculated as

$$\eta = \frac{S_s}{S_e} = \frac{14.99\,\text{mm}}{27.78\,\text{mm}} = 0.54, \tag{12}$$

where S_s is the bridge response calculated from BIS and S_e is the same point bridge response calculated from the finite element model.

The verification factor η is much less than 1. Based on the conversion table,[35] the corresponding adjustment factor Z_v in Eq. (11) is greater than 1, which indicates the actual performance of the bridge is much better than the design. Based on the consideration of load testing, it can be concluded that the actual load-carrying capacity of the tested bridge satisfies the requirement.

5. Conclusions and Recommendations

To provide an effective bridge evaluation method based on field test BILs and BISs, this paper provides a multiple-scenario bridge evaluation method. A four-span continuous bridge evaluation example is introduced at the end of this paper. Major conclusions of this paper include the following aspects:

(1) Based on the current governing code and guidelines in China and the United States, the complete evaluation of a bridge requires bridge design, field inspection, and load testing data. Based on possible combinations of available information, possible application scenarios of BIL- and BIS-based bridge evaluation methods are summarized.
(2) A multiple-scenario bridge evaluation method based on BIL and BIS is proposed. For permit load verification, BIS is used as a bridge metamodel to calculate the heavy vehicle-induced load effect. For performance degradation checking, BIS is used for comparison between previous load testing results. For load-carrying capacity evaluation, BIS is used to account for the favorable effect of the field-testing results.
(3) A four-span continuous bridge field test and evaluation example is introduced at the end of this paper to verify the effectiveness of the proposed method. BILs and BISs of the tested bridge are identified. The evaluation results indicate that the actual performance of the bridge is much better than the design, and BISs give more accurate evaluation results than BILs.

Limitations and future development of this study: The evaluation methods proposed in this paper are based on the linear superposition of BILs and BISs. The superposition principle will no longer apply when the bridge has undergone severe degradation, or when the virtual loads applied to the line of influence are so high that the bridge may have entered a nonlinear state. Therefore, establishing the relationship between load magnitude and bridge response and thus establishing a correction method for bridge evaluation results is an important research field.

Acknowledgments

This research work was jointly supported by the National Natural Science Foundation of China (Grant Nos. 52208147, 52250011 and 52078102), Key Laboratory of Damage Diagnosis for Engineering Structures of Hunan Province, the Fellowship of China Postdoctoral Science Foundation (Grant No. 2022TQ0053), Foundation for High Level Talent Innovation Support Program of Dalian (Grant No. 2022RQ017), and the Fundamental Research Funds for the Central Universities (Grant Nos. DUT22ZD213 and DUT22QN235).

References

1. ASCE (American Society of Civil Engineering), 2021 infrastructure report card: Bridges. Retrieved from infrastructure report card (2021). http://www.infrastructurereportcard.org/cat-item/bridges/.
2. MoT (Ministry of Transport of the People's Republic of China), Statistical bulletin on the development of the transport industry in 2021 (2022). https://xxgk.mot.gov.cn/2020/jigou/zh-ghs/202205/t20220524_3656659.html.
3. Z. W. Chen, Q. L. Cai and J. Li, Stress influence line identification of long suspension bridges installed with structural health monitoring systems, *Int. J. Struct. Stab. Dyn.* **16**(4) (2016) 1640023.
4. X. M. Yang, C. X. Qu, T. H. Yi, H. N. Li and H. Liu, Modal analysis of a bridge during high-speed train passages by enhanced variational mode decomposition, *Int. J. Struct. Stab. Dyn.* **20**(13) (2020) 2041002.
5. A. M. Abdallah, R. A. Atadero and M. E. Ozbek, A state-of-the-art review of bridge inspection planning: Current situation and future needs, *J. Bridge Eng.* **27**(2) (2022) 03121001.
6. S. Alampalli, D. M. Frangopol, J. Grimson, M. W. Halling, D. E. Kosnik, E. O. L. Lantsoght, D. Yang and Y. E. Zhou, Bridge load testing: State-of-the-practice. *J. Bridge Eng.* **26**(3) (2021) 03120002.
7. Y. F. Zhu, W. X. Ren and Y. F. Wang, Structural health monitoring on Yangluo Yangtze River Bridge: Implementation and demonstration, *Adv. Struct. Eng.* **25**(7) (2022) 1431–1448.
8. S. B. Lin and Y. J. Wang, Crack-depth estimation in concrete elements using ultrasonic shear-horizontal waves, *J. Perform. Constr. Fac.* **34**(4) (2020) 04020064.
9. R. R. Hou and Y. Xia, Review on the new development of vibration-based damage identification for civil engineering structures: 2010–2019, *J. Sound Vib.* **491** (2021) 115741.
10. X. M. Yang, C. X. Qu, T. H. Yi, H. N. Li and H. Liu, Modal analysis of a bridge during high-speed train passages by enhanced variational mode decomposition, *Int. J. Struct. Stab. Dyn.* **20**(13) (2020) 2041002.
11. R. J. Sherman, M. H. Hebdon and J. B. Lloyd, Diagnostic load testing for improved accuracy of bridge load rating, *J. Perform. Constr. Fac.* **34**(5) (2020) 04020082.
12. P. Olaszek, M. Lagoda and J. R. Casas, Diagnostic load testing and assessment of existing bridges: Examples of application. *Struct. Infrastruct. Eng.* **10**(6) (2014) 834–842.
13. S. W. Lin, Y. L. Du, T. H. Yi and D. H. Yang, Model updating using bridge influence lines based on an adaptive metamodel global optimization method, *J. Bridge Eng.* **27**(3) (2022) 04022003.
14. TRB (Transportation Research Board), *Primer on Bridge Load Testing* (Washington, DC, 2019).

15. X. Zheng, D. H. Yang, T. H. Yi and H. N. Li, Development of bridge influence line identification methods based on direct measurement data: A comprehensive review and comparison, *Eng. Struct.* **198** (2019) 109539.
16. X. Zheng, D. H. Yang, T. H. Yi and H. N. Li, Bridge influence surface identification method considering the spatial effect of vehicle load, *Struct. Control Health Monit.* **28**(8) (2021) e2769.
17. Y. T. Wei, D. H. Yang, T. H. Yi, H. N. Li, P. Zhou and X. W. Xue, Bridge static influence line identification based on structural dynamic responses under high-speed trains, *Int. J. Struct. Stab. Dyn.* **23**(11) (2023) 2350126.
18. X. Zheng, D. H. Yang, T. H. Yi and H. N. Li, Bridge influence line identification from structural dynamic responses induced by a high-speed vehicle, *Struct. Control Health Monit.* **27**(7) (2020) e2544.
19. E. J. OBrien, M. J. Quilligan and R. Karoumi, Calculating an influence line from direct measurements, *Proc. Inst. Civ. Eng. Bridge Eng.* **159**(1) (2006) 31–34.
20. Z. W. Chen, L. Zhao, W. J. Yan, K. V. Yuen and C. Wu, A statistical influence line identification method using Bayesian regularization and a polynomial interpolating function, *Struct. Control Health Monit.* **29** (2022) e3080.
21. W. J. Yan and K. V. Yuen, A new probabilistic frequency-domain approach for influence line extraction from static transmissibility measurements under unknown moving loads, *Eng. Struct.* **216** (2020) 110625.
22. T. Khuc and F. N. Catbas, Structural identification using computer vision-based bridge health monitoring, *ASCE J. Struct. Eng.* **144**(2) (2018) 04017202.
23. Y. Zhou, S. Zhou, G. W. Hao and J. Zhang, Bridge influence line identification based on big data and interval analysis with affine arithmetic, *Measurement* **183** (2021) 109807.
24. R. Ono, T. M. Ha and S. Fukada, Analytical study on damage detection method using displacement influence lines of road bridge slab. *J. Civil Struct. Health Monit.* **9** (2019) 565–577.
25. M. M. Alamdari, L. L. Ge, K. Kildashti, Y. C. Zhou, B. Harvey and Z. Y. Du, Non-contact structural health monitoring of a cable-stayed bridge: Case study, *Struct. Infrastruct. Eng.* **15**(8) (2019) 1119–1136.
26. M. M. Alamdari, K. Kildashti, B. Samali and H. V. Goudarzi, Damage diagnosis in bridge structures using rotation influence line: Validation on a cable-stayed bridge, *Eng. Struct.* **185** (2019) 1–14.
27. Q. L. Cai, Z. W. Chen, S. Y. Zhu and L. Y. Mo, On damage detection of beam structures using multiple types of influence lines, *Structures* **42** (2022) 449–465.
28. Y. Zhou, S. K. Di, C. S. Xiang, W. R. Li and L. X. Wang, Damage identification in simply supported bridge based on rotational-angle influence lines method, *Trans. Tianjin Univ.* **24**(6) (2018) 587–601.
29. X. Zheng, T. H. Yi, D. H. Yang and H. N. Li, Stiffness estimation of girder bridges using influence lines identified from vehicle-induced structural responses, *J. Eng. Mech.* **147**(8) (2021) 4021042.
30. Z. Sun, D. M. Siringoringo and Y. Fujino, Load-carrying capacity evaluation of girder bridge using moving vehicle, *Eng. Struct.* **229** (2021) 111645.
31. N. B. Wang, W. Shen, C. R. Guo and H. P. Wan, Moving load test-based rapid bridge capacity evaluation through actual influence line, *Eng. Struct.* **252** (2022) 113630.
32. AASHTO (American Association of State Highway and Transportation Officials), *Manual for Bridge Rating Through Load Testing* (AASHTO, Farmington Hills, MI, 1988).

33. AASHTO (American Association of State Highway and Transportation Officials), *The Manual for Bridge Evaluation* (AASHTO, Farmington Hills, MI, 2018).
34. MoT (Ministry of Transport of the People's Republic of China), *JTG 023-1985: Code for Design of Highway Reinforced Concrete and Prestressed Concrete Bridges and Culvert* (China Communications Press, Beijing, 1985).
35. MoT (Ministry of Transport of the People's Republic of China), Highway old bridge load bearing capacity identification methods (for trial implementation) (1988).
36. MoT (Ministry of Transport of the People's Republic of China), *JTG D62-2004: Code for Design of Highway Reinforced Concrete and Prestressed Concrete Bridges and Culverts* (China Communications Press, Beijing, 2004).
37. MoT (Ministry of Transport of the People's Republic of China), *JTG/T-J21-2011 Specifications for Inspection and Evaluation of Load Bearing Capacity of Highway Bridges* (China Communications Press, Beijing, 2011).
38. MoT (Ministry of Transport of the People's Republic of China), *GTJ 3362-2018 Specifications for Design of Highway Reinforced Concrete and Prestressed Concrete Bridges and Culverts* (China Communications Press, Beijing, 2018).

https://doi.org/10.1142/9789819814770_0025

Chapter 25

Seismic Response Control of a Nonlinear Tall Building Under Mainshock–Aftershock Sequences Using Semi-Active Tuned Mass Damper#

Liangkun Wang*, Ying Zhou† and Weixing Shi‡

Department of Disaster Mitigation for Structures, Tongji University, Shanghai 200092, P. R. China

National Key Laboratory of Disaster Reduction in Civil Engineering Tongji University, Shanghai 200092, P. R. China

*wangliangkun@tongji.edu.cn

†yingzhou@tongji.edu.cn

‡swxtgk@tongji.edu.cn

Under the excitation of strong mainshock (MS), the structure will enter the nonlinear stage, whose hysteretic characteristic can be well-simulated by the Bouc–Wen (BW) model. The accompanying aftershock (AS) may increase structural damage and cause the safety problem, therefore, the advanced structural control technology is worthy of being applied to nonlinear structure under MS–AS sequences. The tuned mass damper (TMD) is a popular vibrational absorber, however, it is sensitive to the frequency deviation and may lose its effect for nonlinear dynamic response control in this issue. To protect the nonlinear tall building under MS–AS sequences effectively, a semi-active TMD (STMD) is investigated in this study, which can vary its frequency and damping in real time simultaneously. It can adapt to structural nonlinear behavior in each time segment through varying its stiffness according to the wavelet transform (WT) based algorithm, and meanwhile, enhance the energy dissipation through switching its damping coefficient in real time. A 10-story nonlinear tall building is proposed as a case study and for comparison, a passive TMD (PTMD) is used. Nine MS–AS sequences are elected and nonlinear structural control performance of STMD is highlighted. Results show that STMD can control structural displacement responses and base shear effectively and performs better than PTMD in reducing the plastic development of the structure as well. Further, energy analysis is proposed and it is verified that STMD has an excellent effect in decreasing the structural input energy from MS–AS sequences and dissipating more energy than PTMD. Especially, it is found that a strong AS may cause a larger response

†Corresponding author.

#This chapter appeared previously on the International Journal of Structural Stability and Dynamics. To cite this chapter, please cite the original article as the following: L. Wang, Y. Zhou & W. Shi, *Int. J. Struct. Stab. Dyn.*, **23**, 2340027 (2023), DOI: 10.1142/S0219455423400278.

to the nonlinear structure than MS, while the nonlinear control effect of STMD is still significant in that case.

Keywords: Earthquake mitigation; tuned mass damper; mainshock–aftershock sequences; semi-active control; nonlinear dynamics; energy analysis.

1. Introduction

To protect the structure from earthquake excitation and improve its safety, structural control devices, such as base-isolators[1] and additional energy dissipation dampers,[2–5] have been applied widely in real projects. The tuned mass damper (TMD) is a popular vibrational absorber, which includes a mass, springs and dashpots.[6,7] Due to the advantages of small size, easy installation and little influence on the use function of buildings, TMD has been widely used in earthquake attenuation.[8–11] Brandão and Almeida[12] optimized single and multiple TMDs for structural earthquake protections. Etedali *et al.*[13] proposed the optimal design of TMDs based on the artificial intelligence prediction model. Wang *et al.*[14] controlled the seismic response considering soil–structure interaction using an adaptive TMD. Gholizad and Mohammadi[15] presented a reliable design of TMD through the Monte Carlo simulation. Wang *et al.*[16] proposed a novel magnetic multi-mass damper for multimodal response control, while Rathi and Chakraborty[17] optimized TMDs considering structural uncertainty and stochastic earthquake excitation.

It can be known from the above literature review that a well-tuned passive TMD (PTMD) has a good earthquake mitigation performance. It is because for a linear structure, the resonance will dominate the structural response, while the frequency band of resonance is narrow.[18] The optimized PTMD can control the target resonant frequency effectively.[19] However, a PTMD may lose its control effect when mistuned. Further, under the excitation of a strong earthquake, the structure will enter the nonlinear stage, it becomes more challenging for nonlinear structural control. Domizio *et al.*[20] optimized single and multiple TMDs for nonlinear structures. Wang and Wu[21] investigated the earthquake mitigation performance of TMD for a structure with nonlinear elastic foundation. Quaranta *et al.*[22] controlled a nonlinear structure with linear TMD under pulse-like ground motion. Espinoza *et al.*[23] used a TMD to control torsional responses of nonlinear asymmetrical structures. Mohebbi and Joghataie[24] optimized TMDs for nonlinear structural control based on the distributed genetic algorithm. Apart from linear TMDs,[25–27] the earthquake mitigation effect of hysteretic TMD for nonlinear structure was proposed by Boccamazzo *et al.*,[28] while TMDs for nonlinear base isolation buildings was investigated by Hessabi *et al.*[29]

Comparing the two paragraphs above, the attenuation performance of a PTMD for nonlinear structure is not so significant as linear structure. That is because a nonlinear structure does not have a fixed natural frequency and from another perspective, a PTMD can be regarded as always being in a mistuned stage. Therefore, a semi-active TMD (STMD), which can retune its parameters in real time,[30–32] is worth studying in this issue. Sun and Nagarajaiah[33] proposed an STMD with variable damping and stiffness, while the structural stiffness was changed linearly with time. Woo *et al.*[34] controlled linear and nonlinear buildings using STMD. Wang *et al.* proposed an STMD with variable frequency and damping for a nonlinear system[35] and a base-isolated structure with lead rubber bearing,[36] whose hysteretic characteristics were simulated by the Bouc–Wen (BW) model. Karami *et al.*[37] investigated the nonlinear control effect of STMD for a structure based on the bilinear elastoplastic model. A novel STMD was proposed by Sun *et al.*,[38] and it was applied to control the offshore wind turbine under multi-hazards,[39] mitigate its fatigue damage under real wind and wave conditions,[40] while bi-directional vibration could be found in Refs. 41 and 42.

From the reviewed researches, it can be known that STMD has a better nonlinear structural control effect than PTMD generally. However, in addition to the strong mainshock (MS), the accompanying aftershock (AS) may increase structural damage and cause the safety problem.[43–45] Therefore, structural dynamics and nonlinear behaviors under MS–AS sequences should be considered carefully.[46–48] Poiani *et al.*[49] assessed the AS fragility of reinforced concrete industrial structures. Silwal and Ozbulut[50] proposed an AS fragility assessment for steel structures with self-centering dampers. Ghotbi and Taciroglu[51] presented a conditioning criterion for selecting hazard-consistent AS records. Trapani and Malavisi[52] proposed a seismic fragility assessment under MS–AS sequences using an improved incremental dynamic analysis method. Besides, Trevlopoulos *et al.*[53] presented the assessment of seismic risk through seismic period extension and operational earthquake prediction.

Considering the complexity and randomness of MS–AS sequences, the advanced structural control technology is worthy of being applied to nonlinear structure under MS–AS sequences. However, the research on the implementation of a STMD for a nonlinear structure against MS–AS sequences is absent. Further, the control algorithm will be more meaningful if only based on out-put signals which can be measured directly, as the nonlinear structure and MS–AS sequences are filled with uncertainties. Moreover, because the TMD is a frequency-dependent vibrational absorber, the STMD will be worthwhile to adapt to structural nonlinear behavior in each time segment, and can enhance the energy dissipation as well.

To fill this blank, the seismic response of a nonlinear tall building under MS–AS sequences is controlled by a STMD in this study. The STMD with variable pendulum and eddy current damping, together with its combined variable stiffness and damping strategy, is introduced in Sec. 2, respectively. A nonlinear 10-story tall building based on the BW model is proposed as a case study in Sec. 3, while dynamic equations are presented as well. For comparison, a PTMD optimized to the linear building is used. In Sec. 4, numerical results are shown and discussed in detail, while energy analysis is proposed further. At last, conclusions are summarized in Sec. 5.

2. Electromechanical System and Control Algorithm

2.1. *Semi-active TMD with variable pendulum and damping*

For a better nonlinear control effect, a semi-active TMD (STMD) with variable frequency and damping is proposed in this section, whose electromechanical system is illustrated in Fig. 1.

In Fig. 1, a scaled experimental model of the adaptive TMD with variable pendulum length and eddy current damping is shown and the detailed description can be referred to in Ref. 6. For sake of completeness, it is reviewed briefly here.

Pendulum-type TMD is attractive in real applications of tall buildings, whose natural frequency is only dependent on its pendulum length and is usually implemented on the top story to control the first mode.[54] Therefore, to retune its frequency, it can be seen in Fig. 1 that the pendulum is wrapped around the stepper motor. It will adjust the pendulum length under the guidance of a microcontroller,

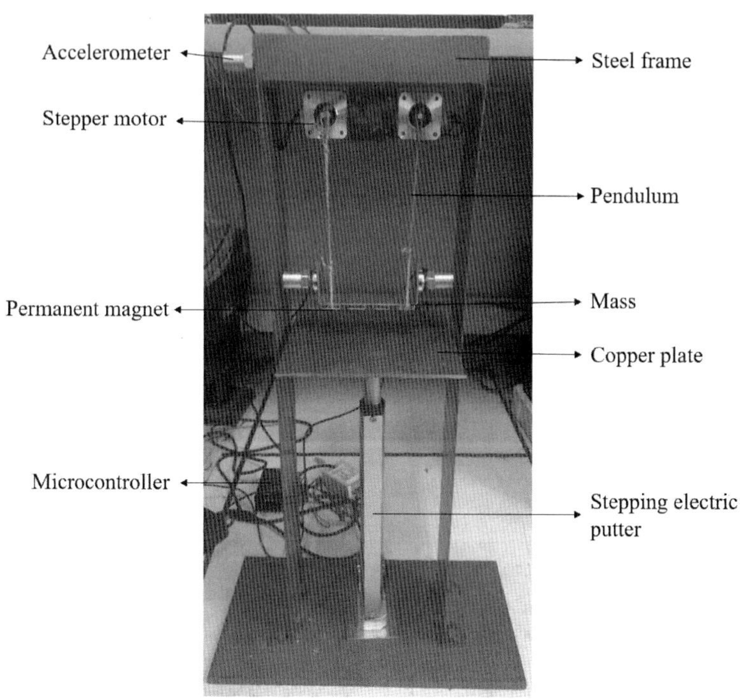

Fig. 1. STMD with variable pendulum length and eddy current damping.[6]

which is the equivalent stiffness of STMD. The scale model in Fig. 1 uses the synchronous rotation of two stepper motors to retune the frequency of STMD.[55] In practical engineering applications, the number of stepper motors and the suspension form can be determined by considering the mass, electric power of the stepper motor and the requirements on the response time of the electromechanical system.

Meanwhile, the STMD in Fig. 1 can change the eddy current damping coefficient in real time as well. Through the relative displacement between the conductor plate (which is copper plate in Fig. 1) and the mass with permanent magnets adsorbed, an eddy current damping force is generated. It was found in Ref. 35 that the eddy current damping was sensitive to the air gap between magnets and conductor plate, while they had a monotonically decreasing relationship. Therefore, a stepping electric putter is presented in Fig. 1 to adjust the air gap and hence damping coefficient. It should be noted that the vertical motion of stepping electric putter should consider the variation of pendulum length as well because for a constant position of conductor plate, the variation of position of mass can also change the air gap and hence eddy current damping. From the above working mechanism of stepper motor and stepping putter, it can be known that the introduced STMD can retune its frequency and damping easily to achieve a better structural control effect under MS–AS sequences.

2.2. *Combined variable frequency and damping control strategy*

Considering the complexity and randomness of MS–AS sequences and uncertainties of the nonlinear structure, it is difficult to know structural modal information exactly in each time segment. Therefore, the control algorithm of STMD will be more meaningful if only based on output signals which can be measured directly to obtain a better robustness. Moreover, as the TMD is a frequency-dependent vibrational absorber, it will be better to adapt to structural nonlinear behavior in real time, and can enhance the energy dissipation as well.

The well-known wavelet transform (WT) is based on the idea of the localization of short time Fourier transform (STFT), designed through an adaptive "time-frequency" window that changes in width with frequency.[18] To overcome the shortcoming of fixed window width of STFT, through WT, the time-frequency distribution of the signal can be focused on analysis, and the vibrational signal can be multi-scale refined through the telescopic translation window to achieve the effect of time subdivision at high frequency and frequency subdivision at low frequency, achieves a better role in capturing and identifying vibrational signals of nonlinear structures.

The displacement signal of the top story is tracked and instantaneous frequency f_N is obtained through WT. Considering that the TMD is focused on the structural first mode and to eliminate noise and other interference in the signal, a preset minimum adjustable frequency $f_{\min}$ and the corresponding maximum $f_{\max}$ are set in the microcontroller first. If the identified frequency f_N is in the adjustable range, the frequency of STMD f_d will be retuned to it; otherwise, it will tuned to the corresponding minimum $f_{\min}$ or maximum $f_{\max}$. Then, the pendulum length L_d will be adjusted according to Eq. (1). Meanwhile, the equivalent stiffness of STMD $k_d(t)$ is retuned as well.

$$L_d = \frac{g}{4\pi^2 f_d^2}. \tag{1}$$

To enhance the energy dissipation of STMD further, a damping switching strategy based on velocity and acceleration signals of the nonlinear structure and STMD is proposed in Eq. (2). The main motivation of this switching algorithm is that the phase difference between them is judged by their velocity difference and the acceleration of the main structure, so that the damping coefficient can be changed to better transfer and absorb the kinetic energy of the main structure and dissipate.

$$\begin{cases} (\dot{x}_d - \dot{x}_p)\ddot{x}_p \leq 0 \geq c_d(t) = 0, \\ (\dot{x}_d - \dot{x}_p)\ddot{x}_p > 0 \geq c_d(t) = c_{\max}, \end{cases} \tag{2}$$

where $\dot{x}_d$ and $\dot{x}_p$ are velocities of STMD and the top floor of the structure, respectively, $\ddot{x}_p$ means the acceleration of the top floor, while $c_d(t)$ and $c_{\max}$ represent semi-active damping coefficient of STMD in real time t and the preset maximum damping coefficient can be adjusted, respectively. It should be noted that after the trial, $c_{\max} = 2c_{\text{opt}}$ in this paper, where c_{opt} is the optimal TMD damping coefficient and will be introduced in Sec. 3.2.

The combined variable stiffness and damping strategy of STMD for nonlinear control is illustrated in Fig. 2. The presented control strategy is mostly based on output signals, very little modal information and no input information are needed, which is extremely suitable for the considered nonlinear structural control under MS–AS sequences.

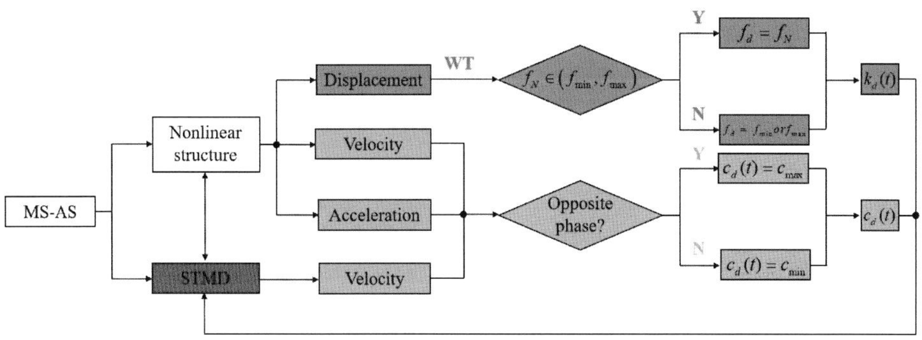

Fig. 2. Combined variable stiffness and damping strategy of STMD for nonlinear control.

3. Dynamic Equations and Case Study

3.1. *Dynamic equations*

Under MS–AS sequences, the nonlinear tall building equipped with an STMD on the top story is shown in Fig. 3.

In Fig. 3, m_N, k_{bwN} and c_N are the mass, nonlinear stiffness based on the BW model and damping coefficient of the Nth floor of the nonlinear tall building, respectively; m_d means the mass of STMD; x_N and x_d represent relative displacements of the Nth floor and STMD, respectively; $\ddot{x}_g$ is the MS–AS sequence.

Therefore, dynamic equations of the nonlinear structure with STMD are presented in Eqs. (3) and (4), respectively.

$$\mathbf{M}\ddot{\mathbf{X}} + \mathbf{C}\dot{\mathbf{X}} + \mathbf{K_e}\mathbf{X} + \mathbf{K_h}\mathbf{v} = -\mathbf{M}^*\ddot{\mathbf{x}}_\mathbf{g}, \tag{3}$$

$$\dot{\mathbf{v}} = (\mathbf{A}\dot{\mathbf{X}} - \beta|\dot{\mathbf{X}}||\mathbf{v}|^{\mu-1}\mathbf{v} - \gamma\dot{\mathbf{X}}|\mathbf{v}|^{\mu})/x_y, \tag{4}$$

where $\mathbf{M}$, $\mathbf{C}$, $\mathbf{K_e}$ and $\mathbf{K_h}$ are structural mass, damping, elastic stiffness and hysteresis stiffness matrices, respectively; $\mathbf{X}$ means the relative displacement vector between floors and $\mathbf{M}^*$ is the diagonal mass matrix; $\mathbf{v}$ is the hysteresis displacement vector of the nonlinear structure, x_y means the yield displacement of each

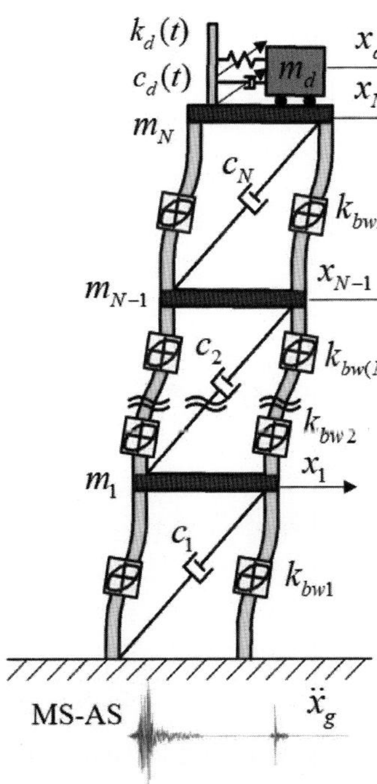

Fig. 3. Nonlinear tall building with STMD under MS–AS sequences.

floor; $\mathbf{A}$, $\boldsymbol{\beta}$, $\boldsymbol{\gamma}$ and $\boldsymbol{\mu}$ represent coefficient matrices for determining the shape of the restoring force curve in the BW model. All these matrices are proposed in Eqs. (5)–(9), respectively.

$$\mathbf{M} = \begin{pmatrix} m_1 & & & & \\ m_2 & m_2 & & & \\ \vdots & \vdots & \ddots & & \\ m_N & m_N & \cdots & m_N & \\ m_d & m_d & \cdots & m_d & m_d \end{pmatrix}, \tag{5}$$

$$\mathbf{C} = \begin{pmatrix} c_1 & -c_2 & & & \\ & c_2 & -c_3 & & \\ & & \ddots & \ddots & \\ & & & c_N & -c_d(t) \\ & & & & c_d(t) \end{pmatrix}, \tag{6}$$

$$\mathbf{K_e} = \begin{pmatrix} \alpha_1 k_1 & -\alpha_2 k_2 & & & \\ & \alpha_2 k_2 & -\alpha_3 k_3 & & \\ & & \ddots & \ddots & \\ & & & \alpha_N k_N & -k_d(t) \\ & & & & k_d(t) \end{pmatrix}, \tag{7}$$

$$\mathbf{K_h} = \begin{pmatrix} (1-\alpha_1)k_1 & -(1-\alpha_2)k_2 & & & \\ & (1-\alpha_2)k_2 & -(1-\alpha_3)k_3 & & \\ & & \ddots & \ddots & \\ & & & (1-\alpha_N)k_N & 0 \\ & & & & 0 \end{pmatrix}, \tag{8}$$

$$\mathbf{M}^* = \mathrm{diag}\begin{pmatrix} m_1 & m_2 & \cdots & m_N & m_d \end{pmatrix}, \tag{9}$$

where α_N is the ratio of the linear restoring force at the Nth floor of the structure to the total restoring force while k_N means its linear stiffness before yielding.

Through the proposed equations, dynamic responses of the nonlinear tall building with a STMD can be calculated by Newmark-β method.

3.2. *Case study*

To verify the nonlinear structural control effect of the presented STMD under MS–AS sequences, a 10-story nonlinear tall building is proposed as a case study in this section.

The mass of the building from the first floor to top is 2.2, 2.0, 1.8, 1.8, 1.8, 1.8, 1.8, 1.8, 1.8 and 1.6 × 10^5 kg, respectively. The inter-story stiffness before yielding is 3.5, 3.2, 3.0, 2.8, 2.8, 2.8, 2.8, 2.8, 2.8 and 2.6 × 10^5 kN/m, respectively. It is assumed that the damping ratio ζ_p for all the modes is 2%. The considered model is a typical shear-type building, then, structural mass and stiffness matrixes can be easily obtained. Rayleigh damping is used to propose the damping matrix, which is shown in the equation of motion. Then, the TMD is added to the equation as well. In the BW model, $\mathbf{A}$, $\boldsymbol{\beta}$, $\boldsymbol{\gamma}$ and $\boldsymbol{\mu}$ are column vectors with values of 1, 0.25, 3 and 1, respectively. α_N is taken as 0.04[20,34,37] and the yield displacement of each floor is 7.3 mm. According to the modal analysis, the first-order frequency of the structure before yielding is 0.98 Hz, whose modal mass participation coefficient is 0.82. As a reference, for the linear structure whose layer stiffness is 0.04 times that of the original structure, the first-order frequency is only 0.20 Hz with the same modal mass participation coefficient.

A PTMD optimized to the pre-yield structure is presented for comparison. It is tuned to the first mode (0.98 Hz) with a modal mass ratio μ of 5%. According to classical TMD optimization formulas in Refs. 56–58, its optimal frequency ratio f_{opt} and damping ratio ζ_{opt} can be obtained through Eqs. (10) and (11).

$$f_{\text{opt}} = \frac{1}{1+\mu}\left(1 - \zeta_p\sqrt{\frac{\mu}{1+\mu}}\right), \tag{10}$$

$$\zeta_{\text{opt}} = \frac{\zeta_p}{1+\mu} + \sqrt{\frac{\mu}{1+\mu}}. \tag{11}$$

As a result, f_{opt} and ζ_{opt} in Eqs. (10) and (11) are 94.82% and 23.73%, respectively, while the frequency of PTMD is 0.93 Hz.

The STMD has the same mass as PTMD. In Fig. 2, to control the structural elastoplastic dynamic response, STMD is expected to cover the frequency range from the pre-yield structure to the post-yield structure. The upper limit of the adjustable frequency is 1.1 times the first-order frequency of the pre-yield structure (0.98 Hz), that is, 1.08 Hz; the lower limit is 0.9 times the first-order frequency of the referenced post-yield structure (0.20 Hz), that is, 0.18 Hz. The STMD can retune itself between (0.18 and 1.08 Hz) to achieve a better nonlinear structural control effect. The lower limit of the adjustable damping coefficient of STMD is 0, while the upper limit is twice the optimized damping coefficient of PTMD.

In the numerical simulation, nine MS–AS sequences with different spectral properties are especially chosen and listed in Table 1, which are downloaded from Ref. 59.

Table 1. Selected MS–AS sequences.[59]

Sequence	Earthquake type	Earthquake name	Year	Station name	Component
1	MS	Coalinga-01	1983	Pleasant Valley P.P.—yard	RSN368_COALINGA.H_H-PVY135
	AS	Coalinga-02			RSN383_COALINGA_A-PVY135
2	MS	Imperial Valley-06	1979	El Centro Array #11	RSN174_IMPVALL.H_H-E11230
	AS	Imperial Valley-07			RSN199_IMPVALL.A_A-E11230
3	MS	Livermore-01	1980	San Ramon—Eastman Kodak	RSN214_LIVERMOR_A-KOD270
	AS	Livermore-02			RSN223_LIVERMOR_B-KOD270
4	MS	Livermore-01	1980	San Ramon Fire Station	RSN215_LIVERMOR_A-SRM340
	AS	Livermore-02			RSN224_LIVERMOR_B-SRM340
5	MS	Northridge-01	1994	Moorpark—Fire Sta	RSN1039_NORTHR_MRP180
	AS	Northridge-04			RSN1681_NORTH142_MRP180
6	MS	Northridge-01	1994	Los Angeles—7-story Univ Hospital (FF)	RSN1007_NORTHR_UNI095
	AS	Northridge-06			RSN1713_NORTH392_UNI180
7	MS	Whittier Narrows-01	1987	Mt Wilson—CIT Seis Sta	RSN663_WHITTIER.A_A-MTW090
	AS	Whittier Narrows-02			RSN715_WHITTIER.B_B-MTW090
8	MS	Irpinia_Italy-01	1980	Calitri	RSN289_ITALY_A-CTR270
	AS	Irpinia_Italy-02			RSN300_ITALY_B-CTR270
9	MS	Chi-Chi_Taiwan	1999	TCU067	RSN1504_CHICHI_TCU067-N
	AS	Chi-Chi_Taiwan-02			RSN2383_CHICHI.02_TCU067E

It should be noted that to make the building completely consume the vibrational energy of MS and investigate its dynamic response under AS, a time interval of 30 s is set between them. Meanwhile, an additional 30 s is added after AS to allow it to freely attenuate vibration. Two typical MS–AS sequences are shown in Fig. 4, which are MS–AS8 and MS–AS9 in Table 1, respectively. It can be found that as for MS–AS8, the peak ground amplitude (PGA) of AS is even larger than MS, while for MS–AS9, the AS is much smaller than MS.

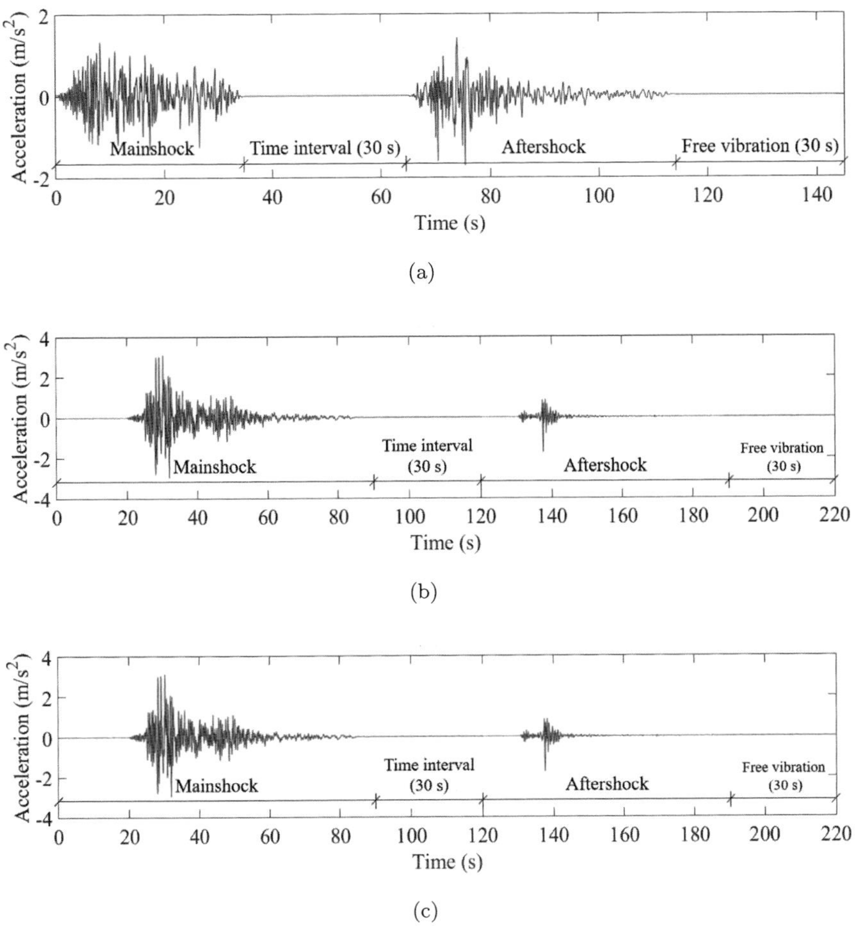

Fig. 4. Two typical MS–AS sequences. (a) MS–AS8 and (b) MS–AS9.

4. Numerical Results and Analysis

4.1. *Dynamic response comparison*

Under nine MS–AS sequences in Table 1, dynamic responses of the structure without TMD, with a PTMD and with a STMD are compared in this section, while maximum and root mean square (RMS) displacements of the top story and maximum inter-story displacements under MS and AS, respectively, are proposed in Tables 2–4 and Fig. 5, respectively.

As can be seen from the numerical simulation results in Tables 2–4 and Fig. 5, the dynamic response of the building varies greatly under the excitation of different MS–AS sequences. Except for the MS–AS8, the dynamic response of the structure under the MS is obviously greater than that of AS. The MS–AS8 also gives us a

Table 2. Comparisons on maximum displacements of the top story under mainshock and aftershock, respectively (cm).

Sequence	Mainshock			Aftershock		
	Without TMD	PTMD	STMD	Without TMD	PTMD	STMD
MS–AS1	12.4468	11.5404	11.1702	2.0201	2.2868	2.2449
MS–AS2	10.8334	9.6568	8.9861	3.8420	4.3858	4.0971
MS–AS3	2.5669	2.2217	2.1139	2.5508	1.8414	1.7663
MS–AS4	2.9218	1.9758	1.5881	1.5567	1.2943	1.4289
MS–AS5	5.4739	4.5355	4.2023	3.2797	3.6895	3.6841
MS–AS6	2.3162	2.1705	2.1343	0.2369	0.4714	0.2833
MS–AS7	0.8624	0.8357	0.8244	0.5727	0.5462	0.5445
MS–AS8	7.4040	7.2770	6.8489	9.6662	8.5290	8.7459
MS–AS9	19.8734	19.5026	18.2198	5.4107	5.5133	4.7102

Table 3. Comparisons on RMS displacements of the top story under mainshock and aftershock, respectively (cm).

Sequence	Mainshock			Aftershock		
	Without TMD	PTMD	STMD	Without TMD	PTMD	STMD
MS-AS1	1.8324	1.8055	1.6817	0.6836	0.9842	0.9321
MS-AS2	2.3762	2.5577	2.3324	1.1414	1.7425	1.5447
MS-AS3	0.6399	0.4941	0.4868	0.5771	0.3934	0.3767
MS-AS4	0.6588	0.4772	0.4391	0.4138	0.2677	0.2573
MS–AS5	1.0778	0.9290	0.8395	0.7642	0.6876	0.7046
MS-AS6	0.5278	0.4640	0.4141	0.0626	0.2649	0.0918
MS-AS7	0.1357	0.0940	0.0932	0.1849	0.1038	0.1016
MS-AS8	1.9199	1.8122	1.6839	1.7229	1.5207	1.4079
MS-AS9	3.5280	3.5004	3.1525	1.4891	2.0541	1.5727

Table 4. Comparisons on structural maximum inter-story displacements under mainshock and aftershock, respectively (mm).

Sequence	Mainshock			Aftershock		
	Without TMD	PTMD	STMD	Without TMD	PTMD	STMD
MS-AS1	26.0780	23.5304	23.4485	3.4834	4.1830	4.3670
MS-AS2	21.9968	19.2251	17.9125	6.4345	6.9431	7.0155
MS-AS3	4.3486	3.8167	3.7785	3.8978	2.6414	2.9006
MS-AS4	4.5752	3.1229	2.3678	2.3395	1.7260	2.1040
MS-AS5	8.8089	9.0545	8.1241	6.2393	6.9794	7.1861
MS-AS6	4.6735	4.2409	4.1767	0.9698	1.1536	0.9056
MS-AS7	2.2308	2.0875	2.0678	1.7751	1.6603	1.6491
MS-AS8	13.6658	12.0243	11.3986	20.6602	16.8485	16.4906
MS-AS9	49.6988	45.7392	43.2938	9.7737	9.8276	8.3547

warning that the damage caused by AS may exceed that of the MS and should be paid close attention to. STMD has a better earthquake reduction effect than PTMD, both in controlling displacement responses of the top floor and the maximum inter-story displacement. Under the excitation of MS9, the structural dynamic response is the greatest and the structure enters the nonlinear stage in the most degree. PTMD cannot change the parameter, thus, its damping effect is relatively weak; while STMD can control displacement responses of the nonlinear building to a great degree and better guarantee its safety. Under the input of AS8, the displacement response of the structure is the largest under all the AS excitations, while STMD also has an excellent control performance, especially on the RMS displacement of the top floor, that is, the fastest vibration attenuation.

For some specific sequences such as MS–AS1, MS–AS2, MS–AS5, and MS–AS6, the dynamic response of the structure under the AS is significantly smaller than that of MS. PTMD and STMD both have a positive control effect for the MS, while having a little negative effect for the AS. Actually, it is also in line with expectations. For example, for linear structures, when the excitation deviates from the resonance interval, the response of the structure will decrease significantly, meanwhile, TMD has almost no control effect or even a slight negative effect. Besides, even for these specific aftershocks, STMD still performs better than PTMD.

To comprehensively evaluate the damping effects of PTMD and STMD on the whole structure under the excitation of the whole MS–AS sequence, the maximum and RMS base shears are taken as indicators for comparison, while numerical results are shown in Table 5 and Fig. 6, respectively. The mathematical expression of base shear F_s can be shown as

$$F_s = \sum_{i=1}^{N} \alpha_i k_i x_{yi} + (1 - \alpha_i) k_i x_{yi} z_i(t), \tag{12}$$

where α_i is the ratio of the linear restoring force at the ith floor to the total restoring force, k_i means the inter-story stiffness of the ith floor, x_{yi} represents the yield

Table 5. Comparisons on maximum and RMS base shears of different control cases ($\times 10^5$ N).

Sequence	Maximum			RMS		
	Without TMD	PTMD	STMD	Without TMD	PTMD	STMD
MS-AS1	24.7493	24.2656	24.3615	3.7316	3.2275	3.1794
MS-AS2	23.9232	23.1638	22.9058	3.7684	3.4125	3.2509
MS-AS3	11.2685	9.8595	9.4781	2.5825	1.7574	1.6276
MS-AS4	11.3199	8.1863	6.6102	2.3103	1.6436	1.5175
MS-AS5	20.1869	18.9322	17.5660	3.9408	3.3099	3.1010
MS-AS6	11.9887	11.4666	11.6922	1.7175	1.2916	1.3316
MS-AS7	6.9244	6.2590	6.1386	0.8182	0.5624	0.5536
MS-AS8	23.1434	22.9153	23.7828	5.5019	4.9783	4.8491
MS-AS9	30.4387	29.9745	30.2327	3.8180	3.6287	3.6059

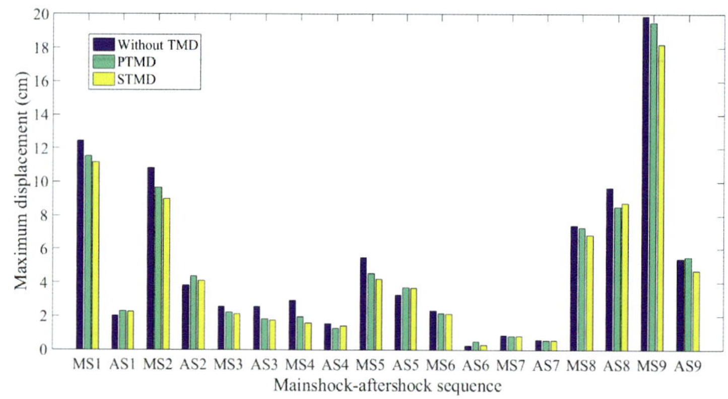

(a)

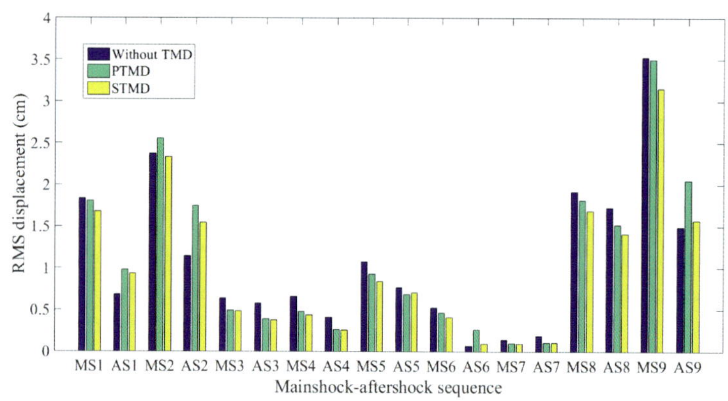

(b)

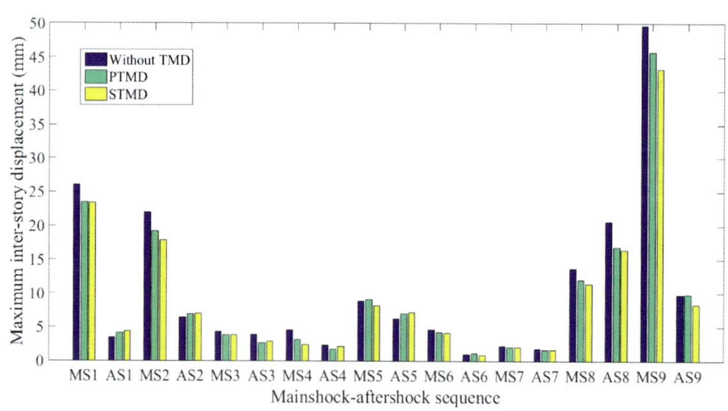

(c)

Fig. 5. Nonlinear dynamic response comparison. (a) Maximum displacement of the top story; (b) RMS displacement of the top story; and (c) Maximum inter-story drift.

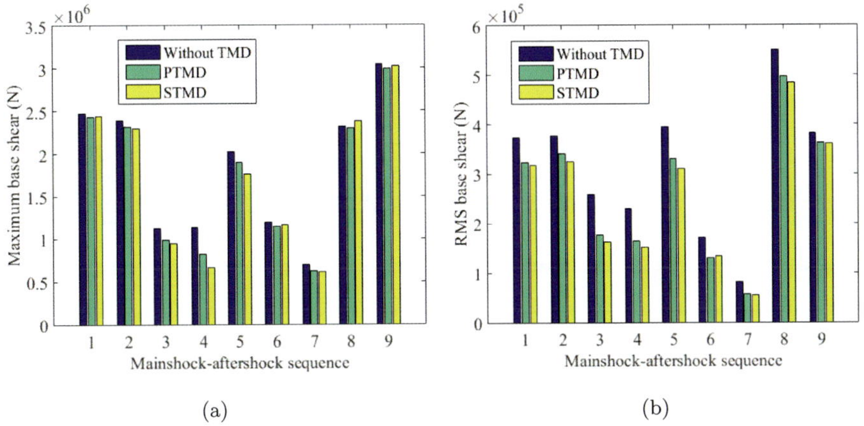

Fig. 6. Comparisons on base shears of different control cases. (a) Maximum and (b) RMS.

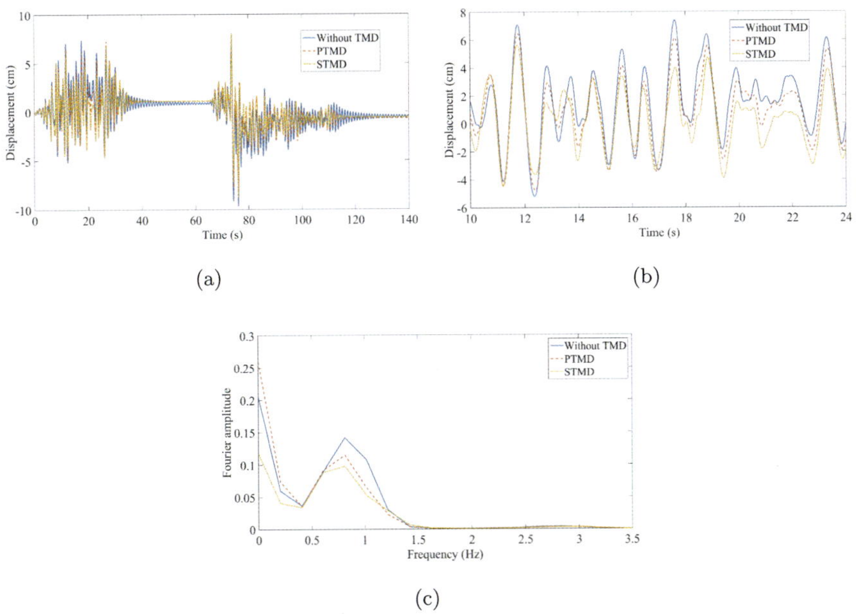

Fig. 7. Displacement response comparison of the top story under MS–AS8. (a) Time history; (b) zoom on time history; and (c) Fourier spectrum.

displacement of the ith floor and $z_i(t)$ is the stress proportional displacement of the ith floor.

In Table 5 and Fig. 6, structural maximum base shear is the maximum under MS–AS9, mainly because of MS9; it has the maximum RMS base shear under

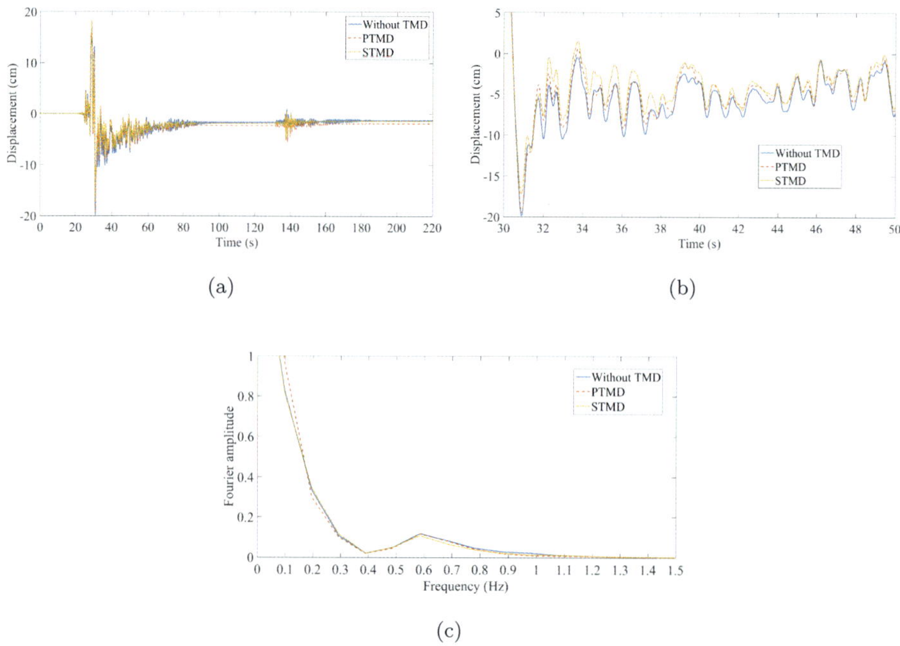

Fig. 8. Displacement response comparison of the top story under MS–AS9. (a) Time history; (b) zoom on time history; and (c) Fourier spectrum.

MS–AS8, mainly because of AS8. Besides, it can be seen that STMD can control base shear effectively as well, the reduction in RMS base shear is more obvious than the maximum and it performs better than PTMD in both maximum and RMS base shears.

4.2. *Time history comparison*

From the previous section, it can be known that the nonlinear tall building has the largest dynamic response under MS–AS8 and MS–AS9 as shown in Fig. 4. To better investigate the nonlinear control effect of STMD in the time domain, these two sequences are chosen as example in this section and displacement response comparisons of the top story are presented in Figs. 7 and 8, respectively.

As can be seen from Fig. 7, the structure has entered the nonlinear stage under the excitation of the MS–AS8, and the dynamic response of the AS section is significantly greater than that of the MS section. It is shown in Fig. 7(a) that STMD has a satisfactory elastoplastic damping effect throughout the whole period. From Fig. 7(b), it can be known that STMD has a satisfactory control effect for the nonlinear structure. Especially, in the range of 20–22 s, the vibration of the structure without TMD cannot return to the equilibrium position, while STMD can control it along the equilibrium position, which shows the effect in controlling the development of structural plasticity. As can be found from Fig. 7(c), STMD has the best control effect in the frequency band of 0.5–1.5 Hz, and its Fourier amplitude of the signal at the frequency of 0 is the smallest, which indicates that the degree of structural plasticity can be most effectively reduced.

As can be seen from Fig. 8, under the excitation of MS–AS9, the dynamic response of the structure in the MS section is very large and the plastic development degree of the structure is the largest, while the peak displacement is more than twice of that under the input of AS8. At the same time, the displacement response of the structure decreases significantly under the excitation of AS9, that is, the dynamic response of the structure is dominated by the MS section under the sequence of MS–AS9. Although STMD can still control the displacement response of the structure most effectively, the reduction is not as obvious as that of the MS–AS8. Therefore, it can be known that STMD is mainly suitable for controlling the nonlinear dynamic response when the plasticity of the structure is small, generally, its damping effect deteriorates with the deepening of the plasticity development degree.

To compare dynamic responses of the whole building, dynamic response comparisons of each story under MS–AS8 and MS–AS9 are presented in Figs. 9 and 10, respectively, where maximum and RMS displacements and maximum inter-story displacement can all be found.

From Figs. 9 and 10, overall, STMD has an excellent damping effect on the whole building. Under the excitation of MS8, the performance of STMD is better than that of PTMD in terms of maximum and RMS displacements. Under the excitation of AS8, STMD is similar to PTMD in controlling the maximum displacement, while it has a better attenuation in controlling the RMS response. Under the input of MS9, STMD has a better control effect than PTMD in both MS and AS sections. In the AS section, PTMD, which cannot adjust its parameters, has an obvious negative effect of amplification vibration on the RMS displacement due to the serious mistuning, while STMD can significantly improve this defect because it can adjust its own frequency and damping in real time. In terms of the maximum inter-story displacement, STMD has the best control performance under the excitation of the two MS–AS sequences. At the same time, under the input of different sequences, either in the MS section or AS section, the maximum inter-story displacement of the structure is different on different floors. The improvement of STMD in controlling the maximum inter-story displacement is beneficial to better control the development of structural plasticity and improve its safety.

To highlight the improvement of STMD in controlling the maximum inter-story displacement, structural hysteretic curve comparisons under MS–AS8 and MS–AS9 are illustrated in Figs. 11 and 12, respectively. It should be noted that according to Figs. 9(c) and 10(c), the floor with the maximum inter-story displacement is considered.

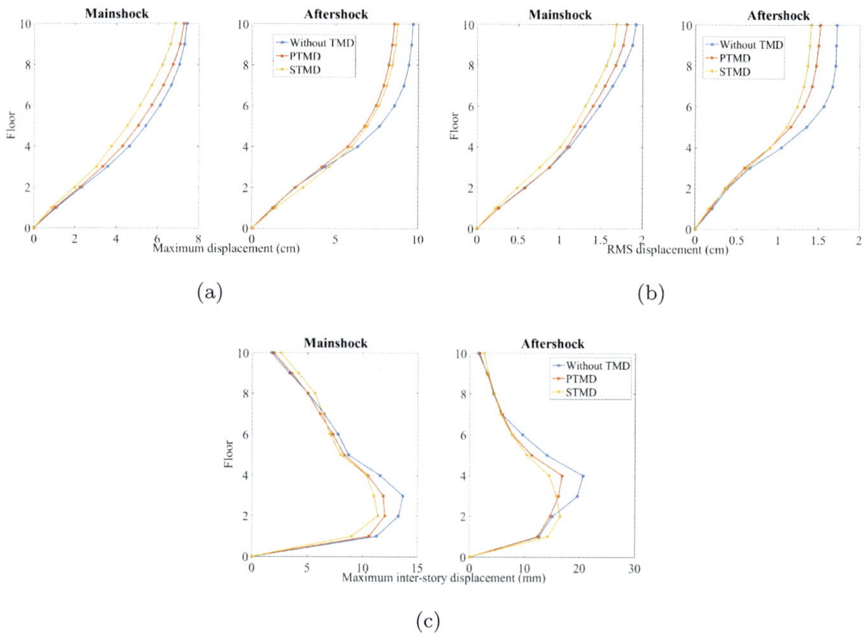

Fig. 9. Dynamic response comparison of each story under MS–AS8. (a) Maximum displacement; (b) RMS displacement; and (c) maximum inter-story displacement.

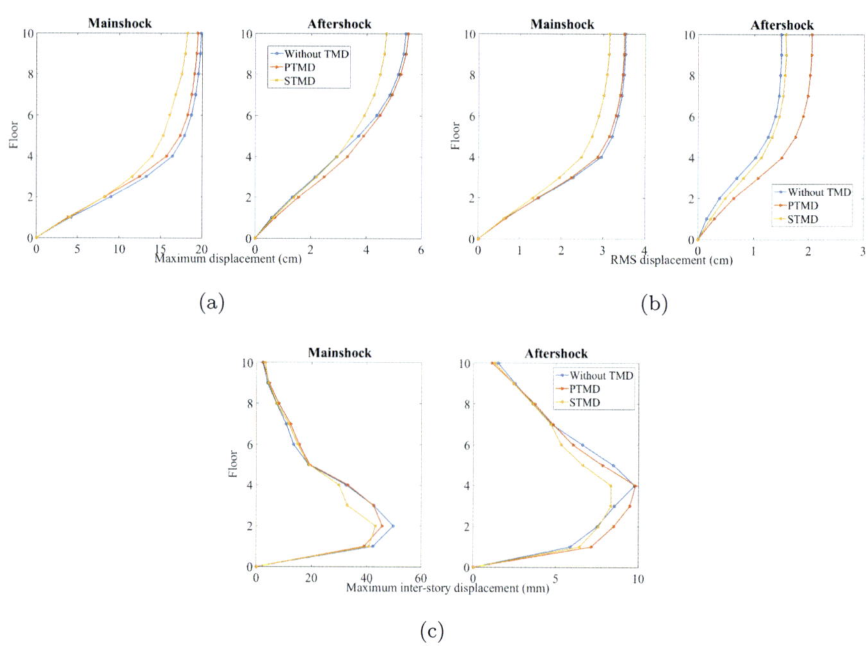

Fig. 10. Dynamic response comparison of each story under MS–AS9. (a) Maximum displacement; (b) RMS displacement and (c) maximum inter-story displacement.

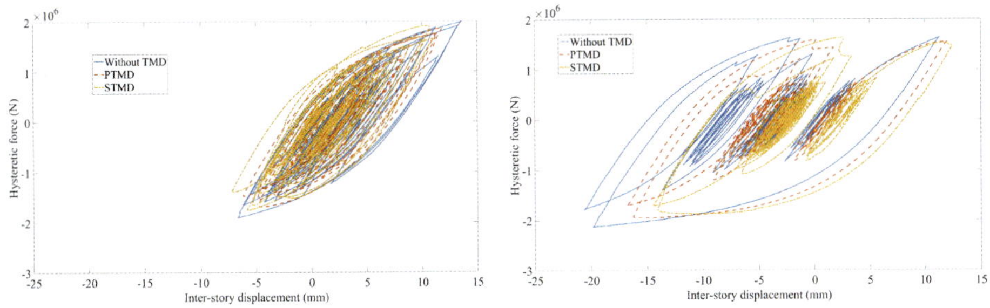

(a) (b)

Fig. 11. Structural hysteretic curve comparison under MS–AS8. (a) The third story under MS8 and (b) the fourth story under AS8.

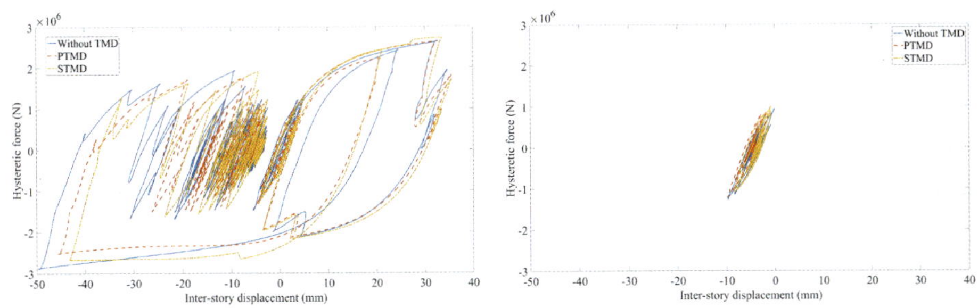

(a) (b)

Fig. 12. Structural hysteretic curve comparison under MS–AS9. (a) The second story under MS9 and (b) the fourth story under AS9.

As can be seen from Figs. 11 and 12, under the effect of AS8, the hysteretic curve of the structure is significantly larger than the MS section, which further highlights the importance of AS consideration and the necessity of adding STMD for nonlinear dynamic response control. However, under the excitation of AS9, the hysteretic curve of the structure is obviously smaller than the MS section. At the same time, STMD can pull the hysteretic curve of the structure towards the origin of displacement, that is, it can effectively control the development of its plasticity and improve the toughness and resilience of the structure.

Finally, take the MS–AS8 as an example, instantaneous parameters (frequency and damping coefficient) of STMD are presented in Fig. 13, respectively.

It can be seen in Fig. 13(a) that according to Sec. 3.2, to protect the structural elastoplastic response, the STMD can vary its frequency between 1.08 Hz (for the elastic structure) and 0.18 Hz (for the nonlinear structure). Meanwhile, it also can be found that the instantaneous frequency is mainly between 1.08 and 0.41 Hz. As

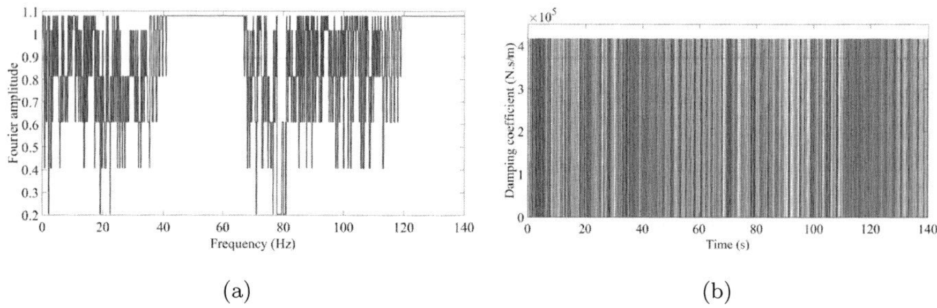

Fig. 13. Instantaneous parameters of STMD under MS–AS8. (a) Frequency and (b) damping coefficient.

a reference, for the linear structure whose inter-story stiffness of the first story is 0.04 times that of the original structure, the first-order frequency is exactly 0.41 Hz. In Fig. 13(b), STMD can switch its own damping coefficient between the minimum and maximum, to achieve a better energy dissipation.

4.3. *Energy analysis*

To further evaluate the energy dissipation of PTMD and STMD on the whole nonlinear tall building under the MS–AS sequence, energy analysis is carried out in this section. Input energy of MS–AS sequence to the structure and as well as dissipated energy by different TMDs are compared and shown in Table 6 and Fig. 14, respectively.

From the above results in Table 6 and Fig. 14, it is obvious that STMD can not only decrease the input energy from the MS–AS sequence to the structure, but also dissipate more energy than PTMD, indicating a perfect performance from the perspective of energy transfer and dissipation. Especially, under the MS–AS9

Table 6. Comparisons on the input energy of MS–AS sequences and dissipated energy by different TMDs ($\times 10^5$ J).

Sequence	Input energy			Dissipated energy	
	Without TMD	PTMD	STMD	PTMD	STMD
MS-AS1	11.0195	10.8554	8.9110	1.5513	2.1095
MS-AS2	6.3857	6.3722	5.2827	1.1169	1.2004
MS-AS3	1.0950	0.9920	0.9602	0.4442	0.5301
MS-AS4	0.8434	0.8317	0.8071	0.4983	0.5238
MS-AS5	6.2041	5.9826	3.3860	1.2355	1.5904
MS-AS6	1.1574	1.1159	0.4900	0.2192	0.2606
MS-AS7	0.2708	0.2622	0.0585	0.0286	0.0301
MS-AS8	14.1829	14.5099	14.8306	3.6529	3.8962
MS-AS9	19.0746	19.0493	18.4742	1.9098	2.9436

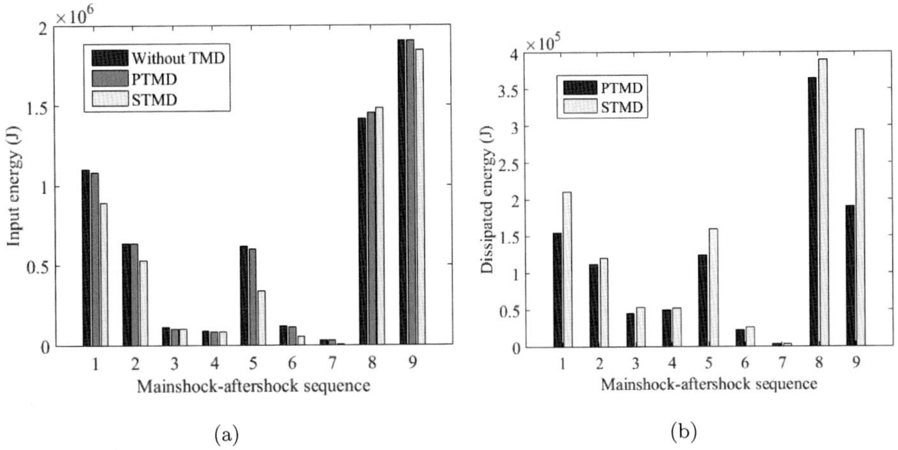

Fig. 14. Energy analysis of the nonlinear tall building. (a) Input energy by MS–AS sequences and (b) dissipated energy by different TMDs.

sequence, the input energy of the STMD case is smaller than the PTMD case, while STMD can even dissipate more energy than PTMD, which shows the superiority in energy dissipation for nonlinear structure.

To highlight the improvement of STMD in energy dissipation, force–displacement comparison and force–velocity comparison of different TMDs are proposed in Figs. 15 and 16, respectively. At the same time, the hysteretic curves of the MS and AS section are shown in different subgraphs, for the convenience of comparison, they are presented in the same coordinate frame.

It can be seen from Figs. 15 and 16 that STMD has a larger curve area than PTMD in both force–displacement curve and force–velocity curve, which indicates a better energy dissipation capacity. STMD can transfer and dissipate more kinetic energy of the nonlinear structure through its own larger stroke and velocity. It also can be found that under the input of MS–AS8, STMD consumes similar energy in the MS segment and AS segment. However, under the excitation of MS–AS9, the energy consumed by STMD under the AS is much less than that in the MS segment, which is related to the intensity of the input AS. In addition, the sharp angle in STMD hysteretic curves is caused by its time-varying stiffness and damping coefficient as shown in Fig. 13.

At last, comparisons on input energy by different MS–AS sequences and dissipated energy by different TMDs in the time domain are presented in Fig. 17.

Comparing Figs. 4 and 17, the platform segment in Fig. 17 is caused by the 30 s time interval preset in Fig. 4. For the MS–AS8, the input energy of the MS and AS segments is similar. For the MS–AS9, the input energy of the MS is significantly greater than that of the AS. As for the ratio between the dissipated energy and

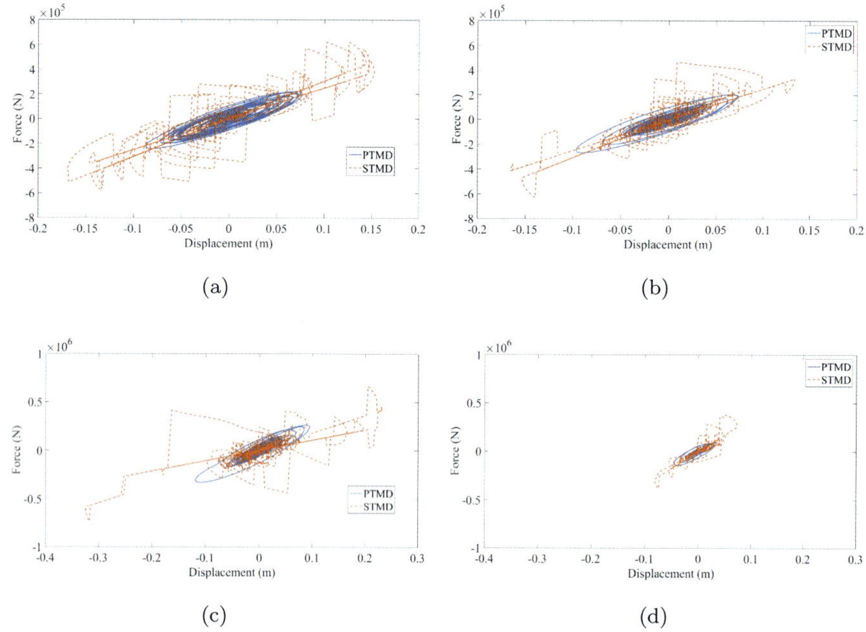

Fig. 15. Force–displacement comparison of different TMDs. (a) MS8; (b) AS8; (c) MS9; and (d) AS9.

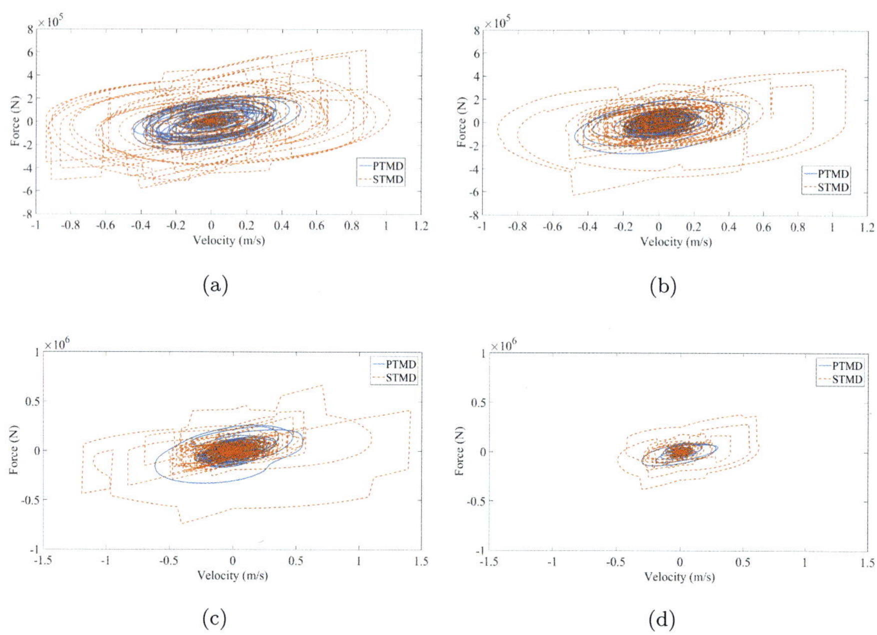

Fig. 16. Force–velocity comparison of different TMDs. (a) MS8; (b) AS8; (c) MS9; and (d) AS9.

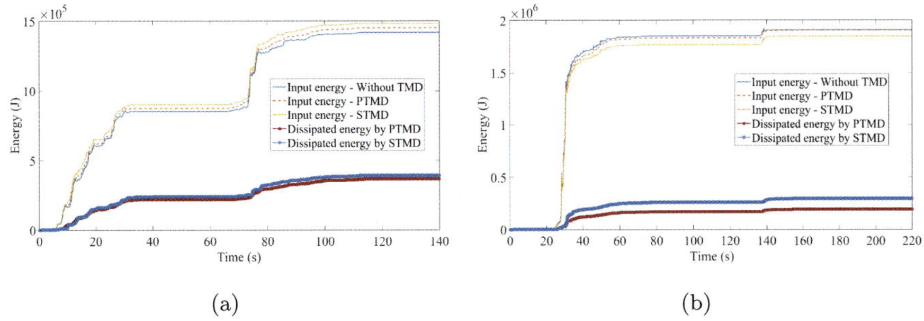

Fig. 17. Comparisons of input energy by different MS–AS sequences and dissipated energy by different TMDs. (a) MS–AS8 and (b) MS–AS9.

input energy, they are 25.18% and 26.27% for PTMD and STMD, respectively for MS–AS8, while they are 10.03% and 15.93%, respectively, for MS–AS9. Therefore, it is obvious that STMD has a better damping effect and energy dissipation than PTMD, and can protect the nonlinear tall building from MS–AS sequence more effectively.

5. Conclusions

Under the input of strong mainshock (MS), the tall building may enter the nonlinear stage, whose hysteretic characteristic can be well-simulated by the Bouc–Wen (BW) model. However, the accompanying aftershock (AS) can increase structural damage and cause the safety problem. Therefore, the advanced structural control technology should be implemented for nonlinear structure under MS–AS sequences. The tuned mass damper (TMD) is a widely applied vibrational absorber, however, a passive TMD (PTMD) is sensitive to the frequency deviation and may lose its effect for nonlinear dynamic response control in this issue.

To protect a nonlinear tall building under MS–AS sequences effectively, a semi-active TMD (STMD) is presented in this study, which can vary its frequency and damping in real time simultaneously. It can adapt to structural nonlinear behavior in each time segment through varying its stiffness according to the wavelet transform (WT) based algorithm, and meanwhile, enhance the energy dissipation through switching its damping coefficient in real time. A 10-story nonlinear tall building is proposed as a case study and a PTMD is used for comparison. Nine MS–AS sequences are elected and nonlinear structural control performance of STMD is highlighted. Further, energy analysis is proposed

From the numerical results proposed in Sec. 4, the following points can be summarized:

(1) The presented combined control strategy for STMD is mostly based on output signals, very little modal information and no input information are needed,

which is extremely suitable for the considered nonlinear structural control under MS–AS sequences.

(2) STMD can control structural displacement responses and base shear effectively and performs better than PTMD in reducing the plastic development of the structure as well. Especially, it is found that a strong AS may cause more damage to the building than the MS, while the nonlinear control effect of STMD is still significant in that case.

(3) STMD can effectively control the development of the structural plasticity and improve its toughness and resilience.

(4) STMD has an excellent effect in decreasing the structural input energy from MS–AS sequences and dissipating more energy than PTMD, which indicates that STMD has a better damping effect and can protect the nonlinear tall building from MS–AS sequence more effectively.

Acknowledgments

The authors are grateful for the financial support received from the National Natural Science Foundation of China (Grant No. 52025083), the National Key Research and Development Program of China (Grant No. 2022YFF0608903), and this study is also sponsored by Shanghai Pujiang Program (22PJ1413600) and supported by the Fundamental Research Funds for the Central Universities (22120220573).

ORCID

Liangkun Wang https://orcid.org/0000-0003-3426-4023
Ying Zhou https://orcid.org/0000-0001-9553-5419
Weixing Shi https://orcid.org/0000-0002-1000-7973

References

1. P. Chen, B. Wang, K. S. Dai and T. Li, Analytical and numerical investigations of base isolation system with negative stiffness devices, *Eng. Struct.* **268** (2022) 114799.
2. B. F. Spencer and S. Nagarajaiah, State of the art of structural control, *J. Struct. Eng. ASCE* **129**(7) (2003) 845–856.
3. J. J. Wang, C. Zhang and Y. Q. Zheng, An inerter-enhanced asymmetric nonlinear energy sink for response mitigation of structures subjected to harmonic and seismic ground excitations, *Struct. Contr. Health Monit.* **29**(12) (2022) e3104.
4. J. J. Wang, C. Zhang, H. B. Li and Z. B. Liu, A vertical-vibro-impact-enhanced track bistable nonlinear energy sink for robust and comprehensive control of structures, *Struct. Contr. Health Monit.* **29**(5) (2022) e2931.
5. J. J. Wang, C. Zhang and Z. B. Liu, Clutching inerters enhanced tuned mass dampers for structural response mitigation under impulsive and seismic excitations, *Struct. Contr. Health Monit.* **29**(2) (2022) e2881.
6. L. K. Wang, S. Nagarajaiah, W. X. Shi and Y. Zhou, Study on adaptive-passive eddy current pendulum tuned mass damper for wind-induced vibration control, *Struct. Des. Tall Spec. Build.* **29**(15) (2020) e1793.

7. S. Nagarajaiah, Adaptive passive, semiactive, smart tuned mass dampers: Identification and control using empirical mode decomposition, hilbert transform, and short-term fourier transform, *Struct. Contr. Health Monit.* **16**(7–8) (2009) 800–841.
8. S. Nagarajaiah and E. Sonmez, Structures with semiactive variable stiffness single/multiple tuned mass dampers, *J. Struct. Eng. ASCE* **133**(1) (2007) 67–77.
9. J. J. Wang, B. Wang, Z. B. Liu, H. B. Li and C. Zhang, Seismic response mitigation of building structures with a novel vibro-impact dual-mass damper, *Eng. Struct.* **215** (2020) 110673.
10. J. J. Wang, C. Zhang, H. B. Li and Z. B. Liu, Experimental and numerical studies of a novel track bistable nonlinear energy sink with improved energy robustness for structural response mitigation, *Eng. Struct.* **237** (2021) 112184.
11. J. J. Wang and Y. Q. Zheng, Development and robustness investigation of track-based asymmetric nonlinear energy sink for impulsive response mitigation, *Eng. Struct.* **286** (2023) 116127.
12. F. Brandão and A. Almeida, Optimum design of single and multiple tuned mass dampers for vibration control in buildings under seismic excitation, *Int. J. Struct. Stab. Dyn.* **22**(8) (2022) 2250078.
13. S. Etedali, Z. K. Bijaem, N. Mollayi and V. Babaiyan, Artificial intelligence-based prediction models for optimal design of tuned mass dampers in damped structures subjected to different excitations, *Int. J. Struct. Stab. Dyn.* **21**(9) (2021) 2150120.
14. L. K. Wang, W. X. Shi and Y. Zhou, Adaptive-passive tuned mass damper for structural aseismic protection including soil-structure interaction, *Soil. Dynam. Earthq. Eng.* **158** (2022) 107298.
15. A. Gholizad and S. Mohammadi, Reliability-based design of tuned mass damper using monte carlo simulation under artificial earthquake records, *Int. J. Struct. Stab. Dyn.* **17**(10) (2017) 1750121.
16. X. Wang, Z. Y. Xu, D. D. Wang, T. Wang, G. Q. Fu and C. J. Lu, Optimization design of the magnetic multi-mass damper for multi-frequency vibration suppression, *Int. J. Struct. Stab. Dyn.* **22**(11) (2022) 2250112.
17. A. K. Rathi and A. Chakraborty, Reliability-based performance optimization of TMD for vibration control of structures with uncertainty in parameters and excitation, *Struct. Contr. Health Monit.* **24**(1) (2017) e1857.
18. L. K. Wang, W. X. Shi, X. W. Li, Q. W. Zhang and Y. Zhou, An adaptive-passive retuning device for a pendulum tuned mass damper considering mass uncertainty and optimum frequency, *Struct. Contr. Health Monit.* **26**(7) (2019) e2377.
19. S. T. Liu, Z. Lu, P. Z. Li, S. W. Ding and F. Wan, Shaking table test and numerical simulation of eddy-current tuned mass damper for structural seismic control considering soil-structure interaction, *Eng. Struct.* **212** (2020) 110531.
20. M. Domizio, H. Garrido and D. Ambrosini, Single and multiple TMD optimization to control seismic response of nonlinear structures, *Eng. Struct.* **252** (2022) 113667.
21. Y. Wang and L. Wu, Effects of tuned mass damper on fixed-fixed 3D nonlinear string resting on nonlinear elastic foundation, *Int. J. Struct. Stab. Dyn.* **17**(4) (2017) 1750047.
22. G. Quaranta, F. Mollaioli and G. Monti, Effectiveness of design procedures for linear TMD installed on inelastic structures under pulse-like ground motion, *Earthq. Struct.* **10**(1) (2016) 239–260.
23. G. Espinoza, J. L. Almazan, F. Benedetti and C. Jara, Torsional balance of nonlinear asymmetrical structures by means of a tuned mass damper, *Struct. Contr. Health Monit.* **26**(11) (2019) e2442.

24. M. Mohebbi and A. Joghataie, Designing optimal tuned mass dampers for nonlinear frames by distributed genetic algorithms, *Struct. Des. Tall Spec. Build.* **21**(1) (2012) 57–76.
25. Z. Zhang and T. Balendra, Passive control of bilinear hysteretic structures by tuned mass damper for narrow band seismic motions, *Eng. Struct.* **54** (2013) 103–111.
26. S. Sgobba and G. C. Marano, Optimum design of linear tuned mass dampers for structures with nonlinear behaviour, *Mech. Syst. Signal Process.* **24** (2010) 1739–1755.
27. E. E. Matta, Lifecycle cost optimization of tuned mass dampers for the seismic improvement of inelastic structures, *Earthq. Eng. Struct. Dyn.* **47**(3) (2018) 714–737.
28. A. Boccamazzo, B, Carboni, G, Quaranta and W. Lacarbonara, Seismic effectiveness of hysteretic tuned mass dampers for inelastic structures, *Eng. Struct.* **216** (2020) 110591.
29. R. M. Hessabi, O. Mercan and B. Ozturk, Exploring the effects of tuned mass dampers on the seismic performance of structures with nonlinear base isolation systems, *Earthq. Struct.* **12**(3) (2017) 285–296.
30. L. K. Wang, S. Nagarajaiah, W. X. Shi and Y. Zhou, Semi-active control of walking-induced vibrations using adaptive tuned mass damper considering human-structure-interaction, *Eng. Struct.* **244** (2021) 112743.
31. S. Nagarajaiah and N. Varadarajan, Short time Fourier transform algorithm for wind response control of buildings with variable stiffness TMD, *Eng. Struct.* **27**(3) (2005) 431–441.
32. X. C. Guan, J. C. Zhang, H. Li and J. P. Ou, Semi-active control for benchmark building using innovative TMD with MRE isolators, *Int. J. Struct. Stab. Dyn.* **20**(6) (2020) 2040009.
33. C. Sun and S. Nagarajaiah, Study on semi-active tuned mass damper with variable damping and stiffness under seismic excitations, *Struct. Contr. Health Monit.* **21**(6) (2014) 890–906.
34. S. Woo, S. Lee and L. Chung, Seismic response control of elastic and inelastic structures by using passive and semi-active tuned mass dampers, *Smart Struct. Syst.* **8**(3) (2011) 239–252.
35. L. K. Wang, W. X. Shi, Y. Zhou and Q. W. Zhang, Semi-active eddy current pendulum tuned mass damper with variable frequency and damping, *Smart Struct. Syst.* **25**(1) (2020) 65–80.
36. L. K. Wang, S. Nagarajaiah, W. X. Shi and Y. Zhou, Seismic performance improvement of base-isolated structures using a semi-active tuned mass damper, *Eng. Struct.* **271** (2022) 114963.
37. K. Karami, S. Manie, K. Ghafouri and S. Nagarajaiah, Nonlinear structural control using integrated DDA/ISMP and semi-active tuned mass damper, *Eng. Struct.* **181** (2019) 589–604.
38. C. Sun, S. Nagarajaiah and A. J. Dick, Family of smart tuned mass dampers with variable frequency under harmonic excitations and ground motions: Closed-form evaluation, *Smart Struct. Syst.* **13**(2) (2014) 319–341.
39. C. Sun, Semi-active control of offshore wind turbines under multi-hazards, *Mech. Syst. Sig. Process.* **99** (2018) 285–305.
40. C. Sun and V. Jahangiri, Fatigue damage mitigation of offshore wind turbines under real wind and wave conditions, *Eng. Struct.* **178** (2019) 472–483.
41. C. Sun, V. Jahangiri and H. Sun, Adaptive bidirectional dynamic response control of offshore wind turbines with time-varying structural properties, *Struct. Contr. Health Monit.* **28**(11) (2021) e2817.

42. V. Jahangiri, C. Sun and F. Kong, Study on a 3D pounding pendulum TMD for mitigating bi-directional vibration of offshore wind turbines, *Eng. Struct.* **241** (2021) 112383.
43. M. Shokrabadi, H. Burton and J. Stewart, Impact of sequential ground motion pairing on mainshock-aftershock structural response and collapse performance assessment, *J. Struct. Eng. ASCE* **144**(10) (2018) 04018177.
44. A. Nikellis and K. Sett, Multihazard Risk Assessment and cost-benefit analysis of a bridge-roadway-levee system, *J. Struct. Eng. ASCE* **146**(5) (2020) 04020050.
45. M. Mirrashid and H. Naderpour, Innovative computational intelligence-based model for vulnerability assessment of rc frames subject to seismic sequence, *J. Struct. Eng., ASCE* **147**(3) (2021) 04020350.
46. D. Ji, W. Wen and C. Zhai, Constant-ductility energy factors of SDOF systems subjected to mainshock-aftershock sequences, *Earthq. Spectra* **37**(2) (2021) 1078–1107.
47. A. Papadopoulos, P. Bazzurro and W. Marzocchi, Exploring probabilistic seismic risk assessment accounting for seismicity clustering and damage accumulation: Part I. Hazard analysis, *Earthq. Spectra* **37**(2) (2021) 803–826.
48. H. Pan and K. Kusunoki, Aftershock damage prediction of reinforced-concrete buildings using capacity spectrum assessments, *Soil. Dynam. Earthq. Eng.* **129** (2020) 105952.
49. M. Poiani, V. Gazzani, F. Clementi and S. Lenci, Aftershock fragility assessment of Italian cast-in-place RC industrial structures with precast vaults, *J. Build. Eng.* **29** (2020) 101206.
50. B. Silwal and O. Ozbulut, Aftershock fragility assessment of steel moment frames with self-centering dampers, *Eng. Struct.* **168** (2018) 12–22.
51. A. Ghotbi and E. Taciroglu, Conditioning criteria based on multiple intensity measures for selecting hazard-consistent aftershock ground motion records, *Soil. Dynam. Earthq. Eng.* **139** (2020) 106345.
52. F. D. Trapani and M. Malavisi, Seismic fragility assessment of infilled frames subject to mainshock/aftershock sequences using a double incremental dynamic analysis approach, *Bull. Earthq. Eng.* **17** (2019) 211–235.
53. K. Trevlopoulos, P. Guéguen, A. Helmstetter and F. Cotton, Earthquake risk in reinforced concrete buildings during aftershock sequences based on period elongation and operational earthquake forecasting, *Struct. Saf.* **84** (2020) 101922.
54. L. K. Wang, W. X. Shi and Y. Zhou, Study on self-adjustable variable pendulum tuned mass damper, *Struct. Des. Tall Spec. Build.* **28**(1) (2019) e1561.
55. L. K. Wang, Y. Zhou and W. X. Shi, Seismic control of a smart structure with semi-active tuned mass damper and adaptive stiffness property, *Earthq. Eng. Resil.* **2**(1) (2023) 74–93.
56. H. Den, *Mechanical Vibrations* (McGraw-Hill, New York, 1985).
57. L. K. Wang, S. Nagarajaiah, Y. Zhou and W. X. Shi, Experimental study on adaptive-passive tuned mass damper with variable stiffness for vertical human-induced vibration control, *Eng. Struct.* **280** (2023) 115714.
58. L. K. Wang, W. X. Shi, Q. W. Zhang and Y. Zhou, Study on adaptive-passive multiple tuned mass damper with variable mass for a large-span floor structure, *Eng. Struct.* **209** (2020) 110010.
59. PEER Ground Motion Database, Pacific Earthquake Engineering Research Center. https://ngawest2.berkeley.edu.

https://doi.org/10.1142/9789819814770_0026

Chapter 26

Towards Probabilistic Robust and Sparsity-Free Compressive Sampling in Civil Engineering: A Review#

Haoyu Zhang, Shicheng Xue, Yong Huang* and Hui Li

Key Lab of Structures Dynamic Behavior
and Control of the Ministry of Education
School of Civil Engineering, Harbin Institute of Technology
Harbin 150090, P. R. China

Key Lab of Smart Prevention and Mitigation of Civil Engineering Disasters
of the Ministry of Industry and Information Technology
Harbin Institute of Technology, Harbin 150090, P. R. China
**huangyong@hit.edu.cn*

Compressive sampling (CS) is a novel signal processing paradigm whereby the data compression is performed simultaneously with the sampling, by measuring some linear functionals of original signals in the analog domain. Once the signal is sparse sufficiently under some bases, it is strictly guaranteed to stably decompress/reconstruct the original one from significantly fewer measurements than that required by the sampling theorem, bringing considerable practical convenience. In the field of civil engineering, there are massive application scenarios for CS, as many civil engineering problems can be formulated as sparse inverse problems with linear measurements. In recent years, CS has gained extensive theoretical developments and many practical applications in civil engineering. Inevitable modelling and measurement uncertainties have motivated the Bayesian probabilistic perspective into the inverse problem of CS reconstruction. Furthermore, the advancement of deep learning techniques for efficient representation has also contributed to the elimination of the strict assumption of sparsity in CS. This paper reviews the advancements and applications of CS in civil engineering, focusing on challenges arising from data acquisition and analysis. The reviewed theories also have applicability to inverse problems in broader scientific fields.

Keywords: Compressive sampling; sparsity; Bayesian inference; generative model; deep learning; structural health monitoring.

*Corresponding author.
#This chapter appeared previously on the International Journal of Structural Stability and Dynamics. To cite this chapter, please cite the original article as the following: H. Zhang *et al.*, *Int. J. Struct. Stab. Dyn.*, **23**, 2340028 (2023), DOI: 10.1142/S021945542340028X.

1. Introduction

1.1. *Overview*

Compressive sampling (CS, also known as compressed sensing or compressive sensing) is a novel data acquisition and processing method that takes advantage of the sparse structure in signals.[1–3] It performs data compression by simply capturing a few linear measurements of the original signal, which is done simultaneously with sampling in the analog domain. By utilizing the sparsity constraint to original signals,[4–8] CS gives rigorous guarantees for stable reconstruction from far fewer measurements (i.e. lower sampling rate) than that required by the Shannon–Nyquist sampling theorem.[9]

The emergence of sparse representation via ℓ_1 minimization which is related to CS dates back to half a century,[10] while CS received rigorous theoretical analysis in 2006 and has since become an active research area.[1–3] Conventional CS can be summarized as finding the sparsest solution of original signals in the transformed domain (e.g. the decomposed representation under pre-defined sparse bases), subject to the constraint of satisfying linear measurements. As a result, the objective function of CS reconstruction consists of both the data-fitting and sparsity-promoting regularization terms. In the past decades, researchers have developed a series of efficient algorithms with theoretical guarantees to solve such a regularized linear inverse problem, leading to the wide application of CS in diverse fields such as imaging and communication.[11–14] In civil engineering, researchers have also made significant contributions to both the theoretical and practical sides of CS.[15]

1.2. *Necessity of CS in civil engineering*

In modern civil engineering systems, massive amounts of data are usually collected for structural diagnosis and assessment, thus calling for effective and reliable data compression approaches. Compared with traditional compression methods which require fully sampled data,[16,17] CS offers a simple linear sensing mechanism in the analog domain, enabling direct data sampling in the compressed form (using special CS sensors), thus greatly relieving the power consumption of high-speed analog-to-digital conversion. Furthermore, CS can also be applied for loss recovery during wireless data transmission, since packet loss of data can be regarded as special data compression.[18] These advantages make CS a powerful approach for reducing the cost of data acquisition, transmission and storage, especially for wireless monitoring platforms.

The underlying theory of CS indicates that if the target object has some practical prior structures (usually, being sparse under specified bases), its information can be stably recovered from very few linear measurements, even if the measured data are contaminated by noise.[1–3,19] In the field of civil engineering, a series of problems can be formulated as a general sparse inverse problem with noisy linear measurements (the nonlinear components can be converted into a linear space by kernel functions, or linearly truncated by the Taylor expansion), where the measured data are often insufficient. For example, in structural health monitoring (SHM), it is important to perform accurate identification of load,[20] modal properties,[21] and damage parameters[22] to facilitate the automatic assessment of structural conditions, while only spatially sparse sensors are available. Similarly, in non-destructive testing (NDT), there is a need to reconstruct meaningful information from a limited number of sensors in order to detect structural damages while reducing the testing time and data collection volume. In addition, the interpolation and extrapolation of load distribution or site property based on very few measured points also pose a significant challenge. All the above problems can be theoretically analyzed and solved in the framework of CS, once we build sparse structures for the target object.

1.3. *Critical issues for application of CS*

Despite the great success of CS in terms of both theoretical advances and algorithmic implementations, it faces some worrisome issues in the scenario of civil engineering.

On the one hand, the majority of CS algorithms are deterministic, meaning they only search for a single "optimal" solution and provide a point estimate to the original data. These approaches will be problematic when the compression ratio (CR) is high and information in the CS measurements is insufficient to precisely determine a proper sparse solution. To address these highly ill-posed problems of CS reconstruction, a Bayesian probabilistic framework can be employed to describe the whole family of plausible solutions based on CS measurements, and the confidence of reconstructed data can be effectively quantified through the posterior uncertainty. This Bayesian framework has the potential to adaptively select more appropriate CS measurements, with the target of reducing the updated posterior uncertainty.[23] Moreover, the deterministic CS algorithms often require a tuning of the coefficient of the sparsity regularization term to ensure a suitable sparse solution, which needs a validation set of data. Advantageously, Bayesian CS treats the coefficient as a "nuisance" hyperparameter, allowing automatically learning it from CS measurements on hand. Through marginalizing over the posterior uncertainty of this hyperparameter, the robustness of CS reconstruction can also be effectively enhanced.[24–26]

On the other hand, due to the large variability from civil engineering problems, the original data may not be sparse enough, or even not be sparse at all, under any pre-defined sparse bases, making conventional CS reconstructions that rely on sparsity constraint fail. Although it is feasible to construct overcomplete dictionaries (in which basis functions are redundant) that could sparsely characterize the target data, by using dictionary learning algorithms for specific datasets,[27,28] sparsity may no longer be the most efficient low-complexity prior structure in the Big Data era. Deep learning-based generative models, which have a powerful ability to capture data features in a much more low-dimensional space, offer an alternative solution to automatically learn a more compact and generalizable representation based on the dataset of interested objects.[29–31] This strategy actively reduces the dimensionality of solution space to that of the low-dimensional learned space; thus, the original data are not required to be sparse at all, alleviating the high requirements of both quantity and quality for CS measurements. Such representations have been demonstrated to exist,[32] and the CS reconstruction based on optimizing the learned representations with nonlinear mappings has also been theoretically guaranteed.[33]

1.4. *Paper organization*

In the remainder of this paper, Sec. 2 introduces the basic concept and theory of CS. Section 3 reviews the advancements of CS utilizing sparsity constraint and their applications in civil engineering, including both the deterministic and Bayesian probabilistic algorithms. Section 4 presents a sparsity-free CS framework based on deep generative models, along with practical case studies of its implementation. Finally, several key summaries and future prospects of CS in civil engineering are discussed in Sec. 5.

2. Critical Concepts and Brief Theory

2.1. *Critical concepts in CS*

Fundamentally, CS is designed to solve some problems of the traditional sampling-and-compression framework in the signal processing field. According to the Shannon–Nyquist sampling theorem which utilizes the band-limitedness of natural signals, the frequency of equally spaced analog-to-digital sampling needs to be higher than twice of the highest frequency component in original signals to ensure the reconstruction without aliasing. When acquiring and transmitting signals with high-frequency components (for example, structural vibration responses under broadband ambient excitations), expensive high-speed sampling is required at first, and then transform coding techniques[34,35] are used to obtain a low-complexity representation of the signal in a transformed domain, where only a few major components and their positions are retained to compress the fully sampled signal, thus ensuring stable signal transmission and efficient storage. This framework not only wastes sampling and coding resources, but also introduces the overhead to transmit the positions of major components.[36]

In CS, the original signal (e.g. the time series or images) is denoted as a discrete vector $\mathbf{x} \in \mathbb{R}^N$. By the knowledge of linear algebra, this vector can be decomposed by a set of basis vectors $\mathbf{\Psi} = [\psi_1, \ldots, \psi_N] \in \mathbb{R}^{N\times N}$, and the corresponding coefficient/representation vector is $\mathbf{s} \in \mathbb{R}^N$:

$$\mathbf{x} = \boldsymbol{\Psi}\mathbf{s}. \tag{2.1}$$

For simplicity, we assume that the basis matrix $\boldsymbol{\Psi}$ is real and orthogonal, thus the original signal $\mathbf{x}$ is equivalent to the representation vector $\mathbf{s} = \boldsymbol{\Psi}^T\mathbf{x}$. $\mathbf{s}$ is K-sparse when it has at most K non-zero elements (the case of interest is $K \ll N$), and belongs to

$$\mathbb{S}_N^K = \{\mathbf{s} \,|\, \mathbf{s} \in \mathbb{R}^N, \|\mathbf{s}\|_0 \leq K\}. \tag{2.2}$$

We call $\mathbf{s}$ compressible (approximately sparse) when it can be well approximated by its first K largest elements. In civil engineering, structural responses are usually dominated by a few oscillators in the frequency domain; structural load distributions in many cases are sparse for discrete loads or can be characterized by a few simple functions for continuous loads. Hence, it is reasonable to consider the signals to be sparse or compressible.

Instead of high-speed sampling the original data at equally spaced points, CS greatly alleviates the power consumption of sampling and coding, by randomly and unequally spaced sampling the original $\mathbf{x}$ with a sub-Nyquist rate, or more generally, by measuring some linear functionals of $\mathbf{x}$, i.e. CS directly samples $\mathbf{y} \in \mathbb{R}^M$ in the compressed form. As shown in Fig. 1, the general compression/sensing process is expressed as

$$\mathbf{y} = \boldsymbol{\Phi}\mathbf{x} + \boldsymbol{\varepsilon} = \boldsymbol{\Phi}\boldsymbol{\Psi}\mathbf{s} + \boldsymbol{\varepsilon} = \boldsymbol{\Theta}\mathbf{s} + \boldsymbol{\varepsilon}, \tag{2.3}$$

where $\boldsymbol{\Phi} \in \mathbb{R}^{M\times N}$ is the measurement matrix, $M < N$; the CR is defined as N/M; $\boldsymbol{\Psi} \in \mathbb{R}^{N\times N}$ is a manually selected sparse basis; $\boldsymbol{\Theta} = \boldsymbol{\Phi}\boldsymbol{\Psi} \in \mathbb{R}^{M\times N}$ is the sensing matrix; and $\boldsymbol{\varepsilon} \in \mathbb{R}^M$ is the measurement error, which is minor for modern sensors. This measurement approach avoids complete sampling, complex coding, and the overhead of storing locations of major components in the traditional

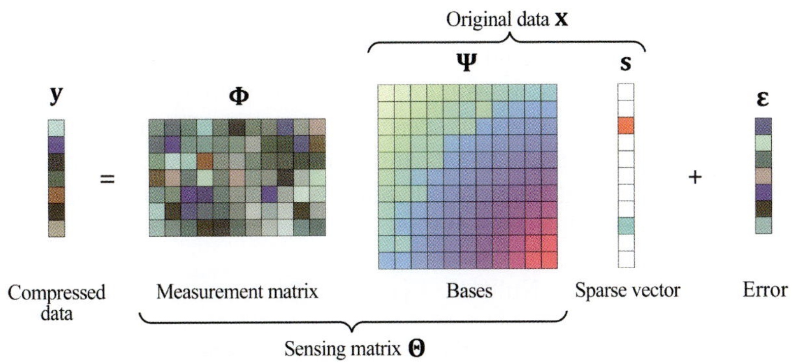

Fig. 1. The general measurement process of CS.

sampling-and-compression framework. It has been implemented on CS sensors,[37–39] with small additional computational burdens.

It is worth noting that the measurement matrix $\mathbf{\Phi}$ is typically non-adaptive, i.e. being independent of the original data $\mathbf{x}$. This "democratic" measurement approach[40,41] is useful when parts of the transmitted data are continuously lost or corrupted by some anomalies (common for wireless transmission), since the missing after compression can be simply treated as a higher CR, while the information of original data is still preserved in each measurement. Researchers have also explored the adaptive measurement for target data to improve the reconstruction performance, such as the rakeness-based CS.[42]

For the decompression/reconstruction of CS, only the noisy incomplete compressed measurements $\mathbf{y}$ and the sensing matrix $\mathbf{\Theta}$ are available to recover the sparse representation vector $\mathbf{s}$, or equivalently, the original data $\mathbf{x}$. Since $M < N$, this problem is ill-posed and mathematically intractable, but the sparsity constraint for $\mathbf{s}$ can help to turn the problem into a solvable one. Utilizing the sparsity prior knowledge to alleviate the ill-posedness of inverse problems has been studied for decades, namely the well-known sparse regression or sparse recovery.[43] CS is characterized by the fact that the bases are user-selected (rather than given) to facilitate the solution of the inverse problem on hand, and provides a rigorous theoretical analysis (rather than a heuristic one) for the algorithms of promoting solution sparsity, which makes CS an active research area with its own specified approaches.[19]

2.2. *Brief theory of recoverability guarantees in CS*

Next, we briefly explore in which situation the linear inverse problem of CS reconstruction can be solved, from the particular to the general case.

2.2.1. *Requirements for stable reconstruction*

We start with the clean measurements $\bar{\mathbf{y}} = \mathbf{\Theta}\mathbf{s}$ without errors. When the $\mathbf{s}$ to be recovered is exactly K-sparse and the positions of its non-zero elements are known, the least squares method can be employed when the number of CS measurements M is not smaller than K. However, positions of non-zero elements are generally unknown. If we desire to uniquely determine any K-sparse vector from clean measurements $\bar{\mathbf{y}}$, it is required that the difference between any two distinct K-sparse vectors is far from the null space of $\mathbf{\Theta}$:

$$\mathrm{Null}(\mathbf{\Theta}) = \{\mathbf{r} \,|\, \mathbf{\Theta}\mathbf{r} = \mathbf{0}\}. \tag{2.4}$$

Otherwise, there will exist $\mathbf{s}_1, \mathbf{s}_2 \in \mathbb{S}_N^K$ satisfying $\mathbf{\Theta}\mathbf{s}_1 - \mathbf{\Theta}\mathbf{s}_2 = \mathbf{\Theta}(\mathbf{s}_1 - \mathbf{s}_2) = \mathbf{0}$, and thus they cannot be distinguished from CS measurements on hand. With the proof in Ref. 44, the sufficient and necessary condition to uniquely recover any K-sparse vector from clean measurements $\bar{\mathbf{y}}$ is that the Null($\mathbf{\Theta}$) does not contain any $\mathbf{r} \in \mathbb{S}_N^{2K}$, which is equivalent to requiring spark($\mathbf{\Theta}$) $> 2K$.[45]

Moreover, when approximately solving for those compressible (approximately sparse) vectors, the requirement to the sensing matrix $\mathbf{\Theta}$ needs to be extended to the null space property (NSP). Briefly, this means that its null space Null($\mathbf{\Theta}$) should also not contain any excessively compressible vectors.[44] The performance of solving for a compressible vector also depends on the extent to which it can be approximated by a sparse vector.

In practice, CS measurements are inevitably contaminated by some errors. Denote the CS reconstruction algorithm as $\Delta : \mathbb{R}^M \to \mathbb{R}^N$, then the reconstructed vector is $\tilde{\mathbf{s}} = \Delta(\mathbf{y})$. In order to reconstruct a K-sparse or compressible vector $\mathbf{s}$ stably, measurement error should not lead to an arbitrarily large reconstruction error on $\mathbf{s}$ (it can be easily generalized to $\mathbf{x}$):

$$\|\Delta(\mathbf{\Theta}\mathbf{s} + \boldsymbol{\varepsilon}) - \mathbf{s}\|_2 \leq C\|\boldsymbol{\varepsilon}\|_2, \tag{2.5}$$

where C is a positive constant. Hence, we need to impose more strict requirements on the sensing matrix $\mathbf{\Theta}$. Candès *et al.*[46,47] proposed the well-known restricted isometry property (RIP) for $\mathbf{\Theta}$. We call $\mathbf{\Theta}$ satisfies the RIP of order K, when for arbitrary K-sparse $\mathbf{s}$, it has

$$(1 - \delta_K)\|\mathbf{s}\|_2^2 \leq \|\mathbf{\Theta}\mathbf{s}\|_2^2 \leq (1 + \delta_K)\|\mathbf{s}\|_2^2, \tag{2.6}$$

where $\delta_K \in (0, 1)$ is a constant. Intuitively, when $\delta_K \to 0$, $\mathbf{\Theta}$ does not change the length of input vectors, i.e. it behaves like an orthogonal matrix. Therefore, the smaller δ_K is, the more similarity between the behaviors of $\mathbf{\Theta}$ and the orthogonal matrix for K-sparse vectors would be, which facilitates the reconstruction. When $\mathbf{\Theta}$ satisfies the RIP of order $2K$, it can maintain the distance between any two distinct K-sparse vectors, which allows us to distinguish sparse solutions under the interference of measurement errors. In addition, the NSP can also be derived from the RIP,[44] which makes $\mathbf{\Theta}$ satisfying the RIP applicable for solving those compressible vectors.

A sensing matrix $\mathbf{\Theta}$ that satisfies the RIP brings a sufficient condition for stable CS reconstruction of exact K-sparse or compressible vectors from noisy measurements. The required order of RIP and the limit for δ_K depend on the used reconstruction algorithm Δ. For reference, some requirements are summarized in Ref. 48.

2.2.2. *Construction of appropriate CS measurements*

The requirements for stable CS reconstruction explained in Sec. 2.2.1 are established on the sensing matrix $\mathbf{\Theta}$; but in practical scenarios, we generally need to select the sparse basis $\mathbf{\Psi}$ after compression, leaving only the measurement matrix $\mathbf{\Phi}$ to be adjusted. In order to construct a matrix $\mathbf{\Theta}$ that satisfies the desired requirements, using random matrices as $\boldsymbol{\Phi}$ can facilitate the construction. For example, when each element of a matrix in $\mathbb{R}^{M \times N}$ is sampled independently and identically from a Gaussian distribution, then with

$$M = \mathcal{O}\left[\frac{K \log(N/K)}{\delta_K^2}\right], \tag{2.7}$$

it will satisfy the RIP of order K with high probability,[49,50] where the constant is δ_K.

The universality of the Gaussian distribution will make $\boldsymbol{\Theta}$ remain Gaussian after multiplying the Gaussian $\boldsymbol{\Phi}$ by $\boldsymbol{\Psi}$, thus automatically makes $\boldsymbol{\Theta}$ satisfy the same RIP. By substituting the order and constant required by the reconstruction algorithm into Eq. (2.7), we can estimate how many Gaussian measurements are sufficient to achieve stable reconstruction, i.e. recoverability guarantees. It also holds for sub-Gaussian matrices.[49,50]

Nonetheless, evaluating the recoverability of an existing sensing matrix $\boldsymbol{\Theta}$ requires computing spark($\boldsymbol{\Theta}$), checking the NSP or the RIP, which is hard and impractical. Some researchers have established guarantees based on the coherence of matrices,[51,52] that is easily computed. Interestingly, the stable reconstruction is also guaranteed when $\boldsymbol{\Phi}$ and $\boldsymbol{\Psi}$ are incoherent.[36] For example, the measurement matrix of randomly deleting rows in $\mathbf{x}$ (i.e. the sub-Nyquist sampling) and the Fourier basis immediately satisfy the incoherence.[4]

Remark 2.1. For the case with complex basis $\boldsymbol{\Psi} \in \mathbb{C}^{N \times N}$ and complex sparse representation $\mathbf{s} \in \mathbb{C}^N$, the main theory in Sec. 2.2 still holds. We can simply treat the real and imaginary parts separately to form a new representation vector $\mathbf{s}' \in \mathbb{R}^{2N}$, and modify the basis accordingly. The zero entries in K-sparse $\mathbf{s}$ let both the real and imaginary parts of the corresponding entries in $\mathbf{s}'$ be zero, which makes $\mathbf{s}'$ being $2K$-sparse and thus does not affect the reconstruction performance. In civil engineering, the commonly used Fourier basis for the frequency analysis is complex.

Remark 2.2. For the selection of sparse basis $\boldsymbol{\Psi}$, dictionary learning methods[27,28] can be used to design overcomplete dictionaries $\mathbf{D} \in \mathbb{R}^{N \times D}$ (with $D > N$ redundant basis vectors) that are more suitable for the target data, where a sparser representation is obtained. This helps to reduce the number of required CS measurements and improve the reconstruction performance. When using dictionaries, it is very important to note that a successful reconstruction on the representation $\mathbf{s}$ does not necessarily equal that of the original data $\mathbf{x}$. Because there may not exist a constant C' that satisfies $\|\mathbf{x}_1 - \mathbf{x}_2\|_2 = \|\mathbf{D}\mathbf{s}_1 - \mathbf{D}\mathbf{s}_2\|_2 \leq C'\|\mathbf{s}_1 - \mathbf{s}_2\|_2$. So, we need to check extra requirements for dictionaries.[53]

3. Developments of Conventional CS with Sparsity Constraint

3.1. *Typical deterministic CS algorithms*

The most straightforward constraint for sparsity is the ℓ_0 norm (i.e. the number of non-zero elements). Based on this, the objective for CS reconstruction of $\mathbf{s}$ is

$$\begin{aligned} \min_{\bar{\mathbf{s}}} \quad & \|\bar{\mathbf{s}}\|_0 \\ \text{s.t.} \quad & \|\mathbf{y} - \boldsymbol{\Theta}\bar{\mathbf{s}}\|_2 \leq \varepsilon_u, \end{aligned} \tag{3.1}$$

where ε_u represents the upper bound of errors, i.e. $\|\boldsymbol{\varepsilon}\|_2 \leq \varepsilon_u$, which need to be tuned.

Solving the ℓ_0 norm regularized problem directly is NP-hard, and a series of greedy strategy-based algorithms[54] were proposed to efficiently solve this problem by iterative approximation to determine the support set (i.e. positions of non-zero entries) and the corresponding element values in $\mathbf{s}$, e.g. orthogonal matching pursuit,[55] iterative hard thresholding,[56] CS matching pursuit,[57] and so on. In the field of civil engineering, some researchers have employed these greedy CS algorithms.[58,59] The convergence of these algorithms is also theoretically guaranteed, but they usually need to preset the number of non-zero elements (opposite to the sparsity degree) as stopping criterion. When the dimensionality of $\mathbf{s}$ is very high or its sparsity degree is relatively low, these greedy algorithms tend to be inefficient and unstable.[51,60]

Considering the efficient mathematical tools in the convex optimization field,[61] more researchers focused on relaxing the ℓ_0 norm to an alternative ℓ_1 norm (the minimum convex upper bound of ℓ_0 norm, as shown in Fig. 2, the vertex of ℓ_1 constrained area is more likely to meet the data-fitting hyperplane) that also promotes sparsity but is convex,[46] or approximating some non-convex norms (e.g. the p norm, $0 < p < 1$) that strongly promote sparsity by reweighting some strongly

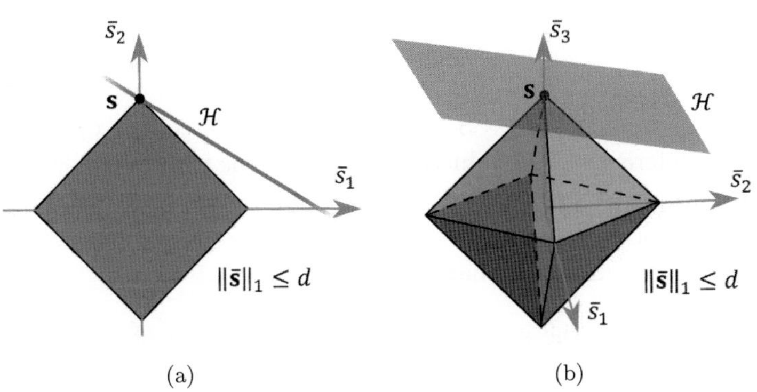

Fig. 2. The ℓ_1 ball with radius d of (a) $\bar{\mathbf{s}} \in \mathbb{R}^2$ and (b) $\bar{\mathbf{s}} \in \mathbb{R}^3$. $\mathcal{H}$ is a hyperplane of solutions fitting $\mathbf{y}$.

convex norms.[62,63] For ℓ_1 minimization-based algorithms, the objective of CS reconstruction has the following unconstrained form:

$$\tilde{\mathbf{s}} = \underset{\bar{\mathbf{s}}}{\operatorname{argmin}}(\|\mathbf{y} - \boldsymbol{\Theta}\bar{\mathbf{s}}\|_2^2 + \rho\|\bar{\mathbf{s}}\|_1), \tag{3.2}$$

where the first and second terms on right side are the data-fitting and sparsity-promoting regularization terms, respectively; the coefficient ρ needs to be tuned to ensure a proper balance between the two terms. Candès[47] has proved that for the noiseless case, the ℓ_0 minimization problem has a unique K-sparse solution when the sensing matrix $\boldsymbol{\Theta}$ satisfies the RIP of order $2K$ with a constant $\delta_{2K} < 1$; the relaxed ℓ_1 minimization problem has the same solution as the ℓ_0 problem, while the required constant is $\delta_{2K} < \sqrt{2} - 1$, which implies that the above convex relaxation is exact. Though more measurements are required to pay for such relaxations, the convex regularizations are more efficient to solve and do not require to preset a sparsity degree as those greedy algorithms.

When using noisy measurements, Candès[47] also provided the bound of reconstruction error. Denote the vector that sets all but the first K largest entries of $\mathbf{s}$ to zero as $\mathbf{s}_K$, it is the best K-sparse solution. With the same condition for noiseless case, the reconstruction error of the solution to relaxed ℓ_1 minimization problem obeys

$$\|\tilde{\mathbf{s}} - \mathbf{s}\|_2 \le C_0\|\mathbf{s} - \mathbf{s}_K\|_1/\sqrt{K} + C_1\varepsilon_u, \tag{3.3}$$

with some constants C_0 and C_1. The reconstruction error depends on the degree to which the representation vector $\mathbf{s}$ can be approximated by a sparse vector $\mathbf{s}_K$, and on the level of errors. This means that using ℓ_1 minimization, we can recover K-sparse vectors exactly from clean measurements, and reconstruct K-sparse vectors stably or compressible vectors approximately from noisy measurements.

There are some classical algorithms to solve the ℓ_1 norm regularized problem, such as basis pursuit denoising,[64] gradient projection for sparse reconstruction,[65] and the least absolute shrinkage and selection operator[66] in statistics that performs data fitting and variable selection simultaneously. Some researchers in civil engineering also studied these CS algorithms based on convex optimization techniques.[18,20,21] Reference 67 introduced how to solve the p norm regularized problem. In addition to the greedy algorithms and the convex relaxation algorithms, there is another combinatorial algorithm[68] that can also solve CS problem, but its model is too specific and the suitable scenarios are limited.

Remark 3.1. Besides the sparsity, for data in the form of non-sparse matrices with low-rankness, the ℓ_0 norm constraint can be replaced by the rank of the reconstructed matrix; being analogous to vector-form data, the corresponding minimum convex relaxation is the nuclear norm of the matrix.[5] In civil engineering, some matrices composed of collected data are reported to present low-rankness.[69,70] Based on this, it is also feasible to reconstruct those non-sparse but low-rank matrices from partially observed elements.

3.2. *Bayesian probabilistic framework for robust CS reconstruction*

Section 3.1 reviews typical deterministic CS algorithms, where a single "optimal" solution is searched for the data to be reconstructed. However, when measurements are highly noisy and incomplete, the inverse problem of CS reconstruction becomes severely ill-posed and is impossible to be solved precisely. To this end, one should describe the whole family of plausible solutions based on CS measurements under a Bayesian probabilistic framework.

In the Bayesian framework, the sparsity prior knowledge is integrated into the prior probability density function (PDF) of the entries in the representation vector (in Bayesian theory, they are named as the parameters of the unknown model $\mathbf{s}$). And the inverse problem of CS reconstruction is treated as the inference of their posterior PDF, which quantifies the updated relative plausibility of different values of the model parameters, with the help of the likelihood function considering available measurements. The framework can be built naturally in the CS reconstruction, since the posterior PDF obtained from Bayes' theorem is consistent with both the likelihood function and prior PDF, which correspond to the data-fitting and sparsity-promoting regularization terms in Eq. (3.2), respectively.

3.2.1. *Basic theory of Bayesian CS*

Ji *et al.*[23] tackled the CS reconstruction problem from a Bayesian perspective for the first time, by employing the sparse Bayesian learning theory.[71] The quantified uncertainty in posterior PDF measures the confidence for reconstructed results. It also provides a method for the optimal design of CS measurements, with the goal of reducing reconstruction uncertainty.[23] The automatic relevance determination prior is assigned to the representation vector, or the model $\mathbf{s}$, which is a sparsity-promoting prior:

$$p(\mathbf{s}\,|\,\boldsymbol{\alpha}) = \prod_{n=1}^{N} p(s_n\,|\,\alpha_n) = \prod_{n=1}^{N} \mathcal{N}(s_n\,|\,0, \alpha_n), \tag{3.4}$$

where $\boldsymbol{\alpha} = (\alpha_1, \ldots, \alpha_N)^T$ represents the independent variance hyperparameters of entries in $\mathbf{s}$. The key to sparsity modelling of $\mathbf{s}$ is that when $\alpha_n \to 0$, it essentially produces a Dirac delta-function at zero for the prior PDF of s_n, and so for its posterior PDF.

In Bayesian CS, subject to the constraint of first two moments, a Gaussian maximum entropy distribution is defined for the prediction error vector $\boldsymbol{\varepsilon}' \sim \mathcal{N}(\boldsymbol{\varepsilon}'\,|\,\mathbf{0}, \beta^{-1}\mathbf{I})$, which contains the measurement error $\boldsymbol{\varepsilon}$ and the modelling error from non-zero compressible components in $\mathbf{s}$. Correspondingly, the likelihood function for CS measurements $\mathbf{y}$ is

$$p(\mathbf{y}\,|\,\mathbf{s},\beta) = \mathcal{N}(\mathbf{y}\,|\,\boldsymbol{\Theta}\mathbf{s},\beta^{-1}\mathbf{I}), \tag{3.5}$$

where β is the precision hyperparameter of model prediction errors.

Using Bayes' theorem, the posterior PDF of $\mathbf{s}$ can be obtained:

$$p(\mathbf{s}\,|\,\mathbf{y},\boldsymbol{\alpha},\beta) = p(\mathbf{y}\,|\,\mathbf{s},\beta)p(\mathbf{s}\,|\,\boldsymbol{\alpha})/p(\mathbf{y}\,|\,\boldsymbol{\alpha},\beta), \tag{3.6}$$

where unknown hyperparameters (in Bayesian theory, they are named as the model class $\mathcal{M}(\boldsymbol{\alpha},\beta)$) exist. To characterize the CS reconstruction of $\mathbf{s}$, its marginal posterior PDF could be produced by using Laplace's asymptotic approximation[72]:

$$p(\mathbf{s}\,|\,\mathbf{y}) = \int p(\mathbf{s}\,|\,\mathbf{y},\boldsymbol{\alpha},\beta)p(\boldsymbol{\alpha},\beta\,|\,\mathbf{y})d\boldsymbol{\alpha}d\beta \approx p(\mathbf{s}\,|\,\mathbf{y},\tilde{\boldsymbol{\alpha}},\tilde{\beta}), \tag{3.7}$$

where $\tilde{\boldsymbol{\alpha}}$ and $\tilde{\beta}$ are the *maximum a posteriori* estimates to these hyperparameters. They are estimated by maximizing $\{\tilde{\boldsymbol{\alpha}},\tilde{\beta}\} = \arg\max p(\boldsymbol{\alpha},\beta\,|\,\mathbf{y}) = \arg\max[p(\mathbf{y}\,|\,\boldsymbol{\alpha},\beta)p(\boldsymbol{\alpha},\beta)]$. For a non-informative prior $p(\boldsymbol{\alpha},\beta)$, as we commonly used, the evidence or marginal likelihood $p(\mathbf{y}\,|\,\boldsymbol{\alpha},\beta)$ is most crucial for the learning of hyperparameters.[23]

Based on the relationship in Eq. (2.1), once we obtain the posterior PDF of sparse representation vector $\mathbf{s}$ through Eq. (3.7), the posterior PDF of original data $\mathbf{x}$ will also be synchronously obtained, which provides the probabilistic reconstruction result.

Several important theoretical advances of Bayesian CS are reviewed in the following. Representative probabilistic reconstruction results will be discussed in Sec. 3.3.3(ii).

Remark 3.2. Learning hyperparameters $\boldsymbol{\alpha}$ and β by maximizing $p(\mathbf{y}\,|\,\boldsymbol{\alpha},\beta)$ automatically involves a trade-off between how well the model $\mathbf{s}$ fits the measured data and how sparse it is, because of Bayesian Ockham's razor[72] at work. There is an information-theoretic interpretation, in which the logarithm of evidence can be expressed as[25]

$$\begin{aligned}\log p(\mathbf{y}\,|\,\boldsymbol{\alpha},\beta) &= \mathbb{E}_{\mathbf{s}\sim p(\mathbf{s}\,|\,\mathbf{y},\boldsymbol{\alpha},\beta)}[\log p(\boldsymbol{y}\,|\,\mathbf{s},\beta)]\\ &\quad - \mathbb{E}_{\mathbf{s}\sim p(\mathbf{s}\,|\,\mathbf{y},\boldsymbol{\alpha},\beta)}\left[\log\frac{p(\mathbf{s}\,|\,\mathbf{y},\boldsymbol{\alpha},\beta)}{p(\mathbf{s}\,|\,\boldsymbol{\alpha})}\right],\end{aligned} \tag{3.8}$$

where the first term is a measure of data fitness; and the second term is the nonnegative information gain from the prior to posterior, which measures increased model complexity. Therefore, by maximizing the (logarithm of) evidence, the learned model class $\mathcal{M}(\tilde{\boldsymbol{\alpha}},\tilde{\beta})$ will produce a model that can explain the CS measurements well and is also sufficiently sparse (i.e. not too complex). This learning method also leads to the benefit that Bayesian CS does not require the laborious tuning of hyperparameters.

Remark 3.3. Compared with the deterministic CS methods in Sec. 3.1, the Bayesian framework has the following benefits: (1) the posterior uncertainty provides the measure of confidence of reconstructed results, which can be used as an objective to design a better measurement strategy of CS; (2) the hyperparameter of sparsity-promoting regularization constraints is automatically determined through maximizing the evidence function, without the need for additional validation data; (3) by integrating the "nuisance" hyperparameter of errors in the quantified posterior PDF out, the robustness of CS reconstruction will be further improved.

3.2.2. *Robustness enhancement for Bayesian CS*

Huang *et al.*[25] found that the reconstruction of Bayesian CS algorithms suffers from a robustness problem: when the number of CS measurements M is smaller than a specified value, the reconstruction sometimes converges to suboptimal solutions with very large reconstruction error and much more non-zero model parameters, as

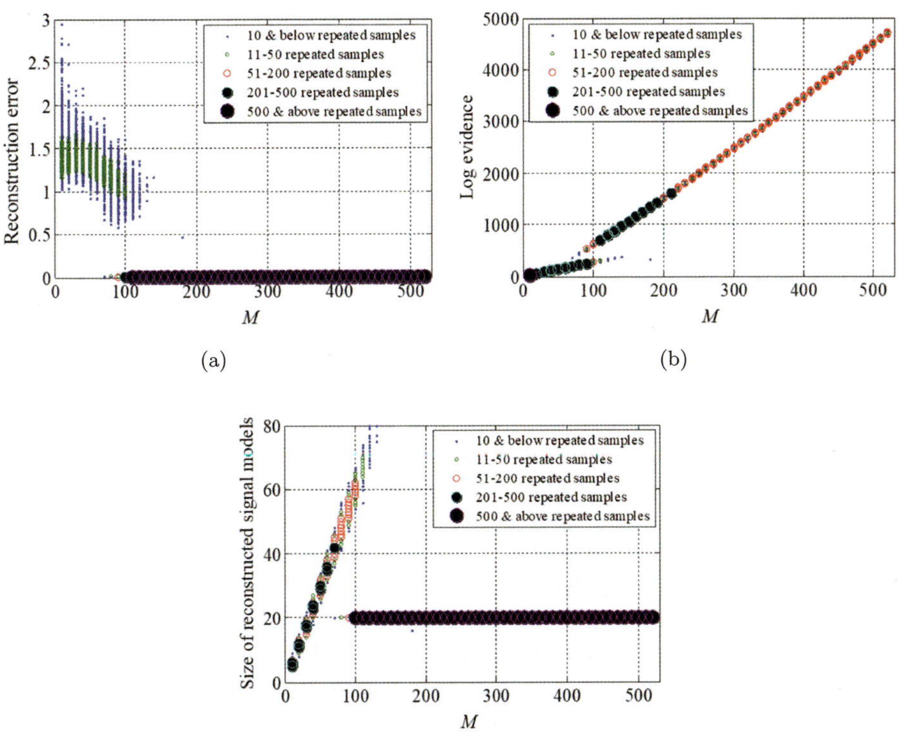

Fig. 3. The robustness problem for reconstruction of Bayesian CS.[25]

shown in Fig. 3. This is because the procedure of learning hyperparameters $\boldsymbol{\alpha}$ and β may get trapped in the local maximums of the evidence $p(\mathbf{y} \mid \boldsymbol{\alpha}, \beta)$.

Several robustness enhancement strategies have been developed, including stochastic optimization over the hyperparameters $\boldsymbol{\alpha}$ which control the model sparsity, and explicit marginalization for the precision hyperparameter β of model prediction errors.[24–26]

Most CS algorithms rely on the assumption that the original data are sparse enough in some bases. For practical applications, such as the SHM signals, the original data are usually approximately sparse with non-zero compressible components. This will also affect the robustness of CS reconstruction. Baron *et al.*[73] utilized a two-state mixture Gaussian model as a prior to capture the approximate sparsity in original data, and implemented Bayesian inference of reconstruction through the belief propagation method. Huang *et al.*[26] found that the reconstruction performance of approximately sparse data is very sensitive to the hyperparameter β. They developed an algorithm with robust treatments to β by explicitly marginalizing it, which was successfully applied to approximately sparse SHM signals from a cable-stayed bridge. In Ref. 26, it is also demonstrated that the quantified posterior uncertainties of their presented algorithm provide a useful tool for the diagnostics of CS reconstruction, since all incorrect reconstructed signal models give much larger average error bars (i.e. mean of diagonal elements of the posterior covariance matrix) than that of the correct reconstruction cases. This is an important benefit of the posterior uncertainty quantification in Bayesian CS.

3.2.3. *Multitask Bayesian CS for data with shared sparse pattern*

In SHM, the structural response signals from adjacent sensors are always very similar, because they are governed by the same few global vibration modes (i.e. presenting joint or shared sparse pattern in the frequency domain). To exploit the redundancy of multiple measurements from adjacent sensors, it is meaningful to develop CS algorithms to jointly learn sparse representations for the relevant tasks by modelling informative relationships within groups of CS measurements $\{\mathbf{y}_l\}_{l=1}^{L}$. This can be regularized by the $\ell_{2,1}$ norm[74,75] of the matrix composed of corresponding representation vectors $\{\mathbf{s}_l\}_{l=1}^{L}$, which constrains the group sparsity. But, it requires tuning of the regularization coefficient.

Multitask Bayesian CS was first proposed by Ji *et al.*,[24] where a shared hyper prior is defined across L tasks $\{\mathbf{y}_l \rightarrow \mathbf{s}_l\}_{l=1}^{L}$, then the shared hyperparameters $\boldsymbol{\alpha}$ are learned based on CS measurements of all tasks, which means that all tasks have the same sparse pattern. To ensure the information shared among multiple tasks is appropriate, one could use some methods (such as the fuzzy C-means in Ref. 76) to classify all tasks into several clusters firstly, so that multitask CS can be utilized for the tasks which belong to the same cluster. Huang *et al.*[77] presented an improved multitask Bayesian CS algorithm to model the data correlation across various relevant tasks, with the application to real SHM signals. By employing multitask Bayesian CS, the diagnosis and corresponding "healing" methods were proposed in Huang *et al.*,[78] which could judge whether the reconstructed result is accurate or not using CS measurements and improve the reconstruction performance when necessary. Wan *et al.*[79] developed an improved complex multitask Bayesian CS algorithm, which applies to sparse representations in the domain of complex numbers.

3.2.4. *Sequential Bayesian CS for data with slowly changing sparse pattern*

Besides the vanilla sparsity, sparse representations of original data may have other special structures, that can also be employed as prior information to improve the CS reconstruction performance. For example, the joint sparse pattern in Sec. 3.2.3, the clustered structure and tree structure of the coefficients under wavelet bases, which have been investigated as an additional prior in Bayesian CS (e.g. Refs. 80 and 81).

Another important structure is the temporally correlated sparsity. In SHM, structural response signals tend to be stationary under ambient excitations, because the condition of structures is relatively stable within a certain time period. Hence, the sparse representation vector should be slowly changing between two consecutive instances, i.e. two sparse representation vectors with a sparse difference. This temporal correlation can be used to aid CS reconstruction at the next time instance, thereby possibly reducing the required number of measurements for accurate reconstruction. Sejdinović *et al.*[82] derived a Bayesian CS framework for dynamically changing sparse signals, and they performed sequential estimation of the posterior PDF via the sequential Monte Carlo sampling. However, the method is not applicable for high-dimensional data, since a large computational cost is required for the stochastic simulation.

Huang *et al.*[83] recently proposed a sequential Bayesian CS algorithm, in which a hierarchical Bayesian model (as shown in Fig. 4) is introduced to incorporate the two sparsity constraints mentioned above.

Their stochastic state-space model is given by

$$\begin{cases} \mathbf{y}_i = \mathbf{\Theta}_i \mathbf{s}_i + \boldsymbol{\varepsilon}_i, \\ \mathbf{s}_i = \mathbf{s}_{i-1} + \mathbf{e}_i, \\ \boldsymbol{\delta}_i = \mathbf{s}_i + \boldsymbol{\epsilon}_i, \end{cases} \tag{3.9}$$

where the subscript $i \geq 1$ represents the ith time instance, and $\mathbf{s}_0$ is a known initial model; the model prediction error is $\boldsymbol{\varepsilon}_i \sim \mathcal{N}(\mathbf{0}, \beta_i^{-1}\mathbf{I})$, which has a precision hyperparameter β_i; $\mathbf{e}_i \sim \mathcal{N}(\mathbf{0}, \text{diag}(\boldsymbol{\lambda}_i))$ is an sparse innovation vector to model the temporal dependence in the evolution from $\mathbf{s}_{i-1}$ to $\mathbf{s}_i$, with hyperparameters $\boldsymbol{\lambda}_i$; a pseudo-data vector $\boldsymbol{\delta}_i = \mathbf{0}$ is introduced, which makes $\boldsymbol{\epsilon}_i = \boldsymbol{\delta}_i - \mathbf{s}_i$ and $\boldsymbol{\epsilon}_i \sim \mathcal{N}(\mathbf{0}, \text{diag}(\boldsymbol{\alpha}_i))$, thus the sparsity constraint for $\mathbf{s}_i$ is imposed to $\boldsymbol{\epsilon}_i$, with hyperparameters $\boldsymbol{\alpha}_i$.

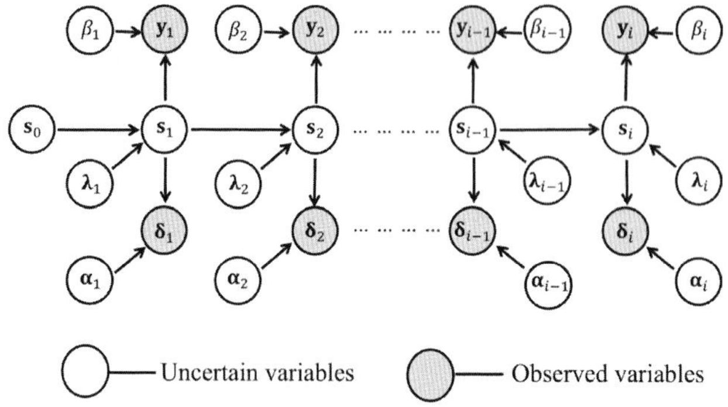

Fig. 4. Graphical representation of the hierarchical model in sequential Bayesian CS.[83]

CS reconstruction is realized by computing the updated posterior PDF based on the data ($\mathbf{y}_{1:i}$ and $\boldsymbol{\delta}_{1:i}$) from the initial time instance to the ith one:

$$\begin{aligned} p(\mathbf{s}_i \,|\, \mathbf{y}_{1:i}, \boldsymbol{\delta}_{1:i}, \boldsymbol{\lambda}_i, \boldsymbol{\alpha}_i, \beta_i) &\propto p(\boldsymbol{\delta}_i \,|\, \mathbf{s}_i, \boldsymbol{\alpha}_i) p(\mathbf{y}_i \,|\, \mathbf{s}_i, \beta_i) p(\mathbf{s}_i \,|\, \mathbf{y}_{1:(i-1)}, \boldsymbol{\delta}_{1:(i-1)}, \boldsymbol{\lambda}_i) \\ &\propto p(\boldsymbol{\delta}_i \,|\, \mathbf{s}_i, \boldsymbol{\alpha}_i) p(\mathbf{y}_i \,|\, \mathbf{s}_i, \beta_i) \int p(\mathbf{s}_i \,|\, \mathbf{s}_{i-1}, \boldsymbol{\lambda}_i) p(\mathbf{s}_{i-1} \,|\, \mathbf{y}_{1:(i-1)}, \boldsymbol{\delta}_{1:(i-1)}) d\mathbf{s}_{i-1}, \end{aligned} \tag{3.10}$$

where the hyperparameters $\{\boldsymbol{\lambda}_i, \boldsymbol{\alpha}_i, \beta_i\}$ can also be learned by

$$\begin{aligned} \{\tilde{\boldsymbol{\lambda}}_i, \tilde{\boldsymbol{\alpha}}_i, \tilde{\beta}_i\} &= \underset{\{\boldsymbol{\lambda}_i, \boldsymbol{\alpha}_i, \beta_i\}}{\operatorname{argmax}} \; p(\boldsymbol{\lambda}_i, \boldsymbol{\alpha}_i, \beta_i \,|\, \mathbf{y}_{1:i}, \boldsymbol{\delta}_{1:i}) \\ &= \underset{\{\boldsymbol{\lambda}_i, \boldsymbol{\alpha}_i, \beta_i\}}{\operatorname{argmax}} \; [p(\mathbf{y}_i, \boldsymbol{\delta}_i \,|\, \mathbf{y}_{1:(i-1)}, \boldsymbol{\delta}_{1:(i-1)}, \boldsymbol{\lambda}_i, \boldsymbol{\alpha}_i, \beta_i) p(\boldsymbol{\lambda}_i, \boldsymbol{\alpha}_i, \beta_i)], \end{aligned} \tag{3.11}$$

where $p(\mathbf{y}_i, \boldsymbol{\delta}_i \,|\, \mathbf{y}_{1:(i-1)}, \boldsymbol{\delta}_{1:(i-1)}, \boldsymbol{\lambda}_i, \boldsymbol{\alpha}_i, \beta_i)$ is the evidence for these hyperparameters. Their analytical solution was derived in Ref. 83 to recursively estimate the posterior PDF of $\mathbf{s}_i$ for each time instance.

3.3. *Applications of CS in civil engineering*

This subsection will review representative scenarios of applying CS in civil engineering, mainly in SHM and NDT, which are inverse problems in nature and have massive data to be processed, as well as accurately modelling load distributions and site properties from severely limited measurement points.

3.3.1. *Structural health monitoring*

Over the last several decades or so, developing the SHM systems for real-time damage diagnosis, condition assessment and early warning of in-service civil infrastructures that suffer from the progressive deterioration or some extreme events has received increasing attentions.[84,85] By placing contact (usually with a low spatial resolution) and non-contact (usually with a low temporal resolution) sensors on the large-scale structures, a massive amount of data need to be collected during the continuous monitoring. In SHM, the problems of handling the high cost of data collection and analyzing data with insufficient resolution can be efficiently tackled by CS techniques.

(i) Reconstruction from compressed data

Continuously collecting massive data brings huge work pressure on the sampling, coding, transmission and storage facilities, especially for wireless sensing devices, and thus there is an urgent need in SHM for efficient data compression. CS provides a simple compression method that is done simultaneously with sampling, and this will greatly save the power consumption of analog-to-digital conversion and coding. Hence, CS is suitable for solving the above problems and guarantees reliable reconstruction.

The stationary vibration response signals of structures under ambient excitations can be decomposed in the modal coordinates as follows:

$$u_n(t) = \sum_m \phi_{m,n} q_m(t), \tag{3.12}$$

where $u_n(t)$ denotes the displacement time history of the structural nth degree of freedom (DoF); $\phi_{m,n}$ denotes the value of the structural mth order mode shape at the nth DoF; $q_m(t)$ denotes the mth order modal response, which mainly corresponds to a sinusoidal function of the vibration of a single DoF system. In general, the structural vibration is dominated by a few active modes, which means that the structural responses (also including acceleration and velocity) can be represented in a more compact fashion in the frequency domain, thus adapting to the sparsity requirement of CS.

Bao *et al.*[86] first explored CS of acceleration response signals of a real bridge, using the ℓ_1 minimization to perform signal reconstruction. As shown in Fig. 5(a), they found that using the Harr wavelet transform as the sparse basis can achieve better performance than that using the Fourier basis. Note that when the original data are known, the compression efficiency of CS is weaker than that of classical compression methods. It is because the compression process of those classical methods is nonlinear, while CS only performs linear measurements. For pure data compression problems, the advantage of CS is that the data sampling and compression are completed simultaneously, which can significantly reduce the power consumption of high-frequency sampling. Huang *et al.*[25] was the first in SHM to use Bayesian CS to compute the posterior PDF of the reconstructed signals, which provides not only the most probable solutions but also their reconstruction confidence. Jayawardhana *et al.*[87] applied the CS reconstructed signals for damage identification. Based on an

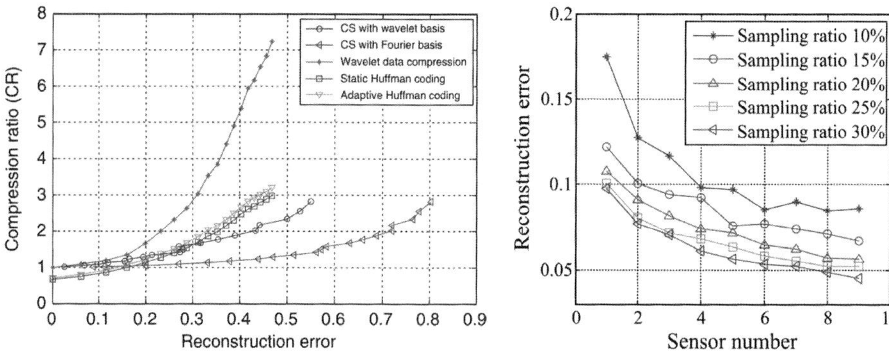

(a) (b)

Fig. 5. (a) Comparison of reconstruction performance between CS and classical data compression methods.[86] (b) The improvement of CS reconstruction performance by using information among relevant sensors.[88]

experimental reinforced concrete structure, they showed that the compression does not compromise the structural damage information contained in original response signals, demonstrating the applicability of CS techniques. Bao *et al.*[88] considered that the response signals in different locations of the same structure are controlled by the same modes and thus have the shared sparse pattern. They described this knowledge by the group sparsity constraint. Using monitoring data of an actual bridge, their experimental results in Fig. 5(b) showed that simultaneous reconstruction of similar signals is more efficient and accurate than that of individually reconstructing each one. Li *et al.*[89] studied CS for monitoring data from piezoelectric ceramic transducers, in which a greedy reconstruction algorithm is used. The effect of different sparsity degrees on the CS reconstruction performance is analyzed based on data from a real concrete tunnel. They found that the reconstruction accuracy is independent of the state of transducers and structural damages. This result broadened the application scenarios of CS in SHM. Kang *et al.*[90] proposed an autoregressive model for post-processing of the CS reconstructed results, and further reduced the reconstruction errors without paying for extra measurements.

Bao *et al.*[91] formalized the forward process of CS as a neural network (NN), and the sparse representation vector was used as network parameters. They proposed to implement the reconstruction by gradient descent methods, so that the advanced graphics processing unit (GPU) hardware optimized for gradient computation can be used to accelerate the CS reconstruction. Chen *et al.*[92] further embedded the iterative process of a classical CS algorithm into the layers of a NN, with the hyperparameters and dictionaries as learnable network parameters that could be trained by specific datasets. This joint training method improved the reconstruction accuracy, which was validated by the monitoring data from rail systems. They showed that the trained network performs reconstruction extremely fast, with the help of advanced GPUs.

Amini *et al.*[93] proposed a workflow to obtain trained dictionaries for a specific dataset of structural vibration responses by dictionary learning, with corresponding deterministic measurement matrices (being deterministic will facilitate the hardware implementation). They also provided effective metrics to quantify the appropriate CR for target datasets. Inspired by the analog-to-information converter based on random demodulation,[38,94] Liang *et al.*[95] designed hardware for directly acquiring data in the compressed form for both active and passive SHM systems, as well as the corresponding software. Unlike the "democratic" measurement, special pre-integrated sampling strategies can effectively improve the signal reconstruction performance,[42,96,97] which is an important direction for developing CS in practical applications. Along this, Zonzini *et al.*[98] explored measurement strategies when non-zero entries of the sparse representation vector are not uniformly distributed, but have localized patterns, which is applicable to structural vibration with well-separated modes. Numerical prior knowledge was established on a structure-specific finite element model, which adapted the sampling strategy according to the second-order statistics of simulated responses. They also designed adaptive bases for the target response signals. A model-assisted framework was realized by exploring the balance among the sampling rate, on-board storage requirement, communication overhead and signal reconstruction accuracy. It has also been verified experimentally that such modification barely affects the accuracy of subsequent modal identification,[99] and damage detection.[100]

Benefitted from the advances in deep learning-based computer vision techniques, recent SHM systems widely use wired surveillance cameras and wireless devices equipped with image sensors (e.g. drones and smartphones[101]) to acquire image data of structures. Image data contain a large number of data points (i.e. pixels); thus, its volume far exceeds that of traditional time series signals, which requires more efficient data compression. This is very important to maximize the working duration of wireless platforms. Yang *et al.*[102] explored the use of discrete cosine transform as the sparse basis for SHM image data. They verified CS reconstruction performance using real image data of pipe structures, concrete structures, etc. Huang *et al.*[83] investigated the application of their sequential Bayesian CS algorithm to the reconstruction of temporally sparse changed image sequences.

(ii) Reconstruction from under-sampled data

In SHM, the presence of under-sampling is more usual than that of compression, both in the temporal and spatial domains. Random under-sampling also adapts to the framework of CS, and the reconstruction is theoretically guaranteed.

Yang *et al.*[21] performed modal identification based on non-uniform, low-rate randomly sampled response signals (i.e. the signals are under-sampled in the temporal

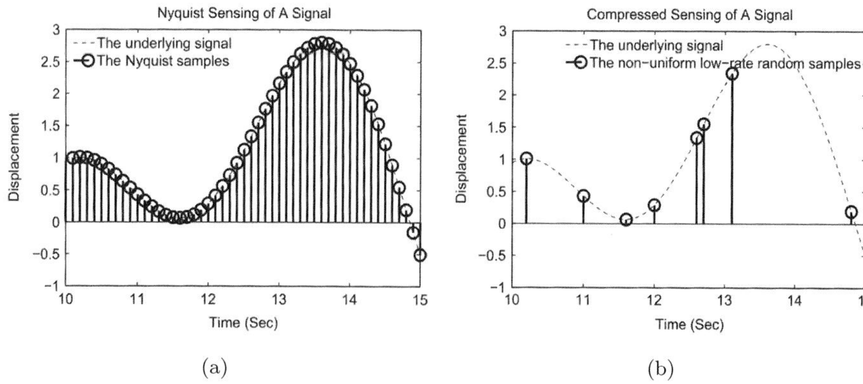

Fig. 6. Illustration of under-sampling in the temporal domain.[21]

domain, as shown in Fig. 6). They first obtained the mode shapes by blind source separation, and then reconstructed the uniform, high-rate modal responses based on CS, from which the frequencies and damping ratios were also accurately identified. Ganesan *et al.*[103] used CS to simultaneously alleviate the need for vibration monitoring at a high sampling rate over long time periods and with a large number of sensors for spatially dense measurement. For the spatial parts, since the mode shapes of simple structures are sparse in the spatial-frequency domain, they applied CS to recover high-resolution mode shapes from only a small number of measurement points. This not only improved the accuracy and resolution of damage identification, but is also important for alleviating the expensive deployment costs, mass effects and possible failures of densely placed contact sensors. Mei *et al.*[104] employed several mobile vehicles as dense measurement points to perform the structural monitoring. The data acquired by the mobile measurement points were both temporally and spatially sparse (under-sampled). It formed a matrix with a large number of unobserved entries. They found that this matrix has low-rankness, which allows matrix completion by CS, then modal identification can be performed based on the reconstructed matrix with dense sampling. Jana *et al.*[105] used a dictionary learning approach to establish data-driven spatial bases for linear time-invariant structural vibration, which allows the application of CS to reconstruct the desired full-field response from a finite number of measurement points (i.e. the spatial under-sampling, as shown in Fig. 7). This is also useful to guide the design of appropriate number and placement locations of contact sensors.

Non-contact vibration measurement usually relies on the image data of structures with high spatial resolution. The continuous motion tracking of structural nodes and components will generate a tremendous amount of data. Bhowmick *et al.*[106] evaluated the optical flow data from images (for vision-based displacement measurement) at a low rate randomly, in both the temporal and spatial domains (as

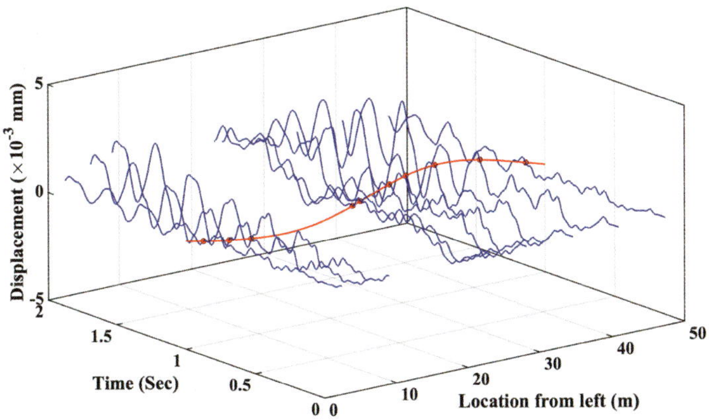

Fig. 7. Illustration of under-sampling in the spatial domain.[105]

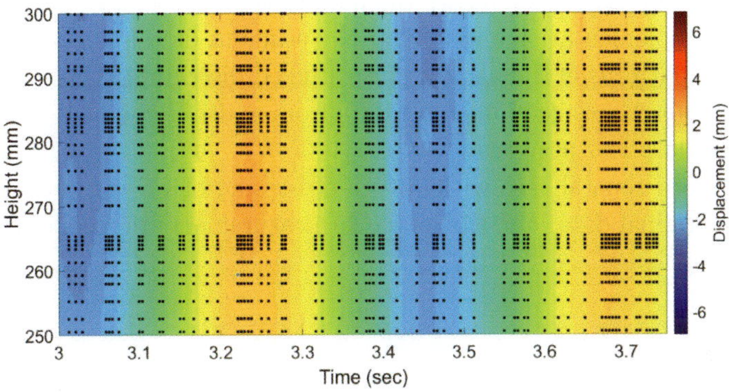

Fig. 8. Illustration of spatial-temporal under-sampling, where black dots represent sampling.[106]

shown in Fig. 8). After under-sampling, they successfully employed CS to reconstruct the full-field displacement time history, by utilizing the low-rankness and sparsity of the original optical flow data.

(iii) Recovery of lost data

In SHM, data loss often occurs during the transmission, especially for wireless monitoring systems, which will severely affect the subsequent condition assessment for structures.[107] Therefore, SHM systems should have the ability to automatically recover lost data. Note that we can also treat the abnormal data recovery as the same loss recovery problem if we remove the identified anomalies, e.g. outliers, from the collected data.

Benefitting from the "democracy" of CS measurement, the data loss that occurs after compression simply corresponds to a higher CR. Therefore, the data loss recovery problem can be solved by CS algorithms. In this context, Bao *et al.*[18] addressed the data loss problem by compressing the original data before transmitting them back. Their results showed that the modal identification accuracy is almost unaffected, using the data that reconstructed by CS algorithms. Zou *et al.*[108,109] provided a hardware implementation of this approach[18] and further introduced redundancy to resist data loss that may occur during transmission. This approach was also applied for the data loss recovery problem when fast relative motion exists between the base station and the wireless sensors.[110] Huang *et al.*[26] investigated the application of their developed robustness-enhanced Bayesian CS algorithm for the loss recovery of SHM signals with approximately sparse representations. Huang *et al.*[77] further applied a multitask version of their Bayesian CS algorithm in SHM, which improves the data loss recovery performance using spatial correlation from the adjacent sensor nodes of the lost node. Amini *et al.*[111] improved the random demodulation method, which makes the measurement process of CS more suitable for hardware implementation and robust to continuous data loss patterns. Based on similar sparse patterns of signals, they also studied distributed CS[112] and discussed the simultaneous reconstruction of lost data from multiple channels and under multiple types of data loss patterns.

For the more general case without pre-possessing of CS compression but with parts of the original data lost, the forward (measurement) process of data loss can be expressed as

$$\mathbf{x}_{\text{Lost}} = \mathbf{m} * \mathbf{x}, \tag{3.13}$$

where $\mathbf{m} \in \mathbb{R}^N$ is a vector with all elements of one, except for the entries where data loss occurs, which are zero; "$*$" denotes Hadamard product with element-wise multiplication. The process above can also be expressed in the matrix multiplication form, and obviously, when continuous data loss occurs, the concentrated pattern of zero entries will make the corresponding sensing matrix not necessary to satisfy the recoverability condition required by CS. In this case, reconstruction can be achieved by leveraging the correlation among relevant data, i.e. using complete data from adjacent sensors as a reference for CS.

Yang *et al.*[113] found that the data matrix composed of multi-channel response signals presents low-rankness, which is desired by the theory of CS. Partially lost channels can be completed/recovered using other multiple channels based on the low-rankness constraint. Simultaneously reconstructing their respective missing parts leads to better computational efficiency. They conducted experiments using seismic responses of building structures and monitored acceleration responses of long-span bridges. The results showed that successful recovery was achieved for single-channel data loss based on the sparsity constraint, as well as multi-channel data loss based on the low-rankness constraint. Tang *et al.*[107] investigated how to sparsely characterize the non-stationary signals (e.g. the seismic responses) by the harmonic wavelet basis. Using data of multiple channels with similar sparse patterns, they proposed to implement the recovery by gradient descent methods under the group sparsity constraint, and the continuously lost data were successfully recovered. Jana *et al.*[114] addressed the data loss problem occurring in cable vibrations that are measured by wireless sensors. They introduced small windows to transform the response of a linear time-varying system into an approximate linear time-invariant system, and successfully achieved loss recovery based on CS algorithms. After that, the time-varying cable force was effectively identified from the recovered responses.

As to data corrupted by abnormal values, Chen *et al.*[115] constructed a Hankel matrix for structural responses with outliers, and found that the matrix can be divided into a clean data matrix with low-rankness and a sparse outlier matrix, which is also applicable to the framework of CS. Based on the two constraints, robust data recovery can be achieved and outliers can be detected at the same time.

For the non-contact monitoring of large structures (by processing SHM image data), Ngeljaratan *et al.*[116] investigated the loss of vision-based vibration measurements due to the failure of target tracking algorithms after anomalies occurring in the image data (e.g. overexposure or motion blurring). Their experimental results demonstrated that CS can be used to accurately recover the lost parts, and that the recovered results can be applied to accurate identification of modal parameters.

(iv) Extension to inverse problems with nonlinear measurements

SHM involves a series of inverse problems that have nonlinear measurements, such as the identifications of structural loads, modal properties and damage parameters. Tackling their ill-posedness that is caused by incomplete measurement points through sparsity constraints has been studied for a long time, but the nonlinear mapping from the target objects to the corresponding measurements is not applicable to the theoretical framework of CS. By deriving the sensitivity matrix[117,118] from small sparse changes in the target objects to measurements (e.g. partial derivatives of the influence line of a load), or constructing dictionaries for sparse representation through some model-assisted methods,[20,119,120] the nonlinear mapping can be approximately linearized and thus be applicable to the CS framework. In addition, for the model updating based on incomplete modal data, Huang *et al.*[22,121] proposed several Bayesian methods where the nonlinear eigenvalue decomposition appears in the joint prior PDF to provide a soft constraint and the initial reference values for sparse modelling are used as the pseudo data. Although modal properties are nonlinearly related to system parameters (e.g. stiffness), a series of coupled linear-in-the-parameter problems are involved in the formulation.[122] This allows a tractable sparse model updating and damage identification using CS.

To ensure the recoverability of target objects, a RIP-based or similar analysis for these approximated forward processes should be performed.[118] However, there is a lack of relevant studies in the field of civil engineering, though sparse constraints do improve the accuracy of inversion. Therefore, we only comment on this extension (in fact, it is a very broad research area) without providing a comprehensive review.

(v) Bypass reconstruction by compressive signal processing

The aim of CS is to reconstruct entire original signals, but in fact we may only focus on a few features of interest. In some special cases, we can directly extract the desired features from CS measurements (termed as compressive signal processing[123]), which not only saves computational resources but also avoids the intermediate signal reconstruction errors. Park *et al.*[124] investigated modal identification using CS measurements of structural vibration responses. Based on the differential equations of vibration of multiple DoF systems, the associated theoretical justification was also provided. This approach allows SHM systems to sample and transmit data in a more efficient manner (e.g. only transmit the extracted features), and helps to minimize power consumption of battery-operated wireless sensors.

(vi) Summary

In general, SHM systems typically require high speed time-domain sampling and dense spatial-domain sampling for effective damage diagnosis and condition assessment, which brings massive data. The occurrence of data loss and anomalies also severely affects the health monitoring. Therefore, the implementation of SHM requires expensive sampling, transmission, storage and pre-processing, which induces a gap to practical applications. As we reviewed above, CS techniques bring the light, making it possible to reconstruct full-field, high-rate responses from spatially sparse, randomly low-rate measurements, as well as the data loss recovery, within a unified framework.

3.3.2. *Non-destructive testing*

Ultrasound, acoustic emission and guided waves are typical NDT techniques used for the detection of structural damage, and they usually acquire high-frequency signals. Capturing these waveforms requires high sampling rates, which generates large amounts of data. In addition, wavefield-based damage detection in the guided wave-based NDT requires a lot of measurement points, which is time-consuming. Hence, CS is also widely studied in NDT. For NDT problems here, the original data $\mathbf{x}$ in Eq. (2.3) are the high-dimensional time-domain signal with a high sampling rate or wavefield; CS measurements $\mathbf{y}$ may be under-sampled in the temporal or spatial domain; the sparse basis $\boldsymbol{\Psi}$ can be manually selected by utilizing fast Fourier transform, wavelet kernels, wave propagation equation, etc.; the over-complete sensing matrix $\boldsymbol{\Theta}$ can be pre-designed as illustrated in Eq. (3.14) later.

For the temporal domain applications, CS was investigated to reconstruct ultrasonic and acoustic emission signals from a few linear measurements, which reduced the burden of data sampling, storage and transmission. Bai *et al.*[58] presented a CS-based ultrasonic signal reconstruction scheme. Several sparse bases and reconstruction algorithms were selected in a comparative study with the traditional discrete wavelet transform-based compression method to verify the practicality of CS for ultrasonic signals. To quantify the uncertainty in the reconstruction of ultrasonic signals, Fuentes *et al.*[125] investigated the application of Bayesian CS, and the data-driven dictionaries that obtained by dictionary learning were adopted. The quantified uncertainties on signal reconstruction grow with the decrease of the number of CS measurements, which helps to determine an appropriate CR for ultrasonic signals. Liu *et al.*[126] applied a greedy algorithm to reconstruct high-rate acoustic emission signals from temporally under-sampled measurements. Their results demonstrated that the reconstructed signals effectively retain key features in the frequency spectrum. Song *et al.*[127] developed an improved sparsity adaptive matching pursuit algorithm to reconstruct acoustic emission signals in rails. Compared with other existing methods, more noticeable damage information was retained in their reconstruction.

As to the spatial domain applications, CS was employed for reconstructing the guided wavefield from measurements that were under-sampled in spatial domain (e.g. Ref. 128). The simplified diagram of this scheme is illustrated in Fig. 9. According to the Lamb wave propagation equation, the element $h_{m,n}^{(\mu)}(f)$ of the sensing matrix $\mathbf{\Theta}$ is computed as

$$h_{m,n}^{(\mu)}(f) = \frac{1}{\sqrt{d_{m,n}}} \exp[-jk_{m,n}^{(\mu)}(f) d_{m,n}], \tag{3.14}$$

where $d_{m,n}$ is the distance between a possible damage and measurement location, $k_{m,n}^{(\mu)}(f)$ refers to the wavenumber of guided wave corresponding to mode μ. After

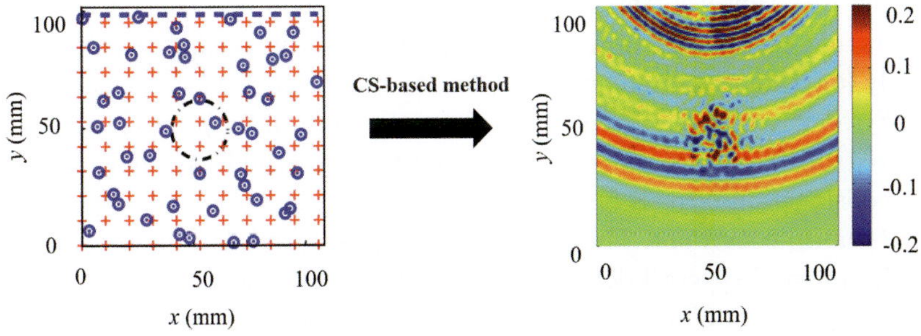

Fig. 9. Simplified diagram of wavefield reconstruction using CS.[128]

constructing the sensing matrix $\mathbf{\Theta}$, the spectral projected gradient algorithm for ℓ_1 norm minimization was employed for sparse inversion to reconstruct the wavefields, which greatly reduced the measurement cost.[128] However, as further reducing the number of measurement points, the quality of the wavefield reconstructed by CS algorithms may be severely degraded. To compensate for this degradation, Esfandabadi *et al.*[129] combined CS and super-resolution convolutional NN to perform the guided wavefield reconstruction. Even when the number of measurement points was reduced to 10% of the total grid points, the wavefield can still be reconstructed with high quality. To further reduce the volume of data for each signal, Sabeti *et al.*[130] investigated the recovery of wavefields from measurements that were under-sampled in both the spatial and temporal domains, which was based on a two-dimensional ultrasonic guided wave model.

Overall, CS techniques provide an effective choice to reduce the volume of data and the consumed time for data acquisition in NDT.

3.3.3. *Structural load and environment modelling*

In addition to the monitoring and inspection of structures, another major issue in civil engineering is the accurate modelling of structural load and environment, e.g. wind load distributions and site properties. Accurate environmental conditions facilitate rational force analysis and reliability analysis. However, the locations of measurement points are usually spatially sparse, leading to the consequence that the interpolation and extrapolation may not be so reliable. In these scenarios, CS also has the potential to play an important role.

(i) CS for the load distribution

Civil structures are often subjected to wind excitation, which is one of the dominant loads for those flexible structures such as long-span bridges and high-rise buildings. Accurate measurement to the distribution of wind field (e.g. the time histories of wind speed and pressure) is of great significance for the design and assessment of civil structures. However, this in turn raises the issues of both the dense placement of sensors and costly collection of tremendous data. Yan *et al.*[131] implemented the randomly low-rate measurement to obtain the compressed wind speed data. Then, their CS reconstruction was constrained by the nonconvex p norm regularization and solved by the alternating direction multiplier method. As shown in Fig. 10, their reconstruction error is obviously smaller than that of the CS algorithm with convex ℓ_1 norm constraint. The reconstructed wind speed data preserved the characteristics of wind spectral density and rose diagram. Bayındır *et al.*[132] performed CS to the lift and drag measurements for von Kármán vortices generated by winding circular and elliptical columns, and the reconstructed measurements still can be used to analyze von Kármán vortices effectively. Wan *et al.*[133] investigated CS

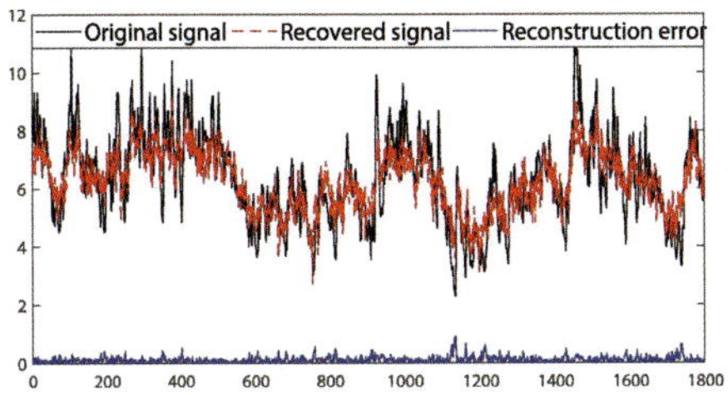

(a)

(b)

Fig. 10. Comparison of reconstruction performance between CS algorithms (a) with convex ℓ_1 norm constraint and (b) with nonconvex p norm constraint.[131] The latter one has a stronger sparsity-promoting effect.

for wind speed data collected from the large-scale spatial structures. Since the wind speed from nearby sensors may be similar despite the time shifts, a dedicated dictionary was designed for sparse representation by taking time shifts into account. Comparing with the linear interpolation method and the CS method using Fourier basis, the error of wind speed that was reconstructed by the CS method with the dedicated dictionary was smaller, especially in cases with high CRs. Pasparakis *et al.*[134] used CS to perform the reconstruction of wind time histories and the extrapolation of wind field in two spatial dimensions, leveraging the sparsity and low-rankness constraints, respectively. The statistics of the underlying stochastic field were accurately estimated from their reconstructed wind data.

The safe operation of marine structures is related to the wave field of seas. The long-term wave records are non-stationary, but can be sparsely characterized by the harmonic wavelet basis. Using adaptive basis, Malara *et al.*[135] explored the use of CS to extrapolate free surface displacements of seas. The results demonstrated that the adaptive bases greatly improved the accuracy of CS reconstruction, and spectral characteristics were effectively retained in the reconstructed free surface displacements. Malara[136] further investigated the extrapolation of free surface displacements based on existing wave pressure head records by CS algorithms, overcoming the limitations of measurement techniques.

(ii) CS for the site property

Modelling of site properties involves the evaluation of natural materials (e.g. soil), which have a high degree of variability. Bayesian CS algorithms are useful to effectively account for these uncertainties. Zhao *et al.*[137] applied Bayesian CS to reconstruct geo-data from discrete measurements with unequal intervals, and computed the posterior PDF of the reconstructed data. Based on this, the derivatives (which are often required in geotechnical engineering) of geo-data were calculated and the associated uncertainties of derivatives were also quantified. Xu *et al.*[138] proposed

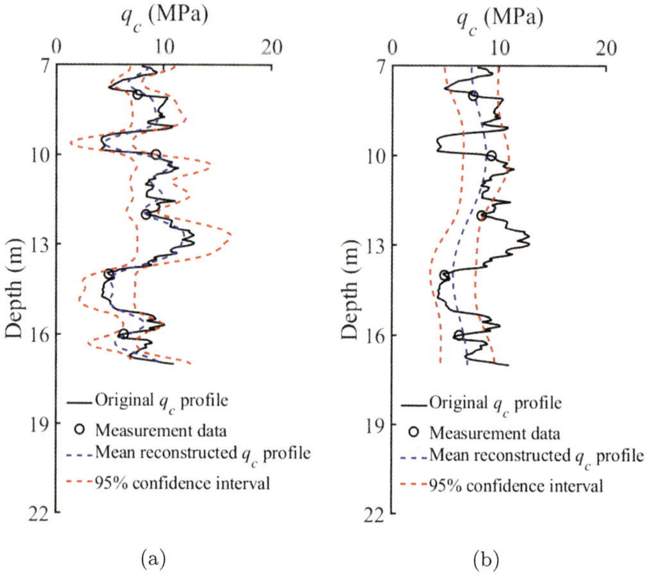

Fig. 11. Interpolated results of extremely sparse geo-data by the (a) collaborative and (b) existing Bayesian CS algorithms.[138] They are probabilistic estimations with associated posterior uncertainties.

a collaborative Bayesian CS algorithm to interpolate extremely sparse geo-data by fusing relevant information from multiple data sources. As demonstrated in Fig. 11, the collaborative Bayesian CS algorithm enabled to interpolate extremely sparse geo-data more properly than existing Bayesian CS. The risk assessment of geotechnical systems requires multiple complex nonlinear simulations and therefore has a huge computational burden. Constructing response surface functions from a few simulation samples is a cost-effective method that has much lower computational complexity. Li *et al.*[139] investigated the reconstruction of the response surface functions by Bayesian CS, and verified the effectiveness of the reconstructed response surface function for slope reliability analysis considering the spatial variability of soil properties.

Site properties are measured by *in situ* testing or laboratory testing. However, the measurements are usually carried out at a few locations due to the time, budget or technical limitations. This raises a question of how to select the most effective sampling locations to obtain as much information as possible about the spatial distribution of site properties. Bayesian CS is also a powerful tool to tackle this problem. Zhao *et al.*[140] explored the determination of efficient sampling locations by uncertainty quantification in Bayesian CS. Using the available measurement data, an updated posterior PDF of the soil property profile was applied to determine next sampling locations based on the information entropy.

In short, CS greatly alleviates the measurement burden required for structural load and environment modelling and therefore has a good potential for real applications.

4. Developments of Emerging Sparsity-Free CS

In the real-world civil engineering applications, the original data may not be necessarily sparse in common transformed domains. Thereby, conventional CS relying on the sparse constraints is no longer applicable as per the theoretical analysis in Sec. 2, and thus may lead to large reconstruction errors. Besides, the nonlinear process of CS reconstruction consumes a large amount of time and computational resources, hindering the real-time processing of CS in civil engineering. To address these limitations, this section provides a review of the efforts on developing fast sparsity-free CS reconstruction algorithms.

4.1. *Generative-model-based CS*

4.1.1. *Motivation*

In recent years, the field of civil engineering has seen significant advances in collecting a large amount of data. These data generally imply the rich information of environmental loads and service behavior of structures. They have the potential to serve as the reference to construct target-specific priors. In this context, we can revisit the orientation of CS: use sparse representations of high-dimensional data as prior constraints for reconstruction. Considering that deep learning models are capable of automatically learning data features, and also proved to outperform traditional hand-designed features in many areas,[141,142] we may employ deep learning to find a more effective data representation than sparse features for constraining the CS reconstruction.

Deep learning-based generative models, that automatically capture low-dimensional latent representation $\mathbf{z} \in \mathbb{R}^Q$ of high-dimensional data, have attracted growing attention (e.g. the well-known generative adversarial network,[29] termed as GAN, and variational autoencoder[30]). By randomly sampling in the low-dimensional latent space, pre-trained generators could draw new realistic samples from the training dataset. Once the generator successfully models the distribution of real data, it can be used as a new prior constraint for CS reconstruction. Such a constraint is established from a more natural assumption: the target data lie near the range of a well pre-trained generator $G_{\boldsymbol{\theta}} : \mathbb{R}^Q \to \mathbb{R}^N$ ($\boldsymbol{\theta}$ is the parameter of pre-trained generator; typically, $Q \ll N$), rather than being sparse under some specified bases. With the active dimensionality reduction effect of $G_{\boldsymbol{\theta}}$, the solution space of CS is explicitly reduced to the low-dimensional latent space, where much less elements need to be determined. Results in Ref. 33 demonstrated that the generative models do automatically construct excellent representations that are more concise than the standard sparse models. Hence, the low-complexity prior automatically modelled by pre-trained generators becomes a competitive alternative to the conventional sparse constraint in CS.

4.1.2. *Requirements and theoretical guarantee for robust recovery*

Generative-model-based CS was first proposed by Bora *et al.*[33] By assuming that the original data lie near the range of a pre-trained generator $G_{\boldsymbol{\theta}}$, this method alleviates the ill-posedness in reconstruction directly. Referring to the restricted eigenvalue condition[143] (REC) in the traditional CS theory that essentially requires all the approximately sparse vectors to be far from the null space of the sensing matrix, Bora *et al.*[33] made a similar recoverability guarantee for the generative-model-based CS. They established a modified REC for the measurement matrix $\boldsymbol{\Phi}$ on the set of natural data $\mathcal{S} \subseteq \mathbb{R}^N$ (where the target data may locate, so called S-REC):

$$\|\boldsymbol{\Phi}(\mathbf{x}_1 - \mathbf{x}_2)\|_2 \geq \gamma\|\mathbf{x}_1 - \mathbf{x}_2\|_2 - \delta, \tag{4.1}$$

where $\mathbf{x}_1$, $\mathbf{x}_2$ are any two distinct vectors in the set $\mathcal{S}$; $\gamma > 0$ is a scale parameter; and δ is an additional slack term, $\delta \geq 0$. Note that since this new approach establishes naturalness constraints directly on the space of original data, only constraints on the measurement matrix are required ($\boldsymbol{\Phi}$ rather than $\boldsymbol{\Theta}$). Under these conditions, we say that the matrix $\boldsymbol{\Phi}$ satisfies S-REC$(\mathcal{S}, \gamma, \delta)$. With this, if two data $\mathbf{x}_1$ and $\mathbf{x}_2$ have significant difference, then the distance between their measurements $\boldsymbol{\Phi}\mathbf{x}_1$ and $\boldsymbol{\Phi}\mathbf{x}_2$ will also be large, allowing us to estimate the original data from CS measurements.

To ensure that the theoretical guarantees are established in the space of data $\mathbf{x}$, rather than in the space of measurements $\mathbf{y}$, the measurement matrix should also preserve norms, i.e. small measurement errors indicate small reconstruction errors. Based on the discussion in Sec. 2, it is known that a random matrix obeying a Gaussian distribution satisfies this property. Here, we denote the target original data as $\mathbf{x}^*$, which unnecessarily exist in $\mathcal{S}$, and the compressed data are $\mathbf{y}^* = \boldsymbol{\Phi}\mathbf{x}^* + \boldsymbol{\varepsilon}$. When the matrix $\boldsymbol{\Phi}$ satisfies S-REC$(\mathcal{S}, \gamma, \delta)$ and preserves norms ($\|\boldsymbol{\Phi}\mathbf{x}\|_2 \leq 2\|\mathbf{x}\|_2$ for every $\mathbf{x} \in \mathbb{R}^N$) with high probability, for $\tilde{\mathbf{x}}$ that approximately minimizes $\|\mathbf{y}^* - \boldsymbol{\Phi}\mathbf{x}\|_2$ over $\mathcal{S}$, with addictive optimization error ϵ:

$$\|\mathbf{y}^* - \boldsymbol{\Phi}\tilde{\mathbf{x}}\|_2 \leq \min_{\mathbf{x}\in\mathcal{S}} \|\mathbf{y}^* - \boldsymbol{\Phi}\mathbf{x}\|_2 + \epsilon, \tag{4.2}$$

Bora *et al.*[33] gave the reconstruction error bound with high probability:

$$\|\tilde{\mathbf{x}} - \mathbf{x}^*\|_2 \leq \left(\frac{4}{\gamma} + 1\right) \min_{\mathbf{x}\in\mathcal{S}} \|\mathbf{x} - \mathbf{x}^*\|_2 + \frac{1}{\gamma}(2\|\boldsymbol{\varepsilon}\|_2 + \epsilon + \delta). \tag{4.3}$$

In the generative-model-based CS, $\mathcal{S}$ should be taken as the range of the pre-trained generator $G_{\boldsymbol{\theta}}$, i.e. $G_{\boldsymbol{\theta}}(\mathbf{z})$. The next problem is how to construct the measurement matrix $\boldsymbol{\Phi}$ that satisfies the above conditions for $G_{\boldsymbol{\theta}}(\boldsymbol{z})$.

For any two different vectors $\mathbf{z}_1$, $\mathbf{z}_2$ in $\mathbb{R}^Q$, the pre-trained generator $G_{\boldsymbol{\theta}}$ is usually a L-Lipschitz function[33]:

$$\|G_{\boldsymbol{\theta}}(\mathbf{z}_1) - G_{\boldsymbol{\theta}}(\mathbf{z}_2)\|_2 \leq L\|\mathbf{z}_1 - \mathbf{z}_2\|_2, \tag{4.4}$$

where L is a positive constant. Bora *et al.*[33] assumed that the input low-dimensional representations $\mathbf{z}$ have bounded norm, simply, a hyper-ball with radius r: $\mathcal{B}^Q(r) = \{\mathbf{z} \,|\, \mathbf{z} \in \mathbb{R}^Q, \|\mathbf{z}\|_2 \leq r\}$ (this restriction is acceptable for the typical generative models,[29,30] which model the latent distribution $p_{\mathbf{z}}(\mathbf{z})$ as a Gaussian with exponentially decaying tails). They proved that for $\alpha < 1$, if the number of CS measurements satisfies

$$M = \Omega\left(\frac{Q}{\alpha^2}\log\frac{Lr}{\delta}\right), \tag{4.5}$$

then the random measurement matrix $\boldsymbol{\Phi}$ with each entry obeying the Gaussian distribution $\mathcal{N}(0, 1/M)$ independently and identically will satisfy S-REC$(G_{\boldsymbol{\theta}}(\mathcal{B}^Q(r)), 1-\alpha, \delta)$ with high probability. Note that the established measurement bound in Eq. (4.5) contains no unknown sparsity parameter, which is more practical than that of conventional CS.

Combining the above results, for any $\mathbf{x}^*$ in $\mathbb{R}^N$, using a random Gaussian matrix $\boldsymbol{\Phi}$ (each entry obeying $\mathcal{N}(0, 1/M)$ independently and identically) with the row number $M = \mathcal{O}[Q\log(Lr/\delta)]$, Bora *et al.*[33] showed that for L-Lipschitz $G_{\boldsymbol{\theta}}$, minimize $\|\mathbf{y}^* - \boldsymbol{\Phi}G_{\boldsymbol{\theta}}(\mathbf{z})\|_2$ in a bounded low-dimensional representation space $\|\mathbf{z}\|_2 \leq r$ to obtain $\tilde{\mathbf{z}}$ by arbitrary algorithms (with the addictive error ϵ in the optimization result), there is the following reconstruction error bound with high probability:

$$\|G_{\boldsymbol{\theta}}(\tilde{\mathbf{z}}) - \mathbf{x}^*\|_2 \leq 6\min_{\|\mathbf{z}\|_2 \leq r}\|G_{\boldsymbol{\theta}}(\mathbf{z}) - \mathbf{x}^*\|_2 + 3\|\boldsymbol{\varepsilon}\|_2 + 2\epsilon + 2\delta, \tag{4.6}$$

where the parameters r (as large as possible for exploring latent space) and δ (as small as possible for well-conditioned $\boldsymbol{\Phi}$) can be set to a form related to L (leading to a constant time losing on M, which is mild), such that roughly $\mathcal{O}(Q\log L)$ measurements are needed to guarantee reconstruction of generative-model-based CS. For D-layer, ReLU-activated generative models, it is proved that $M = \mathcal{O}(QD\log N)$ is sufficient for robust recovery.

Bora *et al.*[33] stated that $\tilde{\mathbf{z}}$ can be solved by any algorithm. Because the deep learning-based generator $G_{\boldsymbol{\theta}}$ is differentiable, they found that the gradient descent can be a good method for the optimization above. As long as the gradient descent method finds a good approximate solution $\tilde{\mathbf{z}}$ to the objective function $\|\mathbf{y}^* - \boldsymbol{\Phi}G_{\boldsymbol{\theta}}(\mathbf{z})\|_2^2$, the output $G_{\boldsymbol{\theta}}(\tilde{\mathbf{z}})$ will be close to the solution that is closest to the original data $\mathbf{x}^*$ in the range of $G_{\boldsymbol{\theta}}$. It was also shown that the multiple random starts technique and Tikhonov regularization can help to avoid local optimums and produce better results. The experimental results presented that generative-model-based CS can achieve the same reconstruction accuracy as the least absolute shrinkage and selection operator[66] even using much fewer measurements, and that the first representation error term in Eq. (4.6) is the dominant error source under a reasonable M, corresponding to the assumption that $\mathbf{x}^*$ lies near the range of $G_{\boldsymbol{\theta}}$.

4.1.3. *Practicality investigation using SHM images for crack segmentation*

Despite the rapid advancements of the deep learning-based generative models, it is still challenging to train generators that can characterize the original data precisely, especially for ultra-high-dimensional data with small training dataset, such as the image data in civil engineering. Limited by the capability of current generative models, since the reconstructed results of generative-model-based CS is constrained to exist in the range of $G_{\boldsymbol{\theta}}$, its CS reconstruction performance saturates and the reconstruction error cannot be reduced to zero even using more measurements, by which those conventional CS algorithms can achieve more accurate reconstruction.[33] Increasing the dimensionality of latent representation may alleviate this problem (i.e. to lower down representation error), but this in turn increases the demand for the number of CS measurements.

The theory established in Sec. 4.1.2 guarantees the reconstruction accuracy in terms of ℓ_2 norm. However, for civil engineering applications, it is sufficient to only identify the features of interest from the data, instead of requiring a full reconstruction. For instance, in structural surface damage detection, the goal is to accurately identify the area of damages instead of analyzing the whole surface. By focusing on the features of interest, the upper bound of reconstruction error has the potential to be effectively reduced (maximizing the use of the assumption that $\mathbf{x}^*$ lies near the range of $G_{\boldsymbol{\theta}}$, or has similar semantics learned by $G_{\boldsymbol{\theta}}$), avoiding the need for very high-quality generators.

Huang *et al.*[32] first explored the generative-model-based CS of SHM image data, with the purpose for crack segmentation, i.e. the feature of interest is crack. They employed a simple deep convolutional GAN[144] to learn the crack features in SHM images (as shown in Fig. 12(a)), which was implemented for CS of compressed SHM images (as shown in Fig. 12(b)), and utilized the similarity of crack regions before and after compression as the reconstruction accuracy metric.

This approach leverages the assumption that the target data are "close" to the range of the pre-trained generator (in term of the target crack feature). This also greatly simplified the training process of the generative model, since it is not required to capture some non-essential background details from image data. In the experiments, the generator trained only for just ten minutes or so outperformed conventional CS algorithms in terms of reconstruction accuracy, calculation speed

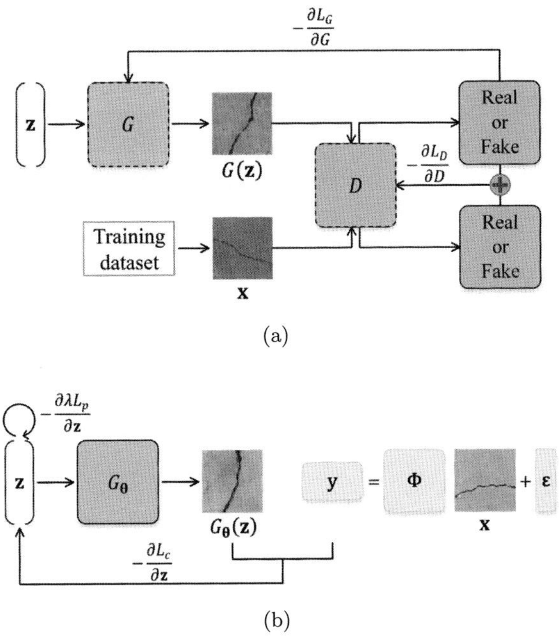

Fig. 12. The (a) training and (b) reconstruction of generative-model-based CS.[32]

and noise robustness. As a reference, the classical image compression algorithm, named JPEG,[145] in the traditional sampling-and-compression framework failed to reconstruct at a CR of 8, while the generative-model-based CS was still able to recover the crack regions from CS measurements at a CR of 64.

Since the reconstructed results are constrained in the data distribution modelled by the pre-trained generator, this implicit constraint on the feature of interest (e.g. crack) allows generative-model-based CS to have a strong noise robustness. The use of advanced GPU for gradient descent computation also leads to a fast reconstruction. Moreover, the method can be applied for the occlusion and motion blur removal if the occlusion and motion blur processes are treated as a linear operation (but not necessarily satisfying the recoverability condition).[32] The super resolution problem in Ref. 33 is also related to this.

More importantly, considering the variability of backgrounds of civil structure images, it is crucial to examine whether the generator G_θ has sufficient generalization capability. Huang *et al.*[32] trained the G_θ using crack images from smooth concrete surface, and tested the CS reconstruction of crack images with complex backgrounds, including steel fatigue cracks, cracks on the surface of post-earthquake building and rough bridge road. The results in Fig. 13 demonstrated that the simply trained G_θ can still accurately reconstruct the crack regions, though the reconstruction of background details, on the other hand, almost fails (maintaining a pattern

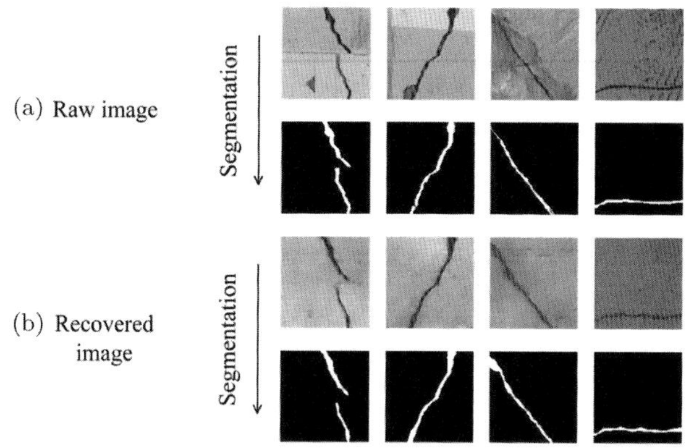

Fig. 13. Illustrative examples of the generalization capability of generative-model-based CS.[32]

similar to that of the training data). This indicates that generative-model-based CS is suitable for practical use, with its good generalization capability.

With all these successful efforts, the application of generative-model-based CS for the structural time series data remains a challenge. For instance, training effective generative models with better performance than well-established sparse representations is an issue.

4.2. *End-to-end reconstruction by deep learning model*

Considering the powerful nonlinear mapping capability of deep learning models, they were employed to learn nonlinear end-to-end mapping relationships from the measurements to target objects. This subsection provides a review on the non-iterative fast reconstruction algorithms following this end-to-end learning idea that does not require reconstruction with intermediate steps, as well as representative applications in civil engineering.

For cases that are consistent with the distribution of training data, the end-to-end non-iterative approach (as shown in Fig. 14) offers undeniable advantage of being accurate and fast, although it lacks interpretability. Dong *et al.*[146] utilized a GAN model for end-to-end CS of SHM vibration signals, achieving fast and accurate reconstruction by considering the consistency of both the time and frequency domains of signals in the loss function. The effect of network hyperparameters on reconstruction performance was also investigated.

In recent years, the end-to-end reconstruction methods have also been widely applied to the recovery of lost data of structural responses, or the prediction of unmeasured data since it can be treated as data loss in spatial domain. Researchers have developed methods adopting various NN architectures, such as the convolutional NN,[148] recurrent NN,[149] GAN,[147,150] graph NN,[151] physics-informed NN,[152]

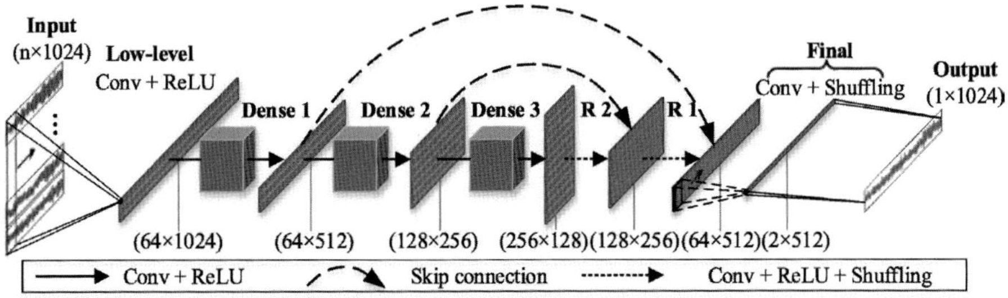

Fig. 14. Illustration of end-to-end reconstruction using trained deep learning models.[147]

and their combinations with time-frequency cross-domain information,[150] data simulated by finite element model,[153] self-attention mechanism,[154,155] multimodal learning for heterogeneous data.[156] Yang *et al.*[157] used a recurrent NN to identify the outliers and used a conditional GAN for repairing data after removing outliers. Pan *et al.*[158] proposed a novel attention-based NN model trained by datasets where the structural seismic responses of measured and unmeasured locations of interest are inputs and outputs, respectively. Experimental results demonstrated that the reconstructed output of trained model can achieve real-time seismic assessment of bridges.

Wang *et al.*[159] discussed the use of a NN model to learn the mapping of data from the online open dataset to the missing part of the target dataset in the same period, when all local records are lost at the same time and cannot be reconstructed using the relationship among sensors. Comerford *et al.*[160] used a NN model to recover the missing stochastic process records for power spectrum estimation. They[161] further used multiple simple NN models to quantify the uncertainties of recovered records. Xiang *et al.*[162] trained super-resolution NN models for the low-resolution crack images, which improved the semantic segmentation accuracy. Moreover, the recovery of other lost data in civil engineering, including the temperature data,[163] weigh-in-motion data[164] and wind pressure monitoring data,[165] was also studied by training end-to-end deep learning models.

In fact, Bora *et al.*[33] also conducted CS experiments by training such end-to-end reconstruction model, and achieved very good reconstruction results based on very few CS measurements. Such an approach is applicable to any inverse problem, not just CS, its straightforward training procedure with very fast reconstruction speed is highly attractive in engineering applications, but also has obvious drawbacks, namely, the often-criticized generalization problem of deep learning models. It requires the target data and forward measurement process to be consistent with the training dataset, i.e. it has no adaptive capability for different measurement strategies, and requires retraining the model when replacing the measurement strategy or processing out-of-distribution data, which induces an additional overhead compared to the method in Sec. 4.1. In addition, it is easily affected by abnormal patterns in data such as dependent noise and outliers.

5. Concluding Remarks and Future Prospects

Compressive sampling (CS) originates in the field of signal processing, but its impact goes far beyond that. It enables efficient sampling by simply measuring some linear functionals of original signals, at a much lower rate than the requirements by the Shannon–Nyquist sampling theorem. More importantly, the CS theory guarantees accurate reconstruction of original signals from noisy incomplete linear measurements, by imposing practical sparsity constraints. This allows rigorous analysis and solution in a unified framework for many inverse problems in civil engineering. Advances in CS bring significant convenience to engineering scenarios that are plagued by the trade-off between low-cost measurements and accurate reconstruction.

Under appropriate bases, it is common to present sparsity for civil engineering objects, including signals and images in SHM and NDT, structural damages, load distributions, and site properties, while they are often severely under-sampled. These inverse problems with sparsity prior drive the developments of CS theories and applications in the field of civil engineering. Specifically, large uncertainties in CS reconstruction from highly noisy and incomplete measurements motivate the use of a Bayesian probabilistic framework in CS. The quantified posterior uncertainty gives insights on the confidence level of reconstructed results and helps to improve the robustness of the whole CS process. Moreover, the pre-defined sparsity may be not necessarily general and efficient, due to the huge variability of civil engineering problems. Powerful deep generative models trained by target datasets automatically capture an efficient low-dimensional representations of target objects. The generative-model-based CS actively reduces the solution space to the low-dimensional latent space, thus not requiring the sparsity at all. With the researchers' continuous efforts, advancements in CS have brought important inspiration and convenient solution to inverse problems in broader scientific fields.

In the future, there are several crucial problems to be tackled for developing a CS-unified civil engineering system.

5.1. *Real-time reconstruction*

In contrast to the simple interpolation reconstruction guaranteed by the Shannon–Nyquist sampling theorem, CS requires some complicated methods to reconstruct the original data. The reconstruction speed of current algorithms is insufficient for real-time monitoring of civil engineering systems, especially for very high-dimensional problems. To speed up CS reconstruction, we can utilize special structures in data, such as group sparsity to facilitate simultaneous reconstruction for multiple relevant data[88] (may be split from the same large-scale data[166]), or tree sparsity to reduce the computational complexity of reconstruction by ignoring parts of unknowns at once.[167] These structures also provide additional information to improve the reconstruction performance.

Furthermore, it is imperative to develop modular and portable software for distributed/parallel computing platforms for the reconstruction, as well as adapt the CS algorithms for advanced GPU hardware, such as algorithms in Sec. 4 that incorporate a pre-trained deep generative model or an end-to-end reconstruction model, whose speed is very easy to scale.

5.2. *Robust reconstruction*

Bayesian CS allows quantification of the posterior uncertainty, which measures confidence for the reconstructed data. The estimated posterior uncertainty provides a useful tool for reconstruction diagnostics without knowledge of the original data. However, the accurate quantification of posterior uncertainties by more effective stochastic simulation methods still needs further exploration in real-world applications.

One important issue of those probabilistic robust Bayesian CS algorithms is the high computational resource consumption because of uncertainty estimation, especially when dealing with high-dimensional objects and large-scale datasets. Research topics include variable-resolution methods that permit progressive refinement of model parameterization, and recursive Bayesian estimation over time using sequential datasets.

Moreover, Bayesian inference for sparse inversion considering nonlinear components requires further investigation.[168] Approximate Bayesian Computation[169] methods have the potential to bypass the difficulty of evaluating analytical formulas of likelihood functions (may be difficult, or even elusive for real-world problems).

5.3. *Transferable reconstruction*

The suitability of assumed prior constraints is crucial for the success of CS reconstruction. A highly transferable constraint among similar inverse problems can greatly facilitate the application of CS, and reduce the implementation cost. Although sparsity is a common choice, it may not be the most efficient one. As reviewed before, the deep learning models (generative or end-to-end models) that learn mappings without sparsity constraints can significantly improve reconstruction performance. However, whether such relationships trained on specific datasets can be sufficiently generalized for civil engineering problems, especially for end-to-end models, remains an open question. To avoid the laborious tuning for deep learning models, the generalization issue may be solved by the broad learning system with the incremental learning capability,[170] allowing real-time updating of network parameters based on the latest data. Transfer learning such as domain adaptation[171] or heterogeneous transfer learning[172,173] can also be applied. These approaches may provide fast reconstruction speed and high generalization ability for new target classes of data.

5.4. *Decentralized reconstruction*

In practice, CS reconstruction is usually performed after the compressed data are transmitted back to a central base station, imposing significantly high computing power requirements on the base station. A more cost-effective and robust decentralized system can be achieved by implementing the CS reconstruction on several edge computing devices. The greedy strategy-based CS algorithms, which is technically easy to implement, demonstrate high reconstruction efficiency and have been applied on low-power edge computing devices with limited recourses.[174] However, other CS reconstruction algorithms with complex computation operators and high computational complexity still have a relatively long way to go, for which model pruning[175] is a feasible choice. In other engineering fields, the edge collaborative CS wireless sensor networks have been investigated, which include the self-developed nodes, adaptive transmission scheduling and other techniques.[176–178]

When we actually do not need the complete original data but only their key features, decentralized edge computing devices can also facilitate to perform the compressive signal processing[123] (avoiding intermediate errors) and transmit the required features directly.

Acknowledgments

The authors acknowledge financial support from National Key Research and Development Program of China [Grant No. 2021YFF0501003] and National Natural Science Foundation of China [Grant Nos. U2139209 and 52078174].

ORCID

Haoyu Zhang https://orcid.org/0009-0004-9620-7344

Shicheng Xue https://orcid.org/0009-0009-2927-0788

Yong Huang https://orcid.org/0000-0002-7963-0720

Hui Li https://orcid.org/0000-0001-9198-3951

References

1. E. J. Candès, Compressive sampling, in *Proc. Int. Congress of Mathematicians*, Vol. 3 (2006), pp. 1433–1452.
2. E. J. Candès, J. K. Romberg and T. Tao, Stable signal recovery from incomplete and inaccurate measurements, *Commun. Pure Appl. Math.* **59**(8) (2006) 1207–1223.

3. D. L. Donoho, Compressed sensing, *IEEE Trans. Inf. Theory* **52**(4) (2006) 1289–1306.
4. E. J. Candès, J. K. Romberg and T. Tao, Robust uncertainty principles: Exact signal reconstruction from highly incomplete frequency information, *IEEE Trans. Inf. Theory* **52**(2) (2006) 489–509.
5. M. Fazel, E. Candes, B. Recht and P. Parrilo, Compressed sensing and robust recovery of low rank matrices, in *Proc. 42nd Asilomar Conf. Signals, IEEE Systems and Computers* (IEEE, 2008), pp. 1043–1047.
6. R. G. Baraniuk and M. B. Wakin, Random projections of smooth manifolds, *Found. Comput. Math.* **9**(1) (2009) 51–77.
7. R. G. Baraniuk, V. Cevher, M. F. Duarte and C. Hegde, Model-based compressive sensing, *IEEE Trans. Inf. Theory* **56**(4) (2010) 1982–2001.
8. M. F. Duarte and Y. C. Eldar, Structured compressed sensing: From theory to applications, *IEEE Trans. Signal Process.* **59**(9) (2011) 4053–4085.
9. C. E. Shannon, Communication in the presence of noise, *Proc. IRE* **37**(1) (1949) 10–21.
10. J. F. Claerbout and F. Muir, Robust modeling with erratic data, *Geophysics* **38**(5) (1973) 826–844.
11. M. Lustig, D. Donoho and J. M. Pauly, Sparse MRI: The application of compressed sensing for rapid MR imaging, *Magn. Reson. Med.* **58**(6) (2007) 1182–1195.
12. L. C. Potter, E. Ertin, J. T. Parker and M. Cetin, Sparsity and compressed sensing in radar imaging, *Proc. IEEE* **98**(6) (2010) 1006–1020.
13. S. C. Li, L. Da Xu and X. H. Wang, Compressed sensing signal and data acquisition in wireless sensor networks and internet of things, *IEEE Trans. Ind. Inform.* **9**(4) (2012) 2177–2186.
14. Z. Gao, L. Dai, S. Han, I. Chih-Lin, Z. Wang and L. Hanzo, Compressive sensing techniques for next-generation wireless communications, *IEEE Wirel. Commun.* **25**(3) (2018) 144–153.
15. I. A. Kougioumtzoglou, I. Petromichelakis and A. F. Psaros, Sparse representations and compressive sampling approaches in engineering mechanics: A review of theoretical concepts and diverse applications, *Probab. Eng. Mech.* **61** (2020) 103082.
16. N. Jayant, Signal compression: Technology targets and research directions, *IEEE J. Sel. Areas Commun.* **10**(5) (1992) 796–818.
17. S. W. Ma, X. F. Zhang, C. M. Jia, Z. H. Zhao, S. Q. Wang and S. S. Wang, Image and video compression with neural networks: A review, *IEEE Trans. Circuits Syst. Video Technol.* **30**(6) (2019) 1683–1698.
18. Y. Q. Bao, H. Li, X. D. Sun, Y. Yu and J. P. Ou, Compressive sampling-based data loss recovery for wireless sensor networks used in civil structural health monitoring, *Struct. Health Monit.* **12**(1) (2013) 78–95.
19. M. Vidyasagar, *An Introduction to Compressed Sensing* (Society for Industrial and Applied Mathematics, 2019).
20. L. Zhang, Y. Liang and L. Yu, Moving force identification based on group Lasso and compressed sensing, *Int. J. Struct. Stab. Dyn.* **22**(15) (2022) 2250170.
21. Y. C. Yang and S. Nagarajaiah, Output-only modal identification by compressed sensing: Non-uniform low-rate random sampling, *Mech. Syst. Signal Process.* **56** (2015) 15–34.
22. Y. Huang and J. L. Beck, Hierarchical sparse Bayesian learning for structural health monitoring with incomplete modal data, *Int. J. Uncertain. Quantif.* **5**(2) (2015) 139–169.

23. S. H. Ji, Y. Xue and L. Carin, Bayesian compressive sensing, *IEEE Trans. Signal Process.* **56**(6) (2008) 2346–2356.
24. S. H. Ji, D. Dunson and L. Carin, Multitask compressive sensing, *IEEE Trans. Signal Process.* **57**(1) (2008) 92–106.
25. Y. Huang, J. L. Beck, S. Wu and H. Li, Robust Bayesian compressive sensing for signals in structural health monitoring, *Comput. Aided Civ. Infrastruct. Eng.* **29**(3) (2014) 160–179.
26. Y. Huang, J. L. Beck, S. Wu and H. Li, Bayesian compressive sensing for approximately sparse signals and application to structural health monitoring signals for data loss recovery, *Probab. Eng. Mech.* **46** (2016) 62–79.
27. K. Kreutz-Delgado, J. F. Murray, B. D. Rao, K. Engan, T. W. Lee and T. J. Sejnowski, Dictionary learning algorithms for sparse representation, *Neural Comput.* **15**(2) (2003) 349–396.
28. J. Mairal, F. Bach, J. Ponce and G. Sapiro, Online dictionary learning for sparse coding, in *Proc. 26th Annual Int. Conf. Machine Learning* (ACM, 2009), pp. 689–696.
29. I. Goodfellow, J. Pouget-Abadie, M. Mirza, B. Xu, D. Warde-Farley, S. Ozair, A. Courville and Y. Bengio, Generative adversarial nets, in *Proc. 27th Int. Conf. Neural Information Processing Systems*, Vol. 2 (2014), pp. 2672–2680.
30. D. P. Kingma and M. Welling, Auto-encoding variational bayes, *arXiv preprint* (2013), arXiv:1312.6114.
31. D. P. Kingma and P. Dhariwal, Glow: Generative flow with invertible 1 × 1 convolutions, in *Proceedings of the 32nd International Conference on Neural Information Processing Systems* (Curran Associates Inc., 2018), pp. 10236–10245.
32. Y. Huang, H. Y. Zhang, H. Li and S. Wu, Recovering compressed images for automatic crack segmentation using generative models, *Mech. Syst. Signal Process.* **146** (2021) 107061.
33. A. Bora, A. Jalal, E. Price and A. G. Dimakis, Compressed sensing using generative models, in *ICML'17: Proceedings of the 34th International Conference on Machine Learning*, (JMLR.org, 2017), pp. 537–546.
34. S. Mallat, *A Wavelet Tour of Signal Processing* (Elsevier, 1999).
35. V. K. Goyal, Theoretical foundations of transform coding, *IEEE Signal Process. Mag.* **18**(5) (2001) 9–21.
36. R. G. Baraniuk, Compressive sensing [lecture notes], *IEEE Signal Process. Mag.* **24**(4) (2007) 118–121.
37. J. A. Tropp, M. B. Wakin, M. F. Duarte, D. Baron and R. G. Baraniuk, Random filters for compressive sampling and reconstruction, in *Proc. IEEE Int. Conf. Acoustics Speech and Signal Processing* (IEEE, 2006), p. III.
38. S. Kirolos, J. Laska, M. Wakin, M. Duarte, D. Baron, T. Ragheb and R. Baraniuk, Analog-to-information conversion via random demodulation, in *Proc. IEEE Dallas/CAS Workshop on Design, Applications, Integration and Software* (IEEE, 2006), pp. 71–74.
39. J. Romberg, Compressive sensing by random convolution, *SIAM J. Imaging Sci.* **2**(4) (2009) 1098–1128.
40. M. A. Davenport, J. N. Laska, P. T. Boufounos and R. G. Baraniuk, A simple proof that random matrices are democratic, *arXiv preprint* (2009), arXiv:0911.0736.

41. J. N. Laska, P. T. Boufounos, M. A. Davenport and R. G. Baraniuk, Democracy in action: Quantization, saturation, and compressive sensing, *Appl. Comput. Harmon. Anal.* **31**(3) (2011) 429–443.
42. M. Mangia, R. Rovatti and G. Setti, Rakeness in the design of analog-to-information conversion of sparse and localized signals, *IEEE Trans. Circuits Syst. I Regul. Pap.* **59**(5) (2012) 1001–1014.
43. T. Hastie, R. Tibshirani and M. Wainwright, *Statistical learning with sparsity: The lasso and generalizations* (CRC press, 2015).
44. Y. Eldar and G. Kutyniok (eds.), *Compressed Sensing: Theory and Applications* (Cambridge University Press, Cambridge, 2012), doi:10.1017/CBO9780511794308.
45. D. L. Donoho and M. Elad, Optimally sparse representation in general (nonorthogonal) dictionaries via $\ell 1$ minimization, *Proc. Natl Acad. Sci. USA* **100**(5) (2003) 2197–2202.
46. E. J. Candes and T. Tao, Decoding by linear programming, *IEEE Trans. Inf. Theory* **51**(12) (2005) 4203–4215.
47. E. J. Candes, The restricted isometry property and its implications for compressed sensing, *C. R. Math.* **346**(9–10) (2008) 589–592.
48. M. A. Davenport and M. B. Wakin, Analysis of orthogonal matching pursuit using the restricted isometry property, *IEEE Trans. Inf. Theory* **56**(9) (2010) 4395–4401.
49. R. G. Baraniuk, M. A. Davenport, R. A. DeVore and M. B. Wakin, A simple proof of the restricted isometry property for random matrices, *Constr. Approx.* **28** (2008) 253–263.
50. R. DeVore, G. Petrova and P. Wojtaszczyk, Instance-optimality in probability with an $\ell 1$-minimization decoder, *Appl. Comput. Harm. Anal.* **27**(3) (2009) 275–288.
51. J. A. Tropp and A. C. Gilbert, Signal recovery from random measurements via orthogonal matching pursuit, *IEEE Trans. Inf. Theory* **53**(12) (2007) 4655–4666.
52. Z. Ben-Haim, Y. C. Eldar and M. Elad, Coherence-based performance guarantees for estimating a sparse vector under random noise, *IEEE Trans. Signal Process.* **58**(10) (2010) 5030–5043.
53. E. J. Candes, Y. C. Eldar, D. Needell and P. Randall, Compressed sensing with coherent and redundant dictionaries, *Appl. Comput. Harmon. Anal.* **31**(1) (2011) 59–73.
54. J. A. Tropp, A. C. Gilbert and M. J. Strauss, Algorithms for simultaneous sparse approximation. Part I: Greedy pursuit, *Signal Process.* **86**(3) (2006) 572–588.
55. Y. C. Pati, R. Rezaiifar and P. S. Krishnaprasad, Orthogonal matching pursuit: Recursive function approximation with applications to wavelet decomposition, in *Proc. 27th Asilomar Conf. Signals, Systems and Computers*, (IEEE, 1993), pp. 40–44.
56. T. Blumensath and M. E. Davies, Iterative thresholding for sparse approximations, *J. Fourier Anal. Appl.* **14** (2008) 629–654.
57. D. Needell and J. A. Tropp, CoSaMP: Iterative signal recovery from incomplete and inaccurate samples, *Appl. Comput. Harmon. Anal.* **26**(3) (2009) 301–321.
58. Z. L. Bai, S. L. Chen, Q. Y. Xiao, L. C. Jia, Y. B. Zhao and Z. M. Zeng, Compressive sensing of phased array ultrasonic signal in defect detection: Simulation study and experimental verification, *Struct. Health Monit.* **17**(3) (2018) 434–449.
59. S. Sawant, S. Banerjee and S. Tallur, Performance evaluation of compressive sensing based lost data recovery using OMP for damage index estimation in ultrasonic SHM, *Ultrasonics* **115** (2021) 106439.

60. J. A. Tropp, Greed is good: Algorithmic results for sparse approximation, *IEEE Trans. Inf. Theory* **50**(10) (2004) 2231–2242.
61. J. A. Tropp, Algorithms for simultaneous sparse approximation. Part II: Convex relaxation, *Signal Process.* **86**(3) (2006) 589–602.
62. R. Chartrand and V. Staneva, Restricted isometry properties and nonconvex compressive sensing, *Inverse Probl.* **24**(3) (2008) 035020.
63. R. Chartrand and W. Yin, Iteratively reweighted algorithms for compressive sensing, *2008 IEEE Int. Conf. Acoustics, Speech and Signal Processing* (IEEE, 2008), pp. 3869–3872.
64. S. S. Chen, D. L. Donoho and M. A. Saunders, Atomic decomposition by basis pursuit, *SIAM Rev.* **43**(1) (2001) 129–159.
65. M. A. Figueiredo, R. D. Nowak and S. J. Wright, Gradient projection for sparse reconstruction: Application to compressed sensing and other inverse problems, *IEEE J. Sel. Top. Signal Process.* **1**(4) (2007) 586–597.
66. R. Tibshirani, Regression shrinkage and selection via the lasso, *J. R. Stat. Soc. Series B (Methodological)* **58**(1) (1996) 267–288.
67. Z. He, A. Cichocki, R. Zdunek and S. Xie, Improved FOCUSS method with conjugate gradient iterations, *IEEE Trans. Signal Process.* **57**(1) (2008) 399–404.
68. G. Cormode and S. Muthukrishnan, Combinatorial algorithms for compressed sensing, in *Proc. 40th Annual Conf. Information Sciences and Systems* (IEEE, 2006), pp. 198–201.
69. Y. Yang and S. Nagarajaiah, Blind denoising of structural vibration responses with outliers via principal component pursuit, *Struct. Control Health Monit.* **21**(6) (2014) 962–978.
70. Y. C. Yang, S. Nagarajaiah and Y. Q. Ni, Data compression of very large-scale structural seismic and typhoon responses by low-rank representation with matrix reshape, *Struct. Control Health Monit.* **22**(8) (2015) 1119–1131.
71. M. E. Tipping, Sparse Bayesian learning and the relevance vector machine, *J. Mach. Learn. Res.* **1** (2001) 211–244.
72. J. L. Beck, Bayesian system identification based on probability logic, *Struct. Control Health Monit.* **17**(7) (2010) 825–847.
73. D. Baron, S. Sarvotham and R. G. Baraniuk, Bayesian compressive sensing via belief propagation, *IEEE Trans. Signal Process.* **58**(1) (2009) 269–280.
74. A. Argyriou, T. Evgeniou and M. Pontil, Multi-task feature learning, in *Proceedings of the 19th International Conference on Neural Information Processing Systems* (MIT Press, 2006), pp. 41–48.
75. G. Obozinski, B. Taskar and M. Jordan, Multi-task feature selection, Statistics Department, UC Berkeley, Technical Report, **2**(2.2) (2006) 2.
76. Q. Wang, P. L. Zhang, C. Meng, H. G. Wang and C. Wang, Multi-task Bayesian compressive sensing for vibration signals in diesel engine health monitoring, *Measurement* **136** (2019) 625–635.
77. Y. Huang, J. L. Beck and H. Li, Multitask sparse Bayesian learning with applications in structural health monitoring, *Comput. Aided Civ. Infrastruct. Eng.* **34**(9) (2019) 732–754.
78. Y. Huang, C. S. Shao, S. Wu and H. Li, Diagnosis and accuracy enhancement of compressive-sensing signal reconstruction in structural health monitoring using multi-task sparse Bayesian learning, *Smart Mater. Struct.* **28**(3) (2019) 035001.

79. H. P. Wan, G. S. Dong, Y. Luo and Y. Q. Ni, An improved complex multi-task Bayesian compressive sensing approach for compression and reconstruction of SHM data, *Mech. Syst. Signal Process.* **167** (2022) 108531.
80. L. Yu, H. Sun, J. P. Barbot and G. Zheng, Bayesian compressive sensing for cluster structured sparse signals, *Signal Process.* **92**(1) (2012) 259–269.
81. L. H. He, H. J. Chen and L. Carin, Tree-structured compressive sensing with variational Bayesian analysis, *IEEE Signal Process. Lett.* **17**(3) (2009) 233–236.
82. D. Sejdinović, C. Andrieu and R. Piechocki, Bayesian sequential compressed sensing in sparse dynamical systems, in *Proc. 48th Annual Allerton Conf. Communication, Control, and Computing (Allerton)* (IEEE, 2010), pp. 1730–1736.
83. Y. Huang, J. L. Beck, H. Li and Y. L. Ren, Sequential sparse Bayesian learning with applications to system identification for damage assessment and recursive reconstruction of image sequences, *Comput. Meth. Appl. Mech. Eng.* **373** (2021) 113545.
84. C. R. Farrar and K. Worden, An introduction to structural health monitoring, *Phil. Trans. R. Soc. A* **365**(1851) (2007) 303–315.
85. Y. Bao, Z. C. Chen, S. Y. Wei, Y. Xu, Z. Y. Tang and H. Li, The state of the art of data science and engineering in structural health monitoring, *Engineering* **5**(2) (2019) 234–242.
86. Y. Q. Bao, J. L. Beck and H. Li, Compressive sampling for accelerometer signals in structural health monitoring, *Struct. Health Monit.* **10**(3) (2011) 235–246.
87. M. Jayawardhana, X. Q. Zhu, R. Liyanapathirana and U. Gunawardana, Compressive sensing for efficient health monitoring and effective damage detection of structures, *Mech. Syst. Signal Process.* **84** (2017) 414–430.
88. Y. Q. Bao, Z. Q. Shi, X. Y. Wang and H. Li, Compressive sensing of wireless sensors based on group sparse optimization for structural health monitoring, *Struct. Health Monit.* **17**(4) (2018) 823–836.
89. H. D. Li, D. Ai, H. P. Zhu and H. Luo, An Orthogonal Matching Pursuit based signal compression and reconstruction approach for electromechanical admittance based structural health monitoring, *Mech. Syst. Signal Process.* **133** (2019) 106276.
90. J. Kang, W. Ren, Y. L. Xie, Y. Zhao and J. F. Wang, An enhanced method to reduce reconstruction error of compressed sensing for structure vibration signals, *Mech. Syst. Signal Process.* **183** (2023) 109585.
91. Y. Bao, Z. Y. Tang and H. Li, Compressive-sensing data reconstruction for structural health monitoring: A machine-learning approach, *Struct. Health Monit.* **19**(1) (2020) 293–304.
92. S. X. Chen, Y. Q. Ni and L. Zhou, A deep learning framework for adaptive compressive sensing of high-speed train vibration responses, *Struct. Control Health Monit.* **29**(8) (2022) e2979.
93. F. Amini, Y. Hedayati and H. Zanddizari, Determining the number of measurements for compressive sensing of traffic-induced vibration data, *Measurement* **152** (2020) 107259.
94. J. N. Laska, S. Kirolos, M. F. Duarte, T. S. Ragheb, R. G. Baraniuk and Y. Massoud, Theory and implementation of an analog-to-information converter using random demodulation, *2007 IEEE Int. Symposium on Circuits and Systems*, (IEEE, 2007), pp. 1959–1962.
95. D. Liang, Q. Han, Y. Cai, K. Yu and Y. Zhang, Compressive sampling system based on random demodulation for active and passive structural health monitoring, *Smart Mater. Struct.* **31**(6) (2022) 065005.

96. M. Mangia, F. Pareschi, V. Cambareri, R. Rovatti and G. Setti, Rakeness-based design of low-complexity compressed sensing, *IEEE Trans. Circuits Syst. I Regul. Pap.* **64**(5) (2017) 1201–1213.
97. T. Loss, M. J. Colbrook and A. C. Hansen, Stratified sampling based compressed sensing for structured signals, *IEEE Trans. Signal Process.* **70** (2022) 3530–3539.
98. F. Zonzini, M. Zauli, M. Mangia, N. Testoni and L. De Marchi, Model-assisted compressed sensing for vibration-based structural health monitoring, *IEEE Trans. Ind. Inform.* **17**(11) (2021) 7338–7347.
99. F. Zonzini, M. Zauli, M. Mangia, N. Testoni and L. De Marchi, HW-oriented compressed sensing for operational modal analysis: The impact of noise in MEMS accelerometer networks, in *Proc. Sensors Applications Symposium* (IEEE, 2021), pp. 1–5.
100. F. Zonzini, A. Carbone, F. Romano, M. Zauli and L. De Marchi, Machine learning meets compressed sensing in vibration-based monitoring, *Sensors* **22**(6) (2022) 2229.
101. G. Karameşe *et al.*, Vision-based dynamic response measurements of structures by using smartphone videos and alternative techniques, *Int. J. Struct. Stab. Dyn.* **23**(7) (2023) 2350069.
102. Y. C. Yang and S. Nagarajaiah, Robust data transmission and recovery of images by compressed sensing for structural health diagnosis, *Struct. Control Health Monit.* **24**(1) (2017) e1856.
103. V. Ganesan, T. Das, N. Rahnavard and J. L. Kauffman, Vibration-based monitoring and diagnostics using compressive sensing, *J. Sound Vib.* **394** (2017) 612–630.
104. Q. P. Mei, N. Shirzad-Ghaleroudkhani, M. Gül, S. F. Ghahari and E. Taciroglu, Bridge mode shape identification using moving vehicles at traffic speeds through non-parametric sparse matrix completion, *Struct. Control Health Monit.* **28**(7) (2021) e2747.
105. D. Jana and S. Nagarajaiah, Data-driven full-field vibration response estimation from limited measurements in real-time using dictionary learning and compressive sensing, *Eng. Struct.* **275** (2023) 115280.
106. S. Bhowmick and S. Nagarajaiah, Spatiotemporal compressive sensing of full-field Lagrangian continuous displacement response from optical flow of edge: Identification of full-field dynamic modes, *Mech. Syst. Signal Process.* **164** (2022) 108232.
107. Z. Y. Tang, Y. Q. Bao and H. Li, Group sparsity-aware convolutional neural network for continuous missing data recovery of structural health monitoring, *Struct. Health Monit.* **20**(4) (2021) 1738–1759.
108. Z. L. Zou, Y. Q. Bao, H. Li, B. F. Spencer and J. P. Ou, Embedding compressive sensing-based data loss recovery algorithm into wireless smart sensors for structural health monitoring, *IEEE Sens. J.* **15**(2) (2014) 797–808.
109. Z. L. Zou, Y. Q. Bao, F. D. Deng and H. Li, An approach of reliable data transmission with random redundancy for wireless sensors in structural health monitoring, *IEEE Sens. J.* **15**(2) (2014) 809–818.
110. Y. Q. Bao, Y. Yu, H. Li, X. Q. Mao, W. F. Jiao, Z. L. Zou and J. P. Ou, Compressive sensing-based lost data recovery of fast-moving wireless sensing for structural health monitoring, *Struct. Control Health Monit.* **22**(3) (2015) 433–448.
111. F. Amini, Y. Hedayati and H. Zanddizari, Exploiting the inter-correlation of structural vibration signals for data loss recovery: A distributed compressive sensing based approach, *Mech. Syst. Signal Process.* **152** (2021) 107473.

112. M. F. Duarte, S. Sarvotham, D. Baron, M. B. Wakin and R. G. Baraniuk, Distributed Compressed Sensing of Jointly Sparse Signals, in *Conference Record of the Thirty-Ninth Asilomar Conference on Signals, Systems and Computers*, (IEEE, 2005), pp. 1537–1541.
113. Y. C. Yang and S. Nagarajaiah, Harnessing data structure for recovery of randomly missing structural vibration responses time history: Sparse representation versus low-rank structure, *Mech. Syst. Signal Process.* **74** (2016) 165–182.
114. D. Jana, S. Nagarajaiah, Y. C. Yang and S. L. Li, Real-time cable tension estimation from acceleration measurements using wireless sensors with packet data losses: Analytics with compressive sensing and sparse component analysis, *J. Civ. Struct. Health Monit.* **12**(4) (2022) 797–815.
115. S. Y. Chen, Y. W. Wang and Y. Q. Ni, Gross outlier removal and fault data recovery for SHM data of dynamic responses by an annihilating filter-based Hankel-structured robust PCA method, *Struct. Control Health Monit.* **29**(12) (2022) e3144.
116. L. Ngeljaratan, M. A. Moustafa and G. Pekcan, A compressive sensing method for processing and improving vision-based target-tracking signals for structural health monitoring, *Comput.-Aided Civ. Infrastruct. Eng.* **36**(9) (2021) 1203–1223.
117. C. D. Zhang and Y. L. Xu, Comparative studies on damage identification with Tikhonov regularization and sparse regularization, *Struct. Control Health Monit.* **23**(3) (2016) 560–579.
118. R. Hou, Y. Xia, Q. Xia and X. Zhou, Genetic algorithm based optimal sensor placement for L1-regularized damage detection, *Struct. Control Health Monit.* **26**(1) (2019) e2274.
119. J. Kang and Z. D. Duan, Modal parameter identification with compressed samples by sparse decomposition using the free vibration function as dictionary, *Smart Struct. Syst.* **25**(2) (2020) 123–133.
120. B. Xu and L. Yu, A novel regularized adaptive matching pursuit for moving force identification using multiple criteria and prior knowledge, *Int. J. Struct. Stab. Dyn.* **23** (2023) 2350117.
121. Y. Huang, J. L. Beck and H. Li, Bayesian system identification based on hierarchical sparse Bayesian learning and Gibbs sampling with application to structural damage assessment, *Comput. Methods Appl. Mech. Eng.* **318** (2017) 382–411.
122. K. V. Yuen, J. L. Beck and L. S. Katafygiotis, Efficient model updating and health monitoring methodology using incomplete modal data without mode matching, *Struct. Control Health Monit.* **13**(1) (2006) 91–107.
123. M. A. Davenport, P. T. Boufounos, M. B. Wakin and R. G. Baraniuk, Signal processing with compressive measurements, *IEEE J. Sel. Top. Signal Process.* **4**(2) (2010) 445–460.
124. J. Y. Park, M. B. Wakin and A. C. Gilbert, Modal analysis with compressive measurements, *IEEE Trans. Signal Process.* **62**(7) (2014) 1655–1670.
125. R. Fuentes, C. Mineo, S. G. Pierce, K. Worden and E. J. Cross, A probabilistic compressive sensing framework with applications to ultrasound signal processing, *Mech. Syst. Signal Process.* **117** (2019) 383–402.
126. C. Liu *et al.*, Acoustic emission signal processing for rolling bearing running state assessment using compressive sensing, *Mech. Syst. Signal Process.* **91** (2017) 395–406.
127. S. Song *et al.*, An improved structural health monitoring method utilizing sparse representation for acoustic emission signals in rails, *IEEE Trans. Instrum. Meas.* **72** (2023) 6500511.

128. O. Mesnil and M. Ruzzene, Sparse wavefield reconstruction and source detection using compressed sensing, *Ultrasonics* **67** (2016) 94–104.
129. Y. Keshmiri Esfandabadi *et al.*, Deep learning for enhancing wavefield image quality in fast non-contact inspections, *Struct. Health Monit.* **19**(4) (2020) 1003–1016.
130. S. Sabeti and J. B. Harley, Spatio-temporal undersampling: Recovering ultrasonic guided wavefields from incomplete data with compressive sensing, *Mech. Syst. Signal Process.* **140** (2020) 106694.
131. J. W. Yan, H. Peng, Y. Yu and Y. Z. Luo, Compressive sensing of wind speed based on non-convex ℓp-norm sparse regularization optimization for structural health monitoring, *Eng. Struct.* **194** (2019) 346–356.
132. C. Bayındır and B. Namlı, Efficient sensing of von Kármán vortices using compressive sensing, *Comput. Fluids.* **226** (2021) 104975.
133. H. P. Wan, G. S. Dong and Y. Z. Luo, Compressive sensing of wind speed data of large-scale spatial structures with dedicated dictionary using time-shift strategy, *Mech. Syst. Signal Process.* **157** (2021) 107685.
134. G. D. Pasparakis, K. R. dos Santos, I. A. Kougioumtzoglou and M. Beer, Wind data extrapolation and stochastic field statistics estimation via compressive sampling and low rank matrix recovery methods, *Mech. Syst. Signal Process.* **162** (2022) 107975.
135. G. Malara, I. A. Kougioumtzoglou and F. Arena, Extrapolation of random wave field data via compressive sampling, *Ocean Eng.* **157** (2018) 87–95.
136. G. Malara, Compressive sampling-based extrapolation of free surface displacement data from pressure measurements, *Ocean Eng.* **266** (2022) 113044.
137. T. Y. Zhao and Y. Wang, Differentiation of discrete data with unequal measurement intervals and quantification of uncertainty in differentiation using Bayesian compressive sampling, *Comput. Geotech.* **122** (2020) 103537.
138. J. B. Xu, Y. Wang and L. L. Zhang, Interpolation of extremely sparse geo-data by data fusion and collaborative Bayesian compressive sampling, *Comput. Geotech.* **134** (2021) 104098.
139. P. P. Li and Y. Wang, Development of an efficient response surface method for highly nonlinear systems from sparse sampling data using Bayesian compressive sensing, *ASCE-ASME J. Risk. Uncertain. Eng. Syst. Part A.-Civ. Eng.* **7**(4) (2021) 4021050.
140. T. Y. Zhao and Y. Wang, Determination of efficient sampling locations in geotechnical site characterization using information entropy and Bayesian compressive sampling, *Can. Geotech. J.* **56**(11) (2019) 1622–1637.
141. Y. Bengio, A. C. Courville and P. Vincent, Unsupervised feature learning and deep learning: A review and new perspectives, *CoRR*, abs/1206.5538, 1(2665), 2012 (2012).
142. Y. LeCun, Y. Bengio and G. Hinton, Deep learning, *Nature* **521**(7553) (2015) 436–444.
143. P. T. Bickel, R. Ya'acov and T. B. Alexandre, Simultaneous analysis of lasso and Dantzig selector, *Ann. Statist.* **37**(4) (2009) 1705–1732.
144. A. Radford, L. Metz and S. Chintala, Unsupervised representation learning with deep convolutional generative adversarial networks, *arXiv preprint* (2015), arXiv:1511.06434.
145. G. K. Wallace, The JPEG still picture compression standard, *IEEE Trans. Consum. Electron.* **38**(1) (1992) xviii–xxxiv.
146. G. S. Dong, H. P. Wan, Y. Luo and M. D. Todd, A fast sparsity-free compressive sensing approach for vibration data reconstruction using deep convolutional GAN, *Mech. Syst. Signal Process.* **188** (2023) 109937.

147. G. Fan, J. Li, H. Hao and Y. Xin, Data driven structural dynamic response reconstruction using segment based generative adversarial networks, *Eng. Struct.* **234** (2021) 111970.
148. G. Fan, J. Li and H. Hao, Lost data recovery for structural health monitoring based on convolutional neural networks, *Struct. Control Health. Monit.* **26**(10) (2019) e2433.
149. S. Jeong, M. Ferguson, R. Hou, J. P. Lynch, H. Sohn and K. H. Law, Sensor data reconstruction using bidirectional recurrent neural network with application to bridge monitoring, *Adv. Eng. Inform.* **42** (2019) 100991.
150. X. M. Lei, L. M. Sun and Y. Xia, Lost data reconstruction for structural health monitoring using deep convolutional generative adversarial networks, *Struct. Health Monit.* **20**(4) (2021) 2069–2087.
151. J. Niu, S. L. Li and Z. L. Li, Restoration of missing structural health monitoring data using spatiotemporal graph attention networks, *Struct. Health Monit.* **21**(5) (2022) 2408–2419.
152. P. Ni, Y. X. Li, L. M. Sun and A. Wang, Traffic-induced bridge displacement reconstruction using a physics-informed convolutional neural network, *Comput. Struct.* **271** (2022) 106863.
153. Y. X. Li, P. Ni, L. M. Sun and W. Zhu, A convolutional neural network-based full-field response reconstruction framework with multitype inputs and outputs, *Struct. Control Health Monit.* **29**(7) (2022) e2961.
154. K. J. Jiang, Q. Han, X. L. Du and P. H. Ni, Structural dynamic response reconstruction and virtual sensing using a sequence to sequence modeling with attention mechanism, *Autom. Constr.* **131** (2021) 103895.
155. G. Fan, Z. Y. He and J. Li, Structural dynamic response reconstruction using self-attention enhanced generative adversarial networks, *Eng. Struct.* **276** (2023) 115334.
156. B. W. Du, L. Y. Wu, L. L. Sun, F. Xu and L. C. Li, Heterogeneous structural responses recovery based on multi-modal deep learning, *Struct. Health Monit.* **22** (2023) 799–813.
157. K. Yang, Y. L. Ding, H. C. Jiang, H. W. Zhao and G. Luo, A two-stage data cleansing method for bridge global positioning system monitoring data based on bi-direction long and short term memory anomaly identification and conditional generative adversarial networks data repair, *Struct. Control Health Monit.* **29**(9) (2022) e2993.
158. Y. Pan, C. E. Ventura and T. Li, Sensor placement and seismic response reconstruction for structural health monitoring using a deep neural network, *Bull. Earthq. Eng.* **20** (2022) 4513–4532.
159. Z. W. Wang, A. D. Li, W. M. Zhang and Y. F. Zhang, Long-term missing wind data recovery using free access databases and deep learning for bridge health monitoring, *J. Wind. Eng. Ind. Aerodyn.* **230** (2022) 105201.
160. L. Comerford, I. A. Kougioumtzoglou and M. Beer, An artificial neural network approach for stochastic process power spectrum estimation subject to missing data, *Struct. Saf.* **52** (2015) 150–160.
161. L. Comerford, M. Beer and N. W. Lu, Revealing prediction uncertainty in artificial neural network based reconstruction of missing data in stochastic process records utilizing extreme learning machines, in *Proc. IEEE Symposium Series on Computational Intelligence* (2017), pp. 1–7.
162. C. Xiang, W. Wang, L. Deng, P. Shi and X. Kong, Crack detection algorithm for concrete structures based on super-resolution reconstruction and segmentation network, *Autom. Constr.* **140** (2022) 104346.

163. H. Liu, Y. L. Ding, H. W. Zhao, M. Y. Wang and F. F. Geng, Deep learning-based recovery method for missing structural temperature data using LSTM network, *Struct. Monit. Maint.* **7**(2) (2020) 109–124.
164. Y. Z. Zhuang, J. C. Qin, B. Chen, C. Z. Dong, C. B. Xue and S. M. Easa, Data loss reconstruction method for a bridge weigh-in-motion system using generative adversarial networks, *Sensors* **22**(3) (2022) 858.
165. Y. Q. Ni and M. Li, Wind pressure data reconstruction using neural network techniques: A comparison between BPNN and GRNN, *Measurement* **88** (2016) 468–476.
166. L. Gan, Block compressed sensing of natural images, *2007 15th Int. Conf. Digital Signal Processing*, (IEEE, 2007), pp. 403–406.
167. W. J. Liang, G. X. Lin and C. S. Lu, Tree structure sparsity pattern guided convex optimization for compressive sensing of large-scale images, *IEEE Trans. Image Process.* **26**(2) (2016) 847–859.
168. F. Filippitzis, M. D. Kohler, T. H. Heaton and J. L. Beck, Sparse Bayesian learning for damage identification using nonlinear models: Application to weld fractures of steel-frame buildings, *Struct. Control Health Monit.* **29**(2) (2022) e2870.
169. J. M. Marin, P. Pudlo, C. P. Robert and R. J. Ryder, Approximate Bayesian computational methods, *Stat. Comput.* **22**(6) (2012) 1167–1180.
170. S. C. Kuok and K. V. Yuen, Model-free data reconstruction of structural response and excitation via sequential broad learning, *Mech. Syst. Signal Process.* **141** (2020) 106738.
171. S. J. Pan and Q. Yang, A survey on transfer learning, *IEEE Trans. Knowl. Data Eng.* **22**(10) (2010) 1345–1359.
172. K. Weiss, T. M. Khoshgoftaar and D. Wang, A survey of transfer learning, *J. Big Data* **3**(1) (2016) 1–40.
173. P. Gardner, L. A. Bull, N. Dervilis and K. Worden, On the application of kernelised Bayesian transfer learning to population-based structural health monitoring, *Mech. Syst. Signal Process.* **167** (2022) 108519.
174. L. Bai, P. Maechler, M. Muehlberghuber and H. Kaeslin, High-speed compressed sensing reconstruction on FPGA using OMP and AMP, *2012 19th IEEE Int. Conf. Electronics, Circuits, and Systems (ICECS 2012)*, (IEEE, 2012), pp. 53–56.
175. M. Żarski, B. Wójcik, K. Książek and J. A. Miszczak, Finicky transfer learning—A method of pruning convolutional neural networks for cracks classification on edge devices, *Comput.-Aided Civ. Infrastruct. Eng.* **37**(4) (2022) 500–515.
176. C. H. Zhao, B. P. Tang, Y. Huang and L. Deng, Edge collaborative compressed sensing in wireless sensor networks for mechanical vibration monitoring, *IEEE Trans. Ind. Inform.* **19** (2023) 8852–8864.
177. C. H. Zhao, B. P. Tang, L. Deng, Y. Huang and Q. K. Li, Multilevel adaptive near-lossless compression in edge collaborative wireless sensor networks for mechanical vibration monitoring, *IEEE Trans. Ind. Electron.* **70** (2023) 11703–11713.
178. C. H. Zhao, B. P. Tang, Y. Huang and H. Fu, Multi-layer joint optimization of packet size and adaptive transmission scheduling of wireless sensor networks for mechanical vibration monitoring, *IEEE Internet Things J.* **10** (2022) 6444–6455.

https://doi.org/10.1142/9789819814770_0027

Chapter 27

Coupling Dual Nonlinearity of Arch Bridge Bearing Capacity Analysis Method#

Jianting Zhou*,§, Shaorui Wang*,¶,§§, Zengwu Liu†,‖, Ligui Yang†,**, Zhilu Wang‡,†† and Yingbin Li†,‡‡

*State Key Laboratory of Mountain Bridge and Tunnel Engineering
School of Civil Engineering, Chongqing Jiaotong University
Chongqing, Nan'an District, 400074, P. R. China

†School of Civil Engineering, Chongqing Jiaotong University
Chongqing, Nan'an District, 400074, P. R. China

‡School of Civil Engineering, Chongqing University
Chongqing, Shapingba District, 400074, P. R. China

§jtzhou@cqjtu.edu.cn
¶ruiruiplace@163.com
‖1754595545@qq.com
**yangligui@cqjtu.edu.cn
††zhlwang@cqu.edu.cn
‡‡281650659@qq.com

In order to realistically restore the mechanical properties of the arch bridge under load and damage processes, a coupled dual nonlinear analysis method is established for the arch-bridge load–carrying capacity by considering the effects of bidirectional force transfer and adaptive section stiffness. First, based on the geometric nonlinear theoretical method of bidirectional force transfer, the macroscopic structural deformation of arch ribs under different loading conditions and load levels is preliminarily calculated, including the bending and axial deformation of each section of the arch, which is used to characterize the development of plasticity. According to deformation or strain, the fiber model analysis method is used to determine the true moment–curvature and axial force–curvature development paths of key sections, and then the elastic–plastic flexural and compressive stiffnesses of each section under different deformation states are derived. The elastic–plastic stiffness is substituted back to the geometric nonlinear equations to obtain the deformation and internal force along the arch ribs considering the cross-sectional plasticity. Finally, the above process is cycled until the equivalence condition is satisfied between the structural internal force/deformation and cross-sectional ones. In order to verify the rationality and accuracy of such a coupling method, a model test is carried out on a stiff skeletal concrete suspension chain line arch with a span of 12 m, whose results match satisfactorily with that of calculated by the coupling dual

§§Corresponding author.

#This chapter appeared previously on the International Journal of Structural Stability and Dynamics. To cite this chapter, please cite the original article as the following: J. Zhou *et al.*, *Int. J. Struct. Stab. Dyn.*, **23**, 2340030 (2023), DOI: 10.1142/S0219455423400308.

nonlinearity method. Compared with the method suggested in the code JTG3362-2018, the proposed method can better predict the development of deflection of different arch rib sections, as well as the actual internal force state. The bending moment calculated as per the code is 7–16% larger than the test result at the ultimate capacity state, while the method of this paper is within 5%.

Keywords: Geometric nonlinearity; material nonlinearity; additional bending moment; bearing capacity; model experiment.

1. Introduction

Arch bridges are widely used in mountainous roads and railroads with complex conditions because of their high stiffness, high load–bearing performance, good seismic performance and low cost.[1,2] In China, the span diameter of arch bridge is increasing, the world's largest steel pipe concrete arch bridge — Pingnan three bridges (Fig. 1), the span diameter has reached 575 m. The largest span diameter-reinforced concrete arch bridge has been completed for the Shanghai-Kunshan Railway Beipanjiang Bridge (Fig. 2), the span diameter is 445 m. The under-construction strong skeleton concrete arch bridge-Tian'e Longtan Special Bridge, the span diameter has reached 600 m.

The arch is a compression–bending structure, and the cross-section has axial force, bending moment and shear force at the same time. Since the length and slenderness of the arch structure is relatively large, the internal force of the arch is mainly based on axial pressure and bending moment. Influenced by the curvature of the axis, the axial force and bending moment of the arch affect each other. In order to ensure the accuracy of the calculation results of the internal forces and displacements of the arch structure, it is important to study the influence of arch geometric nonlinearity.[3,4] At the same time, the influence of material nonlinearity

Fig. 1. Pingnan three bridges.

Fig. 2. Beipanjiang bridge.

cannot be ignored when the arch bridge reaches its ultimate load carrying capacity at high stress level. Therefore, the geometric and material nonlinearities have been paid close attention in the research and design of arch bridges.

For the geometric nonlinearity of arch, scholars at home and abroad have established various nonlinear arch models using static method, energy method and finite element method to solve the arch internal force, deformation and stability and other related problems.[5–11] Timoshenko[12] considered the effect of geometric nonlinearity on the structure, established the buckling analysis model of compressed circular ring and curved rod, and obtained the second-order bending deflection displacement from the physical equation. The differential equations, combined with the boundary conditions of the circular arch to find out the exact characteristic equation of bifurcation instability and critical load of the circular arch. Dinnik[13] used the force method to find out the expression of internal force first and then the expression of displacement according to the physical and geometric equations in the study of circular arch under the action of uniform static water pressure. Deng[14] derived the geometrically nonlinear analytical expressions for the displacement and internal force of the circular arch considering the second-order effect generated by the axial force based on the basic equations of the hydrostatic method, and constructed the finite unit by the direct stiffness method in matrix form, and further carried out the instability analysis by the hydrostatic method. Shen and Ma *et al.*[15] studied the in-plane stability of deep circular arches under uniform radial pressure using small deformation theory and ANSYS, starting from the micro-segment equilibrium equation, and calculated the axial displacement generalized solution. Cheng and Tong[16] used the strain–displacement relationship in finite deformation theory according to the assumption of flat section, completely considered the second-order effects of transverse stress and shear stress, and derived the equilibrium equation for in-plane nonlinear analysis of the arch using the principle of virtual work, which gave the general theory of in-plane bending instability of circular arc arch. The above scholars' research mainly focuses on the analytical solution of the stability problem of circular arch in specific working conditions, but there is no unified theoretical model for the nonlinear internal force and displacement of circular arch under arbitrary working conditions. With the increasing span of arch bridges,[17–19] the accuracy of arch bridge design and construction has put forward new requirements for theoretical calculations. Therefore, when calculating the geometric nonlinear internal forces of the arch, it is necessary to consider both the coupling effect of the vertical deformation of the arch and the horizontal force of the arch foot, and the effect of the coupling effect of the horizontal deformation of the arch and the vertical force of the arch.

The research of some scholars shows that the elastic compression caused by the axial force has a nonnegligible effect on the internal force of the arch. Li *et al.*[20] used the elastic center method to derive the internal force calculation formula of the no-hinged parabolic arch and found that the error of the calculated thrusts reached 30% without considering the influence of elastic compression of the arch ribs under the concentrated load. Zhou *et al.*[21] derived the analytical solution of strain-influenced line for a no-hinged arch with variable cross-sectional suspension based on Ritter's formula and found that the deviation of calculation of axial force with or without increases with the decrease of vector-to-span ratio. Some scholars, on the other hand, have studied the analytical solutions of arch structures from the perspective of approximate integration. Yang *et al.*[22] derived an analytical formula for parabolic arch deflection based on the elastic center method, using the approximate parabolic fitting of suspension lines. Hu *et al.*[23,24] derived the practical analytical formulas for the internal forces of parabolic arch based on the elastic center method and MEXE method without considering axial compression and considering axial compression in two cases. However, the above studies did not consider the effect of arch deformation on the internal force. Some other scholars consider the interaction between the vertical deformation of the arch and the internal force to consider the nonlinear problem of the arch. Chatterjee first considered the effect of the vertical deformation of the arch on the internal force of the arch and found that the internal force of the arch increases after considering the vertical deformation of the arch compared with the elastic theory.[25] He *et al.* considered that the differential equation for the deformation of the arch should account for the effect of the horizontal thrust and the additional bending moment generated by the vertical deformation.[26] Liu and Zhou,[27] Zhang *et al.*[28] and Chen[29] considered the influence of vertical deformation of the arch on the internal force of the arch, derived the differential equation of the arch considering the influence of vertical deformation, and carried out internal force analysis of concrete no-hinged arch and steel pipe concrete no-hinged arch, all of which concluded that the influence of vertical deformation of the arch on the internal force of the arch needs to be considered. However, the above scholars' analysis objects are all parabolic no-hinged arches, and no research has been carried out on suspended chain line no-hinged arches; in addition, for the geometric nonlinearity of the arch, only the influence of vertical deformation is considered in the differential equation of the arch deformation, and the influence of the horizontal deformation of the arch and the additional bending moment caused by the rotation of the arch are not considered. If the influence of the above factors is not considered, the nonlinear internal forces of the arch will show large deviations.

When the load acting on the arch exceeds a certain magnitude, the arch not only has the problem of geometric nonlinearity, but also has material nonlinearity. In terms of member material nonlinearity, many scholars have used numerical methods[30,31] to study the positive section bearing capacity of columns, but it requires iterative calculation, which has the disadvantages of difficult convergence and low computational efficiency and is not convenient for engineering applications. Liu *et al.*[32] analyzed the bearing capacity of stacked eccentric compressed short columns using the strain distribution in the cross section and proposed a theoretical formula for the bearing capacity calculation using the integral method. Li *et al.*[33] analyzed the bearing capacity of concrete short columns based on the concrete and reinforcement principal structure relationship, and solved the section internal force and curvature relationship based on the section strain. Chen *et al.*[34] selected all possible section edge strains of compressed short columns to solve the section internal force and curvature, and derived the analytical equation of the elastic–plastic moment-axis force–curvature correlation of the compressed bar section. Shuang *et al.*[35,36] addressed the problem of reinforcement in the limit state of bearing capacity of compression–bending members, and proposed the method of determining the stress from strain before calculating the internal force of the cross-section, which can be calculated with complete consideration of the concrete and reinforcement intrinsic relationship curves without iterative calculations, and drew the nomogram for dimensionless reinforcement calculation. The above scholars' researches for material nonlinearity are focused on biased short columns, and less research is done at the structural level, especially for arch structures. Zhao *et al.*, Fan *et al.* and Li *et al.*[37–39] proposed the nonlinear dynamic characteristics theory of rotating thick rings, supports and elastic supports based on the elastic theory, and providing theoretical support for the relevant calculations.

This paper integrates the material nonlinearity and geometric nonlinearity calculation methods, proposes the geometric-material dual nonlinearity calculation method of no-hinged arch bearing capacity based on adaptive stiffness, and prepares the dual nonlinearity program based on MATLAB platform to realize the rapid and accurate calculation of internal force of suspension arch under any load level, which solves the difficult problem that the existing methods are difficult to consider both geometric and material dual nonlinearity.

2. Fundamentals of Analytical Method

2.1. *Bending moment expressions for arch sections based on "bidirectional force transfer"*

In order to analyze the internal force and deformation of the nonhinged arch under arbitrary load, a schematic diagram of the mechanical analysis model of the nonhinged arch is established, as shown in Fig. 3.

In the nonhinged arch shown in Fig. 3, the origin of the coordinate system is located at the left arch foot, the calculated span is l, the calculated vector height is f, the vertical reaction force at the left arch foot B is V_B, the horizontal reaction force is H. The moment is M_B. The arbitrary vertical load on the arch is $P_1, P_2, P_j \ldots P_n$. Under the action of arbitrary load, the axis of the arch is deformed, the point m changes to the point m', the vertical deformation is w, the horizontal deformation is δ, the distance from the origin before the deformation of the point m is x_m. Without considering the deformation of the arch axis, the bending moment at any point m in the arch can be expressed as

$$M_m = M_B + V_B x_m - H y_m - \sum_{j=1}^{n} P_j (x_m - x_j). \tag{1}$$

Considering only the effect of vertical deformation of the arch axis, the bending moment at any point m in the arch can be expressed as

$$M_m = M_B + V_B x_m - H(y_m - w) - \sum_{j=1}^{n} P_j (x_m - x_j). \tag{2}$$

On the basis of formula (2), considering the influence of the horizontal deformation δ of the arch axis on the internal force of the arch, the bending moment at any point m in the arch can be expressed as follows: considering the vertical deformation w and horizontal deformation δ of the arch axis, the bending moment at any point m in the arch can be expressed as follows:

$$M_m = M_B + V_B(x_m + \delta) - H(y_m - w) - \sum_{j=1}^{n} P_j (x_m - x_j). \tag{3}$$

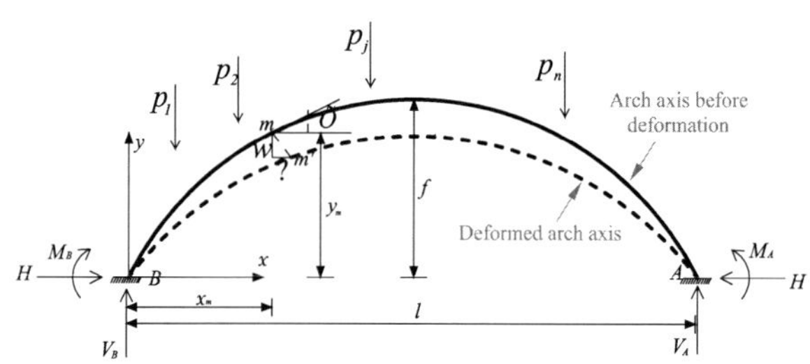

Fig. 3. Calculation diagram of the arch under arbitrary load.

Compared with formulae (2) and (3), formula (1) bending moment at any point of the arch axis does not take into account the influence of the deformation of the arch axis. Using formula (1) to analyze the internal force of small span arch does not cause large deviation. However, with the increase of the arch span diameter, the length–thin ratio is also increasing, and the deformation of the arch under external load increases, so it is necessary to consider the influence of deformation on the internal force of the arch, otherwise it will cause large deviation. Compared with formula (2), formula (3) not only considers the influence of the deformation of the arch axis on the internal force, but also takes into account the coupling effect of the horizontal thrust of the arch foot and the vertical deformation of the arch, as well as the coupling effect of the vertical reaction force and the horizontal deformation of the arch, so formula (3) reflects the internal force state of the arch more accurately.

2.2. *Differential equation for deformation of arch*

In the nonhinged arch mechanics model shown in Fig. 3, the vertical deformation at any point of the arch axis under vertical load is w, and the horizontal deformation is δ, and the equilibrium differential equation is as follows:

$$\frac{d^2\omega}{dx^2} = -\frac{M}{EI\cos\varphi} + \frac{d}{dx}\left(\frac{\alpha Q}{GA}\right) + \left(\frac{N}{EA}\right)' y' + \frac{N}{EA}y'', \tag{4}$$

$$\frac{d\delta}{dx} = \omega' y' - \frac{N}{EA}(1 + y'^2), \tag{5}$$

where M is bending moment of any section of the arch, Q is shear force of any section of the arch, N is axial force of any section of the arch and y' and y'' are the first-order derivative and second-order derivative of the arch axis.

Formulae (4) and (5) are the second-order differential control equations of the internal forces and deformation of the arch and consider the effect of axial force and shear force. Since the main arch length to slenderness ratio of arch bridges is usually greater than 5, the effect of shear on internal forces can be neglected.[3,40]

If the arch axis is the suspension chain line, a coordinate system is established at the left arch foot, see Fig. 1. The suspension chain line equation[41] can be expressed as

$$y = f - \frac{f}{m-1}\left(chk\left(1 - \frac{2x}{l}\right) - 1\right), \tag{6}$$

where $k = \ln(m + \sqrt{m^2 - 1})$, f is calculated vector height, m is arch axis coefficient and l is calculated span diameter.

For formula (6), the first-order derivative function and the second-order derivative function can be expressed as

$$y' = \frac{f}{m-1} \cdot \frac{2k}{l} shk\left(1 - \frac{2x}{l}\right), \tag{7}$$

$$y'' = -\frac{f}{m-1} \cdot \frac{4k^2}{l^2} chk \left(1 - \frac{2x}{l}\right). \tag{8}$$

Substituting formulae (6)–(8) into formulae (4) and (5), formulae (4) and (5) are the second-order differential equations for the internal force and deformation of the arch in the suspended chain line.

When the effect of shear force in the arch is not considered, formula (2) is substituted into formula (4), the second-order differential equation of the arch under any accidental load can be obtained

$$\frac{d^2\omega}{dx^2} = -\frac{1}{EI\cos\varphi}\left[M_B + V_B x - H(y-\omega) - \sum_{j=1}^{n} P_j(x - x_j)\right] + \left(\frac{N}{EA}\right)' y' + \left(\frac{N}{EA}\right)' y''. \tag{9}$$

Next, the $\cos\varphi = \frac{1}{\sqrt{1+y'^2}}$ and $N \approx H/\cos\varphi$ is substituted into formula (9). The result can be expressed as:

$$\frac{d^2\omega}{dx^2} = -\frac{1}{EI\cos\varphi}\left[M_B + V_B x - H(y-\omega) - \sum_{j=1}^{n} P_j(x - x_j)\right] + \frac{Hy''}{EA\cos\varphi}(2 - \cos^2\varphi). \tag{10}$$

Formula (10) contains $H\omega/EI\cos\varphi$ term, and the presence of $\cos\varphi$ makes the explicit analytical solution of formula (10) not available. The difference method can be used to discretize formula (10), as shown in Fig. 4, and formula (10) can be rewritten as

$$\frac{EI\cos\varphi_i}{\Delta x^2}(\omega_{i-1} - 2\omega_i + \omega_{i+1}) + M_B + V_B x_i - H(y_i - \omega_i) - \frac{I}{A} Hy''(2 - \cos^2\varphi_i) = \sum_{j=1}^{n} P_j(x_i - x_j). \tag{11}$$

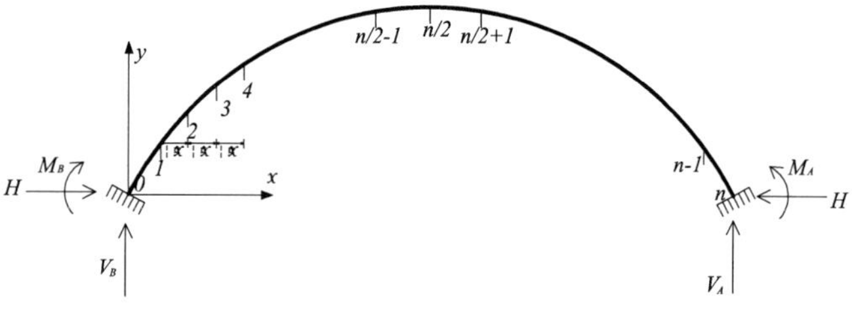

Fig. 4. Schematic diagram of discrete nonhinged arch.

Formula (11) can be constructed for $n+1$ equations, but there are $n+2$ unknowns for $\omega_1, \omega_2, \omega_3 \ldots \omega_{n-1}$ and M_B, H, V_B. An additional equation must be added to solve for the above parameters.

The integration of formula (5) leads to the following equation:

$$\begin{aligned}\delta(x) &= \int_0^x \omega' y' dx - \int_0^x \frac{N}{EA}(1+y'^2)dx + C \\ &= \omega y' \mid_0^x - \int_0^x \omega y'' dx - \frac{H}{EA}\int_0^x \frac{(1+y'^2)}{\cos\varphi}dx + C.\end{aligned} \tag{12}$$

The displacements and angles of rotation of B and A at the foot of the no-hinged arch are 0. At the foot of arch B, $x=0$, $\delta_B = \omega_B = 0$, it is known that $C=0$ in formula (12), then formula (12) can be rewritten as

$$\delta(x) = \omega(x)y'(x) - \int_0^x \omega(x)y''(x)dx - \frac{H}{EA}\int_0^x (1+y'^2)^{3/2}dx. \tag{13}$$

At the foot of the arch A, when $x=l, \delta_A = \omega_A = l$, formula (13) can be transformed into:

$$-\int_0^l w(x)y''(x)dx - \frac{H}{EA}\int_0^l (1+y'^2)^{3/2}dx = 0. \tag{14}$$

Using the Simpson formula, formula (14) can be rewritten as

$$\begin{aligned}&EA\left(4\sum_1^{n/2}\omega_{2i-1}y''_{2i-1} + 2\sum_1^{n/2}\omega_{2i}y''_{2i}\right) \\ &+H\left[\frac{1+y_0'^2}{\cos\varphi_0} + \frac{1+y_n'^2}{\cos\varphi_n} + 4\sum_1^{n/2}\frac{1+y_{2i-1}'^2}{\cos\varphi_{2i-1}} + 2\sum_1^{n/2-1}\frac{1+y_{2i}'^2}{\cos\varphi_{2i}}\right] = 0.\end{aligned} \tag{15}$$

The parameters $\omega_1, \omega_2, \omega_3 \ldots \omega_{n-1}$ and M_B, H, V_B can be obtained by formulae (11) and (15).

The horizontal deformation δ can be solved by formula (13), using the Simpson formula, formula (13) can be rewritten as

$$\begin{aligned}\delta(x_i) = \omega_i y'_i &- \frac{\Delta x}{6}\left(\omega_0 y''_0 + \omega_i y''_i + 4\sum_1^{n/2}\omega_{2i-1}y''_{2i-1} + 2\sum_1^{n/2-1}\omega_{2i}y''_{2i}\right) \\ &- \frac{H}{EA}\cdot\frac{\Delta x}{6}\left[\frac{1+y_0'^2}{\cos\varphi_0} + \frac{1+y_n'^2}{\cos\varphi_n} + 4\sum_1^{n/2}\frac{1+y_{2i-1}'^2}{\cos\varphi_{2i-1}} + 2\sum_1^{n/2-1}\frac{1+y_{2i}'^2}{\cos\varphi_{2i}}\right].\end{aligned} \tag{16}$$

The parameters $\omega_1, \omega_2, \omega_3 \ldots \omega_{n-1}, \delta_1, \delta_2, \delta_3 \ldots \delta_{n-1}$ and M_B, H, V_B of the arch under any load can be obtained by formulae (11), (15) and (16). The above parameters can be used to solve formula (3) to obtain the bending moment of any section under any vertical load with vertical deformation and horizontal deformation of the suspended chain line arch.

2.3. *Cross-sectional adaptive stiffness calculation method*

Taking a single steel pipe and concrete section as an example, the formula for calculating the internal force of the section considering material nonlinearity is derived by using the material intrinsic structure relationship and back-calculating the stress by means of the strain of the section.

The formulas for the calculation of the steel pipe and concrete principal structure relationship are shown in the literature (plus the latest literature) for a circular pipe concrete section shown in Fig. 5, with the steel pipe outer diameter $D = 2r$ and the concrete inner diameter r_c.

Using the geometric relationship of strains in Fig. 6, the height of the pressurized zone can be calculated as

$$x = \frac{D\varepsilon_{s1}}{\varepsilon_{s1} - \varepsilon_{s2}} = \frac{2r\varepsilon_{s1}}{\varepsilon_{s1} - \varepsilon_{s2}}. \tag{17}$$

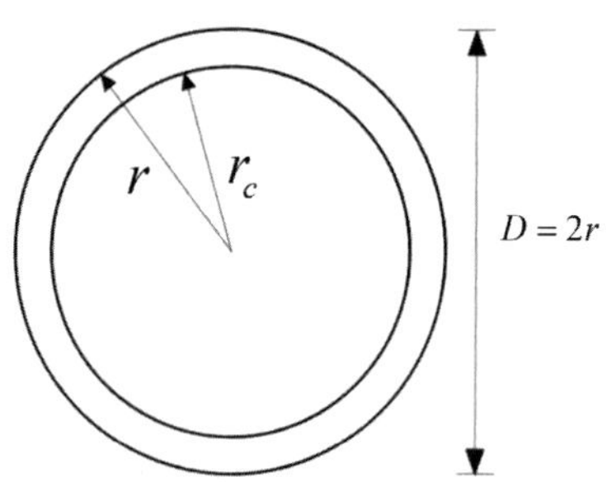

Fig. 5. Schematic diagram of steel pipe concrete section.

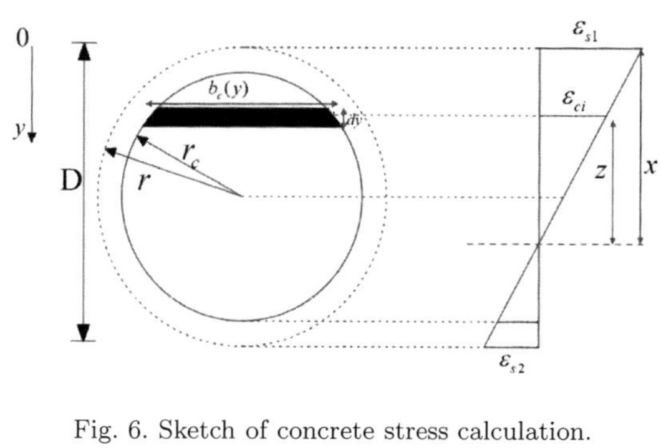

Fig. 6. Sketch of concrete stress calculation.

(i) For the concrete part of the pipe, the height x of the pressure zone is known, then the strain ε_{ci} at any one fiber in the height range of the section can be determined as

$$\varepsilon_{ci} = \frac{z}{x}\varepsilon_{s1} = \frac{x-y}{x}\varepsilon_{s1}. \tag{18}$$

Substituting the strain expression into the concrete principal structure yields the stress $\sigma_{ci}(y)$ at any fiber.

The concrete section width changes along the height, when $t \leq y \leq 2r - t$, satisfying $t = r - r_c$, and $b_c(y) = 2\sqrt{r_c^2 - (r-y)^2}$.

With the computational expressions for stresses, the stresses and stress moments over the micro-area can be integrated to calculate the axial forces and bending moments. The concrete axial force N_c in the pipe, and the bending moment M_c can be expressed as

$$N_c = \int_t^{2r-t} \sigma_c(y)b_c(y)dy, \quad M_c = \int_t^{2r-t} \sigma_c(y)b_c(y)\,|r-y|\,dy. \tag{19}$$

(ii) For the steel tube part of the section, knowing the height x of the pressure zone, the strain ε_{si} at any one fiber within the height range of the cross-section can be determined using the geometric relationship of strains in Fig. 7 as

$$\varepsilon_{si} = \frac{z}{x}\varepsilon_{s1} = \frac{x-y}{x}\varepsilon_{s1}. \tag{20}$$

When the native structure of the steel pipe is known, it is possible to determine the stress $\sigma_{si}(y)$ at any one fiber within the height of the section, the width of the steel pipe section changes with the height of the section, which changes as follows.

$$b_s(y) = \begin{cases} 2\sqrt{r^2-(r-y)^2}, & 0 \leq y \leq t, \\ 2\sqrt{r^2-(r-y)^2} - 2\sqrt{r_c^2-(r-y)^2}, & t \leq y \leq 2r-t, \\ 2\sqrt{r^2-(r-y)^2}, & 2r-t \leq y \leq 2r. \end{cases} \tag{21}$$

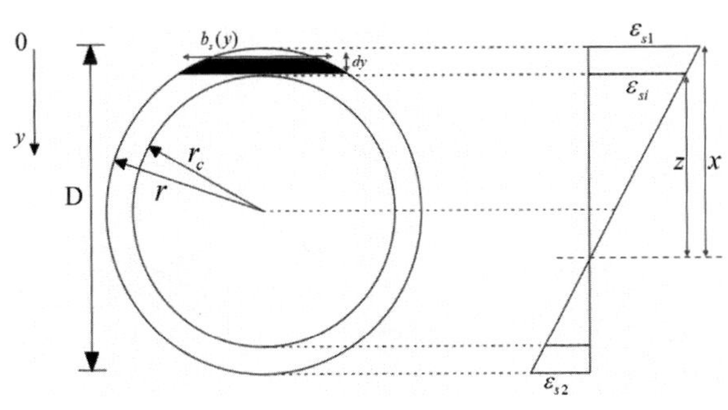

Fig. 7. Steel pipe stress calculation sketch.

With the calculation of stress expressions, it is possible to integrate the stress and stress moment on the micro-area to calculate the axial force and bending moment. According to the force balance relationship, the axial force Ns and bending moment M_s of the steel tube section can be expressed as

$$N_s = \int_0^{2r} \sigma_s(y) b_s(y) dy, \quad M_s = \int_0^{2r} \sigma_s(y) b_s(y) \left| r - y \right| dy. \tag{22}$$

The axial force and bending moment of the concrete and steel pipe are superimposed to obtain the total axial force N and bending moment M on a single concrete section as follows:

$$N = N_c + N_s, \quad M = M_c + M_s. \tag{23}$$

2.4. *Coupled analytical model considering dual nonlinearity*

Under any vertical load, axial force, bending moment and shear force are generated in the no-hinged arch, ignoring the effect of shear force, the axial force causes the arch rib to undergo axial compression deformation ε_{0i}, while the bending moment causes the arch rib to undergo bending deformation, producing curvature ϕ. As shown in Fig. 8, under any vertical load $P_1, P_2, \ldots, P_n$, the arch rib curved section a, b, c, the three points produce vertical deformation ω and horizontal deformation δ, which change from the initial coordinates (x_{i-1}, y_{i-1}), (x_i, y_i), (x_{i+1}, y_{i+1}) to $(x_{i-1} + \delta_{i-1}, y_{i-1} - \omega_{i-1})$, $(x_i + \delta_i, y_i - \omega_i)$, $(x_{i+1} + \delta_{i+1}, y_{i+1} - \omega_{i+1})$.

The deflection deformation caused by the bending moment can be quantified by the deformation curvature ϕ, which can be expressed as

$$\phi_i = \frac{|\omega|}{(1 + \omega_i'^2)^{3/2}}, \tag{24}$$

where ϕ_i is the deformation curvature and w_i is the vertical displacement of the ith node.

The axial deformation ε_{0i} of the arch rib under the action of axial force (positive for compression) can be expressed as

$$\varepsilon_{0i} = \frac{-\Delta s_i}{s_i} \approx \sqrt{\frac{[(x_i + \delta_i) - (x_{i-1} + \delta_{i-1})]^2 + [(y_i + \omega_i) - (y_{i-1} + \omega_{i-1})]^2}{(x_i - x_{i-1})^2 + (y_i - y_{i-1})^2}}, \tag{25}$$

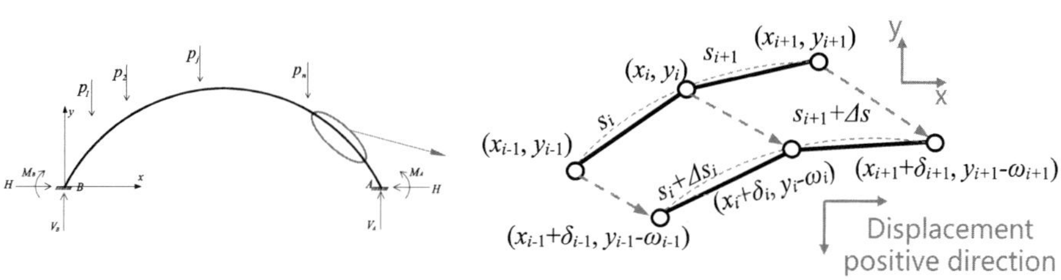

Fig. 8. Deformation of arch ribs.

where s_i is the arch rib i arc length, Δs_i is the arch rib i arc length change, x_i is the arch rib i point horizontal coordinates, y_i is the arch rib i point vertical coordinates, δ_i is the arch rib i point horizontal deformation and ω_i is the arch rib i point vertical deformation.

2.5. *Program implementation of the method*

In order to facilitate the use of the above formula to solve the internal force and deformation of the no-hinged arch of the suspension chain line considering the double nonlinearity, the calculation procedure was prepared using Matlab, and the calculation process is shown in Fig. 9. The process of solving the internal force and deformation of the no-hinged arch of the suspension chain line is as follows.

The key steps are as follows:

(i) Determine the number of segments n according to the size of the span L of the suspension line arch, which can determine $x_0, x_1, x_2 \ldots x_n$ and $\Delta x = L/n$; substitute the geometric parameters m, L, f and $\mathrm{x_i}$ of the suspension line arch into the suspension line equation to obtain the arch axis coordinates y, first-order y' derivatives and second-order y'' derivatives; discretize the cross section.

(ii) Structural geometric nonlinear analysis is performed to obtain the bending moments, axial forces, and deformation values ω of the arch for each section.

(iii) According to the initial assumption of the upper and lower edge strain ε_1 and ε_2 of the section, to solve the strain at any position of the section, the strain is substituted into the material intrinsic relationship to obtain the stress σ_i, and the integral formula is used to solve the bending moment M' and axial force N' of the section in the reverse direction.

(iv) The curvature of any section of the arch can be deduced from the calculation results of step 2, then obtain the strain ε_1 and ε_2, and get the updated bending moment M' and axial force N'. The bending stiffness EI and axial compression stiffness EA of the arch are repeatedly adjusted and calculated iteratively until $M = M'$, $N = N'$.

(v) The deformation value ω of the key section of the output arch rib, the bending moment M and axial force of the section N.

It should be noted that under different vertical loading conditions, the stiffness of each section of the arch rib is affected differently. When the loading force cannot be the structure into the plastic state of the material, the calculated bending moment and displacement are the same as the pure geometric nonlinear calculation results. When the loading force makes the arch structure reach the cross-sectional bearing capacity state, then the arch rib part of the cross-sectional material enters plasticity, and the calculated bending moment of the arch rib is the bending moment at the bearing capacity state, and the maximum value of the arch rib bending moment is the ultimate bending moment at the bearing capacity state.

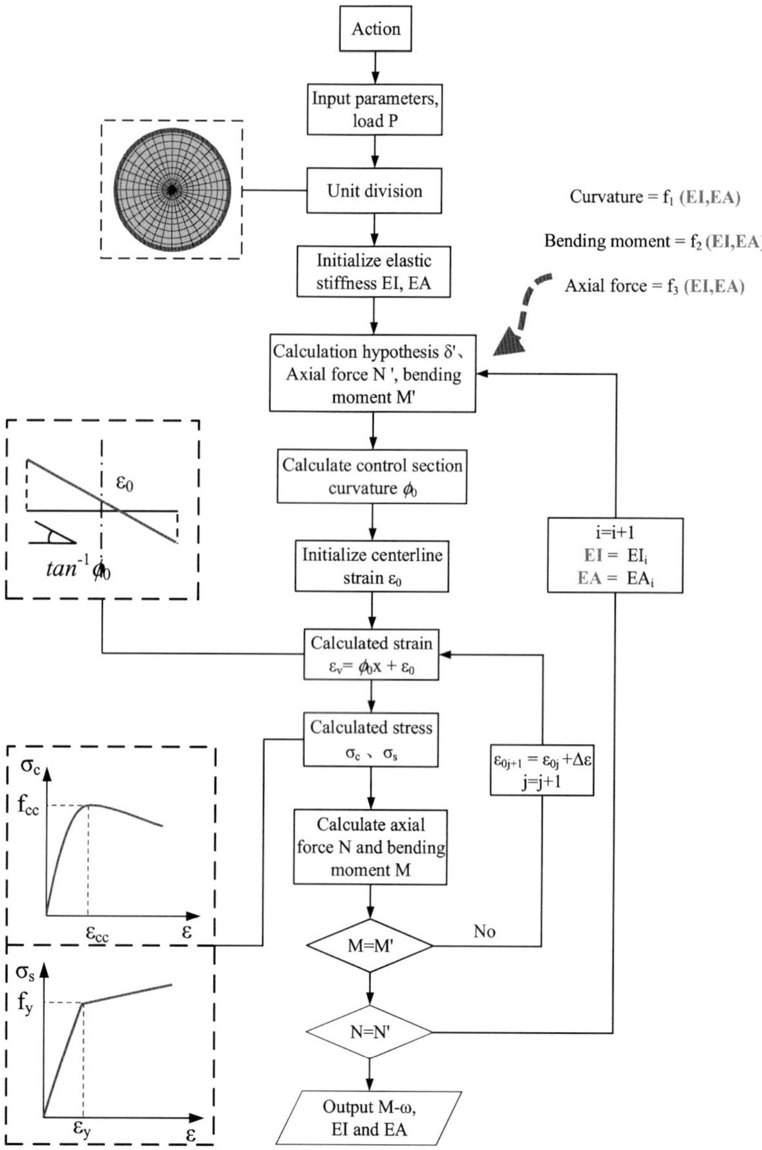

Fig. 9. Flow chart of double nonlinear theory calculation.

3. Whole Process Model Test of Arch Bridge Damage

3.1. *Test overview*

In order to verify the correctness of the above theory, a test of the stiff skeletal concrete arch was carried out. The span diameter of the arch is 12 m, the vector

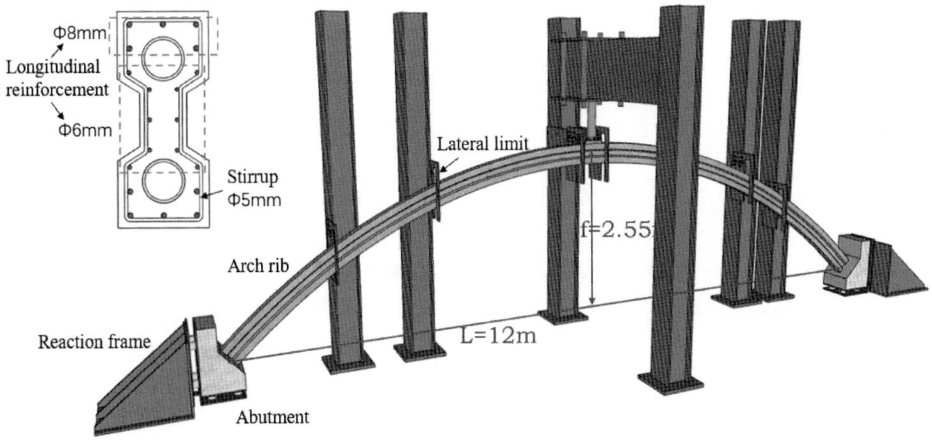

(a) Elevation of test arch

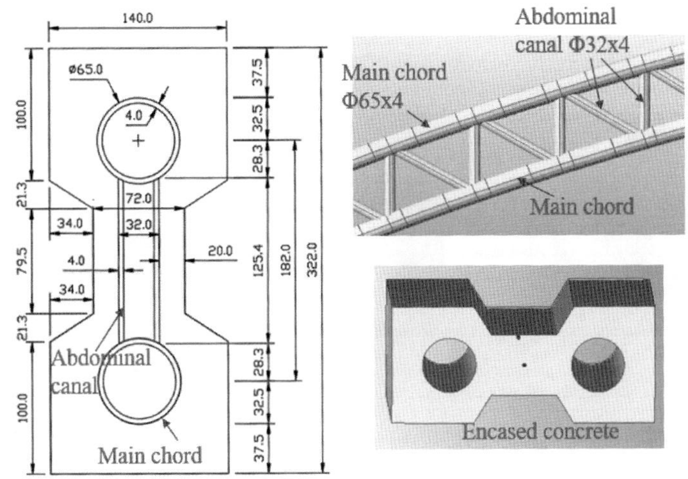

(b) Test arch cross-sectional drawing (unit: mm)

Fig. 10. Elevation of the test arch.

height is 2.55 m, and the arch axis coefficient is 1.8. The main arch steel is Q345, and the concrete inside the pipe is C80. The outer diameter of the pipe is 65 mm, and the wall thickness is 26 mm. The longitudinal reinforcement rate is 2.056%, and the loading method is arch top loading. The elevation of the test arch is shown in Fig. 10.

3.2. *Test setup and test protocol*

The test system is shown in Fig. 11. The test system consists of abutment reaction frame device, jack supporting loading device, arch rib transverse limiting device,

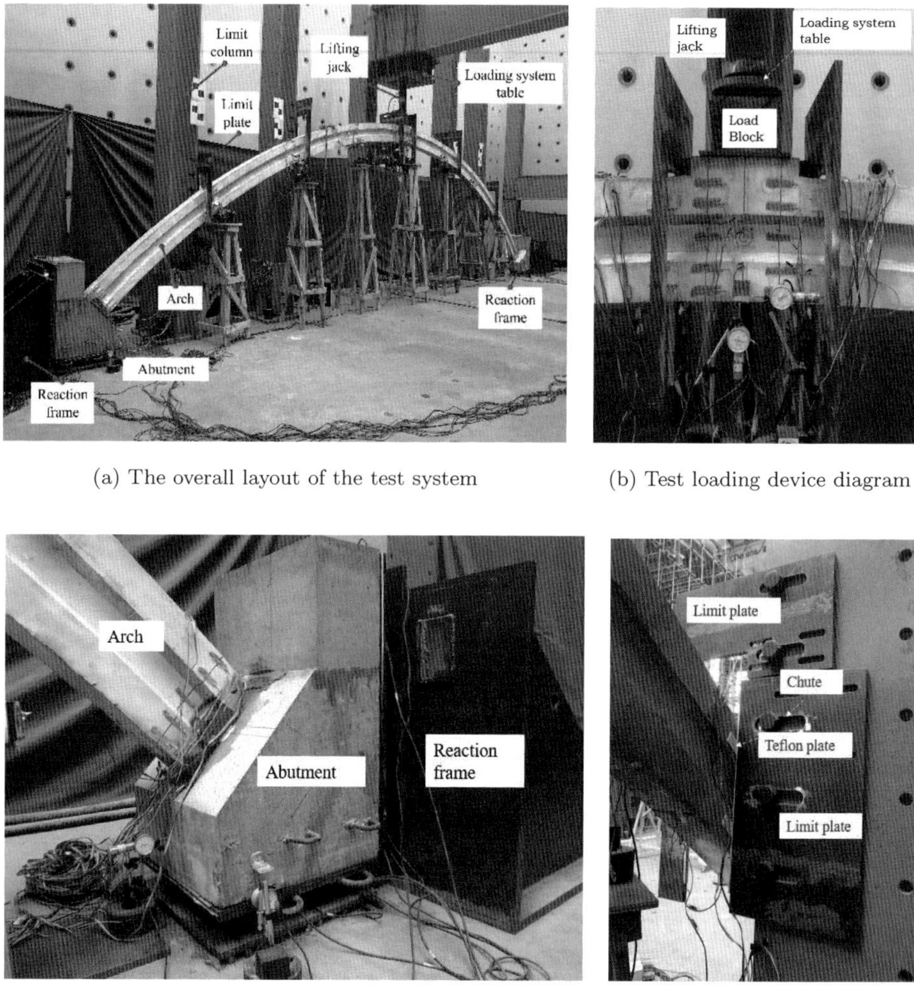

(a) The overall layout of the test system (b) Test loading device diagram

(c) Diagram of abutment and counterforce frame (d) Lateral limit device

Fig. 11. Diagram of loading and limit device.

abutment reaction force testing device, and the test arch containing the abutment. The abutment reaction frame device includes the reaction frame, anchor rods to fix the reaction frame. The reaction frame restricts the sliding of the abutment in the whole process of test arch loading, and pre-pressure is done before the experiment. Arch rib loading device consists of a jack, pressure sensor, loading plate and supporting tool, loading to ensure that the pressure sensor, jack, loading plate, and arch rib loading pad in a vertical straight line. The arch rib transverse limit device consists of limit beam, limit plate, and fixed bolts, which can be adjusted by sliding grooves to ensure that the test arch does not appear out-of-plane deformation in the loading process, and paste teflon plate between the arch and limit plate to reduce the friction between the two. The abutment reaction force testing device consists of steel plate and pressure sensor, the bottom surface of the sensor is fixed with the steel plate, and the top surface of the sensor is in near-smooth contact with the steel plate.

An important innovation of the test in this paper is to obtain the reaction force of the arch foot during the whole process of the test arch by setting pressure sensors at the abutment. The abutment reaction force testing device consists of steel plate and pressure sensor, the bottom surface of the sensor is fixed with the steel plate, four sensors are set for each steel plate, and the top surface of the sensor is in smooth contact with the steel plate. The vertical force testing device is located between the bottom surface of the abutment and the ground, and the horizontal force testing device is located between the back surface of the abutment and the counterforce frame. In order to eliminate the friction between the sensor and the steel plate, a rolling roller with a diameter of 2 cm is placed under the steel plate of the base of the abutment and lubricant is applied on the surface, while two layers of teflon plates with a thickness of 1 mm are added between the steel plate and the abutment and lubricant is applied. The abutment reaction force testing device is shown in Fig. 12.

Vertical force, horizontal thrust, and bending moment at the arch foot can be calculated based on the principle of static equilibrium, as shown in formula (26).

$$
\begin{aligned}
H &= H_1 + H_2 \\
V_B &= V_1 + V_2 \\
M_B &= H_1 \times h_1 - H_2 \times h_2 + V_1 \times v_1 - V_2 \times v_2.
\end{aligned}
$$

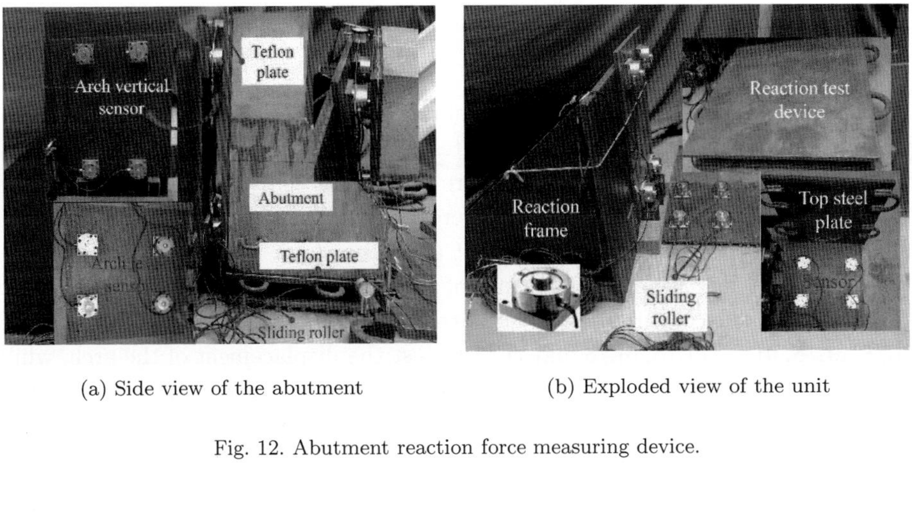

(a) Side view of the abutment (b) Exploded view of the unit

Fig. 12. Abutment reaction force measuring device.

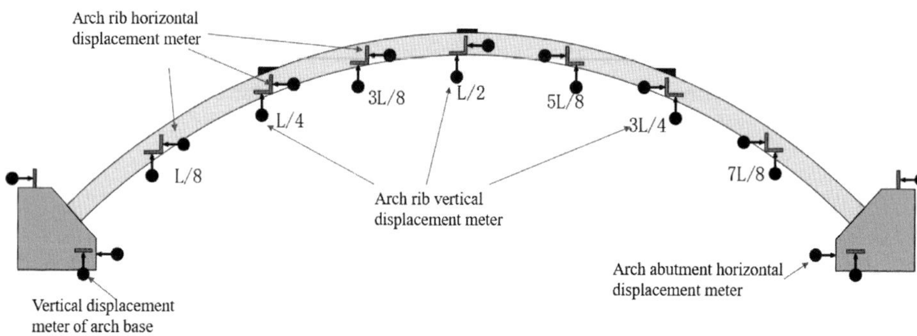

(a) General arrangement of horizontal and vertical displacement meters

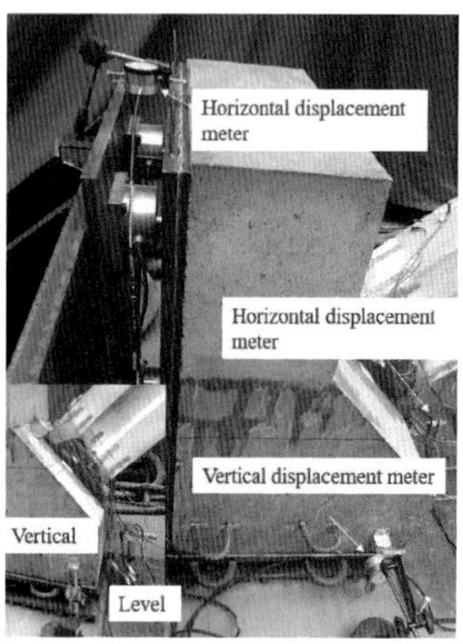

(b) Displacement gauge arrangement at abutment

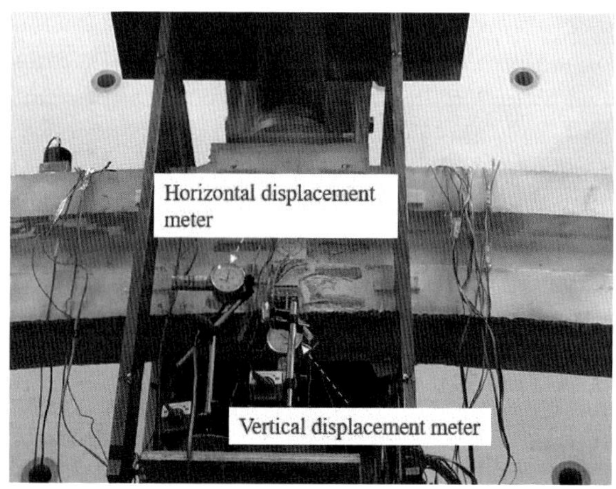

(c) Displacement gauge arrangement at arch top

Fig. 13. Displacement meter arrangement diagram.

In the formula: H_1 and H_2- the combined forces of the upper and lower row pressure sensors for the horizontal force test of the arch foot; V_1 and V_2- the combined forces of the front and rear pressure sensors for the vertical force test of the arch seat; H_1, h_2, v_1, v_2. The height difference and distance between the center of the sensor and the center of the arch foot axis.

To test the displacement and strain of the test arch during loading, horizontal and vertical displacement gauges were arranged at the selected key sections L/8, 2L/8, 3L/8, 4L/8, 5L/8, 6L/8 and 7L/8 to test the displacement of the arch, while horizontal, vertical and out-of-plane displacement gauges were arranged at the arch base, and the displacement gauges arrangement of the test arch (B1, B2 and B3) is shown in Fig. 13. In order to analyze the damage mode of the stiff skeleton arch, the synergistic working performance of the steel pipe and the outsourced concrete, and the influence of different longitudinal reinforcement rates and loading methods on the arch bearing capacity, longitudinal strain gauges were arranged on the surface of the outsourced concrete of the test arch in nine sections, including L/8, L/4, 3L/8, L/2, 5L/8, 3L/4, 7L/8 and both sides of the arch foot section, and nine gauges were arranged in each section. The upper and lower main chord steel pipes are also arranged with strain gauges in the above nine sections, respectively, at the upper and lower edges of the steel pipes, with a total of four gauges per section, and two circumferential strain gauges on the surfaces of the upper and lower chords on both sides of the loading point, in addition to two longitudinal ones. The longitudinal reinforcement is also arranged in nine key sections with the same strain gauges as the outer concrete and steel pipe sections, with six gauges in each section. A total of 90 strain gauges were arranged on the surface of the concrete, 48 strain gauges on the surface of the upper and lower chords, and 12 strain gauges on the surface of the reinforcement, for a total of 150 strain gauges.

The mechanical properties of the steel pipe were tested according to the tensile test procedure for metal materials (GB/T 228.1-2010),[42] and the yield strength of the steel pipe was 359 MPa, the tensile strength was 538 MPa, the modulus of elasticity was 210 GPa, and Poisson's ratio was 0.31. The axial compression tests of concrete specimens were carried out according to the physical and mechanical properties of concrete test methods,[43,44] and the cube compressive strength, prismatic axial compressive strength and modulus of elasticity of the outer concrete were 57.23 MPa, 47.2 MPa and 36.487 GPa, respectively. The yield strength and modulus of elasticity of longitudinal reinforcement of 5 mm diameter are 636.88 MPa and 160 667 MPa, respectively. The yield strength and modulus of elasticity of longitudinal reinforcement of 8 mm diameter are 480.49 MPa and 198 267 MPa, respectively. The yield strength and modulus of elasticity of longitudinal reinforcement of 6 mm diameter are 590.49 MPa and 198 267 MPa, respectively. The yield strength and modulus of elasticity of longitudinal reinforcement diameter 6 mm were 590.09 MPa and 178 867 MPa, respectively. The yield strength and modulus of elasticity of hoop diameter 5 mm were 523.18 MPa and 184 100 Mpa, respectively.

After doing the material performance test and erecting the test arch, the damage test of the test arch was started. In order to ensure the accuracy of the test arch test data in the whole process of the test, preloading test was conducted before the test. The pre-loading can eliminate the inelastic deformation of the test device, as well as check whether the loading device, displacement meter and strain gauge are normal, and whether the structure is effective in transmitting force. The pre-load load is determined by the calculation before the test and does not exceed 30% of the ultimate load. The test object is the arch of concrete wrapped with strong skeleton, because the concrete cracking strain in tension is relatively small, so the pre-load does not exceed 20% of the ultimate load, the formal loading using multiple graded loading, graded load is 5 kN. To ensure that the arch in each level of load test arch deformation is sufficient, each level of load hold load 2 min. when the test arch loading force change is small or force unloading phenomenon, but the displacement change is large, it means that the arch has reached the ultimate bearing capacity state, stop loading.

4. Analysis of Test Results and Method Validation

4.1. *Test phenomena and damage patterns*

The displacement of the test arch foot can be reflected by the abutment, and the transverse, horizontal and vertical displacements of the arch foot in the limit state of bearing capacity are shown in Table 1. The maximum horizontal and vertical displacements of the test arch foot are 1.31 mm, respectively, the displacement in the face of the abutment is small and negligible, and no concrete cracks are seen in the starting surface of the arch at the foot of the arch, which means that the test arch does not flatten and rotate during the whole process of loading, which satisfies the effect of fixed no-hinged arch boundary, and also shows that the abutment has achieved the purpose of no-hinged arch test. In addition, the lateral displacement of both the test arch and the abutment is less than 1 mm, and the deformation outside the arch face can be neglected, so this test can be regarded as an in-plane damage test.

The damage pattern of the test arch is shown in Fig. 14 under the ultimate bearing capacity condition, the cracks are mainly in the left and right arch foot, L/8–2L/8 area, 3L/8–7L/8 area, and at the top of the arch. Among them, only the cracks at the top of the arch are fully developed in the direction of section height, and some cracks almost penetrate the section, and the maximum width of the cracks is 0.514 mm. The distribution range of cracks at the top of the arch is about 200 cm, showing symmetrical distribution pattern, and the cracks below the loading point are vertical, and the cracks on both sides are inclined and distributed from the lower edge of the section, pointing to the upper edge of the arch at about 45°. The longest cracks in the L/8–2L/8 area and the 3L/8–7L/8 area are in the left and right foot, the L/8–2L/8 area, the 3L/8–7L/8 area and the 7L/8 area. The longest cracks in the L/8–2L/8 area and 3L/8–7L/8 area only extend to 1/2 height of the section, and the cracks are perpendicular to the arch axis, the cracks in these two areas show the characteristics of more and denser but not wider, the maximum width of the cracks

Table 1. Test abutmenting displacement (unit: mm).

	Left side abutment			Right side abutment			6L/8
Specimen number	Lateral displacement	Horizontal displacement	Vertical displacement	Lateral displacement	Horizontal displacement	Vertical displacement	Lateral displacement
B1	0.061	1.113	0.82	−0.035	1.31	0.622	0.312

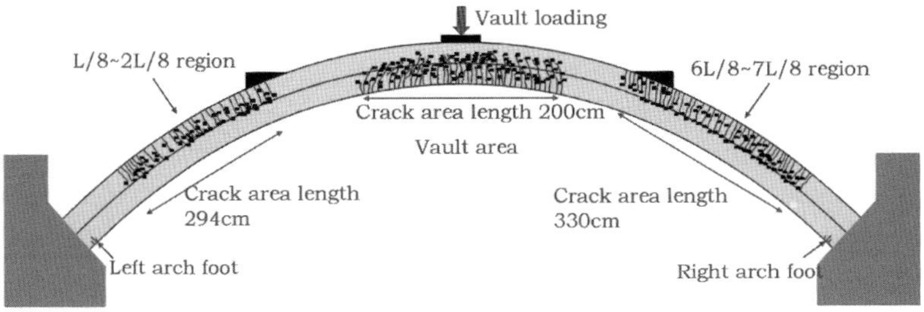

(a) Crack distribution map of damage state

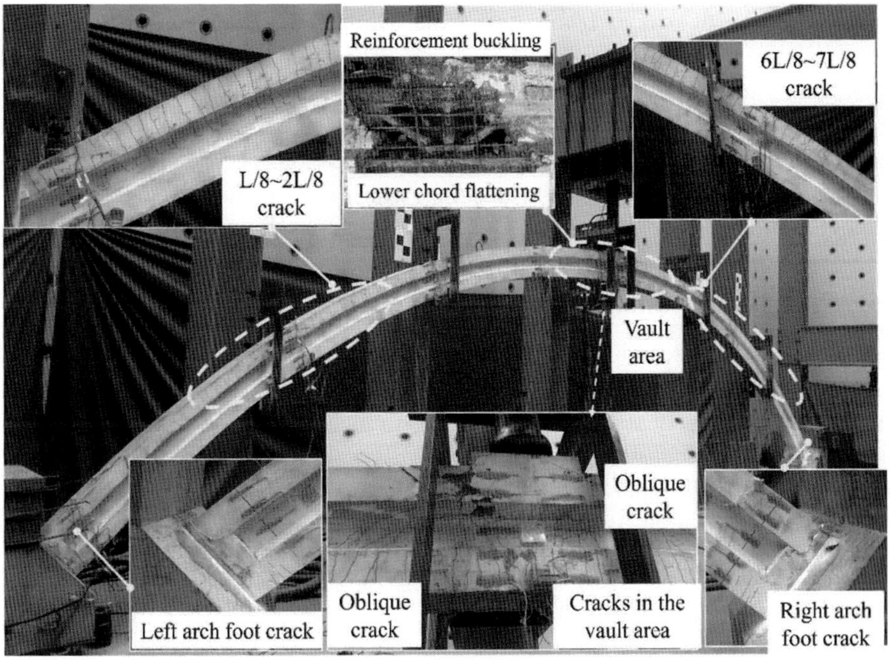

(b) Crack distribution map of damage state

Fig. 14. Test arch damage mode diagram.

are 0.134 mm and 0.089 mm, respectively. When the concrete was dissected at the top of the arch, it was found that the internal upper and lower main chords were deformed and flattened without local buckling, and the reinforcement was buckled under pressure, and the final damage mode was single hinge damage at loading. The cracks in the whole arch of the test arch show symmetrical distribution pattern, the crack distribution lengths in the L/8–2L/8 area and 3L/8–7L/8 area are 294 cm and 330 cm respectively; the crack length in the arch top area is 200 cm, symmetrical about the loading block.

4.2. *Comparison analysis of test value and theoretical value*

Test arch at 170 kN, the lower edge of the lower chord steel pipe strain yielding; at 235 kN, the upper edge of the lower chord steel pipe strain yielding; at 249 kN, the loading at the upper edge of the section of concrete compression cracks, but the arch can continue to bear the force. In favor of safety consideration, it is defined that when the upper edge of the lower chord steel pipe yields, the section appears plastic hinge and reaches the section bearing capacity limit state, and the corresponding arch rib bending moment is the bearing capacity limit bending moment. Through the double nonlinear bearing capacity algorithm proposed in this paper can calculate the arch rib deformation under different loading force, the theoretical value of arch rib deformation of test arch and test value comparison is shown in Fig. 15, the deformation at the maximum section of vertical upward deflection and downward deflection of arch rib is selected to demonstrate, respectively L/4 and arch top (L/2) section. It can be seen from the figure that the theoretical and test values of the test arch deformation agree well, and although deviations occur between them after the hinge, the change pattern is the same, which can predict the development of the arch rib deflection better.

4.3. *Comparative analysis with the current specification*

From the abutment reaction force, test device shown in Fig. 12 of Sec. 2.2 can obtain the abutment reaction force of the test arch, through formula (26) can calculate the horizontal force, vertical force and bending moment at the foot of the arch, and then use the actual measurement of the deformation of each key section of the test arch, through the formula (3) can calculate the bending moment value of each key section of the test arch. When the loading force makes the test arch reach out of

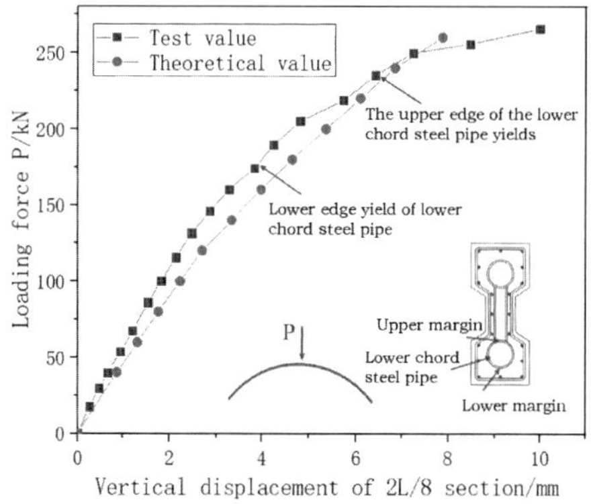

(a) L/4 cross section

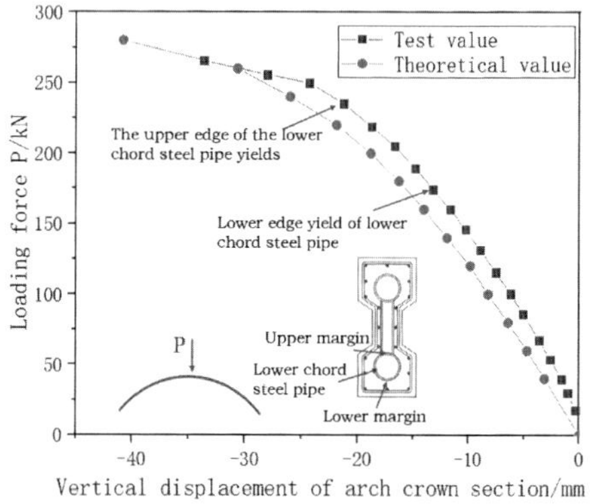

(b) Vault section

Fig. 15. Comparison of deformation of test arch B1.

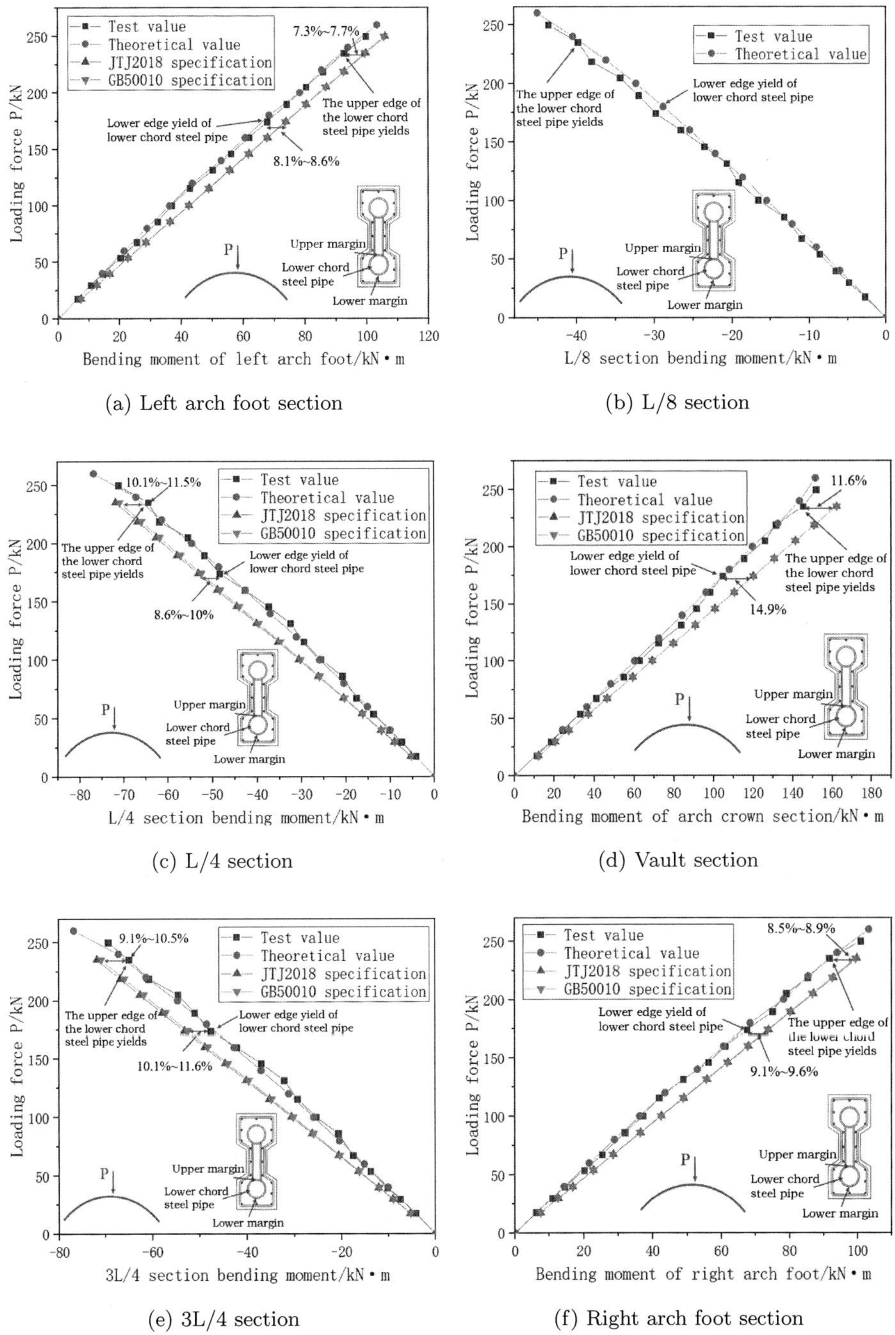

Fig. 16. Test arch bending moment comparison and bearing capacity state and specification comparison.

the hinge state, the state at this time is the section bearing capacity state, and the corresponding most unfavorable section bending moment value is the bearing capacity limit bending moment.

Figure 16 is the comparison of bending moment values of test arch key section under different loading force states, this section not only pairs the theoretical and test values of test arch key section bending moment, but also compares the test arch key section bending moment with different codes to reflect the advanced nature of this paper's method. From the figure, it can be seen that the theoretical value of the double nonlinear bending moment calculation in this paper and the test is in good agreement, although a few loading conditions have slight deviations between the two values, but the development law is consistent, which can effectively predict the internal force development of the whole process of arch force. This indicates that the theoretical values can effectively reflect the real bending moment state of each section of the arch after considering the adaptive stiffness reduction of the arch by the algorithm of this paper.

In addition, the comparison between the key section bending moment and the code value shows that the code (Design Code for Highway-Reinforced Concrete and Prestressed Concrete Bridges and Culverts (JTG 3362-2018), Concrete Structure Design Code (GB50010-2002)) calculated bending moment values are greater than the theoretical (test) bending moment values, and the test arch bending moment values and the code calculated values under different loading forces have different Deviation, in the section bearing capacity state, the normative calculated bending moment value of left arch foot is 7.3–7.7% larger than the theoretical value and test value, L/4 section is 10.1–11.5%, arch top section is 11.6%, 3L/4 is 9.1–10.5%, and right arch foot section is 8.5–8.9%. Through the comparison between the specification value and the test value, it can be seen that the current specification overestimates the arch rib bending moment and cannot accurately reflect the actual internal force state of the arch rib, which cannot provide more accurate guidance for the arch design.

5. Conclusion

Based on the theoretical derivation, a coupled dual nonlinear analysis method is built for the load–carrying capacity of arch bridges by considering the effects of bidirectional force transfer and adaptive sectional stiffnesses, showing good reliability and accuracy compared with the arc model specimens in the whole loading process. The major achievements and findings are summarized as follows:

(i) In terms of geometric nonlinearity, the geometric nonlinear differential equations for internal force and displacement of no-hinged arch in suspension line are derived based on the bending moment equation of "bi-directional force transfer", and the geometric nonlinear calculation method for internal force and displacement of no-hinged arch in suspension line considering bi-directional force transfer is established by using the joint solution of difference method and Simpson's formula.

(ii) For the material nonlinearity, from the section bending and axial deformation, the stress–strain intrinsic structure relationship of material is used to derive the formula for calculating the internal force of the suspension line arch section, revealing the development law of the real moment–curvature and axial force–axial deformation relationship of the arch section, establishing the self-adaptive stiffness calculation method of any section during the whole loading process of the arch, and proposing the stiffness adaptive calculation method of the no-hinged arch. Finally, the geometric and material nonlinearity of the coupled arch forms the geometric-material dual nonlinear calculation method of the no-hinged arch bearing capacity based on the adaptive stiffness.

(iii) Under the arch top loading condition, the stiff skeleton steel pipe concrete arch is a single-hinged damage mode with cracks in multiple places, and the steel pipe stiff skeleton improves the plastic capacity of the arch and enhances the ductile deformation capacity of the arch. After comparison, the theoretical and experimental values of the test arch deformation match well, and the change law is basically the same, although there is a certain deviation between the two after the hinge, but can even predict the development trend of arch rib deflection better.

(iv) The theoretical calculation results of bending moment considering double nonlinearity and the test results of this paper are in good agreement at different cross-sections, and the comparison with the current code shows that the calculated bending moment of the code is 7–16% larger at different cross-sections, which indicates that the theoretical algorithm proposed in this paper is better than the code of Design of Highway Reinforced Concrete and Prestressed Concrete Bridges and Culverts (JTG 3362-2018) and can accurately reflect the actual force and damage of each section of the arch rib. It can accurately reflect the actual force and damage state of each section of the arch rib.

Acknowledgments

This research is jointly funded by the Guangxi Key Research and Development Program (key technology for the construction of extra large span skeletal concrete arch bridges, GuikeAB22036007-8), Chongqing Technology Innovation and Application Development Special Key Project (development and application of intelligent construction and operation and maintenance technology for large span arch bridges in mountainous areas, CSTB2022TIAD-KPX0205), Science and Technology Innovation Project of Chongqing Education Commission "Construction of Twin Cities Economic Circle in Chengdu-Chongqing Region" (new method for intelligent control of the construction of extra-large span steel tube and concrete arch bridges,

KJCXZD2020032), Natural Science Foundation of Chongqing (research on the calculation method of segmental arch formation for large span arch bridges based on the influence of process errors and environmental factors, cstc2021jcyj-msxm2491). This support is gratefully acknowledged.

References

1. B. C. Chen, J. G. Wei, J. Zhou and J. P. Liu, Application status and Prospect of concrete filled steel tubular arch bridges in China, *J. Civil. Eng.* **50**(6) (2017) 12.
2. B. C. Chen and J. P. Liu, Summary of world arch bridge construction and technology development, *J. Transp. Eng.* **20**(1) (2020) 15.
3. H. T. Yin, Y. Y. Geng, H. Y. Yang and Y. Y. Wang, Study on in-plane stability performance of parabolic-shaped steel pipe concrete no-hinged arch with large aspect ratio, *J. Build. Struct.* **38**(S1) (2017) 444–450.
4. M. Ding, J. H. Shi, B. T. Wang, H. Z. Wang, T. Deng, S. H. Luo and X. G. Jiang, Analytic geometric nonlinear circular arch unit, *Eng. Mech.* **38**(7) (2021) 1–8+29.
5. L. Li and K. Lei, Research on influence factors of bearing capacity of concrete-filled steel tubular arch for traffic tunnel, *Symmetry* **14**(1) (2022) 167.
6. K. M. Liu and X. Z. Sun, Bearing capacity of concrete filled steel tube circular arch under the six-point uniformly distributed loading and its engineering application, *Adv. Mater. Sci. Eng.* **2021** (2021) 8288648.
7. Y. F. Zhao, W. W. Xie and L. F. Yang, Elastic modulus reduction method for stable bearing capacity analysis of dumbbell-shaped CFST arches, *J. Phys. Conf. Ser.* **2044**(1) (2021) 012163.
8. C. C. Yuan, Q. Hu, Y. Y. Wang and C. Y. Liu, Out-of-plane stability of fixed concrete-filled steel tubular arches under uniformly distributed loads, *Mag. Concr. Res.* **73**(18) (2020) 945–957.
9. W. Jiang, L. Wang and Y. J. Li, Out-plane stability safety factors of CFST arches using inverse finite element reliability method, *IOP Confe. Ser. Mater. Sci. Eng.* **392**(2) (2018) 022035.
10. Mark Andrew Bradford and Y. L. Pi, Geometric nonlinearity and long-term behavior of crown-pinned CFST arches, *J. Struct. Eng.* **141**(8) (2014), https://doi.org/10.1061/(ASCE)ST.1943-541X.0001163.
11. X. R. Wu, C. Y. Liu, W. Wang and Y. Y. Wang, In-plane strength and design of fixed concrete-filled steel tubular parabolic arches, *J. Bridge Eng.* **20**(12) (2015), https://doi.org/10.1061/(ASCE)BE.1943-5592.0000766.
12. S. Timoshenko, *Theory of Elastic Stability* (McGraw-Hill Book Company Inc, New York, 1961).
13. A. H. Dinnik, *Stability of Arch* (Architectural Engineering Press, Beijing, 1958) (in Chinese).
14. T. Deng, *Geometric Nonlinear Analysis and its Application of Circular Arch Based on Large Displacement Model* (China Agricultural University, Beijing, 2018) (in Chinese).
15. B. Shen and K. J. Ma, In-plane stability of deep circular arch under uniform radial pressure, *Mech. Eng.* **35**(2) (2013) 62–64.
16. P. Cheng and G. S. Tong, A general theory for in-plane nonlinear analysis of circular arches, *Eng. Mech.* **22**(1) (2005) 93–101.
17. J. Zheng and J. Wang, Concrete-filled steel pipe arch bridges in China, *Engineering* **4**(1) (2018) 143–155.

18. J. Zheng, New development trend of long span concrete arch bridges in China, *J. Chongqing Jiaotong Univ. (Nat. Sci. Edn.)* **35**(S1) (2016) 8–11.
19. L. Han, W. Li and B. Reidar, Developmentsand advanced applications of concrete-filled steel tubular (CFST) structures: Members, *J. Constr. Steel Res.* **100** (2014) 211–228.
20. X. P. Li, H. Chen and Y. Zhang, Exact formula for calculating the internal forces of parabolic arches, *Sci. Technol. Eng.* **10**(6) (2010) 1453–1457.
21. Y. Zhou, C. C. Xu, Q. Zhao and X. Z. Wang, Analytical solution of strain-influenced lines for no-hinged arches with variable cross-section suspension lines, *J. Comput. Mech.* 1–6.
22. Y. H. Yang, L. J. Liu, W. G. Sun and T. Z. Hao, Practical analytical solutions for the line of influence of deflection of parabolic no-hinged arches with equal sections, *J. Comput. Mech.* **34**(4) (2017) 480–486.
23. C. F. Hu, W. X. Ren and J. Z. Shang, Improvement study of MEXE correction method based on approximate curve integration, *Xiangtan Univ. (J. Nat. Sci.)* **35**(1) (2013) 63–66+96.
24. C. F. Hu, X. Y. Lu, H. H. Gan and W. X. Ren, Practical analytical solutions for suspended chain line arches based on approximate integration, *J. Central South Univ. (Nat. Sci. Edn.)* **46**(3) (2015) 1058–1065.
25. P. N. Chatteerjee, *On the Deflection Theory of Ribbed Two-Hinged Elastic Arches* (The University of Illinois, Chicago, 1948).
26. X. He, F. He and X. Zhang, Geometric nonlinear analysis of arch bridges deflection theory, *J. China Highway Eng.* **3** (1991) 47–54+67.
27. X. Liu and Y. Zhou, Internal force calculation of long-span concrete no-hinged arch considering nonlinear effect, *J. Hunan Univ. (Nat. Sci. Edn.)* **1** (1987) 29–38.
28. J. Zhang, Q. Zheng and R. Qin, Spline collocation method for internal force analysis of arch bridge with large deflection, *J. Changsha Jiaotong Univ.* **3** (2001) 53–57.
29. J. Chen, Spline moment collocation method for nonlinear internal force analysis of long-span arch, *J. Guangxi Univ. (Nat. Sci. Edn.)* **2** (1998) 181–185.
30. J. T. Jiao, B. Diao, Y. H. Ye and S. C. Li, Bearing capacity and ductility of reinforced concrete Z-shaped columns with positive section, *J. Beijing Univ. Aeronaut. Astronaut.* **167**(1) (2007) 110–114.
31. Q. S. Cui and H. D. Nie, Numerical analysis of the positive section load bearing capacity of HRB500 reinforced concrete Z-shaped section bi-directional compression bent column, *J. Build. Struct.* **35**(S1) (2014) 119–125.
32. H. B. Liu, J. G. Zheng and P. D. Zou, Calculation of eccentric compressive load bearing capacity of short columns with laminated reinforced concrete circular section, *J. Jilin Univ. (Eng. Edn.)* **41**(S2) (2011) 159–163.
33. L. Li, D. H. Zhou, C. Shuang and Y. C. He, Comparison of different calculation methods for second-order reinforcement of reinforced concrete columns, *J. Comput. Mech.* **1–9**(2) (2023) 295–302.
34. X. Chen, D. Zhou, S. Zhang, P. Wang and L. Q. Li, An analytical approach to the elastic-plastic moment-curvature correlation for compression-bending sections, *Eng. Mech.* **31**(11) (2014) 175–182+197.
35. C. Shuang, D. H. Zhou, X. Chen, L. Li and S. W. Lan, A simple method for calculating the second-order bearing capacity of columns with circular sections, *J. Jilin Univ. (Eng. Edn.)* **50**(6) (2020) 2178–2185.

36. C. Shuang, D. H. Zhou, X. Chen, L. Li and S. W. Lan, Graphical algorithm for the positive section load capacity of reinforced concrete bi-directionally biased members, *J. Civil. Eng.* **53**(9) (2020) 77–84.
37. C. H. Zhao, W. H. Pan and Y. Z. Luo, Exact matrix stiffness method for out-of-plane buckling analysis of funicular arches considering warping deformations, *Int. J. Struct. Stab. Dyn.* **23**(1) (2023) 2350010.
38. B. Fan, Z. M. Wang and Q. B. Wang, Nonlinear dynamic characteristics of rotating thick ring on the elastic foundation based on plane stress theory, *Int. J. Struct. Stab. Dyn.* (2023), https://doi.org/10.1142/S021945542450055X.
39. Z. W. Li, M. M. Xie, J. A. Li, Y. H. Li and Y. Q. Yang, Linear and nonlinear vibration of axially loaded timoshenko beam with elastic supports: Effects of transition parameter, *Int. J. Struct. Stab. Dyn.* (2023), https://doi.org/10.1142/S0219455424500676..
40. C. F. Hu and Y. M. Huang, In-plane nonlinear elastic stability of pin-ended parabolic multi-span continuous arches, *Eng. Struct.* **190** (2019) 435–446.
41. Z. W. Liu, J. T. Zhou, Y. X. Wu, X. You and Y. H. Hao, Linear control method for arch ring of oblique-stayed buckle cantilever pouring reinforced concrete arch bridge, *Adv. Civil Eng.* **2021** (2021) 6633717.
42. China Iron and Steel Industry Association, *Tensile Testing of Metallic Materials: Part 1: Room Temperature Test Method: GB/T 228.1-2010* (China Standards Press, Beijing, 2011) (in Chinese).
43. Ministry of Housing and Urban Rural Development of the People's Republic of China, *Standard for test Method of Physical and Mechanical Properties of Concrete: GB/T 50081-2019* (China Construction Industry Press, Beijing, 2019) (in Chinese).
44. L. H. Han, *Concrete Filled Steel Tubular Structures: Theory and Practice*, 3rd ed. (Science Press, Beijing, 2016) (in Chinese).

https://doi.org/10.1142/9789819814770_0028

Chapter 28

Fiber–Based Model for Simulating Inelastic Interaction Buckling of Rectangular CFST Slender Columns with Unequal Wall Thickness#

Tuan Trung Le*, Vipulkumar Ishvarbhai Patel*, Qing Quan Liang†,§, Mizan Ahmed‡ and Phat Huynh*

*School of Computing, Engineering and Mathematical Sciences
La Trobe University, Bendigo, VIC 3552, Australia

†College of Sport, Health and Engineering, Victoria University
PO Box 14428, Melbourne, VIC 8001, Australia

‡School of Civil and Mechanical Engineering, Curtin University
Kent Street, Bentley, WA 6102, Australia
§Qing.Liang@vu.edu.au

Rectangular concrete-filled steel tubular (RCFST) columns can be designed to have unequal wall thickness to maximize their resistance to combined axial and uniaxial bending loads. However, there is very little published research on their local–global interaction buckling behavior. This paper presents an efficient fiber-based simulation model for analyzing the inelastic interaction buckling of eccentrically loaded RCFST slender columns with unequal wall thickness. The model considers the progressive local–global interaction buckling, distributed plasticity, concrete cracking and crushing, second-order effects, and geometric imperfections. The mathematical model is programmed to compute not only the axial load–displacement responses but also the interaction curves of axial load and moment of slender RCFST columns. The simulation modeling is validated by experimentally measured data. The validated computer model is used to parametrically study the interaction local–global buckling responses of RCFST columns while the range analysis is conducted to identify the relative significance of key parameters. It is demonstrated that the proposed computer model accurately simulates experimentally measured responses. Hence, it can be used with confidence in practice to analyze and design RCFST slender columns fabricated with steel plates of unequal thickness. The existing design standard AS/NZS 2327:2017 provides acceptable strength predictions of RCFST slender columns and beam-columns while the procedure specified in Eurocode 4 overestimates their capacities.

§Corresponding author.
#This chapter appeared previously on the International Journal of Structural Stability and Dynamics. To cite this chapter, please cite the original article as the following: T. T. Le *et al.*, *Int. J. Struct. Stab. Dyn.*, **23**, 2340031 (2023), DOI: 10.1142/S021945542340031X.

Keywords: Concrete-filled steel columns; interactive inelastic buckling; local and post-post buckling; simulation modeling; inelastic analysis; range analysis; unequal-thickness box section; AS/NZS 2327:2017; Eurocode 4.

1. Introduction

Rectangular concrete-filled steel tubular (RCFST) columns have been frequently utilized in tall buildings, bridges, and offshore structures because they possess high performance.[1] When a RCFST column is under combined axial and uniaxial bending loads, the concrete below the neutral axis is in tension, thereby contributing little to the capacity of the column.[1] To compensate this, the tension flange thickness can be increased to enhance the strength of the RCFST column. In addition, the use of a thicker compression flange can considerably increase the local buckling strength of RCFST columns. The capacity of a RCFST column fabricated with unequal-thickness steel plates as shown in Fig. 1 is higher than that of the one having an equal wall thickness and the same steel ratio. However, slender RCFST columns having noncompact steel box sections of unequal thickness and a member slenderness ratio between 22 and 120 may undergo the inelastic local–global instability. Both simulation and experimental works have been very limited on the interaction buckling in RCFST slender columns with unequal-thickness section. Therefore, this paper deals with this issue by developing a robust and efficient model for analyzing the inelastic interaction buckling of such composite columns.

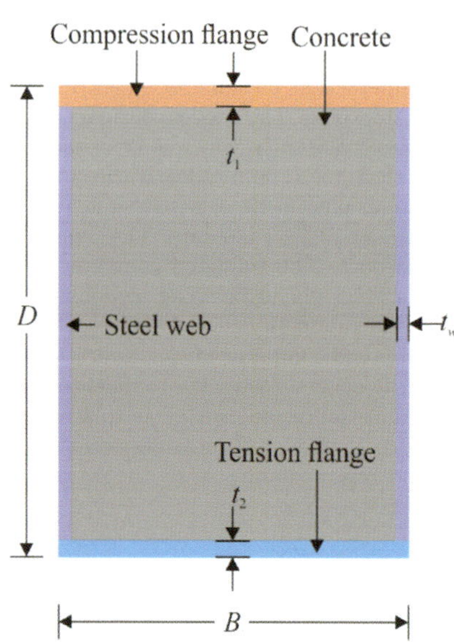

Fig. 1. Cross-section of RCFST column having unequal wall thickness.

There have been extensive studies on the experimentations of RCFST slender columns constructed with steel box section of equal thickness. Shakir-Khalil and Zeghiche[2] undertook experiments on slender full-scale RCFST columns where the width-to-thickness was 24. The tested columns were found to fail by the global buckling without localized-buckling of compact sections. Varma *et al.*[3] reported test results on the response of high strength concrete-filled steel tubular (CFST) square beam-columns where the ratio of width-to-thickness was 48 or 32. It was observed that the compression flange underwent inelastic local buckling before the peak load was reached while the localized buckling of the webs took place in the post-peak regime. Mursi and Uy[4] investigated experimentally the interaction instability of uniaxially loaded slender high-strength square CFST columns with plate slenderness ratios varying from 36 to 56.8. They also tested slender steel box columns constructed with normal strength concrete and high-strength steel under biaxial loads.[5] The results showed that inelastic local buckling might occur before or after the peak load was attained. Further experimental investigations into the responses of high-strength and ultra-high-strength CFST columns with high slenderness ratios were reported by Liu,[6] Perea *et al.*,[7] Liew *et al.*,[8] and Xiong *et al.*[9] The tests indicated that the inelastic local–global interaction buckling took place in RCFST columns with intermediate length, characterized by the interactive and progressive nature.

However, experimental studies of RCFST columns with box section of unequal thickness have been very limited. Tests on the responses of RCFST beams with unequal wall thickness under bending were conducted by Lu *et al.*[10] They reported that the strength and ductility of RCFST beams having unequal wall thickness are higher than those of RCFST beams formed with box section of equal thickness. Zhang *et al.*[11] analytically investigated the sectional properties of RCFST members constructed with four steel plates of unequal thickness. Fu *et al.*[12] undertook experimental studies to determine the responses of twelve RCFST slender columns, which were formed by welding four steel plates with unequal thickness. The influences of the ratio of flange thickness, member slenderness as well as end eccentricity on the response of RCFST columns were investigated. Experimental observations indicated that the failure of RCFST columns tested was induced by the interaction local–global inelastic instability associated with the crushing of concrete. Fu *et al.*[12] suggested that the thickness of the compression flange should be three times the tension flange thickness to achieve economical designs of RCFST slender columns when they are eccentrically loaded.

The inelastic analysis models of RCFST slender columns constructed with box section of equal thickness have been developed by several researchers.[13–22] However, only a few models have incorporated the interaction buckling in the inelastic analysis due to its complexity.[15–22] The walls of a RCFST column loaded uniaxially or biaxially may be under either uniform stress or stress gradient.[18,20] Shanmugam *et al.*[15]

formulated a computer model for the modeling of slender RCFST columns loaded biaxially to failure. Their model considers the strength of post-localized buckling of plates under uniform compressive stress only using the effective width concept. Vrcelj and Uy[16] and Mursi and Uy[17] incorporated the influences of localized instability of steel plates under uniform stress in the nonlinear procedures for slender RCFST columns loaded either uniaxially or biaxially. However, their models do not simulate the progressive failure behavior of steel plates subjected to nonuniform stress in the post-local buckling regime. Fiber element analysis models have been proposed for predicting the inelastic interaction buckling of RCFST slender columns under combined bending and axial loads by Patel *et al.*,[18,19] Liang *et al.*,[20] Phan *et al.*,[21] and Kamil *et al.*[22] The fiber modeling techniques have been demonstrated to simulate well the interactive buckling of flanges and webs under stress gradients combined with the overall column buckling.[18–23]

The only numerical model of RCFST columns having unequal wall thickness was presented by Fu *et al.*[12] who employed ABAQUS to undertake parametric studies. This paper develops a novel approach to account for the progressive local buckling of steel plates with unequal thickness, which is not addressed in the previous studies. The proposed approach provides a more accurate prediction of the behavior of RCFST columns with unequal wall thickness, which is essential for the practical design of such members. The numerical analysis incorporates the effect of localized buckling of unequal walled RCFST column cross-section, which is not considered in other fiber-based models. For this purpose, a fiber-based simulation model is proposed that calculates the deflection responses to eccentric loads and interaction responses to combined bending moment and axial load in RCFST columns with unequal wall thickness. Important features associated with RCFST columns are considered, including the inelastic interaction buckling, distributed plasticity, concrete cracking and crushing, imperfections, and seconder-order effects. The experimental data given by Fu *et al.*[12] is employed to evaluate the accuracy of the simulation model. Extensive numerical analyses are carried out to determine the sensitivity of structural response to various parameters. The range analysis is conducted to determine the relative significance of key parameters on the strengths of RCFST columns. The existing design methods given in AS/NZS 2327a[42] and Eurocode 4[43] are assessed by using the numerical results and test data.

2. Fiber-Based Simulation Model

2.1. *Inelastic sectional analysis*

The inelastic sectional analysis can be performed using the finite element models or fiber element models.[24] In finite element modeling, the column is divided into three-dimensional elements with numerous degrees of freedom along its length. In addition, contact elements are required to model the interaction between the steel

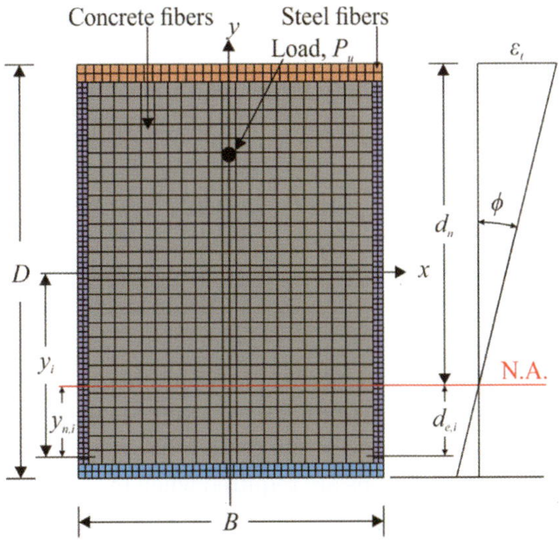

Fig. 2. Discretization of cross-section with unequal wall thickness into fiber elements.

tube and concrete core. Due to a large number of degrees of freedom, the computational cost of nonlinear finite element analysis is significantly higher than that of fiber elements models, which do not require the discretization of the column along its length. Therefore, the simulation model has been developed in this study that utilizes the analysis method of fiber elements, which is a computationally cost–effective and accurate computational modeling technology.[23,25] As shown in Fig. 2, the cross-section of the RCFST column having unequal wall thickness is discretized into many small fiber elements. Steel fibers are assigned with steel constitutive laws while the fibers of concrete follow the concrete stress–strain model. The material constitutive relationships are given in Sec. 3. For a cross-section under combined uniaxial bending moment and axial load, the internal actions of axial force and bending moment are expressed in the mathematical form as

$$P = \sum_{j=1}^{\mathrm{ns}} \sigma_{s,j} A_{s,j} + \sum_{k=1}^{\mathrm{nc}} \sigma_{c,k} A_{c,k}, \tag{1}$$

$$M = \sum_{j=1}^{\mathrm{ns}} (\sigma_{s,j} A_{s,j}) y_j + \sum_{k=1}^{\mathrm{nc}} (\sigma_{c,k} A_{c,k}) y_k, \tag{2}$$

where P stands for the axial force; M denotes the bending moment; $\sigma_{c,k}$ represents the stress at the centroid of concrete fiber k; $\sigma_{s,j}$ denotes the effective stress in the structural steel fiber j; $A_{s,j}$ and $A_{c,k}$ are the areas of the concrete fibers j and k, respectively; x and y represent the centroid coordinates of fibers j and

k, respectively; and nc and ns denote the numbers of concrete and steel fibers, respectively.

To rigorously enforce compatibility in the inelastic modeling of a slender RCFST column, the strain is assumed to linearly distribute within the depth of the box section shown in Fig. 2. This strain profile reflects the fact that the plane sections of the member remain plane after deformation. The longitudinal slip at the interface of the steel box and the filled concrete is taken as zero. However, the proposed method of inelastic analysis allows the effect of localized buckling of steel box in the transverse direction to be accounted for in the strength calculations. The shear deformations are ignored. The distributed plasticity in the cross-section is explicitly modeled by small steel fibers which follow the strain comparability and stress–strain model for structural steel. If the compressive strain of concrete fibers exceeds the specified value of 0.04, the column is assumed to fail so that the analysis is stopped.

The moment–curvature relations are required when modeling the interaction diagram of axial load and bending moment for a RCFST slender column. For the cross-section under axial load (P_a), the inelastic moment–curvature curve is simulated by gradually increasing the curvature; the corresponding moments are computed. For each curvature increment, the solution algorithms[18,20] developed by means of implementing Müller's method[26] are used to iteratively vary the neutral-axis depth to achieve the force equilibrium state. For each trial depth of neutral axis, the corresponding fiber strains and stresses are calculated; the internal action P is determined as the stress resultant. The moment (M) is calculated if the following force equilibrium condition is satisfied:

$$|P_a - P| < \varepsilon_k, \tag{3}$$

where the tolerance for convergence (ε_k) is specified as $\varepsilon_k = 10^{-4}$ by default in the computer program.

It should be noted that the initial localized instability and progressive post-localized buckling of steel box section with unequal thickness are accounted for in the determinations of the moment and axial force as described in Sec. 2.2. The ultimate moment of the cross-section loaded eccentrically is obtained by determining the maximum moment from the simulated moment–curvature curve.

2.2. *Progressive local buckling*

The local instability of the steel plates of a RCFST column is a function of the ratio of width-to-thickness, initial geometric imperfection, residual stress level, steel yield stress, and applied edge stress. The expressions formulated by Liang *et al.*[27] are implemented in the computer simulation program to determine the initial buckling stress of thin plates with different thickness. After the onset of localized instability, the theory of effective widths is used to calculate the post-localized instability strength for steel plates as illustrated in Fig. 3. Liang *et al.*[27] derived the formulas

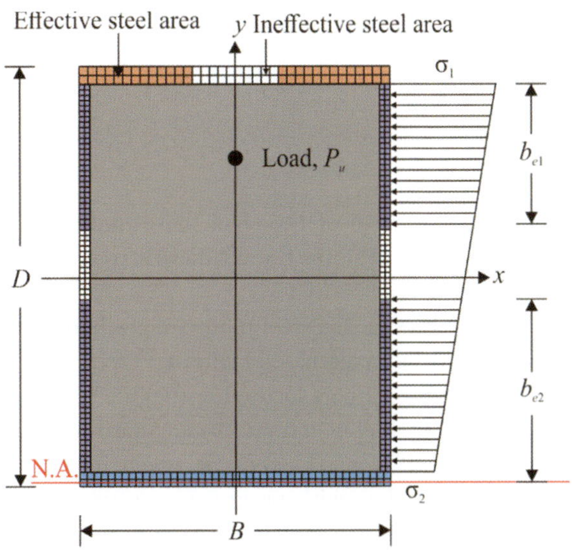

Fig. 3. Effective areas of steel walls with unequal thickness in uniaxially loaded column.

of effective widths for steel plates subjected to stress gradients in RCFST columns under uniaxial and biaxial bending, which are applicable to clamped plates having clear width-to-thickness (b/t) ratios from 30 to 100. The residual stresses due to welding of steel plates as well as localized geometric imperfections have been incorporated in the post-local bucking models.[27] The present mathematical model implements the following expressions derived by Liang *et al.*[27]

$$\frac{b_{e1}}{b} = \begin{cases} 0.2777 + 0.01019\left(\frac{b}{t}\right) - 1.972 \times 10^{-4}\left(\frac{b}{t}\right)^2 \\ \quad +9.605 \times 10^{-7}\left(\frac{b}{t}\right)^3 & (\alpha > 0.0), \\ 0.4186 - 0.002047\left(\frac{b}{t}\right) + 5.355 \times 10^{-5}\left(\frac{b}{t}\right)^2 \\ \quad -4.685 \times 10^{-7}\left(\frac{b}{t}\right)^3 & (\alpha = 0.0), \end{cases} \tag{4}$$

$$\frac{b_{e2}}{b} = (1+\phi)\frac{b_{e1}}{b}, \tag{5}$$

in which $\phi = 1 - \alpha$ and $\alpha = \sigma_2/\sigma_1$, α represents the stress gradient coefficient; σ_2 denotes the minimum edge stress and σ_1 stands for the maximum edge stress. The method of effective strength can also be utilized to calculate the post-localized buckling strength for thin plates. The ultimate strength of thin plates subjected to compressive stress gradient can be determined by the following equation of Liang *et al.*[27]

$$\frac{\sigma_{1u}}{f_{sy}} = (1 + 0.5\phi)\frac{\sigma_u}{f_{sy}} \quad (0 \le \phi < 1.0), \tag{6}$$

where σ_{1u} is the ultimate strength of the steel plate under stress gradient ($\alpha = \sigma_2/\sigma_1$), corresponding to σ_1; σ_u denotes the ultimate strength of the plate stressed uniformly.[27]

The in-plane normal stresses on the steel plate are gradually redistributed based on the theory of effective widths as discussed by Liang[25] to predict the gradual post-localized buckling behavior. Steel fibers located in the ineffective widths are given a zero stress as illustrated in Fig. 3. The interactive localized buckling of compression steel flange and webs with different thickness is taken into account in the fiber model. To model the post-buckling with progressive nature, the ineffective widths of steel box subjected to increasing load are determined by linear integration.[25] After the occurrence of initial localized instability, the stresses on steel fibers are continuously updated to consider the progressive post-local buckling.

2.3. *Inelastic column global buckling*

Slender RCFST columns having slenderness ratios ranging from 22 to 120 and noncompact steel sections undergo the inelastic local–global interaction buckling as observed in experiments.[12] The modeling of inelastic local buckling described in Sec. 2.2 is incorporated in the column global inelastic buckling analysis in the present mathematical model to account for coupled instability. Slender RCFST columns with pin-ended supports and equal end eccentricity (e) as shown in Fig. 4 are treated in this study. The computer model of RCFST columns accounts for geometric and material nonlinearities, geometric imperfections (u_o) initially presented in the column, second-order effects, compatibility, and equilibrium. However, the spread of plasticity along the column length is not considered to significantly save computational time.

The deflected shape of the slender RCFST column as shown in Fig. 4 is represented by a part-sine displacement function as follows:

$$u = u_m \sin\left(\frac{\pi z}{L}\right), \tag{7}$$

in which u_m denotes the lateral displacement that occurs at the column mid-height, and L is the effective length of the slender column.

The curvature φ_m at the mid-height can be computed from the deflection as $\varphi_m = (\pi/L)^2 u_m$. The moment ($M_{\text{ext}}$) acting externally at the column mid-height

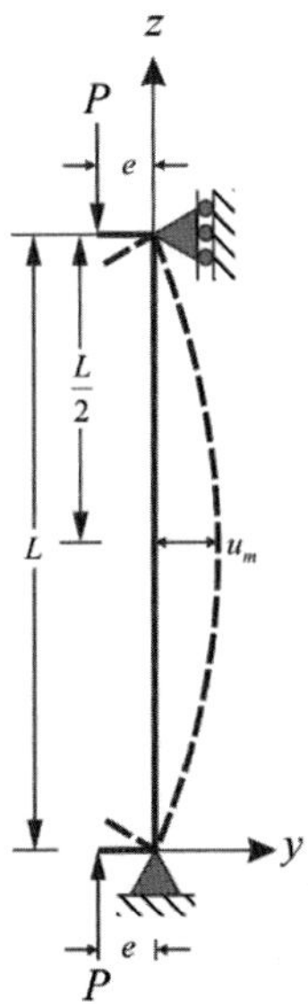

Fig. 4. Pin-ended RCFST slender column loaded eccentrically.

incorporating initial imperfection (u_o), loading eccentricity (e), and the effect of second order is

$$M_{\text{ext}} = P(u_o + e + u_m). \tag{8}$$

To simulate the axial load–displacement curve of RCFST slender columns loaded eccentrically, the proposed method utilizes the displacement-control technique. The lateral displacement u_m at the mid-height of the member is increased in an incremental manner and the corresponding ϕ_m value is computed. For each calculated ϕ_m value, the neutral-axis depth in the cross-section is iteratively varied by means of using iterative solution algorithms[18,28] based on Müller's method[26] to enforce the equilibrium of bending moment at the mid-height. The generalized displacement control method developed by Yang and Shieh[29] may be employed to obtain satisfactory solutions. The displacement compatibility is enforced by the displacement function. The internal action (P), which satisfies the moment equilibrium, is taken as the external load (P_a) applied to the column ends with eccentricity (e). Therefore, the equilibrium conditions can be written as:

$$P_a - P = 0, \tag{9}$$

$$|M - P(u_o + e + u_m)| < \varepsilon_k. \tag{10}$$

The load–deflection modeling process can be terminated by limiting either the deflection or axial load. If the axial load drops to $0.75P_u$ or the deflection exceeds the specified limit, the analysis is terminated.

2.4. *Axial load–moment interaction response*

The inelastic interaction diagrams of axial load and moment (strength envelopes) of slender RCFST columns with unequal wall thickness have not been studied by Fu *et al.*[12] To understand the interaction strengths, computational algorithms have been developed in this study for generating the strength envelopes accounting for inelastic interaction buckling. For the RCFST slender column under an axial load, which is considered as the ultimate axial load (P_u), its ultimate moment (M_u) is calculated as the maximum bending moment ($M_{e,\max}$) applied to the column ends. This means that the maximum loading eccentricity is to be determined. The equilibrium conditions are enforced at the mid-height and expressed as follows:

$$|P_u - P| < \varepsilon_k, \tag{11}$$

$$r_m = |P_u(u_o + u_m) + M_e - M_m| < \varepsilon_k, \tag{12}$$

in which M_e and M_m are the moments at the ends and mid-height, respectively, and r_m stands for the unbalanced bending moment at the column mid-height.

The main steps of the computer algorithms for simulating the strength envelope of a slender RCFST column are illustrated as follows:

Compute P_o by employing the load–deflection modeling procedure
For $P_u = 0$ to $0.9P_o$ step $P_o/10$
Do

$$\phi_m = \phi_m + \Delta\phi_m$$

$$u_m = \phi_m(L/\pi)^2$$

Compute M_m using the moment–curvature modeling procedure
Do
Adjust ϕ_e by the solution method
Compute M_e using the moment–curvature modeling procedure

$$r_m = \text{Abs}[P_u(u_o + u_m) + M_e - M_m]$$

Loop Until $r_m < \varepsilon_k$
If $M_e > M_{e,\max}$ then

$$M_{e,\max} = M_e$$

End if
Loop Until $M_e < 0.98M_{e,\max}$
Next P_u

in which P_o denotes the ultimate axial strength of the concentrically-loaded slender RCFST column, $\Delta\phi_m$ is the curvature increment at the member mid-height, and ϕ_e stands for the curvature at the member ends. If the compressive strain in concrete fibers exceeds the specified limiting value, the loop is exited.

3. Constitutive Laws for Concrete and Steel

3.1. *Confined concrete*

The rectangular steel box of a RCFST column buckles locally outward so that it exerts little lateral pressure on the filled concrete. This confinement effect does not enhance the strength of the filled compressive concrete; however, it remarkably improves the concrete ductility.

Therefore, its effect on the concrete ductility is accounted for in the constitutive laws. The constitutive model of concrete in compression and tension adopted in the present mathematical model is illustrated in Fig. 5. Parts A–D of the curve depicted in Fig. 5 represent the ductility enhancement due to the confinement. The initial modulus of elasticity E_c in both compression and tension is estimated as $E_c = 6900 + 3320\sqrt{\gamma_c f'_c}$ (MPa) specified by ACI Committee 363,[30] in which f'_c denotes the strength of concrete cylinder under compression; γ_c is the coefficient recognizing the column size effect, given by Liang[25] as $\gamma_c = 1.85(D - t_1 - t_2)^{-0.135}$, in which D, t_1, and t_2 are illustrated in Fig. 1.

The Part OA of the material constitutive curve for concrete in compression $(0 \leq \varepsilon_c \leq \varepsilon'_{cc})$ is defined by using the expressions of Mander *et al.*[31] as

$$\sigma_c = \frac{f'_{cc} Q(\varepsilon_c/\varepsilon'_{cc})}{Q - 1 + (\varepsilon_c/\varepsilon'_{cc})^Q}, \tag{13}$$

$$Q = \frac{E_c}{E_c - (f'_{cc}/\varepsilon'_{cc})}. \tag{14}$$

Expressions provided by Liang[25] for Parts AB, BC, and CD as shown in Fig. 5 are

$$\sigma_c = \begin{cases} \gamma_c f'_c & \text{for} \quad \varepsilon'_c < \varepsilon_c \leq \varepsilon'_c, \\ [\beta_c + (0.015 - \varepsilon_c)100(1 - \beta_c)]\gamma_c f'_c & \text{for} \quad 0.005 < \varepsilon_c \leq 0.015, \\ \beta_c \gamma_c f'_c & \text{for} \quad \varepsilon_c > 0.015, \end{cases} \tag{15}$$

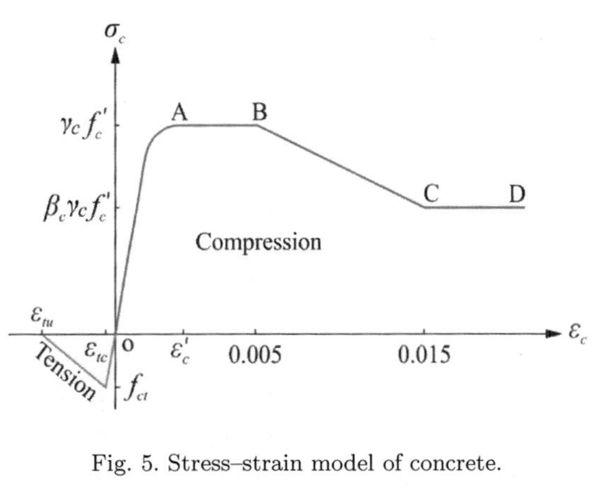

Fig. 5. Stress–strain model of concrete.

in which β_c was given by Liang,[25] accounting for the effect of the confining pressure induced by the tube on the behavior of compressive concrete in the post-peak regime, and is written as

$$\beta_c = \begin{cases} 1.0 & \text{for} \quad \lambda_s \le 24, \\ 1.5 - \lambda_s/48 & \text{for} \quad 24 < \lambda_s \le 48, \\ 0.5 & \text{for} \quad \lambda_s > 48, \end{cases} \tag{16}$$

where $\lambda_s = B_s/t$, $B_s = \max(B, D)$, and t represents the average thickness of the cross-section with unequal wall thickness.

The strain (ε_c') at the maximum stress of unconfined concrete is assumed to increase linearly from 0.002 to 0.003 depending on the concrete strength. As suggested by Liang,[25] if $f_c' \le 28\,\text{MPa}$, $\varepsilon_c' = 0.002$; if $f_c' > 82\,\text{MPa}$, $\varepsilon_c' = 0.003$; if $28 < f_c' \le 82\,\text{MPa}$, the linear interpolation is used to determine ε_c'. For the circular CFST columns, the hoop strain should be considered in the stress–strain modeling of concrete in the post-peak range.[32–34]

The idealized material constitutive model of concrete subjected to tensile stresses is illustrated in Fig. 5. The concrete cracks when its tensile stress is equal to or greater than its tensile strength (f_{ct}), which is estimated as $f_{ct} = 0.6\sqrt{\gamma_c f_c'}$. The tension stiffening of concrete is considered by a linear constitutive relation in the post-cracking regime as shown in Fig. 5. The value of the ultimate strain in tension ε_{tu} is $10\varepsilon_{tc}$, in which ε_{tc} is the cracking strain.

3.2. *Structural steels*

The material constitutive model of structural steels presented by Liang[25] is adopted in the present simulation modeling. This material model shown in Fig. 6 is utilized to simulate the material responses of high-strength, mild, and cold-formed steels in tension and compression. The steel yielding as well as strain-hardening are accounted for as illustrated in Fig. 6, where the rounded part in the range $0.9\varepsilon_{\text{sy}} < \varepsilon_s \le \varepsilon_{\text{st}}$ is expressed by

$$\frac{\sigma_s}{f_{\text{sy}}} = \left(\frac{\varepsilon_s - 0.9\varepsilon_{\text{sy}}}{\varepsilon_{\text{st}} - 0.9\varepsilon_{\text{sy}}}\right)^{\frac{1}{45}}, \tag{17}$$

in which σ_s and ε_s represent the stress and strain in a steel fiber in the longitudinal direction, respectively; ε_s is the steel strain in a fiber, f_{sy} and ε_{sy} stand for the yield stress and strain, respectively. For cold-formed and high-strength steels, the ultimate strain value (ε_{su}) is specified as 0.1; for mild steel, however, it is specified as 0.2 to reflect the good ductility of mild steel.

The surrounding steel tube of an RCFST column exhibits biaxial stress owing to the confinement mechanism. The lateral expansion of concrete under compression induces tension in the surrounding steel tube, which reduces its original yield

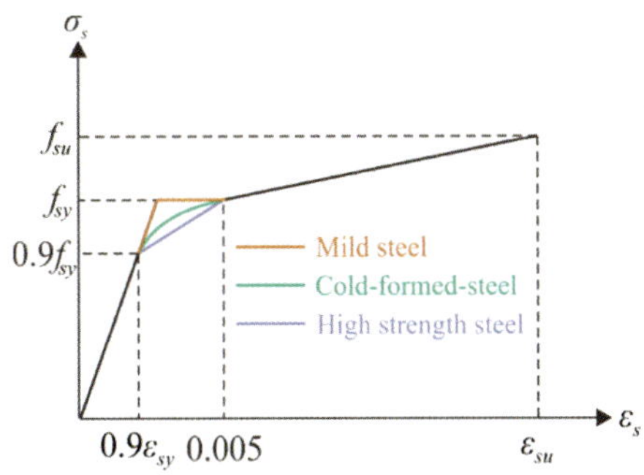

Fig. 6. Material constitutive model of structural steels.

strength. To account for the effects of concrete confinement on the steel tube, the yield strength is multiplied by a factor of 0.9 which is presented in Fig. 6 and Eq. (17).

4. Validation of Simulation Model

4.1. *RCFST beam-columns with unequal tube thickness*

The computer simulation model developed in this research is validated by test data reported by Fu *et al.*[12] Table 1 lists the geometric details in addition to material strengths of the tested RCFST slender columns that were welded by steel plates of unequal thickness. In numerical analyses, the compressive strengths of concrete cubes were reduced to the cylinder concrete compressive strengths by multiplying a factor of 0.85. The loading eccentricities (e) of 20 mm and 40 mm given in Table 1 were measured from the geometric centroid of the cross-sections. They were calculated as 17.3 mm and 37.3 mm measured from the center of gravity of the column cross-sections, respectively. It can be observed from Table 1 that the load–displacement modeling procedure yields good strength predictions. The average value of the predicted strengths ($P_{u,\text{fib1}}$) to the measured values ($P_{u,\text{exp}}$) is 0.93, providing that both coefficient of variation (CoV) and standard deviation (SD) are 0.04. The accuracy of the fiber element model in predicting the ultimate axial strengths is further verified by comparing them with the results obtained from the finite element model ($P_{u,\text{FE}}$) presented by Fu *et al.*[12] in Table 1. The mean value of the predicted ultimate axial strength from the finite element model-to-the fiber element model $P_{u,\text{fib1}}/P_{u,\text{FE}}$ is 0.96, with a standard deviation and coefficient of variation of 0.03. This additional verification further confirms the precise predictions of the ultimate axial strengths by the fiber element model. As shown in Fig. 7, the computer model captures well the experimentally measured load–displacement curves. The comparison of fiber element model predictions and finite element results given by Fu *et al.*[12] is also presented in Fig. 7. The elastic response of the RCFST

Table 1. Comparison of computer simulations and measured strengths of RCFST columns with unequal tube thickness.

Specimen	$B \times D$ (mm)	$t_1 \times t_2$ (mm)	t_w (mm)	L (mm)	e (mm)	$P_{u,\text{exp}}$ (kN)	$P_{u,\text{FE}}$ (kN)	$P_{u,\text{fib1}}$ (kN)	$P_{u,\text{fib2}}$ (kN)	$\frac{P_{u,\text{fib1}}}{P_{u,\text{exp}}}$	$\frac{P_{u,\text{fib2}}}{P_{u,\text{exp}}}$	$\frac{P_{u,\text{fib1}}}{P_{u,\text{FE}}}$	Ref.
B1200-0	110 × 150	7.82 × 7.82	2.92	1 200	0	1 657	1619.3	1 468	—	0.89	—	0.91	
B1200-20	110 × 150	7.82 × 7.82	2.92	1 200	20	1 220	1187.3	1 110	1 124	0.91	0.92	0.93	
B1200-40	110 × 150	7.82 × 7.82	2.92	1 200	40	942	928.2	892	910	0.95	0.97	0.96	
U1200-0	110 × 150	9.68 × 5.88	2.92	1 200	0	1 580	1582.1	1 518	—	0.96	—	0.96	
U1200-20	110 × 150	9.68 × 5.88	2.92	1 200	20	1 220	1216.5	1 203	1 223	0.99	1.00	0.99	
U1200-40	110 × 150	9.68 × 5.88	2.92	1 200	40	994	946.3	952	987	0.96	0.99	1.01	Fu *et al.*[12]
B1600-0	110 × 150	7.82 × 7.82	2.92	1 600	0	1 615	1543.1	1 428	—	0.88	—	0.93	
B1600-20	110 × 150	7.82 × 7.82	2.92	1 600	20	1 196	1125.6	1 055	1 059	0.88	0.89	0.94	
B1600-40	110 × 150	7.82 × 7.82	2.92	1 600	40	948	877.8	844	841	0.89	0.89	0.96	
U1600-0	110 × 150	9.68 × 5.88	2.92	1 600	0	1 573	1526.4	1 512	—	0.96	—	0.99	
U1600-20	110 × 150	9.68 × 5.88	2.92	1 600	20	1 204	1150.1	1 146	1 165	0.95	0.97	1.00	
U1600-40	110 × 150	9.68 × 5.88	2.92	1 600	40	958	891.1	899	922	0.94	0.96	1.01	
Mean										0.93	0.95	0.96	
Standard deviation (SD)										0.04	0.04	0.03	
Coefficient of variation (CoV)										0.04	0.04	0.03	

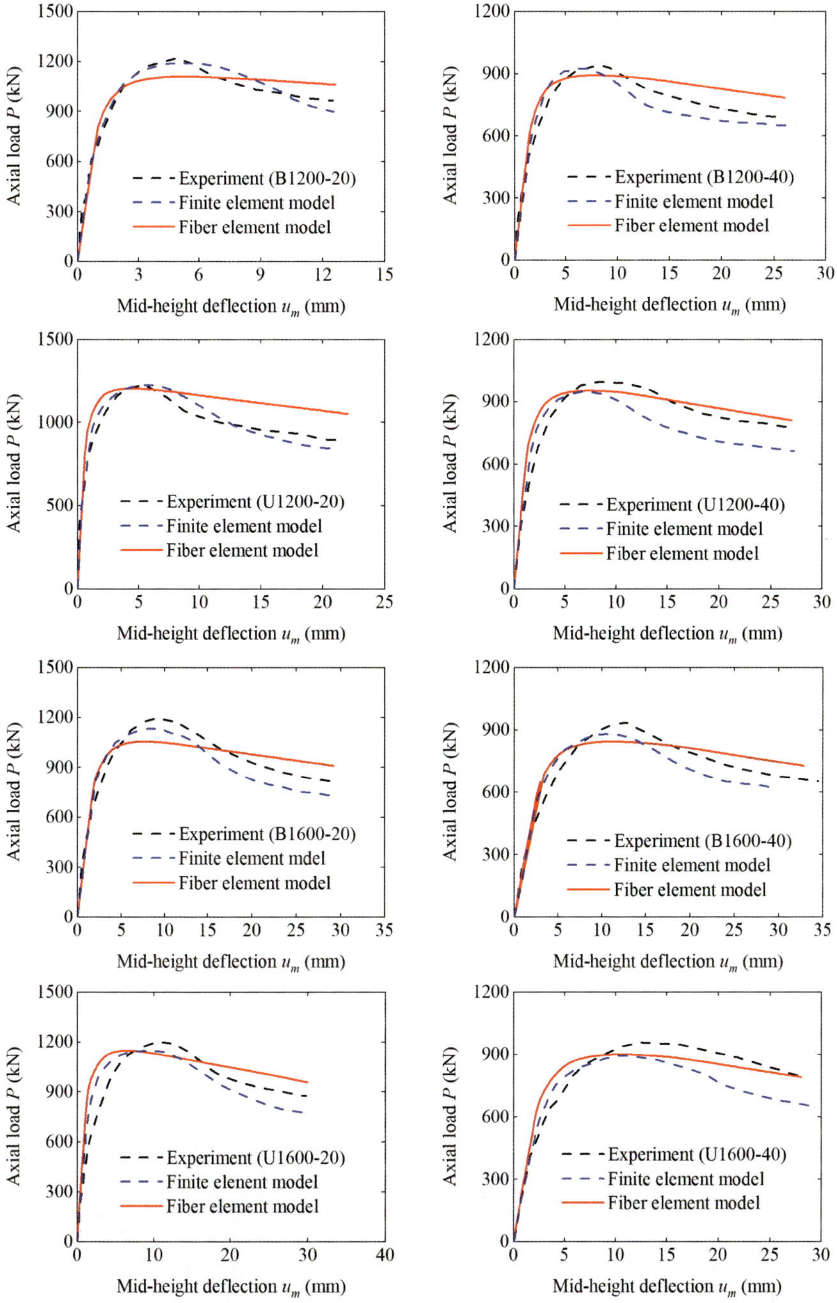

Fig. 7. Load–deflection curves: comparison of experiment and fiber element model for RCFST beam-columns with unequal tube thickness.

Table 2. Comparison of computer simulations and measured strengths of RCFST columns with equal tube thickness.

Specimen	$B \times D \times t$ (mm)	L (mm)	e_x (mm)	e_y (mm)	f'_c (MPa)	f_{sy} (MPa)	E_s (GPa)	$P_{u,\mathrm{exp}}$ (kN)	$P_{u,\mathrm{fib1}}$ (kN)	$\frac{P_{u,\mathrm{fib1}}}{P_{u,\mathrm{exp}}}$	$\frac{P_{u,\mathrm{fib1}}}{P_{u,\mathrm{exp}}}$	$\frac{P_{u,\mathrm{fib2}}}{P_{u,\mathrm{exp}}}$	Ref.
SCH-1	203.7 × 203.9 × 9.96	2 130	0	38	29.9	291	205	1 956	1995.73	1.02	1.01	1.00	
SCH-2	204.0 × 203.3 × 10.01	3 050	0	0	31.1	290	205	2 869	2907.4	1.01	0.99	—	Bridge[35]
SCH-7	152.5 × 152.3 × 6.48	3 050	0	38	31.1	254	205	680	734.0	1.08	1.05	0.97	
R1	80 × 120 × 5	3 210	0	0	37.4	386.3	205	600	600.41	1.00	1.10	—	Shakir-
R2	80 × 120 × 5	3 210	0	24	34.0	386.3	205	393	383.61	0.98	0.98	0.95	Khalil and Zeghiche[36]
S1	149.8 × 149.8 × 4.27	2 700	0	0	31.9	445	200	1 355	1356.99	1.00	1.00	—	Matsui *et al.* [37]
S2	149.8 × 149.8 × 4.27	2 700	0	25	31.9	445	200	847	868.80	1.03	1.02	0.98	
S3	149.8 × 149.8 × 4.27	2 700	0	75	31.9	445	200	552	542.20	0.98	0.98	0.85	
S4	149.8 × 149.8 × 4.27	2 700	0	125	31.9	445	200	383	393.13	1.03	1.02	0.85	
S5	149.8 × 149.8 × 4.27	3 600	0	0	31.9	445	200	1 143	1140.86	1.00	0.99	—	
S6	149.8 × 149.8 × 4.27	3 600	0	25	31.9	445	200	706	734.41	1.04	1.04	1.02	
S7	149.8 × 149.8 × 4.27	3 600	0	75	31.9	445	200	440	464.39	1.06	1.05	0.95	
S8	149.8 × 149.8 × 4.27	3 600	0	125	31.9	445	200	325	346.85	1.07	1.06	0.90	
S9	149.8 × 149.8 × 4.27	4 500	0	0	31.9	445	200	909	895.88	0.99	0.98	—	
S10	149.8 × 149.8 × 4.27	4 500	0	25	31.9	445	200	588	602.78	1.03	1.02	1.01	
S11	149.8 × 149.8 × 4.27	4 500	0	75	31.9	445	200	374	394.64	1.06	1.05	0.96	
S12	149.8 × 149.8 × 4.27	4 500	0	125	31.9	445	200	278	302.98	1.09	1.08	1.04	
C18-1	125 × 125 × 3.2	2 250	0	20.5	94.1	450	200	830	874.28	1.05	1.05	1.04	
C18-3	125 × 125 × 3.2	2 250	0	61.5	94.1	450	200	457	490.57	1.07	1.07	1.06	
C24-0	125 × 125 × 3.2	3 000	0	0	94.1	450	200	1 079	1108.18	1.03	1.02	—	Chung *et al.*[38]
C24-1	125 × 125 × 3.2	3 000	0	20.5	94.1	450	200	664	676.12	1.02	1.02	1.02	
C24-3	125 × 125 × 3.2	3 000	0	61.5	94.1	450	200	375	399.06	1.06	1.06	1.02	
C30-0	125 × 125 × 3.2	3 750	0	0	94.1	450	200	747	814.05	1.09	1.09	—	
CCH1	75 × 75 × 3	1 770	0	3.75	79	450	193.7	414	393.61	0.95	0.95	0.99	
CCH2	65 × 65 × 3	1 770	0	3.25	79	400	189.3	294	263.12	0.89	0.89	0.92	Vrcelj and Uy[39]
CCM1	75 × 75 × 3	1 770	0	3.75	52	450	193.7	343	355.98	1.04	1.04	1.07	
CCm2	65 × 65 × 3	1 770	0	3.25	52	400	189.3	269	242.59	1.01	0.90	0.93	
Mean										1.03	1.02	0.98	
Standard deviation (SD)										0.04	0.05	0.03	
Coefficient of variation (COV)										0.04	0.05	0.03	

columns is accurately predicted. However, the experimentally measured post-peak behavior deviates slightly from the predicted response. It should be noted that the further validation between the results obtained from the finite element model and fiber element model in Fig. 7 shows a good match in the post-peak range. This indicates that the fiber element model predicts well the post-peak response of RCFST columns with unequal tube thickness. Furthermore, in practice a perfect match between simulations and experiments cannot be achieved due to uncertainties in experimental measurements of imperfections and the material properties. A major contributor to the lack of perfect correlation is the variability of concrete. For instance, the strengths of concrete cylinders under compression within a single batch may vary up to ±10%.

The capacity of RCFST slender columns under eccentric loading is taken as the intersection between the line $y = x/e$ and the interaction diagram. The ultimate axial loads ($P_{u,\text{fib2}}$) predicted by the load–moment interaction modeling procedure are provided in Table 1. The average value of $P_{u,\text{fib2}}/P_{u,\text{exp}}$ ratios is 0.95 while SD and CoV are 0.04. This indicates that the developed computational model accurately predicts the load–moment interaction curve.

4.2. *RCFST beam-columns with equal tube thickness*

The proposed fiber-based model is further validated by comparisons of numerical results with test data on RCFST slender beam-columns with equal tube thickness in Table 2. It is observed that the ultimate axial strengths of the beam-columns are accurately predicted by the fiber-based model. However, the ultimate axial strength of Specimen C30-0 is slightly overestimated due to the uncertainty of the actual concrete compressive strength. The mean value of the model-to-experiment $P_{u,\text{fib1}}/P_{u,\text{exp}}$ ratios is 1.03 and both standard deviation and coefficient of variation are 0.07. In addition, the mean value of the model-to-experiment $P_{u,\text{fib2}}/P_{u,\text{exp}}$

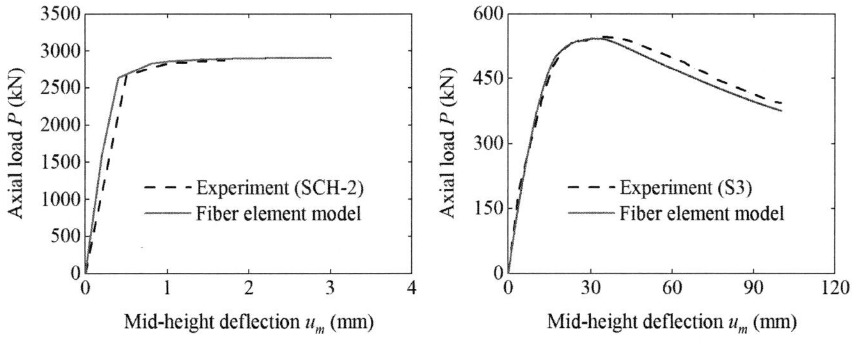

Fig. 8. Load–deflection curves: comparison of experiment and fiber element model for RCFST beam-columns with equal tube thickness.

ratio for load–moment interaction curve is 0.98 with the standard deviation and coefficient variation are 0.03. The predicted load–deflection curves of RCFST slender beam-columns with equal tube thickness are compared with the test results in Fig. 8. Good agreement between the predicted and experimental load–deflection curves is obtained.

5. Parametric Study

The validated computer program was employed to study the influences of flange thickness ratio and member slenderness as well as loading eccentricity on the coupled inelastic instability of slender RCFST columns. The geometry and material parameters employed in the parametric investigations are listed in Table 3. In the parametric study, the steel yield stress of webs was taken as the same as that of the flanges. A value of 200 GPa was assigned to Young's modulus of structural steel. All analyses incorporated the geometric imperfections of $L/1\,000$ initially presented in the columns.

5.1. *Effects of flange thickness ratio*

The inelastic analyses of columns A1–A4 given in Table 3 were undertaken to assess the influences of t_1/t_2 ratio on the inelastic interaction buckling responses. The t_1/t_2 ratios from 1 to 4 were calculated by increasing the compression flange

Table 3. Dimensions and material strengths of RCFST columns.

Column	$B \times D$ (mm)	t_1 (mm)	t_2 (mm)	t_1/t_2	t_w (mm)	e/D	f_{sy} (MPa)	f_{su} (MPa)	f'_c (MPa)	L/r
A1	600 × 700	14	14	1	14	0.1	450	520	80	22
A2	600 × 700	28	14	2	14	0.1	450	520	80	22
A3	600 × 700	42	14	3	14	0.1	450	520	80	22
A4	600 × 700	56	14	4	14	0.1	450	520	80	22
A5	600 × 700	14	14	1	14	0.1	450	520	80	22
A6	600 × 700	14	28	0.5	14	0.1	450	520	80	22
A7	600 × 700	14	42	0.3	14	0.1	450	520	80	22
A8	600 × 700	14	56	0.25	14	0.1	450	520	80	22
A9	700 × 800	20	10	2	8	0.1	420	500	65	22
A10	700 × 800	20	10	2	8	0.1	420	500	65	40
A11	700 × 800	20	10	2	8	0.1	420	500	65	70
A12	700 × 800	20	10	2	8	0.1	420	500	65	100
A13	700 × 800	20	10	2	8	0.1	690	790	100	30
A14	700 × 800	20	10	2	8	0.2	690	790	100	30
A15	700 × 800	20	10	2	8	0.4	690	790	100	30
A16	700 × 800	20	10	2	8	0.6	690	790	100	30
A17	600 × 700	20	8	2.5	14	0.1	450	520	80	22
A18	600 × 700	20	11.39	1.8	12	0.1	450	520	80	22
A19	600 × 700	28	14	2	14	0.1	450	520	40	22
A20	600 × 700	28	14	2	14	0.1	450	520	50	22
A21	600 × 700	28	14	2	14	0.1	450	520	60	22
A22	600 × 700	28	14	2	14	0.1	450	520	70	22

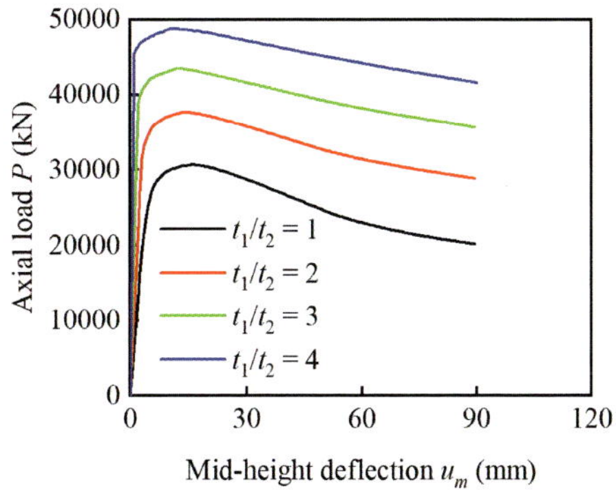

Fig. 9. Effects of flange thickness ratio on axial load–deflection responses.

thickness t_1 for the same tension flange thickness t_2. Figure 9 compares the axial load–displacement relations as a function of t_1/t_2 ratios. It appears from Fig. 9 that the initial flexural stiffness of the member improves as the t_1/t_2 ratio increases. This is because the initial stiffness depends on Young's modulus of steel and concrete; moreover, the contribution of unequal wall thickness to initial bending stiffness increases with increasing the t_1/t_2 ratio. The resistance of the column with unequal wall thickness (t_1/t_2) to axial loading is higher than that of the one having equal wall thickness ($t_1/t_2 = 1$). This is because using a higher t_1/t_2 ratio enhances the localized buckling strength of the flange in compression and the steel area, thereby improving the column strength. In addition, the center of gravity of the column cross-section moves towards the thicker compression flange, which reduces the loading eccentricity, therefore, increasing the capacity of the column. These results indicate that the larger compression flange thickness can effectively be used in an RCFST column loaded uniaxially.

The interaction behavior of axial load and moment in RCFST columns with different flange thickness ratios has not been studied. The beam-columns A1–A4 displayed in Table 3 were simulated and the obtained results are given in Fig. 10(a). It is apparent that the ultimate axial load and maximum bending moment increase as the t_1/t_2 ratio increases by increasing the compression flange thickness t_1 for the same thickness t_2. The impact of t_1/t_2 ratio on the pure bending strength is not significant. Figure 10(b) presents the influences of t_1/t_2 ratio on the capacities of columns A5–A8 (Table 3) by increasing the tension flange thickness t_2 while maintaining compression flange t_1. The pure bending moment increases significantly with increasing the tension flange t_2. It is interesting to note from Fig. 10(b) that the ultimate axial load slightly reduces as t_1/t_2 ratio increases. It can be concluded that the use of a thicker compression flange is beneficial for the columns loaded uniaxially while a thicker tension flange can be used for beams under pure bending without any axial load.

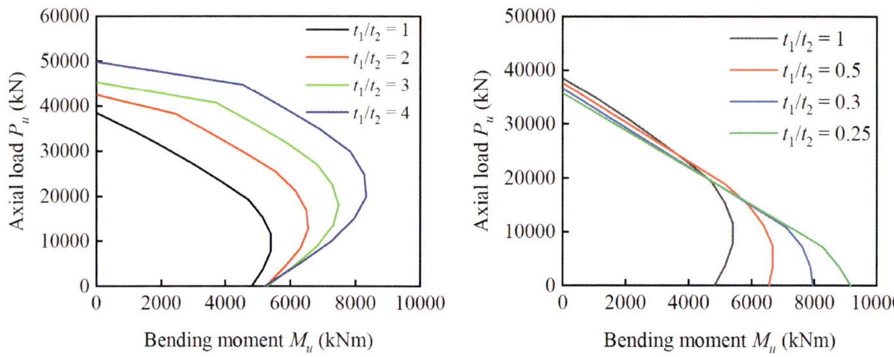

(a) Compression flange thickness effect (same $t_2 = 14\,\text{mm}$)

(b) Tension flange thickness effect (same $t_1 = 14\,\text{mm}$)

Fig. 10. Effects of flange thickness t_1/t_2 ratio on strength envelopes.

5.2. *Effects of member slenderness ratio*

The inelastic strength curves of columns A9–A12 (Table 3) with member slenderness ratios varying from 0 to 120 were predicted by the computer simulation model. Figure 11 presents the column strength curves with varying flange thickness t_1/t_2 ratios, where P_{or} stands for the ultimate axial load of the RCFST column having $t_1/t_2 = 1$. Figure 11 indicates that the ultimate strength increases as the flange thickness t_1/t_2 ratio increases for the same L/r ratio. When the t_1/t_2 ratio is changed from 1 to 4, the ultimate axial load of the column having an L/r ratio of 100 increases by 45%. When designing RCFST columns with a relatively high slenderness ratio, a thicker compression flange can be used to obtain the required capacity.

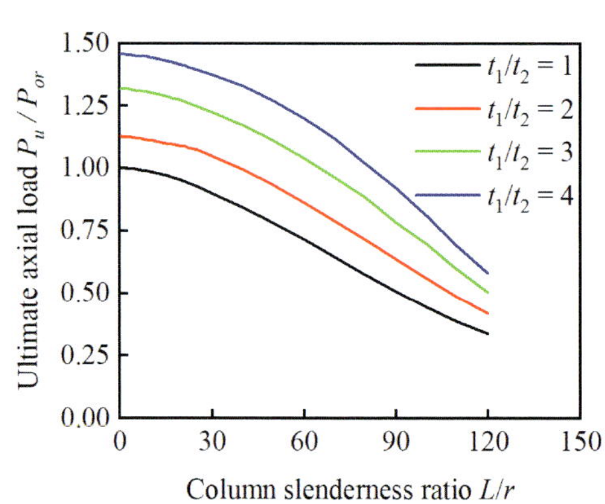

Fig. 11. Effects of flange thickness ratio on column strength curves.

5.3. *Effects of eccentricity ratio*

The computer modeling procedure has been employed to assess the effects of the end eccentricity on the performance of RCFST columns A13–A16 (Table 3) as a function of the flange thickness t_1/t_2 ratio. The results obtained are given in Fig. 12, in which P_{oe} stands for the ultimate axial strength of the cross-section with $t_1/t_2 = 1$. There is a significant reduction in the column's resistance when increasing the e/D ratio. This is caused by the negative effect of the large bending moment at the column ends. The most interesting aspect of these responses shown in Fig. 12 is that for the same e/D ratio, the ultimate load improves as the t_1/t_2 ratio changes from 1 to 4. For e/D ratio of 0.5, a 38% enhancement in the ultimate axial strength is noted if the t_1/t_2 ratio is changed from 1 to 4. It could be concluded that the use of steel box with unequal thickness is beneficial for slender RCFST columns with a high loading eccentricity ratio ($e/D = 0.5$). The strength reduction of RCFST columns located at the corner or side of the building due to large loading eccentricity can be overcome by using an unequal-thickness section with a thicker compression flange. When the RCFST column has a thick compression flange (with $t_1/t_2 = 4$), the strength of the member is shown to increase despite the large end eccentricity. Thus, these research findings indicate that unequal-thickness tube can be used in slender columns with a large loading eccentricity ($0.3 \leq e/D \leq 1.0$) rather than a small eccentricity ($e/D < 0.3$). For the loading with small eccentricity, the conventional CFST columns with equal tube thickness are recommended. In addition, the large deflection at the mid-height generally occurs for the column with a large eccentricity. The increase in mid-height deflection induces tensile stress in the bottom flange. Therefore, it is suggested that the thickness of the tension flange should be larger than that of steel webs.

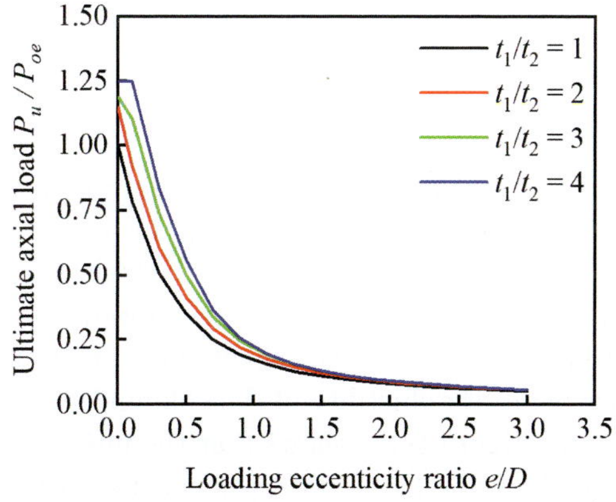

Fig. 12. Effects of end eccentricity on the strength of RCFST columns with different t_1/t_2 ratios.

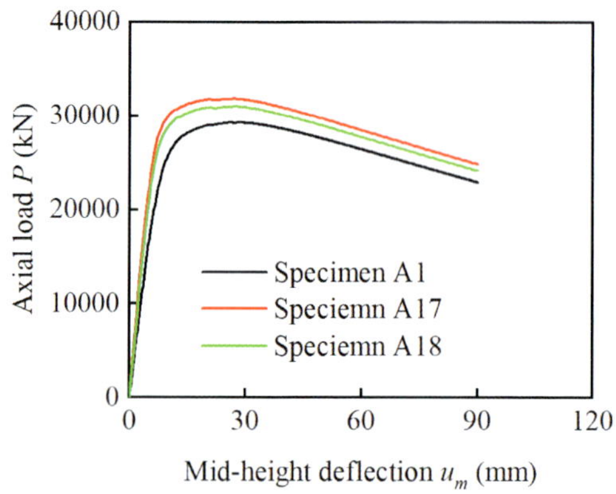

Fig. 13. Performance of RCFST columns designed with varying wall thickness.

5.4. *Economical design of RCFST columns*

As demonstrated in the preceding sections, the thickness of steel box section significantly affects the strengths of RCFST columns. The economical designs of RCFST columns can be achieved by designing steel plates with unequal thickness. To investigate this, columns A17 and A18 listed in Table 3 have been designed by using steel plates with unequal thickness but have the same steel cross-sectional area as that of column A1 with an equal wall thickness. The developed computer model was utilized to simulate their load–deflection responses, which are presented in Fig. 13. It is seen from the Fig. 13 that column A17 has the highest strength and initial bending stiffness; A18 has higher strength and initial bending stiffness than column A1. The ultimate axial load of column A17 is 9% higher than that of column A1. This indicates that column A17 is an economical design that maximizes the strength and bending stiffness of the column for the materials used.

5.5. *Effects of concrete compressive strength*

The effects of concrete compressive strength on the performance of RCFST columns A19–A22 (Table 3) were evaluated using the computer model. The results are presented in Fig. 14. As demonstrated in Fig. 14(a), the ultimate axial strength of the column increases as the concrete strength increases. However, it is important to note that increasing the concrete strength has a slight effect on the initial stiffness of the column. On the other hand, the ductility of the column decreases with an increase in the concrete strength. Figure 14(b) shows that the ultimate bending strength of the column is improved by using higher strength concrete. However, the increase in concrete strength does not have a significant impact on the pure moment capacity of the column.

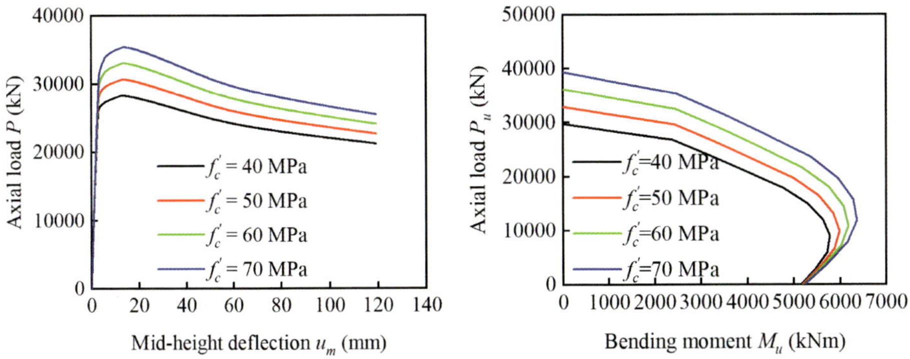

(a) Axial load–deflection responses (b) Strength interaction diagrams

Fig. 14. Effects of concrete strength on the performance of RCFST columns.

6. Range Analysis

The orthogonal design was applied to assess the relative significance of main parameters.[40] The chosen main parameters in the orthogonal design of RCFST columns were t_1/t_2 ratio, slenderness ratio L/r, steel yield strength f_{sy}, and compressive concrete strength f_c'. Each parameter had four levels as presented in Table 4. The columns were arranged based on the orthogonal array, which is illustrated in Table 5. All columns were designed to have a cross-section of 700 × 800 mm and a web thickness of 8 mm. The thickness of the bottom flange t_2 was 10 mm. The loading eccentricity of 80 mm was applied to all columns. The ultimate loads were obtained from the load–deflection modeling procedure and are presented in Table 5.

Range analysis is a statistical method applied to the results from orthogonal design to identify the relative significance of design parameters.[41] The range value (R) of a parameter is calculated as:

$$R_a = \max k_{a,b} - \min k_{a,b}, \tag{18}$$

$$k_{a,b} = \frac{1}{N} \sum_{n=1}^{N} y_{b,n}, \tag{19}$$

Table 4. Four design levels for each parameter.

Design level	t_1/t_2 ratio	L/r ratio	f_{sy} (MPa)	f_c' (MPa)
I-level	1	30	300	40
II-level	2	50	420	60
III-level	3	70	450	80
IV-level	4	90	690	100

Table 5. Ultimate axial loads of RCFST columns from the orthogonal design.

	Parameters				
No.	t_1/t_2 ratio	L/r ratio	f_{sy} (MPa)	f_c' (MPa)	P_u (kN)
B1	2	30	420	60	31 005
B2	4	70	300	80	30 661
B3	1	30	300	40	23 970
B4	3	90	690	40	23 495
B5	1	70	690	60	21 196
B6	3	70	420	100	35 022
B7	2	90	300	100	20 166
B8	1	50	450	100	32 414
B9	4	30	690	100	67 379
B10	2	70	450	40	20 089
B11	3	50	300	60	28 351
B12	4	50	420	40	33 472
B13	4	90	450	60	28 525
B14	2	50	690	80	38 332
B15	1	90	420	80	17 122
B16	3	30	450	80	43 512

where a is the name of the parameter, b is the design level of the parameter; N is the total number of design levels; $y_{b,n}$ is the performance indicator considering the specific parameter at the design level b; and $k_{a,b}$ is the mean value of $y_{b,n}$. The parameter with a higher range value R is a more significant factor that influences the behavior of the design. In the present study, the ultimate axial strength of slender RCFST columns that are eccentrically loaded is used as the performance indicator.

The results obtained by the range analysis are presented in Fig. 15(a). The slenderness ratio is the most important parameter followed by the t_1/t_2 ratio, strength of compressive concrete f_c', and steel yield stress f_{sy}. The yield stress of the steel box is the least significant parameter because the area of the steel box is only 3–8% of the whole cross-sectional area. Figure 15(b) demonstrates the influence of each parameter on the column capacity. Increasing the L/r ratio significantly decreases the load–carrying capacity. On the other hand, increasing the t_1/t_2 ratio, steel yield strength, and concrete strength considerably improves the column capacity. Designing RCFST columns with a high t_1/t_2 ratio is the most effective way to increase the column strength when the column height and section size are constrained due to architecture requirements, as demonstrated by the results of the range analysis presented in Fig. 15(b). Increasing the L/r ratio significantly reduces the column capacity, whereas increasing the t_1/t_2 ratio, steel yield strength, and concrete strength considerably improves the column capacity.

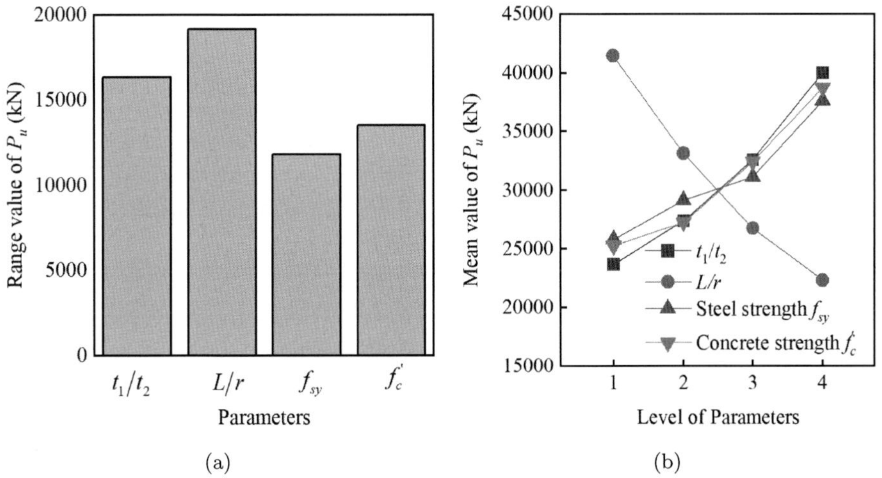

Fig. 15. The results of range analysis: (a) range value; (b) mean value of ultimate capacity.

7. Evaluation of Existing Design Methods

Currently, there are no design standards specifically tailored for RCFST slender columns and beam-columns with unequal steel tube thickness. Therefore, the existing design methods for CFST columns with equal wall thickness given in AS/NZS 2327:2017[42] and Eurocode 4[43] are modified to assess the strengths of RCFST slender columns with unequal steel tube thickness under axial compression and under combined uniaxial bending and axial compression. To facilitate this evaluation, the numerical results and test data presented in preceding sections are used for comparison.

7.1. *AS/NZS* 2327:2017

The ultimate section capacity of RCFST columns under axial compression is calculated as

$$P_{\text{us}}^{\text{AS}} = k_f A_s f_{\text{sy}} + A_c f_c', \tag{20}$$

where k_f is the form factor defined as the ratio of the effective cross-sectional area to the gross area of the rectangular steel tube. The effective widths of the flanges and webs are defined as

$$b_e = b\left(\frac{\lambda_{\text{ep}}}{\lambda_e}\right) \leq b, \tag{21}$$

$$\lambda_e = \frac{b}{t}\sqrt{\frac{f_{\text{sy}}}{250}}, \tag{22}$$

where b is the clear width of the plate; λ_{ep} is compact/noncompact slenderness ratio taken as 64.

The member capacity of a RCFST column under axial compression is determined by

$$P_{\text{uc}}^{\text{AS}} = \alpha_c P_{\text{us}}, \tag{23}$$

where α_c represents the buckling reduction factor for member capacity, defined as

$$\alpha_c = \zeta \left[1 - \sqrt{1 - \left(\frac{90}{\zeta\lambda}\right)^2}\right], \tag{24}$$

$$\zeta = \frac{\left(\frac{\lambda}{90}\right)^2 + 1 + \eta}{2\left(\frac{\lambda}{90}\right)^2}, \tag{25}$$

$$\lambda = \lambda_n + \alpha_a \alpha_b, \tag{26}$$

$$\eta = 0.00326(\lambda - 13.5) \geq 0, \tag{27}$$

$$\lambda_n = 90\lambda_r, \tag{28}$$

$$\alpha_a = \frac{2100(\lambda_n - 13.5)}{\lambda_n^2 - 15.3\lambda_n + 2050}, \tag{29}$$

$$\alpha_b = \begin{cases} 0 & \text{if } k_f < 1, \\ 1 & \text{if } k_f = 1, \end{cases} \tag{30}$$

$$\lambda_r = \sqrt{\frac{P_{\text{us}}}{P_{\text{cr}}}}, \tag{31}$$

$$P_{cr} = \frac{\pi (EI)_{\text{eff}}^{\text{AS}}}{L_e^2}, \tag{32}$$

$$(EI)_{\text{eff}}^{\text{AS}} = E_s I_s + E_c^{\text{AS}} I_c, \tag{33}$$

$$E_c^{AS} = \begin{cases} \rho_c^{1.5} 0.043\sqrt{f_{\text{cmi}}} & \text{if } f_{\text{cmi}} = 0.9(1.2875 - 0.001875 f_c') f_c' \leq 40\,\text{MPa}, \\ \rho_c^{1.5}(0.024\sqrt{f_{\text{cmi}}} + 0.12) & \text{if } f_{\text{cmi}} = 0.9(1.2875 - 0.001875 f_c') f_c' > 40\,\text{MPa}. \end{cases} \tag{34}$$

The section load–moment interaction diagram is calculated using the plastic design procedures. The relation between axial load and bending moment for composite member under combined uniaxial bending and compression is defined as

$$M_{\text{ed}} = \begin{cases} \dfrac{0.9}{k}\left(M - \dfrac{P - \alpha_n P_{\text{us}}^{\text{AS}}}{(\alpha_c - \alpha_n) P_{\text{us}}^{\text{AS}}} M_0\right) & \text{if } P \geq \alpha_n P_{\text{us}}^{\text{AS}}, \\ \dfrac{0.9}{k} M & \text{if } P < \alpha_n P_{\text{us}}^{\text{AS}}, \end{cases} \tag{35}$$

where M_{ed} is the member bending capacity corresponding to compression load P; M is the section bending capacity corresponding to the section compression capacity P;

M_0 is the imperfection moment defined as the intersection between line $y = N_{\text{uc}}$ and the section load–moment interaction diagram; α_n is the factor of interaction curve which is calculated as

$$\alpha_n = \alpha_c(1+\beta_m)/4 \quad \text{with} \quad \beta = 0.66 + 0.44r \geq 0.44, \tag{36}$$

where β is a factor considering the effect of unequal moments at two ends of a column; $r = M_b/M_t$ with M_b and M_t are external bending moment at bottom and top column ends ($M_b = Pe_b$, $M_t = Pe_t$ with e_b and e_t are eccentricities at column ends); k is the coefficient representing the second-order effect, which is calculated as

$$k = \frac{\beta}{1 - P/P^{\text{AS}}_{\text{eff,II}}} \geq 1, \tag{37}$$

$$P^{\text{AS}}_{\text{eff,II}} = \frac{\pi(EI)^{\text{AS}}_{\text{eff,II}}}{L_e^2}, \tag{38}$$

$$(EI)^{\text{AS}}_{\text{eff,II}} = K_o(E_sI_s + K_{c,\text{II}}E^{\text{AS}}_{\text{cm}}I_c) \quad \text{with} \quad K_o = 0.9, K_{c,\text{II}} = 0.2. \tag{39}$$

7.2. *Eurocode* 4

The member capacity of RCFST columns under axial load is defined as

$$P^{\text{EU}}_{\text{uc}} = \chi P^{\text{EU}}_{\text{us}}, \tag{40}$$

$$P^{\text{EU}}_{\text{us}} = A_s f_{\text{sy}} + A_c f'_c, \tag{41}$$

where χ is a reduction factor defined as

$$\chi = \frac{1}{\phi + \sqrt{\phi^2 - \lambda^2}}, \tag{42}$$

$$\phi = 0.5[1 + 0.21(\lambda - 0.2) + \lambda^2], \tag{43}$$

$$\lambda = \sqrt{\frac{P^{\text{EU}}_{\text{us}}}{P_{\text{cr}}}}, \tag{44}$$

$$P_{\text{cr}} = \frac{\pi(EI)^{\text{EU}}_{\text{eff}}}{L_e^2}, \tag{45}$$

$$(EI)^{\text{EU}}_{\text{eff}} = E_sI_s + 0.6E^{\text{EU}}_c I_c, \tag{46}$$

$$E^{\text{EU}}_c = 22\,000[(f'_c + 8)/10]^{0.3} \quad \text{(MPa)}. \tag{47}$$

The section interaction curve is determined by using the plastic design procedure. The load–moment intersection diagram for composite member is calculated as

$$M_{\text{ed}} = \begin{cases} \alpha_M \left[\dfrac{M}{k} - P \dfrac{M_0}{\beta P_{\text{uc}}^{\text{EU}}} \left(1 - \dfrac{P_{\text{uc}}^{\text{EU}}}{P_{\text{cr,II}}} \right) \right] & \text{if } P > 0.1 P_{\text{cr,II}}, \\ \alpha_M \left(M - P \dfrac{M_o}{P_{\text{uc}}^{\text{EU}}} \right) & \text{if } P \leq 0.1 P_{\text{cr,II}}, \end{cases} \tag{48}$$

where M_{ed} is the flexural strength of member corresponding to compression load P; M is the section bending moment capacity under compression load P; M_o is the section bending moment capacity under compressive force $P_{\text{uc}}^{\text{EU}}$; α_M is the reduction factor of steel stress block ($\alpha_M = 0.9$ if $f_{\text{sy}} < 420\text{MPa}$ and $\alpha_M = 0.8$ if $f_{\text{sy}} \geq 420\text{MPa}$). The second-order effects is determined by a factor k determined by

$$k = \frac{\beta}{1 - P/P_{\text{eff,II}}^{\text{EU}}} \geq 1, \tag{49}$$

where Euler's bucking strength $P_{\text{cr,II}}$ is calculated as

$$P_{\text{cr,II}} = \frac{\pi (EI)_{\text{eff,II}}^{\text{EU}}}{L_e^2}, \tag{50}$$

$$(EI)_{\text{eff,II}}^{\text{EU}} = 0.9(E_s I_s + 0.5 E_c^{\text{EU}} I_c). \tag{51}$$

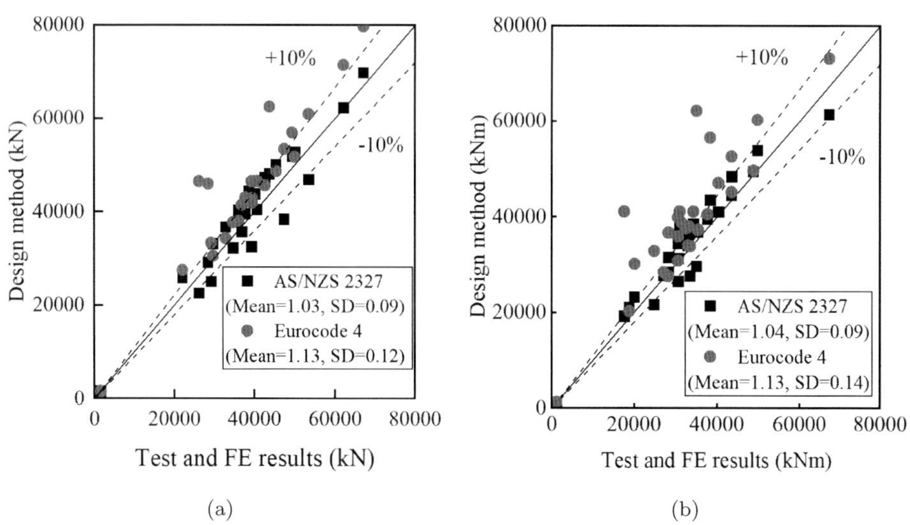

Fig. 16. Verification of AS/NZS 2327:2017[42] and Eurocode 4[43] for (a) axially loaded RCFST slender columns and (b) RCFST slender beam-columns.

7.3. *Validation*

Both Eurocode 4[43] and AS/NZS 2327:2017[42] adopt the same approach for determining the column strength P_{ed} which is the intersection between the line $M = Pe$ and the member load-interaction diagram. The experimental data and numerical results presented in the preceding sections are compared with the predicted strengths from existing design standards. Figure 16 shows that the design method specified in Eurocode 4[43] overestimates the capacity of RCFST slender columns and beam-columns. The design standard AS/NZS 2327:2017,[42] even though provides slightly overestimated strengths, the results are still in acceptable range with the mean of 1.03 and standard deviation of 0.09. Hence, the procedure given in AS/NZS 2327:2017[42] can be used to calculate the strengths of RCFST slender columns and beam-columns with unequal steel tube thickness.

8. Conclusions

A computer simulation model of fiber analysis has been formulated and described in this paper for analyzing the inelastic interaction buckling of slender RCFST columns fabricated by welding steel plates of unequal thickness. The simulation model accounts for not only geometric and material nonlinearities but also progressive inelastic local–global interaction buckling. The agreement between computer simulations and experiments has been shown to be very good. The proposed modeling technique allows inelastic RCFST slender columns with unequal wall thickness to be efficiently analyzed and designed. The assessment of existing design standards AS/NZS 2327:2017 and Eurocode 4 was conducted.

Based on the buckling and range analyses, concluding remarks are provided as follows:

(1) Increasing the compression flange thickness t_1 increases the local buckling strength and steel area, thereby significantly improving the capacities of RCFST columns.
(2) The thick wall compression flange can be used to increase the ultimate axial strengths of RCFST columns while the thickness of the tension flange should be increased to enhance the pure bending strength.
(3) The axial load resistance of RCFST columns decreases with increasing the member slenderness. Slender RCFST columns should be designed to have unequal wall thickness to maximize their resistance to axial and uniaxial bending loads. However, steel box section having equal wall thickness should be used to form short RCFST columns.
(4) It is suggested that unequal-thickness section should be used in slender RCFST columns with the e/D ratios varying from 0.3 to 1.0. The thickness of the compression flange should be greater than that of the tension flange.

(5) The relative significance of parameters examined follows the order: member slenderness, flange thickness ratio, strength of compressive concrete, and yield stress of steel tubes.
(6) The design procedure provided by Eurocode 4 provides overestimated and scattered predictions of the strengths of RCFST slender columns and beam-columns.
(7) The design standard AS/NZS 2327:2017 yields good strength predictions of concentrically and eccentrically loaded RCFST slender columns with unequal steel tube thickness.

ORCID

T. T. Le https://orcid.org/0000-0002-1624-1664
V. I. Patel https://orcid.org/0000-0002-6463-9044
Q. Q. Liang https://orcid.org/0000-0003-0333-2265
M. Ahmed https://orcid.org/0000-0001-5499-3181
P. Huynh https://orcid.org/0000-0001-5291-3731

References

1. Q. Q. Liang, *Analysis and Design of Steel and Composite Structures* (CRC Press, Boca Raton and London, 2014).
2. H. Shakir-Khalil and J. Zeghiche, Experimental behaviour of concrete-filled rolled rectangular hollow-section columns, *Struct. Eng.* **67** (1989) 346–353.
3. A. H. Varma, J. M. Ricles, R. Sause and L.-W. Lu, Experimental behavior of high strength square concrete-filled steel tube beam-columns, *J. Struct. Eng. ASCE* **128**(3) (2002) 309–318.
4. M. Mursi and B. Uy, Strength of concrete filled steel box columns incorporating interaction buckling, *J. Struct. Eng. ASCE* **129**(5) (2003) 626–639.
5. M. Mursi and B. Uy, Behaviour and design of fabricated high strength steel columns subjected to biaxial bending, Part 1: Experiments, *Int. J. Adv. Steel Const.* **2**(4) (2006) 286–315.
6. D. Liu, Behaviour of high strength rectangular concrete-filled steel hollow section columns under eccentric loading, *Thin-Walled Struct.* **42**(12) (2004) 1631–1644.
7. T. Perea, R. T. Leon, J. F. Hajjar and M. D. Denavit, Full-scale tests of slender concrete-filled tubes: Axial behavior, *J. Struct. Eng. ASCE* **139**(7) (2013) 1249–1262.
8. J. R. Liew, M. Xiong and D. Xiong, Design of concrete filled tubular beam-columns with high strength steel and concrete, *Structure* **8** (2016) 213–226.
9. M. X. Xiong, D. X. Xiong and J. Y. R. Liew, Behaviour of steel tubular members infilled with ultra high strength concrete, *J. Const. Steel Res.* **138** (2017) 168–183.
10. F. Lu, S. Li, D. Li and G. Sun, Flexural behavior of concrete filled non-uni-thickness walled rectangular steel tube, *J. Const. Steel Res.* **63**(8) (2007) 1051–1057.
11. Y. Zhang, C.-J. Yu, G.-Y. Fu, B. Chen, S.-X. Zhao and S.-P. Li, Study on rectangular concrete-filled steel tubes with unequal wall thickness, *Steel Compos. Struct.* **22** (2016) 1073–1084.

12. G. Fu, G. Fu, C. Yu, S. Li, F. Wang and J. Yang, Behaviour of rectangular concrete-filled steel tubular slender column with unequal wall thickness, *Eng. Struct.* **236** (2021) 112100.
13. J. F. Hajjar, P. H. Schiller and A. Molodan, A distributed plasticity model for concrete-filled steel tube beam-columns with interlayer slip, *Eng. Struct.* **20**(8) (1998) 663–676.
14. N. Shanmugam and B. Lakshmi, Effect of key parameters on strength of in-filled steel-concrete composite columns, *Int. J. Struct. Stab. Dyn.* **5**(2) (2005) 217–239.
15. N. Shanmugam, B. Lakshmi and B. Uy, An analytical model for thin-walled steel box columns with concrete in-fill, *Eng. Struct.* **24**(6) (2002) 825–838.
16. Z. Vrcelj and B. Uy, Strength of slender concrete-filled steel box columns incorporating local buckling, *J. Const. Steel Res.* **58**(2) (2002) 275–300.
17. M. Mursi and B. Uy, Behaviour and design of fabricated high strength steel columns subjected to biaxial bending, Part II: Analysis and design codes, *Adv. Steel Const.* **2**(4) (2006) 314–354.
18. V. I. Patel, Q. Q. Liang and M. N. S. Hadi, High strength thin-walled rectangular concrete-filled steel tubular slender beam-columns, Part I: Modeling, *J. Const. Steel Res.* **70** (2012) 377–384.
19. V. I. Patel, Q. Q. Liang and M. N. S. Hadi, High strength thin-walled rectangular concrete-filled steel tubular slender beam-columns, Part II: Behavior, *J. Const. Steel Res.* **70** (2012) 368–376.
20. Q. Q. Liang, V. I. Patel and M. N. S. Hadi, Biaxially loaded high-strength concrete-filled steel tubular slender beam-columns, Part I: Multiscale simulation, *J. Const. Steel Res.* **75** (2012) 64–71.
21. D. H. H. Phan, V. I. Patel, Q. Q. Liang, H. Al Abadi and H.-T. Thai, Simulation of uniaxially compressed square ultra-high-strength concrete-filled steel tubular slender beam-columns, *Eng. Struct.* **232** (2021) 111795.
22. G. M. Kamil, Q. Q. Liang and M. N. S. Hadi, Fire-resistance of eccentrically loaded rectangular concrete-filled steel tubular slender columns incorporating interaction of local and global buckling, *Int. J. Struct. Stab. Dyn.* **19**(8) (2019) 1950085.
23. M. Ahmed, Q. Q. Liang, V. I. Patel and M. N. S. Hadi, Local–global interaction buckling of square high-strength concrete-filled double steel tubular slender beam-columns, *Thin-Walled Struct.* **143** (2019) 106244.
24. S. El-Tawil, C. Sanz-Picon and G. Deierlein, Evaluation of ACI 318 and AISC (LRFD) strength provisions for composite beam-columns, *J. Const. Steel Res.* **34**(1) (1995) 103–123.
25. Q. Q. Liang, Performance-based analysis of concrete-filled steel tubular beam–columns, Part I: Theory and algorithms, *J. Const. Steel Res.* **65**(2) (2009) 363–372.
26. D. E. Müller, A method for solving algebraic equations using an automatic computer, *Math. Tables Other Aids Comp.* **10**(56) (1956) 208–215.
27. Q. Q. Liang, B. Uy and J. Y. R. Liew, Local buckling of steel plates in concrete-filled thin-walled steel tubular beam–columns, *J. Const. Steel Res.* **63**(3) (2007) 396–405.
28. V. I. Patel, Q. Q. Liang and M. N. S. Hadi, Numerical analysis of circular concrete-filled steel tubular slender beam-columns with preload effects, *Int. J. Struct. Stab. Dyn.* **13**(3) (2013) 1250065.

29. Y.-B. Yang and M.-S. Shieh, Solution method for nonlinear problems with multiple critical points, *AIAA J.* **28**(12) (1990) 2110–2116.
30. ACI Committee 363, State-of-the-art report on high-strength concrete, *ACI J. Proceed.* **81**(4) (1984) 364–411.
31. J. B. Mander, M. J. N. Priestley and R. Park, Theoretical stress-strain model for confined concrete, *J. Struct. Eng. ASCE* **114**(8) (1988) 1804–1826.
32. M. H. Lai, M. T. Chen, F. M. Ren and J. C. M. Ho, Uni-axial behaviour of externally confined UHSCFST columns, *Thin-Walled Struct.* **142** (2019) 19–36.
33. M. H. Lai, W. Song, X. L. Ou, M. T. Chen, Q. Wang and J. C. M. Ho, A path dependent stress-strain model for concrete-filled-steel-tube column, *Eng. Struct.* **211** (2020) 110312.
34. M. H. Lai, Y. H. Lin, Y. Y. Jin, Q. Fei, Z. C. Wang and J. C. M. Ho, Uni-axial behaviour of steel slag concrete-fill-steel-tube columns with external confinement, *Thin-Walled Struct.* **185** (2023) 110562.
35. R. Q. Bridge, *Concrete Filled Steel Tubular Columns* (School of Civil Engineering, The University of Sydney, Sydney, Australia, 1976), Research Report No. R 283.
36. H. Shakir-Khalil and J. Zeghiche, Experimental behaviour of concrete-filled rolled rectangular hollow-section columns, *Struct. Eng.* **67**(3) (1989) 346–353.
37. C. Matsui, K. Tsuda and Y. Ishibashi, Slender concrete filled steel tubular columns under combined compression and bending, in *Proc. 4th Pacific Structural Steel Conf.*, Singapore, Pergamon, Vol. 3(10) (1995), pp. 29–36.
38. J. Chung, K. Tsuda and C. Matsui, High-strength concrete filled square tube columns subjected to axial loading, in *Seventh East Asia-Pacific Confe. Structural Engineering and Construction*, Kochi, Japan, Vol. 2 (1999), pp. 955–960.
39. Z. Vrcelj and B. Uy, Behaviour and design of steel square hollow sections filled with high strength concrete, *Aust. J. Struct. Eng.* **3**(3) (2002) 153–170.
40. R. K. Roy, *A Primer on the Taguchi Method* (Society of Manufacturing Engineering, 2010).
41. H. Xi, H. Zhang, Y.-L. He and Z. Huang, Sensitivity analysis of operation parameters on the system performance of organic ranking cycle system using orthogonal experiment, *Energy* **172** (2019) 435–442.
42. AS/NZS 2327:2017, *Composite Structures-Composite Steel-Concrete Construction in Buildings* (Standards Australia/Standards New Zealand, Sydney, Australia/Wellington, New Zealand, 2017).
43. Eurocode 4, *Design of Composite Steel and Concrete Structures — Part 1-1: General Rules and Rules for Buildings* (European Committee for Standardization, 2004).

https://doi.org/10.1142/9789819814770_0029

Chapter 29

Floor Seismic Response Spectrum for the Building Equipped with Visco-Elastic Decouplers#

Shi-Li Guo, Yang Xiang* and Guo-Qiang Li

Department of Structural Engineering, Tongji University
Shanghai 200092, P. R. China
**y_xiang@tongji.edu.cn*

By installing the visco-elastic (VE) decouplers between the floor slab and the beam, the floors of a building are isolated from the primary structure. Such a structural system, termed VE decoupled floors, has an increased damping and a reduced inertial force for mitigating the floor seismic response. These features lead to a significantly improved seismic performance of the important facilities resting (or anchored) on floors. This paper focuses on the floor seismic response spectrum (FSRS) of the building equipped with VE decoupled floors. First, the frequency-dependent property of the structure is dealt with, whereas the time-variant stiffness and damping of the VE decoupler are replaced by a couple of stabilized substitutions, and the appropriation of this substitution is numerically validated. Secondly, given the stabilized substitutive model, the FSRS is derived based on the classic complex modal decomposition process and the stochastic dynamics of linear oscillators. In this process, the input seismic excitation is considered via a specific model that accounts for its non-stationary nature. Lastly, a parametrical analysis is conducted to validate the proposed FSRS model and to reveal some special features of the VE decoupled flooring system. It is shown that the FSRS can be better mitigated if a more flexible VE decoupler and a larger mass (isolation) ratio are incorporated. A VE decoupled floor installed at the level of the roof is more effective than the ones installed at the lower levels. Basically, the peak of FSRS can be reduced by 70% if the VE decoupling system is properly designed.

Keywords: Decoupled floor; visco-elastic decoupler; floor seismic response spectrum; complex modal superposition; non-classical damping; non-structural component.

1. Introduction

Recent post-earthquake surveys have demonstrated the importance of a well-protected non-structural system for the resilience of a building.[1–4] From a practical point of view, the building resilience can be achieved via a high-performance

*Corresponding author.
#This chapter appeared previously on the International Journal of Structural Stability and Dynamics. To cite this chapter, please cite the original article as the following: S.-L. Guo, Y. Xiang & G.-Q. Li, *Int. J. Struct. Stab. Dyn.*, **23**, 2340032 (2023), DOI: 10.1142/S0219455423400321.

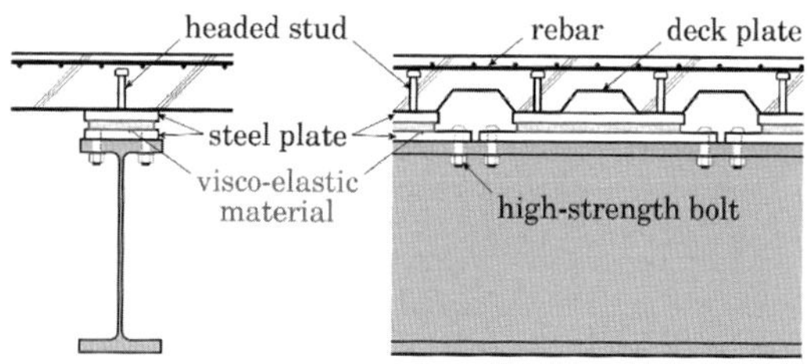

Fig. 1. VE decoupled floor resting on steel beam (Koetaka *et al.*[9]).

structural system.[5,6] By "high-performance", it indicates a simultaneous protection on the primary structure and the secondary elements. Besides, it has an additional constructional advantage compared with conventional structures.[7,8]

Focusing on the target of "high-performance", a typical and promising attempt, termed "steel frame with visco-elastically decoupled floor", was proposed by Koetaka *et al.*[9] The configuration of the proposal is shown in Fig. 1. In this structural system, the floor is decoupled from the steel beam via a flexible visco-elastic (VE) decoupler. The decoupler consists of three parts: (1) The top steel plate to be welded to the steel deck of the floor; (2) the bottom steel plate to be bolted to the steel beam; and (3) the sandwiched VE material to provide vertical stiffness and lateral flexibility. Thanks to the flexibility and damping of the VE decoupler, the floors can be deemed "isolated". The "isolated" floor exhibits a notably reduced acceleration and could protect the acceleration-sensitive building contents from getting damaged. Meanwhile, the reduced floor acceleration yields a mitigated inertial force, leading to a smaller seismic action on the primary structure (steel frame). In addition, the decoupled floor has constructional advantages because the floor slabs can be fully prefabricated, conveniently assembled, and easily demounted.

A series of research have been conducted to reveal the mechanical properties, dynamic behaviors, and seismic response analysis approaches of this novel high-performance structure. First, the static compression characteristics of the VE decoupler, including the short- and long-term cases, are experimentally studied by Koetaka and Sakamoto.[10] The dynamic shearing feature of the VE decoupler, with a vertical compression acting on top, is investigated by Xiang *et al.*[11] Based on the validated material (VE decoupler) model, the structural behavior under seismic excitations was investigated via shaking table tests and numerical simulations. The tests involved single- and two-story scaled structural models (Koetaka *et al.*,[9] Koetaka *et al.*[12]), while the numerical simulation incorporated a wide range of structural parameter (Xiang *et al.*,[13] Xiang and Koetaka[14]). Specifically, a recalibrated complex modal superposition approach is given in Xiang *et al.*,[11] based on which the elastic seismic response of the steel frame with VE decoupled floors can be estimated. According to the preceding researches,[13,14] the peak seismic response of a steel frame equipped with VE decoupled floors could be 50% less than a conventional frame.

Despite the preceding research achievements, the beneficial effect of the VE decoupled floor on mitigating the floor (acceleration) response has not been well quantified. In practice, the seismic response of the VE decoupled floors is needed so as to provide a solid basis for the aseismic design of the non-structural elements. To do so, the most appropriate way is to develop the floor seismic response spectrum (FSRS) for the building, which forms the major task of the current research.

There have been a number of methods for developing the FSRS. The most classic approaches can be dated back to the 1980s,[15–19] when the FSRSs were constructed for systems with and without structure–content interaction in either the time or the frequency domain. In most cases, the frequency approach is much compact than the time history analysis method,[20] but the latter remains the only choice for some special structures such as large-scaled complex ones,[21] base-isolated ones,[22–25] or structures responding in their inelastic stage.[26–29] Recent progresses on the elastic FSRS for classically-damped structures are enhancing the applicability of the FSRS models suggested by design codes.[30–34] Meanwhile, some FSRS models incorporating the inelastic structural behaviors and emphasizing the practical application features are being developed.[35,36]

To propose an appropriate FSRS model for the structures equipped with VE decoupled floors, two major challenges need to be conquered. First, the VE decoupler, in its current form, exhibits a strong frequency-dependency, making its stiffness and damping coefficient fluctuate dramatically under the earthquake excitation. This leads to a time-variant structural model as the stiffness and damping matrices of the structure vary throughout the seismic response process in the time-domain, and prevents the conventional FSRS constructing methods from being adopted. Second, the VE decoupler eliminates the classic nature of the structural damping, and a number of superb approaches for developing the classically damped FSRS are no longer suitable for the situation considered here.

In this paper, the preceding two challenges are tackled and the FSRS for the multi-story building equipped with VE decoupled floors is developed. In doing so, the frequency-dependent property of the building structure is dealt with in Sec. 2, whereas the time-variant stiffness and damping of the VE decoupler are replaced by a couple of stabilized substitutions, and the appropriation of this substitution is numerically validated. Further, theoretical backgrounds and special processes for developing the FSRS are provided in Sec. 3 for the non-classically damped structure discussed herein. Subsequently, a parametrical analysis, which is conducted to validate the proposed FSRS model and to reveal some special features of the VE decoupled flooring system, are given in Sec. 4 before the paper gets closed with conclusions in Sec. 5. In this research, the structure is assumed to behave in its linear elastic stage. Further investigations focusing on its inelastic FSRS are to be reported in the near future.

2. Frequency Dependency and Stabilized Structural Substitution

In this section, the seismic response characteristics of the VE decoupled floors are illustrated. In Secs. 2.1 and 2.2, the floor response-mitigation property of the structure is presented, and the time-variant feature of the mechanical property of the VE decoupler is demonstrated. To make the subsequent theoretical derivations feasible, the stabilized substitution of the structural model is introduced and its rationality is proved in Sec. 2.3.

2.1. *Simulating method and numerical model*

2.1.1. *Structural model*

Without loss of generality, a three-story building structural model is subjected to a realistic earthquake excitation in this section. Two configurations of the structure are considered. First, the floors are rigidly connected to the structure, forming a conventional multi-story frame model. Second, the floors are decoupled from the structure, generating the model shown in Fig. 2(a). Denoting the seismic mass of the jth story by m_j, the mass of the decoupled floor can be expressed by $\nu_j m_j$, where ν_j is the mass ratio between the decoupled floor and its affiliating story. The inter-story stiffness and damping coefficient of the frame is k_j and c_j, respectively, while that of the VE decoupler are $k_{\text{ve},j}$ and $c_{\text{ve},j}$.

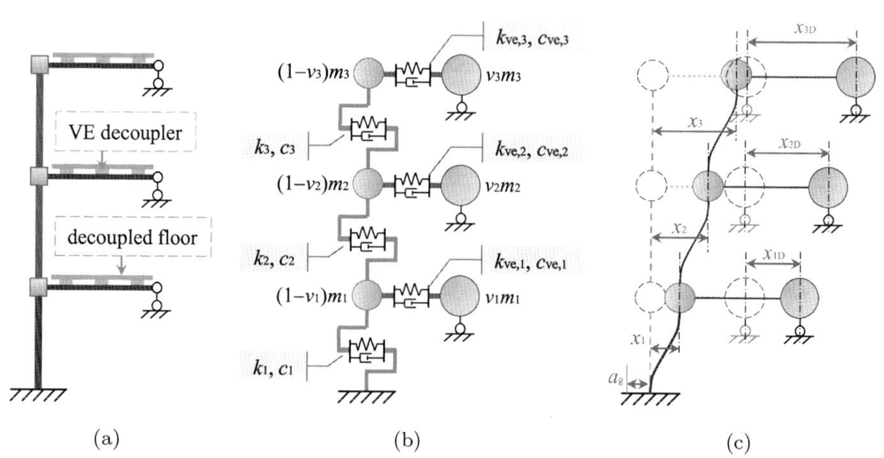

Fig. 2. (a) Frame with the decoupled floors, (b) dynamic model, and (c) displacement denotation.

The dynamic model of the structure equipped with VE decoupled floors is shown in Fig. 2(b), with a dynamic equilibrium of

$$\mathbf{M}\ddot{\mathbf{x}} + \mathbf{C}\dot{\mathbf{x}} + \mathbf{K}\mathbf{x} = -\mathbf{M}\boldsymbol{\iota} \cdot a_g, \tag{2.1}$$

where $\mathbf{M}$, $\mathbf{C}$, and $\mathbf{K}$ are the mass, damping, and stiffness matrices of the structure, $\mathbf{x}$ is the displacement vector, "·" and "··" on top of $\mathbf{x}$ represent its one- and two-order differentials with respect to time. In specific, $\mathbf{x} = [x_1, x_{1\text{D}}, x_2, x_{2\text{D}}, x_3, x_{3\text{D}}]^{\text{T}}$, with the meaning of each element shown in Fig. 2(c). Lastly, a_g is the ground acceleration and $\boldsymbol{\iota}$ is an influential vector with all elements equal to unity. For clarity, $\mathbf{M}$, $\mathbf{C}$, and $\mathbf{K}$ are detailed as follows:

$$\mathbf{M} = \begin{bmatrix} (1-\nu_1)m_1 & & & & & \\ & \nu_1 m_1 & & & & \\ & & (1-\nu_2)m_2 & & & \\ & & & \nu_2 m_2 & & \\ & & & & (1-\nu_3)m_3 & \\ & & & & & \nu_3 m_3 \end{bmatrix}, \tag{2.2}$$

$$\mathbf{C} = \mathbf{C}_0 + \begin{bmatrix} c_{\text{ve},1} & -c_{\text{ve},1} & & & & \\ -c_{\text{ve},1} & c_{\text{ve},1} & & & & \\ & & c_{\text{ve},2} & -c_{\text{ve},2} & & \\ & & -c_{\text{ve},2} & c_{\text{ve},2} & & \\ & & & & c_{\text{ve},3} & -c_{\text{ve},3} \\ & & & & -c_{\text{ve},3} & c_{\text{ve},3} \end{bmatrix}, \tag{2.3}$$

$$\mathbf{K} = \begin{bmatrix} k_1 + k_2 + k_{\text{ve},1} & -k_{\text{ve},1} & -k_2 & & & \\ -k_{\text{ve},1} & k_{\text{ve},1} & 0 & & & \\ -k_2 & 0 & k_2 + k_3 + k_{\text{ve},2} & -k_{\text{ve},2} & -k_3 & \\ & & -k_{\text{ve},2} & k_{\text{ve},2} & 0 & \\ & & k_3 & 0 & k_3 + k_{\text{ve},3} & -k_{\text{ve},3} \\ & & & & -k_{\text{ve},3} & k_{\text{ve},3} \end{bmatrix}. \tag{2.4}$$

Especially, the $\mathbf{C}_0$ in Eq. (2.3) represents the inherent damping of the steel frame, and it is generated via the Rayleigh damping model with a modal damping ratio of $\zeta_1 = \zeta_2 = 0.03$.

2.1.2. *Simulating approach*

As mentioned before, the $k_{\mathrm{ve},j}$ and $c_{\mathrm{ve},j}$ in **K** and **C** varies throughout the seismic responding process due to the frequency-dependency of the VE decoupler. To keep consistent with the preceding researches, we use a Kelvin–Voigt model to compute the values of $k_{\mathrm{ve},j}$ and $c_{\mathrm{ve},j}$.[37]

$$k_{\mathrm{ve},j} = G_0 \frac{1 + ab \cdot \omega_{\mathrm{ve},j}^{2\alpha} + (a+b)\omega_{\mathrm{ve},j}^{\alpha}\cos(\alpha\pi/2)}{1 + a^2 \cdot \omega_{\mathrm{ve},j}^{2\alpha} + 2a \cdot \omega_{\mathrm{ve},j}^{\alpha}\cos(\alpha\pi/2)} \rho_j, \tag{2.5}$$

$$c_{\mathrm{ve},j} = \frac{G_0}{\omega_{\mathrm{ve},j}} \cdot \frac{(b-a)\omega_{\mathrm{ve},j}^{\alpha}\sin(\alpha\pi/2)}{1 + a^2 \cdot \omega_{\mathrm{ve},j}^{2\alpha} + 2a \cdot \omega_{\mathrm{ve},j}^{\alpha}\cos(\alpha\pi/2)} \cdot \rho_j, \tag{2.6}$$

where G_0, a, b, and α are material-dependent parameters, and their values are $G_0 = 39.2\,\mathrm{KPa}$, $a = 5.6 \times 10^{-5}$, $b = 2.1$, $\alpha = 0.558$ (assignment of the parameters are given in Refs. 37 and 38]), ρ_j is the shearing area-to-thickness ratio of the VE decoupler in the jth story, $\omega_{\mathrm{ve},j}$ is the transient deforming (circular) frequency of the VE decoupler under the seismic excitation. The value of $\omega_{\mathrm{ve},j}$ is computed by[38]

$$\omega_{\mathrm{ve},j} = \sqrt{\frac{\sum_{j=1}^{n} |(\dot{\delta}_{\mathrm{ve},j}^{t})^2 - (\dot{\delta}_{\mathrm{ve},j}^{t-j\Delta t})^2|}{\sum_{j=1}^{n} |(\delta_{\mathrm{ve},j}^{t})^2 - (\delta_{\mathrm{ve},j}^{t-j\Delta t})^2|}}, \tag{2.7}$$

where $n = [0.25T_p/\Delta t]$, with T_p being the pre-dominant period of the dynamic response, Δt being the time interval of the dynamic responding process. In Eq. (2.7), $\dot{\delta}_{\mathrm{ve},j}^{t}$ and $\dot{\delta}_{\mathrm{ve},j}^{t-j\Delta t}$ is the shear deforming velocity of the VE decoupler at time t and $t-\Delta t$, respectively, $\delta_{\mathrm{ve},j}^{t}$ and $\delta_{\mathrm{ve},j}^{t-j\Delta t}$ are the accordant deformation. It is worth mentioning that the mechanical model and simulation approach defined in Eqs. (2.5)–(2.7) have been validated by experimental tests conducted by Koetaka *et al.*[9] and Xiang *et al.*[11]

From Eqs. (2.5)–(2.7), the transient value of $\omega_{\mathrm{ve},j}$ would vary during the seismic responding process because of the un-stationary nature of both the input excitation and the output response, i.e. $\dot{\delta}_{\mathrm{ve},j}^{t}$, $\dot{\delta}_{\mathrm{ve},j}^{t-j\Delta t}$, $\delta_{\mathrm{ve},j}^{t}$, and $\delta_{\mathrm{ve},j}^{t-j\Delta t}$. Given the varying $\omega_{\mathrm{ve},j}$ and the frequency-dependency of the VE decoupler, the values of $k_{\mathrm{ve},j}$ and $c_{\mathrm{ve},j}$ would fluctuate throughout the seismic response history, and it would further lead to an unstable **K** and **C** and a special time-variant mechanical property of the structure, as will be demonstrated soon.

2.1.3. *Structural parameters*

In the example illustrated in this section, we assume $k_1 : k_2 : k_3 = 6 : 5 : 3$, $m_1 = m_2 = m_3$, and a fundamental natural period of $T_1 = 0.45\,\mathrm{s}$ for the conventional structural model. For the model with VE decoupled floor, we assume the mass ratios for each story as $\nu_1 = \nu_2 = \nu_3 = 0.8$. The values of ρ_j, which represents the stiffness of the VE decoupler for the jth story, is quantified by

$$\rho_j = \frac{k_{\mathrm{ve},j}^{\mathrm{nor}}}{G_0 \frac{1+ab\cdot\omega_1^{2\alpha}+(a+b)\omega_1^{\alpha}\cos(\alpha\pi/2)}{1+a^2\cdot\omega_1^{2\alpha}+2a\cdot\omega_1^{\alpha}\cos(\alpha\pi/2)}} = \frac{k_{\mathrm{ve},j}^{\mathrm{nor}}}{G_1} = \frac{\kappa_j \cdot k_j}{G_1}, \tag{2.8}$$

where $\omega_1 = 2\pi/T_1$, being the fundamental circular frequency of the structure in its conventional configuration (without VE decoupled floor), G_1 is the nominal storage modulus of the VE decoupler, $k_{\mathrm{ve},j}^{\mathrm{nor}}$ is the nominal stiffness of the VE decoupler in the jth story, and κ_j is the nominal stiffness ratio between the VE decoupler and the frame, $\kappa_j = k_{\mathrm{ve},j}^{\mathrm{nor}}/k_j$. Apparently, the value of ρ_j can be determined as long as κ_j is given. In the subsequent context, we use κ_j to measure the stiffness of the VE decouplers. For the numerical example discussed here, the values of κ_j are assumed to be $\kappa_1 = \kappa_2 = \kappa_3 = 0.15$. It is worth noting that the values of ν_j and κ_j adopted here are within a regular range.[11,13,14]

2.2. *Seismic response characteristic*

Subjecting the structural model, in both the conventional and the decoupled configurations, to an earthquake record plotted in Fig. 3(a), the structural

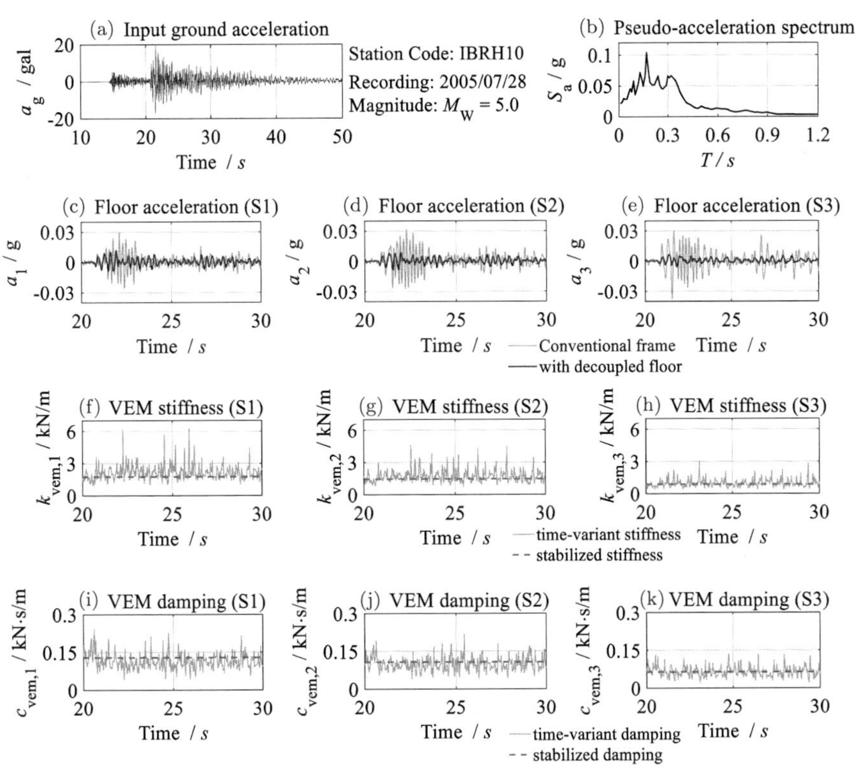

Fig. 3. Input excitation, floor acceleration, and stiffness & damping of VE decoupler.

responses can be obtained by integrating the dynamic equilibrium using the classic Newmark-β method.[39] From Fig. 3(b), the pseudo-acceleration response spectrum of the excitation has a larger amplitude for $T \leq T_1 = 0.45\,\text{s}$. This indicates a pronounced response of the higher-order modes (with modal period shorter than T_1).

The absolute floor accelerations of the conventional case and the floor-decoupled case are shown in Figs. 3(c)–3(e) for story 1 (S1), story 2 (S2), and story 3 (S3), respectively. From the figure, the VE decoupled floors exhibit notably smaller acceleration responses than the conventional flooring system, whereas the peak value of the former is about 70% less than the latter. Such a result implies a dramatic reduction in the amplitude of FSRS of the VE decoupled floors. Regarding the mitigated FSRS, further elaborations are to be given in Secs. 4 and 5.

As mentioned in Sec. 2.1, the structure equipped with VE decoupled floors has a time-variant mechanical property under earthquake excitations because of the frequency-dependency of the VE decoupler. For demonstration, the stiffness of the VE decoupler is plotted in Figs. 3(f)–3(h) for story 1 (S1), story 2 (S2), and story 3 (S3), respectively. From the figure, the fluctuating $k_{\text{ve},j}$ is bounded within a specific range despite its notable time-varying feature exhibited. The damping coefficient of the VE decoupler, $c_{\text{ve},j}$, has a similar feature with $k_{\text{ve},j}$, as shown in Figs. 3(i)–3(k).

In fact, the value of $k_{\text{ve},j}$ and $c_{\text{ve},j}$ mainly oscillates around a nominal value (termed the "stabilized" value hereafter). According to Xiang *et al.*,[11] such a value corresponds to the fundamental circular frequency $\omega_1 = 2\pi/T_1$ of the conventional structure. To get this value, we could use the stable ω_1 to replace the fluctuating $\omega_{\text{ve},j}$ in computing $k_{\text{ve},j}$ and $c_{\text{ve},j}$ and obtain a stabilized $k_{\text{ve},j}$ and $c_{\text{ve},j}$ via Eq. (2.5). For clarity, the values of $k_{\text{ve},j}$ and $c_{\text{ve},j}$ are shown in Figs. 3(f)–3(k).

2.3. *Stabilized substitution of the structural model*

In order to construct a theoretical process for the FSRS of a structure equipped with VE decoupled floors, the time-varying mechanical properties of the structure induced by the fluctuating $k_{\text{ve},j}$ and $c_{\text{ve},j}$ are to be replaced by a set of time-invariant substitutions. From Figs. 3(f)–3(k) and some companioning computations,[11] the stabilized values for $k_{\text{ve},j}$ and $c_{\text{ve},j}$, denoted by $k_{\text{ve},j}^{\text{nor}}$ and $c_{\text{ve},j}^{\text{nor}}$ hereafter, can be appropriately determined by

$$k_{\text{ve},j}^{\text{nor}} = G_0 \frac{1 + ab \cdot \omega_1^{2\alpha} + (a+b)\omega_1^{\alpha}\cos(\alpha\pi/2)}{1 + a^2 \cdot \omega_1^{2\alpha} + 2a \cdot \omega_1^{\alpha}\cos(\alpha\pi/2)} \rho_j, \tag{2.9}$$

$$c_{\text{ve},j}^{\text{nor}} = \frac{G_0}{\omega_1} \cdot \frac{(b-a)\omega_1^{\alpha}\sin(\alpha\pi/2)}{1 + a^2 \cdot \omega_1^{2\alpha} + 2a \cdot \omega_1^{\alpha}\cos(\alpha\pi/2)} \cdot \rho_j. \tag{2.10}$$

Comparing Eqs. (2.9) and (2.10) with Eqs. (2.5) and (2.6), $k_{\text{ve},j}^{\text{nor}}$ and $c_{\text{ve},j}^{\text{nor}}$ correspond to a specific frequency of ω_1. It has been demonstrated by Xiang *et al.*[11] that using $k_{\text{ve},j}^{\text{nor}}$ and $c_{\text{ve},j}^{\text{nor}}$ to replace $k_{\text{ve},j}$ and $c_{\text{ve},j}$ would generate an accurate structural

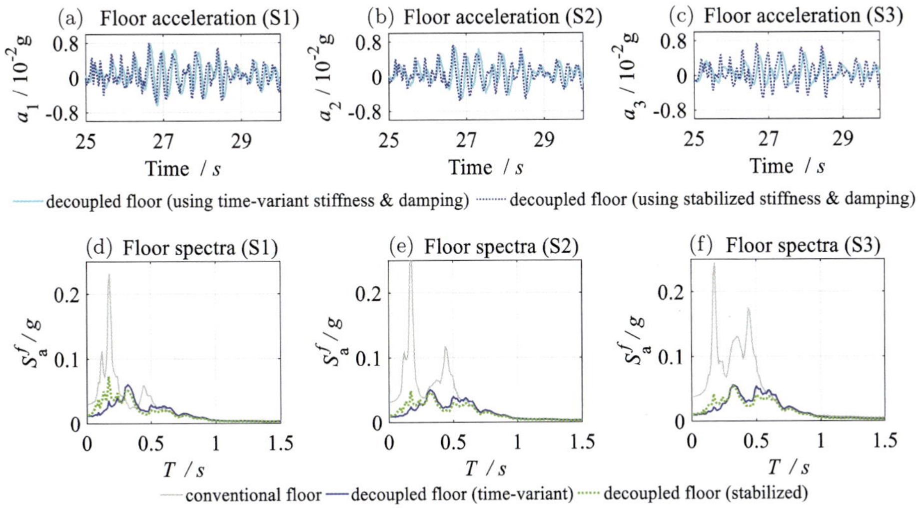

Fig. 4. Floor acceleration responses and FSRSs obtained from time-variant and stabilized (nominal) stiffness and damping of VE decouplers.

displacement response. The values of $k_{\mathrm{ve},j}^{\mathrm{nor}}$ and $c_{\mathrm{ve},j}^{\mathrm{nor}}$ are marked by the dashed lines in Figs. 3(f)–3(k). From the figure, the stabilized substitutions of $k_{\mathrm{ve},j}$ and $c_{\mathrm{ve},j}$ are similar to their realistic time-variant counterparts.

Regarding the estimated floor acceleration, such a substitution may lead to tiny errors, but the error is basically ignorable, as demonstrated by the results shown in Fig. 4. In Figs. 4(a)–4(c), the floor acceleration response histories for story 1 (S1), story 2 (S2), and story 3 (S3), are plotted. Both the accurate results considering the fluctuating $k_{\mathrm{ve},j}$ and $c_{\mathrm{ve},j}$ and the approximate ones generated by the stabilized $k_{\mathrm{ve},j}^{\mathrm{nor}}$ and $c_{\mathrm{ve},j}^{\mathrm{nor}}$ are shown in the figure. The two groups of results are almost identical except for some deviations at several peaks.

More importantly, the FSRS generated by the two groups of floor acceleration histories are shown in Figs. 4(d)–4(f). As expected, the two groups of FSRS are similar with each other, demonstrating the appropriation of the stabilized substitution (using $k_{\mathrm{ve},j}^{\mathrm{nor}}$ and $c_{\mathrm{ve},j}^{\mathrm{nor}}$ to replace $k_{\mathrm{ve},j}$ and $c_{\mathrm{ve},j}$) for deriving the FSRS. Meanwhile, it is shown in Figs. 4(d)–4(f) that the structure equipped with VE decoupled floors have a notably mitigated FSRS, as compared with the conventional scenario.

3. FSRS of the Building with VE Decoupled Floor: Theory

The numerical example in Sec. 2, accompanied by many other computations (Xiang *et al.*[11]), has demonstrated the rationality of using the stabilized nominal stiffness and damping of the VE decoupler in analyzing the structural (including floor) responses. Given the stabilized substitutive model, the FSRS can be derived. The derivation could be conducted based on the classic complex modal decomposition process (Sec. 3.1) and the stochastic dynamics of linear oscillators (Secs. 3.2 and 3.3). In this process, the property of the input seismic excitation is considered via a specific model to be introduced in Sec. 3.4.

3.1. *Model decomposition of non-classical damped system: Revisited*

The structure equipped with VE decoupled floors is a non-classically damped system. For brevity, the modal decomposition of such a structure can be conducted in the state space. The dynamic governing equation of the system is as follows:

$$\mathbf{B}\dot{\mathbf{Z}} + \mathbf{G}\mathbf{Z} = \mathbf{Q}(t), \tag{3.1}$$

where $\mathbf{B}$ and $\mathbf{G}$ are structural characteristic matrices, $\mathbf{Q}(t)$ is the dynamic load on the system, $\mathbf{Z}$ is the response vector which includes both the displacement and the velocity components, $\dot{\mathbf{Z}}$ is the one-order time-differential of $\mathbf{Z}$. Here, $\mathbf{B}$, $\mathbf{G}$, $\mathbf{Q}(t)$, and $\mathbf{Z}$ are detailed as follows:

$$\mathbf{B} = \begin{bmatrix} \bar{\mathbf{0}} & \mathbf{M} \\ \mathbf{M} & \mathbf{C} \end{bmatrix}, \tag{3.2}$$

$$\mathbf{G} = \begin{bmatrix} -\mathbf{M} & \bar{\mathbf{0}} \\ \bar{\mathbf{0}} & \mathbf{K} \end{bmatrix}, \tag{3.3}$$

$$\mathbf{Q}(t) = \begin{bmatrix} \mathbf{0} \\ -\mathbf{M}\iota \cdot a_g \end{bmatrix}, \tag{3.4}$$

$$\mathbf{Z} = \begin{bmatrix} \dot{\mathbf{x}} \\ \mathbf{x} \end{bmatrix}. \tag{3.5}$$

In Eqs. (3.2)–(3.4), $\bar{\mathbf{0}}$ and $\mathbf{0}$ are matrix and vector with all elements equal to zero.

In order to obtain the modal properties of the structure, we assume the system is responding in a specific mode $\mathbf{\Psi}$, and the structural response can be denoted as follows:

$$\mathbf{Z} = \mathbf{\Psi} e^{\gamma t}, \tag{3.6}$$

where γ indicates the vibrating period of the mode. Letting the right-hand side of Eq. (3.1) equal to zero (corresponding to free vibration) and substituting Eq. (3.5) in Eq. (3.1), we have

$$\left(\begin{bmatrix} \bar{\mathbf{0}} & \bar{\mathbf{I}} \\ -\mathbf{K}^{-1}\mathbf{M} & -\mathbf{K}^{-1}\mathbf{C} \end{bmatrix} - \frac{1}{\gamma}\bar{\mathbf{I}} \right) \mathbf{\Psi} = 0, \tag{3.7}$$

based on which the $2N$ eigen-values and the accordant eigen-vectors can be obtained. Here, N is the number of degree-of-freedom of the structure. Denoting the ith and the jth eigen-vectors by $\mathbf{\Psi}_i$ and $\mathbf{\Psi}_j$, respectively, the modal orthogonality would lead us to

$$\begin{cases} \mathbf{\Psi}_i^T \mathbf{B} \mathbf{\Psi}_j = 0, & (i \neq j) \\ \mathbf{\Psi}_i^T \mathbf{G} \mathbf{\Psi}_j = 0, & (i \neq j). \end{cases} \tag{3.8}$$

The natural circular frequency ω_i and modal damping ratio ζ_i of the ith mode are as follows:

$$\omega_i = \left| -\frac{\mathbf{\Psi}_i^T \mathbf{G} \mathbf{\Psi}_i}{\mathbf{\Psi}_i^T \mathbf{B} \mathbf{\Psi}_i} \right|, \tag{3.9}$$

$$\zeta_i = -\mathrm{Re}\left(-\frac{\mathbf{\Psi}_i^T \mathbf{G} \mathbf{\Psi}_i}{\mathbf{\Psi}_i^T \mathbf{B} \mathbf{\Psi}_i} \right) \Big/ \omega_i, \tag{3.10}$$

where $\mathrm{Re}(\cdot)$ means the real part of a complex number.

3.2. *Stochastic response of structure in the state space*

Given the eigen-values and the eigen-vectors, the state space response of the structure can be expressed as follows:

$$\mathbf{Z}(t) = \mathbf{\Psi} \mathbf{V}(t), \tag{3.11}$$

where $\mathbf{\Psi} = \{\mathbf{\Psi}_1, \mathbf{\Psi}_2, \ldots, \mathbf{\Psi}_{2N}\}$ is the collection of the $2N$ eigen-vectors, $\mathbf{V}(t)$ is the modal coordinate vector that shows the contribution of each mode to the global response of the structure. For clarity, the ith element of $\mathbf{Z}(t)$ is computed as follows:

$$Z_i(t) = \sum_{j=1}^{2N} \mathbf{\Psi}_{ij} V_j(t), \tag{3.12}$$

where $\mathbf{\Psi}_{ij}$ is the jth element of eigen-vector $\mathbf{\Psi}_i$, $V_j(t)$ is the modal coordinate of the jth mode. Substituting Eq. (3.11) in Eq. (3.1) would generate

$$\mathbf{B} \mathbf{\Psi} \dot{\mathbf{V}} + \mathbf{G} \mathbf{\Psi} \mathbf{V} = \mathbf{Q}(t). \tag{3.13}$$

By pre-multiplying $\mathbf{\Psi}^{\mathrm{T}}$ with each component in Eq. (3.13) and considering the modal orthogonality, we have

$$\mathbf{\Psi}_j^{\mathrm{T}} \mathbf{B} \mathbf{\Psi}_j \dot{V}_j + \mathbf{\Psi}_j^T \mathbf{G} \mathbf{\Psi}_j V_j = \mathbf{\Psi}_j^{\mathrm{T}} \mathbf{Q}(t). \tag{3.14}$$

Assuming the initial displacement and velocity of the structure are both zero, the value of $V_j(t)$ could be obtained by solving the partial differential equation of Eq. (3.14):

$$V_j(t) = \frac{1}{\boldsymbol{\Psi}_j^{\mathrm{T}}\mathbf{B}\boldsymbol{\Psi}_j}\int_0^t \boldsymbol{\Psi}_j^{\mathrm{T}}\mathbf{Q}(s)e^{\gamma_j(t-s)}\mathrm{d}s$$
$$= \frac{1}{\boldsymbol{\Psi}_j^{\mathrm{T}}\mathbf{B}\boldsymbol{\Psi}_j}\sum_{k=1}^{N}\boldsymbol{\Psi}_{k+N,j}\int_0^t F_k(s)e^{\gamma_j(t-s)}\mathrm{d}s. \quad (3.15)$$

It is reasonable to assume that the mean of the earthquake excitation process is zero, and so is the structural response process. Therefore, as given in Eqs. (3.11), (3.12) and (3.15), the stochastic response of the structure in the state space can be obtained as long as the covariance matrix of $\mathbf{Z}(t)$ is known. To get this matrix, we analyze

$$E[Z_i(t_1)Z_j^*(t_2)] = R_{Z_iZ_j}(t_1,t_2)$$
$$= \sum_{m=1}^{2N}\sum_{n=1}^{2N}\boldsymbol{\Psi}_{im}\boldsymbol{\Psi}_{jn}^* R_{V_mV_n}(t_1,t_2)$$
$$= \sum_{m=1}^{2N}\sum_{n=1}^{2N}\boldsymbol{\Psi}_{im}\boldsymbol{\Psi}_{jn}^* E[V_m(t_1)V_n^*(t_2)], \quad (3.16)$$

where $R_{Z_iZ_j}(t_1,t_2)$ is the correlation function of $Z_i(t)$ and $Z_j(t)$, "*" in the superscript means complex conjugate, $E[\cdot]$ means mathematical expectation, $R_{V_mV_n}(t_1,t_2)$ is the correlation function of the modal coordinates V_m and V_n. Substituting Eq. (3.15) in $R_{V_mV_n}(t_1,t_2)$, after rearrangement, we have

$$R_{V_mV_n}(t_1,t_2) = \frac{1}{\boldsymbol{\Psi}_m^{\mathrm{T}}\mathbf{B}\boldsymbol{\Psi}_m(\boldsymbol{\Psi}_n^{\mathrm{T}}\mathbf{B}\boldsymbol{\Psi}_n^*)}$$
$$\times\sum_{k=1}^{N}\sum_{l=1}^{N}\boldsymbol{\Psi}_{k+N,m}\boldsymbol{\Psi}_{l+N,n}^* A(t_1,t_2), \quad (3.17)$$

where $A(t_1,\ t_2)$ in Eq. (3.17) is detailed as follows:

$$A(t_1,t_2) = \int_0^{t_1}\int_0^{t_2} R_{F_kF_l}(s_1,s_2)\cdot e^{[\gamma_m(t_1-s_1)+\gamma_n^*(t_2-s_2)]}\mathrm{d}s_1\mathrm{d}s_2. \quad (3.18)$$

In Eq. (3.18), $R_{F_kF_l}(s_1,s_2) = E[F_k(s_1)F_l(s_2)]$, and it represents the correlation function of the external force F_k and F_l. Assuming F_k and F_l are stationary stochastic processes, the value of $R_{F_kF_l}(s_1,s_2)$ can be computed by

$$R_{F_kF_l}(s_1,s_2) = \int_{-\infty}^{\infty} S_{F_kF_l}(\omega)\cdot e^{\mathrm{i}\omega(s_2-s_1)}\mathrm{d}\omega, \quad (3.19)$$

where $S_{F_kF_l}(\omega)$ is the cross-spectral density of F_k and F_l.

Putting Eq. (3.19) into Eq. (3.18) and noting that $\lim e^{\gamma_m t} = \lim e^{\gamma_n^* t} = 0$ for $t \to \infty$, we obtain

$$A(t_1, t_2) = \int_{-\infty}^{\infty} S_{F_k F_l}(\omega) \frac{1}{\gamma_m + \mathrm{i}\omega} \cdot \frac{1}{\gamma_n^* - \mathrm{i}\omega} e^{\mathrm{i}\omega(t_2 - t_1)} \mathrm{d}\omega. \tag{3.20}$$

Considering Eqs. (3.20), (3.17), and (3.16), we have

$$\begin{aligned} R_{Z_i Z_j}(t_1, t_2) &= \sum_{m=1}^{2N} \sum_{n=1}^{2N} \boldsymbol{\Psi}_{im} \boldsymbol{\Psi}_{jn}^* R_{V_m V_n}(t_1, t_2) \\ &= \int_{-\infty}^{\infty} \sum_{m=1}^{2N} \sum_{n=1}^{2N} \frac{\boldsymbol{\Psi}_{im} \boldsymbol{\Psi}_{k+N,m}}{\boldsymbol{\Psi}_m^{\mathrm{T}} \mathbf{B} \boldsymbol{\Psi}_m (\gamma_m + \mathrm{i}\omega)} \cdot \frac{\boldsymbol{\Psi}_{jn}^* \boldsymbol{\Psi}_{l+N,n}^*}{(\boldsymbol{\Psi}_n^{\mathrm{T}} B \boldsymbol{\Psi}_n^*)(\gamma_n^* - \mathrm{i}\omega)} \\ &\quad \times \sum_{k=1}^{N} \sum_{l=1}^{N} S_{F_k F_l}(\omega) e^{\mathrm{i}\omega\tau} \mathrm{d}\omega = R_{Z_i Z_j}(\tau), \end{aligned} \tag{3.21}$$

wherein $\tau = t_2 - t_1$.

A stational stochastic process ($Z_i(t)$ or $Z_j(t)$) follows the subsequent rule

$$R_{Z_i Z_j}(\tau) = \int_{-\infty}^{\infty} S_{Z_i Z_j}(\omega) e^{\mathrm{i}\omega\tau} \mathrm{d}\omega. \tag{3.22}$$

A comparison between Eq. (3.21) and (3.22) leads us to

$$S_{Z_i Z_j}(\omega) = \sum_{m=1}^{2N} \sum_{n=1}^{2N} \frac{\boldsymbol{\Psi}_{im} \boldsymbol{\Psi}_{k+N,m}}{\boldsymbol{\Psi}_m^{\mathrm{T}} \mathbf{B} \boldsymbol{\Psi}_m (\gamma_m + \mathrm{i}\omega)} \cdot \frac{\boldsymbol{\Psi}_{jn}^* \boldsymbol{\Psi}_{l+N,n}^*}{(\boldsymbol{\Psi}_n^{\mathrm{T}} \mathbf{B} \boldsymbol{\Psi}_n^*)(\gamma_n^* - \mathrm{i}\omega)} \sum_{k=1}^{N} \sum_{l=1}^{N} S_{F_k F_l}(\omega), \tag{3.23}$$

which shows the cross-spectral density of the response Z_i and Z_j in the state space. Given $S_{Z_i Z_j}(\omega)$ and Eq. (3.22), the covariance matrix of $\mathbf{Z}(t)$ is obtained.

3.3. *Stochastic modeling of excitation for constructing the FSRS*

From Eqs. (3.22) and (3.23), the value of $S_{F_k F_l}(\omega)$ is needed in deriving $R_{Z_i Z_j}(\tau)$. In this section, $S_{F_k F_l}(\omega)$ and the accordant correlation function of F_k and F_l are analyzed.

The dynamic equilibrium of a structure equipped with VE decoupled floor is given in Eq. (2.1). In Eq. (2.1), the structural responses are "relative" values (the ground is taken as the reference object). Since our target is to learn the FSRS, which is related to the absolute floor acceleration, the dynamic equilibrium of Eq. (2.1) should be rewritten as follows:

$$\mathbf{M}\ddot{\mathbf{x}}^t + \mathbf{C}\dot{\mathbf{x}}^t + \mathbf{K}\mathbf{x}^t = \mathbf{C}\iota\dot{x}_g + \mathbf{K}\iota x_g, \tag{3.24}$$

where $\ddot{\mathbf{x}}^t$, $\dot{\mathbf{x}}^t$, and $\mathbf{x}^t$ is the absolute acceleration, velocity, and displacement response of the structure, respectively, $\dot{x}_g$ and x_g are the velocity and displacement of the ground. From Eq. (3.24), the structure is subjected to an equivalent external dynamic action termed

$$\mathbf{F}(t) = \mathbf{C}\iota\dot{x}_g + \mathbf{K}\iota x_g. \tag{3.25}$$

Since it needs to analyze the value of $S_{F_kF_l}(\omega)$ before the $S_{Z_iZ_j}(\omega)$ in Eq. (3.23) can be obtained, we have to consider $R_{F_kF_l}(t_1,t_2) = E[F_k(t_1)F_l(t_2)]$ and conduct the subsequent derivation

$$\begin{aligned} R_{F_kF_l}(t_1,t_2) &= E[F_k(t_1)F_l(t_2)] \\ &= E\left[\left(\sum_{i=1}^{N} C_{ki}\dot{x}_g(t_1) + \sum_{i=1}^{N} K_{ki}x_g(t_1)\right)\right. \\ &\quad \left.\cdot\left(\sum_{i=1}^{N} C_{li}\dot{x}_g(t_2) + \sum_{i=1}^{N} K_{li}x_g(t_2)\right)\right] \\ &= E\left[\sum_{i=1}^{N} C_{ki}\sum_{i=1}^{N} C_{li}\dot{x}_g(t_1)\dot{x}_g(t_2) + \sum_{i=1}^{N} C_{ki}\sum_{i=1}^{N} K_{li}\dot{x}_g(t_1)x_g(t_2)\right. \\ &\quad \left. + \sum_{i=1}^{N} K_{ki}\sum_{i=1}^{N} C_{li}x_g(t_1)\dot{x}_g(t_2) + \sum_{i=1}^{N} K_{ki}\sum_{i=1}^{N} K_{li}x_g(t_1)x_g(t_2)\right]. \end{aligned} \tag{3.26}$$

For brevity, the following denotations are used hereafter

$$\sum_{i=1}^{N} C_{ki} = C_k, \quad \sum_{i=1}^{N} C_{li} = C_l, \quad \sum_{i=1}^{N} K_{ki} = K_k, \quad \sum_{i=1}^{N} K_{li} = K_l. \tag{3.27}$$

Meanwhile, by noting that

$$\begin{cases} R_{\dot{x}_g}(t_1,t_2) = E[\dot{x}_g(t_1)\dot{x}_g(t_2)], \\ R_{x_g}(t_1,t_2) = E[x_g(t_1)x_g(t_2)], \\ R_{\dot{x}_gx_g}(t_1,t_2) = E[\dot{x}_g(t_1)x_g(t_2)], \\ R_{x_g\dot{x}_g}(t_1,t_2) = E[x_g(t_1)\dot{x}_g(t_2)]. \end{cases} \tag{3.28}$$

Equation (3.26) can be rewritten as follows:

$$\begin{aligned} R_{F_kF_l}(t_1,t_2) &= C_kC_lR_{\dot{x}_g}(t_1,t_2) + C_kK_lR_{\dot{x}_gx_g}(t_1,t_2) \\ &\quad + K_kC_lR_{x_g\dot{x}_g}(t_1,t_2) + K_kK_lR_{x_g}(t_1,t_2). \end{aligned} \tag{3.29}$$

Assuming that $\dot{x}_g$ and x_g are both stationary Gaussian stochastic processes, while F_k and F_l are joint stationary stochastic processes, we have

$$R_{F_kF_l}(\tau) = C_kC_lR_{\dot{x}_g}(\tau) + C_kK_lR_{\dot{x}_gx_g}(\tau) + K_kC_lR_{x_g\dot{x}_g}(\tau) + K_kK_lR_{x_g}(\tau), \tag{3.30}$$

$$S_{F_kF_l}(\omega) = \frac{1}{2\pi}\int_{-\infty}^{\infty} R_{F_kF_l}(\tau)e^{-\mathrm{i}\omega\tau}\mathrm{d}\tau, \tag{3.31}$$

$$S_{F_kF_l}(\omega) = C_kC_lS_{\dot{x}_g}(\omega) + C_kK_lS_{\dot{x}_gx_g}(\omega) + K_kC_lS_{x_g\dot{x}_g}(\omega) + K_kK_lS_{x_g}(\omega). \tag{3.32}$$

In Eq. (3.32), $S_{x_g}(\omega)$ and $S_{\dot{x}_g}(\omega)$ is the power spectral density (PSD) of the ground displacement x_g and velocity $\dot{x}_g$, respectively, $S_{\dot{x}_gx_g}(\omega)$ and $S_{x_g\dot{x}_g}(\omega)$ are the cross-spectral density of x_g and $\dot{x}_g$. For a stationary excitation process, it exhibits the following feature

$$\begin{cases} S_{\dot{x}_g}(\omega) = \omega^2 S_{x_g}(\omega) \\ S_{a_g}(\omega) = \omega^4 S_{x_g}(\omega) \\ S_{x_g\dot{x}_g}(\omega) = -S_{\dot{x}_gx_g}(\omega) = \mathrm{i}\omega S_{x_g}(\omega) \end{cases} \tag{3.33}$$

wherein $S_{a_g}(\omega)$ is the PSD of the ground acceleration a_g. Putting Eq. (3.33) into Eq. (3.32) yields

$$S_{F_kF_l}(\omega) = \frac{S_{a_g}(\omega)}{\omega^4}[C_kC_l\omega^2 + K_kK_l + (K_kC_l - C_kK_l)\mathrm{i}\omega]. \tag{3.34}$$

3.4. *Accounting for the non-stationary property of the earthquake excitation*

In the preceding derivations, the excitation and the structural responses are assumed stationary. A realistic seismic input, however, exhibits non-stationary properties that may have a notable influence on the structural response. In this section, the non-stationary feature of the input is accounted for. In specific, we use a developed model to process the pseudo-acceleration spectrum provided in design codes, and to obtain the PSD function of the non-stationary excitations. The adopted model is discussed in Ref. 40.

$$S_{a_g}(\omega) = \frac{4\zeta}{\pi\omega}R^2(\omega,\zeta)\frac{1}{M(\omega,\zeta)f^2(\omega)}, \tag{3.35}$$

where $R(\omega,\zeta)$ is the pseudo-acceleration response spectrum of the considered excitation with a damping ratio of ζ, while $M(\omega,\zeta)$ is a modification factor that accounts for the deviation between a stationary and a non-stationary response process, $f(\omega)$ is termed "peak value factor" for amending the peak value of the estimated response. The values of $M(\omega,\zeta)$ and $f(\omega)$ are computed by Jiang and Hong[40]

$$M(\omega,\zeta) = \frac{t_2 - 0.3t_1}{t_d} + \frac{e^{-2\zeta\omega t_2} - e^{-2\zeta\omega t_1}}{2\zeta\omega t_d} + \frac{1}{pt_d}[1 - e^{-p(t_d+t_1/2-t_2)}], \tag{3.36}$$

$$f(\omega) = \sqrt{2\ln\frac{\omega}{\pi}t_d + \frac{0.5772}{2\ln\frac{\omega}{\pi}t_d}}. \tag{3.37}$$

In compliance with the Chinese seismic design code,[41] the values of t_1, t_2, t_d, and p are given by Ref. 42, as shown in Table 1.

3.5. *Generating the FSRS for the building with VE decoupled floors*

Now we have obtained the $S_{Z_iZ_j}(\omega)$ and the $S_{F_kF_l}(\omega)$, as given in Eqs. (3.23) and (3.34), respectively. Accordingly, the state space response $\mathbf{Z}(t)$ of the structure can be fully determined as long as the adjusted $S_{a_g}(\omega)$ is properly selected, as shown in Eq. (3.35). Regarding the floor response, we could use the (absolute) displacement response to derive the accordant acceleration. Taking the structural model introduced in Sec. 2.1.1 for example, the floor displacement and acceleration responses are as follows:

$$\begin{cases} \bar{Z}_r(t) = Z_{N+r}(t) \\ S_{\ddot{\bar{Z}}_r}(\omega) = \omega^4 S_{Z_{N+r}}(\omega) \end{cases} \tag{3.38}$$

where $Z_{N+r}(t)$ is the $(N+r)$th component of $\mathbf{Z}(t)$, $\bar{Z}_r(t)$ is the displacement response of the rth degree-of-freedom of the structure, $S_{\ddot{\bar{Z}}_r}(\omega)$ and $S_{Z_{N+r}}(\omega)$ are the PSD of $\ddot{\bar{Z}}_r(t)$ and $Z_{N+r}(t)$, respectively. It is worth mentioning that $\ddot{\bar{Z}}_r(t)$, which represents the acceleration response of the rth degree-of-freedom, is the two-order time-differential of $\bar{Z}_r(t)$.

Subjecting an oscillator with circular frequency ω_0 and damping ratio ζ_0 to the floor acceleration response, the PSD of the displacement response of the oscillator is as follows:

$$S_{ur}(\omega) = |H(\omega_0)|^2 S_{\ddot{\bar{Z}}_r}(\omega), \tag{3.39}$$

where $S_{ur}(\omega)$ is the PSD of the response, $H(\omega_0)$ is the frequency response function detailed as follows:

$$H(\omega_0) = \frac{1}{\omega_0^2 - \omega^2 + 2\mathrm{i}\varepsilon_0\omega_0\omega}. \tag{3.40}$$

Assuming the response process of the oscillator has a mean value of zero, the variance of its displacement, σ_{ur}, should be

$$\sigma_{ur}^2 = \int_0^\infty S_{ur}(\omega)\mathrm{d}\omega. \tag{3.41}$$

Table 1. Values of t_1, t_2, t_d, and p.

Site class	t_1/s	t_2/s	t_d/s	p/1/s
I	0.5	5.5	6.69	0.45
II	0.8	7.0	8.41	0.35
III	1.2	9.0	10.92	0.25
IV	1.6	12.0	15.49	0.15

To consider the effect of the non-stationary property of the excitation (represented by $S_{\ddot{Z}_r}(\omega)$), we take the same process given by Eq. (3.36) to amend σ_{ur}:

$$\bar{\sigma}_{ur}^2 = M(\omega_0, \varepsilon_0)\sigma_{ur}^2, \tag{3.42}$$

where $\bar{\sigma}_{ur}$ is the amended σ_{ur}. Additionally, the "peak value factor" given by Eq. (3.37) is adopted to modify the peak value of the estimated response

$$d_{ur}^{\text{Peak}} = f(\omega_0)\bar{\sigma}_{ur}. \tag{3.43}$$

Given the peak displacement response of the oscillator d_{ur}^{Peak}, the accordant pseudo-acceleration response spectrum, or in other words, the FSRS, can be derived via

$$\text{FRS}(\omega_0, \varepsilon_0) = \omega_0^2 d_{ur}^{\text{Peak}}. \tag{3.44}$$

4. FSRS of the Building with VE Decoupled Floor: Numerical Validation and Parametrical Analyses

Given the theoretical expression of the FSRS for VE decoupled floors (see Sec. 3), a numerical validation of the derived solution is presented in this section. For brevity and consistency, we still use a three-story structural model in the analysis. First, the seismic excitations adopted in the numerical computation is introduced (Sec. 4.1). Some important results are shown in Secs. 4.2 and 4.3 to demonstrate two key issues: (1) The rationality of using a stabilized substitutive structural model to replace the time-variant (frequency-dependent) model in deriving the FSRS; (2) the accuracy of the theoretically derived FSRS (compared with the "accurate" ones obtained from time history analyses). Subsequently, the beneficial effect of the VE decoupled floors on mitigating the amplitude of FSRS is presented in Sec. 4.4. Also given in Sec. 4.4 is the effect of some important structural design parameters on the FSRS.

4.1. *Seismic excitations*

Taking a design spectrum in the Chinese seismic design code[41] as the target spectrum, a total number of 50 artificial earthquake excitations are generated. The target spectrum corresponds to a medium-firm constructional site that has a predominant period of 0.35 s. The target spectrum and the response spectra of the artificial excitations are shown in Fig. 5. From the figure, the mean spectrum of the artificial excitations is in compliance with the target one. To be clear, all the spectra in Fig. 5 are 3% damped.

4.2. *Accuracy of FSRS obtained from the substitutive model*

As mentioned in Sec. 2.2, the frequency-dependency of the VE decoupler has led to a time-variant stiffness and damping of the structure. To deal with this feature, a stabilized substitution of the structure (using a nominal stiffness & damping

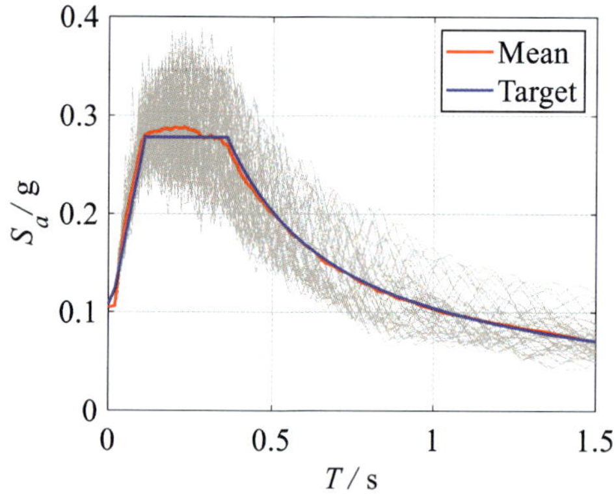

Fig. 5. Target design spectrum and response spectra of artificial excitations.

parameter to replace the time-variant one) is proposed and has been demonstrated effective in reproducing the structural (including floor) response under earthquakes (Sec. 2.3). In this section, the conclusion drawn in Sec. 2.3 is further validated using the 50 earthquake excitations introduced in Sec. 4.1.

The adopted 3-story structural model, as shown in Fig. 2, has a fundamental period of $T_1 = 0.45\,\text{s}$. The mass distribution of the structure is $m_1 : m_2 : m_3 = 1 : 1 : 1$, $\nu_1 = \nu_2 = \nu_3 = 0.7$. The stiffness property of the structure is defined by $k_1 : k_2 : k_3 = 1 : 0.75 : 0.5$, $\kappa_j = k_{\text{ve},j}/k_j = 0.2$ ($j = 1, 2$, and 3). The inherent damping of the structure is set to 3% and the initial damping matrix (see $\mathbf{C_0}$ in Eq. (2.3)) is constructed via the Rayleigh damping model. Subjecting the structure to the 50 earthquake excitations and computing the structural (including floor) response via the time history analysis method, the FSRS is obtained. From the computation, two kinds of structural models are considered: (1) The "accurate" model that accounts for the time-variant property of the structural stiffness and damping; (2) The "substitutive" model that uses the stabilized (nominal) stiffness and damping. The two kinds of structural model yield two groups of FSRSs, termed the "time-variant" one and the "stabilized" one. The mean of the FSRS given by the 50 excitations are plotted in Fig. 6, in which the vertical axis, S_a^f represents the amplitude of the FSRS.

From Fig. 6, the "time-variant" FSRS and its "stabilized" substitution are quite alike, as the error between them is basically negligible. For the FSRS of story 1 (see Fig. 6(a) for S1), the stabilized result is a little bit smaller (deviation less than 5%) than its accurate (time-variant) counterpart at $T \approx 0.35$ s. For the FSRS of story 3 (see Fig. 6(c) for S3), the stabilized result at $T \approx 0.55\,\text{s}$ is larger than the accurate case (deviation on the safe side), and the amplitude of such deviation is not pronounced. Besides these specific scenarios, the "time-variant" FSRS and its "stabilized" substitution are identical. From Fig. 6, using the stabilized substitutive

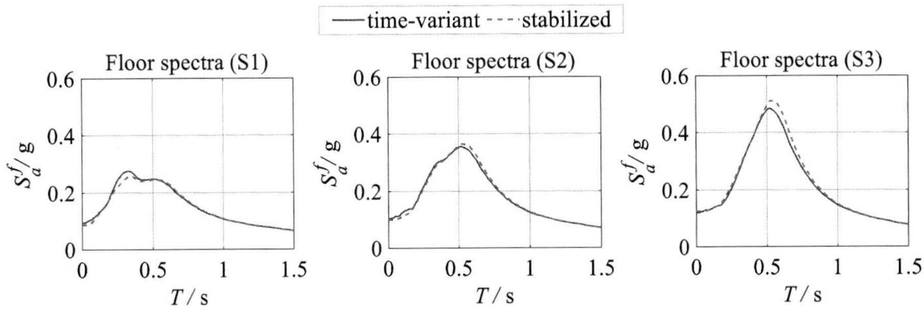

Fig. 6. FSRS obtained from the accurate time-variant structural model and the substitutive stabilized structural model.

structural model to compute the FSRS would generate appropriate estimations that exhibit good accuracies.

4.3. *Accuracy of FSRS obtained from the proposed theoretical approach*

Based on the stabilized substitutive structural model, the FSRS is derived via the theoretical approach given in Sec. 3. For brevity, we still use the structure detailed in Sec. 4.2 to illustrate the accuracy of the theoretical approach. The FSRSs obtained from the "accurate" time history analysis (using the time-variant structural model) and the ones given by the theoretical approach are plotted in Fig. 7. Each curve in the figure represents the mean of the 50 results yielded by the 50 excitations.

From Fig. 7(a), for the 1st story (S1), the theoretical solution and the "accurate" numerical results are alike. Although there is an underestimation at $T \leq T_1 = 0.45$ s and an overestimation at $T \geq T_1 = 0.45$ s, the errors are quite tiny. From Figs. 7(b) and 7(c), the theoretical approach either well predicts or slightly overestimates the amplitude of the FSRS for story 2 (S2) and story (S3). Basically, the deviation between the theoretical and the numerical solutions are induced by two issues: (1) The substitutive model, based on which the theoretical method is established, may generate a slight overestimation of the amplitude of FSRS, as shown in Fig. 6; (2) the process in Sec. 3.4, which is adopted to account for the non-stationary nature of the earthquake excitation, is an approximate one and may lead to some sort of bias. Nonetheless, the results in Fig. 7 demonstrates the good accuracy of the FSRS derived from the theoretical approach established in Sec. 3.

4.4. *FSRS of the building with VE decoupled floor in different design scenarios*

Given the good accuracy of the FSRS obtained from the proposed method, a series of parametrical computation, which incorporates different structural

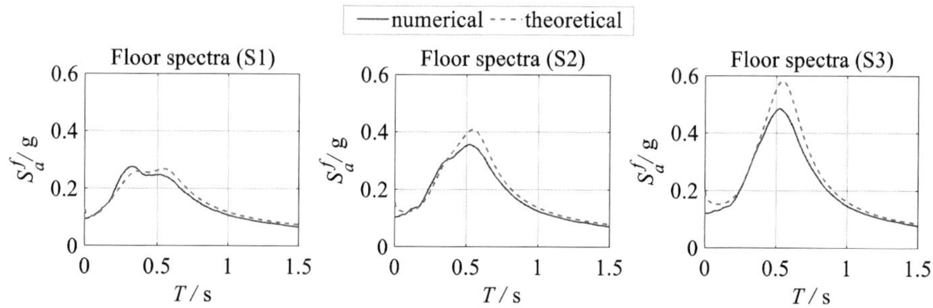

Fig. 7. FSRS obtained from the time history analysis method and from the proposed theoretical approach.

configurations and design parameters, is conducted and the results are presented. Some design suggestions are drawn to provide guidance for the engineering practice.

4.4.1. *FSRS of the building with height-wise varied floor-decoupling configurations*

In Figs. 8(a)–8(g), a total number of seven floor-decoupling configurations are presented. The one in Fig. 8(h) represents a conventional 3-story structure wherein all the floors are rigidly connected to the frame. In this case (taken as the benchmark scenario), the fundamental period of the frame is set to $T_1 = 0.45$ s, while its mass and stiffness is defined as $m_1 : m_2 : m_3 = 1 : 1 : 1$, $k_1 : k_2 : k_3 = 1 : 0.75 : 0.5$. Regarding the floor-decoupling cases, the seven situations given in Figs. 8(a)–8(g) exhibit height-wise differences. In all these seven cases, the stiffness of the VE decoupler (if any) is controlled by $\kappa_j = k_{\mathrm{ve},j}/k_j = 0.2\,(j = 1, 2, \text{ and/or } 3)$, the mass ratio of the decoupled floor (if any) is $\nu_j = 0.7\,(j = 1, 2, \text{ and/or } 3)$. For convenience, the peak of the amplitude of FSRS is denoted by $\mathrm{FSRS}_{\max}$ in the subsequent discussion.

Starting from Fig. 8(g), the frame with a single VE decoupled floor at story 1 exhibits an $\mathrm{FSRS}_{\max}$ of 0.7 g, 1.1 g, and 1.7 g for the 1st, 2nd, and 3rd story, respectively. Compared with the ones in Fig. 8(h), which has an $\mathrm{FSRS}_{\max}$ of 0.7 g, 1.4 g, and 2.0 g for story 1, 2, and 3, Fig. 8(g) situation has a marginal reduction in the amplitude of FSRS. Considering the cases in Figs. 8(e)–8(g), as the height (level) of the decoupled floor rises, the beneficial effect of the decoupled floor increases. For example, in Fig. 8(e), wherein the VE decoupled floor is installed at the roof, the value of $\mathrm{FSRS}_{\max}$ becomes 0.3 g, 0.4 g, and 0.6 g for the 1st, 2nd, and 3rd story, respectively. Such values are notably smaller than the cases in Figs. 8(g) and 8(h).

Comparing Figs. 8(a)–8(d) with Fig. 8(e)–8(g), increasing the quantity of decoupled floor would lead to a more significant reduction in $\mathrm{FSRS}_{\max}$. For brevity, the values of $\mathrm{FSRS}_{\max}$ in these situations are not duplicated since the straightforward

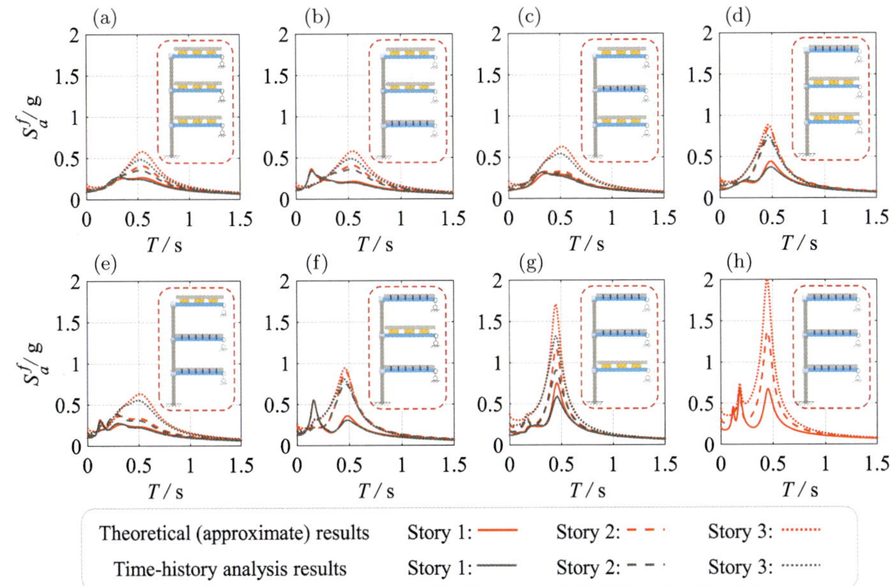

Fig. 8. FSRS of the building with height-wise varied floor-decoupling configurations.

demonstrations are shown in the figure. Especially, the cases in Figs. 8(a)–8(c), and 8(e) have more pronounced effects in controlling the amplitude of FSRS than the other cases, i.e. the ones in Figs. 8(d)–8(g). Specifically, the FSRSs at story 2 and story 3 are much better controlled in the former four cases. This is mainly because the roof of the former four structures is equipped with VE decoupled floors, and such a configuration (roof isolation) is quite effective for mitigating the acceleration response of the structure. Further insights regarding this issue are beyond the scope of this paper, yet it will be reported in the near future elsewhere.

Additionally, in Fig. 8(h), the peak of FSRS ($\mathrm{FSRS_{max}}$) corresponds to $T = 0.45$ s, which equals the fundamental period of the structure ($T_1 = 0.45\,\mathrm{s}$). While in Figs. 8(a)–8(g), the $\mathrm{FSRS_{max}}$ appears at some periods larger than 0.45 s. This is because the VE decoupled floor lengthens the period of the structure, making the peak of FSRS appear at $T > 0.45\,\mathrm{s}$.

4.4.2. *$FSRS_{\max}$ of the building with varied decoupler stiffness and mass ratio*

In this section, the structural model shown in Fig. 2 is adopted to illustrate the effects of κ_j and ν_j ($j = 1, 2,$ and 3) on the FSRS. The structure has a $T_1 = 0.45\,\mathrm{s}$ and a modal damping ratio of 3%. In the floor decoupling scenario, we assume $\kappa_1 = \kappa_2 = \kappa_3 = \kappa$ and $\nu_1 = \nu_2 = \nu_3 = \nu$, and we use $\mathrm{FSRS_{max}}$ to quantify the amplitude of FSRS. The values of κ_j and ν_j considered are $\kappa_j = 0.05 - 0.85$ (with an

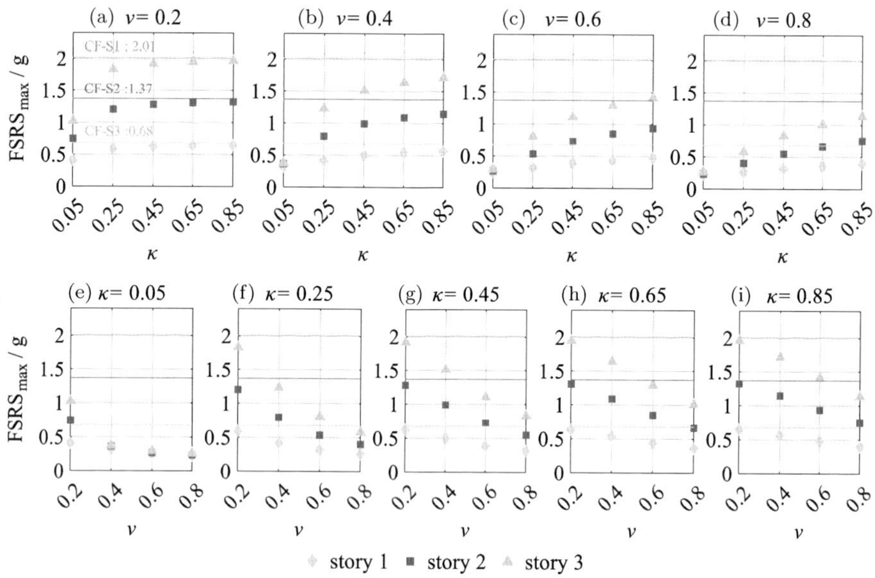

Fig. 9. $FSRS_{max}$ of the building with varied decoupler stiffness and mass ratio (horizontal lines indicate the $FSRS_{max}$ of a conventional frame without VE floor).

interval of 0.2) and $\nu_j = 0.2 - 0.8$ (interval = 0.2). A total number of 20 structural models, installed by VE decoupled floors with varied parameters (different κ_j and ν_j) are analyzed using the theoretical approach developed in Sec. 3. The values of $FSRS_{max}$ are plotted in Fig. 9. In the analysis, the input earthquake excitations are modeled via the method given in Sec. 3.4 whereas the target spectrum of the excitations is the one introduced in Sec. 4.1. In specific, the colored horizontal lines in Fig. 9 represents the values of $FSRS_{max}$ of a conventional frame, which could serve as a benchmark to show the advantage of the VE floors in controlling the magnitude of the FSRS.

From Figs. 9(a)–9(d), the value of $FSRS_{max}$ decreases as the VE decoupler becomes flexible (κ gets smaller). This is rational because a smaller κ corresponds to a larger structural flexibility (longer period) and an enlarged structural damping. From Figs. 9(e)–9(i), the value of $FSRS_{max}$ decreases as the mass of the decoupled floor increases (larger ν). This is also rational because a larger ν indicates that more mass is decoupled (isolated) from the primary structure and the inertial force of the structure is better mitigated. This would lead to a reduced acceleration response. The curves in Fig. 9 show that the FSRS of the top floor (roof) is better controlled (value of $FSRS_{max}$ decreases more significantly) in a VE decoupled flooring system as long as the values of κ and ν are properly selected. This is a favored property because commonly the amplitude of FSRS at the top floor is the largest among all the floors, and it usually requires a more effective measure to mitigate the $FSRS_{max}$ therein.

5. Conclusions

This research focuses on the FSRS of the building equipped with VE decoupled floors. By substituting the frequency-dependent structural model and deriving the dynamic response of the non-classically damped structure, the FSRS of the VE decoupled flooring system is quantified and some important features of such FSRSs are revealed. The main findings are as follows:

(1) A building structure equipped with the VE decoupled floor exhibits some time-varying stiffness and damping under earthquake excitations because of the frequency-dependency of the VE decoupler. Such a time-varying model can be replaced by a substitutive "stabilized" one using a set of "nominal" mechanical features for the VE decouplers. The FSRSs of the structure, generated by the time-varying "accurate" model and the substitutive "approximate" model, are basically identical.

(2) The building structure equipped with VE decoupled floors is non-classically damped, and the FSRS of such a structure could be derived based on the complex modal decomposition method accompanied by a stochastic simulation of the response and excitation in the state space. In such a process, the non-stationary nature of the earthquake excitation should be accounted for. A specific processing method in compliance with the Chinese seismic design spectrum is integrated in this study to consider the non-stationary nature of the excitations.

(3) The derived FSRS, using the preceding theoretical method and the approximate substitutive structural model, exhibits an ideal accuracy. The results of a series of numerical computation demonstrates that the FSRS of the building equipped with VE decoupled floors can be better mitigated if: (1) A more flexible VE decoupler and a larger mass (isolation) ratio is incorporated; (2) the VE decoupled floor is installed at the roof (near top) level. According to the computation, the peak of FSRS can be reduced by 70% if the VE decoupling system is properly designed.

Acknowledgments

This work is supported by the National Key Research and Development Program of China (No. 2022YFC3801904). This financial support is gratefully acknowledged. The consistent guidance to the research on the VE decoupled flooring system from Prof. Yuji Koetaka at Kyoto University, Japan, is gratefully acknowledged.

ORCID

S.-L. Guo https://orcid.org/0009-0006-2135-4395
Y. Xiang https://orcid.org/0000-0002-3589-5436

References

1. R. P. Dhakal, Damage to non-structural components and contents in 2010 Darfield earthquake, *Bull. New Zealand Soc. Earthq. Eng.* **43**(4) (2010) 404–411.
2. M. Reyners, Lessons from the destructive Mw 6.3 Christchurch, New Zealand, earthquake, *Seismol. Res. Lett.* **82**(3) (2011) 371–372.
3. R. S. Vicente, H. Rodrigues, H. Varum, A. Costa and J. A. R. Mendes da Silva, Performance of masonry enclosure walls: Lessons learned from recent earthquakes, *Earthq. Eng. Eng. Vib.* **11**(1) (2012) 23–34.
4. E. Miranda, G. Mosqueda, R. Retamales and G. Pekcan, Performance of nonstructural components during the 27 February 2010 Chile earthquake, *Earthq. Spectra* **28**(1_suppl1) (2012) 453–471.
5. Q. Mao, W. Yuan and J. Wu, Design and experimental study of a multiple-degree-of-freedom vibration absorber by using an inertial actuator, *Int. J. Struct. Stab. Dyn.* (2023) 2450006, DOI: 10.1142/S0219455424500068.
6. Z. Lu, M. Zhou, N. Ma and J. Du, Comparative studies on nonlinear structures with multiple tuned mass damper and multiple tuned impact damper, *Int. J. Struct. Stab. Dyn.* **23**(14) (2023) 2350158.
7. G. Q. Li, Some important concepts on high-performance steel structures and realization, *Prog. Steel Build. Struct.* **22**(5) (2020) 1–18.
8. S. Y. Zhai, K. Cao, G. Q. Li, W. Y. Wang, C. Chen and Q. Liu, A state-of-art review in inter-module connections of column-supported modular steel buildings, *Prog. Steel Build. Struct.* **24**(11) (2022) 1–11.
9. Y. Koetaka, K. Matsumoto and Y. Yano, Seismic response of single-story steel moment frame with isolated floor system, *15th World Conf. Earthquake Engineering*, 24–28 September 2012, Lisbon, pp. 1–8.
10. Y. Koetaka and Y. Sakamoto, Compression and shear characteristics of laminated visco-elastic materials for seismic floor system, *J. Struct. Constr. Eng. (Transact AIJ).* **82**(734) (2017) 533–542 (in Japanese).
11. Y. Xiang, Y. Koetaka and K. Nishira, Steel frame with aseismic floor: From the viscoelastic decoupler model to the elastic structural response, *Earthquake Eng. Struct. Dyn.* **50**(6) (2021) 1651–1670.
12. Y. Koetaka and Y. Sakamoto, Shaking table test of two-story steel moment frame with seismic floor system, *16th World Conf. Earthquake Engineering*, 9–13 January 2017, Santiago, Chile, pp. 1–12.
13. Y. Xiang, Y. Koetaka and N. Okuda, Single-story steel structure with LVEM-isolated floor: Elastic seismic performance and design response spectrum, *Eng. Struct.* **196** (2019) 109314.
14. Y. Xiang and Y. Koetaka, Structural feasibility of incorporating the LVEM-isolated floor in the first story of a two-story steel frame, *Eng. Struct.* **199** (2019) 109686.
15. M. F. Ishac and A. C. Heidebrecht, Generation of lateral and rotational floor response spectra by an alternative approach, *Earthq. Eng. Struct. Dyn.* **10**(1) (1982) 47–58.
16. M. J. Yan, Fast floor response spectra generation technique, *J. Press. Vessel Technol.* **105** (1983) 35–41.

17. T. Igusa and A. D. Kiureghian, Generation of floor response spectra including oscillator-structure interaction, *Earthq. Eng. Struct. Dyn.* **13**(5) (1985) 661–676.
18. A. Asfura and A. D. Kiureghian, Floor response spectrum method for seismic analysis of multiply supported secondary systems, *Earthq. Eng. Struct. Dyn.* **14**(2) (1986) 245–265.
19. L. E. Suarez and M. P. Singh, Floor response spectra with structure–equipment interaction effects by a mode synthesis approach, *Earthq. Eng. Struct. Dyn.* **15**(2) (1987) 141–158.
20. A. Paskalov and S. Reese, Deterministic and probabilistic floor response spectra, *Soil Dyn. Earthq. Eng.* **23**(7) (2003) 605–618.
21. P. Varpasuo, The development of the floor response spectra using large 3D model, *Nucl. Eng. Des.* **192**(2–3) (1999) 229–241.
22. I. Politopoulos, I. Sergis and F. Wang, Floor response spectra of a partially embedded seismically isolated nuclear plant, *Soil Dyn. Earthq. Eng.* **78** (2015) 213–217.
23. F. G. Fan and G. Ahmadi, Floor response spectra for base-isolated multi-storey structures, *Earthq. Eng. Struct. Dyn.* **19**(3) (1990) 377–388.
24. J. W. Jung, H. W. Jang, J. H. Kim and J. W. Hong, Effect of second hardening on floor response spectrum of a base-isolated nuclear power plant, *Nucl. Eng. Des.* **322** (2017) 138–147.
25. S. Li, P. Tan and H. Ma, Improved response spectrum method based on real-complex hybrid modal superposition for base-isolated structures, *Int. J. Struct. Stab. Dyn.* **23**(9) (2023) 2350095.
26. R. A. Medina, R. Sankaranarayanan and K. M. Kingston, Floor response spectra for light components mounted on regular moment-resisting frame structures, *Eng. Struct.* **28**(14) (2006) 1927–1940.
27. M. Oropeza, P. Favez and P. Lestuzzi, Seismic response of nonstructural components in case of nonlinear structures based on floor response spectra method, *Bull. Earthq. Eng.* **8**(2) (2010) 387–400.
28. V. Vukobratović and P. Fajfar, A method for the direct determination of approximate floor response spectra for SDOF inelastic structures, *Bull. Earthq. Eng.* **13**(5) (2015) 1405–1424.
29. C. Petrone, G. Magliulo and G. Manfredi, Floor response spectra in RC frame structures designed according to Eurocode 8, *Bull. Earthq. Eng.* **14**(3) (2016) 747–767.
30. W. Jiang, B. Li, W. C. Xie and M. D. Pandey, Generate floor response spectra: Part 1. Direct spectra-to-spectra method, *Nucl. Eng. Des.* **293** (2015) 525–546.
31. R. Wang, W. C. Xie and M. D. Pandey, Generation of floor and tertiary response spectra of structures under seismic excitations at multiple supports, *Earthq. Eng. Struct. Dyn.* **51**(4) (2022) 853–874.
32. M. Surana, Y. Singh and D. H. Lang, Effect of structural characteristics on damping modification factors for floor response spectra in RC buildings, *Eng. Struct.* **242** (2021) 112514.
33. H. Anajafi, R. A. Medina and C. Adam, Evaluation of the floor acceleration response spectra of numerical building models based on recorded building response data, *J. Earthq. Eng.* **26**(12) (2022) 6378–6402.
34. H. Anajafi and R. A. Medina, Lessons learned from evaluating the responses of instrumented buildings in the United States: The effects of supporting building characteristics on floor response spectra, *Earthq. Spectra* **35**(1) (2019) 159–191.

35. R. J. Merino, D. Perrone and A. Filiatrault, Consistent floor response spectra for performance-based seismic design of nonstructural elements, *Earthq. Eng. Struct. Dyn.* **49**(3) (2020) 261–284.
36. K. Haymes, T. Sullivan and R. Chandramohan, A practice-oriented method for estimating elastic floor response spectra, *Bull. New Zealand Soc. Earthq. Eng.* **53**(3) (2020) 116–136.
37. K. Kasai, M. Teramoto, K. Okuma and K. Tokoro, Constitutive rule for viscoelastic materials considering temperature, frequency, and strain sensitivities, *J. Struct. Constr. Eng.* **543** (2001) 77–86 (in Japanese).
38. Y. H. Huang, T. Kato, A. Wada, M. Iwata, T. Takeuchi and K. Okuma, The dynamic mechanical model of viscoelastic dampers relying on the frequency and temperature, *J. Struct. Constr. Eng.* **516** (1999) 91–98 (in Japanese).
39. A. K. Chopra, *Dynamics of Structures: Theory and Applications to Earthquake Engineering*, 2nd edn. (Prentice Hall, New Jersey, 2000).
40. J. Jiang and F. Hong, Conversion between power spectrum and response spectrum and artificial earthquakes, *Earthq. Eng. Eng. Vib.* **4**(3) (1984) 1–11 (in Chinese).
41. Ministry of Housing and Urban-Rural Development and Administration of Quality Supervision, Inspection and Quarantine of the People's Republic of China. Code for seismic design of buildings. GB50011–2010 (ver 2016), Beijing, China (2016).
42. M. Zhang, Z. Zhang and T. Li, Study on parameters of Clough-Penzien seismic random model corresponding to response spectrum of GB50011-2001, *World Earthq. Eng.* **23**(1) (2007) 56–60 (in Chinese).

https://doi.org/10.1142/9789819814770_0030

Chapter 30

Disk Model Effect for Road Surface Roughness Using Convolution Method[#]

Der-Shen Yang[*,‡,¶], Qianhui Zhang[*], Hao Xu[†,§,¶], Kwesi Sagoe-Crentsil[*] and Wenhui Duan[*]

[*]*Department of Civil Engineering, Monash University Clayton, Victoria 3800, Australia*

[†]*School of Civil Engineering, Chongqing University Chongqing 400045, P. R. China*

[‡]*ted.yang@monash.edu*

[§]*hxu@cqu.edu.cn*

In most related studies on road surface roughness, the vehicle's wheel is often using a contact point model rather than a disk model. This results in neglecting the wheel's size and interaction with the road. Consequently, the vehicle's response may not be genuinely reflected, especially for the massive topic of noise, vibration, and harshness (NVH). Unlike the existing approach targeting the power spectrum, this paper proposes a new convolution method to tackle the disk effect and operates directly on the spatial domain, i.e. road surface roughness. By using a designed periphery function, it can simulate the wheel geometry passing through road surface roughness. The periphery function acts as a filter to the road surface roughness that can filter out smaller oscillations. Some examples involving roughness from ISO 8608 standards were tested. It is shown herein that the proposed method can match the theoretical result (using the geometry method (GM)) not only in the spatial domain but also in the power spectral density (PSD). Since the convolution is performed under the spatial domain, the proposed method can directly apply the disk model to any existing road surface roughness with different spectral compositions in practice. Understanding the disk effect reduces the higher frequency of the vehicle's response depending on roughness severity, which may significantly impact the vehicle design for ride comfort, road surface roughness extraction, bridge health monitoring using the drive-by method, etc.

Keywords: Convolution; road roughness; power spectral density; disk model; point model.

[¶]Corresponding authors.

[#]This chapter appeared previously on the International Journal of Structural Stability and Dynamics. To cite this chapter, please cite the original article as the following: D.-S. Yang *et al.*, *Int. J. Struct. Stab. Dyn.*, **23**, 2340033 (2023), DOI: 10.1142/S0219455423400333.

1. Introduction

Early works on road surface roughness have focused on the runway that affects aircraft operation.[1] Click or tap here to enter text. Considering the random character of the road surface roughness, the power-spectral analysis of the runway was first adopted to simulate the runway height variation.[2] It has been found that the power spectrum can represent the runway characteristics concisely in which the roughness is distributed to the various frequency components.[3] Regarding the power spectral density (PSD), Dodds and Robson completed a description of road surface roughness by a single autocorrelation function evaluated from any longitudinal track.[4] This work was proposed to British Standards Institution (BSI) and was later become ISO 8608 standard (ISO, 1995-09-01).[5,6] Since then, PSD from ISO 8608 has become the standard in simulating different classes of road surface profiles and demonstrated the formation in reporting the measured road roughness data. To adopt the artificial road profile based on ISO 8608, researchers have improved its statistical distribution regarding the random process.[7,8]

Followed by numerous field studies in road profiles, researchers first assumed that the contact model between the vehicle and road surface is a "point model" for characterizing the vehicle response spectrum. Bruscella *et al.* introduced a universal methodology for analyzing discretely sampled sealed bituminous road profile data.[9] They are one of the first studies to measure and digitize the road profile with a laser profilometer in Victoria, Australia. Andren presented a literature survey of such PSD approximations and tested four approximations on longitudinal road profile data from the entire Swedish state road network.[10] Rouillard introduced a numerical approach for decomposing non-stationary random vibration signals into constituent Gaussian elements from collected road profiles.[11] In addition, Loprencipe and Cantisani proposed unified procedures to measure the spectrum and link it to the texture level.[12] The relationships between rolling noise and pavement texture were evaluated, and the surface characteristics as a spectrum in the domain of spatial frequencies (or wavelengths) are shown in Fig. 1.[12–14] More recently, Mucka compared the spectrum parameters of actual roads with ISO 8608 road classes used in a sample of 27 scientific papers. The studies list the road classes corresponding to the actual road conditions, e.g. motorways and local roads.[15]

Considering the complexity of various road roughness, the World Bank proposed the international roughness index (IRI) to measure the roughness that is stable with time.[16] By adopting the "point model" for the vehicle and road surface interaction, the IRI presents the road roughness with slope units (in/mi, m/km, etc.) obtained from the response of a predefined vehicle, also known as a "golden car".[17] Adopting a golden car not only reduces the complexity of a vehicle model but also eliminates the heavy computation task of the vehicle and road interaction. This gives the IRI a huge edge to be commonly adopted in practice, as it is most economically viable for

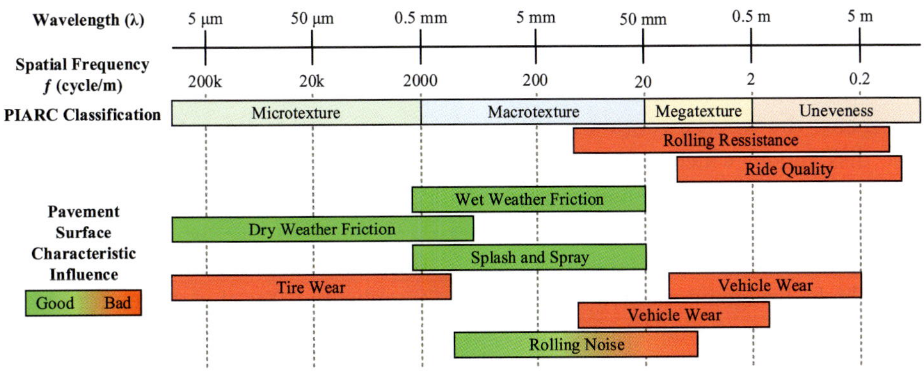

Fig. 1. Pavement surface characteristic influence.[12–14]

road improvement projects.[18,19] For example, Nguyen *et al.* presented the effect of road surface roughness to recommend a safe vehicle speed.[20] In their study, the IRI is adopted to measure the acceleration, and the wheel bumping is also discussed.

Owing to a standardized procedure for measuring IRI, the golden car is simplified by quarter-car model to represent the sprung mass system.[21] The IRI intends to serve as a general pavement condition indicator based on the contact point.[22] Therefore, Kropac and Mucka compared different indicators from standards, e.g. PSD from ISO 8608, IRI, three-band indicators (TBI), and their mutual relationships.[23] The study shows all the indicators are mutually convertible but may increase the uncertainty of converted quantities. Thus, the mechanism of the vehicle's response needs further investigation owing to its complexity, e.g. wheels, suspensions, axles, etc.

To better understand the vehicle's response, researchers have sought to simulate the physical interaction model between the contact point and the vehicle itself. Considering the rail–road vehicle, the wheel–rail contact mechanics model has been studied since the 1900s.[24] Kalker developed various wheel–rail contact theories that can be used to determine the tangential forces and spin moment between the wheel and the rail.[25,26] In addition, many researchers have improved the accuracy while reducing computational costs for calculating the contact mechanics model.[27–30] Considering the motor vehicle, the tire-road interaction will be examined by modeling pneumatic tires and its impact on vehicle dynamics.[31] Different tire simulation models have been purposed, e.g. rigid ring, multibody tire, to study the tire dynamic response for comfort and handling analysis.[32–34] Besides, the vehicle's model has been refined by including the suspension system, especially for investigating ride comfort and rolling resistance.[35] By studying the road profile and its spectrum, the relationship to the vehicle vibration response was characterized.[36–38] It is worth noting that road roughness is considered a major source of excitation for the riding environment of the heavy vehicle driver.[39] Such an unwanted vibration may pose a considerable risk while driving in highway traffic.

In view of the above-mentioned studies, e.g. ride comfort and roughness measurement, the wheel's contact model is often considered a contact "point model." The disk model which considers the contact duration as the wheel interacts with road surface roughness should also be included.[40–42] Chang *et al.* pioneered the disk model by introducing a geometry method (GM) to accommodate the disk model for wheels moving over highway bridges.[43] However, adopting the GM will make the vehicle–bridge interaction analysis more complicated. Xu *et al.* have proposed a refined roughness formula to account for the wheel size effect based on the ISO 8608 standard.[44] Although the refined formula can generate different roughness levels, it can neither consider a non-stationary road roughness nor modify the existing road profile to include the wheel size effect. It is noted that the vehicle's noise, vibration, and harshness (NVH) is a massive topic in vehicle dynamics. The disk effect can reduce the vehicle's response in higher frequency.[6,43,44] This has a significant impact on multiple areas, including road surface roughness extraction,[45–48] riding comfort for the vehicle,[36–38] bridge health monitoring by a passing vehicle,[49–53] etc.

In this study, a new convolution method is proposed. This method can easily tackle the disk geometry effect and adjust any other frequency period over road surface roughness. Considering a rigid wheel passing over the road with rotation, the proposed method can accommodate any type of road roughness from any standard, as it is facilitated under the spatial domain. The layout of the paper is as follows. In Sec. 2, the adopted ISO 8608 standard is introduced. Section 3 compares the response from the point model with the disk model. Section 4 introduces the newly proposed convolution method. This includes the formulation of the periphery function. In Sec. 5, two numerical examples are given to verify the reliability of the proposed method.

2. Road Surface Roughness

To simulate the road roughness, the road profile characterization and PSD fitting presented in ISO 8608 were adopted.[6] The road roughness $r(x)$ is defined as

$$r(x) = \sum_{i=0}^{N} \sqrt{2G_d(n_i)\Delta n} \cdot \cos(n_i x_c + \theta_i), \tag{1}$$

where N denotes the total number of harmonic waves, n_i the ith spatial frequency in the range of 1 to 100 cycle/m with increment $\Delta n = 0.04$ cycle/m, θ the random phase angle from 0 to 1, which is based on normal distribution, x_c the location of the road and the displacement PSD function $G_d(n_i)$ is defined as

$$G_d(n_i) = G_d(n_0)\left(\frac{n_i}{n_0}\right)^{-2}, \tag{2}$$

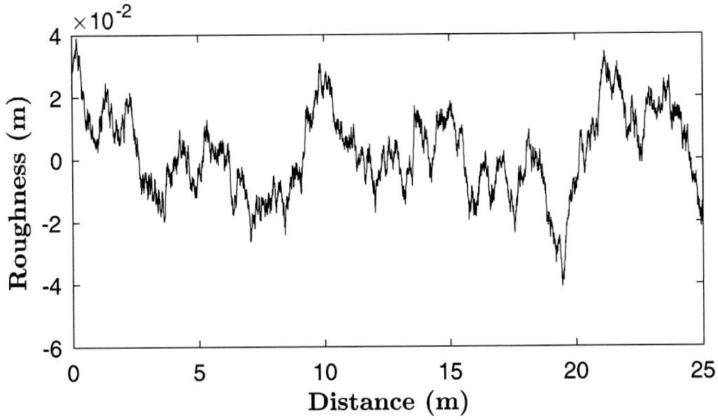

Fig. 2. Profiles of standard road roughness.

where the reference spatial frequency is $n_0 = 0.1$ cycle/m and the $G_d(n_0)$ is determined by the roughness class. This study adopts a higher standard for displacement PSD to seek wider application. For bridge design, it is common to consider the ISO 8608 standard road surface roughness with $G_d(n_0) = 1 \times 10^{-6}\,\text{m}^3$, as shown in Fig. 2. A low roughness is taken as $G_d(n_0) = 0.01 \times 10^{-6}\,\text{m}^3$ and a high roughness as $G_d(n_0) = 16 \times 10^{-6}\,\text{m}^3$.[54,55]

3. Moving Path Using Point and Disk Models

In recent years, many researchers have studied moving load problems, e.g. vehicle–bridge interaction. Regarding road surface roughness, it is important to assume the contact model between the vehicle and the road pavement. Therefore, the point model is widely used despite the lack of accuracy when compared to the disk model. In this section, we compare the moving path of a passing vehicle based on the point model and disk model.

3.1. *Moving path based on point model*

In the moving load problem, the conventional "point model" is commonly adopted due to its simplicity and straightforwardness. Consider a single degree-of-freedom (DOF) test vehicle shown in Fig. 3. The vehicle is represented by a sprung mass of m_v supported by a spring-dashpot unit of stiffness k_v and damping coefficient c_v. Conventionally, the wheel (or tire) of the vehicle is simulated as a point moving over the road surface. The point model becomes very unrealistic when passing rough pavements since the "point" can touch virtually any point of the rough surface, and it tends to exaggerate the high-frequency part of the roughness frequencies. In many cases, the presence of roughness, coupled with higher vehicle speeds, has become

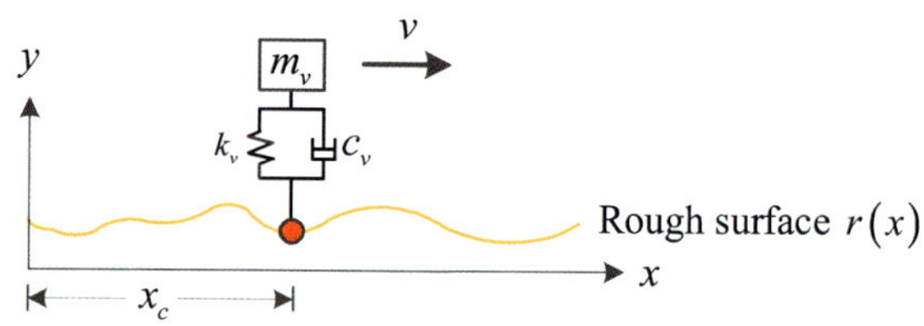

Fig. 3. Single DOF test vehicle using contact point model.

an obstacle to the scanning of bridge frequencies by moving test vehicles, although it is not always unconquerable.[44]

3.2. *Moving path based on disk model*

Followed by the point model, the contact interface between the wheel and the road surface needs to be studied to project the moving path based on the disk model. As shown in Fig. 4, assuming that the original wheel center (denoted as O) moves to coordinate x_c at a certain time step t_c that is known given the speed v of the vehicle. Considering the wheel rotates and passes over the road surface roughness without disk deformation. The moving path can be determined by the wheel's center, typically connected to the vehicle's body. The solid and dashed curves represent the periphery of the rigid wheel directly in contact with the road surface roughness and the smooth surface, excluding the roughness $r(x)$, respectively.

Consider first the special case where the surface roughness is excluded. The periphery of the rigid wheel (as shown by the dashed curve in Fig. 4) is directly in contact with the smooth surface at point C. Hence, the coordinate x_c of the contact point coincides with the coordinate of the disk center.

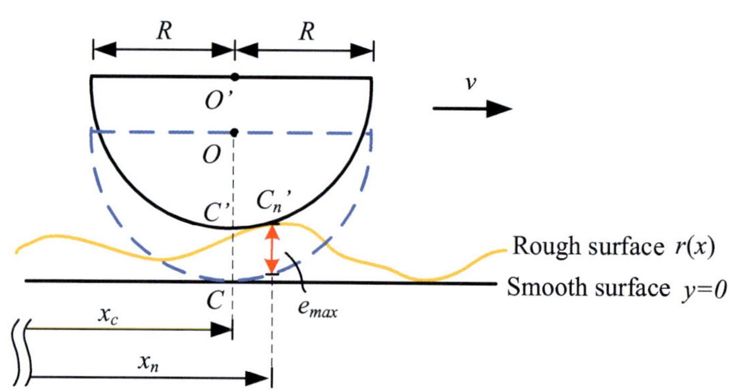

Fig. 4. Contact interface between disk model and road roughness.

Next, consider the more general case with the surface roughness $r(x)$ included. In most previous studies, the radius R of the rigid disk is taken as zero, meaning that the rigid disk is a "point" and is in contact with point C. As such, the elevation of the disk is equal to the height of the surface roughness $r(x_c)$. In this study, however, the wheel is assumed to be of finite size, i.e. with a radius R, rather than as a point. As shown by the solid curve in Fig. 4, the wheel gets in touch with the surface profile at point C_n," rather than at point C, and the elevation of the disk lifted by the surface roughness is $r(x_n)$, rather than the original value of $r(x_c)$.

The following is the existing GM's procedure for determining the contact point position x_c and disk elevation $r(x_n)$ with the disk center O located at coordinate x_c in each time step[43]:

Step 1. Divide the wheel's diameter $2R$ into N intervals in the horizontal direction.

Step 2. For the nth interval, calculate the height r_n of the surface roughness and the height s_n of the wheel periphery above the smooth surface.

Step 3. For the nth interval, calculate the elevation e_n of the wheel lifted by the surface roughness by subtracting the wheel periphery height $p(x_n)$ from the roughness $r(x_n)$, i.e. $e_n = r(x_n) - p(x_n)$. The rigid wheel will be lifted or dropped to a height of $|e_n|$.

Step 4. For all the N intervals, determine the elevation $r(x_c)$ of the whole wheel by calculating the maximum value of the lift e_n, i.e. $r(x_c) = \max(e_n)$. Also, determine the contact point position x_c as the coordinate corresponding to the maximum lift $e_{\max}$. For example, the maximum lift $e_{\max}$ shown in Fig. 4. It represents exactly the elevation or lift of the disk, $r(x_c)$, when it moves to coordinate x_c, and the coordinate x_n corresponding to the maximum lift $e_{\max}$ represents exactly the contact point position. The difference between the locations x_c and x_n is denoted as the shift $(x_n - x_c)$ in the contact point coordinate.

3.3. *Comparison of moving paths using point and disk models*

To compare the moving paths using the point model with the disk model, Fig. 5 shows the results using two different models. From Fig 5(a), it can be seen that the moving path of the disk model is much smoother than the path of the point model because the wheel does not contact each point of the road surface. Moreover, the moving path is smoother as we increase the wheel's radius R from 0.2 m to 0.3 m. Therefore, the moving path of the disk model is essentially the upper bound of the road surface roughness. In addition, Fig. 5(b) shows the moving paths using the point and disk models in the frequency domain. It has shown that the disk model has filtered out some of the higher frequencies. Despite that the use of the disk model or other advanced ones is more realistic, it adds substantial difficulty to analysis and cannot be popularly adopted.

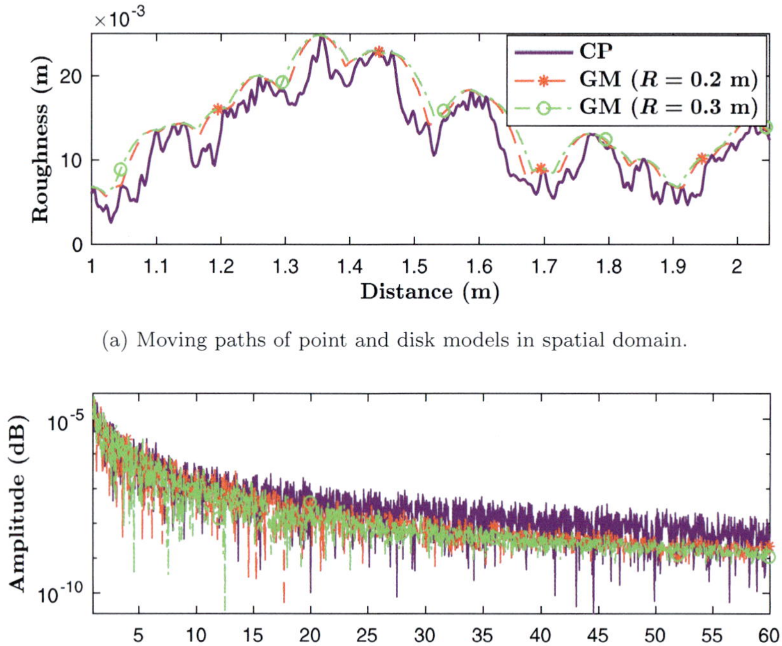

(a) Moving paths of point and disk models in spatial domain.

(b) Moving paths of point and disk models in frequency domain.

Fig. 5. Comparison of moving paths using point and disk models.

4. Disk Effect Using Convolution Method

From the physics point of view, the disk model can better reflect the interaction between the vehicle and road roughness than the point model. While using the GM (shown in Sec. 3.2) can simulate the disk model, it is more complicated than using the point model. In this section, we propose the convolution method by adopting the periphery function. The periphery function will be designed and tested out with different building block functions. Finally, the building block function akin to the surface roughness will be adopted for further investigation.

4.1. *Formulation of disk model*

To simulate the disk effect (or wheel size effect) using the convolution method, one needs to design a function for the wheel to convolute over the surface roughness $r(x_c)$. Section 3.3 has shown that the moving path of the disk model has filtered out some of the higher frequencies when compared to the use of the point model. Thus, a new type of convolution function needs to be proposed.

Judging from the natural shape of the disk model in Fig. 4, the wheel does not touch every point of the surface roughness during its travel. As the wheel moves

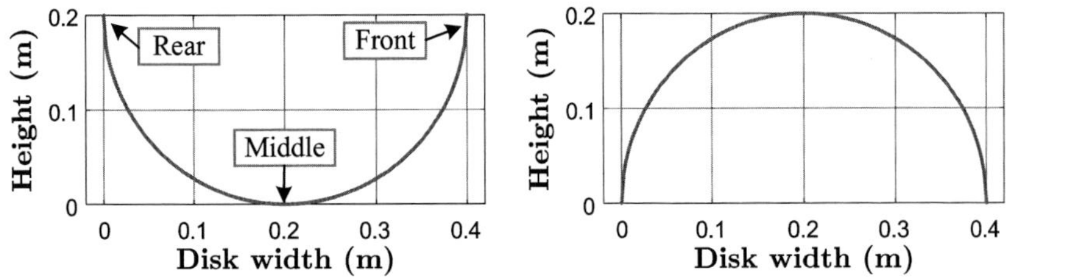

(a) Wheel geometry shape.

(b) Periphery function.

Fig. 6. Comparison of disk geometry and periphery function.

forward, the front and rear parts of the wheel are far away from the ground compared to the middle part of the wheel. The middle part is most likely to contact the ground surface and raise up the whole wheel. Therefore, a special function needs to be used to reflect the natural geometry shape of the disk model for convolution.

Assuming the disk has a radius of $R = 0.2\,\text{m}$, the wheel geometry shape is shown in Fig. 6(a). To convolute over the roughness, the wheel geometry shape is converted as a periphery function $p(x)$ (shown in Fig. 6(b)):

$$p(x_n) = c_g[R - \sqrt{R^2 - (x_n - x_c)^2}], \tag{1}$$

where c_g is the geometry coefficient, x_c the center location of the disk model in the longitudinal direction and x_n the distant variable ranging from $(x_c - R)$ to $(x_c + R)$ based on the disk center x_c. Despite the periphery function considering the wheel's geometry shape, it has overestimated the contribution from the front and rear parts and underestimated the contribution from the middle part of the wheel. Further improvement will be demonstrated.

4.2. *Convolution of the disk model with building block functions*

To verify the feasibility of the convolution method, we test the periphery function with the building block (or essential) functions (shown in Fig. 7) from signal processing. This enables us to study the convolution method under different conditions. By setting the amplitude to the height to $R = 0.2$ m, the essential functions, including impulse function, Heaviside function, and rectangular function, are illustrated in Figs. 7(a)–7(c). Judging from the shape of the essential functions, the roughness function (shown in Fig. 5) can be viewed as a constitution of multiple impulse functions with different amplitudes. Noting the amplitude 2 m for essential functions is exaggerated for surface roughness, as the purpose herein is to demonstrate the proposed method.

By adopting the above essential function with an amplitude of 0.2 m, Fig. 8 compares the results from the GM (outlined in Sec. 3.2) and the proposed convolution method using the periphery function. It can be seen the results from using

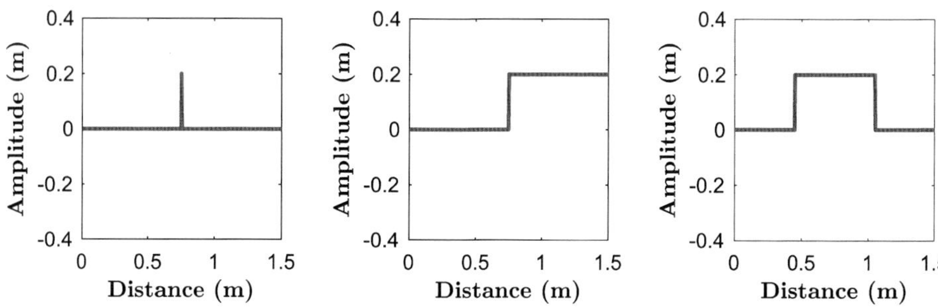

(a) Impulse function. (b) Heaviside function. (c) Rectangular function.

Fig. 7. Essential functions from signal processing.

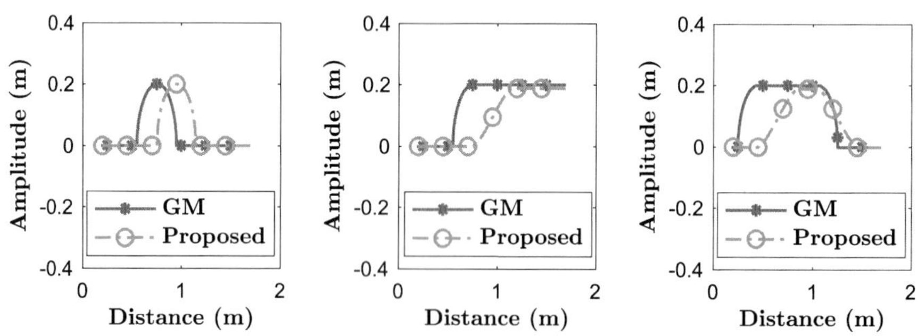

(a) Impulse function. (b) Heaviside function. (c) Rectangular function.

Fig. 8. Comparison of the results from GM and proposed method.

the periphery function fits well with the GM for the impulse function (shown in Fig. 8(a)). However, both results do not match well for Heaviside and Rectangular functions (shown in Figs. 8(b) and 8(c)). The convolution method's results do not reflect the changes in essential functions timely. As the periphery function passes over the plateau in the Heaviside and rectangular functions, the convolution method can only reflect the changes until the whole wheel function passes through.

To further investigate the impulse function, the amplitude is set to 2×10^{-2}, 2×10^{-3}, 2×10^{-4} m, and testing with the GM and proposed method in Fig. 9. As the amplitude reduces from 2×10^{-2} m (shown in Fig. 9(a)) to 2×10^{-4} m (shown in Fig. 9(c)), the duration of the contact between the spike and the wheel decreases. However, the proposed convolution method's results do not decrease the contact duration as we reduce the impulse amplitude. In other words, the transition period of the wheel passing over the impulse function is not reflected properly. It is worth noting that the "contact area" of the wheel decreases, especially for the height of

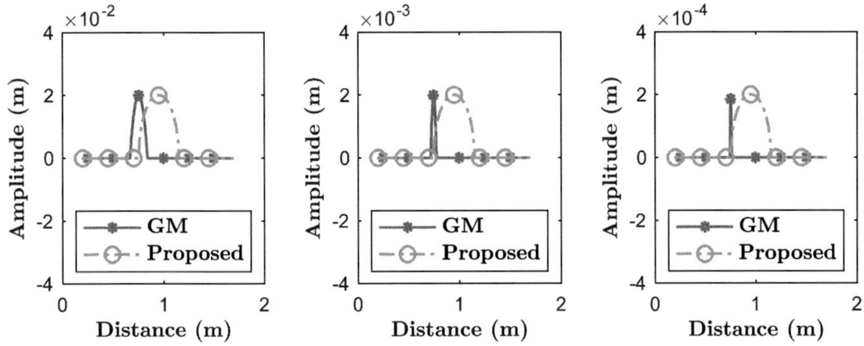

(a) Amplitude: 2×10^{-2} m. (b) Amplitude: 2×10^{-3} m. (c) Amplitude: 2×10^{-4} m.

Fig. 9. Comparison of different amplitudes for impulse function using GM and proposed method.

the impulse much smaller than the wheel's radius. Such a contact area of the wheel should be factored into the periphery function.

4.3. *Modified periphery function for disk model*

In this section, the modified periphery function is introduced for the road surface amplitude r_a, which is much smaller than the radius of the wheel. This is a more typical situation as the surface roughness is very small and does not become an obstacle. The modified function is based on the original periphery function in Sec. 4.1 and includes the concept of the contact area in Sec. 4.2.

To modify the periphery function, there are two steps. First, consider the period outside of contact duration. The non-contact area of the original function is set to zero. Next, consider the period during the wheel in contact with the impulse function, the whole baseline of the contact area shifts to zero. The modified periphery function is shown in Fig. 10. It can be seen the modified function is much narrower than the original function, as it mainly concerns the contact area. Besides, the height of the modified function gradually increases in the contact area. The attribute for the natural shape of the disk model remains (as mentioned in Sec. 4.1).

By adopting the essential function with an amplitude of 2×10^{-4} m, Fig. 11 compares the results from the GM (outlined in Sec. 3.2) and the proposed convolution method using the modified periphery function. It can be seen that the results from using the proposed convolution method fit well with the GM for all essentials. The impulse function's result for the proposed method is properly demonstrated in Fig. 11(a). Moreover, the proposed method is able to reflect the transition period of the periphery function passes over the plateau in Figs. 11(b) and 11(c). Considering it is unlikely to have a plateau in the road surface roughness, our focus will be testing

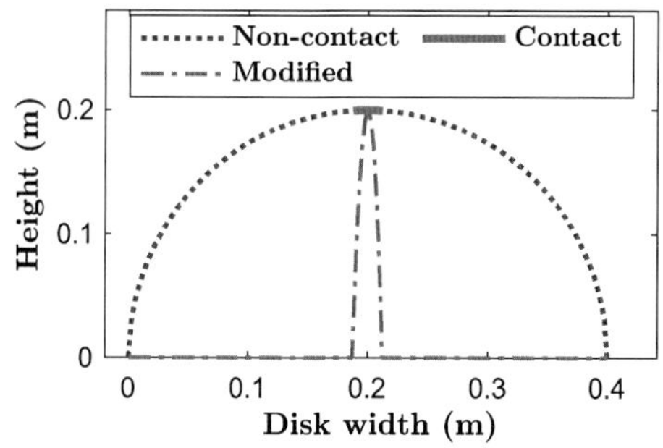

Fig. 10. Comparison of original (contact and non-contact areas) and modified periphery functions (normalized maximum height to 0.2 m).

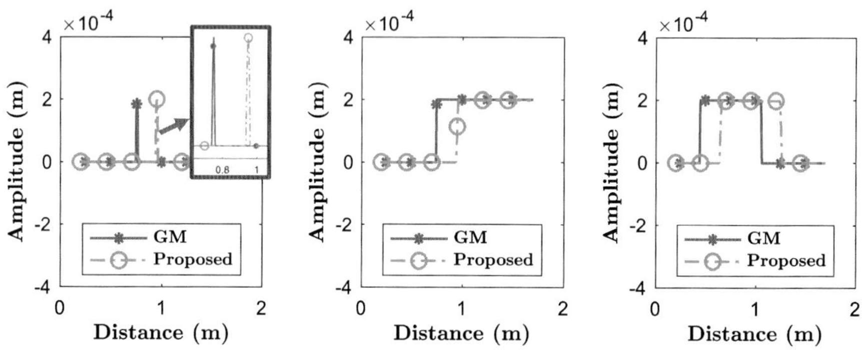

(a) Impulse function. (b) Heaviside function. (c) Rectangular function.

Fig. 11. Comparison of different essential functions using GM and proposed method.

with the impulse function and its extension, e.g. the shah function, for verification in the following section.

5. Example Problems

In this section, the proposed convolution method is verified with two examples. In the first example, the shah function is adopted to test out the proposed method. By using different spatial periods, the shah function can simulate road roughness with various frequencies. In the second example, the ISO standard roughness is adopted. By comparing with the conventional GM, the proposed method is tested out with various severity of surface roughness.

5.1. *Verification with shah function*

Considering the surface roughness as a shah function (also known as Dirac comb), a passing wheel with a radius of 0.2 m and speed of 3 m/s will be adopted to test the

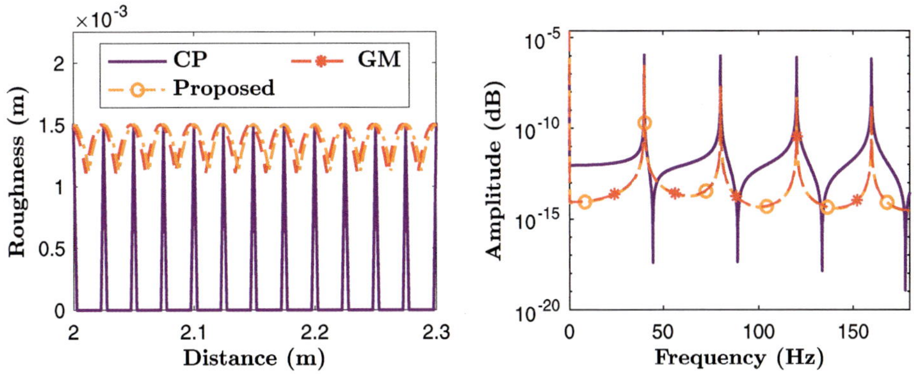

(a) Displacement.

(b) Spectrum.

Fig. 12. Comparison of proposed method with GM and contact point (CP) response.

proposed convolution method and verify it with the geometry approach. It is known the shah function is a periodic function of impulse function. Thus, the proposed method will apply to different periods of the shah function, which can be viewed as a composition of different frequencies of surface roughness. To simulate the shah function, the spatial periodic is first assumed to be 0.025 m with an amplitude of 1.5×10^{-3} m and a length of 25 m, and the time step is discretized into 10,000 steps as the wheel travel through the shah function.

Figure 12 compares the proposed convolution method with the GM and the result from contact point measurement (the same as the shah function). From Fig. 12(a)), it can be seen the moving paths for the wheel using the proposed method and GM have a similar periodic path despite having a different baseline in the spatial domain. To measure the frequency of response, the PSD of each response is given in Fig. 12(b). The peak frequencies for the contact point response (a.k.a. shah function) are still the same as the proposed method and GM despite considering the disk or point models. However, the peak amplitude of the higher frequencies is reduced for the proposed method, and the GM is reduced. This is owing to the shape of the disk model smoothing out some smaller vibrations during its course of moving forward.

To further study the reduction of the peak amplitude in frequency, we consider the shah function with a different spatial period. By using the proposed method and the GM, the peak amplitude from frequency responses is extracted and shown in Fig. 13. In comparing the proposed method to the GM, the reduction of the peak amplitudes matches well between the two methods under various spatial periods of the shah function. Besides, the reduction of the peak amplitude is highly correlated to the higher frequency. It is worth noting that the constitution of all

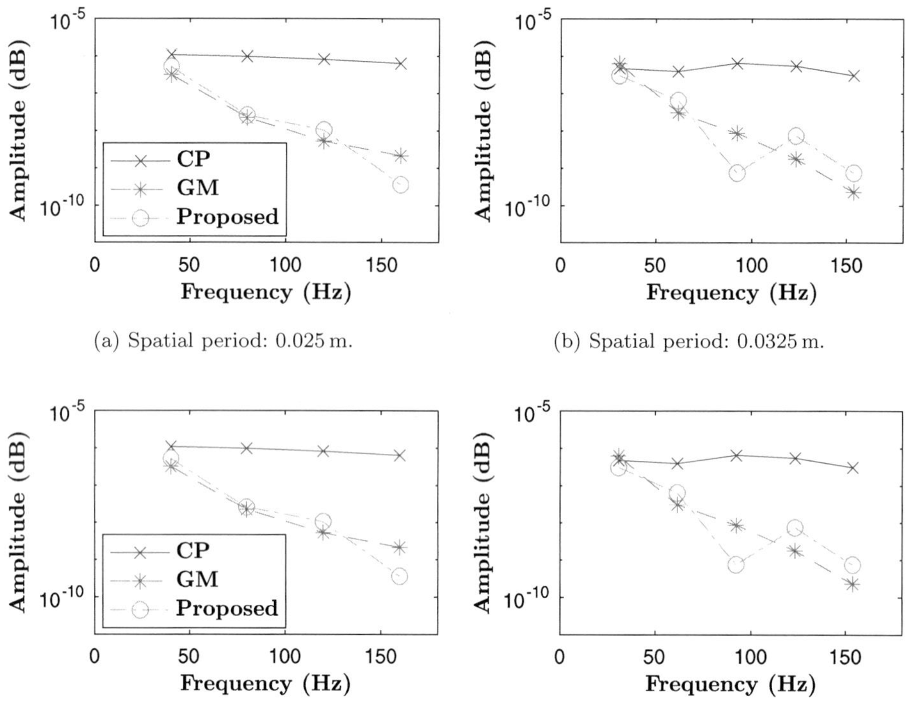

(a) Spatial period: 0.025 m.

(b) Spatial period: 0.0325 m.

(c) Spatial period: 0.04 m.

(d) Spatial period: 0.0475 m.

Fig. 13. Comparison of peak amplitude from PSD with different spatial periods of the shah function.

the frequency spectrums can represent the road surface roughness. The proposed convolution method using a modified periphery function is capable of simulating the disk effect.

5.2. *Verification with ISO 8608 road surface roughness*

To illustrate the reliability of the proposed convolution method, first, we consider a passing wheel with a radius of 0.2 m and speed of 3 m/s (same as Example 5.1) to test on the ISO 8608 standard road surface roughness. With the use of the displacement PSD function in Eqs. (1) and (2), the profile of the standard road roughness can be generated with randomization due to the random phase angle, as shown in Fig. 2.

In order to adopt the modified periphery function, the critical task is to determine the contact and non-contact area of the disk model, as shown in Fig. 10. Thus, the road surface amplitude is crucial to determine. Since the road surface roughness is composed of a random phase angle in the normal distribution and often with bias

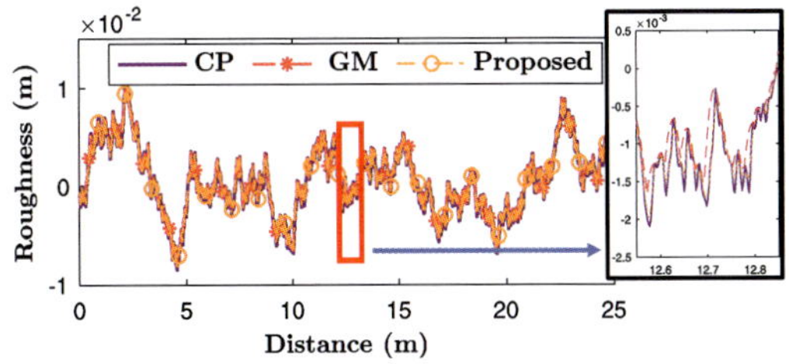

(a) Displacement.

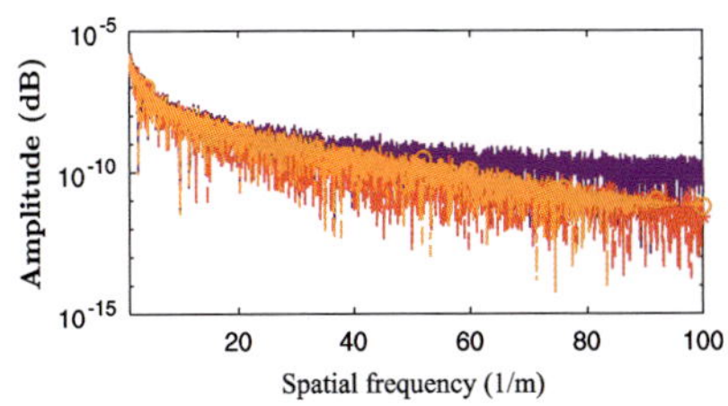

(b) Spectrum.

Fig. 14. Comparison of response using proposed method and GM under standard surface roughness.

error,[6] the road surface amplitude per distance can be calculated as

$$r_a = \sigma / L, \tag{1}$$

where σ is the standard deviation of the road profile (shown in Fig. 2) and$L$ the distance of the road surface roughness. This road surface amplitude can be used as the impulse term to determine the contact and non-contact area in Fig. 10.

Figure 14 compares the response from the proposed method using a modified periphery function and the GM. From Fig. 14(a), it can be seen the result from the proposed method follows the trend of the GM. Both methods are able to restrain small displacement from the surface roughness. More importantly, Fig. 14(b) shows that both methods have a similar reduction rate for the frequency amplitude.

In Fig. 15, the example further extends to different roughness. Referring to Sec. 2, a low (smooth) and a high (poor) road surface roughness is tested out with the same passing wheel. It can be seen the response spectrum from the proposed method and GM matches very well as it passes through the low surface roughness (shown in Fig. 15(a)). As one increases the roughness to high surface roughness, the response spectrum from the proposed method and GM slightly deviates in the higher frequencies (shown in Fig. 15(b)). However, the results from the two methods still match quite well in general. Moreover, considering the wheel effect, the higher surface roughness will likely have a higher amplitude reduction in higher frequency.

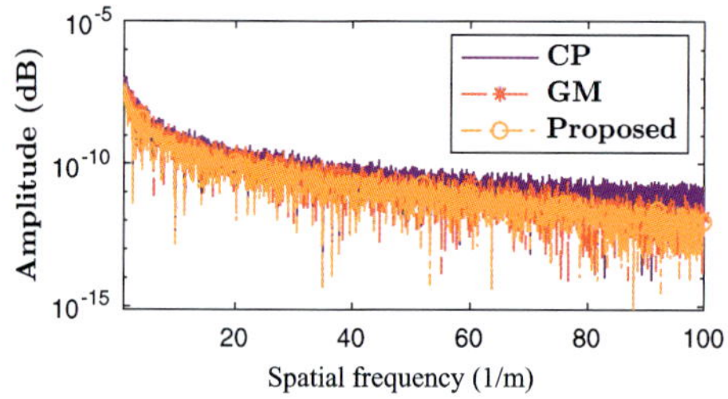

(a) Low surface roughness.

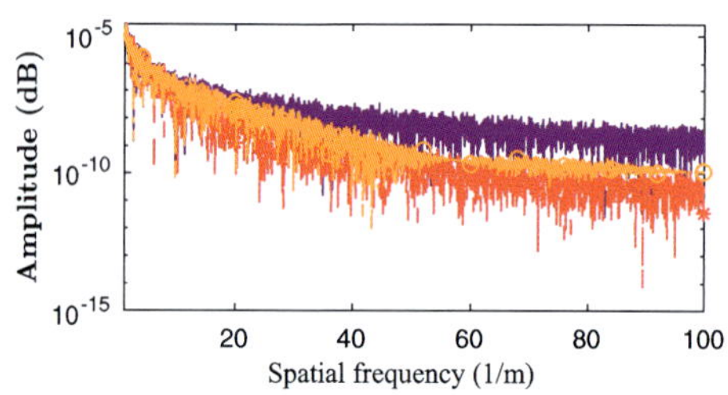

(b) High surface roughness.

Fig. 15. Comparison of response spectrum using proposed method and GM under low and high surface roughness.

It is worth noting that this study only shows the wheel with the same radius passing through different surface roughness. This is owing to the condition of a larger wheel radius passing through a higher road surface roughness being similar to that of a smaller wheel radius passing through a lower road surface roughness. There is no need to show the response of different radii as the case of different surface roughness is demonstrated.

6. Conclusion

In this paper, the convolution method to consider the disk effect is proposed. First, by adopting the periphery function, the proposed method is tested with essential functions of the road surface roughness. Followed by improving the outcome with essential functions, the modified periphery function is formulated. Finally, the proposed convolution method is verified with the traditional GM under different severity of ISO road surface roughness. From the illustrative example, the following conclusions are drawn:

- The proposed method is able to include the disk effect for road surface roughness.

- The proposed method is generic to apply to different types of road surface roughness (including non-stationary behavior), and the displacement and frequency responses are highly correlated to the actual condition.
- The modified periphery function acts like a filter to the road surface roughness while filtering out smaller oscillations. Thus, the proposed method can easily be corporate with different types of filters to suit it needs to filter the unwanted frequency.
- When the surface roughness is low, the frequency amplitude reduction for the disk effect also reduces.
- With the consideration of the contact and non-contact area of the wheel, the result from the modified periphery function prevails over the periphery function.

The illustrative example presented herein demonstrated the capability of the convolution method in considering the disk effect. Further research directions include incorporating contact mechanic models and using deconvolution to the measured response from the wheel, which can assist in capturing the surface roughness in the field.

Acknowledgment

This research study is financial supported by the Australian Research Council (Nos. IH150100006 and DE190101249), the National Natural Science Foundation of China (Grant No. 52208146), and China Postdoctoral Science Foundation (Grant No. 2022M720580).

References

1. J. C. Houbolt, Runway roughness studies in the aeronautical field, *Trans. Am. Soc. Civ. Eng.* **127** (1962) 427–447.
2. J. H. Walls, J. C. Houbolt and H. Press, Some Measurements and Power Spectra of Runway Roughness, *Natl. Advisory Comm. for Aeronautics*, Techical Note 3305, Langley Field (1954).
3. J. C. Houbolt, J. H. Walls and R. F. Smiley, On Spectral Analysis of Runway Roughness and Loads Developed During Taxiing, *Natl. Advisory Comm. for Aeronautics*, Techical Note 3484, Langley Field (1955).
4. C. J. Dodds and J. D. Robson, The description of road surface roughness, *J. Sound Vib.* **31** (1973) 175–183.
5. *BS ISO 8608:1995*: *Mechanical Vibration. Road Surface Profiles. Reporting of Measured Data* (International Organization for Standardization, 1995).
6. *BS ISO 8608:2016*: *Mechanical Vibration. Road Surface Profiles. Reporting of Measured Data* (International Organization for Standardization, British Standards Institute, London, 2016).
7. B. R. Davis and A. G. Thompson, Power spectral density of road profiles, *Veh. Syst. Dyn.* **35** (2001) 409–415.
8. G. Loprencipe and P. Zoccali, Use of generated artificial road profiles in road roughness evaluation, *J. Mod. Transp.* **25** (2017) 24–33.

9. B. Bruscella, V. Rouillard and M. Sek, Analysis of road surface profiles, *J. Transp. Eng.* **125** (1999) 55–59.
10. P. Andrén, Power spectral density approximations of longitudinal road profiles, *Int. J. Veh. Des.* **40**(1–3) (2006) 2–14.
11. V. Rouillard, Decomposing pavement surface profiles into a Gaussian sequence, *Int. J. Veh. Syst. Model. Test.* **4**(4) (2009) 288–305.
12. G. Loprencipe and G. Cantisani, Unified analysis of road pavement profiles for evaluation of surface characteristics, *Mod. Appl. Sci.* **7**(8) (2013) 1–14.
13. J. C. Wambold, C. E. Antle, J. J. Henry and Z. Rado, International PIARC experiment to compare and harmonize texture and skid resistance measurements, *PIARC*, Paris (1995), p. 346.
14. G. Bitelli, A. Simone, F. Girardi and C. Lantieri, Laser scanning on road pavements: A new approach for characterizing surface texture, *Sensors* **12** (2012) 9110–9128.
15. P. Múčka, Simulated road profiles according to ISO 8608 in vibration analysis, *J. Test. Eval.* **46** (2018) 20160265.
16. M. W. Sayers, On the calculation of international roughness index from longitudinal road profile, *Transp. Res. Rec.* (1501) (1995) 1–12.
17. T. D. Gillespie, M. W. Sayers and L. Segel, *Calibration of Response-type Road Roughness Measuring Systems* (Transportation Research Board, Ann Arbor, Michigan, 1980).
18. C. R. Bennett and I. D. Greenwood, *Modelling Road User and Environmental Effects in HDM-4*, Vol. 7 (The Highway Development and Management Series Collection, Birmingham, 2018).
19. J. A. Padilla and O. G. dela Cruz, Methods on calculating the international roughness index: A literature review, in *Sustainable Development Approaches* (Springer International Publishing, 2022), pp. 11–9.
20. X. Nguyen, T. Nguyen and P. H. Tran, The effect of road surface roughness to recommended speed of vehicles, *IOP Conf. Ser.: Mater. Sci. Eng.* **886** (2020) 012014.
21. G. Loprencipe, P. Zoccali and G. Cantisani, Effects of vehicular speed on the assessment of pavement road roughness, *Appl. Sci.* **9** (2019) 1783.
22. M. W. Sayers and S. M. Karamihas, *The Little Book of Profiling: Basic Information About Measuring and Interpreting Road Profiles* (The Regent of the University of Michigan, Ann Arbor, 1998).
23. O. Kropáč and P. Múčka, Indicators of longitudinal road unevenness and their mutual relationships, *Road Mater. Pavement Des.* **8** (2007) 523–549.
24. F. W. Carter, On the action of a locomotive driving wheel, *Proc. R. Soc. Lond. A Contain. Pap. Math. Phys. Character* **112** (1926) 151–157.
25. K. L. Johnson, The effect of a tangential contact force upon the rolling motion of an elastic sphere on a plane, *J. Appl. Mech.* **25** (1958) 339–346.
26. K. E. Zaazaa and A. L. Schwab, Review of Joost Kalker's wheel–rail contact theories and their implementation in multibody codes, in: *7th Int. Conf. Multibody Systems, Nonlinear Dynamics, and Control, Parts A, B and C*, Vol. 4 (ASMEDC, 2009), pp. 1889–1900.
27. J. Piotrowski and H. Chollet, Wheel–rail contact models for vehicle system dynamics including multi-point contact, *Veh. Syst. Dyn.* **43** (2005) 455–483.
28. J. Piotrowski and W. Kik, A simplified model of wheel/rail contact mechanics for non-Hertzian problems and its application in rail vehicle dynamic simulations, *Veh. Syst. Dyn.* **46** (2008) 27–48.

29. MSh Sichani, R. Enblom and M. Berg, Comparison of non-elliptic contact models: Towards fast and accurate modelling of wheel–rail contact, *Wear* **314** (2014) 111–117.
30. B. Liu, S. Bruni and E. Vollebregt, A non-Hertzian method for solving wheel–rail normal contact problem taking into account the effect of yaw, *Veh. Syst. Dyn.* **54** (2016) 1226–1246.
31. H. B. Pacejka, *Tire and Vehicle Dynamics* (Elsevier, Bangalore, India, 2002).
32. A. J. C. Schmeitz, *A Semi-Empirical Three-Dimensional Model of the Pneumatic Tyre Rolling Over Arbitrarily Uneven Road Surfaces* (Delft University of Technology, 2004),
33. P. Lugner, H. Pacejka and M. Plöchl, Recent advances in tyre models and testing procedures, *Veh. Syst. Dyn.* **43** (2005) 413–426.
34. F. Farroni, A. Sakhnevych and F. Timpone, A three-dimensional multibody tire model for research comfort and handling analysis as a structural framework for a multi-physical integrated system, *Proc. Inst. Mech. Eng. D: J. Autom. Eng.* **233** (2019) 136–146.
35. R. S. Sharp and D. A. Crolla, Road vehicle suspension system design a review, *Veh. Syst. Dyn.* **16** (1987) 167–192.
36. A. M. A. Soliman, M. M. S. Kaldas and S. A. Abdallah, Influence of Road Roughness on the Ride and Rolling Resistance for Passenger Car (2013).
37. P. Múčka, G. J. Stein and P. Tobolka, Whole-body vibration and vertical road profile displacement power spectral density, *Veh. Syst. Dyn.* **58** (2020) 630–656.
38. R. Bridgelall, *A Participatory Sensing Approach to Characterize Ride Quality*, In eds. J. P. Lynch, K.-W. Wang and H. Sohn (SAE 2001 World Congress, 2014), p. 90610A.
39. R. Hassan and K. Mcmanus, *Heavy Vehicle Ride and Driver Comfort* (2001).
40. R. Hassan and R. Evans, Road roughness characteristics in car and truck wheel tracks, *Int. J. Pavement Eng.* **14** (2013) 736–745.
41. V. Žuraulis, L. Levulytė and E. Sokolovskij, The impact of road roughness on the duration of contact between a vehicle wheel and road surface, *Transport* **29** (2014) 431–439.
42. J. P. Yang and C.-Y. Cao, Wheel size embedded two-mass vehicle model for scanning bridge frequencies, *Acta Mech.* **231** (2020) 1461–1475.
43. K. C. Chang, F. B. Wu and Y. B. Yang, Disk model for wheels moving over highway bridges with rough surfaces, *J. Sound Vib.* **330** (2011) 4930–4944.
44. H. Xu, M. H. Wang, Z. L. Wang, D. S. Yang, Y. H. Liu and Y. B. Yang, Generation of surface roughness profiles for inclusion in vehicle–bridge interaction analysis and test application, *Int. J. Struct. Stab. Dyn.* (2022).
45. L. Janani, V. Sunitha and S. Mathew, Influence of surface distresses on smartphone-based pavement roughness evaluation, *Int. J. Pavement Eng.* **22** (2021) 1637–1650.
46. N. Abulizi, A. Kawamura, K. Tomiyama and S. Fujita, Measuring and evaluating of road roughness conditions with a compact road profiler and ArcGIS, *J. Traffic Transp. Eng.* (*Engl. Ed.*) **3** (2016) 398–411.
47. L. Kumar, T. Tallam and C. Naveen Kumar, Assessment of ride quality and road roughness by measuring the response from a vehicle mounted android smartphone, *IOP Conf. Ser.: Earth Environ. Sci.* **982** (2022) 012062.
48. K. Feng, M. Casero and A. González, Characterization of the road profile and the rotational stiffness of supports in a bridge based on axle accelerations of a crossing vehicle, *Comput.-Aided Civ. Infrastruct. Eng.* (2023).

49. Y. B. Yang and J. P. Yang, State-of-the-art review on modal identification and damage detection of bridges by moving test vehicles, *Int. J. Struct. Stab. Dyn.* **18** (2017) 1850025.
50. A. Malekjafarian, R. Corbally and W. Gong, A review of mobile sensing of bridges using moving vehicles: Progress to date, challenges and future trends, *Structures* **44** (2022) 1466–1489.
51. T. J. Matarazzo *et al.*, Crowdsourcing bridge dynamic monitoring with smartphone vehicle trips, *Commun. Eng.* **1** (2022) 29.
52. D. S. Yang and C. M. Wang, Bridge damage detection using reconstructed mode shape by improved vehicle scanning method, *Eng. Struct.* **263** (2022) 114373.
53. H. Xu, Y. H. Liu, M. Yang, D. S. Yang and Y. B. Yang, Scanning and separating vertical and torsional–flexural frequencies of thin-walled girder bridges by a single-axle test vehicle, *Thin-Walled Struct.* **182** (2023) 110266.
54. K. C. Chang, F. B. Wu and Y. B. Yang, Effect of road surface roughness on indirect approach for measuring bridge frequencies from a passing vehicle, *Interact. Multisc. Mech.* **3** (2010) 299–308.
55. J. P. Yang and W. C. Lee, Damping effect of a passing vehicle for indirectly measuring bridge frequencies by EMD technique, *Int. J. Struct. Stab. Dyn.* **18** (2018) 1850008.

https://doi.org/10.1142/9789819814770_0031

Chapter 31

A Synthetic Integrated Wireless Aerodynamic Evaluation System for Train–Bridge System#

Simin Zou*,†, Xuhui He*,†,‡,§ and Hanfeng Wang*,†

*School of Civil Engineering, Central South University
Changsha 410075, P. R. China

†National Engineering Laboratory for High-Speed Railway Construction
Changsha 410075, P. R. China

‡Hunan Provincial Key Laboratory for Disaster Prevention and Mitigation
of Rail Transit Engineering Structure
Changsha 410075, P. R. China
§xuhuihe@csu.edu.cn

The development of high-speed railway networks and the increased running speeds of high-speed trains (HSTs) have made the aerodynamic interference between HSTs and their surrounding environments increasingly important. Compared with a traditional wind tunnel test, systematically understanding the aerodynamic characteristics of HSTs involves relatively more stringent requirements, highlighting the need to develop experimental methods and technologies with enhanced dynamic performance. Central South University (CSU) developed a wireless data acquisition system, named as the in-model sensory and wireless data acquisition — remote control and processing system (ISWDA-RCPS), which can operate onboard a novel moving train and infrastructure rig. The system was developed to meet current wind tunnel data collection needs, and it avoids the physical cables used in conventional devices, which are extremely susceptible to induced noise. The system accepts inputs from various sensors and transfers the data wirelessly to an access point outside a wind tunnel's test section. To analyze the feasibility of the ISWDA-RCPS concerning its sensing capabilities and wireless communications, we conduct experiments in multiple operating conditions. Finally, pressure measurements are acquired from a moving Fuxing HST model at different points and used to analyze the aerodynamic behavior of the model.

Keywords: High-speed railway; train; bridge; moving-train and infrastructure rig; data acquisition; aerodynamic characteristics.

§Corresponding author.
#This chapter appeared previously on the International Journal of Structural Stability and Dynamics. To cite this chapter, please cite the original article as the following: S. Zou, X. He & H. Wang, *Int. J. Struct. Stab. Dyn.*, **23**, 2340034 (2023), DOI: 10.1142/S0219455423400345.

1. Introduction

Recent developments in railway technology have made new high-speed railways an increasingly important means of transportation globally. Chinese express railway construction best reflects these current trends and prospects.[1,2] Given present trends, the railway network has been expanding into Europe and the rest of the world.[3,4] Furthermore, as train operating speeds gradually increase, high-speed passenger trains (300–400 km/h) and maglevs (600 km/h) in high-speed rail (HSR) systems are becoming more prevalent.[5–7] However, these increased running speeds may introduce issues that threaten operating safety.[8]

Unlike aircraft, high-speed trains (HSTs) face many operational conditions derived from infrastructures and the environment, including bridges, tunnels, wind/noise barriers, and forest zones. These are significant factors that limit the safety and stability of train systems; in addition, vast differences exist among railway conditions worldwide. Recent incidents in China and the USA should not be ignored in transportation infrastructure, such as the oscillation of the Humen Bridge and the Verrazano-Narrows Bridge which were driven by wind,[9,10] the wind-induced vibration affects transport safety, bridge strength and fatigue resistance.[11] The primary task is to ensure safety and comfort; thus, research on the aerodynamic problems of trains and bridges should be highly prioritized.[12–14]

Ideally, HSTs should be tested under natural atmospheric conditions. However, the costs of performing experiments on a real HST are substantial, and such tests are subject to regulation.[15] Hence, wind tunnel tests are indispensable to modern transportation and civil engineering.[16,17] Today, no transport scheme or long-span bridge designer would consider committing to executing an advanced design without first measuring its aerodynamic properties, stability, and controllability in a wind tunnel.[18,19] However, the particular aerodynamic characteristics of HSTs and surrounding infrastructure result in a higher demand for aerodynamic research through wind tunnel test technologies and methods.

In many previous wind tunnel experiments, the similarity criteria basically meet the engineering research needs; thus, static wind tunnel modelling methods have gained momentum and been used to guide mobility measurements in engineering practice.[20,21] In aerodynamics research, the most familiar force and pressure measurement methods are used to test the aerodynamic characteristics of structures.[22–24] Recently, some scholars have established methods to compare the results obtained from pressure distribution data with the data derived from force measurements, resulting in a coordinated correction approach for the pressure and force tests.[25,26] However, pressure measurements have advantages regarding efficiency and detailed aerodynamic characteristic analyses over force measurements.[27–29]

Although static wind tunnel modelling methods are still in wide use,[30–33] due to the complexity and particularity of the aerodynamic characteristics of HSTs and infrastructure systems, some problems still exist: (1) the ground boundary layer effect needs to be eliminated to ensure that the aerodynamic signals faithfully capture the loading distribution[34]; (2) high-speed relative movement effects include unsteady aerodynamic flows[35]; (3) the correct aerodynamic interference between a train and infrastructure must be determined[36]; (4) the interference induced by the inertial force is hard to be eliminated[26]; (5) long cables in typical systems suffer from electromagnetic interference, are cumbersome, are generally inflexible. A few moving train experiments have been carried out worldwide in recent years,[37–39] and popularity has increased dramatically. The upsurge has been driven by many powerful railway market forces and increasing aerodynamic problems as train speeds increase. Due to the lack of high-quality experimental wind fields and requirement of an experimental method closer to actual "high-speed railway engineering", researchers are increasingly taking advantage of moving train experiments and improved experimental facilities to better study the aerodynamic characteristics of train and bridge systems in wind tunnels.[40–43] Although advancement in relative movement has been implemented in a moving train–bridge system, the moving model experimental technology still needs improvement.

To further address these issues, a new moving train rig was developed in a wind tunnel at Central South University (CSU), and an experimental train was launched on a bridge to acquire the necessary testing conditions. To further evaluate the aerodynamic characteristics of the moving train, an integrated wireless test system — the in-model sensory and wireless data acquisition — remote control and processing system (ISWDA-RCPS) — was developed and implemented; the system incorporates radio frequency (RF) wireless telemetry, in-model embedded electronics, and sensor techniques. The sensor signals are digitized onboard the model and transmitted using microwave telemetry. The data are collected and displayed in real-time on a standard personal computer using software customized for this application. Compared with conventional wind tunnel instruments, the ISWDA-RCPS has outstanding wireless transmission performance, no inertial forces, and a relatively low cost. Moreover, the system provides a larger application base and can fit into smaller high-speed moving objects. The reliability and robustness of this system have been demonstrated through long-term operations in a wind tunnel environment. Both these tests and a comparison of the collected data with those collected by conventional methods verified the good performance of the ISWDA-RCPS.

2. CSU-WT1 Wind Tunnel

The CSU wind tunnel design was carefully prepared and intended to be both an energy-saving and well-testing approach, therefore, a closed-loop was adopted.[44] The wind tunnel was built by railway workers and installed in the National Engineering Laboratory for Construction Technology of High-speed Railways at Central South University. This installation, whose schematic is shown in Fig. 1, is believed to be the only railway-owned wind tunnel in China that experiences regular use.

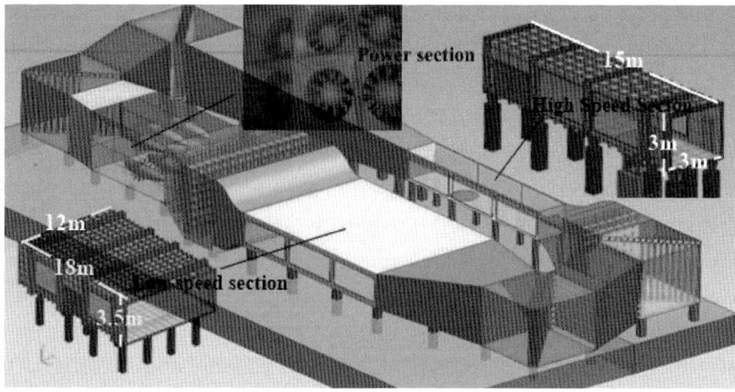

Fig. 1. CSU-WT1 wind tunnel.[45]

The CSU-WT1 wind tunnel consists of two test sections in which the model is kept: a contraction section, a settling section before the test section, and a diffuser after the test section. A fan located after the diffuser creates the wind. The wind tunnel is equipped with six large fans (3 × 2) that push air from the left side (shown in purple). The pushed air passes through turning vanes and then through a contraction zone to streamline the flow in the relatively large test section (12 × 3.5 m). The right side has a square section with dimensions of 3 × 3 m followed by a diffuser zone with varying dimensions. To meet the demand for high-speed railway applications, the area with the highest wind speed (close to 90 m/s in the square section) is called the high-speed section. The high experimental wind speeds and the large cross-test section of the CSU-WT1 wind tunnel are suitable for improving and modifying new experimental technology and facilities because of its diameter, which can be used to test railroad trains efficiently.

3. A Moving-Train and Infrastructure Facility

As the field of railway aerodynamics has matured, the motivation for and focus on improving the realism of experimental investigations has increased. No unified understanding exists because rail traffic and infrastructure vary across different environments. A moving-train and infrastructure facility has been established to address open-air pressure pulses; aerodynamic loads on trains, bridges, tunnels, and tracks (ballast pick-up); crosswind effects; passing train events (transient aerodynamic effects); train-infrastructure interactions (loads on trains and pressure wave phenomena); and slipstream effects. This experimental system can correct the relative motion between a train and the surface at high speeds and any infrastructure or ground configuration. A schematic diagram of a moving train rig in the CSU-WT1 wind tunnel is shown in Fig. 2. The moving train and infrastructure facility is intended to help determine and understand the abovementioned aerodynamic problems. It benefits from higher train speeds than most current

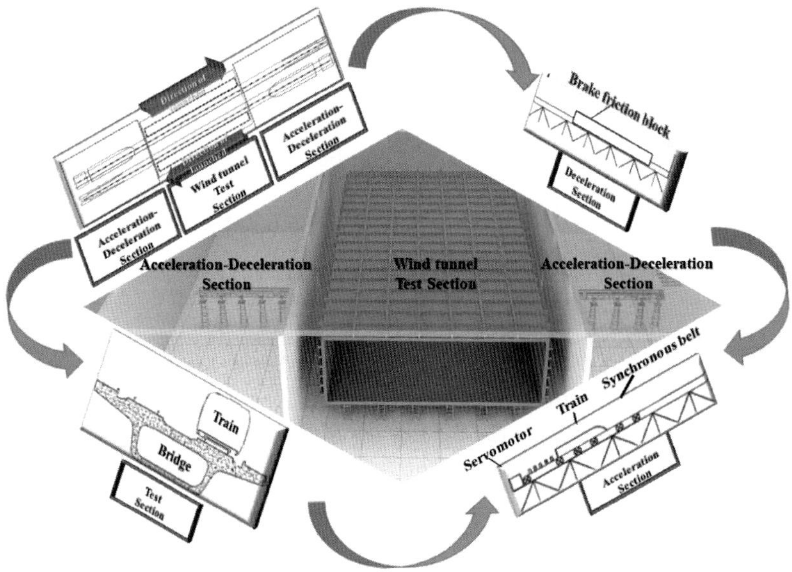

Fig. 2. Schematic diagram of the moving-train and infrastructure facilities in the CSU-WT1 wind tunnel.

moving-model systems, its extra-wide wind tunnel cross-section, and a flexible approach for changing trains and infrastructure models.

The main components include the wind tunnel train testing model, a high-strength synchronous conveyor belt, a hysteresis mechanism, a test guide rail, a wind tunnel test section and measurement and control systems. The high-speed motor and high-strength synchronous conveyor belt are efficiently coordinated and easily controlled. This configuration allows for the effective control of train model speeds. The high-speed motor and high-strength synchronous conveyor rotate together, and the quality of the train model can determine the initial speed of this model. Furthermore, the initial acceleration distance of the train model required by the system is relatively short, which is conducive to the reuse and development of the currently constructed wind tunnel laboratory.

The facility can generally achieve high-speed train tests within favorable measurement times. The test section covers the entire cross-section of the wind tunnel, and the range of the infrastructure length can be consistent with the test section to maximize the use of the section width of the wind tunnel and realize long-distance measurements. Moreover, the terrain advantage of the wind tunnel also allows tests to be conducted in the low-speed section of the wind tunnel. The test model is arranged vertically in accordance with the incoming flow direction. To enable the best use of the width of the wind tunnel test section, the acceleration and deceleration sections of the high-speed moving train device are outside the wind tunnel.

The motion system uses inertia as the high-speed driving power of the model, the scale ratio of the model is 1:8–1:30, the mass of the model is 10–30 kg, and the model is driven from accelerating and sliding to testing to decelerating to stopping on a 34 m long track. The synchronous conveyor belt is used as a power source; the vehicle is fixed on the belt, and the conveyor belt can be accelerated to simulate movement within a very short period. The conveyor belt runs to the end of the acceleration section; then, the model vehicle leaves the conveyor belt and is launched forward along the track at a high start-up speed. Moreover, to ensure the vehicle's stability in the wind tunnel, the vehicle chassis is equipped with a rail-holding device. Furthermore, adding a wind tunnel perpendicular to the path of the moving train model allows crosswinds to be generated. The crosswinds simulated by such an experimental setup benefit from an accurate incoming yaw profile and can critically simulate the transient natures of crosswinds that HSTs experience during real-world operations. Thus, there is clear motivation and potential to improve the measurement capabilities of moving-model experiments, including pressure and force measurements, the interaction effects between trains and infrastructures (mainly bridges), and the simplicity with which realistic exposure to crosswinds can be modelled.

4. The ISWDA-RCPS

4.1. *Overview*

Figure 3 depicts the overall setup of the system and the general locations of the system components in relation to the model. To conduct systematic tests on the

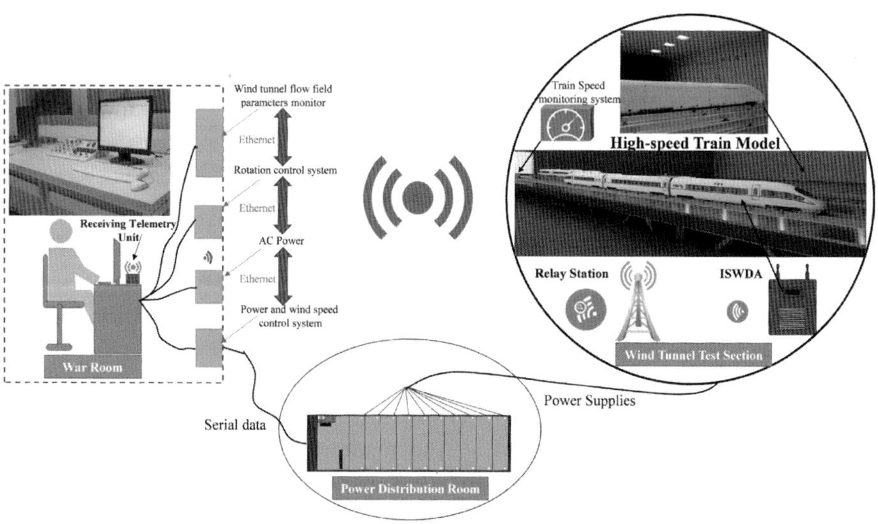

Fig. 3. System setup for determining the aerodynamic characteristics of a moving train in the CSU wind tunnel.

aerodynamic characteristics of an HST and measure the aerodynamic interference effects between trains and infrastructures, a novel embedded electronics package (ISWDA) was designed and fabricated. As shown in Fig. 3, the test model runs continuously at high speed after the system is armed, and ISWDA is triggered simultaneously. Once ISWDA is triggered, data obtained from the model are transmitted through ISWDA to a relay station in the wind tunnel test section. Additionally, the digitized data are sent serially from the wind tunnel test section to a telemetry receiving unit in the war room (a remote location on the other side of the CSU-WT1 wind tunnel used for an expanded system setup, resulting in a complete ISWDA-RCPS experimental analysis system).

The ISWDA-RCPS provides substantial testing methods, acquisition capability, and working pattern improvements compared to traditional wind tunnel test equipment. By integrating wireless transmission into the ISWDA-RCPS, the restrictions imposed by traditional wired technologies are eliminated, and the system's efficiency is enhanced. On the one hand, real-time aerodynamic data can be obtained via wireless technology without spatial limitations, and on the other hand, the received data can be effectively processed and managed through the system. The acquisition devices automatically achieve signal modulation and collection using built-in microprocessors and customized software.

The ISWDA-RCPS (Fig. 3) consists of three main components: an in-model scanning system, a data retransfer system and a control and management system. The front end of the test cycle is the in-model scanning system. A complete in-model scanning system requires capturing aerodynamic data from the moving train model and the target signals. The data retransfer system controls data transmission between the in-model scanning and control and management systems. The control and management system consists of one or more PC-compatible computers running dedicated software for data collection and instruction issuance.

A customized wireless communication network is used in the system to enable leapfrog communications. Each device has an independent address; the data retransfer system has the highest-level address, while the in-model scanning system and the control and management system have different address levels according to their positions in the complete network. Thus, to avoid command conflicts in the network, each device responds only to commands from its superior device after initialization. The in-model scanning system can also return data directly to the control and management system; thus, it can respond to commands from superior and subordinate devices.

In-model scanning system. Each in-model scanning system is self-contained and includes a series of sensors, a transceiver, a sampling circuit board, and a battery. The system is designed to accommodate different types of sensors. In this study, the in-model scanning system includes only wind pressure transducers; however, the system could easily be modified to include acceleration, displacement, force, or strain sensors. In general, sensors that generate a maximum of 16-bit digital or analog signals are compatible with the current hardware after proper signal conditioning.

Data retransfer system. The data retransfer system receives information from the in-model scanning system and sends it to the control and management system. However, the data retransfer system forwards sensor data and transmits commands from the control and management system to the in-model scanning system. Based on individual test requirements, the data retransfer system can be installed inside or outside the wind tunnel. A mounted data retransfer system working at 5 GHz controls the bidirectional communications between the control and management system and the data retransfer system.

Control and management system. Because large amounts of data must be collected and the train model needs to be triggered, the most obvious choice for control and management is an industrial personal computer (IPC), either a desktop or a laptop depending on the need for portability. During the commissioning stage for each case, tablet or advanced handheld smart terminals can be used for smaller amounts of data, and real-time pneumatic signals can be displayed on these devices for experimental convenience. The current configuration consists of a Lenovo IPC with an Intel i7-4790 processor; this can be considered a moderate processing speed requirement because the computer also must control the moving train rig. Slower devices could be used, but they would reduce the system's data transfer rate. The specific data collection software is written in C++ for the MS Windows operating system.

4.2. *Characteristics and composition of the ISWDA-RCPS*

The ISWDA-RCPS mainly consists of up to seven separate units distributed around the wind tunnel test section; these units communicate with the war room via a wireless network. An architecture summarizing the role of each component is presented in Fig. 4. These units comprise two main aspects: the ISWDA mechanism in the test model and the outside RCPS. Four unique components are contained in the ISWDA module: a wind pressure sensor system, a data acquisition system, a transceiver, and a battery pack. Each unit is in a metal box and embedded in the test model. The data path flows from the wind pressure sensor system to the data acquisition system and then to the transceiver. At this point, the data are received in the wind tunnel test section and sent to a higher-level supervisory computer with customized software. Internally, the test model design reserves space to allow several independent systems to be housed in the train and tested during wind tunnel operation. The test model was produced using 3D printing technology and possesses a composite outer covering, which allows data to be transmitted through the train with

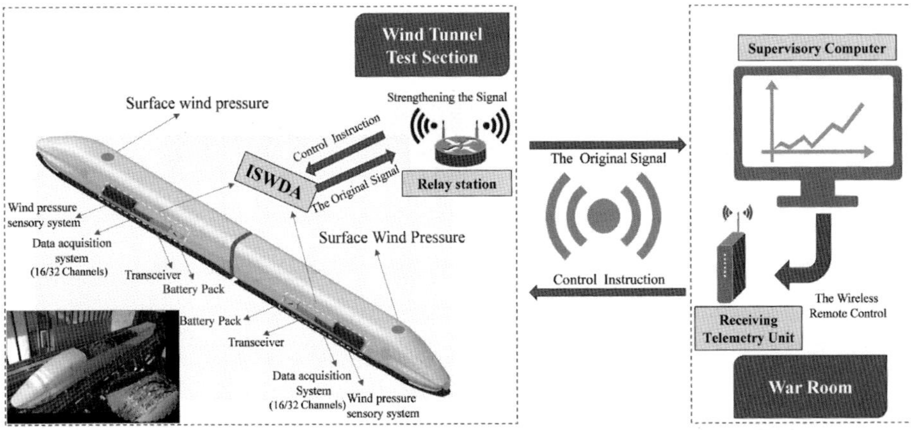

Fig. 4. The topology of the ISWDA-RCPS.

no external antenna and enables the ideal situation with no cables or wires external to the model.

The wind pressure sensory system (WPSS) is the most fundamental and vital element in the overall system. The proposed WPSS consists of numerous pressure sensors distributed at points of interest on the surface of an HST. The WPSS sensors selected for this test are commercially available, are small, and have relatively low costs. An embedded integration of sensor systems that meet the imposed sensor resolution and physical constraint requirements was designed and fabricated in-house. The WPSS consists of the basic components described above, several pressure transducers with signal conditioning circuits, and several pressure sensing ports, all enclosed in an aluminum metallic casing. Both the 16- and 32-channel modules of the WPSS can fit into the interior space of the train. To test the stability of the sensors without considering the effects of temperature variations, tests were performed inside the empty wind tunnel chamber at a near-constant temperature $T = 25°\text{C}$. The pressure remained stable throughout the experiment. No filtering was used, and the signal was obtained at a sampling rate of 2 000 Hz (Fig. 5).

The data acquisition system (DAS) is an in-model unit customized for HST wind tunnel test applications. This unit significantly improves the signal-to-noise ratio because the cables linked to the WPSS are very short (inches compared to tens or hundreds of feet). In a typical system, expensive instrumentation amplifiers and special-purpose cables offset the electromagnetic interference imposed on long cables. Examples of the internal circuit boards utilized for these DAS modules are shown in Fig. 6. The data acquisition boards include a signal conditioning circuit to increase the output resolution of the dynamic pressure measurements. The 16- and 32-channel DAS units are identical except for the number of channels. As depicted, various voltage signals can be connected to the in-model DAS. It

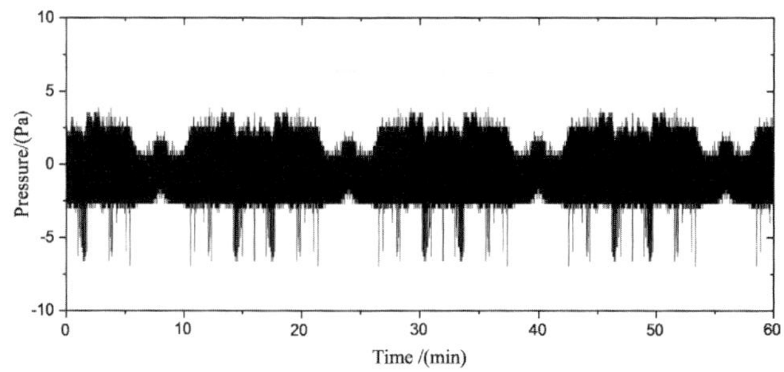

Fig. 5. Pressure data were measured in the empty wind tunnel chamber experiment at a constant temperature.

Fig. 6. A DAS internal circuit board.

is possible because of each unit's ability to route multiplexed analog signals to a programmable gain amplifier, which allows the DAS to change ranges and optimize the output resolution. The collected analog signals are sent to a 16-bianalog-to-digitalal (A/D) converter, where the signals are digitized and transmitted or stored within the logging unit in solid-state memory.

The transceiver is an onboard microcontroller that creates a half-duplex communication link between the DAS and the upper computer. This transceiver is capable of transmitting information at up to 54 Mbps over a distance of approximately 50 m. Using the same microprocessor and clock frequency in the relay stations and telemetry receiving units makes it easier to achieve good synchronization, resulting in higher transmission speeds. The transceiver communicates with the acquisition system using a standard serial transmission protocol.

The built-in battery is a 12-V Li-ion battery with a capacity of 2.2 Ah. This battery weighs 850 g and is one of the heaviest components of the ISWDA mechanism because it is the electricity source that provides power for all the model components. When fully charged, the battery allows the ISWDA module to run continuously for more than 18 h.

The units described above constitute the in-model subsystem (ISWDA), which forms the front-end target portion of the ISWDA-RCPS for acquisition and information transmission. The ISWDA mechanism is the core of the ISWDA-RCPS; it equips wind tunnel tests with characteristic aerodynamic data and supports communications between the test section and the war room. The in-model subsystem also prepares for future HST model performance studies by measuring other indicators, which is possible due to the flexible way in which the components are wired; this subsystem will be used in future studies to estimate the safety and stability of trains and to provide deep-level estimations of the dynamic responses of trains and infrastructure. It will include four main research systems: safety, noise, vibration, and harshness (NVH) systems, as well as an aerodynamic property evaluation module. Such studies focus on evaluating loads and responses under different crosswind environments. Benefiting from the customizability and flexible assembly of the WPSS and the moving train model rig, a range of safety and comfort features associated with HSTs under strong winds, such as wheels and rails, and the vibration, acceleration, and displacement of trains and bridges will be studied in this laboratory in the future. Furthermore, the equipment can be combined with hybrid simulation technology to obtain more specific visual details and used to analyze mechanisms and evaluate the relationship between the HST and infrastructure under crosswinds. Some of the plans for developing this system include further reducing the size of the embedded electronics, improving the sensor accuracy, increasing the telemetry data transfer rate, increasing the number of data channels, and improving the battery power to enable completely wireless applications, among other upgrades.

The following section specifies the units outside the model that (together with the units described above) form the complete ISWDA-RCPS.

The relay station is a data switching center that receives information from the in-model units and sends it to the telemetry receiving units. The relay station is utilized in application evaluations under extreme circumstances to provide timely and effective information. The relay station acts as a data receiver and a transmitter that can transmit commands from the telemetry receiving units to the in-model transceiver. As mentioned above, the in-model transceiver can send data directly to the telemetry receiving units. However, the transceiver might be limited due to special test scenarios, and unexpected results may be encountered during the test, such as occasional RF signal dropout. This dropout occurs due to the multipath environment in the test section and the needs to transmit RF signals through not only the outer skin of the model but also the wind tunnel walls on the floor and both sides of the test section. The RF dropout was proven to be a function of the model attitude, and the signal loss was limited to a small range of pitches. Due to range and reliability considerations, the relay station must respond quickly to inputs while simultaneously parsing the telemetry receiving unit and the upper computer responses rapidly and continuously.

The operation of the entire system is controlled by the computer's data collection software and the firmware in the relay station and in-model units. The system works on a unidirectional link between the upper computer and the ISWDA module because of the half-duplex characteristics of the utilized transceivers. A typical transmission sequence, starting with the ISWDA mechanism being ready to transmit and the upper computer being ready to receive, is shown in Fig. 4.

4.3. *Implementation and validation of the ISWDA-RCPS*

As mentioned before, the current application of the system involves pressure transducers as the main sensors. The system digitizes the analogue signals from the pressure transducers through the A/D converter. Therefore, the data collected by the system consist of voltage signals that must be converted to pressure values using a calibration curve. Calibration curves were constructed for the WPSS currently in use.

To test the stability and accuracy of the ISWDA-RCPS, extensive laboratory experiments were carried out. The experiments were performed using two different facilities at CSU: a signal generator and a low-speed, blowdown-type wind tunnel with an open test section. In both approaches, measurements could be taken directly at the same measurement point, or the pressure signals could be transferred through tubes toward the pressure transducers, where the pressure is converted into an electrical signal. In these processes, a 'scannivalve'-type pressure scanner was selected for a pressure comparison. Two measurement systems were allocated to each point, and measurements were carried out simultaneously at all the points.

The in-house-designed and manufactured device with a loudspeaker is shown in Fig. 7(a). The loudspeaker was covered with a transparent flat polyvinyl chloride (PVC) plate and placed in an airtight space to eliminate air leakage or pressure when the sound signal was created. The loudspeaker produced sounds known in advance (i.e. before performing the experiments). These sounds included sine curves with various characteristics and various noise signals. The results of the experiment are shown in Fig. 7(b). The pressures of the two systems remained stable throughout the experiment. No filtering process was used, and the signal was obtained at a sampling rate of 625 Hz. The ISWDA-RCPS data tracked nicely with the pressure scanner data, yielding less than a 10% extreme error for all channels (the ISWDA-RCPS

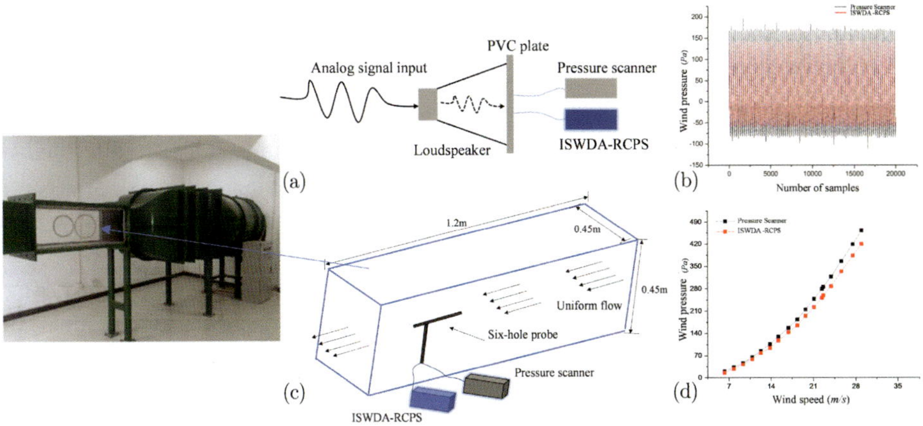

Fig. 7. Pressure data compared to the conventional wind tunnel measurement equipment results: (a) the experimental setup for generating pressure signals using a loudspeaker; (b) the time history of the wind pressure; (c) the experimental setup for generating pressure signals using a wind tunnel; (d) the mean wind pressures measured by the probe at different wind speeds.

data are shown as a solid red line), which is the result of the different apertures used by the ISWDA-RCPS and the pressure scanner.

Figure 7(c) shows a schematic view of the CSU-2 wind tunnel at CSU. This low-speed blowdown-type wind tunnel is operable in open-circuit mode with a blown-air configuration. The test section is 0.45 m high, 0.45 m wide, and 1.2 m long. A honeycomb, screens and a nozzle enhance the airflow's uniformity and reduce the turbulence intensity at the test section inlet. The turbulence intensity at the inlet cross-section is less than 0.5%. A six-hole probe was placed in the center of the wind tunnel section and connected to two systems, the ISWDA-RCPS and the pressure scanner. The goal was to acquire data for every test condition and compare the results of the data collected by the ISWDA-RCPS to those collected by the corresponding "conventional device." During this test, the wind speed varied from 6.2 to 31 m/s. The configuration of the data collection system for this test was similar to that in the loudspeaker test. Again, the data recordings were synchronized, and the data were displayed in real-time in both raw and engineering units. Figure 7(d) depicts the pressure data acquired under each wind speed condition in the wind tunnel for the same six-hole probe. These data represented one orifice at the top of the probe head and were compared to one co-positioned port on the opposite side. The figure shows that these pressure data tracked nicely with the pressure scanner measurements. A closer inspection shows that the mean wind pressure had larger bias errors at higher wind speeds caused by different apertures and air outlet pipe diameters. More discussions on these measurements will be presented in subsequent publications.

5. Aerodynamic Characteristics of an HST Running on a Bridge

5.1. *Train–bridge system*

As mentioned above, the aerodynamic characteristics of an HST crossing a bridge were experimentally determined at the CSU-1 wind tunnel. This unique facility is dedicated to identifying, understanding, and solving high-speed railway aerodynamics problems. The train model prototype was a Fuxing Chinese HST with a locomotive (leading vehicle) 27.15 m long, 3.36 m wide, and 3.81 m high. Its geometry scale ratio was set to 1:16.8. The dimensions of the cars were 1 616 mm long, 227 mm high, and 200 mm wide, as shown in Fig. 8. The selected bridge model type was a supported box girder — the most widely used type for high-speed railway lines. In the wind tunnel test, the model dimensions also had a scale ratio of 1:16.8, and its length, width, and height were 12 m, 1.408 m, and 0.216 m, respectively. In this experiment, which "piggybacked" on the adopted scale ratio with a particular model, several independent systems could be housed in the train and tested during this wind tunnel test procedure.

5.2. *Signal triggering*

A stand-alone data system was developed to record and display real-time data. This system was remotely triggerable by the control and management system to

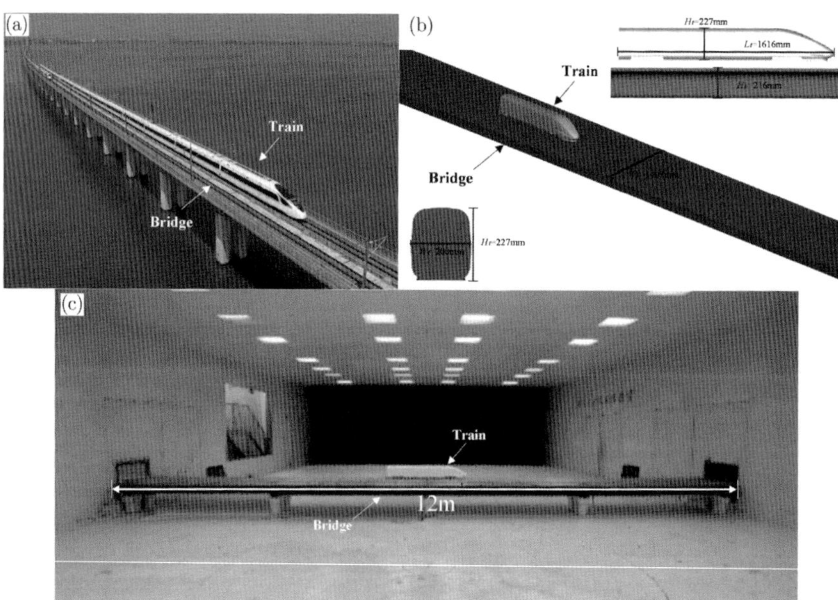

Fig. 8. Train–bridge system: (a) prototype of a train and bridge, (b) scale model of the train and bridge deck, and (c) train and bridge in the wind tunnel test section.[43]

ensure that each dataset was synchronized to the moving train rig data for post-test analysis.

To ensure that sudden and fast events not triggered directly by the experimentalist could be recorded in full, the system was equipped with an automatic triggering system within the control interface. The control display interface features three different modes for triggering data acquisition. In both modes, the sampling rate can be adjusted independently for each pair of channels. The three triggering modes are as follows:

(1) Manually triggered from the user interface for one channel. This mode allows for manual triggering at different times and sampling rates on the 32 pairs of channels.
(2) Manually triggered from the user interface of one channel; this action also automatically triggers all the remaining channels simultaneously.
(3) Automatic triggering of all 16 and 32 channels using an electrical input.

5.3. *Repeatability of the ISWDA-RCPS*

Because the ISWDA-RCPS adopts independently developed wireless wind pressure test technology, it was necessary to conduct repeated tests to ensure the stability and effectiveness of the test results and to perform an accurate analysis. After static and dynamic calibration inside and outside the laboratory, the test equipment was evaluated in a wind tunnel test for HSTs. These repeatability tests needed to be conducted at high speeds.

Figure 9 shows a cross-section of the vehicle–bridge system and the layout of the measuring points (T1–T16) on the high-speed moving train model. The mean and standard deviation pressure values were considered to represent accuracy and repeatability, respectively:

$$\bar{x} = \frac{1}{2}\sum_{i-1}^{n} x_i, \tag{1}$$

$$\sigma = \sqrt{\frac{\sum_{i=1}^{n}(x_i - \bar{x})^2}{n-1}}, \tag{2}$$

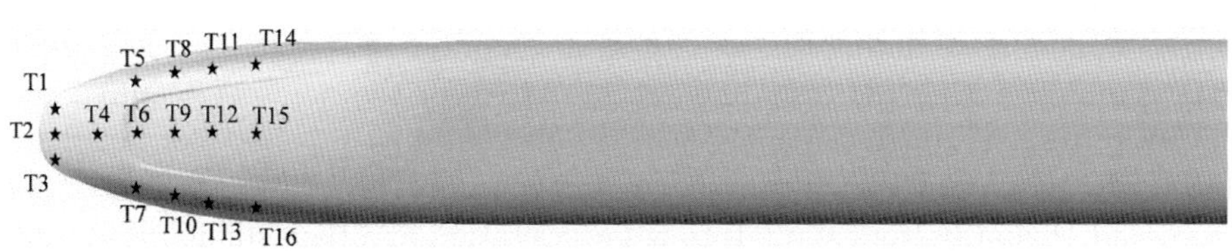

Fig. 9. The layout of the measuring points.

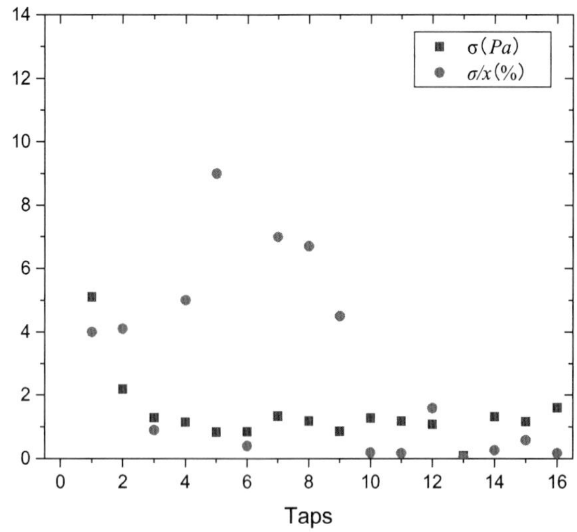

Fig. 10. Relative standard error.

where $x_1, x_2, \ldots, x_i$ are the data obtained during a certain run and n represents a set of measurement data acquired from each measuring point.

Figure 10 shows the uncertainties derived from 15 repeatability tests associated with these measurements based on the standard error and mean statistics (σ is the standard deviation; $\bar{x}$ is the arithmetic mean) for each point measured during a moving train model test, in which 16 pressure taps were installed on the train.

The ratio of the standard error to the average value never exceeded 10%, indicating that the repeatability of the test results was excellent, the test device had a good and stable working state, and the obtained model surface wind pressure had good accuracy.

5.4. *Experimental results*

To test the aerodynamic characteristics of the train during high-speed movements under a crosswind, the main sections of the train model were selected for testing. See Fig. 11(a) for the section measuring points 1–15 distribution. The pressure values at different taps were obtained by comparing the pressures to those measured by sensors on the train surface. Figure 11(b) shows the wind pressure-time history data on the train surface during the test; these are representative plots of the pressure data acquired during the tests and are indicative of the datasets acquired from the entire test. Through the analysis of the results of each period, we found that during the process of vehicle movement, the wind pressure on the vehicle surface proceeded obviously from static through acceleration and then stabilized until the braking deceleration caused it to reach a stable static value. This approach allowed for the extraction and analysis of the steady wind pressure period during the moving

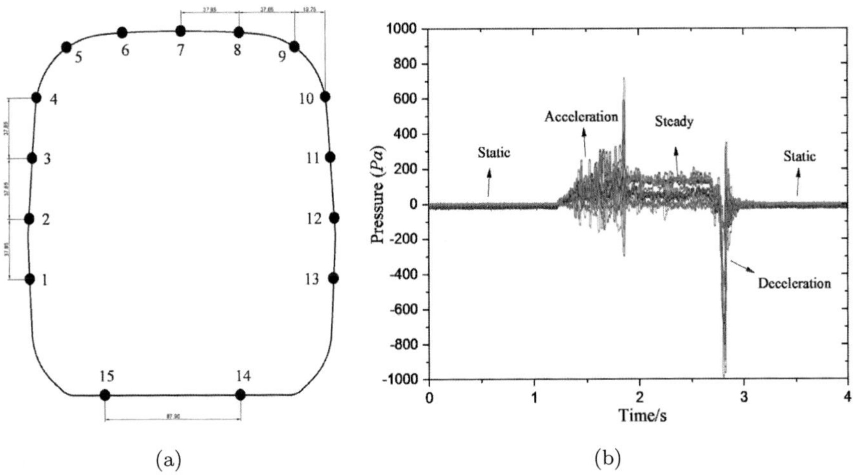

Fig. 11. (a) Measurement points on the center cross-section of the train and (b) wind pressure on the train model.

process, enabling us to obtain the wind pressure distribution on the train surface. The changes in the pressure curve over time reveal the aerodynamic characteristics of the train inside the wind tunnel test section, which were highly influenced by train movement. However, the ISWDA-RCPS can meet the imposed HST surface wind pressure test requirements.

To study the influence of a crosswind on the surface pressure of the train, we selected a certain section of the train model surface for measuring the point layout. During the test, the crosswind speed was controlled by the wind tunnel. The layout of the 15 measuring points is shown in Fig. 11(a). The train surface pressure was usually expressed as a nondimensional pressure coefficient during this test.

To study the influence of the aerodynamic characteristics of a vehicle under different wind conditions, aerodynamic characteristics tests were carried out under three different conditions: in case 1, the wind speed was 0 m s^{-1} with a train speed of $32 \text{ m} \cdot \text{s}^{-1}$, i.e. the wind tunnel did not open; in case 2, the wind speed was $6 \text{ m} \cdot \text{s}^{-1}$ with a train speed of $32 \text{ m} \cdot \text{s}^{-1}$; in case 3, the wind speed was $6 \text{ m} \cdot \text{s}^{-1}$, and the train was stationary on the bridge, i.e. the train speed was $0 \text{ m} \cdot \text{s}^{-1}$. The distribution of the average wind pressure coefficient on the train's surface is shown in Fig. 10. To further analyze the influence of crosswinds on vehicles, two types of speeds were used as reference dimensions for a comparative analysis according to different conditions. As shown in Fig. 12(a), the absolute value of the train surface wind pressure affected by a crosswind with a speed of $6 \text{ m} \cdot \text{s}^{-1}$ was greater than that under no wind, indicating that crosswinds significantly impact the train surface wind pressure. As shown in Fig. 12(b), when the crosswind speed was used as the reference dimension to compare the impact of vehicle movement,

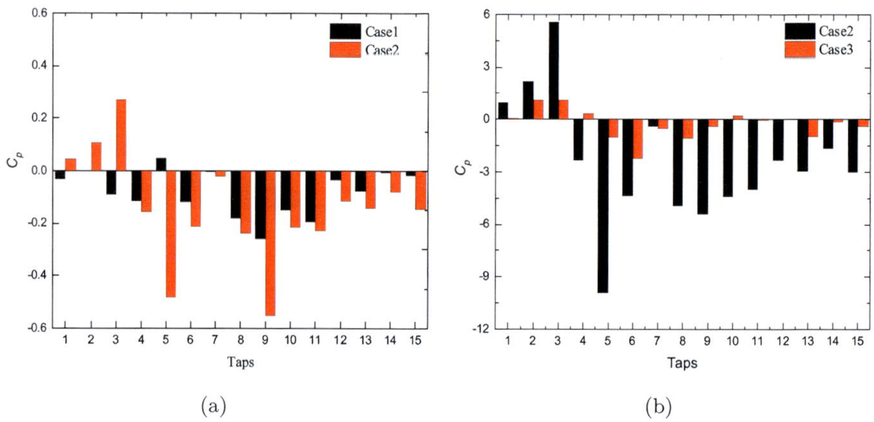

Fig. 12. Wind pressure distribution on the train: (a) the effect of crosswinds alone and (b) the effect of crosswinds when moving.

the effect on a vehicle moving at high speed was much greater than on a stationary vehicle. Similar change rules were captured in other wind tunnel tests. Overall, the tests showed that a static model could not be used to simulate the aerodynamic characteristics of a train at high speeds, and it is difficult to simulate the wind shear, turbulence, and aerodynamic interference between structures under actual operating conditions.

6. Conclusion

Understanding the aerodynamic characteristics of an HST in different environments can help to directly evaluate the drag reduction, safety, and stability of the train, and pressure distribution measurement is essential. Many methods for measuring pressure distributions have been proposed in previous studies, but no method has been found that is portable, inexpensive, and can be used to measure pressure distributions during train model operations on infrastructure. However, all these measurements can be obtained by placing the ISWDA-RCPS in the train model. Our experiments showed that the ISWDA-RCPS could reliably measure pressure measurements in a novel moving train rig and that the tested aerodynamic characteristics vary under different train conditions. The system can be used to identify these aspects, allowing the aerodynamic characteristics of the HST induced by various infrastructures and environments to be assessed.

Acknowledgments

The study was funded by the National Natural Science Foundations of China (Grant No. 51925808, Grant No. U1934209) and the Tencent Foundation (Xplorer Prize

2021), Key Project, (Grant No. 2021-Key-04-2) and China Postdoctoral Science Foundation (Grants No. 2022M713516).

References

1. X. H. He, T. Wu, Y. F. Zou, Y. F. Chen, H. Guo and Z. W. Yu, Recent developments of high-speed railway bridges in China, *Struct. Infrastruct. Eng.* **13**(12) (2017) 1584–1595.
2. B. Gao, G. W. Li and Z. H. Zhen, *High-Speed Rail: An analysis of the Chinese Innovation System* (World Scientific Publishing Company, Singapore, 2021).
3. N. Akcay, Turkey-China relations within the concept of the New Silk Road Project, *Bölgesel Araştırmalar Dergisi* **1**(3) (2017) 73–96.
4. F. Xu, *The Belt and Road: The Global Strategy of China High-Speed Railway* (Springer, New York, 2018).
5. Q. Gu, B. Yuan, O. Ozcelik, R. J. Chang, W. Guo, H. M. Zhou, J. Y. Lin, T. Wang and J. G. Zhang, A novel moving load integration method for real-time hybrid shaking table test of high-speed maglev vehicle-bridge interaction system, *Int. J. Struct. Stab. Dyn.* **22**(10) (2022) 2241013.
6. W. M. Zhai, P. F. Liu, J. H. Lin and K. Y. Wang, Experimental investigation on vibration behaviour of a CRH train at speed of 350 km/h, *Int. J. Rail Transp.* **3**(1) (2015) 1–16.
7. G. X. Deng, Y. Peng, C. G. Yan and B. Wen, Running safety evaluation of a 350 km/h high-speed freight train negotiating a curve based on the arbitrary Larangian-Eulerian method, *Proc. Inst. Mech. Eng. F J. Rail Rapid Transit.* **235**(9) (2021) 1143–1157.
8. H. Q. Tian, Review of research on high-speed railway aerodynamics in China, *Transp. Safety Environ.* **1**(1) (2019) 1–21.
9. D. S. Nino and A. Luongo, Nonlinear aeroelastic in-plane behavior of suspension bridges under steady wind flow, *Appl. Sci.* **10**(5) (2020) 1689.
10. M. Zhang, F. Xu and H. Yu, A simplified model to evaluate peak amplitude for vertical vortex-induced vibration of bridge decks, *Int. J. Mech. Sci.* **192** (2021) 106145.
11. M. Hosomi, H. Kobayashi and Y. Nitta, Fatigue strength design for vortex-induced oscillation and buffeting of a bridge, *J. Wind Eng. Ind. Aerodyn.* **67–68** (1997) 227–237.
12. W. Tao, P. Lou, Y. F. Zou, C. Z. Cai, X. H. He and S. Jiang, Effects of curved wind barrier on the aerodynamic characteristics of a train–bridge system and its static wind load, *Int. J. Struct. Stab. Dyn.* **22**(10) (2022) 2241006.
13. Z. Y. Yao, N. Zhang, M. C. Zhu and X. D. Li, Dynamic interaction analysis and running safety assessment of the wind-train–bridge system considering the moving train's aerodynamic coupling with crosswind, *Int. J. Struct. Stab. Dyn.* **23**(15) (2023) 2350173.
14. J. C. Wen, Q. Li, L. Zhao, Q. Huang and Z. G. Lu, Aerodynamic characteristics of express freight train on bridges based on wind tunnel tests, *Int. J. Struct. Stab. Dyn.* **22**(10) (2022) 2241005.
15. L. Pii, E. Vanoli, F. Polidoro, S. Gautier and A. Tabbal, A full scale simulation of a high speed train for slipstream prediction, in *Proc. Transport Research Arena*, 14–17 April 2014, Paris, France.
16. J. E. Cermak, Wind-tunnel development and trends in applications to civil engineering, *J. Wind Eng. Ind. Aerodyn.* **91**(3) (2003) 355–370.

17. V. T. Cang, Transportation engineering technology: A review, *Int. J. Mech. Eng. Appl.* **1**(3) (2015) 76–79.
18. A. Kareem and Y. Tamura, *Advanced Structural Wind Engineering* (Springer, Japan, 2013).
19. R. Zhou, Y. J. Ge, Y. X. Yang, Y. L. Du and L. H. Zhang, Aerodynamic performance evaluation of different cable-stayed bridges with composite decks, *Steel Compos. Struct.* **34**(5) (2020) 699–713.
20. R. H. Scanlan and J. J. Tomko, Airfoil and bridge deck flutter derivatives, *J. Eng. Mech. Div.* **976** (1971) 1717–1737.
21. G. Diana, D. Rocchi, T. Argentini and S. Muggiasca, Aerodynamic instability of a bridge deck section model: Linear and nonlinear approach to force modelling, *J. Wind Eng. Ind. Aerodyn.* **98** (2010) 363–374.
22. E. Hjorth-Hansen, A. Jakobsen and E. Strømmen, Wind buffeting of a rectangular box girder bridge, *J. Wind Eng. Ind. Aerodyn.* **42**(1) (1992) 1215–1226
23. A. L. Smith, D. J. Mee, W. J. T. Daniel and T. Shimoda, Design, modelling and analysis of a six component force balance for hypervelocity wind tunnel testing, *Comput. Struct.* **79**(11) (2001) 1077–1088.
24. Q. S. Yang, D. Y. Li, Y. Hui and S. S. Law, Estimation of extreme wind pressure coefficient in a zone by multivariate extreme value theory, *Wind Struct.* **31**(3) (2010) 197–207.
25. X. Chen and A. Kareem, Validity of wind load distribution based on high frequency force balance measurements, *J. Struct. Eng.* **131**(6) (2005) 984–987.
26. L. Yan, L. D. Zhu and R. G. J. Flay, Comparison of force-balance and pressure measurements on deck strips on a stationary bridge model, *J. Wind Eng. Ind. Aerodyn.* **164** (2017) 96–107.
27. M. Matsumoto, H. Shirato, K. Araki, T. Haramura and T. Hashimoto, Spanwise coherence characteristics of surface pressure field on 2-D bluff bodies, *J. Wind Eng. Ind. Aerodyn.* **91**(1–2) (2003) 155–163.
28. F. Cheli, R. Corradi, D. Rocchi, G. Tomasini and E. Maestrini, Wind tunnel tests on train scale models to investigate the effect of infrastructure scenario, *J. Wind Eng. Ind. Aerodyn.* **98**(6–7) (2010) 353–362.
29. F. Cheli, F. Ripamonti, D. Rocchi and G. Tomasini, Aerodynamic behaviour investigation of the new EMUV250 train to cross wind, *J. Wind Eng. Ind. Aerodyn.* **98**(4–5) (2010) 189–201.
30. K. H. Yu, X. H. He, C. Z. Cai, L. Yan and Y. F. Zou, Aerodynamic characteristics of trains on a viaduct with non-uniform cross-section under crosswinds by wind tunnel tests, *Adv. Struct. Eng.* **24**(9) (2021) 1769–1781.
31. L. Li, T. H. Liu, Z. J. Guo, W. H. Li and Y. T. Xia, On the effect of rail-end slope in train aerodynamics under crosswind, *Veh. Syst. Dyn.* **60**(6) (2022) 1888–1908.
32. M. Suzuki and Y. Hibino, Field tests and wind tunnel tests on aerodynamic characteristics of train/vehicles under crosswinds, *Quart. Rep. RTRI* **57** (2016) 55–60.
33. A. Orellano and M. Schober, Aerodynamic performance of a typical high-speed train, in *Proc. 4th WSEAS Int. Conf. Fluid Mechanics and Aerodynamics*, Elounda, Greece, 21–23 August 2006, pp. 18–25.
34. H. Kwon, Y. Park, D. Lee and M. Kim, Wind tunnel experiments on Korean highspeed trains using various ground simulation techniques, *J. Wind Eng. Ind. Aerodyn.* **89** (2001) 1179–1195.

35. S. M. Zou, X. H. He and H. F. Wang, Numerical investigation on the crosswind effects on a train running on a bridge, *Eng. Appl. Comput. Fluid Mech.* **14**(1) (2020) 1458–1471.
36. M. Wang, X. Z. Li, J. Xiao, Q. Y. Zou and H. Q. Sha, An experimental analysis of the aerodynamic characteristics of a high-speed train on a bridge under crosswinds, *J. Wind Eng. Ind. Aerodyn.* **177** (2018) 92–100.
37. C. J. Baker, Train aerodynamic forces and moments from moving model experiments, *J. Wind Eng. Ind. Aerod.* **24** (1986) 227–251.
38. Q. Yang, J. Song and G. Yang, A moving model rig with a scale ratio of 1/8 for high speed train aerodynamics, *J. Wind Eng. Ind. Aerod.* **152** (2016) 50–58.
39. F. Dorigatti, M. Sterling, C. J. Baker and A. D. Quinn, Crosswind effects on the stability of a model passenger train-A comparison of static and moving experiments, *J. Wind Eng. Ind. Aerod.* **138** (2015) 36–51.
40. M. Bocciolone, F. Cheli, R. Corradi, S. Muggiasca and G. Tomasini, Crosswind action on rail vehicles: Wind tunnel experimental analyses, *J. Wind Eng. Ind. Aerod.* **96** (2008) 584–610.
41. Y. Li, P. Hu, Y. L. Xu, M. Zhang and H. Liao, Wind loads on a moving vehicle-bridge deck system by wind-tunnel model test, *Wind Struct.* **19** (2014) 145–167.
42. H. Xiang, Y. Li, S. Chen and C. Li, A wind tunnel test method on aerodynamic characteristics of moving vehicles under crosswinds, *J. Wind Eng. Ind. Aerod.* **163** (2017) 15–23.
43. X. Z. Li, M. Wang, J. Xiao, Q. Y. Zou and D. J. Liu, Experimental study on aerodynamic characteristics of high-speed train on a truss bridge: A moving model test, *J. Wind Eng. Ind. Aerod.* **179** (2017) 26–38.
44. X. H. He, Y. F. Zou, H. F. Wang, Y. Han and K. Shi, Aerodynamic characteristics of a trailing rail vehicles on viaduct based on still wind tunnel experiments, *J. Wind Eng. Ind. Aerod.* **135** (2014) 22–33.
45. X. H. He and S. M. Zou, Advances in wind tunnel experimental investigations of train–bridge systems, *Tunn. Undergr. Space Technol.* **118** (2021) 104157.
46. L. X. Huang and Y. Fang, Convergence analysis of wireless remote iterative learning control systems with dropout compensation, *Math. Probl. Eng.* **2013** (2013) 609284.
47. C. Peng, Y. Song, X. P. Xie, M. Zhao and M. R. F., Event-triggered output tracking control for wireless networked control systems with communication delays and data dropouts, *IET Control. Theory Appl.* **10**(17) (2016) 2195–2203.

https://doi.org/10.1142/9789819814770_0032

Chapter 32

Bridge Frequency Identification Using Cross-Power Spectra of Vehicle Vibrations from Multiple Moving Speeds#

Xuzhao Lu

College of Civil Engineering, Tongji University
1239 Siping Road, Shanghai 200092, P. R. China
luxuzhao_1992@tongji.edu.cn

Chul-Woo Kim *, Kai-Chun Chang † and Zhuoran Han ‡

Department of Civil and Earth Resources Engineering
Kyoto University, Kyoto 615-8540, Japan
**kim.chulwoo.5u@kyoto-u.ac.jp*
†*chang.kaichun.4z@kyoto-u.ac.jp*
‡*han.zhuoran.78r@st.kyoto-u.ac.jp*

Limin Sun

College of Civil Engineering, Tongji University
1239 Siping Road, Shanghai 200092, P. R. China
lmsun@tongji.edu.cn

In recent years, a rapid bridge health monitoring technology has been developed using an instrumental moving vehicle. Using recorded vehicle vibration data, bridge frequencies are identified for bridge health monitoring or finite element model updating. Target bridge frequencies with significant amplitudes in the vehicle's vibration frequency spectra are expected to be found. However, in the coupled vehicle–bridge interaction (VBI) system, bridge vibration-relevant vehicle dynamics might not be noticeable. The bridge frequency would be difficult to identify because of the potential influence of road roughness. To resolve this difficulty, a novel bridge frequency identification method is proposed to mitigate the negative effects of road roughness. First, theoretical derivations are done to ascertain the VBI system dynamic characteristics considering road surface roughness. Our findings showed that the road roughness-relevant vehicle dynamics are closely related with the traveling speed, whereas the bridge frequency remains approximately constant. Theoretical investigations indicated that cross-power spectra between vehicle dynamics at multiple moving speeds are effective to mitigate the negative effects of road roughness. Presumably, it is feasible to identify the target bridge frequency from the cross-power spectra. Both the dynamic characteristics of the VBI system and the effectiveness of the proposed method for bridge frequency identification were examined using finite element

*Corresponding author.
#This chapter appeared previously on the International Journal of Structural Stability and Dynamics. To cite this chapter, please cite the original article as the following: X. Lu *et al.*, *Int. J. Struct. Stab. Dyn.*, **23**, 2340035 (2023), DOI: 10.1142/S0219455423400357.

simulations and laboratory experiments. Compared to existing methods, the proposed method is widely applicable to real-world situations and difficulties.

Keywords: Bridge frequency identification; cross-power spectra; multiple moving speeds; vehicle–bridge interaction (VBI) system.

1. Introduction

Bridge frequency identification is an important task for bridge health monitoring and finite element model updating. For bridge health monitoring, through changes in bridge frequencies, one can detect changes in bridge properties and therefore make a diagnosis about the bridge service status.[1] In finite element model updating, a bridge frequency can be a key observation used to evaluate the discrepancy between a finite element model and an actual bridge. To identify bridge frequencies, a conventional method is to install sensors on a target bridge.[2] However, for the huge numbers of short-span bridges in urban transport systems, a monitoring system using conventional methods requires great investment in huge amounts of sensors, and great amounts of human labor for data analysis, which is presumably unsustainable. To reduce investment and labor particularly, an alternative method is to use a moving vehicle to identify the bridge frequency based on a vehicle–bridge interaction (VBI) system.[3–5] In a VBI system, a bridge vibrates under vehicular loading. Moreover, the bridge vibration would excite vehicle vibration through interaction between the bridge surface and the vehicle tires. Bridge frequencies, specifically the vertical acceleration of bridges, have been proven theoretically as identifiable from the spectra of vehicle vibrations. Therefore, the structural integrity of a bridge is identifiable. Compared to conventional methods, a method with a moving vehicle would be rapid and require fewer resources.[6]

Generally, a target bridge frequency would be identified from vehicle spectra by choosing a frequency with a large amplitude. Nevertheless, that choice can be difficult in practice because of possible interference from uneven road surfaces.[7] When a vehicle passes over a bridge, the uneven road surface (i.e. road roughness) will excite the vehicle. In fact, the vehicle acceleration spectrum has higher amplitudes at frequencies corresponding to vibrations excited by road surface irregularities than those of the bridge.[8] Although the road roughness condition for a specific road is deterministic, its spatial frequencies and corresponding amplitudes are always unknown. Therefore, for the dynamic responses of the vehicle in a single trip, one can hardly determine whether a chosen frequency is the bridge frequency, a road roughness-relevant frequency, or a frequency induced by noise.

Malekjafarian and O'Brien[9] applied strong external excitation to a bridge to reduce the negative effects of road roughness. Their findings indicated that excessive bridge vibrations were transmitted to the running vehicle, making it possible to extract the vibration corresponding to the bridge frequency from the vehicle vibration. However, realizing this approach requires powerful actuators, which might be available for laboratory experiments. This approach can be realized only with great difficulty for practical applications.

Yang *et al.*[7,10] and Jian *et al.*[11] specifically examined the modification of moving vehicle systems. A common method is to use several specific single-degree-of-freedom (SDOF) trailers (or the front and rear axles) following a tractor. By calculating the difference in time-domain or frequency-domain responses between these trailers, the influence of bridge roughness can be eliminated theoretically. However, it is also true that an ideal SDOF trailer is difficult to achieve. Moreover, modes unrelated to bridge vibrations, corresponding to the rocking and pitching of SDOF trailers, appear in the vehicle vibration spectrum, thereby making the identification of bridge frequencies difficult. Furthermore, the second strategy requires that moving vehicles follow strict regulations for the moving speed and trajectory, which is also challenging to realize in practical engineering situations.

Leaving aside physical difficulties related to the renovation of bridges and vehicles, more advanced signal-processing methods have been proposed. Bases of a Fourier series were transformed into other spaces.[12–14] It was expected that insignificant non-significant frequency in a frequency spectrum would become significant in another space. Yang and Chen[15] proposed the use of the Stochastic Subspace Identification (SSI) method, which treats road roughness as a random excitation source. The statistical concept was also used by Liu *et al.*[16] They proposed a machine learning method to identify bridge damage. Data-driven approaches have also been reported recently. Cheema *et al.*[19] proposed a novel data-driven approach by integrating a nonlinear dimensionality-reduction technique using Uniform Manifold Approximation and Projection (UMAP) together with a non-parametric clustering technique using Hierarchical Density-Based Spatial Clustering of Applications with Noise (HDBSCAN). Kaur *et al.*[17] proposed an unsupervised approach using only the measured responses from vehicles passing over a bridge. Nevertheless, most of these methods give no plain physical explanation of the vehicle dynamics in new spaces, which can be questionable for engineering applications, especially when material or geometric properties vary among bridges.

For this study, to identify the bridge frequency, we first analyzed the theoretical dynamic responses of the VBI system while particularly addressing the potential relation between road roughness and vehicle dynamics. This idea was highlighted by Agostinacchio, Ciampa, and Olita,[18] who showed that when a vehicle moves on a road, the frequency of the vehicle vibration excited by road irregularities is proportional to the vehicle speed: i.e. the discrete Fourier transform (DFT) curves in the spectrum would scale with the moving speed. Our theoretical analysis clarifies the relationship between the vehicle vibration frequency and the vehicle, bridge, and road roughness properties. It was found that by changing the vehicle speed, the frequency of the road-roughness-induced vehicle vibrations would change proportionally. Therefore, in this study, after confirming the relation between vehicle vibration and road roughness in a VBI system, a new strategy with multiple vehicles assigned at different speeds was proposed. This strategy of multiple vehicles was inspired by the research reported by Ghaleroudkhani *et al.*,[20] who showed that the averaged DFTs of the dynamic motion from multiple vehicles can identify the bridge frequency effectively. Based on the theoretical analysis, this study calculates the cross-spectra of vehicle acceleration for multiple vehicles. Compared to Refs. 7, 9, and 10, this method does not need to strictly regulate the multiple vehicles' moving speed and trajectories, which is supposed to be more practical in real-world applications. Besides, in this study, we also found that the surfaces of the wheels were uneven and would cause periodical vehicle vibrations. The uneven wheel surface is another kind of unwanted vibration source but has rarely been investigated and solved in past studies. Nagayama *et al.*[21] used a similar method for bridge frequency identification, but they put little emphasis on the theoretical derivation that can explain the essential mechanism underlying this method. This study was conducted to bridge this gap. Numerical simulations and laboratory experiments were also undertaken for the validation of this method.

Theoretical derivation of this method on the VBI system is presented in Sec. 2, including derivation of the dynamic responses of the VBI system and the mechanism of the cross-power spectrum. A relevant numerical analysis is presented in Sec. 3. Validation work with laboratory experiments is presented in Sec. 4.

2. Theoretical Analysis

2.1. *Dynamics of the VBI system considering road roughness*

When a beam is a simply supported bridge with constant cross-section properties, and when the vehicle is simplified as a SDOF system consisting of a spring and a sprung mass as shown in Fig. 1, then the undamped dynamic equation of the vehicle in the VBI system is shown in Eq. (1)[3] as follows:

$$m_v \ddot{u}_v(t) + k_v[u_v(t) - u_w(t)] = 0, \tag{1}$$

$$u_w(t) = u_b(x,t)|_{x=vt} + R(x)|_{x=vt}, \tag{2}$$

where $u_v(t)$ denotes the vertical displacement of the sprung mass. Also, $\ddot{u}_v(t)$ represents the vehicle's vertical acceleration. $u_w(t)$ stands for the vertical displacement of the contact point. m_v and k_v respectively signify the vehicle mass and the spring stiffness. $R(x)$ denotes the road roughness at the location x; and v stands for the vehicle speed.

The dynamic equation of the bridge in VBI system is shown as follows:

$$\begin{aligned} &\bar{m}\ddot{u}_b(x,t) + EIu_b''''(x,t) \\ &\quad = \{k_v[u_v(t) - (u_b(x,t)|_x = vt + R(x)|_{x=vt})] - m_v g\}\delta(x - vt). \end{aligned} \tag{3}$$

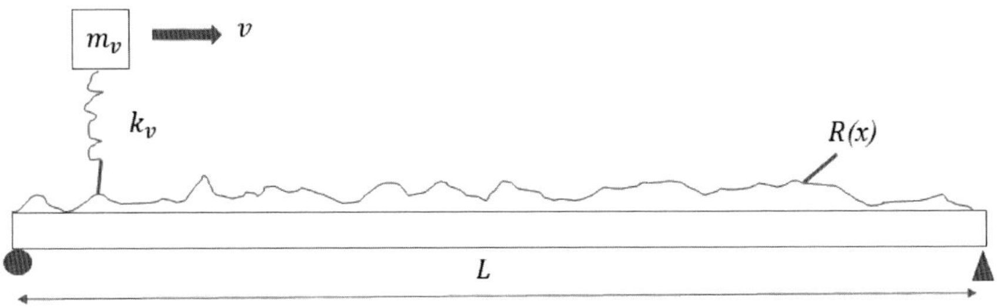

Fig. 1. Single degree of freedom (SDOF) VBI system.

In that equation, $\bar{m}$ stands for the unit mass, E expresses the elastic modulus, and I denotes the moment of inertia. g signifies the acceleration of gravity, δ is the delta function, where the prime denotes differentiation with respect to coordinate x, and the dot denotes differentiation with respect to time t.

To resolve the equations above, Eq. (2) is substituted into Eq. (1) to obtain Eq. (4).

$$m_v \ddot{u}_v(t) + k_v [u_v(t) - u_b(x,t)|_{x=vt} - R(x)|_{x=vt}] = 0. \tag{4}$$

As given in Eq. (5), vehicle displacement u_v can be decomposed into three components: Vehicle-relevant component $u_{v,1}$ satisfying the equation of motion in Eq. (6a), bridge-relevant component $u_{v,2}$ satisfying the equation of motion in Eq. (6b), and roughness-relevant component $u_{v,3}$ satisfying the equation of motion in Eq. (6c).

$$u_v(t) = u_{v,1}(t) + u_{v,2}(t) + u_{v,3}(t), \tag{5}$$

$$m_v \ddot{u}_{v,1}(t) + k_v u_{v,1}(t) = 0, \tag{6a}$$

$$m_v \ddot{u}_{v,2}(t) + k_v [u_{v,2}(t) - u_b(x,t)|_{x=vt}] = 0, \tag{6b}$$

$$m_v \ddot{u}_{v,3}(t) + k_v [u_{v,3}(t) - R(x)|_{x=vt}] = 0. \tag{6c}$$

Equation (6a) presents the free vibration of the vehicle, for which the relevant frequency is defined as presented below.

$$\omega_v = \sqrt{\frac{k_v}{m_v}}. \tag{6d}$$

Moving $R(x)|_{x=vt}$ in Eq. (6c) to the right-hand side yields Eq. (6e).

$$m_v \ddot{u}_{v,3}(t) + k_v u_{v,3}(t) = k_v R(x)|_{x=vt}. \tag{6e}$$

Equation (6e) represents a forced vibration of the vehicle by the equivalent external force $k_v R(x)|_{x=vt}$ related to road roughness. The road roughness can be assumed as a superposition of simple spatial cosine waves as presented as follows:

$$R(x)|_{x=vt} = \sum_{i=1}^{N} A_i \cos(\Omega_i x). \tag{6f}$$

Therein, Ω_i represents the spatial frequency of the road roughness. In this study, we focus on frequency identification. The random spatial phase theoretically shifts the road roughness profile along the spatial coordinate but does not change the spatial frequency, and therefore it is omitted in Eq. (6f) and the following discussions without losing much generality. It is noteworthy that Eqs. (6a) and (6c) are uncoupled equations that can be solved directly with vibration components having frequency of ω_v or Ω_i.

Equation (6b) consists of the bridge vibration $u_b(x,t)|_{x=vt}$. It is a coupled equation that should be solved in combination with Eq. (3). As reported by Yang *et al.*,[3] when the vehicle mass is much less than the bridge mass, the term of $k_v(u_v(t) - u_b(x,t)|_{x=vt})$ in Eq. (3) is negligible in comparison with $m_v g$. Therefore, the bridge vibration shown in Eq. (3) is divisible into the following two components.

$$u_b(x,t) = u_{b,1}(x,t) + u_{b,2}(x,t). \tag{7}$$

Therein, $u_{b,1}(x,t)$ denotes the bridge dynamics attributable to moving force $m_v g$; $u_{b,2}(x,t)$ denotes the bridge dynamics attributable to a time-varying force $k_v R(x)|_{x=vt}$.

Equation (3) can be transformed further into a superposition of two sub-equations shown in Eqs. (8a) and (8b).

$$\bar{m}\ddot{u}_{b,1}(x,t) + EIu''''_{b,1}(x,t) = -m_v g\delta(x - vt), \tag{8a}$$

$$\bar{m}\ddot{u}_{b,2}(x,t) + EIu''''_{b,2}(x,t) = k_v[-R(x)|_{x=vt}]\delta(x - vt). \tag{8b}$$

Equation (8a) denotes a moving force $m_v g$ traveling across the bridge. Also, Eq. (8b) denotes a time-varying force $k_v R(x)|_{x=vt}$ moving across the bridge. For a linear dynamic system, the solution to Eq. (8a) is expressed in terms of the modal shapes, $\phi_n(x)$, and associated modal coordinates, $q_{b,1,n}(t)$, as follows:

$$u_{b,1}(x,t) = \sum_{n} \phi_n(x) q_{b,1,n}(t). \tag{9}$$

The nth modal equation corresponding to Eq. (8a) can then be obtained as follows:[3]

$$\ddot{q}_{b,1,n} + \omega_{b,n}^2 q_{b,1,n} = \frac{-2m_v g}{\bar{m}L} \sin\frac{n\pi vt}{L}. \tag{10a}$$

The solution to Eq. (10a) yields Eq. (10b) as follows:

$$q_{b,1,n}(t) = \frac{\Delta_{st,n,1}}{1 - S_{n,1}^2}\left[\sin\left(\frac{n\pi vt}{L}\right) - S_n \sin\omega_{b,n}t\right], \tag{10b}$$

for which the following variables are used:

$$\Delta_{st,n,1} = \frac{-2m_v gL^3}{n^4\pi^4 EI}, \tag{11a}$$

$$S_{n,1} = \frac{n\pi v}{L\omega_{b,n}}, \tag{11b}$$

$$\omega_{b,n} = \frac{n^2\pi^2}{L^2}\sqrt{\frac{EI}{\bar{m}}}. \tag{11c}$$

Similarly, the nth modal equation corresponding to Eq. (8b) is obtainable as follows:

$$\ddot{q}_{b,2,n} + \omega_{b,n}^2 q_{b,2,n} = \frac{-2k_v \sum_{i=1}^{N} A_i \cos(\Omega_i vt)}{\bar{m}L}\sin\frac{n\pi vt}{L}. \tag{12}$$

The solution to Eq. (12) is shown as in Eq. (13):

$$\begin{aligned} q_{b,2,n}(t) &= \frac{\Delta_{st,n,2}}{1 - S_{n,2,1,i}^2}\left[\sum_{i=1}^{N} A_i\{\sin(\Omega_i vt + \omega_n t) - S_{n,2,1,i}\sin\omega_{b,n}t\}\right] \\ &\quad - \frac{\Delta_{st,n,2}}{1 - S_{n,2,2,i}^2}\left[\sum_{i=1}^{N} A_i\{\sin(\Omega_i vt - \omega_n t) - S_{n,2,2,i}\sin\omega_{b,n}t\}\right], \end{aligned} \tag{13}$$

where

$$\omega_n = \frac{n\pi v}{L}, \tag{14a}$$

$$\Delta_{st,n,2} = -\frac{k_v L^3}{n^4\pi^4 EI}, \tag{14b}$$

$$S_{n,2,1,i} = \frac{\Omega_i v + \omega_n}{\omega_{b,n}}, \tag{14c}$$

$$S_{n,2,2,i} = \frac{\Omega_i v - \omega_n}{\omega_{b,n}}. \tag{14d}$$

Therefore, by substituting Eq. (10b) onto Eq. (9), one can obtain $u_{b,1}(x,t)$ as follows:

$$u_{b,1}(x,t) = \sum_n \frac{\Delta_{st,n,1}}{1 - S_{n,1}^2}\left[\sin\left(\frac{n\pi vt}{L}\right) - S_n \sin\omega_{b,n}t\right]\sin\frac{n\pi x}{L}. \tag{15a}$$

Similarly,

$$u_{b,2}(x,t) = \sum_n \left\{ \begin{array}{l} \left[\sum_{i=1}^{N} \frac{\Delta_{st,n,2}}{1 - S_{n,2,1,i}^2} A_i \{ \sin(\Omega_i vt + \omega_n t) - S_{n,2,1,i} \sin \omega_{b,n} t \} \right] \\ - \left[\sum_{i=1}^{N} \frac{\Delta_{st,n,2}}{1 - S_{n,2,2,i}^2} A_i \{ \sin(\Omega_i vt - \omega_n t) - S_{n,2,2,i} \sin \omega_{b,n} t \} \right] \end{array} \right\}$$
$$\times \sin \frac{n\pi x}{L}. \tag{15b}$$

Because the two components $u_{b,1}$ and $u_{b,2}$ dominate bridge dynamics, by the bridge-relevant component $u_{v,2}$ in the vehicle response in Eq. (5) can also be decomposed into two components (Eq. (16)): $u_{v,2,1}$, which satisfies the equation of motion in Eq. (16a) having $u_{b,1}$ term and $u_{v,2,2}$, which satisfies the equation of motion in Eq. (16b) having $u_{b,2}$ term. They are as follows:

$$m_v \ddot{u}_{v,2,1}(t) + k_v [u_{v,2,1}(t) - u_{b,1}(x,t)|_{x=vt}] = 0, \tag{16a}$$

and

$$m_v \ddot{u}_{v,2,2}(t) + k_v [u_{v,2,2}(t) - u_{b,2}(x,t)|_{x=vt}] = 0, \tag{16b}$$

where

$$u_{v,2}(t) = u_{v,2,1}(t) + u_{v,2,2}(t). \tag{16c}$$

Substituting Eqs. (9) and (10b) into Eq. (16a), one can obtain the solution[3] to Eq. (16a) using the Duhamel integral as presented as follows:

$$u_{v,2,1}(t) = \sum_n [A_{1,n} \cos(2\omega_n t) + A_{2,n} \cos(\omega_v t) + A_{3,n} \cos(\omega_{b,n} t - \omega_n t)$$
$$+ A_{4,n} \cos(\omega_{b,n} t + \omega_n t)]. \tag{17}$$

In that equation, $A_{1,n}$, $A_{2,n}$, $A_{3,n}$, and $A_{4,n}$ are constant values determined by $\omega_{b,n}$, ω_v, and ω_n. Similarly, by substituting Eqs. (15b)–(16b), one can solve the following Duhamel integral.[22]

$$u_{v,2,2}(t) = \omega_v \int_0^t \sum_n \left\{ \begin{array}{l} \left[\sum_{i=1}^{N} \frac{\Delta_{st,n,2}}{1 - S_{n,2,1,i}^2} A_i \{ \sin(\Omega_i v\tau + \omega_n \tau) - S_{n,2,1,i} \sin \omega_{b,n} \tau \} \right] \\ - \left[\sum_{i=1}^{N} \frac{\Delta_{st,n,2}}{1 - S_{n,2,2,i}^2} A_i \{ \sin(\Omega_i v\tau - \omega_n \tau) \\ - S_{n,2,2,i} \sin \omega_{b,n} \tau \} \right] \end{array} \right\}$$
$$\times \sin \frac{n\pi v\tau}{L} \sin \omega_v (t - \tau) d\tau. \tag{18}$$

Thereby, one obtains the solution as follows:

$$u_{v,2,2}(t) = \sum_n \sum_i [B_{1,n} \cos(\Omega_i vt - 2\omega_n t) + B_{2,n} \cos(\Omega_i vt) + B_{3,n} \cos(\Omega_i vt + 2\omega_n t)$$
$$+ B_{4,n} \cos(\omega_v t) + B_{5,n} \cos(\omega_{b,n} t - \omega_n t) + B_{6,n} \cos(\omega_{b,n} t + \omega_n t)], \tag{19}$$

Table 1. Vibration components and frequencies in vehicle dynamics.

	Frequency	Relation to moving speed
Vehicle frequency	ω_v	None
Road roughness frequency	$\omega_{R,i}v$	Strong
Driving force frequency	$2\omega_n = 2n\pi v/L$	Strong
Coupling bridge-driving force	$\omega_{b,n} - \omega_n$	Weak
frequency	$\omega_{b,n} + \omega_n$	Weak
Coupling road roughness-driving force	$\Omega_i v - 2\omega_n$	Strong
frequency	$\Omega_i v + 2\omega_n$	Strong

where $B_{1,n}$, $B_{2,n}$, $B_{3,n}$, $B_{4,n}$, $B_{5,n}$, and $B_{6,n}$ are constant values determined by $\omega_{b,n}$, ω_v, Ω_i and ω_n.

2.2. *Application of cross power spectra to bridge frequency identification*

2.2.1. *Relation of vibration components to the moving speed*

All the vibration components (derived from Eqs. (6a), (6c), (17), and (19)) related to vehicle dynamics are presented in Table 1. The relation between their frequencies and moving speeds can be evaluated with three levels. The vehicle frequency, i.e. ω_v, is independent of the moving speed. Its relation is therefore denoted as "None." The road roughness frequency $\Omega_i v$, the driving force frequency $2\omega_n = 2n\pi v/L$, and the frequency of the coupling driving force and road roughness $\Omega_i v \pm 2\omega_n$, are proportional to the moving speed and are therefore denoted as having a "Strong" relation. For the frequency of the coupling bridge and driving force vibration $\omega_{b,n} \pm \omega_n$ of mid-span and short-span highway bridges in a speed limit, the first driving force frequency is much lower than the first bridge frequency. Paultre *et al.*[23] investigated the fundamental frequencies of 898 highway bridges and concluded that, for most of these bridges, the first natural frequency varies between 2 Hz and 5 Hz. Furthermore, as He and Zhu[24] reported based on their research findings, for typical moving speeds of 36, 72, and 108 km/h, the ratio of the first driving-force frequency to the first natural frequency of the bridge ranges from 0 to 0.15 (i.e. $\omega_n \ll \omega_{b,n}$). Therefore, the change in the moving speed is expected to have only a slight influence on the coupling bridge-driving force frequency. The relation is designated as "Weak". Moreover, under a common speed, $\omega_{b,n} \pm \omega_n$ can be treated approximately as $\omega_{b,n}$.

2.2.2. *Cross-power spectra*

For generality, one can assume two signals as presented below.

$$x_0(t) = A_0 \cos \omega t + e_0(t), \tag{20}$$

$$x_1(t) = A_1 \cos(\omega t + \varphi) + e_1(t). \tag{21}$$

These signals have the same frequency, but they have mutually differing amplitudes and phases. Furthermore, the noises $\{e_0(t)\}$ and $\{e_1(t)\}$ are uncorrelated zero-mean stationary processes: i.e. independent variables. The cross-correlation function between $\{x_0(t)\}$ and $\{x_1(t)\}$ can be expressed as follows.[25]

$$\begin{aligned} R_{x_0x_1}(\tau) &= E\{x_0(t)x_1(t+\tau)\} \\ &= E\{A_0A_1\cos\omega t\cos[(\omega(t+\tau)+\varphi)] + A_0e_1(t)\cos\omega t \\ &\quad + A_1e_0(t)\cos[(\omega(t+\tau)+\varphi)] + e_0(t)e_1(t)\} \\ &= A_0A_1E\{\cos\omega t\cos[(\omega(t+\tau)+\varphi)]\} = A_0A_1R_{x_\omega x_\omega}(\tau). \end{aligned} \tag{22}$$

Therein, $E\{\cdot\}$ stands for the mathematical expectation, τ is the time lag and $R_{x_\omega x_\omega}(\tau)$ is the auto-correlation of $x_\omega(t)$. Also, $e_0(t)$ and $e_1(t)$ are independent variables; $e_0(t)$ and $e_1(t)$ are uncorrelated with $\cos\omega t$. Therefore, $E\{e_0(t)e_1(t)\} = 0$, $E\{e_1(t)\cos\omega t\} = 0$, and $E\{e_0(t)\cos[(\omega(t+\tau)+\varphi)]\} = 0$. $x_\omega(t)$ can be expressed as follows:

$$x_\omega(t) = \cos\omega t. \tag{23}$$

Following the Wiener–Khinchin theorem, the Fourier transform of the $R_{x_\omega x_\omega}(\tau)$ is the power spectral density (PSD) of $x_\omega(t)$.

$$\mathcal{F}(R_{x_\omega x_\omega}(\tau)) = P_{x_\omega x_\omega}(f). \tag{24}$$

In that equation, $\mathcal{F}(\cdot)$ is the Fourier transform operator; $P_{x_\omega x_\omega}$ denotes the PSD function. Then substituting Eq. (22) into Eq. (24), one can obtain the following cross-power spectrum.

$$\mathcal{F}(R_{x_0x_1}(\tau)) = A_0A_1\mathcal{F}(R_{x_\omega x_\omega}(\tau)) = A_0A_1P_{x_\omega x_\omega}(f). \tag{25}$$

It is noteworthy that the illustration of $P_{x_\omega x_\omega}(f)$ would show a peak only at the frequency corresponding to ω.

2.2.3. *Application of cross-power spectra for bridge frequency identification*

This study relies on the assumption that the vehicle frequency ω_v does not closely approximate the bridge frequency $\omega_b \pm \omega_n$ or the road roughness frequency $\Omega_i v$. Consequently, in the time domain, the components are mutually uncorrelated. If a vehicle passes over a bridge twice with different speeds v_0 and v_1, then the frequency of components denoted as "Strong" would change proportionally with the moving speed. Therefore, on the second vehicle trip, these "Strong" components would become uncorrelated to those on the other trip. With the calculation of the cross-correlation function between vehicle dynamics in these two trips, as indicated in Eq. (22), these "Strong" components can be eliminated from calculation results. The frequency of the "None" and "Weak" components would be approximately constant. These components would be reserved in the cross-correlation function. Consequently, considering the potential uncorrelated measuring noise, the cross-correlation function of the vehicle dynamics can be derived as presented below.

$$R_{u_{v0}u_{v1}}(\tau) = R_{\omega_v\omega_v}(\tau) + R_{\omega_b\omega_b}(\tau). \tag{26}$$

Similar to the derivation of Eq. (25), the cross-power spectrum exhibit peaks at frequencies corresponding to vehicle frequency ω_v and bridge frequency ω_b. The potential negative influences from road roughness and measuring noise can thereby be greatly reduced. The cross-power spectrum derivations in this and the last sections could apply to vehicle displacement and acceleration responses because the latter response is the second derivative of the former and keeps the same frequency in the same cosine functions with modulated amplitude (the coefficients of the cosine functions). We use vehicle acceleration responses for bridge frequency identification, hereafter, as it is more easily measured in practice.

3. Numerical Validation

As described hereinafter, three finite element models of a VBI system are built.[26] The road roughness in the first and second models is a single sinusoid curve. The roughness in the third model is a combination of multiple sinusoid curves simulating random road roughness. These three models were designed for the following validations.

(1) Characteristics of the road roughness-relevant bridge dynamics.
(2) The road roughness-relevant vehicle vibration frequency is proportional to the moving speed.
(3) The effectiveness of the proposed method for bridge frequency identification considering road roughness.

3.1. *Numerical models*

All these models are simple VBI systems consisting of a simply supported bridge and a sprung-mass vehicle. The first model is designed to validate the theoretically derived dynamic responses of the simple VBI system. Since the driving-force frequency ω_n is always rather small in comparison with the road-roughness frequency $\Omega_i v$, it is supposed to be hard to distinguish the road-roughness relevant vehicle acceleration between $u_{v,3}(t)$ (in Eq. (6c), the vehicle vibration components with the frequency $\Omega_i v$) and $u_{v,2,2}(t)$ (in Eq. (18), the vehicle vibration components with the frequency $\Omega_i v \pm 2\omega_n$). Considering that $u_{v,2,2}(t)$ is directly derived from Duhamel integral of Eq. (15b), the first model aimed at Eq. (15b), with a focus on the presented proportional relationship between the amplitude of road-roughness frequency $\Omega_i v$ and the amplitude of the road-roughness relevant bridge vibration frequency $\Omega_i v \pm \omega_n$. This simulation would first confirm the existence of road-roughness-relevant bridge vibration, and secondly, validate the derived proportional relationship in Eq. (15b).

It should be noted that in most of the existing research,[27–29] the road roughness condition is neglected or set as a good condition, and the influence of road roughness on bridge and vehicle dynamics is very limited and hard to be distinguished. Therefore, for the first model, a specific VBI system model was designed to make the road roughness-relevant bridge responses significant. The bridge span was 20 m long and the first bending frequency was 1 Hz. The vehicle mass was set at 21.07 kg, which is approximately 1.8% of the bridge mass. The VBI system was simulated following Ref. 30.

The VBI system described by Yang *et al.*[3] was simulated as the second and third models. The bridge and vehicle parameters are presented respectively in Tables 2 and 3. Two-dimensional VBI models are built with the commercial software package ABAQUS,[31] as presented in Fig. 2. This is a 2D model and the simply supported beam was simulated with shell elements. The road roughness was simulated with a thin layer of shell elements that are attached to the top of the beam. The heights of these elements are set with different values according to the roughness condition. The vehicle is simulated as a sprung mass block that connects to a wheel with a spring. The mass block and the wheel were simulated with shell elements. The wheel remains in contact with the road roughness. The analytical time step was set as 0.002 s with the Hilbert–Hughes–Taylor method. Simulation of the interaction between the wheel and the road roughness follows the penalty method reported from an earlier study.[26]

3.2. *Road roughness condition*

The first and the second model use the simple road roughness designed as a sinusoid wave as follows:

$$R(x) = A \times \sin(4\pi x) \quad (\text{unit} : \text{m}), \tag{27}$$

Table 2. Bridge parameters.

Span length (m)	25
Unit mass (kg/m)	4 800
EI ($\text{GN} \cdot \text{m}^2$)	3.3
1st bending frequency (Hz)	2.08

Table 3. Vehicle parameters.

Mass (kg)	1 200
Spring stiffness (kN/m)	500
1st frequency (Hz)	3.25

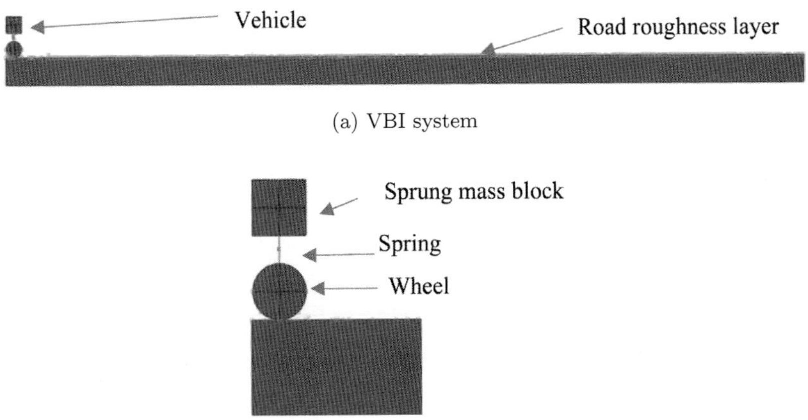

(a) VBI system

(b) Vehicle

Fig. 2. Numerical model of a VBI system.

where A is the amplitude and x is the longitudinal coordinate. It can be calculated that the spatial frequency is 2 cycles/m.

In the third model, according to ISO-8608,[32] the road roughness was simulated as a periodically modulated random process. It is specified by the PSD function, as shown in Eq. (28).

$$S_r(\omega_s) = A_r(\omega_s/\omega_{s0})^{-2}. \tag{28}$$

Therein, $S_r(\omega_s)$ represents the PSD in m^2/cycle/m, A_r is the roughness coefficient in m^2/cycle/m, ω_{s0} denotes the discontinuity frequency equal to $1/2\pi$ (cycle/m), and ω_s stands for the spatial frequency in cycle/m. From the PSD, a surface profile was generated with a random number φ distributed uniformly between 0 and 2π. The road profile is given in its discrete form as Eq. (29).

$$R(x) = \sum_{k=1}^{N} \left(4A_r \left(\frac{2\pi k}{L_c \omega_{s0}}\right)^{-2} \frac{2\pi}{L_c}\right)^{1/2} \cos(\omega_{sk}x - \varphi_k). \tag{29}$$

In that equation,

$$\omega_{sk} = k\Delta\omega_s, \Delta\omega_s = \frac{2\pi}{L_c} \tag{30}$$

and L_c represents the bridge span length. For this study, the road roughness condition was assumed to be "good." N was set as 20. The spatial frequency ω_{sk} ranges from 0.1 cycle/m to 2 cycle/m.

3.3. *Validations*

For the first model, three amplitudes A of the road roughness were set as 0.0057 m (A), 0.0114 m (2A), and 0.0228 m (4A). The moving speed was 10 m/s, and

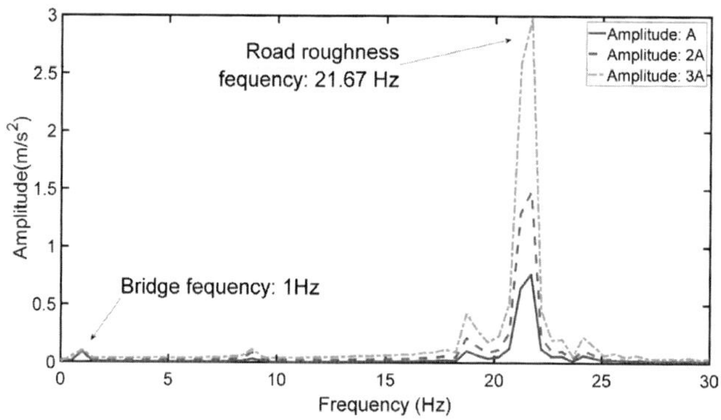

Fig. 3. Acceleration frequency spectra of mid-span beam in the first model.

the road roughness-relevant bridge vibration frequency should be around 20 Hz (= 2 cycle/m × 10 m/s), which can be observed in the bridge mid-span acceleration spectrum in the finite element simulation (as shown in Fig. 3). Moreover, it can be found in Fig. 3 that the amplitude (the peak at 21.67 Hz) of the road roughness-relevant bridge vibration is proportional to the road roughness amplitude. Therefore Eq. (15b) is numerically validated. Further, it can be confirmed that the change in road roughness amplitude would influence the amplitude of the road-roughness relevant bridge vibration, but would not change the frequency value, and therefore would not influence the target identification result of frequency.

For the second model, three moving speeds were assigned to the vehicle: 5 m/s, 7.5 m/s, and 10 m/s. The vehicle dynamics of the second model with these three moving speeds are presented respectively in Figs. 4(a)–4(c). The corresponding frequency spectra are portrayed in Figs. 4(d)–4(f). The frequency spectra show that the bridge frequency and the vehicle frequency remain almost constant under different vehicle speeds. More importantly, the frequencies which are dominant because of the road roughness-relevant vehicle vibration are observed around 10 Hz (= 2 cycle/m × 5 m/s), 15 Hz ($= 2\frac{\text{cycle}}{\text{m}} \times 7.5\frac{\text{m}}{\text{s}}$), and 20 Hz (= 2 cycle/m × 10 m/s) as expected, thereby validating the derived linear relation between the vehicle speed and road roughness-relevant vehicle vibration.

In the third model, the frequency spectra[33] of vehicle dynamics at three speeds under random road roughness are exhibited in Fig. 5(a). The amplitudes of the bridge frequency and the vehicle frequency are significant because the road roughness condition is good. From the cross-power spectra (portrayed in Fig. 5(b)), it can be ascertained that the frequency corresponding to the peak is 2.1 Hz, which is very close to the first-bending frequency of the bridge (2.08 Hz). Furthermore, the vehicle frequency and the bridge frequency are approximately equal. Although another significant peak is expected to exist at the vehicle frequency, it is difficult to distinguish them for this finite element model.

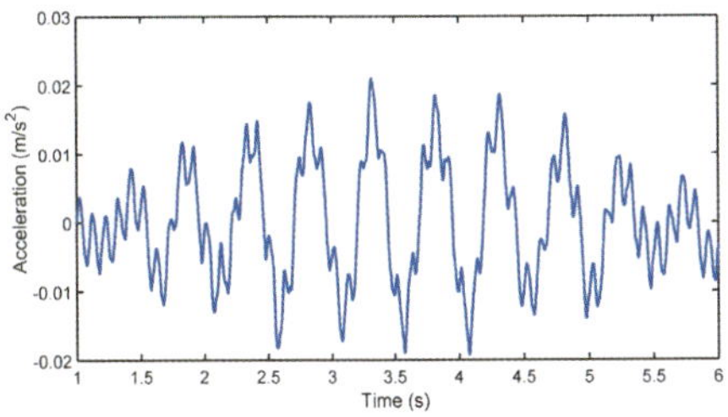

(a) [Acceleration time history at 5 m/s]

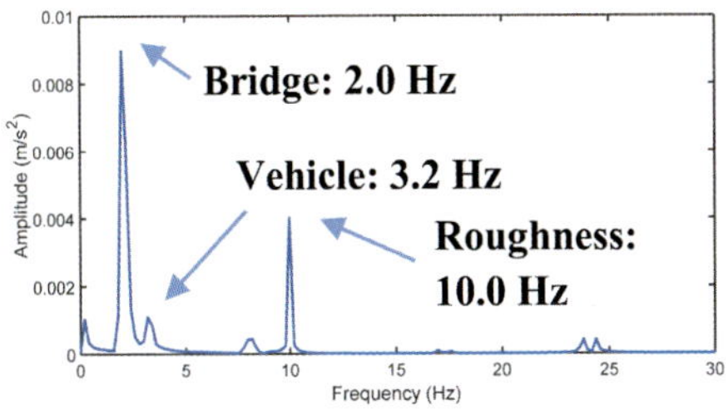

(d) [Frequency spectrum at 5 m/s]

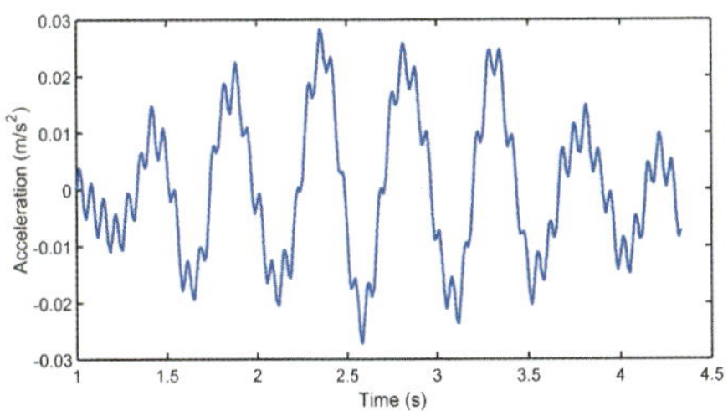

(b) [Acceleration time history at 7.5 m/s]

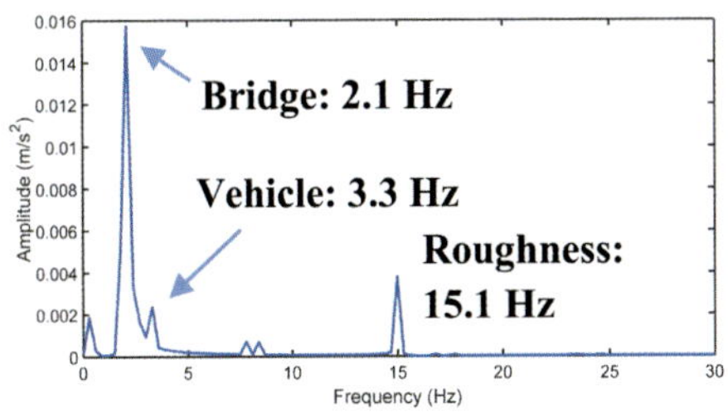

(e) [Frequency spectrum at 7.5 m/s]

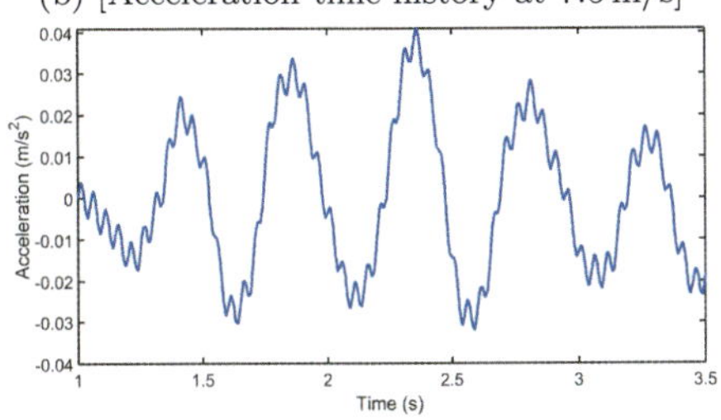

(c) [Acceleration time history at 10 m/s]

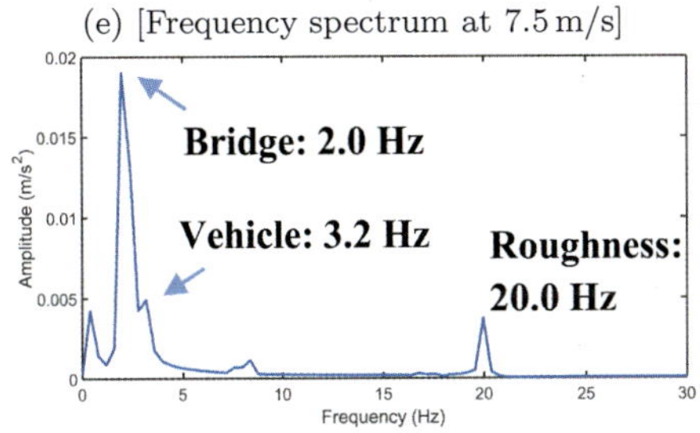

(f) [Frequency spectrum at 10 m/s]

Fig. 4. Vehicle dynamics of the second model at three moving speeds.

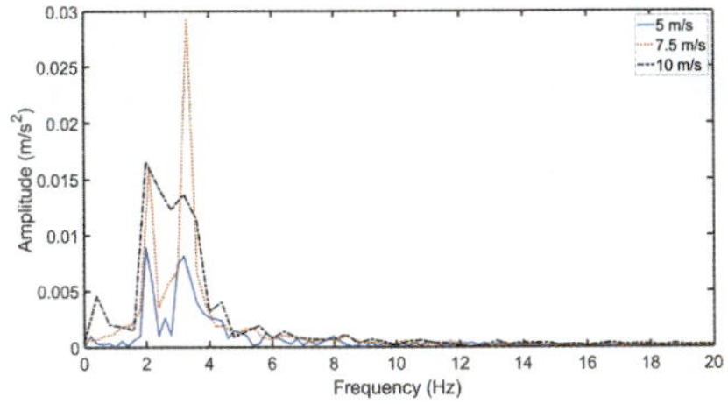

(a)

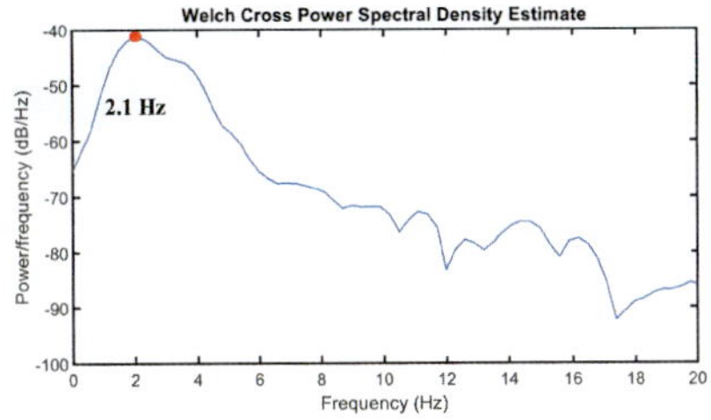

(b)

Fig. 5. Vehicle dynamics of the third model. (a) Frequency spectra at three speeds and (b) cross-power spectrum between 5 m/s and 7.5 m/s.

4. Experiments

Laboratory experiments were conducted to validate the derived dynamic characteristics of VBI systems, especially to ascertain the road roughness-relevant vehicle dynamics and the effectiveness of the proposed bridge frequency identification method.

4.1. *Experiment setup*

The VBI system used for experimentation consists of three beams for the bridge and a vehicle model. These three beams are simply supported beams (Fig. 6). The span length of the middle beam is 5.4 m. The remaining two spans are respectively intended for acceleration and deceleration spans. The acceleration span was installed to keep the vehicle traveling at a constant speed on the middle beam. All these beams were designed with a constant H-type cross-section. The middle beam properties are portrayed in Table 4.

Figure 7 shows the vehicle as a sprung-mass system that moves along a pair of trails installed on the bridge. The rail surface has been shaped to emulate a very good road roughness condition. The exact height of the trail surface is assumed to be unknown. A servomotor and a fixed pulley were installed at each end of the bridge. To form a driving system, the motor was connected to the vehicle with a nylon rope. Accelerometers were installed at the bridge span center and on the sprung mass. The sampling frequency of the acceleration measurement was 200 Hz. The vehicle sprung mass was 11.83 kg. The respective frequencies of the vehicle, the bridge, and their corresponding frequencies in the VBI system (by parking the vehicle on the span center of the bridge and conducting a free vibration test) are presented in Table 5.

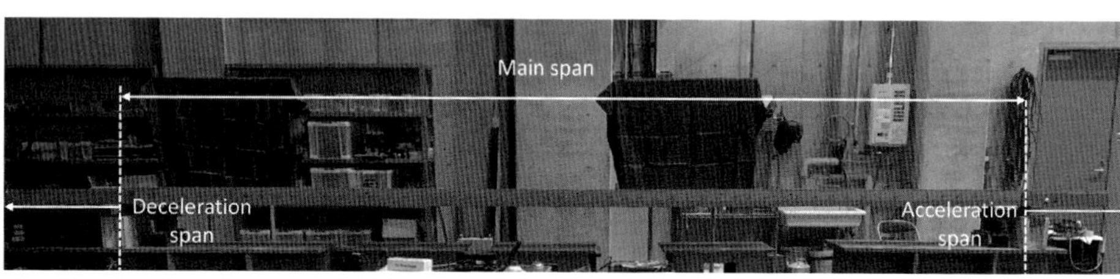

Fig. 6. Beams used for experimentation.

Table 4. Middle beam properties.

Property	Value
Span length (m)	5.4 m
Young's modulus	2.1×10^{11} N/m^2
Density	7 850 kg/m^3

Fig. 7. Moving vehicle.

Table 5. Vehicle and bridge frequencies during free vibration tests.

	Individual frequency (Hz)	VBI frequency (Hz)
Vehicle	6.39	6.65
Bridge	3.59	3.38

Since this study focuses on the potential influence of road roughness and noise, the frequency of the vehicle, as prior knowledge, was designed to be much higher than the bridge's first frequency to avoid any potential interference effect. The experiment considers five vehicle speeds: 0.32 m/s, 0.43 m/s, 0.59 m/s 0.80 m/s, and 1.00 m/s. Considering that in the actual situation, short and medium-span bridges whose first natural bending frequency is close to the test beam have a span ranging from 30 m to 35 m, the vehicle speed of 1.0 m/s in the experiment can be converted to 22.8 km/h for a 30-meter span and 26.6 km/h for a 35-meter span.

It is noteworthy that the wheel surface is not smooth. In fact, it represents another kind of road roughness. It was tested that when the vehicle moves at a constant speed, such an uneven wheel surface would also induce a periodic vibration in the vehicle with a specific frequency. The vehicle wheel diameter is 0.0318 m. When the vehicle speed is 1 m/s, the vehicle vibration frequency induced by the uneven wheel surface can be calculated as follows:

$$f_{\text{wheel}} = \frac{1\,\text{m/s}}{\pi \times 0.0318\,\text{m}} = 10.01\,\text{Hz}. \tag{31}$$

4.2. *Road roughness-relevant components in vehicle dynamics*

Frequency spectra of accelerations for the bounce motion of the vehicle and for the bridge under five speeds are depicted in Fig. 8. A dominant peak (denoted as "Roughness" meaning the road roughness-relevant component) can be found for both the vehicle vibration spectrum and the bridge vibration spectrum. Furthermore, the dominant frequency varies with the vehicle speed. Vibrations corresponding to this frequency might be related to the uneven vehicle wheel surface.

Table 6 shows that the experimentally obtained and theoretical results mutually match well, demonstrating that the varying dominant frequency in Fig. 8 with vehicle speed might result from the uneven vehicle wheel surface. As presented in Fig. 9, a similar trend of frequency variation with vehicle speed is apparent under different vehicle masses. Therefore, one can infer that the road roughness-relevant component is induced by the uneven wheel surface of the vehicle. It should be noted that the uneven wheel surface is another kind of inconsistency of moving trajectory of multiple vehicles. For actual engineering, methods proposed by the previous research might be hard to eliminate the interference because the surfaces of the tire between multiple vehicles, or the front and rear axles, might be different.

4.3. *Bridge frequency identification with the cross-power spectrum*

Before identifying the bridge frequency from vehicle vibrations, the frequency spectra of the bridge acceleration at the span center are analyzed. From the frequency spectra of the beam obtained under speeds of 0.42 m/s, 0.59 m/s, 0.80 m/s, and 1.00 m/s, with prior knowledge that the bridge's first natural frequency is in the range of 3 Hz and 4 Hz, frequencies with the highest amplitude in each spectrum are presented in Table 7. Table 7 shows that the identified frequency decreases concomitantly with increasing vehicle speed, which might correspond to $\omega_b - \omega_n$. However, the frequency at 0.80 m/s increased, which might correspond to $\omega_b + \omega_n$.

When the vehicle moves at 0.32 m/s, the frequency of road roughness-relevant components in the bridge and vehicle vibrations are in close resonance to the bridge's first natural frequency. Also, the amplitude corresponding to the frequency is so high as to shadow the bridge's first natural frequency. Therefore, from the frequency spectrum of the bridge acceleration (Fig. 8(b)), identifying the bridge frequency directly with no prior knowledge is difficult. For the reason presented above, the 0.32 m/s case is not discussed further for these analyses.

To illustrate some benefits of the proposed method over the general "individual-run auto-spectrum" method, this study first identifies the bridge frequency directly from the frequency spectra of vehicle vibrations. Data taken from the 20 trips of the vehicle on the bridge for each speed are investigated.

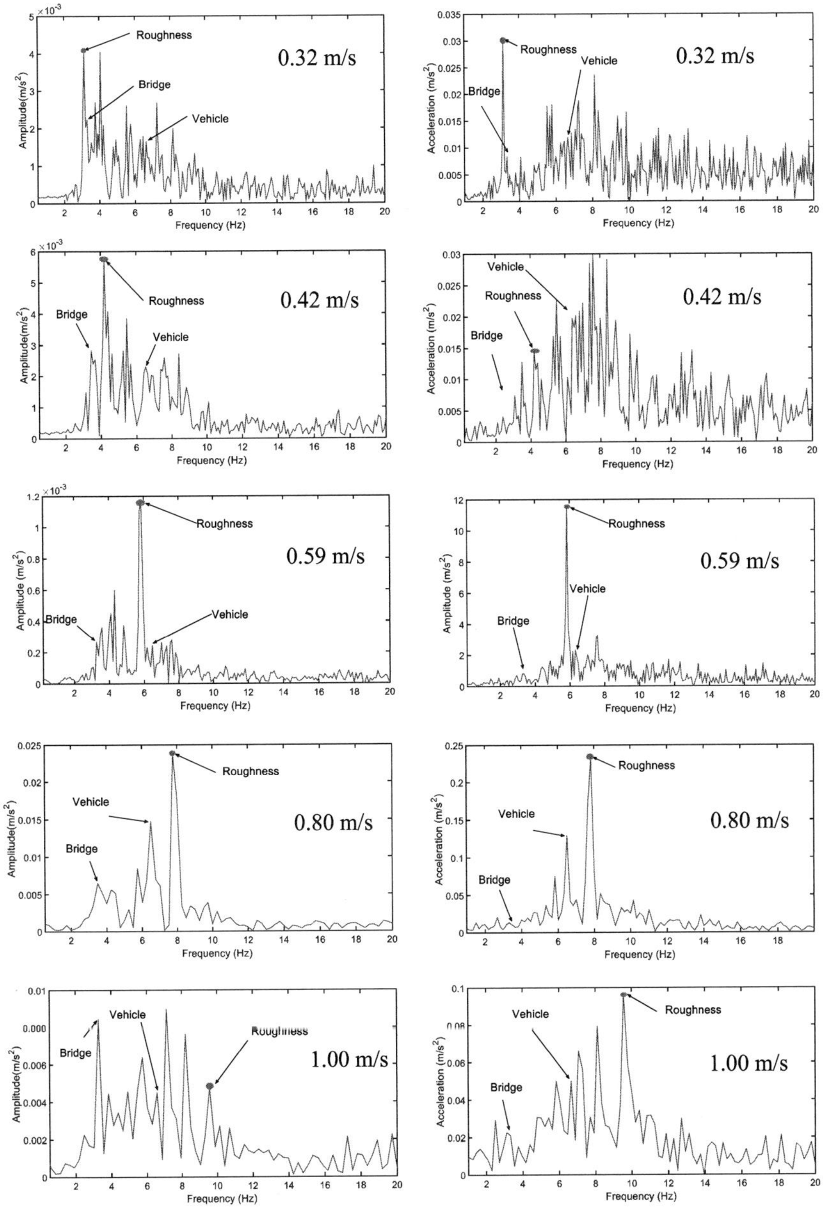

(a) Bridge dynamics

(b) Vehicle dynamics

Fig. 8. Fourier amplitude of vehicle and bridge accelerations at five vehicle speeds (red solid circles denoted as "Roughness" show the frequency caused by the road roughness-relevant vehicle vibration).

Table 6. Road roughness-relevant vehicle vibration frequency at five speeds.

Speed (m/s)	Frequency (Hz)		
	Experiment	Theoretical result	Error (%)
0.32	3.1	3.2	4.68
0.43	4.1	4.4	6.06
0.59	5.8	6.1	5.83
0.80	7.7	8.1	5.23
1.00	9.6	10.1	5.27

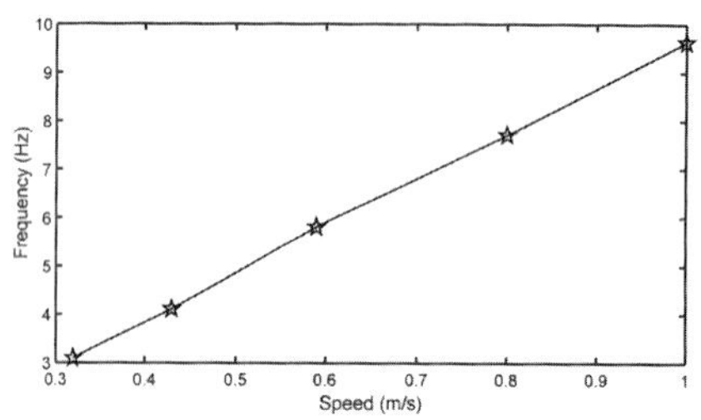

(a) 11.83 kg

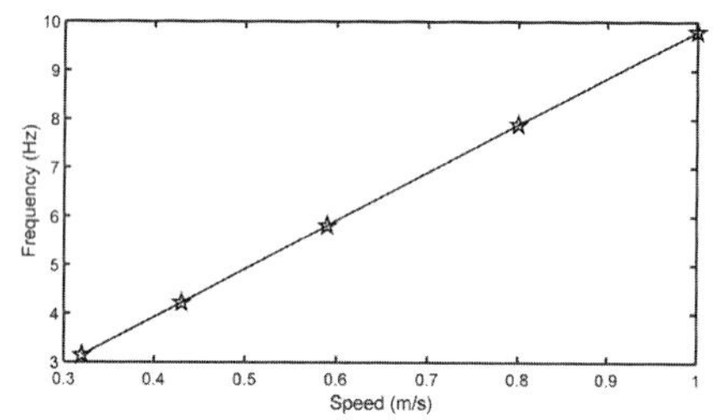

(b) 15.82 kg

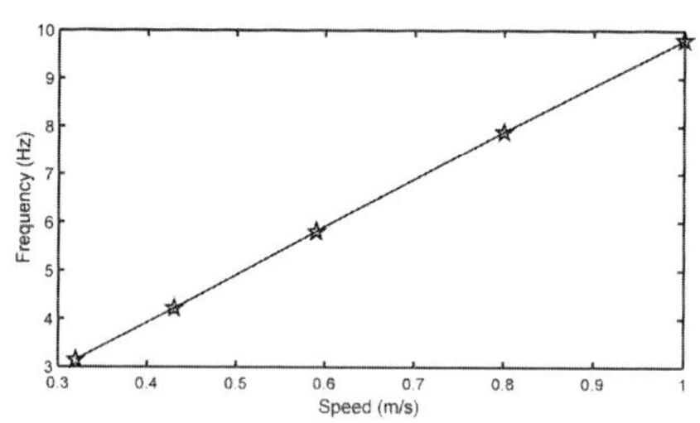

(c) 21.68 kg

Fig. 9. Relation between moving speed and the roughness-relevant vehicle vibration frequency.

Table 7. Identified frequencies taken from Fourier spectra of bridge acceleration.

Speed	0.42 m/s	0.59 m/s	0.80 m/s	1.00 m/s
Frequency	3.5 Hz	3.5 Hz	3.7 Hz	3.3 Hz

Taking the VBI system dynamics at 0.80 m/s as an example, those Fourier spectra of bridge vibration and vehicle vibration are presented in Fig. 10. In bridge vibration frequency spectra, the frequency corresponding to the most significant peak (F-MSP) between 3 Hz and 4 Hz remains almost constant along the 20 trips, as

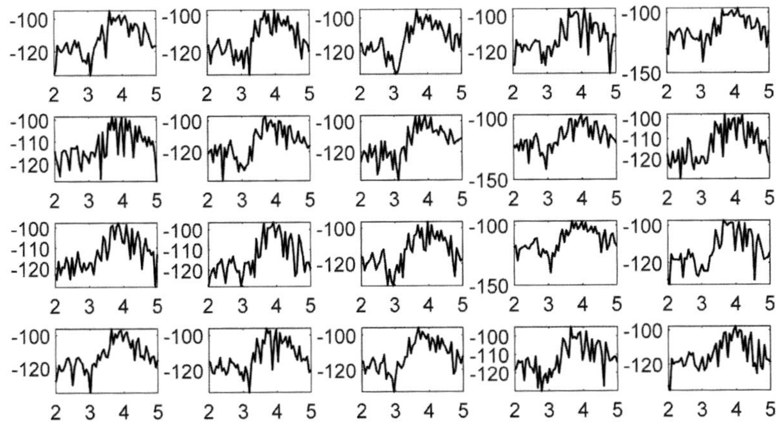

(a)

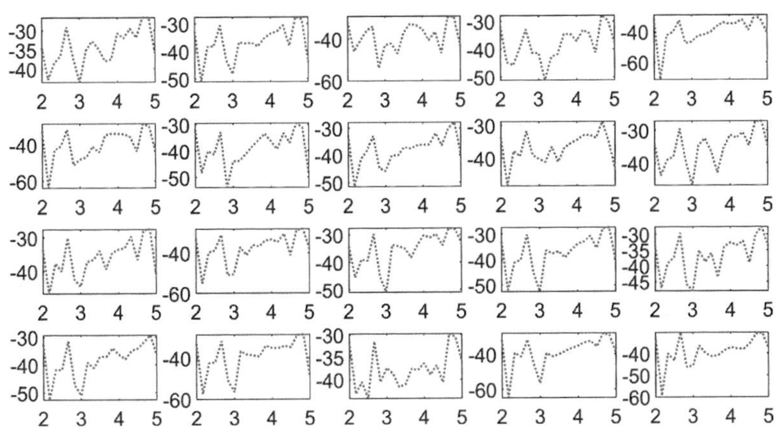

(b)

Fig. 10. PSD of bridge and vehicle accelerations from 20 trips under vehicle speed of 0.80 m/s. (a) PSD of bridge acceleration (horizontal axis, frequency (Hz); vertical axis, PSD (dB)) and (b) PSD of vehicle acceleration (horizontal axis, frequency (Hz); vertical axis, PSD (dB)).

shown in Fig. 10(a). The F-MSP variation between 3.6 Hz and 3.7 Hz demonstrates that, with bridge vibration, the bridge-relevant component is always more important than the road roughness-relevant components and noises. However, for the vehicle vibration, as shown in Fig. 10(b), the bridge vibration-relevant components are not as significant as that shown in Fig. 10(a). From multiple trips with the same running speed, the F-MSP in the vehicle spectra is identified as multiple values 3.2 Hz, 3.3 Hz, 3.5 Hz, 3.7 Hz, or 3.8 Hz. It is supposed that the F-MSP might be corresponding to road roughness or measurement noise rather than bridge vibration. Consequently, for a single trip, it is difficult to ascertain whether an F-MSP is

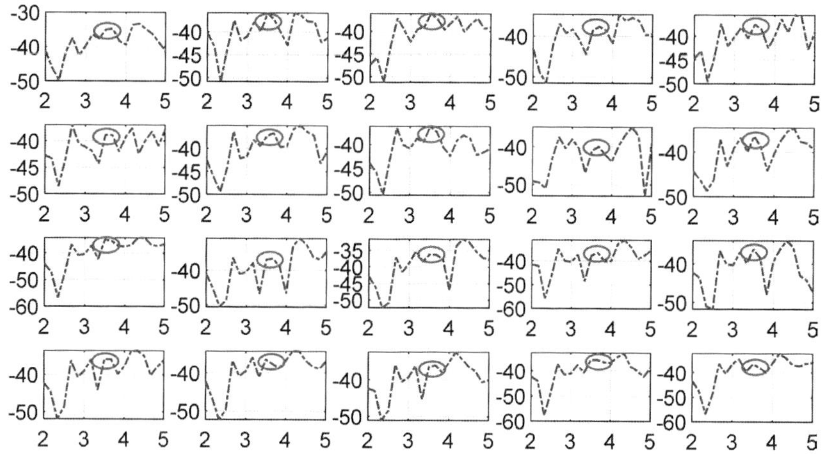

Fig. 11. Cross-power spectra of vehicle acceleration between 0.43 m/s and 0.80 m/s (the significant peaks, which denotes dominant frequencies of the bridge vibration, between 3.5 Hz and 3.667 Hz are marked with ellipses)(horizontal axis, frequency (Hz); vertical axis, PSD (dB)).

a bridge frequency, a road roughness-related frequency, or a frequency induced by noise without the other information from the other trips.

To identify the bridge frequency more clearly, 20 cross-power spectra of vehicle acceleration between 0.42 m/s and 0.80 m/s are investigated as presented in Fig. 11, which shows that the F-MSP in the cross-power spectra is almost constant. Dominant frequencies between 3.5 Hz and 3.7 Hz, which were supposed to be the target first frequency of the bridge, are observed. To emphasize the validity of the cross-power spectra, the vehicle auto-power spectra are overlapped with cross-power spectra, as presented in Fig. 12 (marked with circles). In the range of 3–4 Hz of the cross-power spectra, the F-MSPs have the same value in multiple trips. In comparison to Fig. 10(b), the bridge frequency can be more intuitively identified in the cross-power spectra. Furthermore, two frequencies between 3.5 Hz and 3.7 Hz are closely related, respectively, to the bridge frequency when the vehicle speeds are 0.42 m/s and 0.80 m/s.

Averaged cross-power spectra between other speeds are estimated and presented in Fig. 13. There are obviously significant F-MSPs in the range of 3 Hz and 4 Hz in these spectra of vehicle acceleration. Significant frequencies among all these cross-power spectra are presented in Table 8. Table 9 presents the summarized bridge frequencies under different speeds. A comparison of these bridge frequencies of bridge vibrations (Table 7) and cross-power spectra (Table 9) shows that the errors are few, which validates the proposed method as effective to identify a bridge frequency.

Moreover, the identified frequencies at 0.42 m/s and 0.59 m/s (Table 9) turned out to have equal values. In fact, the frequencies identified from bridge vibration spectra at these two speeds are nearly equal, making it difficult to distinguish them

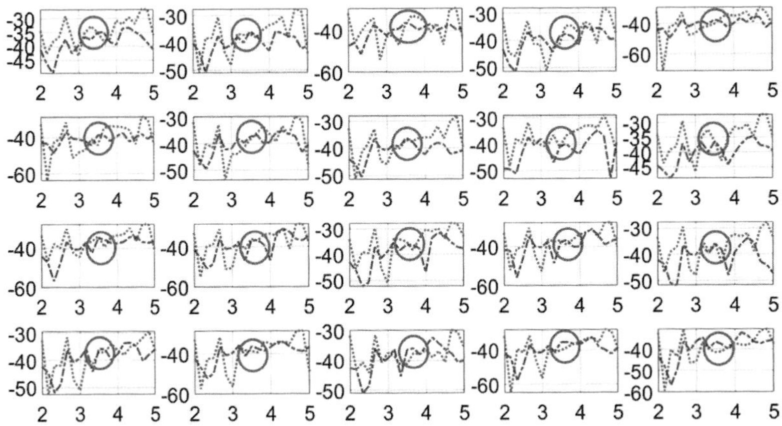

Fig. 12. Cross-power spectra from accelerations measured under vehicle speeds of 0.43 m/s and 0.80 m/s (blue dashed lines) and the PSD spectra from acceleration at the speed of 0.80 m/s (red dot lines) (Th comparison between these two spectra in the range of 3–4 Hz are marked with circles) (horizontal axis, frequency (Hz); vertical axis, PSD (dB)).

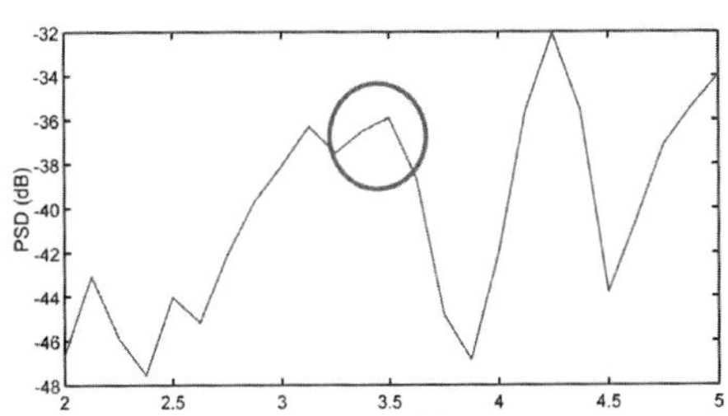

(a) 0.43 m/s and 0.59 m/s

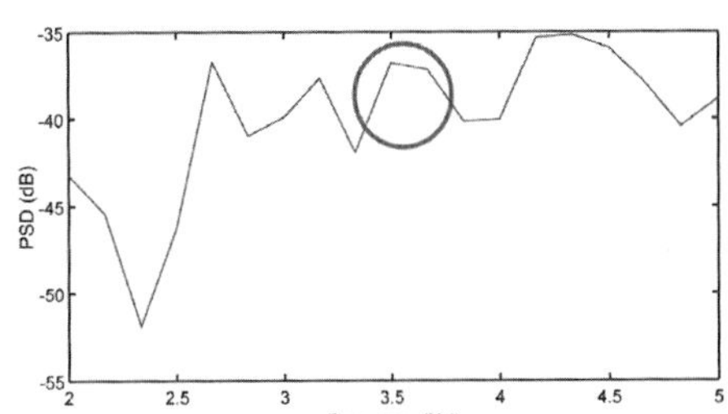

(b) 0.43 m/s and 0.80 m/s

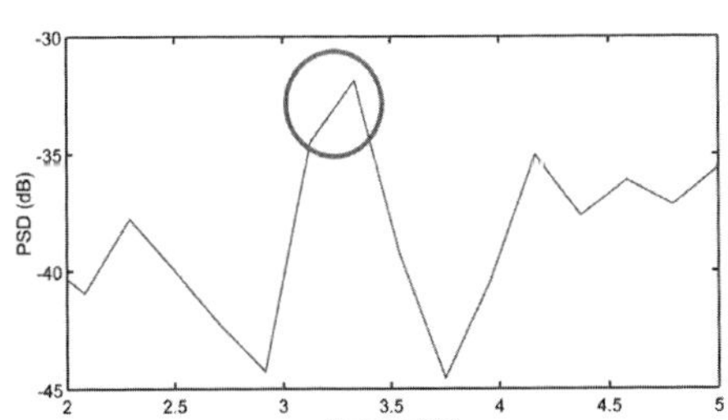

(c) 0.43 m/s and 1.00 m/s

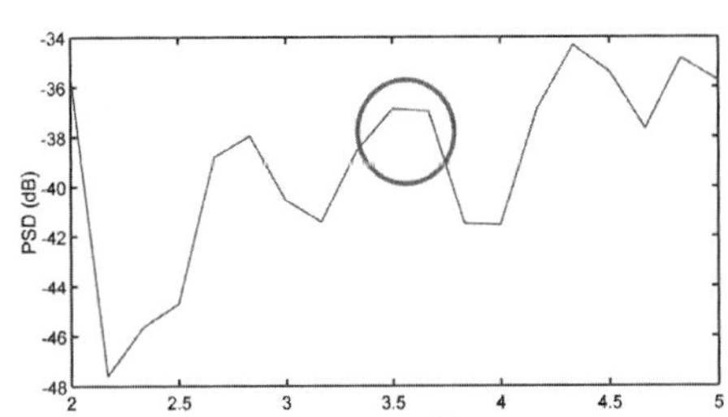

(d) 0.59 m/s and 0.80 m/s

Fig. 13. Averaged cross-power spectra from accelerations different under vehicle speeds. (The circles marked the dominate frequencies. Values of these frequencies are listed in Table 8).

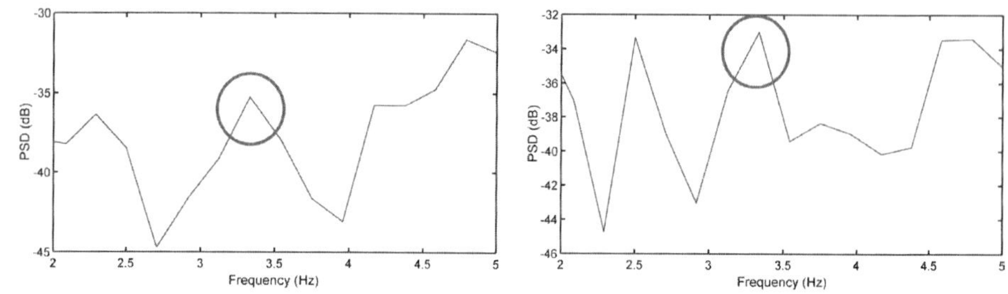

(e) 0.59 m/s to 1.00 m/s (f) 0.80 m/s to 1.00 m/s

Fig. 13. (*Continued*)

Table 8. Identified frequencies of cross-power spectra.

	0.43 m/s	0.59 m/s	0.80 m/s	1.00 m/s
0.43 m/s	—	3.50 Hz	3.50 Hz/3.67 Hz	3.33 Hz
0.59 m/s	3.50 Hz	—	3.50 Hz/3.67 Hz	3.33 Hz
0.80 m/s	3.50 Hz/3.67 Hz	3.50 Hz/3.67 Hz	—	3.33 Hz
1.00 m/s	3.33 Hz	3.33 Hz	3.33 Hz	—

Table 9. Bridge frequency identified from cross-power spectra under different speeds.

Speed	0.42 m/s	0.59 m/s	0.80 m/s	1.00 m/s
Frequency	3.50 Hz	3.50 Hz	3.67 Hz	3.33 Hz
Error	−0.57%	1.19%	−0.46%	1.93%

from one another. More importantly, the influence of the moving speed on changes in the bridge frequency is presumed to be very slight for these two low speeds. Therefore the bridge frequency in the VBI system was identified as approximately 3.50 Hz. Comparison with the tested bridge frequency (3.59 Hz, presented in Table 5) with free vibrations demonstrates that the frequency identification achieves high precision.

5. Conclusions

A novel method was proposed to identify a bridge frequency by reducing the potential negative influence of road roughness using the cross-power spectrum of the acceleration responses collected from multiple moving instrumented vehicles. The theoretical dynamics of a VBI system considering road roughness was first derived from dynamic equations. Components of vehicle dynamics were analyzed focusing on the relation between their frequencies and the moving speed. Subsequently, another theoretical derivation was conducted to illustrate the mechanism of the cross-power spectrum for reducing the influence of road roughness. Finite element simulations and several laboratory experiments were used to validate the proposed method's derived relation and effectiveness. The salient conclusions obtained from this study are presented below.

(1) In a VBI system, road roughness acts as a time-varying moving force on the bridge. Theoretically, the road roughness-relevant bridge vibration would also excite vehicle vibration. Furthermore, road roughness would directly induce a kind of vehicle vibration.
(2) The vehicle dynamics consists of vibration components corresponding to the bridge frequency, vehicle frequency, driving force frequency, coupling bridge-driving force frequency, and coupling road roughness-driving force frequency. The bridge frequency and the vehicle frequency remain almost constant at different moving speeds; the other frequencies are linearly correlated to the moving speed.
(3) The potential negative influence of road roughness can be eliminated, theoretically, by calculating the cross-power spectra among vehicle accelerations at different moving speeds. Vehicle dynamic components for which frequencies are linearly correlated to the moving speed will be filtered out from the cross-power spectrum. Components corresponding to the constant bridge frequency can then be identified.
(4) Results of finite element simulations and laboratory experiments will validate the theoretically-derived relation and demonstrate the effectiveness of the proposed method for the bridge frequency. Specifically in the experiments, although the bridge frequency identified from the bridge vibration spectra varies at different moving speeds, the proposed method can also capture the bridge frequency from cross-power spectra of vehicle dynamics, thereby reducing the potential influence of road roughness and noise.

This method is presumed to be useful for practical engineering. It presents the benefits of methods using vehicle-borne instruments over the general method using sensors installed on bridges. Moreover, it provides an advanced solution to bridge-frequency identification with clear physical interpretations, which can be expected to be generally applicable. Also, the proposed method is easier to implement in practice in comparison with some existing methods because this method does not need to regulate strictly the multiple vehicles' speeds and trajectories. Nevertheless, some conditions restricting the utility of the proposed method are as follows: (1) The necessity for prior knowledge of the vehicle frequency, the non-correlated assumption of frequencies between multiple vehicles, the bridge, and road roughness; (2) the bridge frequency and the vehicle frequency must not be too similar; and (3) multiple trips for each speed are recommended to achieve precise results, which might also be time-consuming; (4) potential effect from structural and vehicle damping. Solutions to these restrictions will be sought through future research.

Acknowledgments

A part of this work is supported as a JSPS Bilateral Joint Research Project by Grant No. JPJSBP120217405, which is greatly appreciated. The first author would like to thank the International Postdoctoral Exchange Fellowship Program founded by Office of China Postdoc Council (OCPC).

ORCID

Xuzhao Lu https://orcid.org/0000-0003-0215-2756
Chul-Woo Kim https://orcid.org/0000-0002-2727-6037
Kai-Chun Chang https://orcid.org/0000-0002-3097-0898
Zhuoran Han https://orcid.org/0009-0007-4322-1101
Limin Sun https://orcid.org/0000-0002-6370-6559

References

1. A. Elhattab, N. Uddin and E. J. O'Brien, Drive-by bridge frequency identification under operational roadway speeds employing frequency independent underdamped pinning stochastic resonance (FI-UPSR), *Sensors* **18**(12) (2018) 4207.
2. P. Asadollahi and J. Li, Statistical analysis of modal properties of a cable-stayed bridge through long-term wireless structural health monitoring, *J. Bridge Eng.* **22**(9) (2017) 04017051.
3. Y. B. Yang, C. W. Lin and J. D. Yau, Extracting bridge frequencies of the dynamic response of a passing vehicle, *J. Sound Vib.* **272**(3–5) (2004) 471–493.
4. C. W. Kim, R. Isemoto, P. J. McGetrick, M. Kawatani and E. J. O'Brien, Drive-by bridge inspection from three different approaches, *Smart Struct. Syst.* **13** (2014) 775–796.
5. A. Malekjafarian, P. J. McGetrick and E. J. O'Brien, A review of indirect bridge monitoring using passing vehicles, *Shock Vib.* **2015** (2015) 286139.
6. A. Malekjafarian, R. Corbally and W. Gong, A review of mobile sensing of bridges using moving vehicles: Progress to date, challenges and future trends, *Structures* **44** (2022) 1466–1489.
7. Y. B. Yang, Y. C. Li and K. C. Chang, Using two connected vehicles to measure the frequencies of bridges with rough surface: A theoretical study, *Acta Mech.* **223** (2012) 1851–1861.
8. Y. Zhan and F. T. K. Au, Bridge surface roughness identification based on vehicle–bridge interaction, *Int. J. Struct. Stab. Dyn.* **19** (2019) 1950069.
9. A. Malekjafarian and E. J. O'Brien, On the use of a passing vehicle for the estimation of bridge mode shapes, *J. Sound Vib.* **397** (2017) 77–91.
10. Y. B. Yang, X. Q. Mo, K. Shi, Z. L. Wang, H. Xu and Y. T. Wu, Contact residue for simultaneous removal of vehicle's frequency and surface roughness in scanning bridge frequencies using two connected vehicles, *Int. J. Struct. Stab. Dyn.* **21**(13) (2021) 2171006.

11. X. Jian, Y. Xia and L. Sun, An indirect method for bridge mode shapes identification based on wavelet analysis, *Struct. Control Health Monit.* **27** (2020) e2630.
12. C. Tan, A. Elhattab and N. Uddin, "Drive-by" bridge frequency-based monitoring utilizing wavelet transform, *J. Civ. Struct. Health Monit.* **7** (2017) 615–625.
13. C. Tan and N. Uddin, Hilbert transform based approach to improve extraction of "drive-by" bridge frequency, *Smart Struct. Syst.* **25**(3) (2020) 265–277.
14. N. Jin, Y. B. Yang, E. G. Dimitrakopoulos, T. S. Paraskeva and L. S. Katafygiotis, Application of short-time stochastic subspace identification to estimate bridge frequencies of a traversing vehicle. *Eng. Struct.* **230** (2021) 111688.
15. Y. B. Yang and C. W. Lin, Vehicle–bridge interaction dynamics and potential applications, *J. Sound Vib.* **284** (2005) 205–226.
16. J. Liu, S. Chen, M. Bergés, J. Bielak, J. H. Garrett, J. Kovačević and H. Y. Noh, Diagnosis algorithms for indirect structural health monitoring of a bridge model via dimensionality reduction, *Mech. Syst. Signal Process.* **136** (2020) 106454.
17. P. Cheema, M. Makki Alamdari, K. C. Chang, C. W. Kim and M. Sugiyama, A drive-by bridge inspection framework using non-parametric clusters over projected data manifolds, *Mech. Syst. Signal Process.* **180** (2022) 109401.
18. K. Kaur, M. Makki Alamdari, E. Atroshchenko, K. C. Chang and C. W. Kim, Unsupervised learning-based framework for indirect structural health monitoring using adversarial autoencoder, *J. Sound Vib.* **550** (2023) 117598.
19. M. Agostinacchio, D. Ciampa and S. Olita, The vibrations induced by surface irregularities in road pavements — a Matlab® approach, *Eur. Transp. Res. Rev.* **6** (2014) 267–275.
20. N. Shirzad-Ghaleroudkhani, Q. Mei and M. Gül, Frequency identification of bridges using smartphones on vehicles with variable features, *J. Bridge Eng.* **25** (2020) 04020041.
21. T. Nagayama, A. P. Reksowardojo, D. Su and T. Mizutani, Bridge natural frequency estimation by extracting the common vibration component from the responses of two vehicles, *Eng. Struct.* **150** (2017) 821–829.
22. R. W. Clough and J. Penzien, *Dynamics of Structures* (McGraw-Hill Companies, New York, 1975).
23. P. Paultre, O. Chaallal and J. Proulx, Bridge dynamics and dynamic amplification factors — a review of analytical and experimental findings, *Can. J. Civil Eng.* **19**(2) (1992) 260–278.
24. W. Y. He and S. Zhu, Moving load-induced response of damaged beam and its application in damage localization, *J. Vib. Control.* **22**(16) (2016) 3601–3617.
25. H. Von Storch and W. Z. Francis, *Statistical Analysis in Climate Research* (Cambridge University Press, 2002).
26. X. Lu, C. W. Kim and K. C. Chang, Finite element analysis framework for dynamic vehicle–bridge interaction system based on ABAQUS, *Int. J. Struct. Stab. Dyn.* **20**(3) (2020) 2050034.
27. Y. B. Yang, J. D. Yau, S. Urushdaze and T. Y. Lee, Historical review on resonance and cancellation of simply supported beams subjected to moving train loads: From theory to practice, *Int. J. Struct. Stab. Dyn.* (2023) 2340008, https://doi.org/10.1142/S0219455423400084.
28. E. J. O'Brien, S. Wilson, J. Keenahan and Y. Ren, A Bayesian approach to the estimation of road profile and bridge damage from a fleet passing vehicle measurements, *Int. J. Struct. Stab. Dyn.* (2023) 2450040, https://doi.org/10.1142/S0219455424500408.

29. J. P. Yang and C. C. Wang, Analytical formulations for amplifier-enhanced VBI systems, *Int. J. Struct. Stab. Dyn.* (2023) 2471001, https://doi.org/10.1142/S02194554247100019.
30. Y. B. Yang and K. C. Chang, Extraction of bridge frequencies from the dynamic response of a passing vehicle enhanced by the EMD technique, *J. Sound Vib.* **322** (4–5) (2009) 718–739.
31. M. Smith, *ABAQUS/Standard User's Manual, Version 6.9* (Dassault Systèmes Simulia Corp., 2009).
32. International Organization for Standardization, ISO 8608:2016. Mechanical vibration — Road surface profiles — Reporting of measured data, Standard.
33. The MathWorks Inc., MATLAB version: 9.11.0 (R2021b), Natick, Massachusetts: The MathWorks Inc (2021), https://www.mathworks.com.

https://doi.org/10.1142/9789819814770_0033

Chapter 33

On Time-Domain Solutions to the Free Vibration of SDOF Systems with Linear Hysteretic Damping[#]

Wenjun Gao[*,‡] and Xilin Lu[†,§]

[*]*Department of Disaster Mitigation for Structures*
College of Civil Engineering, Tongji University
Shanghai, P. R. China

[†]*State Key Laboratory of Disaster Reduction in Civil Engineering*
Tongji University, Shanghai, P. R. China
[‡]*2014Joker@tongji.edu.cn*
[§]*lxlst@tongji.edu.cn*

This article focuses on the free vibration of single-degree-of-freedom (SDOF) systems with linear hysteretic damping and investigates the consistency and convergency of the solution to the free vibration in time domain. The criteria associated with the unit-impulse response function at $t = 0$ are suggested to check the consistency of possible solutions. Exhaustive and systematic derivations are presented to clarify the confusing results in the literature. Subsequently, the integration in the Hilbert term proves to be nonconvergent when an exponential term ($e^{\bar{s}t}$) is assumed to be the general form of displacement solutions. Hence, it can be concluded that the exact solution to the free vibration of SDOF systems with linear hysteretic damping is still an open question. At last, a numerical verification method is proposed to assess the accuracy of possible solutions.

Keywords: Linear hysteretic damping; free vibration; unit-impulse response function; single-degree-of-freedom system; exact solution.

1. Introduction

Linear hysteretic damping, also called complex stiffness damping,[1] complex damping,[2] hysteretic damping[3] or structural damping,[4] is introduced based on large quantities of the experimental data and observations resulting from solid material tests, especially for construction materials. Most material constitutive tests show

[‡]Corresponding author.
[#]This chapter appeared previously on the International Journal of Structural Stability and Dynamics. To cite this chapter, please cite the original article as the following: W. Gao & X. Lu, *Int. J. Struct. Stab. Dyn.*, **23**, 2340036 (2023), DOI: 10.1142/S0219455423400369.

that the energy losses of solid materials per strain cycle are approximately proportional to the square of strain amplitude and nearly independent of strain rate over a considerable frequency range ($0.01 \sim 100\,\mathrm{Hz}$).[5]

Indeed, the classic viscous damping model that introduces damping forces related to velocities definitely leads to the frequency-dependent energy loss in the steady state of harmonic excitation. Recent discussions can be found in the literature.[6] For instance, $m\ddot{u} + c\dot{u} + ku = p\sin(\hat{\omega}t)$, where m, c, k, t, p, $\hat{\omega}$, u, $\dot{u}$ and $\ddot{u}$ are the mass, viscous damping coefficient, stiffness, time, amplitude of the excitation force, circular frequency of the excitation force, displacement, velocity and acceleration, respectively. In the steady state, the displacement amplitude ρ can be determined using the expression $\rho(k,\xi,p,\beta) = (p/k)/\sqrt{(1-\beta^2)^2 + (2\xi\beta)^2}$, where $\beta = \hat{\omega}/\omega$, $\omega = \sqrt{k/m}$, $\xi = c/(2m\omega)$ are the frequency ratio, undamped circular frequency and viscous damping ratio, respectively. According to the expression of $\rho(k,\xi,p,\beta)$, it can yield three different steady-state systems with the same ρ, k, ξ but different p, β as illustrated in Fig. 1(a), where the different excitation-force amplitudes are denoted as p_1, p_2, p_3 and the corresponding frequency ratios are denoted as β_1, β_2, β_3. Thus, the energy losses of the three viscously damped systems indicated in Fig. 1(a) can be calculated by $W_{\mathrm{d},i} = \pi k\rho_0^2(2\xi\beta_i)$, $i = 1,2,3$. The values of $W_{\mathrm{d},i}$ ($i = 1,2,3$) are equal to the areas of the ellipses generated by the damping force-displacement relationship in the steady state, as shown in Fig. 1(b). Although the vibration amplitudes are exactly the same, the energy losses are different. This feature violates the phenomenon observed in most material constitutive tests, especially for construction materials. Certainly, more than three steady-state systems under different values of p and β can be marked in Fig. 1(a), but the marked three cases sufficiently clarify the fact. By contrast, following a similar derivation process, the complex stiffness damping model[1] leads to the same energy losses regardless of the different frequency ratios [denoted as $\tilde{\beta}_1$, $\tilde{\beta}_2$, $\tilde{\beta}_3$ in Fig. 1(c)] and excitation-force amplitudes [denoted as $\tilde{p}_1$, $\tilde{p}_2$, $\tilde{p}_3$ in Fig. 1(c)] as long as the damped systems have the same complex-stiffness damping ratio ζ, stiffness k and vibration amplitude $\tilde{\rho}_0$ in the steady state of harmonic excitation.

These characteristics of the complex stiffness model are quite suitable for building structures whose fundamental modes usually fall in low frequencies,[7] regardless of more complicated damping mechanisms suitable for composite structures[8,9] or thermoelastic vibrations.[10] Although rate-independent linear damping may lead to similar structural performance in seismic analyses of building structures as rate-dependent damping does when their energy loss factors are close to each other, it has been systematically demonstrated that different damping mechanisms truly affect the optimal form in structural optimization problems involving dynamics.[11,12]

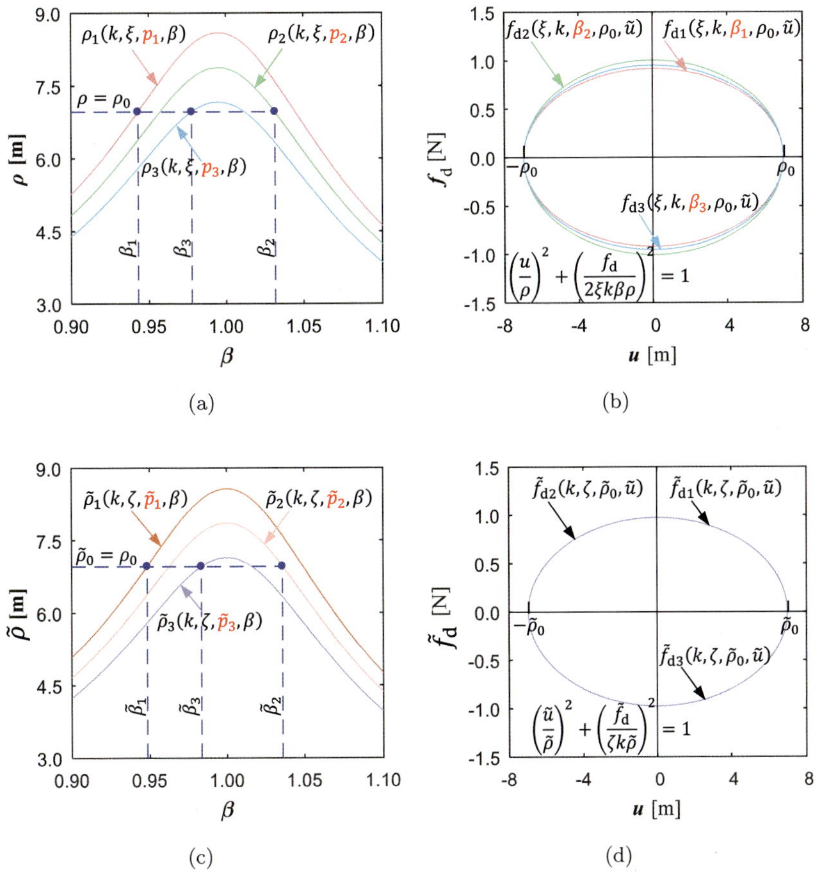

Fig. 1. Vibration amplitude-frequency relationships and energy losses in different damping models. (a) Vibration amplitude-frequency relationships in viscously damped SDOF systems. (b) Energy losses in viscously damped SDOF systems. (c) Vibration amplitude-frequency relationships in SDOF systems with the complex stiffness damping model. (d) Energy losses in SDOF systems with the complex stiffness damping model.

In 1995, Inaudi *et al.* suggested a set of differential equations for systems with linear hysteretic damping.[13] This damping model outweighs previous models in terms of allowing a consistent Fourier transform between the frequency-domain and time-domain equations. Although its convenience and consistency are widely recognized, the linear hysteretic damping model[13] indeed causes difficulties in solving the corresponding free vibration in time domain. The time-domain analytical solution to free vibration with the linear hysteretic damping model is of paramount importance. It provides further insights into the decay mechanism of vibration amplitude and damped fundamental frequency. Also, the analytical solution helps to establish a theoretical model in damping tests utilizing dynamic properties of free vibration. In addition, it can be viewed as a basic but reliable reference result when other sophisticated rate-independent damping models need to be investigated. Nevertheless, no literature has reported the analytical solution to the free vibration problem so far.

In this study, the criteria associated with the unit-impulse response function at $t = 0$ are suggested to check the consistency of possible solutions. Then, the detailed demonstrations on the divergence issues in solving the free vibration of SDOF systems with the linear hysteretic damping model are presented. Also, a simple numerical verification method is proposed as valid and effective tool to assess the accuracy of possible solutions. This study provides a mathematically distinguished path to clarifying the underlying divergence nature in the linear hysteretic damping model.[13]

2. SDOF Systems with Linear Hysteretic Damping

The differential equation of motion for an SDOF system with linear hysteretic damping can be written in the following form,[13]

$$m\ddot{u}(t) + ku(t) - \eta k\hat{u}(t) = f(t), \tag{1}$$

where m, k, η, $\ddot{u}(t)$, $u(t)$ and $f(t)$ are the mass, stiffness, hysteretic damping coefficient, acceleration, displacement and external force in the SDOF system, respectively; $\hat{u}(t)$ is denoted as the Hilbert transform result of $u(t)$ according to the Ref. 14 and also called the Hilbert term in this study. The transform from $u(t)$ to $\hat{u}(t)$ is expressed as follows:

$$\hat{u}(t) = H[u(t)] = \frac{1}{\pi}\int_{-\infty}^{+\infty} \frac{u(\tau)}{t-\tau} d\tau, \tag{2}$$

where $H[\cdot]$ denotes the Hilbert transform. Obviously, the Hilbert transform of a time-domain signal still leads to a time-domain signal. Note that the integral range in Eq. (2) is $(-\infty, +\infty)$ and the Hilbert term $\hat{u}(t)$ needs the displacement information over the entire time domain. Therefore, the system represented by Eq. (2) is a noncausal system and cannot be physically implemented.[15] This noncausality is also inherited by the system in Eq. (1), which means that the SDOF system in Eq. (1) cannot be a physical model.

When the external force $f(t)$ in Eq. (2) is constantly equal to zero, the differential equation of free vibration is yielded as

$$\ddot{u}(t) + \omega^2 u(t) - \eta\omega^2\hat{u}(t) = 0, \tag{3}$$

where ω is the circular frequency of the SDOF system and $\omega = \sqrt{k/m}$.

Sun *et al.* claim the time-domain solution to the displacement in Eq. (3) can be expressed as follows[16,17]:

$$u(t) = \mathrm{e}^{-\lambda_{\mathrm{H}} t}[A_1 \cos(\omega_{\mathrm{H}} t) + A_2 \sin(\omega_{\mathrm{H}} t)] = \rho \cdot \mathrm{e}^{-\lambda_{\mathrm{H}} t} \cdot \cos(\omega_{\mathrm{H}} t - \phi) \tag{4}$$

$$\begin{cases} \omega_{\mathrm{H}} = \sqrt{\frac{1}{2}\omega^2 + \frac{1}{2}\sqrt{\omega^4 + \eta^2\omega^4}} \\ \lambda_{\mathrm{H}} = \frac{\eta\omega^2\sqrt{2}}{2\sqrt{\omega^2 + \sqrt{\omega^4 + \eta^2\omega^4}}} \\ A_1 = u(0) \\ A_2 = \frac{1}{\omega_{\mathrm{H}}}(\dot{u}(0) + \lambda_{\mathrm{H}} u(0)) \\ \rho = \sqrt{A_1^2 + A_2^2} \\ \phi = \tan^{-1}\left(\frac{A_2}{A_1}\right), \end{cases} \tag{5}$$

where $u(0)$ and $\dot{u}(0)$ are the initial displacement and velocity of the SDOF system. Also note that ω_{H} and λ_{H} in Eq. (5) can be rewritten as

$$\begin{cases} \omega_{\mathrm{H}} = \frac{\eta\omega}{\sqrt{2(\sqrt{1+\eta^2} - 1)}} \\ \lambda_{\mathrm{H}} = \omega\sqrt{\frac{1}{2}(\sqrt{1+\eta^2} - 1)} \end{cases} \tag{6}$$

and

$$\lambda_{\mathrm{H}}^2 + \omega_{\mathrm{H}}^2 = \omega^2\sqrt{1+\eta^2} \tag{7}$$

$$2\lambda_{\mathrm{H}}^2 = \omega^2(\sqrt{1+\eta^2} - 1) \tag{8}$$

$$\lambda_{\mathrm{H}}\omega_{\mathrm{H}} = \frac{1}{2}\eta\omega^2. \tag{9}$$

When $\eta \to 0$, it yields

$$\frac{\omega_{\mathrm{H}}}{\lambda_{\mathrm{H}}} = \frac{1 + \sqrt{1+\eta^2}}{\eta} \approx \frac{2}{\eta}. \tag{10}$$

3. Unit-Impulse Response

By performing the Fourier transform on Eq. (1), the vibration of the SDOF system with the linear hysteretic damping can be expressed in frequency domain as follows[13]:

$$-m\varpi^2 U(\varpi) + kU(\varpi) + \mathrm{i}\eta k \cdot \mathrm{sgn}(\varpi) U(\varpi) = F(\varpi) \tag{11}$$

$$\begin{cases} U(\varpi) = \Gamma[u(t)] = \displaystyle\int_{-\infty}^{+\infty} u(t) \mathrm{e}^{-\mathrm{i}\varpi t} \mathrm{d}t \\ F(\varpi) = \Gamma[f(t)] = \displaystyle\int_{-\infty}^{+\infty} f(t) \mathrm{e}^{-\mathrm{i}\varpi t} \mathrm{d}t \\ \Gamma[-\eta k \hat{u}(t)] = \displaystyle\int_{-\infty}^{+\infty} -\eta k \hat{u}(t) \mathrm{e}^{-\mathrm{i}\varpi t} \mathrm{d}t = \mathrm{i}\eta k \cdot \mathrm{sgn}(\varpi) \Gamma[u(t)], \end{cases} \tag{12}$$

where $\Gamma[\cdot]$ denotes the Fourier transform; i represents an imaginary unit; the symbol, ϖ, indicates that the domain of definition has been transformed into frequency domain by performing the Fourier transform.

The transfer function $H_{\mathrm{H}}(\varpi)$ for Eq. (11) is expressed as

$$H_{\mathrm{H}}(\varpi) = \frac{1}{-m\varpi^2 + k + \mathrm{i}\eta k \cdot \mathrm{sgn}(\varpi)}. \tag{13}$$

Then, according to dynamics theories,[1] the inverse Fourier transform of the transfer function $H_H(\varpi)$ results in the unit-impulse response function of the SDOF system with the linear hysteretic damping

$$h_{\mathrm{H}}(t) = \frac{1}{2\pi} \int_{-\infty}^{+\infty} H_{\mathrm{H}}(\varpi) \mathrm{e}^{\mathrm{i}\varpi t} \mathrm{d}\varpi. \tag{14}$$

It turns out to be very difficult to obtain the integration result of Eq. (14). As far as we know, there is no literature reporting that integration result. Fortunately, when $t = 0$, the integrand in Eq. (14) can be extremely simplified, leading to the analytical result proposed by Hu[18] as follows:

$$h_{\mathrm{H}}(0) = \frac{1}{2\pi} \int_{-\infty}^{+\infty} H_{\mathrm{H}}(\varpi) \mathrm{d}\varpi = -\frac{1}{2\sqrt{mk}} \sqrt{\frac{-1 + \sqrt{1+\eta^2}}{2(1+\eta^2)}}. \tag{15}$$

Hu[18] also pointed out that $h_{\mathrm{H}}(0) < 0$. This conclusion is also reported in the Ref. 17. It is worth mentioning that $h_{\mathrm{H}}(0) < 0$ indicates a nonzero displacement before the affection of a unit impulse, according to Eq. (15), which violates physics in the real world.

Another approach to derive the unit-impulse response function of an SDOF system with the linear hysteretic damping is based on the definition and utilizes the displacement solution to the free vibration.

Provided that the initial displacement and velocity are 0 and $1/m$ (where m is the mass), respectively, in the displacement solution, the SDOF system exactly possesses a unit momentum. Then, the displacement response of the SDOF system under these specific conditions is definitely the same as its unit-impulse response function. For example, the displacement solution to an undamped SDOF system is written as[1]

$$u(t) = u(0) \cdot \cos(\omega t) + \frac{\dot{u}(0)}{\omega} \sin(\omega t), \tag{16}$$

where ω is the circular frequency of the SDOF system and $\omega = \sqrt{k/m}$. When $u(t) = 0$ and $\dot{u}(t) = 1/m$, the displacement solution in Eq. (16) becomes the unit-impulse response function $h(t)$ of the undamped SDOF system and can be expressed as

$$h(t) = \frac{1}{m\omega} \sin(\omega t). \tag{17}$$

Another example is the SDOF system with linear viscous damping. In this case, the general displacement solution is written as[1]

$$u(t) = \mathrm{e}^{-\xi\omega t}\left[u(0) \cdot \cos(\omega_{\mathrm{D}} t) + \frac{\dot{u}(0) + \xi\omega \cdot u(0)}{\omega_{\mathrm{D}}} \sin(\omega_{\mathrm{D}} t)\right], \tag{18}$$

where ξ is the viscous damping coefficient and $\omega_{\mathrm{D}} = \omega\sqrt{1-\xi^2}$. Again, by specifying $u(t) = 0$ and $\dot{u}(t) = 1/m$, the displacement solution in Eq. (18) becomes the unit-impulse response function $h_D(t)$ of the SDOF system with linear viscous damping and can be expressed as[1]

$$h_{\mathrm{D}}(t) = \mathrm{e}^{-\xi\omega t} \cdot \frac{1}{m\omega_{\mathrm{D}}} \sin(\omega_{\mathrm{D}} t). \tag{19}$$

Hence, using the general displacement solution of an SDOF system is a valid and reliable way to yield the exact expression of the unit-impulse response function. However, applying this golden rule to Eq. (4), which is viewed as the displacement solution to Eq. (3) that represents an SDOF system with the linear hysteretic damping in the Refs. 16 and 17 the final result is definitely contradictory to Eq. (15). Apparently, substituting $u(t) = 0$ and $\dot{u}(t) = 1/m$ into Eq. (4), it yields

$$\tilde{h}_{\mathrm{H}}(t) = \mathrm{e}^{-\lambda_{\mathrm{H}}\omega t} \cdot \frac{1}{m\omega_{\mathrm{H}}} \sin(\omega_{\mathrm{H}} t). \tag{20}$$

According to Eq. (20), $\tilde{h}_{\mathrm{H}}(0) = 0$ which violates the result in Eq. (15) where $h_{\mathrm{H}}(0) < 0$.

The purpose of Sec. 3 is to verify Eq. (4) in terms of its unit-impulse response. According to Eq. (4), the resulting unit-impulse response is expressed by Eq. (20). However, Eq. (20) yields a zero value when $t = 0$ and therefore contradicts the verification condition, namely, Eq. (15), which indicates that the exact unit-impulse response must achieve a negative value when $t = 0$. Hence the verification result proves that Eq. (4) is not the exact time-domain solution to the displacement in Eq. (3).

4. Divergence of the Hilbert Term

To investigate the inconsistency between Eqs. (15) and (20) further, it is worth reviewing the solving process of Eq. (3), specifically the convergence of the Hilbert term $\hat{u}(t)$. In this way, the obstacles in the solving process can be fully unveiled. Also, the failure of Eq. (4) as the exact solution to Eq. (3) is essentially validated.

Normally, solving the second-order linear differential equation in Eq. (3) assumes that it has a particular solution in the form of $\mathrm{e}^{\bar{s}t}$, where $\bar{s}$ is a complex number and denoted as $\bar{s} = \chi + \mathrm{i}\delta$ (χ and δ represent the real part and imaginary part respectively, and i is an imaginary unit). Then, the crucial problem is that whether the designated form, i.e. $\mathrm{e}^{\bar{s}t}$, can render the Hilbert term $\hat{u}(t)$, namely, the integral in Eq. (2), convergent. To address this issue, substituting $u(t) = \mathrm{e}^{\bar{s}t}$ into Eq. (2), and thus the Hilbert term $\hat{u}(t)$ is calculated as

$$\hat{u}(t) = H[\mathrm{e}^{\bar{s}t}] = \frac{1}{\pi}\int_{-\infty}^{+\infty}\frac{\mathrm{e}^{\bar{s}\cdot\tau}}{t-\tau}\mathrm{d}\tau = -\frac{1}{\pi}\int_{-\infty}^{+\infty}\frac{\mathrm{e}^{\bar{s}(\tau-t)}\cdot\mathrm{e}^{\bar{s}t}}{\tau-t}\mathrm{d}(\tau-t) = \mathrm{e}^{\bar{s}t}G \tag{21}$$

$$G = -\frac{1}{\pi}\int_{-\infty}^{+\infty}\frac{\mathrm{e}^{\bar{s}\kappa}}{\kappa}\mathrm{d}\kappa, \tag{22}$$

where Eq. (21) performs a variable substitution, i.e. $\kappa = \tau - t$.

Then, the convergence of G in Eq. (22) determines the convergence of the Hilbert term $\hat{u}(t)$. Then, the integral in Eq. (22) is rewritten as a contour integral as follows:

$$\begin{aligned} G &= \lim_{R,1/r\to+\infty} -\frac{1}{\pi}\left(\int_{\Gamma_N} + \int_{\Gamma_P}\right)\frac{\mathrm{e}^{\bar{s}\cdot\bar{\kappa}}}{\bar{\kappa}}\mathrm{d}\bar{\kappa} = \lim_{R,1/r\to+\infty}\frac{1}{\pi}\left(\int_{\mathrm{C}_R} + \int_{\mathrm{C}_r}\right)\frac{\mathrm{e}^{\bar{s}\cdot\bar{\kappa}}}{\bar{\kappa}}\mathrm{d}\bar{\kappa} \\ &= \lim_{R,1/r\to+\infty}\frac{1}{\pi}\left(\int_{\mathrm{C}_R} + \int_{\mathrm{C}_r}\right)\frac{\mathrm{e}^{\chi\bar{\kappa}}}{\bar{\kappa}}\cdot\mathrm{e}^{\mathrm{i}\delta\cdot\bar{\kappa}}\mathrm{d}\bar{\kappa}, \end{aligned} \tag{23}$$

where the integration paths, i.e. Γ_N, Γ_P, C_r and C_R and their orientations are illustrated in Fig. 2. The radii of the external and inner semicircles in Fig. 2 are R and r, respectively.

Since the integrand $\mathrm{e}^{\bar{s}\cdot\bar{\kappa}}/\bar{\kappa}$ in Eq. (23) is fully analytic along the integration paths, which form a closed curve, and their neighborhood, the following result can be

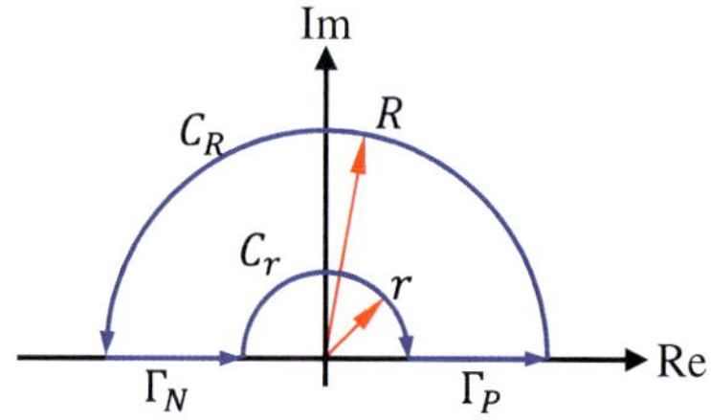

Fig. 2. Integration paths and directions in Eq. (23).

easily obtained using the Cauchy integral theorem (also called the Cauchy-Goursat integral theorem).

$$\oint \frac{\mathrm{e}^{\bar{s}\cdot\bar{\kappa}}}{\bar{\kappa}} \mathrm{d}\bar{\kappa} = \left(\int_{\Gamma_N} + \int_{\mathrm{C}_r} + \int_{\Gamma_P} + \int_{\mathrm{C}_R} \right) \frac{\mathrm{e}^{\bar{s}\cdot\bar{\kappa}}}{\bar{\kappa}} \mathrm{d}\bar{\kappa} = 0. \tag{24}$$

By inspection of Eq. (23), one necessary condition for the convergence of G is that the integral along both C_r and C_R must be convergent simultaneously. Clearly, when $r \to 0$, the integration of the integrand $\mathrm{e}^{\bar{s}\cdot\bar{\kappa}}/\bar{\kappa}$ along C_r converges to $-\mathrm{i}\pi$ as follows:

$$\lim_{r\to 0} \int_{\mathrm{C}_r} \frac{\mathrm{e}^{\bar{s}\cdot\bar{\kappa}}}{\bar{\kappa}} \mathrm{d}\bar{\kappa} = -\mathrm{i}\pi \tag{25}$$

the detailed demonstrations for Eq. (25) are listed in Appendix A.1. Also note that the convergence of Eq. (25) do not rely on the value of δ, which is the imaginary part of $\bar{s}$.

When $R \to +\infty$, the integration of the integrand $\mathrm{e}^{\bar{s}\cdot\bar{\kappa}}/\bar{\kappa}$ along C_R is calculated as

$$\lim_{R\to+\infty} \int_{\mathrm{C}_R} \frac{\mathrm{e}^{\bar{s}\cdot\bar{\kappa}}}{\bar{\kappa}} \mathrm{d}\bar{\kappa} = \lim_{R\to+\infty} \int_0^{\pi} \frac{\mathrm{e}^{\bar{s}R\mathrm{e}^{\mathrm{i}\theta}}}{R\mathrm{e}^{\mathrm{i}\theta}} \mathrm{d}\left(R\mathrm{e}^{\mathrm{i}\theta}\right) = \lim_{R\to+\infty} \int_0^{\pi} \mathrm{i}\mathrm{e}^{\bar{s}R\mathrm{e}^{\mathrm{i}\theta}} \mathrm{d}\theta, \tag{26}$$

where Eq. (26) performs a variable substitution, i.e. $\bar{\kappa} = R\mathrm{e}^{\mathrm{i}\theta}$, considering the integration path C_R. However, the calculation result in Eq. (26) definitely fails to converge. The detailed demonstrations can be found in Appendix A.2.

The establishment of Eq. (4) requires that G in Eq. (21) converges to $-\mathrm{i}$, which ensures that the Hilbert term $\hat{u}(t)$ has an explicit integration result, i.e. $\hat{u}(t) = -\mathrm{i}\mathrm{e}^{\bar{s}t}$. To judge the convergence of G in Eq. (21), combining Eqs. (23), (24) and (25) yields

$$G = \lim_{R,1/r\to+\infty} \frac{1}{\pi} \left(\int_{\mathrm{C}_R} + \int_{\mathrm{C}_r} \right) \frac{\mathrm{e}^{\bar{s}\cdot\bar{\kappa}}}{\bar{\kappa}} \mathrm{d}\bar{\kappa} = -\mathrm{i} + \frac{1}{\pi} \left(\lim_{R\to+\infty} \int_{\mathrm{C}_R} \frac{\mathrm{e}^{\bar{s}\cdot\bar{\kappa}}}{\bar{\kappa}} \mathrm{d}\bar{\kappa} \right). \tag{27}$$

Clearly, the convergence of G depends on the second term in Eq. (27). Since the second term, i.e. Eq. (26), has proved to be nonconvergent, G in Eq. (27) is also nonconvergent and cannot converge to $-\mathrm{i}$, meaning that Eq. (4) cannot establish.

Thus, it can be concluded that Eq. (4) is not the displacement solution to Eq. (3) that represents an SDOF system with the linear hysteretic damping.

5. Numerical Verification

Another approach that can be utilized to check the displacement solution in Eq. (4) is to compare the Hilbert term resulting from Eqs. (3) and (4) with that according to Eq. (2) which gives the mathematical definition.

By specifying the values of $u(0)$ and $\dot{u}(0)$ in Eq. (4), it yields two displacement solution candidates, which are denoted as $u_{c1}(t)$ and $u_{c2}(t)$, respectively, as follows.

$$u_{c1}(t) = \mathrm{e}^{-\lambda_{\mathrm{H}}t}[A_1 \cos(\omega_{\mathrm{H}}t) + A_2 \sin(\omega_{\mathrm{H}}t)] \quad (-\infty < t < +\infty) \tag{28}$$

$$u_{c2}(t) = \begin{cases} 0 & (t < 0) \\ \mathrm{e}^{-\lambda_{\mathrm{H}}t}[A_1 \cos(\omega_{\mathrm{H}}t) + A_2 \sin(\omega_{\mathrm{H}}t)] & (t \geq 0). \end{cases} \tag{29}$$

The difference between $u_{c1}(t)$ and $u_{c2}(t)$ lies in their values over the range $t < 0$. Clearly, $u_{c1}(t)$ has nonzero values when $t < 0$, whereas $u_{c2}(t)$ is established as a piecewise function to ensure it always gets zero when $t < 0$.

Then, substituting the displacement solution candidate, $u_{c1}(t)$ or $u_{c2}(t)$, into the first two terms of Eq. (3), the residual term that replaces the original term, $\hat{u}(t)$, can be calculated as follows:

$$\tilde{u}_{c1}(t) = \frac{1}{\eta\omega^2}(\ddot{u}_{c1}(t) + \omega^2 u_{c1}(t)) \quad (-\infty < t < +\infty) \tag{30}$$

or

$$\tilde{u}_{c2}(t) = \begin{cases} 0 & (t < 0) \\ \dfrac{1}{\eta\omega^2}(\ddot{u}_{c2}(t) + \omega^2 u_{c2}(t)) & (t \geq 0). \end{cases} \tag{31}$$

When $t < 0$, Eq. (30) loses its physical meaning, since an SDOF system in the real world should keep stationary before free vibration. This drawback is overcome by Eq. (31), but it brings in a nondifferential point at $t = 0$.

On the other hand, substituting $u_{c1}(t)$ and $u_{c2}(t)$ respectively into Eq. (2) to perform the Hilbert transform, it yields

$$\hat{u}_{c1}(t) = H[u_{c1}(t)] = \frac{1}{\pi}\int_{-\infty}^{+\infty} \frac{u_{c1}(\tau)}{t - \tau} d\tau, \tag{32}$$

$$\hat{u}_{c2}(t) = H[u_{c2}(t)] = \frac{1}{\pi}\int_{-\infty}^{+\infty} \frac{u_{c2}(\tau)}{t - \tau} d\tau. \tag{33}$$

If the perfect agreement exists either between $\tilde{u}_{c1}(t)$ obtained by Eq. (30) and $\hat{u}_{c1}(t)$ obtained by Eq. (32) or between $\tilde{u}_{c2}(t)$ obtained by Eq. (31) and $\hat{u}_{c2}(t)$ obtained by Eq. (33), it means that Eq. (4) indeed provides a true solution to Eq. (3). Otherwise, Eq. (4) cannot be viewed as the solution to Eq. (3).

For instance, by specifying $u(0) = 0$, $\dot{u}(0) = 15$, $\eta = 0.15$, $k = 16$ and $m = 1$, it yields examples of $\tilde{u}_{c1}(t)$, $\tilde{u}_{c2}(t)$. The corresponding $\hat{u}_{c1}(t)$ and $\hat{u}_{c2}(t)$ are approximately computed using a numerical integration approach at a time interval of 0.01 from -2 to 15. The numerical estimation results are denoted as $\hat{u}_{Hc1}(t)$ and $\hat{u}_{Hc2}(t)$. Then, $\tilde{u}_{c1}(t)$ and $\hat{u}_{Hc1}(t)$ are compared in Fig. 3(a), while $\tilde{u}_{c2}(t)$ and $\hat{u}_{Hc2}(t)$ are compared in Fig. 3(b). Apparently, $\tilde{u}_{c1}(t)$ does not fully overlap $\hat{u}_{Hc1}(t)$ in Fig. 3(a), nor does $\hat{u}_{c2}(t)$ fully overlap $\hat{u}_{Hc2}(t)$ in Fig. 3(b). This mismatch provides another

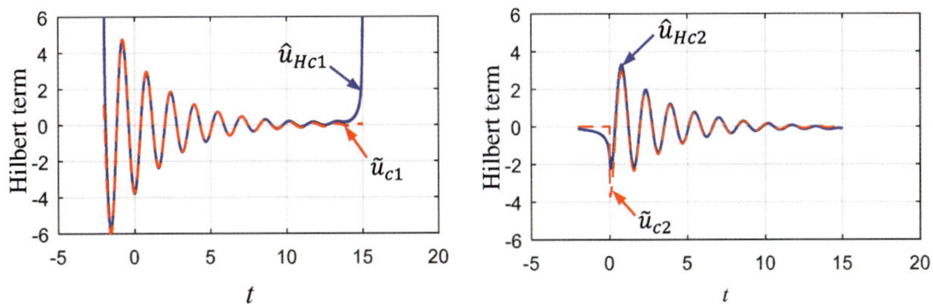

(a) Comparison between $\tilde{u}_{c1}(t)$ and $\hat{u}_{Hc1}(t)$

(b) Comparison between $\tilde{u}_{c2}(t)$ and $\hat{u}_{Hc2}(t)$

Fig. 3. Comparison of the residual terms and numerical results of the Hilbert terms.

compelling evidence for the conclusion that Eq. (4) cannot be viewed as the solution to Eq. (3).

6. Discussions

In fact, substituting $u_{c1}(t)$ [Eq. (28)] into the differential equation of free vibration in Eq. (1) and replacing $\hat{u}(t)$ with $\tilde{u}_{c1}(t)$ [Eq. (30)], it leads to a new SDOF system with special damping model as follows:

$$m\ddot{u}(t) + 2\lambda_{\mathrm{H}} m\dot{u}(t) + m(\lambda_{\mathrm{H}}^2 + \omega_{\mathrm{H}}^2)u(t) = f(t), \tag{34}$$

where λ_{H} and ω_{H} refer to Eq. (6); $\lambda_{\mathrm{H}}^2 + \omega_{\mathrm{H}}^2$ refers to Eq. (7); m is the mass; $f(t)$ represents the imposed external force.

To compare with the classic differential equation of free vibration using the linear hysteretic damping in Eq. (3), Eq. (34) is reformed into

$$\ddot{u}(t) + \omega^2 u(t) + 2\lambda_{\mathrm{H}}^2 u(t) + 2\lambda_{\mathrm{H}}\dot{u}(t) = 0. \tag{35}$$

By comparing Eqs. (35) and (3), it is found that the linear hysteretic damping term, $-\eta\omega^2\hat{u}(t)$, in Eq. (3) is replaced by a new damping model, which is expressed as $2\lambda_{\mathrm{H}}^2 u(t) + 2\lambda_{\mathrm{H}}\dot{u}(t)$ in Eq. (35). Clearly, this damping model is significantly different from the linear hysteretic damping term in Eq. (3) due to its relations to both velocity and displacement. The damping force in Eq. (35) can be calculated as

$$\begin{aligned} f_{\mathrm{d}}(t) &= 0 \quad (t < 0) \\ f_{\mathrm{d}}(t) &= 2m(\lambda_{\mathrm{H}}^2 u(t) + \lambda_{\mathrm{H}}\dot{u}(t)) \quad (t \geq 0) \\ &= m\omega^2\left(\sqrt{1+\eta^2} - 1\right)u(t) + 2m\omega\sqrt{\frac{1}{2}\left(\sqrt{1+\eta^2} - 1\right)} \cdot \dot{u}(t) \\ &= \tilde{\eta} \cdot ku(t) + m\omega\sqrt{2\tilde{\eta}} \cdot \dot{u}(t), \end{aligned} \tag{36}$$

where $\tilde{\eta} = \sqrt{1+\eta^2} - 1$ and $k = m\omega^2$.

Substituting the solution, Eq. (4), into the case of free vibration expressed by Eq. (35), the damping force can be simplified as

$$\begin{aligned} f_{\mathrm{d}}(t) &= -m\eta\omega^2 \cdot \rho \cdot \mathrm{e}^{-\lambda_{\mathrm{H}} t} \cdot \sin(\omega_{\mathrm{H}} t - \phi) \quad (t \geq 0) \\ &= -\eta k \cdot \rho \cdot \mathrm{e}^{-\lambda_{\mathrm{H}} t} \cdot \cos\left(\omega_{\mathrm{H}} t - \phi - \frac{\pi}{2}\right), \end{aligned} \tag{37}$$

where ρ and ϕ refer to Eq. (5).

The idealized SDOF system in Eq. (34) can also be transformed into frequency domain

$$-m\varpi^2 U(\varpi) + kU(\varpi) + (k\tilde{\eta} + 2\mathrm{i} \cdot \lambda_{\mathrm{H}} m\varpi)U(\varpi) = F(\varpi), \tag{38}$$

where $F(\omega)$ is the Fourier transform of external excitations, and $\tilde{\eta} = \sqrt{1+\eta^2} - 1$. Again, by comparing Eq. (38) and the frequency domain expression for the SDOF system with the linear hysteretic damping, Eq. (11), their differences are obvious and prove that the SDOF system in Eq. (34) is not the same as the system in Eq. (1). Hence, the exact solution to Eq. (35) is not the exact solution to Eq. (3).

Since λ_{H} is always positive, there exists the Fourier transform of $f_{\mathrm{d}}(t)$ in Eq. (37) as follows.

$$\Gamma[f_{\mathrm{d}}(t)] = -\eta k\rho \cdot \Gamma[\Psi_{\mathrm{e}}(t) \cdot \sin(\omega_{\mathrm{H}} t - \phi)] = -\frac{\eta k\rho}{2\pi} \cdot \Gamma[\Psi_{\mathrm{e}}(t)] * \Gamma[\sin(\omega_{\mathrm{H}} t - \phi)], \tag{39}$$

where

$$\Psi_{\mathrm{e}}(t) = \begin{cases} 0 & (t < 0) \\ \mathrm{e}^{-\lambda_{\mathrm{H}} t} & (t \geq 0) \end{cases} \tag{40}$$

$$\Gamma[\Psi_{\mathrm{e}}(t)] = \frac{1}{\lambda_{\mathrm{H}} + \mathrm{i}\varpi} = \frac{1}{\lambda_{\mathrm{H}}^2 + \varpi^2} - \mathrm{i}\frac{\varpi}{\lambda_{\mathrm{H}}^2 + \varpi^2} \tag{41}$$

$$\begin{aligned} \Gamma[\sin(\omega_{\mathrm{H}} t - \phi)] &= -\pi \sin\phi \cdot (\delta(\varpi - \omega_{\mathrm{H}}) + \delta(\varpi + \omega_{\mathrm{H}})) \\ &\quad - \mathrm{i}\pi \cos\phi \cdot (\delta(\varpi - \omega_{\mathrm{H}}) - \delta(\varpi + \omega_{\mathrm{H}})). \end{aligned} \tag{42}$$

Herein, $\delta(\cdot)$ represents the Dirac delta function. While the Fourier transform of $\hat{u}_{c2}(t)$ in Eq. (33) can be calculated as

$$\Gamma[\hat{u}_{c2}(t)] = \Gamma\left[u_{c2}(t) * \frac{1}{\pi t}\right] = \Gamma[u_{c2}(t)] \cdot \Gamma\left[\frac{1}{\pi t}\right] = -\mathrm{i} \cdot \mathrm{sgn}(\varpi) \cdot \Gamma[u_{c2}(t)], \tag{43}$$

where

$$\Gamma[u_{c2}(t)] = \Gamma[\rho \cdot \Psi_{\mathrm{e}}(t) \cdot \cos(\omega_{\mathrm{H}} t - \phi)] = \frac{\rho}{2\pi} \cdot \Gamma[\Psi_{\mathrm{e}}(t)] * \Gamma[\cos(\omega_{\mathrm{H}} t - \phi)] \tag{44}$$

$$\begin{aligned} \Gamma[\cos(\omega_{\mathrm{H}} t - \phi)] &= \pi \cos\phi \cdot (\delta(\varpi - \omega_{\mathrm{H}}) + \delta(\varpi + \omega_{\mathrm{H}})) \\ &\quad + \mathrm{i}\pi \sin\phi \cdot (-\delta(\varpi - \omega_{\mathrm{H}}) + \delta(\varpi + \omega_{\mathrm{H}})). \end{aligned} \tag{45}$$

Since $\Gamma[\rho \cdot \Psi_{\mathrm{e}}(t) \cdot \sin(\omega_{\mathrm{H}} t - \phi)] \neq \Gamma[\hat{u}_{\mathrm{c}2}(t)]$ according to Eqs. (39)–(43), it can be concluded that $\hat{u}_{\mathrm{c}2}(t) \neq \rho \cdot \Psi_{\mathrm{e}}(t) \cdot \sin(\omega_{\mathrm{H}} t - \phi)$. This result also proves that Eq. (4) cannot provide an exact solution to Eq. (3), though it is the exact solution to Eq. (35).

7. Conclusions

This study investigates time-domain solutions to SDOF systems with the linear hysteretic damping. The primary contributions are concluded in the following aspects.

- An effective and convenient approach is suggested to verify possible time-domain solutions to the free vibration of SDOF systems with the linear hysteretic damping. This approach utilizes the exact unit-impulse response of the SDOF system at $t = 0$, see Eq. (15), as a necessary condition to check possible solutions. Also, a simple but effective method is suggested to obtain the unit-impulse response of possible solutions via specifying initial displacement and velocity to yield a unit momentum in the SDOF system. Finally, the unit-impulse response of possible solutions is compared with that of the exact solution at $t = 0$.
- Another method proposed to verify possible solutions to Eq. (3) is to compare a possible solution-induced Hilbert term, see Eq. (30) or (31), with the numerical integration result based on the definition expression, i.e. Eq. (2).
- This study rigorously demonstrates that the assumed exponential form ($\mathrm{e}^{\bar{s}t}$), fails to enable the Hilbert term $\hat{u}(t)$ in Eq. (2) to yield a convergent result, in the process of solving the second-order linear differential equation, namely, Eq. (3). As a result, the exact solution to the free vibration of SDOF systems with linear hysteretic damping is still an open problem so far.

Acknowledgments

This study is supported by National Natural Science Foundation of China (Grant No. 52208195).

Appendix A.1. Proofs

A.1. *Demonstration of Eq.* (25)

Proof. The integrand in Eq. (25) can be further partitioned into two parts as follows:

$$\lim_{r\to 0}\int_{\mathrm{C}_r}\frac{\mathrm{e}^{\bar{s}\cdot\bar{\kappa}}}{\bar{\kappa}}\mathrm{d}\bar{\kappa} = \lim_{r\to 0}\int_{\mathrm{C}_r}\frac{\mathrm{e}^{\bar{s}\cdot\bar{\kappa}}}{\bar{s}\cdot\bar{\kappa}}\mathrm{d}(\bar{s}\cdot\bar{\kappa}) = \lim_{r\to 0}\int_{\hat{\mathrm{C}}_r}\frac{\mathrm{e}^{\bar{z}}}{\bar{z}}\mathrm{d}\bar{z} = \lim_{r\to 0}\int_{\hat{\mathrm{C}}_r}\frac{1}{\bar{z}} + \Upsilon(\bar{z})\mathrm{d}\bar{z}, \tag{A.1}$$

where $\bar{z} = \bar{s} \cdot \bar{\kappa}$; the integration path $\hat{C}_r$ is obtained by rotating C_r at an angle of α around the coordinate origin of the complex plane; α represents the orientation angle of $\bar{s}$ in the complex plane.

The second integrand $\gamma(\bar{z})$ in Eq. (A.1) can be expanded as follows:

$$\Upsilon(\bar{z}) = 1 + \sum_{n=2}^{+\infty} \frac{\bar{z}^{n-1}}{n!}. \tag{A.2}$$

Obviously, $\gamma(\bar{z})$ is analytic at $\bar{z} = 0$ and $\gamma(0) = 1$. Hence, there always exists a sufficiently small value of $|\bar{z}|$ to make $|\gamma(\bar{z})| \leq 2$ hold, resulting in the following inequalities when $r \to 0$.

$$\int_{\hat{C}_r} \Upsilon(\bar{z})\mathrm{d}\bar{z} \leq \int_{\hat{C}_r} |\Upsilon(\bar{z})|\mathrm{d}\bar{z} \leq 2\int_{\hat{C}_r} \mathrm{d}l = 2\pi r \tag{A.3}$$

$$-2\pi r = -2\int_{\hat{C}_r} \mathrm{d}l \leq -\int_{\hat{C}_r} |\Upsilon(\bar{z})|\mathrm{d}\bar{z} \leq \int_{\hat{C}_r} \Upsilon(\bar{z})\mathrm{d}\bar{z}. \tag{A.4}$$

Thus, according to the Sandwich Theorem, Eqs. (A.3) and (A.4) definitely lead to a zero value for the second part in Eq. (A.1), namely,

$$\lim_{r\to 0} \int_{\hat{C}_r} \Upsilon(\bar{z})\mathrm{d}\bar{z} = 0. \tag{A.5}$$

Furthermore, the first part in Eq. (A.1) can be easily calculated as follows:

$$\lim_{r\to 0} \int_{\hat{C}_r} \frac{1}{\bar{z}}\mathrm{d}\bar{z} = \lim_{r\to 0} \int_{\hat{C}_r} \frac{1}{re^{\mathrm{i}\theta}}\mathrm{d}(re^{\mathrm{i}\theta}) = \lim_{r\to 0} \int_{\pi+\alpha}^{\alpha} \mathrm{i}\mathrm{d}\theta = -\mathrm{i}\pi. \tag{A.6}$$

By combining the results in Eqs. (A.5) and (A.6), Eq. (25) is readily established. □

Appendix A.2. Divergence of Eq. *(26)*

Proof.

The integrand in Eq. (26) can be further expanded as follows:

$$\begin{aligned}
\mathrm{i}e^{\bar{s}Re^{\mathrm{i}\theta}} &= \mathrm{i}e^{SRe^{\mathrm{i}(\theta+\alpha)}} \\
&= \mathrm{i}e^{[SR\cdot\cos(\theta+\alpha)+\mathrm{i}\cdot SR\cdot\sin(\theta+\alpha)]} \\
&= \mathrm{i}e^{SR\cdot\cos(\theta+\alpha)}e^{\mathrm{i}\cdot SR\cdot\sin(\theta+\alpha)} \\
&= \mathrm{i}e^{SR\cdot\cos(\theta+\alpha)}[\cos(SR\cdot\sin(\theta+\alpha)) + \mathrm{i}\cdot\sin(SR\cdot\sin(\theta+\alpha))] \\
&= -e^{SR\cdot\cos(\theta+\alpha)}\sin(SR\cdot\sin(\theta+\alpha)) + \mathrm{i}e^{SR\cdot\cos(\theta+\alpha)}\cos(SR\cdot\sin(\theta+\alpha)),
\end{aligned} \tag{A.7}$$

where $\bar{s} = \chi + \mathrm{i}\delta$, $S = |\bar{s}| = \sqrt{\chi^2 + \delta^2}$ and α represents the orientation angle of $\bar{s}$ in the complex plane.

The convergence of Eq. (26) requires that the integrations for both the real and imaginary parts of $\mathrm{i}\cdot\exp(\bar{s}R\mathrm{e}^{\mathrm{i}\theta})$ must converge simultaneously when $R \to \infty$, considering that these two parts cannot be mixed or added together. Therefore, it should individually investigate each part. For brevity, only the real part is exhaustively analyzed herein. The integration for the real part of $\mathrm{i}\cdot\exp(\bar{s}R\mathrm{e}^{\mathrm{i}\theta})$ can be rewritten as follows:

$$\begin{aligned} J_\theta\Big|_{\substack{0\\R\to+\infty}}^{\pi} &= \lim_{R\to+\infty}\int_0^{\pi} -\mathrm{e}^{SR\cdot\cos(\theta+\alpha)}\sin(SR\cdot\sin(\theta+\alpha))\mathrm{d}\theta \\ &= \lim_{R\to+\infty}\int_{\alpha}^{\pi+\alpha} -\mathrm{e}^{SR\cdot\cos\varphi}\sin(SR\cdot\sin\varphi)\mathrm{d}\varphi \\ &= J_\varphi\Big|_{\substack{\alpha\\R\to+\infty}}^{\pi+\alpha} \end{aligned} \tag{A.8}$$

where $\varphi = \theta + \alpha$ for convenience.

To reveals the divergence of J_φ in Eq. (A.8), the issue is separately discussed according to the quadrant of $\bar{s}$ in the complex plane, as illustrated in Fig. A.1, also employing the notation $x = \sin(\theta+\alpha)$ so that $\cos(\theta+\alpha) = \sqrt{1-x^2}$ or $-\sqrt{1-x^2}$.

Case 1:

When $\bar{s}$ lies in the first quadrant where $\chi > 0$, $\delta > 0$ and $\alpha > 0$, J_φ in Eq. (A.8) can be calculated as

$$\begin{aligned} J_\varphi\Big|_{\substack{\alpha\\R\to+\infty}}^{\pi+\alpha} &= J_\varphi\Big|_{\substack{\alpha\\R\to+\infty}}^{\pi/2} + J_\varphi\Big|_{\substack{\pi/2\\R\to+\infty}}^{\pi-\alpha} + J_\varphi\Big|_{\substack{\pi-\alpha\\R\to+\infty}}^{\pi} + J_\varphi\Big|_{\substack{\pi\\R\to+\infty}}^{\pi+\alpha} \\ &= J_\varphi\Big|_{\substack{\alpha\\R\to+\infty}}^{\pi/2} + J_\varphi\Big|_{\substack{\pi/2\\R\to+\infty}}^{\pi-\alpha} \end{aligned} \tag{A.9}$$

where due to the symmetry, the third and fourth terms in Eq. (A.9) can be eliminated because of

$$J_\varphi\Big|_{\substack{\pi-\alpha\\R\to+\infty}}^{\pi} + J_\varphi\Big|_{\substack{\pi\\R\to+\infty}}^{\pi+\alpha} = 0. \tag{A.10}$$

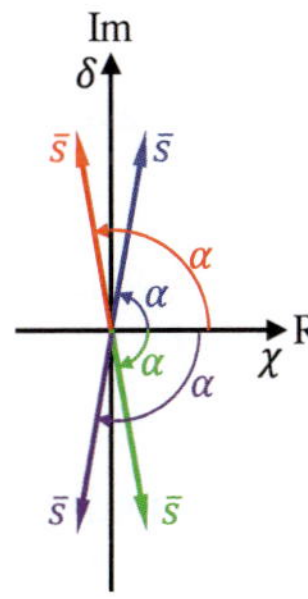

Fig. A.1. Orientations of $\bar{s}$ in the complex plane.

The residual terms in Eq. (A.9) are further expressed as

$$J_\varphi\Big|_{\substack{\alpha\\R\to+\infty}}^{\pi/2} = J_x\Big|_{\substack{\sin\alpha\\R\to+\infty}}^{1} = \lim_{R\to+\infty}\int_{\sin\alpha}^{1} -\mathrm{e}^{SR\sqrt{1-x^2}}\sin(SR\cdot x)\mathrm{d}x \tag{A.11}$$

$$J_\varphi\Big|_{\substack{\pi/2\\R\to+\infty}}^{\pi-\alpha} = J_x\Big|_{\substack{1\\R\to+\infty}}^{\sin\alpha} = \lim_{R\to+\infty}\int_{1}^{\sin\alpha} -\mathrm{e}^{-SR\sqrt{1-x^2}}\sin(SR\cdot x)\mathrm{d}x. \tag{A.12}$$

In Eq. (A.11), when $R\to+\infty$, $SR\sqrt{1-x^2}$ becomes infinite but $\sin(SR\cdot x)$ oscillates over the range $[-1,1]$. As a result, Eq. (A.11) is nonconvergent. In Eq. (A.12), when $R\to+\infty$, $\exp(-SR\sqrt{1-x^2})\to 0$ and $\sin(SR\cdot x)$ is bounded. Therefore, Eq. (A.12) is convergent. Since only the first term in Eq. (A.9) is nonconvergent, it can be concluded that J_θ in Eq. (A.8) is nonconvergent.

Case 2:

When $\bar{s}$ lies in the second quadrant where $\chi<0$, $\delta>0$ and $\alpha>0$, J_φ in Eq. (A.8) can be calculated as

$$\begin{aligned} J_\varphi\Big|_{\substack{\alpha\\R\to+\infty}}^{\pi+\alpha} &= J_\varphi\Big|_{\substack{\alpha\\R\to+\infty}}^{\pi} + J_\varphi\Big|_{\substack{\pi\\R\to+\infty}}^{2\pi-\alpha} + J_\varphi\Big|_{\substack{2\pi-\alpha\\R\to+\infty}}^{3\pi/2} + J_\varphi\Big|_{\substack{3\pi/2\\R\to+\infty}}^{\pi+\alpha} \\ &= J_\varphi\Big|_{\substack{2\pi-\alpha\\R\to+\infty}}^{3\pi/2} + J_\varphi\Big|_{\substack{3\pi/2\\R\to+\infty}}^{\pi+\alpha} \end{aligned} \tag{A.13}$$

where due to the symmetry, the first and second terms in Eq. (A.13) can be eliminated because of

$$J_\varphi\Big|_{\substack{\alpha\\R\to+\infty}}^{\pi} + J_\varphi\Big|_{\substack{\pi\\R\to+\infty}}^{2\pi-\alpha} = 0. \tag{A.14}$$

The residual terms in Eq. (A.13) are further expressed as

$$J_\varphi\Big|_{\substack{2\pi-\alpha\\R\to+\infty}}^{3\pi/2} = J_x\Big|_{\substack{-\sin\alpha\\R\to+\infty}}^{-1} = \lim_{R\to+\infty}\int_{-\sin\alpha}^{-1} -\mathrm{e}^{-SR\sqrt{1-x^2}}\sin(SR\cdot x)\mathrm{d}x \tag{A.15}$$

$$J_\varphi\Big|_{\substack{3\pi/2\\R\to+\infty}}^{\pi+\alpha} = J_x\Big|_{\substack{-1\\R\to+\infty}}^{-\sin\alpha} = \lim_{R\to+\infty}\int_{-1}^{-\sin\alpha} -\mathrm{e}^{SR\sqrt{1-x^2}}\sin(SR\cdot x)\mathrm{d}x. \tag{A.16}$$

Following the reasoning process mentioned for Eqs. (A.11) and (A.12), it is found that Eq. (A.15) is convergent whereas Eq. (A.16) is nonconvergent. Since only the fourth term in Eq. (A.13) is nonconvergent, it can be concluded that J_θ in Eq. (A.8) is nonconvergent.

Case 3:

When $\bar{s}$ lies in the third quadrant where $\chi < 0$, $\delta < 0$ and $\alpha < 0$, J_φ in Eq. (A.8) can be calculated as

$$\begin{aligned} J_\varphi\Big|_{\substack{\alpha \\ R\to+\infty}}^{\pi+\alpha} &= J_\varphi\Big|_{\substack{\alpha \\ R\to+\infty}}^{-\pi/2} + J_\varphi\Big|_{\substack{-\pi/2 \\ R\to+\infty}}^{-\pi-\alpha} + J_\varphi\Big|_{\substack{-\pi-\alpha \\ R\to+\infty}}^{0} + J_\varphi\Big|_{\substack{0 \\ R\to+\infty}}^{\pi+\alpha} \\ &= J_\varphi\Big|_{\substack{\alpha \\ R\to+\infty}}^{-\pi/2} + J_\varphi\Big|_{\substack{-\pi/2 \\ R\to+\infty}}^{-\pi-\alpha} \end{aligned} \tag{A.17}$$

where due to the symmetry, the third and fourth terms in Eq. (A.17) can be eliminated because of

$$J_\varphi\Big|_{\substack{-\pi-\alpha \\ R\to+\infty}}^{0} + J_\varphi\Big|_{\substack{0 \\ R\to+\infty}}^{\pi+\alpha} = 0. \tag{A.18}$$

The residual terms in Eq. (A.17) are further expressed as

$$J_\varphi\Big|_{\substack{\alpha \\ R\to+\infty}}^{-\pi/2} = J_x\Big|_{\substack{\sin\alpha \\ R\to+\infty}}^{-1} = \lim_{R\to+\infty}\int_{\sin\alpha}^{-1} -\mathrm{e}^{-SR\sqrt{1-x^2}}\sin(SR\cdot x)\mathrm{d}x \tag{A.19}$$

$$J_\varphi\Big|_{\substack{-\pi/2 \\ R\to+\infty}}^{-\pi-\alpha} = J_x\Big|_{\substack{-1 \\ R\to+\infty}}^{\sin\alpha} = \lim_{R\to+\infty}\int_{-1}^{\sin\alpha} -\mathrm{e}^{SR\sqrt{1-x^2}}\sin(SR\cdot x)\mathrm{d}x. \tag{A.20}$$

Again, following the reasoning process mentioned for Eqs. (A.11) and (A.12), it is found that Eq. (A.19) is convergent whereas Eq. (A.20) is nonconvergent. Since only the second term in Eq. (A.17) is nonconvergent, it can be concluded that J_θ in Eq. (A.8) is nonconvergent.

Case 4:

When $\bar{s}$ lies in the fourth quadrant where $\chi > 0$, $\delta < 0$ and $\alpha < 0$, J_φ in Eq. (A.8) can be calculated as

$$\begin{aligned} J_\varphi\Big|_{\substack{\alpha \\ R\to+\infty}}^{\pi+\alpha} &= J_\varphi\Big|_{\substack{\alpha \\ R\to+\infty}}^{0} + J_\varphi\Big|_{\substack{0 \\ R\to+\infty}}^{-\alpha} + J_\varphi\Big|_{\substack{-\alpha \\ R\to+\infty}}^{\pi/2} + J_\varphi\Big|_{\substack{\pi/2 \\ R\to+\infty}}^{\pi+\alpha} \\ &= J_\varphi\Big|_{\substack{-\alpha \\ R\to+\infty}}^{\pi/2} + J_\varphi\Big|_{\substack{\pi/2 \\ R\to+\infty}}^{\pi+\alpha} \end{aligned} \tag{A.21}$$

where due to the symmetry, the first and second terms in Eq. (A.21) can be eliminated because of

$$J_\varphi\Big|_{\substack{\alpha \\ R\to+\infty}}^{0} + J_\varphi\Big|_{\substack{0 \\ R\to+\infty}}^{-\alpha} = 0. \tag{A.22}$$

The residual terms in Eq. (A.21) are further expressed as

$$J_\varphi\Big|_{\substack{-\alpha \\ R\to+\infty}}^{\pi/2} = J_x\Big|_{\substack{\sin\alpha \\ R\to+\infty}}^{1} = \lim_{R\to+\infty}\int_{\sin\alpha}^{1} -\mathrm{e}^{SR\sqrt{1-x^2}}\sin(SR\cdot x)\mathrm{d}x \tag{A.23}$$

$$J_\varphi\Big|_{\substack{\pi/2 \\ R\to+\infty}}^{\pi+\alpha} = J_x\Big|_{\substack{1 \\ R\to+\infty}}^{\sin\alpha} = \lim_{R\to+\infty}\int_{1}^{\sin\alpha} -\mathrm{e}^{-SR\sqrt{1-x^2}}\sin(SR\cdot x)\mathrm{d}x. \tag{A.24}$$

Again, following the reasoning process mentioned for Eqs. (A.11) and (A.12), it is found that Eq. (A.23) is nonconvergent whereas Eq. (A.24) is convergent. Since only the third term in Eq. (A.21) is nonconvergent, it can be concluded that J_θ in Eq. (A.8) is nonconvergent.

Above all the four cases, J_θ in Eq. (A.8) fails to converge, meaning that the integration for the real part of $\mathrm{i}\cdot\exp(\bar{s}R\mathrm{e}^{\mathrm{i}\theta})$, see Eq. (A.7), cannot achieve a convergent result. Hence, Eq. (26) proves to be nonconvergent. □

References

1. R. W. Clough and J. Penzien, *Dynamics of Structures*, 2nd edn. (McGraw-Hill Publishing Company, New York, 1993).
2. J. Q. Zhu and M. Zhu, Calculation of complex damping response spectra from earthquake records, *Earthq. Eng. Eng. Vib.* **20**(2) (2000) 19–23 (in Chinese).
3. Q. L. Liu, X. Y. Fu and X. Z. Xiong, A mode superposition method for seismic calculation of hybrid structures based on hysteretic damping assumption, *China Civ. Eng. J.* **54**(8) (2021) 101–110 (in Chinese).
4. R. Bishop, The treatment of damping forces in vibration theory, *J. R. Aeronaut. Soc.* **59**(539) (1955) 738–742.
5. Z. He, Z. Liao and X. Wang, Some notes on theory of complex damping, *Earthq. Eng. Eng. Vib.* **22**(1) (2002) 1–6 (in Chinese), doi: 10.13197/j.eeev.2002.01.001.
6. A. Güllü, E. Yüksel and E. Altintaş, Closed-form demonstration and shake table verification of damping effect on the seismic energy, *Int. J. Struct. Stab. Dyn.* **22**(12) (2022) 2250121.
7. N. Satake, K. Suda, T. Arakawa, A. Sasaki and Y. Tamura, Damping evaluation using full-scale data of buildings in Japan, *J. Struct. Eng.* **129**(4) (2003) 470–477.
8. V. Gunasekaran, S. Gulhane, S. Gupta, J. Pitchaimani, V. Rajamohan and G. Manickam, Structural-acoustic response analysis of variable stiffness laminates with inherent material damping, *Int. J. Struct. Stab. Dyn.* **22**(12) (2022) 2250127.
9. Q. Dai, Y. Liu, Z. Qin and F. Chu, Damping and frequency response characteristics of functionally graded fiber-reinforced composite cylindrical shells, *Int. J. Struct. Stab. Dyn.* **22**(9) (2022) 2250107.
10. Y. Ge and A. Sarkar, Thermoelastic damping in vibrations of small-scaled rings with rectangular cross-section by considering size effect on both structural and thermal domains, *Int. J. Struct. Stab. Dyn.* **23**(3) (2023) 2350026.
11. W. Gao, F. Wang and O. Sigmund, Systematic design of high-Q prestress micro membrane resonators, *Comput. Methods Appl. Mech. Eng.* **361** (2020) 112692.
12. D. Høj, F. Wang, W. Gao, U. B. Hoff, O. Sigmund and U. L. Andersen, Ultra-coherent nanomechanical resonators based on inverse design, *Nat. Commun.* **12** (2021) 5766.
13. J. A. Inaudi and J. M. Kelly, Linear hysteretic damping and the Hilbert transform, *J. Eng. Mech.* **121** (1995) 626–632.
14. S. L. Hahn, *Hilbert Transforms Signal Processing* (Artech House, Boston, 1996).
15. N. Makris, On the physical meaning of time-domain constitutive models with complex parameters, *Meccanica* **55**(3) (2020) 453–467.
16. P. Sun, H. Yang, W. Zhao and Z. Wang, Time domain calculation method based on hysteretic damping model, *Eng. Mech.* **36**(6) (2019) 13–20 (in Chinese).
17. P. Sun, H. Yang and Y. Zhao, Time-domain calculation method of improved hysteretic damped system based on frequency-dependent loss factor, *J. Sound Vib.* **488** (2020) 115658, doi: 10.1016/j.jsv.2020.115658.
18. H. Hu, Structural damping model and system dynamic response at time domain, *Chin. J. Appl. Mech.* **10**(1) (1993) 34–44 (in Chinese).

https://doi.org/10.1142/9789819814770_0034

Chapter 34

A Physics-Data Hybrid Framework to Develop Bridge Digital Twin Model in Structural Health Monitoring#

Li-Feng Qin, Wei-Xin Ren* and Chuan-Rui Guo

College of Civil and Transportation Engineering
Shenzhen University, Shenzhen 518060
Guangdong Province, P. R. China
*renwx@szu.edu.cn

Digital twin aims to create a virtual model for a physical structure by combining measurement data in structural health monitoring. The most important feature is to achieve the physical structure-monitoring data synchronization. For this purpose, a physics-data hybrid framework to develop the bridge digital twin model in structural health monitoring is proposed in the paper. The physical base is firstly formed by the finite element model of the digital representation for the physical bridge that can fully incorporate both structural geometry and structural state. The data base is then built by all measurement data of the monitored bridge. By defining the context that is common to both physical base and data base, the mirror relationship between physical base and data base for the specified context is formulated. To achieve the best matching of the mirror relationship by minimizing process, the digital twin model in terms of the specified context can be developed. In such a way, the proposed framework integrates physical knowledge and data intelligence into one model. A demonstration of a simulated simply supported beam is provided to show how the digital twin model is developed by using proposed physics-data hybrid framework. It is found that the generated digital twin model is consistent with the current structural state of the beam. The presented physics-data hybrid framework helps in clearer understanding of the realization of digital twin model in structural health monitoring, providing a new perspective for smart bridge solutions.

Keywords: Digital twin; data base; bridge health monitoring; structural dynamics.

1. Introduction

As an essential part of the infrastructure, bridges play an irreplaceable role in ensuring travel for humans and transportation safety. While bridge safety accidents

*Corresponding author.
#This chapter appeared previously on the International Journal of Structural Stability and Dynamics. To cite this chapter, please cite the original article as the following: L.-F. Qin, W.-X. Ren & C.-R. Guo, *Int. J. Struct. Stab. Dyn.*, **23**, 2340037 (2023), DOI: 10.1142/S0219455423400370.

have occurred frequently with the increase in operation time, the operation and maintenance of bridges have become the primary challenge of bridge engineering.[1] Structural health monitoring technologies monitor the operational status in real time by placing sensors on structures, and make assessment and early warning of safety conditions.[2,3] With the rapid development of structural health monitoring, a data challenge emerges due to insufficient capability in managing, processing and interpreting extensive monitoring datasets to extract useful information.[4] On the other hand, Internet of things, Big data, Artificial intelligence, Cloud computing, and other new technologies have been widely used. Therefore, it is necessary to provide more smart solutions to ensure bridge health.

One emerging area of research that addresses this challenge is the creation of "digital twin". As a virtual representation of the physical infrastructure, the digital twin can be updated in near real-time as new data is collected and provides feedback into the physical structure.[5] By generating a reliable digital twin model, it is possible to utilize a variety of services, such as prediction, control, assessment and optimization of the actual object's status. The digital twin was once listed in the top ten strategic technologies as an important enabling technology to realize digital transformation and intelligent upgrade.[6,7]

The digital twin concept was first introduced as "digital equivalent to a physical product" by Professor Grieves in 2003, which included three main parts: physical components, virtual models, and their connections of data and information.[8] NASA gave the first detailed definition of digital twin: an integrated multiphysics, multiscale, probabilistic simulation of an as-built vehicle or system that uses the best available physical models, sensor updates, and fleet history to mirror the life of its corresponding flying twin.[9] Since then, the digital twin has been a hot spot in a variety of fields including manufacturing, power, machinery, aircraft, medical, automobile, aerospace, and smart city.[10] While there is no consolidated view on what digital twin is, Tao *et al.*[11] presented a five-dimensional digital twin model and established its standard system. Bao *et al.*[12] proposed an approach of modeling and operations for product digital twins, process digital twins and operation digital twins in the context of manufacturing and the interoperation modes between these digital twins were explained accordingly. Jiang *et al.*[13] discussed the digital twin framework and application criteria in smart manufacturing. Wang *et al.*[14] presented a digital twin reference model for rotating machinery fault diagnosis and corresponding model construction requirements were discussed. Worden *et al.*[15] proposed a rigorous representation theory of digital twin across a number of domains, including pure mathematics, computer science, engineering and physics. While these studies in the digital twin field have focused on the explicit frameworks, which applications faced challenges to support different integration levels.

At present, the digital twin is mainly applied in the industrial field, through building a digital twin model of the product, completing the design and optimization of the product, realizing resource deployment and intelligent production in order to improve production efficiency.[16] Qian *et al.*[17] proposed the framework of digital twin workshop system, which enables to achieve optimal workshop production management and control. Tao *et al.*[18] proposed a new method with big data for product design, manufacturing, and service driven by digital twin, so as to drive product design, manufacturing, and service to be more efficient, smart, and sustainable. Shang guan *et al.*[19] proposed a physical-virtual convergence approach using digital twin technology for a satellite system's fault diagnosis and operation monitoring. Lu *et al.*[20] established a digital twin and application framework for electromechanical products to realize the mechanism of product design iterative optimization, fault diagnosis and prediction. For industry, the digital twin is more concerned with geometric model, dynamic display and data presentation. On the contrary, in structures, researchers are more concerned with the mechanical properties of the structure and the safety state. Kang *et al.*[21] provided a bridge digital health multimedia service, which can update the characteristic parameters of the structure through real-time operation data collected by sensors so that the digital twin bridge and the actual bridge can be consistent. Ye *et al.*[22] established a railway bridge digital twin model with BIM technology and realized the visualization of real-time monitoring data. Mohammadi[23] provided a three-dimensional digital representation of the bridge via both unmanned aerial vehicle photogrammetry and terrestrial laser scanning, which can provide safer, cheaper, more reliable, and less distributive bridge inspections. Ritto and Rochinha[24] explored the digital twin method integrating physics-based models with machine learning and applied to the damage detection of wind turbine structure. Chakraborty and Adhikari[25] introduced time scales to establish a multi-time scale digital twin framework for dynamic system. Considering the uncertainty of a structure, Li *et al.*[26] used the dynamic Bayesian network to construct a digital twin model to predict the probability of crack growth. Tsialiamanis *et al.*[27] proposed a digital twin framework combining the stochastic finite element method and conditional generative adversarial networks to predict the probability distribution of uncertainty parameters.

The digital twin is a general concept that has been widely used in many areas. Even though the digital twin has received a significant attention from many aspects, the definition of digital twin is not so unique depending on application areas. There is no systematic, clear and realizable reference framework on how to generate a full digital twin model. Each approach has its own norms, with little crossover. In nature, the digital twin emphasizes more on the integration of data and physics, as well as the consistency and synchronization of virtual and physical spaces. The digital twin model is the foundation and core of digital twin theories and technologies. Although most studies in the field of digital twins have focused on manufacturing and proposed explicit architectures, the concept of digital twin is very different between infrastructures and industrial products. The most important feature of digital twin in the structural health monitoring is to achieve the physical structure-monitoring data synchronization. A full digital twin model as the foundation for structural safety assessment in structural health monitoring has not been realized. To build a digital twin model, the essential issues of physical base, data base and mirroring between two models as well as the best matching between two models have to be clarified.

With this regard, this paper is aimed at presenting a physics-data hybrid framework to develop the bridge digital twin model combining both physical base generated by the finite element model as the digital representation of physical bridge and data base generated by the extensive bridge health monitoring data. The key issues are clarified including digital representation of physical structure, digital base for multi-source monitoring data, context, mirror implementation and achieving the best matching between digital physical model and digital base. Through the full combination of the physical model and information data, the interpretability and expressibility of the results are enhanced and synchronized which lays an important foundation for the subsequent safety assessment and life prediction of the bridge.

2. Physics-Data Hybrid Framework to Develop Bridge Digital Twin Model

In structural health monitoring, the digital twin should focus more on the structural safety performance, which is the key difference from other fields. As the environment and load of a given bridge structure are extremely complex, this means that the physics of bridge structure is unlikely to be known at all scales and in all contexts. This makes its digital twin model complex and even difficult to implement. Therefore, based on the "context" view proposed by Worden *et al.*,[15] the current physics-data hybrid framework to develop the bridge digital twin model is proposed, which consists of five main parts as follows: (1) The physical base generated by the finite element model that is the digital representation of the physical bridge. (2) The data base generated by multi-source monitoring data of the monitored bridge. (3) The context common to both physical base and data base. (4) The mirror relationship between physical base and data base for the specified context. (5) The best matching to achieve the mirror relationship. As a general framework, this framework can take more factors into account in the physical base, fuse multi-source monitoring data in the data base, and even create more complex contexts and mirror relationships to realize a multi-layered digital twin model, providing guidance for the digital twin in bridge health monitoring.

2.1. *Digital representation of the physical structure*

To develop the bridge digital twin model in bridge health monitoring aimed at evaluating the operational safety of monitored physical bridge, the first issue is to achieve the digital representation of the physical bridge or target bridge as shown in Fig. 1 to form the digital physical base. At present, many advanced technologies such

Fig. 1. Physical bridge.

as 3D laser scanning, photogrammetry, CAD, and building information modeling (BIM) are used to establish the three-dimensional digital model of the physical structure to restore the real three-dimensional scene.

The latest 3D laser scanning tools can get accurate measurements and a good image of the structure. Using precision tripod or unmanned aerial vehicle (UAV) mounted laser scanners, any structure can be mapped to mm accuracies. In such a way, the digital data are collected remotely without the need to physically touch anything. The advanced 3D laser scanning is very quick and can easily record several million data points in a minute and upload the information to the cloud. The 3D digital model of any physical structure can be generated in the built environments by using the latest 3D laser scanning hardware and software technologies. The illustration of the 3D digital model of a physical bridge generated by 3D laser scanning is as shown in Fig. 2. Alternately, Building Information Model (BIM) is a technology for

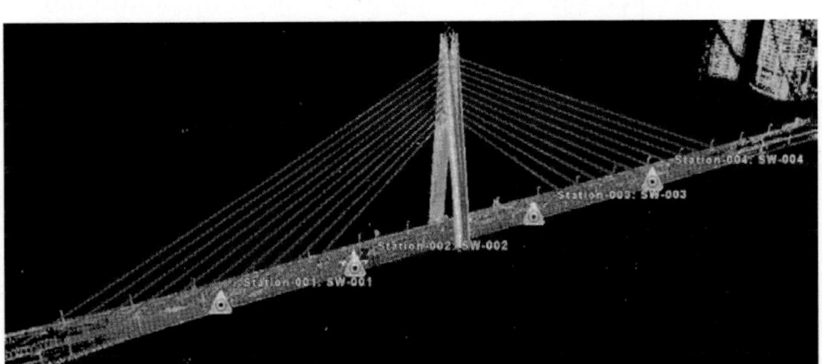

Fig. 2. The 3D digital model of a physical bridge generated by 3D laser scanning.

Fig. 3. The 3D digital model of a physical bridge generated by BIM.

digital, object-oriented modeling of structures. It is more sophisticated and detailed compared to regular 3D modeling, though it does rely on the geometric information provided by CAD. BIM platforms have been developed for effective usage of IT tools to meet the need for digital prototyping and visualization in planning, design, construction and maintenance. Figure 3 illustrates a digital information model of a physical bridge generated by the BIM platform.

Either the point cloud 3D digital model generated by 3D laser scanning or the 3D digital model generated by BIM provides a large amount of digital geometrical information. They are considered as the digital twins in many areas. It is true to a certain extent. However, such 3D digital models only contain the "shape" information that is the geometry of the physical structure. In fact, a physical structure is essentially designed to bear the dedicated loads such as dead loads, live loads and occasional loads like wind, impact and earthquake loads by using dedicated materials. Namely, a physical structure contains not only the "shape" information, but also the "state" information. Such "state" information is the physical information of the physical structure including the material properties and a short cycle predictive structural performance analysis under applied loads and operational environments to determine the structural state which is safe or not. The point cloud model and BIM fall short of the digital twin concept that can continuously update the state of the physical structure. They lack features essential for structural health monitoring.

The finite element method is a popular numerical method for solving structural analysis problems to produce a stale solution of a physical structure under loading. In the finite element method, a continuous physical structure is discretized into a finite number of small elements so that the working state of the continuous structure is estimated by assembling the similar properties of discretized elements per node. The discretized 3D finite element model is indeed the 3D digital model of a physical structure containing both geometric information and material information to facilitate structural analysis, providing an effective mean for the digital representation of

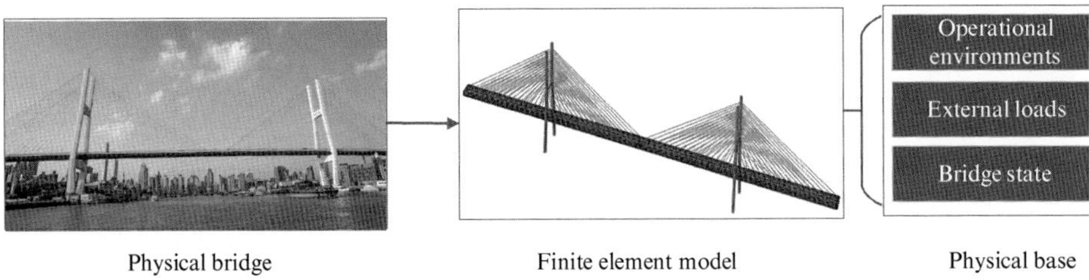

Fig. 4. Physical base build by digital representation of the physical bridge.

the physical structure. Such a structural finite element model meets all the requirements for containing both "shape" information and "state" information of a physical structure where structural material properties, operational environments, dedicated loads, structural analysis and performance evaluation can be easily incorporated. Therefore, the finite element model is considered as a good choice to be the digital representation of the physical bridge to build the physical base as shown in Fig. 4 in current proposed physics-data hybrid framework to develop bridge digital twin model for structural health monitoring.

From the perspective of structural health monitoring, the bridge's health state is affected by many factors. The change of operational environment directly causes the state change of bridge structure. Wind load and vehicle load are the primary external loads carried by bridges during their service, leading to a change in the bridge state. Therefore, it is necessary to establish its twin model by considering the effects of environments and loads. For this reason, the digital representation of physical bridge can be characterized by three items: bridge state S, environment E and external load V. The state S is the physical objects of interests referred to structural responses. Environment E could be considered as the all-natural environments and exterior geographical conditions of the physical bridge. External load V is the total loads carried by the bridges. Assuming that a vector $\underline{s}(t)$ represents the bridge state S at an instantaneous time t, one yields

$$\underline{s}(t) = \{s_1(t), s_2(t) \cdots s_{Ns}(t)\}. \tag{1}$$

Since the physical bridge's state cannot be represented by only one feature, it is represented by Ns states. The state vector $\underline{s}(t)$ consists of a set of Ns instantaneous responses which completely characterize its state. The specific number of features required to represent the bridge's state needs to be determined according to the actual situation and demand.

Similar to the physical bridge's state S, the environment E and external load V are represented by the environment vector $\underline{e}(t)$ and external load vector $\underline{v}(t)$ that consist of a set of environmental variables and external load variables, respectively, at an instantaneous time t as follows:

$$\underline{e}(t) = \{e_1(t), e_2(t) \cdots e_{Ne}(t)\}, \tag{2}$$

$$\underline{v}(t) = \{v_1(t), v_2(t) \cdots v_{Nv}(t\}. \tag{3}$$

It is well known that environmental variables and external load variables of the physical bridge will result in the change of the bridge state, i.e. a change in such a variable will evoke a difference in the state $\underline{s}(t)$. Once the physical bridge is represented by the digital finite element model, the specified state S can be determined by the known environment E and external load V. The changes in the operational environments such as temperature and wind can be incorporated with the finite element model as a kind of applied load on the physical bridge. The crack, corrosion and scouring of the physical bridge can be considered as a kind of element stiffness reduction in the finite element model.

2.2. *Data base of the monitored bridge*

Since recent 20 years, the structural health monitoring (SHM) has gained considerable attention in civil engineering, especially in bridge engineering. In general, a SHM process involves the sensor implementation on the monitored bridge, data acquisition and transmission over time using periodically sampled structural response measurements from an array of sensors, the extraction of damage-sensitive features from these measurements, and the statistical analysis of these features to determine the current bridge performance state. Subsequently, the structural health monitoring refers to a technology that allows for recording the structural behavior over a long duration to assess performance, identify anomalies, and detect damage to ensure the integrity, reliability, and safety of the bridges. For the long-term SHM, the big data are continuously generated by this process where data are periodically updated information regarding the ability of the structure to perform its intended function. Such big data constitutes the complete "data base" for the monitored physical bridge.

The data base is simply understood as a general information infrastructure. Its main function is to provide unified and standardized digital service support for various industries and services. At the same time the data base requires the data to be able to aggregate, store, and even generate intelligent services based on data, similar to human brain. According to this definition, the data base usually needs a series of processing on the original data to realize digital operation. For example, the data from different sources are transformed into a uniform specification to achieve consistent expression of multi-source heterogeneous data. The multi-data association is established through target differentiation, entity recognition and information association to achieve the multi-sensor data fusion. The data preprocessing is used to reduce the noise of original data and improve the data quality. The ability is provided for the users with optimization and decision-making reference through

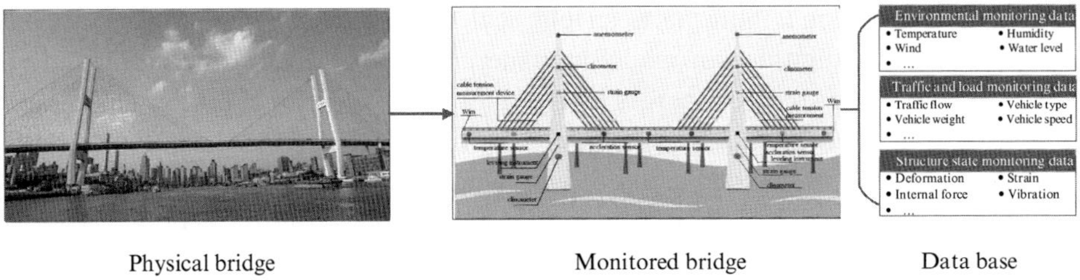

Fig. 5. Data base from structural health monitoring system.

artificial intelligence and machine learning. However, due to technical limitations the current data base is often used for data storage and primary artificial intelligence.

Structural health monitoring system can provide a large amount of data for digital twin, but these data present multi-source heterogeneous characteristic. The physical variables that need to be monitored for the bridge can be roughly divided into three categories: environmental monitoring, operational load monitoring and structural state monitoring. The environmental monitoring refers to the bridge site's natural environment and geographical conditions such as temperature, humidity, wind, water level, scouring and earthquake. The operational load refers to the load generated by traffic on the bridge, in which monitoring objects include traffic flow, vehicle type, speed and axle load. The structural state monitoring refers to the bridge's static and dynamic response monitoring, mainly including stress, strain, displacement, acceleration and internal force. All these monitoring data constitute the data base D as shown in Fig. 5 for the digital twin. In other words, the data base D for the bridge digital twin is that all data nodes grow on the base. According to the classification, the data base D obtained by the structural health monitoring system is the set of environmental monitoring data, operational load monitoring data and structural state monitoring data.

The data base is a link to realize the interconnection between the physical bridge and real-time monitored data. If such a link is achieved, a bridge digital twin model of physics-data hybrid can be developed in structural health monitoring. The key is the feature extraction where the feature is common to both physical base generated by digital representation of the physical bridge and data base generated by monitoring data of the monitored bridge.

2.3. *Context*

As the original monitoring data or data base D has the "5V" characteristics, namely, volume, velocity, variety, value and veracity, data preprocessing, the data fusion, data analysis and data storage, etc. are usually required to extract the feature from monitoring data, and provide data preparation for establishing mirror relationship to realize the best matching between physical base and data base. As mentioned above, the bridge digital twin model in structural health monitoring is difficult to represent the whole structure across its entire range of possible behavior, as the cost is too high and the existing technology is hard to realize. Matching the required features according to the dedicated requirements can be a feasible way to develop the bridge digital twin model in structural health monitoring. In general, Worden *et al.*[15] gave the definition of context C as follows:

$$C = \{e_i^C \in E, v_j^C \in V, s_k^C \in S; i, j, k\}. \tag{4}$$

It can be seen that Context C is defined as the execution environment and load conditions required for research purposes and the parameters to be predicted. The subset $\{e_i^C\}$, $\{v_i^C\}$ and $\{s_k^C\}$ are referred to as the environmental context, the load context, and the response or predictive context of the structure. Depending on the purpose of developed bridge digital twin model, the fundament of context selection is that the selected contexts are common to both physical base generated by digital representation of the physical bridge and data base generated by monitoring data of the monitored bridge. In such a way, a bridge digital twin model that matches the contexts of interest can be developed in the physics-data hybrid framework. To facilitate the implementation, the digital twin herein is not to establish a global model of the structure or system in all physical contexts. In fact, it may be the case that different bridge digital twin models are possible to cover the dedicated contexts of interest completely.

In structural health monitoring, the context of the structural state mainly includes the static characteristics and dynamic characteristics. The structural static context may refer to deflection, rotation, stress, strain, etc. The structural dynamic context includes dynamic responses, natural frequency, modal shape, and modal strain energy. Context C may be either direct context or indirect context common to both physical base and data base. For the direct deflection context as example, the deflection context is the direct monitoring measurement data in the data base, while it is also the structural state contact in the physical base that can be calculated by the finite element model-digital representation of the physical bridge. The indirect context can be extracted and associated through feature extraction, system identification, machine learning, multi-data fusion, and other methods. For the structural dynamic characteristics (natural frequency and more shape) as example, they can be indirectly determined by system identification through the direct dynamic acceleration measurement data in the data base, while those in the physical base can be also determined by the eigenvalue calculation through the finite element model.

Note that the changing environmental conditions and external loads will strongly affect the physical bridge's responses. By relying on the huge data link provided by the health monitoring system, through clustering analysis, dictionary calibration, intelligent algorithms and other ways, it is possible to establish an intelligent fusion mechanism of multi-source heterogeneous big data which fully combines the environment, load and structural state. Namely, to form a context that can comprehensively reflect the key characteristics incorporated with the environment, load and structural state in both physical base and data base are the key issues to develop the bridge digital twin model in structural health monitoring.

2.4. *Mirror relationship*

Once the dedicated context is extracted, it is necessary to establish the mirror relationship between the actual observation and the prediction in the physical base of this specific context. As mentioned above, the actual observation in the data base obtained from all measurements of the physical bridge in the structural health monitoring system includes the environments, loads and response, which is expressed as a set

$$\mathrm{X}^c(t) = \{\underline{e}^c(t), \underline{v}^c(t), \underline{r}^c(t)\}, \tag{5}$$

where superscript C represents the specific "context" of interest, while $\underline{e}^c(t)$, $\underline{v}^c(t)$ and $\underline{r}^c(t)$ are the observation vectors of the environment, load and response for a given context. If the bridge is regarded as a system, the input of the system is environmental information and load information, and the output of the system is the response of the bridge. Therefore, the structure S response $\underline{r}^c(t)$ for a specific context C can be expressed as

$$\underline{r}^c(t) = S[\underline{e}^c(t), \underline{v}^c(t)]. \tag{6}$$

With regard to the prediction of the physical bridge in the physical base, similarly, the prediction $\hat{X}^c(t)$ for a specific context C can be expressed as a vector set

$$\hat{X}^c(t) = \{\underline{\hat{e}}^c(t), \underline{\hat{v}}^c(t), \underline{\hat{r}}^c(t)\}, \tag{7}$$

where vectors $\underline{\hat{e}}^c(t)$ and $\underline{\hat{v}}^c(t)$ are the operational environments and applied loads on the physical bridge, while vector $\underline{\hat{r}}^c(t)$ refers to predicted structural responses in the physical base. As the physical bridge is herein digitally represented by the finite element model, obviously, the structural responses at the same time scale can be easily predicted by performing the finite element analysis with the operational environments and applied loads on the physical bridge, which is expressed as

$$\underline{\hat{r}}^c(t) = M^{PC}[\underline{\hat{e}}^c(t), \underline{\hat{v}}^c(t)], \tag{8}$$

where M^{PC} refers to the physical structure represented by digital physical model for a specific context C. It can be seen that vectors $\underline{\hat{e}}^c(t)$ and $\underline{\hat{v}}^c(t)$ are the input of the physical bridge, while vector $\underline{\hat{r}}^c(t)$ denotes the output responses of the physical bridge at the same temporal scale for a context C.

Referring to the mirror definition given by Worden et al.,[15] a mirror relationship d on a given context C can be defined by the actual observation in the data base and the prediction in the physical base in terms of environments, loads and responses as shown below:

$$d^c(\hat{\underline{e}}^c(t), \underline{e}^c(t)) \le \varepsilon_1, \tag{9}$$

$$d^c(\hat{\underline{v}}^c(t), \underline{v}^c(t)) \le \varepsilon_2, \tag{10}$$

$$d^c(\hat{\underline{r}}^c(t), \underline{r}^c(t)) \le \varepsilon_3, \tag{11}$$

where ε_i in which $i = 1, 2, 3$ is a user-defined threshold, depending on the cost of digital twin modeling and user requirements. To some extent, users can decrease accuracy to reduce cost, and also increase cost to enhance accuracy. Only the observations in the data base and the predictions in the physical base satisfy the above mirror conditions, a bridge digital twin model M^C for the specified context can be built out of physics-data hybrid in the structural health monitoring. As such a bridge digital twin model always incorporates the environmental, load and response variables, it may be a multi-scale and multi-physics model. Once the physical base and data base of the physical bridge are built in the structural health mentoring, to achieve the extraction of the dedicated content, a bridge digital twin model can be developed in terms of the best matching of the mirror relationships.

2.5. *Best matching of mirror relationships*

After the mirror relationships are determined, the most important thing is how to achieve the best matching of such mirror relationships between physical base and data base. By calculating the distance d^c between the prediction and observation, if the result d^c meets the user-defined threshold in the mirror relationship, it is indicated that the physical base and data base have achieved the best matching. This process can be denoted by

$$f(\theta) = \min \|\hat{X}^c(t) - \mathrm{X}^c(t)\|, \tag{12}$$

where θ are the parameters of physical structure. $\hat{X}^c(t)$ denotes the predictions in the physical base for the specific context C, while $\mathrm{X}^c(t)$ denotes the observations in the data base for the specific context C.

To realize the minimization, some optimization technologies are needed for short cycle predictive analysis of process outcomes. The optimization algorithm may be either single-objective or multi-objective. Due to the big data in the structural health monitoring, some more advanced methods in deep learning are required to be incorporated. This process is further denoted by the symbol $M^{DL}(\hat{X}^c(t), \mathrm{X}^c(t))$ where the superscript DL represents methods of deep learning to optimize parameters.

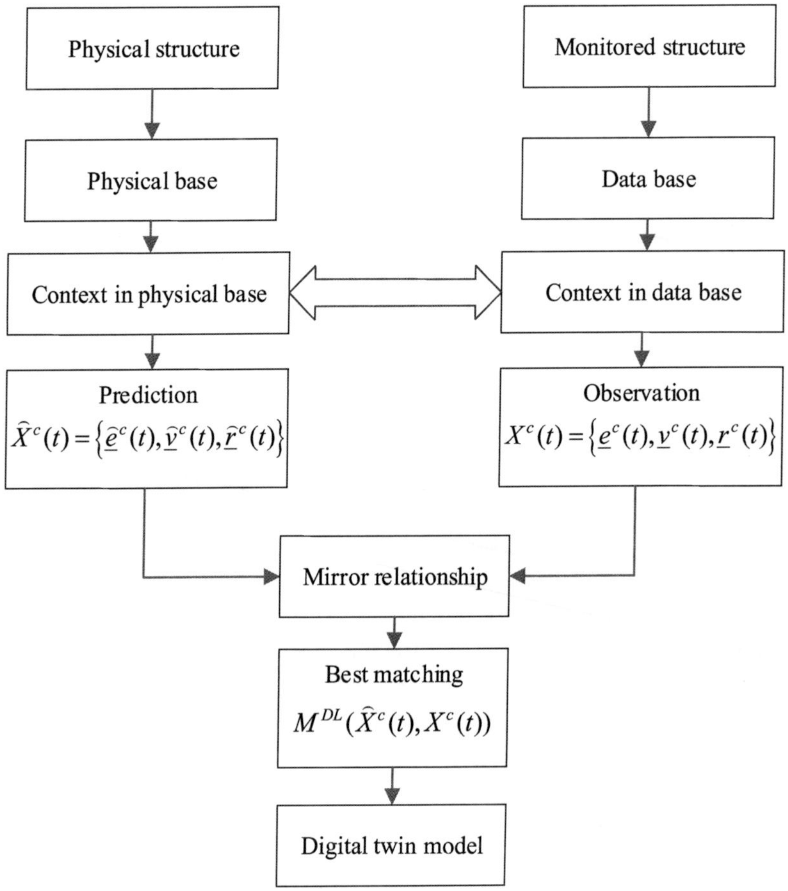

Fig. 6. A flowchart of physics-data hybrid framework to develop bridge digital twin model in structural health monitoring.

The model that has achieved the best matching between physical base and data base for the specified context is regarded as the digital twin model. A flowchart of the proposed physics-data hybrid framework to develop bridge digital twin model in structural health monitoring is as shown in Fig. 6.

This digital twin conceptual framework fully incorporates both physical structure and monitoring data information, through the physical structure digitally represented by the finite element model, digital base for multi-source monitoring data extracted context common to both physical structure and monitoring data, mirror relationship implementation, the best matching of mirror relationship between physical base and data base. With such a built digital twin model, the structural state assessment and intelligent maintenance of the monitored bridge can be subsequently carried out.

3. A Demonstration of Physics-Data Hybrid Framework to Develop Digital Twin Model

For complex engineering applications, it is hard to have a clear understanding of the mechanism of bridge fatigue damage and safety performance degradation caused by temperature, vehicle, wind and scour, etc. Building a general bridge digital twin model in all the contexts is unrealistic based on current technology. At present, in the aspect of structural health monitoring, most of the research studies have been done on structural state and safety assessment based on the change in structural dynamic characteristics. It is the most practicable way to develop the digital twin model mainly considering the dynamic properties of the physical bridge. As the structural dynamic parameters are the structural inherent characteristics of the bridge, the structural frequency is one of the main dynamic indexes reflecting the bridge state. To take the first five natural frequencies as the specific context, a simply supported beam is herein implemented to demonstrate how a digital twin model is developed through the proposed physics-hybrid framework.

A simulated simply supported beam is as shown in Fig. 7. The beam parameters are as follows: $L = 6$ m, density $\rho = 2\,500\,\mathrm{kg/m^3}$, elastic modulus $E = 3.2 \times 10^{10}$ Pa, section size $A = 0.25\,\mathrm{m} \times 0.2\,\mathrm{m}$ and inertia moment of cross-section $I = 1.66 \times 10^{-4}\,\mathrm{m^4}$. Considering that the structure has been operated over a long time, a damage at 3.6 m from the left end is assumed to simulate the current state of the beam. Such a damage is referred to as the stiffness reduction due to structural performance deterioration.

Assuming that the beam is a linear system, neglecting the effects of shear deformation and rotary inertia, the beam can be physically represented by the equation of motion as

$$\rho A \frac{\partial^2 u(x,t)}{\partial t^2} + c\frac{\partial u(x,t)}{\partial t} + EI\frac{\partial^2 u(x,t)}{\partial x^2} = 0, \tag{13}$$

where ρ is the density of material, A is the cross-section area, c is the viscous damping, EI is the bending stiffness of the beam, and $u(x,t)$ is the dynamic displacement field that is a function of the position of the independent variable x and time t. The

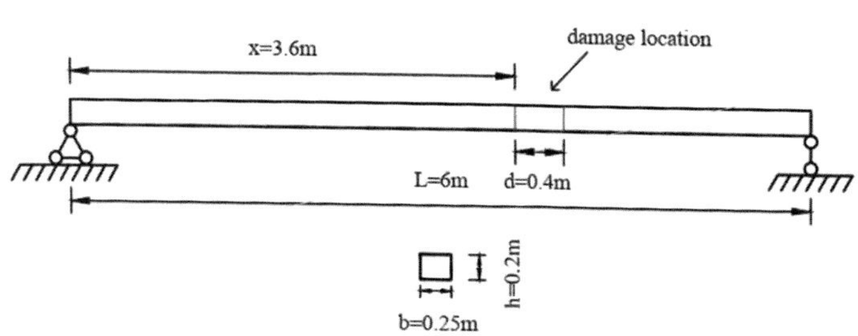

Fig. 7. A simply supported beam.

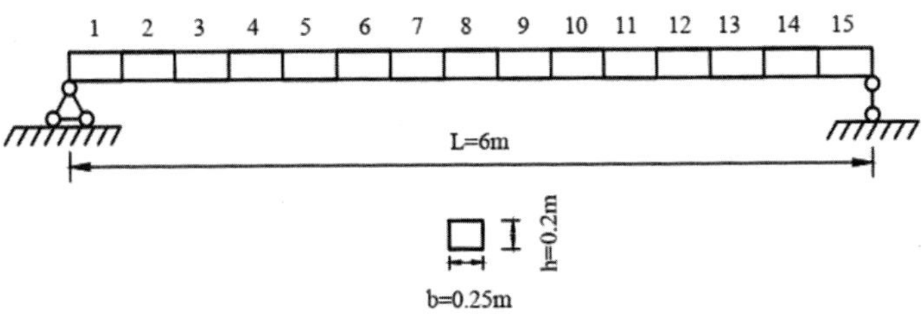

Fig. 8. Finite element model of the beam.

Table 1. The first five natural frequencies (Hz) of the beam.

Natural frequency	1	2	3	4	5
Physical base	8.99	35.92	80.63	142.93	222.53
Data base	7.78	33.24	79.14	128.04	146.22

nth natural frequency of a simply supported undamping beam can be calculated by

$$w_n = n^2\pi^2\sqrt{\frac{EI}{mL^4}} \quad n = 1, 2, \ldots, \tag{14}$$

where m is the density per unit length.

To digitally simulate the initial state of the physical beam, a digital physical model represented by the finite element model is as shown in Fig. 8. It is established by dividing the beam into 15 beam elements equally. As the digital finite element model of the beam includes both geometrical and material information, the operational and applied loads can be incorporated to predict the performance state or responses so that the physical base of the beam can be built. The first five natural frequencies representing the initial state of the physical beam can be calculated either from Eq. (14) or from the established finite element model as shown in Table 1.

With regard to building data base of the physical beam, to simulate the "monitored beam", the bending stiffness of the beam at element 10 is assumed to be reduced by 50% as shown in Fig. 9. For such a simulated "monitored bridge", several accelerometers may be mounted on the node points of the physical beam to measure the dynamic acceleration responses of the data base.

By using the time-history acceleration response measurements, the natural frequencies of the beam at current state can be indirectly identified through the modal analysis in either frequency-domain or time-domain. The identified first five natural frequencies of the simulated "monitored bridge" are listed in Table 1. It can be found that the natural frequencies do have a bit difference between physical base and data base.

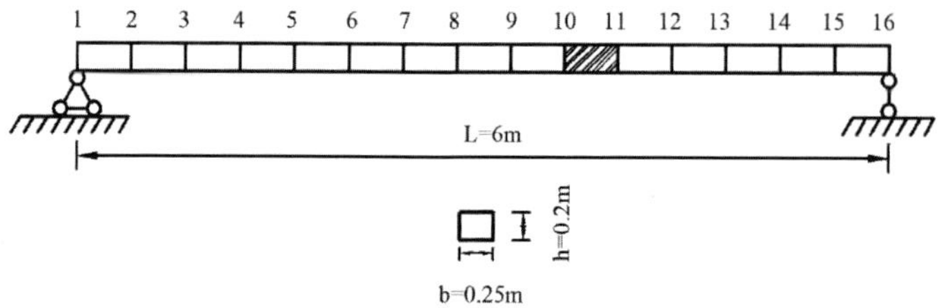

Fig. 9. Simulated "monitored bridge" (current state of the physical bridge).

Note that most of the external factors such as environment and load information are not included in either physical base or data base for such a simulated simply supported beam, as it is just a simple demonstration of how the digital twin model is built through the proposed physics-hybrid framework in the structural health monitoring.

Once both physical base and data base are built, as discussed above, the first five natural frequencies that are common to the physical base and the data base are chosen as a subset of the specific context in this demonstration as follows:

$$C = \{x_i \in S, \quad i = 5\}, \tag{15}$$

where x_i refers to the ith natural frequency, while S refers to the structural state. More exactly, the contexts in the data base and the physical base are

$$x_i^C = \{7.78\,\text{Hz}, 33.24\,\text{Hz}, 79.14\,\text{Hz}, 128.04\,\text{Hz}, 146.22\,\text{Hz}\}, \tag{16}$$

$$\hat{x}_i^C = \{8.99\,\text{Hz}, 35.92\,\text{Hz}, 80.63\,\text{Hz}, 142.93\,\text{Hz}, 222.53\,\text{Hz}\}, \tag{17}$$

where x_i^C denotes the first five natural frequencies in the data base generated by the dynamic acceleration response measurements of the monitored physical structure through the modal parameter identification, while $\hat{x}_i^C$ denotes the first five natural frequencies in the physical base calculated by the finite element analysis through the finite element model that digitally represents the physical structure.

Thereafter, the mirror relationship between the natural frequencies as the specified context C can be set up as follows:

$$d^c(\hat{x}_i^C, x_i^C) = \frac{|\hat{x}_i^C - x_i^C|}{x_i^C} \times 100\% \le \varepsilon_T, \quad \varepsilon_T = 10\%, \tag{18}$$

where a user-defined threshold is given to be $\varepsilon_T = 10\%$. Calculating the distance or error d^c between $\hat{x}_i^C$ and x_i^C can determine whether the mirror relationship is satisfied or not.

As a digital twin model of the physical structure, the best matching of the specified context between physical base and data base should be achieved to minimize the distance under the dedicated mirror relationship as follows:

$$\min\|\hat{x}_i^C - x_i^C\| \quad i = 5. \tag{19}$$

It can be seen that the optimization algorithm is necessary to calculate the optimal parameters of the physical structure to achieve the best matching. In the proposed physics-data hybrid framework to develop the digital twin model in structural health monitoring, the digital representation of the physical bridge is the finite element model. In the case of structural finite element model involved, the finite element model updating is a feasible way to achieve the best matching between physical base and data base.

The structural finite element model ensures that each parameter has a practical physical meaning. The finite element model updating process of a structure involves the construction objective function, selection of the updated physical parameters and optimization algorithm. As a result, the structural physical parameters θ to be updated have to be determined before the finite element model updating. In the current simple demonstration, two updated physical parameters of the beam are chosen. They are the elastic modulus θ_1 that refers to the material property and inertia moment θ_2 at the damage location that refers to the section geometry of the physical beam.

In general, the structural finite element model involves a huge number of elements and parameters so that the continuous updating of the finite element model is time-consuming as so many iterations are required. To mobilize the finite element model updating process and optimization iteration, the response surface-based finite element model updating approach is herein adopted. With the two selected structural parameters, the following quadratic polynomial response surface is used to replace the finite element model

$$\hat{y}^c = \beta_0 + \sum_{i=1}^{m} \beta_i \theta_i + \sum_{i=1}^{m} \sum_{j \geq 1}^{m} \beta_{ij} \theta_i \theta_j, \tag{20}$$

where $\beta_0, \beta_i, \beta_{ij}$ are the regression coefficients. θ_i denotes the structural physical parameters. $\hat{y}^c$ represents the response value at the specified structural physical parameters θ_i. In this case, $m = 2$.

The central composite design is adopted to determine the sample points. The least-square regression is then performed to fit the response surface to the sampled data. The typical contracted response surfaces are as shown in Fig. 10.

To check the accuracy of the constructed response surface model, the root mean squared error (RSME) value is used as follows:

$$\mathrm{RMSE}(\hat{y}_i^{RS} \hat{y}_i^C) = \sqrt{\frac{1}{n} \sum_{i=1}^{n} (\hat{y}_i^C - \hat{y}_i^{RS})^2}, \tag{21}$$

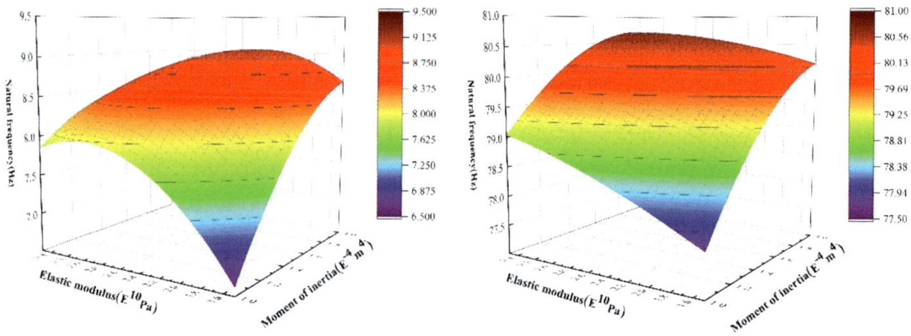

(a) The first natural frequency

(b) The third natural frequency

Fig. 10. Typical response surfaces.

Table 2. RMSE values for each mode.

Mode	1	2	3	4	5
RMSE	0.0028	0.0002	0	0.0090	0.0028

Table 3. The digital twin model in a specific frequency context (Hz).

Context	$\hat{X}^c$ in physical base	X^c in data base	Digital twin model	ε_i
f_1	8.99	7.78	7.78	0
f_2	35.92	33.24	33.25	0.03%
f_3	80.63	79.14	79.15	0.01%
f_4	142.93	128.04	128.09	0.04%
f_5	149.14	146.22	146.07	0.10%

where $\hat{y}_i^{RS}$ is the response value of the response surface model, while $\hat{y}_i^C$ is the predicted response of the confirmation samples from the finite element model. The RSME values for each mode are listed in Table 2. It can be found that all RMSE values are close to zero, which demonstrates that the created response surfaces have a good accuracy.

The established mirror relationship that is the physics-data synchronization can be regarded as the objective function in the finite element model updating process. The optimization algorithm used to minimize the objective function is a fmincon method in MATLAB. Once the best matching of the objective function is achieved, the digital twin model of the physical beam is developed. The first five natural frequencies of the built digital twin model for the simulated beam are as shown in Table 3.

The natural frequencies in the physical base actually represent the initial state of the physical structure, while the natural frequencies in the data base represent the current state of the physical structure or monitored structure. It can be found from Table 3 that the first five natural frequencies of the developed digital twin model for the beam are almost consistent with those in the data base, indicating that such a digital twin model can represent the current state of physical structure in term of a specific context. Namely, the digital twin model built in the structural health monitoring can provide the structural state in the real time. The further damage detection, structural assessment and state prediction can be therefore carried out with the developed digital twin model.

4. Conclusions

The most important feature of digital twin in the structural health monitoring is to achieve the physical structure-monitoring data synchronization. For this purpose, a physics-data hybrid framework to develop the bridge digital twin model in structural health monitoring is proposed in the paper. The physical structure is digitally represented by the finite element model to fully incorporate both structural geometry and structural state to form the physical base. The corresponding data base is built by combining measurement data of the physical structure in structural health monitoring. By defining the context that is common to both physical base and data base, the mirror relationship between physical base and data base for the specified context can be conceived. To achieve the best matching of the mirror relationship, the digital twin model that incorporated both physical structure and monitored structure can be developed. A simulated simply supported beam is implemented to demonstrate how the digital twin model is developed by using proposed physics-data hybrid framework. It is found that the generated digital twin model is consistent to the actual beam in terms of the specified context.

The concept and implementation of proposed physics-data hybrid framework to develop the digital twin model in structural health monitoring is clear and simple. This framework integrates physical knowledge and data intelligence into one model. The digital twin model built in the structural health monitoring can reflect the structural updated state in the real time. Subsequently, the damage detection, structural assessment and structural state prediction can be further performed with such a developed digital twin model.

The actual bridge digital twin will be faced with a large amount of data, many influencing factors, and other problems. The presented physics-data hybrid framework helps in clearer understanding of the realization of digital twin model in structural health monitoring, providing a new perspective for smart bridge solutions. Considering the varying operating conditions and the complex coupling effects on the actual monitored bridge, the construction of a multi-factor coupling digital twin model, including multi-data, uncertainty quantification, verification and validation, is still challenging in the structural health monitoring.

Acknowledgments

Financial support from projects of Shenzhen Science and Technology Innovation Program under grant Nos. JSGG20210802093207022, KQTD20180412181337494, ZDSYS20201020162400001 and GJHZ20200731095802007 is acknowledged.

ORCID

Li-Feng Qin https://orcid.org/0009-0009-2762-9355
Chuan-Rui Guo https://orcid.org/0000-0002-7141-2972

References

1. X. H. Zhou and X. G. Zhang, Thoughts on the development of bridge technology in China, *Engineering* **5**(6) (2019) 1120–1130.
2. Y. F. Duan *et al.*, Advanced finite element model of Tsing Ma bridge for structural health monitoring, *Int. J. Struct. Stab. Dyn.* **11**(2) (2011) 313–344.
3. L. Ren *et al.*, Structural health monitoring system developed for Dalian stadium, *Int. J. Struct. Stab. Dyn.* **16**(4) (2016) 1640018.
4. X. J. Ye *et al.*, A deep learning-based method for automatic abnormal data detection: Case study for bridge structural health monitoring, *Int. J. Struct. Stab. Dyn.* **23** (2023) 2350131, doi: 10.1142/s0219455423501316.
5. F. Tao and Q. L. Qi, Make more digital twins, *Nature* **573**(7775) (2019) 490–491.
6. B. Brenner and V. Hummel, Digital twin as enabler for an innovative digital shopfloor management system in the ESB logistics learning factory at Reutlingen - University, *Procedia Manuf.* **9** (2017) 198–205.
7. S. Yun, J. H. Park and W. T. Kim, Data-centric middleware based digital twin platform for dependable cyber-physical systems, *9th Int. Conf. Ubiquitous Future Netw.* (2017), pp. 922–926.
8. M. Grieves and J. Vickers, Digital twin: Mitigating unpredictable, undesirable emergent behavior in complex system, in *Transdisciplinary Perspectives on Complex Systems* (Springer, Cham, Switzerland, 2017).
9. E. Glaessgen and D. Stargel, The digital twin paradigm for future NASA and U.S. air force vehicles, *53rd AIAA/ASME/ASCE/AHS/ASC Structures, Structural Dynamics and Materials Conference*, Honolulu, 2012.
10. F. Tao *et al.*, Five-dimension digital twin model and its ten applications, *Comput. Integr. Manuf. Syst.*, **25**(01) (2019) 1–18.
11. F. Tao *et al.*, Digital twin driven prognostics and health management for complex equipment, *CIRP Ann-Manuf. Technol.* **67**(1) (2018) 169–172.
12. J. S. Bao, D. S. Guo, J. Li and J. Zhang, The modelling and operations for the digital twin in the context of manufacturing, *Enterp. Inf. Syst.* **13**(4) (2019) 534–556.
13. H. F. Jiang *et al.*, How to model and implement connections between physical and virtual models for digital twin application, *J. Manuf. Syst.* **58** (2021) 36–51.
14. J. J. Wang *et al.*, Digital twin for rotating machinery fault diagnosis in smart manufacturing, *Int. J. Prod. Res.* **57**(12) (2019) 3920–3934.

15. K. Worden *et al.*, On digital twins, mirrors and virtualisations: frameworks for model verification and validation, *ASCE-ASME J. Risk Uncertainty Eng. Syst. Part B Mech. Eng.* **6**(3) (2019) 030902.
16. F. Tao *et al.*, Digital twin in industry: State-of-the-Art, *IEEE Trans. Ind. Inform.* **15**(4) (2019) 2405–2415.
17. W. Qian *et al.*, Multidimensional data modeling and model validation for digital twin workshop, *J. Comput. Inf. Sci. Eng.* **21** (2021) 031005-1.
18. F. Tao *et al.*, Digital twin-driven product design, manufacturing and service with big data, *Int. J. Adv. Manuf. Technol.* **94** (2018) 3563–3576.
19. D. S. Shangguan, L. P. Chen and J. W. Ding, A digital twin-based approach for the fault diagnosis and health monitoring of a complex satellite system, *Symmetry-Basel* **12**(8) (2020) 1307.
20. Y. Lu, X. L. Qiu and Y. Xing, Digital twin-based operation simulation system and application framework for electromechanical products, *Int. Conf. Computer, Control and Robotics* (2021), pp. 146–150.
21. J. S. Kang, K. Chung and E. J. Hong, Multimedia knowledge-based bridge health monitoring using digital twin, *Multimed. Tools Appl.* **80**(26–27) (2021) 34609–34624.
22. C. Ye *et al.*, A digital twin of bridges for structural health monitoring, *12th International Workshop on Structural Health Monitoring* (Stanford University, 2019).
23. M. Mohammadi *et al.*, Quality evaluation of digital twins generated based on UAV photogrammetry and TLS: Bridge case study, *Remote Sens.* **13**(17) (2021) 3499.
24. T. G. Ritto and F. A. Rochinha, Digital twin, physics-based model, and machine learning applied to damage detection in structures, *Mech. Syst. Signal Proc.* **155** (2021) 107614.
25. S. Chakraborty and S. Adhikari, Machine learning based digital twin for dynamical systems with multiple time-scales, *Comput. Struct.* **243** (2021) 106410.
26. C. Li *et al.*, Dynamic Bayesian network for aircraft wing health monitoring digital twin, *AIAA J.* **55**(3) (2017) 1–12.
27. G. Tsialiamanis *et al.*, On generative models as the basis for digital twins, *Data-Centric Eng.* **2** (2021) e11.

https://doi.org/10.1142/9789819814770_0035

Chapter 35

Adaptive Excitation Frequency Matching and Acoustic-Feature-Based Automatic Hammering Inspection System[#]

Huang Huang[*,‡], Xi Huang[†,§], Zhishen Wu[*,¶], Mohamad Najib Alhebrawi[†,||] and Zhishui Liang[*,**]

[*]*Southeast University, International Institute for Urban Systems Engineering Nanjing 211189, P. R. China*

[†]*Ibaraki University, Department of Urban and Civil Engineering Hitachi 316-8511, Japan*

[‡]*hhly9527@gmail.com*
[§]*22nd302r@vc.ibaraki.ac.jp*
[¶]*zswu@seu.edu.cn*
[||]*najeebhub@gmail.com*
[**]*zs_liang@seu.edu.cn*

Hammering tests are employed for detecting surface and internal damage; however, their accuracy depends on the experience and expertise of inspectors. Because automation techniques can address the limitations of manual hammering tests, researchers have developed automatic hammering inspection systems to replace or assist conventional hammering inspection. Although several studies on automatic hammering inspection have been reported, several challenges hinder the practical application of their results. The relative hammering frequency and identification algorithms for complex and diverse damage are critical factors in implementing developed techniques. In a previous study, an automatic hammering inspection device for rapid and automated damage detection in practical applications was developed. This device can remotely adjust the hammering angle and force to increase detection accuracy and efficiency. Furthermore, an artificial-intelligence-enhanced damage identification method was developed to identify fine concrete cracks accurately using acoustic data. This study developed an automatic hammering inspection system with an adaptive matching function and automatic damage identification to increase the accuracy and efficiency of damage detection. First, by imitating human auditory judgment, an adaptive matching algorithm for the excitation frequencies of different damage types was developed to determine the optimal excitation frequency. Second, the relationships between acoustic features and damage information were analyzed. Finally, two concrete specimens with artificial damage were tested using the proposed system, with the relationships between the acoustic features in the time and frequency domains, as well as the damage location and depth, investigated

[¶]Corresponding author.
[#]This chapter appeared previously on the International Journal of Structural Stability and Dynamics. To cite this chapter, please cite the original article as the following: H. Huang *et al.*, *Int. J. Struct. Stab. Dyn.*, **23**, 2340038 (2023), DOI: 10.1142/S0219455423400382.

experimentally. The results show that the proposed system can accurately identify surface and internal damage distributions up to a depth of 100 mm at 20 mm intervals.

Keywords: Automated hammering inspection; acoustic-feature-based damage identification; adaptive excitation frequency matching.

1. Introduction

Recently, the prompt detection, localization, and quantification of structural damage have gained increasing importance, attracting significant attention to damage inspection, durability evaluation, and performance prediction of existing structures.[1–3] For regular structural inspections of bridges, tunnels, and other structures, hammering tests are simple inspection methods adopted worldwide. Generally, successfully implementing a hammering test requires inspectors to understand the setup of a hammer to achieve optimal results and evaluate damage based on acoustic responses. However, inadequate training can lead to incorrect conclusions. Furthermore, the monitoring of structures located in dangerous areas is often difficult. Therefore, an automatic inspection system that can replace or assist conventional hammering inspections is crucial.

Hammering inspections characterize concrete material qualities, such as strength and crack formation. Extensive research has been conducted employing automatic hammering inspection to detect damage to various structures (e.g. peeling failures of the outer walls of high-rise buildings,[4] delamination of concrete bridge decks,[5] and tunnel damage[6]). However, existing methods can only detect severe surface damage, while the relationships between acoustic features and damage information remain unclear.

Two key factors are known to affect the performance of automatic inspection systems — automation degree of the inspection device and efficiency and accuracy of the inspection results. In our previous study, an automatic hammering device consisting of an automatic hammer operating at different excitation frequencies, a six-degree-of-freedom robotic arm, and a wireless directional microphone was developed.[7] The developed device ensured that the hammering position and obtained acoustic data were sufficiently standardized. Furthermore, an artificial-intelligence-enhanced approach was developed to identify fine concrete cracks accurately, achieving the successful identification of crack widths of 0.2, 0.4, and 0.6 mm.[8]

For an automatic inspection system, the inspection device must be calibrated according to the damage type. For example, different excitation frequencies are required to obtain effective sound data at different damage depths. Additionally, the acoustic features corresponding to different damage depths vary. Consequently, an inspection device may yield unsatisfactory performance.[9–12] Therefore, developing an automatic detection system that can independently determine the appropriate frequencies and algorithms to apply according to different damage types is crucial.

Moreover, an automatic hammering inspection system with an adaptive matching function was developed in this study. In the proposed method, the sound features of the spectral centroids related to intact and damaged concrete specimens are used as training data for determining the correlation between sound features and damage. During training, adaptive matching is configured to generate optimal excitation frequencies. The practicability of the proposed system for identifying and localizing compound and internal damage was investigated through experiments.

2. Theoretical Background of Hammering Tests and Sound Features

In this section, the properties and requirements of hammering inspection are discussed, followed by the theoretical background on sound feature extraction and the adaptive matching algorithm for automatic hammering inspection.

2.1. *Requirements for hammering inspection*

For hammering inspection, investigations of the presence or absence of damage, as well as damage localization and quantitative analysis, are crucial. Many studies on different hammering inspection devices and methods have been conducted to achieve accurate detection results. Asano *et al.*[13] developed an impact inspection method based on the dropping of a steel ball. The maximum amplitude and peak frequency of the excited sound was correlated with the dimension and depth of defects. Miyoshi *et al.*[14] proposed a continuous hammering test method using a rotational hammering device. Based on their experimental results, the relationship between damage

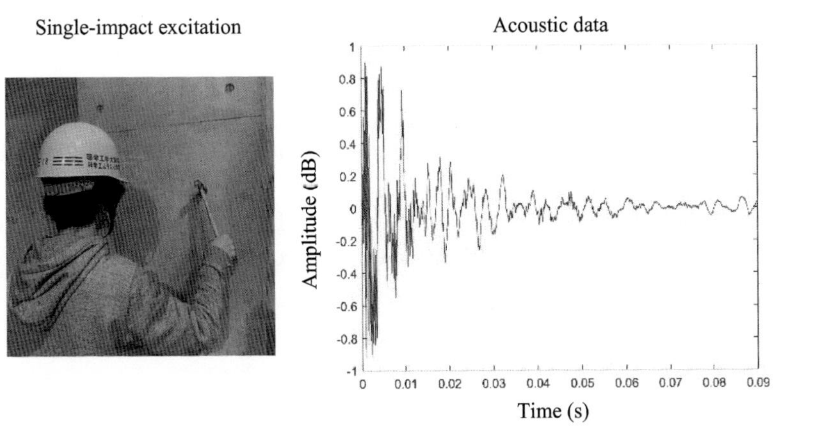

(a) Single-impact excitation (b) Acoustic data

Fig. 1. (a) Conventional hammering inspection and (b) single-impact hammering excited sound data.

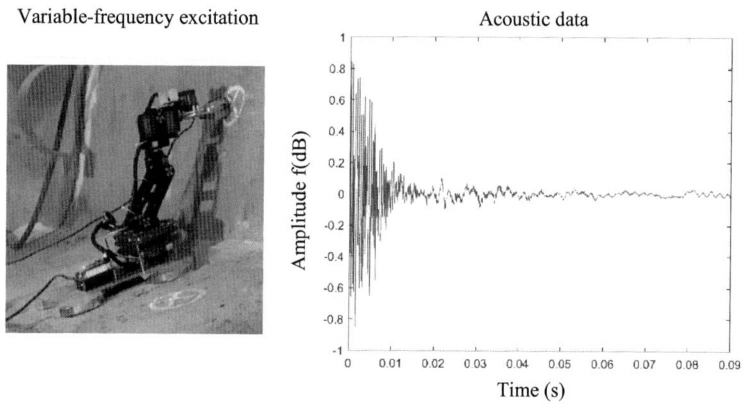

(a) Variable-frequency excitation

(b) Acoustic data

Fig. 2. (a) Automatic hammering inspection and (b) variable-frequency hammering excited sound data.

location and size was confirmed by comparing changes in frequency characteristics and maximum sound pressure. Additionally, Nakayama *et al.*[15] studied the effects of concrete strength and sound features and reported that changes in the hammering position are among the causes of sound feature changes. Despite the convenience of these methods, the random selection of impact forces and frequencies by inspectors (see Fig. 1) hinders their practicality and applicability, thus highlighting the need for improvements in the accuracy and efficiency of the hammering inspection method.

To identify surface, internal, and composite damage in complex environments, an identification method based on variable-frequency hammering has been proposed, as shown in Fig. 2.[12] Experimental results revealed that the acoustic data generated by variable-frequency impact hammers could be used to identify multiple types of damage in concrete structures with an accuracy of 97.8%, higher than that obtained using single-frequency hammering. In acoustic analysis, frequency features are extracted for sound feature analysis. In this study, the frequency features under different excitation frequencies were compared and analyzed to define accurate sound frequency features and optimal excitation conditions.

2.2. *Frequency features of sound data*

The peak frequency of a sound data series is a characteristic parameter commonly used as a sound feature in the frequency domain. However, sound data extracted from onsite hammering tests contain multiple peak frequencies, which appear as a frequency band rather than a single peak frequency. Therefore, accurately evaluating integrated responses in the frequency band is crucial.

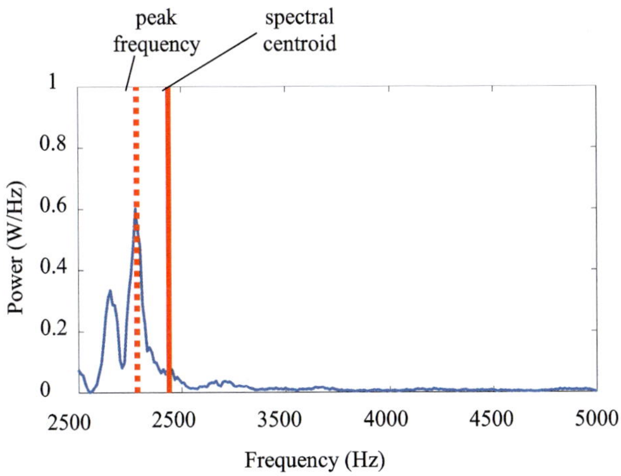

Fig. 3. Peak frequency and spectral centroid of sound data from a hammering test.

The spectral centroid is another sound frequency feature used to express the center of a frequency band and magnitude information. The individual centroid of a spectral frame is defined as the sum of the magnitudes multiplied by the frequency divided by the sum of the magnitudes as follows:

$$\text{spectral centroid} = \frac{\sum_{k=1}^{N} kF[k]}{\sum_{k=1}^{N} F[k]}, \tag{1}$$

where $F[k]$ denotes the amplitude corresponding to bin k in the discrete Fourier transform spectrum.

In Fig. 3, the dotted red line represents the peak amplitude of the extracted frequencies, whereas the solid red line represents the spectral centroid of the frequency band from 2 500 to 5 000 Hz. Because sound frequency features are related to the excitation frequency, if information in a wide frequency band can be excited during hammering tests, the related damage information can be evaluated over a wide frequency range. In the following subsection, additional information regarding excitation frequencies and damage identification is provided.

2.3. *Relationships between excitation frequencies and sound features*

Sound features are strongly dependent on excitation frequencies. The excitation condition of a single impact can excite sound features only in a narrow frequency band. Plots of changes in extracted sound features and excitation frequencies are presented in Fig. 4. The extracted sound features for excitation frequencies of 2, 5, and 10 Hz are similar and concentrated in narrow frequency bands. In contrast to the single-impact method, variable-frequency excitation successfully identifies all

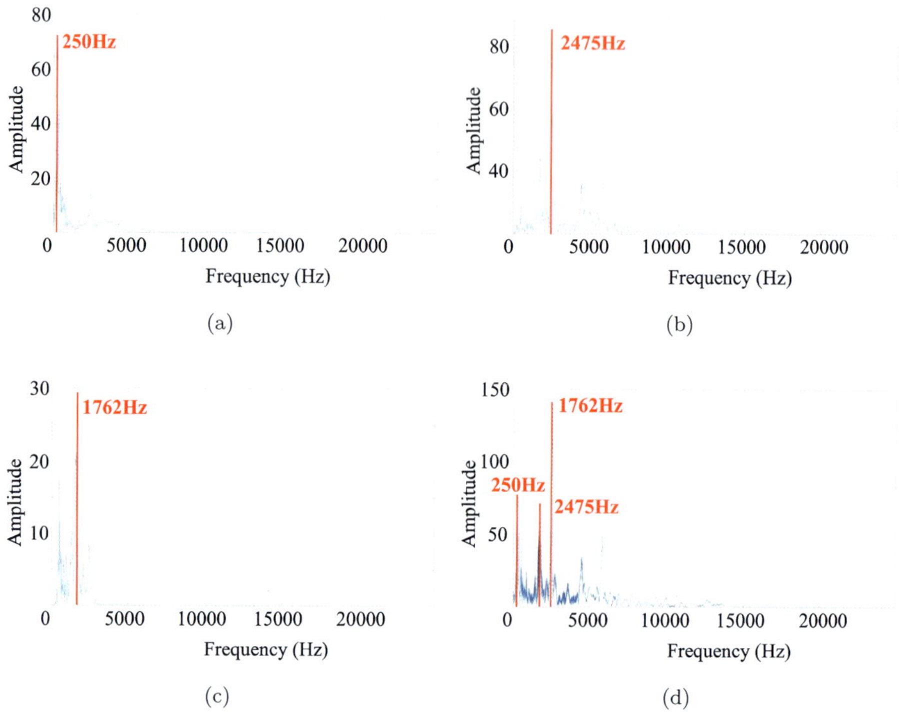

Fig. 4. Sound features for excitation frequencies of (a) 2, (b) 5, and (c) 10 Hz and (d) variable-frequency excitation from 2–10 Hz.

peak frequencies under single-impact excitation. Furthermore, it provides additional frequency information suitable for selecting the optimal excitation frequency for complex damage identification.

3. Adaptive Excitation Frequency Matching and Acoustic-Feature-Based Automatic Hammering Inspection System

3.1. *Imitating human behavior for automatic hammering inspection*

As shown in Fig. 5, a well-trained inspector excels at employing sounds to assess damage during hammering tests. Acoustic features are auditory attributes that convey crucial information regarding damage types. Despite having been considered to occupy multidimensional vectors, different excitation conditions are expected to cause different sound responses corresponding to acoustic features. Therefore, identifying possible damage types correlating with optimal excitation frequencies remains a significant challenge. Moreover, developing a function that adaptively

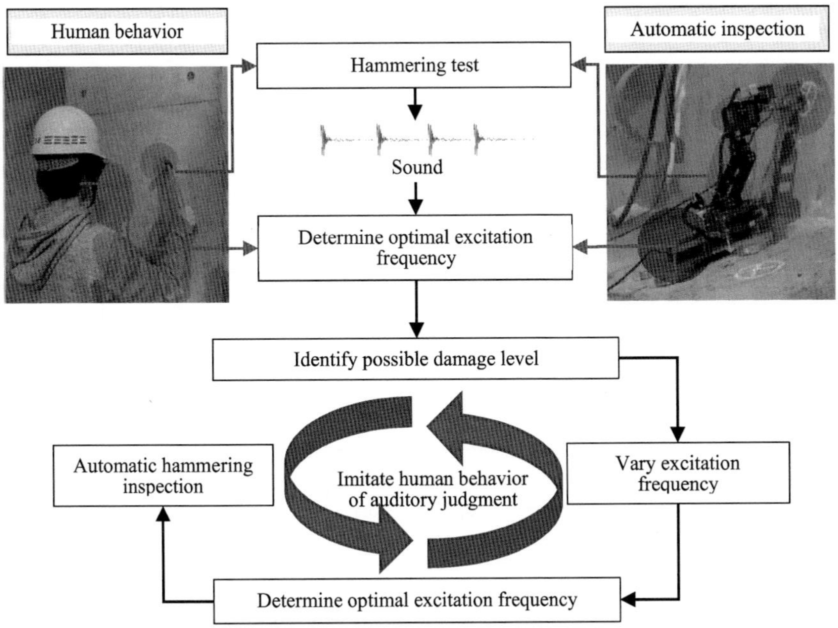

Fig. 5. Adaptive adjustment of the excitation frequency for automatic hammering inspection.

adjusts the excitation frequency is crucial for automatic hammering inspection. This study proposes an adaptive approach to adjust the excitation frequency based on human auditory judgment.

3.2. *Principle of adaptive excitation frequency matching*

To develop a function for the adaptive adjustment of excitation frequencies, sound data were analyzed using different excitation frequencies and damage types from published studies.[7–9] A cross-correlation model was used to reveal damage-type correlations in different excitation frequency ranges. Acoustic features such as the peak frequency, spectral centroid, and spectra in different frequency bands were observed to have generic acoustic correlations, which can be used as training data to learn the correlations between excitation frequencies and damage types. Figure 6 illustrates the principle underlying the adaptive matching of optimal excitation frequencies based on possible damage types.

3.3. *Method for adaptive excitation frequency matching*

In this study, to determine the correlations between sound features and damage, the sound features of the spectral centroids corresponding to intact and damaged concrete specimens were used as training data. To consider the effects of damage on

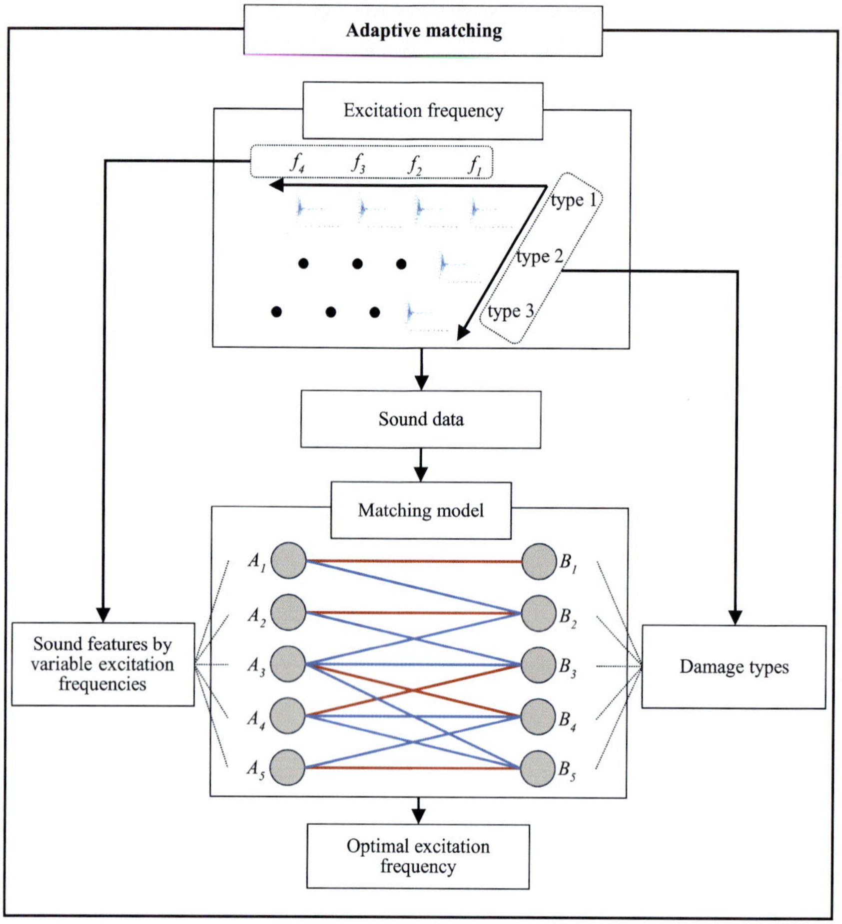

Fig. 6. Principles of adaptive excitation frequency matching.

different sound features (spectral centroid values), sound features were normalized using the zero-mean normalized cross-correlation function, which was extended by weighting each frequency component as follows:

$$\boldsymbol{S}(\boldsymbol{A}, \boldsymbol{u}, \boldsymbol{x}) = \frac{\sum_{k \in K} u_{(k)}(A_{(k)} - \bar{A})(x_{(k)} - \bar{x})}{\sqrt{\sum_{k \in K} u_{(k)}(A_{(k)} - \bar{A})^2}\sqrt{\sum_{k \in K} u_{(k)}(x_{(k)} - \bar{x})^2}}, \tag{2}$$

where x is a vector consisting of the amplitude of each frequency component obtained from the damage to be evaluated, A is the frequency template vector obtained from the training data used to compare and evaluate the new evaluation object, K is the index set of frequency components utilized to evaluate the similarity between x and A, and u is the weight of each frequency component associated with index set K and represents the characteristic vector of the hammering test. The range of the function $\boldsymbol{S}(\boldsymbol{A}, \boldsymbol{u}, \boldsymbol{x})$ is defined as $[-1, 1]$, where a value close to one indicates excellent similarity between the frequency component and damage type.

3.4. *Proposed system*

Based on the theoretical background presented in Sec. 2, an adaptive excitation frequency matching approach was developed to increase the accuracy of automatic hammering inspection. A flowchart of the proposed method is presented in Fig. 7. Particularly, adaptive matching for the optimal excitation frequency and acoustic-feature-based damage identification is divided into two stages. First, pre-detection is performed according to three excitation frequencies ranging from low to high. Subsequently, matching is implemented according to the correlations between the sound characteristics of the three excitation frequencies and different damage types.

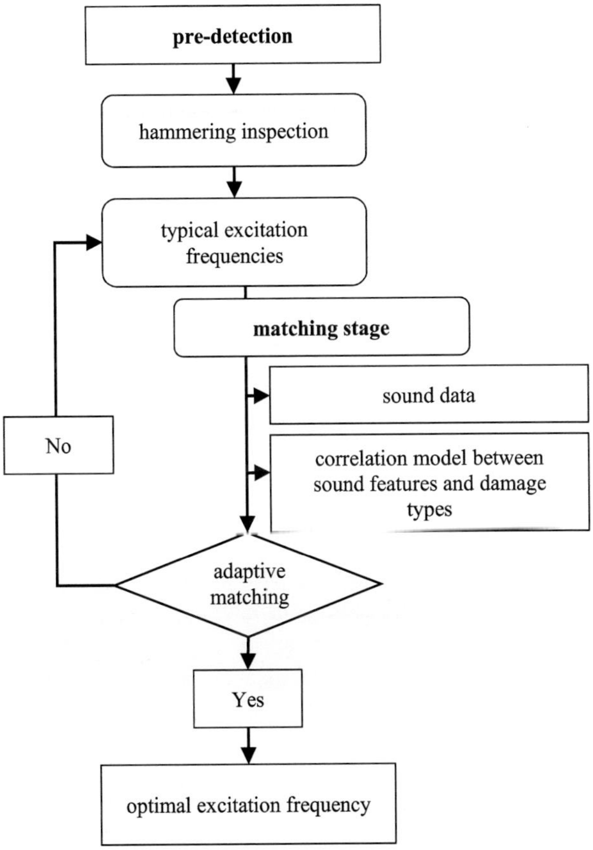

Fig. 7. Flowchart of the proposed adaptive excitation frequency matching approach.

Finally, the optimal excitation frequency is obtained by narrowing the range of excitation frequencies.

4. Experimental Setup and Procedure

4.1. *Detection of internal damage*

To simulate the internal damage of concrete specimens, foam was buried 60 mm under a concrete specimen at a depth of 100 mm, as shown in Fig. 8. The foam covered the distance from 200–260 mm from the left side of the concrete. The concrete specimen had dimensions of 400 (length) $\times$120 (height) $\times$120 (width) mm and was divided into two parts, as shown in Fig. 8. Hammering was performed from left to right in 20 mm intervals with frequencies ranging from 2 to 10 Hz, resulting in the formation of 20 hammering sections.

4.2. *Detection of surface and internal damage*

To verify the applicability of the proposed method, inspection tests were performed on a concrete beam with a length of 420 mm, width of 60 mm, and height of 60 mm. The concrete beam was divided into 21 areas, each 20 mm wide, with the damage type in each area classified into four categories. In Fig. 9, "health," "surface damage," "internal damage," and "compound damage" refer to no damage, surface

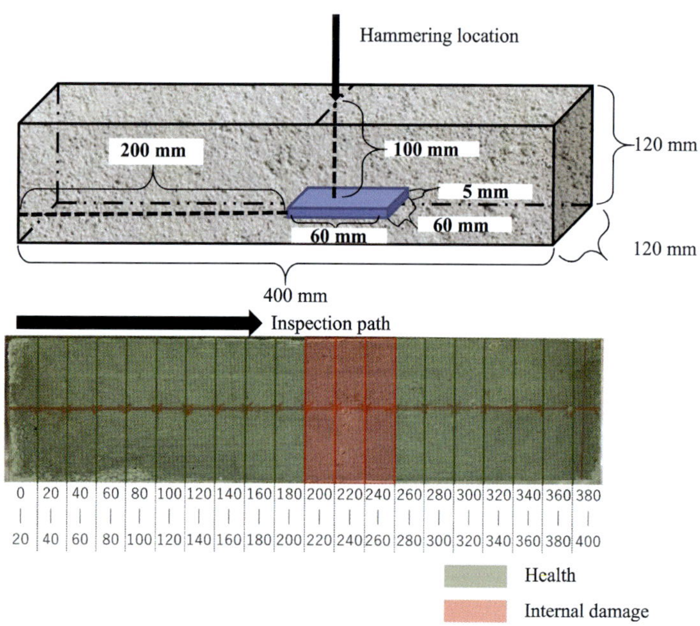

Fig. 8. Schematic of a concrete specimen with internal damage.

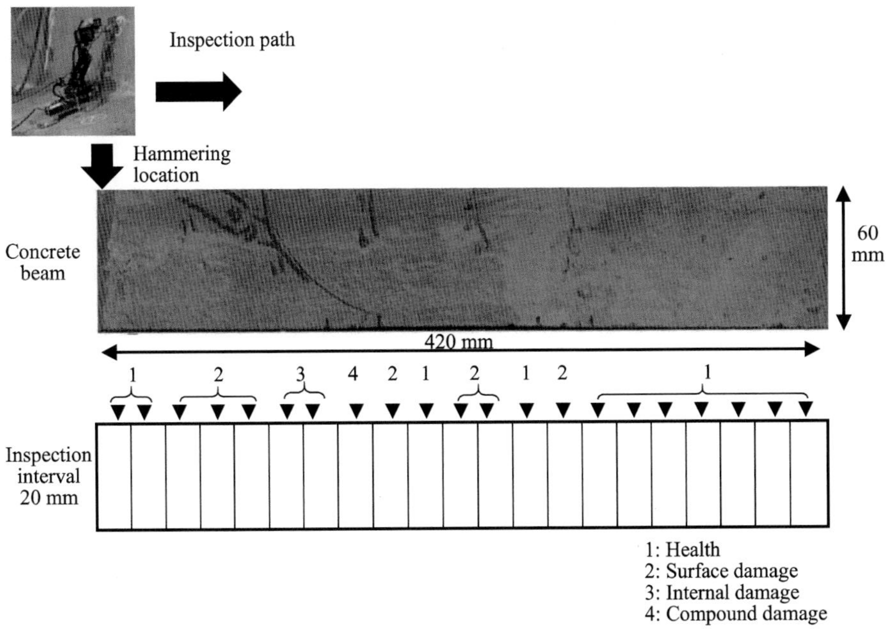

Fig. 9. Schematic of concrete areas under four damage classifications.

cracks, internal cracks, and both surface and internal cracks, respectively. The automatic hammering device inspected each area along the beam direction. Sound data were obtained by hammering the concrete at frequencies ranging from 2 to 10 Hz.

5. Experimental Results and Discussion

5.1. *Adaptive excitation frequency matching results*

Figure 10 compares the variable-frequency-excited peak frequencies and spectral centroids in the "healthy" and "damaged" areas, respectively, based on data from the detection of internal damage. The solid red lines represent the spectral centroids, and the dotted red lines represent the peak frequencies. As indicated, the excited spectral centroids describe the frequencies of signal components with large magnitudes in a spectrum and reflect the distribution of the power spectrum information related to sound data.

5.2. *Sound-feature-based damage identification results obtained using variable-frequency hammering*

Figure 11 presents the relationship between the excitation frequency and spectral centroid of the "healthy" area obtained using the variable-frequency hammering

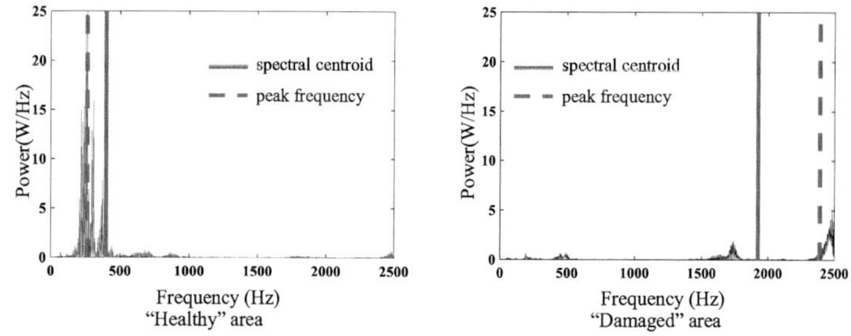

(a) 0–2500 Hz

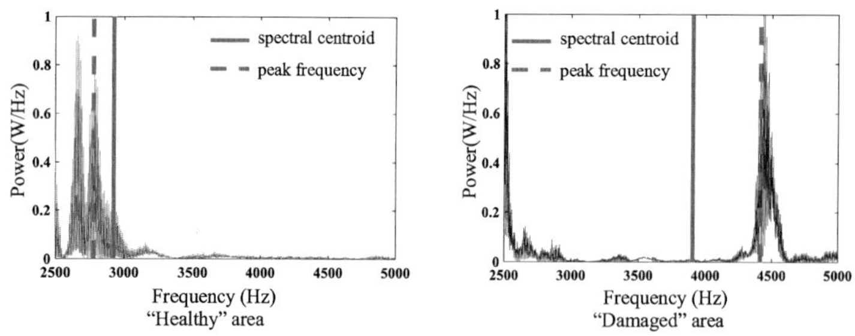

(b) 2500–5000 Hz

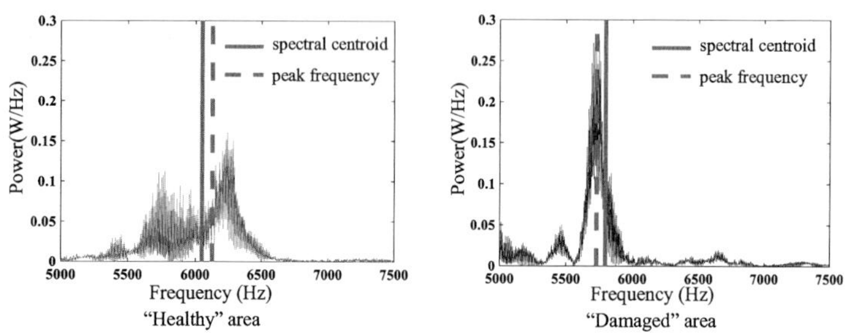

(c) 5000–7500 Hz

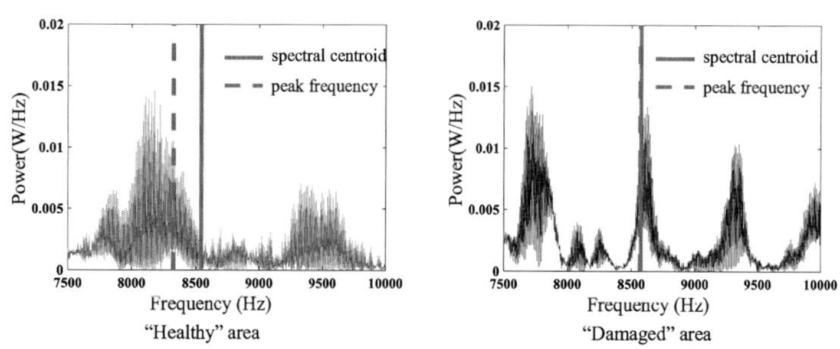

(d) 7500–10000 Hz

Fig. 10. Variable-frequency-excited peak frequencies and spectral centroids in "healthy" and "damaged" areas in the bands (a) 0–2500 Hz, (b) 2500–5000 Hz, (c) 5000–7500 Hz, and (d) 7500–10000 Hz.

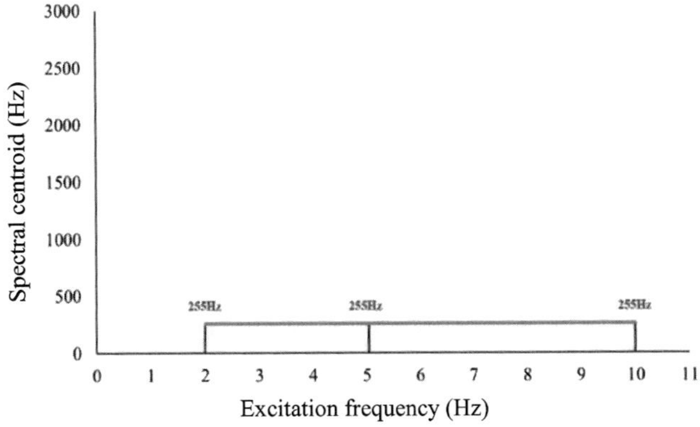

Fig. 11. Relationship between the excitation frequency and spectral centroid (healthy area).

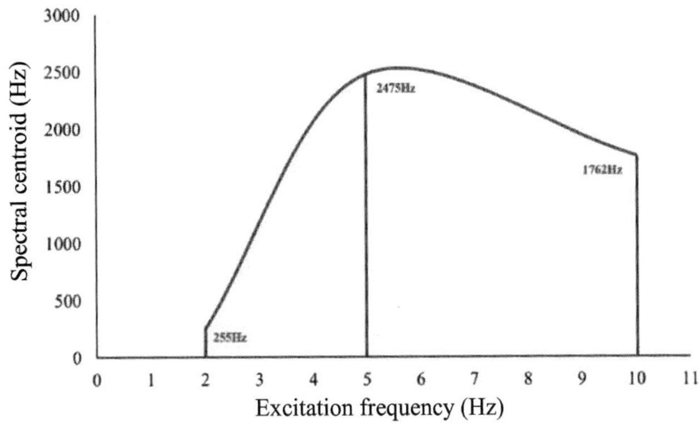

Fig. 12. Relationship between the excitation frequency and spectral centroid (damaged area).

method based on data from the detection of internal damage. The sound features are relatively stable; thus, the detection results are unaffected by excitation frequency optimization.

By contrast, in Fig. 12 — also derived from the detection of internal damage — the relationship between the excitation frequency and spectral centroid for the "damaged" area reveals different spectral responses under different excitation frequencies. Damage can be localized correctly by leveraging this relative fluctuation.

5.3. *Sound-feature-based damage localization results obtained using variable-frequency hammering*

5.3.1. *Internal damage experiment results*

In Fig. 13, the "health" and "internal damage" areas are marked in red, and the blue line represents the sound features extracted using the variable-frequency hammering

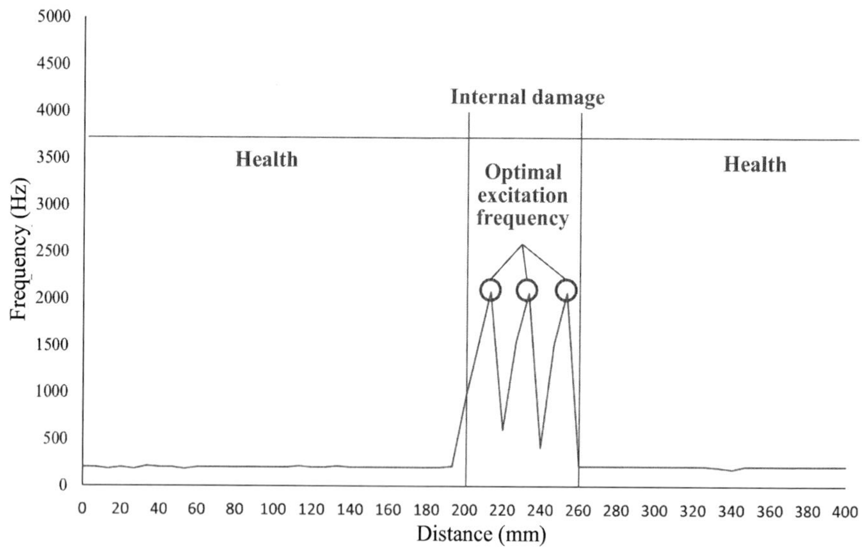

Fig. 13. Relationship between the excitation frequency and spectral centroid (internal damage areas).

method. The plot changes abruptly at approximately 200–260 mm, with the location of this change being consistent with the actual damage location. Therefore, the proposed method is suitable for internal damage detection in concrete structures.

5.3.2. *Compound damage experiment results*

The changes observed in the sound features extracted using the variable-frequency hammering method are presented in Fig. 14. As shown at the top of the figure, each hammering point corresponds to a 20 mm wide area. The "health" area spans 0–40 mm and 300–420 mm; the "surface damage" area covers 40–60 mm, 80–100 mm, 200–220 mm, and 260–280 mm; the "internal damage" area covers 100–160 mm; and the "compound damage" area extends from 100 to 160 mm. Red markers and lines highlight the detection outcomes at the optimal excitation frequency.

Each inflection point (denoted by a red circle) aligns with an observable crack in the actual specimen. The frequency within the "healthy" area remains relatively consistent. Conversely, the frequency variations in the "compound damage" area indicate extensive cracks reaching depths of 60 mm, effectively representing the damage distribution. The detection results at the optimal excitation frequency provide a precise representation of this distribution. For example, the three cracks on the right are similar. However, the spectral centroid on the left is higher than that on the right, possibly corresponding to the shapes and sizes of the cracks. Different cracks or different spectral centroids from dominant frequencies can cause changes owing to quantitative factors, such as damage depth or size.

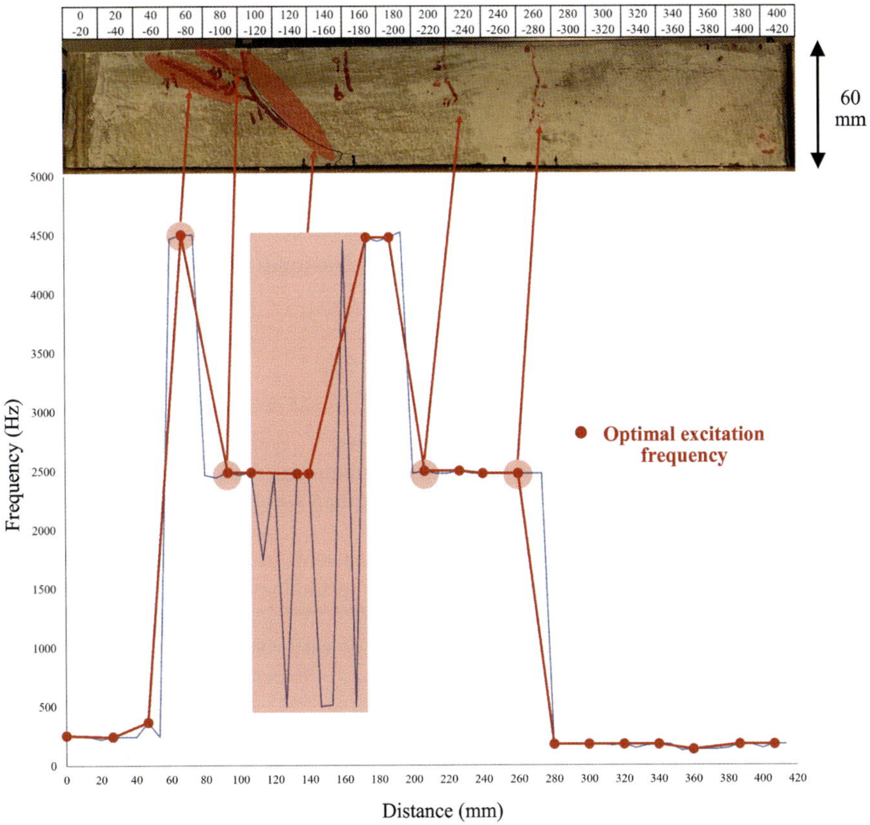

Fig. 14. Relationship between the excitation frequency and spectral centroid ("compound damage" areas).

6. Conclusions

The applicability of sound-feature-based damage identification was investigated using an automatic hammering inspection technique. The enhancement of adaptive frequency matching and the ability to detect compound damage were verified through laboratory experiments. The following conclusions were drawn:

(1) An adaptive matching algorithm was designed to emulate human auditory discernment, focusing on excitation frequencies and various damage types. This algorithm facilitates the determination of optimal excitation frequencies—ascertained from fixed excitation frequencies of 2, 5, and 10 Hz for internal damage with an approximate depth of 60 mm.
(2) An analysis of the correlation between acoustic characteristics and damage types revealed that peak frequencies and spectral centroids can serve as effective acoustic indicators to differentiate damage types.

(3) The experimental results confirmed the effectiveness of the proposed system, which is capable of accurately identifying internal and compound damage up to a depth of 100 mm in 20 mm intervals.

Acknowledgments

This research was funded by the National Key Research and Development Program of China (Grant No. 2019YFC1511103) and the National Natural Science Foundation of China (Grant No. 52178115).

References

1. R. Raišutis R. Kažys, L. Mažeika, E. Žukauskas, V. Samaitis, L. Draudvilienė and A. Vladišauskas, An adjustment-free NDT technique for defect detection in multilayered composite constructions using ultrasonic guided waves, *Int. J. Struct. Stab. Dyn.* **14**(8) (2014) 1440025.
2. M. Hussin, T. H. T. Chan, S. Fawzia and N. Ghasemi, Identifying the prestress force in prestressed concrete bridges using ultrasonic technology, *Int. J. Struct. Stab. Dyn.* **20**(10) (2020) 2042014.
3. H. Lee, H. W. Park and H. Sohn, Pipe defect visualization and quantification using longitudinal ultrasonic modes, *Int. J. Struct. Stab. Dyn.* **14**(5) (2014) 1440008.
4. J. Y. L. Kasahara, H. Fujii, A. Yamashita and H. Asama, Weakly supervised approach to defect detection in concrete structures using hammering test, in *Proc. 2019 IEEE 8th Global Conf. on Consumer Electronics (GCCE)* (IEEE, 2019), pp. 973–974.
5. G. Zhang, R. S. Harichandran and P. Ramuhalli, An automatic impact-based delamination detection system for concrete bridge decks, *NDT & E Int.* **45** (2012) 120–127.
6. H. Fujii, A. Yamashita and H. Asama, Defect detection with estimation of material condition using ensemble learning for hammering test, in *Proc. 2016 IEEE International Conference on Robotics and Automation (ICRA)*, 2016, Stockholm, Sweden, pp. 3847–3854.
7. X. Huang, H. Huang and Z. Wu, Development of an automated hammer testing system for concrete structures, *Int. J. Sci. Manage. Stud.* **6** (2023) 32–45.
8. M. N. Alhebrawi, H. Huang and Z. Wu, Artificial intelligence enhanced automatic identification for concrete cracks using acoustic impact hammer testing, *J. Civil Struct. Health Monit.* **13** (2023) 469–484.
9. X. Huang, H. Huang and Z. Wu, Development of a variable-frequency hammering method using acoustic features for damage-type identification, *Appl. Sci.* **13**(3) (2023) 1329.
10. H. Fujii, A. Yamashita and H. Asama, Automated diagnosis of material condition in hammering test using a boosting algorithm, in *Proc. 2014 IEEE Int. Workshop on Advanced Robotics and its Social Impacts* (IEEE, 2014), pp. 101–107.
11. A. Watanabe, J. Even, L. Y. Morales and C. Ishi, Robot-assisted acoustic inspection of infrastructures-cooperative hammer sounding inspection, *J. Robot. Soc. Japan* **33**(7) (2015) 548–554.

12. L. Zhang, H. Ji, L. Liu and J. Zhao, Time–frequency domain characteristics of acoustic emission signals and critical fracture precursor signals in the deep granite deformation process, *Appl. Sci.* **11**(17) (2021) 8236.
13. K. Goda, H. Kosugi, S. Aoyama and A. Kobayashi, Quantitative evaluation for thickness and cavity of concrete models by impact acoustic method, *Trans. Jpn. Soc. Irr. Drain. Reclam. Eng.* **71** (2003) 73–82.
14. N. Matsui, F. Kuratani, T. Yoshida and Y. Hasebe, An automatic detection method for concrete defects based on self-organizing map for rotary hammering test, in *Proc. Int. Conf. Motion and Vibration Control* (The Japan Society of Mechanical Engineers, 2020), p. 10054.
15. T. Kamada, M. Asano, M. Kunieda and K. Rokugo, Quantitative nondestructive evaluation of defects in concrete surface layer by impact acoustics methods, *Doboku Gakkai Ronbunshu* **2002**(704) (2002) 65–79.

https://doi.org/10.1142/9789819814770_0036

Chapter 36

A Small-Deformation Rate-Independent Continuous-Flow Model for Elasto-Plastic Frames Allowing Rapid Fatigue Predictions in Metallic Structures[#]

Dominic Jarecki, Bensingh Dhas, Arun Srinivasa*
and J. N. Reddy

Department of Mechanical Engineering
Texas A&M University College Station
Texas 77845, USA
*arun-r-srinivasa@tamu.edu

Fatigue analysis in metallic frame structures can be challenging due to associated computational costs; if localized plasticity is involved, then the approach of three-dimensional (3D) continuum plasticity models for direct computation of stresses will be infeasible for the analysis of cyclic loading that would need to be modeled in medium- to high-cycle fatigue and vibratory fatigue applications. This difficulty is particularly accentuated in architected structures, for which high-resolution 3D finite element analysis (FEA) would be prohibitively expensive. In this work, we propose an alternative approach based on the use of novel elasto-plastic frame model with continuous flow (i.e. no sharp yield function) for modeling 3D frame and lattice structures. Rather than splitting the strains (as is done in classical plasticity) we split the deformation measures, extension, curvature and twist, into elastic and plastic components and postulate a rate type evolution rule for the plastic variables in terms of the stress resultants (axial force, bending moment, and torque). The combination of structural models together with the use of elasto-plastic operator split to solve the resulting boundary value problem allows for much faster determination of localized plasticity than continuum models can provide. The use of a continuous transition from elastic to rate-independent plasticity (as opposed to an abrupt change with classical plasticity models) allows us to capture localized microplasticity and determine resulting fatigue progression using a cycle-count-free, plastic work-based approach, formulated in terms of the curvatures and resultants.

*Corresponding author.
#This chapter appeared previously on the International Journal of Structural Stability and Dynamics. To cite this chapter, please cite the original article as the following: D. Jarecki *et al.*, *Int. J. Struct. Stab. Dyn.*, **23**, 2340039 (2023), DOI: 10.1142/S0219455423400394.

We demonstrate that (a) the model is able able to reproduce the response of 3D FEA with very few elements and (b) the model has the ability to rapidly predict the fatigue life under variable amplitude combined loading with relatively few frame elements.

Keywords: Low-cycle fatigue; rate-independent plasticity; micro-architected materials; geometrically-exact beam theory; smooth-yielding elastic-plastic constitutive model.

1. Introduction

Modeling the initiation and evolution of plastic deformation complex frame-like structures (such as those possible by three-dimensional (3D) printing) — at varying length scales and for a variety of disparate applications — has been a challenge extensively examined in the literature, with a particular focus on approximating key aspects of the deformation behavior without overly-onerous computational costs. The aim of this paper is to develop an approach based on a novel rate-independent continuous flow model for elasto-plastic frames that significantly reduces computational cost and allows for rapid evaluation of fatigue and failure response with only slight loss of accuracy compared with full finite element analysis (FEA) calculations.

For example, fractal frame structures like Menger Sponge-type micro-architectures have been considered for application in next-gen biomedical implants (see Refs. 1 and 2) — their porous structure promotes superior osseointegration, and their tailorable stiffness allows better matching to that of bone, helping prevent stress-shielding and separation of the implant from the surrounding bone. However, simulation of even a single unit cell of these structures is quite challenging; recent works have required anywhere from 300 000 tetrahedral elements[3] to 2 million[4] for simulation of $M = 3$ sponges.

In the civil engineering literature, knowledge and localization of plastic deformation within a structure has been examined extensively through limit analysis, in particular via the concept of a plastic hinge.[5–8] Such plastic hinge models (or more generally "lumped plasticity models") possess the flexibility to handle both hardening and softening cases (e.g. Refs. 9 and 10), and is extremely useful for gross failure analysis, this approach is severely limited for predicting fatigue where accumulation of localized microplasticity (even in the nominally elastic regime) is the determining factor before full cross-sectional yielding.

There are several beam and rod models in ABAQUS that have some form of plastic deformation capabilities, however, since they assume elastic response until full cross-sectional yielding they are not suitable for modeling the evolution of *microplastic* deformations in the beam; this precludes their use fatigue modeling under cyclic load that are crucial for bioimplants. Examination of this issue is included in Secs. 4.2 and 4.5.

For example, Tabatabaei *et al*[11] employed a plastic-hinge-type formulation to create a computationally-efficient model of micro-architected structures; they allow plastic hinge formation to be determined during simulation via a von Mises-like yield criterion stated in terms of cross-sectional stress-resultants (see Eqs. (45) and (46) in Ref. 11). Similarly Challamel[12] has considered the dynamic shakedown of structures using just a one degree of freedom plasticity model.

Geometrically-exact beam theories (based on the use of direct moment–curvature and twist relationships rather than 3D elasticity) have been explored in the literature for large, nonlinear deformations of elastic rods and shells for some time,[13] with early works exploring elastic-plastic deformations in shells due to Simo and Kennedy[14] and beams due to Saje*et al.*[15–17] Recent efforts have produced models suitable for use with elastic-plastic analysis in terms of stress resultants for large deformation of beams,[18] and elastic-visco-plastic deformations.[19] However, all the models described implement an elasto-plastic (EP) transition that is abrupt — the material is fully elastic until a certain limit in the space of moments and forces is reached after which flow ensues. Though quite suitable for static failure analysis, discrete jumps from elastic to (permanent) plastic behavior (rather than smooth variation between the states) will not lend itself to capturing fatigue progression under cyclic deformation.

The situation is particularly challenging for lifetime-damage estimation under cyclic loading. This estimation has been accomplished through means of a *cycle count*.[20] If, in addition to the lengthy EP simulation over structures with complex substructures such as a Menger sponge, a cycle count at every critical point is required, understanding the damage mechanics of multiple instances of an architected structure being designed is undoubtedly out of reach.

Time-domain analysis of complex structures using a complete elastic-plastic FEA followed by a cycle count at all points through the body is often computationally intractable for cases of interest, thus workarounds are often used. Remaining in the time-domain, notch correction (see Refs. 21–23) and constitutive modeling (for instance using the Garud[24] or Jiang*et al*[25–27] kinematic hardening plasticity models) of a linear-elastic FEA followed by a cycle count limited to points anticipated to present the most damage is one approach often encountered. Alternative approaches have included analyzing the progression of fatigue from the frequency-domain[28–31]; by posing the problem as a statistical analysis of failure resulting from vibration, a dramatically faster calculation is possible, but at the cost of both accuracy (for some cases) and generality.

Following the work of Rajagopal and Srinivasa,[32] Wang*et al*[33] formulated a smooth-yielding plasticity model for the case of a one-dimensional (1D) element stated in terms of moment and curvature, but only considered yielding resulting from bending for a single beam. In order to develop a physics-based approach to fatigue, Mozafari*et al*[34–36] extended this continuous flow plasticity model which allows fatigue damage to be accumulated through the use of total energy dissipated, without a cycle count.

The key features of the model developed here are:

(1) A novel approach to elasto-plastic behavior of beams and frames directly in terms of stress resultants and center-line kinematics. The yield behavior is not restricted to just the nodes but throughout the structure.
(2) A systematic procedure for the evaluation of structural parameters for different beam cross-sections.
(3) Verification and validation of the results for a variety of load cases.
(4) Application to fatigue loading predictions for combined loading conditions.

Section 2 describes the equations governing the elastic-plastic continuous flow (EPCF) which forms the core of this work, as well the important particulars of relating the generalized forces to the flow potential (Sec. 2.2) and to the fatigue damage progression (Sec. 2.3). Section 3 describes some salient details of the numerical approximation, in particular highlighting the variety of algorithms that can be successfully used with the EPCF model. Finally, Sec. 4 provides an extensive array of results to demonstrate the capabilities of the EPCF model. Beginning with a demonstration of how the required material parameters can be fit using simple results determined from cantilever beam experiments, Secs. 4.2–4.4 show how the models predictions extend well to much more complicated geometries. Section 4.5 shows how the generalization of accumulated microplastic work (AMW) fatigue parameter described by Mozafari*et al*[34] to beam elements can produce results of excellent accuracy, but at significantly reduced computational cost.

2. Theory

2.1. *Elastic-plastic continuous flow model for beams*

It is well known that when considering the yielding behavior in rods under bending and torsion (or combinations), the moment–curvature relationship does not exhibit a sharp yield point. This occurs for two reasons:

(1) The geometry; in an EP beam yielding begins at the outer fibers and moves towards the center so that at intermediate loads that are below the full section yielding loads, there is a mix of purely elastic and plastic regions; commercial finite element method (FEM) codes using classical J2 plasticity can reasonably capture this effect.
(2) The material; under axial loading, the whole cross-section yields at the same time, nonetheless, axial stress–strain curves reveal behavior that is rarely well-captured by "perfect" elastic-plastic deformation. As can be seen in Fig. 3(a), ABAQUS predicts a sharp transition in axial deformation, though this is not typically observed experimentally (see, for instance, the fits and plastic transition parameters determined in Mozafari*et al*[36]).

Thus, we anticipate a clearly visible "gradual transition" from purely elastic to full plastic response in such structures; if this is not captured, plastic deformation in the structural element will be significantly under-represented. The EPCF model possesses requisite flexibility to capture both sharp transition (if needed) and gradual transition through increasing microplastic deformation.

Our aim is to model this behavior at a structural level without having to integrate across the cross-section at every time step. Thus, we consider a beam in its reference configuration with its neutral axis oriented along the x-direction. Under deformation, we denote displacements of the beam along the x- (axial), y-, and z- (transverse) directions by u, v, and w, respectively. The twist angle of the beam is denoted by ϕ. For small deformations the strain and curvature measures of the beam can be written as, $\mathbf{q} = [\epsilon, \kappa_x, \kappa_y, \kappa_z]^T$, where the strains are related to the displacement as

$$\epsilon = \frac{du}{dx}, \quad \kappa_x = \frac{d\phi}{dx}, \quad \kappa_y = \frac{d^2w}{dx^2}, \quad \kappa_z = \frac{d^2v}{dx^2}. \tag{2.1}$$

As in small strain plasticity, we introduce the additive splitting of the generalized strain $\mathbf{q} = \mathbf{q}_e + \mathbf{q}_p$.

The generalized load experienced by the beam is denoted by load vector $\mathbf{Q} = [n, m_x, m_y, m_z]^T$, where n is the axial force, m_x is the axial moment or torque, and m_y and m_z are the bending moments around the other two axes.[a] The generalized load vector $\mathbf{Q}$ is related to the generalized strain measures using the following relationship:

$$\mathbf{Q} = \mathbb{C}(\mathbf{q} - \mathbf{q}_p). \tag{2.2}$$

The generalized modulus $\mathbb{C}$ is a 4×4 matrix in which the diagonal elements are the section moduli under axial deformation, torsion, and bending, $(\mathrm{EA}, \mathrm{GJ}, \mathrm{EI}_z, \mathrm{EI}_y)^T$, and off-diagonal terms are zero if no phenomena such as warping are considered.

As the frame deforms the loading path traces a curve in the four-dimensional (4D) generalized strain space $\mathbf{q}$. Following Mozafari*et al*[34] and Jarecki*et al*[37] at every point in the generalized strain space, we will introduce a field of directions represented by 4D unit vectors $\mathbf{N}$ that represent the "loading direction" so that as long the generalized strain rate makes a positive angle with $\mathbf{N}$, plastic flow is possible while no plastic flow occurs if the loading direction is negative, i.e.

$$\dot{\mathbf{q}}_p \begin{cases} = 0, & \text{if } \dot{\mathbf{q}} \cdot \mathbf{N} \leq 0, \\ \neq 0, & \text{otherwise,} \end{cases} \tag{2.3}$$

Thus, *in this model, there is no sharp transition between elastic and plastic behavior.* Instead, there is always plastic flow during loading.

[a]Since this is based on Euler–Bernoulli beam theory, the shear forces do no work and so are not included in the list.

Following Rajagopal and Srinivasa[32] it can be shown that the continuity of the stress as a function of time for different load paths, and the associated flow rule of classical plasticity can be generalized by specifying that the plastic generalized curvatures evolve as

$$\dot{\mathbf{q}}_p = \beta \mathbf{N}, \tag{2.4}$$

where β is magnitude of the plastic strain rate. The value of β controls the transition from purely elastic to fully plastic flow.

To see this, we take the time derivative of (2.2) and obtain

$$\dot{\mathbf{Q}} = \mathbb{C}(\dot{\mathbf{q}} - \dot{\mathbf{q}}_p) = \mathbb{C}(\dot{\mathbf{q}} - \beta \mathbf{N}). \tag{2.5}$$

In particular, $\beta = 0$ implies no plastic flow and so (2.5) reduces to a purely elastic response.

On the other hand, if we set

$$\beta = \frac{\dot{\mathbf{q}} \cdot \mathbb{C}\mathbf{N}}{\mathbf{N} \cdot \mathbb{C}\mathbf{N}}, \tag{2.6}$$

then (2.5) implies that $\dot{\mathbf{Q}} \bullet \mathbf{N} = 0$ and fully plastic flow ensues in the direction $\mathbf{N}$. In other words, we obtain a plastic hinge.

The central idea here is to choose a constitutive relation for β that (a) satisfies (2.3), (b) ensures rate-independent behavior, and (c) ensures that the response smoothly transitions from elastic to fully plastic flow when the fully plastic yielding of the cross-section is reached.

In order to achieve this, we define a "gross" or "fully plastic" flow function $F(\mathbf{Q}) \leq 0$ such that $F(\mathbf{Q}) = 0$ represents section yielding. We then specify the flow direction $\mathbf{N}$ to be

$$\mathbf{N} = \frac{\frac{\partial F}{\partial \mathbf{Q}}}{\left\| \frac{\partial F}{\partial \mathbf{Q}} \right\|}. \tag{2.7}$$

Finally, we specify the constitutive relation for the plastic strain rate magnitude β to be

$$\dot{\mathbf{q}}_p = [1 + \tanh\left(\alpha(\mathbf{N})F(\mathbf{Q})\right)] \left\langle \frac{\mathbb{C}\mathbf{N} \bullet \dot{\mathbf{q}}}{\mathbf{N} \bullet \mathbb{C}\mathbf{N}} \right\rangle \mathbf{N}, \tag{2.8}$$

where, as previously, $\langle \bullet \rangle$ denotes Macaulay brackets. Note that the tanh function ensures a smooth transition from 0 to 1. The transition function $\alpha(\mathbf{N})$ controls the transition from microplastic to bulk flow in the frame. Its dependence on the flow direction $\mathbf{N}$ is vital to account for the ability to model the response due to combined loading as we shall see later. Small values of α represent a very gradual transition from microplastic to section yield such as that encountered in bending or torsion while large values of α represent a more abrupt transition.

This completes the description of the smooth-yielding model developed for beams; the only remaining challenge is to determine a suitable means of determining how applied moments and forces push the model towards elastic-plastic behavior. The form of the yield function (or flow potential, in the context of this work) in terms of stress resultants is a nontrivial matter, examined at some length in the literature (see, in particular, Secs. 1.4 and 1.5 from Ref. 38; for a more recent examination, see Ref. 18). The specific form employed in this paper is discussed in the following Sec. 2.2.

2.2. *Relating the flow potential to the section yielding conditions*

To exploit the material constants used with the 3D stress–strain formulation of the model, we need a means of relating the generalized forces $\mathbf{Q}$ to the yield stress of the material and the cross-section of the beam. For our purposes, the plastic limit moduli, well-reported in the structural literature (see, for instance, Ref. 39 for torsion and Ref. 40 for bending), are suitable.

The flow function is given by a von Mises-like yield condition

$$F(\mathbf{Q}) = f(\mathbf{Q}) - \sigma_y = \left[(\mu_n n)^2 + (\mu_x m_x)^2 + (\mu_y m_y)^2 + (\mu_z m_z)^2\right]^{1/2} - \sigma_y. \quad (2.9)$$

In Eq. (2.9), σ_y denotes the yield stress and the μ parameters provide a bridge between stress and the generalized loading.

As mentioned previously, significant effort has been expended to determine a form for the yield function/flow potential that is suitable for use with stress resultants. Causing particular issue is the fact that the generalized-load yield surface — in addition to being geometry-dependent — is likely not a hypersphere, or even a hyper-ellipse.

Herrnböck*et al*[18] for instance, suggest fits to the following form for a flow potential in terms of stress resultants:

$$\left|\frac{n_1}{n_{1y}}\right|^{\alpha} + \left|\frac{n_2}{n_{2y}}\right|^{\alpha} + \left|\frac{n_3}{n_{3y}}\right|^{\gamma} + \left|\frac{m_1}{m_{1y}}\right|^{\delta} + \left|\frac{m_2}{m_{2y}}\right|^{\delta} + \left|\frac{m_3}{m_{3y}}\right|^{\xi} - 1 = 0, \quad (2.10)$$

where the n_i are forces, m_i are moments, and α, γ, δ, ξ are potentially distinct exponents. Mollica and Srinivasa[41] have shown a systematic procedure for obtaining flow potentials of any degree of complexity. Such an approach can be tailored to a wide range of response possibilities but at the cost of increasing the amount of curve fitting. Though use of specific carefully-tailored flow potentials would undoubtedly improve accuracy. However, in order to illustrate the efficacy of our approach, we demonstrate in this paper that the EPCF model can determine suitable fatigue predictions, even when only using the simplest models with very minimal curve fitting using a hyper-ellipsoid flow potential.

With a specific form chosen posited for the flow potential, we can now determine the surface normal (Eq. (2.7)) as

$$\mathbf{N} = \frac{1}{\sqrt{(\mu_n n)^2 + (\mu_x m_x)^2 + (\mu_y m_y)^2 + (\mu_z m_z)^2}} \begin{Bmatrix} \mu_n^2 n \\ \mu_x^2 m_x \\ \mu_y^2 m_y \\ \mu_z^2 m_z \end{Bmatrix}. \tag{2.11}$$

We report the forms of the limit moduli for two cross-sections, though of course, numerous others are available in the literature, or can be derived using techniques of the theory of elasticity. (Note that for all cases, we take $\mu_n = 1/A$.) For a beam of rectangular cross-section oriented along the x-axis, defined by height h oriented along the y-axis and width w oriented along the z-axis, we have

$$\mu_x = \frac{2\sqrt{3}}{a^3\left(\frac{b}{a} - \frac{1}{3}\right)}, \quad \mu_y = \frac{4}{hw^2}, \quad \mu_z = \frac{4}{h^2 w}, \tag{2.12}$$

where μ_x corresponds to torsion and μ_y, μ_z to bending, and $a = \max(h, w)$ and $b = \min(h, w)$. For the particular case of a square cross-section, the simpler expressions $\mu_x = 3\sqrt{3}/h^3, \mu_y = \mu_z = 4/h^3$ are obtained.

For a beam oriented along the x-axis possessing an elliptical cross-section, described by major and minor axis $(a + b)$ and $(a - b)$, respectively, we have the following:

$$\mu_x = \frac{3\sqrt{3}}{2\pi a^3\left(1 - 4.5\left(\frac{b}{a}\right)^2 + 4\left(\frac{b}{a}\right)^3\right)}, \quad \mu_y = \frac{3}{4(a+b)(a-b)^2},$$
$$\mu_z = \frac{3}{4(a+b)^2(a-b)}. \tag{2.13}$$

Assuming that the major axis is oriented along the y-axis and the minor axis along the z-axis. Obviously, the bending parameters are flipped if the opposite arrangement is true. For the particular case of a circular cross-section (radius R) the simpler expressions $\mu_x = 3\sqrt{3}/(2\pi R^3), \mu_y = \mu_z = 3/(4R^3)$ are obtained.

Note that for non-circular cross-sections, if significant torsional deformation is anticipated, then the generalized modulus $\mathbb{C}$ should be modified to include warping; these constants alone will not compensate for this effect with an unmodified modulus.

2.3. *Damage accumulation based on accumulated microplastic work for fatigue failure criterion in smooth-yielding model*

Mozafari*et al*[36] determine progress to fatigue in terms of AMW, using the following expression:

$$\eta_{\text{stress}}(t) = \int_0^t \boldsymbol{\sigma} : \dot{\boldsymbol{\varepsilon}}^P dt, \tag{2.14}$$

where : denotes the tensor contraction. When AMW (in units of energy per volume) reaches critical value η_f, fatigue failure is predicted to occur. We suggest an analogous expression in terms of force resultants

$$\eta_{\text{FR}}(t) = \int_0^t \mathbf{Q} \boldsymbol{\cdot} \dot{\mathbf{q}}^P dt. \tag{2.15}$$

Here we have AMW in terms of individual beam elements, in units of energy per *length.*

Below, we compare the AMW calculations for simple loads (axial, bending, and torsion) of a cantilever beam of length L oriented with its neutral axis along the x-axis and constant, symmetric cross-section of area A, which should allow us to accurately approximation η_{stress} from η_{FR}.

Axial Deformation. For the small strain approximation ($\varepsilon_{xx} \approx du/dx$), both AMW parameters are related by cross-sectional area (as should be expected given the units)

$$\eta_{\text{FR}}(t) = \int \text{EA}\left(\frac{du}{dx} - \varepsilon^P\right) \dot{\varepsilon}^P dt = A\left(\eta_{\text{stress}}(t)\right).$$

Bending Deformation. We limit ourselves to a beam with a square (height, h) or circular (radius, r) cross-section; For the loading described, the maximum stress will occur at $z = \pm h/2$ and $z = \pm r$ for the square and circular cross-sections, respectively.

For the case in which the cantilever beam so described is loaded with a transverse load on the neutral axis, the maximum uniaxial stress is given by

$$\sigma_{xx} = \frac{Mz}{I} = -zE\left(\frac{d^2w}{dx^2} - \kappa_z^p\right).$$

Recognizing that for the beam loaded in this way subject to the small strain approximation, we have $\varepsilon_{xx}^P \approx z\kappa_z^P$, we can relate the AMW in terms of stress with that determined using the generalized force and curvature relationships upon substitution for the particular form of the moment of inertia

$$\eta_{\mathrm{FR}}(t) = \int \mathrm{EI}\left(\frac{d^2w}{dx^2} - \kappa_z^p\right)\dot{\kappa}_z^P dt. \tag{2.16}$$

In the case of a square cross-section (where $I = (h^4/12) = (A/3)(z^2)$)

$$\eta_{\mathrm{FR}}(t) = \frac{A}{3}\left(\int zE\left(\kappa_z - \kappa_z^P\right)(z\dot{\kappa}_z^P)dt\right) = \left(\frac{A}{3}\right)\eta_{\mathrm{stress}}(t).$$

In the case of a circular cross-section (for which $I = (\pi r^4/4) = (A/4)(z^2)$)

$$\eta_{\mathrm{FR}}(t) = \frac{A}{4}\left(\int zE\left(\kappa_z - \kappa_z^P\right)(z\dot{\kappa}_z^P)dt\right) = \left(\frac{A}{4}\right)\eta_{\mathrm{stress}}(t).$$

Torsional Deformation. We follow the same process for a cantilever loaded by a twisting moment, T around the x-axis which produces angular rotation ϕ. The maximum uniaxial stress in this same cantilever is given by

$$\sigma_{xy} = \frac{Mr}{J} = rG\left(\frac{d^2\phi}{dx^2} - \kappa_x^p\right).$$

Recognizing that for the beam loaded in this way subject to the small strain approximation, we have $\varepsilon_{xy}^P \approx r\kappa_x^P$, we can relate the AMW in terms of stress with that determined using the generalized force and curvature relationships upon substitution for the particular form of the polar moment of inertia

$$\eta_{\mathrm{FR}}(t) = J\int G\left(\kappa_x - \kappa_x^p\right)\dot{\kappa}_x^P dt.$$

In the case of a square cross-section (where $I = (h^4/6) = (2A/3)(z^2)$)

$$\eta_{\mathrm{FR}}(t) = \frac{2A}{3}\left(\int zG\left(\kappa_x - \kappa_x^p\right)(z\dot{\kappa}_x^P)dt\right) = \left(\frac{2A}{3}\right)\eta_{\mathrm{stress}}(t).$$

In the case of a circular cross-section (for which $J = (\pi r^4/2) = (A/2)(z^2)$)

$$\eta_{\mathrm{FR}}(t) = \frac{A}{2}\left(\int zG\left(\kappa_x - \kappa_x^p\right)(z\dot{\kappa}_x^P)dt\right) = \left(\frac{A}{2}\right)\eta_{\mathrm{stress}}(t).$$

Note that we must multiply these proportionality factors by (1/2) in this case, however, due to the fact that shear multiplication terms $\sigma_{ij}\varepsilon_{ij}^p, i \neq j$ appear twice in a tensor contraction.

The foregoing discussion suggests the following form for η, used throughout this work:

$$\eta(t) = \frac{1}{A}\int \left(n\dot{\varepsilon}^P + \phi_{\mathrm{bend}}(m_y\dot{\kappa}_y^P + m_z\dot{\kappa}_z^P) + \phi_{\mathrm{tors}}(T\dot{\kappa}_x^P)\right)dt, \tag{2.17}$$

where for a square cross-section $\phi_{\mathrm{bend}} = \phi_{\mathrm{tors}} = 3$, and for a circular cross-section $\phi_{\mathrm{bend}} = \phi_{\mathrm{tors}} = 4$. This form is suitable for simple loadings, but undoubtedly becomes inaccurate for more complex ones; nonetheless, as demonstrated in the next section, we can still use this to leverage the model from Mozafari*et al*[34–36] for excellent fatigue life predictions, but at significantly lower computational cost.

3. Numerical Approximation

3.1. *Elastic-plastic boundary value problem*

Assuming no body force, the generalized force balance equations for an Euler–Bernoulli beam oriented along the x-axis (or rotated to lie along this direction) can be written as

$$\int_{x_a}^{x_b} \left[n\frac{du}{dx} + m_x\frac{d\phi}{dx} + m_y\frac{d^2w}{dx^2} + m_z\frac{d^2v}{dx^2} \right] dx = 0. \tag{3.1}$$

Selecting Lagrange test function ψ for axial displacement and small rotation due to torsion, and Hermite cubic φ for the transverse displacements and their first derivative, we substitute the individual equations contained in (2.2) into Eq. (3.1); partitioning (3.1) with respect to distinct virtual displacements (denoted below with a δ before the appropriate test function), we have

$$\int_{x_a}^{x_b} \left[\mathrm{EA}\left(\frac{d\psi}{dx} - \epsilon_p\right)\right] \frac{d\delta\psi}{dx} dx = 0, \quad \int_{x_a}^{x_b} \left[\mathrm{GJ}\left(\frac{d\psi}{dx} - (\kappa_x)_p\right)\right] \frac{d\delta\psi}{dx} dx = 0, \tag{3.2}$$

for axial and torsional deformation on a given element from x_a to x_b and

$$\int_{x_a}^{x_b} \left[\mathrm{EI}_z\left(\frac{d^2\varphi}{dx^2} - (\kappa_z)_p\right)\right] \frac{d^2\delta\varphi}{dx^2} dx = 0,$$

$$\int_{x_a}^{x_b} \left[\mathrm{EI}_y\left(\frac{d^2\varphi}{dx^2} - (\kappa_y)_p\right)\right] \frac{d^2\delta\varphi}{dx^2} dx = 0, \tag{3.3}$$

for bending about both transverse axes.

Numerically, we can accurately and expediently solve the flow rule and transverse force balance over the finite element (FE) mesh using operator split: that is, any given load increment is split into two steps, (i) an elastic step — freezing the plastic curvature — and (ii) a plastic step — freezing the plastic rate.

To describe the formulation over a single reference element oriented along the x-axis from x_a to x_b, we introduce discrete nodal variables u_j, ϕ_n, v_J, and w_N representing the axial displacement, twist, and the transverse displacements and bending rotations along the y- and z-directions, respectively. The axial displacement u_j and twist ϕ_n are interpolated using Lagrange interpolation functions, denoted below by ψ. To maintain continuity for both the transverse displacement and the bending rotation (related to the transverse displacement by the first derivative for an Euler–Bernoulli beam), we employ Hermite cubic interpolation functions nodal variables v_J and w_N. The form for Lagrange and Hermite interpolation functions as well as the particulars of Gauss-point numerical integration are well-known see e.g. Ref. 42.

For the deformations described by this model, we have

$$\begin{aligned}
\left(\int_{x_a}^{x_b}\left[\mathrm{EA}\frac{d\psi_i}{dx}\right]\frac{d\psi_j}{dx}dx\right)u_j &= \int_{x_a}^{x_b}[\mathrm{EA}\epsilon_p]\frac{d\psi_i}{dx}dx,\\
\left(\int_{x_a}^{x_b}\left[\mathrm{GJ}\frac{d\psi_m}{dx}\right]\frac{d\psi_n}{dx}dx\right)\phi_n &= \int_{x_a}^{x_b}[\mathrm{GJ}(\kappa_x)_p]\frac{d\psi_m}{dx}dx,\\
\left(\int_{x_a}^{x_b}\left[\mathrm{EI}_z\frac{d^2\varphi_I}{dx^2}\right]\frac{d^2\varphi_J}{dx^2}dx\right)v_J &= \int_{x_a}^{x_b}[\mathrm{EI}_z(\kappa_z)_p]\frac{d^2\varphi_I}{dx^2}dx,\\
\left(\int_{x_a}^{x_b}\left[\mathrm{EI}_y\frac{d^2\varphi_M}{dx^2}\right]\frac{d^2\varphi_N}{dx^2}dx\right)w_N &= \int_{x_a}^{x_b}[\mathrm{EI}_y(\kappa_y)_p]\frac{d^2\varphi_M}{dx^2}dx,
\end{aligned} \tag{3.4}$$

where the lower-case indices shown above take the values of 1 and 2 indices per element, and the upper-case indices (associated with the Hermite cubic test functions) take the values 1 through 4 indices per element. We can state this concisely (treating the right-hand side as a plastic forcing contribution) as $[K_{ij}](\Delta_j) = (F_\mathrm{p})_i$, where, now in this case, j varies through the 12 DOFs associated with a single FE.

In Eq. (3.4), we take plastic curvatures as values from the previous load increment, $\mathbf{q}_p^n$; this constitutes the first update step of the algorithm, in which the current values of the curvature, that is $\mathbf{q}^{n+1}$, are obtained. The algorithm selected below is justified noting that for almost all inelastic processes, the loading vector and flow direction do not change much within a load increment.[43]

3.2. *Update of plastic variables*

We state an algorithm described in Ref. 37 in a form suitable for use with the moment–curvature model here as Algorithm 1. The plastic update step takes the configurational variables at the $(n+1)$th step and the plastic variables at the nth step to produce the increment in plastic variables $\Delta\mathbf{q}_p$. The change in the plastic variables within each load step needs to be updated consistent with the yield condition and flow rule. We denote the nth load step flow direction by $\mathbf{N}^n$. Considering Eq. (2.4) we arrive at the following condition on plastic strain rate β:

$$\beta = \left[1 + \tanh\left(\alpha(\mathbf{N}^n)F(\mathbf{q}^{n+1}, \mathbf{q}_p^{n+1})\right)\right]\left\langle\frac{\mathbb{C}\mathbf{N}^n\cdot(\mathbf{q}^{n+1} - \mathbf{q}^n)}{\mathbf{N}^n\bullet\mathbb{C}\mathbf{N}^n}\right\rangle. \tag{3.5}$$

3.3. *Determination of model structural parameters*

Since this is a frame model, the response is governed not only by the type of loading, but also by the cross-sectional geometry. For a given geometry, the following features

Result: Computes $\mathbf{q}^{n+1}$, Given $\mathbf{q}^n$

START: $\Delta\mathbf{q}_p = 0$, ERROR $= 1$, TOL $= 10^{-6}$, $i = 0$, imax;
$\mathbf{q} \longleftarrow \mathbf{q}^{n+1}$;

while ERROR $>$ TOL and $i <$ imax **do**
  $\mathbf{q}_p \longleftarrow \mathbf{q}_p^n + \Delta\mathbf{q}_p$;
  $\Delta\mathbf{q} \longleftarrow \mathbf{q} - \mathbf{q}^n$;
  $\phi = ||\Delta\mathbf{q}_p||$;

  Compute the loading direction $\mathbf{N} = \mathbf{N}(\mathbf{q}, \mathbf{q}_p)$;
  if $L := \mathbb{C}\mathbf{N} \bullet \Delta\mathbf{q} / \mathbb{C}\mathbf{N} \bullet \mathbf{N} > 0$ **then**
    $\beta = \beta(\mathbf{q}, \mathbf{q}_p)$;
    $\theta = -\frac{\partial \beta}{\partial \mathbf{q}_p} \bullet \mathbf{N}$;
    $d\mathbf{q}_p = \frac{(\beta + \theta\phi)L}{1 + \theta L}\mathbf{N}$;
  else
    $d\mathbf{q}_p = 0$;
  end
  ERROR $\longleftarrow ||d\mathbf{q}_p - \Delta\mathbf{q}_p||$;
  $\Delta\mathbf{q}_p \longleftarrow d\mathbf{q}_p$;

  $i \longleftarrow i + 1$
end

$(\mathbf{q}^{n+1}, \mathbf{q}_p^{n+1}) \longleftarrow (\mathbf{q}, \mathbf{q}_p)$;
$n \longleftarrow n + 1$

Algorithm 1: Picard iteration performed at each Gauss point of EP BVP; $\mathbf{Q}_i^n$ denotes the temporary generalized forcing values available after iteration i.

of the elastic-plastic deformation were considered essential to the proper modeling of the fracture behavior.

(1) Initial slope of force–displacement curves (Section moduli, EA, EI, GJ);
(2) Perfect plasticity limit forces/moments for single-load cases (Yield stress, σ_y);
(3) Elastic-plastic behavior in transition region (Transition parameter, α).

The determination of the first two categories is straightforward, following directly from the geometry and standard experimental procedures; alpha can be modified for each single-loading case separately until a suitable match is obtained. Figure 1 shows this process for the case of pure axial loading.

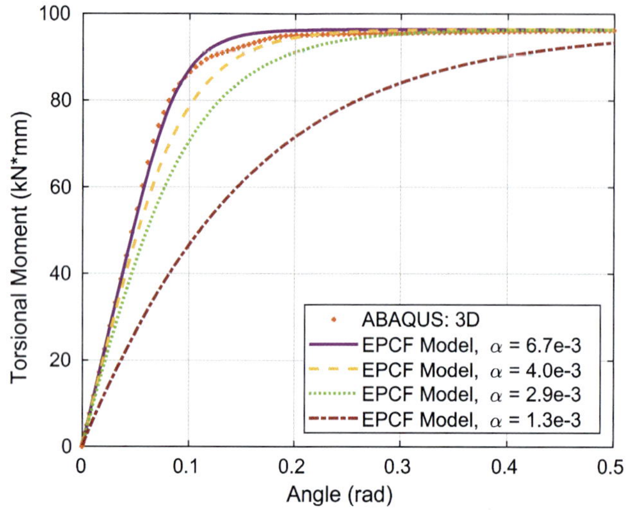

Fig. 1. Even if the underlying material has a sharp transition from elastic to plastic, the structural response under bending and torsion will have a smooth transition to plastic response. The EPCF model can be tuned to better approximate the EP transition region via judicious selection of α. This is crucial for fatigue calculations based on damage accumulation.

Given force–displacement and moment–curvature experimental data (or data from a high-fidelity simulation), we suggest the following methods to assess the accuracy of the EP beam model and to obtain replicable fits to the given material parameters.

Method 1: Residuals. The two residuals below can be employed to assess accuracy of the continuous-yielding EP beam model

$$\begin{aligned} R_1 &= \sum_{i=1}^{4} ||\max(Q_i) - \max(Q_i^{\exp})||, \\ R_2 &= \sum_{i=1}^{4} \sum_{q_i(t_j)=q_i(t_0)}^{q_i(t_f)} \Big|\Big|\kappa_{qQ}(t_j, q_i, Q_i) - \kappa_{qQ}^{\exp}(t_j, q_i, Q_i)\Big|\Big|, \end{aligned} \tag{3.6}$$

where values with the superscript "exp" denote quantities obtained experimentally or numerically from a (presumed) higher fidelity model (such as a full 3D continuum-based simulation). For the first residual perfect plasticity is assumed (so that a plateau value is actually attained); the greater the value of R_1, the greater the difference between the numerically predicted plateau value for a given element of the generalized load vector Q_i and the experimental value $Q_i^{\exp}$. For the second residual, a forward difference approximation of the numerical (EPCF) curvature, $\kappa_{qQ}(t_j, q_i, Q_i)$ is compared with the experimentally determined values, $\kappa_{qQ}^{\exp}(t_j, q_i, Q_i)$, in the EP transition region, indexed along generalized displacement values from (user-selected) t_i to t_f; the curvature measure is given as

$$\kappa_{qQ}(t_j, q_i, Q_i) = \frac{(q_i(t_{j+1}) - q_i(t_j))\,|Q_i(t_{j+1}) - Q_i(t_j)|}{(Q_i(t_{j+2}) - 2Q_i(t_{j+1}) - Q_i(t_{j+1}))^{3/2}}. \tag{3.7}$$

No residual is included for the initial (elastic) slope — it is determined from standard experiments, and no modification is required for use here — or for the yield stress — once a plateau value is identified for one of the force resultant of interest, the yield stress follows from the plastic limit moduli.

Method 2: Slope Intersection. For the case of the elasto-plastic beam and frame response, we identify four characteristic features of the response (see Fig. 2 as well as the discussion in Refs. 44 and 45, in which this approach is applied to nonlinear models of biological fibers). (1) The initial slope which represents the elastic stiffness D_0, (2) the final slope D_∞ which represents the flow behavior after full section yielding and is related to the flow function g hardening characteristics, (3) the intersection point (Q_0, q_0) between the initial and final slope lines (related to the form of the flow function) and (4) the point (Q_1, q_0) corresponding to the drop in the stress due to continuous flow (related to the transition parameter α). We note that this method reduces to standard definitions of yield if the structure has a sharp yield point; it also separates the "transient region" from the full section yield thus obviating the need for complex hardening rules to account for this phenomenon. This approach provides an easily-replicable method for fitting data in an EP simulation.

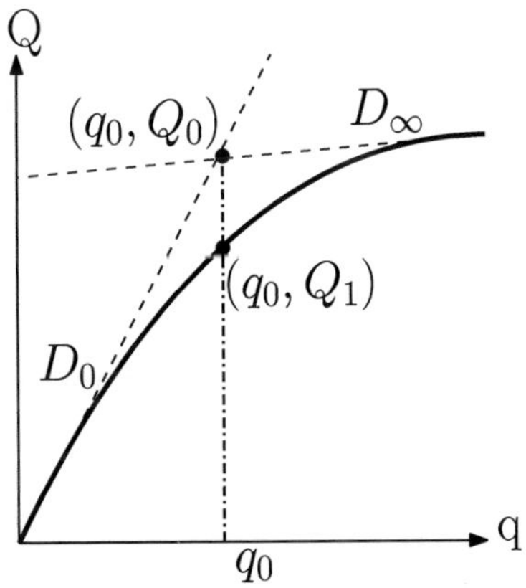

Fig. 2. Slope intersection method for fitting our elastic-plastic deformation model to data. The key measurements that will be used for the parameter fitting are (1) the initial slope D_0, (2) the final slope D_∞, (3) the point of intersections (Q_0, q_0) and the point (Q_1, q_0) that represents deviation from the sharp yield.

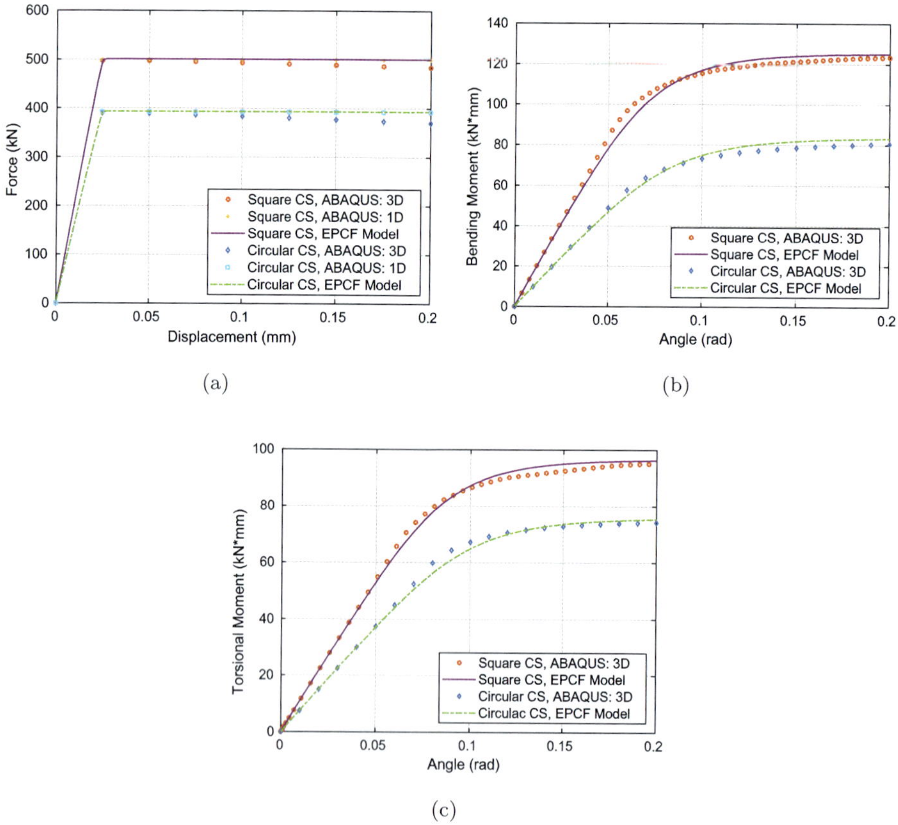

Fig. 3. Single load (a) axial, (b) bending and (c) torsion simulation used to fit α parameter in EPCF model for rectangular and circular cross-sections.

In the case of the EPCF model, the smoothness parameter α selected should be the one that minimizes error at the point on the stress–strain (or generalized force–displacement) curve directly beneath the intersection point of the bilinear representation generated from the two slopes. Figure 2 shows what this fitting method looks like graphically.

Figure 3 below shows single load simulations, representative of the type of data that can be used to fit the material parameters needed for the EPCF beam elements. For each of the different load types, different α values are determined; the values obtained for the square and circular cross-sections under the loading conditions considered above are listed in Table 1.

We will set the transition function $\alpha(\mathbf{N})$ to be simply linear in the direction of loading, i.e.

$$\alpha(\mathbf{N}) = \sum_{I=1}^{3} \alpha_i N_i. \tag{3.8}$$

Table 1. List of the transition function α values determined for pure loading cases; all values are shown in units of 1/MPa. These are used in Eq. (3.8).

Cross-section	α_{ax}	α_{bend}	α_{tors}
Square	0.1	0.00571	0.00667
Circle	0.1	0.00571	0.00667

With the α_i being obtained from curve fitting the "pure" loading cases: α_1 being fit to axial loading (see Fig. 3(a)) α_2 being fit to pure bending (see Fig. 3(b)) and α_3 being fit to pure torsion (see Fig. 3(c)). The values of α_i for square and circular cross-sections can be seen in Table 1.

4. Results

4.1. *Monotonic combined loading studies*

The predictions of the EPCF model was compared to FE simulations performed in ABAQUS. The design of the structures used is shown in Fig. 4.

One end of the beam was fully constrained, and loads were applied on the other end of the beam. Every node/element at the end surface was coupled to a single reference point and individual and combined loads such as vertical displacement, and rotations were imposed on that point.

Figure 5(a) shows the results obtained for loading via an off-axis displacement (i.e. flexure, not pure bending); the EPCF model captures the behavior predicted in ABAQUS fairly well here.

Figures 5(b)–5(e) shows the comparison between the smooth-yielding model for the tension-torsion combined loading and bending-torsion case, respectively, for both square and circular cross-sections. Notice the prominent mismatch in generalized load plateau height for the case of torsion in the square cross-section; this is

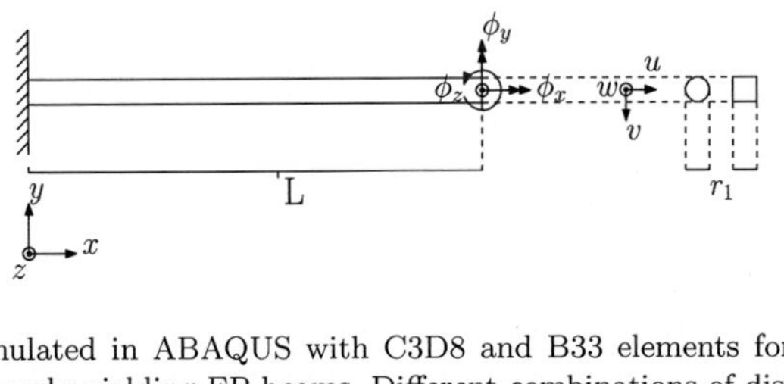

Fig. 4. Cantilever simulated in ABAQUS with C3D8 and B33 elements for comparison against predictions of continuously-yielding EP beams. Different combinations of displacements and rotations were applied to the beam and the tip deflection versus the applied load was compared (see also Fig. 5).

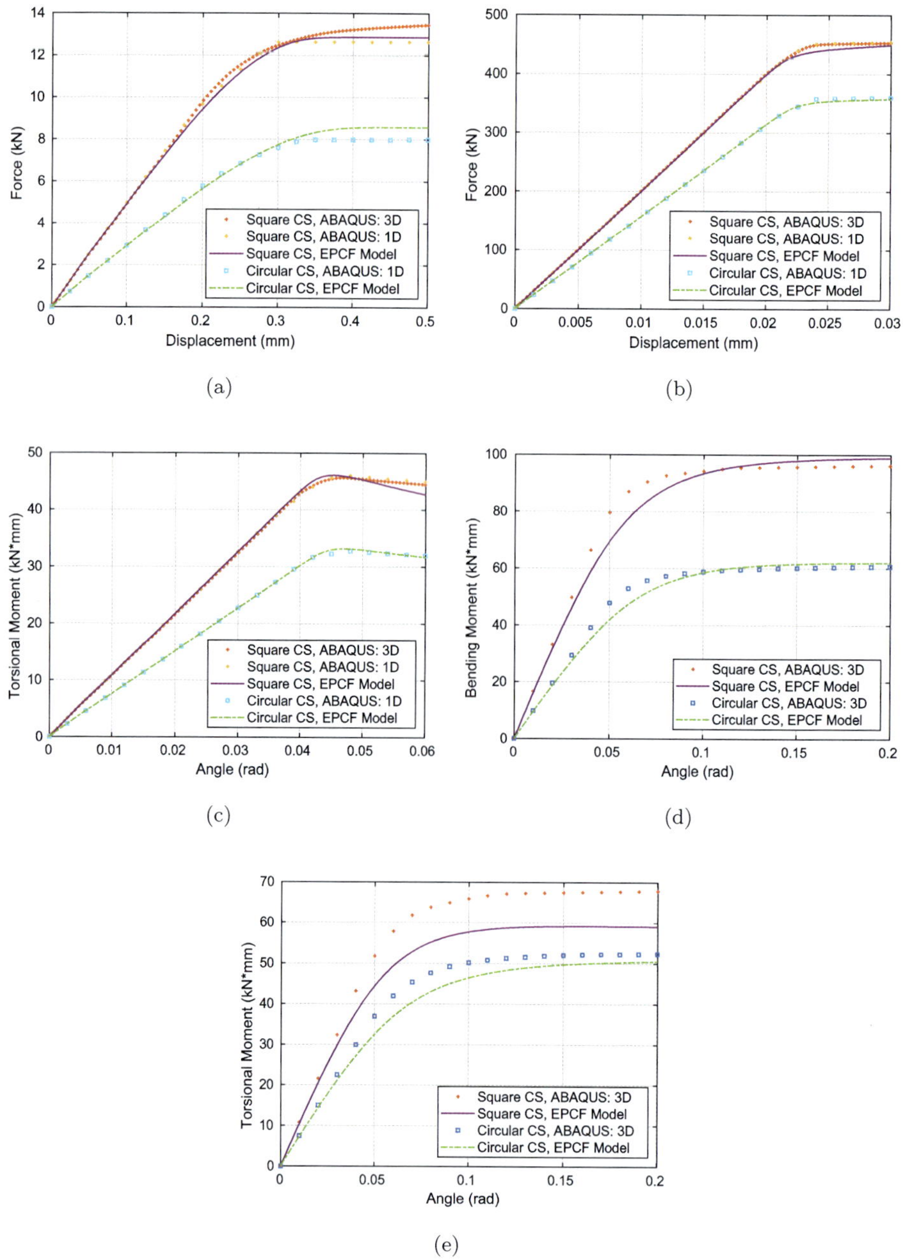

Fig. 5. Results of simultaneous monotonous end loading of a cantilever beam: comparison between ABAQUS 3D and the EPCF model. (a) Force–displacement response due to transverse loading load (b) Force–displacement response due to simultaneous axial load and twist (c) Torque versus twist response due to simultaneous axial load and twist (d) Moment versus transverse rotation response due to combined bending and torsion (e) Torque versus twist angle response due to combined bending and torsion

undoubtedly due to the omission of warping in the EPCF formulation. Future modifications to include warping will likely capture the final generalized load plateau for non-circular cross-sections under torsion much more accurately.

4.2. *Monotonic and cyclic loading of beams under various end loadings*

Once the smoothness parameter has been appropriately selected using single-load input as described previously, the calibrated results extend well to handle cyclic deformations. A comparison between cyclic deformations predicted by ABAQUS against those from the EPCF model is shown below in Fig. 6.

As noted before, the sharp transition of J2 plasticity (and the piecewise jump in behavior in the hardening model used in ABAQUS) will not capture the continuous

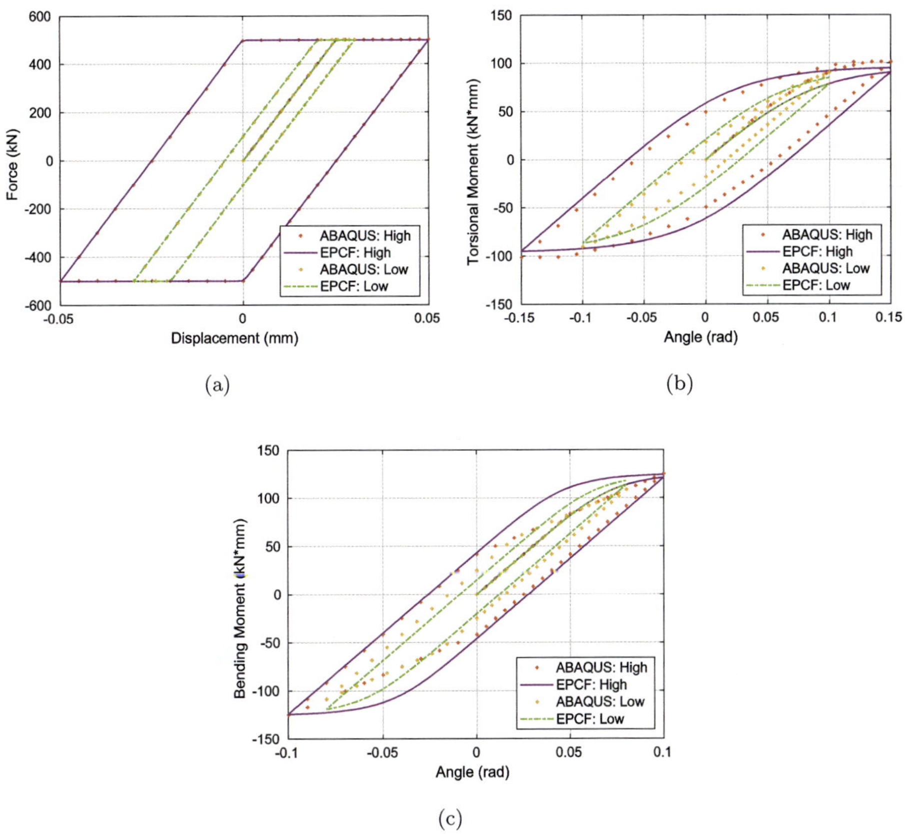

Fig. 6. Cyclic deformation comparison between ABAQUS and the EPCF model for single load (a) Tension, (b) Torsion and (c) Bending. The continuous-yield behavior of the EPCF model allows it to compute fatigue damage from the AMW, which would not be possible with current commercial code.

accrual of plasticity (and associated damage from microplastic work), so — though simulations using models in this class are useful as a comparison — they would not be suitable for the AMW fatigue calculation described here.

4.3. *Deformation of frames*

In this section, we show the efficacy of the new beam model in the simulation of the response of more complex structures. We begin with the consideration of two-dimensional (2D) and 3D portal frames. We compare the model response with that of an ABAQUS 3D model with a square-shaped cross-section for every vertical support (centerline length $L_2 = 11.5$ mm) and horizontal beam (centerline length $L_1 = 9$ mm). The width w and height h of the cross sections were 1 mm. The end surfaces (3D ABAQUS) and end nodes (EPCF) of the vertical support in the frame were constrained. The nodes/elements at the shared sections between a support and a beam at a top corner of a frame were coupled to a reference node and loads were applied to that assigned node accordingly.

Figure 7(c) shows the region where the displacement was applied to deform the 3D frame. The size of the applied loading region was selected to be big enough to eliminate any erroneous localized deformation while still being small enough to simulate a point-wise loading condition on a frame.

The 3D eight node brick element C3D8 element was used to simulate both beams and frames. Notably, the geometry and applied load in the 2D frame were in 2D but a 3D element was used to simulate it when the square-shaped cross-section of the frame was considered.

Three loading cases were considered; for both the 2D and 3D portal frames, applied displacement of $u_{\text{in}} = 5$ mm was used (Fig. 7), giving one case for the 2D frame, one for the 3D frame in which both applied displacements were in the same direction, and one final case for the 3D frame in which the applied displacements were in the opposite direction (push–pull).

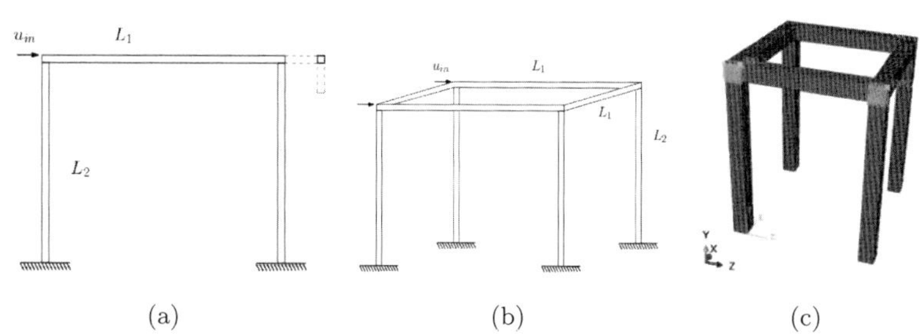

Fig. 7. Structures simulated in ABAQUS with C3D8 elements for comparison against predictions of continuous-flow EP beam elements. (a) 2D portal frame, (b) 3D portal frame, (c) FE mesh used for 3D portal frame in ABAQUS, with regions of load application highlighted.

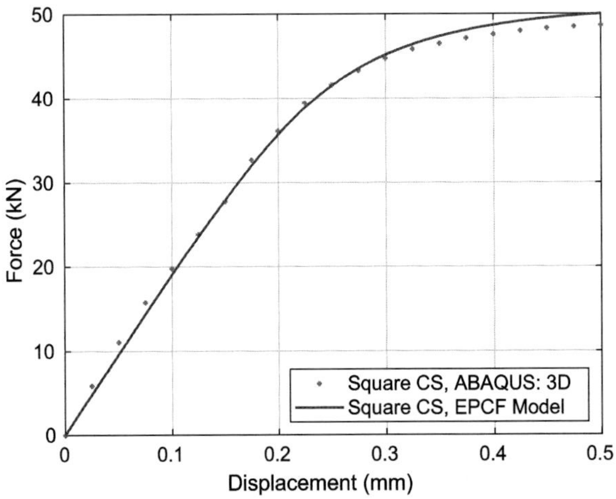

Fig. 8. Load displacement at point of application on 2D portal frame.

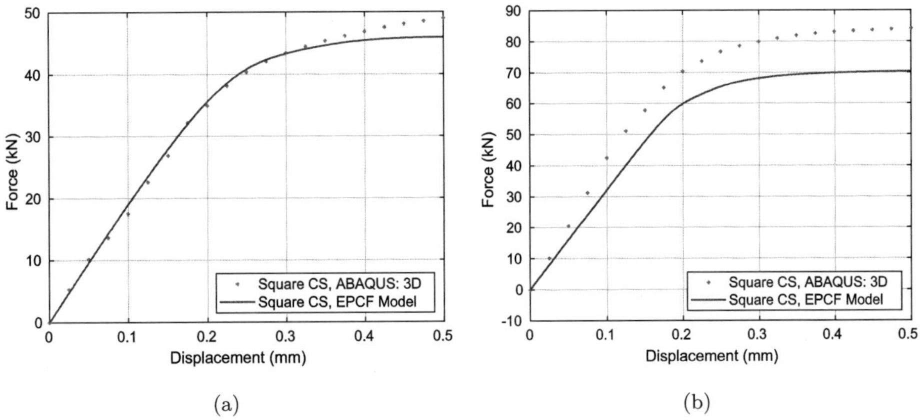

Fig. 9. Load displacement at point of application of 3D portal frame; (a) applied displacements in the same direction, (b) applied displacements in alternate directions. The discrepancy in Fig. 23 is due to warping which is not accounted for in the current formulation.

The force–displacement curves are presented and compared well to the analytical results as shown in Figs. 8 and 9, with only the push–pull case exhibiting discrepancies with the result predicted by ABAQUS.

The push–pull 3D frame plateau loads predicted by the EPCF model and ABAQUS differed somewhat (Fig. 9(b)); it is likely that this is due to torsion. For the push–pull case, the vertical support beams twist around the central axis somewhat, and, because the cross-section employed is non-circular, warping is anticipated. Modifications to the EPCF model to include warping would likely address this. Nonetheless, we are able to effectively produce results with significantly reduced computational effort that reasonably represent the anticipated deformation results of complex loadings and structures, using only data gleaned from material parameters, geometry, and simple fits to pure load cases, which was our intended result.

4.4. *Deformation of micro-architected geometries*

In this section, we demonstrate how the EPCF model (with judicious choice of section moduli parameters) can be used to model the elastic-plastic deformation in complex micro-architected geometries.

We consider the recent work of Kushwaha*et al.*[4] The fractal Menger sponge geometry (created by repeating, progressively smaller, self-similar removals of a cube from an original solid cube) has been evaluated as a potential design for prostheses, for two reasons principally: first, the tailorable architecture allows a better match between the elastic modulus of bone and the prosthesis, and second, the porous structure promotes superior osseointegration as the surrounding bone heals and accepts the prosthesis. As an engineering problem, however, the Menger sponge presents a challenge; clearly predictive analysis must be performed before prostheses with these novel designs can be cleared for use in humans, but meshing itself is a challenge. Prediction of the likely fatigue life is computationally prohibitive.

Kushwaha*et al*[4] performed numerical simulation of various Menger sponge structures using ANSYS mechanical APDL; their mesh convergence analysis revealed that they required about 2 million quadratic tetrahedral elements (SOLID 187). We aim to replicate the qualitative behavior they observed, but with dramatically fewer elements.

We referred to Kushwaha's provided materials, using an elastic modulus of 1.55 GPa and a Poisson ratio of 0.3. Taking the specific yield stress of 40.70 for the solid specimen (denoted L0; see Kushwaha, Table 3) and the reported density (1.25 g cm^{-3}), we have a yield stress of 50.875 kPa.

From Kushwaha's description of the Menger sponge compressive tests, and their reported boundary conditions for the base of the sponge in their ANSYS simulations, we fix displacements and rotations in for all nodes on the base of our selected meshes. Determining the particulars of the compressive loading, however, is not as clear. The experimental loading used a central pillar to press the specimen, partially distributing this load via a top plate (see Fig. 5 in Ref. 4), but the relative size of the pillar is not reported. Assuming that the top plate would not distribute the load uniformly but that greatest contact would be experienced at the center of the specimen (and in the absence of any details on their chosen numerical implementation) we employed a loading profile windowed with the square root of a sine function centered over the largest (centermost) cavity.

The Menger sponge is symmetric; arbitrarily, we orient the bottom part of the sponge in the xy-plane (and the bottom part of the 2D reduction in the x-plane) at $z = 0$. Denoting side length by L (equal to 5 cm, to match the stated geometry in Kushwaha*et al*[4] we have selected the following form for the boundary condition at the top face ($z = L$) of the 2D reduction (Sierpinski carpet):

$$w_{\text{app}}(x) = \sqrt{\sin\left(\frac{\pi x}{L}\right)}, \tag{4.1}$$

where w_{app} denotes the maximum applied displacement boundary condition in the z-direction (gradually approached as a ramp loading from zero to maximum over the top surface). Similarly, for the 3D simulation, we employ the loading boundary condition

$$w_{\text{app}}(x, y) = \sqrt{\sin\left(\frac{\pi x}{L}\right)\sin\left(\frac{\pi y}{L}\right)}. \tag{4.2}$$

For both the 2D and 3D simulations, we reduced the computational load by trying to enforce symmetry down the center-line (2D) or center-planes (3D). For the 2D case, we maintained $u = \theta_y = \theta_z = 0$ at the centerline ($x = L/2$); for the 3D case, we enforced $u = v = \theta_y = \theta_z = 0$ at the center-planes ($x = L/2$ and $y = L/2$).

To attempt replication of the behavior of solid regions of the Menger sponge with 1D beam elements, section moduli corresponding to elements large enough along the cross-section that they would touch surrounding elements are used; the parameters relating the flow potential to the generalized forces were calculated without modification (see Table 2).

The deformations predicted by the EPCF model for symmetric loadings of the $M = 0, 1$ (Fig. 10) and $M = 2, 3$ (Fig. 11) 2D sponges are shown below; visually, they appear similar to the results obtained by Kushwaha*et al*using ANSYS APDL, most notably with the high stress regions occurring in the same locations.

Though the material they tested was non-metallic (PLA), we show in Fig. 12 that the EPCF model produces specific (density-normalized) compressive stress plateaus that seem reasonably consistent with the experimentally-predicted specific compressive stress values at small strains, replicating the behavior they describe

Table 2. List of section moduli and flow potential parameters used for all levels of Sierpinski carpet simulation (consistent discretization allows the same values to be used for each). Note that, in this case, symmetry causes $\text{EI}_x = \text{EI}_y$ and $\mu_y = \mu_z$, so only one value is reported for bending.

EA (kN)	EI (kN/mm)	GJ (kN/mm)	$\mu_n(\text{mm}^{-2})$	$\mu_y(\text{mm}^{-3})$	$\mu_x(\text{mm}^{-3})$
1.33e3	9.49e1	7.27e1	1.17	5.04	6.55

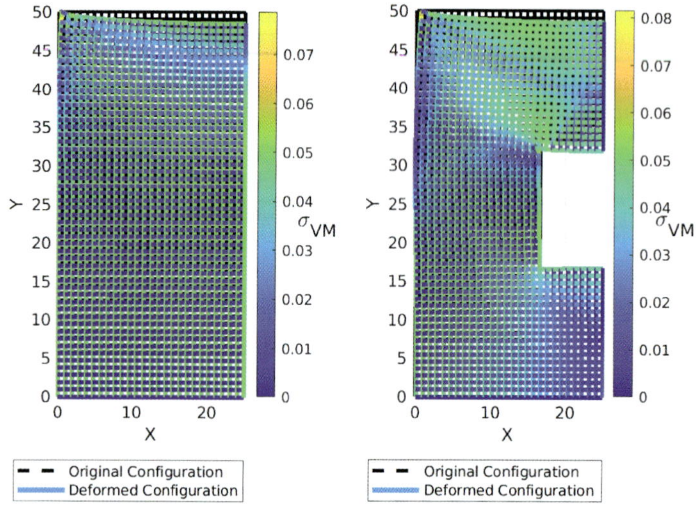

(a) (b)

Fig. 10. Deformation of (a) $M = 0$ and (b) $M = 1$ Sierpinski carpets under centrally-loaded compressive force using EPCF model with 2860 and 2774 frame elements, respectively. Displayed von Mises stresses are in MPa. (Compare deformations with result from ANSYS APDL with 2 million tetrahedral elements in Kushwaha*et al*[4]).

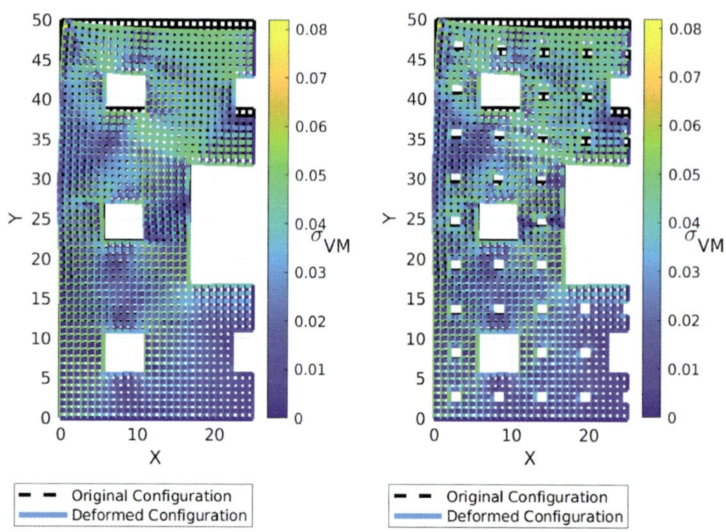

(a) (b)

Fig. 11. Deformation of (a) $M = 2$ and (b) $M = 3$ Sierpinski carpets under centrally-loaded compressive force using EPCF model with 2564 and 2496 frame elements, respectively. Displayed von Mises stresses are in MPa. (Compare deformations with result from ANSYS APDL with 2 million elements in Kushwaha*et al*[4]).

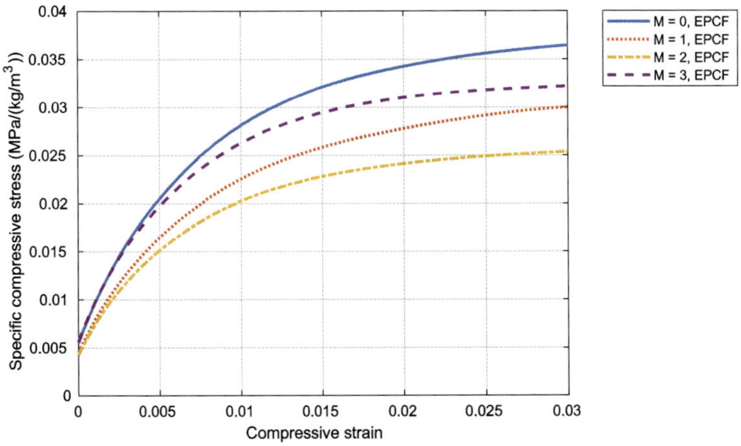

Fig. 12. Specific compressive stress (stress/density) results determined using EPCF model 2D sponge simulations, for 3D printed PLA Menger sponge specimens of increasing Menger sponge level; plateau values for $M = 0$, 1, and 2 compare favorably against Fig. 1(c) in Ref. 4, but more detailed comparison is not possible due to the use of PLA, as no loading–unloading data was provided.

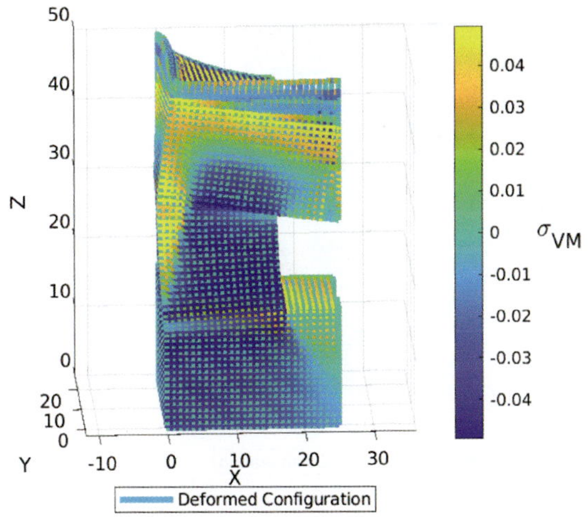

Fig. 13. Deformation of $M = 1$ Menger sponge under centrally-loaded compression using EPCF model with 94734 frame elements. Displayed von Mises stresses are in MPa. Pictured deformations are scaled by a factor of 5 for visibility.

(elastic loading region followed by a plateau in the compressive stress). In particular, Kushwaha*et al*[4] state that "It can be concluded that there exist two distinctive compressive strain regimes for all the levels of the structure: (1) linear elastic region and (2) a plateau region". Figure 11 shows qualitatively similar behavior, in particular showing close matches of the plateau stress with much less computational

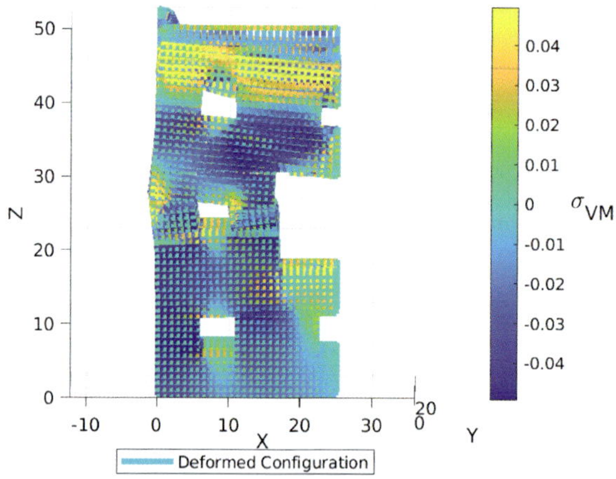

Fig. 14. Deformation of $M = 2$ Menger sponge under centrally-loaded compression using EPCF model with 74976 frame elements. Displayed von Mises stresses are in MPa. Pictured deformations are scaled by a factor of 5 for visibility.

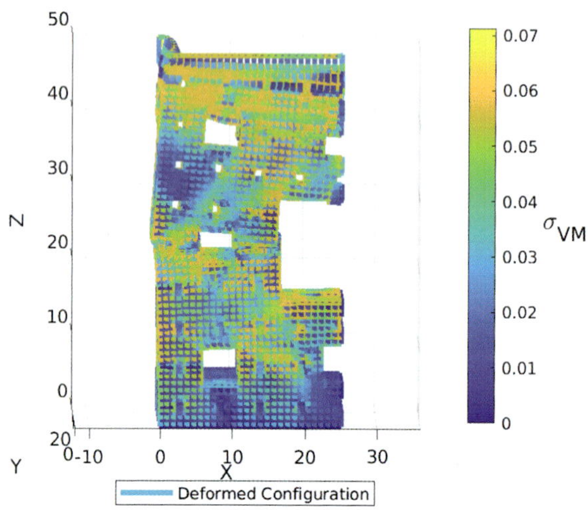

Fig. 15. Deformation of $M = 3$ Menger sponge under centrally-loaded compression using EPCF model with 65572 frame elements. Displayed von Mises stresses are in MPa. Pictured deformations are scaled by a factor of 5 for visibility.

resources and effort. To address the 2D reduction of the 3D model, the effective densities used to determine the specific stress are modified from the 3D values as 1.11, 0.988, and 0.7023 kg/m^3 for the $M = 1$, 2, and 3 carpets, respectively, (determined via subtracting empty portions from the 2D sponge and multiplying the resulting fractional value times the bulk density of 1.25 kg/m^3).

Only the plateau associated with the $M = 3$ sponge disagrees strongly with the experimentally-predicted plateau, with the EPCF predicting a much higher compressive stress than observed; referring to the supplementary videos associated with Kushwaha*et al*[4] we see that, for the $M = 3$ case, the thin geometry associated with the numerous holes in the sponge leads to early buckling, a failure mode that is not accounted for in the small-deformation EPCF model, thus it is to be expected that the EPCF would anticipate higher stress before failure. Additionally, we note from the supplementary video that the initial loading does not appear to deform the structure much, and the initial strains recorded in the plot may be due only to cross-head movement.

While using a 2D analog of the Menger sponge can produce comparative results significantly faster, the EPCF model is, of course, suitable for use with 3D simulations also. Using a tessellation density that is consistent between the $M = 1, 2,$ and 3 sponges (with the same density along the cross-section as that of the 2D Sierpinski carpet meshes), we performed a simulation of a compressive loading of three Menger sponges.

The locations of maximum von Mises stress predicted by the EPCF model compare favorably with those obtained in Kushwaha*et al*[4] as can be seen below in Figs. 16 and 17. We should note that the location of the maximum stresses is probably the extent of the comparison possible in this case, since the result shown in Kushwaha*et al*[4] is for larger deformation than would be accurate to simulate under the assumptions of the EPCF model. (Our maximum compressive strain is

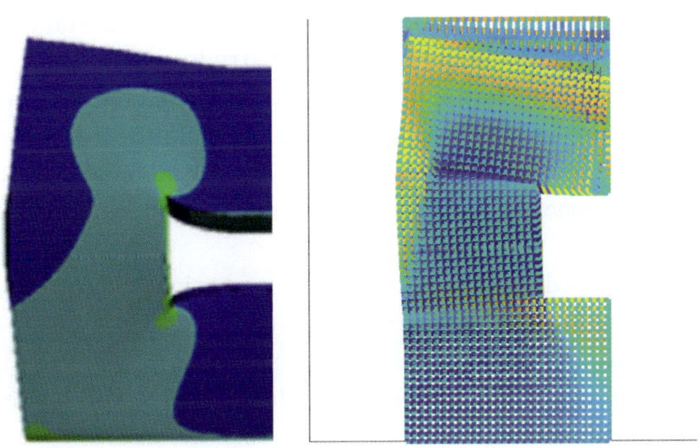

(a) (b)

Fig. 16. Comparison of (b) EPCF deformation (scaled by a factor of 5) predictions with (a) fully-deformed result from ANSYS APDL with 2 million elements in Kushwaha*et al*[4] for $M = 1$ Menger sponge. Note that the ANSYS model has much larger strains at its final state (large deformation model), so the stresses shown should not be identical.

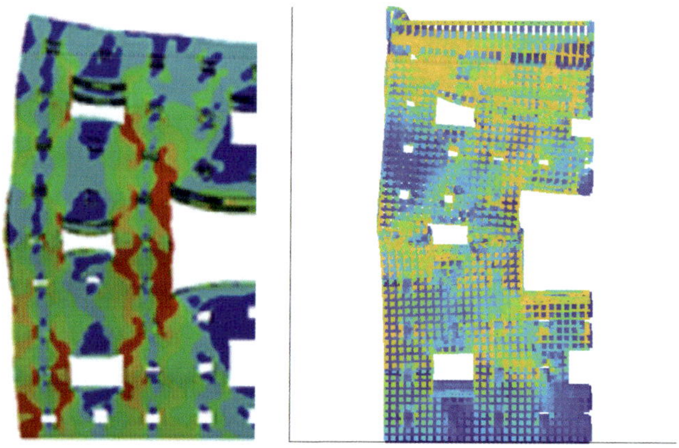

(a) (b)

Fig. 17. Comparison of (b) EPCF deformation (scaled by a factor of 5) predictions with (a) fully-deformed result from ANSYS APDL with 2 million elements in Kushwaha*et al*[4] for $M = 3$ Menger sponge. Despite the different color-mapping scheme selected, it is evident that peak stresses are predicted in the same location for each model.

half that in their simulation, but as can be seen in Fig. 12, the plateau stress is still approached, even at this reduced strain input.) Additionally, uncertainty about the form of the loading boundary conditions used in their ANSYS simulation likely influences any discrepancy between the results.

With the ability to predict elastic-plastic deformations in complex structures but at greatly reduced effort from both the engineer and the computer, the simulation of lengthy cyclic histories for this kind of structure is in reach. Though large deformations are not suitable for use with the EPCF model as described in this work, we are not interested in those for the purposes of fatigue analysis. (With this limitation in mind, care must be taken when using this model to avoid cases in which, even though macroscopic strains are small, locally large strains would be produced.)

The next section demonstrates how the physically based AMW fatigue parameter can be used with this model to predict fatigue without addition of a significant computational penalty (as compared to commonly-used methods such as cycle counts, for which determination of the elastic-plastic history is only the beginning of the fatigue analysis). This can in turn be used to leverage the structural calculation capabilities demonstrated here.

4.5. *Application to physically based fatigue predictions*

Dogbone specimens of various standardized shapes are used to replicate the stress states that appear in objects of engineering importance in the controlled environment of the laboratory. Direct simulation of the nonlinear, elastic-plastic stress–strain state that occurs in the body during deformation is, traditionally, quite computationally expensive, and requires careful meshing.

A recent work by Mozafari*et al*[36] demonstrates the performance of the smooth-yielding model combined with the AMW fatigue parameter by predicting the fatigue life associated with multiple dogbone specimens. As can be seen in Fig. 12 of Ref. 36, defining the geometry accurately requires quite a few elements (612, 1545, and 5604, for the cases shown); we are able to produce comparable fatigue predictions with only 8 elements.

We reference the extensive Al 7075-T651 experimental results provided in Zhao and Jiang's excellent 2008 paper[46]; as described in detail in this work, multiple specimen types were used to perform tension, torsion, and tension-torsion tests. Using the specimen geometry taken together with the material parameters (E = 0.717e5 MPa, $\nu = 0.306$, YS = 501 MPa), we determine the section moduli and the parameters that relate generalized load and yield (i.e. the μ-parameters). As an initial guess for the AMW at failure, we use the value reported by Mozafari*et al*[36] for Al 7075-*T6* ($\eta_f = 2\,850$ MJ/m^3) as the fatigue failure criterion, recognizing that the difference in material processing (and, potentially, surface finish) may require adjustments to the actual eta$_f$ value that would be appropriate for use.

The complete structural parameters used by the EPCF model for this effort are reported in Tables 3 and 4. As noted in previous sections, the smoothness parameter, α, associated with the EPCF beam element under pure torsion, pure bending, and pure axial deformation, is distinct. Using nonlinear FEA simulations with perfect plasticity as a baseline, however, would result in a continuous-flow parameter α parameter that is too large to accurately accrue fatigue damage from AMW. Referencing Mozafari*et al*[36] — in which the ξ value ($\approx 1/\alpha$) is determined

Table 3. List of section moduli used for Al 7075-T651 fatigue predictions, with the relevant specimen (and experimental runs from Zhao and Jiang 2008) noted.

Specimen	EA (kN)	EI (kN/mm)	GJ (kN/mm)
Solid cylinder (2a)	7.68e6	6.55e7	5.02e7
Tubular (4a), (5c)–(5f)	9.91e6	6.05e8	4.63e8

Table 4. List of flow potential parameters used for Al 7075-T651 fatigue predictions, with the relevant specimen (and experimental runs from Zhao and Jiang 2008) noted.

Specimen	μ_n(mm^{-2})	μ_y(mm^{-3})	μ_x(mm^{-3})
Solid cylinder (2a)	0.00933	0.00376	0.00415
Tubular (4a), (5c)–(5f)	0.00723	0.00103	0.00113

to be 107 MPa for the case of Al-7075 T6 — we select values near the inverse of their value for both tension and torsion, in particular, $\alpha_n = 8e-3$ MPa^{-1} (for tension), and $\alpha_x = 10e-3$ MPa^{-1} (for torsion).

Multiple strain control loading paths (including proportional, exactly 90° out-of-phase, and two out-of-phase Lissajous curves) are described in Fig. 4 of Zhao and Jiang[46]; it is easy to convert these into load paths in displacement. We use the various displacement load paths as control input for an EPCF simulation set up as follows:

- Half of the specimen midsection geometry (from $x = L/2$ to $x = L$, where L depends on the particular specimen) was simulated;
- Eight nodes were used to discretize the length of the geometry into frame elements (following appropriate convergence analysis);
- Symmetry was enforced on one end ($x = L/2$); it was treated as fixed relative to the end under displacement control;
- Displacement was prescribed on the other end ($x = L$), according to the appropriate waveforms for tension and torsion channels provided in Zhao and Jiang.[46]

The EPCF simulations performed against the fatigue data collected in Table 2(a) in Ref. 46 were fully-reversed axial tension; Table 4(a) in Ref. 46 contained data from fatigue experiments performed in fully-reversed torsion. Table 5(d) in Ref. 46 contained data collected under 90° out-of-phase loading, and Tables 5(d) and 5(e) in Ref. 46 contained loading paths defined by Lissajous curves. No functional definition was provided for the Lissajous curves, but we fit the curves shown as follows; for 5(e), we specified axial channel u_{in} and shear channel ϕ_{in} as

$$u_{\text{in}} = \frac{\Delta u_{\text{in}}}{2}\sin(t),$$

$$\phi_{\text{in}} = \frac{\Delta \phi_{\text{in}}}{2}\sin\left(2(8\pi)t - \frac{\pi}{1.9}\right),$$

where and for 5(f)

$$u_{\text{in}} = \frac{\Delta u_{\text{in}}}{2}\sin(t),$$

$$\phi_{\text{in}} = \frac{\Delta \phi_{\text{in}}}{2}\sin\left(4(8\pi)t - \frac{\pi}{2.3}\right),$$

where $\Delta u_{\text{in}}/2$ and $\Delta \phi_{\text{in}}/2$ denote the amplitudes for the prescribed displacements. Using these load paths, we are able to replicate the fatigue predictions under tension and tension-torsion with high accuracy, for most points staying within a factor of 2 (see Fig. 18), with only the 90° out-of-phase results (circular, EPCF, 5d in Fig. 18) presenting an issue.

Experimental results were taken from Tables 2, 4, and 5 in Ref. 46, omitting — for this work — nonzero mean cases. Future modification to the EPCF model

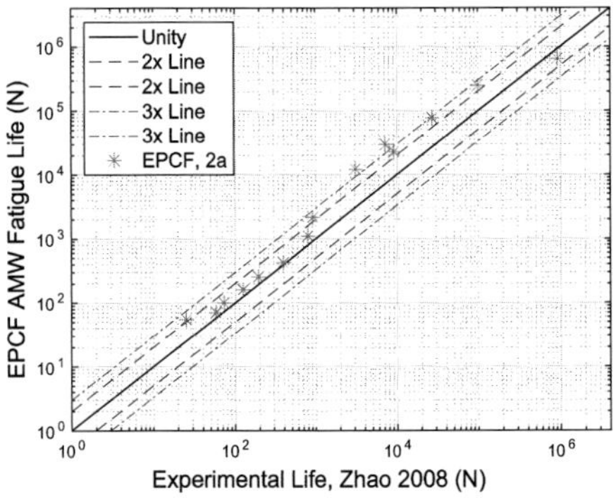

(a)

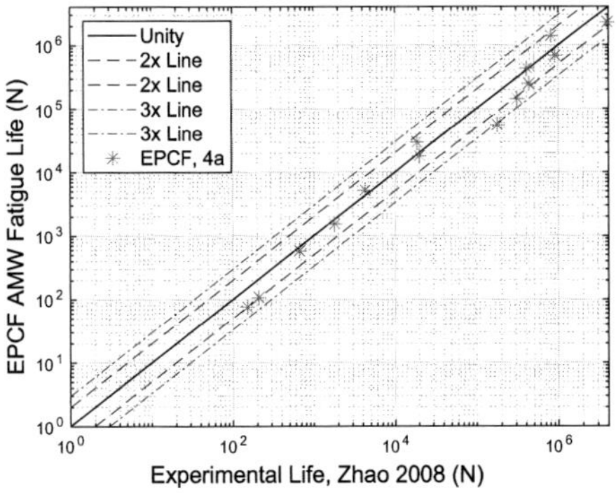

(b)

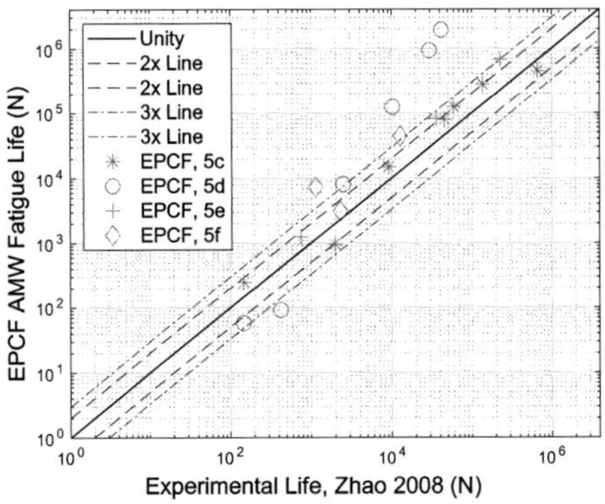

(c)

Fig. 18. Fatigue predictions produced by EPCF compared to experimental results of Zhao and Jiang[46]: (a) under tension, with zero mean stress (fatigue data from Table 2(a)); (b) under torsion with zero static axial load. Fatigue data taken from Table 4(a); and (c) Under tension-torsion with four different loading paths (fatigue data from Tables 5(c)–5(f) in Zhao and Jiang[46]).

to include kinematic hardening should allow general cyclic cases to be accurately handled.

5. Conclusion

This work demonstrates that a moment–curvature-based EPCF frame model can provide full-structural plasticity calculations in a computationally-efficient and accurate manner, replicating results obtainable by 3D FEA but with significantly lower degrees of freedom and greatly-reduced meshing efforts, particularly for complex micro-structures.

This makes it possible to carry out cycle-count-free fatigue calculations via use of the AMW, something which classical J2 plasticity models (even with frame elements) available in commercial software are not capable of. This exciting development brings full-structural, time-domain fatigue calculations closer to the hands of the design engineer.

However the current approach has very simple cross sectional kinematics — the CS rotates rigidly. This is unrealistic and results in early plasticity (due to stiffer elastic response). The results can be further improved by considering cross sectional warping (such as by using an approach similar to that developed by Payette and Reddy[47]).

Acknowledgments

A. R. Srinivasa and Dominic Jarecki gratefully acknowledge the support of TDA, Inc. in conducting this research. J. N. Reddy and A. R. Srinivasa would also like to acknowledge National Science Foundation CMMI grant No. 1952873 for their support in the performance of this work. The authors are pleased to acknowledge the discussions with Dr. Archana Arbind (Indian Institute of Technology, Kharagpur, India) for some aspects of the formulation. The paper is dedicated to Professor Yeong–Bin Yang on his 70th birthday. Professor Yang's research and professional leadership has enormously benefited the structural mechanics community all over the world, and we wish him good health and happiness in the years to come.

References

1. M. Zhao, H. Qing, Y. Wang, J. Liang, M. Zhao, Y. Geng, J. Liang and B. Lu, Superelastic behaviors of additively manufactured porous NiTi shape memory alloys designed with Menger sponge-like fractal structures, *Mater. Des.* **200** (2021) 109448.
2. X. Zhang, F. Yang, B. Liu and J. Deng, Design of Menger sponge fractal structural NiTi as bone implants, *Model. Simul. Mater. Sci. Eng.* **29**(8) (2021) 084001.
3. N. Viet, N. Karathanasopoulos and W. Zaki, Effective stiffness, wave propagation, and yield surface attributes of Menger sponge-like pre-fractal topologies, *Int. J. Mech. Sci.* **227** (2022) 107447.
4. B. Kushwaha, K. Dwivedi, R. S. Ambekar, V. Pal, D. P. Jena, D. R. Mahapatra and C. S. Tiwary, Mechanical and acoustic behavior of 3D-printed hierarchical mathematical fractal Menger sponge, *Adv. Eng. Mater.* **23**(4) (2021) 2001471.
5. D. W. White, Plastic-hinge methods for advanced analysis of steel frames, *J. Constr. Steel Res.* **24**(2) (1993) 121–152.
6. O. Bayrak and S. A. Sheikh, Plastic hinge analysis, *J. Struct. Eng. ASCE* **127**(9) (2001) 1092–1100.
7. J. Y. R. Liew, H. Chen, N. E. Shanmugam and W. F. Chen, Improved nonlinear plastic hinge analysis of space frame structures, *Eng. Struct.* **22**(10) (2000) 1324–1338.
8. M. H. Scott and G. L. Fenves, Plastic hinge integration methods for force-based beam–column elements, *J. Struct. Eng. ASCE* **132**(2) (2006) 244–252.
9. G. Cocchetti and G. Maier, Elastic–plastic and limit-state analyses of frames with softening plastic-hinge models by mathematical programming, *Int. J. Solids Struct.* **40**(25) (2003) 7219–7244.

10. D. Ehrlich and F. Armero, Finite element methods for the analysis of softening plastic hinges in beams and frames, *Comput. Mech.* **35**(4) (2005) 237–264.
11. M. Tabatabaei and S. N. Atluri, Simple and efficient analyses of micro-architected cellular elastic-plastic materials with tubular members, *Int. J. Plast.* **99** (2017) 186–220.
12. N. Challamel, Dynamic analysis of elastoplastic shakedown of structures, *Int. J. Struct. Stab. Dyn.* **5**(02) (2005) 259–278.
13. P. M. Pimenta and T. Yojo, Geometrically exact analysis of spatial frames, *Appl. Mech. Rev.* **46**(11) (1993) 118–128.
14. J. C. Simo and J. Kennedy, On a stress resultant geometrically exact shell model. Part V. Nonlinear plasticity: Formulation and integration algorithms, *Comput. Methods Appl. Mech. Eng.* **96**(2) (1992) 133–171.
15. M. Saje, I. Planinc, G. Turk and B. Vratanar, A kinematically exact finite element formulation of planar elastic-plastic frames, *Comput. Methods Appl. Mech. Eng.* **144**(1–2) (1997) 125–151.
16. M. Saje, G. Turk, A. Kalagasidu and B. Vratanar, A kinematically exact finite element formulation of elastic–plastic curved beams, *Comput. Struct.* **67**(4) (1998) 197–214.
17. D. Zupan and M. Saje, Finite-element formulation of geometrically exact three-dimensional beam theories based on interpolation of strain measures, *Comput. Methods Appl. Mech. Eng.* **192**(49–50) (2003) 5209–5248.
18. L. Herrnböck, A. Kumar and P. Steinmann, Geometrically exact elastoplastic rods: Determination of yield surface in terms of stress resultants, *Comput. Mech.* **67**(3) (2021) 723–742.
19. O. Weeger, D. Schillinger and R. Müller, Mixed isogeometric collocation for geometrically exact 3D beams with elasto-visco-plastic material behavior and softening effects, *Comput. Methods Appl. Mech. Eng.* **399** (2022) 115456.
20. ASTM, E1049–85–Standard practices for cycle counting in fatigue analysis (2017).
21. A. Ince and G. Glinka, A numerical method for elasto-plastic notch-root stress strain analysis, *J. Strain Anal. Eng. Des.* **48**(4) (2013) 229–244.
22. A. Ince, G. Glinka and A. Buczynski, Computational modeling of multiaxial elasto-plastic stress–strain response for notched components under non-proportional loading, *Int. J. Fatigue* **62** (2014) 42–52.
23. A. Ince and G. Glinka, Innovative computational modeling of multiaxial fatigue analysis for notched components, *Int. J. Fatigue* **82** (2016) 134–145.
24. Y. Garud, *Prediction of Stress-Strain Response under General Multiaxial Loading* (ASTM International, 1982).
25. Y. Jiang and H. Sehitoglu, Modeling of cyclic ratchetting plasticity, part I: Development of constitutive relations, *J. Appl. Mech. Trans. ASME* **63**(3) (1996) 720–725.
26. Y. Jiang and H. Sehitoglu, Modeling of cyclic ratchetting plasticity, part II: Comparison of model simulations with experiments, *J. Appl. Mech. Trans. ASME* **63**(3) (1996) 726–733.
27. Y. Jiang and J. Zhang, Benchmark experiments and characteristic cyclic plasticity deformation, *Int. J. Plast.* **24**(9) (2008) 1481–1515.
28. D. Benasciutti and R. Tovo, Cycle distribution and fatigue damage assessment in broad-band non-gaussian random processes, *Probab. Eng. Mech.* **20**(2) (2005) 115–127.

29. T. Dirlik and D. Benasciutti, Dirlik and tovo-benasciutti spectral methods in vibration fatigue: A review with a historical perspective, *Metals* **11**(9) (2021) 1333.
30. M. Bonte, A. de Boer and R. Liebregts, Determining the von Mises stress power spectral density for frequency domain fatigue analysis including out-of-phase stress components, *J. Sound Vib.* **302**(1–2) (2007) 379–386.
31. D. Jarecki, A. R. Srinivasa, N. Iyyer and P. Thamburaja, Spectral fatigue prediction with a unified strain-based approach to handle localized plastic excursions, *Mech. Syst. Signal Process,* Under review (2023).
32. K. R. Rajagopal and A. R. Srinivasa, Inelastic response of solids described by implicit constitutive relations with nonlinear small strain elastic response, *Int. J. Plast.* **71** (2015) 1–9.
33. Z. Wang, A. R. Srinivasa, K. R. Rajagopal and J. N. Reddy, Simulation of inextensible elasto-plastic beams based on an implicit rate type model, *Int. J. Non-Linear Mech.* **99** (2018) 165–172.
34. F. Mozafari, P. Thamburaja, A. R. Srinivasa and N. Moslemi, A rate independent inelasticity model with smooth transition for unifying low-cycle to high-cycle fatigue life prediction, *Int. J. Mech. Sci.* **159** (2019) 325–335.
35. F. Mozafari, P. Thamburaja, A. R. Srinivasa and S. Abdullah, Fatigue life prediction under variable amplitude loading using a microplasticity-based constitutive model, *Int. J. Fatigue* **134** (2020) 105477.
36. F. Mozafari, P. Thamburaja, N. Moslemi and A. R. Srinivasa, Finite-element simulation of multi-axial fatigue loading in metals based on a novel experimentally-validated microplastic hysteresis-tracking method, *Finite Elem. Anal. Des.* **187** (2021) 103481.
37. D. Jarecki, A. R. Srinivasa and N. Iyyer, Rate-independent elasto-plastic materials—a brief history and some new developments, *Int. J. Adv. Eng. Sci. Appl. Math.* **13**(1) (2021) 3–17.
38. W. J. Stronge and T. Yu, *Dynamic Models for Structural Plasticity* (Springer, 1993).
39. M. Bruneau, C.-M. Uang and R. Sabelli, *Ductile Design of Steel Structures* (McGraw-Hill Education, 2011).
40. R. G. Budynas and A. M. Sadegh, *Roark's Formulas for Stress and Strain* (McGraw-Hill Education, 2020).
41. F. Mollica and A. R. Srinivasa, A general framework for generating convex yield surfaces for anisotropic metals, *Acta Mech.* **154**(1) (2002) 61–84.
42. J. N. Reddy, *Introduction to the Finite Element Method* (McGraw-Hill Education, New York, 2019).
43. J. C. Simo, Algorithms for static and dynamic multiplicative plasticity that preserve the classical return mapping schemes of the infinitesimal theory, *Comput. Methods Appl. Mech. Eng.* **99**(1) (1992) 61–112.
44. N. A. Kazerooni, A. Srinivasa and A. Freed, Orthotropic-equivalent strain measures and their application to the elastic response of porcine skin, *Mech. Res. Commun.* **101** (2019) 103404.
45. A. D. Freed and K. Rajagopal, A promising approach for modeling biological fibers, *Acta Mech.* **227** (2016) 1609–1619.
46. T. Zhao and Y. Jiang, Fatigue of 7075-T651 aluminum alloy, *Int. J. Fatigue* **30**(5) (2008) 834–849.
47. G. Payette and J. Reddy, A nonlinear finite element framework for viscoelastic beams based on the high-order Reddy beam theory, *J. Eng. Mater. Technol.* **135**(1) (2013) 011005.

https://doi.org/10.1142/9789819814770_bmatter

Author Index